TREATISE ON
INVERTEBRATE PALEONTOLOGY

Prepared under the Guidance of the
Joint Committee on Invertebrate Paleontology

Paleontological Society — *Society of Economic Paleontologists and Mineralogists* — *Palaeontographical Society*

Directed and Edited by

RAYMOND C. MOORE

Part F
COELENTERATA

BY FREDERICK M. BAYER, HILBRAND BOSCHMA, HORACIO J. HARRINGTON, DOROTHY HILL, LIBBIE H. HYMAN, MARIUS LECOMPTE, EUGENIA MONTANARO-GALLITELLI, RAYMOND C. MOORE, ERWIN C. STUMM, AND JOHN W. WELLS

GEOLOGICAL SOCIETY OF AMERICA
and
UNIVERSITY OF KANSAS PRESS

1956

Reprinted 1963

Library of Congress Catalogue Card
Number: 53-12913

Printed in the U.S.A. by
THE UNIVERSITY OF KANSAS PRESS
Lawrence, Kansas

Address all communications to The Geological Society of America, 419 West 117 Street, New York 27, N.Y.

The *Treatise on Invertebrate Paleontology* has been made possible by (1) a grant of funds from The Geological Society of America through the bequest of Richard Alexander Fullerton Penrose, Jr., for preparation of illustrations and partial defrayment of organizational expense; (2) contribution of the knowledge and labor of specialists throughout the world, working in co-operation under sponsorship of The Palaeontographical Society, The Paleontological Society, and The Society of Economic Paleontologists and Mineralogists; and (3) acceptance by the University of Kansas Press of publication without cost to the Societies concerned and without any financial gain to the Press.

TREATISE ON INVERTEBRATE PALEONTOLOGY

Directed and Edited by
RAYMOND C. MOORE

PARTS

The indicated Parts (excepting the first and last) are to be published at whatever time each is ready. All may be assembled ultimately in bound volumes. In the following list, already published Parts are marked with a double asterisk (**) and those in press or nearing readiness for press are marked with a single asterisk (*). Each is cloth bound with title in gold on the cover. Copies are available on orders sent to the Geological Society of America at 419 West 117th Street, New York 27, N.Y., at prices quoted, which very incompletely cover costs of producing and distributing them but on receipt of payment the Society will ship copies without additional charge to any address in the world.

The list of contributing authors is subject to change.

(A)—INTRODUCTION.
B—PROTISTA 1 (chrysomonads, coccolithophorids, diatoms, etc.).
*C—PROTISTA 2 (foraminifers).
**D—PROTISTA 3 (radiolarians, tintinnines) (xii+195 p., 1050 figs.). $3.00.
**E—ARCHAEOCYATHA, PORIFERA (xviii + 122 p., 728 figs.). $3.00.
**F—COELENTERATA (this volume).
**G—BRYOZOA (xii + 253 p., 2,000 figs.). $3.00.
H—BRACHIOPODA.
I—MOLLUSCA 1 (chitons, scaphopods, gastropods).
J—MOLLUSCA 2 (gastropods).
K—MOLLUSCA 3 (nautiloid cephalopods).
*L—MOLLUSCA 4 (ammonoid cephalopods).
M—MOLLUSCA 5 (dibranchiate cephalopods).
N—MOLLUSCA 6 (pelecypods).
O—ARTHROPODA 1 (trilobitomorphs).
**P—ARTHROPODA 2 (chelicerates, pycnogonids) (xvii + 181 p., 565 figs.). $3.50.
Q—ARTHROPODA 3 (ostracodes).
R—ARTHROPODA 4 (branchiopods, cirripeds, malacostracans, myriapods, insects).
S—ECHINODERMATA 1 (cystoids, blastoids, edrioasteroids, etc.).
T—ECHINODERMATA 2 (crinoids).
U—ECHINODERMATA 3 (echinozoans, asterozoans).
**V—GRAPTOLITHINA (xvii + 101 p., 358 figs.). $3.00.
W—MISCELLANEA (worms, conodonts, problematical fossils.
(X)—ADDENDA (index).

CONTRIBUTING AUTHORS

AGER, D. V., Imperial College of Science and Technology, London, England.
AMSDEN, T. W., Oklahoma Geological Survey, Norman, Okla.
ARKELL, W. J., Sedgwick Museum, Cambridge University, Cambridge, Eng.
BAIRSTOW, LESLIE, British Museum (Natural History), London, Eng.
BARKER, R. WRIGHT, Shell Development Co., Houston, Tex.
BASSLER, R. S., U.S. National Museum, Washington, D.C.
BATTEN, ROGER, L., University of Wisconsin, Madison, Wisconsin.
BAYER, FREDERICK M., U.S. National Museum, Washington, D.C.
BELL, W. CHARLES, University of Texas, Austin, Tex.
BERDAN, JEAN M., U.S. Geological Survey, Washington, D.C.
BOARDMAN, R. S., U.S. Geological Survey, Washington, D.C.
BOSCHMA, H., Rijksmuseum van Natuurlijke Historie, Leiden, Netherlands.

BOWSHER, ARTHUR L., U.S. Geological Survey, Washington, D.C.

BRAMLETTE, M. N., Scripps Institution of Oceanography, La Jolla, California.

BRANSON, CARL C., University of Oklahoma, Norman, Okla.

BROWN, D. A., University of Otago, Dunedin, N.Z.

BULMAN, O. M. B., Sedgwick Museum, Cambridge University, Cambridge, Eng.

CAMPBELL, ARTHUR S., St. Mary's College, St. Mary's College, Calif.

CARPENTER, FRANK M., Biological Laboratories, Harvard University, Cambridge, Mass.

CASEY, RAYMOND, Geological Survey of Great Britain, London, Eng.

CHAVAN, ANDRÉ, Chantemerle à Seyssel (Ain), France.

COLE, W. STORRS, Cornell University, Ithaca, N.Y.

COOPER, G. ARTHUR, U.S. National Museum, Washington, D.C.

COX, L. R., British Museum (Natural History), London, Eng.

CURRY, DENNIS, Pinner, Middlesex, Eng.

†DAVIES, L. M., Edinburgh, Scot.

DECHASEAUX, C., Laboratoire de Paléontologie à la Sorbonne, Paris, France.

DURHAM, J. WYATT, Museum of Palaeontology, University of California, Berkeley, Calif.

EAMES, F. E., Anglo-Iranian Oil Company, London, Eng.

ELLIOTT, GRAHAM F., Kensington, Eng.

EMERSON, W. K., American Museum of Natural History, New York, N.Y.

EXLINE, HARRIET, Rolla, Mo.

FELL, H. BARRACLOUGH, Victoria University College, Wellington, N.Z.

FISHER, D. W., New York State Museum, Albany, N.Y.

FLOWER, ROUSSEAU H., New Mexico Institute of Mining and Technology, Socorro, N.Mex.

FRIZZELL, DONALD L., Missouri School of Mines, Rolla, Mo.

FURNISH, WILLIAM M., State University of Iowa, Iowa City, Iowa.

GARDNER, JULIA, U.S. Geological Survey, Washington, D.C.

GEORGE, T. NEVILLE, Glasgow University, Glasgow, Scot.

GLAESSNER, M. F., University of Adelaide, Adelaide, S.Austral.

HAAS, FRITZ, Chicago Natural History Museum, Chicago, Ill.

HANNA, G. DALLAS, California Academy of Sciences, San Francisco, Calif.

HÄNTZSCHEL, WALTER, Geologisches Staatsinstitut, Hamburg, Ger.

HARRINGTON, H. J., University of Kansas, Lawrence, Kans.

HASS, WILBERT H., U.S. Geological Survey, Washington, D.C.

HATAI, KOTORA, Tohoku University, Sendai, Japan.

HAWKINS, H. L., Reading University, Reading, Eng.

HEDGPETH, JOEL, Scripps Institution of Oceanography, University of California, La Jolla, Calif.

HENNINGSMOEN, GUNNAR, Paleontologisk Museum, University of Oslo, Oslo, Norway.

HERTLEIN, L. G., California Academy of Sciences, San Francisco, Calif.

HESSLAND, IVAR, Geologiska Institutet, University of Stockholm, Stockholm, Swed.

HILL, DOROTHY, University of Queensland, Brisbane, Queensl.

HOLTHUIS, L. B., Rijksmuseum van Natuurlijke Historie, Leiden, Netherlands.

HOWE, HENRY V., Louisiana State University, Baton Rouge, La.

HOWELL, B. F., Princeton University, Princeton, N.J.

HYMAN, LIBBIE H., American Museum of Natural History, New York, N.Y.

JAANUSSON, VALDAR, Paleontological Inst., Uppsala, Sweden.

KAMPTNER, ERWIN, Naturhistorisches Museum, Wien, Aus.

KEEN, MYRA, Stanford University, Stanford, Calif.

KESLING, ROBERT V., Paleontological Museum, University of Michigan, Ann Arbor, Mich.

KING, RALPH H., University of Kansas, Lawrence, Kans.

KNIGHT, J. BROOKES, Sarasota, Florida.

KUMMEL, BERNHARD, Museum of Comparative Zoology, Harvard University, Cambridge, Mass.

LA ROCQUE, AURÈLE, Ohio State University, Columbus, Ohio.

LAUBENFELS, M. W. DE, Oregon State College, Corvallis, Ore.

LECOMPTE, MARIUS, Institut Royal des Sciences Naturelles, Bruxelles, Belg.

LEONARD, A. BYRON, University of Kansas, Lawrence, Kans.

LEVINSON, S. A., Humble Oil & Refining Company, Houston, Tex.

LOCHMAN-BALK, CHRISTINA, New Mexico Institute of Mining and Geology, Socorro, N.Mex.

LOEBLICH, A. R., JR., U.S. National Museum, Washington, D.C.

LOEBLICH, HELEN TAPPAN, U.S. Geological Survey, Washington, D.C.

LOHMAN, KENNETH E., U.S. Geological Survey, Washington, D.C.

LOWENSTAMM, HEINZ A., California Institute of Technology, Pasadena, Calif.

LUDBROOK, N. H., Department of Mines, Adelaide, S.Austral.

MARWICK, J., New Zealand Geological Survey, Wellington, N.Z.

MELVILLE, R. V., Geological Survey and Museum, London, Eng.

MILLER, ARTHUR K., State University of Iowa, Iowa City, Iowa.

MONTANARO-GALLITELLI, EUGENIA, Istituto di Geología e Paleontología, Università, Modena, Italy.

MOORE, RAYMOND C., University of Kansas, Lawrence, Kans.

MORRISON, J. P. E., U.S. National Museum, Washington, D.C.

MUIR-WOOD, H. MARGARET, British Museum (National History), London, Eng.

NEWELL, NORMAN D., American Museum of Natural History (and Columbia University), New York, N.Y.

OKULITCH, VLADIMIR J., University of British Columbia, Vancouver, B.C.

OLSSON, AXEL A., Coral Gables, Florida.

PALMER, KATHERINE VAN WINKLE, Paleontological Research Institution, Ithaca, N.Y.

PECK, RAYMOND E., University of Missouri, Columbia, Mo.

PETRUNKEVITCH, ALEXANDER, Osborn Zoological Laboratory, Yale University, New Haven, Conn.

POULSEN, CHR., Universitetets Mineralogisk-Geologiske Institut, København, Denm.

POWELL, A. W. B., Auckland Institute and Museum, Auckland, N.Z.

PURI, HARBANS, Florida State Geological Survey, Tallahassee, Fla.

RASETTI, FRANCO, Johns Hopkins University, Baltimore, Md.

REGNÉLL, GERHARD, Paleontologiska Institution, Lunds Universitets, Lund, Swed.

REHDER, HARALD A., U.S. National Museum, Washington, D.C.

REICHEL, M., Bernouillanum, Basel University, Basel, Switz.

RICHTER, EMMA, Senckenberg Natur-Museum, Frankfurt-a.-M., Ger.

RICHTER, RUDOLF, Universität Frankfurt-a.-M., Frankfurt-a.-M., Ger.

ROBERTSON, ROBERT, Stanford University, Stanford, Calif.

SCHINDEWOLF, O. H. Geologisch-paläontologisches Institut der Universität Tűbingen, Tűbingen, Ger.

SCHMIDT, HERTA, Senckenbergische Naturforschende Gesellschaft, Frankfurt-a.-M., Ger.

SCOTT, HAROLD W., University of Illinois, Urbana, Ill.

SHAVER, ROBERT, University of Mississippi, University, Miss.

SIEVERTS-DORECK, HERTHA, Stuttgart-Möhringen, Ger.

SMITH, ALLYN G., Berkeley, Calif.

SOHN, I. G., U.S. Geological Survey, Washington, D.C.

†SPENCER, W. K., Crane Hill, Ipswich, Suffolk, Eng.

STAINBROOK, MERRILL A., Brandon, Iowa.

STEHLI, FRANK, California Institute of Technology, Pasadena, Calif.

STENZEL, H. B., University of Houston, Houston, Tex.

STEPHENSON, LLOYD W., U.S. Geological Survey, Washington, D.C.

STØRMER, LEIF, Paleontologisk Institutt, University of Oslo, Oslo, Nor.

STUBBLEFIELD, C. J., Geological Survey and Museum, London, Eng.

STUMM, ERWIN C., Museum of Paleontology, University of Michigan, Ann Arbor, Mich.

SWAIN, FREDERICK M., University of Minnesota, Minneapolis, Minn.

SYLVESTER-BRADLEY, P. C., University of Sheffield, Sheffield, Eng.

TASCH, PAUL, University of Wichita, Wichita, Kans.

TEICHERT, CURT, U.S. Geological Survey, Federal Center, Denver, Colo.

THOMPSON, M. L., University of Kansas, Lawrence, Kans.

THOMPSON, R. H., University of Kansas, Lawrence, Kans.

Tiegs, O. W., University of Melbourne, Melbourne, Victoria, Austral.

Tripp, Ronald P., Glasgow, Scot.

Ubaghs, G., Université de Liège, Liège, Belg.

Vokes, H. E., Johns Hopkins University, Baltimore, Md.

Wanner, J., Scheidegg (Allgäu), Bayern, Ger.

Weir, John, Glasgow University, Glasgow, Scot.

Weller, J. Marvin, University of Chicago, Chicago, Ill.

Wells, John W., Cornell University, Ithaca, N.Y.

Whittington, H. B., Museum of Comparative Zoology, Harvard University, Cambridge, Mass.

Williams, Alwyn, Queens University of Belfast, Belfast, N.Ire.

Wills, L. J., Romsley, Eng.

†Withers, T. H., Bournemouth, Eng.

Wright, C. W., Kensington, Eng.

†Wrigley, Arthur, Norbury, London, Eng.

Yochelson, Ellis L., U.S. Geological Survey, Washington, D.C.

Yonge, C. M., University of Glasgow, Glasgow, Scot.

Youngquist, Walter L., University of Kansas, Lawrence, Kans.

†—Deceased.

EDITORIAL PREFACE

The aim of the *Treatise on Invertebrate Paleontology,* as originally conceived and consistently pursued, is to present the most comprehensive and authoritative, yet compact statement of knowledge concerning invertebrate fossil groups that can be formulated by collaboration of competent specialists in seeking to organize what has been learned of this subject up to the mid-point of the present century. Such work has value in providing a most useful summary of the collective results of multitudinous investigations and thus should constitute an indispensable text and reference book for all persons who wish to know about remains of invertebrate organisms preserved in rocks of the earth's crust. This applies to neozoologists as well as paleozoologists and to beginners in study of fossils as well as to thoroughly trained, long experienced professional workers, including teachers, stratigraphical geologists, and individuals engaged in research on fossil invertebrates. The making of a reasonably complete inventory of present knowledge of invertebrate paleontology may be expected to yield needed foundation for future research and it is hoped that the *Treatise* will serve this end.

The *Treatise* is divided into parts which bear index letters, each except the initial and concluding ones being defined to include designated groups of invertebrates. The chief purpose of this arrangement is to provide for independence of the several parts as regards date of publication, because it is judged desirable to print and distribute each segment as soon as possible after it is ready for press. Pages in each part will bear the assigned index letter joined with numbers beginning with 1 and running consecutively to the end of the part. When the parts ultimately are assembled into volumes, no renumbering of pages and figures is required.

The outline of subjects to be treated in connection with each large group of invertebrates includes (1) description of morphological features, with special reference to hard parts, (2) ontogeny, (3) classification, (4) geological distribution, (5) evolutionary trends and phylogeny, and (6) systematic description of genera, subgenera, and higher taxonomic units. In general, paleoecological aspects of study are omitted or little emphasized because comprehensive treatment of this subject is being undertaken in a separate work, prepared under auspices of a committee of the United States National Research Council. A selected list of references is furnished in each part of the *Treatise.*

Features of style in the taxonomic portions of this work have been fixed by the Editor with aid furnished by advice from the Joint Committee on Invertebrate Paleontology representing the societies which have undertaken to sponsor the *Treatise*. It is the Editor's responsibility to consult with authors and co-ordinate their work, seeing that manuscript properly incorporates features of adopted style. Especially he has been called on to formulate policies in respect to many questions of nomenclature and procedure.

The subject of family and subfamily names is reviewed briefly in a following section of this preface, and features of *Treatise* style in generic descriptions are explained.

A generous grant of $35,000 has been made by the Geological Society of America for the purpose of preparing *Treatise* illustrations. Administration of expenditures has been in charge of the Editor and most of the work by photographers and artists has been done under his direction at the University of Kansas, but sizable parts of this program have also been carried forward in Washington and London.

FORM OF ZOOLOGICAL NAMES

Many questions arise in connection with the form of zoological names. These include such matters as adherence to stipulations concerning Latin or Latinized nature of words accepted as zoological names, gender of generic and subgeneric names, nominative or adjectival form of specific names, required endings for some family-group names, and numerous others. Regulation extends to capitalization, treatment of particles belonging to modern patronymics, use of neo-Latin letters, and approved methods for converting diacritical marks. The magnitude and complexities of nomenclature problems surely are enough to warrant the complaint of those who hold that zoology is the study of animals rather than of names applied to them.

CLASSIFICATION OF ZOOLOGICAL NAMES

In accordance with the "Copenhagen Decisions on Zoological Nomenclature" (London, 135 p., 1953), zoological names may be classified usefully in various ways. The subject is summarized here with introduction of designations for some categories which the *Treatise* proposes to distinguish in systematic parts of the text for the purpose of giving readers comprehension of the nature of various names together with authorship and dates attributed to them.

CO-ORDINATE NAMES OF TAXA GROUPS

Five groups of different-rank taxonomic units (termed *taxa,* sing., *taxon*) are discriminated, within each of which names are treated as co-ordinate, being transferrable from one category to another without change of authorship or date. These are: (1) Species Group (subspecies, species); (2) Genus Group (subgenus, genus); (3) Family Group (tribe, subfamily, family, superfamily); (4) Order/Class Group (suborder, order, subclass, class); and (5) Phylum Group (subphylum, phylum). In the first 3 of these groups, but not others, the author of the first-published valid name for any taxon is held to be the author of all other taxa in the group which are based on the same nominate type and the date of publication for purposes of priority is that of the first-published name. Thus, if author A in 1800 introduces the family name X-idae to include 3 genera, one of which is *X-us;* and if author B in 1850 divides the 20 genera then included in X-idae into subfamilies called X-inae and Y-inae; and if author C in 1950 combines X-idae with other later-formed families to make a superfamily X-acea (or X-oidea, X-icae, etc.); the author of X-inae, X-idae and X-acea is A, 1800, under the Rules. Because taxonomic concepts introduced by authors B and C along with appropriate names surely are not attributable to author A, some means of recording responsibility of B and C are needed. This is discussed later in explaining proposed use of *"nom. transl."*

The co-ordinate status of zoological names belonging to the species group is stipulated in Art. 11 of the present Rules; genus group in Art. 6 of the present Rules; family group in paragraph 46 of the Copenhagen Decisions; order/class group and phylum group in paragraphs 65 and 66 of the Copenhagen Decisions.

ORIGINAL AND SUBSEQUENT FORMS OF NAMES

Zoological names may be classified according to form (spelling) given in original publication and employed by subsequent authors. In one group are names which are entirely identical in original and subsequent usage. Another group comprises names which include with the original subsequently published variants of one sort or another. In this second group, it is important to distinguish names which are inadvertent changes from those constituting intentional emendations, for they have quite different status in nomenclature. Also, among intentional emendations, some are acceptable and some quite unacceptable under the Rules.

VALID AND INVALID NAMES

Valid names. A valid zoological name is one that conforms to all mandatory provisions of the Rules (Copenhagen Decisions, p. 43-57) but names of this group are divisible into subgroups as follows: (1) *"inviolate names,"* which as originally published not only meet all mandatory requirements of the Rules but are not subject to any sort of alteration (most generic and subgeneric names); (2) *"perfect names,"* which as they appear in original publication (with or without precise duplication by subsequent authors) meet all mandatory requirements and need no correction of any kind but which nevertheless are legally alterable under present Rules (as in changing the form of ending of a published class/order-group name); (3) *"imperfect names,"* which as originally published and with or without subsequent duplication meet mandatory requirements but contain defects such as incorrect gender of an adjectival specific name (for example, *Spironema recta* instead of *Spironema rectum*) or incorrect stem or form of ending of a family-group name (for example, Spironemidae instead of Spironematidae); (4) *"transferred names,"* which are derived by valid emendation from either of the 2nd or 3rd subgroups or from a pre-existing transferred name (as illustrated by change of a family-group name from -inae to -idae or making of a superfamily name); (5) *"improved names,"* which include necessary as well as somewhat arbitrarily made emendations allowable under the Rules for taxonomic categories not now covered by regulations as to name form and alterations that are distinct from changes that distinguish the 4th subgroup (including names derived from the 2nd and 3rd subgroups and possibly some alterations of 4th subgroup names). In addition, some zoological names included among those recognized as valid are classifiable in special categories, while at the same time belonging to one or more of the above-listed subgroups. These chiefly include (7) *"substitute names,"* introduced to replace invalid names such as junior homonyms; and (8) *"conserved names,"* which are names that would have to be rejected by application of the Rules except for saving them in their original or an altered spelling by action of the International Commission on Zoological Nomenclature in exercising its plenary powers to this end. Whenever a name requires replacement, any individual may publish a "new name" for it and the first one so introduced has priority over any others; since newness is temporary and relative, the replacement designation is better called substitute name rather than new name. Whenever it is considered desirable to save for usage an otherwise necessarily rejectable name, an individual cannot by himself accomplish the preservation, except by unchallenged action taken in accordance with certain provisions of the Copenhagen Decisions; otherwise he must seek validation through ICZN.

It is useful for convenience and brevity of distinction in recording these subgroups of valid zoological names to introduce Latin designations, following the pattern of *nomen nudum, nomen novum,* etc. Accordingly, the subgroups are (1) *nomina inviolata* (sing., *nomen inviolatum,* abbr., *nom. inviol.*); (2) *nomina perfecta* (sing., *nomen perfectum,* abbr., *nom. perf.*); (3) *nomina imperfecta* (sing., *nomen imperfectum,* abbr., *nom. imperf.*); (4) *nomina translata* (sing., *nomen translatum,* abbr., *nom. transl.*); (5) *nomina correcta* (sing., *nomen correctum,* abbr., *nom. correct.*); (6) *nomina substituta* (sing., *nomen substitutum,* abbr., *nom. subst.*); (7) *nomina conservata* (sing., *nomen conservatum,* abbr., *nom. conserv.*).

Invalid names. Invalid zoological names consisting of originally published names that fail to comply with mandatory provisions of the Rules and consisting of inadvertent changes in spelling of names have no status in nomenclature. They are not available as replacement names and they do not preoccupy for purposes of the Law of Homonomy. In addition to *nomen nudum,* invalid names may be distinguished as follows: (1) *"denied names,"* which consist of originally published names (with or without subsequent duplication) that do not meet mandatory requirements of the Rules; (2) *"null names,"* which comprise unintentional alterations of names; and (3) *"vain or void names,"* which consist of invalid emendations of previously published valid or invalid names. Void names do have status in nomenclature, being classified as junior synonyms of valid names.

Proposed Latin designations for the indicated kinds of invalid names are as follows: (1) *nomina negata* (sing., *nomen negatum,* abbr., *nom. neg.*); (2) *nomina nulla* (sing., *nomen nullum,* abbr., *nom. null.*); (3) *nomina vana* (sing., *nomen vanum,* abbr., *nom. van.*). It is desirable in the *Treatise* to identify invalid names, particularly in view of the fact that many of these names (*nom. neg., nom null.*) have been considered incorrectly to be junior objective synonyms (like *nom. van.*), which have status in nomenclature.

SUMMARY OF NAME CLASSES

Partly because only in such publications as the *Treatise* is special attention to classes of zoological names called for and partly because new designations are now introduced as means of recording distinctions explicitly as well as compactly, a summary may be useful. In the following tabulation valid classes of names are indicated in boldface type, whereas invalid ones are printed in italics.

Definitions of Name Classes

nomen conservatum (nom. conserv.). Name otherwise unacceptable under application of the Rules which is made valid, either with original or altered spelling, through procedures specified by the Copenhagen Decisions or by action of ICZN exercising its plenary powers.

nomen correctum (nom. correct.). Name with intentionally altered spelling of sort required or allowable under the Rules but not dependent on transfer from one taxonomic category to another ("improved name"). (*See* Copenhagen Decisions, paragraphs 50, 71-2-a-i, 74, 75, 79, 80, 87, 101; in addition, change of endings for categories not now fixed by Rules.)

nomen imperfectum (nom. imperf.). Name that as originally published (with or without subsequent identical spelling) meets all mandatory requirements of the Rules but contains defect needing correction ("imperfect name"). (*See* Copenhagen Decisions, paragraphs 50-1-b, 71-1-b-i, 71-1-b-ii, 79, 80, 87, 101.)

nomen inviolatum (nom. inviol.). Name that as originally published meets all mandatory requirements of the Rules and also is uncorrectable or alterable in any way ("inviolate name"). (*See* Copenhagen Decisions, paragraphs 152, 153, 155-157).

nomen negatum (nom. neg.). Name that as originally published (with or without subsequent identical spelling) constitutes invalid original spelling and although possibly meeting all other mandatory requirements of the Rules, is not correctable to establish original authorship and date ("denied name"). (*See* Copenhagen Decisions, paragraph 71-1-b-iii.)

nomen nudum (nom. nud.). Name that as originally published (with or without subsequent identical spelling) fails to meet mandatory requirements of the Rules and having no status in nomenclature, is not correctable to establish original authorship and date ("naked name"). (*See* Copenhagen Decisions, paragraph 122.)

nomen nullum (nom. null.). Name consisting of an unintentional alteration in form (spelling) of a previously published name (either valid name, as *nom. inviol., nom. perf., nom. imperf., nom. transl.;* or invalid name, as *nom. neg., nom. nud., nom. van.,* or another *nom. null.*) ("null name"). (*See* Copenhagen Decisions, paragraphs 71-2-b, 73-4.)

nomen perfectum (nom. perf.). Name that as originally published meets all mandatory requirements of the Rules and needs no correction of any kind but which nevertheless is validly alterable ("perfect name").

nomen substitutum (nom. subst.). Replacement name published as substitute for an invalid name, such as a junior homonym (equivalent to "new name").

nomen translatum (nom. transl.). Name that is derived by valid emendation of a previously published name as result of transfer from one taxonomic category to another within the group to which it belongs ("transferred name").

nomen vanum (nom. van.). Name consisting of an invalid intentional change in form (spelling) from a previously published name, such invalid emendations having status in nomenclature as junior objective synonyms ("vain or void name"). (*See* Copenhagen Decisions, paragraphs 71-2-a-ii, 73-3.)

Except as specified otherwise, zoological names accepted in the *Treatise* may be understood to be classifiable either as *nomina inviolata* or *nomina perfecta* (omitting from notice *nomina correcta* among specific names) and these are not discriminated. Names which are not accepted for one reason or another include junior homonyms, a few senior synonyms classifiable as *nomina negata* or *nomina nuda,* and numerous junior synonyms which include both objective *(nomina vana)* and subjective (all classes of valid names) types; effort to classify the invalid names as completely as possible is intended.

NAME CHANGES IN RELATION TO GROUP CATEGORIES

SPECIFIC AND SUBSPECIFIC NAMES

Detailed consideration of valid emendation of specific and subspecific names is unnecessary here because it is well understood and relatively inconsequential. When the form of adjectival specific names is changed to obtain agreement with the gender of a generic name in transferring a species from one genus to another, it is never needful to label the changed name as a *nom. transl.* Likewise, transliteration of a letter accompanied by a diacritical mark in manner now called for by the Rules (as in changing originally published *brőggeri* to *broeggeri*) or elimination of a hyphen (as in changing originally published *cornu-oryx* to *cornuoryx* does not require "*nom. correct.*" with it. Revised provisions for emending specific and subspecific names are stated in the report on Copenhagen Decisions (p. 43-46, 51-57).

GENERIC AND SUBGENERIC NAMES

So rare are conditions warranting change of the originally published valid form of generic and subgeneric names that lengthy discussion may be omitted. Only elimination of diacritical marks of some names in this category seems to furnish basis for valid emendation. It is true that many changes of generic and subgeneric names have been published, but virtually all of these are either *nomina vana* or *nomina nulla*. Various names which formerly were classed as homonyms are not now, for two names that differ only by a single letter (or in original publication by presence or absence of a diacritical mark) are construed to be entirely distinct. Revised provisions for emendation of generic and subgeneric names also are given in the report on Copenhagen Decisions (p. 43-47).

Examples in use of classificatory designations for generic names as previously given are the following, which also illustrate designation of type species, as explained later.

Kurnatiophyllum THOMSON, 1875 [**K. concentricum;* SD GREGORY, 1917] [=*Kumatiophyllum* THOMSON, 1876 *(nom. null.)*; *Cymatophyllum* THOMSON, 1901 *(nom. van.)*; *Cymatiophyllum* LANG, SMITH & THOMAS, 1940 *(nom. van.)*].

Stichophyma POMEL, 1872 [**Manon turbinatum* RÖMER, 1841; SD RAUFF, 1893] [=*Stychophyma* VOSMAER, 1885 *(nom. null.)*; *Sticophyma* MORET, 1924 *(nom. null.)*].

Stratophyllum SMYTH, 1933 [**S. tenue*] [=*Ethmoplax* SMYTH, 1939 (*nom. van. pro Stratophyllum*); *Stratiphyllum* LANG, SMITH & THOMAS, 1940 (*nom. van. pro Stratophyllum* SMYTH) (*non Stratiphyllum* SCHEFFEN, 1933)].

Placotelia OPPLIGER, 1907 [**Porostoma marconi* FROMENTEL, 1859; SD DELAUBENFELS, herein] [=*Plakotelia* OPPLIGER, 1907 *(nom. neg.)*].

Walcottella DELAUB., *nom. subst.*, 1955 [*pro Rhopalicus* SCHRAMM., 1936 (*non* FÖRSTER, 1856)].

Cyrtograptus CARRUTHERS, 1867 [*nom. correct.* LAPWORTH, 1873 (*pro Cyrtograpsus* CARRUTHERS, 1867), *nom. conserv.* proposed BULMAN, 1955 (ICZN pend.)]

FAMILY-GROUP NAMES; USE OF "*NOM. TRANSL.*"

The Rules now specify the form of endings only for subfamily (-inae) and family (-idae) but decisions of the Copenhagen Congress direct classification of all family-group assemblages (taxa) as co-ordinate, signifying that for purposes of priority a name published for a unit in any category and based on a particular type genus shall date from its original publication for a unit in any category, retaining this priority (and authorship) when the unit is treated as belonging to a lower or higher category. By exclusion of -inae and -idae, respectively reserved for subfamily and family, the endings of names used for tribes and superfamilies must be unspecified different letter combinations. These, if introduced subsequent to designation of a subfamily or family based on the same nominate genus, are *nomina translata,* as is also a subfamily that is elevated to family rank or a family reduced to subfamily rank. In the *Treatise* it is desirable to distinguish the valid emendation comprised in the changed ending of each transferred family group name by the abbreviation "*nom. transl.*" and record of the author and date belonging to this emendation. This is particularly important in the case of superfamilies, for it is the author who introduced this taxon that one wishes to know about rather than the author of the superfamily as defined by the Rules, for the latter is merely the individual who first defined some lower-rank family-group taxon that contains the nominate genus of the superfamily. The publication of the author containing introduction of the superfamily *nomen translatum* is likely to furnish the information on taxonomic considerations that support definition of the unit.

Examples of the use of "*nom. transl.*" are the following.

Subfamily STYLININAE d'Orbigny, 1851

[*nom. transl.* EDWARDS & HAIME, 1857 (*ex* Stylinidae D'ORBIGNY, 1851)]

Superfamily ARCHAEOCTONOIDEA Petrunkevitch, 1949

[*nom. transl.* PETRUNKEVITCH, herein (*ex* Archaeoctonidae PETRUNKEVITCH, 1949)]

Superfamily CRIOCERATITACEAE Hyatt, 1900

[*nom. transl.* WRIGHT, 1952 (*ex* Crioceratitidae HYATT, 1900)]

FAMILY-GROUP NAMES; USE OF "*NOM. CORRECT.*"

Valid emendations classed as *nomina correcta* do not depend on transfer from one category of family-group units to another but most commonly involve correction of the stem of the nominate genus; in addition, they include somewhat arbitrarily chosen modification of ending for names of tribe or superfamily. Examples of the use of "*nom. correct.*" are the following.

Family STREPTELASMATIDAE Nicholson, 1889

[*nom. correct.* WEDEKIND, 1927 (*ex* Streptelasmidae NICHOLSON, 1889, *nom. imperf.*)]

Family PALAEOSCORPIIDAE Lehmann, 1944

[*nom. correct.* PETRUNKEVITCH, herein (*ex* Palaeoscorpionidae LEHMANN, 1944, *nom. imperf.*)]

Family AGLASPIDIDAE Miller, 1877

[*nom. correct.* STØRMER, herein (*ex* Aglaspidae MILLER, 1877, *nom. imperf.*)]

Superfamily AGARICIICAE Gray, 1847

[*nom. correct.* WELLS, herein (*ex* Agaricioidae VAUGHAN & WELLS, 1943, *nom. transl. ex* Agariciidae GRAY, 1847)]

FAMILY-GROUP NAMES; USE OF "*NOM. CONSERV.*"

It may happen that long-used family-group names are invalid under strict application of the Rules. In order to retain the otherwise invalid name, appeal to ICZN is needful. Examples of use of *nom. conserv.* in this connection, as cited in the *Treatise*, are the following.

Family ARIETITIDAE Hyatt, 1874

[*nom. correct.* HAUG, 1885 (*pro* Arietidae HYATT, 1875), *nom. conserv.* proposed ARKELL, 1955 (ICZN pend.)]

Family STEPHANOCERATIDAE Neumayr, 1875

[*nom. correct.* FISCHER, 1882 (*pro* Stephanoceratinen NEUMAYR, 1875, invalid vernacular name), *nom conserv.* proposed ARKELL, 1955 (ICZN pend.)]

FAMILY-GROUP NAMES; REPLACEMENTS

Family-group names are formed by adding letter combinations (prescribed for family and subfamily but not now for others) to the stem of the name belonging to genus (nominate genus) first chosen as type of the assemblage. The type genus need not be the oldest in terms of receiving its name and definition, but it must be the first-published as name-giver to a family-group taxon among all those included. Once fixed, the family-group name remains tied to the nominate genus even if its name is changed by reason of status as a junior homonym or junior synonym, either objective or subjective. According to the Copenhagen Decisions, the family-group name requires replacement only in the event that the nominate genus is found to be a junior homonym, and then a substitute family-group name is accepted if it is formed from the oldest available substitute name for the nominate genus. Authorship and date attributed to the replacement family-group name are determined by first publication of the changed family-group name.

The aim of family-group nomenclature is greatest possible stability and uniformity, just as in case of other zoological names. Experience indicates the wisdom of sustaining family-group names based on junior subjective synonyms if they have priority of publication, for opinions of different workers as to the synonymy of generic names founded on different type species may not agree and opinions of the same worker may alter from time to time. The retention similarly of first-published family-group names which are found to be based on junior objective synonyms is less clearly desirable, especially if a replacement name derived from the senior objective synonym has been recognized very long and widely. To displace a much-used family-group name based on the senior objective synonym by disinterring a forgotten and virtually unused family-group name based on a junior objective synonym because the latter happens to have priority of publication is unsettling. Conversely, a long-used family-group name founded on a junior objective synonym and having priority of publication is better continued in nomenclature than a replacement name based on the senior objective synonym. The Copenhagen Decisions (paragraph 45) take account of these considerations by providing a relatively simple procedure for fixing the desired choice in stabilizing family-group names. In conformance with this, the *Treatise* assigns to contributing authors responsibility for adopting provisions of the Copenhagen Decisions.

Replacement of a family-group name may be needed if the former nominate genus is transferred to another family-group. Then the first-published name-giver of a family-group assemblage in the remnant taxon is to be recognized in forming a replacement name.

FAMILY-GROUP NAMES; AUTHORSHIP AND DATE

All family-group taxa having names based on the same type genus are attributed to the author who first published the name for any of these assemblages, whether tribe, subfamily, or family (superfamily being almost inevitably a later-conceived taxon). Accordingly, if a family is divided into subfamilies or a subfamily into tribes, the name of no such subfamily or tribe can antedate the family name. Also, every family containing differentiated subfamilies must have a nominate *(sensu stricto)* subfamily, which is based on the same type genus as that for the family, and the author and date set down for the nominate subfamily invariably are identical with those of the family, without reference to whether the author of the family or some subsequent author introduced subdivisions.

Changes in the form of family-group names of the sort constituting *nomina correcta,* as previously discussed, do not affect authorship and date of the taxon concerned, but in publications such as the *Treatise* it is desirable to record the authorship and date of the correction.

ORDER/CLASS-GROUP NAMES; USE OF *"NOM. CORRECT."*

Because no stipulation concerning the form of order/class-group names is given yet by the Rules, emendation of all such names actually consists of arbitrarily devised changes in the form of endings. Nothing precludes substitution of a new name for an old one, but a change of this sort is not considered to be an emendation. Examples of the use of *"nom. correct."* as applied to order/class-group names are the following.

Order DISPARIDA Moore & Laudon, 1943

[*nom. correct.* MOORE, 1952 (*ex* Disparata MOORE & LAUDON, 1943)]

Suborder FAVIINA Vaughan & Wells, 1943

[*nom. correct.* WELLS, herein (*ex* Faviida VAUGHAN & WELLS, 1943)]

Suborder FUNGIINA Verrill, 1865

[*nom. correct.* WELLS, herein (*ex* Fungiida DUNCAN, 1884, *ex* Fungacea VERRILL, 1865)]

TAXONOMIC EMENDATION

Emendation has two measurably distinct aspects as regards zoological nomenclature. These embrace (1) alteration of a name itself in various ways for various reasons, as has been reviewed, and (2) alteration of taxonomic scope or concept in application of a given zoological name, whatever its hierarchical rank. The latter type of emendation primarily concerns classification and inherently is not associated with change of name, whereas the other type introduces change of name without necessary expansion, restriction, or other modification in applying the name. Little attention generally has been paid to this distinction in spite of its significance.

Most zoologists, including paleozoologists, who have signified emendation of zoological names refer to what they consider a material change in application of the name such as may be expressed by an importantly altered diagnosis of the assemblage covered by the name. The abbreviation *"emend."* then may accompany the name, with statement of the author and date of the emendation. On the other hand, a multitude of workers concerned with systematic zoology think that publication of *"emend."* with a zoological name is valueless because more or less alteration of taxonomic sort is introduced whenever a subspecies, species, genus, or other assemblage of animals is incorporated under or removed from the coverage of a given zoological name. Inevitably associated with such classificatory expansions and restrictions is some degree of emendation affecting diagnosis. Granting this, still it is true that now and then somewhat radical revisions are put forward, generally with published statement of reasons for changing the application of a name. To erect a signpost at such points of most significant change is worth while, both as aid to subsequent workers in taking account of the altered nomenclatural usage and as indication that not-to-be-overlooked discussion may be found at a particular place in the literature. Authors of contributions to the *Treatise* are encouraged to include records of all specially noteworthy emendations of this nature, using the abbreviation "*emend.*" with the name to which it refers and citing the author and date of the emendation.

In Part G (Bryozoa) and Part D (Protista 3) of the *Treatise,* the abbreviation "*emend.*" is employed to record various sorts of name emendations, thus conflicting with usage of "*emend.*" for change in taxonomic application of a name without alteration of the name itself. This is objectionable. In Part E (Archaeocyatha, Porifera) and later-issued divisions of the *Treatise,* use of "*emend.*" is restricted to its customary sense, that is, significant alteration in taxonomic scope of a name such as calls for noteworthy modifications of a diagnosis. Other means of designating emendations that relate to form of a name are introduced.

STYLE IN GENERIC DESCRIPTIONS

DEFINITION OF NAMES

Most generic names are distinct from all others and are indicated without ambiguity by citing their originally published spelling accompanied by name of the author and date of first publication. If the same generic name has been applied to 2 or more distinct taxonomic units, however, it is necessary to differentiate such homonyms, and this calls for distinction between junior homonyms and senior homonyms. Because a junior homonym is invalid, it must be replaced by some other name. For example, *Callopora* HALL, 1851, introduced for Paleozoic trepostome bryozoans, is invalid because GRAY in 1848 published the same name for Cretaceous-to-Recent cheilostome bryozoans, and BASSLER in 1911 introduced the new name *Hallopora* to replace HALL's homonym. The *Treatise* style of entry is:

Hallopora BASSLER, 1911 [*pro Callopora* HALL, 1851 (*non* GRAY, 1848)].

In like manner, a needed replacement generic name may be introduced in the *Treatise* (even though first publication of generic names otherwise in this work is avoided). The requirement that an exact bibliographic reference must be given for the replaced name commonly can be met in the *Treatise* by citing a publication recorded in the list of references, using its assigned index number, as shown in the following example.

Mysterium DELAUBENFELS, *nom. subst.* [*pro Mystrium* SCHRAMMEN, 1936 (ref. 40, p. 60) (*non* ROGER, 1862)] [**Mystrium porosum* SCHRAMMEN, 1936].

For some replaced homonyms, a footnote reference to the literature is necessary. A senior homonym is valid, and in so far as the *Treatise* is concerned, such names are handled according to whether the junior homonym belongs to the same major taxonomic division (class or phylum) as the senior homonym or to some other; in the former instance, the author and date of the junior homonym are cited as:

Diplophyllum HALL, 1851 [*non* SOSHKINA, 1939] [**D. caespitosum*].

Otherwise, no mention of the existence of a junior homonym is made.

CITATION OF TYPE SPECIES

The name of the type species of each genus and subgenus is given next following the generic name with its accompanying author and date, or after entries needed for definition of the name if it is involved in homonymy. The originally published combination of generic and trivial names for this species is cited, accompanied by an asterisk (*), with notation of the author and date of original publication. An exception in this procedure is made, however, if the species was first published in the same paper and by the same author as that containing definition of the genus which it serves as type; in such case, the initial letter of the generic name followed by the trivial name is given without repeating the name of the author and date, for this saves needed space. Examples of these 2 sorts of citations are as follows:

Diplotrypa NICHOLSON, 1879 [**Favosites petropolitanus* PANDER, 1830].

Chainodictyon FOERSTE, 1887 [**C. laxum*].

If the cited type species is a junior synonym of some other species, the name of this latter also is given, as follows:

Acervularia SCHWEIGGER, 1819 [**A. baltica* (=**Madrepora ananas* LINNÉ, 1758)].

It is judged desirable to record the manner of establishing the type species, whether by original designation or by subsequent designation, but various modes of original designation are not distinguished.

Original designation of type species. The Rules provide that the type species of a genus or subgenus may be recognized as an original designation if only a single species was assigned to the genus at the time of first publication (monotypy), if the author of a generic name employed this same name for one of the included species (tautonymy), if one of the species was named *"typus," "typicus,"* or the like, if the original author explicitly indicated the species chosen as the type, or if some other stipulations were met. According to convention adopted in the *Treatise,* the absence of any indication as to manner of fixing the type species is to be understood as signifying that it is established by original designation, the particular mode of original designation not being specified.

Subsequent designation of type species; use of "SD" and "SM." The type species of many genera are not determinable from the publication in which the generic name was introduced and therefore such genera can acquire a type species only by some manner of subsequent designation. Most commonly this is established by publishing a statement naming as type species one of the species originally included in the genus, and in the *Treatise* fixation of the type species in this manner is indicated by the letters "SD" accompanied by the name of the subsequent author (who may be the same person as the original author) and the date of publishing the subsequent designation. Some genera, as first described and named, included no mentioned species and these necessarily lack a type species until a date subsequent to that of the original publication when one or more species are assigned to such a genus. If only a single species is thus assigned, it automatically becomes the type species and in the *Treatise* this subsequent monotypy is indicated by the letters "SM." Of course, the first publication containing assignment of species to the genus which originally lacked any included species is the one concerned in fixation of the type species, and if this named 2 or more species as belonging to the genus but did not designate a type species, then a later "SD" designation is necessary. Examples of the use of "SD" and "SM" as employed in the *Treatise* follow.

Hexagonaria GÜRICH, 1896 [**Cyathophyllum hexagonum* GOLDFUSS, 1826; SD LANG, SMITH & THOMAS, 1940].

Muriceides STUDER, 1887 [**M. fragilis* WRIGHT & STUDER, 1889; SM WRIGHT & STUDER, 1889].

SYNONYMS

Citation of synonyms is given next following record of the type species and if 2 or more synonyms of differing date are recognized, these are arranged in chronological order. Objective synonyms are indicated by accompanying designation "(obj.)," others being understood to constitute subjective synonyms. Examples showing *Treatise* style in listing synonyms follow.

Calapoecia BILLINGS, 1865 [**C. anticostiensis;* SD LINDSTRÖM, 1833] [=*Columnopora* NICHOLSON, 1874; *Houghtonia* ROMINGER, 1876].

Staurocyclia HAECKEL, 1882 [**S. cruciata* HAECKEL, 1887] [=*Coccostaurus* HAECKEL, 1882 (obj.); *Phacostaurus* HAECKEL, 1887 (obj.)].

A synonym which also constitutes a homonym is recorded as follows:

Lyopora NICHOLSON & ETHERIDGE, 1878 [**Palaeopora? favosa* M'COY, 1850] [=*Liopora* LANG, SMITH & THOMAS, 1940 (*non* GIRTY, 1915)].

Some junior synonyms of either objective or subjective sort may take precedence desirably over senior synonyms wherever uniformity and continuity of nomenclature are served by retaining a widely used but technically rejectable name for a generic assemblage. This requires action of ICZN using its plenary powers to set aside the unwanted name and validate the wanted one, with placement of the concerned names on appropriate official lists. In the *Treatise* citation of such a conserved generic name is given in the manner shown by the following example.

Tetragraptus SALTER, 1863 [*nom. correct.* HALL, 1865 (*pro Tetragrapsus* SALTER, 1863), *nom. conserv.* proposed BULMAN, 1955, ICZN pend.] [**Fucoides serra* BRONGNIART, 1828 (=*Graptolithus bryonoides* HALL, 1858].

ABBREVIATIONS

Some authors' names and most stratigraphic and geographic names are abbreviated in order to save space. General principles for guidance in determining what names should be abbreviated are frequency of repetition, length of name, and avoidance of ambiguity. Abbreviations used in this division of the *Treatise* are explained in the following alphabetically arranged list.

Abbreviations

Abhandl., Abhandlungen
Abstamm., Abstammung
Abt., Abteilung, -en
Acad., Academia, Académie, Academy
Accad., Accademia
Advanc., Advancement
Afhandl., Afhandling, -ar
Afr., Africa, -n
Agric., Agriculture, -al
Akad., Akademie
Ala., Alabama
Alb., Albian
Algonk., Algonkian
Am., America, -n
Anis., Anisian
Ann., Annual, Annals
Antarct., Antarctica
Anz., Anzeiger
Apt., Aptian
Arch., Archiv, -es
Arct., Arctic
Arg., Argentina
Ariz., Arizona
Ark., Arkansas
Årsskr., Årsskrift
art., article, -s
Artinsk., Artinskian
Ashgill., Ashgillian
AsiaM., Asia Minor
Assoc., Association
Atl., Atlantic
Aus., Austria
Austral., Australia
Auvers., Auversian
Avd., Avdelning, -ar
Bd., Band, Bände
Baj., Bajocian
Balcomb., Balcombian
Balt., Baltic
Barr., BARRANDE, JOACHIM
Bath., Bathonian
Bayer., Bayerischen
B.C., British Columbia
Beförd., Beförderung
Beil., Beilage
Beitr., Beitrag, Beiträge
Belg., Belgium, -que
Bih., Bihang
Bill., BILLINGS, ELKANAH
Biol., Biological
Blainv., BLAINVILLE, H. M. D. DE
Bodenforsch., Bodenforschung, -en
Boh., Bohemia
Böhm., Böhmischen
Bol., Bolivia
Bol., Boletím, -ín
Boll., Bollettino
Braz., Brazil
Br.I., British Isles
Bull., Bulletin, -s
C., Central
Calif., California
Cam., Cambrian
Camp., Campanian
Can., Canada
Canad., Canadian
Carb., Carboniferous
Carib., Caribbean
Carn., Carnian
Carpath., Carpathian Mountains
Cenom., Cenomanian
Centralbl., Centralblatt
Chamisso-E., CHAMISSO & EYSENHARDT
Chazy., Chazyan
Chester., Chesteran
circ., circular
Cl., Classe
Claiborn., Claibornian
Clinton., Clintonian
Coll., Collection, -s; College
Com., Comité
Comm., Committee
Comp., Comparative
Congr., Congress
Conn., Connecticut
Contr., Contribution, -s
Coral., Corallian
cosmop., cosmopolitan
Couvin., Couvinian
Cret., Cretaceous
Czech., Czechoslovakia
d., das, der, die
Dan., Danian
Danm., Danmarks
Defr., DEFRANCE, M. J. L.
Denkschr., Denkschriften
Denm., Denmark
deutsch., deutschen
Dev., Devonian
Div., Division
Dobr., DOBROLYUBOVA, T. A.
Downton., Downtonian
Duch.-M., DUCHASSAING & MICHELOTTI
Dyb., DYBOWSKI, W. N.
E., East
Ehr., EHRENBERG, C. G.
Eichw., EICHWALD, C. E. VON
Ellesm., Ellesmereland
emend., emendation
Ems., Emsian
Eng., England
Eoc., Eocene
Esch., ESCHCHOLTZ
espan., español, -a
Est., Estonia
Étal., ÉTALLON, A.
etc., et cetera
Eth., ETHERIDGE, ROBERT, JR.
Eth.-F., ETHERIDGE, ROBERT, JR., & FOORD, A. H.
Eu., Europe
Exped., Expedition
f., för, für
fasc., fascicle
Fenton-F., FENTON, C. L., & FENTON, M. A.
fig., figure, -s
fis., fisiche
Fla., Florida
Foren., Forening
Fr., France
Frasn., Frasnian
From., FROMENTEL, E. DE
Ga., Georgia
Gargas., Gargasian
G.Brit., Great Britain
Gegen., GEGENBAUER
geogn., geognostische
Geol. (geol.), Geological, -ischen, -iska, -y
Ger., Germany
Gesell., Gesellschaft
Givet., Givetian
Goldf., GOLDFUSS, G. A.
Gotl., Gotland
Gshel., Gshelian

Hall-C., HALL, JAMES, & CLARKE, J. M.
Hall-W., HALL, JAMES, & WHITFIELD, R. P.
Handb., Handbuch
Handl., Handlingar
Helvet., Helvetiae
Hett., Hettangian
Hist., History, -ia
Hkl., HAECKEL, E.
Hung., Hungary
I., Island, -s
ICZN., International Commission on Zoological Nomenclature
Ill., Illinois
Imp., Imperial
Ind., Indiana
Ind.O., Indian Ocean
IndoPac., Indo-Pacific
indukt., induktive
Inst., Institut, Institute, Institution, Instituto, Instituut, Instituciò, Institutet
Inv., Investigations
Ire., Ireland
ital., italiana
Jaarb., Jaarboek
Jackson., Jacksonian
Jahrb., Jahrbuch
Jahresber., Jahresbericht
Jahresh., Jahreshefte
Jahrg., Jahrgang
Jones-Hill, JONES, O. A., & HILL, DOROTHY
Jour., Journal
Jur., Jurassic
k., kaiserlich
k.k., kaiserlich königlich
Kans., Kansas
Keys., KEYSERLING, A.
kgl., koeniglich
Kimm., Kimmeridgian
Kinderhook., Kinderhookian
Kl., Klasse
Kon., KONINCK, L. G. DE
Kük., KÜKENTHAL, WILLY
Ky., Kentucky
L., Land, Lower
Lab., Laboratory, -ies
Ladin., Ladinian
Lamx., LAMOUROUX, J. V. F.
Lancefield., Lancefieldian
Landesanst., Landesanstalt
Landesmus., Landesmuseum
Lang-S., LANG, W. D., & SMITH, STANLEY
Lang-S.-T., LANG, W. D., SMITH, STANLEY, & THOMAS, H. D.
lat., lateral
Leuck., LEUCKART
Lief., Lieferung, -en
Lind., LINDSTRÖM, G.
Linn., Linnean
Lit., Literary
livr., livraison
Llandeil., Llandeilian
Lockport., Lockportian
Lond., Londinian
long., longitudinal
Lonsd., LONSDALE, WILLIAM
Ludlov., Ludlovian
Lusit., Lusitanian
Lut., Lutetian
M., Middle
m., meter, -s
mm., millimeter, -s
Maastr., Maastrichtian
MacGill., MACGILLIVRAY, P. H.
Mag., Magazine
Mähr., Mährisch.
MalayArch., Malay Archipelago
Manch., Manchuria
Mass., Massachusetts
Mat. (mat.), matimatiche, Matematik
math., mathematische
Md., Maryland
M.Edw., MILNE-EDWARDS, HENRI
M.Edw.-H., MILNE-EDWARDS, HENRI, & HAIME, JULES
Meddel., Meddelelser
Meded., Mededeelingen
Medit., Mediterranean
Meek-W., MEEK, F. B., & WORTHEN, A. H.
Mem., Memoir, -ia, -s
Mém., Mémoire, -s
Mich., MICHELIN, J. L. H.
Mich., Michigan
Micr., Microscopical
Midl., Midland
Midway., Midwayan
mijnw., mijnwezen
Miller-G., MILLER, S. A., & GURLEY, W. F. E.
Min., Mineralogie
Minn., Minnesota
Mio., Miocene
Misc., Miscellaneous
Miss., Mississippi, -an
Mitt., Mitteilungen
Mo., Missouri
Mohawk., Mohawkian
Mon., Monograph
Mont.Gall., MONTANARO-GALLITELLI, EUGENIA
Moore-J., MOORE, R. C., & JEFFORDS, R. M.
Mus., Musée, Museo, Museum
N., North
Namur., Namurian
nat., natural, -i, naturelle, -s
Natl., National
naturf., naturforschende
Naturgesch., Naturgeschichte
Naturh., Naturhistorie
Naturv., Naturvetenskap
naturwiss., naturwissenschaftlich
N.Car., North Carolina
Nebr., Nebraska
Nederl., Nederlandsch, -e
Neocom., Neocomian
Nev., Nevada
Nich., NICHOLSON, H. A.
Nich.-E., NICHOLSON, H. A., & ETHERIDGE, ROBERT, JR.
Nich.-L., NICHOLSON, H. A., & LYDEKKER, R.
Nich.-T., NICHOLSON, H. A., & THOMSON, J.
N.J., New Jersey
no., number, -s, numéro, -s, número, -s
Nor., Norian
Norf., Norfolk
N.S.W., New South Wales
N.Y., New York
N. Z., New Zealand
N.Zem., Nova Zembla
oberrhein., oberrheinischen
obj., objective
Obshch., Obshchestva
Ok., OKULITCH, V. J.
Okla., Oklahoma
Oligo., Oligocene
Onond., Onondagan
Orb., D'ORBIGNY, A. D.
Ord., Ordovician
p., page, -s
Pa., Pennsylvania
Pac., Pacific
Paläont., Paläontologia, -ie, -ische
Palaeontogr., Palaeontographica, -al
Paleont., Paleontologia, -ica, -ical, -y
Paleontogr., Paleontographica
pend., pending
Penn., Pennsylvanian
Perm., Permian
Petersb., Petersburg
Philip., Philippines
Philos., Philosophical
pl., plate, -s
Pleisto., Pleistocene
Plio., Pliocene
Pol., Poland
Pont., Pontian
Port., Portuguese, Portugal
Portl., Portlandian
P.R., Porto Rico
Precam., Precambrian
preuss., preussische
Proc., Proceedings
Prof., Professional
pt., part, -s
Pub., Publication
Quart., Quarterly
Que., Quebec
Queensl., Queensland
Quenst., QUENSTEDT, F. A.
Quoy-G., QUOY, J. R. C., & GAIMARD, J. P.
Raf., RAFINESQUE, C. S.
Raur., Rauracian
Rec., Recent
reconstr., reconstructed, -ion
Reichsanst., Reichsanstalt
Reichsmus., Reichsmuseum
Rend., Rendiconti, Rendus
Rept., Report, -s
Riv., Rivista
Rom., ROMINGER, CARL
Roy., Royal, -e

Rupel., Rupelian
Russ. (russ.), Russia, russisch
S., South, Sea
Sandb., SANDBERGER, GUIDO
Scand., Scandinavia, -n
S.Car., South Carolina
Schind., SCHINDEWOLF, O. H.
Schloth., SCHLOTHEIM, E. F.
Schr., Schrift, en
Sci., Science, Scienza, Scienze, Scientific
Scot., Scotland
sec., section, -s
Selsk., Selskabs
Senon., Senonian
Sequan., Sequanian
ser., series, serial
Serv., Serviço
sh., shale
Sib., Siberia
Sil., Silurian
Sinem., Sinemurian
Sitzungsber., Sitzungsberichte
Skr., Skrifter
Smith-T., SMITH, STANLEY, & TREMBERTH, R.
Soc., Society, Société
Somali., Somaliland
Soshk., SOSHKINA, E. D.
Soshk.-D.-P., SOSHKINA, E. D., DOBROLYUBOVA, T. A., & PORFIRIEV, V.
sp., species
Spec., Special
Spitz., Spitzbergen
Sta., Station
Steinm., STEINMANN, G.
Stuck., STUCKENBERG, A.
Supp., Supplement
Sver., Sveriges
Swed., Sweden, Swedish
Switz., Switzerland
tang., tangential
Tasm., Tasmania
Term.-T. TERMIER, HENRI, & TERMIER, GENEVIÈVE
Tert., Tertiary
Tex., Texas
Thomson-N., THOMSON, J., & NICHOLSON, H. A.
Thuring., Thuringian
Tithon., Tithonian
Tolm., TOLMACHEV, I. P.
Tournais., Tournaisian
Trans., Transactions
transv., transverse
Traut., TRAUTSCHOLD, H.
Trav., Travaux
Trias., Triassic
trop., tropical, tropics
Turon., Turonian
T.Woods., TENISON-WOODS, J. E.
U. Upper
u., und
Ulr., ULRICH, E. O.
U.S., United States
USA, United States (America)
Undersøg., Undersøgelse
Undersök., Undersökning
Univ., Université, University, Universitets, Universidade, Università
Urgon., Urgonian
USSR, Union of Soviet Socialist Republics
v., volume, -s
Va., Virginia
Valang., Valanginian
Ver., Verein, -s
Vererb., Vererbungslehre
Verh., Verhandlung, -en, Verhandelingen
Vict., Victoria
Vidensk., Videnskabs
Vindob., Vindobonian
Visé., Viséan
Vt., Vermont
W., West
Waag.-W., WAAGEN, WILLIAM, & WENTZEL, JOSEPH
Walc., WALCOTT, C. D.
Wdkd., WEDEKIND, R.
Wenlock., Wenlockian
Wetensch., Wetenschappen
White-W., WHITE, C. A., & WHITFIELD, R. P.
Wilcox., Wilcoxian
Winch., WINCHELL, A.
Winch.-S., WINCHELL, N. H., & SCHUCHERT, CHARLES
Wiss. (wiss.), Wissenschaften, -liche
Wright-S., WRIGHT, E. P., & STUDER, T.
Württ., Württemberg
Wyo., Wyoming
Yabe-H., YABE, H., & HAYASAKA, I.
Yabe-S., YABE, H., & SUGIYAMA, T.
Yorks., Yorkshire
Zap., Zapiski
Zeitschr., Zeitschrift
Zentralbl., Zentralblatt
Zool., Zoology, Zoologia, Zoological, Zoologiska, Zoologi, Zoologie, Zoologische

REFERENCES TO LITERATURE

Each part of the *Treatise* is accompanied by a selected list of references to paleontological literature consisting primarily of recent and comprehensive monographs available but also including some older works recognized as outstanding in importance. The purpose of giving these references is to aid users of the *Treatise* in finding detailed descriptions and illustrations of morphological features of fossil groups, discussions of classifications and distribution, and especially citations of more or less voluminous literature. Generally speaking, publications listed in the *Treatise* are not original sources of information concerning taxonomic units of various rank but they tell the student where he may find them; otherwise it is necessary to turn to such aids as the *Zoological Record* or NEAVE's *Nomenclator Zoologicus*. References given in the *Treatise* are arranged alphabetically by authors and accompanied by index numbers which serve the purpose of permitting citation most concisely in various parts of the text; these citations of listed papers are inclosed invariably in parentheses and are distinguishable from dates because the index numbers comprise no more than 3 digits. Ordinarily, index numbers for literature references are given at the end of generic or family diagnoses.

SOURCES OF ILLUSTRATIONS

At the end of figure captions an index number is given to supply record of the author of illustrations used in the *Treatise*, reference being made to an alphabetically arranged list of authors' names which follows. The names of authors, but generally not individual publications, are cited. Illustrations consisting of exact copies of previously published figures (except for possible change of scale) are distinguished by the use of an asterisk (*) with the index number, and previously unpublished illustrations are marked by the letter "n" (signifying "new") with the index number; all other indications of the sources of illustrations are construed to mean "after" the cited author or authors, that is, embodying some degree of change. Addition of the abbreviation "mod." denotes appreciable alteration of the source figure.

RAYMOND C. MOORE

PART F

COELENTERATA

By Frederick M. Bayer, Hilbrand Boschma, Horacio J. Harrington, Dorothy Hill, Libbie H. Hyman, Marius Lecompte, Eugenia Montanaro-Gallitelli, Raymond C. Moore, Erwin C. Stumm, and John W. Wells

CONTENTS

INTRODUCTION

By Raymond C. Moore

As defined in this volume, the phylum Coelenterata comprises all of the so-called diploblastic lower invertebrates belonging to the Eumetazoa, that is, the numerically and structurally disparate groups named Cnidaria and Ctenophora. Excluded are the Porifera or sponges, which are classified as structurally distinct from Eumetazoa and designated as Parazoa. The coelenterates differ from other Eumetazoa, all of which have bilateral symmetry, in being characterized by primary radial or biradial symmetry derived from the larval stage and retained throughout ontogeny. Thus, the cnidarians and ctenophores taken together are definable as radiate forms or Radiata (Cuvier, 1816; *non* Lamarck, 1809), whereas remaining eumetazoans are the bilateral forms or Bilateria. Another distinction is that among the Radiata (or Coelenterata) no body cavities exist between the outer cell layer (ectoderm, epidermis) and inner cell layer (endoderm, gastrodermis), for these layers are in contact or connected by more or less structureless substance of jelly-like nature termed mesogloea; in most Bilateria, on the other hand, such spaces are developed and on account of this character they are often termed Coelomata.

The Cnidaria are distinguished mainly by possession of stinging capsules (nematocysts) and are characterized by ability to move their body parts muscularly. The Ctenophora lack nematocysts and retain the ciliary locomotion of the larval stage (planula) but are relatively advanced in having more definite organization of the digestive system and in development of an aboral sensory region. The common features and differences of the 2 groups are appropriately recognized by classifying them as subphyla of the phylum Coelenterata.

The contributors to this Part of the *Treatise* are not to be understood as committed individually to features of classification which have been adopted for the volume as a whole; this is my responsibility; but each author is granted the privilege of expressing his own views, and his conformance to the general plan of systematic organization of the text does not imply full endorsement necessarily. Some features of arrangement reflect effort on the part of the editor to make clearly and readily distinguishable the authorship of respective subdivisions, several of which do not coincide with ideal major segments adapted for placement in sequence conforming to zoological classification.

OUTLINE OF CLASSIFICATION

A tabular summary of main divisions recognized in the phylum Coelenterata is desirable near the beginning of the volume and because a number of departures from customary classification are introduced, these call for some explanation.

First, decision to include ctenophores with other radiate Eumetazoa in a single phylum called Coelenterata and division of this assemblage into 2 subphyla named Cnidaria and Ctenophora is based on considerations given in preceding paragraphs.

Next, the conclusion that the polypoid and medusoid types of organization, instead of representing a more or less unexplainable "alternation of generations," constitute the product of evolutionary differentiation in which the polypoid form is a persistent early-growth form and the medusoid a normal adult type of coelenterate leads to interpretation of the medusoids as initial types of Coelenterata. This is consonant with the paleontological record, which includes numerous Lower Cambrian and even some Precambrian medusoid fossils. Consequently, the simplicity of hydroid forms is not accepted as basis for placing them in first position among various types of coelenterates. Precedence is assigned to early medusoids.

Main Divisions of the Coelenterata

Cnidaria *(subphylum). Precam.-Rec.*
- Protomedusae *(class). Precam.-Ord., ?Sil., ?Penn.*
  - Brooksellida *(order). Precam.-Ord., ?Sil., ?Penn.*
- Dipleurozoa *(class). L.Cam.*
  - Dickinsoniida *(order). L.Cam.*
- Scyphozoa *(class). Cam.-Rec.*
  - Scyphomedusae *(subclass). ?L.Cam., U.Jur.-Rec.*

Stauromedusida *(order)*. *Rec.*
Carybdeida. *?U.Jur., Rec.*
Coronatida. *?L.Cam., U.Jur.-Rec.*
Semaeostomatida. *?U.Jur., Rec.*
Lithorhizostomatida. *U.Jur.*
Rhizostomatida. *?U.Jur., Rec.*
?Scyphomedusae incertae sedis. *L.Cam.*
Conulata *(subclass)*. *Cam.-Trias.*
Conulariida *(order)*. *Cam.-Trias.*
Conchopeltina *(suborder)*. *M.Ord.*
Conulariina. *Cam.-Trias.*

Hydrozoa *(class)*. *Cam.-Rec.*
Trachylinida *(order)*. *?L.Cam., ?M.Jur., U.Jur.-Rec.*
Trachymedusina *(suborder)*. *?L.Cam., ?M.Jur., L.Cret.-Rec.*
Narcomedusina. *U.Jur.-Rec.*
Hydroida *(order)*. *Cam.-Rec.*
Gymnoblastina *(suborder)*. *?Trias.-?Cret., Eoc.-Rec.*
Calyptoblastina. *Cam.-Rec.*
Spongiomorphida. *Trias.-Jur.*

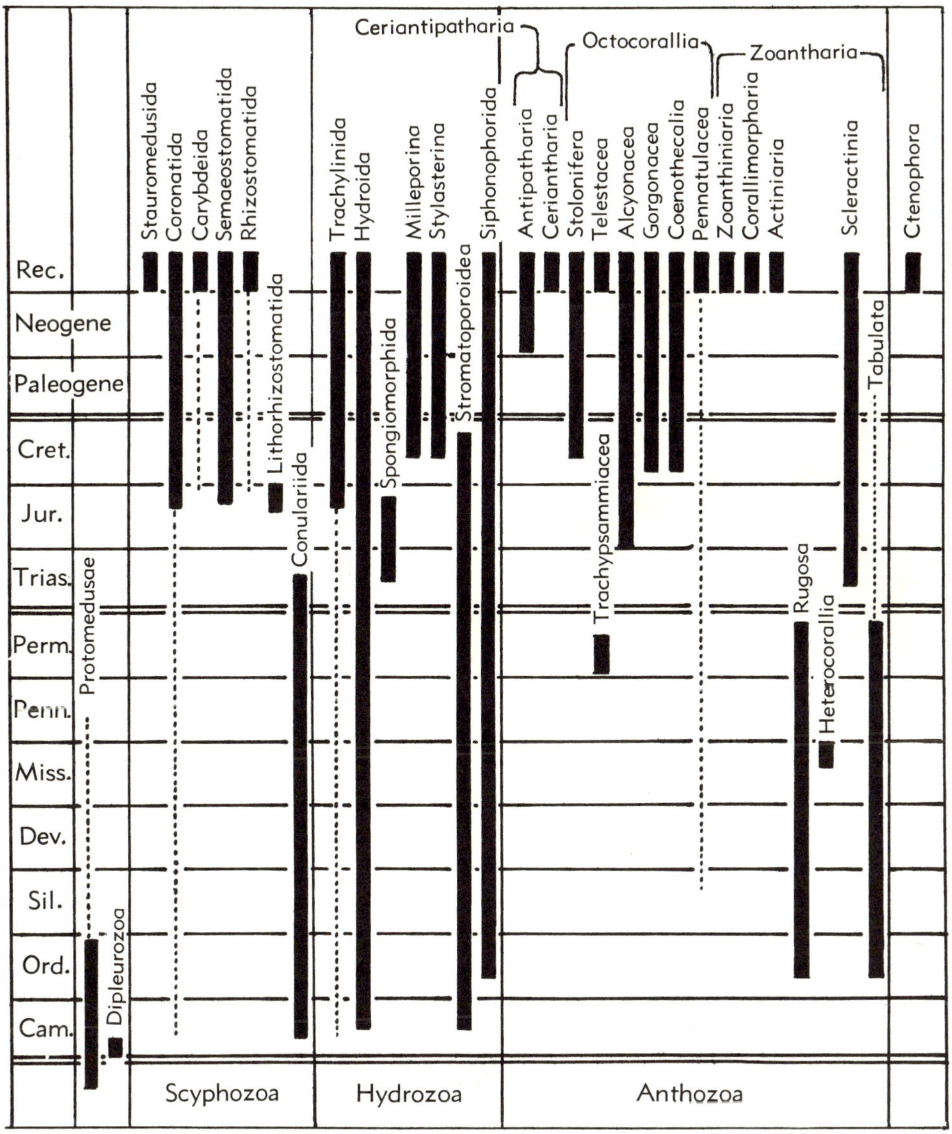

FIG. 1. Geologic distribution of main divisions of Coelenterata (MOORE, n).

Milleporina. *U.Cret.-Rec.*
Stylasterina. *U.Cret.-Rec.*
Stromatoporoidea. *Cam.-Cret.*
Siphonophorida. *Ord.-Rec.*
Calycophorina *(suborder). Rec.*
Physophorina. *Rec.*
Rhizophysaliina. *Rec.*
Chondrophorina. *Ord.-Rec.*
Anthozoa *(class). M.Ord.-Rec.*
Ceriantipatharia *(subclass). Mio.-Rec.*
Antipatharia *(order). Mio.-Rec.*
Ceriantharia. *Rec.*
Octocorallia *(subclass). ?Sil., Perm.-Rec.*
Stolonifera *(order). Cret.-Rec.*
Telestacea. *Rec.*
Alcyonacea. *L.Jur.-Rec.*
Trachypsammiacea. *Perm.*
Coenothecalia. *Cret.-Rec.*
Gorgonacea. *Cret.-Rec.*
Scleraxonia *(suborder). Cret.-Rec.*
Holaxonia. *Cret.-Rec.*
Pennatulacea. *?Sil., Rec.*
Sessiliflorae *(suborder). ?Sil., Rec.*
Subselliflorae. *Cret.-Rec.*
Incertae sedis. *Sil.-Cret.*
Zoantharia *(subclass). Ord.-Rec.*
Zoanthiniaria *(order). Rec.*
Corallimorpharia. *Rec.*
Actiniaria. *Rec.*
Rugosa. *Ord.-Perm.*
Streptelasmatina *(suborder). Ord.-Perm.*
Columnariina. *Ord.-Perm.*
Cystiphyllina. *Ord.-Dev.*
Heterocorallia *(order). Carb.*
Scleractinia. *M.Trias.-Rec.*
Astrocoeniina *(suborder). M.Trias.-Rec.*
Fungiina. *M.Trias.-Rec.*
Faviina. *M.Trias.-Rec.*
Caryophylliina. *Jur.-Rec.*
Dendrophylliina. *U.Cret.-Rec.*
Tabulata *(order). M.Ord.-Perm., ?Trias.-?Eoc.*
Ctenophora *(subphylum). Rec.*

Geological distribution of the main divisions of coelenterates is shown graphically in Figure 1.

ACKNOWLEDGMENTS

The completion of preparatory work needed for bringing the contents of this volume on Coelenterata to readiness for publication has depended on tireless labor by many individuals other than myself, and part of this at least is of such magnitude as to make the omission of stated acknowledgments unallowable. Normally there is no reason to express thanks to an author for materials published under his name, because the writing is at his initiative and, where financial remuneration is not involved (none going to *Treatise* authors), he receives intangible credit measured by the value of his publication. This does not apply to several segments of *Treatise* Part F on Coelenterata, and accordingly it is appropriate to thank DOROTHY HILL and JOHN W. WELLS for adding materially to work already done in producing their major chapters by acceptance of responsibility for organizing lesser units. Very special appreciation is expressed to HORACIO J. HARRINGTON because work done by him has served to avoid many months of delay in publication which without his help would have been certain. Until June, 1955, contributions by authors assigned to prepare medusoid coelenterates and conulariids, both arranged at the beginning of the *Treatise* project, were awaited, places for them being held open; finally, when it became evident that these could not be obtained, Dr. HARRINGTON agreed to aid me in substituting for these delinquent authors. His ability and energy have been combined to achieve accomplishment of the mission in incredibly short time. In this connection, it is appropriate to thank ROBERT VOSPER, Director of Libraries at the University of Kansas, and members of the library staff who helped obtain nearly 150 needed interlibrary loans of books sought simultaneously for studies of the literature on medusoids and conulariids. Also, appreciation may be expressed to Chancellor F. D. MURPHY and other administrative officers of the University of Kansas for approval of appointment of Dr. HARRINGTON as Research Professor assigned mainly to assisting me on the *Treatise*.

H. DIGHTON THOMAS, of the British Museum (Natural History), furnished very helpful criticism and advice pertaining to part of this volume.

Work on many illustrations contained in *Treatise* Part F, as in others, has been done by NANCY-LOU PATTERSON and JANE SHACKELFORD WHITMORE, artists, and ROBERT O. FAY, in charge of the photographic laboratory. Figures in the chapters by HILBRAND BOSCHMA and F. M. BAYER, however, were prepared by these authors or under their direction. K. E. CASTER very kindly furnished several photographs of Cambrian medusoids from Australia which had been obtained from R. G. SPRIGG, and J. W. WELLS supplied a photograph of the remarkable specimen of *Conchopeltis* which is reproduced in the chapter on Scyphozoa. Deserving acknowledgment and thanks for this and previously issued *Treatise* volumes are the skill and unusual co-operation of members of the University of Kansas Press, especially C. K. HYDER, editor; T. C. RYTHER, superintendent of printing; and J. J. VERBURG, foreman. Many other persons who are not named individually also have shared in making this book available.

CNIDARIA—GENERAL FEATURES

By Dorothy Hill and J. W. Wells

CONTENTS

To the ancients, outstanding features of the animals now collectively known as cnidarian coelenterates were their ability to sting and resemblance of their growth forms to those of plants, whence the names Acalephae, Cnidae, and later Zoophyta. Eighteenth-century workers noted their tentacles, whence were derived the terms "polyps" and Polypi (from their analogy with cuttlefish) or Radiata (from their evident radial symmetry). Early in the 19th century, their characteristic gastrovascular cavity **(coelenteron)** was seen to distinguish them from other radiate animals such as echinoderms, and hence the name Coelenterata was introduced. The Ctenophora, originally included in the Coelenterata, have been separated as a distinct phylum by many zoologists but here are classed as a subphylum.

NATURE OF CNIDARIA

The Cnidaria are free-swimming or sedentary animals, lacking definite head and centralized nervous system, and having fundamentally biradial structure, and a body wall of 2 layers of cells, an outer **ectoderm** and an inner **endoderm,** between which is a primarily structureless layer, the **mesogloea.** They are particularly characterized by **tentacles,** and stinging or adhesive bodies (**nematocysts,** Fig. 2). Within the body wall is a single cavity, the coelenteron, acting as a digestive tube with only a mouth opening, and complicated in some groups by partitions or canals. Digestion is partly intracellular; the nervous system consists of a diffuse network of cells (Fig. 3), and the muscular system is simple. In sexual reproduction the egg develops into an ovoid ciliated larva **(planula),** which usually becomes attached and grows into some type of polyp. Asexual reproduction is by various modes of budding. Cnidarians generally are regarded as representative of the primitive stock of metazoans characterized by acquired specializations.

The Cnidaria are remarkable for polymorphism, signifying the variety of form that a species may present. This includes 2 types, **polyp** and **medusa,** derivable one from the other. A polyp is sedentary, in shape an elongated cylinder attached to or resting on the substratum at the aboral end, with the mouth and tentacles at the free (oral) end. A medusa is a free-swimming, bowl-shaped animal with marginal tentacles, a very short oral-aboral axis, and great thickening of the mesogloea. Dominance of the types varies in the classes. Among Hydrozoa, both polypoid and medusoid forms may occur in the same species, the medusoid form preponderates in the Scyphomedusae, and the class Anthozoa is entirely polypoid (Fig. 4). Where both polypoid and medusoid forms occur in a species, the polyps reproduce only asexually, budding off medusae that reproduce sexually. The polyp may be regarded as a persistent larval stage and the medusa as the completely evolved cnidarian.

ORIGIN AND EVOLUTION OF CLASSES

The general evolutionary pattern of major groups of Coelenterata is speculative, for it is only hinted at in the structural series of living forms and has found little or no confirmation from the fossil record. The classical picture of the origin and evolution of the phylum assumes (1) that the coelenterates are the nearest living representatives of the ancestral stock of all other metazoan groups and (2) that of the classes, the Hydrozoa are most primitive structurally and arose first. The more distant origin of the phylum is thought to lie in a ciliated polarized free-swimming bottom-feeder provided with ectoderm and endoderm, largely because all coelenterate groups possess a planula larval stage having these attributes. This hypothetical protocnidarian has been termed metagastraea. Further developments followed one or the other of 2 theoretical lines.

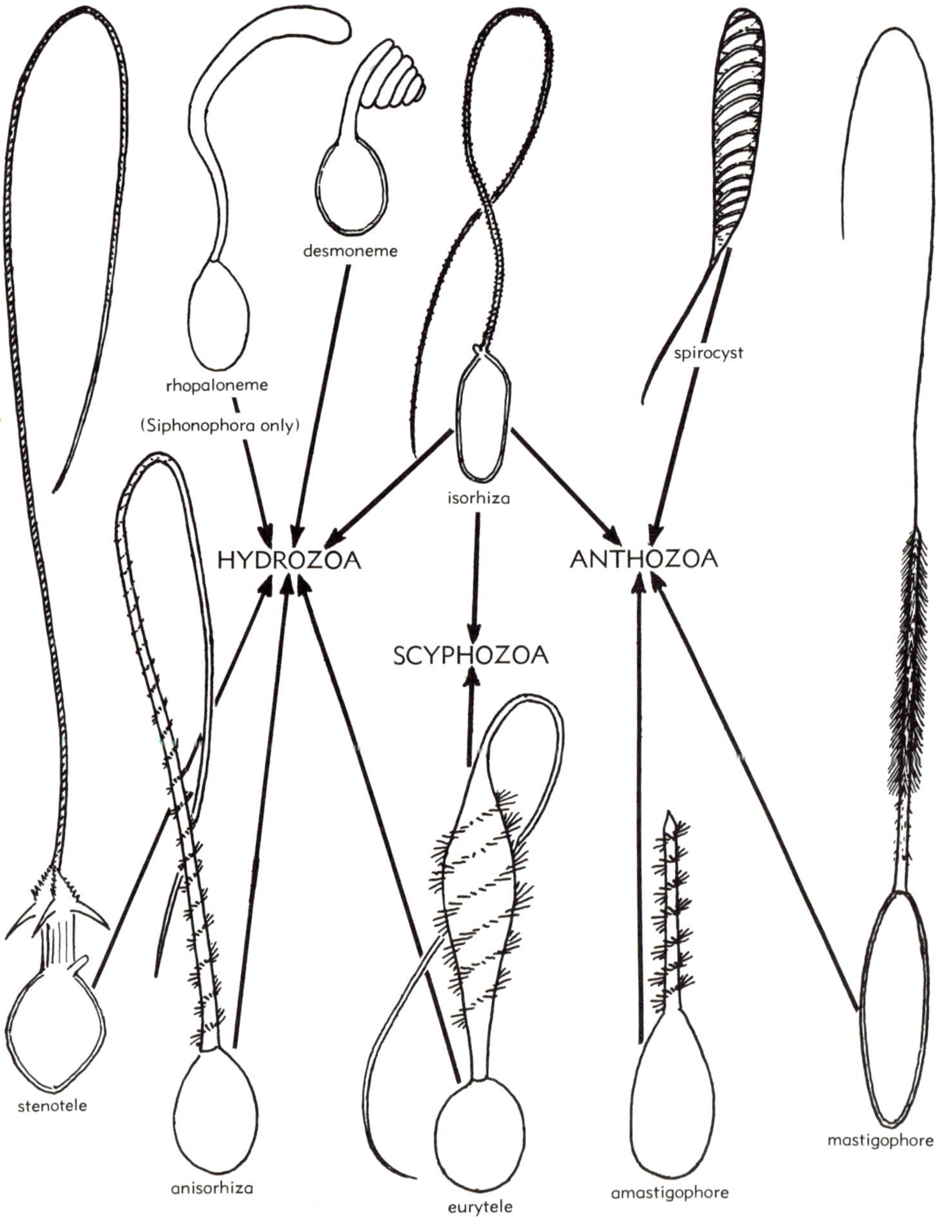

Fig. 2. Types of nematocysts and their occurrence in main groups of coelenterates (nematocysts much enlarged) (Hill & Wells, n).

One theory (Fig. 5*A*) postulates that attachment of the metagastraea at its aboral pole gave rise to a hydra-like polyp, from which hydroid colonies arose by asexual budding and that by development of polymorphism, some polyps became modified into medusae specialized for pelagic life and sexual reproduction.

Difficulties arise with this assumption, and many investigators consider more credible the hypothesis that ancestral coelenterates were primitive medusae derived from protocnidarians by development of tentacles and resembling the actinula larva of living Trachylinida (Fig. 5*C*). The hydrozoans could have arisen from such medusae if these first acquired the habit of asexual budding and then developed a prolonged attached stage as a polypoid colony that produced medusae by budding. This would account for the asexual nature of hydrozoan polypoids and the retention of a fully developed medusa stage, and it points to trachylines as the most primitive living coelenterates. The Scyphozoa (Scyphomedusae) could have originated from the same trachylinid actinula stage as the Hydrozoa by exploitation of the medusoid condition, in which the principal changes were develop-

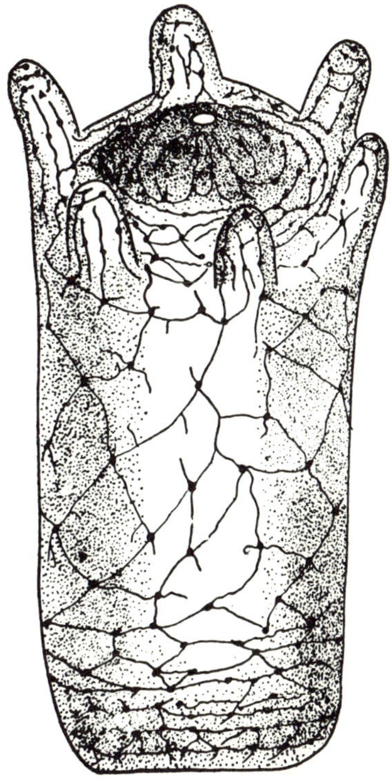

Fig. 3. Nervous system of *Hydra* (reticulate connection of nerve fibers doubtful) (Hadzi).

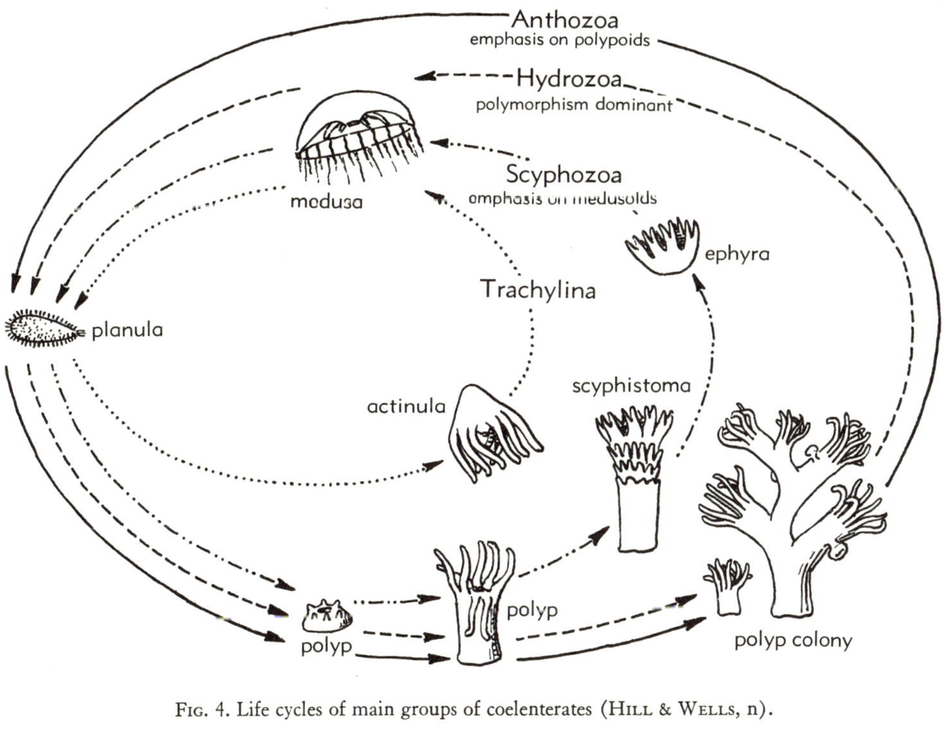

Fig. 4. Life cycles of main groups of coelenterates (Hill & Wells, n).

ment of mesenteries associated with increase in size. In another direction from the trachylinid stem, emphasis on the polypoid stage, with increase in size and development of mesenteries, could have given rise to the Anthozoa, in which all traces of the medusa stage are absent, and the polypoid stage, originally a larval stage, dominates.

A different evolutionary scheme has been suggested by HADZI, who would derive the coelenterates from the rhabdocoele flatworms, an order of the class Turbellaria, phylum Platyhelminthes, more or less by retrogression (Fig. 4*C*). Several workers have pointed out that there is no real difference in the structure of anemones and planarians as regards general body layers. On this plan, the anthozoans represent the primary group and first to evolve, the scyphozoans being derived from them, and hydrozoans constituting the most highly modified group. HADZI's hypothesis is open to the objection that if Anthozoa are considered to be the primary coelenterates, the derivation of medusae from them becomes unthinkable.

Unfortunately, although the paleontological record (Fig. 1) provides much information on evolution of some coelenterate orders

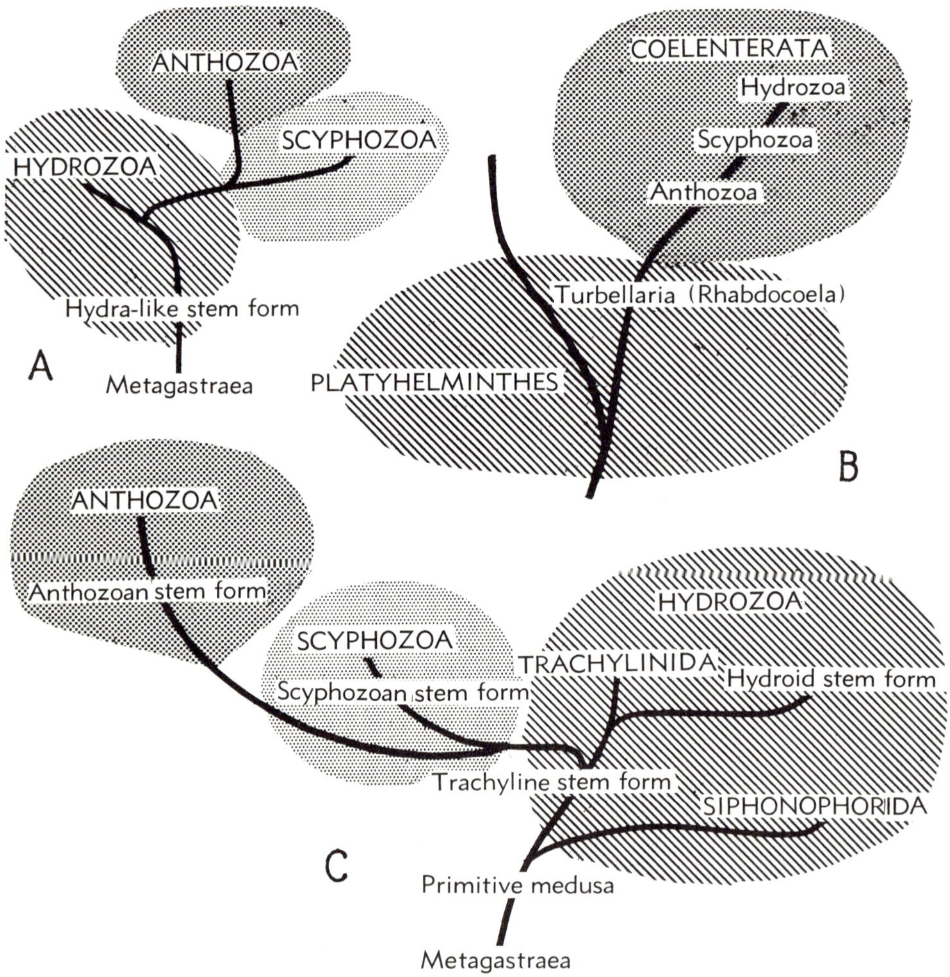

FIG. 5. Postulated phylogenetic relationship of main divisions of the coelenterates. *A,* Development through *Hydra*-like stem form with polymorphic differentiation of free-swimming medusae from attached polyps. *B,* Development through rhabdocoele flatworms, as postulated by HADZI. *C,* Development through primitive medusa and trachyline stem form (HILL & WELLS, n).

with skeletons, it provides no confirmation of any of these speculative lines of development of the great classes. As pointed out subsequently, separation of the classes of Cnidaria probably was completed in the unknown time before beginning of the Cambrian.

A skeleton, calcitic in some, aragonitic, horny or chitinophosphatic in others, has developed at different times in different coelenterate lines, each being a secondary development consequent on establishment of the polypoid stage and not necessarily indicating relationship. This sporadic appearance of a major structural type seems to characterize the coelenterates, among which it has occurred in about half of the main groups.

GEOLOGICAL IMPORTANCE

The living coelenterates are marine except for a few fresh-water hydroids, and fossils are known only from marine sediments. Although some soft-bodied forms are known from rare fossil impressions left in mud or sand, cnidarian fossils are dominantly the remains of groups with calcareous skeletons; many orders lacking skeletons are not known as fossils.

The oldest cnidarian recorded is a doubtful scyphomedusa from late Precambrian rocks of the Grand Canyon in Arizona; hydrozoans, scyphomedusae, and possibly anthozoans occur sporadically in Cambrian rocks; hydrozoan fossils, especially stromatoporoids, are more numerous in Paleozoic than in Mesozoic or Tertiary strata; Scyphomedusae are rare throughout; anthozoans became important in the Middle Ordovician, and have retained their importance into Recent time (Fig. 1).

Coelenterates live in all marine environments today, although, like most animals, they are more numerous on continental shelves than elsewhere. Forms with skeletons are sessile and benthonic but a few horny hydrozoans attach themselves to floating weed. Forms with calcareous skeletons seem on the whole to prefer a hard substratum, but forms with horny skeletons may attach themselves also on muddy bottoms. Calcareous anthozoans and hydrozoans in the past, as today, are the most important inhabitants of organic reef environments; many of them act and have acted as framework organisms; they are and have been more prolific in warm waters than in cold.

Phylum COELENTERATA Frey & Leuckart, 1847

[Originally included Porifera and Ctenophora] [=Zoophyta LINNÉ, 1758 (*partim*); Radiata LAMARCK, 1802 (*partim*); Actinozoa DE BLAINVILLE, 1831 (*partim*); Nematozoa BBONN, 1831; Nematophora HUXLEY, 1852 (*partim*), Coelentera LANKESTER, 1877; includes Cnidaria HAECKEL, 1881 (*non* Cnidaires M.EDW.-H., 1857); Cnidarea DELAGE & HÉROUARD, 1901; Enterocoela GRABAU, 1913 (*non* HUXLEY, 1875)]

Multicellular animals with biradial or radiobilateral symmetry; body wall with cells arranged in 2 layers (ectoderm and endoderm) connected by a structureless mesogloea containing cells chiefly of ectodermal origin; within the body wall is a single cavity (enteron) having a single opening (mouth) for ingestion and egestion and commonly being divided by radial folds and partitions; digestion partly intracellular; nervous system comprised of a network of cells; no respiratory, excretory, or circulatory systems. *Precam.-Rec.*

Subphylum CNIDARIA Hatschek, 1888

[Actinozoa DE BLAINVILLE, 1830; *non* Cnidaires MILNE-EDWARDS & HAIME, 1857 (=Anthozoa *hic*); *nec* Cnidaria VERRILL, 1865 (=Anthozoa *hic*); *nec* Cnidaria HAECKEL, 1881 (=Coelenterata *hic*)]

Mainly distinguished by possession of stinging capsules (nematocysts) and well-developed muscular action. Commonly with power of budding, by which either individuals or colonies may be formed; sexual reproduction typically producing an ovoid uniformly ciliated larva (planula); polymorphism characteristic, chief types being fixed cylindrical polyps and free cup-shaped medusae; some groups with endo- or exoskeleton of calcitic, aragonitic, horny or chitinophosphatic nature.[1] *Precam.(Algonk.)-Rec.*

[1] Includes 3 main classes (Hydrozoa, Scyphozoa, Anthozoa) in addition to the 2 assemblages (Protomedusae, Dipleurozoa) designated by CASTER and HARRINGTON & MOORE as classes (p. F21, F24). The modern meaning of the term Scyphozoa is quite different from the original usage (GÖTTE, 1887), which embraced the Anthozoa and Ctenophora but consistent use during many decades for the acraspedote medusae supports retention of the name. As defined in the *Treatise*, the Scyphozoa include subclasses called Scyphomedusae and Conulata (p. F28).

MORPHOLOGY OF LIVING COELENTERATES

By Libbie H. Hyman

CONTENTS

CHARACTERISTICS OF COELENTERATES

The phylum Coelenterata, preferably termed Cnidaria after removal of the sponges and ctenophores, exemplifies a grade of animal structure characterized in two important ways, (1) in being composed of tissues not constituted into organs, and (2) in the arrangement of its parts on a radially symmetrical plan (Fig. 6). The body consists essentially of a solid wall inclosing a simple or more or less subdivided central cavity (coelenteron) that is closed at one end (aboral) and open by the mouth at the other end (oral). The wall is composed of an outer epithelium (ectoderm or epidermis), an inner epithelium (endoderm or gastrodermis) bounding the central cavity or cavities, and an intermediate tissue (mesogloea or mesenchyme) varying in thickness and construction. Other tissues present are a nervous network and strata of muscular fibers; these occur between the epithelia or at their bases.

The cnidarian body is constructed on the radial plan; that is, all parts are symmetrically arranged around the central oral-aboral axis, and any plane through the body that includes the whole of this axis will divide the body into halves that are identical with each other regardless of the direction of the plane. Radial symmetry is generally believed to be the original morphological plan of the phylum; it is altered to biradial and bilateral symmetry in many members of the class Anthozoa, but such alteration is commonly considered to be secondary (Fig. 7). In biradial symmetry the parts are arranged symmetrically not only with reference to the oral-aboral axis but also with reference to 2 planes at right angles to each other, arbitrarily called sagittal and transverse planes. The halves obtained by section through the oral-aboral axis and the sagittal plane are identical with each other but not identical with the halves obtained by section through the axis and the transverse plane although these are also identical with each other. In bilateral symmetry parts are arranged symmetrically with reference to one plane only, the median sagittal plane. A section along this plane in the anteroposterior axis will yield identical halves, and this is not possible along any other plane of section. Symmetry observed in coelenterates is illustrated by diagrammatic transverse sections (Fig. 8).

Other prominent structural features of cnidarians are the tentacles and the nematocysts. The tentacles are long, slender, hollow or solid extensions of the wall that surround the oral end in one or more cycles. They are wanting in very few members of the phylum. The nematocysts or stinging capsules (not cells) are secreted bodies containing a coiled thread (actually a tube) that, under certain conditions of stimulation, is emitted to the exterior by turning inside out. Such everted threads penetrate prey and stupefy it by means of a toxic fluid of unknown nature emitted from the tip of the thread. All cnidarians are provided with nematocysts which are in fact diagnostic of the living members of the phylum.

The phylum illustrates to a marked degree the phenomenon of polymorphism,

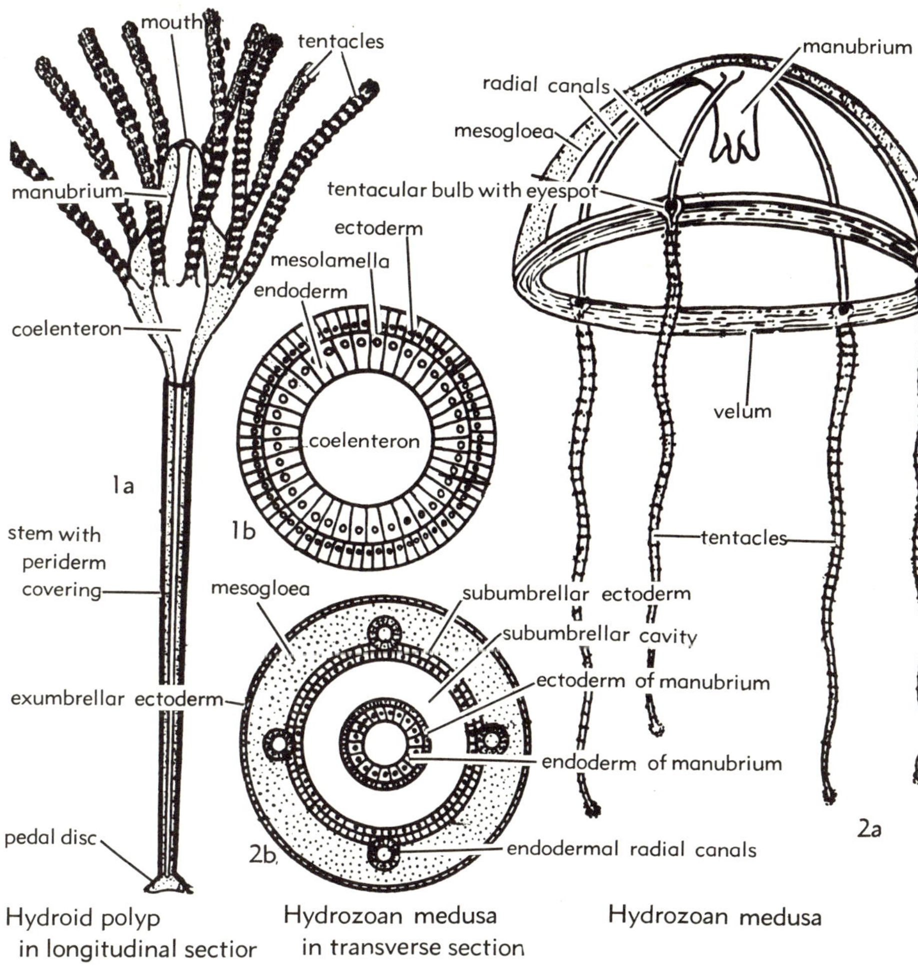

FIG. 6. Diagrams of coelenterate structure. *1,* Hydroid polyp; *1a,* partly in longitudinal section showing periderm-covered stalk with basal attachment and tentacle-encircled hydranth bearing the mouth and containing an undivided central digestive cavity (coelenteron); *1b,* transverse section through hydranth (much enlarged) showing cell layers. 2. Hydrozoan medusa; *2a,* somewhat oblique and partly sectional view of umbrella-shaped body with pendent tentacles, the mouth being located at end of a tubular extension (manubrium) at the summit of the vault on the concave surface; *2b,* transverse section above end of the manubrium. (Modified from L. H. HYMAN, *Invertebrates: Protozoa through Ctenophora,* McGraw-Hill, New York.)

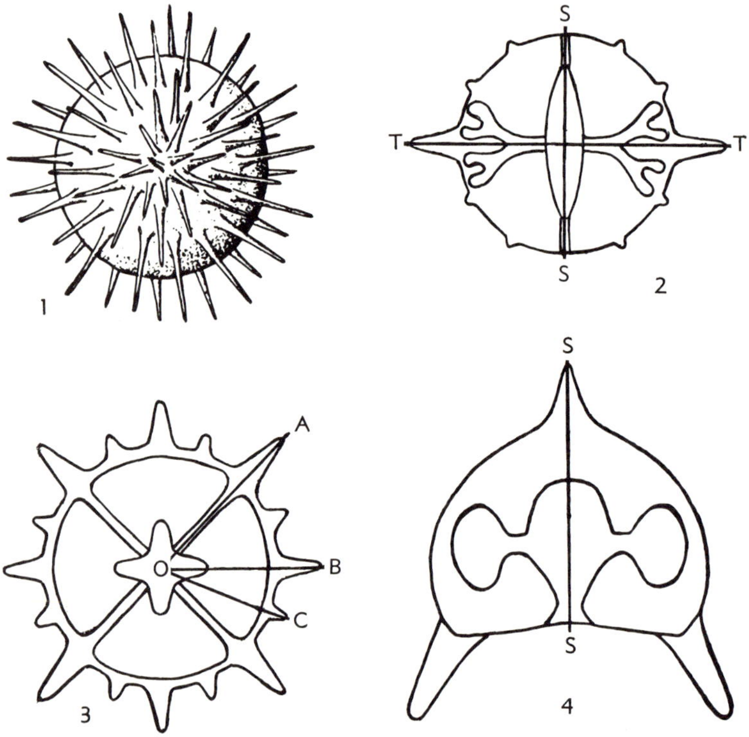

FIG. 7. Symmetry in invertebrate animals. *1,* Spherical symmetry. *2,* Biradial symmetry, characterized by lack of identity of halves produced by section along sagittal and transverse diameters. *3,* Radial symmetry of tetramerous type, characterized by identical nature of halves produced by section along any 2 diameters at right angles to each other. *4,* Bilateral symmetry, with differentiation of dorsal and ventral surfaces. *AO,* perradius; *BO,* interradius; *CO,* adradius; *SS,* sagittal axis; *TT,* transverse axis (modified from L. H. HYMAN, *Invertebrates: Protozoa through Ctenophora,* McGraw-Hill, New York).

which is to say, that it expresses itself in a variety of morphologies that are, however, reducible to 2 plans, the polypoid and the medusoid. The polypoid cnidarian has a columnar form, attached at the aboral end, free at the oral end, which is directly encircled by tentacles (Fig. 6). The intermediate layer of the wall is relatively thin, varying from a thin cement to a stratum of connective tissue. The medusoid type, free-swimming, has the shape of a bowl or saucer with tentacles around the rim and mouth at the end of a tubular projection depending from the center of the concavity of the bowl (Fig. 6,*2a*). The intermediate layer is here very thick, composed of a gelatinous material, whence the common name of jellyfish applied to medusoids. The terms craspedote and acraspedote used with reference to medusae refer to the presence or absence of a circular muscular shelf extending inward from the bowl edge.

The polypoid type exhibits a high degree of asexual reproduction and tends to form colonies fixed to a substratum (Fig. 9). Such colonies require support and hence are provided with some sort of skeleton, either external or internal or both. The form and nature of this skeleton varies in the different groups of Cnidaria. The medusoid type is solitary, although a few forms are capable of budding off other medusae like themselves, and, being free-swimming, hence supported by the water, does not require and among living forms is not provided with any skeleton. Sexual reproduction commonly results in the formation of a swimming larva that in general settles and develops into some variant of the polypoid type. The medusoid type, when present in

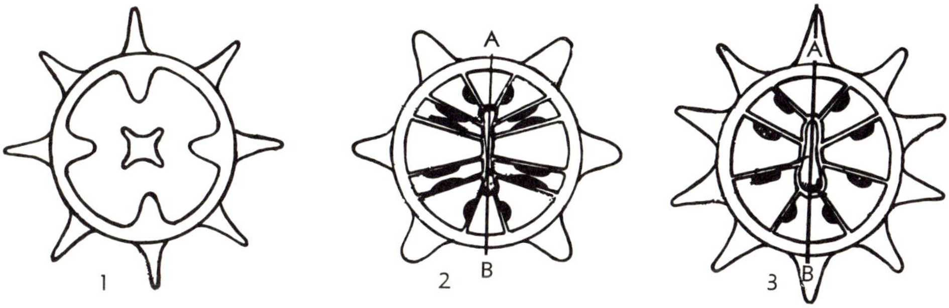

FIG. 8. Diagrammatic transverse sections of coelenterates illustrating symmetry. *1,* Hydrozoan or scyphozoan medusa, showing tetramerous radial symmetry. *2,* Anthozoan (anemone) showing hexamerous biradial symmetry. *3,* Anthozoan (octocoral) showing octamerous radiobilateral symmetry. *AB,* plane of bilateral symmetry (HYMAN).

the life cycle, arises by asexual budding from the polypoid.

The existing cnidarians are almost exclusively marine and widely distributed throughout the oceans, although most abundant on the continental shelves. The polypoid types are benthonic, with the exception of a few genuine pelagic forms besides others that habitually live attached to floating seaweeds. The medusoid types are pelagic, either neritic or oceanic, and some are limited to the deeper waters of the oceans.

CLASSIFICATION OF CNIDARIA

The Cnidaria (HATSCHEK, 1888) may be defined as radial, biradial, or radiobilateral Metazoa composed of tissues, without definite organs; parts organized around an oral-aboral axis, lacking cephalization; with one internal cavity, the coelenteron, simple or subdivided, closed aborally, open orally by the mouth surrounded (with few exceptions) by one or more circles of tentacles; with nematocysts as a defensive and food-capturing device; solitary or colonial, fixed or free-swimming; colonies produced by budding processes, generally provided with a skeleton; highly polymorphic, attached phases polypoid, free-swimming phases medusoid; sexual reproduction generally with a free larva that attaches to become a polypoid type; almost exclusively marine; ?Precambrian; Cambrian to Recent.

The Cnidaria are divided into 3 classes:[1] Hydrozoa, Scyphozoa, and Anthozoa.

The Hydrozoa are polymerous or tetramerous radially symmetrical cnidarians, that occur as polyps or medusae or both; coelenteron simple or branched but not provided with a stomodaeum, or with septal projections or with nematocyst-bearing structures; oral end of the polyp elongated into a manubrium; medusa of the craspedote type without rhopalia.

The Scyphozoa are solitary acraspedote tetramerous radially symmetrical medusae or medusa-like polypoids in which the coelenteron is subdivided in larva or adult into a central space and 4 interradial pockets by 4 projecting interradial septa; the coelenteron is armed with nematocyst-bearing tentacles (gastric filaments); the margin of the medusae bears sense organs (rhopalia) in the form of reduced and altered tentacles; the life cycle includes a polypoid larva that differs from a hydrozoan polyp in its tetramerous symmetry and subdivision of the coelenteron by 4 longitudinal septa and that develops directly into the adult medusae or gives off medusae by transverse fission.

The Anthozoa are exclusively polypoid cnidarians with hexamerous, octamerous, or polymerous biradial or radiobilateral symmetry; the oral end is expanded radially into an oral disc bearing hollow tentacles; a stomodaeum is present, usually provided with one or more siphonoglyphs; the coelenteron is more or less subdivided by endodermal septa some or all of which are fastened to the stomodaeum; parts of septa not fastened to the stomodaeum are edged with a cnidoglandular filament; all traces of a medusoid stage are absent; the members are solitary or colonial and commonly provided with skeletons.

[1] See classification given on p. F2-F4 for main divisions of Coelenterata which are recognized in this volume of the *Treatise*.—EDITOR.

HYDROZOA

MORPHOLOGICAL FEATURES

The Hydrozoa typically occur as radially symmetrical polypoid forms, either solitary or colonial, mostly the latter. The ordinary hydrozoan polyp consists of the attached **base,** the slender erect stem or **stalk,** and the elongated terminal expansion or **hydranth** that bears the mouth and tentacles. Colonies arise by formation of stem buds that grow out into branches and by basal outgrowths **(stolons)** that send up new stems; the latter develop terminal hydranths and repeat the budding and branching processes. These colonies are generally stiffened and protected by an external skeleton in the form of a chitinous tube. In the orders Milleporina and Stylasterina, however, the polyp colony occupies a massive calcareous exoskeleton permeated by the stems of the polyps in the form of tubes. The living part of a hydrozoan colony is hollow, containing a continuous coelenteron. Hydrozoan polyps show a good deal of polymorphism, often occurring in several types within the same colony.

FIG. 9. Hydroid colony of calyptoblastic type *(Campanularia),* illustrating sympodial growth (HYMAN).

Many hydrozoans also have a medusoid stage, budded off asexually to swim about as a free medusa. In such case the polypoid stage is wholly asexual and the medusa is the sexual form, producing sex cells that develop into the polypoid stage by way of a free larva, termed **planula.** However, the medusoid stage often shows more or less reduction or abortion, finally being reduced to an altered polyp **(gonozooid)** that produces the sex cells directly. In one group of Hydrozoa, the order Trachylina, on the contrary, the polyp stage is greatly reduced or absent and the free medusa dominates.

In the past, the hydrozoan life cycle that consists of an asexual polypoid phase budding off sexually reproducing medusae was regarded as typical and was interpreted as an alternation of asexual and sexual generations (metagenesis). It is more consistent with the facts about other invertebrate groups and a more physiological interpretation to regard the polypoid phase as a persistent juvenile stage that has continued to multiply asexually, whereas the medusoid phase is the definitive adult. It follows as a necessary corollary to this interpretation that the medusoid type is the original and primitive coelenterate.

DIVISIONS

The Hydrozoa are classified as follows:[1]

Order Hydroida, the hydroids. Polypoid generation well developed, solitary or colonial, mostly provided with a chitinous exoskeleton; with free medusae or aborted structures representing medusae present on the polypoids; medusae provided with ocelli or ectodermal statocysts or both.

Order Milleporina, the millepores. Hydroid colony occupying the surface layer of a massive calcareous exoskeleton with pores through which the polyps protrude; polyps dimorphic; protective polyps with capitate tentacles; with free but degenerated medusae, formed in special cavities.

Order Stylasterina. Similar to Milleporina, but protective polyps small, without tentacles; without free medusae; sex products formed on aborted medusae in special cavities.

[1] See footnote, p. F13.

Order Trachylina, the trachyline medusae. Hydrozoa with only the medusoid stage well developed; hydroid stage reduced or wanting; sense organs of medusae are statocysts and tentaculocysts with endodermal statoliths.

Order Siphonophora, the siphonophores. Hydrozoa existing as floating or free-swimming highly polymorphic colonies without skeleton, composed of an assortment of polypoid and medusoid individuals attached to a stem; without free medusae.

Order Chondrophora. Floating colonies consisting of polymorphic members attached to the underside of a disciform expansion; with free medusae.

SCYPHOZOA

MORPHOLOGICAL FEATURES

The Scyphozoa or Scyphomedusae are cnidarians primarily of the medusoid type which reaches a climax in this class. They have the same general appearance as the medusae of the Hydrozoa but are commonly much larger and of more complicated construction. Like the hydromedusae, the Scyphomedusae are based on a plan of tetramerous radial symmetry; that is, their main parts are arranged symmetrically in 4 or some multiple of 4 around the central oral-aboral axis of symmetry. This axis, therefore, falls at the intersection of 2 principal (but identical) planes of symmetry crossing at right angles. These 2 planes are those on which the mouth angles and gastric pockets lie. The 4 radii thus indicated are called **perradii** and the 4 radii halfway between them are termed **interradii;** when necessary, radii halfway between adjacent per- and interradii are referred to as **adradii.** There are obviously 4 perradii and interradii but 8 adradii. Section along the 4 perradii or the 4 interradii results in 4 morphologically identical quadrants; this is the essence of tetramerous symmetry.

The highly gelatinous body **(bell** or **umbrella)** varies in shape from a goblet or trumpet to a dome, bowl, or saucer; it bears tentacles on the margin at 4 or more radii or in indefinite number, further complicated assemblages of sensory organs called **rhopalia.** The bell margin is typically scalloped on the plan of 4 and the tentacles and rhopalia are definitely related to the scallops. A rhopalium comprises an equilibratory organ (tentaculocyst), chemoreceptive sensory patches, and in some Scyphozoa eye spots. The scyphozoan bell lacks the velum characteristic of the hydrozoan medusa, that is, a circular muscular shelf extending inward from the bell margin, although a slight indication of this structure is present in some Scyphozoa. This difference leads to the names **craspedote medusae** and **acraspedote medusae** for hydro- and scyphomedusae, respectively.

From the center of concavity of the bell depends the beginning part of the digestive tract in the form of a quadrangular tube **(manubrium)** opening at its free end by the 4-cornered mouth. The mouth corners, commonly drawn out into frilly **oral lobes** or **arms,** define the perradii. The manubrium is lined with endoderm, not as earlier supposed with inturned ectoderm; hence, as in the Hydrozoa, a pharynx or stomodaeum is lacking in the Scyphozoa. The manubrium leads into the general coelenteron located in the substance of the bell; as characteristic of the Scyphozoa, in contrast with hydrozoan medusae, the coelenteron is subdivided into a central stomach and 4 gastric pockets located on the perradii produced by 4 partitions or septa projecting into the interior along the interradii. As a further characteristic of Scyphozoa, the free edges of the 4 partitions bear numerous tentacle-like projections, termed **gastric filaments,** that are armed with nematocysts and participate in the capture of prey. Definite gastric pockets and interradial partitions are lacking in adults of the higher orders of Scyphozoa, being confined to the larval stages, but gastric filaments are invariably present. From the periphery of the central stomach or its pockets gastrovascular canals, commonly much branched, are given off to the bell margin where a ring canal may or may not be present.

Another distinctive feature of the Scyphozoa, wanting in hydrozoan medusae, is the presence between the bell margin and the manubrium of a deep depression, the **subumbrellar funnel,** pushed into the partition in each interradius. This fundamental feature again is present in the adults of only the lower orders, being confined to the larval stage of the higher ones.

Strong muscles, running in circular and radial directions, are present; especially noticeable is the coronal muscle, a strong circular band on the subumbrellar side near the periphery. By its contractions the bell pulsates, enabling the medusa to swim and keep afloat. The nervous system consists of the usual plexus in the base of both epithelial layers, without cephalic concentration, but ganglionic concentrations of nervous tissue occur in connection with each rhopalium and these rhopalial ganglia are important in controlling activities of the animal.

REPRODUCTION

Adult Scyphozoa reproduce exclusively by the sexual method; the gonads are borne on the sides of the interradial partitions where present, or in the floor of the stomach periphery. The early larva (planula) attaches and develops into a polypoid type that reproduces extensively by asexual methods. It differs considerably from the hydroid polyp in having a trumpet-shaped body attached by an adhesive basal disc and 4 subumbrellar funnels and 4 interradial partitions. It is thus, like the adult, definitely tetramerous, unlike the hydrozoan polyp. Consequently, tetramerous radial symmetry is found to be fundamental to the class Scyphozoa. The scyphozoan polyp produces medusae by transverse fission of its summit; such juvenile medusae differ considerably from the adult in appearance and gradually develop the adult morphology.

OCCURRENCE AND CLASSIFICATION[1]

As the Scyphozoa are devoid of skeletal elements one would not expect any preservation as fossils. However, a few impressions left by them in favorable situations exist. The Scyphozoa are and always have been exclusively marine.

The Scyphozoa or Scyphomedusae are defined as acraspedote tetramerous medusae or medusa-like polypoids in which the coelenteron bears the gonads and gastric filaments armed with nematocysts; the coelenteron in adult or polypoid young is subdivided by 4 interradial partitions into a central space and 4 perradial pockets; the subumbrella is indented by 4 interradial pits (subumbrellar funnels) sunk into the partitions when these are present; complicated marginal sense bodies (rhopalia) are present; life history includes a polypoid tetramerous young provided with interradial septa and subumbrellar funnels; this develops directly into the adult or gives off medusae by transverse fission. The class is divided into 4 orders:

Order Stauromedusae or Lucernariida. Adult as a solitary attached polypoid with goblet- or trumpet-shaped body, provided with interradial partitions, subumbrellar funnels, and 8 marginal groups of tentacles; not represented as fossils.

Order Carybdeida or Cubomedusae. Free medusae of cuboidal or quadrangular shape with 4 tentacles or groups of tentacles at the corners and 4 rhopalia between the corners; tentacles borne on a flattened bladelike base, the pedalium; with interradial gastrovascular septa and subumbrellar funnels.

Order Coronatae. Free medusae with scalloped bells, characterized by a circular coronal groove; tentacles borne singly on pedalia, alternating with rhopalia; with interradial partitions and subumbrellar funnels.

Order Semaeostomeae. Typical flat or saucer-like free medusae with generally 8 or 16 rhopalia borne in niches between the lappets; rhopalia alternating with one to many tentacles; mouth angles drawn out into long frilly lobes; interradial partitions, gastric pockets, and subumbrellar funnels wanting in the adult, but present in the polypoid young.

Order Rhizostomeae. Tentacles wanting; oral lobes branched into 8 thick gelatinous arms that obliterate the mouth and contain numerous small mouths; otherwise as in Semaeostomeae.

ANTHOZOA

MORPHOLOGICAL FEATURES

The Anthozoa are strictly polypoid cnidarians in which all trace of a medusoid stage has been lost from the ontogeny. The broad morphological alterations from the radial to the tetramerous type already seen at work in the hydrozoan medusa and in both hydroid and medusoid phases of the Scyphozoa are further evolved in the Anthozoa,

[1] See footnote, p. F13, and p. F27-F66.

resulting in biradial and radiobilateral types of symmetry. Tetramerous symmetry easily passes into biradial symmetry. Whereas in tetramerous symmetry the 4 perradii have identical ends and planes along them divide the body into identical halves or quadrants, in biradial symmetry the 2 ends of one pair of perradii differ from the 2 ends of the other pair of perradii. Consequently, a plane along each pair of perradii will produce identical halves but those from one pair will not be identical with those from the other pair (Fig. 8). The 2 pairs of perradii are arbitrarily called **sagittal** and **transverse axes.** Bilateral symmetry is easily derived from either the tetramerous or the biradial type by the development of unlikeness along the sagittal axis.

The bilateral symmetry found in the Anthozoa is of the radiobilateral type, that is, it retains features of radial symmetry in the form of radiating partitions and hence is clearly an alteration of radial symmetry. It is therefore quite unlike the bilateral symmetry of the true bilateral animals which do not retain radial features, although presumably derived from a radial ancestor. Consequently HADZI's attempt to derive the Anthozoa from the flatworms appears poorly founded. The symmetry of the majority of the Anthozoa is best referred to as radiobilateral.

The anthozoan polyp differs in several ways from the hydrozoan polyp. It is relatively short and stout, with the oral end flattened into a disc bearing a central mouth encircled by one or more circlets of tentacles. As foreshadowed in the Scyphozoa, the coelenteron is no longer a simple cavity but is divided up into chambers by longitudinal partitions, the **septa** or **mesenteries.** The mouth leads into the coelenteron by way of an elongated pharynx or **stomodaeum** probably lined by inturned ectoderm. This opens freely below into the general coelenteron. Some of the septa just mentioned reach from body wall to pharynx, hence are termed perfect or complete septa; others that do not reach the pharynx are termed imperfect or incomplete. The free edges of all mesenteries are provided with a sinuous thickening called the septal **filament** armed with glands and nematocysts that reminds of the gastric filaments of the Scyphozoa. As all septa have free edges basal to the pharynx it is clear that all of the coelenteric chambers are in communication. The anthozoan pharynx commonly bears one or more **siphonoglyphs.** These are flagellated grooves by which water currents are directed into the interior. The presence of the siphonoglyphs defines the sagittal axis, and the mouth is also commonly elongated in this axis.

Another difference between the anthozoan and hydrozoan polyp is the much greater thickness of the intermediate layer between the 2 body epithelia in the former. This intermediate layer is invariably cellular and consists of some form of connective tissue. Thus, the standard concept of the coelenterates as 2-layered animals fails with regard to the Anthozoa and is best abandoned as of no comparative morphological value.

The anthozoan polyp may be solitary, in which case it is usually temporarily attached by a basal disc or other type of basal expansion. Most generally, however, anthozoans are colonial and the colonies are then supported by skeleton of varied nature, commonly developed strongly. Further, as in hydrozoan colonies, the polyps of anthozoan colonies are often di- or polymorphic.

The muscular system is mostly well developed, especially in the form of retractor muscles along the septa by which the body can be shortened quickly. The arrangement of these retractor muscles is of great systematic importance among the existing anthozoans. The nervous system takes the form of the usual network, not concentrated into anything resembling a center. The sex cells are produced along the septa and develop into a swimming larva of the planula type that settles and develops directly into a polypoid.

OCCURRENCE AND CLASSIFICATION[1]

The Anthozoa are exclusively marine and those with calcareous skeletons are well-preserved as fossils, furnishing, indeed, many types no longer in existence.

The Anthozoa are divisible into 2 great groups or subclasses, the Octocorallia with octamerous radiobilateral symmetry and the Zoantharia with various types of biradial or radiobilateral symmetry.

[1] See footnote, p. F13, and chapters on Anthozoa.

SUBCLASS OCTOCORALLIA

The octocoral (or alcyonarian) polyp differs from all other anthozoan polyps in several respects. It is provided with 8 pinnate tentacles symmetrically arranged. The coelenteron is subdivided by 8 complete single septa that extend from the body wall to the pharynx into 8 chambers, each of which continues into a tentacle. The tentacles therefore alternate with the septa. The pharynx is provided with one siphonoglyph (or sulcus) and the side of the polyp that bears the siphonoglyph is arbitrarily termed ventral, or better sulcal. On its sulcal face, that is, the side toward the siphonoglyph, each septum bears a strong longitudinal muscle band, the retractor, by which the polyp can be contracted and its oral end drawn into the interior. The 8 symmetrically arranged tentacles and septa confer on the alcyonarian polyp a superficial appearance of octamerous radial symmetry (whence the name Octocorallia) but the presence of one siphonoglyph and the arrangement of the septal muscles modify this symmetry in the direction of bilaterality. There is but one plane of symmetry, that which bisects the siphonoglyph, leaving 4 septa in one half and 4 in the other.

Below the pharynx, the septa are free and gradually diminish in breadth to the polyp base, being finally reduced to mere ridges. Generally, only the 2 asulcal septa, that is, those opposite the siphonoglyph, reach the base. The free septal edge is thickened into a cord, the septal filament, mostly of digestive nature, but the filaments of the 2 asulcal septa differ from the other 6, being heavily flagellated and serving to create an upward current, whereas at the siphonoglyph the current runs inwards.

The existing octocorals are colonial. In simplest groups, the colony runs over a substratum in the form of single polyps erected from a stolon or mat. Most Octocorallia, however, occur as lobed or branching colonies in which only the oral part of the polyps protrudes; the basal parts are imbedded in a common flesh or coenenchyme and are connected with each other by gastrodermal (endodermal) tubes that permeate the coenenchyme. New polyps arise as a bulge from the gastrodermal tube, and this bulge then sprouts tentacles and forms a mouth and pharynx by epidermal invagination. The coenenchyme is covered externally with epidermis and within consists of a gelatinous mesenchyme containing amoeboid cells.

The polyps may be dimorphic, consisting of 2 sorts, the ordinary polyps or zooids and the siphonozooids. The latter are reduced in all structures except the siphonoglyph, which is strongly developed. The siphonozooids do not feed but serve to drive a water current through the colony.

The octocorallian (alcyonarian) colony is supported by skeleton which is generally a product of the cells of the mesenchyme, and is either calcareous or horny in nature. It may consist of separate calcareous spicules or of spicules fused by a calcareous cement or of amorphous calcareous substance or of calcareous spicules coated by or united by horn or of strands and lamellae of horn.

The nature of the skeleton more or less determines the different orders. Four of the 6 orders here recognized are represented by fossil remains, but in general the Octocorallia are not of paleontological importance.

Order Stolonifera. Polyps arising singly from a creeping base, not partly imbedded in a coenenchymal mass; may be provided with warty calcareous spicules secreted by the cells of the mesenchyme. The best-known member of this order is the red organ-pipe coral (*Tubipora*), in which each polyp is inclosed in a skeletal tube of fused calcareous spicules. The ensemble of these tubes resembles the pipes of a pipe-organ. The tubes are further united at intervals by transverse platforms. Curiously enough, *Tubipora* is not known as a fossil.

Order Telestacea. A small group, exemplified by *Telesto*, in which the colonies occur as simple or branched stems arising from a creeping base; not represented as fossils.

Order Alcyonacea. Comprises the fleshy types of octocorals, known as soft corals, in which the polyps are partly imbedded in a common massive or lobed coenenchyme. The skeleton consists of separate warty calcareous spicules secreted by the cells of the mesenchyme and commonly present in great profusion; these spicules are scantily represented as fossils.

Order Coenothecalia. Also a small group, represented by fossils and by one existing genus, the blue coral, *Heliopora*, with a massive calcareous skeleton composed of fibers of aragonite fused into lamellae. The skeleton contains numerous blind tubes at right angles to the surface; these occur in 2 sizes, larger containing a polyp, and smaller containing gastrodermal tubes, united inside a thin superficial coenenchyme. As the colony grows, the living parts are pushed outward by the formation of transverse calcareous partitions and hence only the periphery of the mass is alive.

Order Gorgonacea. Conspicuous colonial animals of tropical and subtropical waters, including the sea whips, sea feathers, sea fans and other types collectively known as gorgonians or horny corals. Typically, the stems of gorgonians are supported by a horny axial rod which is covered over by a thin layer of coenenchyme containing the usual warty calcareous spicules and bearing the small or minute polyps. The red coral, *Corallium,* differs from other gorgonians in having a purely calcareous axis of cemented spicules. Gorgonian spicules may be preserved as fossils.

Order Pennatulacea. Fleshy colonies composed of the elongated body of the primary axial polyp and of numerous secondary polyps that spring laterally from the primary polyp. The coenenchyme as in other octocorals contains calcareous spicules that may be scantily preserved as fossils.

SUBCLASS ZOANTHARIA

This subclass includes all existing Anthozoa in which the number and arrangement of the septa and number and structure of the tentacles are otherwise than in the Octocorallia. The group is therefore more variable in morphology than is the subclass Octocorallia.

The cylindrical polyp is solitary or colonial and with or without a skeleton, here invariably an exoskeleton, secreted by the ectodermal surface of the animal. The oral end of the polyp forms an oral disc bearing hollow tentacles in one or more circlets or in radiating rows or differentiated into oral and marginal circlets. A well-developed pharynx is present, usually provided with 1 or 2 siphonoglyphs but in some zoantharians without them. The coelenteron is divided into compartments by septa, of which both the complete type reaching the pharynx and the incomplete type not reaching the pharynx are generally present. The septa are coupled, that is, in symmetrical correspondence on the 2 sides of the plane of symmetry that bisects the corners of the elongated mouth. In most individuals the septa are also paired, that is, occur in 2's close together. Thus we get the terms **exocoel** for the coelenteric chambers between pairs of septa and **endocoel** for the narrow chamber between the two members of a pair. Each septum is provided upon one face with a longitudinal retractor muscle and the arrangement of these retractors is of considerable importance among existing anthozoans. Usually the retractors of paired septa face each other, thus located on the side towards the endocoel, except for the pair of septa attached to a siphonoglyph where the retractors are on the exocoelic side. This fact distinguishes the septa of a siphonoglyph, which therefore are called directives. As in Octocorallia, the free edges of the septa are provided with filaments of glandular nature and also armed with nematocysts. Gonads are borne on the septa just back of the filaments.

The arrangement of the septa varies in the class but most commonly, as in anemones and stony corals, is hexamerous; that is, the pairs of septa occur in cycles of 6 or some multiple of 6. In a perfect hexamerous zoantharian, there are 2 siphonoglyphs, one at each end of the laterally flattened pharynx, 6 pairs of complete septa, that is, 2 pairs of directives, and 2 nondirective pairs (primaries) on each side; in the 6 exocoels between the complete pairs occur 6 incomplete pairs, also called secondaries; in the 12 exocoels so formed, 12 pairs of smaller incomplete septa, or tertiaries, and so on. As the animal grows, additional incomplete pairs arise. The hollow tentacles are related to the coelenteric chambers; typically there is one tentacle to each exocoel and one to each endocoel. Consequently the tentacles occur in the same cycles as the coelenteric chambers and increase in number with the formation of new septa.

Where 2 siphonoglyphs are present, the symmetry is of the hexamerous biradial type and the plane of symmetry is that which bisects the 2 siphonoglyphs. If but one siphonoglyph is present, the symmetry is radiobilateral.

The Zoantharia comprise 5 existing orders, of which one is provided with calcareous exoskeleton and hence of great paleontological significance.

Order Actiniaria. This order includes the sea anemones, familiar animals of the seashore, solitary polyps with a tentaculate oral end commonly of great beauty. As this group is devoid of skeleton it is not represented in fossil form.

Order Zoanthidea. A small group of mostly colonial forms, resembling small anemones, but differing from the latter decidedly in septal arrangement and other structural details. As the zoanthids lack skeleton they are of no paleontological importance.

Order Antipatharia. Also a small group, known as the black or thorny corals. They form slender branching plantlike colonies, having, like gorgonians, a hornlike skeletal axis covered with a thin coenenchyme bearing the small polyps. The type of skeleton is not readily preservable and hence this group has few fossil representatives.

Order Ceriantharia. The cerianthids appear like long cylindrical anemones but differ from these in having the tentacles differentiated into smaller oral and larger marginal sets and in an entirely different arrangement of septa all of which are perfect, reaching the pharynx. The soft body without skeleton is naturally not preservable in a fossil state.

Order Scleractinia. Zoantharians with hexamerous symmetry and calcareous hard parts; they are described in detail in a later section.

Extinct Orders. The extinct **Rugosa, Heterocorallia,** and **Tabulata,** are not discussed here because they are described in other sections of this volume.

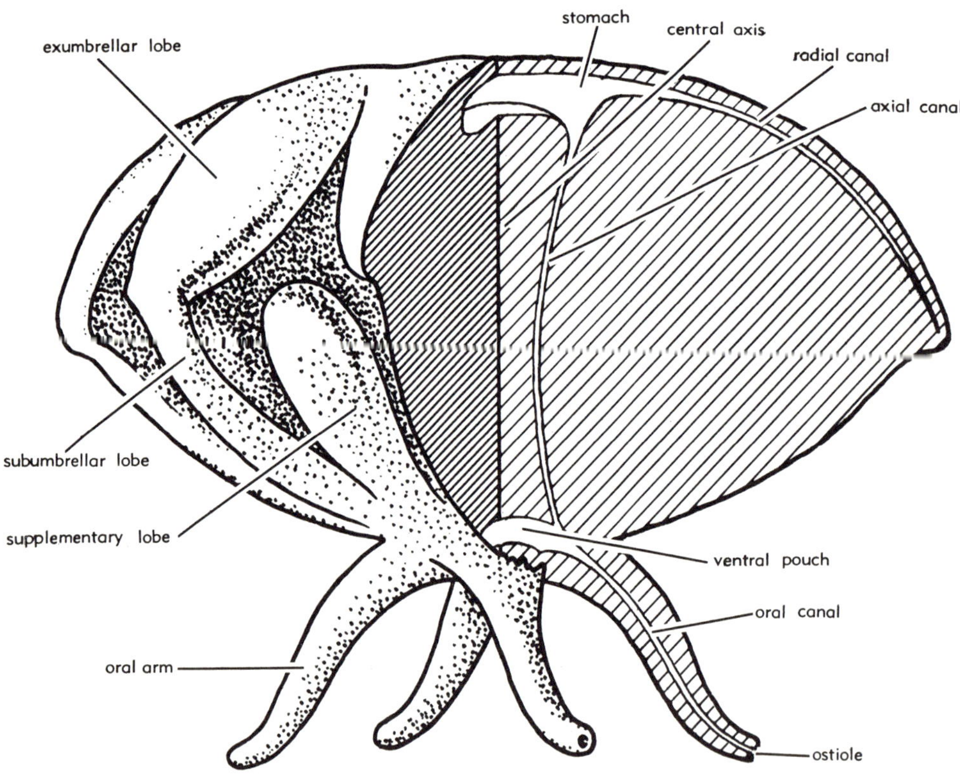

FIG. 10. Reconstruction of *Brooksella* showing morphology of external features and nature of internal structures (HARRINGTON & MOORE, n).

PROTOMEDUSAE

By H. J. Harrington and R. C. Moore

The class Protomedusae was introduced by Caster in 1945 to accommodate the peculiar Cambrian medusoid remains described by Walcott in 1896 under the names *Brooksella* and *Laotira,* referred by him to the class Scyphomedusae. These remains, preserved in cherty nodules 1 to 10 cm. in diameter—the famous "star cobbles" of the Middle Cambrian Conasauga shale of Coosa Valley, Alabama—show great variability in external shape. The simplest and most symmetrical specimens are subellipsoidal bodies displaying swollen, melon-like radial segments, sharply defined by deep radial grooves. The number of segments is highly variable, between 4 and 15 or more, and though some specimens display tetramerous symmetry, many others have an odd number of segments, 5 or 7 being common. The segments or lobes are usually irregular and those of the presumed upper or exumbrellar surface are typically wider than the corresponding ones of the lower surface. In addition, supplementary lobes are seen in many specimens, intercalated between the principals, usually inserted between the narrower subumbrellar segments. Walcott offered the plausible suggestion that the supplementary lobes acted as additional oral arms. Appendages interpreted as true oral arms, normally 4 or 5, are preserved on the subumbrellar side of a few specimens.

Walcott's reconstruction of the internal structure, based both on exfoliated specimens and polished sections, shows a central, disc-shaped aboral stomach giving rise to radial canals, one in each lobe, and to vertical axial canals connecting the stomach with a smaller, disc-shaped ventral cavity ("buccal stomach"). This cavity gives rise to radial canals, one in each oral arm, which appear to open to the exterior in a single ostiole at the distal extremity of the arms (Fig. 10). Gonads have not been recognized. Probably they were located in the central axis of the body below the aboral stomach.

The peculiar and highly variable lobation of the body, and particularly the presence of supplementary incomplete lobes, is a characteristic of primary importance which distinguishes the class from all other Coelenterata. Such features as lack of marginal tentacles and absence of central mouth, which Walcott regarded as indicating a rhizostomatid alliance, were in all probability developed independently and have no phylogenetic significance.

The class Protomedusae seems restricted to the late Proterozoic and early Paleozoic. In addition to the Middle Cambrian forms originally described by Walcott, remains assignable to this class are now known from Algonkian rocks of Arizona, Upper Cambrian strata of Wyoming and Lower Ordovician beds of Sweden and France. A reference to *Laotira* sp. from the Pennsylvanian of Sinai Peninsula, Egypt, by Couyat & Fritel, has never been substantiated by figures or descriptions.

Class PROTOMEDUSAE Caster, 1945

Highly variable, primitive Coelenterata, with bodies swollen into radial pouches divided by deep sulci, commonly with incomplete supplementary lobes intercalated between principals; marginal sensory organs and tentacles absent; central stomach aboral, discoidal, giving rise to simple radial canals, one in each lobe, and to vertical axial canals connecting stomach with ventral cavity; ring canal absent; central mouth absent; oral arms 4 or ?5, each with mesial canal springing from ventral cavity and with ?single ostiole at distal end. *Precam.(Algonk.), M. Cam.-Ord., ?Penn.*

Order BROOKSELLIDA Harrington & Moore, nov.

Characters of class. *Precam.(Algonk.), M.Cam.-Ord., ?Penn.*

Family BROOKSELLIDAE Walcott, 1898

Body ellipsoidal to discoidal, deeply lobate with 4 to 15 or more irregular melon-like segments; exumbrellar lobes wide, fused at mid-field, continued ventrally on subum-

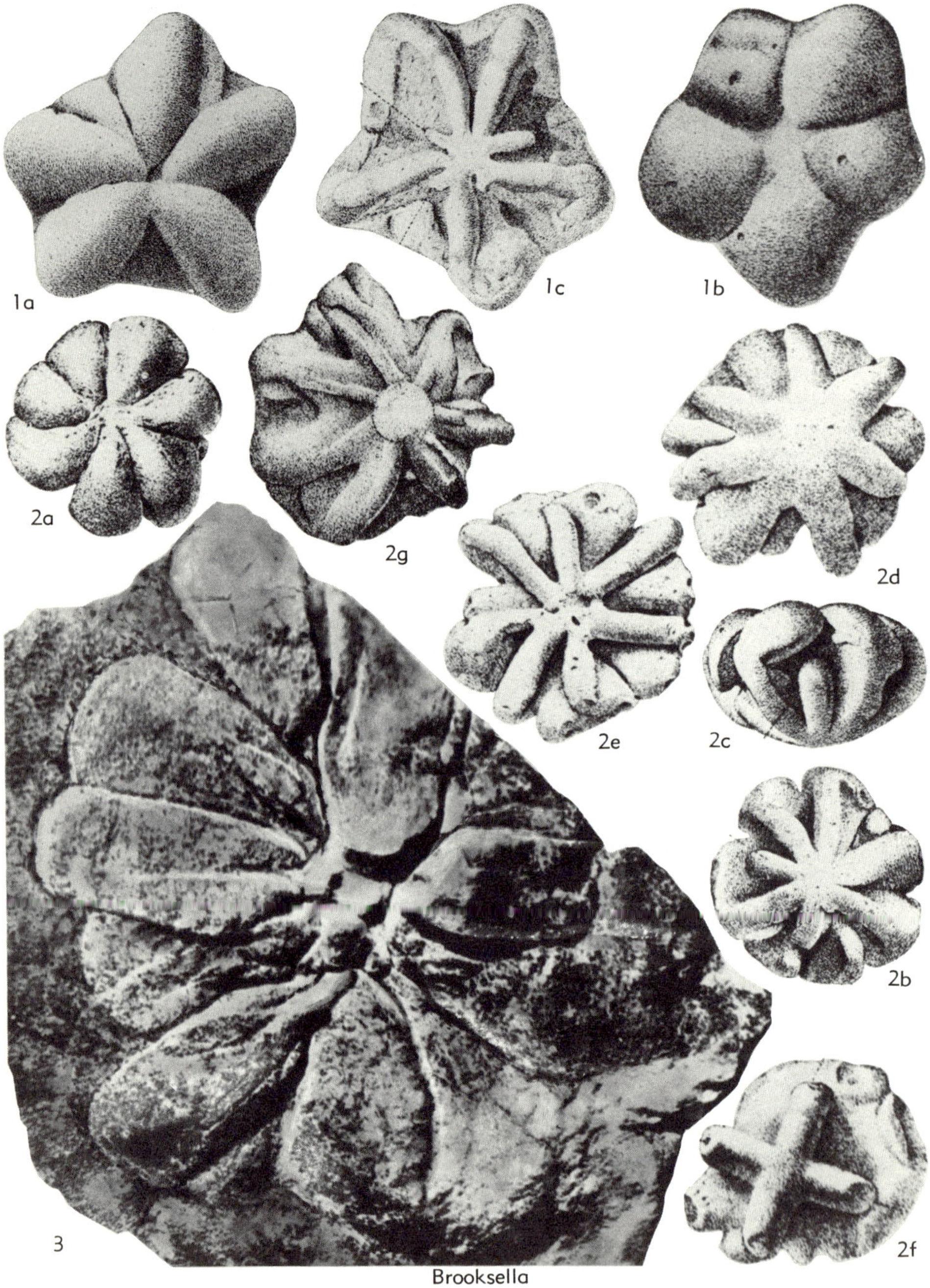

FIG. 11. Precambrian and Cambrian species of *Brooksella* from North America. *1, B. cambria,* M.Cam., Ala.; *1a,b,* exumbrellar views, ×1; *1c,* subumbrellar view, ×1 (WALCOTT). *2, B. alternata,* M. Cam., Ala.; *2a,d,* exumbrellar views; *2b,e-g,* subumbrellar views; *2c,* side view; all×1 (WALCOTT). *3, B. canyonensis,* Precam., Ariz.; holotype, ×1 (VAN GUNDY).

brella in narrow bands fused centrally; supplementary lobes, when present, seemingly fused internally along axial region of body. *Precam.(Algonk.), M.Cam.-Ord., ?Penn.*

Brooksella WALCOTT, 1896 [*nom. conserv.* proposed HARRINGTON & MOORE, 1955 (ICZN pend.)] [**B. alternata* (=*Laotira cambria* WALCOTT, 1896)] [?=*Staurophyton* MEUNIER, 1891 (**S. bagnolensis*), L.Ord., Fr.; *Laotira* WALCOTT, 1896]. Characters of family. *Precam.(Algonk.), M.Cam.-U. Cam.*, USA; *Ord.*, Swed.-?Fr.; *?Penn.*, Egypt.——FIG. 11,*2*. **B alternata*, M.Cam., Ala.; *2a-c*, exumbrellar, subumbrellar, and side views; *2d,e*, exumbrellar and subumbrellar views of another specimen; *2f*, oral arms of 3rd specimen; *2g*, subumbrellar view of 4th specimen; all ×1 (WALCOTT).——FIG. 11,*1*; 12,*1*. *B. cambria* (WALCOTT), M.Cam.-U.Cam., USA (type of *Laotira*) (11,*1*, M. Cam., Ala.; 12,*1*, U.Cam., Wyo.); 11,*1a,b*, exumbrellar views of 2 specimens, ×1; 11,*1*, subumbrellar view, ×1 (WALCOTT); 12,*1*, exumbrellar impression, ×0.65 (CASTER).——FIG. 12,*2*, *?B. cambria* (WALCOTT), L.Ord., Fr. (type of *Staurophyton bagnolensis* MEUNIER); ×1 (MEUNIER) ——FIG. 11,*3*. *B. canyonensis* BASSLER, Precam. (Algonk.), Ariz.; holotype, ×1 (VAN GUNDY).——FIG. 12,*3*. *B. silurica* (VON HUENE), Ord., Swed.; holotype, ×1 (VON HUENE).

REFERENCES

Bassler, R. S.

(1) 1941, *A supposed jellyfish from the Precambrian of the Grand Canyon:* U.S. Natl. Mus., Proc., v. 89, no. 310H, p. 519-522, 1 pl.

Caster, Kenneth E.

(2) 1942, *A laotirid from the Upper Cambrian of Wyoming:* Am. Jour. Sci., v. 240, p. 104-112, pl. 1.

Couyat, J. & Fritel, P.

(3) 1912, *Sur des empreintes (Méduses, Algues) recueillies dans le Carbonifère des environs de Suez:* Comptes Rendus Acad. Sci. Paris, v. 155, p. 795.

Huene, F. von

(4) 1904, *Geologische Notizen aus Oeland und Dalarne, sowie über eine Meduse aus dem Untersilur:* Zentralbl. f. Mineralogie, etc., Jahrg. 1904, p. 450-461, fig. 1-6.

Meunier, Stanislas

(5) 1891, *Staurophyton bagnolensis Stan. Meun. Nouveau fossile des Grès armoricains de Bagnoles (Orne):* Le Naturaliste, 13 année, 2me. ser., p. 134, 1 fig.

Van Gundy, C. E.

(6) 1951, *Nankoweap group of the Grand Canyon Algonkian of Arizona:* Geol. Soc. Am., Bull., v. 62, p. 963-969, pl. 1.

Walcott, Charles D.

(7) 1896, *Fossil jellyfishes from the Middle Cambrian terrane:* U.S. Natl. Mus., Proc., v. 18, p. 611-615, pl. 1.

(8) 1898, *Fossil Medusae:* U.S. Geol. Survey, Mon. 30, ix+201 p., 47 pl.

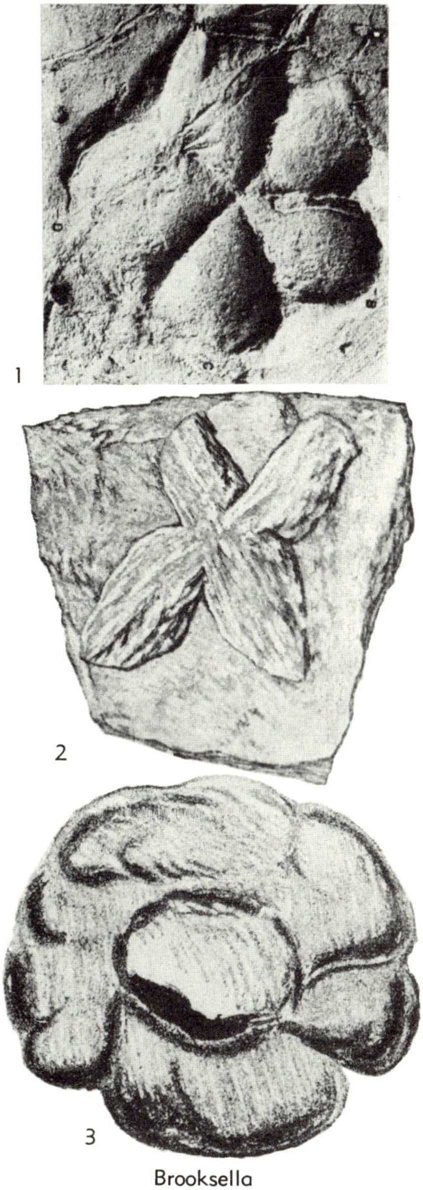

FIG. 12. Cambrian and Ordovician species of *Brooksella. 1, B. cambria*, U.Cam., Wyo.; exumbrellar impression, ×0.65 (CASTER). *2, ?B. cambria*, L. Ord., Fr.; exumbrellar view, ×1 (MEUNIER). *3, B. silurica*, Ord., Swed.; holotype, ×1 (VON HUENE).

DIPLEUROZOA

By H. J. Harrington and R. C. Moore

The class Dipleurozoa was introduced (1) to accommodate some very distinctive Lower Cambrian fossils from southern Australia described by Sprigg in 1947 under the generic name *Dickinsonia*. They are exceptionally neat impressions of small to medium-sized discoidal bodies of elliptical outline resembling a medusoid umbrella with radiating segments and marginal tentacles, but characterized by marked bilateral symmetry and differentiated extremities. A short median furrow along the main axis of the elliptical body gives off numerous simple, diverging lateral segments, shorter and more numerous at one extremity of the ellipse. The margin is scalloped into very short, rounded lappets, each bearing a simple tentacle (Fig. 13).

Sprigg suggested that the remains could represent a new and extinct class of Coelenterata or that, on account of their bilateral symmetry, they may belong in the Siphonophorida. Though bilateral symmetry is a common feature among the siphonophores, there is nothing to suggest that the remains under consideration represent the float or pneumatophore of a colonial coelenterate. The impressions indicate that the living animal had a consistent gelatinous body more or less homogeneous in structure, comparable to the bells of the medusae. One specimen shows what are here regarded as tentacles springing out from the marginal lappets; these structures were interpreted by Sprigg as chitinous radii, but their relation to the marginal lappets makes this view very implausible and strengthens the case for tentacles. These same relations—each tentacle springing out from the mesial part of a lappet make it highly improbable that they could represent submarginal dactylozooids of a colonial coelenterate comparable to the living genera *Velella* or *Porpita*.

Dipleurozoans seem to be restricted to the Lower Cambrian and are known exclusively from South Australia.

Class DIPLEUROZOA Harrington & Moore, 1955

Primitive, specialized Coelenterata with bell-shaped body of elliptical outline, displaying marked bilateral symmetry and differentiated extremities; median furrow along main axis of elliptical body giving off numerous flat, simple, diverging lateral segments separated by narrow grooves reaching periphery; margin scalloped into very short lappets, each bearing simple tentacle. *L.Cam.*

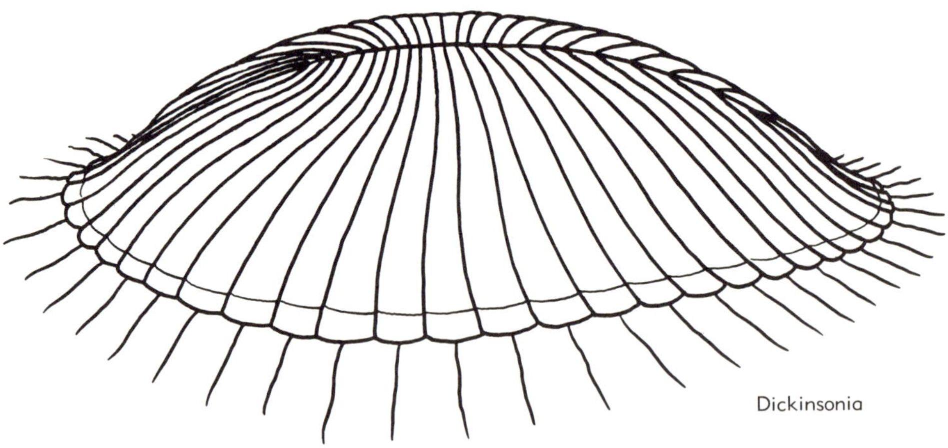

Fig. 13. *Dickinsonia spriggi* Harrington & Moore, oblique view (reconstr.) showing slightly scalloped margin with narrow lobes each bearing a short filiform tentacle. The bilateral symmetry and differentiation in character of the axial extremities are depicted, approx. ×1 (Harrington & Moore, n).

Order DICKINSONIIDA Harrington & Moore, 1955

Characters of class. *L.Cam.*

Family DICKINSONIIDAE Harrington & Moore, 1955

Body elliptical to ovoid in outline; median longitudinal furrow giving off 70 to 140

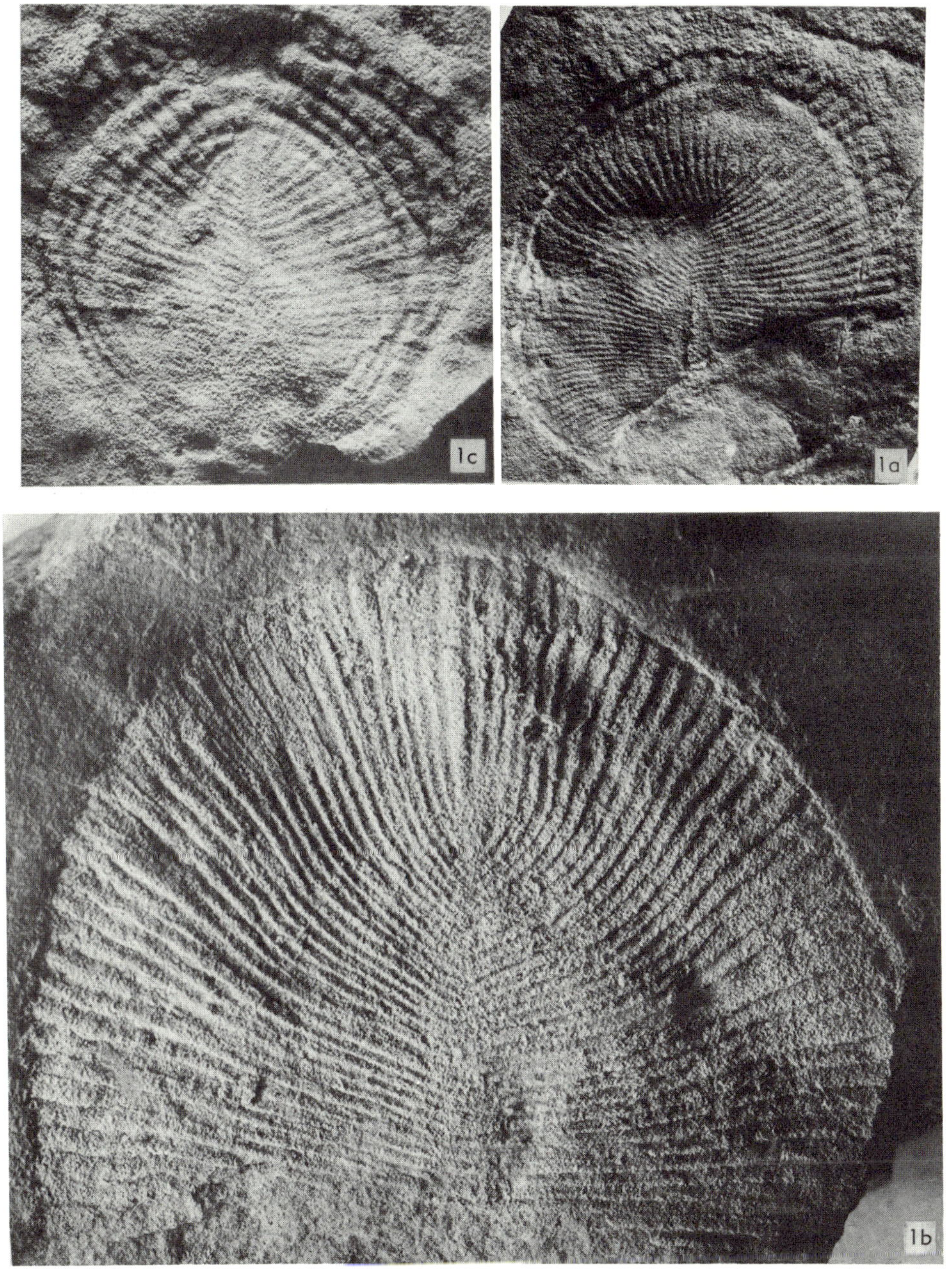

FIG. 14. *Dickinsoni costata* SPRIGG, from Lower Cambrian of South Australia; *1a,* ×1.25; *1b,* ×0.9; *1c,* ×1 (SPRIGG).

FIG. 15. *Dickinsonia minima* SPRIGG, from Lower Cambrian of South Australia; *1a,* ×0.8; *1b,c,* ×1 (SPRIGG).

FIG. 16. *Dickinsonia spriggi* HARRINGTON & MOORE, from Lower Cambrian of South Australia, holotype.

simple diverging lateral segments; segments at one extremity (?anterior) of elliptical body more numerous, narrower and less oblique than at opposite (?posterior) end. *L.Cam.*, S.Austral.

Dickinsonia Sprigg, 1947 [**D. costata*]. Characters of family. *L.Cam.*, S.Austral.——Fig. 14,*1*. **D. costata*; *1a*, holotype, ×0.9; *1b*, another specimen, ×1.25; *1c*, 3rd specimen, ×1 (Sprigg).——Fig. 15, *1*. *D. minima* Sprigg; *1a*, holotype, ×0.8; *1b*, another specimen, ×1; *1c*, 3rd specimen, ×1 (Sprigg).——Fig. 16. *D. spriggi* Harrington & Moore; holotype, ×1 (Sprigg).

REFERENCES

Harrington, H. J., & Moore, R. C.

(1) 1955, *Kansas Pennsylvanian and other jellyfishes:* Kansas Geol. Survey, Bull. 114, pt. 5, p. 153-163, pl. 1-2.

Sprigg, R. G.

(2) 1947, *Early Cambrian(?) jellyfishes from the Flinders ranges, South Australia:* Roy. Soc. S.Australia, Trans., v. 71, pt. 2, p. 212-224, pl. 5-8.

(3) 1949, *Early Cambrian "jellyfishes" of Ediacara, South Australia, and Mt. John, Kimberley District, Western Australia:* Same, v. 73, pt. 1, p. 72-99, pl. 9-21.

SCYPHOZOA

By R. C. Moore and H. J. Harrington

CONTENTS

Class SCYPHOZOA Götte, 1887

[emend. Haeckel, 1891, excluding Anthozoa, Ctenophora]

Mostly tetramerous (Fig. 17) free-swimming or attached Cnidaria lacking hard parts or protected by a thin chitinophosphatic periderm and typically characterized by presence of 4 interradial endodermal septa associated with longitudinal muscles of ectodermal origin and in some extinct forms with mineralized septa in this position. *Cam.-Rec.*

The Scyphozoa are exclusively marine coelenterates which include both attached (polypoid) and free-swimming (medusoid) forms. Among living representatives of the class, medusoids are overwhelmingly the more prominent, but fossils which here are assigned to the Scyphozoa mostly lack any obvious resemblance to modern free-swimming types. This generalization is dependent on revised classification that recognizes the widely distributed, not rare fossils called conulariids as belonging to the Scyphozoa, for otherwise the statement would be untrue and one could only say that the paleontological record of Scyphozoa is lamentably small. It becomes necessary to give reasons for adding the long-extinct much-debated conulariids to the scyphozoans; also, the addition inevitably affects some names and their application.

During several decades zoologists have employed either or both of the names Scyphozoa and Scyphomedusae, introduced by HAECKEL (1891) and LANKESTER (1877), respectively, as applied to the so-called true jellyfishes, that is, the nonveiled (acraspedote) types. The older name, Acraspeda (GEGENBAUR, 1856), has not gained favor in spite of its appropriateness. In the classification adopted in the *Treatise,* the most generally accepted name, Scyphozoa, is recognized for the class and at the same time Scyphomedusae is employed in its customary taxonomic application but defined as a subclass. Another subclass named Conulata is introduced to contain extinct organisms which now are assigned position in the Scyphozoa; this subclass is divided into orders, one of which comprises the Conulariida. Diagnoses of the class and subclasses are given here as a basis of the discussion which follows; systematic descriptions of lower-rank divisions are reserved for sections devoted to Scyphomedusae and Conulata.

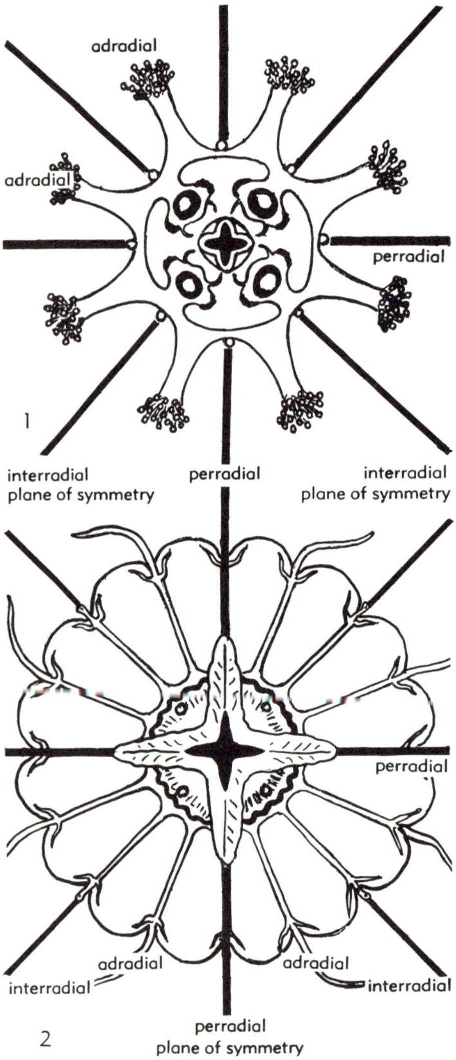

FIG. 17. Planes of symmetry in Scyphozoa. *1,2,* Subumbrellar views of a stauromedusid and a semaeostomatid showing position of perradial and interradial planes of symmetry; planes or lines halfway between these planes are termed adradial (modified from L. H. HYMAN, *Invertebrates: Protozoa through Ctenophora,* McGraw-Hill, New York).

Subclass SCYPHOMEDUSAE Lankester, 1881

Mostly tetramerous medusae and medusalike polypoids, termed acraspedote because lacking true veil, although some possess a very similar structure (velarium); coelenteron in larva or adult partly divided by 4 interradial endodermal septa; except in a few groups, subumbrella indented by 4 deep ectoderm-lined funnels in position of the interradial septa and associated with longitudinal muscles of ectodermal origin; margin bearing reduced tentacles or rhopalia (tentaculocysts with endodermal statoliths); mesogloea cellular, containing fibers and loose amoeboid cells; polypoid larva (scyphistoma) with tetramerous symmetry defined by 4 septa dividing coelenteron, larva growing directly into adult or producing medusae by transverse fission. *Cam.-Rec.*

Subclass CONULATA Moore & Harrington, nov.

Mainly tetramerous cone-shaped to elongate pyramidal or subcylindrical organisms bearing a thin chitinophosphatic periderm which commonly bears transverse and longitudinal markings but may be smooth, some bearing an attachment disc at the pointed aboral extremity but in part probably living unattached; oral end may be protected by inward bent edges of periderm; oral margin with tentacles at least in some forms and probably in all. *M.Cam.-Trias.*

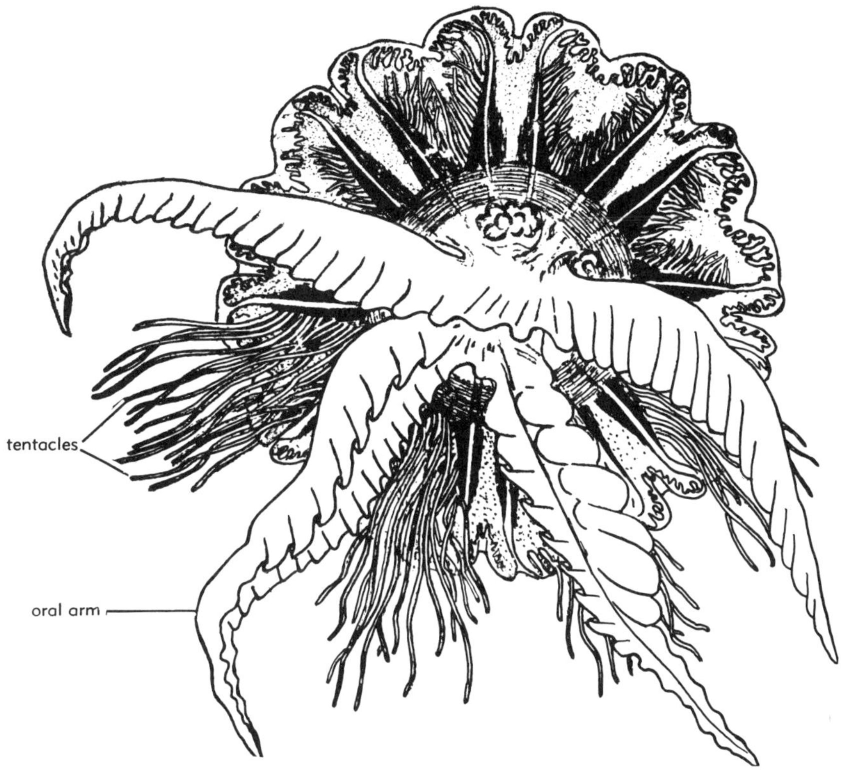

FIG. 18. Oblique view of subumbrellar surface of a modern scyphomedusan (*Cyanea*, order Samaeostomatida), showing prominent oral arms and areas of grouped tentacles; body parts are symmetrically repeated around the bell. *Cyanea arctica* attains a diameter of 2 m. or more, being distinguished as the largest known coelenterate (after L. H. HYMAN, *Invertebrates: Protozoa through Ctenophora*, McGraw-Hill, New York).

DISTINGUISHING CHARACTERS OF SCYPHOZOA

The outstanding feature of the Scyphozoa is strong development of tetrameral radial symmetry throughout life from larval to adult stages and in both polypoid and medusoid states. The body is divisible into identical quadrants by either of 2 sets of planes which intersect at right angles along the axis extending from center of the mouth to the aboral pole (Fig. 17). One set of such planes of symmetry can be placed so as to pass through the mouth arms or laterally outstretched angles and internally to divide each of 4 gastric pouches into halves; by definition these planes are termed perradial. The other set of planes of symmetry belongs halfway between the perradial planes so as to pass through inner angulations of the mouth and coincide with septa that bound the gastric pouches; these planes are designated as interradial. Any radial line or structure located halfway between adjacent perradial and interradial planes is called adradial but none of these define symmetry. In nearly all scyphozoans body parts are symmetrically repeated around the longitudinal axis in the number of 4 or multiples of 4 (Figs. 17, 18); a very few forms having scyphozoan attributes otherwise exhibit a hexameral plan and constitute exceptions.

BASIS FOR CLASSIFICATION

The recognition of divisions assigned rank of subclasses within the Scyphozoa, as indicated by the diagnoses just given, needs

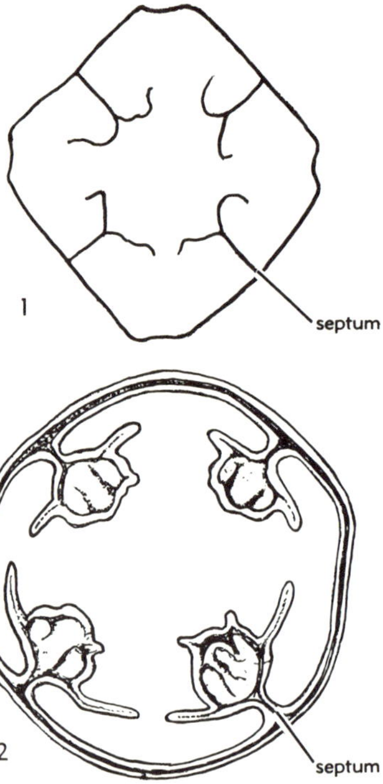

FIG. 19. Transverse sections of a Silurian conulate (*1, Eoconularia loculata,* ×15) and a modern scyphomedusan (*Craterolophus thetys,* ×1) showing similarity in development of bifurcate interradial septa formed of chitinophosphatic substance in the fossil but of soft tissue (endoderm) in living genera (KIDERLEN).

explanation. The conclusion that conulariids are coelenterates having closest affinities with scyphozoans is not new. KIDERLEN (1937) may be credited with marshaling most clearly the evidence of structural resemblances of conulariids and living scyphozoans which points to placement of both in proper phylogenetic relationships.

The Scyphomedusae, as defined to comprise living types of Scyphozoa and a very few extinct jellyfishes which have nearly identical structural features, are distinguished primarily by features of tetrameral symmetry which include partial compartmentation of the coelenteron by 4 interradially placed septa associated in some orders with deep funnel-shaped infolding of ectoderm in the oral region and development of longitudinal ectoderm-derived musculature that is so placed as to cause bifurcation of the septa. The scyphomedusan polypoid becomes anchored by a secretion which hardens into a chitinous substance, although the free-swimming adult lacks chitinous external parts.

The Conulata are characterized by a chitinophosphatic thin periderm that covers and protects the soft parts. Excepting some forms which are considered to represent an antecedent evolutionary stage, they exhibit fundamental tetrameral symmetry and interradial septal structures which are interpreted as identical in relationship to ectodermal infolding and musculature observed in modern Scyphomedusae. This is illus-

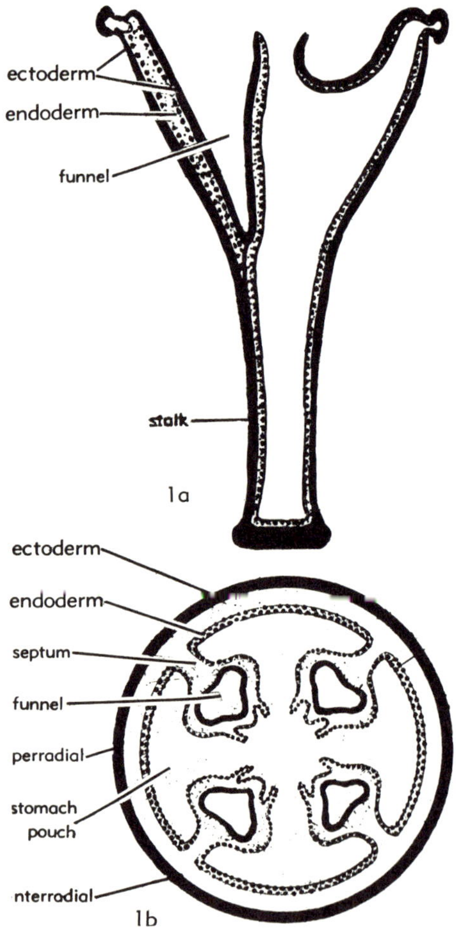

FIG. 20. Longitudinal and transverse sections of *Haliclystus* (Stauromedusida), Rec., showing ectoderm-lined funnels located in positions of the interradial septa; the longitudinal section is partly through an interradius (at left) and partly through a perradius (at right) (modified from L. H. HYMAN, *Invertebrates: Protozoa through Ctenophora,* McGraw-Hill, New York).

trated by transverse sections of representatives of the Conulata and Scyphomedusae (Fig. 19). The forked pattern of the septa that project into the interior of *Conularia loculata* exactly corresponds to the bifurcated septa of living scyphomedusans such as *Cratelolophus* and *Haliclystus* (Fig. 20, *1b*).

COMPARISON OF SCYPHOMEDUSAE AND CONULATA

The body form of polypoid scyphozoans ranges from slender subcylindrical through steep-sided conical or pyramidal shapes to trumpet-like and relatively low broadly flared types. The medusoid form may be goblet-shaped with length greater than width, broadly rounded, cuboid, or low saucer-shaped. Some Scyphozoa live in fixed position throughout life, being attached to seaweed or other foreign objects or the sea bottom by a chitinous secretion. Others are attached temporarily and able to re-attach the body to some support. A large majority of the living scyphozoans are free-swimming, movement being effected by muscular pulsations of peripheral parts of the jellyfish dome or bell.

MUSCULATURE

Musculature is almost wholly ectodermal. Especially noteworthy is the occurrence of longitudinal muscles that run along the interradial septa; these also are ectodermal and are associated in many living scyphozoans with very deep funnel-shaped infoldings of the ectoderm that extend from the surface near the mouth to vicinity of the aboral pole (Fig. 20). Similar longitudinal muscles are inferred to have characterized the Conulata, serving for inward pull on interradial edges of the periderm in order to close the aperture and, according to Bouček (1939), possibly functioning like the coronal muscles of modern Scyphomedusae for pulsatory swimming movements. Existence of such scyphozoan-type musculature in the Conulata rests on the wholly plausible but hardly provable interpretation of the median line and associated internal ridge or septum of conulariids as the place of longitudinal muscle attachment which was steadily shifted in adapertural direction during growth of the animal (Fig. 43). The interradial position of the septa and their inwardly bifurcate form in many Scyphomedusae are matched by the interradial placement of peridermal septa in many Conulata, at least one fossil form (*Conularia*

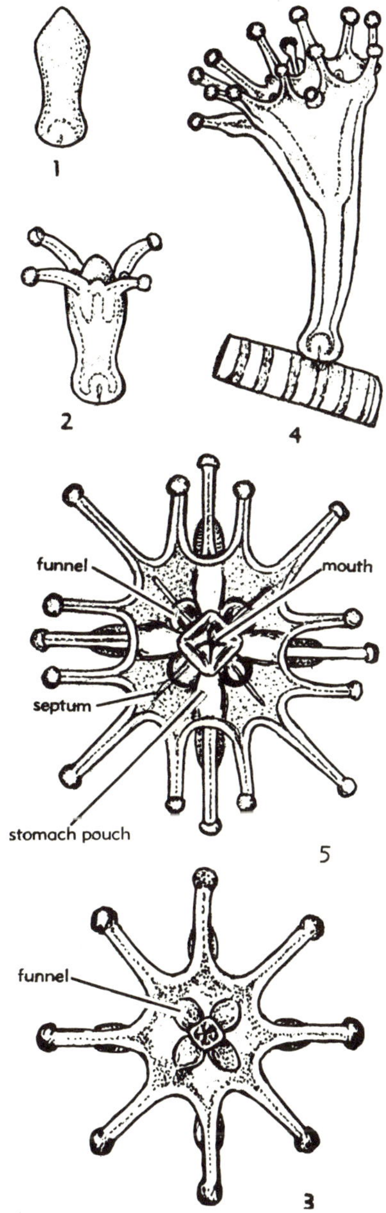

Fig. 21. Development of tentacles in ontogeny of a stauromedusid polyp *(Haliclystus)*, showing *(1-5)* successive stages with 0, 4, 8, and 16 tentacles (after L. H. Hyman, *Invertebrates: Protozoa through Ctenophora*, McGraw-Hill, New York).

loculata) from Silurian rocks showing clearly marked inward bifurcations of these septa (Fig. 19,*1*). Taken in conjunction with other evidence, this similarity strongly supports classification of the Conulata among the Scyphozoa.

TENTACLES

Both Scyphomedusae and Conulata bear tentacles. In many genera of the scyphomedusans these are evident in larval or polypoid stages of growth (Fig. 21) but greatly reduced or lacking in adults; in others they are well developed throughout life, being either hollow or solid, strongly muscled and generally armed with abundant stinging cells (Fig. 18). In stauromedusid polyps, for example, at first 2 tentacles appear, then 4, 8, 16, and so on; the first 4 tentacles are developed in perradial position and the next 4 interradially (Fig. 21). Adult Stauromedusida may have hundreds of tentacles grouped on 8 adradial lobate extensions of the oral surface (Fig. 17,*1*). The occurrence of tentacles at the apertural margin in Conulata is hardly open to doubt, but traces of these are actually preserved, so far as known, only in some specimens of *Conchopeltis,* of Middle Ordovician age from New York (Fig. 22). Originally described as a gastropod, this form is proved by its clearly defined tetrameral symmetry and possession of

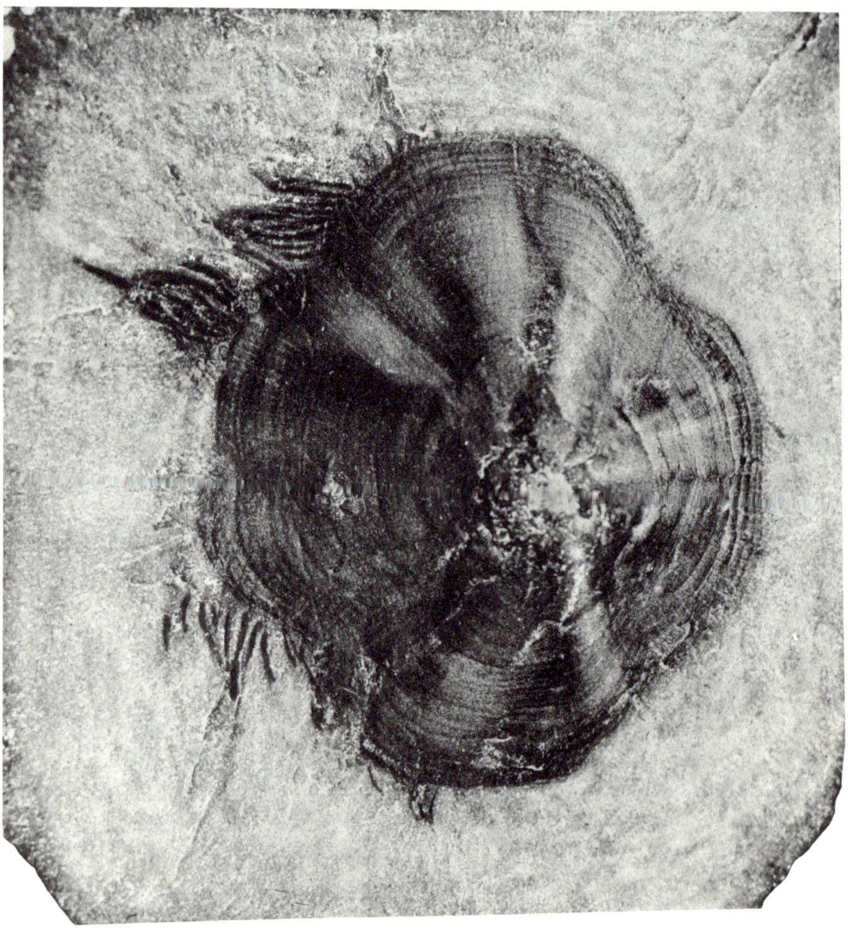

FIG. 22. *Conchopeltis alternata* WALCOTT, Middle Ordovician (Trenton), from Trenton Falls, N.Y., ×1.5, an unusually broad low conulariid with very thin periderm, apical view of specimen in Cornell University collections showing terminal parts of several tentacles that project beyond margin of the aperture. Another somewhat smaller specimen (not figured) also exhibits preservation of tentacles (J. W. WELLS photograph).

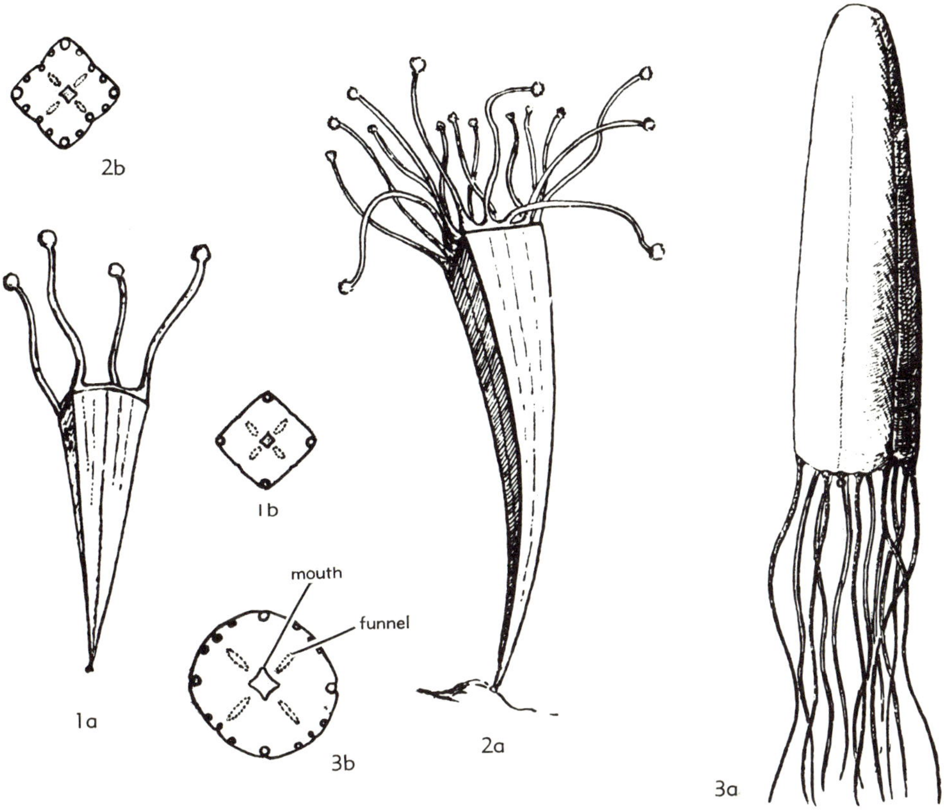

FIG. 23. Reconstructions of conulariids showing inferred number and location of tentacles, based on linear thickenings of the periderm interpreted to originate from attachment of tentacle muscles. *1a,b, Conularia cambria,* ×1, and *2a,b, Archaeoconularia fecunda,* ×0.7, are represented in attached position; *3a,b, Exoconularia consobrina,* ×0.7, is depicted as a free-swimming medusoid, its apical extremity sealed over by a transverse diaphragm (schott) (KIDERLEN).

peripheral tentacles to belong with the Scyphozoa. KIDERLEN (1937) has made reconstructions of tentacles arranged around the aperture of various types of Conulata, basing interpretation partly on peridermal traces such as longitudinal lines and furrows presumed to mark locations of tentacle muscle attachments; these reconstructions are very reasonable, especially in view of their correspondence to successive stages in tentacle development among modern Scyphomedusae (Fig. 23).

PERIDERM

The nature of the periderm of typical Conulata deserves attention. On first consideration, skeletal parts consisting predominantly of calcium phosphate seem utterly foreign to the Coelenterata, for all well-determined members of the phylum either lack hard parts entirely or only possess ability to secrete calcium carbonate or a chitinous substance. This consideration, however, is merely negative and not at all decisive. Modern stauromedusids may have a fixed attachment formed by chitin. The absence of phosphatic mineral salts associated with the chitin formed by some Scyphozoa and other kinds of living coelenterates does not rule out classification of extinct organisms as coelenterates merely because, like the Conulata, they are shown to be characterized by a chitinophosphatic covering. Definitive testimony must be of another sort. Accordingly, examination with open mind admits readily the possibility of recognizing a kind of coelenterate which differs from others in this character. It may be noted in passing that statoliths secreted by living Scyphozoans are composed of calcium sulphate and calcium phosphate.

The chief observational features relating to the periderm of Conulata are its composition, extreme thinness, indication of original flexibility at least in areas adjacent to the aperture, evidence of its wholly ectodermal origin, variety of generally very fine external markings, nature of attachment and structural adaptation for freedom from attachment. Each of these will be reviewed briefly.

Composition. Study of the periderm of several species belonging to conulate genera as seen in thin sections or polished sections has established the presence of microscopic laminae parallel to the surface. They consist of chitin, chitinophosphatic substance of seemingly variable phosphatic content, and essentially pure calcium phosphate. The minerals phosphorite and kolophanite have been reported (Bouček & Ulrich, 1929). Published analyses of chemical composition show figures such as 66.6 and 70.9 per cent $CaPO_4$ for Ordovician species, and a recently made analysis of peridermal fragments from a Pennsylvanian conulariid indicates approximately 96 per cent $CaPO_4$ (indefiniteness being due to extremely minute quantity). Reddish to purplish brown, bluish gray, and white coloration of well-preserved specimens is interpreted to signify original pigment because the colors have been observed to show correlation with specific characters and to be independent of the matrix in which specimens occur.

Thinness. The periderm of individuals belonging to the Conulata is thin, in some species as little as 0.056 to 0.070 mm. in thickness. Averages of the minimal and maximal measurements of periderm thickness of 7 species (maxima including thickened areas at angles) are 0.23 and 0.37 mm., respectively, and the average represented by these figures is 0.30 mm. This thinness accords with use of the term periderm for the protective covering of Conulata, rather than exoskeleton or shell.

Flexibility. The flexibility of the conulate periderm is attested by the common occur-

Fig. 24. Conulate specimens indicating flexibility of the periderm. *1, Exoconularia pyramidata,* M.Ord., Czech., ×0.7, showing indented but unfractured periderm (Kiderlen). *2, E. consobrina,* U.Ord., Fr.(Normandy), ×1; apertural end with corner obliquely sectioned showing that the periderm is folded on itself as well as bent inward to effect closure of the aperture (Kowalski). *3a,b, Archaeoconularia slateri,* M.Ord., Scot., ×1; oblique and apertural views showing plication of periderm (Reed).

rence of distorted but unbroken fossils. Some are merely flattened, whereas others show irregular indentations crossing the periderm without sign of associated fractures (Fig. 24,*1*). The inwardly bent apertural edges of many conulariids, some showing a complicated pattern of infolding along creases, furnish patent evidence of periderm flexibility in this region, for no specimens have been found that show sign of hingement or seam in positions of abrupt bending (Fig. 24,*2,3*). Furthermore, it is clear that the location of apertural inflections of the periderm must have shifted progressively away from the apex as growth added to length. Generally this growth seems to have been at nearly constant rate and although some specimens give indication (by irregularities in crowding of external features) of periodic halts or slower growth rates from time to time, none of these is associated with trace of localized prolonged periderm bending back and forth. Hence, it is safe to assert that in spite of the appearance of rigidity given by specimens having relatively coarse transverse ribs, the periderm of Conulata was flexible.

Ectodermal origin. The chitinophosphatic protective covering of conulariids was not secreted by endodermal layers of the body as might be possible if soft tissues extended over its outer surface. Rather, it was built from the outside inward and evidently, therefore, a product of secretion by ectodermal cells. Proof of this consists of the manner in which the exterior surface on any side of the conulariid furnished a place of attachment for inarticulate brachiopods, small edrioasteroids, other conulariids, and the like (Fig. 25). The nature of the scars clearly indicates that the host conulariid was alive when its periderm became a fixation site for other invertebrates, for the covering left behind by a dead conulariid would be incapable of localized distortion required to fit the shape of the attached organism. Thus, there is small room for doubt as to the truly external nature of the periderm. It follows that surface markings such as transverse ribs, tubercles, and like features are initial products of ectodermal secretion as growth in the apertural region proceeds, the smooth interior and inwardly built peridermal septa being the result of continued secretion by

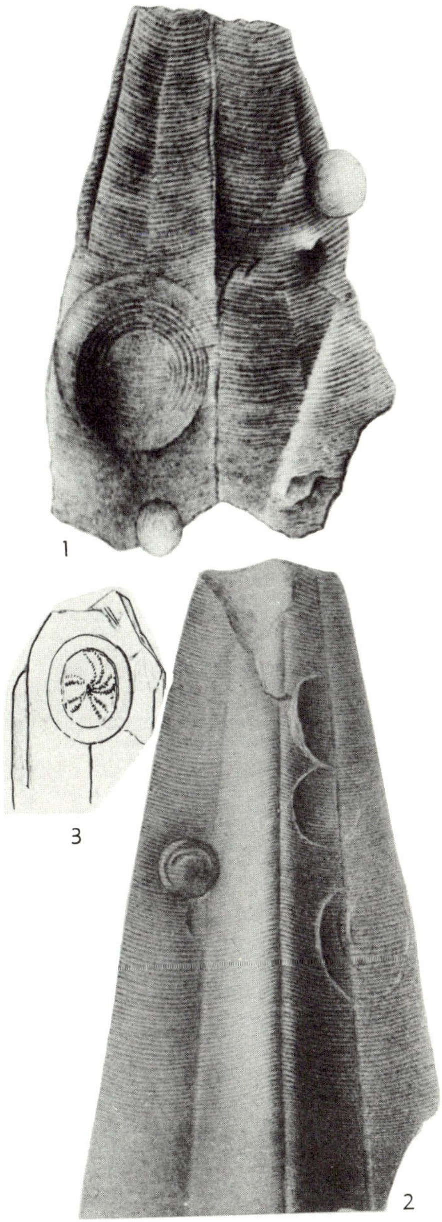

FIG. 25. Exterior of the conulate periderm as place of anchorage for other organisms. *1, Conularia continens,* M.Dev., N.Y., ×1; specimen with 2 small attached inarticulate brachiopod shells and scar left by a much larger inarticulate (HALL). *2, C. undulata,* M.Dev., N.Y., ×1; specimen also showing attachment scars on 2 faces (HALL). *3, Agelacrinites,* an edrioasteroid, on a specimen of *Exoconularia consobrina,* M.Ord., Czech., ×1 (BARRANDE).

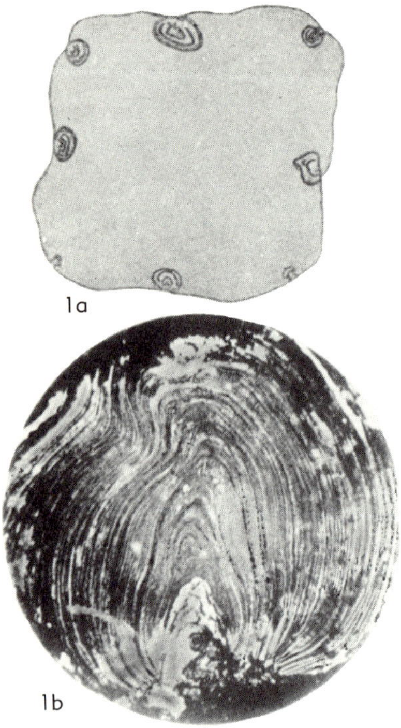

Fig. 26. Localized thickening of conulate periderm. *1a, Transverse section of Ctenoconularia* sp., Dev., S.Am.(Bol.), ×1, showing unusual inward growth of periderm at mid-line of faces and at corners; *1b*, one of these thickenings showing laminated structure, ×50 (Knod).

the ectoderm of chitinophosphatic deposits. The laminated structure and local deflection of the laminae to join with internal septa accord with this interpretation (Fig. 26). The ectodermal nature of the periderm and septa has special significance in connection with the correlation of Conulata with Scyphomedusae, because a common distinguishing character of the latter is deep invagination of the ectoderm layer in positions of the 4 septa and development of ectodermal muscles that interrupt or divide the septa.

FIXATION

Discovery of numerous young conulariids in attached position and study of their mode of attachment (Ruedemann, 1898) throws interesting light on what was probably the prevailing mechanism for fixation of Conulata in early growth stages. This consists of a minute subcircular chitinophosphatic disc which may have fine radial striae on its lower or proximal surface and bear a pair of rings that define the margins or marginal impressions of an inner and outer inverted bowl above the disc (Fig. 28). The disc and bowls probably were not held in place by sucker-like action but by relatively firm cementation of the chitin to the substratum, thus paralleling the chitinous attachment of some modern stauromedusans which cannot move about after place of growth is determined by early fixation. The inverted bowls or cups of the conulariid attachment doubtless served a buttressing function.

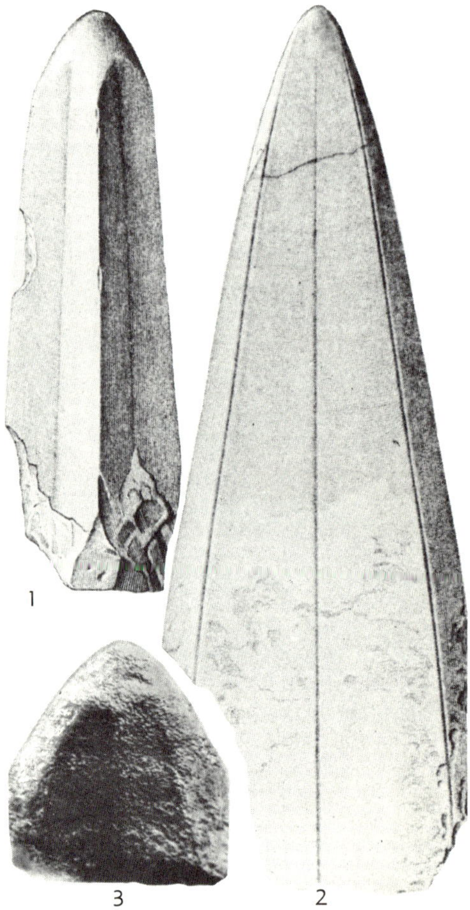

Fig. 27. Smoothly rounded proximal extremity of unattached conulariids. *1, Anaconularia anomala*, M.Ord., Czech., ×1 (Barrande). *2. Exoconularia consobrina*, M.Ord., Czech., ×1 (Barrande). *3, E. pyramidata*, U.Ord., Fr.(Normandy), ×1 (Kowalski).

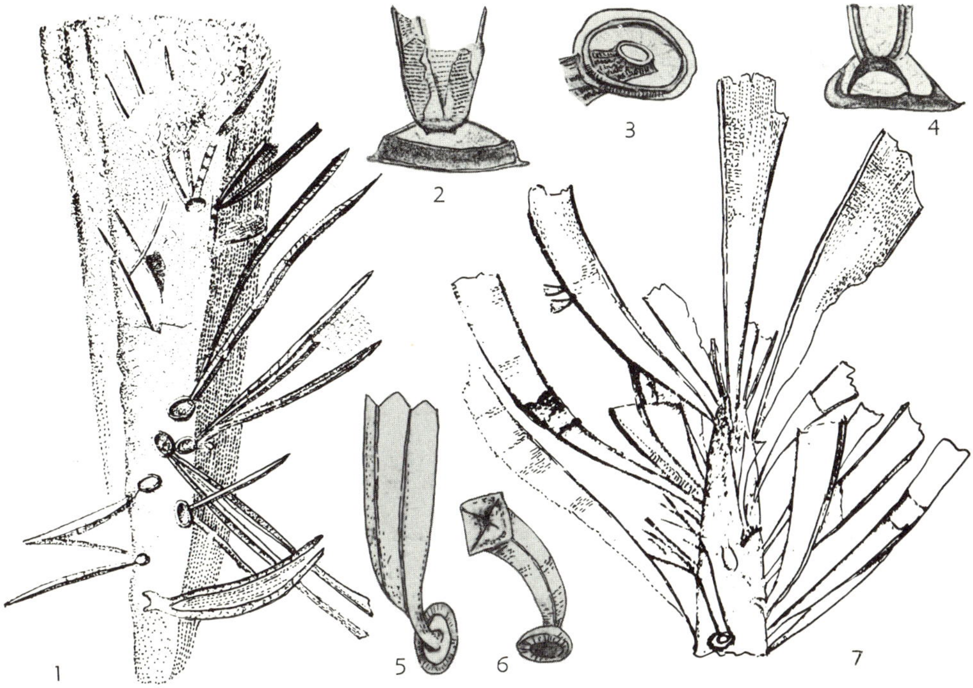

FIG. 28. Attachment of conulariids. *1-7, Sphenothallus angustifolius,* M.Ord., N.Y.; *1,* adult with several young individuals fixed to its periderm, some clearly showing fixation discs but generally with only the linear thickenings at corners of their periderm preserved, ×2; *2-4,* fixation discs, *4* showing longitudinal section, ×10; *5,6,* juvenile individuals, *5* showing lappet-like distal prolongation of faces and *6* showing aperture closed by inbending of these lappets, ×8; *7,* holotype of *S. angustifolius* consisting of several individuals flattened to leaflike form, one (upper left) with attached young specimens, ×1 (all after RUEDEMANN).

UNATTACHED MODE OF LIFE

That many conulariids, if not most of them, were not attached during adult stages is indicated by the frequently observed occurrence of a rounded smooth apical end of mature-sized fossils. The smooth surface, convex toward the apex, is the exposed part of secondarily secreted periderm joined peripherally to the 4 side walls in the manner of a nautiloid septum (Fig. 27). Because longitudinal ridges or walls directed radially inward occur in some conulariids and are termed septa, the curved transverse peridermal structure is called diaphragm (*cloison,* Fr.; *Schott,* Ger.). The uniform adapical convexity of the diaphragm, its smooth confluence with the lateral walls, and common occurrence not far from the original apex indicate that secretion of a diphragm was not for the purpose of closing off an open end produced by accidental injury but a normal adaptation for breaking free from an early-life anchorage (Figs. 24,*1*; 27). Fracture of the flexible, elastic periderm is unlikely and unless resorption and weakening of a band of the side walls were effected as an antecedent to breaking free and building the diaphragm, it is difficult to understand how the change from attached to free-living took place. In any case, no recorded observations support the conjecture that free-living conulariids lay inertly on the sea bottom, resting on any of their 4 sides; instead, it seems reasonable to postulate a free-swimming medusoid existence, as illustrated by a reconstruction drawn by KIDERLEN (1937) (Fig. 23, *3a*). The possibility of increased buoyancy by development of a small gas chamber next to the apical diaphragm is worthy of mention, although the thin periderm of the conulariids surely could be carried without much trouble by the animal in swimming.

In comprehensive view of the characters found to distinguish the conulariids and allied forms on the one hand and modern Scyphomedusae on the other, association of all them in the class Scyphozoa seems most appropriate. Accordingly, they are so classified here. Distinctions that serve to separate the Conulata from the Scyphomedusae are as far-reaching, however, as their resemblances are important. This is taken into account in defining each group as a subclass.

REFERENCES

Barrande, Joachim

(1) 1867, *Ordre des Ptéropodes:* Système silurien du centre de la Bohême (Prague), v. 3, pt. 1, xv+179 p., 16 pl.

Bouček, Bedřich

(2) 1939, *Conularida:* in SCHINDEWOLF, O. H., Handbuch der Paläzoologie, Borntraeger (Berlin), Band 2A, Lief. 5, p. A113-A131, fig. 1-13.

———, & Ulrich, F.

(3) 1929, *Étude sur la coquille du genre Conularia Miller:* Vestník Státního Geologického Ústavu Československé republiky (Prag.), v. 5, p. 2-3.

Hyman, L. H.

(4) 1940, *Invertebrates: Protozoa through Ctenophora:* McGraw-Hill (New York), ix+726 p., 221 fig. (Scyphozoa, p. 497-538, fig. 161-174).

Kiderlen, H.

(5) 1937, *Die Conularien: über Bau und Leben der ersten Scyphozoa:* Neues Jahrb. f. Mineralogie, Beil.-Band 77, Abt. B, p. 113-169, fig. 1-47.

Mayer, A. G.

(6) 1910, *The Scyphomedusae:* in Medusae of the world, Carnegie Inst. Washington, Pub. 109, v. 3, p. 499-735, pl. 56-76, fig. 328-428.

Ruedemann, Rudolf

(7) 1898, *The discovery of a sessile Conularia:* New York State Geol. (Albany), Ann. Rept. 15, p. 701-720, pl. 1-5.

Termier, Henri, & Termier, Geneviève

(8) 1948, *Position systématique et biologie der conulaires:* Revue Scientifique (Paris), année 86, fasc. 12, no. 3300, p. 711-722, fig. 1-25.

SCYPHOMEDUSAE

By H. J. Harrington and R. C. Moore

CONTENTS

MORPHOLOGY

The Scyphomedusae are solitary, free-swimming animals consisting of a gelatinous bell either cuboidal or pyramidal, goblet,- trumpet-, dome-, bowl-, or saucer-shaped, typically displaying tetramerous symmetry, though in some cases arranged upon an hexameral plan. They range in size from a few to 50 cm. in diameter, but some species may attain dimensions of 2 m., being the largest known coelenterates.

The upper, aboral or convex side of the bell is termed **exumbrella;** the lower, oral or concave side is known as **subumbrella.** In the Coronatida, the exumbrellar surface is subdivided by a horizontal circular groove, the **coronal groove** (Fig. 30). The margin of the bell is usually scalloped into **lappets** and may bear a **velarium,** analogous to the velum of the hydrozoan medusae, but arising as an extension of the subumbrella. From the center of the subumbrella there hangs down a short, quadrangular tube, the **manubrium,** typically opened at its lower extremity in a 4-cornered **mouth.** The mouth angles are commonly drawn into short **oral lobes,** but in the Semaeostomatida they extend into long, curtain-like projections, called **oral arms** (Figs. 18, 33). In the Rhizostomatida the edges of the long, frilled, branched oral lobes, termed **mouth arms,** fuse so as to obliterate the mouth opening but contain instead hundreds or thousands of minute suctorial mouths or **ostioles.** The 4 corners of the mouth or the 4-branched mouth arms mark off the 2 main planes of tetramerous symmetry. The 4 umbrellar radii, placed 90° apart, upon which lie the 4 corners of the mouth, are termed **perradii;** the 4 radii at 45° position between the perradii are called **interradii;** and those located midway between these are called **adradii.**

The manubrium leads aborally into a gastric cavity which typically contains 4 soft-tissue partitions or **septa,** projecting along the interradii part way into the interior; these septa divide the cavity into a **central stomach** and 4 perradial **stomach pouches.** The free inner edge of the septa bear numerous tentacle-like **gastric filaments.** The septa are absent in the adult Semaeostomatida and Rhizostomatida, but commonly present in their scyphistoma larvae. The stomach pouches give rise to **radial canals,** which may be numerous and branched, leading to a **ring canal** running along the bell margin. In the orders Stauromedusida, Carybdeida and Coronatida, the subumbrellar surface bears 4 deep interradial, funnel-like pits, the **subumbrellar funnels,** which probably aid in respiration. In the Stauromedusida, Carybdeida and Coronatida, the **gonads** are borne on both sides of the gastric septa as elongated bodies projecting into the stomach pouches; the total number of gonads is 8, 2 in each pouch. In the Semaeostomatida and Rhizostomatida, which lack septa, the gonads (usually 4) occur in the floor of the gastric pouches.

Tentacles and **sensory organs** arise from the bell margin, usually to the number of 4 or of some multiple thereof. The tentacles are borne either between sensory organs, in niches between lappets, on the lappets, or on the subumbrellar surface. Tentacles may be solid or hollow, single or in clusters, and may be very numerous; they are absent in the Rhizostomatida. In the Carybdeida and Coronatida they are borne on gelatinous basal expansions, the **pedalia.** The sensory organs or **rhopalia** are located either on the sides of the bell, on pedalia, or in niches between lappets; they are essentially tentaculocysts acting as organs of equilibrium. In many genera they are provided with **ocelli,** sensitive to light. Nematocysts are abundantly distributed on the tentacles, oral lobes, oral arms, and mouth arms.

DISTRIBUTION

Twelve genera of fossil medusae are now assigned to the Scyphomedusae, but with 4 exceptions, the reference to this subclass is somewhat tentative and provisional. Two doubtful, very imperfectly known genera of uncertain family relationships have been described from the Lower Cambrian of South Australia. The remainder are mostly Upper Jurassic forms from Solnhofen, Germany, and species from Upper Cretaceous and Eocene strata of Italy, Albania, Poland, and Greece.

CLASSIFICATION

The number of fossil genera in each category is tabulated after the corresponding taxa. A question mark preceding a numeral (as ?9) indicates number of tentatively referred genera.

Main Divisions of Scyphomedusae

Scyphomedusae *(subclass)* (4+?9). *?L.Cam., U.Jur.-Rec.*
  Stauromedusida *(order). Rec.*
    Tesseranthidae. *Rec.*
    Eleutherocarpidae. *Rec.*
    Cleistocarpidae. *Rec.*
  Carybdeida *(order)* (?1). *?U.Jur., Rec.*
    Carybdeidae (?1). *?U.Jur., Rec.*
  Coronatida *(order)* (1+?4). *?L.Cam., U.Jur.-Rec.*
    Periphyllidae (1). *U.Jur.-Rec.*
      Periphyllinae. *Rec.*
      Paraphyllininae (1). *U.Jur.-Rec.*
    Ephyridae. *Rec.*
    Atorellidae. *Rec.*
    Collaspididae (?4). *?L.Cam., ?U.Jur., ?U.Cret., ?Eoc., Rec.*
  Semaeostomatida *(order)* (2). *?U.Jur., Rec.*
    Pelagiidae. *Rec.*
    Cyaneidae. *Rec.*
    Ulmariidae. *Rec.*
    ?Semaeostomitidae (1). *U.Jur.*
    ?Eulithotidae (1). *U.Jur.*
  Lithorhizostomatida *(order)* (1). *U.Jur.*
    Rhizostomitidae (1). *U.Jur.*
  Rhizostomatida *(order)* (?1). *?U.Jur., Rec.*
    Cassiopeidae. *Rec.*
    Cepheidae. *Rec.*
    Catostylidae. *Rec.*
    Leptobrachiidae (?1). *?U.Jur., Rec.*
    Rhizostomatidae. *Rec.*
    Archirhizidae. *Rec.*
?Scyphomedusae incertae sedis (2). *L.Cam.*

SYSTEMATIC DESCRIPTIONS

Subclass SCYPHOMEDUSAE Lankester, 1881

[=Acraspeda GEGENBAUR, 1856; Scyphozoa HAECKEL, 1891 *(partim)*]

A diagnosis of this subclass appears in a preceding section which states distinguishing features of the Scyphozoa and discusses division of the class into subclasses Conulata and Scyphomedusae (p. F27-F38). *?L. Cam., U.Jur.-Rec.*

Order STAUROMEDUSIDA Haeckel, 1880

[*nom. correct.* HARRINGTON & MOORE, herein (*pro* Stauromedusae HKL., 1880)] [=Lucernaridae JOHNSTON, 1847 *(partim)*; Lucernaria CLARK, 1863]

Bell pyramidal-conical, flaring at oral end into broad concave subumbrella, tapering aborally into stalk, commonly attached to objects by aboral apex; margin usually with 8 adradial lobes bearing tentacles, commonly hollow, knobbed and in clusters; rhopalia absent or replaced by 8 anchors with ocelli or lithocysts; 4 long interradial septa; 4 large and deep subumbrellar funnels typically present; mouth cruciform, simple. *Rec.*

Genera belonging to this order are classified into 3 families, none of which contains known fossil representatives. They are: Tesseranthidae HAECKEL, 1880, pelagic, free-swimming medusae without lobes and with solid, nonknobbed tentacles; Eleutherocarpidae CLARK, 1863, sessile medusae with 4 simple perradial stomach pouches; and Cleistocarpidae CLARK, 1863, sessile medusae with 4 perradial stomach pouches divided into 2 spaces. Twelve genera known, including *Lucernaria* O.F.MÜLLER, 1776, and *Haliclystus* CLARK, 1863. Stauromedusida are inhabitants of cold seas; sessile forms found in bays and coastal waters.

Order CARYBDEIDA Claus, 1886

[*nom. correct.* MAYER, 1910 (*pro* Charybdeida CLAUS, 1886)] [=Marsupialidae L. AGASSIZ, 1862; Cubomedusae HAECKEL, 1880]

Bell cubical with interradial angles and perradial sides; margin without lappets; vel-

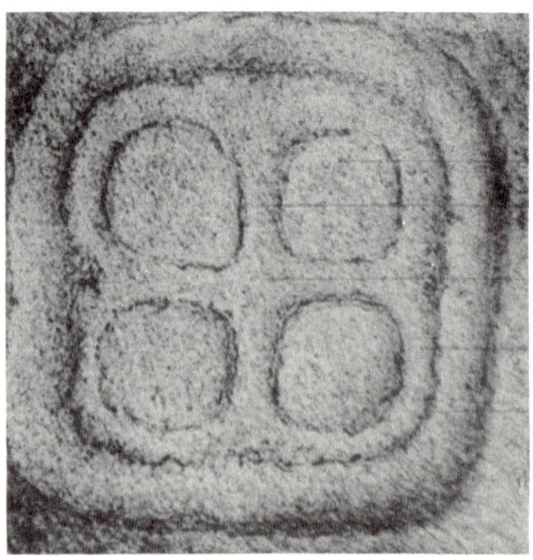

FIG. 29. Carybdeida: *Quadrimedusina quadrata,* Jur., Ger.; holotype, ×0.7 (HAECKEL).

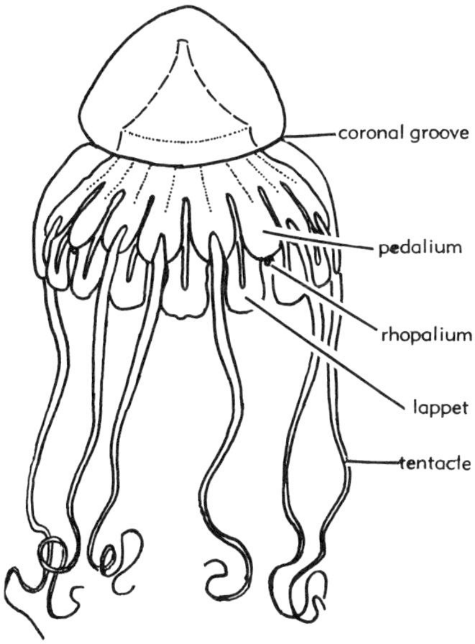

FIG. 30. Coronatida: *Periphylla hyacinthina,* one of the most common living members of the order, found in deep waters of all oceans. It has a purple bell with 16 lappets at the margin, 12 bearing tentacles (L. H. HYMAN, *Invertebrates: Protozoa through Ctenophora,* McGraw-Hill, New York).

arium present; 4 interradial hollow tentacles, single or in clusters, bordered by spatuliform pedalia; 4 perradial knob-shaped rhopalia located within niches upon sides of bell; 4 interradial septa; subumbrellar funnels absent; mouth cruciform, simple; oral lobes very short. *?U.Jur., Rec.*

Inhabitants of the warmer waters of all present oceans, found mostly in the open sea but also swarming in harbors and bays.

Family CARYBDEIDAE Gegenbaur, 1856

Characters of order. *?U.Jur., Rec.*

Carybdea PÉRON & LESUEUR, 1809 [**C. marsupialis*]. *Rec.* Other living genera: *Procharybdis* HAECKEL, 1880; *Tamoya* F. MÜLLER, 1859; *Tripedalia* CONANT, 1897; *Chiropsalmus* L.AGASSIZ, 1862; *Chirodropus* HKL., 1880.

?Quadrimedusina HARRINGTON & MOORE, 1955 [**Medusites quadratus* HAECKEL, 1869]. Subumbrellar impression of square outline, with rounded angles; margin entire; submarginal groove present (?ring canal); central area with 4 radial grooves (?perradial canals) reaching submarginal groove at mid-distance between angles; resulting quadrants with single, large subcircular marking. *U.Jur.,* Ger.——FIG. 29. **Q. quadrata* (HAECKEL); holotype, ×0.7 (HAECKEL).

Order CORONATIDA Vanhöffen, 1892

[*nom. correct.* HARRINGTON & MOORE, herein (*pro* Coronatae VANHÖFFEN, 1892] [=Discomedusae HAECKEL, 1866 (*partim*); Coronata MAAS, 1903 (*non* JAEKEL, 1918)]

Bell conical, dome-shaped or discoidal; umbrella divided by horizontal coronal groove; margin cleft into lappets alternating with pedalia; 4 to 32 rhopalia and 4 to 32 solid tentacles borne by alternating pedalia; 4 subumbrellar funnels; mouth cruciform, simple; manubrium short. *?L.Cam., U.Jur.-Rec.*

Living forms are typical inhabitants of the deeper parts of the ocean but surface forms in warmer waters are also known (Fig. 30).

Family PERIPHYLLIDAE Claus, 1886

[=Peromedusae HAECKEL, 1880 (*partim*)]

Coronatida with 4 rhopalia, 4 or more lappets and 4 or more pedalia bearing tentacles. *U.Jur.-Rec.*

Subfamily PERIPHYLLINAE Claus, 1886

[*nom. transl.* BIGELOW, 1909 (*ex* Periphyllidae CLAUS, 1886)]

Rhopalia interradial; tentacles 4 to 28; lappets 8 to 32. *Rec.*

Periphylla STEENSTRUP, 1837 [**P. hyacinthina*]. *Rec.* Other living genera: *Pericolpa* HAECKEL, 1880; *Periphyllopsis* VANHÖFFEN, 1900; *Nauphontopsis* FEWKES, 1885.

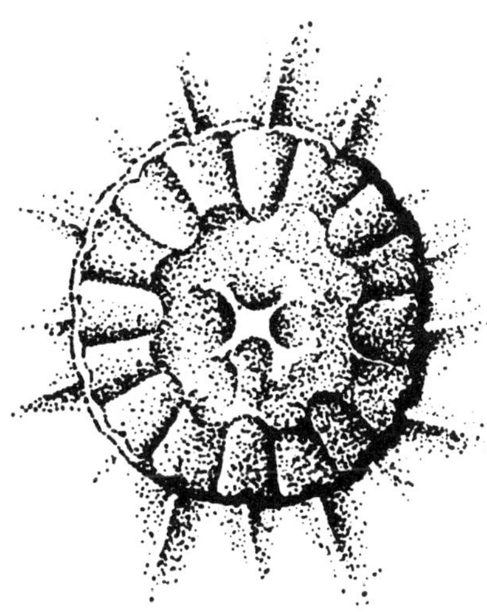

FIG. 31. Coronatida: *Epiphyllina distincta,* U.Jur., Ger.; holotype, ×3 (KIESLINGER).

Subfamily PARAPHYLLININAE Maas, 1903

[*nom. transl.* Bigelow, 1909 (*ex* Paraphyllinidae Maas, 1903)]

Rhopalia perradial; lappets 16; tentacles 12 (4 interradial, 8 adradial). *U.Jur.-Rec.*

Paraphyllina Maas, 1903 [**P. intermedia*]. *Rec.* Only living genus.

Epiphyllina Kieslinger, 1939 [*pro Paraphyllites* Maas, 1906 (*non* Hyatt, 1900)] [**Paraphyllites distinctus* Maas, 1906]. Mid-field smooth; pedalia 16, wide, rhombic, alternating along periphery with 16 narrow crescentic lappets; tentacles 12, 4 carried by interradial, 8 by adradial pedalia. Differs from *Paraphyllina* in having 8 crescentic ad

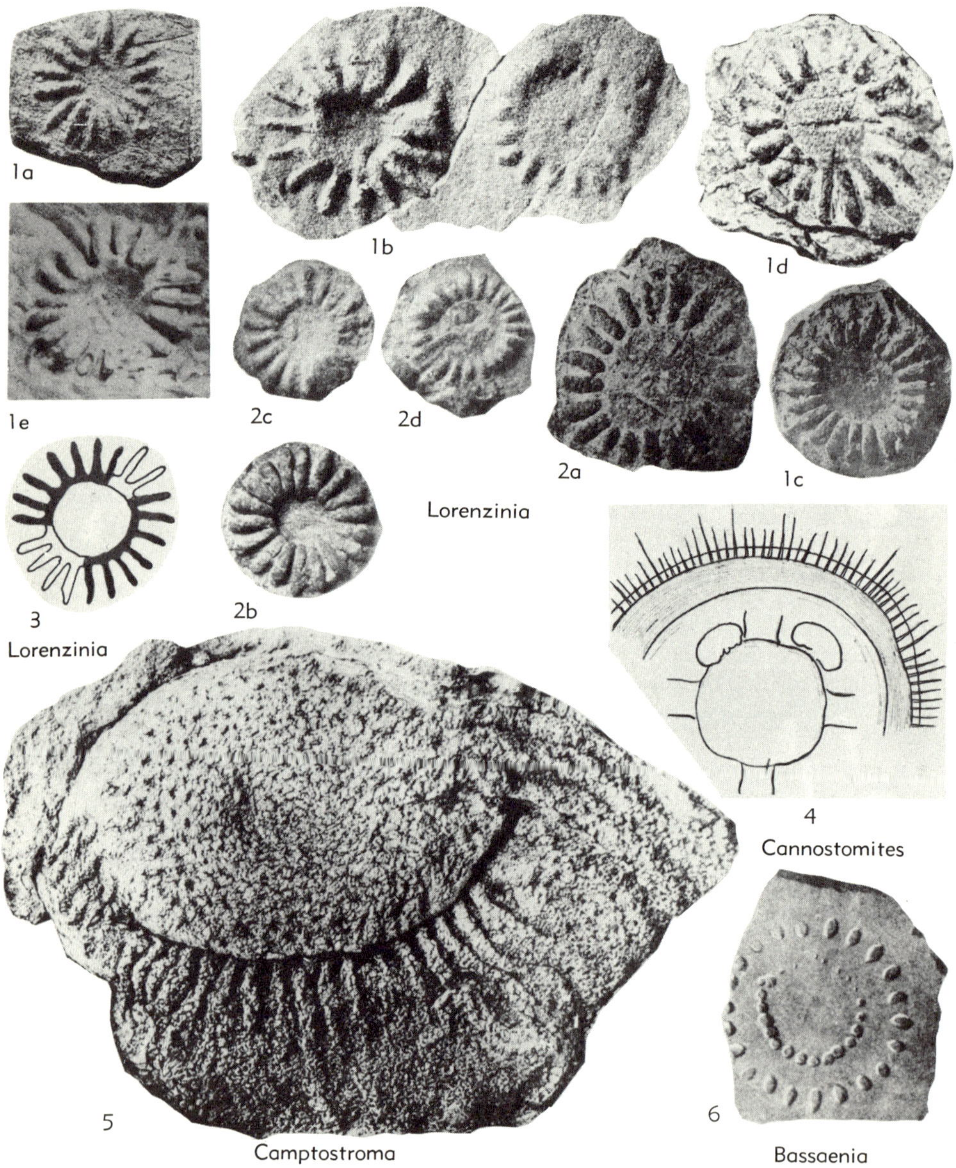

Fig. 32. Coronatida: L.Cam.-Eoc. fossil forms. *1, Lorenzinia apenninica,* ?Cret.-Eoc., Eu.; *1a,* holotype, ×1; *1b-e,* other specimens, all ×0.7 except *1c,* ×0.5 (Gortani and others). *2, L. carpathica,* Eoc., Eu.; *2a,* holotype, ×0.5; *2b-2d,* other specimens, ×0.7, ×0.5 (Zuber and others). *3, L. kulczynskii,* Eoc., Pol.; holotype, ×0.7 (Kuznian). *4, Cannostomites multicirratus,* U.Jur., Ger.; holotype, ×0.2 (Kieslinger). *5, Camptostroma roddyi,* L.Cam., Pa.; holotype, ×1 (Ruedemann). *6, Bassaenia moreae,* U.Cret., Greece; holotype, ×0.5 (Renz).

radial gonads placed 45° apart instead of grouped in pairs on both sides of the 4 interradii. *U.Jur.*, Ger.——FIG. 31. **E. distincta* (MAAS); holotype, ×3 (KIESLINGER).

Family EPHYRIDAE Haeckel, 1880

[*ex Ephyra* HKL., 1880 (=*Palephyra* HKL., 1880)] [=Ephyridae+Linergidae HKL., 1880; Ephyropsidae CLAUS, 1883]

Coronatida with 4 perradial and 4 interradial rhopalia; 8 or more tentacles and 16 or more lappets. *Rec.*

Although *Ephyropsites jurassicus* VON AMMON, 1906, from Solnhofen, formerly was placed in this family, probably it is an exumbrellar impression of *Rhizostomites admirandus* (Lithorhizostomatida); accordingly, no fossil representatives of the family are now recognized.

Palephyra HAECKEL, 1880 [**P. antigua*] [=*Ephyra, Zonephyra* HKL., 1880]. *Rec.* Other living genera: *Nausithoe* KÖLLIKER, 1853; *Linuche* ESCHSCHOLTZ, 1829.

Family ATORELLIDAE Vanhöffen, 1902

Coronatida with 6 rhopalia. *Rec.*

Atorella VANHÖFFEN, 1902 [**A. subglobosa*]. *Rec.* Only known genus.

Family COLLASPIDIDAE Haeckel, 1880

[*nom. correct.* HARRINGTON & MOORE, herein (*pro* Collaspidae HKL., 1880, *ex Collaspis* HKL., 1880=*Atolla* HKL., 1880); HAECKEL's term Collaspidae was accepted as family name by MAYER, 1910, and having priority, is valid according to Copenhagen Decisions (1953)] [=Atollidae BIGELOW, 1909]

Coronatida with more than 8 sense organs alternating with an equal number of tentacles; lappets twice as numerous as tentacles. *?L.Cam., ?U.Jur., ?U.Cret., ?Eoc., Rec.*

Atolla HAECKEL, 1880 [**A. wyvillei*] [=*Collaspis* HKL., 1880]. *Rec.* Only living genus.

?Lorenzinia GABELLI, 1900 [**L. apenninica*]. Body discoidal, convex; mid-field smooth, concave; margin with 16 to 24 prominent, outstretched, elongated subrectangular lobes (?pedalia), sharply separated from each other, raised at proximal end. *?Cret.*, Italy; *Eoc.*, Italy-Cyprus-Pol.-Albania. This problematic "form genus" may be a coronatid medusa. Abundant impressions with 11 to 13 irregular lobes, found associated together in some Eocene strata of Guipuzcoa, Spain, have been referred to *L. apenninica* but actually they belong to a different problematic organism which is not a medusoid.——FIG. 32,*1*. **L. apenninica*, Eoc., Eu.; *1a*, holotype, Italy, ×1 (GORTANI); *1b,c*, other specimens from Italy, ×0.7, ×0.5; (*1b*, SIMONELLI; *1c*, AGRISANI; *1d*, specimen from Greece, ×0.7 (MITZOPOULOS); *1e*, specimen from Cyprus, ×0.7 (RENZ).——FIG. 32,*2*. *L. carpathica* (ZUBER), Eoc., Eu.; *2a*, holotype. Pol., ×0.5 (ZUBER); *2b*, another specimen from Poland, ×0.7 (KUZNIAR); *2c,d*, specimens from Albania, ×0.5 (SAXL).——FIG. 32,*3*. *L. kulczynskii* (KUZNIAR), Eoc., Pol.; holotype, ×0.7 (KUZNIAR).

?Bassaenia RENZ, 1925 [**Lorenzinia (B.) moreae*]. Impressions consisting of an external ring of 22 separated, elliptical protuberances and an internal ring with an equal number of separated, smaller knobs in the radii of the larger ones. *U.Cret.*, Greece. (There is no apparent reason to regard this problematic form as a "subgenus" of *Lorenzinia*, as done both by RENZ and KIESLINGER. Its assignment to the Collaspididae is very questionable.)——FIG. 32,*6*. **B. moreae*; holotype, ×0.5 (RENZ).

?Cannostomites MAAS, 1902 [**C. multicirratus*]. Bell with strongly developed ring muscle; marginal lappets numerous (?136), somewhat irregular; tentacles relatively short, one at each notch between lappets; mid-field with large central ?stomach of circular outline, truncating 4 wide radial bands (?oral arms); quadrants between radial bands bearing reniform depression adjacent to ?stomach, regarded as gonads. Known by a single subumbrellar impression. *U.Jur.*, Ger.—FIG. 32,*4*. **C. multicirratus;* holotype, ×0.2 (KIESLINGER).

?Camptostroma RUEDEMANN, 1933 [**C. roddyi*]. Body discoidal, outline elliptical due to distortion, upper side strongly convex, underside concave; upper side with large, circular, convex, rugose mid-field, separated by deep annular groove from outer ring of numerous (?52 to 60) moderately long, narrow radial lobes, commonly somewhat curved near distal extremity; lobes and underside with tiny rounded spots in a polygonal network of fine lines. RUEDEMANN interpreted the rugosity of the mid-field as due to a reticulate meshwork of tiny spicules surrounding round pores and monticules, and that of the lobes and underside as due to a polygonal network of small platelike spicules. Accordingly, he regarded the fossil as a colonial hydrozoan. Accepted here is the view of KIESLINGER, who interpreted the rugosity as due to shrinkage of a gelatinous mass during dessication and placed the genus among the coronatids. *L.Cam.*, Pa.; *L.Ord.*, Ger.——FIG. 32.*5* **C. roddyi*, L.Cam., Pa.; holotype, ×1 (RUEDEMANN). [Other species: *?C. germanicum* HUNDT, L. Ord. (Tremadoc), Ger.; known from single, very obscure impression.]

Order SEMAEOSTOMATIDA L.Agassiz, 1862

[*nom. correct.* HARRINGTON & MOORE, herein (*pro* Semaeostomae L.AGASSIZ, 1862)] [=Semostomae HAECKEL, 1880; Discomedusae HKL., 1880 (*partim*); Semaeostomata VANHÖFFEN, 1888; Semaeostomae MAAS, 1907]

Bell discoidal, without coronal furrow and without pedalia; margin cleft into lappets; tentacles hollow, arising from subumbrella, lappets or niches between lappets;

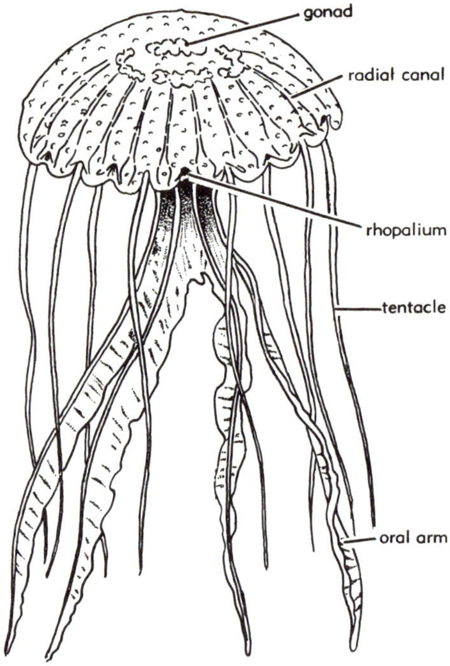

FIG. 33. Semaeostomatida: *Pelagia*, a leading modern representative which inhabits open oceans; oral arms long (L. H. HYMAN, *Invertebrata: Protozoa through Ctenophora*, McGraw-Hill, New York).

rhopalia typically 8 or 16, arising from niches between lappets; radial septa absent; mouth central, simple, with the 4 perradial angles developed into large gelatinous oral arms. ?U.Jur., Rec.

Semaeostomatids are inhabitants of coastal waters of all oceans in all latitudes, commonly occurring in great numbers.

Family PELAGIIDAE Gegenbaur, 1856

Semaeostomatida with 8 or 16 rhopalia (4 perradial, 4 interradial, and when present, 8 adradial); 8 or more tentacles arising from niches between lappets; 16 to 64 lappets; simple lenticular stomach giving rise to completely separated, radiating pouches; ring canal absent; gonads occupying 4 interradial folds in wall of subumbrella; mouth cruciform. *Rec.*

Pelagia PÉRON & LESUEUR, 1809 [**Medusa noctiluca* FORSKÅL, 1775]. *Rec.*——FIG. 33. *Pelagia* sp. (HYMAN). Other living genera: Chrysaora PÉRON & LESUEUR, 1809; *Dactylometra* L.AGASSIZ, 1862; *Kuragea* KISHINOUYE, 1902; *Sanderia* GÖTTE, 1886.

Family CYANEIDAE L.Agassiz, 1862

Semaeostomatida with 8 or 16 rhopalia; tentacles in 8 or 16 clusters or arising diffusely in a wide annulus from subumbrella; stomach giving rise to radiating peripheral pouches and these to numerous branching nonanastomosing blind canals in lappets; ring canal absent; gonads in 4 complexly folded interradial outpocketings of wall of subumbrella; oral arms curtain-like. *Rec.*

Cyanea PÉRON & LESUEUR, 1809 [**Medusa capillata* LINNÉ, 1746]. *Rec.* (Fig. 18). Other living genera: *Desdemona* L.AGASSIZ, 1862; *Drymonema* HAECKEL, 1880; *Patera* LESSON, 1843.

Family ULMARIIDAE Haeckel, 1880

[*nom. correct.* HARRINGTON & MOORE, herein (*pro* Ulmaridae HAECKEL, 1877, *ex Ulmaris* HKL., 1880 =*Discomedusa* CLAUS, 1877); HAECKEL's term Ulmaridae was accepted as family name by MAYER, 1910, and having priority, is valid according to Copenhagen Decisions (1953)] [=Flosculidae + Ulmaridae HKL., 1880]

Semaeostomatida with 8 or 16 rhopalia; lappets 32 to 64; tentacles 24 to numerous; radial canals simple or branched, some anastomosing; ring canal present; gonads 4, interradial; oral arms curtain-like. *Rec.*

Discomedusa CLAUS, 1877 [**D. lobata*] [=*Ulmaris, Umbrosa* HAECKEL, 1880]. *Rec.* Several other living genera, including *Aurellia* PÉRON & LESUEUR, 1809, and *Phacellophora* BRANDT, 1835.

?Family SEMAEOSTOMITIDAE Harrington & Moore, nov.

[=Lithosemaeiden HAECKEL, 1874 (invalid vernacular name not based on generic name)]

Bell with numerous (?112) lappets and an equal number of tentacles; stomach giving rise to 4 perradial pouches; radial canals 16, simple (4 perradial, 4 interradial, 8 adradial); ring canal present; gonads 4, interradial; oral arms 4, very long. *U.Jur.*

Semaeostomites HAECKEL, 1874 [**S. zitteli*]. Tentacles simple, arising from niches between lappets; gonads large, centrifugal; 16 subtriangular adradial markings between distal ends of stomach pouches and gonads; mouth opening indistinct. Known from single specimen. *U.Jur.*, Ger.——FIG. 34. **S. zitteli*; holotype, ×1.2 (WALCOTT).

?Family EULITHOTIDAE Kieslinger, 1939

[=Eulithoten HAECKEL, 1869 (invalid vernacular name)]

Bell dome-shaped, with 16 lappets, 8 clusters of tentacles and 8 rhopalia; gonads 16; oral arms 4, short. *U.Jur.*

Eulithota HAECKEL, 1869 [**E. fasciculata* (=*Solnhofenistomites stechowi* KUHN, 1839, obj.)]

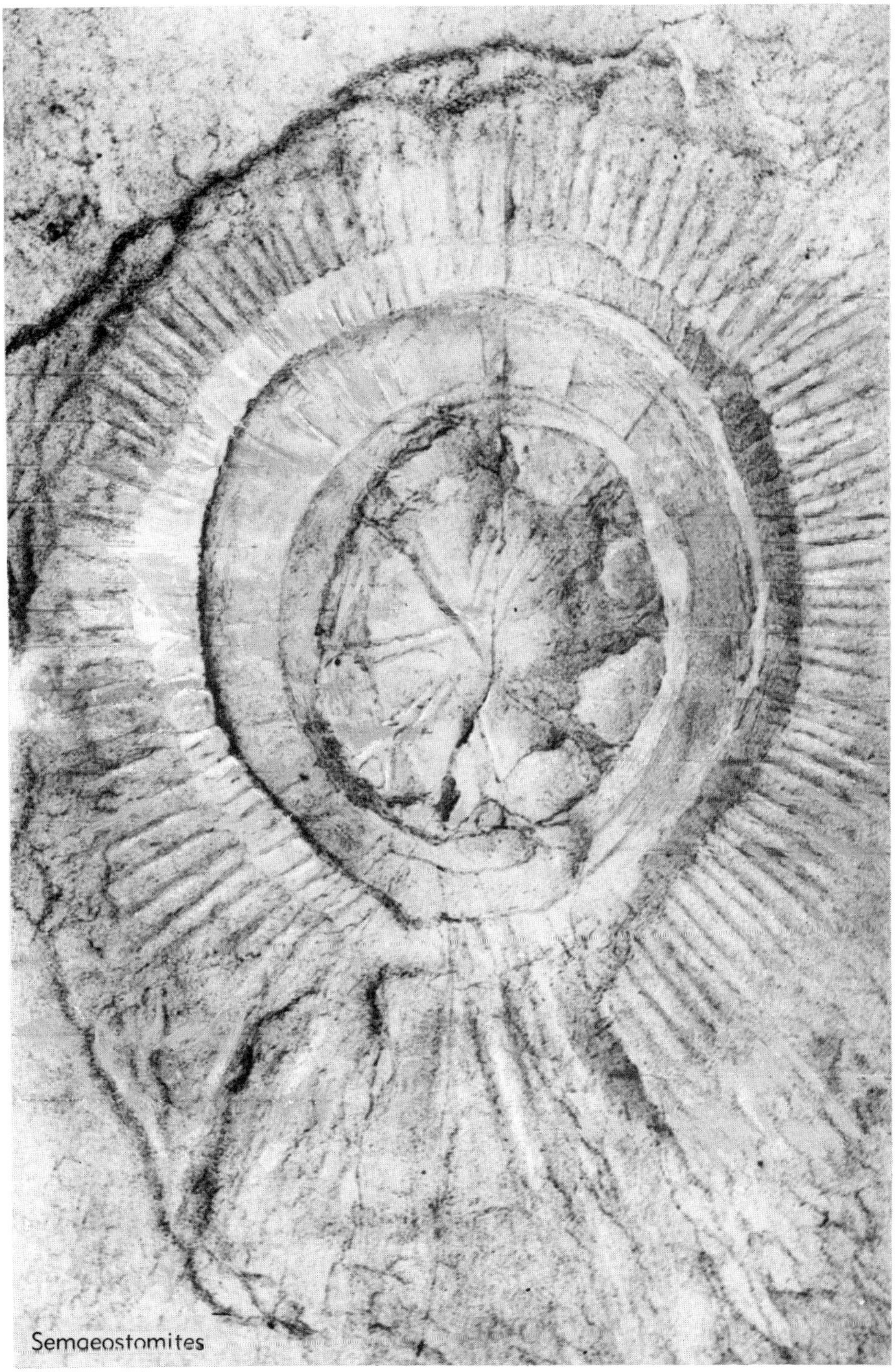

FIG. 34. Semaeostomatida: *Semaeostomites zitteli,* U.Jur., Ger.; holotype, ×1.2 (WALCOTT).

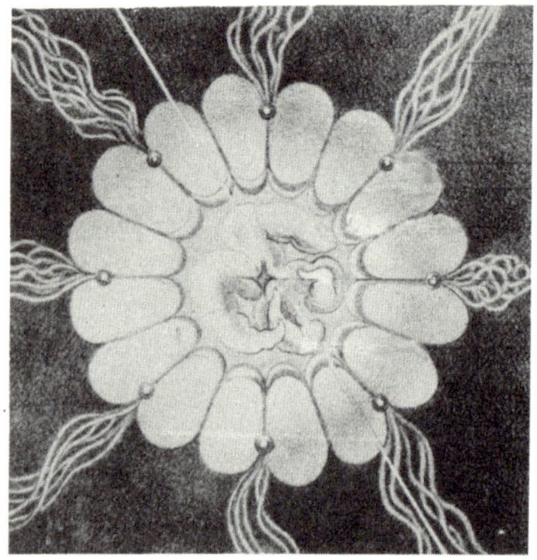

FIG. 35. Semaeostomatida: *Eulithota fasciculata*, subumbrellar surface (reconstr.), ×0.7 (HAECKEL).

[=*Solnhofenistomites* KUHN, 1938 (obj.)]. Tentacles in 8 clusters arising from niches between lappets; knob at base of each cluster interpreted as rhopalium; radial canals 16; gonads crescentic; mouth opening indistinct (?cruciform); faint indication of 4 simple, short oral arms. Known from single specimen. *U.Jur.*, Ger.——FIG. 35. **E. fasciculata;* subumbrellar surface (reconstr.), ×0.7 (HAECKEL).

Order LITHORHIZOSTOMATIDA von Ammon, 1886

[*nom. transl. et correct.* HARRINGTON & MOORE, herein (*ex* Lithorhizostomeae VON AMMON, 1886)]

Bell dome-shaped, with 16 lappets, 8 clus-lappets; rhopalia 8 (4 perradial, 4 interradial); tentacles short, ?ramified, in 8 clusters, arising from exumbrellar surface; pedalia absent; gonadic sacs 4, reniform to subtriangular, with individual opercula; ring canal present; ring muscle powerful; mouth cruciform, functional, with nonfunctional secondary branches; oral arms absent. *U.Jur.*

This order is introduced to accommodate the famous Solnhofen *Rhizostomites admirandus* which, since HAECKEL'S original description, has commonly been referred to the order Rhizostomatida. The presence of a functional cruciform mouth (Fig. 36,*2*), absence of mouth arms and presence of 8 clusters of tentacles precludes reference to the Rhizostomatida without a drastic and unwarranted redefinition of this order. The clusters of tentacles, borne by the exumbrella, cannot be homologized with the tentaculiform extensions of the marginal lappets of *Lobonema* MAYER, the only known rhizostomatid with such organs. None of the excellently preserved specimens shows traces of oral arms. Long mouth arms, however, are present in *Leptobrachites trigonobrachius,* known from a single laterally compressed specimen from Solnhofen and regarded by KIESLINGER as a crushed individual of *R. admirandus.* There is nothing, however, to support this view, as practically the only features that can be distinguished in the poorly preserved specimen are the mouth arms (8 according to HAECKEL, 5 according to BRANDT). None of the diagnostic characters of *R. admirandus* are displayed. CASTER suggested that *R. admirandus* could be a trachylinid medusa, interpreting as exumbrellar the impressions regarded by all other authors as subumbrellar. When all the Solnhofen material originally described as belonging to different genera and species but shown by KIESLINGER to be different "preservation aspects" of *R. admirandus* is taken into account, CASTER'S ingenious interpretation does not resist a critical analysis.

Family RHIZOSTOMITIDAE Harrington & Moore, nov.

Characters of order. *U.Jur.*

Rhizostomites HAECKEL, 1866 [**R. admirandus* (=*R.? lithographicus* HKL., 1866; *Hexarhizites insignis* HKL., 1870; *Myogramma speciosum* MAAS, 1902; *M. speciosissimum* VON AMMON, 1908; *Ephyropsites jurassicus* VON AMMON, 1908] [=*Hexarhizites* HKL., 1870; *Myogramma* MAAS, 1902; *Ephyropsites* V. AMMON, 1908]. (1) SUBUMBRELLAR IMPRESSIONS OF SPECIMENS WITH RADIAL MUSCLES RELAXED (type of *Rhizostomites admirandus*). Mouth cruciform, functional (Fig. 36,*2*), each branch bifurcating into nonfunctional adradial grooves which give rise to 4 insert interradial lobes (forming an "iron-cross") and 4 exsert intercalary triangular perradial lobes; central portion of each lobe (insert and exsert) swollen into a low, rounded knob; oral arms absent; mouth area framed by rugose ring bearing 4 pads, one at distal end of each of the 4 interradial lobes; pads reniform to subtriangular, with inner concentric marking (?gonadic sacs and opercula); central area surrounded by annular zone, smooth adcentrally, concentrically striated along outer half, separated by sharply marked narrow band (?ring canal) from outer annular zone (ring muscle) which is concentrically striated; margin circular, cleft into 128 short, round lappets with 8 deeper perradial and interradial notches bearing small knobs regarded as rhopalia. (2) SUBUMBRELLAR

IMPRESSIONS OF SPECIMENS WITH RADIAL MUSCLES CONTRACTED (type of *Myogramma speciosum*). Differs from preceding in having pinnatisected marking (contracted radial muscles) in rings around mouth area; margin of bell contracted into 8-lobed rosette disclosing the tentacles borne by exumbrella. (3) EXUMBRELLAR IMPRESSION OF SPECIMENS WITH RADIAL MUSCLES RELAXED (type of *Ephyropsites jurassicus*). Central area circular, smooth, separated by sharp narrow ring from annular zone bearing 16 radial lobes (8 large, subtrapezoidal, 8 narrower, parallel-sided) marked off by fine raised striae and incompletely split by a less conspicuous radial line; each subtrapezoidal lobe bears a cluster of ?ramified tentacles located at ⅔ the distance from center of umbrella to periphery; each smaller lobe bears, at approximately same location, a short transverse stria from which springs outward a pair of short radial lines. (4) EXUMBRELLAR IMPRESSION OF SPECIMENS WITH RADIAL MUSCLES CONTRACTED (type of *Myogramma speciosissimum*). Differs from preceding in having pinnatisected radial markings ending at boundary of central area which is not sharply defined; the 16 principal radial striae continue onto this area for half the distance to center and gradually disappear leaving a central smooth mid-field. (5) *Hexarhizites insignis,* known from a single impression, is regarded as the subumbrellar surface (radial muscles relaxed) of an abnormal individual with 12 oral lobes. *U.Jur.,* Ger.——FIGS. 36-39. **R. admirandus;* 36,*1*, subumbrellar surface, radial muscles relaxed, ×0.28; 36,*2*, *ibid.*, another specimen, ×0.58 (both from BRANDT); 37, *ibid.*, a 3rd specimen, ×0.37 (WALCOTT); 38,*1*, *ibid.*, a 4th specimen, ×0.3; 39,*2*, *ibid.*, a 5th specimen, ×0.5

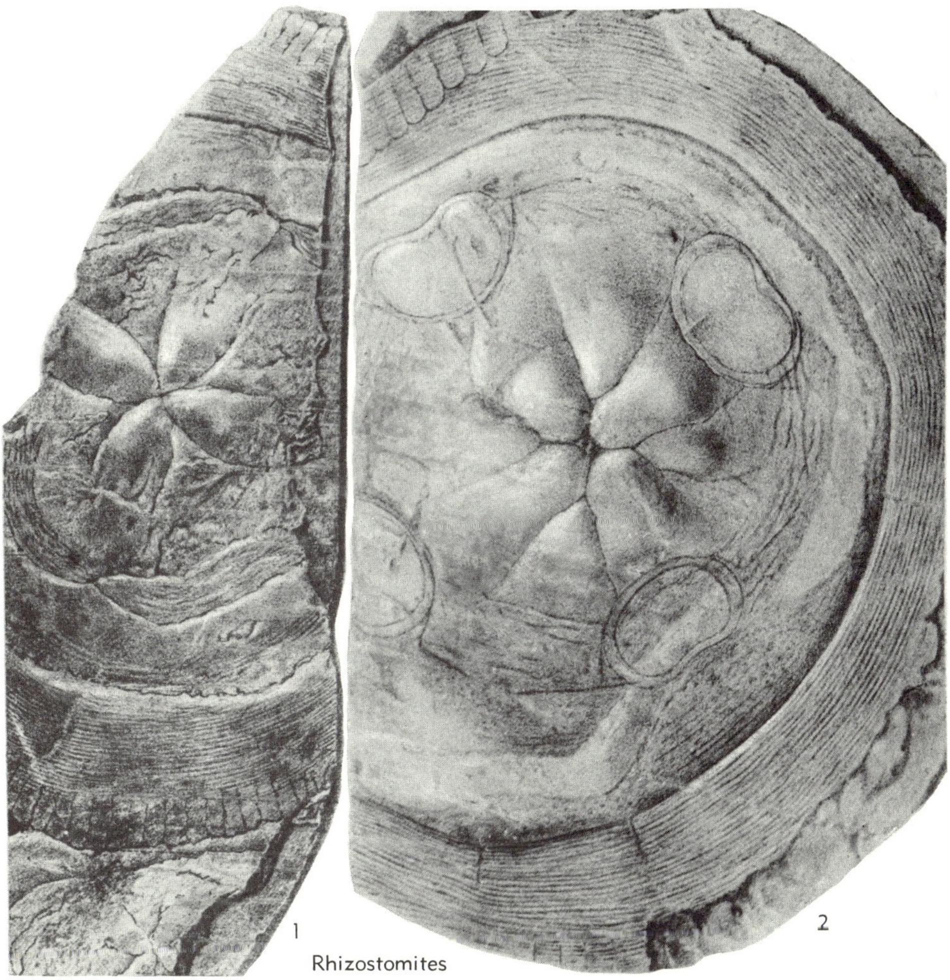

FIG. 36. Lithorhizostomatida: *Rhizostomites admirandus,* U.Jur., Ger.; *1,2,* specimens showing subumbrellar surface with radial muscles relaxed, ×0.28, ×0.52 (BRANDT).

both from VON AMMON); 39,*1*, subumbrellar surface, radial muscles contracted ("*M. speciosum*"), ×0.33 (MAAS); 38,*2*, exumbrellar surface, radial muscles relaxed ("*E. jurassicus*"), ×0.7; 38,*3*, exumbrellar surface, radial muscles contracted ("*M. speciosissimum*"), ×0.17 (both from VON AMMON).

Order RHIZOSTOMATIDA Cuvier, 1799

[*nom. correct.* HARRINGTON & MOORE, herein (*pro* Rhizostomae CUVIER, 1799)] [=Rhizostomidae ESCHSCHOLTZ, 1829; Rhizostomeaea L.AGASSIZ, 1862; Rhizostomata VANHÖFFEN, 1888]

Bell dome-shaped or discoidal, without coronal furrow, without pedalia and without marginal tentacles (except in *Lobonema* MAYER, in which the lappets are prolonged in tentaculiform organs); margin cleft into 8 or more lappets; rhopalia 8 or more; central mouth absent; numerous ostioles borne upon 8 adradial, fleshy, branchlike mouth arms which arise from center of subumbrella. *?U.Jur., Rec.*

Rhizostomatids are inhabitants of tropical seas and the majority of genera confined to the warm waters of the Indo-Pacific region.

Family CASSIOPEIDAE Claus, 1883

[=Rhizostomata pinnata VANHÖFFEN, 1888; Arcadomyaria MAAS, 1903].

Rhizostomatida with 8 linear, pinnately or complexly branching mouth arms. *Rec.*

Cassiopea PÉRON & LESUEUR, 1809 [**Medusa andromeda* FORSKÅL, 1775] [=*Polyclonia* HAECKEL, 1880]. *Rec.* Other genus: *Toreuma* HKL., 1880.

Family CEPHEIDAE Claus, 1883

[=Chaunostomidae CLAUS, 1883; Rhizostomata dichotoma VANHÖFFEN, 1888; Radiomyaria MAAS, 1903]

Rhizostomatida with 8 separate mouth

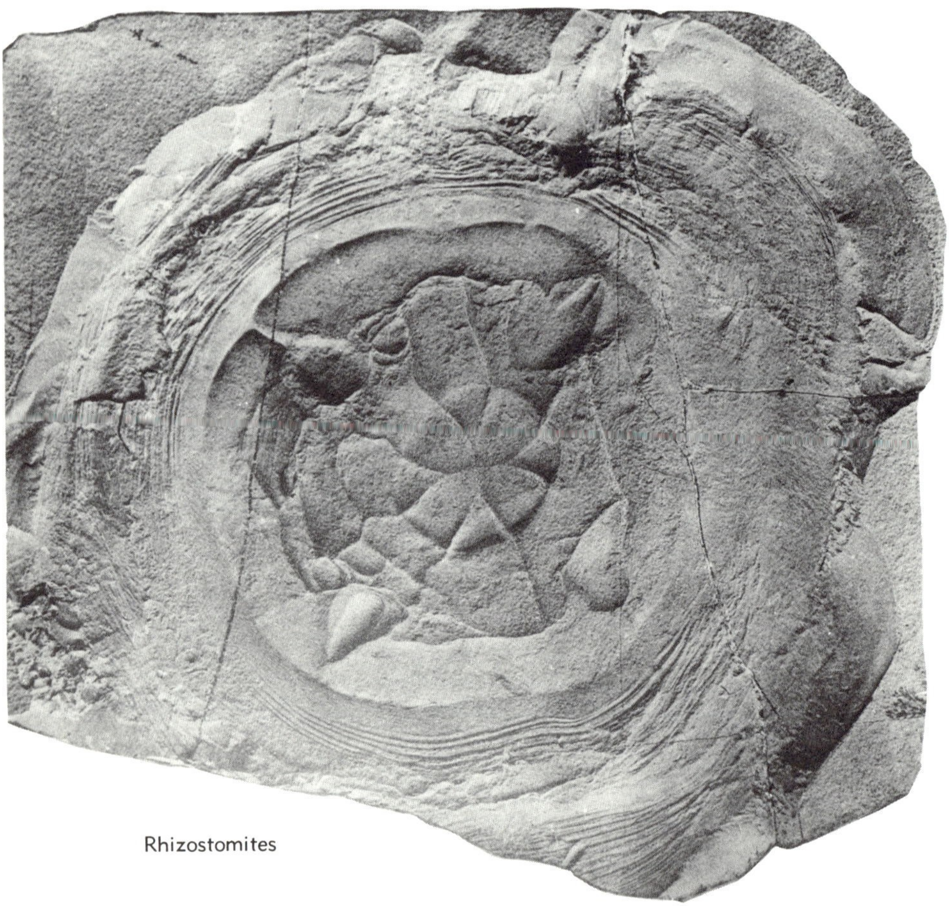

FIG. 37. Lithorhizostomatida: *Rhizostomites admirandus*, U.Jur., Ger.; subumbrellar surface, radial muscles relaxed (WALCOTT).

FIG. 38. Lithorhizostomatida: *Rhizostomites admirandus,* U.Jur., Ger.; *1,* subumbrellar surface, radial muscles relaxed; *2,* exumbrellar surface, radial muscles relaxed, ×0.7; *3,* exumbrellar surface, radial muscles contracted, ×0.17 (VON AMMON).

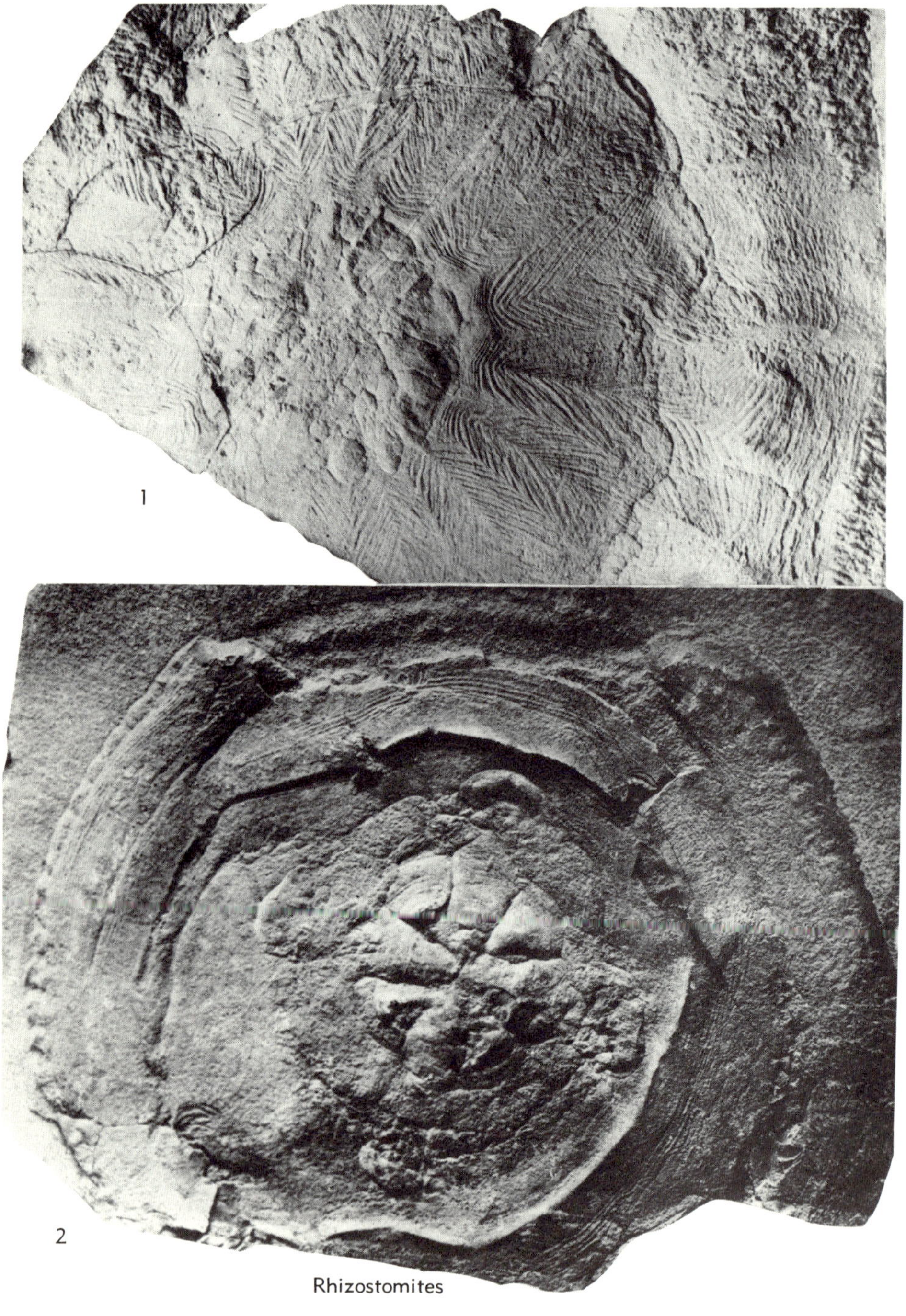

Fig. 39. Lithorhizostomatida: *Rhizostomites admirandus,* U.Jur., Ger.; *1,* subumbrellar surface, radial muscles contracted, ×0.33 (Maas); *2,* subumbrellar surface, radial muscles relaxed, ×0.5 (von Ammon).

arms the lower end of which give rise to 2 expanded leaflike side walls, the outer edges of which give rise to secondary branches and bear frilled ostioles; scapulets absent. *Rec.*

Cephea PÉRON & LESUEUR, 1809 [**Medusa octostyla* FORSKÅL, 1775] [=*Stylorhiza* HAECKEL, 1880; *Netrostoma, Halipetasus* SCHULTZE, 1898; *Microstylus, Perirhiza* KISHINOUYE, 1902]. *Rec.* Other living genera: *Polyrhiza, Cotylorhiza* L.AGASSIZ, 1862.

Family CATOSTYLIDAE Claus, 1883

[=Lychnorhizidae CLAUS, 1883; Rhizostomata triptera + R. trigona VANHÖFFEN, 1888; Rhizostomata triptera MAAS, 1903]

Rhizostomatida in which the lower parts of the 8 separate mouth arms are 3-winged or Y-shaped in cross section, being expanded in a ventral and 2 dorsal lamellae; frilled ostioles borne upon the 3 lamellae; scapulets absent. *Rec.*

Catostylus L.AGASSIZ, 1862 [**C. mosaicus*] [=*Rhacopilus*, L.AGASSIZ, 1862; *Toxoclytus* L.AGASSIZ, 1863; *Crambessa* HAECKEL, 1869; *Loborhiza* VANHÖFFEN, 1888]. *Rec.* Several other genera, including *Versura* HKL., 1880; *Lychnorhiza* HKL., 1890; *Lobonema* MAYER, 1910.

Family LEPTOBRACHIIDAE Claus, 1883

[=Rhizostomata lorifera VANHÖFFEN, 1888]

Rhizostomatida with very elongate, narrow, lashlike mouth arms; lower parts of arms 3-winged in cross section; ostioles developed upon and near the angles; upper parts of mouth arms very short and partially fused to arm disc by a series of arches spanning from one arm to another. *?U.Jur., Rec.*

Leptobrachia BRANDT, 1838 [**Rhizostoma leptotus* CHAMISSO & EYSENHARDT, 1821] [=*Leonura* HAECKEL, 1880]. *Rec.* Other living genera: *Thysanostoma* L.AGASSIZ, 1862; *Lorifera* HKL., 1880.

?**Leptobrachites** HAECKEL, 1869 [**L. trigonobrachius* (=*Pelagiopsis leuckarti* BRANDT, 1871, obj.)]. [=*Pelagiopsis* BRANDT, 1871 (obj.)]. Based on single, obscure specimen from Solnhofen. Bell dome-shaped; margin cleft into lappets; pedalia and tentacles absent; mouth arms very long. According to HAECKEL there are 48 lappets and 8 3-sided prismatic mouth arms. BRANDT accepted only 10 lappets and 5 mouth arms. VON AMMON regarded the fossil as a laterally crushed specimen of *Rhizostomites admirandus*, a view opposed by CASTER, *U.Jur.*, Ger.——FIG. 40. **L. trigonobrachius;* holotype, ×0.3 (BRANDT).

Family RHIZOSTOMATIDAE Claus, 1883

[*nom. correct.* HARRINGTON & MOORE, herein (*pro* Rhizostomidae CLAUS, 1883)] [=Stomolophidae CLAUS, 1883; Rhizostomata scapulata VANHÖFFEN, 1888]

Rhizostomatida in which each of the 8 mouth arms bears a pair of wing-shaped outgrowths (scapulets) which arise from dorsal side of each arm near its point of origin from arm disc; frilled ostioles developed upon upper and outer sides of scapulets and upon lower parts of mouth arms. *Rec.*

Rhizostoma CUVIER, 1799 [**R. pulmo*] [=*Pilema* HAECKEL, 1880]. *Rec.* Other living genera: *Eupilema, Rhopilema* HKL., 1880; *Stomolophus* L. AGASSIZ, 1862.

Family ARCHIRHIZIDAE Haeckel, 1880

[=Rhizostomata simplicia VANHÖFFEN, 1888]

Rhizostomatida with 8 simple, separate, unbranched mouth arms; 16 radial canals give rise to side branches; ostioles without appendages, on ventral side of mouth arms. *Rec.* Family of doubtful validity because represented by 3 living genera seen only by HAECKEL.

Archirhiza HAECKEL, 1880 [**A. primordialis*]. *Rec.* Other living genera: *Haplorhiza, Cannorhiza* HKL., 1880.

FIG. 40. Rhizostomatida: *Leptobrachites trigonobrachius,* U.Jur., Ger.; holotype, ×0.3 (BRANDT).

?SCYPHOMEDUSAE INCERTAE SEDIS

Pseudorhopilema SPRIGG, 1949 [**P. chapmani*]. Based on isolated impressions interpreted as subumbrellar mid-field. Mid-field slightly convex, with central groove giving rise to system of dichotomously branched primary, secondary and ?tertiary grooves, regarded as gastric canals at base of mouth arms. *L.Cam.*, S.Austral.——FIG. 41,*1*. **P. chapmani;* holotype ×0.7 (SPRIGG).

Pseudorhizostomites Sprigg, 1949 [**P. howchini*]. Based on isolated impressions interpreted by Sprigg as subumbrellar mid-field. Mid-field convex, with cruciform grooves, each branching simply once; slight suggestion of secondary dichotomous branching at end of one or more of 8 subradial grooves. *L.Cam.*, S.Austral.——Fig. 41,*3*. **P. howchini*; holotype, ×0.7 (Sprigg).——Fig. 41,*2*. *P.* sp.; *2a,b*, ×0.9 (Sprigg).

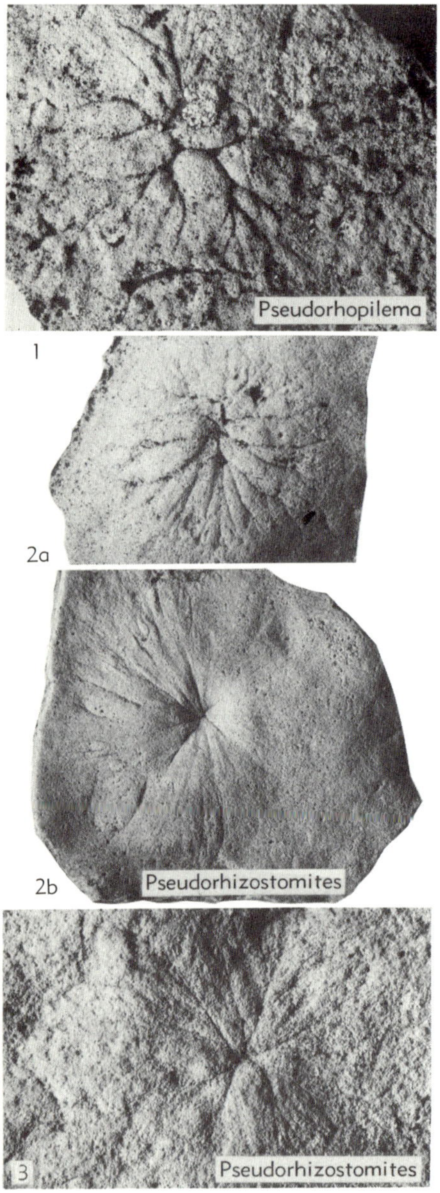

Fig. 41. Scyphomedusae incertae sedis: Lower Cambrian fossils from South Australia. *1, Pseudorhopilema chapmani;* holotype, ×0.7 (Sprigg). *2, Pseudorhizostomites* sp.; *2a,b*, specimens, ×0.9 (Sprigg). *3, Pseudorhizostomites howchini;* holotype, ×0.7 (Sprigg).

REFERENCES

Ammon, L. von

(1) 1886, *Über neue Exemplare von jurassischen Medusen:* Abhandl. Kgl. Bayer. Akad. Wiss., Math.-Nat. Kl., Band 15, p. 120-165, 3 fig., pl. 1-2.

(2) 1908, *Über eine coronate Qualle (Ephyropsites jurassicus) aus dem Kalkschiefer:* Geognostische Jahreshefte, Band 19, p. 169-186, pl. 3.

Angrisani, Cecilia

(3) 1917, *Sopra un avanzo di Attinia del calcareo dei dintorni di Sepino in provincia di Campobasso:* Rend. R. Accad. Sci. Fis. Mat. Napoli, fasc. 1-2, anno 55, p. 54-59, 2 fig.

Brandt, A.

(4) 1871, *Über fossile Medusen:* Mém. Acad. Imp. Sci. St. Petersbourg (7), 16, no. 11, p. 1-28, pl. 1-2.

Desio, A.

(5) 1923, *Sopra una Lorenzinia del Flysch dei dintorni di Firenze:* Riv. ital. Paleont., anno 29, fasc. 1-2, p. 7-10, pl. 1.

Fucini, Antonio

(6) 1908, *La Pania di Corfino:* Soc. Geol. Ital., Boll., tomo 27, p. 61-97, fig. 1.

(7) 1912, *Studi geologici sul Promontorio Argentario:* Ann. Univ. Toscana, tomo 32, Mem. 4, 57 p., 1 fig.

Gabelli, L.

(8) 1900, *Sopra un interesante impronta medusoide:* Il pensiero aristotelico nella sci. moderna (Bologna), tomo 1, no. 2, p. 74, fig. 1.

Gortani, M.

(9) 1920, *Osservazzioni sulle impronte medusoidi del Flysch (Lorenzinia e Atollites):* Riv. ital. Paleont., anno 26, fasc. 3-4, p. 56-72, pl. 2-3.

Haeckel, Ernst

(10) 1866, *Über zwei neue fossile Medusen aus der Familie der Rhizostomiden:* Neues Jahrb. f. Min. etc., Jahrg. 1866, p. 257-282, pl. 5-6.

(11) 1869, *Über die fossilen Medusen der Jurazeit:* Zeitschr. f. wiss. Zoologie, Band 19, p. 538-562, pl. 42.

(12) 1874, *Über eine sechszählige fossile Rhizostome und eine vierzählige fossile Semaeostomae:* Jenaische Zeitschr. f. Naturwiss., Band 8 (Neue Folge, Band 1), p. 308-330, pl. 10-11.

Harrington, H. J., & Moore, R. C.

(13) 1955, *Kansas Pennsylvanian and other jellyfishes:* Kansas Geol. Survey, Bull. 114, pt. 5, p. 153-163, pl. 1-2.

Hundt, Rudolf

(14) 1938, *Camptostroma aus dem Phycodensschichten Ostthüringens:* Zentralbl. f. Min. etc., Jahrg. 1938, Abt. B, p. 477-479, 1 fig.

Kieslinger, A.

(15) 1924, *Medusae fossiles:* Fossilium Catalogus. I. Animalia. Pars 26, p. 1-20.

(16) 1939, *Scyphozoa:* in SCHINDEWOLF, O. H., Handb. der Paläozoologie, Band 2A, Lief 5, p. A69-A109, fig. 1-42.

(17) 1939, *Revision der Solnhofen Medusen:* Palaeont. Zeitschr., Band 21, p. 287-296.

Kuhn, Oskar

(18) 1939, *Eine neue Meduse (Hydromeduse) aus dem Oberjura von Solnhofen:* Zool. Anzeiger, Band 122, no. 11-12, p. 307-312, fig. 1-3.

Kuzniar, Wictor

(19) 1911, *Kilka problematvznych skamielin z fliszu karpakiego:* Kosmos (Lember), v. 36, p. 517-524, 1 pl.

Llueca, F. G.

(20) 1927, *Noticia sobre el hallazgo de la Lorenzinia apenninica de Gabelli en el Eoceno de Guipuzcoa:* Bol. Soc. Espan. Hist. Nat., p. 45-56.

Maas, Otto

(21) 1902, *Über Medusen aus dem Solnhofen Schiefer:* Palaeontographica, Band 48, p. 293-315, pl. 22-23.

(22) 1906, *Über eine neue Medusengattung aus dem lithographischen Schiefer:* Neues Jahrb. f. Min. etc., Jahrg. 1906, Band 2, p. 90-99, fig. 1-4.

Mayer, A. G.

(23) 1910, *The Scyphomedusae:* in Medusae of the world, Carnegie Inst. Washington, Pub. 109, v. 3, p. 499-735, pl. 56-76, fig. 328-428.

Mitzopoulos, Max

(24) 1939, *Ein Medusen-Vorkommen im Eozänflysch des Peloponnes:* Praktika Akademias Athenon (Athens), v. 14, p. 258-259, 1 pl.

Neviani, Antonio

(25) 1926, *Medusa giurassica di Solenhofen (Baviera), Rhizostomites admirandus Haeckel:* Atti Accad. Nuovi Lincei, tomo 19, pt. 1, p. 54-60, 1 pl.

Renz, C.

(26) 1925, *Problematische Medusenabdrücke aus der Olonos-Pindos-zone des Westpeloponnes:* Verh. naturf. Gesell. Basel, Band 36, p. 220-233, fig. 1.

(27) 1930, *Ein Medusenvorkommen im Altterträr der Insel Cypern:* Eclogae Geol. Helvet., v. 23, p. 295-300, fig. 1.

Ruedemann, Rudolf

(28) 1933, *Camptostroma, a Lower Cambrian floating hydrozoan:* U.S. Natl. Mus., Proc., v. 82, no. 2954, art. 13, p. 1-8, pl. 1-4.

Ruggieri, G.

(29) 1943, *Impronta probabile di Medusa in un'argilla dell'Appennina forlivese:* Soc. Geol. Ital., Boll., tomo 61, no. 3, p. 26-31, fig. 1.

Saxl, Ottilie

(30) 1917, *Über ein Juravorkommen bei Skutari im Albanien:* Jahrb. k. k. geol. Reichsanst., Jahrg. 1916, Band 66, Heft 2, p. 281-292, fig. 1.

Simonelli, Vittorio

(31) 1905, *Intorno alcune singolari paleoicniti del Flysch apenninico:* Mem. R. Accad. Sci. Bologna, ser. 6, tomo 2, no. 3-4, p. 261-273, fig. 1.

Sprigg, R. G.

(32) 1947, *Early Cambrian(?) jellyfishes from the Flinders ranges, South Australia:* Roy. Soc. S.Australia, Trans., v. 71, pt. 2, p. 212-224, pl. 5-8.

(33) 1949, *Early Cambrian "jellyfishes" of Ediacara, South Australia, and Mt. John, Kimberley District, Western Australia:* Same, v. 73, pt. 1, p. 72-88, pl. 9-21.

Walcott, Charles D.

(34) 1898, *Fossil Medusae:* U.S. Geol. Survey, Mon. 30, ix+201 p., 47 pl.

Zuber, Rudolf

(35) 1910, *Eine fossile Meduse aus dem Kreideflysch der ostgalizischen Karpathen:* Verhand. k. k. geol. Reichsanst., no. 2, p. 57-59, fig. 1.

CONULATA

By R. C. Moore and H. J. Harrington

CONTENTS

A diagnosis of the subclass Conulata, accompanied by discussion of characters which are considered to support classification of this extinct group of invertebrates as a division of the Scyphozoa, has been given in the section entitled "Scyphozoa." Here, attention is directed to characters of the single order (Conulariida) that at present is assigned to the subclass. The conulariids are a varied but relatively stable assemblage having well-defined morphological attributes. We exclude from Conulata and also from Coelenterata fossils classed as belonging to the Serpulitidae, although Kiderlen (1937) and some other authors have interpreted the chitinophosphatic, generally somewhat compressed elongate tubes with thickened longitudinal rods incorporated in the periderm as a simple type of early conulate, that is, close relative of the Conulariida. Howell (1949) has pointed out that *Serpulites* Blumenbach, 1803, introduced as a name for Mesozoic fossils found near Hanover, Germany, is clearly distinct from *Serpulites* in the sense used by Murchison (1839) for designation of Silurian fossils, this application of the name (erroneously attributed to MacLeay) being ascribable clearly to Murchison. For Paleozoic fossils which have been referred to *Serpulites* (as *S. angustifolius, S. gracilis*) Howell has published the new generic name *Tubelelloides*. This genus and the Middle Cambrian forms called *Urotheca* Matthew, 1899 (*non* Cocteau & Bibron, 1843), renamed *Tubulella* by Howell (1949), and various others (as *Coleoprion, Coleolus,* etc.), are considered to be worms and not at all suggestive of any known kinds of coelenterates. This conclusion certainly is acceptable in the light furnished by contemplation of such an attenuated threadlike form as *Tubulella flagellum* (Matthew) (see Howell, 1949, pl. 1, fig. 10).

Order CONULARIIDA Miller & Gurley, 1896

[*nom. correct.* Moore, in Moore, Lalicker & Fischer, 1952 (*pro* Conularida Miller & Gurley, 1896)]

Characters of subclass. *M.Cam.-L.Trias.*

MORPHOLOGY

The periderm of conulariids typically has a steep-sided pyramidal form. From an attachment disc (*appareil fixateur, Haftscheibe*) (Fig. 27), the 4 sides or faces of the

(*Explanation of Fig 42, continued from facing page.*)

surface, at right mold of opposite face on matrix, enlarged (17*). *8, Diconularia clavus,* M.Sil., Eng.; transverse rows of tubercles at corner, ×10 (29*). *9, Conularia cribristriata,* M.Dev., N.Y.; facial ornament, ×10 (20*). *10, C. continens,* M.Dev., N.Y.; facial ornament, ×10 (20*). *11, C. africana,* Dev., S.Am. (Bol.); middle part of face, ×5 (Knod*). *12, Pseudoconularia dalecarliae,* Sil., Swed.; part of face, ×10 (21*). (Note: Inasmuch as conulariid specimens are illustrated both with apertural extremity directed upward and downward, presumably corresponding to orientations in attached and free-swimming conditions, respectively, no arbitrary orientation of figured specimens is adopted here and where the direction toward the aperture is not self-evident, this is indicated by an arrow.)

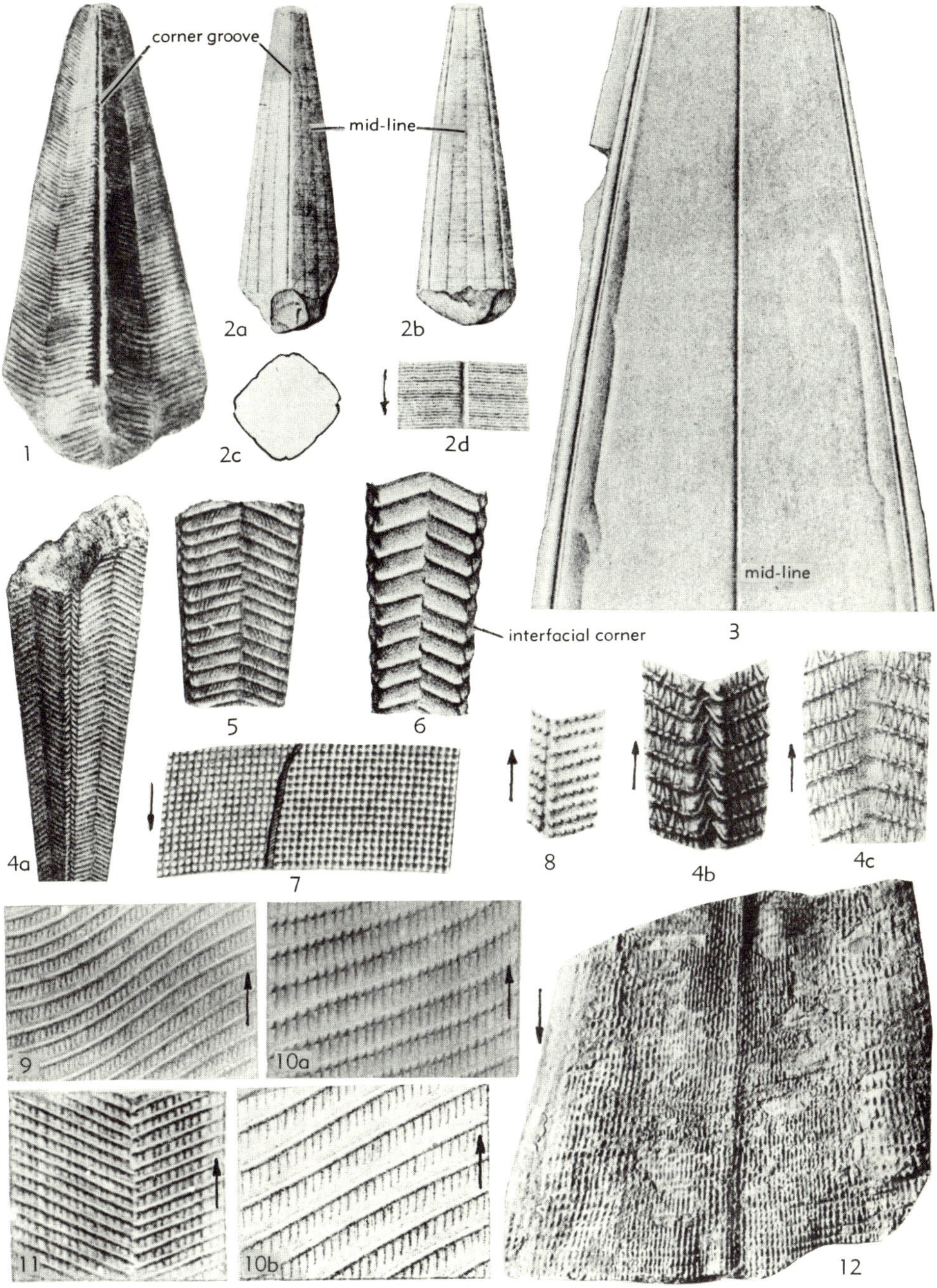

FIG. 42. Morphological features of Conulariida. *1, Paraconularia tubericosta,* L.Carb., Eng., ×1 (32*). *2, Archaeoconularia fecunda,* M.Ord., Czech.; *2a,b,* corner and facial views, ×1; *2c,* transverse section, ×1; *2d,* part of face, ×4 (17*). *3, A. insignis,* M.Ord., Czech.; part of face showing prominent mid-line and corner furrows, ×1 (17*). *4, Paraconularia warthi,* Perm., India; *4a,* corner view, ×1; *4b,c,* corner and mid-line of face, enlarged (33*). *5, P. tenuistriata,* Perm., India; face, ×1 (33*). *6, P.? laevigata,* Perm., India; face, ×1 (33*). *7, Exoconularia exquisita,* M.Ord., Czech.; facial ornament, part of left showing

(Continued on facing page.)

pyramidal form diverge toward the aperture with angles between them measuring normally about 90 degrees. The corner angles typically are furrowed by a groove, but in some genera this interfacial hollow is not developed (Fig. 42). This part of the periderm commonly is somewhat thickened and experience indicates that characters of the corner region have special significance in classification of conulariids (SINCLAIR, 1952).

Along the mid-line of each face is a more or less distinct longitudinal furrow, ridge, or pigmented line that divides the face into halves; this line is a prominent feature of many fossils, being accentuated by offsets in transverse ridges or rows of tubercles on the faces (Fig. 42,*5,6*) and by development internally of a longitudinal ridge or septum (Fig. 19). The strength and mode of expression of this facial mid-line (*ligne pariétale, Mittellinie*) has classificatory importance secondary to that of periderm corner characters. The central part of each half of a face may bear a distinct longitudinal line (lateral line, *Seitenlinie*) which is interpreted by KIDERLEN (1937) to be the trace of tentacular muscle attachment, and close to the mid-line on either side may be another pair of lines (*Nebenlinien*) which mark the position of septal ridges on the interior of the periderm, as in *Metaconularia* (Fig. 42,*2a,b*).

The surface of the faces somewhat rarely is smooth, except for cross wrinkles. Generally, very fine to moderately coarse transverse ribs or rows of tubercles extend from corners of the periderm to the mid-line, be-

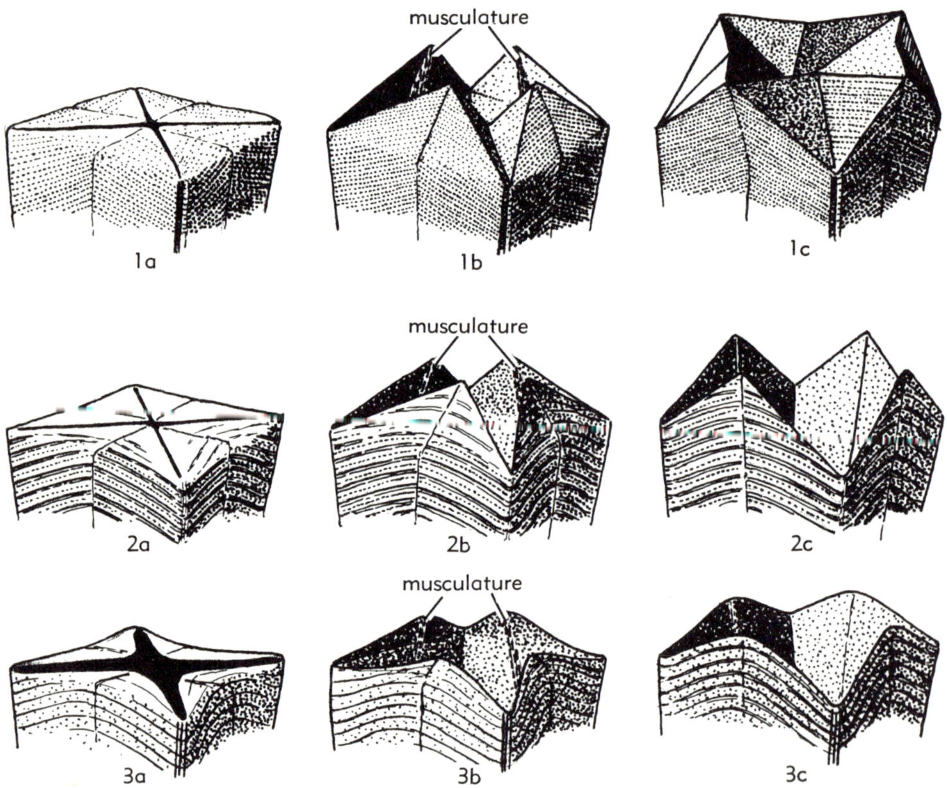

FIG. 43. Apertural region of conulariids showing types of closure by inward bending of the flexible periderm. *1a-c,* Plicated closure, characterized by evenness of apertural margin with sharp folds extending inward from corners, diagrams showing closed, partly open, and widely open positions. *2a-c,* Triangular lappet type of closure, distinguished by adapertural extensions of faces that fit neatly together when pulled inward, showing closed, partly open, and widely open positions. *3a-c,* Lobate lappet type, like *2a-c* except for shorter and marginally curved form of lappets. Inferred musculature for closing aperture is shown in *1b, 2b, 3b* (28n).

ing confluent or offset in position at the mid-line (Fig. 42). These transverse markings tend to slope or curve adaperturally more or less strongly. Interspaces may be devoid of markings or crossed by closely spaced barlike ridges; especially in forms having well-developed tubercles or pustules along the ribs or in transverse rows, these tend to be connected from row to row across the interspaces. In *Pseudoconularia,* accentuation of longitudinal arrangement of tubercles gives predominance of it over transverse features of surface sculpture (Fig. 42,*12*). Characters of the sculpture usually are reliable for discrimination of species, but they have subordinate value for differentiation of genera.

The aperture of conulariids may be partially or entirely closed by inward bending of distal parts of the faces. Three types of closure are distinguished, one consisting of plication of the transversely nearly straight apertural margin (Fig. 43,*1a-c*), and the others consisting of inbending of lappet-like triangular or semicircular extensions of the faces (Fig. 43,*2a-c, 3a-c*). Although many fossils show characters of apertural closure, the extreme thinness of this obviously very flexible part of the periderm, possibly more chitinous than phosphatic, may account for common lack of preservation.

A smoothly rounded, adapically convex diaphragm (*Schott, cloison*) constitutes the proximal extremity of some specimens which have broken free from an earlier attached location (Fig. 28).

CLASSIFICATION

Division of the Conulariida into at least 2 groups assigned rank as suborders is justified by present knowledge. These consist of nearly all described forms in one assemblage, here termed Conulariina, and *Conchopeltis* representing the other. This gives rise to highly disparate divisions, as must be recognized, yet distinction which sets *Conchopeltis* well removed from other conulariids seems necessary. Whether morphological characters of *Conulariella* are likewise different enough from those of common types of conulariids to call for placing it in a separate suborder is doubtful. The classification adopted here is essentially that proposed by Sinclair in 1952. Data recorded in the systematic descriptions on stratigraphic and geographic distribution of genera are subject in varying degree to additions and corrections, because thorough study of specimens and the literature has not been possible in the time allotted for preparing this section. A summary of supra-generic divisions now recognized in the Conulata and the recorded stratigraphic range of each is given in the following tabulation. Numbers of genera defined at present are indicated.

Main Divisions of Conulata

Conulata *(subclass)* (20). *M.Cam.-L.Trias.*
 Conulariida *(order)* (20). *M.Cam.-L.Trias.*
  Conchopeltina *(suborder)* (1). *M.Ord.*
   Conchopeltidae (1). *M.Ord.*
  Conulariina *(suborder)* (19). *M.Cam.-L.Trias.*
   Conulariellidae (1). *M.Cam.-L.Ord.*
   Conulariidae (17). *U.Cam.-Perm.*
    Conulariinae (9). *U.Cam.-Perm.*
    Paraconulariinae (4). *M.Ord.-L.Perm.*
    Ctenoconulariinae (4). *M.Ord.-M.Dev., ?L.Miss.*
   Conulariopsidae (1). *L.Trias.*

Suborder CONCHOPELTINA Moore & Harrington, nov.

Broad low conical forms with strong tetrameral symmetry, periderm very thin and presumably consisting mainly of a chitinous substance (no calcium phosphatic mineralization observed); numerous tentacles along apertural margin which lacks any sign of inward bending to effect closure. *M.Ord.*

Family CONCHOPELTIDAE Moore & Harrington, nov.

Characters of suborder. *M.Ord.*

Conchopeltis Walcott, 1876 [**C. alternata*]. Faces gently convex with broadly lobate margins, separated laterally by wide and shallow interfacial furrows; width of cone approximately 3 times greater than height; surface of periderm marked by faint wavy growth lines concentric with margin and sharply defined fine radial striae. *M.Ord. (Trenton.),* N.Y., Minn.——Figs. 22, 44. **C. alternata,* N.Y.; 22, specimen with tentacles preserved, Cornell Univ. coll., ×1.5 (34n); 44,*1a,c,* internal and external molds of holotype, ×0.9; 44,*1b,* side view of holotype, ×0.9 (23*).

Suborder CONULARIINA Miller & Gurley, 1896

[*nom. transl. et correct.* MOORE & HARRINGTON, herein (*ex* Conularida MILLER & GURLEY, 1896)]

Steeply pyramidal, generally 4-sided forms with thin chitinophosphatic periderm, faces bearing faint to strong transverse lirae or ribs more or less curved adaperturally and continuous or offset along mid-line of faces, which may be defined as a furrow or ridge or pigmented line or which is not evident

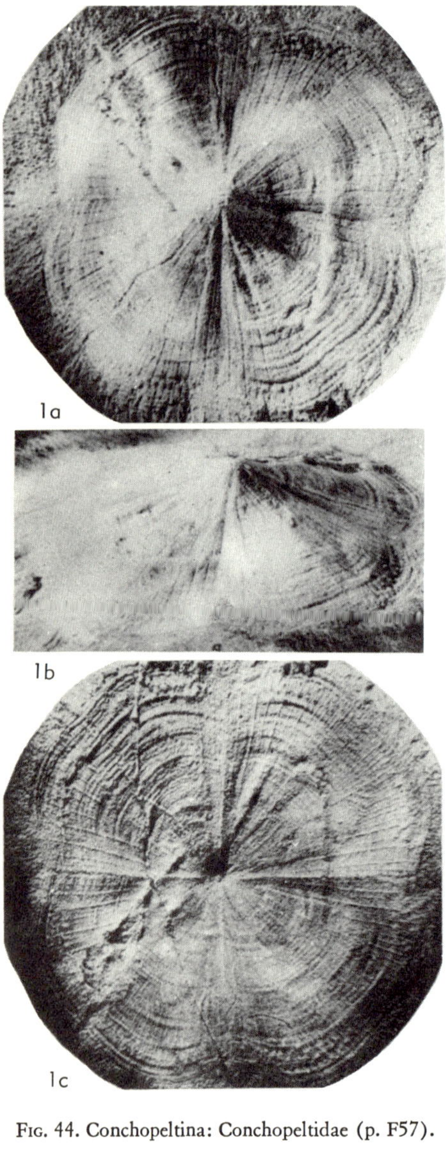

FIG. 44. Conchopeltina: Conchopeltidae (p. F57).

FIG. 45. Conulariina: Conulariellidae (p. F58).

at all; interfacial corners rounded or depressed in a more or less accentuated groove; apex with fixation disc, at least in well-preserved young specimens; aperture partly or entirely closable by inwardly bent edges of faces. *M.Cam.-Trias.*

Family CONULARIELLIDAE Kiderlen, 1937

Flattened trumpet-shaped forms with rectangular cross section having 2 opposite sides several times longer than others; aperture straight-sided, without sign of inward bending of faces; faces marked by transverse ribs that do not curve adaperturally; without mid-lines on faces or internal septa; interfacial corners grooved. *M.Cam.-L.Ord.*

Conulariella BOUČEK, 1928 (**Conularia robusta* BARRANDE, 1867). Differs from *Conularia* and other Conulariidae in rectilinear transverse ribs on faces and from several genera in absence of internal longitudinal ridges (septa). *M.Cam.-L. Ord.*, Eu.——FIG. 45,*1*. **C. robusta* (BARRANDE), Czech.; *1a,b*, side and transv. sec., ×1 (*1a*, 18*; *1b*, 28n).

Family CONULARIIDAE Walcott, 1886

Steep-sided pyramidal forms, mostly with quadrangular cross section, faces generally marked by evident transverse lirae or ribs which are deflected adaperturally in crossing the face but longitudinal sculpture may dominate or the surface may be smooth except for small wrinkles; angles at corners typically thickened somewhat, with or without distinct furrow; mid-line of faces defined by narrow groove or ridge that in some genera corresponds to internal septa and it may be accompanied by 1 or 2 distinct longitudinal lines on either side, or mid-line may not be distinguishable as

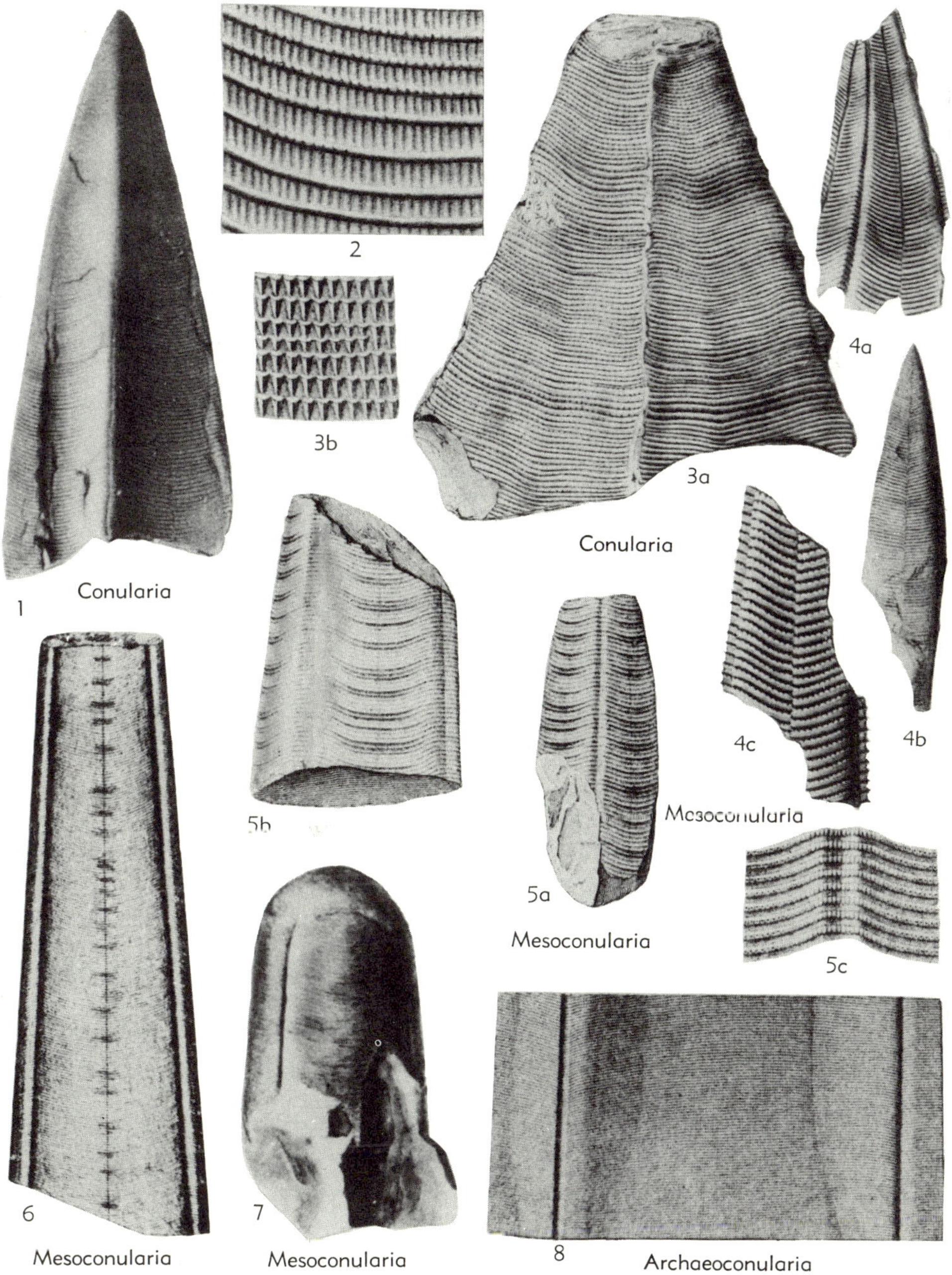

FIG. 46. Conulariina: Conulariidae (p. F60-F61).

morphological feature; aperture closable by inward bending of facial periderm and apical extremity may consist of smoothly rounded diaphragm (*Schott*). *U.Cam.-Perm.*

Subfamily CONULARIINAE Walcott, 1886

[*nom. transl.* SINCLAIR, 1952 (*ex* Conulariidae WALCOTT, 1886)]

Corners of periderm indented by longitudinal furrow which does not interrupt transverse sculpture and is not accompanied by distinct thickening. *U.Cam.-Perm.*

Conularia SOWERBY, 1821 [**C. quadrisulcata* (=*C. sowerbyi* DE BLAINVILLE, 1825; *C. cancellata* SANDBERGER, 1847) (*non C. quadrisulcata* SLATER, 1907=*C. tubericosta* SANDB., 1847)] [=*Plectoconularia* BOUČEK, 1939 (obj.)]. Transverse ribs well defined, closely spaced, finely tuberculate, with interspaces crossed by bars or striae; facial mid-line not marked surficially by groove or ridge nor produced internally as septal ridge. *U.Cam.-Perm.*, world-wide.——FIG. 23,*1a,b*, *C. cambria* WALCOTT, U.Cam., N.Am.; reconstr., ×1 (22*).——FIGS. 25,*1*; 42,*10*. *C. continens* HALL, M.Dev., N.Y.; 25,*1*, specimen with attached inarticulate brachiopods and attachment scars, ×1 (20*); 42,*10*, facial ornament, ×10 (20*).——FIG. 25,*2*. *C. undulata* CONRAD, M.Dev., N.Y.; specimen with attachment scars of brachiopods, ×1 (20*).——FIG. 42,*9*. *C. cribristriata* HALL, M.Dev., N.Y.; facial ornament, ×10 (20*).——FIGS. 42,*11*; 46,*1*. *C. africana* SALTER, Dev., S.Am.(Bol.-Brazil); 42,*11*, facial orna-

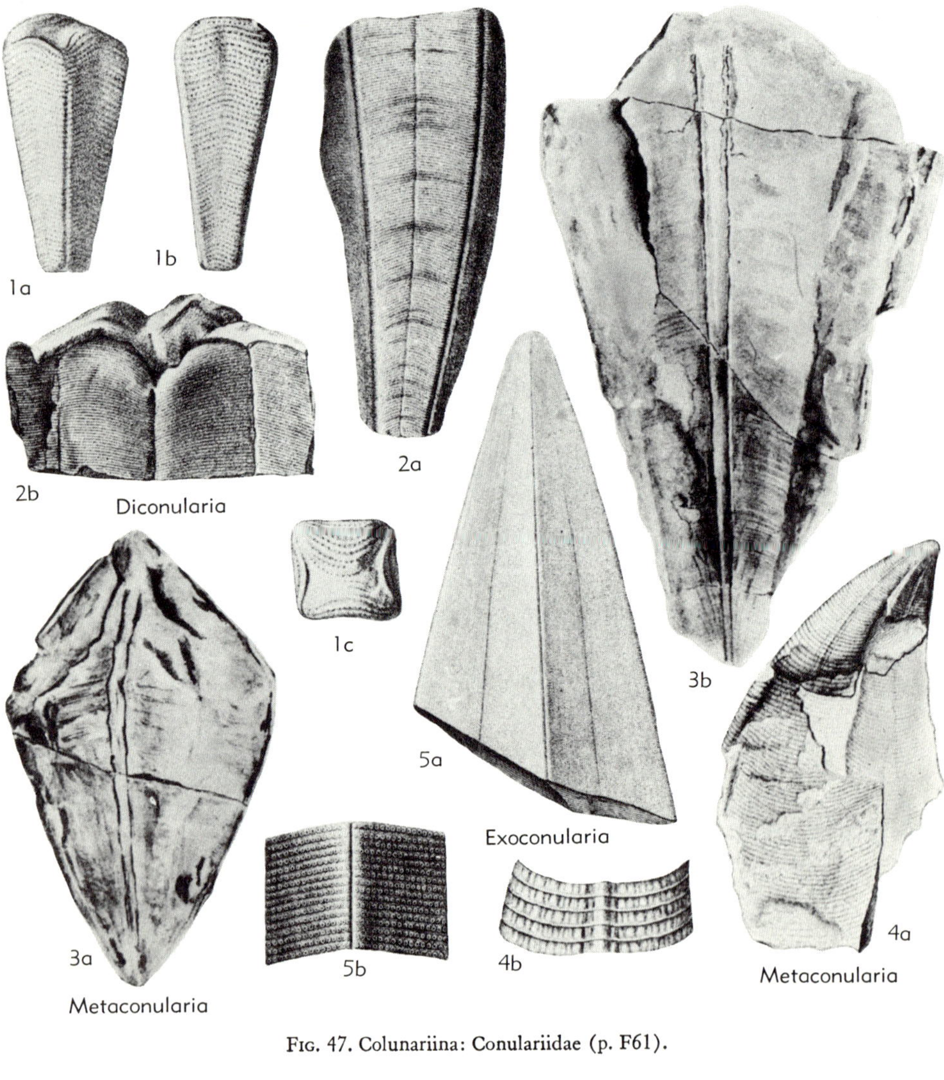

FIG. 47. Colunariina: Conulariidae (p. F61).

ment, ×5 (KNOD*); 46,*1*, side view, ×1 (19*).——FIG. 46,*3*. *C. niagarensis* HALL, M.Sil., Ind.; *3a*, side view of mold, ×1; *3b*, surface ornament, enlarged (20*).——FIG. 46,*2*. *C. congregata* HALL, U. Dev., N.Y.; surface ornament, ×10 (20*).

Archaeoconularia BOUČEK, 1939 [**Conularia insignis* BARRANDE, 1867]. Small to medium-sized, periderm very thin, faces with prominent median groove and adapertural part extended as acute triangular lappet that may be bent inward for closure of aperture; surface marked by fine transverse threads gently arched adaperturally and crossed by faint longitudinal striae; apical diaphragm observed in some species. *L.Ord.-Sil.*, Eu.-N.Am.——FIGS. 42,*3*; 46,*8*. **A. insignis* (BARRANDE), Ord., Czech.; 42,*3*, type, ×1; 46,*8*, part of face, ×2 (17*).——FIG. 24,*3a,b*. *A. slateri* (REED), U.Ord.(Ashgill.), Scot.; *3a,b*, side and apertural views showing closure by infolded periderm, ×1 (29*).——FIGS. 23,*2a,b;* 42,*2a-d*. *A fecunda* (BARRANDE), Ord., Czech.; 23,*2a,b*, reconstr., ×0.7 (22*); 42,*2a-c*, side views and transv. sec., ×1; 42,*2d*, part of face, ×4 (17*).

Mesoconularia BOUČEK, 1939 [**Conularia fragilis* BARRANDE, 1867]. Faces showing more or less distinct mid-line but no secondary longitudinal lines, transverse ribs distinct. *L.Dev.-Penn.*, Eu.-Afr.-N. Am.-S.Am.——FIG. 46,*5*. **M. fragilis* (BARRANDE), L.Dev., Czech.; *5a,b*, side views of 2 specimens, ×1; *5c*, part showing corner groove and transverse ribs, enlarged (17*).——FIG. 46,*4*. *M. ulrichana* (CLARKE), Dev., Brazil(Parana); ×1 (19*).——FIG. 46,*6*. *M. roeperi* (MILLER & GURLEY), Penn., Pa.; ×0.7 (27*).——FIG. 46,*7*. *M. globosa* (SLATER), L.Carb., Eng.; specimen with apical diaphragm, ×1 (32*).

Diconularia SINCLAIR, 1952 [**Conularia micronema* MEEK, 1871]. Corners simple, with furrow or not furrowed; faces bearing close-spaced transverse ribs marked by prominent subcircular pustules; mid-line more or less defined on surface and generally associated with low septal ridge on inner side of periderm. *M.Ord.-Penn.*, N.Am.-Eu.——FIG. 47,*2*. **D. micronema* (MEEK), L.Miss., Ohio.; *2a*, holotype, ×1; *2b*, another specimen with periderm infolded for closure of aperture, ×2 (26*).——FIGS 42,*8;* 47,*1*. *D. clavus* (REED), M.Sil., Eng.,; 42,*8*, ornamentation, ×10; 47,*1a,b*, corner and side views, ×1; 47,*1c*, aperture showing inwardly bent rounded lappets, ×1 (all 29*).

Metaconularia FOERSTE, 1928 [**Conularia aspersa* LINDSTRÖM, 1884]. Corners with rather inconspicuous shallow furrow; periderm of faces very thin, surface marked by nearly horizontal transverse rows of minute granules which also may be aligned vertically; mid-line of faces inconstantly marked by faint groove or not defined externally, on inner side enclosed by a pair of well-developed septal ridges which in compressed specimens may distort outer surface so as to produce longitudinal linear elevations; aperture closed by extensions of facial periderm in form of triangular lappets. *M. Ord.-U.Sil.*, Eu.-N.Am.——FIG. 47,*3*. **M. aspersa* (LIND.), U.Sil.(Ludlov.), Eng.; *3a*, aperture closed by triangular lappets, ×1; *3b*, side of another larger specimen, ×1 (32*).——FIG. 47,*4*. *M. solitaria* (BARRANDE), M.Ord., Czech.; *4a,b*, flattened specimen, ×1, and detail of surface, enlarged (17*).

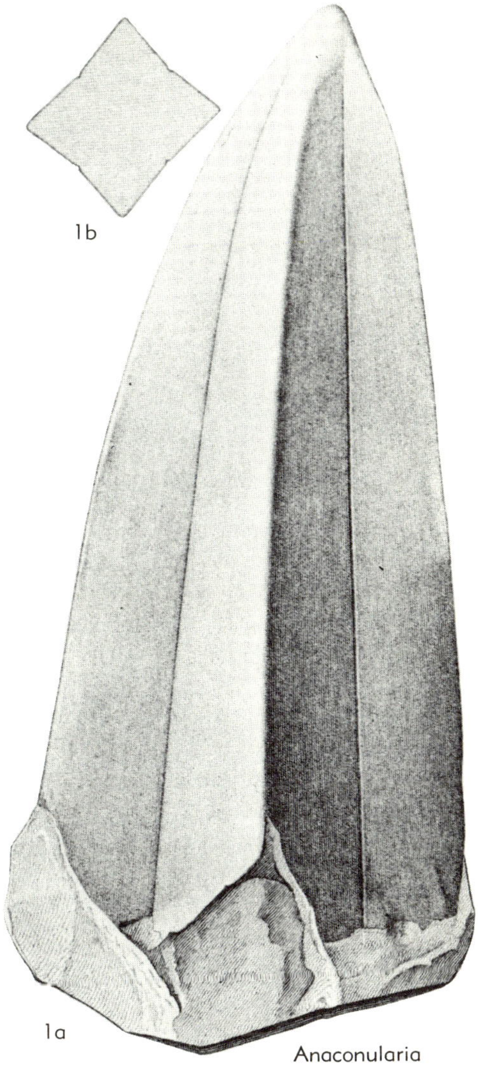

FIG. 48. Conulariina: Conulariidae (p. F62).

Exoconularia SINCLAIR, 1952 [**Conularia exquisita* BARRANDE, 1867]. Faces with accessory longitudinal lines varying in position in different species, surface bearing transverse rows of pustules. *L.Ord.-U.Ord.*, Eu.-N.Am.——FIG. 42,*7;* 47,*5*. **E. ex-*

FIG. 49. Conulariina: Conulariidae (p. F63-F64).

quisita (BARRANDE), M.Ord., Czech.; 42,*7*, ornament, enlarged; 47,*5a,b*, side view, ×1, and part of surface showing rows of tubercles, enlarged (17*).——FIGS. 24,*1*; 28,*3*. *E. pyramidata* (HOENINGHAUS), M.Ord., Czech.; 24,*1*, distorted but unbroken specimen indicating flexibility of periderm, ×0.7 (22); 28,*3*, specimen with schott, ×1 (25).——FIGS. 23,*3*; 24,*2*; 25,*3*. *E. consobrina* (BARRANDE), M.Ord.-U.Ord., Czech.-Fr.; 23,*3*, reconstr. of free-swimming medusoid adult, ×0.7 (22*); 24,*2*, apertural region with oblique section showing infolded periderm, ×1 (25*); 25,*3*, part of surface with attached edrioasteroid *(Agelacrinites)*, ×0.75 (22*).

Anaconularia SINCLAIR, 1952 [**Conularia anomala* BARRANDE, 1867]. Corners without trace of furrows, being elevated in rounded or sharp ridge. *M.Ord.*, Eu.——FIGS. 28,*1*; 48,*1*. *A. anomala* (BARRANDE), M.Ord., Czech.; 48,*1a,b*, side view and transv. sec., ×1 (17*). 28,*1*, specimen with schott, ×1 (17*).

Palaenigma WALCOTT, 1886 [**pro Tetradium*

SCHMIDT, 1874 (*non* DANA, 1846)] [**Tetradium wrangeli* SCHMIDT, 1874]. *L.Ord.*, Eu.

sima BARRANDE, 1867]. Faces with mid-line defined by low but distinct ridge but without acces-

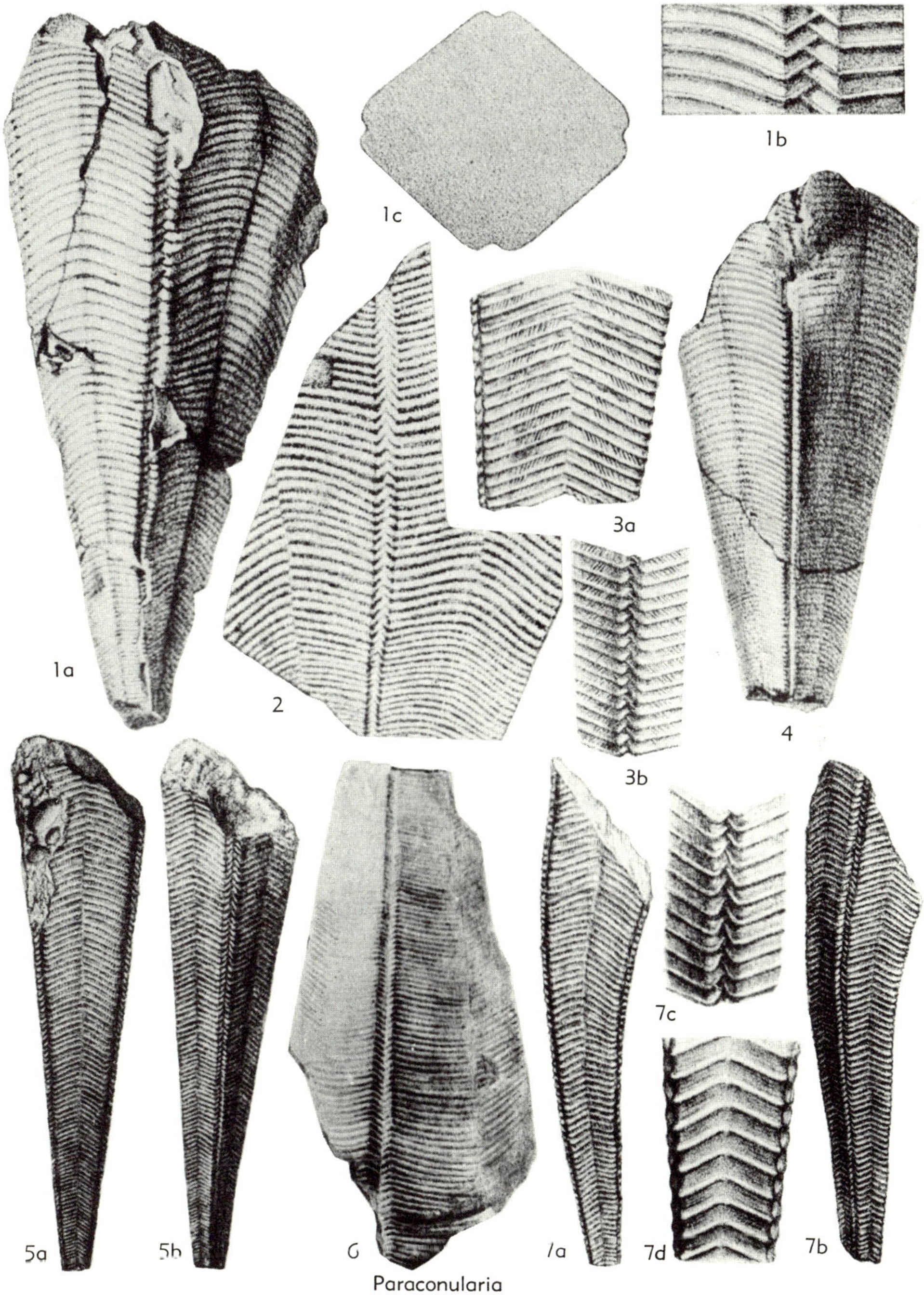

FIG. 50. Conulariina: Conulariidae (p. F65).

ing of closely spaced longitudinal rows of vertically elongate tubercles. *Sil.*, Eu.-N.Am.——Fig. 49,*1*. **P. grandissima* (Barrande), Czech.; *1a,b*, side view of 2 specimens, ×1 (17*).——Fig. 49,*2*. *P. dalecarliae* Hessland, Swed.; *2a,b*, part of surface, ×15, ×10 (21*).

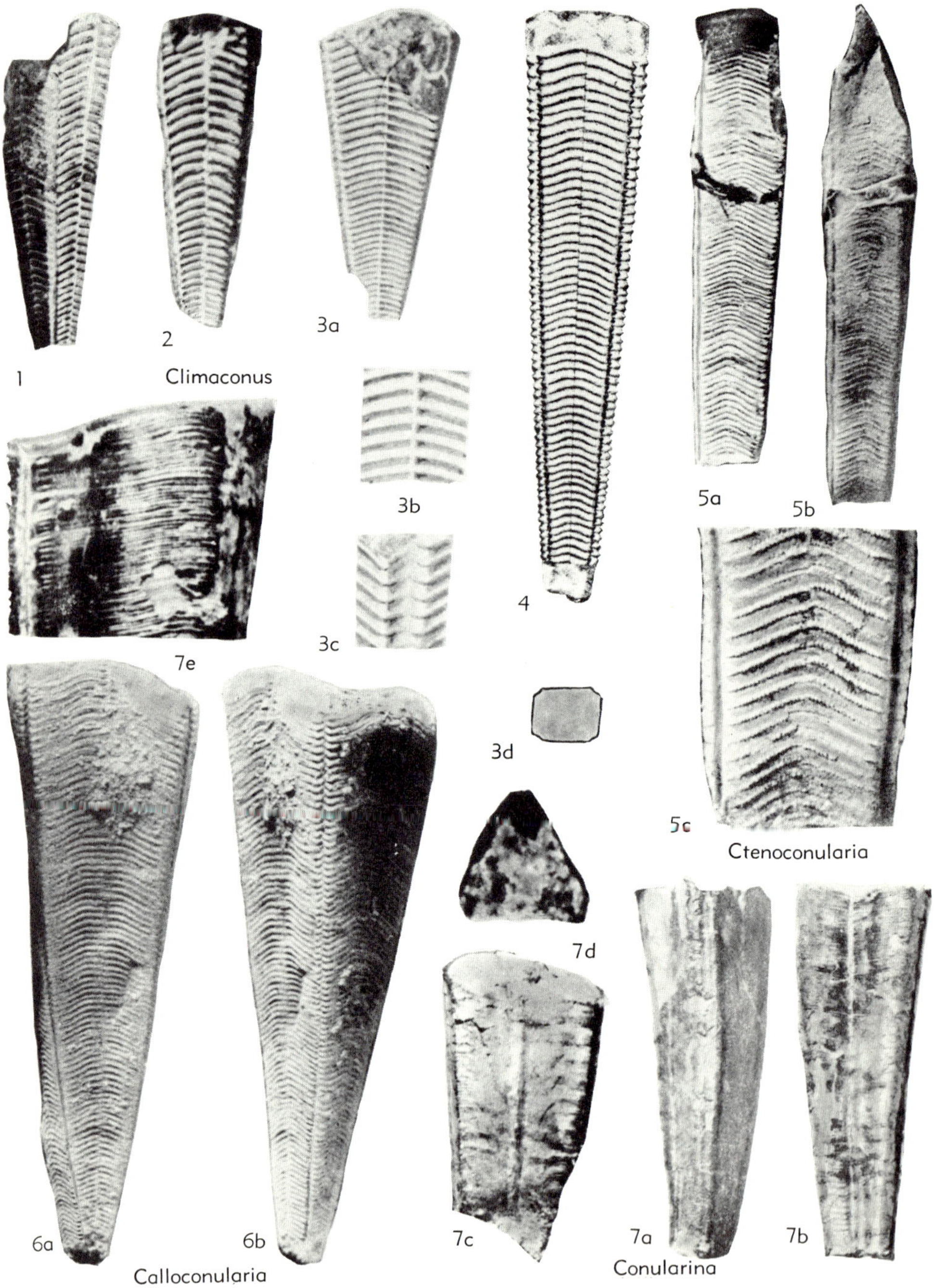

Fig. 51. Conulariina: Conulariidae (p. F65).

Subfamily PARACONULARIINAE Sinclair, 1952

Corners of periderm sharply inflected, with surface sculpture interrupted at corner furrows, transverse ribs in alternating or opposed positions along edges of furrows. *M. Ord.-L.Perm.*

Paraconularia SINCLAIR, 1940 [**Conularia inaequicostata* DEKONINCK, 1883]. Transverse ribs moderately strong, faintly tuberculate, abruptly bent adaperturally at their terminations on edges of corner furrows; mid-line on faces indicated only by slight deflection of ribs along it. *M.Sil.-L.Carb.*, Eu.-N.Am.——FIG. 50,*1*. **P. inaequicostata* (KON.), L.Carb(Tournais.), Belg.; *1a-c*, ×1 and enlarged (24*).——FIG. 50,*2*. *P. blairi* (MILLER & GURLEY), L.Miss.(Kinderhook.), Mo.; ×1 (27*).——FIG. 50,*6*. *P. tubericosta* (SANDBERGER), L.Carb., Eng.; ×1 (form identified by SLATER as *Conularia quadrisulcata*) (SLATER).——FIG. 50,*3*. *P. tenuistriata* (M'COY), Perm., India; ×1 (33*).——FIG. 50,*4*. *P. irregularis* (KON.), L.Carb. (Visé.), Belg.; ×1 (24*).——FIG. 50,*5*. *P. worthi* (WAAGEN), Perm., India; *5a,b*, ×1 (33*).——FIG. 50,*7*. *P. laevigata* (MORRIS), Perm., India; *7a,b*, ×1; *7c,d*, enlarged (33*).

Calloconularia SINCLAIR, 1952 [**C. strimplei*]. Transverse ribs low and crowded, joined by longitudinal bars in interspaces. *Penn.*, N.Am.—— FIG. 51,*6*. **C. strimplei*, U.Penn.(Missourian), Okla.; *6a,b*, ×1 (31*).

Eoconularia SINCLAIR, 1943 [**Conularia loculata* WIMAN, 1894]. Rather small, corner furrows with slightly thickened periderm; mid-line of faces not strongly marked but internally a thin distally bifurcate septum reaches far into interior; surface smooth except for cross wrinkles. *M.Ord.-M.Sil.*, Eu.-N.Am.——FIG. 19,*1*. **E. loculata* (WIMAN), M.Sil., Gotl.; transv. sec. showing bifurcate interradial septa, ×15 (35*).

Neoconularia SUGIYAMA, 1942 [**Conularia rectangularis* HAYASAKA, 1920]. Faces flat and smooth except for transverse irregular wrinkles, without sign of mid-line; cross section rectangular (known only from steinkerns). *L.Perm.*, Japan.

Subfamily CTENOCONULARIINAE Sinclair, 1952

Corners more or less furrowed and strengthened by internal carinae or other thickening of periderm. *M.Ord.-M.Dev., ?L.Miss.*

Ctenoconularia SINCLAIR, 1952 [**C. obex*]. Very slender, with shallow corner furrows marking position of strong internal carinae; mid-line of faces marked by interruption of transverse ribs, which comprise internal as well as external elevations of periderm due to localized thickening, ribs interrupted at margins of corner furrows, with pustulose crests but unlike *Conularia s.s.*, interspaces lack cross bars. *M.Ord.-M.Dev., ?L.Miss.*——FIG. 51,*5*. **C. obex*, M.Ord.(Blackriveran), Minn.; *5a,c*, ×2.5; *5b*, ×7 (41*).——FIG. 51, *4*. *?C. greenei* (MILLER & GURLEY). L.Miss.(Keokuk), Ind.; ×1 (27).

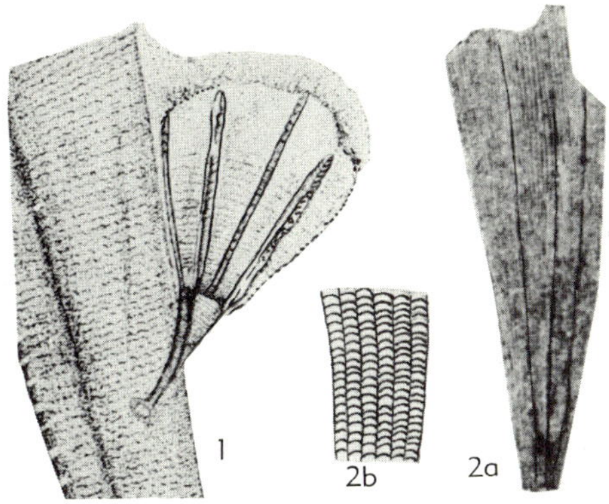

FIG. 52. *Sphenothallus angustifolius* (p. F65).

Climaconus SINCLAIR, 1952 [**Conularia quadrata* WALCOTT, 1879]. Small forms with square to rectangular cross sections; faces with strong smooth transverse ribs that along mid-line meet in opposed or alternating position, mid-line appears externally as ridge with adaperturally diminished height in each interspace and having straight or zigzag appearance, not associated with internal septal ridge; corner furrows may have transverse wrinkles. *M.Ord.-U.Ord.*, N.Am.-Eu.——FIG. 51, *2*. **C. quadratus* (WALCOTT), M.Ord., N.Y.; ×2 (31*).——FIG. 51,*1*. *C. batteryensis* (TWENHOFEL), U.Ord., Que.; ×2 (31*).——FIG. 51,*3*. *C. bottnicus* (HOLM), M.Ord., Swed.; *3a,d*, ×1.5; *3b,c*, ×3 (31*).

Conularina SINCLAIR, 1942 [**Conularia triangulata* RAYMOND, 1905]. Slender; faces with mid-line expressed as low ridge flanked by shallow depressions that form low ridges internally; transverse section triangular or quadrangular; surface marked by fine transverse striae crossed commonly by fine oblique folds near shoulders of corner furrows. *M.Ord.*, N.Am.——FIG. 51,*7*. **C. triangulata* (RAYMOND), Vt.; *7a-d*, ×2; *7e*, ×7.5 (31*).

Sphenothallus HALL, 1847 [**S. angustifolius*; SD MOORE & HARRINGTON, herein (=*Conularia gracile* HALL, 1847)] [=*Hallotheca* BOUČEK, 1936 (obj.); *Glyptoconularia* SINCLAIR, 1952 (type, *Conularia gracilis* HALL, 1847)]. Corners strengthened by pronounced thickening of periderm and faces with mid-line produced inward as carina; surface with longitudinal ornamentation. *M.Ord.*, N.Am.——FIG. 27,*7*; 52,*1*,*2*. **S. angustifolius*, N.Y.; 27,*7*, holotype, ×1; 52,*1*, young individual attached to adult, ×2 (30*); 52,*2a,b*, holotype of *Conularia gracilis*, ×1, and ornament, enlarged (20*).

Family CONULARIOPSIDAE Sugiyama, 1942

Corners of pyramid raised, lacking furrows; faces marked by coarse sinuous longitudinal ridges and by widely spaced transverse folds which may be adventitious. *L. Trias.*

Conulariopsis SUGIYAMA, 1942 [**C. quadrata*]. Characters of family. *L. Trias.*, Japan.

REFERENCES

Barrande, Joachim

(1) 1867, *Ordre des Ptéropodes:* Système silurien du centre de la Bohême (Prague), v. 3, pt. 1, xv+179 p., 16 pl.

Bouček, Bedřich

(2) 1928, *Revise Ceskych Paleozoictych Konularii (Revision de conulaires paléozoiques de la Bohême):* Palaeontographica Bohemiae, (Prague), v. 11.

(3) 1939, *Conularida:* in SCHINDEWOLF, O.H., Handbuch der Paläozoologie, Borntraeger (Berlin), Band 2A, Lief. 5, p. A113-A131, fig. 1-13.

Clarke, J. M.

(4) 1913, *Fosseis devonianos de Paraná:* Serv. Geol. Min. (Rio de Janeiro), Mon. 1, xx+353 p., 27 pl., text figs.

Holm, G.

(5) 1893, *Sveriges Kambrisk-Siluriska Hyolithidae och Conulariidae:* Sver. Geol. Undersok., (C) no. 112 (Stockholm).

Howell, B. F.

(6) 1949, *New hydrozoan and brachiopod and new genus of worms from the Ordovician Schenectady formation of New York:* Wagner Free Inst. Sci. (Philadelphia), Bull., v. 24, no. 1, p. 1-8, pl. 1-2.

Kiderlen, H.

(7) 1937, *Die Conularien, über Bau und Leben der ersten Scyphozoa:* Neues Jahrb. f. Mineralogie, Beil.-Band 77, Abt. B, p. 113-169, fig. 1-47.

Kowalski, J.

(8) 1935, *Les conulaires; quelques observations sur leur structure anatomique:* Soc. Sci. Nat. Ouest France, Bull., ser. 5, tome 5, p. 281-293, pl. 12, fig. 1-3.

Knight, J. B.

(9) 1937, *Conchopeltis Walcott, an Ordovician genus of Conularida:* Jour. Paleont., v. 11, p. 186-188, pl.

Lindström, G.

(10) 1884, *On the Silurian Gastropoda and Pteropoda of Gotland:* K. Svenska Vetenskapsakad. Handl., Bind 19 (6).

Ruedemann, Rudolf

(11) 1898, *The discovery of a sessile Conularia:* New York State Geol. (Albany), Ann. Rept. 15, p. 701-720, pl. 1-5.

Sinclair, G. W.

(12) 1940, *A discussion of the genus Metaconularia with descriptions of new species:* Roy. Soc. Canada, Trans., ser. 3, v. 34, sec. 4, p. 101-121, pl. 1-3.

(13) 1942, *The Chazy Conularida and their congeners:* Carnegie Mus. Ann., v. 28, art. 10, p. 219-240, pl. 1-3.

(14) 1944, *Notes on the genus Archaeoconularia and Eoconularia:* Roy. Soc. Canada, Trans., ser. 3, v. 4, p.

(15) 1952, *A classification of the Conularida:* Chicago Nat. Hist. Mus., Fieldiana, Geology, v. 10, no. 13, p. 135-145, fig. 56.

Slater, I. L.

(16) 1907, *A monograph of British Conulariae:* Palaeontograph. Soc. (London), Mon. 61, p. 1-40, pl. 1-5.

SOURCES OF ILLUSTRATIONS

Explanation of the use of index numbers for citing sources of illustrations is given in the Editorial Preface.

(17) **Barrande, Joachim**
(18) **Bouček, Bedřich**
(19) **Clarke, J. M.**
(20) **Hall, James**
(21) **Hessland, Ivar**
(22) **Kiderlen, H.**
(23) **Knight, J. B.**
(24) **Koninck, L. G. de**
(25) **Kowalski, J.**
(26) **Meek, F. B.**
(27) **Miller, S. A., & Gurley, W. F. E.**
(28) **Moore, R. C., & Harrington, H. J.**
(29) **Reed, F. R. C.**
(30) **Ruedemann, Rudolf**
(31) **Sinclair, G. W.**
(32) **Slater, I. L.**
(33) **Waagen, Wilhelm**
(34) **Wells, J. W.**
(35) **Wilman, Carl**

HYDROZOA—GENERAL FEATURES

By Dorothy Hill and J. W. Wells

GENERAL NATURE AND TYPES OF HYDROZOA

Hydrozoans are either polymorphic, with both polypoid and medusoid forms, or monomorphic and medusoid. Their symmetry is tetramerous or polymerous and radial—any 2 diameters at right angles dividing the animal into similar halves. Their coelenteron is without a stomodaeum or gullet, and it is not divided by projecting ridges or partitions. The mesogloea does not contain cells that have wandered in from either ectoderm or endoderm. Tentacles may be hollow or solid and sex cells ripen in the ectoderm. The oral end of the polyp is elongated, and characteristic of the hydrozoan medusoid is the **velum** or shelf on the inner rim of the umbrella. Exoskeletons occur in some orders and these are either horny or calcareous.

The class may be divided into 7 orders (Trachylinida, Hydroida, Spongiomorphida, Milleporina, Stylasterina, Stromatoporoidea, Siphonophorida) which differ in degree of polymorphism and relative importance of the polypoid and medusoid generations. The trachylinids, regarded by some as the most primitive, are medusae, which as a whole lack a polypoid generation; if one is present, it is minute and attached. The siphonophorids, like the trachylinids, never develop a skeleton but they exhibit marked polymorphism. Because of the diversity of modified medusoid or polypoid individuals of several kinds which compose the swimming or floating siphonophorid colonies, they are regarded commonly as the most specialized group of hydrozoans, although some zoologists consider the polypoid and medusoid kinds to be below the grade of individuals and thus interpret the siphonophorids as the most primitive coelenterates.

Among hydroids the polypoid generation is dominant and usually colonial; the polyps may bud off free medusae, or there may be structures resembling abortive medusae. In the milleporines and stylasterines, which secrete calcareous skeletons, the polypoid colonies have 2 kinds of retractile polyps, whereas the very small medusae perish after a few hours of life during which sex cells are shed. The stromatoporoids and spongiomorphids are extinct orders, but they also seem to have comprised colonies of polyps.

The graptolites, which formerly often were classed as an extinct group of hydrozoans, now are considered to be hemichordates allied to the living pterobranchs.

Hydrozoans with calcareous skeletons are invariably sessile benthonic forms, which seem to flourish better in warm than cold waters. Milleporines and stylasterines are common in Tertiary and Recent reef environments, and the stromatoporoids were characteristic of many Paleozoic and Mesozoic organic reefs. Siphonophores and trachylines are floaters and swimmers, as are the medusoid generations of other orders. The hydroids, being dominantly polypoid, are mostly sessile, and are either benthonic or attached to floating weeds. Most are colonial; a few, such as *Hydra,* are solitary. The latter genus and a few medusoid hydroids are unusual in that they live in fresh water.

Class HYDROZOA Owen, 1843

[Originally included hydroids only; extended by Huxley, 1859, to include siphonophores and trachylines but not milleporines] [=Anoecia Huxley, 1852; Ectocarpen O.Hertwig & R.Hertwig, 1879 (*sine* Ctenophora); Aphacellae Lendenfeld, 1884; Hydromedusae Zittel, 1900 (*non* Vogt, 1851); Hydrozoaria Delage & Hérouard, 1901]

Coelenterates with tetrameral or polymeral radial symmetry; polymorphic, with both polypoid and medusoid forms, or exclusively medusoid; enteron without stomodaeum or nematocysts, and not divided by projecting ridges and partitions; mesogloea noncellular; sex cells ripening in the epidermis; oral end of polyp elongated into a hydranth; medusae almost invariably with velum; exoskeleton, if present, generally chitinous, rarely calcareous. *L.Cam.,-Rec.*

TRACHYLINIDA

By H. J. Harrington and R. C. Moore

CONTENTS

The hydrozoan medusae are all morphologically alike, but important differences in their ontogenetic development divide them into 2 natural groups, the Trachylinida and the medusae of the Hydroida. The trachylinids are characterized by a reduction or absence of the polypoid generation, having typically a planula-actinula-medusa life history. In many genera the tentacles of the actinula larva become those of the medusa, and as the bell margin continues to grow after the tentacles have become fully developed, they are "left behind," deeply rooted in the jelly-like mesogloea and springing from the exumbrellar surface of the bell. Though the trachylinids and the medusae of the Hydroida have sense organs structurally similar, the lithocysts of the former contain endodermal lithocytes and statoliths and are, therefore, not comparable with the ectodermal sense organs of the latter. The trachylinids are best regarded as medusiform actinulae which are not homologous with the medusae of the Hydroida, having attained independently a medusiform appearance. Their ontogenetic development indicates that they are the most primitive of living coelenterates.

MORPHOLOGY

The trachylinid medusae are solitary, free-swimming animals of small to moderate size, consisting of a gelatinous bell similar to that of the Scyphomedusae (Fig. 53). As in that subclass, the upper side of the body is termed **exumbrella,** while the lower side is termed **subumbrella.** The margin of the bell may be either circular or scalloped into **lappets.** An extension of both the exumbrella and the subumbrella, known as **velum,** forms a muscular annular band partly closing the subumbrellar concavity. From the center of the subumbrella there hangs down a cylindrical or quadrangular tube, the **manubrium,** which in some genera may be found at the end of a long projection of the subumbrella, called **pseudomanubrium,** containing extensions of the radial canals. The manubrium opens at its lower extremity into a mouth and leads aborally into a central gastric cavity or **stomach.** The manubrium is absent in the Narcomedusina where the mouth opens directly into the capacious stomach. The gastric cavity is typically outpocketed into 4, 8, or 16 radial **stomach pouches.** These give rise to **radial canals,** commonly 4, 6, or 8, leading into a circular **ring canal** running along the margin of the bell which, in its turn, may give rise to numerous **blind centripetal canals** extending toward the summit of the bell. In some genera, however, the ring canal is absent and the radial canals have blind distal ends. The **gonads** occur either in folds or pendent sacs beneath the radial canals or in the floor of the stomach pouches.

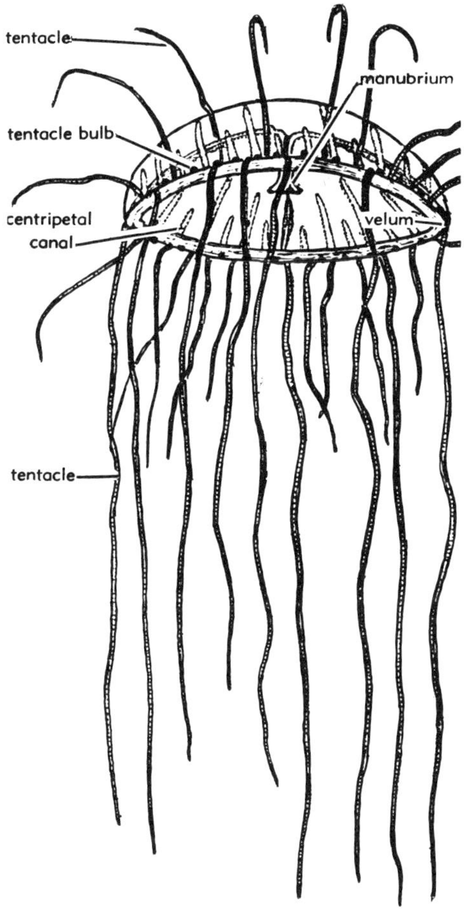

FIG. 53. A living trachylinid, *Olindias,* of the suborder Trachymedusina, showing inverted bowl-shaped form and extended manubrium (pseudomanubrium) at the center of the under (subumbrellar) surface. In addition to long pendent tentacles distributed around the margin, relatively stiff shorter ones project upward (L. H. HYMAN, *Invertebrates: Protozoa through Ctenophora,* McGraw-Hill, New York).

The **tentacles,** of the same or greater number than the radial canals, commonly spring from the exumbrellar surface and may bear **adhesive discs** near their tips. The endodermal core of the tentacles normally continue into the jelly as **tentacular roots,** and the ectoderm of their inner faces may extend on the exumbrella to the bell margin forming thick bands, the **peronia,** which divide the margin into lappets. The **otoporpae** are tracts similar to the peronia, located on the exumbrellar side above the lithocysts. Ocelli are generally absent but **lithocysts** (organs of equilibrium) of endodermal origin are commonly present, located between the **tentacle bases** and of the same or greater number than the radial canals. **Nematocysts** (nettling or stinging organs) are abundantly distributed on the tentacles, peronia, otoporpae, bell margin and, in certain forms, in streaks ascending the exumbrella.

The radial canals, tentacles, sense organs, gonads, and other structures are typically arranged upon a plan of tetramerous symmetry. The 4 main radii, corresponding to the 4 basic radial canals and tentacles, are termed **perradii;** the 4 radii at 45° position between the perradii are called **interradii;** those located midway between perradii and interradii are known as **adradii.**

DISTRIBUTION

Very few fossil records of unquestionable trachylinid medusae are known. These belong in the genus *Kirklandia* from the Lower Cretaceous of Texas and Germany. In addition, 7 genera are now tentatively placed in this order, 5 of them having uncertain family relationships. With the exception of a Lower Cambrian form from South Australia, they are European genera ranging in age from Middle Jurassic to Early Cretaceous.

CLASSIFICATION

The number of fossil genera in each of the following categories is tabulated after the corresponding taxa. A question mark preceding a numeral (as ?7) indicates number of genera tentatively classified.

Main Divisions of Trachylinida

Trachylinida *(order)* (1+?7). *?L.Cam., ?M.Jur., U.Jur.-Rec.*
- Trachymedusina *(suborder)* (1+?1). *?L.Cam., ?M.Jur., U.Jur.-Rec.*
  - Olindiadidae. *Rec.*
  - Petasidae. *Rec.*
  - Limnocnididae. *Rec.*
  - Pectyllididae. *Rec.*
  - Kirklandiidae (1). *?M.Jur., L.Cret.*
  - Trachynematidae (?1). *?L.Cam., Rec.*
  - Geryoniidae. *Rec.*
- Narcomedusina *(suborder)* (?1). *?U.Jur., Rec.*
  - Solmariidae. *Rec.*
  - Cuninidae (?1). *Rec.*
  - Aeginidae. *Rec.*

?Trachylinida incertae sedis (5). *L.Cam.-L.Cret.*

SYSTEMATIC DESCRIPTIONS

Order TRACHYLINIDA Haeckel, 1877

[*nom. correct.* PEARSE, 1936 (*pro* Trachylinae HKL., 1877)]

Hydrozoid medusae having sense organs consisting of statocysts and tentaculocysts of ectodermal origin; tentacles usually inserted above bell margin, springing from exumbrellar surface; with reduced or absent polypoid generation, commonly developing from actinula larvae either directly or by budding. *?L.Cam., ?M.Jur., U.Jur.-Rec.*

Living trachilinids are inhabitants of the open ocean of world-wide distribution, more frequent in warmer seas than in cold waters. Exceptions are 2 fresh-water genera (Fig. 53).

Suborder TRACHYMEDUSINA Haeckel, 1866

[*nom. correct.* HARRINGTON & MOORE, herein (*pro* Trachymedusae HKL., 1866)] [=Trachomedusae HKL., 1879]

Margin of bell entire, not cleft into lappets; radial canals 4, 6, or 8; gonads commonly developed upon radial canals; tentacles usually solid. *?L.Cam., ?M.Jur., L. Cret.-Rec.*

Family OLINDIADIDAE Haeckel, 1879

[*nom. correct.* HARRINGTON & MOORE, herein (*pro* Olindiadae HKL., 1879)]

Some or all tentacles projecting from sides of bell above margin, not grouped in clusters, with adhesive discs; gonads linear, saclike or folded upon the 4 or 6 radial canals. *Rec.*

Olindias F.MÜLLER, 1861 [**O. sambaquiensis*]. *Rec.*——FIG. 53. *O.* sp., Bermuda (HYMAN). Several other living genera.

Family PETASIDAE Haeckel, 1879

Tentacles without adhesive disc; radial canals 4; gonads linear or saclike; lips 4. *Rec.*

Petasus HAECKEL, 1879 [**P. atavus*]. *Rec.* Several other living genera, including the fresh-water *Craspedacusta* LANKESTER, 1880.

Family LIMNOCNIDIDAE Mayer, 1910

Tentacles numerous, hollow, not in clusters, without adhesive disc; lithocysts numerous, on exumbrellar side of velum; radial canals 4 or 6; gonads diffuse in ectoderm of stomach wall; mouth round. *Rec.*

Limnocnida GÜNTHER, 1893 [**L. tanganjicae*]. Rec. Only known genus (fresh-water).

Family PECTYLLIDIDAE Haeckel, 1879

[*nom. correct.* HARRINGTON & MOORE, herein (*pro* Pectillidae HKL., 1879)] [*ex Pectyllis* HKL., 1879 (=*Ptychogastria* ALLMAN, 1878)] [=Ptychogastridae MAYER, 1910]

Numerous more or less isolated clusters of tentacles, some with adhesive disc; numerous free lithocysts; radial canals 8; stomach 8-lobed; lips 4. *Rec.*

Ptychogastria ALLMAN, 1878 [**P. polaris* (=*Pectyllis arctica* HAECKEL, 1879)] [=*Pectyllis, Pectiss, Pectanthis* HKL., 1879]. *Rec.* Only known genus.

Family KIRKLANDIIDAE Caster, 1945

Subumbrellar mid-field lobate, typically with 8 adradial sulci bifurcating near periphery, originating 4 perradial and 4 interradial large, petalate insert lobes (stomach pouches) and 8 adradial, smaller, triangular exsert lobes; radial canals 8 (4 perradial, 4 interradial), each with bulging outpocketing genital sac containing paired gonads; mouth quadrate, functional; manubrium very short; oral arms absent; peripheral area imperfectly known; ?tentacles ?8, rodlike, covered with ovate pustules (?nettling structures); exumbrellar mid-field typically with 8 radiating shallow sulci. *?M.Jur., L.Cret.*

Kirklandia CASTER, 1945 [**K. texana*]. Characters of family. *?M.Jur.,* Ger., *L.Cret.,* USA.——FIG. 54,*1*. **K. texana*, L.Cret., Tex.; *1a*, holotype, subumbrellar surface; *1b*, paratype, subumbrellar surface; *1c*, subumbrellar surface of another specimen; *1d*, paratype, exumbrellar surface; *1e*, side view of another specimen; all ×0.75; *1f*, ?nettling structures of tentacle, ×3.5 (CASTER).——FIG. 55. *?Kirklandia* sp., M.Jur., Ger.; *1*, subumbrellar surface; *2*, *ibid.*, another specimen, both ×1 (LÖRCHER).

Family TRACHYNEMATIDAE Gegenbaur, 1856

[*nom. correct.* HARRINGTON & MOORE, herein (*pro* Trachynemidae GEGEN., 1856)] [*ex Trachynema* GEGEN., 1856 (=*Rhopalonema* GEGEN., 1856)] [=Aglauridae HAECKEL, 1879]

Tentacles without adhesive disc; radial canals 8, simple; ring canal simple. *?L. Cam., Rec.*

Rhopalonema GEGENBAUR, 1856 [**R. velatum* (=*Trachynema ciliatum* GEGEN., 1856)] [=*Trachynema* GEGEN., 1856; *Calyptra* LEUCKART, 1856]. *Rec.* Numerous living genera.

?Beltanella SPRIGG, 1947 [**B. gilesi*]. Exumbrellar impression circular; outer ring (?velum) concentrically striated, narrow; inner field (?stomach) large, circular, with 2 central concentric grooves (?oral structures); intermediate annular zone crossed by 4 paired grooves 90° apart (?radial canals), with 2 evenly spaced circular markings in each quadrant (?gonads); ring canal and tentacles absent. *L.Cam.*, S.Austral. SPRIGG compared the only known impression with the living genus *Rhopalonema*.——FIG. 56. **B. gilesi; 1a,* reconstruction of umbrella with section removed, ×0.5; *1b,* diagrammatic sketch of subumbrellar surface, ×0.55 (SPRIGG); *2,* holotype, ×0.55 (SPRIGG).

Family GERYONIIDAE Eschscholtz, 1829

Radial canals 4 or 6; gonads flat, expanded, leaflike; stomach upon gelatinous peduncle; ring canal giving rise to blindly ending centripetal canals. *Rec.*

Geryonia PÉRON & LESUEUR, 1809 [**G. proboscidalis*]. *Rec.* Other living genus: *Liriope* LESSON, 1843.

Suborder NARCOMEDUSINA Haeckel, 1879

[*nom. correct.* HARRINGTON & MOORE, herein (*pro* Narcomedusae HKL., 1879)]

FIG. 54. Trachymedusinaa: *Kirklandia texana,* L.Cret., Tex.; *1a-e,* ×0.75; *1f,* ×3.5 (CASTER).

FIG. 55. Trachymedusina: *?Kirklandia* sp., M.Jur., Ger.; subumbrellar side of 2 specimens, ×1 (LÖRCHER).

Margin of bell scalloped by peronia; gonads borne in floor of stomach. *?U.Jur., Rec.*

Family SOLMARIIDAE Haeckel, 1879

[*nom. correct.* HARRINGTON & MOORE, herein (*pro* Solmaridae HKL., 1879)] [=Peganthidae HAECKEL, 1879]

Stomach circular, simple, without stomach pouches; saccules may arise from subumbrellar floor of stomach; all other parts of gastrovascular system lacking. *Rec.*

Solmaris HAECKEL, 1879 [**S. godeffroyi*]. *Rec.* Other living genus: *Pegantha* HKL., 1879.

Family CUNINIDAE Broch, 1929

Central stomach giving rise to simple, undivided pouches; marginal scallops, peronia and tentacles of same number (8 or more) as pouches. *?U.Jur., Rec.*

Cunina ESCHSCHOLTZ, 1829 [**C. globosa*]. *Rec.* Other living genera: *Solmissus, Cunoctantha* HAECKEL, 1879.

?Acalepha BEYRICH, 1849 [**A. deperdita*] [=*Craspedonites* HAECKEL, 1865 (obj.); *Trachynemites* HKL., 1869 (obj.)]. Subumbrellar impression with 2 smooth concentric outer rings, inner rosette with 8 radial segments and central circular area; oral arms absent; tentacles ?absent. HAECKEL interpreted the central area as stomach and mouth, the groove along the periphery of rosette as ring canal, the internal outer ring as the umbrellar margin and the external ring as a furrow in the matrix. BRANDT regarded the 2 outer rings as annular folds in the umbrella developed during entombment and the inner structure as a central stomach of circular outline giving rise to 8 undivided stomach pouches. MAYER accepted

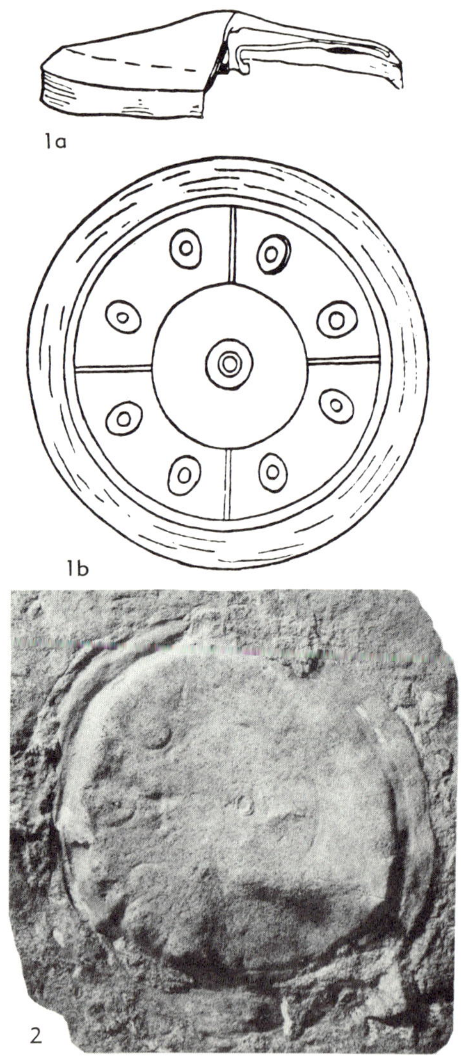

Beltanella

FIG. 56. Trachymedusina: *Beltanella gilesi*, L.Cam., S.Austral.; *1*, section and subumbrellar side (reconstr.); *2*, holotype, ×0.55 (SPRIGG).

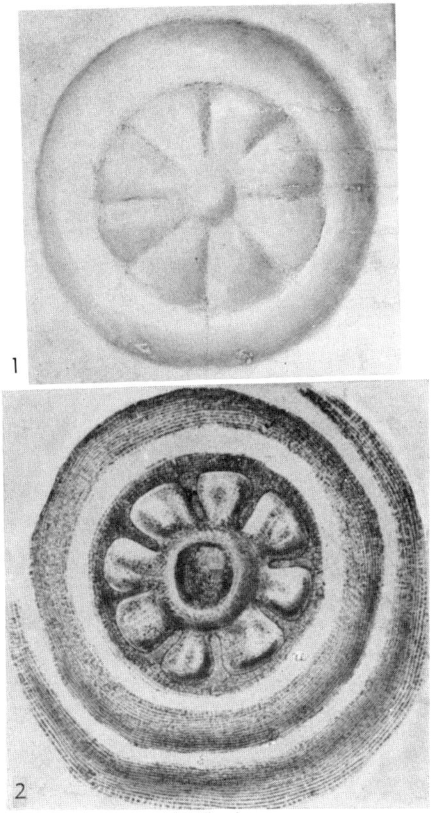

FIG. 57. Narcomedusina: *Acalepha deperdita,* U.Jur., Ger.; 2 specimens, ×0.7, ×0.25 (HAECKEL, WALCOTT).

BRANDT's interpretation and compared the fossils with the living genus *Cunoctantha. U.Jur.,* Ger.——FIG. 57. **A. deperdita; 1,* plaster impression, according to HAECKEL, ×0.7 (HAECKEL); *2,* plaster impression of a different specimen, according to BRANDT, ×0.25 (WALCOTT).

Family AEGINIDAE Haeckel, 1879

[=Cunanthidae HKL., 1879]

Central stomach giving rise to bifurcate pouches; 2 or 4 pouches between successive tentacles. *Rec.*

Aegina ESCHSCHOLTZ, 1829 [**A. citrea*]. *Rec.* Several living genera, including *Aeginopsis* BRANDT, 1836, and *Solmundella* HAECKEL, 1879.

?TRACHYLINIDA INCERTAE SEDIS

Acraspedites HAECKEL, 1869 [**Medusites antiquus* HKL., 1865]. Subumbrellar impression with 2 smooth, concentric outer rings, large circular central area and 2 intermediate rings with 8 radial lobes. HAECKEL interpreted the central area as stomach and mouth surrounded by 8 gonads (inner lobes), the radial grooves marking off the 8 outer lobes as radial canals, the periphery of the outer lobes as the subumbrellar margin cleft into 8 lappets and bearing a ring canal, the internal outer ring as a fold of the umbrella developed during entombment, and the external ring as a furrow in the matrix. Accepted here is the view of CASTER, who regarded the 8 larger lobes as stomach pouches separated by narrow sulci (not radial canals), bearing genital sacs (inner lobes). *U.Jur.,* Ger.——FIG. 58. **A. antiquus* (HAECKEL); holotype, ×0.375 (HAECKEL).

Atollites MAAS, 1902 [**A. zitteli*]. Body discoidal, of circular outline; mid-field with small, circular, smooth depression (?mouth), encircled by slightly convex annular zone crossed by 12 to 14 raised,

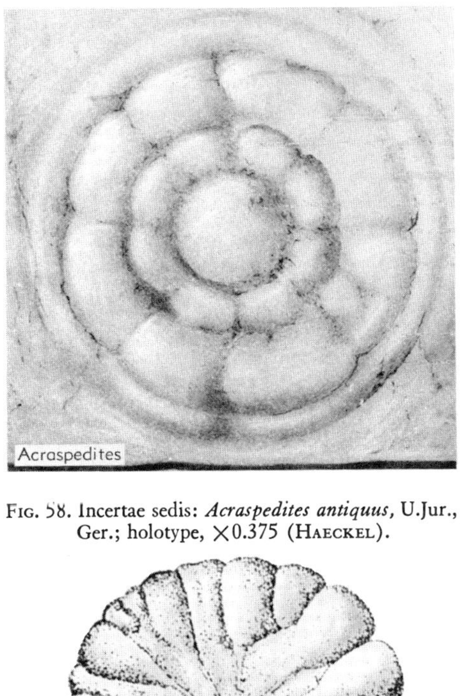

FIG. 58. Incertae sedis: *Acraspedites antiquus,* U.Jur., Ger.; holotype, ×0.375 (HAECKEL).

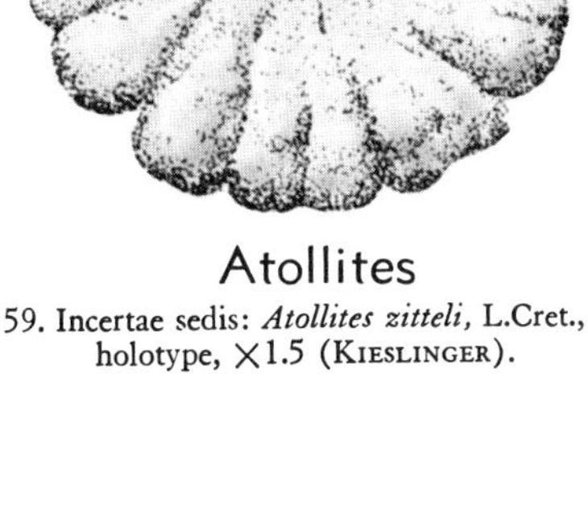

FIG. 59. Incertae sedis: *Atollites zitteli,* L.Cret., Ger.; holotype, ×1.5 (KIESLINGER).

narrow radial bands (?radial canals); external zone with 12 to 14 pyriform lobes, wider and thicker at periphery, sharply separated by deep radial sulci, some partly split by supplementary radial grooves. Both MAAS and KIESLINGER regarded *Atollites* as a collaspidid scyphomedusa. As here interpreted, the remains represent the central subumbrellar area of a trachylinid medusa displaying 12 to 14 radial canals bearing centrifugal genital sacs. *L.Cret.*, Ger.——FIG. 59. **A. zitteli;* holotype, ×1.5 (MAAS).

Ediacaria SPRIGG, 1947 [**E. flindersi* (=*Tateana inflata* SPRIGG, 1949; *Cyclomedusa radiata* SPRIGG, 1949] [=*Tateana* SPRIGG, 1949]. (1) SUBUM-

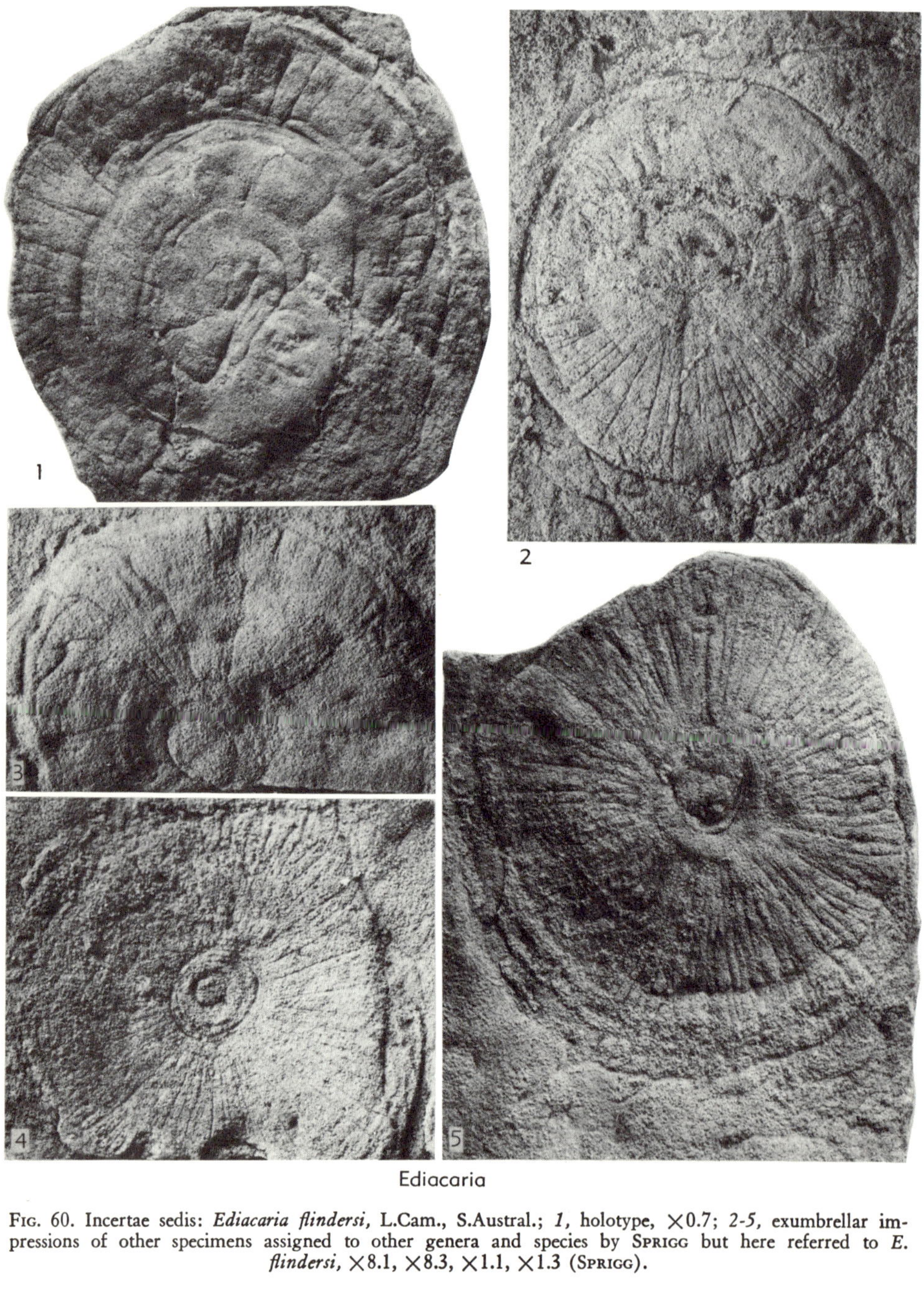

Ediacaria

FIG. 60. Incertae sedis: *Ediacaria flindersi*, L.Cam., S.Austral.; *1*, holotype, ×0.7; *2-5*, exumbrellar impressions of other specimens assigned to other genera and species by SPRIGG but here referred to *E. flindersi*, ×8.1, ×8.3, ×1.1, ×1.3 (SPRIGG).

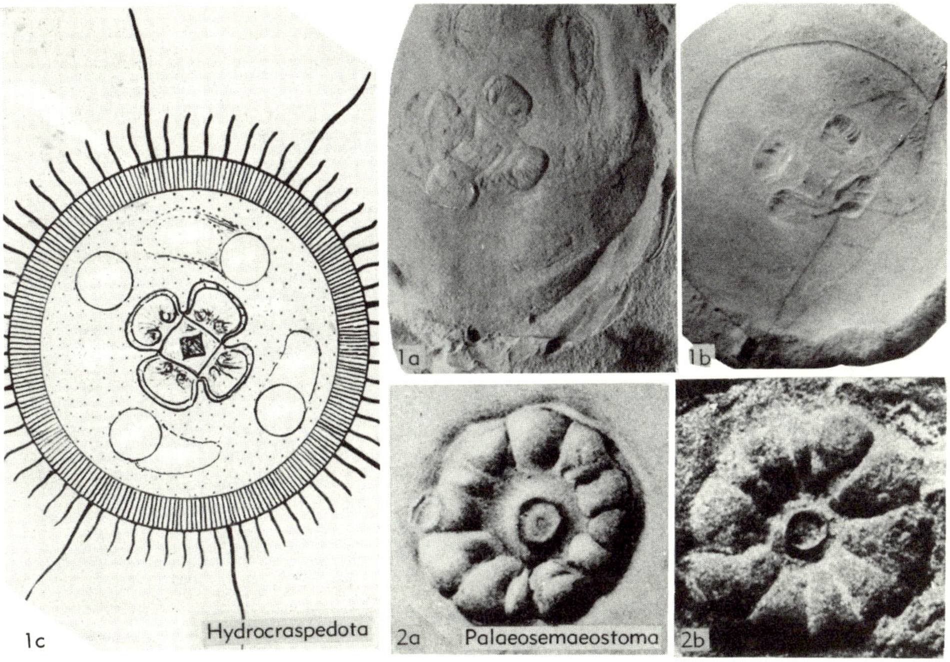

FIG. 61. Incertae sedis: Jurassic trachylinids. *1, Hydrocraspedota mayri,* U.Jur., Ger.; *1a,b,* holotype and young specimen, ×0.35; *1c,* subumbrellar surface and tentacles (reconstr.), ×0.25 (KOLB). *2, Palaeosemaeostoma geryonides,* M.Jur., Ger.; holotype and another specimen, ×1, ×0.5 (VON HUENE, KIESLINGER).

BRELLAR IMPRESSION (type of *E. flindersi*). Circular; margin ?entire, without notches; outer annular zone with numerous radial grooves, irregularly distributed, of unequal length and depth, some bifurcating centripetally, crossed by few irregular concentric rugae; intermediate annular zone somewhat inflated, with few, indistinct radial grooves; imperfectly separated from central mid-field by annular groove; mid-field with ?ring of ?8 small circular knobs (only 4 displayed) and asymmetrical triangular lobe attached to center and spreading outward. (2) EXUMBRELLAR IMPRESSION (type of *Tateana inflata* and of *Cyclomedusa radiata*). Circular, with circular central area (?stomach) marked by one or several concentric grooves and wide outer ring with numerous radial grooves, irregularly distributed and increasing in number toward periphery by implantation and bifurcation, crossed by several coarse concentric rugae. SPRIGG interpreted the radial grooves as radial canals, the outermost concentric ruga as epimarginal ring canal, the triangular lobe in the subumbrellar mid-field as a flattened manubrium, and the small circular knobs as outpocketing genital sacs. As here interpreted, the outer annular zone of the subumbrellar surface represents the impression of a velum with numerous radial shrinkage furrows due to desiccation during entombment, the ?8 central knobs are regarded as centripetal gonads (?located upon radial canals, traces of which may be represented by the radial grooves of the intermediate zone), and the triangular lobe as the impression of a flattened, campanulate manubrium. *Tateana inflata* and *Cyclomedusa radiata* are here regarded as exumbrellar impressions of *Ediacaria flindersi*. *L.Cam.*, S.Austral.——FIG. 60. **E. flindersi; 1,* holotype, ×0.7; *2,* exumbrellar impression (type of *Tateana inflata*), ×8.1; *3,* another exumbrellar impression (type of *T. inflata*), ×8.3; *4,* exumbrellar impression (type of *Cyclomedusa radiata*), ×1.1; *5,* another exumbrellar impression (type of *C. radiata*), ×1.3 (SPRIGG).

Hydrocraspedota KOLB, 1951 [**H. mayri*]. Margin circular, entire; velum narrow, with 600 to 800 fine radial striae (?impressions of statocysts); gonads 4, elliptical, perradial, centrifugal (observed only in adult specimens); mid-field with central quadrate mouth, short manubrium and 4 perradial reniform impressions of mouth tentacles, each with whirl-like bunch of raised striae at both ends; marginal tentacles ?not preserved. KOLB compared the impressions with the living freshwater genus *Craspedacusta*. *U.Jur.*, Ger.——FIG. 61,*1*. **H. mayri; 1a,* holotype, mature individual, ×0.3; *1b,* young individual, ×0.35; *1c,* reconstructions of subumbrellar surface, mature individual, ×0.25 (KOLB).

Palaeosemaeostoma RÜGER, 1933 [*pro Medusa* WAGNER, 1932 (*non* LINNÉ, 1758)] [**Medusina geryonides* VON HUENE, 1901 (=*Medusa gorgonoides* WAGNER, 1932)]. Body depressedly convex; ?subumbrellar surface with rosette of 11 or 12 pillowy sectors sharply defined by deep, narrow grooves; center of disc with depressed circular (?mouth) area encircled by sharp, raised narrow collar; lobation extending around border of disc; ?exumbrellar side with central hollow stem. *M.Jur.*, Ger.——FIG. 61,*2*. **P. geryonides* (v.HUENE); *2a*, holotype, ×1 (VON HUENE); *2b*, another specimen, ×0.5 (KIESLINGER).

REFERENCES

Beyrich, E.

(1) 1849, *Über Acalepha deperdita:* Zeitschr. deutsch. geol. Gesell., Band 1, p. 437.

Caster, Kenneth E.

(2) 1945, *A new jellyfish (Kirklandia texana Caster) from the Lower Cretaceous of Texas:* Paleontogr. Americana, v. 3, p. 168-220, pl. 1-5, fig. 1-8.

Haeckel, Ernst

(3) 1865, *Über fossile Medusen:* Zeitschr. f. wiss. Zoologie, Band 15, p. 504-514, pl. 39.

(4) 1869, *Über die fossilen Medusen der Jurazeit:* Zeitschr. f. wiss. Zoologie, Band 19, p. 538-562, pl. 42.

Huene, F. von

(5) 1901, *Kleine paläontologische Mitteilungen: I. Medusina geryonides:* Neues Jahrb. f. Min. etc., Jahrg. 1901, p. 1-12, fig. 1-2.

Kiderlen, Helmut

(6) 1935, *Die Doggermedusen der Schwabischen Alb:* Jahresber. u. Mitt. oberrhein. geol. Ver., Neue Folge, Band 24, p. 100-108, fig. 1-5.

Kieslinger, A.

(7) 1924, *Meduse fossiles:* Fossilium Catalogus. I. Animalia. Pars 26, p. 1-20.

(8) 1939, *Scyphozoa:* in SCHINDEWOLF, O. H., Handb. der Paläozoologie, Band 2A, Lief 5, p. A69-A109, fig. 1-42.

Kolb, Anton

(9) 1951, *Hydrocraspedota mayri n. g. n. sp., eine Hydromedusae aus den Plattenkalk von Pfalzpaint:* Geol. Blätter f. Nordost-Bayers u. angrenzenden Gebiete, Band. 1, p. 113-116, pl. 5-6, fig. 1.

Kuhn, Oskar

(10) 1937, *Eine fossile Meduse aus dem Opalinus-Ton von Franken:* Centralbl. f. Min. etc., Band 8, p. 334-336, fig. 1.

Lörcher, E.

(11) 1931, *Eine neue fossile Qualle aus dem Opalinusschichten und ihre paläogeographische Bedeutung:* Jahresb. u. Mitt. oberrhein. geol. Ver., Neue Folge, Band 20, p. 44-48, pl. 1.

Maas, Otto

(12) 1902, *Über Medusen aus dem Solnhofen Schiefer:* Palaeontogr., Band 48, p. 293-315, pl. 22-23.

Rüger, L.

(13) 1933, *Medusen aus dem württembergischen unteren Dogger und Bemerkungen über die Stammegeschichte der Medusen:* Jahresb. u. Mitt. oberrhein. geol. Ver., Neue Folge, Band 22, p. 39-43.

——— & Rüger-Haas, P.

(14) 1926, *Palaeosemaeostoma geryonides v. Huene sp., eine sessile Meduse aus dem Dogger von Wehigen in Württemberg und Medusina liasica n. sp., eine coronatenähnliche Meduse aus dem mittleren Lias von Hechingen in Württemberg:* Sitzber. Heidelberg. Akad. Wiss., Math.-Nat. Kl., Jahrg. 1925, Abhandl. 15, p. 1-22, pl. 1-2, fig. 1-2.

Sprigg, R. G.

(15) 1947, *Early Cambrian (?) jellyfishes from the Flinders ranges, South Australia:* Roy. Soc. S.Australia, Trans., v. 71, pt. 2, p. 212-224, pl. 5-8.

(16) 1949, *Early Cambrian "jellyfishes" of Ediacara, South Australia, and Mt. John, Kimberley District, Western Australia:* Same, v. 73, pt. 1, p. 72-99, pl. 9-21.

Wagner, G.

(17) 1932, *Beobachtungen am Meeresstrand:* Aus der Heimat, Band 45, p. 161-172.

Walcott, Charles D.

(18) 1898, *Fossil Medusae:* U. S. Geol. Survey, Mon. 30, ix+201 p., 47 pl.

MEDUSAE OF THE HYDROIDA

By H. J. Harrington and R. C. Moore

CONTENTS

The medusae of the Hydroida obviously do not constitute a separate order of Coelenterata, as they represent the completely evolved, sexual generation sprung by budding from the asexual polyps. However, the hydroid generation from which many medusae presumably came has not been definitely identified. Moreover, the hydroid and medusoid generations may evolve independently in such manner that very similar medusae are budded from quite different hydroid types and vice versa. These difficulties make it necessary to use a special classification for the medusae of the Hydroida side by side with that developed for the hydroids, and it must be apparent that the medusoid classification becomes imperative and unavoidable when the corresponding polyps are unknown. The alternative would be to classify such medusae as incertae sedis within the Hydroida until their polyps are identified, but it must be obvious that in this case the fossil medusae would remain forever in such a category.

MORPHOLOGY

The medusae of the Hydroida are morphologically similar to the trachylinids, consisting of a gelatinous bell typically dome-, bowl-, or saucer-shaped (Fig. 62). As in the trachylinids, the upper side of the bell is termed **exumbrella,** the lower side being known as **subumbrella.** The margin of the bell is always circular and unscalloped. A **velum** is typically present, as well as a **manubrium,** which in some genera is found

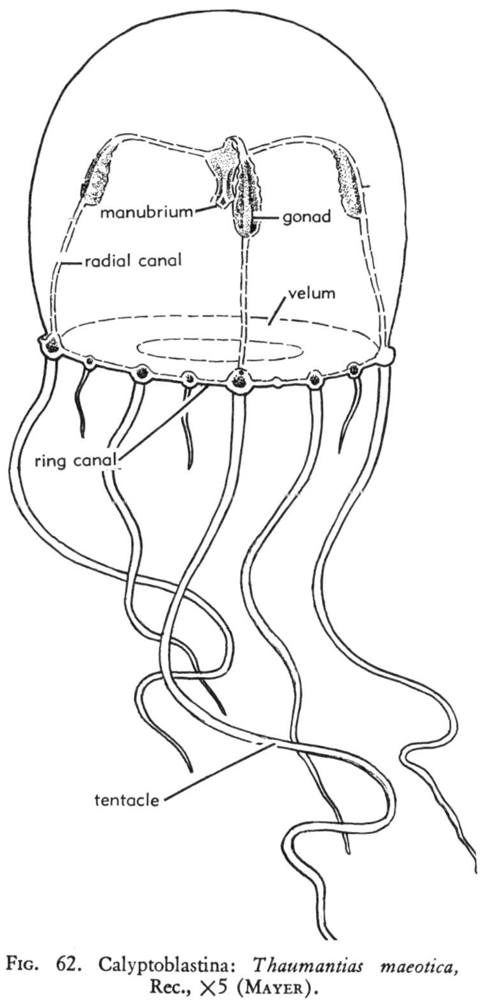

Fig. 62. Calyptoblastina: *Thaumantias maeotica,* Rec., ×5 (Mayer).

at the end of a long **pseudomanubrium.** As in the trachylinids, the manubrium opens at its lower end in a **mouth** and leads aborally into a central gastric cavity or **stomach.** The stomach gives rise to **radial canals,** typically 4 or some multiple thereof, leading into a circular **ring canal** running along the margin of the bell. The **gonads** are found either encircling the manubrium or upon the radial canals.

The bell margin bears **tentacles,** typically 4 or some multiple thereof. In some genera they are very numerous, while in others only 2, 1, or none are developed. The **tentacle bases** commonly bear **ocelli** sensitive to light, and the subumbrellar side of the velar base may bear **statocysts** (sense organs of equilibrium) of ectodermal origin. **Nematocysts** are abundantly distributed on the manubrium, bell margin, exumbrella, and tentacles.

The medusae of the Hydroida usually display marked tetramerous symmetry. The terms **perradii, interradii,** and **adradii** are used in the same sense as applied to the Trachylinida.

DISTRIBUTION

Fossil medusae of the Hydroida seem exceedingly rare. None has been unquestionably identified as belonging in this order and only 2 forms, one from the Lower Cambrian of South Australia, the other from the Upper Pennsylvanian of Nebraska, are here tentatively classified as such.

CLASSIFICATION

The medusae of the Hydroida have received collective designations which, of course, have no status in coelenterate classifications, being synonyms of other widely used terms. The name Hydromedusae, employed by MAYER in 1910, as well as the older names Cryptocarpae ESCHSCHOLTZ, 1829, Gymnopthalmae FORBES, 1848, and Craspedota GEGENBAUR, 1856, are comprehensive designations covering all the nonscyphozoan medusae and, therefore, include the Trachylinida, here classed separately. The more restricted term Leptolinae HAECKEL, 1879, on the other hand, is exactly equivalent to the order Hydroida and the 2 recognized suborders of Leptolinae (Anthomedusae HAECKEL, 1879, and Leptomedusae HAECKEL, 1866) are the exact counterparts of the suborders Gymnoblastina and Calyptoblastina, respectively, of the Hydroida. Clearly, then, both classifications are equivalent down to the level of suborders, and it is only in the taxa of family and lesser rank that important differences appear. It is debatable which of the 2 parallel classifications should be regarded as the natural one and receive preference, but the fact that the medusoid classification applies to completely evolved sexual forms while the hydroid classification is based on asexual polyps regarded as persistent larval stages, suggests that the former deserves the priority currently given to the latter.

The number of known fossil genera in each category is recorded after the corresponding taxa. The question mark (as ?2) indicates tentative classification.

Main Divisions of Hydroid Medusae

Hydroida (Leptolinida) *(order)* (?2). *L.Cam., ?Penn., Rec.*
- Gymnoblastina (Anthomedusina) *(suborder)* (?1). *?Penn., Rec.*
  - Sarsiidae. *Rec.*
  - Cladonematidae. *Rec.*
  - Oceaniidae (?1). *?Penn., Rec.*
- Calyptoblastina (Leptomedusina) *(suborder)* (?1). *?L.Cam., Rec.*
  - Thaumantiadidae (?1). *?L.Cam., Rec.*
  - Eucopidae. *Rec.*
  - Aequoreidae. *Rec.*

SYSTEMATIC DESCRIPTIONS

Medusae of the Order HYDROIDA

[=Leptolinae HAECKEL, 1879]

Veiled medusae developed through alternation of generations from Gymnoblastina or Calyptoblastina hydroids; tentacles grow outward from bell margin after bell has developed; ocelli and otocysts, when present, of ectodermal origin. *?L.Cam., ?Penn., Rec.*

The medusae of the Hydroida are inhabitants of coastal waters, rarely seen in the open ocean far from land, for they cannot maintain themselves in situations unsuited to the growth of their hydroids.

Medusae of the Suborder GYMNOBLASTINA

[=Oceanidae GEGENBAUR, 1856 (*partim*); Anthomedusae HAECKEL, 1879]

Medusae with gonads in the ectoderm of the manubrium; marginal otocysts and sensory clubs absent. *?Penn., Rec.*

Family SARSIIDAE Forbes, 1848

[*nom. correct.* HARRINGTON & MOORE, herein (*pro* Sarsiadae FORBES, 1848)] [=Codomidae HAECKEL, 1879]

Gonads ringlike, encircling the manubrium; radial canals 4 to 6, simple; tentacles simple; mouth without oral tentacles and without prominent lips. *Rec.*

Sarsia LESSON, 1843 [**S. tubulosa*]. *Rec.* Numerous other genera.

Family CLADONEMATIDAE Gegenbaur, 1856

[*nom. correct.* HARRINGTON & MOORE, herein (*pro* Cladonemidae GEGENBAUR, 1856)]

Marginal tentacles feathered or branched; radial canals 4 or more, simple or branched; gonads ringlike, encircling stomach or segregated upon interradial and adradial sides of manubrium or developed in brood pouch above stomach. *Rec.*

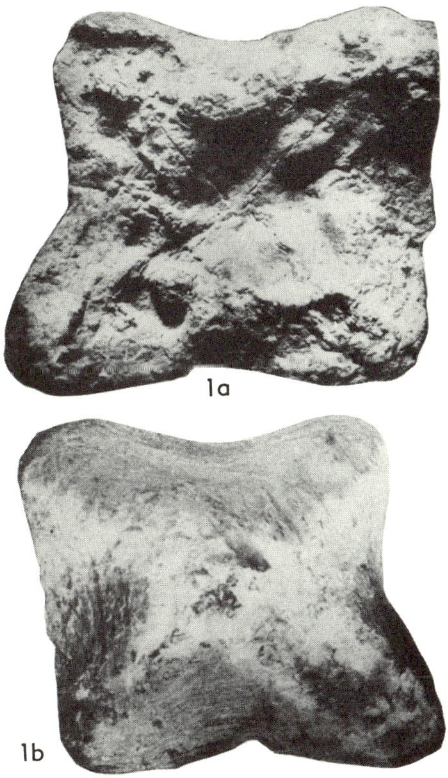

FIG. 63. Gymnoblastina: *Crucimedusina walcotti,* U.Penn., Nebr.; *1a,b,* subumbrellar and exumbrellar sides of holotype, ×0.34 (BARBOUR).

Cladonema DUJARDIN, 1843 [**C. radiatum*]. *Rec.* Several other living genera.

Family OCEANIIDAE Eschscholtz, 1829

Marginal tentacles single or in clusters; simple; gonads 4 or 4 pairs, isolated upon interradial or adradial sides of stomach; mouth with 4 lips. *?Penn., Rec.*

Oceania PÉRON & LESUEUR, 1809 (*sensu* KÖLLIKER, 1853) [**O. armata* KÖLLIKER, 1853]. *Rec.* Numerous other genera.

?Crucimedusina HARRINGTON & MOORE, 1955 [*Medusina walcotti,* BARBOUR, 1914]. Bell pyramidal-conical; subumbrella slightly concave, quadrangular in outline, with interradial rounded angles and perradial concave sides; mouth cruciform, large; exumbrella pyramidal-conical, with 4 rounded ridges meeting at apex and 4 slightly concave sides; surface marked by concentric rugae. *U.Penn.,* Nebr. The fossils may be compared with the living genus *Bougainvillia* LESSON, 1836.——FIG. 63. **C. walcotti* (BARBOUR); *2a,b,* holotype, subumbrellar and exumbrellar views, ×34 (BARBOUR).

Medusae of the Suborder CALYPTOBLASTINA

[=Leptomedusae HAECKEL, 1866]

Medusae with gonads upon radial canals; lithocysts, when present, of ectodermal origin. *?L.Cam., Rec.*

Family THAUMANTIADIDAE Gegenbaur, 1856

[*nom. correct.* HARRINGTON & MOORE, herein (*pro* Thaumantiadae GEGENBAUR, 1856)] [=Cannotidae HAECKEL, 1879]

Lithocysts present; gonads upon or extending outward over radial canals, not confined exclusively within walls of stomach. *?L.Cam., Rec.*

Thaumantias ESCHSCHOLTZ, 1829 (*sensu* MAYER, 1910) [**Laodicea cellularia* A.AGASSIZ, 1865]. *Rec.* ——FIG. 62. *T. maeotica,* Rec.; ×5 (MAYER). Numerous other genera.

?Protodipleurosoma SPRIGG, 1949 [**P. wardi*]. Exumbrellar impression circular; outer ring (?velum) concentrically striated, wide; mid-field large, circular, with ovoid, unevenly constricted lobate central area (?stomach) and irregularly developed radial grooves (?radial canals) branched nondichotomously; primary grooves strong, reaching submarginal groove (?ring canal), giving rise to shorter grooves by branching near base. *L.Cam.,*

FIG. 64. Calyptoblastina: *Protodipleurosoma wardi*, L.Cam., S.Austral.; holotype, ×0.8 (SPRIGG).

S.Austral. SPRIGG compared the fossil with the living genus *Dipleurosoma* BOECK, 1866.——FIG. 64. **P. wardi*; holotype, ×0.8 (SPRIGG).

Family EUCOPIDAE Gegenbaur, 1856

Lithocysts present; radial canals less than 8; gonads developed upon radial canals. *Rec.*

Eucope GEGENBAUR, 1856 (*non* L.AGASSIZ, 1862) [**Thaumantias globosa* FORBES, 1848]. *Rec.* Numerous other genera.

Family AEQUOREIDAE Eschscholtz, 1829

Otocysts present; radial canals 8 or more. *Rec.*

Aequorea PÉRON & LESUEUR, 1809 [**A. forskalea*]. *Rec.* Several other genera.

REFERENCES

Barbour, E. H.

(1) 1914, *Medusina walcotti, a Carboniferous jellyfish:* Am. Jour. Sci., ser. 4, v. 38, no. 228, p. 505-506, fig. 1-2.

Harrington, H. J. & Moore, R. C.

(2) 1955, *Kansas Pennsylvanian and other jellyfishes:* Kansas Geol. Survey, Bull. 114, pt. 5, p. 153-163, pl. 1-2.

Kieslinger, A.

(3) 1924, *Medusae fossiles:* Fossilium Catalogus. I. Animalia. Pars 26, p. 1-20.

(4) 1939, *Scyphozoa:* in SCHINDEWOLF, O. H. Handb. d. Palaeozoologie, Band 2A, Lief 5, p. A69-A109, fig. 1-42.

Sprigg, R. G.

(5) 1949, *Early Cambrian "jellyfishes" of Ediacara, South Australia, and Mt. John, Kimberley District, Western Australia:* Roy. Soc. S.Australia, Trans., v. 73, pt. 1, p. 72-88, pl. 9-21.

Walcott, Charles D.

(6) 1898, *Fossil Medusae:* U. S. Geol. Survey, Mon. 30, ix+201 p., 47 pl.

HYDROIDA AND SPONGIOMORPHIDA

By Dorothy Hill and J. W. Wells

CONTENTS

HYDROIDA

MORPHOLOGY

The solitary hydroid polyp, divisible into the attached base, stalk, and main body region or **hydranth,** is attached either by a **pedal disc** or by rootlike holdfasts (Fig. 6). In colonial forms, stolons or rhizomes extended over the substratum, as either a branching or anastomosing tangle, the anchoring **hydrorhiza** being fused into a peridermal mat with peridermal spines projecting up among the polyp bases; from this arises either the **hydrocaulus** (or main stalk) (Figs. 9, 65) of an arborescent colony, or the subequal single stalks **(caulomes)** of a matlike colony. The hydrocaulus is usually simple but may be fascicled (polysiphonic) (Fig. 66), that is, covered by stolons from the hydrorhizae or by branches from below. Four types of budding occur: (1) **hydrorhizal,** in which polyps spring singly from stolons of the hydrorhiza or mat or from a group of parallel stolons projecting vertically as a bundle (rhizocaulome); (2) **monopodial with terminal hydranths** (Fig. 67), giving rise in many gymnoblasts to racemose branching in which the stems of all buds have a growing point just below the terminal hydranth for production of one new hydranth; (3) **monopodial with terminal growing points** (Fig. 68), in higher calyptoblasts with tip of the hydrocaulus and of each branch a growing point, just behind which new zooids appear; and (4) **sympodial** (Fig. 69), the hydranth stems in lower calyptoblasts having no growth zone, each successive bud overlapping its parent hydranth so as to produce **cymose branching.**

Stolons, caulomes, and hydrocaulus of many forms are enclosed in a chitinous or in some forms a calcareous **periderm** transversely wrinkled near branching points. This is the only part known fossilized. In tubularian gymnoblasts, the periderm does not cover any part of the hydranth or gonads (Fig. 6,*1a*), but in calyptoblasts, a chitinous cuplike **hydrotheca** (Fig. 70) may protect the hydranth and a gonotheca the gonads; where protective zooids of different size and shape from the hydranths are developed, **dactylothecae,** usually much smaller than the hydrothecae, form over them. These thecae are marked off from the stems by a constriction or **shelf** (diaphragm) of periderm. The hydrothecae may be sessile and may be closed by a lid **(operculum)** with one to 4 parts.

The medusoids of the Hydroida are small animals without periderm and therefore unlikely to be found as fossils. They differ from the Scyphozoa but resemble the siphonophores and trachylines in having a velum or inner shelf around the rim of the umbrella. They are solitary and free-swimming, with form of a gelatinous bowl- or bell-shaped umbrella from whose rim and concave subumbrellar surface the principal parts depend (Fig. 6, *2a*). Most of them

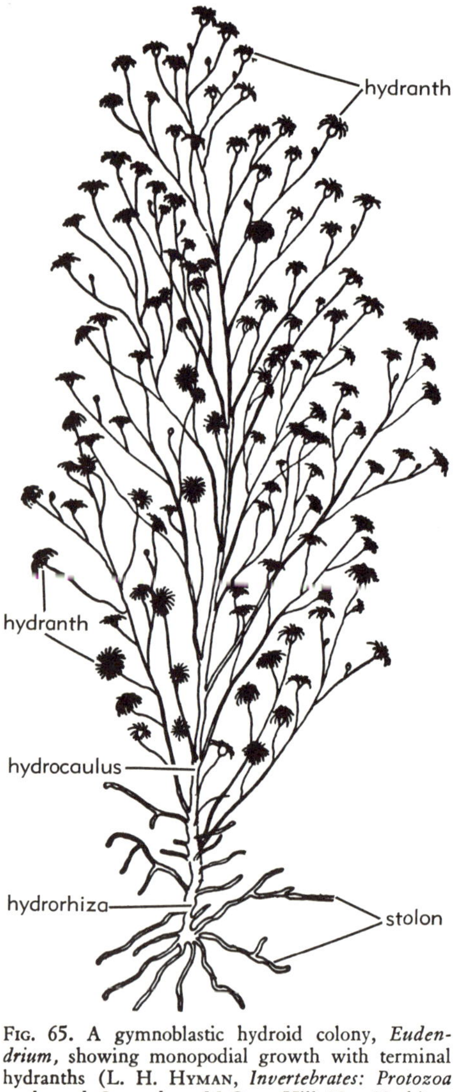

Fig. 65. A gymnoblastic hydroid colony, *Eudendrium,* showing monopodial growth with terminal hydranths (L. H. Hyman, *Invertebrates: Protozoa through Ctenophora,* McGraw-Hill, New York).

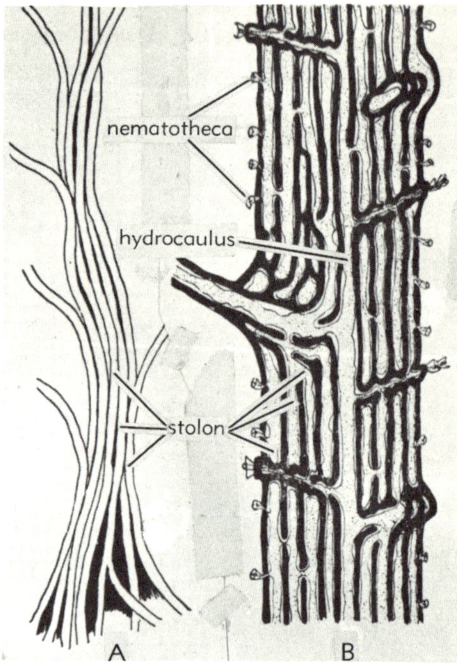

Fig. 66. Structures in hydroid colony growth. *A, Eudendrium,* with simple type of fascicled stem composed of parallel stolons. *B, Plumularia,* with complicated type of fascicled stem, with true hydrocaulus in center encircled by stolons (L. H. Hyman, *Invertebrates: Protozoa through Ctenophora,* McGraw-Hill, New York).

have perfect tetrameral radial symmetry. From the center of the subumbrellar surface hangs the **manubrium,** which is a cylindrical or quadrangular tube opening directly into the gastrovascular cavity and bearing the mouth. The manubrium commonly is surrounded by a ring of tentacles heavily loaded with nematocysts at their outer ends. The gastrovascular cavity contracts into 4 **radial canals** (in some forms, into 6 or 8 radial canals), which radiate to the margin and open into a ring or **marginal canal.** The margin is unscalloped and bears most of the tentacles and sense organs, the tentacles being usually in multiples of 4. The base of each tentacle is commonly enlarged into an **ocellar bulb,** bearing an ocellus or other sensory patches such as **statocysts.**

DISTRIBUTION

Living hydroids are almost exclusively marine, occurring typically in littoral and

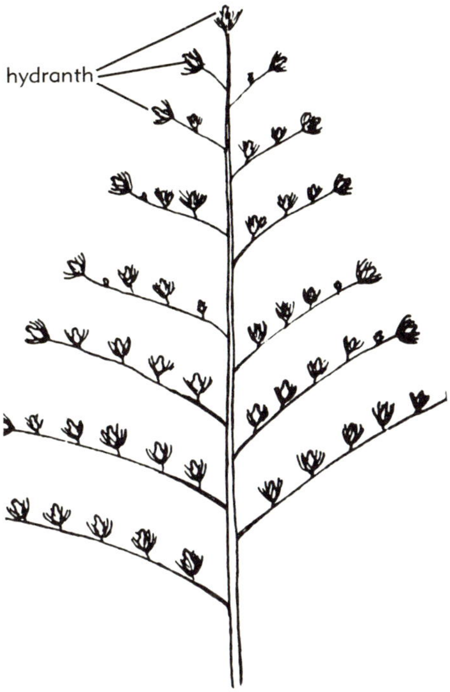

FIG. 67. Monopodial type of hydroid colony with terminal hydranths, *Pennaria* (L. H. HYMAN, *Invertebrates: Protozoa through Ctenophora,* McGraw-Hill, New York).

shallow offshore waters, but some, like the plumularians, go down to depths of 550 fathoms or more. Most prefer a hard substratum but some are adapted to soft bottoms, and many of these are found down to 2,750 fathoms. The colony form is susceptible to environmental changes. The group is most abundant in temperate and cold zones. The hydroid medusae are typically pelagic and planktonic animals of the shore and shallow seas. All are strictly carnivorous. Only a few of the many families are known as fossils.

Gymnoblastine hydroids are known doubtfully from Jurassic strata of the Mediterranean region and certainly from the Cretaceous and Tertiary of Europe. Fossils assigned to this suborder are also found in Lower Cambrian rocks of Australia. Calyptoblastines are recorded from the Middle Cambrian and Lower Ordovician of Victoria, Upper Cambrian of Texas, Middle Ordovician of Virginia and New York, and Middle Devonian of Michigan, as well as from Mesozoic and Tertiary rocks.

Order HYDROIDA Johnston, 1836

[=Sertulariens AUDOUIN & M.-EDWARDS in LAMARCK, 1828; Exoarier RAPP, 1829 *(partim)*; Nudibrachiata FARRE, 1837; Polypiaria GRAY, 1840; Hydrozoa OWEN, 1843; Hydroidea DANA, 1846; Hydraria M.EDW.-H., 1850; Diplomorpha CARUS, 1863 *(partim)*]

Polypoid generation dominant, the few solitary species nude; colonial genera with exoskeleton generally horny, weakly calcareous in some; budding off free medusae or with structures resembling abortive medusae. Medusae velate, with ocelli and ectodermal statocysts. *Cam.-Rec.*

The order includes 3 suborders: Eleutheroblastina, Gymnoblastina, and Calyptoblastina, all defined by ALLMAN, 1871; of these the few Eleutheroblastina are solitary, naked forms (*Hydra*) and are not found fossil.

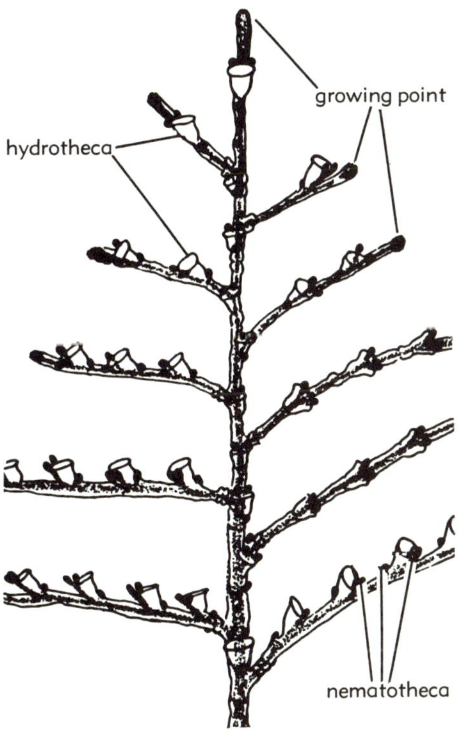

FIG. 68. Monopodial type of hydroid colony with terminal growing points, *Plumularia* (L. H. HYMAN, *Invertebrates: Protozoa through Ctenophora,* McGraw-Hill, New York).

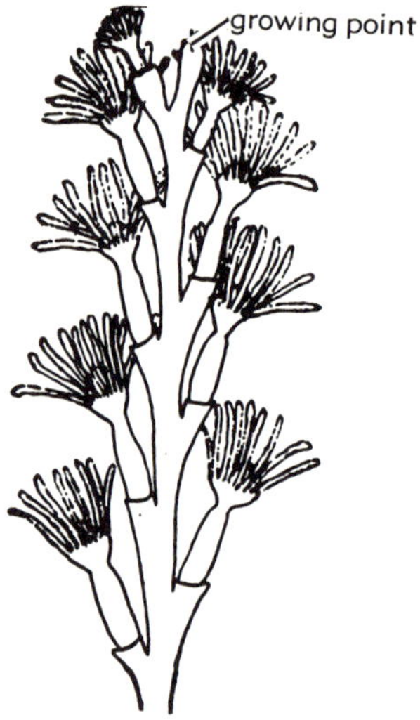

FIG. 69. Branch end of *Halecium* showing sympodial growth (L. H. HYMAN, *Invertebrates: Protozoa through Ctenophora,* McGraw-Hill, New York).

Suborder GYMNOBLASTINA Allman, 1871

[=Tubularina EHR., 1834; Athecata HINCKS, 1868; Anthomedusae HKL., 1896]

Colonial; hydranths without hydrothecae; gonophores naked; exoskeleton of hydrorhizae and caulomes horny, or in some calcareous. Free medusae tall, bell-like, with ocelli but lacking statocysts; gonads borne on the manubrium. *?Trias., ?Jur., ?Cret., Eoc.-Rec.*

Family HYDRACTINIIDAE Agassiz, 1862

[*nom. correct.* HINCKS, 1868 (*ex* Hydractinidae AGASSIZ, 1862)] [Incl. ?Syringosphaeridae DUNCAN, 1879 (as order); ?Heterastrididae FRECH, 1890; ?Ellipsactinidae POLJAK, 1936; ?Heterastridiidae KÜHN, 1939]

Encrusting, caulomes rising from a chitinous or calcareous mat of tangled hydrorhizae with periderm; caulomes in vertical tubes in the peridermal mat. *?Trias., ?Jur., ?Cret., Eoc.-Rec.*

Hydractinia BENEDEN (1841, *genus caelebs*), 1844 [**Alcyonium echinatum* FLEMING, 1828] [=*Echinochorium* HASSALL, 1841; *Synhydra* QUATREFAGES, 1843; *Synhydre* BENEDEN, 1844; *Hydractinea* ALLMAN, 1864; *?Kerunia* MAYER-EYMAR, 1899 (symbiotic with hermit crab); *Cyclactinia, Poractinia* VINASSA, 1901; *?Hydrocorella, Janaria, Hydrissa* STECHOW, 1921; *?Hydroconophora, Hydrokerion* GREGORY, 1931]. Periderm horny or calcareous; encrusting mat laminated and with spines projecting upward among the polyp bases so as to leave pores in the mat; commonly encrusting gastropod shells. *?Jur.-?Cret.,* Eu.; *Eoc.-Rec.*——FIG. 71,*1*. *H. echinata* (FLEMING), Rec., Eu.; *1a,* colony on shell occupied by hermit crab, ×1; *1b,c,* transv. secs. of chitinous spine protruding from base of colony, near base and summit of spine (chitin oblique shaded), ×50; *1d,* long. sec. of spine showing proximal part of hydranths developed from coenosarc covering spine, ×37.5 (ALLMAN); *1e,* long. sec. of part of base, ×30 (9).

?Cycloporidium PARONA, 1909 [**C. tubiforme*]. *U. Cret. (Cenom.),* Italy.

?Ellipsactinia STEINMANN, 1878 [**E. ellipsoidea*]. Hemispherical or elliptical; thick, finely tuberculate laminae with few, irregular connections. *Trias.-Cret.,* Eu.-Afr.——FIG. 71,*2*. **E. ellipsoidea,* U.Jur., Aus.; *2a,* part of surface, ×8; *2b,c,* median sec. ×1, ×10 (14).

?Heterastridium REUSS, 1865 [**H. conglobatum;* SD DIENER, 1921] [=*Stoliczkaria, Syringosphaera* DUNCAN, 1878]. Spherical or nodular, commonly with minute cephalopod shell as nucleus; laminae and pillars finely perforate, the perforations of successive laminae lying above one another, forming fine, radial unwalled tubules (?dactylopores); with large radial cylindrical unwalled, discontinuous tubes (?gastropores, ?gonopores); without tabulae; surface tuberculations markedly stellate with skeletal tissue below them much thickened but not astrorhizal. GERTH has suggested that these forms were pelagic. *Trias.,* Eu.-Asia-Timor.——FIG. 72. [**H. conglobatum*] [=*H. pachystylum*

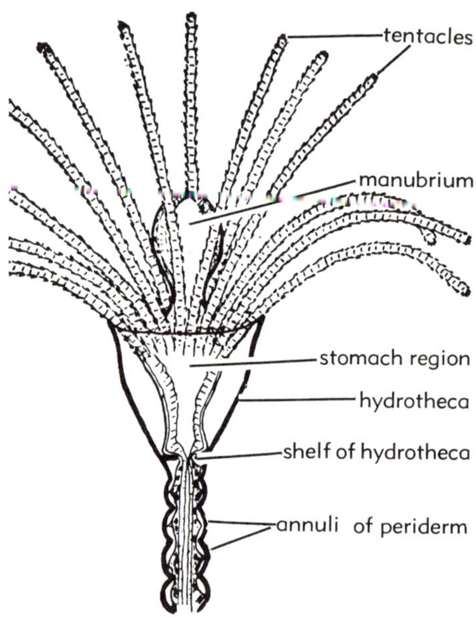

FIG. 70. Calyptoblastine hydranth, *Obelia,* with single filiform circlet of tentacles (L. H. HYMAN, *Invertebrates: Protozoa through Ctenophora,* McGraw-Hill, New York).

FRECH], *U.Trias.*, Aus.; *a,b,* transv. and long secs., ×5 (STEINMANN).

?**Philippinactinia** YABE & SUGIYAMA, 1939 [**P. hasimotoi*]. *Oligo.*, Philip.

?**Stromactinia** VINASSA, 1901 [**S. triassica*]. *U. Trias.*, Carpath.

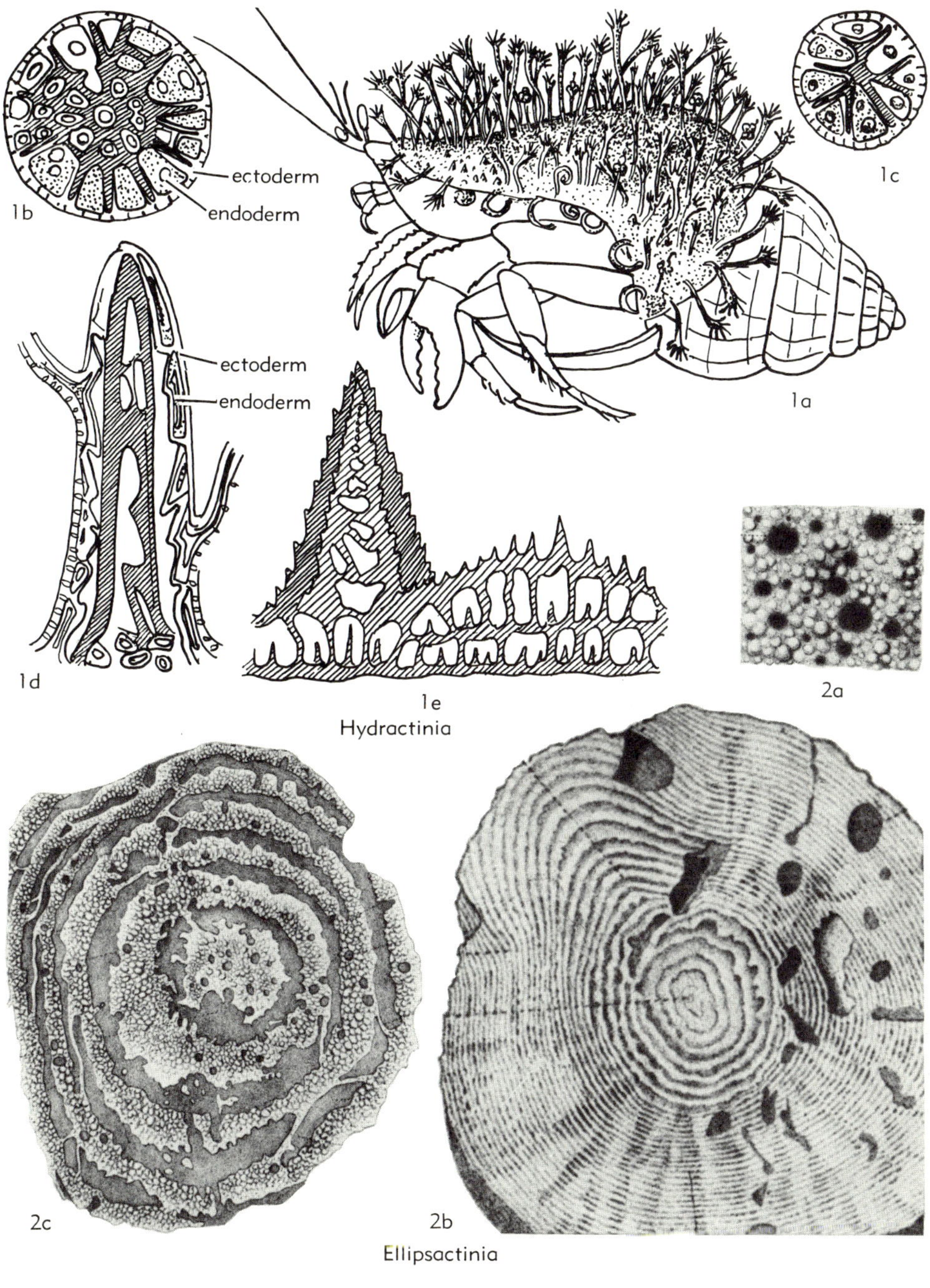

FIG. 71. Hydractiniidae. *1a-e, H. echinata,* Rec., showing living colony, ×1, and sections of chitinous skeleton, enlarged. *2a-c, E. ellipsoidea,* U.Jur., surface enlarged, and sections.

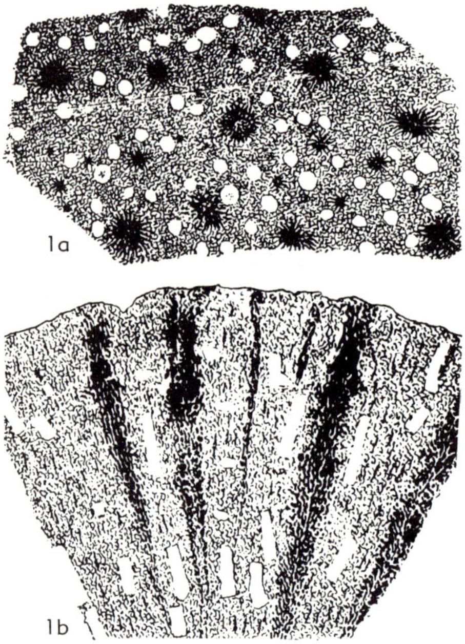

FIG. 72. Hydractiniidae. *1a,b, H. pachystylum,* U. Trias., sections, ×5.

ANTHOMEDUSAE

The medusoid generations of Gymnoblastina, many of which are not correlated reliably with known polypoids, are collectively designated Anthomedusae. The fossil record of the group is treated in sections of this volume entitled "Medusae of the hydroids" and "Medusae incertae sedis and unrecognizable forms."

Suborder CALYPTOBLASTINA Allman, 1871

[=Sertulariadae FLEMING, 1828; Sertularina EHR., 1834; Thecaphora HINCKS, 1868; Leptomedusae HAECKEL, 1866 (often used for calyptoblastean medusae)]

Hydranths provided with hydrothecae; gonophores enclosed in gonothecae; enclosing periderm may be chitinous; free medusae bowl- or saucer-shaped, usually with statocysts and gonads borne on radial canals. *Cam.-Rec.*

Several fossil forms have been described that may pertain to this suborder, but they cannot safely be referred to any living family.

Family UNCERTAIN

Archaeocryptolaria CHAPMAN, 1919 [**A. skeatsi;* SD D. THOMAS, herein]. Hydrocaulus slender, more or less erect or slightly flexuous. Hydrothecae cylindrical or elongate-conical, narrowing very slightly toward base; adnate or attached to hydrocaulus for some distance; aperture circular, lip slightly everted. Periderm chitinous, coarsely wrinkled or scaly. Gonothecae elongate pyriform and incurved to axis. *M.Cam.,* Vict.-Tasm.; *U. Cam.,* Tex.; *M.Ord.,* Va.——FIG. 73,*3.* **A. skeatsi;* M.Cam., Austral., ×2.5 (10).

Archaeolafoea CHAPMAN, 1919 [**A. longicornis*]. Hydrocaulus slender, flexuous, with few lateral branches. Hydrothecae long conical, narrowing slightly toward base, apertures circular; borne equally on sides of hydrocaulus and branches at angles of 15 to 50 degrees. Periderm transversely wrinkled. Gonothecae small, elongate pyriform. *M.Cam.,* Vict.-Tasm.-N.Y.; *U.Cam. (Lancefield.),* Vict., *Ord.,* N.Y.——FIG. 73,*1.* **A. longicornis, M.Cam.,* Vict.-Tasm.; *1a,* colony, ×2.5; *1b,* gonotheca, ×12.5 (10).

Protohalecium CHAPMAN & THOMAS, 1936 [**P. hallianum*]. Hydrocaulus short, repeatedly branching, with shallow hydrothecae on stalks. *M.Cam.,* Vict.-Tasm.

Sphenoecium CHAPMAN & THOMAS, 1936 [**Sphenothallus filicoides* CHAPMAN, 1917; SD D. THOMAS, herein]. Hydrocaulus with short stipe or pedicle with wedge-shaped hydrothecae arranged radially or serially. *M.Cam.,* Vict.-Tasm.

Palaeokylix EISENACK, 1932 [**P. chitinosus*]. Slender chitinous pipes with irregular swellings; simply

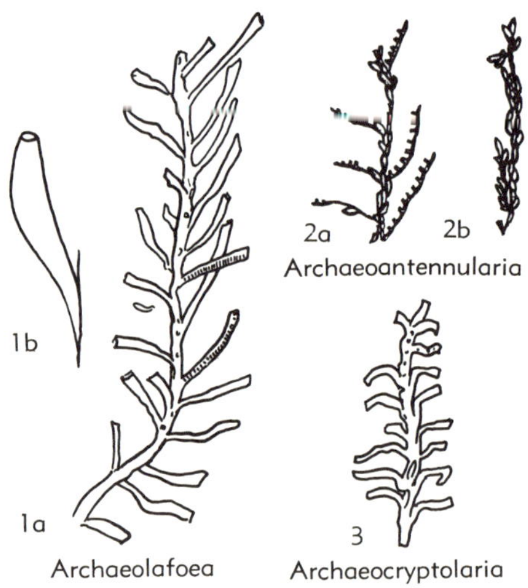

FIG. 73. Calyptoblastina. *1a,b, A. longicornis,* M. Cam., colony, ×2.5, and gonotheca, ×12.5. *2a,b, A. byersi,* M.Dev., ×2. *3, A. skeatsi,* M.Cam., ×2.5.

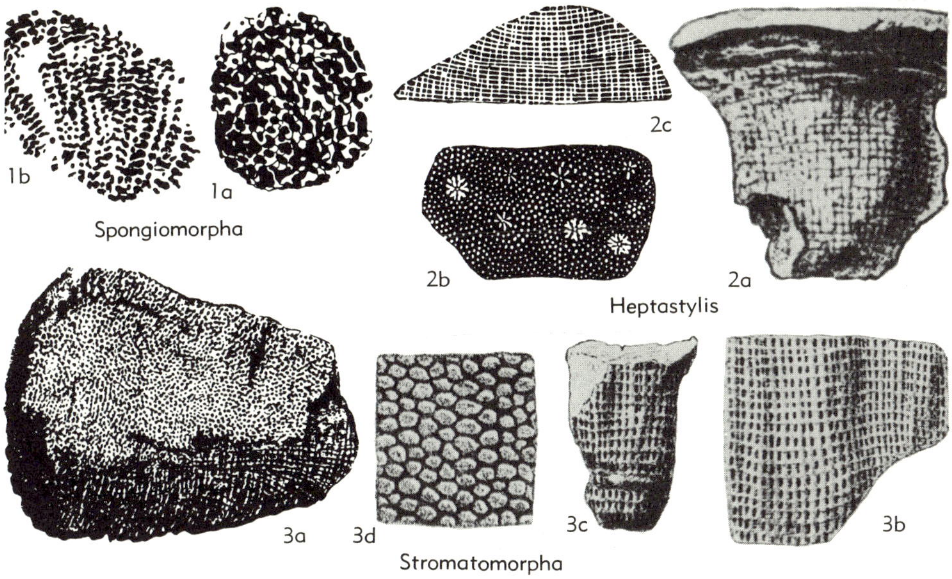

FIG. 74. Spongiomorphidae. *1a,b, S. acyclica,* Trias., sections, ×4. *2a-c, H. stromatoporoides,* Trias., exterior, ×2, and secs., ×4. *3a-d, S. stylifera,* Trias.; *3a,* ×1; *3b,c,* ×2; *3d,* ×6.

branching, ending in short wide theca (6). *Ord.,* Balt.

Palaeotuba EISENACK, 1934 [**P. polycephala*]. Hydrocaulus dichotomously branching, branches widening gradually to terminal thecae (6). *Ord.,* Balt.

Cylindrotheca EISENACK, 1934 [**C. profunda*]. Hydrocaulus simply branched, with long cylindrical theca (6). *Sil.,* Czech.

Mesokylix EISENACK, 1935 [**M. hydra*]. Hydrocaulus branched, thecae shallow, wide terminal cups (6). *M.Jur.,* Ger.

Archaeoantennularia DECKER, 1952 [**A. byersi*]. Branches radiating in all directions from central axis, which consists in part of large gonangia and in part of hydrothecae. *M.Dev.,* Mich.——FIG. 73, *2. *A. byersi;* ×2 (11).

?**Kristinella** RICHTERS, 1913 [**K. monilifera*]. *Cret.,* Ger.

LEPTOMEDUSAE

The term Leptomedusae refers to the medusoid generations of Calyptoblastina, many of which are not correlated with their polypoid generation. Tentatively classed with Leptomedusae is *Protodipleurosoma* SPRIGG, from Lower Cambrian rocks of South Australia. Reference to this and other medusoid fossils is contained in sections of Part F entitled "Medusae of the hydroids" and "Medusae incertae sedis and unrecognizable forms."

SPONGIOMORPHIDA

MORPHOLOGY

The spongiomorphs seem to be more closely allied to the stromatoporoids than to any other group. When first described, they were referred to the scleractinian corals on account of the apparent grouping of 6 vertical pillars (?trabeculae) about a 7th. They possess none of the real characters of anthozoan skeletons, however, whereas all their characteristic features are found within the Hydrozoa. Thus, their vertical pillars and concentric retiform laminae are found in the hydractinians and stromatoporoids, and they lack the ?gastropores of the sphaeractiniids and possess ?astrorhizal structures not found in the hydractinians. They might be referred to the Stromatoporoidea, although the regular astrorhizal structures are somewhat more specialized than in that group. They are characteristic of the Triassic, particularly of warm waters of the Tethys. They have also been recorded from the Jurassic of Japan and northern Africa.

Order SPONGIOMORPHIDA Alloiteau, 1952

Stromatoporoid-like hydrozoans in which stellate ?astrorhizal vertical tubules may develop from the radial grouping of 6 vertical pillars about a 7th. *Trias.-Jur.*

Family SPONGIOMORPHIDAE Frech, 1890

Massive, radial pillars dominant, united by annular thickenings or by horizontal bars, which may merge to form concentric but perforate laminae; latilaminae not developed; 6 pillars may be grouped starwise as a vertical, tubular, astrorhizal space around a 7th; tabulae may be developed. *Trias.-Jur.*

Spongiomorpha FRECH, 1890 [**S. acyclica;* SD DIENER, 1921] [=*Heptastylopsis* FRECH, 1890]. Horizontal bars not merged to form concentric laminae; tabulae numerous. *Trias.*, Eu.-Asia-N.Am.-N.Afr.——FIG. 74,*1*. **S. acyclica*, Eu.; *1a,b*, tang. and radial secs., ×4 (12).

Heptastylis FRECH, 1890 [**H. stromatoporoides*]. Horizontal bars merged to form well-developed concentric laminae; no tabulae; with stellate ?astrorhizal groupings of pillars. *Trias.*, Eu.-N.Am.; *Jur.*, Japan-N.Afr.——FIG. 74,*2*. **H. stromatoporoides*, Trias., Eu.; *2a*, exterior of side, ×2; *2b,c*, tang. and radial secs., ×4 (12).

Stromatomorpha FRECH, 1890 [**S. stylifera;* SD DIENER, 1921] [=*Cylicopsis* LEMAITRE, 1935]. Vertical pillars thickened ringwise at regular intervals, the rings of neighboring pillars united by porous concentric laminae. *Trias.*, Eu.-N.Am.; *Jur.*, Asia-N.Afr.—FIG. 74,*3*. **S. stylifera*, Trias., Eu.; *3a*, exterior, ×1; *3b,c*, radial surfaces, ×2; *3d*, transv. sec., ×6 (12).

GENERA OF UNCERTAIN SYSTEMATIC POSITION

Delheidia DOLLFUSS, 1898 [**Tubipora proxima*]. *Oligo.*, Belg.

Graptogeitonia RUEDEMANN, 1934 [**G. adherens*]. *Ord.*, N.Y.

Jillua KRUMBECK, 1913 [**J. tubifera*]. Encrusting hydrorhiza of unknown hydrozoan. *U.Trias.*, E. Indies.

Lovčenipora GIATTINI, 1903 [**L. vinassai*]. *U. Trias.-Jur.*, Eurasia.

Plumalina HALL, 1858 [**Filicites? plumaria* HALL, 1843]. *U.Dev.(Senecan)*, N.Y.

Rhizostromella PARONA, 1909 [**R. appenina*]. *U. Cret.(Cenom.)*, Italy.

Roseoporella SPRIESTERSBACH, 1935 [**R. rhenana*]. *Dev.*, Ger.

Stoliczkiella CARTER, 1888 [**S. theobaldi*]. *?Cret.*, Karakorum.

Tarrichnium WANNER, 1938 [**T. balanocrini*]. Traces of hydrorhizae of unknown hydrozoans in crinoid columns. *Mio.*, E.Indies.

GENERIC NAMES REFERRED TO HYDROZOA BUT BELONGING TO OTHER GROUPS

Aphrostoma GÜRICH, 1906 (alga or stromatolite), *L.Carb.*, Belg.

Aprutinopora PARONA, 1909 (bryozoan), *Cret.*, Italy.

Bradyzoum KÜHN, 1939 (Foraminifera) [=*Bradya* STACHE, 1873 (*non* BOEK, 1873)]. *U.Cret.*, Eu.

Chondrostroma GÜRICH, 1906 (alga or stromatolite), *L.Carb.*, Belg.

Cladocoropsis FELIX, 1907 (alga), *U.Jur.*, Jugoslavia.

Corynoides NICHOLSON, 1867 (graptolite), *Sil.*, Scot.

Cylindrohyphasma STEINMANN, 1878 (?sponge), *M.Carb.*, Moscow.

Malacostroma GÜRICH, 1906 (alga or stromatolite), *L.Carb.*, Belg.

Marinduqueia YABE & SUGIYAMA, 1939 (?sponge), *Eoc.*, Philip.

Millarella CARTER, 1888 (*?Parkeria*), *U.Cret.*,Eng.

Palaeocoryne DUNCAN & JENKYNS, 1869 (bryozoan), *Miss.*, Scot.

Parastroma DOUVILLÉ, 1927 (rudistid), *U.Cret.*, Cuba.

Parkeria CARPENTER, 1870 (Foraminifera), *Cret.*, Eng.

Porosphaera STEINMANN, 1878 (sponge), *Cret.*, Eu.

Ptychostroma GÜRICH, 1906 (alga or stromatolite), *L.Carb.*, Belg.

Spongiostroma GÜRICH, 1906 (alga or stromatolite), *L.Carb.*, Belg.

Thalamospongia D'ORBIGNY, 1850 (probably sponge) [=*Thalamosmilia* DE FROMENTEL, 1860; *Thalaminia* STEINMANN, 1878; *Thalamina* WAAGEN & WENTZEL, 1887]. *Jur.-L.Cret.*, Fr.

Thaumatostroma WEISSERMEL, 1926 (stromatolite), *Eoc.*, SW.Afr.

Websteria M.EDW.-H., 1850 (bryozoan), *Eoc.*, Eng.

REFERENCES

Publications cited in the following list are judged to be helpful in furnishing additional information concerning Hydrozoa (especially hydroids and spongiomorphs) and in offering a guide to more extensive literature. The index numbers enclosed in parentheses in the column at left are employed in the text for identification of the publications.

Allman, G. J.

(1) 1871-72, *A monograph of the gymnoblastic or tubularian hydroids:* Roy. Soc. (London), xxii+450 p., 23 pl., 77 fig.

Chapman, Frederick, & Thomas, D. E.

(2) 1936, *The Cambrian hydroids of the Heathcote and Monegeeta districts:* Proc. Roy. Soc. Vict., v. 48, p. 193-212, pl. 14-17.

Frech, F.

(3) 1890, *Die Korallenfauna der Trias, I. Die Korallen der juvavischen Triasprovinz:* Palaeontographica, Bd. 37, 116 p., 21 pl.

Gerth, H.

(4) 1915, *Die Heterastridien von Timor:* Paläont. von Timor, Lief. 2, Abt. 4, p. 61-69, pl. 42.

Kühn, O.

(5) 1927, *Zur Systematik und Nomenklatur der Stromatoporen:* Neues Jahrb. Mineralogie, Geologie, Paläontologie, Abt. B, Centralbl., p. 546-551.

(6) 1939, *Hydrozoa:* in SCHINDEWOLF, O. H., Handbuch der Paläozoologie, Borntraeger (Berlin), Bd. 2A, Lief. 5, p. A1-A68.

LeMaitre, Dorothée

(7) 1935, *Études paléontologiques sur le Lias du Maroc, Description des spongiomorphides et des algues:* Service des Mines [Maroc], Notes et Mém., no. 34, 58 p., 12 pl.

Steinmann, G.

(8) 1878, *Über fossile Hydrozoan aus den Familie der Corynidae:* Palaeontographica, Bd. 25, p. 101-124, pl. 12-14.

SOURCES OF ILLUSTRATIONS

Explanation of the use of index numbers for citing sources of illustrations is given in the Editorial Preface.

(8) **Carter, H. J.**
(10) **Chapman, Frederick**
(11) **Decker, C. E.**
(12) **Frech, F.**
(13) **Hill, Dorothy, & Wells, J. W.**
(14) **Steinmann, G.**

MILLEPORINA AND STYLASTERINA

By HILBRAND BOSCHMA[1]

CONTENTS

INTRODUCTION

The Milleporina and Stylasterina, which now are recognized as groups of hydrozoans that secrete a calcareous skeleton, were classified among scleractinian Anthozoa before the structure of their soft parts became known. The affinity of *Millepora* with hydroids was first demonstrated by AGASSIZ (1) in 1858 and similarly that of the stylasterines by MOSELEY (30, 31) about 20 years later. For a long time the 2 groups were considered to be so closely allied as to warrant classifying them together in an order called Hydrocorallina but HICKSON (18, 19) defined the Milleporina (1901) and Stylasterina (1906) as independent orders distinct from other hydrozoans. The structure of the soft parts, however, indicates that each may find a place among the hydroids. Thus, BROCH's (8) classification of the Hydroida assigns the groups with a calcareous skeleton to the suborder Athecatae, arranging the Milleporidae into the section Capitata containing forms with short capitate tentacles that bear a terminal knob, and the Stylasteridae into the section Filifera containing forms with elongate tapering filiform tentacles. Both divisions include numerous families that lack calcareous hard parts. Recent tendency is to recognize the Hydroida, Milleporina, and Stylasterina as separate orders of hydrozoan polyps (23).

The Milleporina differ from the Stylasterina not only in shape of the tentacles but also in the manner of their sexual reproduction, for milleporines produce free-swimming medusae, whereas sexual individuals of the stylasterines are reduced to gonophores.

According to current definition, the Milleporina incorporate 2 family assemblages that probably are not at all closely allied; these are the Milleporidae and Axoporidae. In

[1] The author is much indebted to RAYMOND C. MOORE for editorial assistance in preparing this chapter.

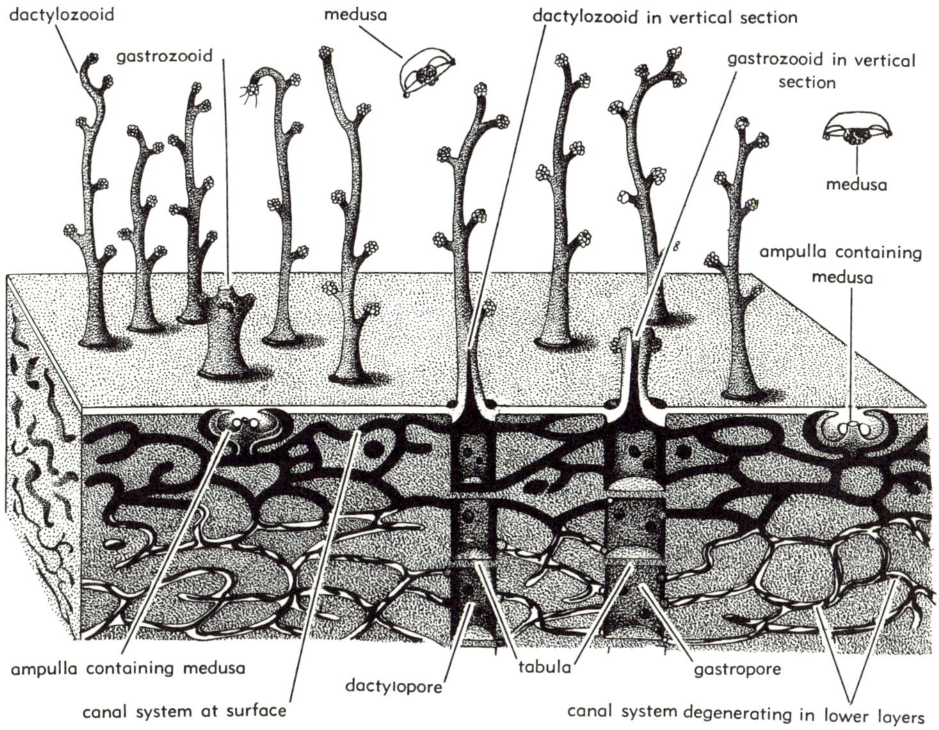

FIG. 75. Diagrammatic sketch of part of a *Millepora* colony showing soft parts and structure of hard parts, enlarged (41 mod.).

1939, KÜHN (26) included 4 genera (*Millepora, Milleaster, Axopora, Diamantopora*) in the Milleporidae, no other family being then recognized. The dimorphic polyps of *Millepora* are enclosed in tubular cavities of the colonial skeleton, which are differentiated as gastropores and dactylopores, both lacking axial structures. *Milleaster* seems not to be significantly different. The genus *Axopora* is characterized by the occurrence of gastropores that bear an axial column (gastrostyle) and by absence of dactylopores. The presence of gastrostyles is a feature suggesting the Stylasterina, but genera of this order have both gastropores and dactylopores. The structure of *Diamantopora* and *Axoporella* resembles that of *Axopora*. These genera comprise a group that probably deserves the rank of a separate order, distinct from both the Milleporina and Stylasterina, but for the present they are classed as a family (Axoporidae) of the Milleporina (5, 6).

MILLEPORINA

Order MILLEPORINA Hickson, 1901

Gastrozooids short, cylindrical, with 4 short tentacles; dactylozooids elongate, with 5 to 7 tentacles; tentacles capitate; sexual generation of free-swimming medusae. *U. Cret.-Rec.*

MORPHOLOGICAL FEATURES

SOFT PARTS

Recent species of the genus *Millepora* are the only available forms for description of the soft parts of Milleporina. These show that the living part of colonies **(coenosarc)** is restricted to superficial layers where gas-

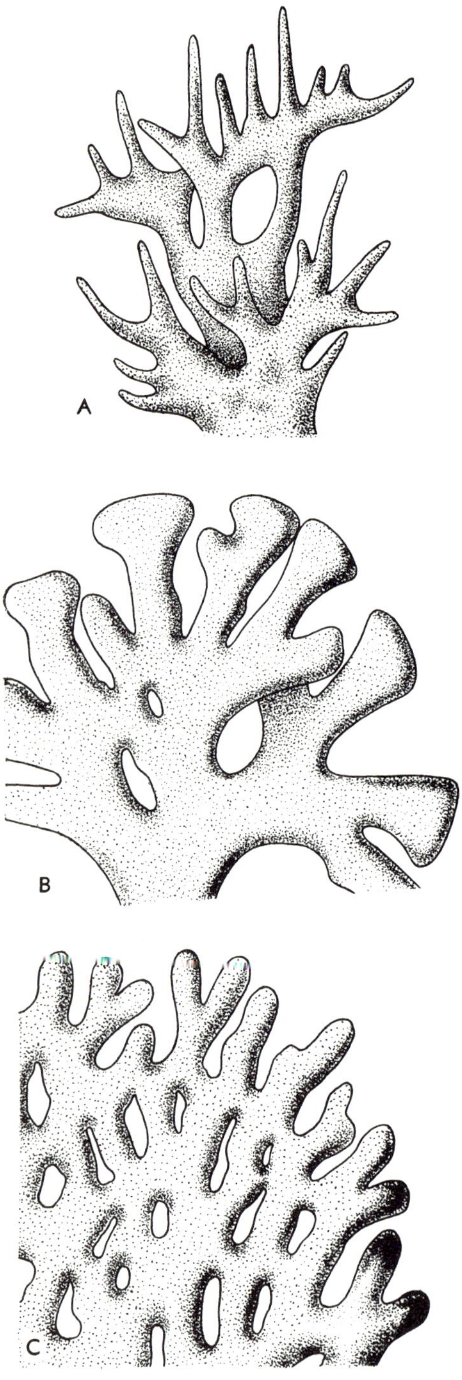

FIG. 76. Growth forms of *Millepora* illustrated by distal parts of colonies belonging to Recent species: *A, M. alcicornis* LINNÉ, W. Indies; *B, M. tenera* BOSCHMA, Molluccas; *C, M. dichotoma* FORSKÅL, Red Sea; all ×1 (35n).

tropores and dactylopores have formed their most recent tabulae (Fig. 75). In this area an intricate meshwork of minute canals occupies small pores that extend between trabeculae of the skeleton, tissue of the canal system being alive near the surface of the colony but degenerating in layers below the most recently secreted tabulae.

Two kinds of polyps occur in tubular openings perpendicularly to the surface of a colony, **gastrozooids** lodged in relatively large cavities termed gastropores, and **dactylozooids** situated in somewhat narrower tubes called dactylopores (Fig. 75 shows hardly perceptible difference in width of gastropores and dactylopores). Both types of polyps connect with the **canal system** at their bases. The gastrozooids are short cylindrical polyps provided with 4 capitate tentacles (round knob at tip) that surround a short conical hypostome with central circular mouth. The dactylozooids are arranged in circular rows with 5 to 7 individuals around a gastrozooid, forming what is termed a **cyclosystem.** Where the polyps are unusually crowded, arrangement in cyclosystems becomes obscured and only the presence of numerous small pores (dactylopores) associated with fewer large pores (gastropores) can be discerned. The dactylozooids are long slender mouthless polyps, which bear 5 to 7 short capitate tentacles. The sexual individuals occur in the upper part of special pores **(ampullae)** consisting of rounded cavities in which sexual generation produces small medusae that eventually become free-swimming. These are hydromedusae of reduced structure, for they lack velum, tentacles, and radial canals. The sexual cells develop on the manubrium, which is a tubular projection of the mouth.

HARD PARTS

Colonies of *Millepora* commonly build large variously shaped calcareous structures, which are upstanding where growth conditions are favorable but laterally extended incrustations where the environment is unfavorable. Some species secrete broad solid plates, whereas others produce branching forms with pointed or rounded tips (Figs. 76*A,C*) or laterally compressed broad tips (Fig. 76*B*), all branching mostly in a single plane to form fan-shaped colonies.

The surface of the calcareous skeleton shows numerous small subcircular or irregularly polygonal openings of **gastropores** surrounded by the slightly smaller apertures of **dactylopores** of similar shape (Fig. 77,*1b*). The outline of these pores locally is definitely stellate in colonies of various species (Fig. 77,*1a*). Gastropores and dactylopores may be arranged in cyclosystems, as described previously, or such a pattern may not be evident. The **coenosteum** between the pores appears very finely porous because of the very minute cavities between trabeculae composing the skeleton, as well shown in sections (Fig. 77,*1c*).

The rounded cavities occupied by ampullae are rare in *Millepora,* although they are known in many different species (2, 3). Generally these openings are greater in width than height and are covered by part of the skeleton that rises slightly above surrounding areas of the surface. The reticulate mass of trabeculae overlying an ampulla has a characteristic shape in each species. Colonies of *Millepora* are dioecious, for the relatively small male ampullae and slightly larger female ampullae are confined to different colonies.

The genus *Milleaster* incrusts gastropod shells or grows in subramose colonies (Fig. 77,*2*). Relatively large subrounded pores (gastropores) confluent with the surface are surrounded by more numerous smaller stellate orifices (dactylopores) located at the summit of tubercle-like elevations and by many minute rounded openings (cavities between trabeculae) that lack raised margins. In at least one species (*M. interglacialis* RICHARDS), however, the stellate pores are not elevated.

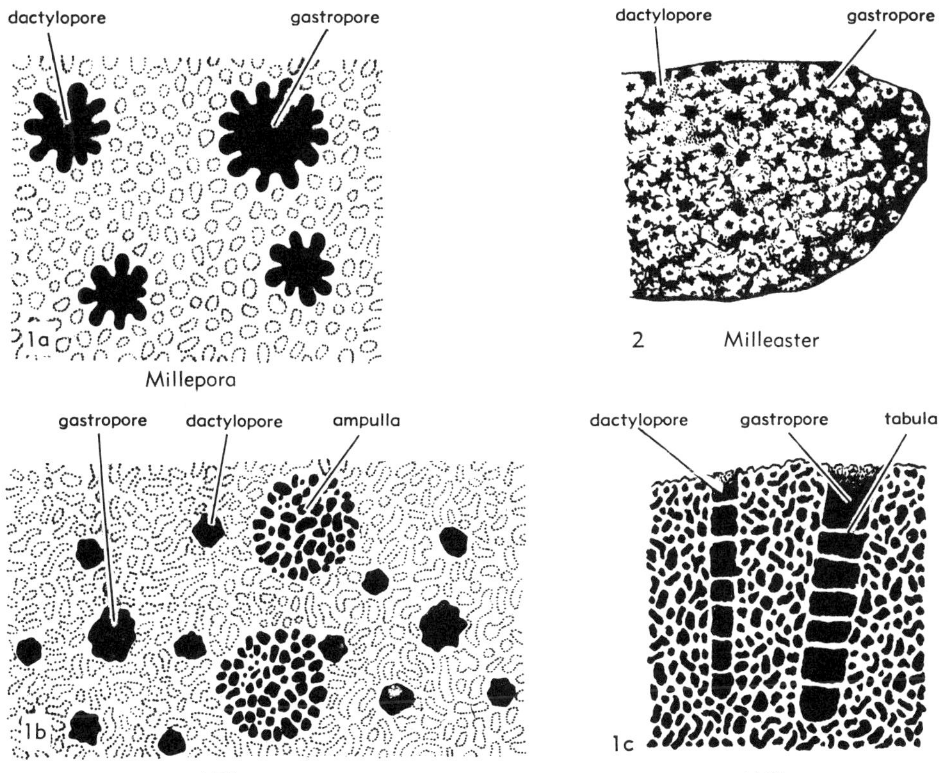

FIG. 77. Milleporidae. *1a, M. complanata,* Rec., ×36. *1b,c, M. tenera,* Rec., ×20. *2, M. incrustans,* Mio., ×8.

The skeleton of *Axopora* contains only one kind of pores, seemingly gastropores, which are regularly distributed circular or somewhat polygonal openings (Fig. 78,*1b*). The central part of these pores contains a longitudinally grooved pillar **(gastrostyle),** which seems to be built by several firmly united parallel rods (Figs. 78,*1b,2*).

In *Diamantopora,* the style is described as solid, without longitudinal grooves (Fig. 78,*4*), and in *Axoporella* it bears numerous small spines, like those of the gastrostyles in Stylasterina (Fig. 78,*3*).

OCCURRENCE

Modern species of *Millepora* are found commonly on coral reefs, generally at depths not exceeding 30 m., which seems to be correlated with dependence of the colonies on symbiotic unicellular algae (zooxanthellae) that need light for their processes of assimilation.

Among many fossils assigned to *Millepora,* HICKSON (17) judges that few really belong to the genus. BOSCHMA (5) concludes that *M. tornquisti* BOSCHMA [=*M. cylindrica* TORNQUIST (*non* REUSS)], from Eocene rocks of Madagascar, can be recognized almost surely as a species of *Millepora,* and the reported occurrence (34) of *M. alcicornis* LINNÉ in Pleistocene deposits of the Canal Zone can be accepted. Possibly the form described as *M. parva* NIELSEN, from the Danian at Faxe, Denmark, also belongs to the genus, but this is not certain.

SYSTEMATIC DESCRIPTIONS

Family MILLEPORIDAE Fleming, 1828

[*nom. correct.* AGASSIZ, 1858 (*pro* Milleporadae FLEMING, 1828)]

Colonial skeleton with gastropores associated with smaller more numerous dactylopores; gastrostyles and dactylostyles lacking. *U.Cret.-Rec.*

Millepora LINNÉ, 1758 [**M. alcicornis;* SD M.EDW.-H., 1850] [=*Palmipora* BLAINV., 1830]. Colony arborescent, composed of upstanding plates, or incrusting. Gastropores and dactylopores circular to irregularly polygonal at orifices, exceptionally stellate; walls not extended above surface; 5 to 7 dactylopores commonly arranged around each gastropore forming indistinct cyclosystems. Ampullae in the coenosteum forming little or no bulge at the surface, covered by somewhat loose-textured plates. *U.Cret.(Dan.)-Rec.*——FIG. 77, *1b,c. M. tenera* BOSCHMA, Rec.; *1b,* surface of specimen from C.Pac.(Yap) showing 2 cyclosystems and 2 ampullae, ×20; *1c,* long. sec. of specimen from E.Indies, ×20 (35*).——FIG. 77,*1a. M. complanata* LAMARCK, Rec., W.Indies; surface showing stellate pores, ×36 (35*).

Milleaster ULR., 1904 [**M. incrustans*]. Colony incrusting, some with a few short branches. Gastropores with irregularly circular apertures, walls not raised above surface; dactylopores with star-shaped orifices on summits of papillae; cyclosystems not evident. Ampullae unknown. *Mio.-Pleisto.*——FIG. 77,*2.* **M. incrustans,* Mio., Md.; surface, ×8 (35*).

Family AXOPORIDAE Boschma, 1951

Gastropores with distinct gastrostyle, dactylopores lacking. *Eoc.-Oligo.*

Axopora M.EDW.-H., 1850 [**Geodia pyriformis* MICH., 1847]. Colony incrusting, forming lumpy protuberances or arborescent. Gastropores with circular opening; gastrostyles longitudinally grooved. Ampullae unknown. *Eoc.-Oligo.*——FIG. 78,*1a,b. A. solanderi* (DEFR.), Eoc., Fr.; *1a,* colony, ×1; *1b,* surface, ×20 (*1a,* 38; *1b,* 35*).——FIG. 78,*2. A. michelini* DUNCAN, Oligo., Eng.; long. sec. of gastropore, enlarged (37).

Diamantopora WEISSERMEL, 1913 [**D. lotzi*]. Like *Axopora* but gastropores with irregularly polygonal apertures and gastrostyles without longitudinal grooves. *Eoc.*——FIG. 78,*4.* **D. lotzi,* Eoc., S.Afr.; transv. sec. showing gastropores and canals of coenosteum (black), and gastrostyles and trabeculae (white), ×20 (35*).

Axoporella BOSCHMA, 1954 [**A. kolosváryi*]. Like *Axopora* but gastrostyles bearing many small spines. *Eoc.*——FIG. 78,*3.* **A. kolosváryi,* Eoc., Hung.; gastrostyle fragments, ×100 (35*).

GENERA ERRONEOUSLY ASSIGNED TO MILLEPORINA

Arachnopora WOODS, 1878.
Millestroma GREGORY, 1898.
Milleporidium STEINM., 1903.
Promillepora DEHORNE, 1920.
Thaumatostroma WEISSERMEL, 1926.

STYLASTERINA

Order STYLASTERINA Hickson & England, 1905

Gastrozooids short cylindrical, with filiform tentacles or lacking tentacles; dactylozooids small finger-shaped, without tentacles; sexual generation by gonophores that remain attached to colony. *U.Cret.-Rec.*

MORPHOLOGICAL FEATURES

COLONIAL FORM

The shape of stylasterine colonies exhibits great differences. Some species are incrusting, others subramose with short lumpy branches, and still others have an arborescent form with very many slender branches. Some colonies are bushy, with branches extending in all directions, and many are fan-shaped, with branches disposed chiefly or wholly in a single plane. In these last-mentioned flabelliform colonies, slight differences in the nature of the opposite nearly flat surfaces serve to distinguish one as anterior (generally somewhat concave) and the other as posterior (mostly gently convex). In some genera cyclosystems are confined to the anterior surface of the colony and in others they tend to be more numerous on this surface. The ampullae occur in greater number on the posterior surface and in some species they are restricted to this surface. An irregularly bushy shape characterizes some colonies in which a polychaete worm occupies a tube formed by thin layers of the stylasterine skeleton; nearly all known specimens of *Conopora* display these features, but a few colonies that lack an associated worm have a fairly regular flabelliform shape.

SOFT PARTS

The coenosarc generally extends more deeply below the surface of stylasterine genera than in *Millepora*. It consists chiefly of a complicated network of canals that connect with the gastrozooids, dactylozooids, and ampullae. The canals have a membranous wall consisting of an outer (ectoderm) layer

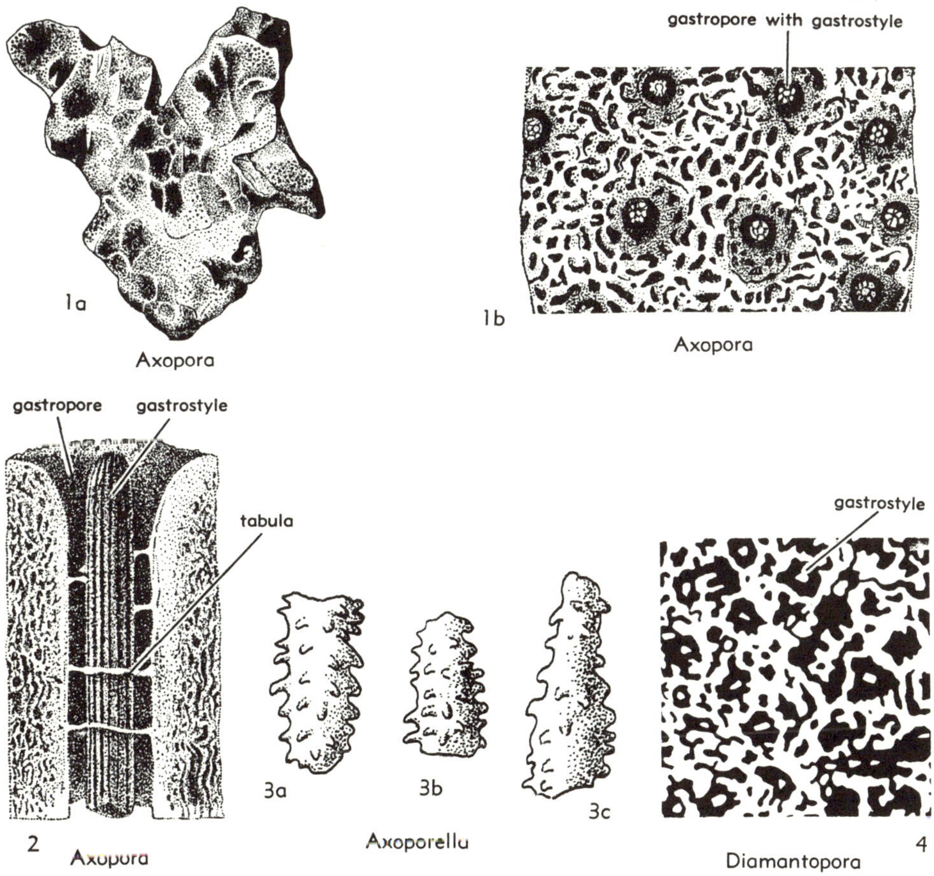

FIG. 78. Axoporidae. *1a,b, A. solanderi,* Eoc., ×1, ×20. *2, A. michelini,* Oligo., enlarged. *3a-c, A. kolosvaryi,* Eoc., gastrostyle fragments, ×100. *4, D. lotzi,* Eoc., ×20.

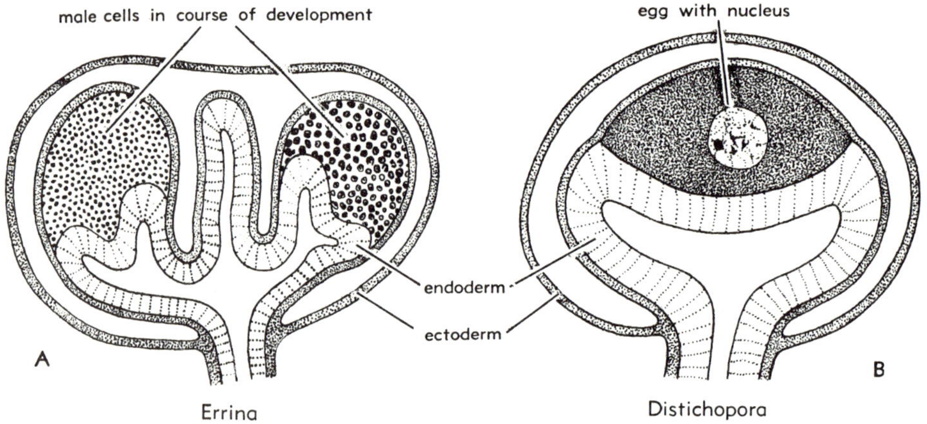

FIG. 79. Longitudinal sections of gonophores showing structure. *A*, Male gonophore of *Errina*, ×160. *B*, Female gonophore of *Distichopora*, ×160.

and an inner (endoderm) layer, which are joined to corresponding parts of the polyps and ampullae.

The gastrozooids are hydroid polyps of simple structure having a short cylindrical shape and a conical domelike hypostome that bears a round or four-cornered mouth at the top. A single whorl of filiform tentacles occurs at the base of the hypostome. The tentacles lack globular extremities differentiated from the stalklike lower part and thus are not capitate. Some stylasterine species possess long tapering tentacles (Fig. 84,*1b*) and others have short stubby ones (Fig. 81,*1b*), among which a few show slightly swollen extremities, although none are really capitate; in the gastrozooids of some genera tentacles are lacking.

The dactylozooids in all Stylasterina consist of polyps that have undergone strong reduction. They lack tentacles and indeed are themselves like simple tentacles; evidently they perform a tentacular function. The dactylozooids of *Pliobothrus* are simple elongate-conical mouthless bodies that are attached basally to the bottom of dactylopores. Those of *Sporadopora* differ only in being joined sideways to the lower part of dactylopores. In *Errina* and *Errinopsis* one side of the dactylozooids is attached to the median part of a grooved spine, leaving a free finger-shaped part that projects outward from the grooved spine (Fig. 84,*1b*). The dactylozooids of *Paraërrina* resemble those of *Errina* in shape, adhering to one side of the dactylopore and attached also to the dactylostyle. The free extremity of dactylozooids, as well shown in genera of the Stylasterinae (Fig. 81,*1b*), projects inward toward the center of the cyclosystem.

The **gonophores** of the Stylasterina are sexual individuals that remain attached to the colonies. Most species are dioecious, some colonies bearing only male gonophores and others having only female ones, but both types have been observed in a single colony of *Distichopora* and *Stylaster*. The gonophores develop in separate cavities (ampullae) and their highly simplified structure suggests that they belong to the styloid type of hydrozoan polyps with the endoderm enlarged into a spadix. The spadix may bifurcate or be divided into several parts, the top portion bearing the genital cells that in the male develop into sperm cells (Fig. 79*A*) and in the female into egg cells (Fig. 79*B*). The spadix and genital cells are covered by a layer of ectoderm, around which is another ectoderm layer that envelops the whole gonophore and covers the ampulla in a position next beneath the calcareous wall of the ampulla. Eggs are fertilized within the ampullae, developing there into planulae, which are larvae consisting of a solid inner mass of endoderm covered by a layer of ectoderm. After an ampulla has fulfilled its function, the cavity becomes filled by coenosteum traversed by the canal system as in adjacent parts of the skeleton.

HARD PARTS

The skeleton secreted by stylasterine colonies is distinctly harder than that of members of the Milleporina and it has narrower pores between the calcareous elements. The somewhat brittle coenosteum of the milleporines is composed of trabeculae that are not closely joined, and the surface of colonies appears opaque. The surface of most colonies of the Stylasterina is more or less glossy.

The gastropores normally are tubes of cylindrical shape, which (as in *Stylaster*) may widen in funnel-shaped manner near the orifice. Fairly commonly they are divided by a girdle of inwardly projecting spines or a calcareous ridge into a proximal "ventral chamber" (or **gastrostyle chamber**) and a distal more or less flaring open chamber (Fig. 80,*2*) with margins not pronouncedly elevated above the general surface of the colonial skeleton or, less commonly, with a raised lip on one side directed

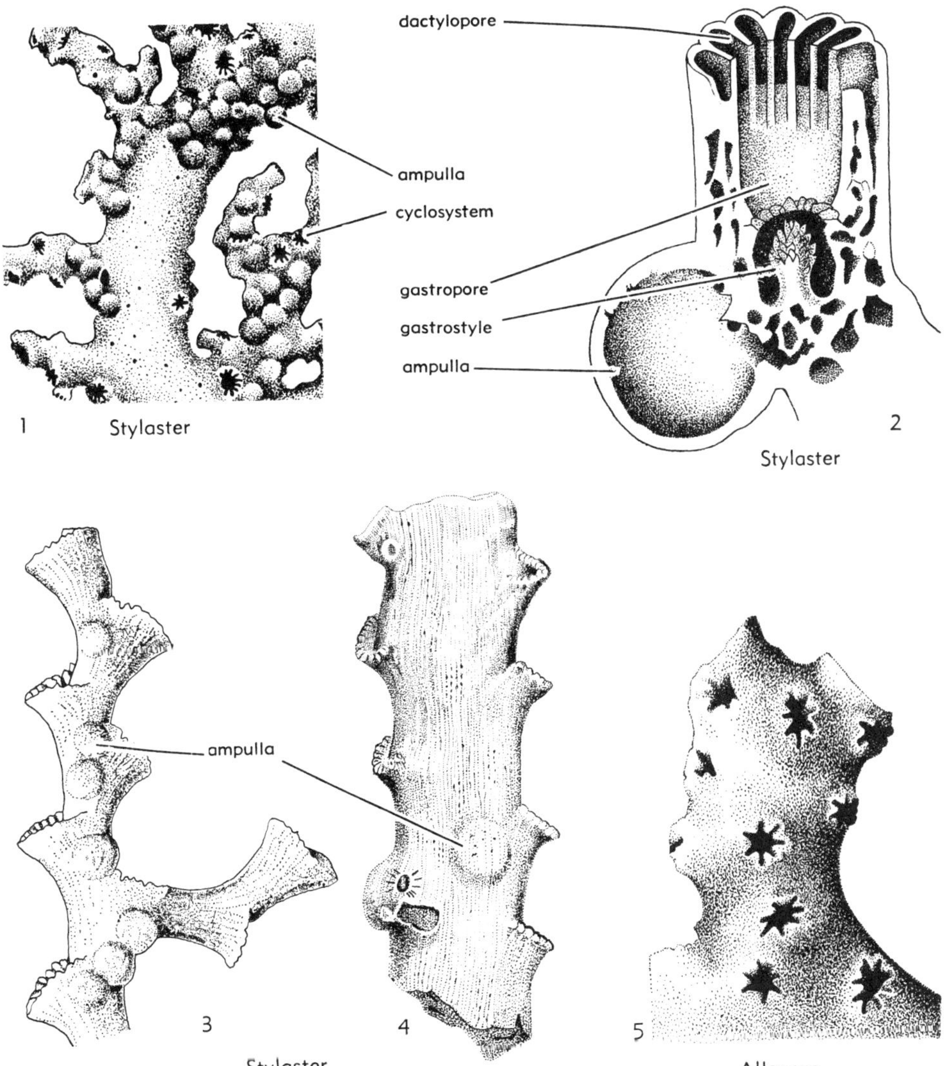

FIG. 80. Stylasteridae. *1, S. rossus,* Rec., ×5. *2, S. densicaulis,* Rec., ×9. *3, S. microstriatus,* Rec., ×10. *4, S. mooraboolensis,* Mio., ×8. *5, A. norvegica,* Rec., ×6 (p. F99-F100).

toward the apex of the colony (Figs. 84,*1a;* 85,*2a,b*). Some gastropores (as in *Pliobothrus* and *Sporadopora*) show the presence of **tabulae.**

The dactylopores are tubular openings in the skeleton narrower than the gastropores, and like the gastropores, they tend to widen toward the orifice. In some genera (as *Astya*), the dactylopores form a very regular circular row of openings around a gastropore, but in others arrangement is less definite. The margins of dactylopores may be even with the surface of the colony or they may be raised as small papillae (Fig. 85,*1b*), and some are characterized by projection of a grooved spine that represents upward continuation of the dactylopore (Fig. 83,*1b;* 84,*1c*).

Gastrostyles, which are axial structures projecting upward from the floor of gastropore tubes, occur in many genera of the Stylasterina but are lacking in others. On their entire surface they bear small spines.

An axial structure in dactylopores of many genera, consisting of a longitudinal row of spicules, is designated as a dactylostyle.

Arrangement of gastropores and dactylopores in cyclosystems, with a varying number of dactylopores grouped around a gastropore, is an important structural character of stylasterine colonies. Radially arranged ridges between the dactylopores (pseudosepta) give the cyclosystems of some Stylasterina the appearance of the calice of a scleractinian coral. Cyclosystems may be scattered indiscriminately over the whole surface of the colony (as in *Allopora, Stylantheca, Conopora*) or may tend to be more abundant on the anterior surface than the posterior surface. In *Stylaster,* cyclosystems are chiefly or entirely restricted to lateral surfaces of the branches; in *Congregopora* they tend to show a linear arrangement on the branches; in *Stenohelia, Crypthelia* (except *C. trophostega* FISHER), and *Astya,* they are almost confined to the anterior surface of the colony. Although the cyclosystems generally are circular in outline, they may be distinctly oval with longer axis disposed longitudinally (as in *Stylaster bellus*) or transversely *(S. gemmascens)* with respect to the branches or horseshoe-shaped (Figs. 80,*3,4*). The pores of *Distichopora* are disposed in lines, rather than cyclosystems, a median row of gastropores being flanked on each side by a row of dactylopores (Fig. 83,*2b*).

The characters of ampullae are various. In *Stylaster* they occur normally on the posterior surface as well as on the anterior surface of the colony, although in some species they are concentrated on a single surface. The ampullae are more or less spherical cavities in the skeleton, generally with a roof that forms a well-marked bulge on the surface of the colony (Figs. 80,*1-4*), but expression of their occurrence in this manner may be very weak or even lacking entirely if the ampullae are buried completely in the coenosteum. The covering roof may have a smooth outer surface or it may be marked by ridges, tubercles, or spines. A small opening in the basal part or on the top can be observed in some ampullae. Those enclosing female gonophores are generally much larger than ampullae that serve for lodgment of male gonophores. Each sort occurs singly or in clusters and ampullae may be closely packed in the coenosteum between cyclosystems (as in *Stylantheca*).

OCCURRENCE

Recent species of Stylasterina are known to live in shallow waters, especially in the tropics, but they occur also in deeper seas (for example, *Crypthelia* to a depth of 2,745 m.). They are known from nearly all parts of the world, including the northern Atlantic and Pacific and waters adjacent to Antarctica.

The oldest fossils that are referred definitely to the Stylasterina occur in Upper Cretaceous (Danian) rocks of northwestern Europe. These include 7 genera and subgenera *(Conopora, Astya, Errina, Inferiolabiata, Pliobothrus, Sporadopora, Congregopora),* of which all but the last-named range to Recent. Species of 3 genera make first known appearance in Tertiary strata, *Stylaster* and *Distichopora* occurring in the Eocene and *Allopora* being recorded from the Oligocene. Eight genera seem to be confined to the Recent.

SYSTEMATIC DESCRIPTIONS

Family STYLASTERIDAE Gray, 1847

Characters of the order. *U.Cret.-Rec.*

Subfamily STYLASTERINAE Gray, 1847

[*nom. transl.* STECHOW, 1921 (*ex* Stylasteridae GRAY, 1847)]

Pores arranged in distinct cyclosystems. *U.Cret.(Dan.)-Rec.*

Stylaster GRAY, 1831 [**Madrepora rosea* PALLAS, 1866; SD M.EDW.-H., 1850] [=*Cyclopora* VERRILL, 1866 (*non* RAF., 1815); *Deontopora* HALL, 1893; *Eustylaster* BROCH, 1914]. Colony arborescent, generally flabelliform, branches slender, rarely coalescent. Cyclosystems on young branches alternating on left and right sides but a few on flat surfaces in older parts of colonies. Gastropores more or less cylindrical, commonly with gastrostyle chamber and bearing a gastrostyle; each surrounded by 6 to 18 dactylopores, which bear dactylostyles. Ampullae scattered singly or grouped in clusters on the flat surfaces, where they form subhemispherical bulges or rarely are not elevated above the surface. *Eoc.-Rec.*——FIG. 80,*1*. **S. roseus* (PALLAS), Rec., W.Indies; colony with female ampullae, ×5 (35*).——FIG. 80,*3*. *S. microstriatus* BROCH, Rec., Japan; part of branch, ×10 (35*).——FIG. 80,*4*. *S. mooraboolensis* (HALL), Mio., Vict., Austral.; part of branch, ×8 (35*).

Allopora EHR., 1834 [**A. oculina* (=*Millepora norvegica* GUNNERUS, 1768)] [=*Cryptaxis* REUSS, 1865 (*non* LOWE, 1855); *Cryptaxiella* KÜHN,

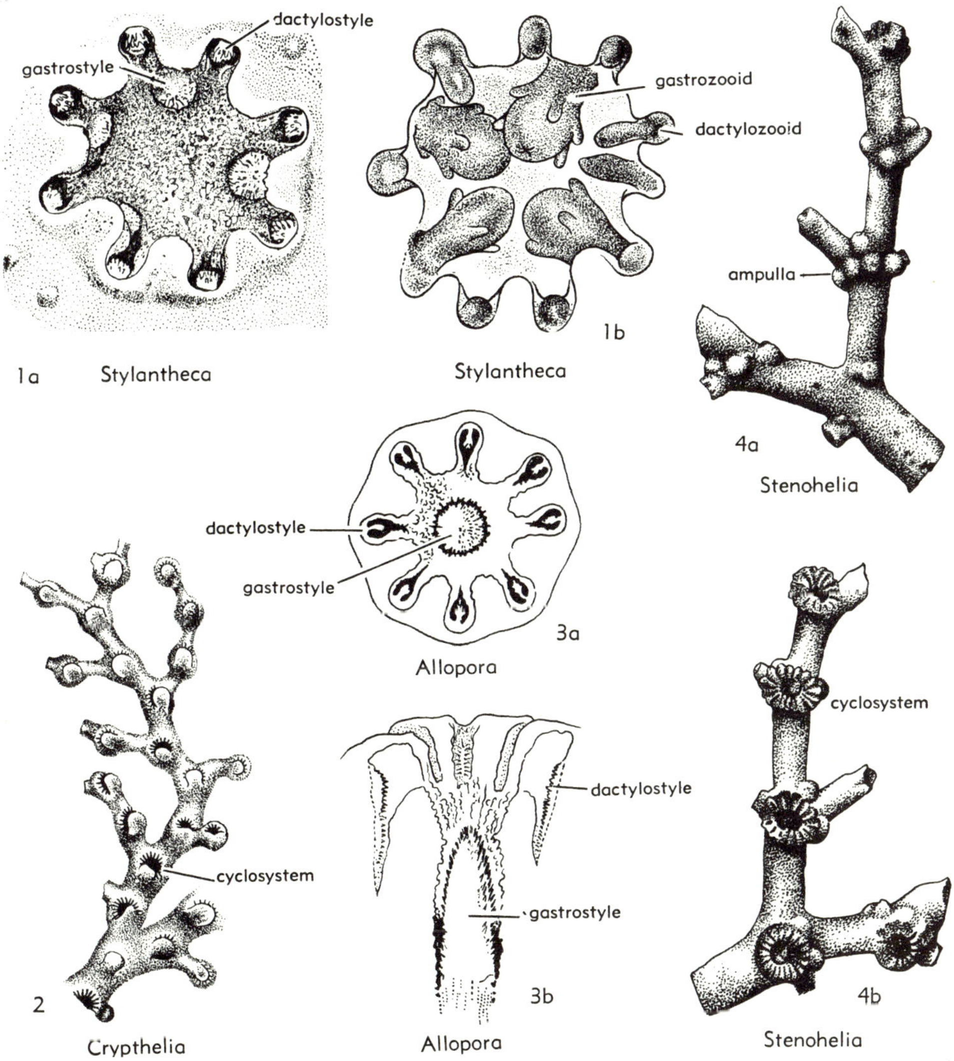

FIG. 81. Stylasteridae. *1a,b, S. porphyra,* Rec., ×30. *2, C. affinis,* Rec., ×2. *3a,b, A. verrilli,* Rec., ×30. *4a,b, S. profunda,* Rec., ×6 (p. F100).

1939]. Colony with rather thick branches, arborescent, flabelliform, lobed, or forming lumpy incrustations. Cyclosystems scattered on all surfaces. Structure as in *Stylaster*. Ampullae buried in coenosteum, with little or no swelling at surface, rarely hemispherical. *Oligo.-Rec.*——FIG. 80,*5*. **A. norvegica* (GUNNERUS), Rec., Norway; part of branch, ×6 (35*).——FIG. 81,*3a,b*. *A. verrilli* DALL, Rec., N.Pac.; *3a,* surface view of cyclosystem, ×30 (40); *3b,* long. sec., ×30 (40).

Stenohelia KENT, 1870 [**Allopora maderensis* JOHNSON, 1862]. Colony arborescent, flabelliform, branches slender, not coalescent. Cyclosystems as in *Stylaster* but all directed toward anterior surface. Gastropores surrounded by 8 to 12 dactylopores. Ampullae hemispherical, prominent, on posterior surface, commonly in groups behind cyclosystems. *Rec.*——FIG. 81,*4a,b*. *S. profunda* MOSELEY, Rec., Kermadec I.; *4a,b,* posterior and anterior surfaces, ×6 (35*).

Stylantheca FISHER, 1931 [**S. porphyra*]. Colony encrusting. Like *Allopora* in structure but cyclosystems generally with 3 to 7 gastrostyles; 8 to 12 dactylopores around gastropores. Ampullae in coenosteum, not bulging surface. *Rec.*——FIG. 81, *1a,b*. **S. porphyra,* Rec., N.Pac.(Calif.); *1a,b,* cyclosystems with 8 dactylostyles around 4 gastrostyles and 10 dactylozooids around 4 gastrozooids, ×30 (40).

Crypthelia M.EDW.-H., 1849 [**C. pudica*] [=*Endhelia* M.EDW.-H., 1849; *Endohelia (nom. van.), Cryptohelia* M.EDW.-H., 1857]. Colony flabelliform, branches slender, rarely coalescent. Cyclosystems only on anterior surface (except in *C. trophostega* FISHER) partly covered by marginally adherent lid; gastropores without gastrostyles, surrounded by 10 to 25 dactylopores without dactylostyles; ventral chamber of gastropore separated by diaphragm with circular opening. Ampullae in coenosteum near cyclosystems or in lid, commonly marked by distinct swellings. *Rec.*——FIG. 81,*2*. *C. affinis* MOSELEY, Rec., W.Indies; part of branch, ×2 (31).——FIG. 82,*1a*. *C. stenopoma* HICKSON & ENGLAND, Rec., E. Indies; upper view of cyclosystem, ×21 (42).——FIG. 82,*1b*. *C. pachypoma* HICKSON & ENGLAND, Rec., E.Indies; oblique view of cyclosystem, ×21 (42).

Conopora MOSELEY, 1879 [**C. tenuis*]. Colony branching irregularly, rarely flabelliform, not coalescent. Cyclosystems generally on all sides but in flabelliform types on anterior surface; 3 to 16 dactylopores around gastropore, interconnecting with lower part of gastropore; no gastrostyles or dactylostyles. Ampullae subhemispherical swellings, mostly on surface turned away from cyclosystems. *U.Cret.(Dan.)-Rec.*——FIG. 82,*2c*. *C. major* HICKSON & ENGLAND, Rec., Mauritius; part of branch with cyclosystems and ampullae, ×9 (36).——FIG. 82,*2a*. **C. tenuis,* Rec., Mauritius; long. sec. of cyclosystem showing openings to dactylopore in lower part of gastropore, ×19 (36).——FIG. 82,*2b*. *C. pauciseptata* BROCH, Rec., Antarct.; 3 cyclosystems on anterior surface of stem, ×15 (35*).

Astya STECHOW, 1921 [*pro Astylus* MOSELEY, 1879 (*non* LAPORTE, 1836)] [**Astylus subviridis* MOSELEY, 1879]. Colony flabelliform, branches slender, not coalescent. Cyclosystems all directed toward anterior surface; about 20 dactylopores around gastropore, which has ventral chamber separated by diaphragm with crescentic opening. No gastrostyles or dactylostyles. Ampullae in circular rows around cyclosystems on anterior surface, surface slightly bulged. *U.Cret.(Dan.)-Rec.*——FIG. 82,*3*. **A. subviridis* (MOSELEY), Rec., SW.Pac.; part of branch, ×3 (44).

Congregopora NIELSEN, 1919 [**C. nasiformis*]. Colony arborescent to flabelliform. Cyclosystems on all sides of branches, tending to linear arrangement; 2 or 3 dactylopores united with a gastropore, forming 3-cornered opening surrounded by slightly raised wall, other isolated dactylopores scattered over surface. No gastrostyles or dactylostyles. Ampullae seemingly hemispherical. *U.Cret. (Dan.)*.——FIG. 82,*4a-c*. **C. nasiformis,* Dan., Denm.; *4a,* colony, ×1 (45); *4b,c,* parts of branches, ×6, ×8 (35*).

Subfamily DISTICHOPORINAE Stechow, 1921

Pores in linear rows, with median row of gastropores flanked on each side by a row of dactylopores. *Eoc.-Rec.*

Distichopora LAMARCK, 1816 [**Millepora violacea* PALLAS, 1766]. Colony flabelliform, with branches generally flattened antero-posteriorly. No cyclosystems, pores being arranged in rows at lateral margins of branches or rarely (*D. irregularis* MOSELEY) on flat surface of branches also; pores deep; gastropores with long gastrostyle, dactylostyles absent. Ampullae forming slight to hemispherical bulges on anterior and posterior surfaces, generally in clusters or united in solid mass. *Eoc.-Rec.*——FIG. 83,*2a*. **D. violacea* (PALLAS), Rec., "South Sea"; colony with many ampullae, ×1 (39).——FIG. 83,*2b*. *D. serpens* BROCH, Rec., SW. Pac.; side view of branch, ×20 (36).

Subfamily ERRININAE Hickson, 1912

[*nom. correct.* STECHOW, 1921 (*ex* Errinina HICKSON, 1912)] [=Sporadoporidae GREGORY, 1930]

Pores not in distinct cyclosystems; if in linear rows, gastropores are somewhat widely separated from dactylopores. *U.Cret. (Dan.)-Rec.*

Errina GRAY, 1835 [**Millepora aspera* LINNÉ, 1767] [=*Labiopora* MOSELEY, 1879]. Colony arborescent, flabelliform to irregular bushy, branches generally not coalescent. Pores not arranged in distinct cyclo-

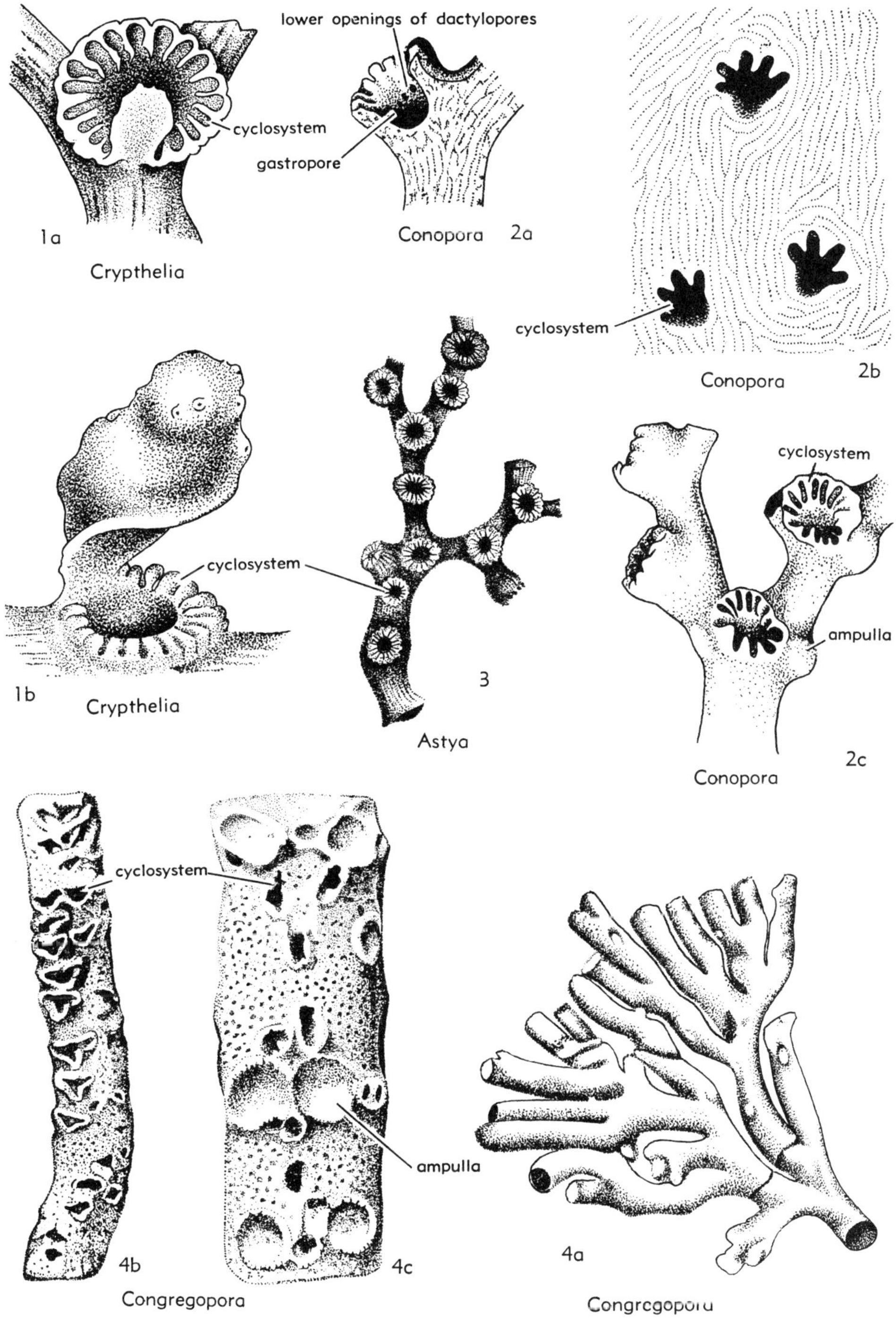

Fig. 82. Stylasteridae. *1a, C. stenopoma,* Rec., ×21. *1b, C. pachypoma,* Rec., ×21. *2a, C. tenuis,* Rec., ×19. *2b, C. pauciseptata,* Rec., ×15. *2c, C. major,* Rec., ×9. *3, A. subviridis,* Rec., ×3. *4a-c, C. nasiformis,* Dan., ×1, ×6, ×8 (p. F100).

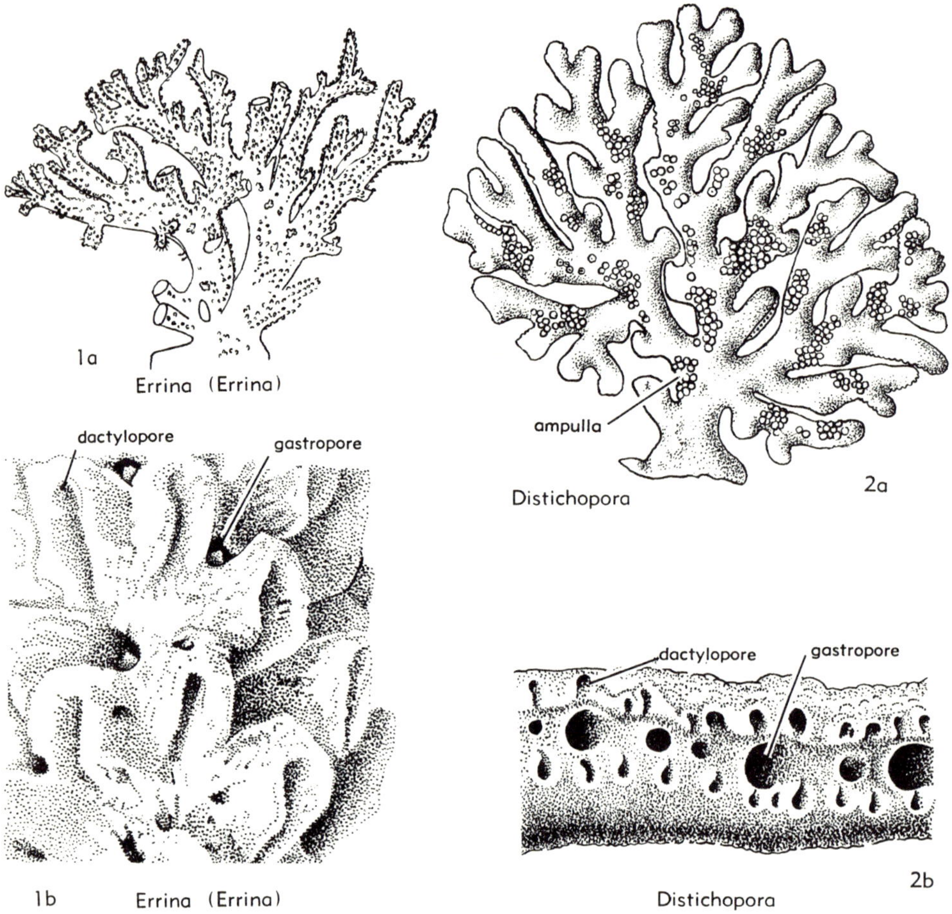

FIG. 83. Stylasteridae. *1a, E. (E.) novaezelandiae,* Rec., *ca.* ×1.2. *1b, E. (E.) antarctica,* Rec., ×16. *2a, D. violacea,* Rec., ×1. *2b, D. serpens,* Rec., ×20 (p. F100-F102).

systems; gastropores with well-developed gastrostyle, not surrounded by a wall. Dactylopores without dactylostyles. Ampullae embedded in coenosteum or making slight to hemispherical bulge at surface. *U.Cret.(Dan.)-Rec.*

E. (Errina) [=*Euerrina* BROCH, 1942]. Dactylopores at base and in groove of a grooved spine with opening directed toward base of colony, some dactylopores without spines scattered over surface. *U.Cret.(Dan.)-Rec.*——FIG. 83,*1a*. *E. (E.) novaezelandiae* HICKSON, Rec., N.Z.; colony, slightly enlarged (41).——FIG. 83,*1b*. *E. (E.) antarctica* (GRAY), Rec., Antarct.; side of part of branch, ×16 (30).

E. (Inferiolabiata) BROCH, 1951 [**Errina labiata* MOSELEY, 1879] [=?*Lepidopora* POURTALÈS, 1871; *Acanthopora* MOSELEY, 1876 (*non* ORB., 1849); *Spinipora* MOSELEY, 1879 (*non* AGASSIZ, 1846); *Labiata* BROCH, 1942 (*non* FABRICIUS, 1823)]. Grooves of grooved spines turned toward top of branches. Some species (*"Lepidopora"*) have gastropores with lip at lower margin present both on anterior and posterior surfaces of colony; dactylopores in a continuous line on lateral margins of branches and scattered among gastropores. *U.Cret.(Dan.)-Rec.*——FIG. 84,*1a*. *E. (I.) glabra* (POURTALÈS), Rec., W.Indies; anterior surface of branchlet, ×6 (46).——FIG. 84,*1b,c*. *E. (I.) echinata* MOSELEY, Rec., S.Atl.; *1b,* gastrozooid and 4 dactylozooids with connecting canals, ×15; *1c,* side of part of branch, ×7 (both 44).

Errinopora FISHER, 1931 [**Errina pourtalesii* DALL, 1884] [=*Protoërrina* BROCH, 1935]. Colony dendritic or flabelliform. Cyclosystems rarely distinct, with 5 to 7 dactylopores around a gastropore, dactylopores commonly scattered irregularly among gastropores, which are not surrounded by wall; dactylopores with grooved spines, groove oriented

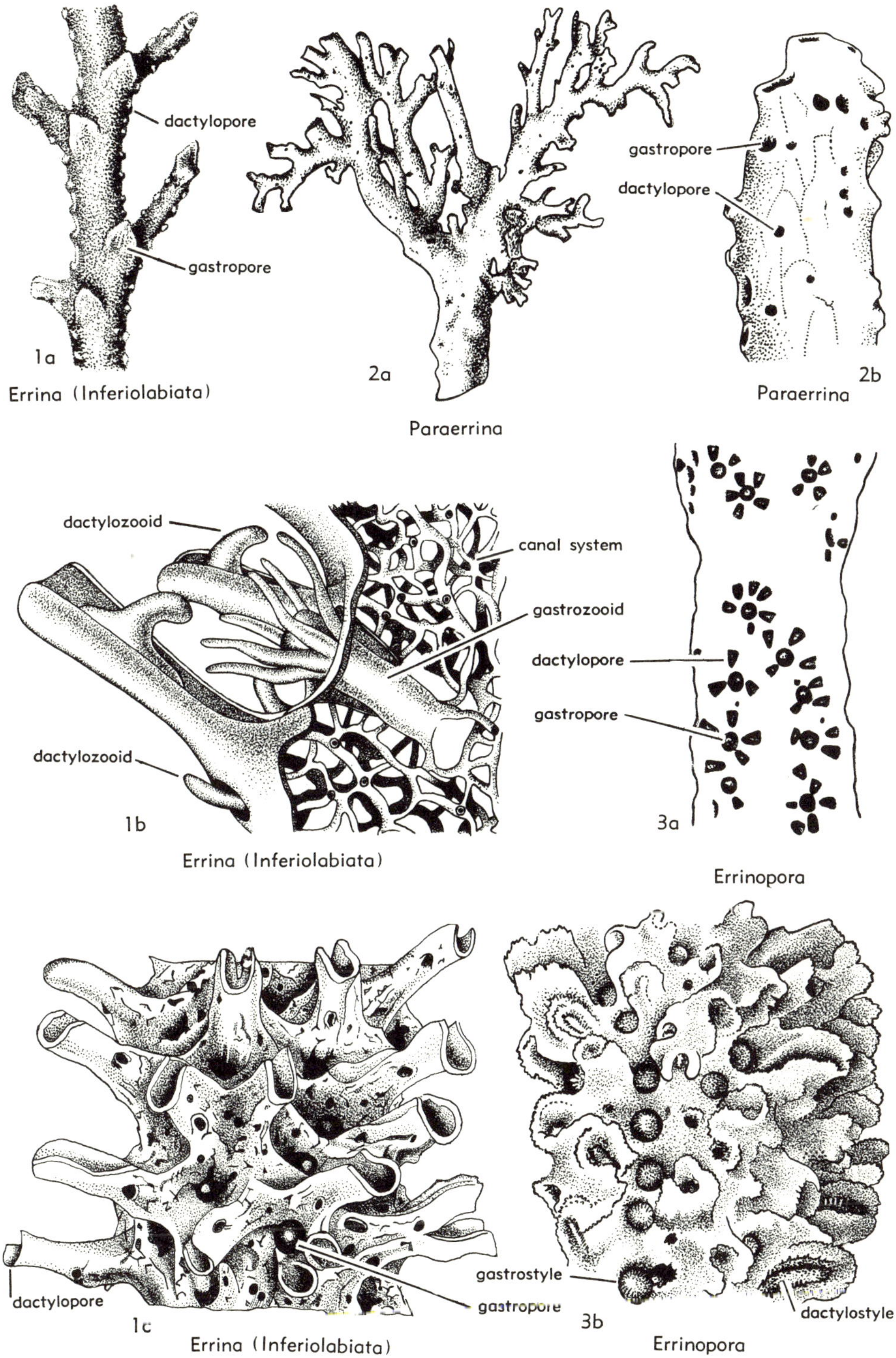

FIG. 84. Stylasteridae. *1a, E. (I.) glabra,* Rec., ×6. *1b,c, E. (I.) echinata,* Rec., ×15, ×7. *2a,b, P. decipiens,* Rec., ×1, ×20. *3a, E. stylifera,* Rec., ×5.5; *3b, E. pourtalesii,* Rec., ×15 (p. F102-F104).

variously. Gastrostyles and dactylostyles present, latter prolonged in grooves of spines. Ampullae form slight swellings or hemispherical bulges at surface. *Rec.*——FIG. 84,*3a*. *E. stylifera* (BROCH), Rec., Okhotsk Sea; side of part of branch showing cyclosystems, ×5.5 (36).——FIG. 84,*3b*. **E. pourtalesii* (DALL), Rec., NE.Pac.(Calif.); side of part of branch, ×15 (40).

Paraërrina BROCH, 1942 [**P. decipiens*]. Like *Errinopora* but gastropores and dactylopores surrounded by low wall only in young branches, opening flush with surface in older parts of colony; no grooved spines. Ampullae form slight swellings to hemispherical bulges. *Rec.*——FIG. 84,*2a,b*. **P. decipiens*, Rec., Mauritius; *2a*, colony, ×1 (36); *2b*, end of branchlet, ×20 (36).

Errinopsis BROCH, 1950 [**E. reticulum*]. Like *Errina* but flabelliform, branches strongly coalescent, forming a network. Ampullae in coenosteum, not forming bulges at surface. *Rec.*——FIG. 85,*4*. **E. reticulum*, Rec., Antarct.; part of colony, ×1 (36).

Phalangopora KIRKPATRICK, 1887 [**P. regularis*]. Colony flabelliform, branches not coalescent. Gastropores generally in single row on anterior and posterior surfaces of branches, with feebly developed lip at lower margin; dactylopores in linear row on lateral margins of branches and scattered among gastropores. Gastrostyles and dactylostyles absent. Ampullae form hemispherical bulges on flat sides of colony. *Rec.*——FIG. 85,*2a,c*. **P. regularis*, Rec., Mauritius; *2a,b*, anterior surface and lateral view of part of branch, ×12; *2c*, part of colony, ×1.5 (both 43).

Pliobothrus POURTALÈS, 1868 [**P. symmetricus*; SD KÜHN, 1928]. Colony ramose, more or less regularly flabelliform, with flattened cylindrical branches. Gastropores without wall or gastrostyle; dactylopores opening at top of round tubercles, without dactylostyles. Ampullae in coenosteum, not visible externally. *U.Cret.(Dan.)-Rec.*——FIG. 85, *1a,b*. **P. symmetricus*, Rec.,N.Atl., W.Indies; *1a*, colony, ×1 (36); *1b*, part of branch, ×6 (46).

Sporadopora MOSELEY, 1879 [*pro Polypora* MOSELEY, 1876 (*non* M'COY, 1844)] [**Polypora dichotoma* MOSELEY, 1876]. Colony consisting of stout vertical stems with some irregular branches, more or less flabelliform, flattened in anterior-posterior direction. Gastropores and dactylopores opening nearly flush with surface; gastrostyles present, dactylostyles absent. Ampullae in coenosteum, not bulging at surface. *U.Cret.(Dan.).-Rec.*——FIG. 85,*3a*. **S. dichotoma* (MOSELEY), Rec., S.Atl.; part of colony, ×0.5 (44).——FIG. 85,*3b*. *S. mortenseni* BROCH, Rec., SW.Pac.(N.Z.); part of branch, ×20 (36).

?**Steganopora** HICKSON & ENGLAND, 1905 [*non Steginopora* ORB., 1853] [**S. spinosa* (?=*Pliobothrus spinosus*, *fide* BROCH, 1942)]. Branch circular, tapering toward end. Gastropores opening flush with surface, dactylopores with orifice at top of long narrow tube; gastrostyles and dactylostyles absent. Ampullae hemispherical. *Rec.*

GENERA ERRONEOUSLY ASSIGNED TO STYLASTERINA

Aphyllacis REUSS, 1872.
Cylindropora WOODS, 1880.
Leptobothrus HALL, 1893.
Palassopora GREGORY, 1930.

REFERENCES

Publications cited in the following list are judged to be most helpful in furnishing additional information concerning the milleporines and stylasterines and in offering a guide to more extensive literature. The index numbers enclosed in parentheses in the column at left are employed in the text for identification of the publications.

Agassiz, Louis

(1) 1858, *The animals of Millepora are hydroid acalephs and not polyps:* Am. Jour. Sci. Arts., ser. 2, v. 26, p. 140-141.

Boschma, H.

(2) 1948, *The species problem in Millepora:* Zool. Verh. Mus. Leiden, no. 1, 115 p., 15 pl., 13 fig.

(3) 1950, *Further notes on the ampullae of Millepora:* Zool. Meded. Mus. Leiden, v. 31, p. 49-61, pl. 1-6, fig. 1-3.

(4) 1951, *Notes on Stylasterina (Hydrocorallia):* Proc. K. Nederl. Akad. Wetensch. (Amsterdam), ser. C, v. 54, p. 451-458, fig. 1-2.

(5) 1951a, *Notes on Hydrocorallia:* Zool. Verh. Mus. Leiden, no. 13, 49 p., 2 pl., 6 fig.

(6) 1954, *De familie Axoporidae:* Versl. K. Nederl. Akad. Wetensch. (Amsterdam), v. 63, p. 99-103, fig. 1-2.

Broch, Hj.

(7) 1914, *Stylasteridae:* Danish Ingolf Exped., v. 5, no. 5, 27 p., 5 pl., 6 fig.

(8) 1924, *Hydroida:* in KÜKENTHAL, WILLY, Handbuch der Zoologie, de Gruyter & Co. (Berlin & Leipzig), v. 1, p. 422-458, fig. 381-421.

(9) 1936, *Untersuchungen an Stylasteriden (Hydrokorallen), Teil 1:* Skr. Norske Videnskapsakad. (Oslo), Mat.-Naturv. Kl., 1936, no. 8, 103 p., 13 pl., 32 fig.

(10) 1942, *Investigations on Stylasteridae (hydrocorals):* Same, 1942, no. 3, 113 p., 6 pl., 38 fig.

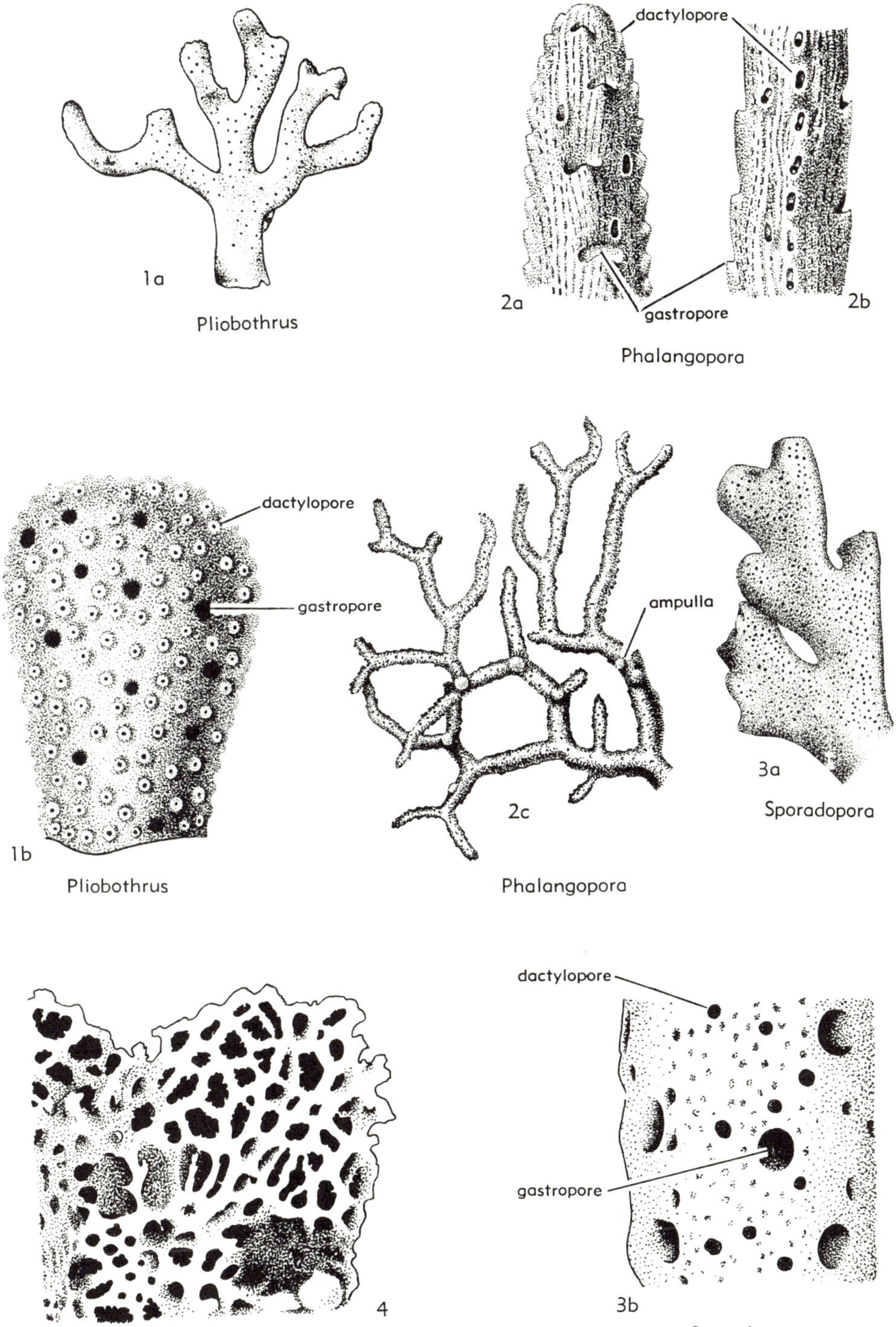

FIG. 85. Stylasteridae. *1a,b, P. symmetricus,* Rec., ×1, ×6. *2a-c, P. regularis,* Rec.; *2a,b,* ×12; *2c,* ×1.5. *3a, S. dichotoma,* Rec., ×0.5. *3b, S. mortenseni,* Rec., ×20. *4, E. reticulum,* Rec., ×1.

(11) 1947, *Stylasteridae (hydrocorals) of the John Murray Expedition to the Indian Ocean, 1933-34:* Sci. Repts., v. 8, no. 2, p. 305-316, pl. 1, fig. 1-4.

(12) 1950, *Stylasteridae (hydrocorals) from southern seas:* Discovery Repts. (London), v. 26, p. 33-46, pl. 2-6, fig. 1-12.

Ellis, J., & Solander, D.

(13) 1786, *The natural history of many curious and uncommon zoophytes:* (London), xii+206 p., 64 pl.

England, H. M.

(14) 1926, *Development of gonophores of the Stylasteridae:* Proc. Zool. Soc. London, 1926, p. 265-283, fig. 1-23.

Fisher, W. K.

(15) 1931, *Californian hydrocorals:* Annals Mag. Nat. Hist., ser. 10, v. 8, p. 391-399, pl. 15-17.

(16) 1938, *Hydrocorals of the North Pacific Ocean:* Proc. U. S. Natl. Mus., v. 84, p. 493-554, pl. 34-76.

Hickson, S. J.

(17) 1898, *Did Millepora occur in Tertiary times?:* Zool. Anz., v. 21, p. 70-71.

(18) 1901, [*Classification of Hydrozoa*]: in Delage, Y., & Hérouard, E., Traité de Zoologie Concrète (Paris), v. 2, pt. 2 (Les Coelentérés), p. 153.

(19) 1906, *Coelenterata and Ctenophora:* Cambridge Nat. Hist. (London), v. 1, p. 243-424, fig. 124-184.

(20) 1912, *On the hydrocoralline genus Errina:* Proc. Zool. Soc. London, 1912, p. 876-896, pl. 94-96.

——— & England, H. M.

(21) 1905, *The Stylasterina of the Siboga Expedition:* Siboga Exped. Mon. 8, 26 p., 3 pl.

(22) 1909, *The Stylasterina of the Indian Ocean:* Trans. Linn. Soc., ser. 2, v. 12 (Zool.), p. 345-354, pl. 44.

Hyman, L. H.

(23) 1940, *The invertebrates, Protozoa through Ctenophora:* McGraw-Hill (New York), xii+726 p., 221 fig.

Kirkpatrick, R.

(24) 1887, *Description of a new genus of Stylasteridae:* Annals Mag. Nat. Hist., ser. 5, v. 19, p. 212-214, pl. 8.

Kühn, O.

(25) 1928, *Hydrozoa:* Fossilium Catalogus, I, Animalia, ed. C. Diener (Berlin), pt. 36, 114 p.

(26) 1939, *Hydrozoa:* in Schindewolf, O. H., Handbuch der Paläozoologie (Berlin), Bd. 2A, p. A3-A68.

Milne-Edwards, H.

(27) 1860, *Histoire naturelle des coralliaires ou polypes proprement dits:* (Paris), v. 3, 560 p.

——— & Haime, Jules

(28) 1850-54, *A monograph of the British fossil corals:* Palaeontograph. Soc. Mon. (London), p. lxxxv+322 p., 72 pl.

(29) 1857, *Histoire naturelle des coralliaires ou polypes proprement dits:* (Paris), v. 2, 633 p., 31 pl.

Moseley, H. N.

(30) 1879, *On the structure of the Stylasteridae, a family of the hydroid stony corals:* Philos. Trans. Roy. Soc. London, ser. B, v. 169, p. 425-503, pl. 34-44.

(31) 1880, *Report on certain hydroid, alcyonarian, and madreporarian corals procured during the voyage of H. M. S. Challenger in the years 1873-76:* Rept. Sci. Results Challenger, Zool. v. 2, pt. 7, p. 11-101, pl. 1-14 (Hydrocorallinae).

Nielsen, K. B.

(32) 1919, *En hydrocoralfauna fra Faxe og bemaerkninger om Danien'ets geologiske stilling:* Danm. geol. Undersøg., ser. 4, v. 1, no. 10, 66 p., 2 pl.

Pourtalès, L. F. de

(33) 1871, *Illustrated catalogue of the Museum of Comparative Zoology at Harvard College, IV, Deep-sea corals.* (Cambridge), 93 p., 8 pl.

Vaughan, T. W.

(34) 1919, *Fossil corals from Central America, Cuba, and Porto Rico, with an account of the American Tertiary, Pleistocene, and Recent coral reefs:* U. S. Natl. Mus., Bull. 103, p. 189-524, pl. 68-152.

SOURCES OF ILLUSTRATIONS

Explanation of the use of index numbers for citing sources of illustrations is given in the Editorial Preface.

(35) **Boschma, H.**
(36) **Broch, Hj.**
(37) **Duncan, P. M.**
(38) **Edwards, H. M., & Haime, Jules**
(39) **Ellis, J., & Solander, D.**
(40) **Fisher, W. K.**
(41) **Hickson, S. J.**
(42) **Hickson, S. J., & England, H. M.**
(43) **Kirkpatrick, R.**
(44) **Moseley, H. N.**
(45) **Nielsen, K. B.**
(46) **Pourtalès, L. F. de**

STROMATOPOROIDEA

By Marius Lecompte

CONTENTS

INTRODUCTION

Stromatoporoids are an extinct group of organisms, exclusively colonial in mode of growth, which secreted a calcareous skeleton having the form of irregular rounded masses, relatively thin sheetlike expansions, and branched or unbranched subcylindrical structures. Some individual colonies exhibit thickness of more than 1 m. and width ranging to nearly 2 m. Others are small, with greatest dimensions less than 1 cm. Fossils recognized as belonging to the group are distributed from Cambrian to Cretaceous, but they are most abundant in some Silurian and Devonian formations, where they may be important reef-builders. Most stromatoporoids show a tendency to split in layers parallel to the surface of the colony, and they break less readily and evenly in directions transverse to these layers. The surface of some colonies is very smooth and devoid of visible markings, whereas that of others is characterized by presence of many small moundlike elevations. These monticules, called mamelons, may be regular or very irregular in distribution, and they may bear radially diverging small channels which subdivide outward.

Biological placement of the stromatoporoids is not firmly established, as indicated by the fact that various authors have suggested their assignment among algae, foraminiferal protozoans, sponges, hydrozoans, some other branch of the coelenterates, and even bryozoans. Conceivably, they represent an independent phylum of invertebrates, not known by students of living animals, but this seems unlikely. The consensus of specialists is that the stromatoporoids are coelenterates which may be segregated as an order of hydrozoans or treated as a class somewhat closely related to the hydrozoans. They are designated here as an order of the Hydrozoa.

MORPHOLOGY

GENERAL FEATURES OF THE SKELETON

FORM OF COLONIES

The skeletal form of stromatoporoid colonies is variable—massive, laminar (sheetlike), and dendroid (Fig. 86,*1-3*)—and as with corals, form of growth may be influenced by environment, although this is not always easy to determine. Some genera and species, nevertheless, are distinguished by their adoption of a massive growth habit or a laterally expanded laminar form. The dendroid colonies generally have been judged to belong to a special group unrelated to the others, some authors going so far as to exclude them from stromatoporoids altogether. The main reason for this somewhat strange interpretation is the presence in dendroid forms of a prominent axial canal, which is construed to be restricted to these dendroid colonies, but in fact it corresponds to structures very common in massive and laminar stromatoporoids; these are the axial canals of astrorhizae, to be described. As noted in the discussion of classification, almost every family of stromatoporoids contains one or more dendroid growth forms directly allied to a massive genus. Although the validity of generic distinction of some of these related forms is doubtful, they are provisionally recognized as separate.

Stromatoporoid colonies generally are free, but some form incrusting growths on other organisms or are intergrown zonally with them, especially where stromatoporoids occur with massive and laminar tabulate corals or less commonly with dendroid tabulates like *Trachypora* (Fig. 87). A more intimate sort of association is observed with the fasciculate tubular organism known since 1841 under the name of *Caunopora* PHILLIPS, which really is a *Syringopora* (Fig. 88). Such intergrowth is found only in Paleozoic rocks to which this commensal tabulate is restricted; it has not the taxonomic value supposed by BARGATZKY (1881) when he introduced the generic name *Diapora,* because it is found in various genera (as *Clathrodictyon, Stromatopora, Stromatoporella, Actinostroma,* and *Rosenella*).

SIZE

The size of colonies is a notably variable character, although it may have some bear-

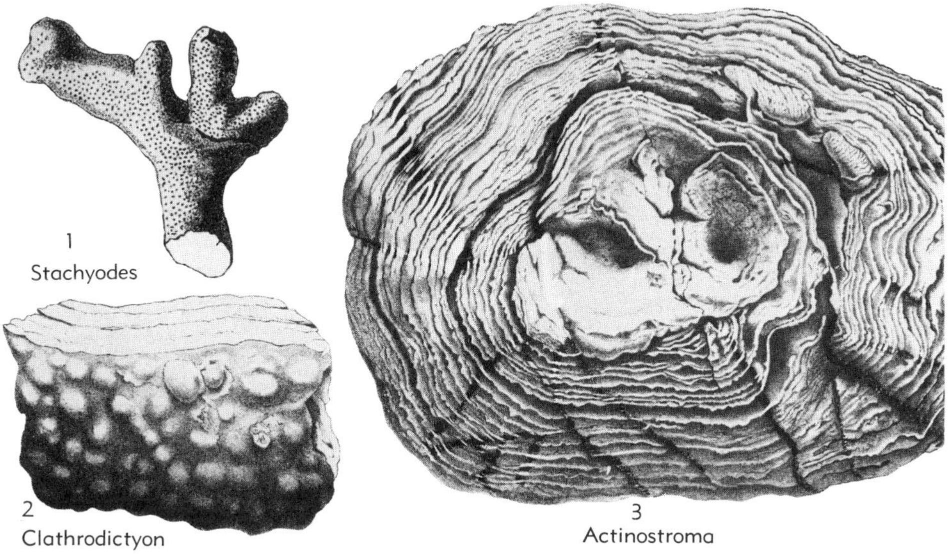

FIG. 86. Stromatoporoid colonies, showing some features of growth form and structure. *1, Stachyodes verticillata* (M'COY), Dev., Ger., a ramose form, ×1.5 (50*). *2, Clathrodictyon variolare* (ROSEN), Sil., Est., showing prominent mamelons on surface and latilaminae in section of colony, ×1 (50*). *3, Actinostrom clathratum* NICH., M.Dev., Eng., basal surface of colony showing many lamellae, ×0.5 (50*).

ing on definition of a species or even of a genus. For instance, *Actinostroma* is distinguished by commonness of colonies having unusually large size, whereas *Clathrodictyon* and *Stromatoporella* generally are represented by rather small laminar colonies. Mesozoic stromatoporoids generally are diminutive, as compared with Paleozoic forms.

SKELETAL STRUCTURE

CHIEF ELEMENTS

The underside of a colony, whether stalked or not, may be covered by an epitheca. The main mass of a colony consists typically of transverse elements, termed **lamellae** and vertical (radial) elements,

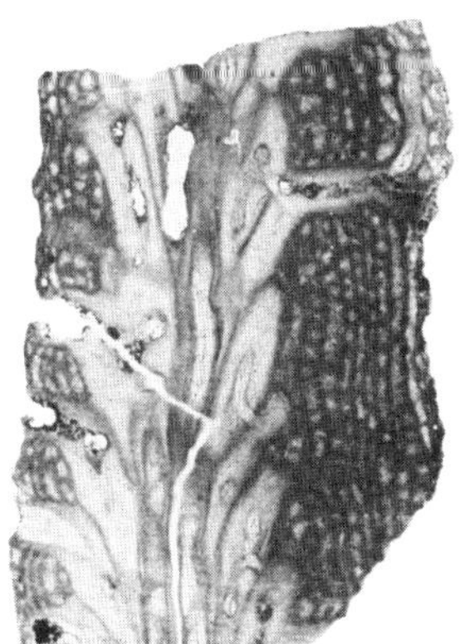

FIG. 87. Longitudinal section of *Trachypora circulipora* KAYSER, a tabulate coral, encrusted by *Stromatoporella* sp., M.Dev.(Givet.), Belg., ×6 (49*).

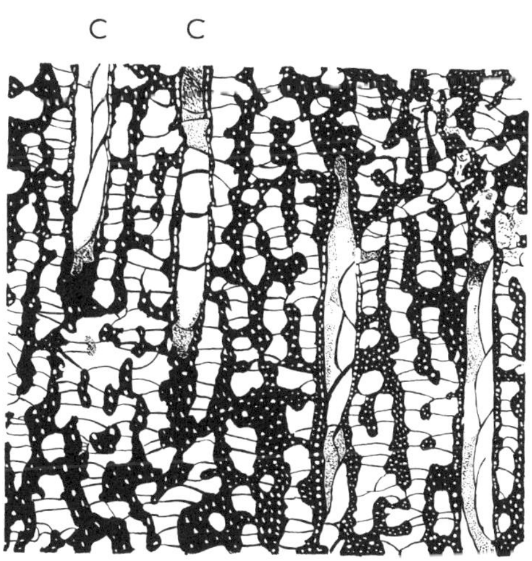

FIG. 88. Intimate intergrowth of a stromatoporoid *(Stromatopora hüpschii)* with the tabulate coral *Syringopora*, M.Dev., Ger., long. sec., ×12. Tubes of the coral, which generally have been described as *Caunopora*, are marked by "C" on the figure (49n).

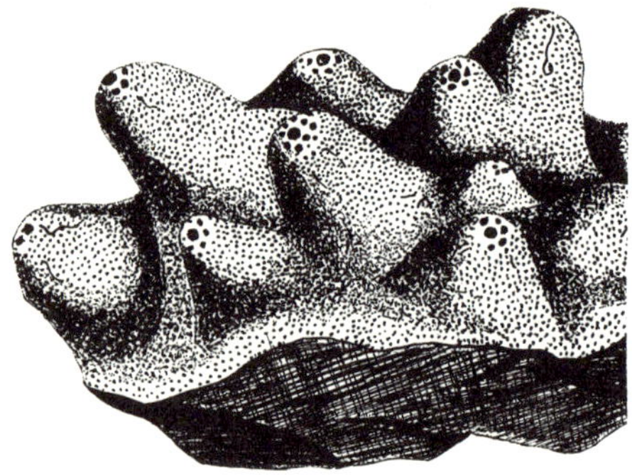

FIG. 89. Part of stromatoporoid colony (*Stromatopora polyostiolata,* M.Dev., Ger.) characterized by unusually prominent mamelons with openings of astrorhizal canals at their summits, ×1.5 (50*).

termed **pillars.** Lamella, which signifies a diminutive thin sheet or fine lamina, is employed for the microscopic transverse layers of stromatoporoids. The upper surface of various species belonging to different genera bears more or less prominent rounded or pointed elevations, termed **mamelons,** and at the center of these is the opening of an astrorhizal axial canal which trends subvertically downward from the surface (Fig. 89). The **astrorhizae,** which comprise star-shaped systems of outwardly branching canals, are clearly defined (Fig. 90) or not discernible, according to whether the surface corresponds to a stage within a lamella or between successive lamellae in building the colonial skeleton. In like manner and depending on the particular structure of the colonial skeleton **(coenosteum),** general

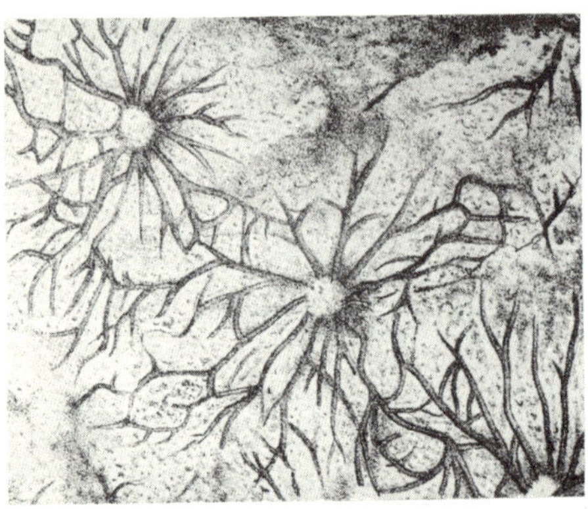

FIG. 90. Stellate pattern of surficial laterally extended branches of astrorhizal canals, seen on slightly weathered specimen of *Stromatoporella eifeliensis* (BARGATZKY), M.Dev., Ger., ×1 (50*).

ornamentation of the surface is tuberculate or reticulate.

The coenosteum of stromatoporoids is essentially trabecular in structure. In most forms, the skeletal elements consist of well-differentiated lamellae and pillars approximately perpendicular to each other (Fig. 91,*1*), but in some the skeleton has an open or close reticulate texture formed by undifferentiated elements (Fig. 91,*2*).

LATILAMINAE

Vertical growth of some colonies proceeds in a periodic manner which results in building fairly regular, moderately thick composite sheets having lateral continuity. These layers, consisting of several individual lamellae, split apart readily along their interface contacts; they are termed latila-

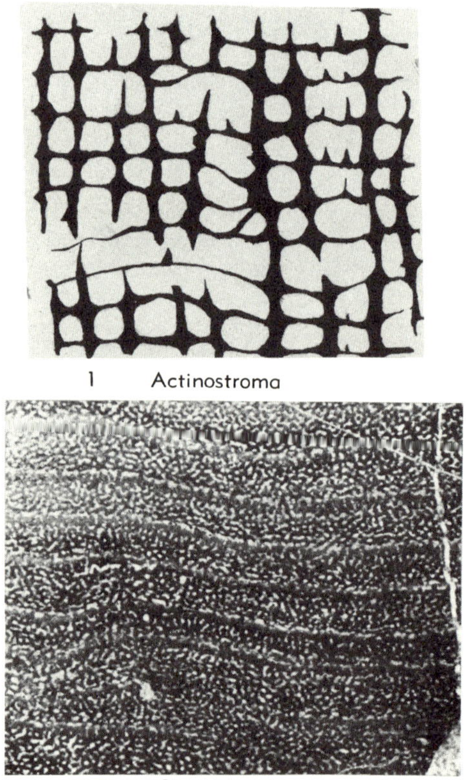

FIG. 91. Lamellae and pillars of stromatoporoid skeleton shown by longitudinal sections. *1, Actinostroma clathratum* NICH., Dev., Eng., ×18, with thin lamellae and relatively stout pillars (50). *2, Stromatopora concentrica* GOLDF., M.Dev., Belg., ×3, with less regular structure of pillars, ×3 (49*).

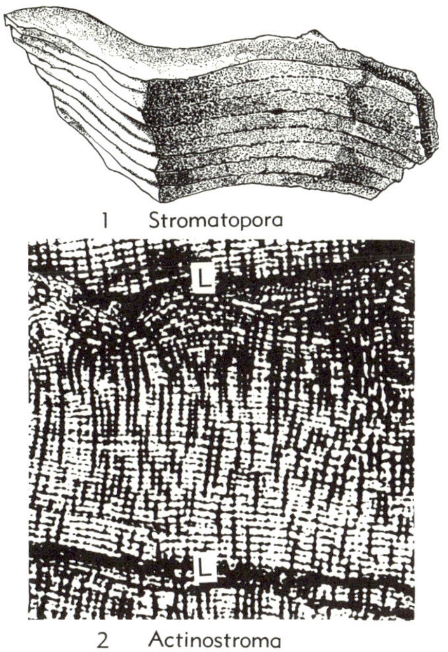

Fig. 92. Latilaminae. *1,* Naturally fractured segment of colony of *Stromatopora concentrica* Goldf., M. Dev., Ger., showing clearly marked thick layered structure denoting latilaminae, ×1 (50). *2,* Longitudinal section of *Actinostroma devonense* Lecompte, U.Dev., Belg., showing pillars and lamellae and indicating hiatus in growth between successive latilaminae (boundaries marked by "L"), ×3 (49n).

minae (Fig. 92,*1*). The presence of latilaminae is not confined to any particular species or genus, but they are best illustrated by some colonies of *Stromatopora concentrica.* Latilamination occurs in stromatoporoids composed of reticulate tissue, as well as in those having well-differentiated skeletal elements. Among the latter, vertical pillars generally are restricted to an individual latilamina, but it happens that some extend without interruption across boundaries of these layers. Boundaries between latilaminae may be marked by increased density or consolidation of the lamellae, by localized intensification of pigment, by a change in structure, such as replacement of reticulate tissue in the upper part of zones of *Stromatopora concentrica* by well-defined pillars or by greater development of astrorhizae at the base of zones.

A structure which sometimes is confused with this regular latilamination is an irregular zonation produced by more or less extensive interruptions in growth, such as are common in tabulate corals. These interruptions are indicated by changes in orientation of skeletal elements and astrorhizal systems, and not uncommonly also by intercalation of a thin layer of sediment (Fig. 92,*2*).

The origin of latilamination has not yet been explained precisely or satisfactorily. Generally, it is thought to denote periodic halts in growth, or alternating acceleration and retardation of growth. The very regular periodicity which characterizes it is not observed in all stromatoporoid colonies of a given deposit.

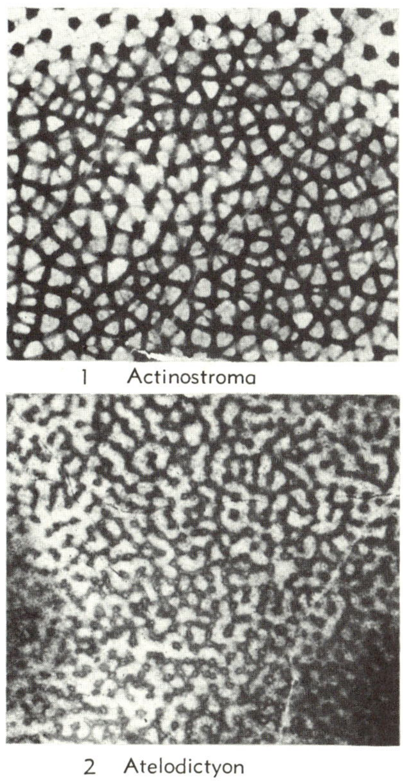

Fig. 93. Meshes of lamellae as seen in tangential sections. *1, Actinostroma devonense* Lecompte, U. Dev., Belg., showing delicate rodlike horizontal elements enclosing subtriangular open spaces, somewhat resembling structure of a hexactinellid sponge, ×12 (49*). *2, Atelodictyon fallax* Lecompte, M. Dev., Belg., showing irregular vermiform pattern of openings in lamella, ×12 (49*).

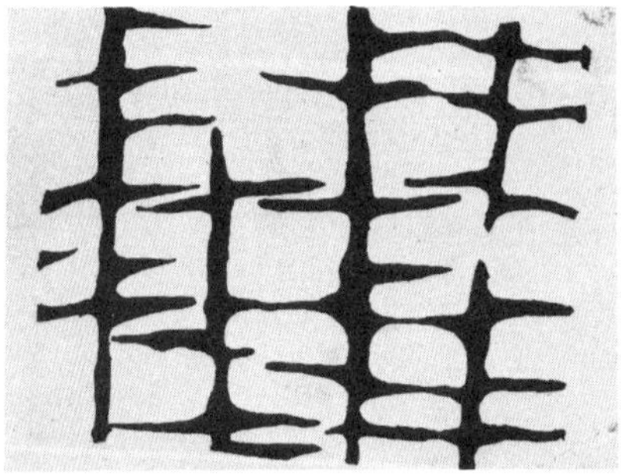

1 Actinostroma

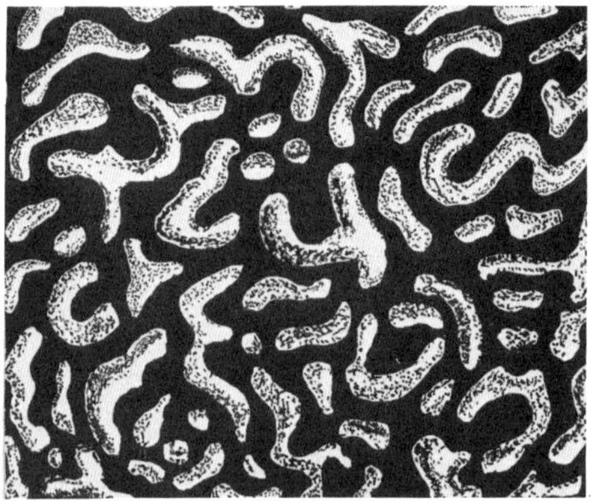

2 Clathrodictyon

Fig. 94. Structure of lamellae and pillars. *1, Actinostroma intertextum* Nich., M.Sil., Eng., showing uneven spacing of lamellae, ×10 (52). *2, Clathrodictyon fastigiatum* Nich., Sil., Eng., tang. sec. in plane of lamella, ×30 (50).

LAMELLAE

As compared with the layers termed latilaminae, the lamellae are very thin, for very many of them occur within a single latilamina; also, lamellae are evident in numerous species of stromatoporoids which lack latilamination. Each lamella consists of a calcareous fibrous network which intersects or impinges on pillars, or separates two contiguous stages of pillars. In genera such as *Actinostroma,* the meshes of the lamellar network have polygonal form resembling the arrangement of siliceous spicules of some hexactinellid sponges (Fig. 93,*1*); in others, the meshes are partly open on one or more sides, making a meandriform pattern as in *Atelodictyon* (Fig. 93,*2*), or a vermiculate design, as in *Clathrodictyon* (Fig. 94,*2*); by thickening, the network becomes a nearly continuous compact sheet which here and there is pierced by pores, as in *Stromatoporella.* Among some forms, especially of Silurian age, the meshes of the lamellae join the pillars at variable levels (Fig. 94,*1*), but in a large majority of species, the lamellae maintain nearly constant levels so as to define regular floors or platforms, arranged one above another (Fig. 91,*1*). In genera such as *Actinostroma, Atelodictyon,* and *Syringostroma,* the lamellae generally are very thin, but in *Hermatostroma, Stromatopora, Stromatoporella,* and others, they are relatively thick. An evolutionary trend toward thickening of lamellar structures seems to exist in some lines, as among Syringostromatidae, but variation in thickness may be influenced by ecologic factors or it may reflect specific differences.

In many genera, lamellae disappear locally or generally. This gives rise to more or less uninterrupted tabular-bearing interpillar spaces extending vertically or sinuously through parts of a colony. Numerous authors have interpreted such structures incorrectly as zooidal tubes.

Lamellae are homogeneous in some genera (Fig. 92,*2*), but cross sections of them in others reveal a dark axial line which has led some authors to conclude that the lamellae are bipartite (Fig. 95,*1,2*). This axial line may be light, instead of dark, as a result of bleaching. Tripp (1932) considers such structure to be a fundamental character, because it offers resemblance to some living athecate hydroids. The presumed bifoliate stromatoporoid lamellae are interpreted by him as the combined roof of one layer and floor of the next following layer, and this he terms the "closed type" *(Stromatoporella, Clathrodictyon).* Simple unipartite lamellae, interpreted as having had only a coenosarc cover, represent the so-called "open type" *(Actinostroma, Stromatopora, Parallelopora).* Distinction of these two types lacks the definiteness supposed by Tripp, for axially divided lamellae occur in some specimens of *Actinostroma* and invariably are present in its Mesozoic descendants, *Actinostromaria* (Fig. 95,*1*). Moreover, proposed classification on this basis disregards important relationships indicated by other characters. The dark axis actually is a microstructural feature consisting of aligned centers of calcification,

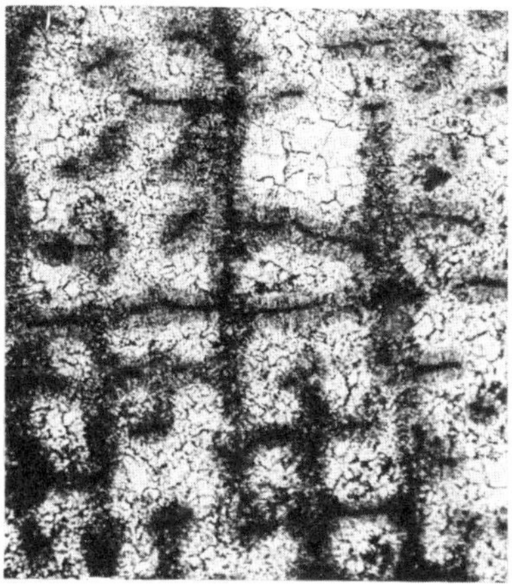

1 Actinostromaria

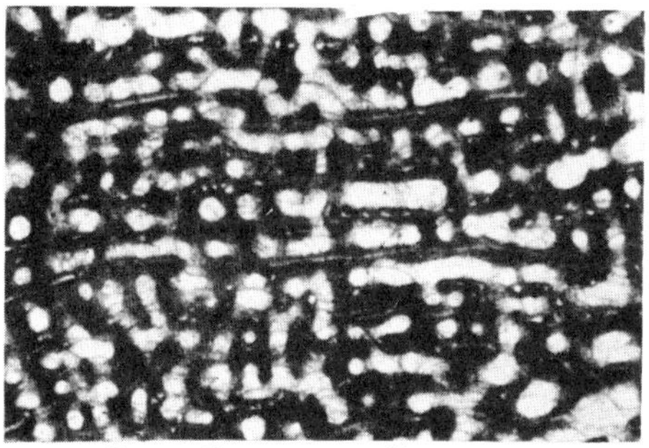

2 Stromatoporella

FIG. 95. Structure of lamellae and pillars. *1, Actinostromaria stellata* CHALMAS, U.Cret.(Cenoman.), Fr., showing dense axial line in pillars and lamellae, long. sec., ×12 (49n). *2, Stromatoporella pertabulata* LECOMPTE, U.Dev.(Frasn.), Belg., long. sec. showing thin clear axial line in several lamellae, ×12 (49*).

and the term "primary lamella" frequently applied to it, like "secondary cover" for clear tissue enclosing it, manifestly lacks histological significance. The existence of an axial canal, as inferred by NICHOLSON (1886), is not supported by observation, for the dark axis does not represent a passageway filled by sediment but is a concentration of carbonaceous pigment within calcareous substance which is identical to the enclosing clear secretion.

PILLARS

The radial or vertical skeletal elements of the stromatoporoid coenosteum in many species are termed pillars. They are classed as continuous if they extend through a number of interlamellar spaces, or discontinuous pillars may falsely suggest existence of continuous pillars, as commonly in *Hermatostroma* and *Syringostroma,* but examination shows clearly that the pillars are interrupted by the lamellae (Fig. 96). Existence of an axial canal in the pillars, which has been reported in some genera, is as unfounded as that in lamellae; the annular appearance of pillars in tangential sections of some species (as *Stromatoporella*) is produced by invagination of a lamella in the plane of the section, for the lamellae may be bent sharply around extremities of the pillars.

RETICULATE TISSUE

Stromatoporoids classed as having reticulate coenosteal structure are distinguished by lack of well-differentiated lamellae and pillars, for the transverse and radial skeletal elements tend to be intergrown intimately. The reticulate tissue is of two types, one characterized by greater distinctness of transverse elements, and the other by dominance of the radial or vertical elements, but there are also intermediate forms which are not referable definitely to either type.

In the type species of *Clathrodictyon,* the transverse and radial elements join without break in continuity to form horizontally elongated vesicles having sufficient regularity of arrangement in successive layers to provide an appearance of lamellation (Fig. 97). This sort of structure, which may be called subreticulate, grades through an intermediate series into types having lamellae and pillars more and more sharply distinct but without loosening of their close bond-

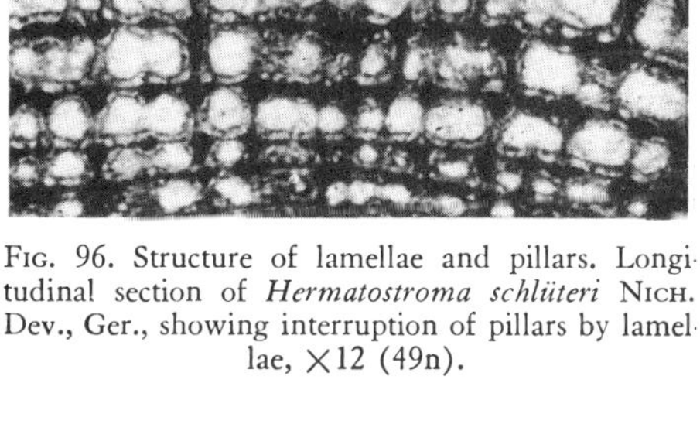

FIG. 96. Structure of lamellae and pillars. Longitudinal section of *Hermatostroma schlüteri* NICH. Dev., Ger., showing interruption of pillars by lamellae, ×12 (49n).

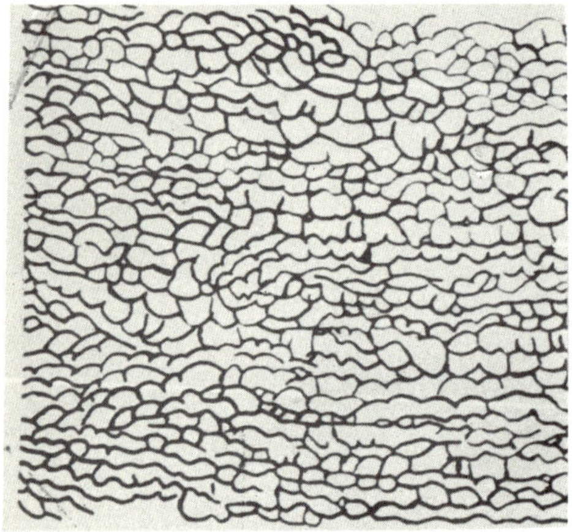

FIG. 97. Structure of lamellae and pillars. Pattern shown by longitudinal section of *Clathrodictyon vesiculosum* NICH. & MURIE, M.Sil.(Clinton.), *Ohio*, ×15 (20).

ing together. Transverse sections of the vertical elements of specimens belonging to this group commonly exhibit a vermiculate appearance (Fig. 94,*2*). *Stromatopora,* and in lesser degree some forms of *Parallelopora,* show network meshes elongated vertically and more or less disarranged, so that lamellation is not discernible; chambers between the vertical elements are traversed by thin tabulae (Fig. 88). The reticulate tissue of these forms also is not strictly definable as a structural type distinct from that characterized by well-differentiated lamellae and pillars, for in *Stromatopora concentrica* the reticulate tissue develops typical pillars in upper parts of the latilaminae, and in other species of this genus, skeletal elements are clearly separate and mutually perpendicular.

Straight or curved thin oblique partitions in horizontal chambers are observed in all types of tissue. These partitions are interpreted as astrorhizal septa, because the extent of their occurrence is correlated with degree of ramification exhibited by astrorhizae.

Thickening of skeletal structures is so great in some colonies, that the chambers are considerably reduced in size and nearly closed. This is especially common in *Stromatoporella* and most pronounced in *S. obliterata.*

ASTRORHIZAE

Structural features. The most remarkable structures of the stromatoporoids are the features named astrorhizae by CARTER (1880). At the surface of the coenosteum, an astrorhiza consists of a stellate system of furrows which lack walls of their own; the furrows spread radially from a circular opening which is located typically on the summit of a mamelon or less commonly on its flank and they tend to branch proceeding outwardly (Figs. 89, 90). A central opening seems to be lacking in the astrorhizae of some Silurian and Mesozoic species. Distally the furrows ramify into the normal network of the coenosteum and become confluent with neighboring systems (Fig. 98,*1a*); laterally also they communicate freely with chambers of the coenosteum. Among forms having interpillar tabulate chambers, these spring upward from lateral canals of astrorhizae, and they may join different levels of such canals (Figs. 98,*1b,2*). The whole structure of the coenosteum is in fact in close relation with the radiation and ramification of astrorhizal canals.

In many stromatoporoids, the astrorhizae are superposed exactly at each lamellar stage, so that the center of any astrorhizal system is occupied by a vertical canal extending from base to summit of the colony, or through one or more latilaminae, or through some lesser thickness (Fig. 98,*3*). In some species the astrorhizae are not superposed, but vary in placement from one growth stage of a colony to another; this irregularity seems to be much less common than has been supposed, however. Thus, reported lack of superposed astrorhizae in *Actinostroma* and use of this character as criterion for distinction of this genus from *Actinostromaria* are erroneous.

Approximate uniformity of spacing of astrorhizae on the surface of hemispherical colonies in spite of increase of this surface by growing is realized by subdivision of the astrorhizal systems in their upward growth (Fig. 99).

The diameter of the vertical canals and that of their radially outspreading branches are highly variable, as is also the rate of diminution in size of the canals away from astrorhizal centers; these measurements have value only for distinction of species. On the other hand, some genera are characterized by the very prominent development of these structures and others by their delicate dimensions. This variation, combined with differently pronounced closeness of normal lamellation, explains the weak or strong contrast between astrorhizal structures and surrounding tissue as seen in vertical sections. Thus, the remarkably delicate astrorhizae of *Actinostroma* have escaped the notice of many observers; also, reported absence of astrorhizae in *Hermatostroma* is an error which may be ascribed to the near-matching in size of the astrorhizal canals and the coarse meshes of associated coenosteal tissue.

The occurrence of astrorhizae in dendroid stromatoporoids has not been perceived generally, and because of their presumed absence, HEINRICH (1914) excluded this group from true stromatoporoids. Their existence, however, hardly admits doubt. Each colony has a single astrorhizal system consisting of a large axial canal and delicate branches which, as in massive stromatoporoids, are disposed parallel to the lamellation.

The ramifications of astrorhizal systems may be superposed during their vertical development, or they may alternate in position. Mamelons are invariably associated with the superposed type, either continuous-

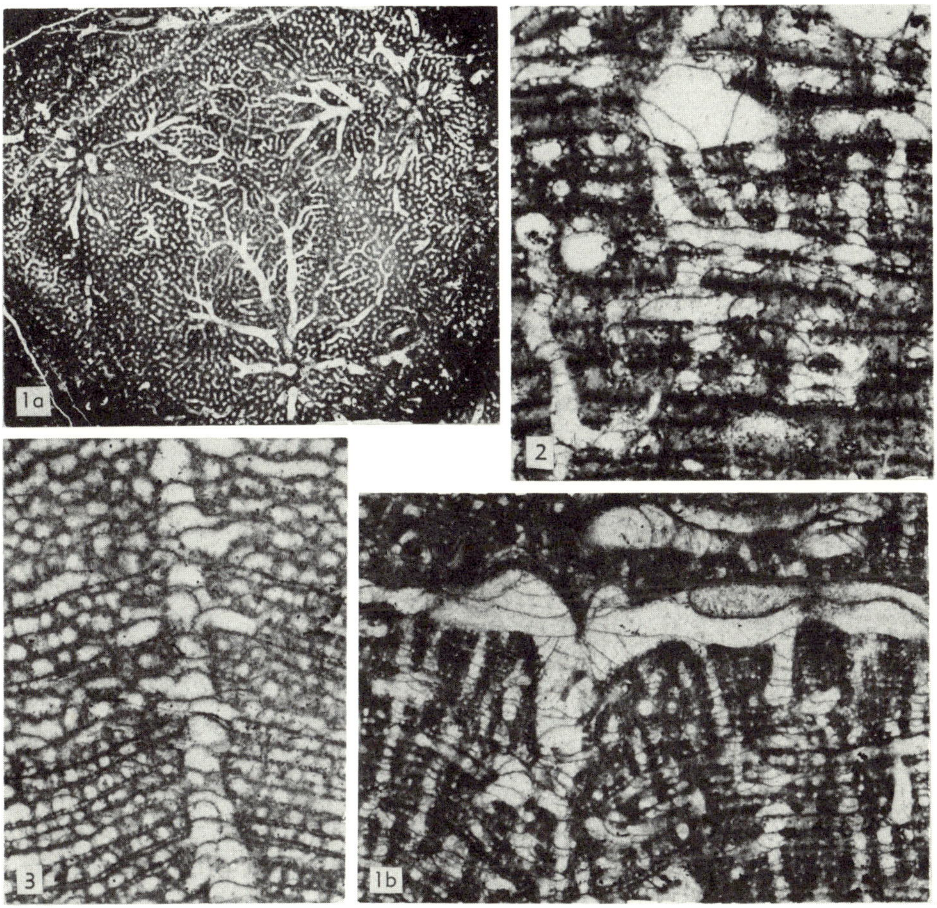

FIG. 98. Structure of astrorhizae. *1, Trupetostroma tenuilamellatum* LECOMPTE, U.Dev.(Frasn.), Belg., *1a,* tang. sec. showing well-defined astrorhizal canals, ×3; *1b,* long. sec. through an astrorhiza showing tabulate vertical canals intersecting lateral canals, ×12 (49*). *2, T. thomasi arduennensis* LECOMPTE, M.Dev. (Givet.), Belg., long. sec. showing tabulate vertical canals enclosing horizontal ones between them ×12 (49*). *3, T. ruedemanni* LECOMPTE, U.Dev.(Frasn.), Belg., long. sec. through a vertical astrorhizal canal, ×12 (49*).

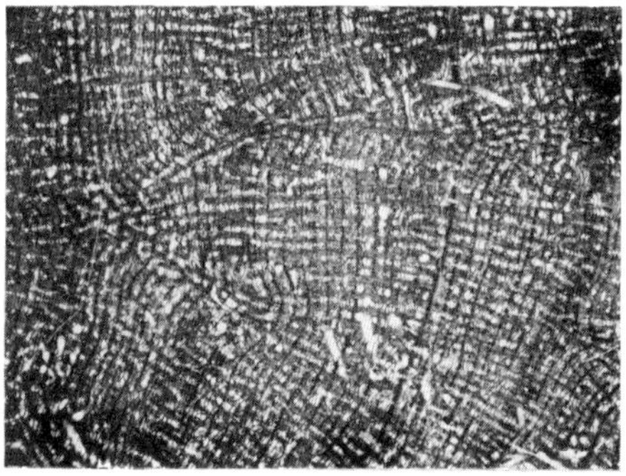

FIG. 99. Structure of astrorhizae. Longitudinal section of *Hermatostroma perseptatum* LECOMPTE, U. Dev.(Frasn.), Belg., showing subdivision of vertical astrorhizal canals that proceeds as colony develops, ×3 (49*).

ly or intermittently, but more or less rapid compensation in upward building of the surface characterizes the other type.

The axial canal and its branches are intersected obliquely by straight or curved tabulae (Fig. 98,*2,3*). These tabulae occur in all species, but exhibit a high degree of variability; generally they are more abundant in astrorhizae with canals of relatively large diameter. Their mode of origin and their function are unexplained.

Among Mesozoic stromatoporoids derived from *Parallelopora,* especially the Milleporellidae and Milleporidiidae, an important change occurs in attitude of the branching astrorhizal passageways. While these run in essentially horizontal position in Paleozoic forms, they tend in later types to become vertical. The Milleporellidae show star-shaped groups of the canals still present on horizontal surfaces but there is no axial canal, the center of the astrorhizal system being occupied by a close-packed bundle of canals which partly spread out radially at certain interlamellar levels, subdividing and passing into the tabulae-bearing vertical chambers (zooidal tubes of authors) between the pillars. Such structure is well shown by *Steinerella mecosola.* In *Milleporidium,* the branching astrorhizal systems have disappeared, the well-defined large tabulate tubes which replace them in the cortical zone passing downward into narrow tabulate chambers of normal type.

This progressive transformation of the astrorhizal systems parallels evolution of the pillars into vertical lamellae which tend to form enclosed tubes with close-spaced tabulae that take the place of vanished horizontal lamellae. Indeed, the general trend of stromatoporoid skeletal evolution seems to be a progressive development of tabulate vertical chambers.

Significance. The function of the stromatoporoid astrorhizal systems has not been determined. They cannot well be interpreted as equivalent to incurrent or excurrent canals of sponges, which are distributed irregularly and are unrelated to particular morphological features of the sponge skeleton. Among stromatoporoids, the secretion of coenosteal tissue was manifestly influenced by the features called astrorhizae, branches of which penetrate all of the skeletal meshwork in the manner of a hydrorhiza. Concordant with minute morphological details of the skeleton, they are seen to be particularly developed as a carpet at the base of latilaminae in Paleozoic forms and also next above a hiatus which defines a surface of necrosis. Their role is evidently one that contributes to building the colonial skeleton, therefore.

CARTER (1880), who first used the term "astrorhiza," interpreted these structures as stellate systems of stolons corresponding to the hydrorhyzae of *Hydractinia echinata,* and with this homology TRIPP (1929) agreed very nearly in showing close resemblance between astrorhizae and the spines of *Hydractinia.* NICHOLSON (1886) accepted the views of CARTER and extended comparison to the anastomosing branchwork of coenosarcal canals in the coenosteum of *Millepora.* He concluded that primary stolons of coenosarc occupied the axial canals and stolonal branches filled the astrorhizal furrows, but found no reason to think that any of the canals housed zooids. KÜHN (1939), on the contrary, has expressed judgment that the axial canals are the only structures recognizable as zooidal tubes, and more precisely, definable as emplacements of blastostyles.

The existence of zooidal tubes in stromatoporoid colonies has been a subject of much controversy. NICHOLSON asserted their occurrence in all stromatoporoids, pointing out that in the Actinostromatidae they are

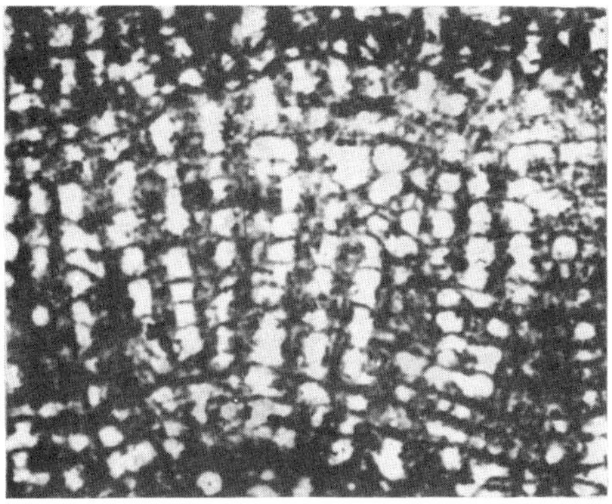

FIG. 100. Supposed zooidal tubes in stromatoporoids. Laterally directed passageways shown in the central part of this longitudinal section of *Parallelopora paucicanaliculata* LECOMPTE, M.Dev.(Givet.), Belg., seem not to be zooidal tubes, ×12 (49*).

represented by the normal, poorly differentiated interpillar spaces and that in the Stromatoporidae they consist of sporadic tabulae-bearing tubes. The presence of zooidal tubes is contested by HEINRICH (1914), TRIPP (1929), YABE & SUGIYAMA (1930), PARKS (1935), and KÜHN (1939), although none offers explanation of the tabulate vertical chambers which especially characterize the Stromatoporidae and related forms. It is very difficult to interpret these chambers as zooidal tubes in forms such as the Devonian *Parallelopora paucicanaliculata* (Fig. 115) and the Mesozoic *Stromatoporellina* and *Steinerella,* in which all interpillar spaces bear tabulae. Vertical sections of these stromatoporoids show features strikingly similar to sections of *Hydractinia echinata* (Figs. 100, 101) and especially *H. circumvestiens,* as remarked by NICHOLSON. Between the spines of *Hydractinia* with their irregular large horizontal prolonga-

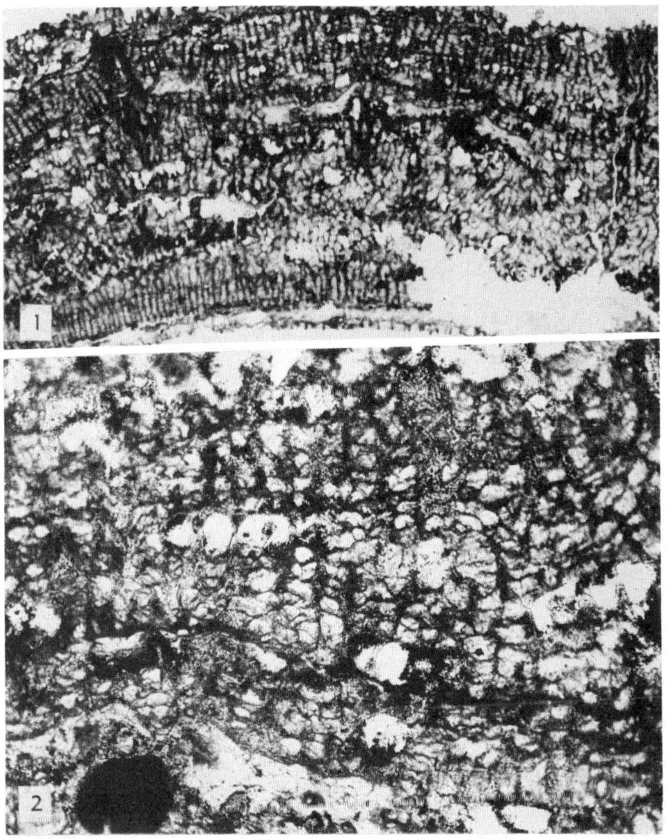

FIG. 101. Structure of modern hydrozoan, *Hydractinia echinata* (FLEMING), North Sea, as shown by longitudinal sections; *1a,* ×3; *1b,* ×50 (49n).

tions which are strikingly like the branched astrorhizal systems of stromatoporoids, the entire coenosteum consists of parallel narrow vertical chambers traversed by close-spaced thin tabulae. In *H. echinata,* these chambers are mostly delimited by trabecular pillars but some are tubular. They connect with the large horizontal canals which become confluent with the furrows on flanks of the spines. At the surface they furnish passageway for hydranths or coenosarcal stolons, and at junctions of 2 or 3 of them, zooids grow upward. NICHOLSON states that in *H. circumvestiens* the vertical tubules, which he interprets as zooidal tubes, are most strongly developed.

MICROSTRUCTURE OF SKELETAL ELEMENTS

The microstructure of stromatoporoid skeletal elements has not yet been studied adequately. The most striking presently known feature is contrast between microstructure of the Paleozoic and Mesozoic forms. Skeletal elements of the latter are finely fibrous wherever observed, but among Paleozoic genera, this structure is found only in *Amphipora.*

PALEOZOIC FORMS

Two essential types of skeletal elements are recognized among Paleozoic stromatoporoids on the basis of microstructure: the *compact* type and the *cellular* or *vacuolar* type.

Compact type. The compact type is observed in the Actinostromatidae, Clathrodictyidae, and Labechiidae. The only structural feature of *Clathrodictyon* and some species of *Actinostroma* is occurrence of a dark axis in skeletal elements, which, as previously noted, has been interpreted erroneously as evidence of a bipartite construction of the lamellae and canaliculation of the pillars. The axis is recognizable even where pigment is lacking, for microcrystalline substance along this line is differently oriented, as shown by examination of thin sections under crossed nicols. A weakly fibrous structure indicated by some specimens may not be an original character, because some degree of recrystallization seems to affect most of these fossils. Such alternation probably accounts for a false cellular microstructure in certain examples of *Actinostroma,* which seems to be only granulation of the recrystallized calcite.

Cellular type. The cellular type of skeletal elements predominates in Paleozoic stromatoporoids. It prevails in the Stromatoporidae, Stromatoporellidae, and Syringostromatidae. The differentiation of subtypes of structure in this group, which has been attempted, such as regular porous, spongy, and irregular vacuolar, lacks value because these terms mostly reflect only conditions of preservation, the pores being enlarged progressively by alteration which transfers calcite from secreted fibrils to crystalline fillings of adjacent minute spaces. It happens that microstructure may not be recognizable at all, and this is especially common in *Stromatoporella.* Nevertheless, a certain systematic variation is observed: for instance, the cellules of the skeletal fiber are generally smaller, more regular, and better aligned in *Stromatoporella* than *Stromatopora* (Figs. 102,*3*,*4*). In the series *Syringostroma-Trupetostroma-Hermatostroma* of the Syringostromatidae, an evolutionary trend is seen in an increased regularity of the cellules, in their elongation, and in their alignment in single row along margins of the skeletal elements (Figs. 96; 102,*2*).

A tubular structure which has been recorded by NICHOLSON and others in various species (as *Stromatoporella granulata, S. eifeliensis, Parallelopora ostiolata, P. goldfussi, Stachyodes verticillata*) is nonexistent, for identification of it rests on misinterpretation of sections in one plane. Actually, these forms exhibit a microreticulate structure defined by a fine dark-pigmented meshwork as it is particularly clear in the type of *Parallelopora ostiolata* (Fig. 102,*1*), but according to state of preservation, this may appear as a fine longitudinal striation, as commonly in *Stachyodes.* No trace of a canal system filled by sediment is found; the pigment underlining the reticulation is purely carbonaceous and occurs in secreted calcium carbonate. This reticulate structure may result from an alteration which enlarges the original cellular alveoles at expense of surrounding fibrillar tissue. In any case, it is not confined to particular genera and within a single genus it may coexist with normal cellular structure, being observed in forms other than those noted by NICHOLSON (such as *Stromatopora, Syringostroma, Trupetostroma, Hermatostroma*). In *H. beuthii* one may see marginal cellules enclosed by dark vertical and transverse fibers in a pattern exactly corresponding to that of the microreticulate structure of *Parallelopora ostiolata*.

Meshed-fiber type. A rather distinct type of skeletal microstructure not previously mentioned is found in the Paleozoic genus *Synthetostroma*. The lamellae of this form are composed of intricately intermingled fibers which, combined with interlamellar partitions, produce a complicated pattern in vertical sections.

MESOZOIC FORMS

Fibrous type. Except for *Dehornella* and *Ceraostroma*, which have a compact type of skeletal microstructure, Mesozoic stromatoporoids are characterized by fibrous structure. Two sorts have been distinguished by STEINER (1932) and YABE & SUGIYAMA (1935): one termed radial, consisting of fibers arranged perpendicularly to the axis of a skeletal element; and the other called fasciculate or "water-jet," comprising fiber

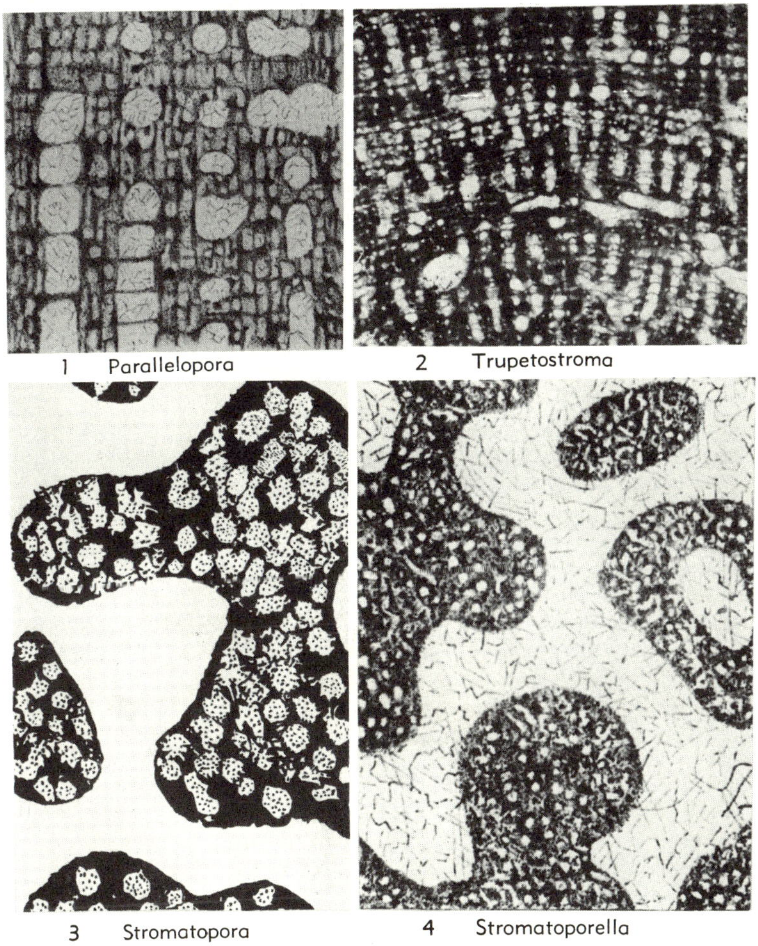

FIG. 102. Microstructure of stromatoporoids. *1, Parallelopora ostiolata* BARGATZKY, Dev., Ger., long. sec. showing microreticulate structure, ×48 (50*). *2, Trupetostroma tenuilamellatum* LECOMPTE, U.Dev. (Frasn.), Belg., long. sec. showing regularly arranged cellules, ×12 (49*). *3, Stromatopora carteri* NICH., M.Sil.(Wenlock.), Eng., tang. sec. showing closely spaced cellules in skeleton, ×48 (50). *4, Stromatoporella granulata* NICH., M.Dev.(Hamilton), N.Y., tang. sec. showing features resembling those of *3*, ×48 (50*).

bundles which open more or less shortly outward somewhat from the axis, with or without a central dark line. Actually, the arrangement of fibers is much more varied; the fibrosity may be complex, and in some, highly diffuse; and centers of calcification, generally marked by a concentration of dark pigment, may not be juxtaposed in continuous alignment along the axis but are quite dissociated, each with its own aureole of fibers.

Dehornella exhibits an aberrant sort of concretionary microcrystalline structure, for parts of skeletal elements, generally along the axis, show single extinction under crossed nicols. This structure is like that of *Ceraostroma* described by Kühn (1926) and interpreted by him to denote a primitive chitinous composition, marking transition from calcareous hard parts of typical stromatoporoids to the chitinous hard parts of various living hydrozoans.

RELATIONSHIPS OF PALEOZOIC AND MESOZOIC TYPES

The contrasts, as well as some similarities, in the nature of skeletal microstructure observed in Paleozoic and Mesozoic stromatoporoids call for elucidation. Steiner's (1932) hypothesis that the Paleozoic forms originally possessed a fibrous structure and that this has been lost by secondary alteration is difficult to conceive, especially as regards species which exhibit the cellular type of microstructure. It is hard to understand how recrystallization could produce such regularly arranged cells of nearly constant size in some genera, and how it could explain the evolutionary progression recognized in the family Syringostromatidae. The change in structure observed at the close of Paleozoic time must be interpreted as having subordinate taxonomic significance, because most of the Mesozoic genera manifestly belong in families which contain Paleozoic forms, and in all other characters they are closely similar or identical to older genera.

The classification and interrelations of microstructural characters are too little established by present knowledge to permit using them as a basis for taxonomy of the stromatoporoids. Most authors, in agreement with Nicholson, recognize microstructure as having value for defining genera, or less commonly, species. Heinrich (1914), neglecting reservations expressed by Nicholson, made this the prime basis of general classification and, restricting the order very arbitrarily, divided Paleozoic genera into two families: Actinostromatidae, with homogeneous pillars; and Stromatoporidae, with porous or tubular structure. This too comprehensive grouping takes no account of textural characters and by associating distantly related forms lacks utility for study of phylogeny. According to Steiner, these 2 Paleozoic groups continue into the Mesozoic, characterized by new structural elements, the Actinostromatidae acquiring fibrous radial structure with a dark axis, and the Stromatoporidae assuming a fasciculate fibrous structure. This concept leads her to place *Siphostroma,* which has textural features indubitably close to the Stromatoporidae, in the Actinostromatidae, and to exclude from this family *Burgundia,* which is very close to *Clathrodictyon.* Moreover, a number of Mesozoic genera are too poorly described to provide information on microstructure. From all viewpoints, it is judged premature to undertake classification of the stromatoporoids on microstructural characters.

SYSTEMATIC POSITION OF STROMATOPOROIDS

Lack of definiteness in conclusions based on comparison of structural features of stromatoporoid hard parts and those of living organisms, supplemented by uncertainty as to the significance of astrorhizae, which are the most particular character of this group, has led authors to explore affinities with different main divisions of the invertebrates, especially Foraminifera among the Protozoa, the Porifera, and Hydrozoa among the Coelenterata.

RELATIONS TO FORAMINIFERA

Classification of stromatoporoids among Foraminifera has been advocated especially by Hickson (1934) and Parks (1935), basing opinions chiefly on certain resemblances of *Actinostroma* to the foraminifer *Gypsina plana*. The arguments advanced relate only to secondary characters, however, or are not relevant, such as (1) analogy in appearance of reticulate structure seen in tangential sections, which is of little significance; (2) interpretation of coarse vertical astrorhizal canals of *Actinostroma* as passages for protrusion of pseudopodia; (3) explanation of lamellae in *Stromatoporella* as consequent on alignment of chambers, with pillar-supported interspaces serving only for separation of their rows; (4) existence of a basal layer of chambers different from others in the colony, corresponding to the microspheric or megalospheric proloculus of Foraminifera; (5) occurrence in some stromatoporoids of tubes called caunopores, assumed to represent one of the alternating generations observed among foraminifers; (6) variability in size, form, and arrangement of chambers, as observed both in stromatoporoids and Foraminifera. Lack of validity of most of these points is self-evident. For example, no reason is known for presumption that a certain layer of stromatoporoid chambers is equivalent to a foraminiferal proloculus; the so-called caunopore tubes are identified with reasonable certainty as belonging to tabulate corals; and variation in patterns of chambers in itself means nothing. How can the closed chambers of Foraminifera, compared with the open meshes of stromatoporoid tissue, signify taxonomic affinity? To imagine that the open meshes represent degeneration finds support in no observations, and it runs counter to what is known of evolutionary trends among stromatoporoids, in which the essentially trabecular skeletal tissue tends to a lamellar development with formation of closed tubes. Finally, the essentially spiral architectural plan of Foraminifera is wholly unknown in the stromatoporoids, and not to be found among them are foramina like those of *Gypsina*.

RELATIONS TO SPONGES

The hypothesis of affinity between Stromatoporoidea and Porifera has been supported by many, chiefly on the basis of analogy, plausible at first sight, between the astrorhizae of stromatoporoids and the water canals of sponges. HEINRICH (1914) and TWITCHELL (1929) especially held this view. An astrorhiza, for the authors, suggests an ingeniously modified rhagon type of sponge, but the astrorhizae are perfectly concordant with morphological peculiarities of the stromatoporoids, including a lamellar disposition, which is not observed in the Porifera. Actually, the coenosteum of stromatoporoids contains no space untouched by the highly ramified astrorhizal systems and (in some forms) by the tabulate inter-pillar spaces inrooted on these. Secretion of the skeleton is controlled so plainly and minutely by the astrorhizal structures that these cannot be interpreted as a simple system of water passageways.

The absence of spicules in the coenosteum of stromatoporoids is an impassable obstacle to classification with sponges, and this has not been overlooked by authors. TWITCHELL has drawn argument from a strange specimen of *Stromatopora centrota* which shows abundant broken skeletal elements, and these he regards as fragments of spicules. Judging from the published figures, this interpretation is highly dubious, and if correct, there is still need to determine whether the spicules really belong to the stromatoporoid or to a parasitic sponge. At all events, and contrary to TWITCHELL's assertion, the described specimen is very exceptional.

Some stromatoporoids having structure interpreted to represent a sponge in contracted state are recorded by TWITCHELL. The supposed contracted condition (as in a cited specimen of *Syringostroma densum*) exhibits none of the compensations of skeletal elements which should be found, and the observed structures are in perfect agreement with the recognized evolution of the family.

RELATIONS TO HYDROZOANS

Most authors have subscribed to views of CARTER (1877) in interpreting the stromatoporoids as ancestors of modern hydrozoans. Among these authors are STEINMANN (1878), NICHOLSON (1886), WAAGEN & WENTZEL (1887), DEHORNE (1920), KÜHN (1927, 1939), TRIPP (1929), and STEINER (1932). For reasons which follow, the view is also adopted here.

(1) The astrorhizae, which constitute so characteristic a peculiarity of the stromatoporoids, find satisfactory explanation only

according to this hypothesis. Unlike the canals of sponges, which are more or less complex systems deeply penetrating tissues in discordant manner, the astrorhizae are concordant with skeletal structure and control details of its development. An astrorhizal layer marks the beginning of new growth after an accidental hiatus or completion of a latilamina. The principal branches which ramify distally are joined laterally by a network of secondary canals extending to all chambers defined by the skeletal fibers. Such structures are closely similar to the hydrorhizae of the Hydrozoa. The evolution which astrorhizae undergo in Paleozoic and especially in Mesozoic time, leads to structures which diverge somewhat from observed features of the hydrorhizae, but nevertheless maintain resemblance to features shown by some species of *Hydractinia.*

(2) The existence of zooidal tubes, in part of the Stromatoporoidea at least, has been cited by NICHOLSON and others as a determining argument, if not sole one, for classifying these organisms as hydrozoans. Doubt that these tubes exist or firm denial of them has led some authors to reject the hypothesis of affinity between stromatoporoids and hydrozoans. This rejection of presumed affinities on such basis is not justified, however, for by no means all living hydrozoans possess zooidal tubes. Among stromatoporoids, it may well be that zooids could develop only on the superficial network of coenosarc, leaving no trace in hard parts of the coenosteum. As admitted by NICHOLSON, without doubt this was the case in the families Actinostromatidae and Clathrodictyidae, which are characterized by skeletal structure very closely similar to that of certain modern athecate hydroids.

The zooidal nature of the narrow vertical tabulate chambers of trabecular construction among Stromatoporidae, Stromatoporellidae, Syringostromatidae, and more or less tubular in Stromatoporidae, Milleporellidae, Milleporidiidae, has not been proved. This is difficult to understand, both in the Stromatoporellidae where these structures are so little developed and sporadic, and in certain species of *Parallelopora,* in the Stromatoporidae, and in the derived Mesozoic families where these structures comprise all tissue between the astrorhizae. But as already stated, some hydractinians have essentially comparable structures. NICHOLSON (1886, p. 68) has drawn attention to *Hydractinia circumvestiens* (Pliocene, England) which has a coenosteum traversed by many vermiform tubules, approximately vertical and parallel; these tubules are interrupted at various levels by irregular, horizontally disposed small chambers, which bend upward at nearly regular lateral intervals toward large pillars (spines) of cribriform texture that may be pierced by an axial canal. The tubules must have served for lodgement of zooids or for passage of coenosarcal stolons. NICHOLSON emphasized the close resemblance of this form with the Stromatoporidae rather than the Actinostromatidae. A vertical section of the chitinous skeleton of *H. echinata* (Fig. 101) shows a comparable structure, which also is strikingly like some species of *Parallelopora* and the Mesozoic Milleporellidae, although the closed tubes of the latter are less developed than in *H. circumvestiens.* New studies of the coenosarcal structure in relation to the more or less tubular coenosteum of *Hydractinia* are needed to throw light on skeletal changes observed in Paleozoic Stromatoporidae and their Mesozoic descendants. Without rejecting possible existence of zooidal tubes in the stromatoporoids, it seems preferable to designate the structures on which their presence is inferred, simply as vertical tabulate chambers; it is interesting to note that they are related closely to the astrorhizae.

(3) Stromatoporoids exhibit some peculiarities which are observed commonly among coelenterates but are unknown in the sponges or foraminifers. These include frequent interruptions in growth, latilaminar structure, and the tendency to build thick deposits of the sort characteristically associated with reefs.

EVOLUTION

The problem of distinguishing and tracing evolutionary trends in stromatoporoid skeletal structure is almost untouched, mainly because appropriate means for investigation have been lacking. Study of thin sections is insufficient because it permits observation of structures only at somewhat widely spaced planes. The method of progressive grinding and making peels or photomicrographs of polished surfaces at closely spaced intervals, has not yet been employed, and this alone seems to be available as a source of needed information. Some applications of this method which we have made to study of the skeleton have not been pursued far enough to permit definitive conclusions, but a few essential features may be indicated.

If attention is directed to families which persist into Mesozoic time, one may perceive plainly an evolutionary tendency toward disappearance of transverse lamellae with concurrent development of radial elements; the horizontal trabecular structures gradually are replaced by tabulae and finally these are the only transverse skeletal elements of the colony. The first step of this evolution appears in *Stromatoporella,* in which a few vertical chambers of moderate height traversed by tabulae are seen (Fig. 95,*2*). These structures are slightly better defined in the Syringostromatidae, even though genera of this family retain well-developed lamellae (Fig. 100). The tendency is emphasized in the Stromatoporidae, especially *Parallelopora,* which leads to the Milleporellidae and the Mesozoic Milleporidiidae. In the Middle Devonian (Givetian) *P. paucicanaliculata,* lamellation already approaches disappearance and all chambers between the pillars are intersected by tabulae (Fig. 100); these chambers have been described by many authors as zooidal tubes.

Increasing dominance of vertical elements and replacement of horizontal trabeculae by tabulae are associated with a tendency of the pillars to develop into vertical lamellae, which evolve progressively toward the form of closed tubes. The first stages of this must be sought in study of Paleozoic forms. Evolution shown by *Parallelopora paucicanaliculata* is already such that successive transversely ground surfaces show a practically continuous network and only rarely a few isolated pillars. The tendency is accelerated and more generalized in Mesozoic, even among Actinostromatidae. In *Actinostromaria* (Jur.-Cret.), the lamellation is still well marked but discontinuous in all species. Some forms, such as *A. salevensis* show by grinding a somewhat persistent network indicating a lamellar and tubular evolution of the vertical elements. The lamellae are not replaced by thin tabulae, but they exhibit important changes. Lamellae bend upward in the form of pillars and pillars turn outward into lamellae, so that relations of these two skeletal elements differ wholly from those of their ancestor *Actinostroma.*

Among the Stromatoporidae, these evolutionary tendencies which were already well defined in Paleozoic time, became accentuated in the Mesozoic. They reached their maximum in a descendant of *Parallelopora* called *Stromatoporellina* (Jur.), in which lamellation has disappeared almost entirely, tubulation of the coenosteum is advanced, and all vertical chambers are traversed by tabulae. This genus leads to the Milleporellidae (Jur.-Cret.), in which tubulation is accentuated and lamellae definitely have disappeared. The Milleporidiidae (Jur.-Cret.), derivatives of this family, include the end products of stromatoporoid evolution; they are characterized by perfectly developed tabulate tubes and in the cortical part of colonies larger and more cylindrical tubes seem to represent remains of differentiated astrorhizae. Moreover, skeletal evolution shown by the last-mentioned families is accompanied by a transformation of the astrorhizal system.

CLASSIFICATION

The various systems of classification proposed for the stromatoporoids reflect large uncertainties in assessing taxonomic value of the diverse morphological features of this group, often interpreted without consideration of phylogeny or notion of affinities with living hydrozoans. Particularly unsatisfactory are classifications which have been made on basis of Paleozoic forms alone. The work of chief contributors is noted chronologically.

WORK OF NICHOLSON

The essential starting point and most fundamental work on stromatoporoids consist of studies recorded by NICHOLSON

(1886). Based on penetrating knowledge of the morphology of these fossils, he divided stromatoporoid genera known to him in two main groups: (1) forms which he assigned to the family Actinostromatidae, characterized by absence of zooidal tubes; and (2) forms included in the Stromatoporidae and Labechiidae, provided with zooidal tubes. The first group was judged to be related rather closely to the hydractinians, whereas the second was interpreted to stand near hydrozoans of the "family Milleporidae" (now referred to the order Milleporina—not the family Milleporidiidae, composed of Mesozoic stromatoporoids). Although Nicholson's plan was accepted and developed by later workers such as Dehorne (1920) and Steiner (1932), no good reason is seen for admitting passage from the Stromatoporidae and Labechiidae, characterized by single-type zooids, to the Milleporina, which have dimorphic zooids of coenosarcal type unknown in any stromatoporoid. Moreover, the skeletal structure of the Stromatoporidae and the families derived from it is much more like the partly tubular coenosteum of *Hydractinia* than colonial skeletons of *Millepora*. Summary of Nicholson's classification is given in the following table.

Nicholson's Classification of Stromatoporoids

Hydractinoidea. Forms without zooidal tubes, related to hydractinians; Actinostromatidae, Idiostromatidae.

Milleporoidea. Forms with zooidal tubes, related to milleporids; Stromatoporidae, Labechiidae.

STUDIES SINCE 1900

Chief contributions to understanding of the relationships to one another of the various kinds of discovered stromatoporoids and efforts to produce useful classification of them are contained in publications by Heinrich, Dehorne, Kühn, Tripp, Steiner, Yabe & Sugiyama, and Lecompte. The conclusions of these investigators may be summarized and compared.

All Paleozoic genera were divided by Heinrich (1914) into 2 families, defined on the basis of skeletal microstructure, but Mesozoic stromatoporoids were not considered. Defects of this treatment have been pointed out in the discussion of microstructure. Megascopic features of several Mesozoic genera clearly indicate assignment to the Actinostromatidae and Stromatoporidae, despite the fact that their microstructure differs from Heinrich's diagnosis for these families, which follows.

Heinrich's Classification of Stromatoporoids

Actinostromatidae. Microstructure of skeletal elements homogeneous.

Stromatoporidae. Microstructure of skeletal elements porous or tubular.

Essential features of Nicholson's classification were adopted by Dehorne (1920), but she added the family Burgundiidae, regarded as intermediate between Actinostromatidae and Stromatoporidae. The Burgundiidae seem too closely related to the Clathrodictyidae to permit acceptance of this classification; also Dehorne's Stromatoporidae include Mesozoic forms (erroneously ascribed to *Stromatopora* and *Stromatoporella*) which clearly belong to other families. An outline of her classification follows.

Dehorne's Classification of Stromatoporoids

Actinostromatidae; Actinostromatinae, Clathrodictyinae.

Burgundiidae.

Stromatoporidae; Stromatoporinae, Stromatoporellinae.

A classification of the stromatoporoids was published by Kühn (1927, 1939). Study of their evolutionary trends and phylogenetic relationships led to the conclusion that the Stromatoporidae are the ancestors of modern hydroids by descent through families named Disjectoporidae and Stromatoporinidae. The Disjectoporidae, however, include genera which do not seem to form a link between the Paleozoic Stromatoporidae and Mesozoic genera assigned by Kühn to the Stromatoporinidae; the Disjectoporidae comprise a side branch. Also, because macrostructure of the Stromatoporidae and Stromatoporinidae is the same and because evolution of Mesozoic forms is exactly in line with that of Paleozoic genera, skeletal microstructure is not a valid criterion for separation of these families. Kühn judged that the Actinostromatidae constitute beginning of an independent line which includes the Labechioidea, interpreted as an order. This seems unacceptable because the

Labechioidea have structural characters which are essentially the same as those of Stromatoporoidea. KÜHN's arrangement of divisions is indicated below.

Kühn's Classification of Stromatoporoids

Stromatoporoidea *(order), Cam.-Cret.*
Actinostromatidae, *Cam.-Cret.* *Clathrodictyidae, Cam.-Perm.*
Stromatoporidae, *Ord.-Perm.*—> Disjectoporidae, *Perm.* —>
Stromatoporinidae, *Trias.-Cret.*—> Hydroidea *(order), Rec.*
Labechioidea (*order,* possibly derived from Actinostromatidae), *Ord.-Dev.*
Sphaeractinoidea (*order,* possibly derived from Disjectoporidae), *Jur.-Cret.*

According to TRIPP (1929), all stromatoporoids have characters closely approaching those of athecate hydroids (gymnoblasts), for the same two types of structures are found in both; the pronounced lamellar structure of the hydroid *Podocoryne* is matched by equivalents in the Clathrodictyidae, Stromatoporellidae, and Syringostromatidae, and the predominant continuous vertical skeletal elements of *Hydractinia* are seen also in the Actinostromatidae, Stromatoporidae, Milleporellidae, and Milleporidiidae. In addition, the partly tubular and tabulate skeleton of hydractinians and that of fossil forms can be cited as sign of relationship which needs elucidation. TRIPP made valuable contribution to the study of systematic relationships of the stromatoporoids, but his definition of groups, given below, is no improvement over that of HEINRICH.

Tripp's Classification of Stromatoporoids

GROUP 1. Closed-tissue forms, with bipartite lamellae.
Skeletal fiber massive, as in *Clathrodictyon.*
Skeletal fiber porous, as in *Stromatoporella.*
GROUP 2. Open-tissue forms, with simple lamellae.
Skeletal fiber massive, as in *Actinostroma.*
Skeletal fiber porous, as in *Stromatopora* and *Parallelopora.*

Important studies by STEINER (1932) led to the conclusion that phylogenetic relationships of the stromatoporoids were not determinable, and accordingly she sought to erect a classification based on structural characters and distinguished by lack of any morphological hiatus. Placement and subdivision of the stromatoporoids by STEINER, as outlined below, satisfactorily indicate relationships to living Tubularia on one hand and Hydrocoralla on the other.

Steiner's Classification of Stromatoporoids

Tubularia.
Stromatoporoidea.
Hydractinoidea; Ellipsactinidae, Actinostromatidae.
Unnamed division; Siphostromatidae, Burgundiidae.
Milleporoidea; Stromatoporidae, Milleporellidae.
Hydrocoralla.

Study of Mesozoic stromatoporoids by YABE & SUGIYAMA (1935) are basis for a classification of these forms in three families. As given below, this is incorporated in classification now adopted.

Yabe & Sugiyama's Classification of Stromatoporoids

Stromatoporidae. All skeletal elements generally trabecular, interspaces tabulate; with or without astrorhizae.
Milleporellidae. Vertical skeletal elements dominant over lamellar and tending to form closed tabulate tubes; astrorhizae weak or lacking.
Milleporidiidae. Vertical skeletal elements persistent, lamellar or tending to form closed tubes; tabulate interspaces include large ones of zooidal nature and small ones; horizontal elements lamellar, much reduced; no astrorhizae.

CLASSIFICATION ADOPTED IN TREATISE

Classification published in papers by LECOMPTE (1951, 1952) is adopted in the *Treatise.* As far as seems practicable in the state of present knowledge, this is defined by considerations of phylogeny and attention is given to features of skeletal structure, including microstructure in manner supplementing other characters. Division of the stromatoporoids in 2 main groups, as proposed by NICHOLSON and followed by others, cannot be maintained, for structures presumed to show relationship of these fossils with millepores do not exist. Also, oversimplified division in large groups based on microstructure, as in classifications by HEINRICH and TRIPP, is unnatural.

Analysis of older classifications and restudy of many genera have led to rejection of some proposed family units and different allocations of some genera. Thus, *Burgundia* is interpreted to belong to the Clathrodictyidae and the family Burgundiidae proposed by DEHORNE is not recognized; similarly the Siphostromatidae of Steiner's classification is not accepted because *Siphostroma* seems clearly to belong to the Milleporellidae. *Ellipsactinia* and the Ellipsactiniidae are judged not to belong with the stromatoporoids, and *Sphaeractinia,* which Steiner included in this family, is assigned to the Actinostromatidae. The classification of Yabe & Sugiyama is adopted with only minor alterations for Mesozoic stromatoporoids; this includes recognition of the family Milleporidiidae, based on *Milleporidium,* which Steiner failed to distinguish from genera of the Milleporellidae.

The subcylindrical to dendroid colonies, which some authors (Heinrich, 1914; Kühn, 1939) have excluded from the Stromatoporoidea, call for discussion because of question about their relation to the massive and laminar types of growth. Nicholson (1886) grouped the 3 genera of dendroid form known to him in the family Idiostromatidae; of these, *Idiostroma* is here assigned to the Syringostromatidae, *Stachyodes* to the Stromatoporidae, and *Amphipora* to uncertain classificatory status. This distribution is in agreement with Ripper's (1937) conclusion that all or nearly all dendroid stromatoporoids belong to genera or at least to families composed mainly of massive forms. Dendroid (*) and massive (**) pairs having almost identical skeletal structure include **Actinostromarianina* (U.Jur.) and ***Actinostromaria* (Jur.-Cret.); **Dendrostroma* (Dev.) and ***Stromatoporella* (Sil.-Carb.); **Parksia* (U.Jur.) and ***Stromatopora* (Ord.-Perm.); **Stachyodes* (M.Dev.-U.Dev.) and ***Parallelopora* (Dev.--?Perm.); **Idiostroma* (Sil.-Dev.) and ***Hermatostroma* (Sil.-Dev.). The presence of a single large axial canal in dendroid colonies has served to mislead many authors; actually this is simply an astrorhizal axial canal from which branches extend into lamellar portions of the coenosteum precisely as in massive colonies. The family Aulaceridae includes Ordovician and Silurian genera of dendroid form, characterized by exceptionally large size of the axial canal and its division into bulbous chambers by strongly arched tabulae. Here belong *Beatricea* and seemingly also *Cryptophragmus.* They are classified among the stromatoporoids, even though their structural peculiarities and stratigraphic placement indicate that they are a very early, surprisingly divergent assemblage.

The families recognized in the *Treatise* section on Stromatoporoidea are listed below with record of their known stratigraphic distribution and statement of the number of genera included in each (shown by figures enclosed by parentheses).

Divisions of Stromatoporoidea

Stromatoporoidea *(order)* (67). *Cam.-Cret.*
- Actinostromatidae *(family)* (7). *Cam.-Cret.*
- Clathrodictyidae (6). *Cam.-Jur.*
- Stromatoporellidae (4). *Sil.-Carb., ?Trias.*
- Syringostromatidae (6). *Sil.-Jur.*
- Stromatoporidae (12). *Ord.-Cret.*
- Disjectoporidae (4). *Perm.-U.Trias.*
- Milleporellidae (4). *Jur.-Cret.*
- Milleporidiidae (2). *Jur.-Cret.*
- Labechiidae (10). *Ord.-Cret.*
- Aulaceridae (6). *Ord.-Sil.*
- Uncertain family (1). Sil.-Jur.

PALEOECOLOGY

Paleoecologic relationships of the stromatoporoids have been studied little, perhaps partly because taxonomy of the group is mostly unfamiliar to geologists and partly because these fossils are most abundant in calcareous rocks where observations of their occurrence are difficult.

The environments to which stromatoporoids seem best suited are about the same as those preferred by the corals. Both are adapted to reef building, but the stromatoporoids seem to tolerate a muddy bottom less well than many corals. Massive and globular colonies tend to occur in rather pure limestone, whereas laterally expanded sheet-like colonies commonly are found in impure limestones or in interstratified marly and clayey beds; in such a setting, interpreted to signify less shallow seas, rather rare laminar stromatoporoids are found associated with coralline bioherms in Devonian sections of the Ardennes region. Like the tabulate corals but in much smaller degree, the coenosteum of stromatoporoid colonies exhibits variations, especially in thickness of the skeletal elements, development of lamellae and pillars, and in spreading apart of these. This variability partly is an expression of local environment.

Paleozoic stromatoporoids contributed to

building reefs much more commonly than the Mesozoic forms. The Silurian and Devonian rocks of many regions contain numerous stromatoporoid-bearing bioherms and biostromes, some of which have considerable thickness and lateral extent. Remarkable are such masses in the Silurian of Gotland, attaining thickness of 10 to 20 m. and horizontal dimensions of several hundred meters; these are distributed in crinoidal limestones and marls. Locally, reeflike deposits consist exclusively of stromatoporoids, but generally tabulate corals and compound rugose corals are associated with them; also, algae of *Solenopora* type, especially *Pilotrix,* are intergrown with the coelenterate colonies and serve to bond fragments together (HADDING, 1941). Globular algae such as *Sphaerocodium* and *Spongiostroma* occur in beds below and above the reefs but not in them. The abundance of broken, more or less worn fragments of corroded stromatoporoid colonies and the presence of the algae are cited by HADDING as evidence of very shallow water, but as shown by the fine sediment associated in the reefs, evidently they were not formed close to a more or less rugged coast but in relatively quiet offshore waters, probably between 5 and 50 m. deep.

In Devonian strata of the Ardennes, bioherms and biostromes composed mainly or exclusively of stromatoporoids occur as lenticular bodies interfingering with surrounding calcareous sediment. Associated tetracorals, tabulates, and algae *(Sphaerocodium, Girvanella)* are quantitatively less important but individually of large size. Conversely, in reefs composed mainly of corals, invariably small, remains of stromatoporoids generally are absent, but if present, they are very thin laminar growths. Massive stromatoporoids of Devonian time were able to grow and build reefs in turbulent waters; corals living in this environment formed large colonies but below the zone of strong wave action their skeletons were small and flattened. Analysis of geologic sections containing reefs built by stromatoporoids, which brings to notice many disconformities, common occurrence of ripple marks, and local deposits of broken, worn reef organisms, indicates shallow-water environment.

Mesozoic stromatoporoids have very little importance as reef builders, even though they are associated locally with reef-formers such as the hexacorals, rudistid clams, and nerineids, and they occur commonly in oölitic limestones which denote a neritic, shallow-water environment. There are few algae, unless the Chaetetidae found in many places belong here at least partly. Chiefly important from paleoecological viewpoints seems to be the association of stromatoporoids with hexacorals.

SYSTEMATIC DESCRIPTIONS

Order STROMATOPOROIDEA Nicholson & Murie, 1878

Hydrozoans distinguished by a calcareous coenosteum basically of trabecular nature, generally with distinct transverse elements (lamellae) and vertical or radial elements (pillars) or with varied sorts of reticulate structure evolving lamellar elements that tended to become tubular; exclusively colonial, massive, sheetlike, or dendroid in growth. A characteristic structure possessed by all but a few genera is termed astrorhiza; it consists typically of a central canal directed normal to the surface and of laterally diverging, outwardly branching furrows on the surface or canals below the surface. Surface with or without tubercles (mamelons). *Cam.-Cret.*

Family ACTINOSTROMATIDAE Nicholson, 1886

[*nom. correct.* LECOMPTE, herein (*pro* Actinostromidae NICH., 1886)]

Colonies massive, flattened, or rarely dendroid. Coenosteum composed of well-differentiated trabecular skeletal elements; lamellae consisting of a hexactinellid network formed by processes joined at regular or irregular levels to more or less continuous vertical pillars. Astrorhizae narrow. Microstructure of skeletal elements massive or fibrous. *Cam.-Cret.*

Actinostroma NICHOLSON, 1886 [**A. clathratum*] [=*?Actinostromella* BOEHNKE, 1915; *Gerronostroma* YAVORSKY, 1931]. Skeletal elements essentially perpendicular to one another; lamellar network commonly formed by complex meshes connected at regular or variable levels to pillars. Skeletal

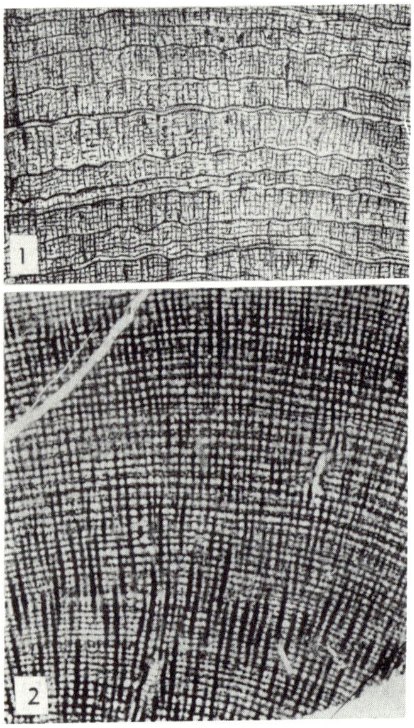

FIG. 103. Actinostromatidae. *1, Actinostroma astroites,* M.Sil., ×10. *2, A. devonense,* U.Dev., ×3.

fibers compact (20). *Cam.-L.Carb.,* cosmop. (This genus generally has been distinguished from *Actinostromaria* by its lack of superposed astrorhizae, but this character is not reliable, since they do occur in some species of *Actinostroma.* Mesozoic species assigned to *Actinostroma* actually belong to *Actinostromaria,* as shown by their fibrous skeletal structure.)——FIGS. 92,*2;* 93,*1;* 103,2. *A. devonense* LECOMPTE, U.Dev. (L.Frasn.), Belg.; long., tang., long. secs., ×3 (49).——FIGS. 86,*3;* 91,*1.* **A. clathratum,* M.Dev., Eng.; 86,*3,* exterior of base, ×0.5; 91,*1,* long. sec., ×12 (50*,50).——FIG. 103,*1. A. astroites* ROSEN, M.Sil. (Wenlock.), Eng.; long. sec., ×10 (50*).——FIG. 94,*1. A. intertextum* NICH., M.Sil., Eng.; long. sec., ×10 (52).

Actinostromaria CHALMAS *in* DEHORNE, 1920 [**A. stellata*]. Coenosteum like that of *Actinostroma* but laminae very discontinuous and more delicate than pillars, which also may be interrupted; the lamellae constitute a simple, regular network, generally with quadrangular but also pentagonal meshes, all in the same plane and many opening on one or more sides so as to produce a meandriform pattern. Pillars commonly are formed by upward turning of lamellar processes. Astrorhizae highly branched, disposed in vertical systems formed by several canals arranged in bundles. Skeletal elements characterized by prominent dark axis surrounded by perpendicular crystalline fibers (7). *Jur.-Cret.,* Eu.-N.Am.-Afr.——FIGS. 95,*1;* 104,*3.* **A. stellata,* Cret. (Cenom.), Fr.; 95,*1,* long. sec., ×12; 104,*3a,b,* long and tang. secs. ×12 (49n all);——FIG. 104,*1. A. salevensis* DEHORNE, Cret., Fr.; tang sec., ×12 (49n).

Actinostromarianina LECOMPTE, 1952 [**Stromatopora milleporoides* DEHORNE, 1920]. A dendroid form with massive base and cylindrical branches: coenosteum as in *Actinostromaria,* pillars bending sharply from axis, occupied by a narrow astrorhizal canal, to meet surface perpendicularly. Astrorhizae feebly developed (18). *U.Jur.,* Eu.——FIG. 104,*4.* **A. milleporoides* (DEHORNE), U.Jur., Rumania; *4a,b,* long. secs. ×3, ×12 (49n).

Atelodictyon LECOMPTE, 1951 [**A. fallax*]. Coenosteum laminar or globular; pillars typically superposed, with appearance of continuity, but restricted in some to interlamellar spaces; lamellae in a subhexactinellid network, or showing in tangential section cellular to chainlike structure. Astrorhizae in vertical systems. Fibers compact. *M.Dev.,* Belg. (Superficially, coenosteum structure resembles that of forms commonly assigned to *Clathrodictyon,* which have pillars not subordinate to lamellae, however.)——FIGS. 93,*2;* 104,2. **A. fallax,* M. Dev.(Couvin.), Belg.; 93,*2,* tang. sec., ×12; 104, *2,* long. sec., ×12 (49*).

Pseudolabechia YABE & SIGUYAMA, 1930 [**P. granulata*]. Colonies laminar. Coenosteum like *Actinostroma,* but distinguished by horizontal lamellar processes not in regular superposed layers and in having vertical columns formed by considerably thickened, upward converging pillars, and marked on the upper surface by large tubercles. Astrorhizae lacking (46). *Ord.-Sil.,* Eu.-Asia.

?**Circopora** WAAGEN & WENTZEL, 1887 [**C. faveolata*], *Perm., Jur.,* Eu.-Asia.

?**Sphaeractinia** STEINMANN, 1878 [**S. diceratina*]. *Jur.-Cret.,* Eu. Imperfectly known, possibly synonym of *Actinostromaria.*

Family CLATHRODICTYIDAE Kühn, 1939

[*nom. correct.* LECOMPTE, herein (*pro* Clathrodictyonidae KÜHN, 1939)]

Coenosteum essentially composed of discontinuous pillars and lamellae intimately associated in manner forming subvesicular tissue, or in varying degrees differentiated. Astrorhizae present. Coenosteal elements compact or microcrystalline. *Cam.-Jur.*

Clathrodictyon NICHOLSON & MURIE, 1878 [**C. vesiculosum*] [=*Stictostroma* PARKS, 1936 (*partim*)]. Typically laminar but may be globular or massive, with skeletal elements poorly differentiated to distinct; pillars restricted to interlamellar spaces, but closely associated with lamellae, which are the predominant structures; interlamellary septa may be abundant. Astrorhizae present or absent (22). *Cam.-Dev.*, cosmop., *Perm.*, Japan. As currently defined, this genus contains groups showing 2 sorts of structures: forms like the type species, with vesicular or subvesicular tissue lacking clear differentiation of pillars and lamellae (Fig. 97); and forms having well-defined lamellae and interlamellar pillars which may be bifurcate or hollow cone-shaped (Fig. 105,*2*). Some authors hold that one form merges with the other through progressive straightening of lamellae and separation of pillars, but the second group possibly is generically distinct.——Fig. 97. **C. vesiculosum*, L.Sil.(Clinton.), Ohio; long. sec., ×10 (50).——Fig. 105,*3*. *C. striatellum* Orb., M.Sil.(Wenlock), Eng.; long. sec., ×24 (50).——Fig. 105,*2*. *C. amygdaloides* Lecompte, M.Dev.(Couvin.), Belg.; long. sec. ×12 (49*). Also, Figs. 86,*2*; 94,*2*.

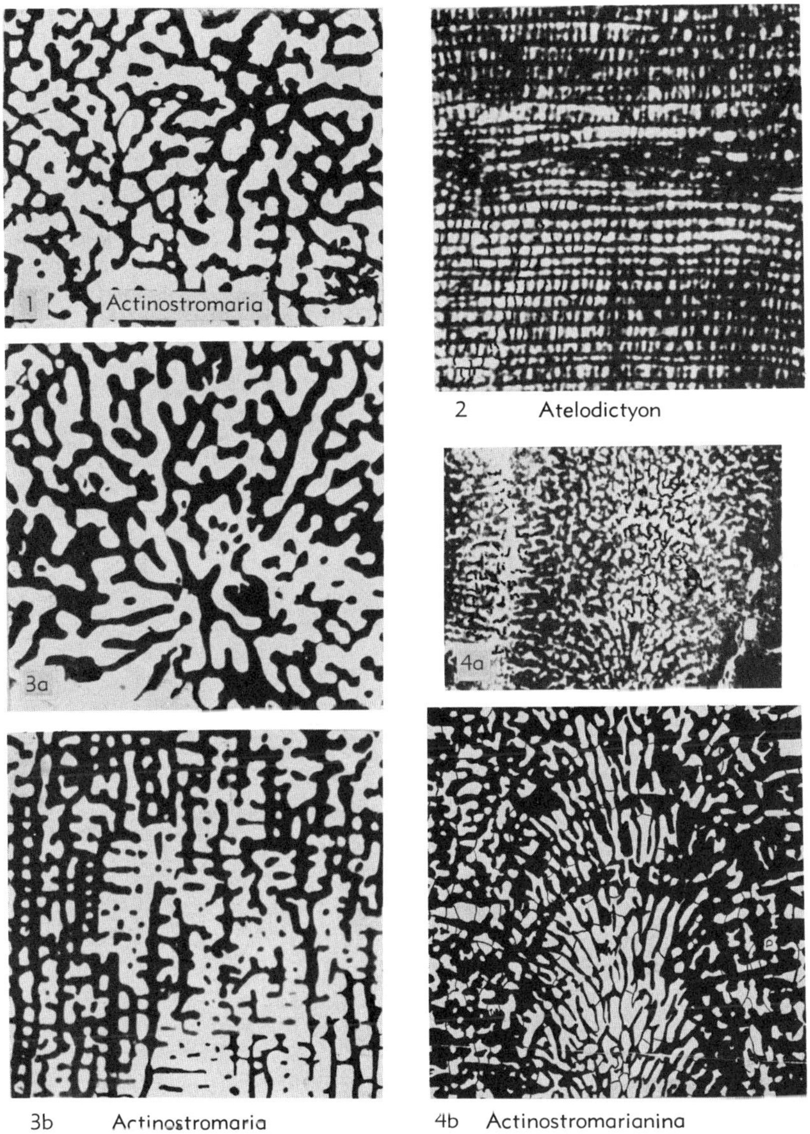

Fig. 104. Actinostromatidae. *1*, *A. salevensis*, Cret., ×12. *2*, *A. fallax*, M.Dev., ×12. *3a,b*, *A. stellata*, Cret., ×12. *4a,b*, *A. milleporoides*, U.Jur., ×3, ×12.

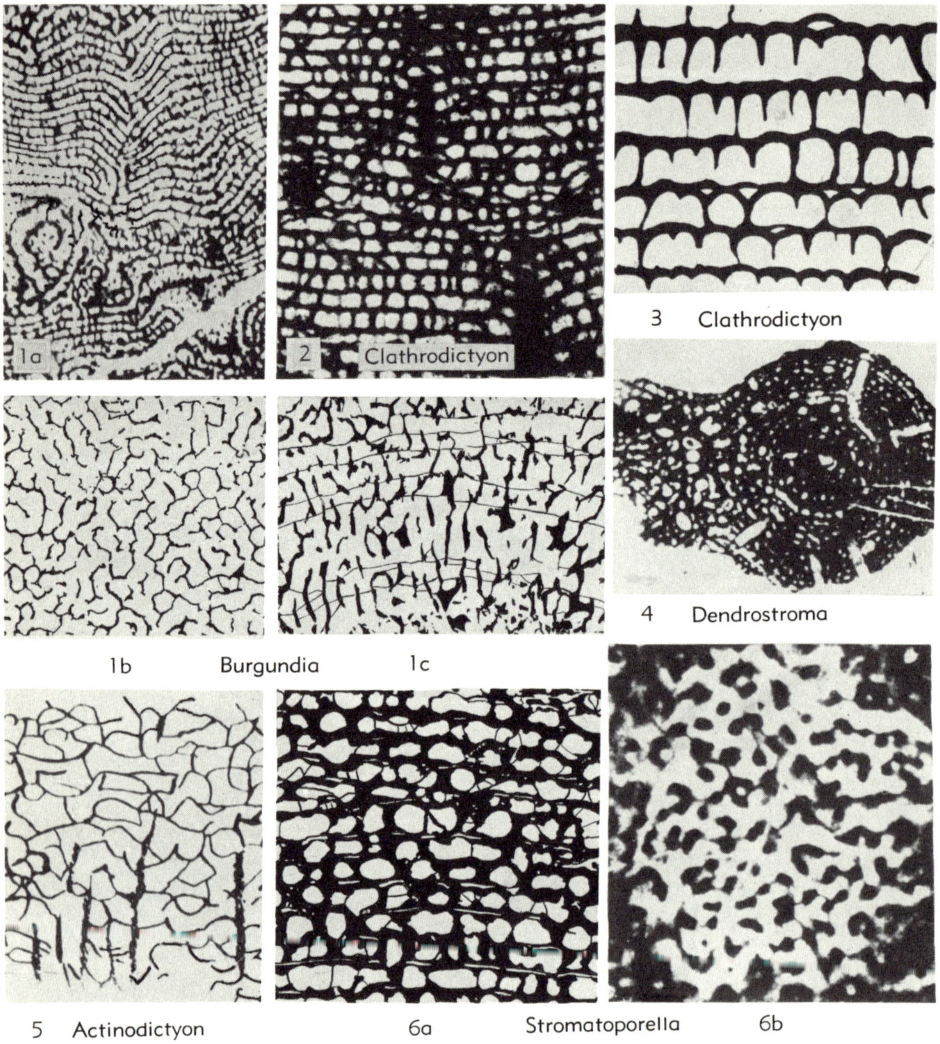

FIG. 105. Clathrodictyidae, Stromatoporellidae. *1a,b, B. trinorchii,* U.Jur., ×3, ×12. *2, C. amygdaloides,* M.Dev., ×12. *3, C. striatellum,* M.Sil., ×24. *4, D. oculatum,* Dev., ×3. *5, A. canadense,* M.Sil., ×10. *6a,b, S. granulata,* M.Dev., ×12.

Actinodictyon PARKS, 1909 [**A. canadense;* SD BASSLER, 1915]. Coenosteum composed of subvesicular tissue traversed by evenly spaced large spongy pillars (24). *M.Sil.-U.Dev.,* Eu.-N.Am.——FIG. 105,*5.* **A. canadense,* M.Sil., Hudson Bay; long. sec. ×10 (51).

Burgundia CHALMAS *in* DEHORNE, 1915 [**B. trinorchii*] [=?*Circoporella* HAYASAKA, 1917]. Growth form laminar, spheroidal, domed, or irregular, with surface marked by mamelons; predominant are well-defined continuous lamellae, more or less wavy and separated from one another in different specimens or parts of the same colony, or partly lenticular; pillars mostly short and straight, bifurcate at top or expanded at one or both extremities, restricted to a single interlamellar space; some pillars long and crooked, unbranched or branched to anastomosing, not restricted to a single interlamellar space; these two types may alternate zonally. Astrorhizae large, in vertical systems with branches abruptly terminated in interlamellar spaces. Lamellae in tangential sections show well-marked reticulate or meandriform structures. Skeletal elements compact, microcrystalline, or granular; a dark or clear filiform axis may be recognized in lamellae and rarely in pillars (6). *Jur.,* Eu.-Afr.-Asia. This genus is closely allied to forms of *Clathrodictyon* having well-differentiated skeletal elements; it is distinguished by partly disordered, branched or anastomosed pillars extending through 2 or more lamellae. So-called zooidal tubes noted by some authors, which rarely occur in sections, seemingly are parts of astrorhizae.——FIG. 105,*1*. **B. trinorchii* U.Jur., Fr.; *1a,b,* long. and tang. secs., ×3; *1c,* long. sec., ×12 (49n).

Stylodictyon NICHOLSON & MURIE, 1878 [**Syringopora columnaris* NICH., 1875]. Coenosteum very dense, formed of well-differentiated lamellae and interlamellar pillars traversed by nearly equidistant stout columns between which curved laminae occur in alternate open and close-spaced groups (22). *Dev.,* N.Am.-Russ.

?**Clavidictyon** SUGIYAMA, 1939 [**C. columnare*]. Colonies columnar, dendroid, or bundled; surface granulose; axial portion composed of reticulate tissue but peripheral part having well-defined lamellae joined by interlamellar pillars. Astrorhizae lacking (37). *M.Sil.,* Japan.

?**Pseudostylodictyon** OZAKI, 1938 [**P. poshanense*]. Columns much thicker and more widely spaced than in *Stylodictyon* and formed by bulging of lamellae only. Astrorhizae seem to be lacking (23). *Ord.,* China.

Family STROMATOPORELLIDAE Lecompte, 1951

Skeletal elements thick, well differentiated, comprising continuous, regularly spaced lamellae and discontinuous pillars confined to interlamellar spaces but superposed in some forms. Straight or tortuous vertical chambers bearing tabulae are poorly developed. Astrorhizae strong. Skeletal elements have porous microstructure. *Sil.-Carb., ?Trias.*

Stromatoporella NICHOLSON, 1886 [**Stromatopora granulata* NICH., 1873] [=?*Clathrocoilona* YAVORSKY, 1931; *Stictostroma* PARKS, 1936 *(partim)*]. Generally laminar, free, with basal epitheca, or encrusting; skeletal elements fairly well individualized and nearly equal in importance; pillars typically limited to interlamellar spaces, with tendency to superposition in some; vertical tabulated chambers few and short. Astrorhizal systems well developed, commonly associated with mamelons. Skeletal elements cellular (20). *Sil.-Carb.,* cosmop. None of the Mesozoic forms described under this name seemingly belong to the genus.——FIG. 105, *6*. **S. granulata* (NICH.), M.Dev.(Couvin.), Belg.; *6a,b,* long. and tang. secs., ×12 (49); also FIG. 102,*4*. Other species, FIGS. 87, 90, 95,*2*.

Dendrostroma LECOMPTE, 1952 [**Idiostroma oculatum* NICHOLSON, 1886]. Dendroid colonies with coenosteum like that of *Stromatoporella,* with thick, well-differentiated pillars and lamellae, the pillars being limited to an interlamellar space. Large axial astrorhizal canal and its lateral branches intersected by partitions (17). *Dev.,* Eu.——FIG. 105,*4*. *D. oculatum* (NICH.), Dev., Ger.; long. sec., ×3 (49n).

Synthetostroma LECOMPTE, 1951 [**S. actinostromoides*]. Free or incrusting colonies. Much-folded lamellae of complex structure, with parallel or intertwined fibrils, and pillars limited to interlamellar spaces but superposed. Astrorhizae very strongly developed, intersected by numerous partitions (17). *M.Dev.,* Belg.——FIG. 106,*3*. **S. actinostromoides,* M.Dev.(Givetian), Belg.; long. sec., ×12 (49*).

?**Lithopora** TORNQUIST, 1900 [**L. koeneni*]. Structure like *Stromatoporella* but the sparse vertical chambers seemingly lack tabulae. Microstructure unknown. *M.Trias.,* Italy.

Family SYRINGOSTROMATIDAE Lecompte, 1951

[*nom. correct.* LECOMPTE, herein (*pro* Syringostromidae LECOMPTE, 1951)]

Well-differentiated continuous pillars and lamellae, thin in early forms but gradually thicker in later ones; vertical tabulated chambers more or less numerous and well developed. Microstructure cellular or reticulate in Paleozoic genera, uniformly in early forms but later tending to marginal localization; fibrous in Mesozoic forms. Skeletal elements with dark axis. *Sil.-Jur.*

Syringostroma NICHOLSON, 1875 [**S. densa;* SD NICH., 1886]. Free or incrusting colonies with coenosteum built of more or less continuous, strong pillars of straight or slightly sinuous form, and reticulate, generally thin, continuous lamellae which show axial lines. Vertical tabulated chambers slightly to well developed. Astrorhizae generally well developed. Skeletal fibers alveolar (19). *Sil.-Dev.,* Eu.-Asia-Austral.-N.Am.——FIG. 107,*2*. *S. centrotum* GIRTY, Sil., N.Y.; long. sec., ×10 (51*).——FIG. 106,*2*. *S. baccatum* LECOMPTE, M. Dev.(Givet.), Belg.; long. sec., ×12 (49*).

Hermatostroma NICHOLSON, 1886 [**H. schlüteri*]. Skeletal elements thick, subequal and mutually perpendicular; interlamellar chambers well delimited, typically rounded, and commonly divided by partitions; vertical tabulated chambers more or less numerous. Astrorhizae generally prominent. Skeletal elements with marginal cellular or canaliculo-cellular structure (20). *Sil.-Dev.,* cosmop.——107,*4*. **H. schlüteri.* M.Dev., Ger.; tang. sec., ×20 (49); also FIG. 96.——FIG. 99. *H. perseptatum* LECOMPTE, U.Dev.(L.Frasn.), Belg.; long. sec., ×3 (49*).

Idiostroma WINCHELL, 1867 [**L. caespitosum;* SD BASSLER, 1915]. Dendroid or fasciculate colonies; pillars superposed and lamellae distinct; large astrorhizal axial canal and lateral branches intersected by partitions; vertical chambers tabulated. Skeletal elements with cellular marginal structure (45). *Sil.-Dev.,,* N.Am.-Eu.-Asia-Austral.——Fig. 107,*3*. *I. roemeri* Nich., Dev., Ger.; transv. sec., ×3 (49*)

Syringostromina Lecompte, 1952 [**Stromatopora choffati* Dehorne, 1920]. Moderately large, irregular subconcretionary colony with coenosteum as in *Syringostroma;* thin, closely spaced, continuous, parallel pillars are intersected by filiform, widely separated lamellae, between which the vertical chambers are divided by numerous, very thin tabulae; in tangential section, the lamellar system shows an extremely fine, complete or meandriform network. Astrorhizae very strongly developed, superposed, with scanty, irregular branches. Skeletal elements longitudinally striate or with trellis structure (18). *U.Jur.,* Eu.——Fig. 106,*4*. **S. choffati* (Dehorne), U.Jur.(Lusit.), Port.; tang. and long. secs., ×12 (49n).

Trupetostroma Parks, 1936 [**T. warreni*]. Compact lamellae pierced by circular pores corresponding to chambers and marking evolutionary tendency toward thickening; strong pillars superposed; vertical tabulated chambers very well developed, connected with astrorhizae. Cellules of skeletal elements tend toward marginal localization.

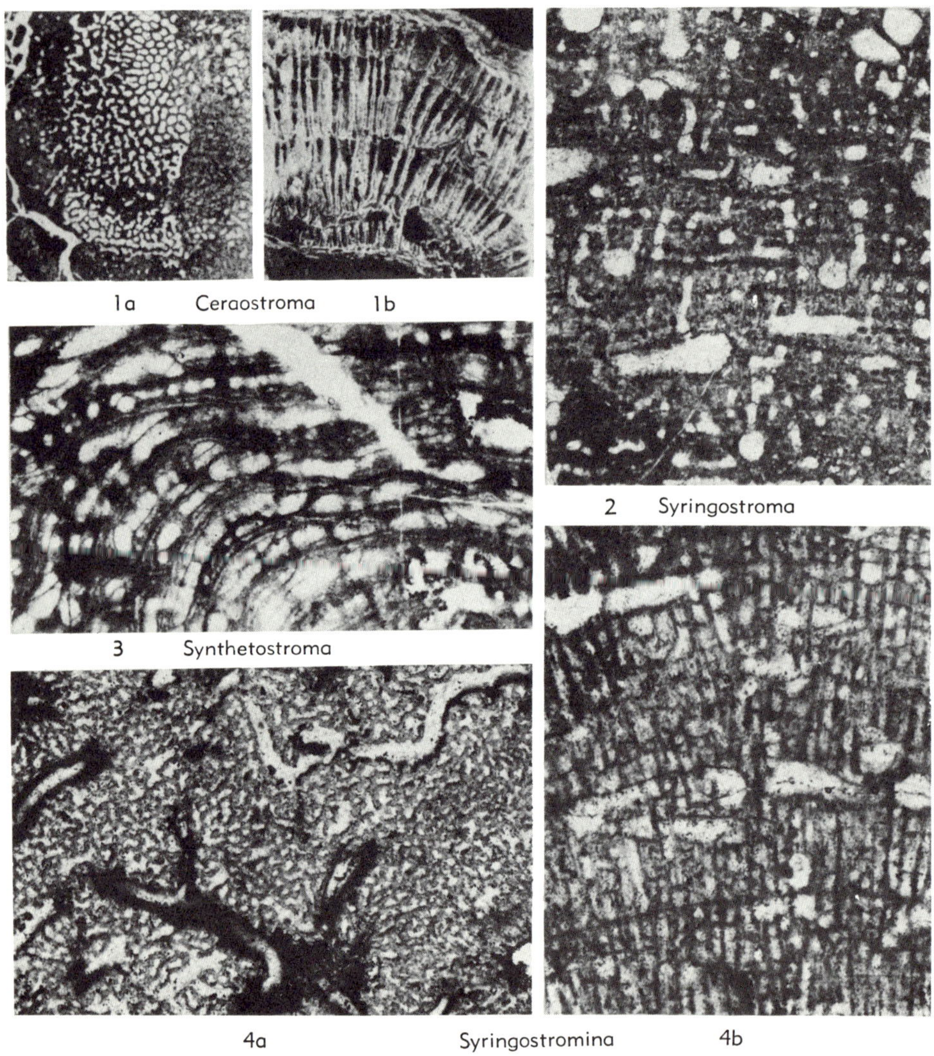

Fig. 106. Stromatoporellidae, Syringostromatidae. *1a,b, G. steinmanni*, U.Jur., ×2.8. *2, S. baccatum,* M.Dev., ×12. *3, S. actinostromoides,* M.Dev., ×12. *4a,b, S. choffati*, U.Jur., ×12.

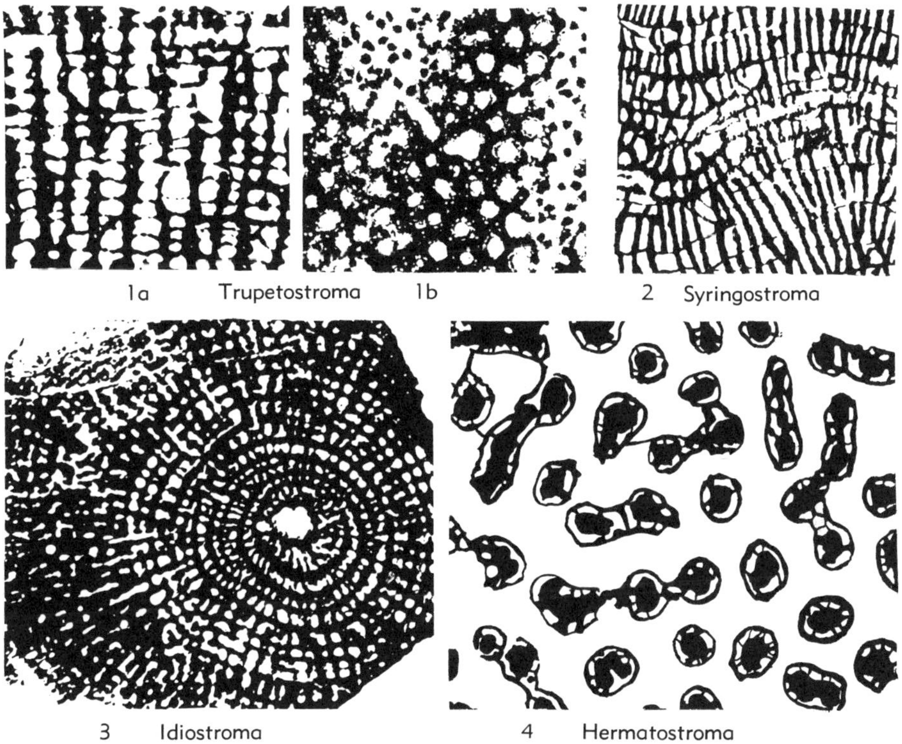

FIG. 107. Syringostromatidae. *1a,b*, *T. warreni*, Dev., ×10. *2*, *S. centrotum*, Sil., ×10. *3*, *I. roemeri*, Dev., ×3. *4*, *H. schlüteri*, M.Dev., ×20.

Astrorhizae prominent (27). *Dev.*, N.Am.-Eu.——FIG. 107,*1*. **T. warreni*, Dev., NW.Can.; *1a,b*, long. and tang. secs., ×10 (51*).——FIG. 98,*1*. *T. tenuilamellatum* LECOMPTE, U.Dev.(L.Frasn.), Belg.; *1a,b*, tang. and long. secs., ×3, ×12 (49*); also FIG. 102,*2*.——FIG. 98,*3*. *T. ruedemanni* LECOMPTE, U.Dev.(L.Frasn.), Belg.; long. sec., ×12 (49*).

?**Ceraostroma** KÜHN, 1926 [**C. steinmanni*]. Colony incrusting, skeleton probably chitinous originally and secondarily calcified. Continuous pillars and wide-spaced lamellae enclosing extended vertical chambers, which bear thin partitions. Astrorhizae well developed (13). *U.Jur.*, Eu.——FIG. 106,*1*. **C. steinmanni*, U.Jur.(Tithon.), Aus.; *1a,b*, long. and tang. secs., ×2.8 (48*).

Family STROMATOPORIDAE Nicholson, 1886

[=Coenostromidae WAAGEN & WENTZEL, 1887]

Colonial form massive, laminar, incrusting, or dendroid. Coenosteum has more or less open or close reticulate structure, lamellation tending to be lost in the reticulation of tissue or to be supplanted by pillars which in certain forms evolve in vertical lamellae. Vertical tabulated chambers numerous. Astrorhizae more or less well developed. Microstructure cellular or fibrous. *Ord.-Cret.*

Stromatopora GOLDF., 1826 [**S. concentrica*] [=*Coenostroma* WINCHELL, 1867; ?*Prisciturben* KUNTH, 1870; *Pachystroma* NICH. & MURIE, 1878; *Rosenia* WAAGEN & WENTZEL, 1887]. Growth massive or laminar, and in some, latilaminar. Coenosteum may be differentiated in strong vertical pillars and stout simple (not bifid) lamellae, but it shows a general tendency toward a closely reticulate structure with progressive loss of lamellation. Interpillar tabulated chambers present. Skeletal elements cellular. Astrorhizae more or less prominent (20). *Ord.-Perm.*, cosmop.——FIGS. 91,*2*; 108,*2*. **S. concentrica*, U.Dev.(L.Frasn.), Belg.; long. secs., ×3, ×12 (49*,49); also FIG. 92, *1*.——FIG. 109,*1*. *S. pachytexta* LECOMPTE, M.Dev., Ger.; long. sec., ×3 (49*). Other species, FIGS. 88, 89, 102,*3*.

Dehornella LECOMPTE, 1952 [**Stromatoporella hydractinoides* DEHORNE, 1918]. Small laminate incrusting form, entire surface covered by conical mamelons with opening of axial canal of stellate astrorhizae at summits; coenosteum beneath mamelons composed of continuous pillars, tortuous and very irregular in thickness, joined by uneven cross bars; between mamelons, continuous lamellae are fairly distinct and very irregular pillars are discontinuous and commonly restricted to an interlamellar space. Pillars cut transversely reveal a pig-

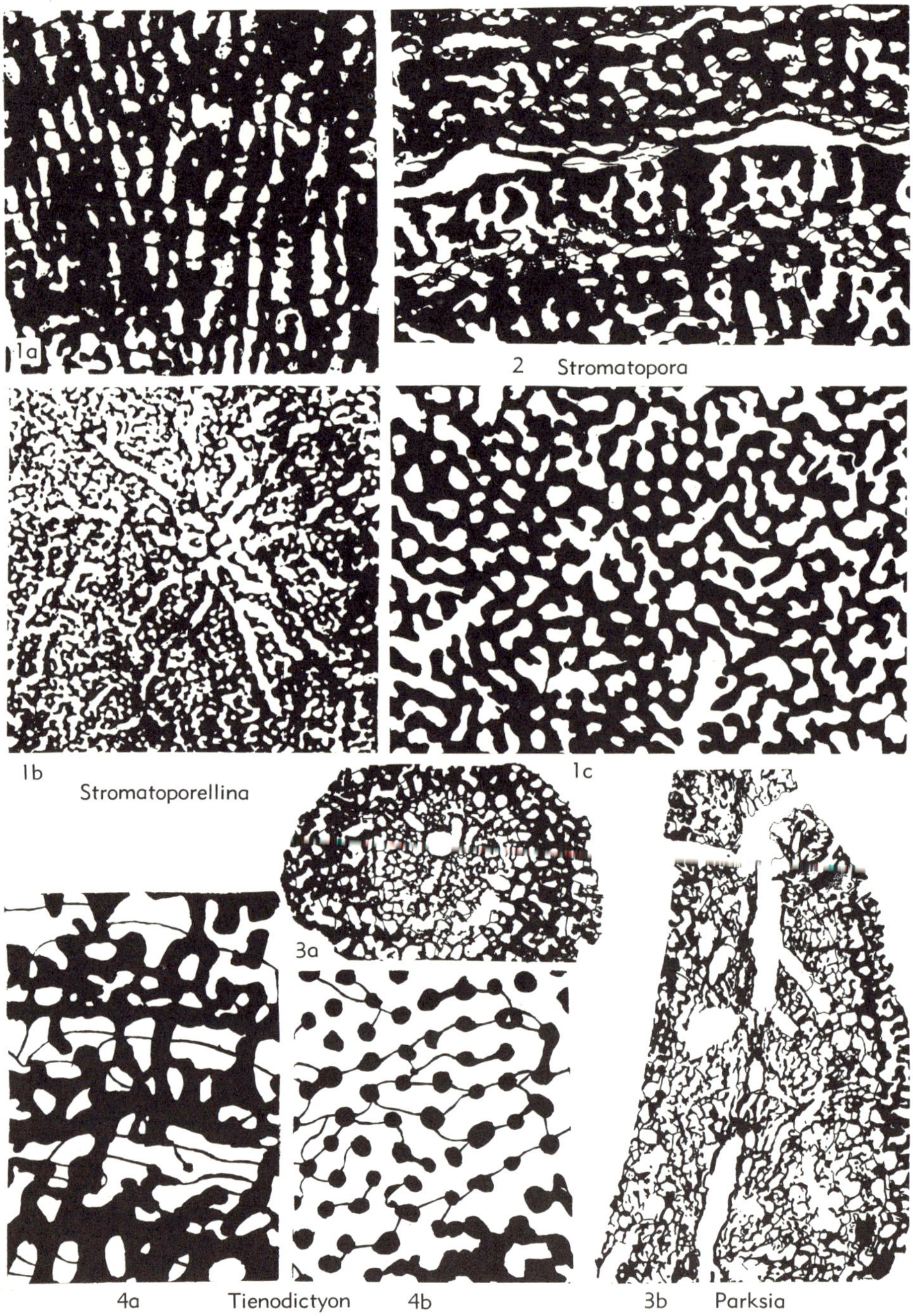

FIG. 108. Stromatoporidae. *1a-c, S. haugi,* U.Cret.; *1a,c,* ×12; *1b,* ×3. *2, S. concentrica,* M.Dev., ×12. *3a,b, P. douvillei,* U.Jur., ×6. *4a,b, T. zonatum,* Dev., ×12.

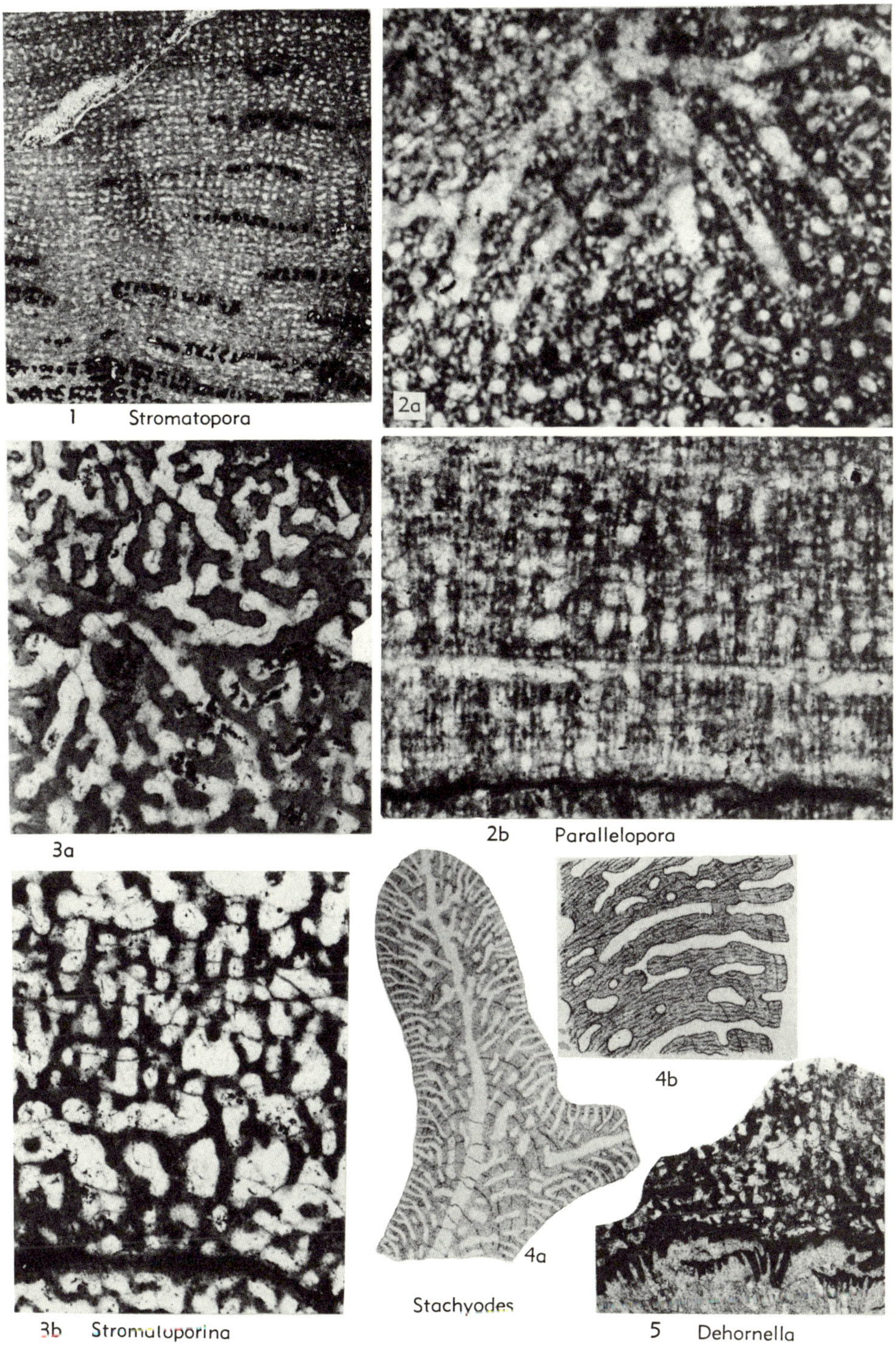

FIG. 109. Stromatoporidae. *1, S. pachytexta,* M.Dev., ×3. *2a,b, P. ostiolata,* M.Dev., ×20. *3a,b, S. choffati,* U.Jur., ×12. *4a,b, S. verticillata,* Dev., ×2, ×12. *5, D. hydractinoides,* U.Jur., ×6.

FIG. 110. Stromatoporidae. *Stachyodes radiata,* U. Dev., ×3.

mented core which undergoes extinction in polarized light, possibly indicating substitution of chitin (18). *U.Jur.,* SW.Eu.——FIG. 109,*5.* **D. hydractinoides,* U.Jur.(Lusitan.), Port.; long. sec., ×6 (49n).

Parallelopora BARGATZKY, 1881 [**P. ostiolata;* SD NICH., 1891]. Massive free colonies with loosely reticulate coenosteum, pillars predominating and lamellation tending to disappear; the stout, more or less continuous pillars are straight or tortuous, and they show tendency to develop in vertical lamellae evolving in closed tubes. All interpillar spaces bearing thin tabulae. Skeletal elements essentially cellular or in some reticulate. Astrorhizae strongly developed (1). *Dev.-?Perm.,* cosmop.——FIG. 109,*2.* **P. ostiolata,* M.Dev., Ger.; *2a,b,* tang. and long. secs., ×20 (49*); also FIG. 102,*1.*——FIG. 100. *P. paucicanaliculata* LECOMPTE, M.Dev. (Givet.), Belg.; long. sec., ×12 (49*).

Parksia LECOMPTE, 1952 [**Stromatopora douvillei* DEHORNE, 1918]. Delicate dendroid colonies with entirely reticulate coenosteum of *Stromatopora* type. Axial astrorhizal canals very large, with coarse lateral branches dipping obliquely and conformably to lamination. Microstructure fibrous, without pigmented dark axes; the fibers, which are longitudinally oriented at axis, bend toward periphery so as to produce wavy extinction under crossed nicols (18). *U.Jur.,* N.Afr.——FIG. 108,*3.* **P. douvillei* (DEHORNE), U.Jur., S.Tunisia; *3a,b,* transv. and long. secs., ×6 (49n).

Plassenia YABE & SUGIYAMA, 1931 [**P. alpina*]. Massive colonies with wavy, perforate lamellae and interlamellar spaces connected by lamellar or rarely columnar vertical partitions, the lamellar ones being irregularly folded and laterally variously joined to one another dividing the interlamellar spaces into numerous minute chambers elongated in various directions; extremely thin tabulae occur uncommonly. Vertical astrorhizae bear lateral branches at irregular intervals. Tangential sections show an irregular network with subcircular meshes, but more commonly elongated or curved in various directions. *U.Jur.,* Eu.

Stachyodes BARGATZKY, 1881 [**Stromatopora (Caunopora) verticillata* M'COY, 1851] [=*Sphaerostroma* GÜRICH, 1896]. Dendroid, fasciculate, or submassive colonies with compact or reticulate coenosteum more or less differentiated in distinct pillars especially in peripheral parts; lamellae filiform, intersecting all tissue. Large axial astrorhizal canals crossed by many straight, oblique, bent, or vesiculose tabulae; adventitious canals distributed, conformably to lamellation. Skeletal elements commonly showing fine longitudinal striae but probably essentially microreticulate (1). *M.Dev.-U.Dev.,* Eu.-Austral.——FIG. 109,*4.* **S. verticillata* (M'COY), Dev., Ger.; *4a,b,* long. sec., ×2, ×12 (50*). Also FIG. 86,*1.* ——FIG. 110. *S. radiata* LECOMPTE, U.Dev.(L.Frasn.), Belg.; long. sec., ×3 (49*).

Stromatoporellata BAKALOV, 1910 [**S. mammilaris*]. *U.Trias.,* Bulgaria.

Stromatoporellina KÜHN, 1927 [**Stromatoporella haugi* DEHORNE, 1917] [=*Epistromatopora, Parastromatopora* YABE & SUGIYAMA, 1935]. Massive mamelonate or branched colonies, coenosteum of *Parallelopora* type, with continuous subparallel pillars, straight or slightly bent, rarely trabecular, mostly lamellate and showing strong tendency to produce adjoined tubes, partly meandriform, of a single type crossed by thin regular tabulae. Astrorhizae well developed, with many tortuous branches, distally ramified; few horizontal canals appear in longitudinal sections. Skeletal elements lacking dark axis are composed of microcrystalline fibers in fine poorly defined bundles (14). *Jur.-Cret.,* Eu.-Afr.-Asia. (Since this form is allied to *Parallelopora,* rather than *Stromatoporella,* as presumed by KÜHN, the generic name is very ill chosen. *Stromatoporellina* embraces forms defined by YABE & SUGIYAMA [1935] as *Epistromatopora,* with granules concentrated in axes of skeletal elements, and *Parastromatopora,* with oblique fibers.)——FIG. 108,*1.* **S. haugi* (DEHORNE), U.Cret., Fr.; *1a,c,* long. and tang. secs., ×12; *1b,* tang. sec., ×3 (49n).

Stromatoporina KÜHN, 1927 [**Stromatopora tornquisti* DENINGER, 1906]. Coenosteum strongly latilaminar with structure like *Stromatopora*, lamellation remaining evident in the reticulation of tissue; pillars irregular, showing no tendency to form tubes; all intercolumnar spaces bearing thin tabulae. Astrorhizae present. Microstructure fibrous, fasciculate, or radial, without dark axis (14). *Trias.-Cret.*, Eu.-Asia. The type species not having been studied, description is based on DEHORNE's (1920) species: *Stromatopora choffati, S. romanica,* and *S. arrabidensis.*——FIG. 109,*3*. *S. choffati* (DEHORNE), U.Jur.(Lusitan.), Port.; *3a,b,* tang. and long. secs., ×12 (49n).

Stromatorhiza BAKALOV, 1906 [**Thamnaraea? granulosa* KOBY, 1888]. Coenosteum entirely reticulate, as in typical *Stromatopora,* but lacking tabular partitions and with homogeneous skeletal elements bearing a strong dark axis. *U.Jur.*, Eu.

Tienodictyon YABE & SUGIYAMA, 1941 [**T. zonatum*]. Colonies massive, with zonal structure and finely granulose surface; strong lamellae of fine trabecular structure separate the zones, which in lower parts consist of reticulate tissue with meshes partly intersected by tabulae, and in upper parts mainly comprise subvertical pillars joined to lamellae by their expanded ends and connected by thin partitions in lamellar disposition. Astrorhizae readily visible in longitudinal sections. *Dev.*, China.——FIG. 108,*4*. **T. zonatum; 4a,b,* long. and tang. secs., ×12 (55).

?**Stromatostroma** BAKALOV, 1910 [**S. triasicum*]. *U.Trias.*, Bulgaria.

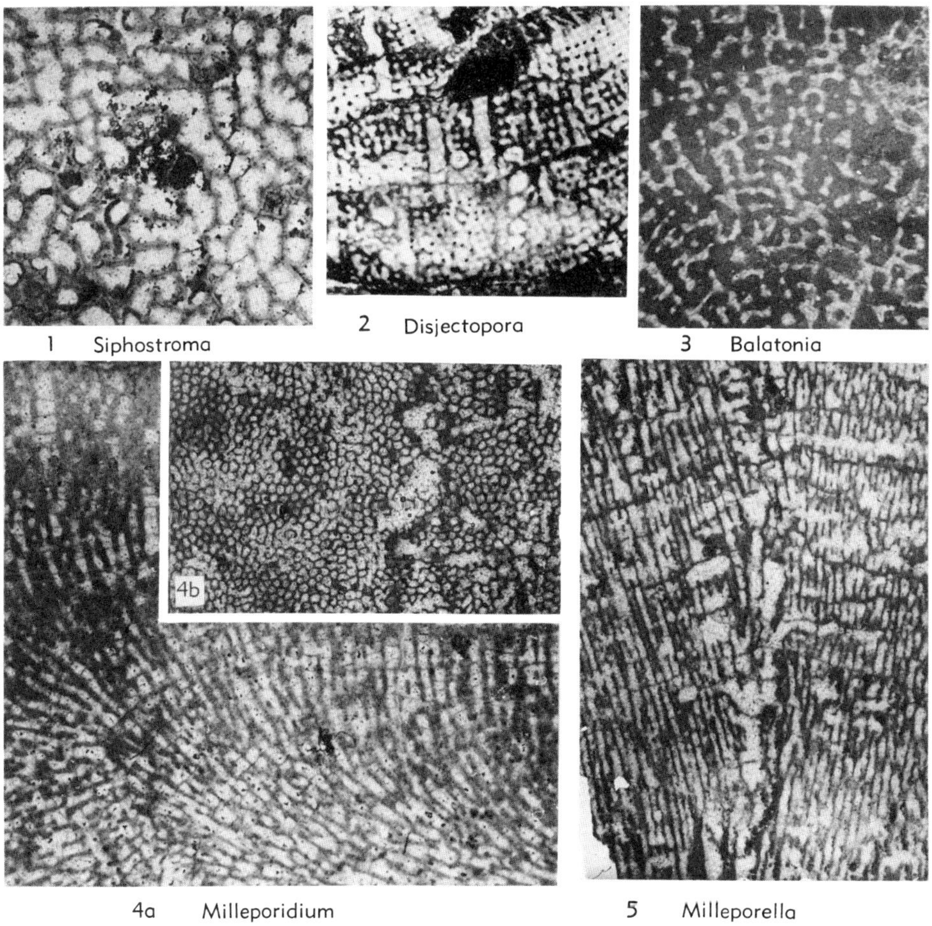

FIG. 111. Disjectoporidae, Milleporellidae, Milleporidiidae. *1, S. arzieri,* L.Cret., ×12. *2, D. japonica,* M. Perm., ×50. *3, B. kochi,* U.Trias., ×7. *4, M. remesi,* U.Jur., ×12. *5a,b, M. adriatica,* L.Eoc., ×6.

Family DISJECTOPORIDAE Tornquist, 1901

Massive forms composed of regular or irregular trabecular tissue, traversed by strong, continuous vertical astrorhizae with tabulated canal and stellate, generally short branches. *Perm.-U.Trias.*

Disjectopora WAAGEN & WENTZEL, 1887 [**D. milleporaeformis*]. Coenosteum composed of trabeculae forming a rectangular network. Astrorhizae with large axial canals, bearing tabulae, horizontal branches generally narrow (42). *Perm.-Trias.*, Asia.——FIG. 111,*2*. *D. japonica* YABE & SUGIYAMA, M.Perm., Japan; long. sec., ×50 (55*).

Balatonia VINASSA, 1908 [**B. kochi*]. Spheroidal colony composed of vermiform elements, in central part arranged loosely in somewhat regular layers, but close and radially disposed toward periphery. Astrorhizae with straight axial tubes, invariably tabulate, and few branches, restricted to outer layers. *U.Trias.*, Eu.——FIG. 111,*3*. **B. kochi*, U.Trias., Hung.; long. sec., ×7 (54*).

Carta STECHOW, 1921 [*pro Carterina* WAAGEN & WENTZEL, 1887 (*non* BRADY, 1884)] [**Carterina pyramidata* WAAG.-W., 1887] [=*Arduorhiza* WAAG.-W., 1889; *Carterinula* STRAND, 1926]. Large massive coenosteum composed of an irregular network of trabeculae traversed in all directions by many continuous astrorhizae with strong axial canals and large branches irregularly ramified intersected by trabecular pseudosepta (33). *Perm.*, Asia.

Irregulatopora WAAGEN & WENTZEL, 1887 [**I. undulata*]. Massive or laminate colonies, coenosteum composed of irregularly arranged trabeculae forming an irregular meshwork. Astrorhizae irregular, with trabecular tabulae (42). *Perm.*, Asia.

Family MILLEPORELLIDAE Yabe & Sugiyama, 1935

Massive or incrusting forms, coenosteum lacking lamellae; vertical elements partly or entirely lamellar, showing strong tendency to form closed tubes of uniform diameter, intersected by tabulae. Astrorhizae still very apparent are grouped in vertical bundles, extending radially sideward into coenosteal tissue; they are related to narrow vertical tabulate chambers. *Jur.-Cret.*

Milleporella DENINGER, 1906 [**M. sardoa*] [=*Istriactis* CHALMAS, 1883]. Subglobular colonies with mamelonate surface; trabecular or lamellar pillars tend to form incomplete or closed tubes joined by filiform cross pieces disposed in lamellar form; tangential sections show a fine fibrous network of regular polygonal meshes, mostly open in meandriform or hieroglyphic patterns. Astrorhizae in close vertical bundles, laterally digitate with uneven arrangement and extent; in tangential sections, they appear as stellate groups of twisted, branching, anastomosed passages which disappear in tissue without appreciable diminution in size. Structure microcrystalline (8). *Jur.-Cret.*, Eu.——FIGS. 111,*5*; 112,*3*. *M. adriatica* (CHALMAS), L.Eoc., Italy; long. and tang. secs., ×6 (49n).

Siphostroma STEINER, 1932 [**S. arzieri*]. Colony with wavy laminae; pillars generally continuous but locally interrupted and irregularly distributed, partly lamellar and tending to form tubes bearing thin, weakly convex tabulae; in expanded interspaces, pillars are absent, tabulae being replaced by subvesicular dissepiments; lamellar platforms few but thick or reduced to a single space or formed by bending of pillars. Tangential sections show a trabecular net of mostly open meshes intersected by irregular vague canal-like spaces, possibly astrorhizal. Pillars commonly have a dark axis with perpendicularly placed fiber bundles around it; locally, pillars are formed of clearly separate, superposed calcification centers and their fibrous aureoles. Tabulae consist of adjoined dark granules without fibrous covering (34). *L.Cret.*, Eu.—FIGS. 111,*1*; 112,*4*. **S. arzieri*, L.Cret.(L.Valang.), Switz.; 111,*1*, long. sec., ×12; 112,*4a*, tang. sec., ×12; 112,*4b*, long. sec., ×82.5 (49n).

Steinerella LECOMPTE, 1952 [**Stromatopora mecosola* STEINER, 1932]. Coenosteum composed of parallel vertical elements, partly trabecular, mainly differentiated in lamellae which tend to form imperfectly or entirely closed tubes with thin tabulae. Astrorhizae very strong, grouped in vertical bundles which partly bend horizontally with upward and downward bifurcations that pass into narrow tabulate intercolumnar chambers; in tangential sections, they form stellate groups with axes marked by many large recurved tubes. Microstructure fibrous, fasciculate, without dark axes (18). *L.Cret.*, Eu.——FIG. 112,*1*. **S. mecosola*, L.Cret.(Neocom.), Switz.; *1a,b*, long. and tang. secs., ×6; *1c*, long. sec., ×12 (49n).——FIG. 112,*2*. *S. loxola* (STEINER), L.Cret.(M.Valang.), Switz.; *2a-c*, long. secs., ×12, ×12, ×55 (49n).

Tosastroma YABE & SUGIYAMA, 1935 [**T. tokunagai*]. Like *Siphostroma* but having platforms developed intermittently at the same level (47). *U.Jur.*, Japan.

Family MILLEPORIDIIDAE Yabe & Sugiyama, 1935

Coenosteum entirely of tubular structure, lacking lamellation. Relatively large, uniform, tabulate tubes are distributed somewhat regularly among small ones comprising the bulk of tissue. Astrorhizae lacking. *Jur.-Cret.*

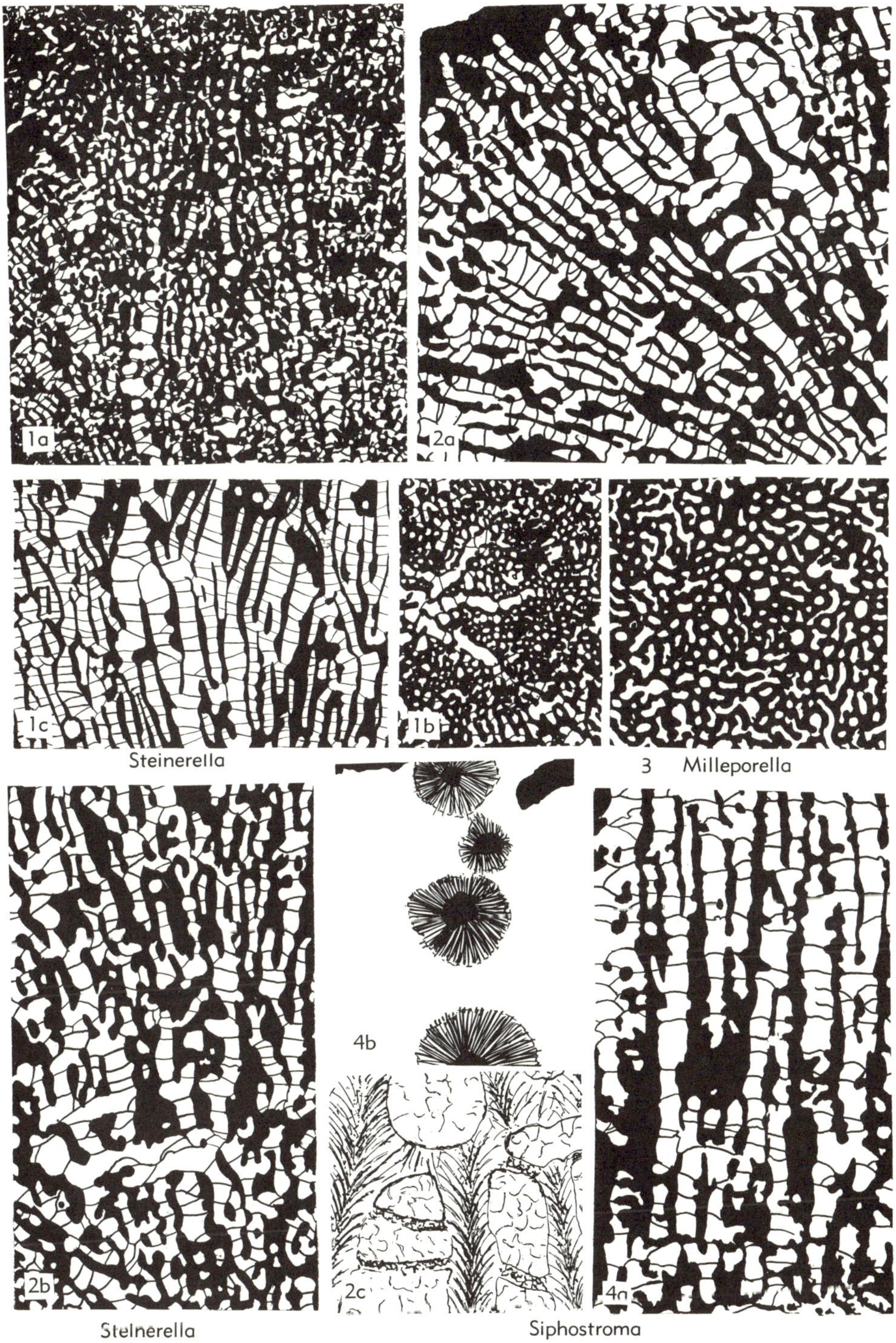

FIG. 112. Milleporellidae. *1a-c, S. mecosola,* L.Cret.; *1a,b,* ×6; *1c,* ×12. *2a-c, S. loxola,* L.Cret.; *2a,b,* ×12; *2c,* ×55. *3, M. remesi,* U.Jur., ×6. *4a,b, S. arzieri,* L.Cret., ×12, ×82.5.

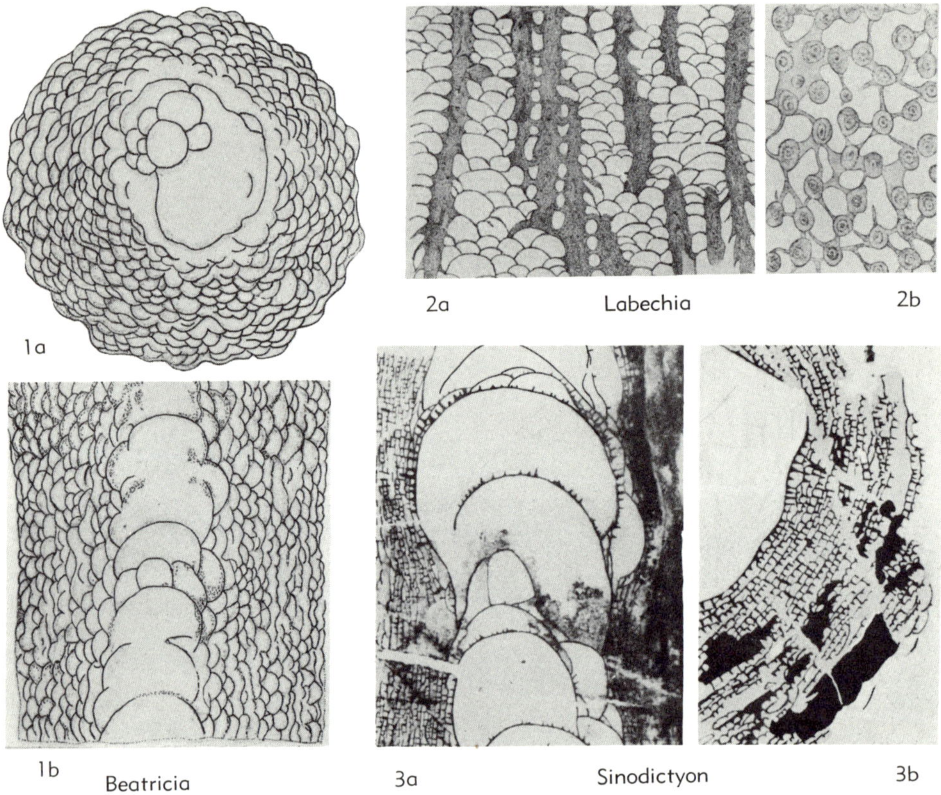

Fig. 113. Labechiidae, Aulaceridae. *1a,b, B. nodulosa,* U.Ord., ×2. *2a,b, L. conferta,* M.Sil., ×10, ×12. *3a,b, S. columnare,* Ord., ×4.

Milleporidium Steinmann, 1903 [**M. remesi*]. Colonies dendroid. Large tabulate tubes, formed from small ones of the axial region, are located in the cortical region. The meandriform pattern of the network seen in transverse sections indicates that union in tubes formed by vertical elements is imperfect; irregular canal-like structures suggest astrorhizal structures, although these are lacking (36). *Jur.-Cret.,* Eu.-Afr.——Fig. 111,*4;* 112,*3.* **M. remesi,* U.Jur.(Tithon.), Aus.; 111,*4,* long. sec., ×12; 112,*3,* tang. sec., ×12 (both 49n).

Myrioporina Kühn, 1939 [*pro Myriopora* Volz, 1904 (*non* Reuss, 1846)] [**Myriopora verbeeki* Volz, 1904]. Coenosteum composed of alternating layers of vermicular and rectilinear elements, the latter containing rounded tubes surrounded by a cyclic system of pores; pillars partly differentiated in tubes (16). *?U.Jur.,* Sumatra.

Family LABECHIIDAE Nicholson, 1885

Massive or laminar colonies with transverse skeletal elements typically composed of vesicular tissue, rarely of horizontal tabulae which invariably lack lamellar arrangement; vertical elements are laminae or strong radially disposed pillars, spiniform or continuous. Astrorhizae present rarely. *Ord.-Carb.*

Labechia M.Edw.-H., 1851 [**Monticularia conferta* Lonsdale, 1839]. Surface covered with tubercles, coalesced in some, making vermiform crests. Coenosteum composed of strong pillars with compact microstructure, joined by thin transverse plates which commonly form a vesicular tissue. Astrorhizae lacking. *Ord.-Carb.,* Eu.-Asia.-N.Am.——Fig. 113,*2.* **L. conferta* (Lonsdale), M.Sil.(Wenlock), Eng.; *2a,b,* long. and tang. secs., ×10, ×12 (50*).

Aulocerium Parks, 1909 [**A. savagei*]. Coenosteum entirely composed of vesicular tissue, lacking pillars but pierced by vertical tubes containing strongly concave tabulae (24). *Sil.,* N.Am.——Fig. 114,*3.* **A. savagei,* M.Sil., Ill.; long. sec., ×10 (51).

Dermatostroma Parks, 1910 [**Labechia papillata* James, 1878]. Thin incrusting colonies covered by closely spaced papillae; coenosteum lacking pillars, formed of thin leaves enclosing irregular chambers (24). *Ord.*, N.Am.

Labechiella YABE & SUGIYAMA, 1930 [**Labechia serotina* NICHOLSON, 1890]. Like *Labechia* but pillars more regularly arranged and transverse elements nearly horizontal or slightly concave. Astrorhizae present (46). *Sil.-Dev.*, Eu.-Asia.——FIG. 114,*2*. **L. serotina* (NICH.), Dev., Eng.; long. sec., ×12 (50).

Lophiostroma NICHOLSON, 1890 [**Labechia? schmidti* NICH., 1890] [=*Chalazodes* PARKS, 1908]. Colonies laminar, surface bearing prominent pointed or blunt mamelons; coenosteum formed by very wavy lamellae, with close-spaced conical elevations in position of the mamelons and forming pillars (20). *Sil.-?Perm.*, N.Am.-Eu.-Asia.

Ludictyon OZAKI, 1938 [**L. vesiculatum*]. Dendroid colonies composed of vesicles growing centrifugally; surface of vesicles covered by numerous very short pillars (23). *Ord.*, China.

Rosenella NICHOLSON, 1886 [**R. macrocystis*]. Massive or laminar growths largely composed of lenticular or cystose lamellae, which bear rudimentary spiniform pillars on their upper surface, not reaching to next higher lamina (20). *Sil.-L.Carb.*, N. Am.-Eu.-Asia.——FIG. 114,*4*. **R. macrocystis*, M. Sil. (Wenlock), Gotl.; long. sec., ×12 (50).

Rosenellina RADUGUIN, 1936 [**R. wellenformis*]. *L.Sil.*, Russ.

Stromatocerium HALL, 1847 [**S. rugosum*]. Like *Labechia* but with pillars replaced by vertical lamellae. *Ord.-?L.Carb.*, N.Am.-Eu.-Asia.

Stylostroma GORSKY, 1938 [**S. crassum*]. Irregu-

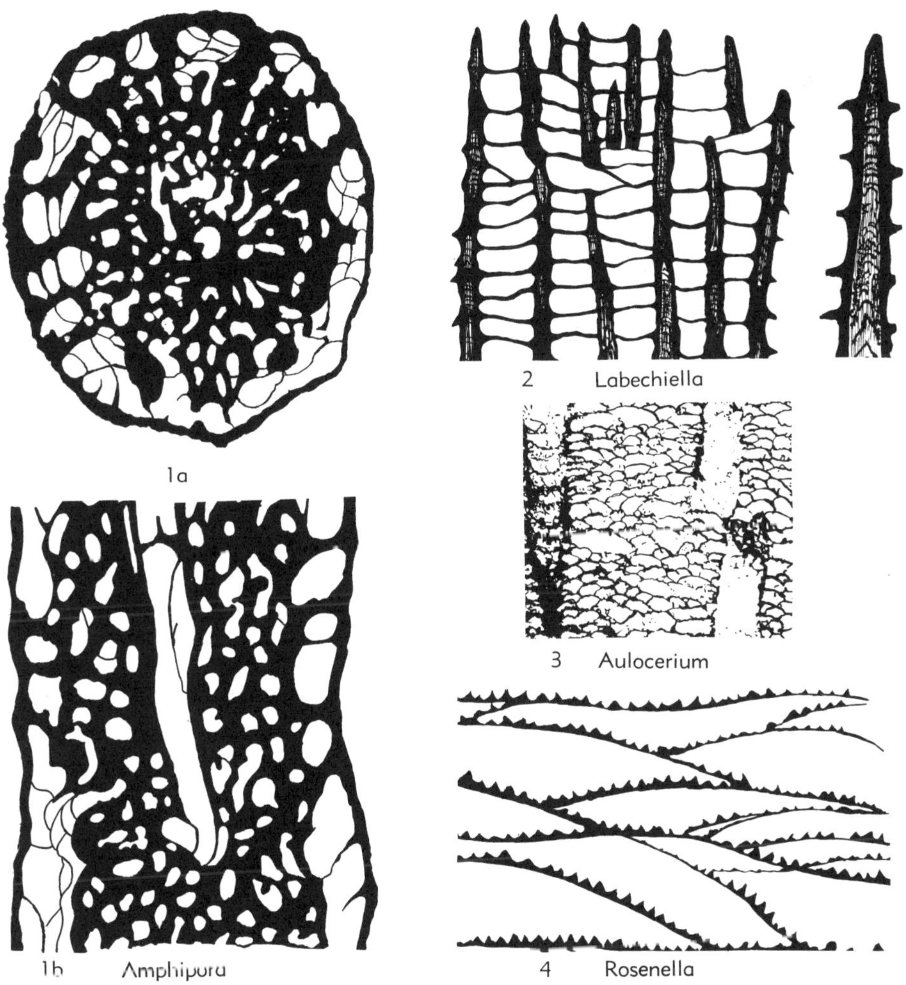

FIG. 114. Labechiidae, Incertae sedis. *1a,b*, *A. ramosa*, Dev., ×12. *2*, *L. serotina*, Dev., ×12. *3*, *A. savagei*, M.Sil., ×10. *4*, *R. macrocystis*, M.Sil., ×12.

larly hemispherical or laminate colonies with incomplete subvesicular lamellae strongly waved, intersected by stout pillars in fan-shaped arrangement. Possibly identical to *Labechia* (9). *U.Dev.-L.Carb.*, Russ.

Family AULACERIDAE Kühn, 1927

Branching colonies with a large axial canal divided into bulbous chambers by strongly arched tabulae, surrounded by vesicular tissue, with or without intersecting pillars or canals. *Ord.-Sil.*

Aulacera PLUMMER, 1843 [**Beatricea undulata* BILLINGS, 1857]. Colonies distinguished from *Beatricea* only by lack of pillars and longitudinally folded surface. *U.Ord.-Sil.*, cosmop.

Beatricea BILLINGS, 1857 [**B. nodulosa*]. Cylindrical unbranched colonies, surface wrinkled or covered with mamelons; a large axial canal is divided by strongly arched tabulae and surrounded by concentric vesicular tissue, generally prominent, and traversed in some forms by radially disposed pillars. *Ord.-Sil.*, N.Am.——FIG. 113,*1*. **B. nodulosa*, U.Ord., Ky.; *1a,b*, transv. and long. secs., ×2 (50).

Sinodictyon YABE & SUGIYAMA, 1930 [**S. columnare*]. Fasciculate or laminar colonies with large axial canal divided into bulbous chambers by highly arched tabulae, bearing spines on their upper surface; peripheral zone composed of continuous closely spaced lamellae which are locally vesiculose and joined by short pillars or rarely by pillars extending through several layers (46). *Ord.*, Manchuria.——FIG. 113,*3*. **S. columnare; 3a,b*, long. and transv. secs., ×4 (55).

Thamnobeatricea RAYMOND, 1931 [**T. parallela*]. Like *Beatricea* but branching and with much reduced vesicular area (29). *M.Ord.*, N.Am.

?**Cryptophragmus** RAYMOND, 1914 [**C. antiquatum*]. Dendroid colonies with large wall-enclosed axial canal, divided into bulbous chambers by irregularly spaced, strongly arched partitions; inner side of canal wall bears a deposit of vesicular tissue. Peripheral tissue composed of thin, weakly separated calcareous leaves traversed by many radially disposed circular openings (28). *M.Ord.*, N.Am.

?**Cladophragmus** RAYMOND, 1931 [**C. bifurcatus*] (29). *M. Ord.*, N.Am.

Family INCERTAE SEDIS

Amphipora SCHULTZ, 1883 [**Caunopora ramosa* PHILLIPS, 1841]. Delicate dendroid colonies, with or without a large axial canal; thin rectilinear or vesiculose platforms may divide the canal or be lacking; coenosteum essentially reticulate, with more or less continuous pillars present in some, tending to form a marginal row of large vesicles; chambers with or without dissepiments. Skeletal elements generally bear a strong dark axis surrounded by radial fibers. *Sil.-Jur.*, Eu.-N.Am.-Afr.-Asia-Austral.——FIG. 114,*1*. **A. ramosa* (PHILLIPS), Dev., Eng.; *1a,b*, transv. and long. secs., ×12 (50). Certain Mesozoic branching forms described as *Stromatopora douvillei* DEHORNE belong to this genus; they lack an axial canal and marginal vesicles, closely resembling some Paleozoic stromatoporoids.

REFERENCES

Publications cited in the following list are judged to be most helpful in furnishing additional information concerning the stromatoporoids and in offering a guide to more extensive literature. The index numbers enclosed in parentheses in the column at left are employed in the text for identification of the publications.

Bargatzky, A.

(1) 1881, *Die Stromatoporen des rheinischen Devons:* Dissertation zur Erlagerung der Doctorwürde bei der hohen philosophischen Fakültät der rhein.-Friedr. Wilhelms Univ. zu Bonn, 78 p.

Boehnke, K.

(2) 1915, *Die Stromatoporen der nordischen Silurgeschiebe in Norddeutschland und in Holland:* Palaeontographica, Bd. 61, p. 147-190, pl. 16-18, fig. 1-35.

Carter, H. I.

(3) 1877, *On the close relationship of Hydractinia, Parkeria and Stromatopora:* Annals Mag. Nat. Hist. (London), ser. 4, v. 19, p. 44-76.

(4) 1879, *On the mode of growth of Stromatopora including the commensalism of Caunopora:* Same, ser. 5, v. 4, p. 101-106.

(5) 1880, *On Stromatopora dartingtonensis with tabulation in the larger branches of the Astrorhiza:* Same, ser. 5, v. 6, p. 339-347, pl. 18.

Dehorne, Y.

(6) 1915, *Sur un actinostromidé du Cénomanien:* Comptes rendus Acad. Sci. Paris, v. 161, p. 733-735.

(7) 1920, *Les Stromatoporoïdés des terrains secondaires:* Mém. explicatif carte geologique de France, 170 p., 17 pl., 31 fig.

Deninger, K.

(8) 1906, *Einige neue Tabulaten und Hydrozoen*

aus mesozoischen Ablagerungen: Neues Jahrb. Mineralogie, Paläontologie, Geologie: Bd. 1, p. 61-70, pl. 1-3.

Gorsky, I. I.

(9) 1938, *Some stromatoporoids from Palaeozoic beds of Novaya Zemlya:* Trans. Arctic Inst., v. 101, p. 14-31.

Heinrich, M.

(10) 1914, *Ueber den Bau und das System der Stromatoporoidea:* Neues Jahrb. Mineralogie, Paläontologie, Geologie, Abt. B, Centralbl. 1914, p. 732-736.

(11) 1914a, *Studien in den Riffkalken des rheinischen oberen Mitteldevons:* Dissertation zur Erlagerung der Doctorwürde bei der hohen philosophischen Fakültät der rhein.-Friedr. Wilhelms Univ. zu Bonn, 58 p., 3 pl.

Hickson, S. J.

(12) 1934, *On Gypsina plana and on the systematic position of the stromatoporoids:* Quart. Journ. Micr. Sci. London, v. 76, p. 433-480, pl. 26-27, fig. 1-13.

Kühn, O.

(13) 1926, *Eine neue Hydrozoen aus dem Stramberger Jura:* Sitzungsber. Akad. Wiss. Wien, Math.-Naturwiss. Kl., Bd. 135, p. 413-424.

(14) 1927, *Zur Systematik und Nomenklatur der Stromatoporen:* Neues Jahrb. Mineralogie, Paläontologie, Geologie, Abt. B, Centralbl. 1927, p. 546-551.

(15) 1928, *Fossilium Catalogus; I. Animalia, Pars 36, Hydrozoa:* (Berlin), 114 p.

(16) 1939, *Hydrozoa:* in Schindewolf, O. H., Handbuch der Paläozoologie, Bd. 2A, 68 p., 96 fig.

Lecompte, M.

(17) 1951-52, *Les stromatoporoïdes du Dévonien moyen et supérieur du bassin de Dinant:* Inst. Roy. Sci. Nat. Belg., pt. 1 (1951), mém. 116, p 1-215, pl. 1-35; pt. 2 (1952), mém. 117, p. 216-359, pl. 36-70.

(18) 1952, *Revision des stromatoporoïdes mesozoïques des collections Dehorne et Steiner:* Inst. Roy. Sci. Nat. Belg., Bull. 28, no. 53, 39 p., 2 pl.

Nicholson, H. A.

(19) 1875, *Descriptions of Amorphozoa from the Silurian and Devonian formations:* Ohio Geol. Survey, v. 2 (Geol., Paleont., pt. 2), p. 245-255, pl. 24.

(20) 1886-92, *A monograph of the British stromatoporoids:* Palaeontograph. Soc. Mon. (London), 234 p., 29 pl.

(21) 1886-90, *On some new or imperfectly known species of stromatoporoids:* Annals Mag. Nat. Hist. (London), ser. 5, v. 17 (1886), p. 225-238, pl. 6-8; v. 18 (1886), p. 8-22, pl. 1-2; v. 19 (1887), p. 1-17, pl. 1-3; ser. 6, v. 7 (1890), p. 309-328, pl. 8-10.

——— & Murie, J.

(22) 1878, *On the minute structure of Stromatopora and its allies:* Jour. Linn. Soc. London, v. 14, p. 187-246, pl. 1-4, fig. 1-5.

Ozaki, K.

(23) 1938, *On some stromatoporoids from the Ordovician limestone of Shantung and South Manchuria:* Jour. Shanghai Sci. Inst., ser. 2, v. 2, p. 205-223, pl. 23-34.

Parks, W. A.

(24) 1907-10 [*Ordovician and Silurian stromatoporoids*]: Univ. Toronto Studies, Geol. Ser., no. 4 (1907), p. 137-172, pl. 1-6; no. 5 (1908), p. 175-240, pl. 7-15; no. 6 (1909), p. 243-292, pl. 16-20; no. 7 (1910), p. 295-344, pl. 21-25.

(25) 1933, *New species of stromatoporoids, sponges, and corals from the Silurian strata of Baie des Chaleurs:* Same, no. 33, 40 p., 8 pl.

(26) 1935, *Systematic position of the stromatoporoids:* Jour. Paleont., v. 9, p. 18-29, pl. 6-7.

(27) 1936, *Devonian stromatoporoids of North America:* Univ. Toronto Studies, Geol. Ser., no. 39, 125 p., 19 pl.

Raymond, P. E.

(28) 1914, *A Beatricea-like organism from the Middle Ordovician:* Can. Geol. Survey, Mus. Bull. 5, (Geol. Ser., no. 21), p. 1-18.

(29) 1931, *Notes on invertebrate fossils, with descriptions of new species:* Bull. Mus. Comp. Zool., Harvard College, Geol. Ser., v. 9, no. 9, p. 165-213, pl. 1-5.

Ripper, E. A.

(30) 1933-37, *The stromatoporoids of Lilydale limestone:* Proc. Roy. Soc. Vict., new ser., v. 45 (1933), p. 152-164, fig. 1-2; v. 49 (1937), p. 178-200, pl. 8-9, fig. 1-4.

(31) 1937, *On the stromatoporoids of the Buchan district, Victoria:* Same. v. 50, p. 11-38, pl. 2-5.

(32) 1938, *Notes on the Middle Palaeozoic stromatoporoid faunas of Victoria:* Same, v. 50, p. 221-243, fig. 1-4.

Stechow, E.

(33) 1921, *Neue Genera und Species von Hydrozoen und anderen Evertebraten:* Archiv Naturgesch., Abt. A, Bd. 87, p. 248-265.

Steiner, A.

(34) 1932, *Contribution à l'étude des stromatopores secondaires:* Bull. Lab. Géol. Univ. Lausanne, v. 50, 117 p., 14 pl.

Steinmann, G.

(35) 1878, *Ueber fossile Hydrozoen aus der Familie der Coryniden:* Palaeontographica, Bd. 25, p. 101-124.

(36) 1903, *"Milleporidium," eine Hydrocoralline aus dem Tithon von Stramberg:* Beitr.

Paläont. Geol. Oester.-Ungarns und Orients, Bd. 15, p. 1-8.

Sugiyama, T.
(37) 1939, *Geological and geographical distribution of stromatoporoids in Japan with notes on some interesting forms:* Jubilee Pub., Comm. 60th Birthday Prof. H. YABE, v. 1, p. 328-456, pl. 1-2.

Tornquist, A.
(38) 1901, *Ueber mesozoische Stromatoporiden:* Sitzungsber. K. Preuss. Akad. Wiss. Berlin, Bd. 47, p. 1115-1123.

Tripp, K.
(39) 1929, *Untersuchungen über den Skelettbau von Hydractinien zu einer vergleichenden Betrachtung der Stromatoporen:* Neues Jahrb. Mineralogie, etc., Beil.-Band, Abt. B, Bd. 62, p. 467-508, pl. 21-30, fig. 1-45.
(40) 1932, *Die Baupläne der Stromatoporen:* Paläont. Zeitschr., Bd. 14, p. 277-293, fig. 1-22.

Twitchell, G. B.
(41) 1929, *The structure and relationships of the true stromatoporoids:* Am. Midl. Naturalist, v. 11, p. 270-307.

Waagen, Wilhelm
(42) 1887, *Salt-Range fossils, Part 7, Coelenterata, Amorphozoa, Protozoa:* Palaeont. Indica, ser. 13, v. 1, p. 925-998.

—— & Wentzel, Joseph
(43) 1886, *Salt-Range fossils, Part 6, Productus Limestone fossils, Coelenterata:* Palaeont. Indica, ser. 13, v. 1, p. 835-924.

Wentzel, Joseph
(44) 1890, *Ueber fossile Hydrocorallinen (Stromatopora und ihre Verwandten):* Lotos, v. 9, p. 1-24, pl. 1-3.

Winchell, Alexander
(45) 1887, *Stromatoporidae, their structure and zoological affinities:* Proc. Am. Assoc. Advanc. Sci. 15th Meet. (Buffalo), for 1886, p. 91-99.

Yabe, H. & Sugiyama, T.
(46) 1930, *On some Ordovician stromatoporoids from South Manchuria, North China and Chôsen (Corea), with notes on two European forms:* Sci. Rept. Tohoku Imp. Univ. (Sendai), ser. 2, v. 14, pt. 1, p. 47-62, pl. 1-7, fig. 1-5.
(47) 1935, *Jurassic stromatoporoids from Japan:* Same, ser. 2, v. 14, p. 135-191, pl. 1-32, fig. 1-32.

SOURCES OF ILLUSTRATIONS

Explanation of the use of index numbers for citing sources of illustrations is given in the Editorial Preface.

(48) **Kühn, O.**
(49) **Lecompte, Marius**
(50) **Nicholson, H. A.**
(51) **Parks, W. A.**
(52) **Ripper, E. A.**
(53) **Steiner, A.**
(54) **Vinassa de Regny, P. E.**
(55) **Yabe, H., & Sugiyama, T.**

SIPHONOPHORIDA

By H. J. Harrington and R. C. Moore

CONTENTS

MORPHOLOGY

The siphonophorids are highly specialized polymorphic hydrozoan colonies consisting of different kinds of modified medusoids and polypoids, which in association are adapted for a planktonic or nektonic existence at or near the surface of seas in all parts of the globe (Figs. 115-117). Many of them, like the Portuguese man-of-war *(Physalia),* possess a float.

The polypoid individuals are classifiable in three categories: gastrozooids (siphons), dactylozooids (palpons), and gonozooids. (1) The **gastrozooids** are the only members of a colony capable of ingesting food. They have a polypoid form but are provided with only a single long hollow tentacle that commonly bears lateral contractile branches **(tentilla)** that terminate in a complex bunch of nematocysts. (2) The **dactylozooids** differ from gastrozooids in absence of a mouth and in having unbranched basal tentacles. Highly modified dactylozooids consist simply of a hollow tentacle-like body which may be associated with gonophores, being then termed **gonopalpons.** (3) The **gonozooids** or reproductive individuals may have polypoid form, resembling gastrozooids even to the extent of possessing a mouth, but they lack tentacles. Commonly they consist of branched stalks called **gonodendra,** which bear clusters of gonophores and may also have gonopalpons.

The medusoid individuals comprise 4 types: swimming bells (nectophores), bracts (hydrophyllia), gonophores, and floats (pneumatophores). (1) The **nectophores** have the form of a true medusa with velum, 4 radial canals, and a ring canal, but lacking mouth, tentacles, and sense organs. (2) The **hydrophyllia** are thick leaf-like gelatinous bodies containing a simple or branched radial canal. (3) The **gonophores** are dioecious sexual members of the colony which may have a medusa-like form and in the Chondrophorina may be set free. (4) The **pneumatophores** are inverted bells consisting of an external exumbrellar wall and an internal subumbrellar wall (air sac), between which is a gastrovascular space that may be divided into chambers by transverse partitions.

All individuals of a colony are budded from a **stem (coenosarc)** which may be long and tubular or expanded radially into a disc fused to the wall of the float. The individuals typically are grouped in clusters **(cormidia),** each of which contains closely associated different kinds of medusoids and polypoids. In siphonophorids having a float and long stem, the upper part of the colony with nectophores is termed the **necto-**

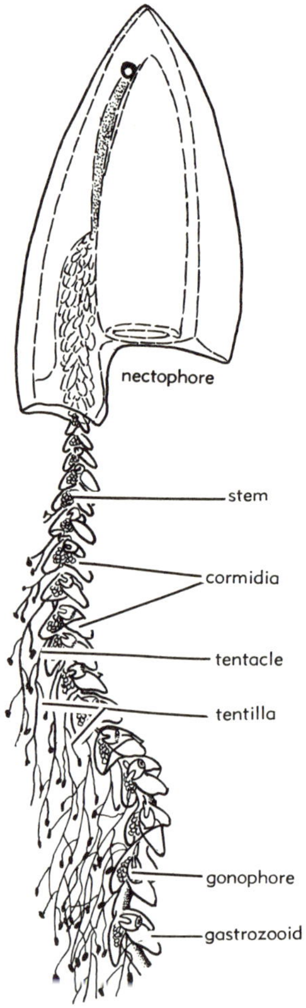

FIG. 115. Calycophorina: *Muggiaea,* Rec., showing swimming bell (nectophore) and trailing stem with cormidia, ×3.3 (L. H. HYMAN, *Invertebrates: Protozoa through Ctenophora,* McGraw-Hill, New York).

some, and the lower part provided with cormidia is termed the **siphonosome** (Fig. 116).

DISTRIBUTION

Fossil siphonophorids are exceedingly rare. Only 5 genera are now recognized as included in this group, all belonging to the suborder Chondrophorina and all represented exclusively by the impression of their floats. Two of these, from the Devonian of New York and Germany, are classed among the Velellidae. The remaining 3 genera belong in the Porpitidae and range in age from Middle Ordovician to Late Devonian, being known from scarce remains found in North America, Australia, and Europe.

CLASSIFICATION

The number of known fossil genera in each category is recorded in the following tabulation.

Main Divisions of Siphonophorida

Siphonophorida *(order)* (5). *M.Ord.-Rec.*
- Calycophorina *(suborder)*. *Rec.*
  - Sphaeronectidae. *Rec.*
  - Diphyidae. *Rec.*
  - Prayidae. *Rec.*
  - Hippopodiidae. *Rec.*
- Physophorina *(suborder)*. *Rec.*
  - Forskaliidae. *Rec.*
  - Agalmatidae. *Rec.*
  - Physophoridae. *Rec.*
  - Anthophysidae. *Rec.*
  - Rhodaliidae. *Rec.*
- Rhizophysaliina *(suborder)*. *Rec.*
  - Rhizophysidae. *Rec.*
  - Physaliidae. *Rec.*
- Chondrophorina *(suborder)* (5). *M.Ord.-Rec.*
  - Velellidae (2). *U.Dev.-Rec.*
  - Porpitidae (3). *M.Ord.-Rec.*

Order SIPHONOPHORIDA Eschscholtz, 1829

[*nom. correct.* HARRINGTON & MOORE, herein (*pro* Siphonophora ESCH., 1829)]

Highly polymorphic, free-swimming or floating colonies composed of several types of polypoid and medusoid individuals attached to a stem or disc; polypoids without oral tentacles; medusoids never developing into complete medusae and rarely free. *M. Ord.-Rec.*

Siphonophorids are typical pelagic animals, inhabitants of the surface of the seas in all parts of the world, more abundant in warmer waters. A few forms live in moderate depths.

Suborder CALYCOPHORINA Leuckart, 1854

[*nom. correct.* HARRINGTON & MOORE, herein (*pro* Calycophorae LEUCK., 1854)]

Siphonophorida without float; summit of colony consisting of one or more nectophores. *Rec.* (Fig. 115).

Genera belonging to this suborder are classified in 4 families, none of which contains known fossil representatives. They are: Sphaeronectidae HUXLEY, 1859, with only one nectophore; Diphyidae ESCHSCHOLTZ, 1829, with 2, or temporarily 3 or 4 nectophores, the oldest definitive one differing from those subsequently formed in the presence of somatocyst; Prayidae KÖLLIKER, 1853, with 2, or temporarily 3 or 4 nectophores, all structurally alike; and Hippopodiidae KÖLLIKER, 1853, with many nectophores, all structurally alike. Many living genera, including *Sphaeronectes* HUXLEY, 1859; *Diphyes* CUVIER, 1817; *Praya, Galeolaria* DE BLAINVILLE, 1834; *Hippopodius, Abyla* QUOY & GAIMARD, 1827.

Suborder PHYSOPHORINA Eschscholtz, 1829

[*nom. correct.* HARRINGTON & MOORE, herein (*pro* Physophorae ESCH., 1829)] [=Physonectae HAECKEL, 1888]

Siphonophorida with apical float devoid of pore, succeeded by length of stem commonly divided into nectosome and siphonosome; forms with enlarged float have reduced or absent nectosome and cormidia crowded together beneath pneumatophore; cormidia forming large and complex groups comprising one or more dactylozooids each with unbranched tentacle, one gastrozooid with branched tentacle, one or more bracts and clusters of gonophores commonly on gonodendra with gonopalpons; cormidia never set free as eudoxids; gonophores never free-swimming. *Rec.* (Fig. 116).

Genera belonging to this suborder are classified in 5 families, none of which contains known fossil representatives. These are: Forskaliidae HAECKEL, 1888, with long stem, well-developed nectosome with elongated nectophores encircling coenosarc in many rows, and long tubular siphonosome; Agalmatidae (*nom. correct.* herein, *pro* Agalmidae) BRANDT, 1935, with normal nectosome, normal or reduced siphonosome and nectophores in 2 alternating rows, closely pressed together; Physophoridae ESCHSCHOLTZ, 1829, with enlarged float, much-shortened stem and crown of dactylozooids above siphonosome; Anthophysidae BRANDT, 1835, with greatly enlarged float, nectophores absent and cormidia crowded together beneath float; and Rhodaliidae HAECKEL, 1888, with bell-like evaginated portion of the float (aurophore) lying

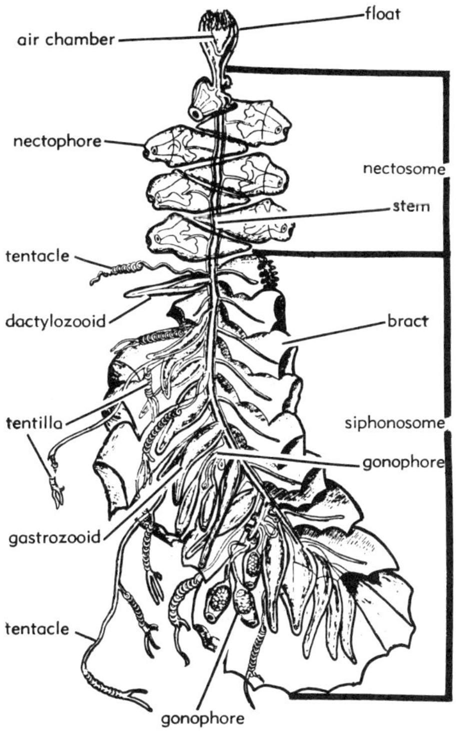

FIG. 116. Physophorina: *Agalma,* Rec., showing small apical float above stem with swimming bells (nectosome) (L. H. HYMAN, *Invertebrates: Protozoa through Ctenophora,* McGraw-Hill, New York).

among a circle of nectophores and opposite zone of proliferation. Several living genera, including *Forskalia* KÖLLIKER, 1853; *Agalma* ESCHSCHOLTZ, 1825; *Stephanomia* PÉRON & LESUEUR, 1807; *Rhodalia, Nectalia* HAECKEL, 1888; *Physophora* FORSKÅL, 1775; *Anthophysa* BRANDT, 1835; *Dromalia* BIGELOW, 1911.

Suborder RHIZOPHYSALIINA Chun, 1882

[*nom. correct.* HARRINGTON & MOORE, herein (*pro* Rhizophysaliae CHUN, 1882)]

Siphonophorida with large hollow float, provided with apical pore; nectophores and bracts absent; cormidia simplified. *Rec.*

The genera of this suborder are classified in 2 families, none of which contains known fossil representatives. They are: Rhizophysidae BRANDT, 1835, with oval float containing gas gland, succeeded by long and stout stem bearing succession of simple cor-

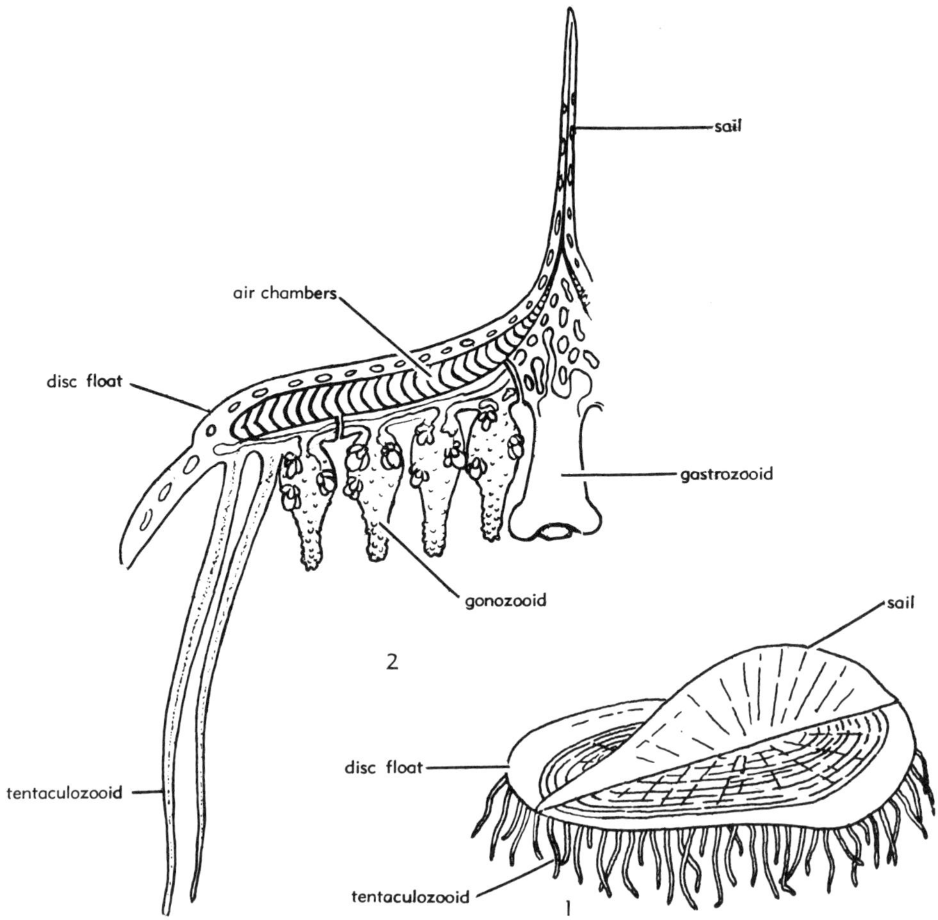

FIG. 117. Chondrophorina: *Velella,* a highly specialized modern siphonophorid with erect crest or sail; *1,* entire specimen; *2,* section across part of colony (L. H. HYMAN, *Invertebrates: Protozoa through Ctenophora,* McGraw-Hill, New York).

midia, commonly composed of one or more gastrozooids with tentacle and one gonodendron bearing gonopalpons and clusters of gonophores; and Physaliidae BRANDT, 1835, with oval, crested, contractile float, and stem shortened to budding coenosarc on ventral surface of float from which there hang down several large and small dactylozooids with tentacles, bunches of gastrozooids without tentacles and many-branched gonodendra bearing gonopalpons, clusters of gonophores and special gelatinous zooids of unknown function. Few known genera, including *Rhizophysa* PÉRON & LESUEUR, 1807, *Epibula* HAECKEL, 1888, and *Physalia* LAMARCK, 1801.

Suborder CHONDROPHORINA Chamisso & Eysenhardt, 1821

[*nom. correct.* HARRINGTON & MOORE, herein (*pro* Chondrophorae CHAMISSO-E., 1821)] [=Disconectae HAECKEL, 1888]

Siphonophorida with stem shortened to flat coenosarc forming firm rounded or oval disc together with float; float containing many concentric air chambers; underside of float with large central gastrozooid encircled by gonozooids with mouth, medusiform gonophores and marginal tentacle-like dactylozooids; gonophores freed as medusae (formerly regarded as separate genus named *Chrysomitra*). *M.Ord.-Rec.* (Fig. 117).

Family VELELLIDAE Brandt, 1835

Float with erect rounded crest; tentilla absent. *U.Dev.-Rec.*

Velella LAMARCK, 1801 [**Medusa velella* LINNÉ, 1758]. *Rec.* (FIG. 117). Only living genus.

Plectodiscus RUEDEMANN, 1916 [**P. molestus*]. Disc elliptical in outline and with compressed conical shape, external membrane ?chitinous; upper surface with numerous concentric sulci giving a corrugated appearance; radial knobs or bosses may be present, obscurely arranged in 3 or 4 cycles; central part of disc bearing diagonal keel, flattened against surface (indistinct in type species). CASTER's detailed analysis of the remains shows that they are pneumatophores of a velellid siphonophorid. *U.Dev.*, N.Y.——FIG. 118,*1*. **P. molestus;* holotype, upper surface, ×0.6 (RUEDEMANN).——

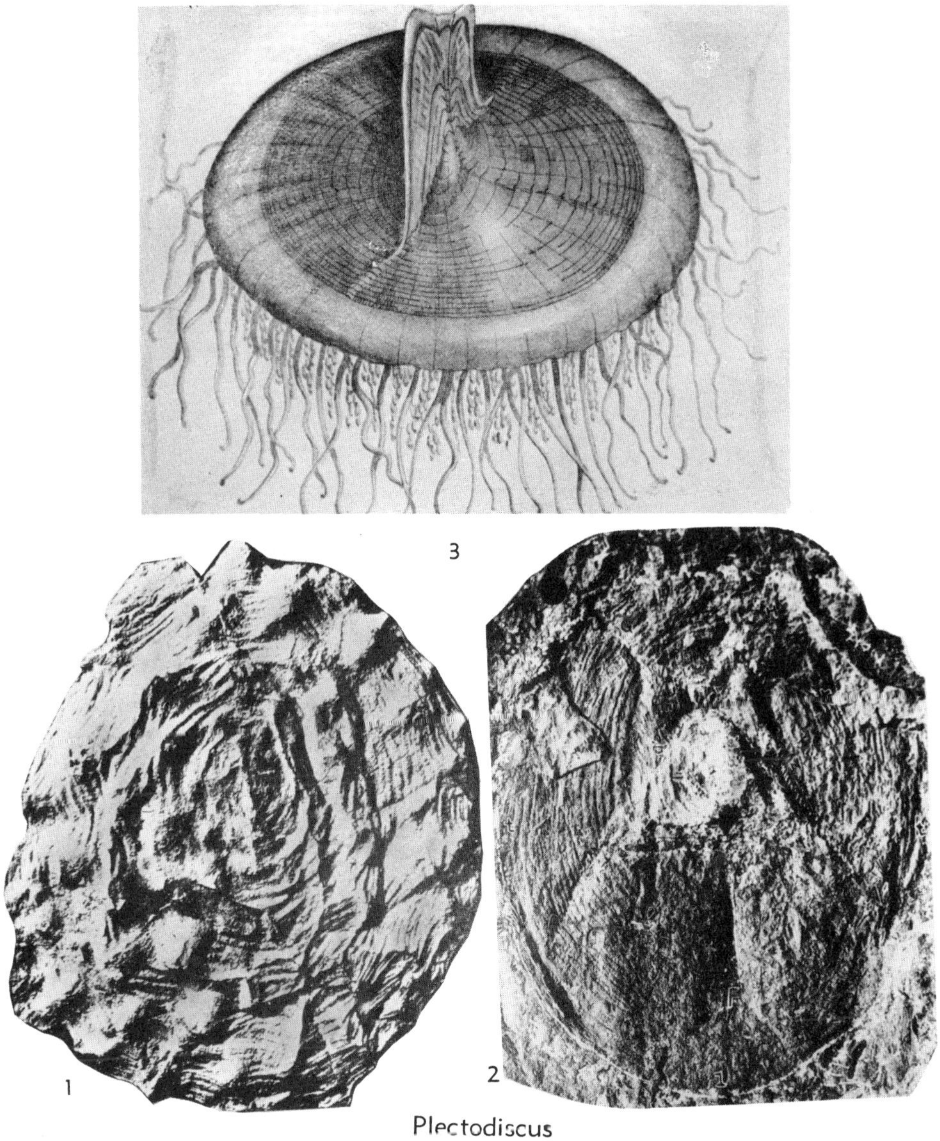

FIG. 118. Chondrophorina: Upper Devonian fossils of velellid type from New York. *1*, *Plectodiscus molestus,* holotype, upper surface, ×0.6 (RUEDEMANN). *2,3*, *P. cortlandensis,* holotype, upper surface, and oblique view from above (reconstr.), ×0.6 (CASTER).

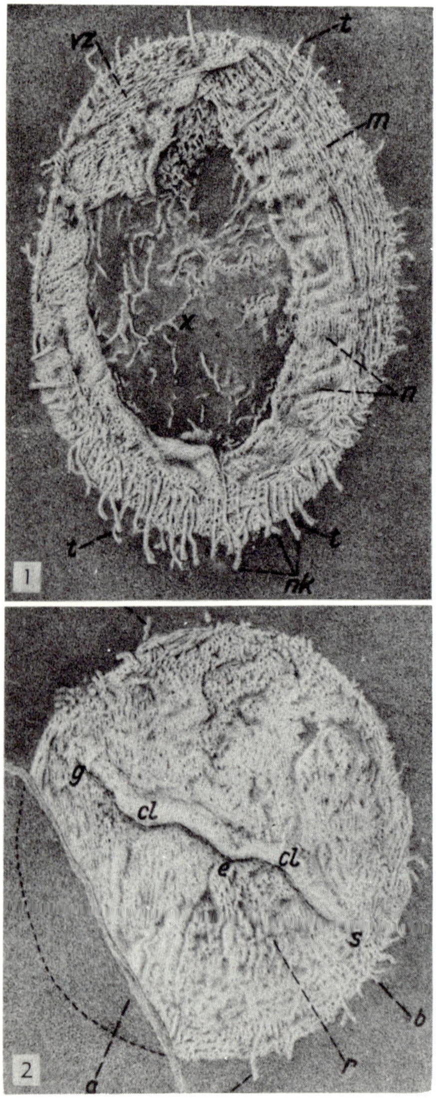

FIG. 119. Chondrophorina: *Paleonectris,* L.Dev. *1,2,* exumbrellar and subumbrellar views of 2 specimens, ×1 (RAUFF).

FIG. 118,*2,3. P. cortlandensis* CASTER; *2,3,* holotype, upper surface and oblique view from above (reconstr.), ×0.6 (CASTER).

Paleonectris RAUFF, 1939 [**P. discoidea*]. Body discoidal, of circular to elliptical outline; exumbrellar surface rugose, with obscure, irregular radial ornamentation (?radial muscular fibres) and thickened, smooth, sinuous band (?flattened "sail") crossing disc along a diameter; a few tentacles protrude for short distance beyond margin; subumbrella with numerous submarginal tentacles (?tentacle-like dactylozooids) having slightly thickened tips; one specimen, which seems to have lost all other elements, has a gaping hole in the center surrounded by irregular ring concentrically striated (?ring muscles); another specimen bears numerous numerous small, branched outgrowths (flattened against disc), regarded as gonostyles. Accepted here is RAUFF's view that the remains represent pneumatophores of a velellid siphonophore. *L.Dev.,* Ger.——FIG. 119,*1,2. *P. discoidea; 1,2,* exumbrellar and subumbrellar views of 2 specimens, ×1 (RAUFF).

Family PORPITIDAE Brandt, 1835

[*non* Porpitidae MOORE & JEFFORDS, 1945 (*ex Porpites* SCHLOTHEIM, 1820, rugose coral)]

Corm high vaulted or flattened disclike, without crest or sails. *M.Ord.-Rec.*

Porpita LAMARCK, 1801 [**Medusa porpita* LINNÉ, 1758]. *Rec.* Other living genus: *Porpema* HAECKEL, 1888.

Discophyllum HALL, 1847 [**D. peltatum*]. Body discoidal, elliptical in outline; upper surface crossed by about 80 simple radiating bands of rounded surface, increasing in width toward periphery; interspaces narrower than bands, with fine median raised line; bands and interspaces crossed by numerous fine concentric lines. RUEDEMANN's interpretation of the remains as pneumatophores of a porpitid siphonophorid is very plausible. *M.Ord.,* N.Y.——FIG. 119,*1. *D. peltatum; 1a,b,* holotype and paratype, ×1.15 (WALCOTT).

Palaeoscia CASTER, 1942 [**P. floweri*]. Thin, delicate disc (?chitinous in life) of circular outline; porelike depression at center from which 16 grooves (in 8 pairs) radiate for 1/7 the distance to periphery; very indistinct transverse lines between some radial grooves; remainder of disc with broad concentric corrugations. Accepted here is CASTER's view that the remains represent pneumatophores of the immature stage of a porpitid siphonophorid. *U.Ord.,* Ohio.——FIG. 119,*2. *P. floweri; 2a,* holotype ×1.15; *2b,* paratype, ×2.8 (CASTER).

Paropsonema CLARKE, 1902 [**P. cryptophya*]. Body compressedly convex to discoidal, of circular outline; upper surface with 2 or 3 concentric cycles of 52 to 60 radial bands looking like knotted cords, separated by smooth, narrower bands bifurcating at inner boundary of each cycle; main radial bands bearing regular succession of concentric rows of pores; center of disc occupied by small circle from which fine lines radiate; tentacles (not preserved in type species) very numerous and delicate, some stronger than others and ?grooved (?hollow); lower surface marked by irregular folds and wrinkles. Both FUCHS and RUEDEMANN interpreted the remains as pneumato-

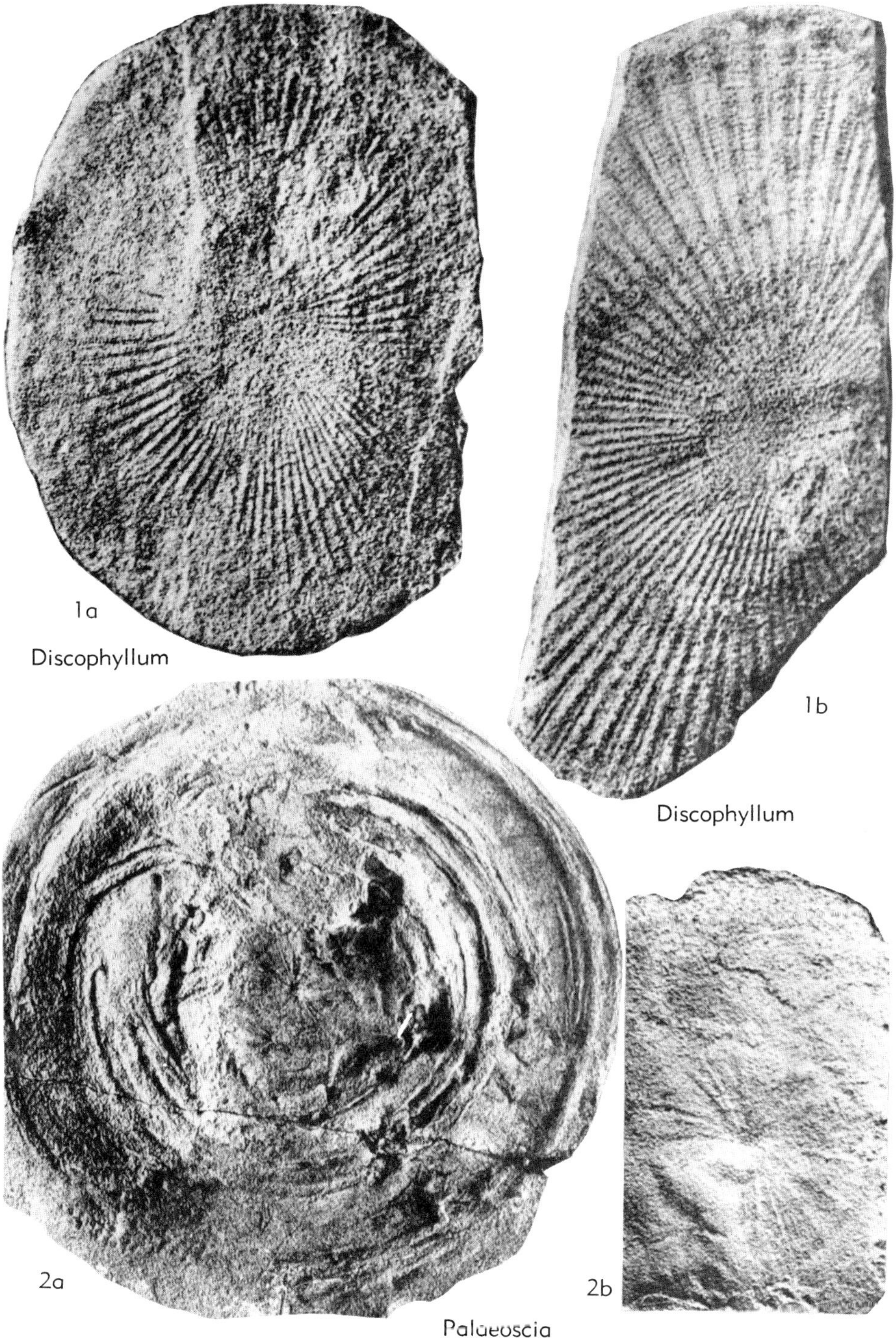

FIG. 120. Chondrophorina: Ordovician fossils from New York and Ohio. *1, Discophyllum peltatum,* M.Ord., N.Y.; *1a,b,* holotype and paratype, ×1.15 (WALCOTT). *2, Palaeoscia floweri,* U.Ord., Ohio; *2a,* holotype, ×115; *2b,* paratype, ×2.8 (CASTER).

phores of a porpitid siphonophorid. The ?hollow tentacles, preserved only in *P. mirabile* (CHAPMAN), originally referred to *Discophyllum*, may represent large dactylozooids similar to the marginal tentacle-like dactylozooids of *Porpita* and the smaller tentacles either gonopalpons or smaller feelers. *Sil.*, S.Austral.; *U.Dev.*, N.Y.——FIG. 121, *1,2*. **P. cryptophya*, U.Dev., N.Y.; *1*, lower surface, ×0.5 (CLARKE); *2*, upper surface of another specimen, ×0.5 (RUEDEMANN).——FIG. 121,*3*. *P. mirabile* (CHAPMAN), Sil., S.Austral.; holotype, upper surface, ×0.3 (CHAPMAN).

REFERENCES

Caster, Kenneth E.

(1) 1942, *Two new siphonophores from the paleozoic:* Paleontogr. Americana (Ithaca, N.Y.), v. 3, p. 56-90, pl. 1-2, fig. 1-6.

Chapman, F.

(2) 1926, *New or little known fossils in the National Museum. Pt. XXX: A Silurian jellyfish:* Roy. Soc. Victoria, Proc., v. 39, p. 13-17, pl. 1-2.

Clarke, John M.

(3) 1902, *Paropsonema cryptophya, a peculiar echinoderm from the intumescens zone (Portage beds) of western New York:* N.Y. State Mus., Bull. 39, v. 8, p. 172-178, pl. 5-9.

Hall, James

(4) 1847, *Corals of the Hudson River group:* Paleont. N.Y., v. 1, p. 247.

Kieslinger, A.

(5) 1924, *Medusae fossiles:* Fossilium Catalogus. I. Animalia, Pars 26, p. 1-20.

(6) 1939, *Scyphozoa:* in SCHINDEWOLF, O. H., Handb. d. Paläozoologie, Band 2A, Lief 5, p. A69-A109, fig. 1-42.

Rauff, H.

(7) 1939, *Paleonectris discoidea Rauff, eine siphonophoridae Meduse auf dem rheinischen Unterdevon nebst Bemerkungen zur umstrittnen Brooksella rhenana Kink.:* Palaeont. Zeitsch., Band 21, p. 194-213.

Ruedemann, Rudolf

(8) 1916, *Note on Paropsonema cryptophya Clarke and Discophyllum peltatum Hall:* N.Y. State Mus., Bull. 189, p. 22-27, pl. 1-2.

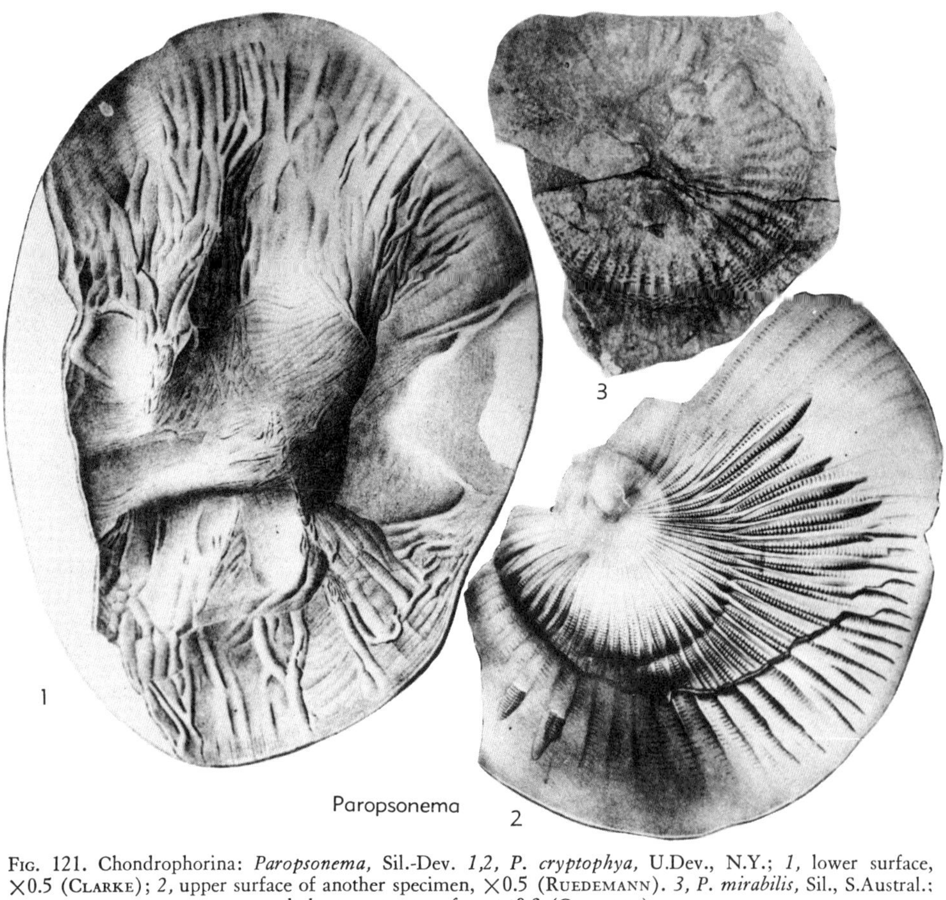

FIG. 121. Chondrophorina: *Paropsonema*, Sil.-Dev. *1,2, P. cryptophya*, U.Dev., N.Y.; *1*, lower surface, ×0.5 (CLARKE); *2*, upper surface of another specimen, ×0.5 (RUEDEMANN). *3, P. mirabilis*, Sil., S.Austral.; holotype, upper surface, ×0.3 (CHAPMAN).

MEDUSAE INCERTAE SEDIS AND UNRECOGNIZABLE FORMS

By H. J. Harrington and R. C. Moore

This section suffers acutely from the label "hypothetical," as it deals with the less distinct and more doubtful fossil records attributed to medusae. Some are, with a degree of probability that amounts to certainty, impressions left by such organisms, displaying enough distinctive features to deserve generic recognition but wanting in essential diagnostic characters to warrant their classification among the trachylinids, leptolinids, or scyphomedusae. These represent, therefore, "genera incertae sedis" within the medusoid assemblage. Others are less distinct medusoid impressions displaying such a paucity of relevant diagnostic features as to make impracticable any generic assignment. These represent "species incertae sedis," referred by their authors mostly to the "genera" *Medusites* and *Medusina,* which, however, have no status in coelenterate classification. *Medusites* Germar, 1825 [**M. capillaris*] [=*Lumbricaria* Goldfuss, 1826], was originally proposed for some problematic bodies now regarded as worm coprolites, having no resemblance to medusoid remains. *Medusina* Walcott, 1898 (type species *Spatangopsis costata* Torell, 1870; SD Caster, 1945), originally proposed "as a generic term to include all species of fossil medusae whose generic characters can not be determined"(!), is a synonym of *Spatangopsis* Torell, 1870 [**S. costata*]. Clearly, then, the medusoid forms referred to *Medusites* are incorrectly assigned to this genus and have no valid generic designation, while those referred to *Medusina,* none of which resembles *S. costata,* cannot be placed in *Spatangopsis,* and, therefore, also lack valid generic designation.

Lastly, a 3rd group comprises unrecognizable forms inadequately described or lacking cogent diagnostic characters, and forms which, though having a vague medusoid aspect, are now regarded as belonging to other organisms, animal or vegetal, or as inorganic impressions.

GENERA INCERTAE SEDIS

Cyclomedusa Sprigg, 1947 [**C. davidi*]. Margin circular, without notches, lappets or tentacles; ?exumbrella with prominent concentric grooves and

Fig. 122. *Cyclomedusa,* L.Cam., S.Austral. *1, C. gigantea,* holotype, ×0.95 (Sprigg). *2,3, C. davidi; 2,* holotype, ×1.3; *3,* another specimen, ×0.9 (Sprigg).

Fig. 123. *Duodecimedusina*, L.Dev., S.Am.; U.Penn., N.Am. *1, D. typica*, U.Penn., Kans.; *1a,b*, holotype, ?subumbrellar surface and side views showing pedestal, ×1 (King). *2, D. ulrichi*, L.Dev., Bol.; ?subumbrellar surface and pedestal, ×1 (Ulrich).

rugae which may extend to margin, and very numerous, fine radial striae not continuous into mid-field, which may contain raised central knob; ?ring canal present. *L.Cam.*, S.Austral. Of the 3 species originally described by Sprigg, *C. radiata* is here regarded as a synonym of *Tateana inflata* Sprigg (=*Ediacaria flindersi* Sprigg, p. F74). Whether the remaining 2 are also synonyms of *Ediacaria* is impossible to decide at present.——Fig. 122,*1*. *C. gigantea* Sprigg; holotype, ×0.95 (Sprigg).——Fig. 122,*2,3*. **C. davidi; 2*, holotype, ×1.3; *3*, another specimen, ×0.9 (Sprigg).

Duodecimedusina King, 1955 [**D. typica*]. Body subdiscoidal; lobes 12, regular, marked off by radial grooves disappearing at mid-field; margin cleft into 12 rounded lappets; mid-field smooth or with small central circular depression. [All observed specimens occur at summit of a cylindrical pedestal consisting of rock matrix (not part of fossil).] *L.Dev.* (Bol.), *U.Penn.* (USA).——Fig. 123,*1*. **D. typica*, Penn.(Missouri), Kans.; *1a,b*, holotype, ?subumbrellar and side view, ×1 (King, in Harrington & Moore).——Fig. 123,*2*. *D. ulrichi* King, L.Dev., Bol.; holotype, ?subumbrellar and side views, ×1 (Ulrich).

Fig. 124. *Madigania*, L.Cam., S. Austral.; *M. annulata*, holotype, ×0.4 (Sprigg).

Madigania Sprigg, 1949 [**M. annulata*]. Margin circular, without notches, lappets or tentacles; ?exumbrellar surface with numerous, conspicuous concentric grooves and rugae; radial striae absent; mid-field smooth or with raised central knob; ?ring canal present. *L.Cam.*, S.Austral.——Fig. 124. **M. annulata;* holotype, ×0.4 (Sprigg).

Paramedusium Gürich, 1930 [**P. africanum*]. Body discoidal, of circular outline; ?upper surface with flat, very delicate radial ribs becoming coarser, irregular and anastomosing near periphery; 2 concentric grooves in distal 3rd of disc. *?Cam.*, SW. Afr.——Fig. 125. **P. africanum;* holotype, ×0.25 (Kieslinger).

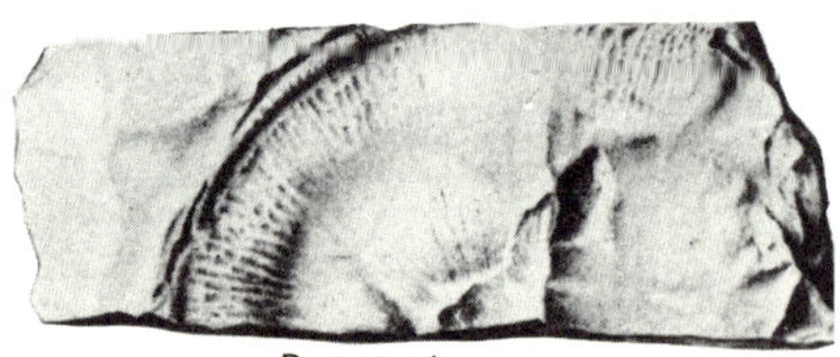

Fig. 125. *Paramedusium*, ?Cam., SW.Afr.; *P. africacum*, holotype, ×0.25 (Kieslinger).

Peytoia Walcott, 1911 [**P. nathorsti*]. Umbrella with 32 lobes marked off by radial grooves, 4 large lobes 90° apart, 7 narrower, in each quadrant; outer extremities of lobes forming 32 short, round lappets; inner extremities with 2 short, broad points projecting inward into mid-field; mid-field subquadrate, inner extremities of large lobes at center of each side; tentacles and ring canal absent. *M.Cam.*, Can. Walcott interpreted the 3 known specimens as subumbrellar impressions and

FIG. 126. *Peytoia,* M.Cam., Can.; *P. nathorsti,* holotype, probably an exumbrellar impression, ×1 (WALCOTT).

the mid-field as a central quadrate mouth, but the specimens are probably best regarded as exumbrellar impressions.——FIG. 126. **P. nathorsti;* holotype, ×1 (WALCOTT).

Protolyella TORELL, 1870 [**P. princeps* (=*Medusites favosus* NATHORST, 1881)] [=*Hydromedusites* FRECH, 1897]. Body discoidal; margin subcircular; ?subumbrellar surface with large, circular, smooth central area and outer annular zone crossed by numerous radial grooves either simple, bifurcating, irregular and anastomosing or bearing "string-of-pearls" nodules (?gonads). *L.Cam.,* Swed.-S.Austral.-?USA.; *M.Cam.,* Czech. This "form genus" is here extended to cover the L.Cam. remains from S.Austral. described by SPRIGG under *Medusina;* the bifurcating radial striation of *"M." filamentus,* interpreted by SPRIGG as marginal tentacles, are here regarded as radial grooves of a specimen in which the impression of the periphery is not preserved. *Camptostroma resseri* RUEDEMANN, L.Cam., N.Y., very different from the type species *C. roddyi,* is also placed in this genus. KIESLINGER regarded it as a synonym of *Protolyella ("Medusina") bohemica,* but this seems too extreme a view.——FIG. 127,*6.* **P. princeps,* L.Cam., Swed.; holotype, ×1 (WALCOTT).——FIG. 127,*5. P. radiata* (LINNARSSON), L.Cam., Swed.; holotype, ×1 (WALCOTT).——FIG. 127,*1. P. asteroides* (SPRIGG), L.Cam., S.Austral., ×0.95 (SPRIGG).——FIG. 127,*2. P. mawsoni* (SPRIGG), L.Cam., S.Austral.; ×0.9 (SPRIGG).——FIG. 127,*7. P. filamenta* (SPRIGG), L.Cam., S.Austral.; holotype, ×1.3 (SPRIGG).——FIG. 127,*4. P. resseri*

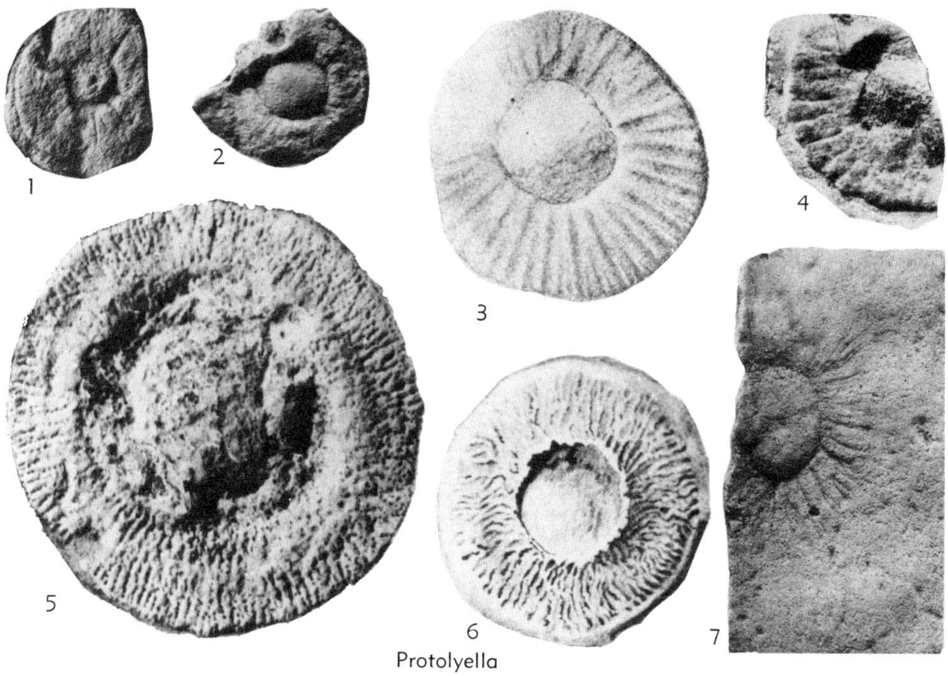

FIG. 127. *Protolyella,* Cam., Eu.-N.Am.-Austral. *1, P. asteroides,* L.Cam., Austral., ×0.95 (SPRIGG). *2, P. mawsoni,* L.Cam., Austral., ×0.9 (SPRIGG). *3, P. bohemica,* M.Cam., Czech., ×0.7 (WALCOTT). *4, P. resseri,* L.Cam., N.Y., ×0.5 (RUEDEMANN). *5, P. radiata,* L.Cam., Swed., ×1 (WALCOTT). *6, P. princeps,* L.Cam., Swed., ×1 (WALCOTT). *7, P. filamenta,* L.Cam., Austral., ×1.3 (SPRIGG).

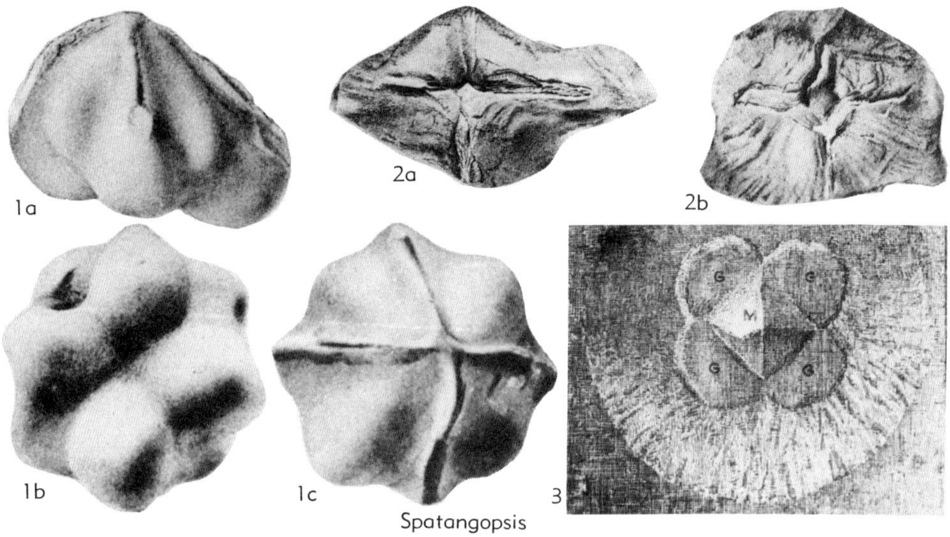

FIG. 128. *Spatangopsis costata,* L.Cam., Swed.; *1a-c,* steinkern, ×1; *2a,b,* subumbrellar impressions, ×0.5; *3,* same, another specimen, ×0.7 (WALCOTT).

(RUEDEMANN), L.Cam., N.Y.; holotype, ×0.5 (RUEDEMANN).——FIG. 127,*3*. *P. bohemica* (GORTANI), M.Cam., Czech.; holotype, ×0.7 (WALCOTT).

Spatangopsis TORELL, 1870 [**S. costata* (=*Agelacrinus lindströmi* LINNARSSON, 1871] [=*Medusina* WALCOTT, 1898 (obj., type species *Spatangopsis costata* TORELL, 1870; SD CASTER, 1945)]. Body discoidal; margin subcircular; exumbrella smooth; subumbrella with numerous radial grooves and large cruciform mouth. One specimen shows large semicircular interradial marking adjacent to mouth, regarded as gonads. Specimens considered as internal molds (steinkerns) of gastric cavity are pyramidal bodies with rounded base or with 4 to 5 rounded basal projections (?roof of gastric cavity), sides bounded by 4 to 5 sharp ridges meeting at apex and triangular areas between ridges flat or with elongated rounded projections (?casts of genital sacs). *L.Cam.,* Swed.——FIG. 128. **S. costata; 1a-c,* side, upper, and lower and side views of a gastric steinkern, ×1; *2a,b,* subumbrellar impressions, ×0.5; *3,* subumbrellar impression of another specimen showing supposed gonads, ×0.7 (all from WALCOTT).

SPECIES INCERTAE SEDIS

"Discophyllum plicatum" BARROIS, 1891. Body small, discoidal, of circular outline, with sharply marked small, circular, ?depressed central area and ?34 to ?40 simple radial ribs, exceptionally bifurcating, separated by narrower, shallow grooves; ribs with obscure, irregular transverse striae. *L.Ord.,* Fr. BARROIS identified the specimens with *Actinophyllum plicatum* PHILLIPS, L.Dev., Eng., regarded the genus as a synonym of *Discophyllum* HALL (here considered as a porpitid siphonophorid) and interpreted the remains as isolated metamers of *Anthomorpha* BORNEMANN (Archaeocyatha). STRAW opposed BARROIS's view, regarded *Actinophyllum* and *Discophyllum* as different genera, and held the French specimens to be specifically (?and generically) distinct from *A. plicatum*. As here interpreted, the French specimens do not seem to belong in either *Actinophyllum* or *Discophyllum;* though possibly medusoid, their relations cannot be established at present.——FIG. 129,*1a,b*. *"D. plicatum," 2,* specimens, ×1 (BARROIS).

"Medusina" atava POHLIG, 1892. Impression circular, with a broad, outer annular zone, concentrically striated, a narrow, smooth inner ring, and a large central mid-field with indistinct traces of 4 petalate lobes 90° apart and 4 smaller, subrhombic exsert lobes. *Perm.,* Ger.——FIG. 129,*3*. *"M." atava,* holotype, ×1 (WALCOTT).

"Medusina" boulangeri VAN STRAELEN, 1926. Known from a single ?exumbrellar impression. Umbrella convex; margin entire, circular, without tentacles; outer annular zone narrow, smooth, separated from large central zone by circular groove; large central area with ?30 irregular radiating grooves, some bifurcating, regarded by VAN STRAELEN as shrinkage furrows due to desiccation during entombment; center of mid-field damaged. *L.Carb.,* Belg.——FIG. 129,*6*. *"M." boulangeri; 6a,b,* holotype, side and ?exumbrellar views, ×0.25 (VAN STRAELEN).

"Medusina" eisfeldensis ADAM, 1950. Known from single specimen. Bell convex; mid-field about 1/3 the diameter of ?exumbrella, with central knob, indistinctly separated from outer annular zone bearing ?12 obscure radial lobes, vaguely defined by shallow radial sulci; margin scalloped into short, round lappets. *L.Trias.*, Ger.——FIG. 129,*4*. *"M." eisfeldensis*, holotype, ×0.5 (ADAM).

FIG. 129. Medusae species incertae sedis. *1, "Discophyllum plicatum,"* L.Ord., Fr., ×1 (BARROIS). *2, "Medusites bicinctus,"* U.Jur., Ger.; ×0.65 (HAECKEL). *3, "Medusina" atava*, Perm., Ger.; ×1 (WALCOTT). *4, "Medusina" eisfeldensis*, L.Trias., Ger.; ×0.5 (ADAM). *5, "Medusites" staurophorus*, U.Jur., Ger.; ×1 (HAECKEL). *6, "Medusina" boulangeri*, L.Carb., Belg.; ×0.25 (VAN STRAELEN). *7, "Medusina" liassica*, L.Jur., Ger.; ×0.8 (RÜGER-HAAS). *8. "Medusina" reiflingensis*, Trias., Aus.; ×1 (KIESLINGER).

"Medusina" liassica Rüger-Haas, 1925. Known from single specimen. ?Exumbrellar impression with small, buttonlike central knob encircled by shallow groove from which 10 shallow sulci radiate, giving rise to 10 subequal lobes ending at periphery in short, rounded lappets. *L.Jur.*, Ger. ——Fig. 129,*7*. *"M." liassica;* holotype, ×0.8 (Rüger-Haas).

"Medusina" lincki Hundt, 1940. ?Subumbrellar impressions 5 to 7 cm. in diameter; margin circular, entire; outer annular zone with 4 or 5 coarse concentric rugae; mid-field large, convex, with cruciform ?mouth. *U.Perm.*, Ger. Very poorly illustrated.

"Medusina" reiflingensis Kieslinger, 1925. Known from single specimen. ?Subumbrellar impression circular, with narrow outer annular zone and large circular mid-field bearing several irregular radial grooves, small central circular marking and 2 broad protuberances, asymmetrically located near periphery of mid-field, regarded by Kieslinger as gonads. *Trias.*, Aus.——Fig. 129,*8*. *"M." reiflingensis,* holotype, ×1 (Kieslinger).

"Medusina" sp., Adam, 1950. Known from single specimen. Small impression with large, smooth, concave mid-field and 16 raised, outer lobes, of elongate quadrangular shape, sharply separated by deep, broad grooves. *M.Trias.*, Ger. The poorly preserved specimen may be compared with *Lorenzinia gabelli.*

"Medusites" bicinctus Haeckel, 1869. Known from single, obscure specimen. Subumbrellar impression circular, with entire margin, 2 smooth outer rings of subequal width, separated by circular groove, and large circular mid-field with indistinct center and 8 radial grooves marking off 8 radial lobes (4 larger 90° apart, one narrower in each quadrant). Caster suggested that *"M." bicinctus* could represent an immature stage of *Rhizostomites admirandus,* but the specimen is too poorly preserved to draw any sound conclusion. *U.Jur.*, Ger.——Fig. 129,*2*. *"M." bicinctus,* holotype, ×0.65 (Haeckel).

"Medusites" staurophorus Haeckel, 1869. Known from single obscure specimen. Margin circular, entire; outer ring narrow, smooth; inner field large, circular, crossed by 4 hyperbolic low ridges marking off a central cruciform excavation (?mouth). *U.Jur.*, Ger.——Fig. 129,*5*. *"M." staurophorus,* holotype, ×1 (Haeckel).

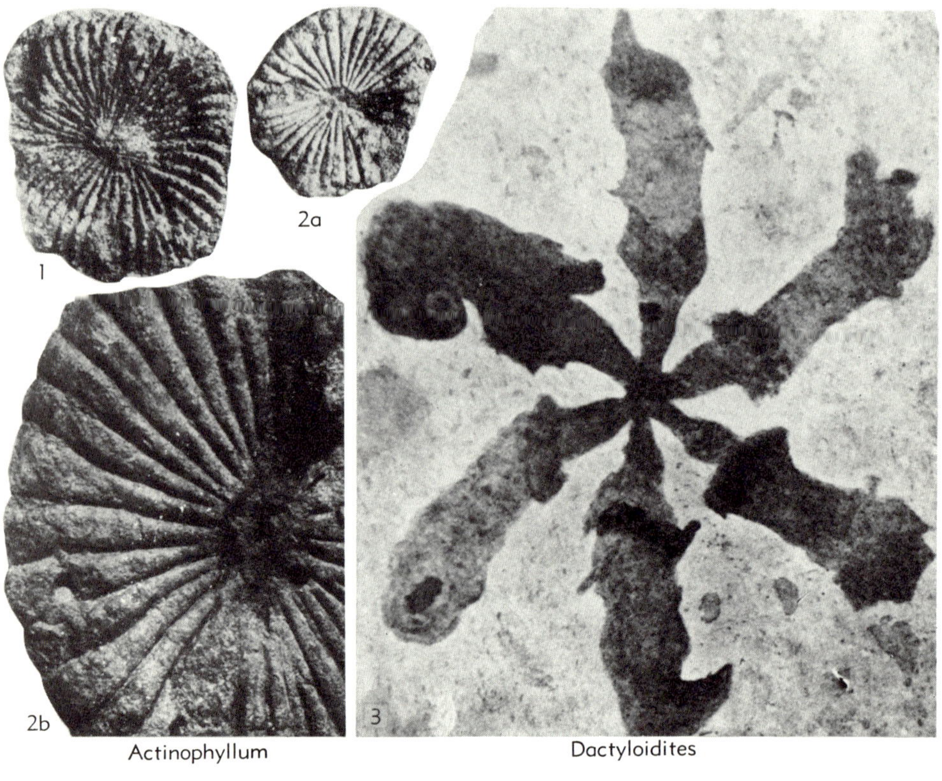

Fig. 130. *Actinophyllum* and *Dactyloidites*. *1*, *A. spinosum,* L.Dev., Eng.; ×2 (Straw). *2*, *A. plicatum,* L.Dev., Eng.; *2a,b*, ×1, ×2.5 (Straw). *3*, *D. asteroides,* L.Cam., N.Y.; ×1 (Walcott).

UNRECOGNIZABLE FORMS AND FORMS ATTRIBUTED TO MEDUSAE BUT NOW REGARDED AS BELONGING TO OTHER ORGANISMS OR AS INORGANIC

Actinophyllum PHILLIPS, 1848 [**A. plicatum*]. Thin, nearly flat circular disc, with small, indistinct, ill-defined central area from which 36 to 56 ribs radiate to periphery; ribs flat, bifurcating near inner extremity and commonly again farther on; surface of ribs either smooth or with tiny spines arranged in alternating series along the middle; interspaces groovelike; ribs reaching margin and continuing on underside of body for half the distance toward center, where they curve upward and disappear leaving a large circular central opening through which internal surface of upper side is displayed. *L.Dev.(Downton.)*, Eng. PHILLIPS originally regarded this problematic form as a calcareous alga resembling the living *Acetabularia;* both M'COY and SALTER suggested its identity with *Spongarium* MILNE-EDWARDS, 1839, U. Sil.(U.Ludlov.), Eng., but STRAW showed that the 2 forms are different, *Spongarium* being almost paper thin, having one concave, smooth surface and the opposite convex and striated (not ribbed); BRODIE regarded the form as a coral; BARROIS argued its identity with *Discophyllum* HALL, 1847, M.Ord., N.Y., and regarded the fossils as metamers of *Anthomorpha* BORNEMANN (archaeocyathid); STRAW opposed this view and went back to PHILLIPS' idea; KIESLINGER placed the form among the medusae incertae sedis. The weight of the evidence suggests that *Actinophyllum* is a calcareous alga.——FIG. 130,*1*. *A. spinosum* STRAW; holotype, ×2 (STRAW).——FIG. 130,2. **A. plicatum*; *2a,b*, holotype, ×1, ×2.5 (STRAW).

Aequorfossa NEVIANI, 1925 [**A. farnesinae*]. *Plio.*, Italy. Known from single obscure impression consisting of 2 concentric circular grooves, regarded by NEVIANI as representing a leptolinid medusa (Aequoridae). Accepted here is the view of SACCO, who regarded the impression as inorganic.

Brooksella rhenana KINKELIN, 1903. *Dev.*, Ger. R. & E. RICHTER have shown that this impression belongs to *Pleurodictyum regulare* MAURER (coral).

Dactyloidites HALL, 1886 [**D. bulbosus* (=*Buthotrepis ? asteroides* FITCH, 1850)]. Starlike remains consisting of a thin carbonaceous film, having 4 to 7 radiating, widely separated lobes, typically expanded and bulbous near distal extremities. *L. Cam.*, N.Y., *?M.Cam.*, Vt. FITCH regarded the fossils as algae; WALCOTT placed them in the Brooksellidae (Protomedusae), a view accepted both by KIESLINGER and CASTER; RUEDEMANN reverted to FITCH's idea and suggested that they could represent remains of algae that lived attached to the sea bottom. The carbonaceous nature of the remains supports their algal origin and makes their medusoid affinities very implausible.——FIG. 130,*3*. **D. asteroides* (FITCH), L.Cam., N.Y.; a typical specimen, ×1 (WALCOTT). *?D. edsoni* RUEDEMANN, 1934, M.Cam., Vt., possibly does not belong in this genus.

Medusichnites MATTHEW, 1891. No type. *Cam.*, Can. Inorganic markings and ?trails.

Medusites GERMAR, 1825 [**M. capillaris*] [=*Lumbricaria* GOLDFUSS, 1826]. *Paleoz.*, Ger. Not a medusoid; ?worm coprolites.

"Medusites" armoricanus PENEAU, 1946. *L.Ord.*, Fr. Unrecognizable.

"Medusites" circularis HAECKEL, 1869. *U.Jur.*, Ger. Unrecognizable.

"Medusites" cretaceus KNER, 1866. *Cret.*, Pol. Unrecognizable.

"Medusites" couffoni PENEAU, 1946. *L.Ord.*, Fr. Unrecognizable.

"Medusites" dubius PENEAU, 1946. *L.Ord.*, Fr. Problematic form of unknown affinities; not a medusoid.

"Medusites" helgolandicus BRANDT, 1871. *Cret.*, Helgoland (Ger.). Unrecognizable.

"Medusites" latilobatus VON AMMON, 1886. *?U.Cret.*, Ger. Not a medusoid; ?sponge.

"Medusites" porpitina HAECKEL, 1869. *U.Jur.*, Ger. Unrecognizable.

Palaegina HAECKEL, 1869 [**P. gigantea*]. *U.Jur.*, Ger. Cephalopod.

Papilionata SPRIGG, 1947 [**P. eyrei*]. *L.Cam.*, S. Austral. Problematic form of unknown affinities; not a medusoid.

"Protoniobia" SPRIGG, 1949 [**P. wadea*]. Known from single, very small impression consisting of a few prominent annular undulations with ring of tiny nodules near center; margin with 4 subcircular attached structures, of different size, arranged in incomplete hexagonal pattern, showing concentric sculpture similar to "parent" impression. *L.Cam.*, W.Austral. SPRIGG regarded the impression as representing a leptolinid medusa (Anthomedusina) with 4 attached buds. Here interpreted as a concretion, inorganic.——FIG. 131. *"*P. wadea"*; holotype, ×9 (SPRIGG).

Trianisites RAFINESQUE, 1821 [**T. cliffordi*]. *Ord.*, Ky. Not a medusoid; ?algae.

FIG. 131. *"Protoniobia wadea,"* L.Cam., W.Austral.; ×9 (SPRIGG).

REFERENCES

Adam, K.D.

(1) 1950, *Erster Medusen Nachweis in der Germanischen Trias:* Neues Jahrb. f. Min. etc., Mon. Band, Heft 11, p. 330-341, fig. 1-4.

Ammon, L. von

(2) 1886, *Über neue Exemplare von jurassischen Medusen:* Abhandl. Kgl. Bayer. Akad. Wiss., Math.-Nat. Kl., Band 15, p. 120-165, pl. 1-2, fig. 1-3.

Barrois, Charles

(3) 1891, *Memoir sur la faune du grès armoricain:* Soc. Geol. du Nord. de la France, Ann. 19, p. 134-237, pl. 1-5.

Brandt, A.

(4) 1871, *Über fossile Medusen:* Mem. Acad. Imp. Sci. St. Petersbourg (7), 16, no. 11, p. 1-18, pl. 1-3, fig. 1-2.

Fitch, Asa

(5) 1850, *Rocks and soils of Washington County (New York):* N.Y. State Agric. Soc., Trans., v. 9, p. 863.

Gortani, M.

(6) 1920, *Osservazzioni sulle impronte medusoidi del Flysch (Lorenzinia e Atollites):* Riv. ital. Paleont., anno 26, fasc, 3-4, p. 56-72, pl. 1-3.

Gürich, G.

(7) 1933, *Die Kuibis-Fossilien der Nama-formation von Sudwest-Afrika. Nachträge und Zusätze:* Paläont. Zeitschr., Band. 15, Heft. 2-3, p. 137-154, fig. 1.

Haeckel, Ernst

(8) 1869, *Über die fossilen Medusen der Jurazeit:* Zeitschr. f. wiss. Zoologie, Band 19, p. 554-561, pl. 42.

Hall, James

(9) 1886, *Note on some obscure organisms in the roofing slates of Washington County, N.Y.:* N.Y. State Mus., 39th Ann. Rept., p. 160, pl. 11.

Harrington, H. J. & Moore, R. C.

(10) 1955, *Kansas Pennsylvanian and other jellyfishes:* Kansas Geol. Survey, Bull. 114, pt. 5, p. 153-163, pl. 1-2.

Hundt, Rudolf

(11) 1940, *Eine Medusina aus dem Kieselschiefer des Ostthüringer Llandovery:* Zentralbl. f. Min. etc., Abt. B, p. 25-29, fig. 1-2.

(12) 1940, *Eine Meduse (Medusina lincki nov. sp.) des oberen Zechsteins Ostthüringens und ihr Einbettungsraum:* Beitr. geol. Thüringen, Band 6, Heft 1, p. 11-16, pl. 1.

Kieslinger, A.

(13) 1924, *Medusae fossiles:* Fossilium Catalogus. I. Animalia. Pars 26, p. 1-20.

(14) 1925, *Eine Meduse aus der alpinen Trias:* Neues Jahrb. f. Min. etc., Abt. B, Band 51, p. 494-502, pl. 1.

(15) 1939, *Scyphozoa:* in Schindewolf, O. H., Handb. d. Paläozoologie, Band 2A, Lief 5, p. A69-A109, fig. 1-42.

Kner, J.

(16) 1865, *Über eine Meduse in Feuerstein:* Sitzungsber. K. Akad. Wiss., Band. 52, p. 480.

Linnarsson, J. G. O.

(17) 1871, *Geognostiska och Paleontologiska Jakttagelser öfver Eophytonsandstenen i Verstergötland:* K. Svenska Vetenskapsakad. Handl., v. 9, no. 7, p. 1-16, pl. 1-2.

Nathorst, A. G.

(18) 1881, *Im aftryck af Medusor i Sveriges Kambriska Lager:* Same, v. 19, no. 1, p. 1-34, pl. 1-6.

(19) 1910, *Ein besonders instruktives Exemplar unter der Medusenabdrucken aus dem Kambrischen Sandstein bei Lugnås:* Sver. Geol. Undersök., Afhandl. (C), Aarsbok 3 (1909), no. 11, p. 1-7, pl. 1.

Neviani, Antonio

(20) 1925, *Di una nuova medusa fossile appartenente alle Aequoridae (Craspedote) rinvenuta nelle argille classiche della Farnesina presso Roma:* Atti Pont. Accad. Nuovi Lincei, tomo 78 (6), p. 148-156, fig. 1.

Peneau, J.

(21) 1946, *Étude sur l'ordovicien inférieur (areniglen — grès armoricain) et sa faune (specialement au Anjou):* Soc. Études Sci. Angers, B. an., 74-76, p. 39-94, 1 pl.

Phillips, John

(22) 1848, *Palaeontological appendix to Professor John Phillips' Memoir on the Malvern Hills compared with the Paleozoic districts of Abberley:* in Phillips, J., & Salter, J. W., Mem. Geol. Survey G.Brit., v. 2, pt. 1, p. 331-386, pl. 30.

Pohlig, H.

(23) 1892, *Altpermische Saurierfährten, Fische und Medusen:* Festschr. zum 70 Geburtst. Rudolf Leuckarts, Leipzig, p. 58-64, fig. 2, pl. 7.

Pompeckj, J. F.

(24) 1896, *Die Fauna des Cambriums von Tejrovic und Skrey in Böhmen:* Jahrb. k. k. geol. Reichsanst., Band 45 (1895), p. 501, pl. 14.

Ruedemann, Rudolf

(25) 1934, *Paleozoic plankton of North America:* Geol. Soc. Am., Mem. 2, 141 p., 26 pl., 6 fig.

Rüger, L., & Rüger-Haas, P.

(26) 1926, *Palaeosemaeostoma geryonides v. Huene sp., eine sessile Meduse aus dem Dogger von Wehingen in Württemberg und Medusina liasica nov. sp., eine coronatenähnliche Meduse aus dem mittleren Lias von Hechingen in Württemberg:* Sitzungsber. Heidelberg. Akad. Wiss., Math.-Nat. Kl., Jahrg. 1925, Abhandl. 15, p. 1-22, pl. 1-2.

Sprigg, R. G.

(27) 1947, *Early Cambrian(?) jellyfishes from the Flinders ranges, South Australia:* Roy. Soc. S.Australia, Trans., v. 71, pt. 2, p. 212-224, pl. 5-8.

(28) 1949, *Early Cambrian "jellyfishes" of Ediacara, South Australia, and Mt. John, Kimberley District, Western Australia:* Same, v. 73, pt. 1, p. 72-99, pl. 9-21.

Straw, S. H.

(29) 1926, *Some notes on the genus Actinophyllum:* Mem. & Proc. Manchester Lit. & Philos. Soc., v. 70, p. 133-139, 1 pl.

Torell, Otto

(30) 1870, *Petrifacta suecana formationis cambricaea:* Lunds Univ. Årsskrift 1869, pt. 2, no. 8, p. 1-14.

Ulrich, Arnold

(31) 1893, *Paleozoische Versteinerungen aus Bolivien:* Neues Jahrb. f. Min. etc., Abt. B, Band 8, p. 5-116, pl. 1-5.

Van Straelen, Victor

(32) 1926, *Sur les premiers restes de Meduses trouvés dans le calcaire carbonifère de Belgique:* Acad. Roy. Belgique, Bull., Cl. Sci., 5me. ser., v. 12, p. 952-956, 1 pl.

Walcott, Charles D.

(33) 1898, *Fossil Medusae:* U. S. Geol. Survey, Mon. 30, ix + 201 p., 47 pl.

(34) 1911, *Middle Cambrian holothurians and medusae:* Smithsonian Misc. Coll., v. 57, no. 3, p. 41-68, pl. 8-13.

Zelisko, J. V.

(35) 1937, *Einige Bemerkungen über das Vorkommen von frei schwimmenden Medusen in der kambrischen Formation:* Zentralbl. f. Min., etc., Abt. B, p. 205-208.

ANTHOZOA—GENERAL FEATURES

By J. W. Wells and Dorothy Hill

Anthozoans are exclusively polypoid, solitary or colonial, with no trace of a medusoid stage in their life cycles, but they retain the free-swimming planula larval stage. They are the only coelenterate group with a **stomodaeum**, a tube extending down from the mouth into the gastrovascular cavity. The stomodaeum may have 1 or 2 grooves **(siphonoglyphs)** lying in a so-called dorsoventral plane which is vertical with respect to the oral surface. The gastrovascular cavity is radially partitioned with respect to the oral-aboral axis by endodermal **mesenteries** extending inward from the inner surface of the column wall; and when these are attached in part to the stomodaeum they are said to be complete or perfect. The **tentacles** are invariably hollow and almost uniformly retractile. Some groups are characterized by a spicular or trabecular calcareous or horny skeleton **(corallum)**.

Anthozoans are exclusively marine, and include a wide variety of common marine animals referred to as sea anemones, corals, soft corals, sea fans, sea pens, sea feathers, and so on.

KINDS OF ANTHOZOANS

The anthozoans are grouped in 3 subclasses: Ceriantipatharia, Octocorallia, and Zoantharia.

CERIANTIPATHARIA

The Ceriantipatharia include 2 orders: Antipatharia (colonial) and Ceriantharia (solitary). The colonies of the former are slender and branching, consisting of a horny skeletal axis covered by a thin coenosarc bearing polyps. Antipatharian polyps generally have only 6 simple tentacles, and 6, 10, or 12 complete mesenteries with very feeble musculature, 6 of which are considered primary, with 2 siphonoglyphs. Ceriantharian polyps lack any skeleton, and have 2 rings of tentacles. The numerous unpaired mesenteries are all complete, arranged in one cycle consisting of the 6 protocnemes in couples on either side of the single siphonoglyph, with many additional couples added serially and only in the dorsal intermesenterial space (Fig. 132,*c*). The **cerinula** larval stage of the ceriantharians resembles the antipatharian polyp, and both orders have 6 protocnemes in couples, of which the first-formed central transverse one is fertile. The subclass is virtually unknown in the fossil state.

OCTOCORALLIA

The Octocorallia are exclusively colonial in habit, with a calcareous spicular or horny (gorgonin) skeleton. The polyps have 8 pinnate tentacles and 8 complete mesenteries, which are directed ventrally from a directive pair said to be dorsal (Fig. 132,*d*). The subclass is known scatteringly by fossils from the Mesozoic onward.

ZOANTHARIA

The Zoantharia include solitary and colonial polyps, with calcareous trabecular skeletons in some orders. The polyps have simple or divided tentacles but never pinnate cyclically arranged ones and are especially distinguished by paired mesenteries. Basically there are 8 complete mesenteries and almost invariably 2 additional lateral couples, making a total of 6 pairs (Fig. 132, *e*), 2 of them being the dorsal and ventral directive pairs; generally additional cyclically arranged pairs are present. New mesenterial pairs are inserted according to several different plans (Fig. 132,*f,g,h*). The subclass is represented paleontologically from the Ordovician (or possibly earlier by a supposed anemone from the Cambrian) to the present day. It includes 6 orders separated largely on the presence or absence of a skeleton and by the mesenterial arrangement and development. These consist of the Zoanthiniaria, Rugosa, Heterocorallia, Corallimorpharia, Scleractinia, Actiniaria, and possibly the extinct Tabulata.

The Zoanthiniaria (zoanthids) are weakly colonial, anemone-like forms lacking a skeleton. The mesenteries are in a single cycle with 2 directive pairs, of which only the ventral one is complete, and new pairs of mesenteries beyond the first 6 are inserted only in the ventrolateral exocoeles on either side of the ventral directive pair (Fig. 132,*f*). They are unknown as fossils, but are believed by some to be representative of the most primitive zoantharians.

The Rugosa are an extinct group known only from Paleozoic rocks, and their relationships to other orders of the subclass are still speculative. The mesenterial arrangement and development can only be inferred from the mode of septal insertion, and this is capable of several interpretations. Many investigators are inclined to see closest affinities of the Rugosa in the Zoanthiniaria, from which they would differ only by having a calcareous exoskeleton and insertion of new septa (and mesenterial pairs) not only in exocoeles on either side of the cardinal septum (ventral mesenterial pair), as in the zoanthinarians, but also in spaces between the alar and counter-lateral septa (between lateral mesenterial pairs), that is, in 4 instead of 2 primary exocoeles (Fig. 132,*g*). Relationship to the endocoelactarian actiniarian anemones (Carlgren, 1918) and to the scleractinian corals has also been claimed (Schindewolf, 1942). The question is far from settled.

The Heterocorallia are a small, extinct Carboniferous group of doubtful affinities.

The orders Corallimorpharia, Scleractinia, and Actiniaria include the living sea anemones and corals, all of them lacking a skeleton except the Scleractinia, which are distinguished from the corallimorphs largely by this criterion. The mesenterial plan is practically identical in all and differs from that of the Zoanthiniaria and ?Rugosa in that the mesenterial pairs beyond the first cycle of 6 are inserted in dorsoventral order in all 6 primary exocoeles (hexactiniarian), rather than in 2 or 4 (Fig. 132,*h*), and in some groups septa may also develop in the endocoeles of the lateral pairs. The Scleractinia alone have a fossil record, from the Middle Triassic to Recent.

The Tabulata are a heterogeneous extinct Paleozoic group whose relations with other anthozoans are uncertain.

EVOLUTION

Three lines of anthozoan development are generally recognized: (1) ceriantipatharian, (2) octocorallian, and (3) zoantharian (Fig. 132).

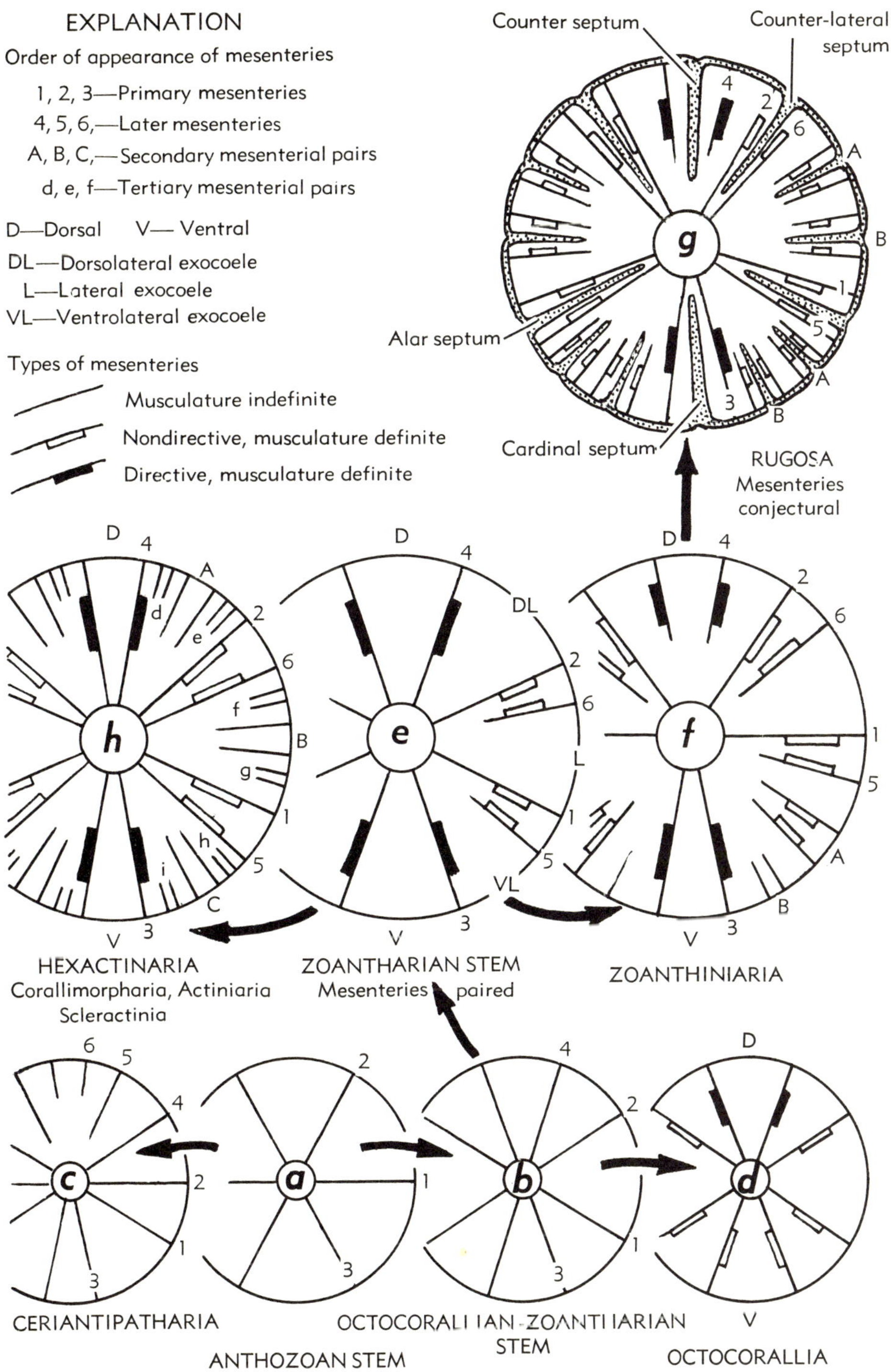

Fig. 132. Relations and mesenterial arrangements of principal anthozoan groups.

Antipatharians seems to represent a very old, primitive group related to the cerianthids. The cerinula larva of the latter is like the antipatharian polyp; both have simple tentacular and mesenterial arrangements and both have 6 mesenteries, of which the central transverse couple is largest. Cerianths diverge by development of additional couples in the dorsal intermesenterial space. The ancestral anthozoan type may have been a solitary polyp, something like the antipatharian polyp with 6 mesenteries in couples and 6 tentacles (Fig. 132,*a*), and by addition of a dorsal directive couple —the last to arise in ontogeny of the other groups—the basic octamerous plan of the other subclasses arose.

The ancestor of the octocorallian and zoantharian lines is considered by some to have been a skeletonless polyp with 8 mesenteries having very weakly developed musculature and 8 tentacles (Fig. 132,*b*), a form leading toward the Octocorallia. From the same stock perhaps arose the basic zoantharian type by the addition in the adult stage of all but the most primitive forms of 4 incomplete mesenteries for the total of 12 in 6 pairs (Fig. 132,*e*). Additional pairs beyond these, arising in the ventrolateral exocoeles, led to the zoanthiniarians (Fig. 132, *f*). The Rugosa may have been derived from these by the development of mesenterial pairs in the lateral as well as in the ventrolateral exocoeles (Fig. 132,*g*) and by appearance of an exoskeleton. The hexactiniarian corals and anemones (Corallimorpharia, Scleractinia, and Actiniaria) added mesenterial pairs in all 6 primary exocoeles, cyclically, in a dorsoventral order (Fig. 132, *h*). They are generally believed to have developed from another branch of the zoantharian stem, although some have tried to see the Rugosa as ancestral to the hexactiniarians, in a direct trend from insertion in 2 exocoeles (zoanthids) through insertion in 4 (rugosans) to insertion in all 6 (hexactiniarians).

Class ANTHOZOA Ehrenberg, 1834

[Less hydroids] [=Endoaria RAPP, 1829; Actinozoa DE BLAINVILLE, 1830 *(partim)*; Actinoidea DANA, 1846; Corallaria MILNE-EDWARDS & HAIME, 1850; Oecioa HUXLEY, 1852; Coralliaria M.-EDW.-H., 1857 *(partim)*; Cnidaires M.EDW.-H., 1857; Actinoida GOSSE, 1860; Polycyclia, Monocyclia *(partim)* BRONN, 1860; Coralligena HUXLEY, 1864; Cnidaria VERRILL, 1865; Coralla HAECKEL, 1866; Scyphopolypi, Scyphozoa *(partim)* GÖTTE, 1887; Anthozoariae DELAGE & HÉROUARD, 1901]

Exclusively polypoid, solitary or colonial, mostly sedentary coelenterates. Oral end expanded as oral disc with centrally located mouth surrounded by peristome, around which are one or more rings of hollow tentacles. Stomodaeum strongly developed, leading from mouth into gastrovascular cavity, generally with 1 or 2 siphonoglyphs. Gastrovascular cavity biradially partitioned into compartments by complete or incomplete mesenteries, some of which bear endodermal gonads. Some groups characteristically with horny or calcareous spicular endoskeleton or calcareous exoskeleton. Exclusively marine. *Ord.-Rec.*

Representatives of this class are divided into three subclasses:

I. Mesenteries unpaired:
   A. Tentacles simple....Ceriantipatharia
   B. Tentacles pinnate............Octocorallia
II. Mesenteries paired..................Zoantharia

REFERENCES

Carlgren, O.

(1) 1918, *Die Mesenterienanordnung der Halcuriiden:* Lund Univ. Årsskr., Neue Folge, Avd. 2, v. 14, 37 p., 1 pl., 25 fig.

Clark, A. H.

(2) 1913, *Cambrian holothurians:* Am. Naturalist, v. 47, p. 488-507.

Delage, Y., & Hérouard, E.

(3) 1901, *Les Coelentérés:* in DELAGE & HEROUARD, Traité de zoologie concrète (Paris), tome 2, pt. 2, x+848 p., 72 pl., 1102 fig.

Dollfus, G.

(4) 1875, *Note sur les empreintes attribuable à un actinie (?Palaeactis vetula) dans les schistes cambriens des Moitiers-d'Allonne:* Soc. nat. Sci., nat.-math. Cherbourg, Mém. 9 (sér. 2), p. 224-232, pl. 3.

Hyman, L. H.

(5) 1940, *The invertebrates: Protozoa through Ctenophora:* McGraw-Hill (New York), xii + 726 p., 221 fig.

Kiesling, A.

(6) 1939, *Scyphozoa:* in SCHINDEWOLF, O. H., Handbuch der Paläozoologie, Borntraeger (Berlin), Band 2A, Lief. 5, p. A69-A109, fig. 1-42.

Pax, F.

(7) 1924, *Hexacorallia:* in KÜKENTHAL's Handbuch der Zoologie, Band 1, p. 770-901, figs.

Ruedemann, Rudolf

(8) 1934, *Paleozoic plankton of North America:* Geol. Soc. America, Mem. 2, vii + 141 p., 26 pl., 6 fig.

Schindewolf, O. H.

(9) 1942, *Zur Kenntnis der Polycoelien und Plerophyllen; eine Studie über den Bau der "Tetrakorallen" und ihre Beziehung zu den Madreporarien:* Reichsamt Bodenforsch., Abh., Neue Folge, Heft 204, 324 p., 36 pl., 155 fig.

Walcott, C. D.

(10) 1911, *Middle Cambrian holothurians and medusae:* Smithson. Misc. Coll., v. 57, no. 3, p. 41-68, pl. 8-13.

CERIANTIPATHARIA

By J. W. Wells and Dorothy Hill

Subclass CERIANTIPATHARIA van Beneden, 1898

[=Hexacorallia Götte, 1902 (*non* Haeckel, 1866)]

Included here are 2 strongly divergent but related orders (1) Antipatharia, the "black corals," which are colonial forms characterized by a thorny skeletal axis of horny gorgonin, with small polyps having only 6 tentacles and 6 to 12 complete unpaired mesenteries, and (2) Ceriantharia, comprising relatively large solitary skeletonless anemone-like polyps with many tentacles arranged in 2 circles and an indefinite number of unpaired mesenteries. The larval stage of the ceriantharians resembles the antipatharian polyps, and both orders differ from other anthozoans in the very weak or indefinite musculature of the mesenteries and in the insertion of new mesenterial couples only in the dorsal intermesenterial space. *Mio.-Rec.*

Order ANTIPATHARIA Milne-Edwards & Haime, 1857

[=Antipathina Ehrenberg, 1834; Antipathidae Gray, 1840; Antipathacea Dana, 1846; Zoantharia cauliculata Milne-Edwards & Haime, 1850; Zoantharia sclérobasiques (Antipatharia) M.Edw.-H., 1857; Hexactina Bronn, 1860; Anticorallia Haeckel, 1896; Antipathidea Bourne, 1900]

Colonial, branching, with thorny horny axial skeleton surrounded by fleshy coenosarc bearing small polyps (Fig. 133,*2*). Polyps with 6 simple or 8 branched tentacles and 6, 10 or 12 complete coupled but unpaired mesenteries and having a single siphonoglyph. *Mio.-Rec.*

These corals are mostly non-reef dwellers and occur in depths of 1 to 1,900 fathoms. About 20 genera are known, only one of which has been found in the fossil state.

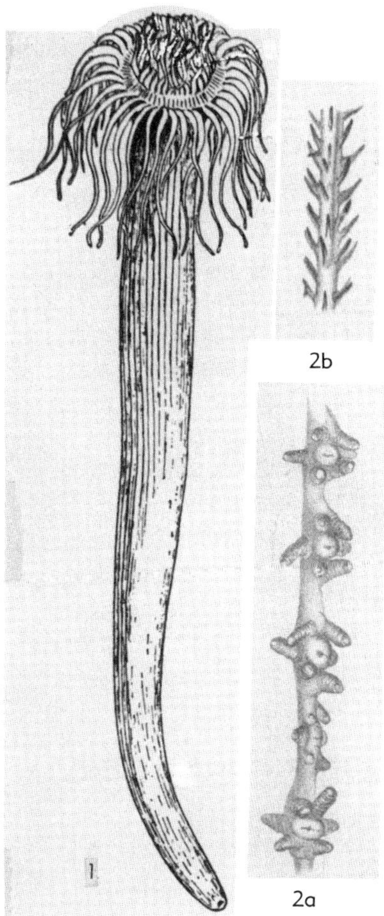

Fig. 133. Representatives of Ceriantipatharia. *1*, *Cerianthus* sp., Rec. (order Ceriantharia), ×1. *2*, *Antipathella subpinnata*, Rec., Medit., 135 fathoms; *2a*, polyps, ×5.5; *2b*, axis, ×26 (order Antipatharia) (*1*, L. H. Hyman, *Invertebrates: Protozoa through Ctenophora*, McGraw-Hill, New York; *2*, after Brook).

Leiopathes HAIME, 1849 [**Antipathes glaberrima* ESPER, 1792]. *Mio.*, Italy; *Rec.*, Medit.

Order CERIANTHARIA Perrier, 1893

[=Cerianthidae MILNE-EDWARDS & HAIME, 1851; Paranemata BRONN, 1860; Tetractiniae LANKESTER, 1877; Heterocoralla HAACKE, 1879 (*non* Heterocorallia SCHINDEWOLF, 1941); Cerianthiae HERTWIG, 1882; Cerianthidea BOURNE, 1900]

Solitary, lacking skeleton. Column elongate, without pedal disc on aboral end. Oral disc with many simple tentacles arranged in 2 rings. Mesenteries complete, coupled but unpaired, in a single cycle. New couples added serially only in dorsal intermesenterial space. *Rec.*

A small group of semiburrowing anemones (Fig. 133,*1*), unknown in the fossil state, unless some vertical cavities in shales and sandstones are traces of cerianthid burrows.

OCTOCORALLIA[1]

By FREDERICK M. BAYER[2]

WITH SECTION ON TRACHYPSAMMIACEA

By EUGENIA MONTANARO-GALLITELLI

CONTENTS

[1] Published with permission of the Secretary of the Smithsonian Institution.
[2] Associate Curator, Division of Marine Invertebrates, U.S. National Museum.

INTRODUCTION

Although the octocorals are not extensively known in the fossil record, their remains have been discovered in deposits as old as the Cretaceous, with doubtful records back to the Silurian. Forms with a solid, calcified axis, such as various species of *Corallium, Isis, Primnoa* and related genera, are noticed more frequently than those that are less durably constructed. By far the largest number of species are of a more transient nature; their only calcified skeletal elements, in the form of minute spicules, are so small as readily to escape detection in the rocks should they not be dissolved before realizing their fossil destiny.

Spicules belonging to octocorals are found in modern marine sediments and occasionally have been observed as fossils. From the amount of calcareous material that octocorals are thought to contribute to reef deposits, it is reasonable to expect their spicules as fossils in larger numbers than actually have been reported. Very possibly the seemingly hopeless problem of identifying such isolated spicular remains has prompted many investigators to ignore them when observed in samples. Effort to identify some octocoral fragments is indeed hopeless, but accurate generic determination of others can be made on the basis of a single sclerite. Many genera have specialized spicules which are absolutely characteristic, but, unfortunately, most of them also have simple sclerite forms that could belong to any of a dozen genera classified in 2 or 3 orders. Since the spicules are apt to be dissociated, it is impossible to determine examples of generalized types. Specific identifications, depending upon the arrangement and combination of spicules, would be possible only with ideally well-preserved material.

None of the standard paleontological works gives more than a brief account of the octocorals, and the systematic treatments of the Recent forms, not written for paleontologists, are of use only to specialized systematists. The aim of this work is to present brief diagnoses of the important genera, both Recent and fossil, with illustrations of their characteristic skeletal elements, as a review of the diverse features of alcyonarian hard parts, for the use of paleontologists in the analysis of marine deposits, bottom sediments, and similar samples.

Previously published illustrations of most species figured in this section of the volume on coelenterates were not found to be adaptable for purposes of the *Treatise;* so new drawings have been prepared in nearly all instances. Furthermore, species chosen to illustrate the genera are not always the genotypes, sometimes because the genotype was thought not to portray the genus to best advantage, and sometimes simply because specimens of it were not available to me for illustration.

ACKNOWLEDGMENTS

I could not in good conscience present the following work without taking notice of the assistance so willingly given by friends and colleagues. The Editor, R. C. Moore, has been most helpful and tolerant of seemingly inexcusable delays. I am grateful to F. A. Chace, Jr., for much good advice, especially in the field of nomenclature; to Helen Duncan, Julia Gardner, J. Brookes Knight, Alfred R. Loeblich, Jr., and John W. Wells, for assistance of many kinds; to Eugenia Montanaro-

GALLITELLI for preparing the section on Trachypsammiacea, with which she is quite familiar; to IRA B. HANSEN, EDITH E. MORTENSEN, HARDING B. OWRE, and J. Q. TIERNEY for criticisms of the manuscript; and to L. B. ISHAM, staff illustrator of the Department of Geology of the U.S. National Museum, for much advice in regard to the figures. I also wish to thank the authorities of the Smithsonian Institution and my supervisors at the United States National Museum for permitting me to undertake what has proved to be an unreasonably time-consuming project.

MORPHOLOGY

Like other Anthozoa, members of the subclass Octocorallia are exclusively polypoid coelenterates. Unlike hydroids, they have a mouth invaginated to form a tubular **pharynx,** and a **gastrovascular cavity** radially partitioned by thin, noncalcareous septa, sometimes called **mesenteries** (Fig. 134). The **autozooids** are of a remarkably uniform type throughout the subclass. They have 8 pinnately branched **tentacles,** which alternate with 8 septa, all complete—i.e., with their upper ends extending from body wall to pharynx. The pharynx, which is a compressed tube, has along one of its narrower sides a strongly ciliated longitudinal tract known as the **siphonoglyph** (or sulcus), which maintains a circulation of water into the gastrovascular cavity. The side of the polyp bearing the siphonoglyph is referred to as "ventral" in older literature, but modern authors call it the **sulcal side** and the opposite the **asulcal.** Longitudinal retractor muscles are situated on the sulcal faces of the septa. The latter, attached along their centrifugal edges to the body wall, join with the oral disc above and with the pharynx along their distal inner edges, thus completely partitioning the upper part of the gastrovascular cavity into 8 blind sacs ending in the tentacles. Below the inner end of the pharynx the edges of the septa hang free and slope rapidly out toward the body wall, where they continue downward as 8 ridges, of which the asulcal pair extends farthest into the gastrovascular cavity. The free septal edges are thickened to form usually convoluted cords called **filaments,** of which the asulcal pair are longest and strongly ciliated. The remaining 6, lacking in some polyps, are shorter and densely glandular.

The type of individual just described is found in all orders of Octocorallia. In certain groups modified individuals may occur also, in which event the basic polyp type is called the autozooid. Modified polyps, called **siphonozooids,** with one rudimentary tentacle or none at all, may be present. In the Pennatulacea, the primary or axial polyp is given a special name, oözooid, and another type of nontentaculate polyp may be represented by the so-called mesozooid. These modified kinds of polyps have little importance from a paleontological standpoint, and accordingly a detailed account of their anatomy is omitted.

The regular octamerous symmetry of the octocorallian autozooid, so obvious externally, is shown by microanatomical investigation to have a basic bilaterality superimposed upon it by the presence of the single siphonoglyph, the modified asulcal pair of septal filaments, and the arrangement of the retractor muscles. By virtue of these features an autozooid may be divided longitudinally into identical halves through only one plane—that which divides the sulcal and the asulcal pairs of septa and consequently the siphonoglyph. Its symmetry is thus radiobilateral.

The upper portion of the autozooid, called **anthocodia** (Fig. 134), is usually retractile, either within the general coenenchymal mass or into the stiffened lower part of the polyp itself, called **anthostele.** In the former case, the polyps in contraction are visible as small pores or stellate apertures, whereas in the latter a distinct **calyx** (verruca) always remains visible as a hillock on the rind (Fig. 134). The tentacles are either folded over the oral disc or introverted into the gastrovascular cavity when the polyp withdraws.

In some instances (Stolonifera, Telestacea), the polyps spring from a basal sheetlike or bandlike stolon and stand as erect, branched or unbranched tubes with walls commonly supported by spicules, but usually the lower portions of the polyps are imbedded in a thick, gelatinous substance containing spicules, the **coenenchyme,** together forming colonies of diverse architecture. The polyps occupy pits in the colonial coenenchyme, but unlike the stony corals, these pits are not marked by radial skeletal elements alternating in position with the septa; however, separate calcareous spicules may be so oriented around the base of the polyp and in its walls as to reflect its octamerous symmetry. The coenenchyme is penetrated by a complex system of tubules,

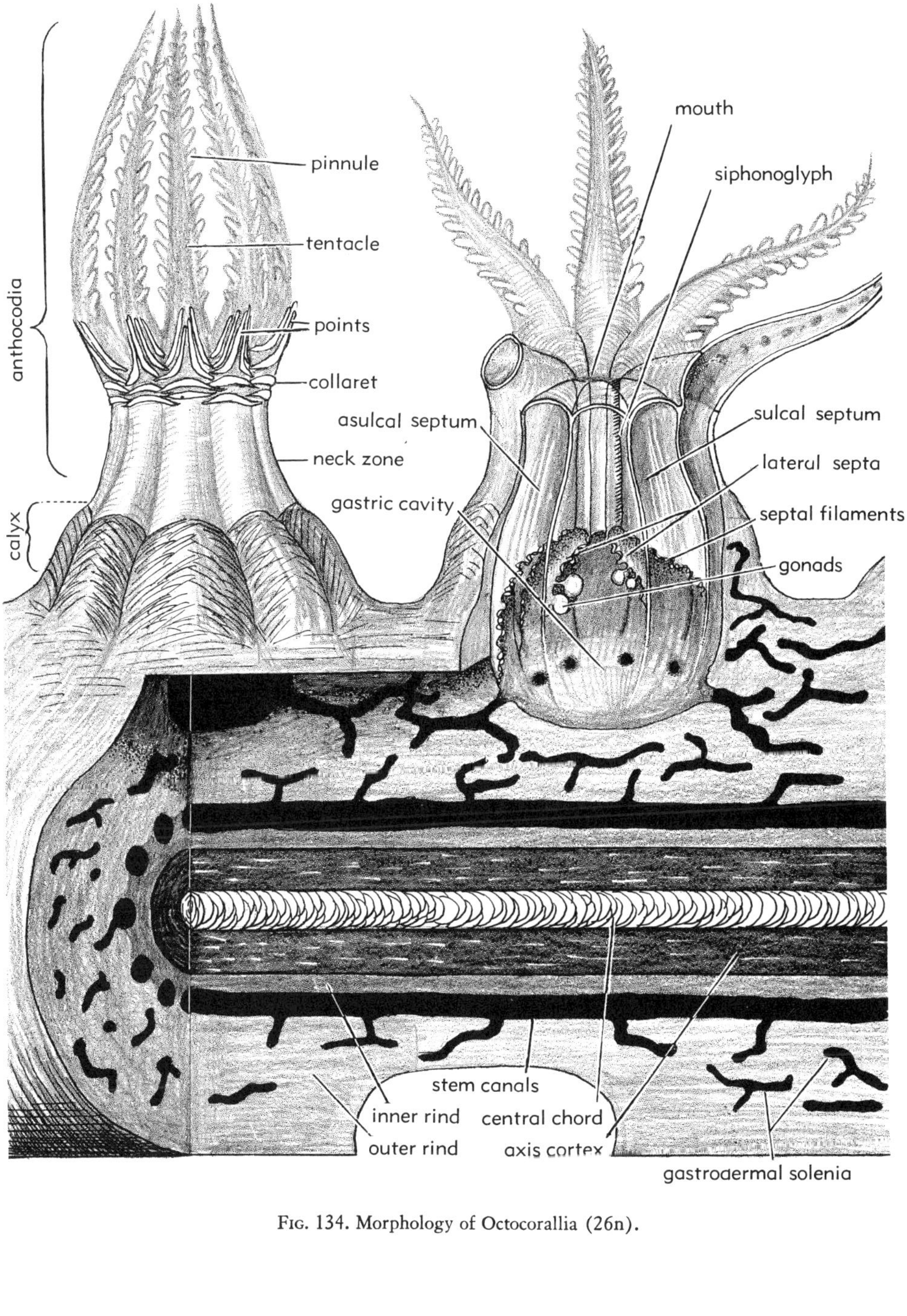

FIG. 134. Morphology of Octocorallia (26n).

the **solenia,** which are lined with gastrodermis and serve to interconnect neighboring polyps.

Commonly, sclerites are distributed in more or less symmetrical arrangement at the base of each tentacle and in the body wall between the septa. Below these tracts, the heavier spiculation of the anthostelar region may be symmetrically disposed, but in very many species this is not so.

In several octocorallian families the anthocodial spiculation is so arranged as to fold over the contracted tentacles and close the aperture of the anthostele like a lid, which commonly is recessed deeply within the calycinal orifice. In the Holaxonia, this structure has been utilized, especially by KÜKENTHAL and his students, to distinguish families (for example, Muriceidae, with operculum; Plexauridae and Acanthogorgiidae, both without operculum). The fact of the matter is that the spicules which make up an operculum are almost universally present and differ mainly in their degree of development. Thus it is impossible to distinguish between these families on a basis of the operculum alone, and keys making use of that feature are unreliable.

The opercula of paramuriceids, plexaurids, gorgoniids, and most other holaxonians consist of a few to many spindles arranged *en chevron* upon and immediately below the tentacle bases; the proximal spicules are ordinarily arranged transversely and form a ring (collaret or crown) around the polyp, upon which the 8 triangular spicular segments (points) rest (Fig. 134). These sclerites are chiefly interseptal in position. In some families, such as the Gorgoniidae and the Telestidae, there may also be some spicules in the body wall along the septa between any interseptal spicules that may be present. In Primnoidae with an operculum consisting of 8 platelike sclerites (all genera but *Primnoides*), the spicules are septally situated and therefore lie between the bases of the tentacles. This operculum, analogous to that of other families, is consequently homologous with the septal rods of telestids and gorgoniids, rather than with the functional opercular points of paramuriceids and plexaurids.

It is not only the arrangement of spicules upon the polyps that is of systematic importance, but also the size and form of the spicules, their distribution in various layers and zones of the coenenchyme, and in some groups even their color.

The basic spicule type, found in all orders (except Coenothecalia, which lack spicules), is the simple spindle, a monaxial rod which is more or less pointed at both ends. In the Pennatulacea they are usually 3-flanged and quite smooth, some being rather blunt. In other orders they are more or less strongly sculptured by spines, complicated warts, or other protuberances. If the spindles are relatively smooth, blunt and short, they may be referred to as rods, but if long, slender and sharp, as needles. Commonly they combine to form twins and cross-shaped deposits, and by various modifications produce scales, plates, thorn scales, clubs, capstans, dumbbells and other forms. Each of these may, in turn, be further specialized; such distinctive modifications will be discussed under the families and genera in which they occur.

Like other coelenterates, the Octocorallia consist of 2 epithelial layers—an outer epidermis derived from ectoderm, and an inner gastrodermis derived from endoderm—separated by a connective substance called the **mesogloea.** The latter is very thin in some places, such as the tentacles and septa, although elsewhere, as in the fleshy colonial tissue between the polyps, it is thick and usually packed with calcareous skeletal elements. The epidermis is a flat, cuboidal or columnar epithelium or syncytium, and consists of supporting, glandular and sensory elements in an outer stratum that generally secretes a thin cuticle, nervous cells in a middle stratum, and muscular cells (which may be imbedded in the mesogloea below) in the innermost stratum. Specialized stinging structures **(nematocysts)** are usually abundant in the surface stratum. The gastrodermis is constructed upon much the same plan but is more commonly found to be syncytial; nematocysts are restricted to the septal filaments. Intracellular symbiotic algae **(zooxanthellae)** are extremely abundant in many species.

The mesogloea is a jelly-like material containing amoeboid cells derived from the epidermis. Some of these become cnidoblasts, producing nematocysts, which ultimately make their way to the outer surface. Others become scleroblasts, producing calcareous sclerites or a horny material, called **gorgonin,** which is a protein that has a lower sulphur content than true horn and commonly contains bromine and iodine united to tyrosin (11). In the anthocodia the mesogloea is thin and in many forms lacking spicules entirely; it constitutes the inner support for the septa, on the sulcal side of which it is raised into longitudinal plates where the retractor muscles are inserted.

The mesogloeal scleroblasts, which produce spicules, account for all of the skeletal elements of the coenenchyme and for certain axial structures. In the Gorgonacea, the axial structures are produced in 2 ways: (1) by more or less complete consolidation of spicules in the medullar region through the agency of horny or calcareous deposits (Scleraxonia); or (2) by the secretion of a horny, or calcareous and horny, central rod by the cells of a tubular upgrowth from the original basal plate of epidermis (Holaxonia). The latter is thus an external structure over which the colony is spread, as it were; whereas the former, made up of mesogloeal spicules and in many species penetrated by gastrodermal solenia, is an internal one. The axes of Scleraxonia and Holaxonia, although analogous in their supporting function, are produced in different ways and are not homologous.

It is characteristic of the Octocorallia that new polyps are asexually produced not by fission but by gemmation from the solenial system of the basal stolon or of the coenenchyme. The colonies so produced are of exceedingly diverse form. In one type, simple tubular polyps spring from a creeping stolonic ribbon or sheet (Fig. 135,*1*). There is rarely much development of mesogloea in the zooid walls, which lack solenial tubes and consequently do not produce daughter polyps. Most members of the Stolonifera follow this plan. In *Tubipora,* however, the spicules of the polyp walls fuse together inseparably and the numerous polyps, bound together by transverse, platform-like secondary stolons, eventually produce colonies of considerable mass. As the polyps ascend, they partition off the lower ends of their rigid tubes by double tabulae, consisting of a funnel-like lower part and an upper, lenticular cap, placed on a level with transverse platforms. The tubes are not filled in below and, for a time at least, living tissue remains within them. New polyps arise from the basal stolon and from the horizontal platforms (Fig. 135,*2*). In the Telestacea, in which the body walls of the polyps have become permeated by solenial tubes, new polyps may arise from the walls of older individuals, thus producing arborescent colonies of moderate size. The sequence of budding seems to be quite constant and forms the basis for separation of the several genera and species comprising the order. A schema of the ramification as seen in *Telesto* is given in Fig. 135,*3*, and the variation displayed by other genera is shown in Fig. 137, *1a,3a,4a*. In the order Alcyonacea, the mesogloea of the polyp walls has developed into a common colonial coenenchyme, in which the endoderm-lined gastric cavities of the polyps are imbedded, and through which ramifies a complex network of solenial tubes from which new polyps arise (Fig. 135,*4*). In certain families, the mesogloea separating the long, closely fascicled primary gastric cavities is more or less reduced and may contain few or no spicules. The resultant colonies are arborescent and, in some cases, closely approximate the appearance of gorgonaceans. In the latter order, however, the gastric cavities of all polyps are uniformly short. In the suborder Scleraxonia, they penetrate the branches only as deep as the medullar zone, into which a few gastrodermal canals may extend. In the Holaxonia, the medullar zone is not a densely spicular coenenchyme but instead a horny axis cylinder (with more or less calcareous matter) into which, of course, no solenia can penetrate (Fig. 135,*6*). In these forms, the cortical coenenchyme, now better termed rind, is comparatively thin and the colonies are usually much branched and arborescent in appearance. In some Scleraxonia, the medullar zone becomes thoroughly consolidated through fusion of the included spicules, and a distinct axis results. These always can be distinguished from the axes of holaxonians by their spicular structure, which is clearly revealed in thin section. In both Scleraxonia and Holaxonia the axis may be either continuous or divided into horny nodes and calcareous internodes. The horny nodes of the articulate scleraxonians invariably contain numerous spicules imbedded in a spongy matrix of gorgonin, whereas those of the holaxonians are pure, translucent gorgonin.

The axis in all Holaxonia is made up of a basic **central chord** (Fig. 134) laid down at the growing branch tips by an axis epithelium, and a secondary cortical layer deposited around the central chord everywhere below the growing tip by an axis epithelium of a different character, which continues to increase the girth of the axis as long as the colony grows. The secondary thickening of the axis appears to be accomplished in 2 distinctly different ways. In the first, the secondary axis epithelium produces scales or fusiform bundles of horny gorgonin more or less completely permeated by calcareous matter, which are closely imbricated or otherwise tightly bound together in roughly concentric layers. In the second, the epithelium deposits columns, irregular prisms or rays of chiefly calcareous material radiating outward from the central chord. Each of the radial elements appears to be the work of a group of epithelial axoblasts acting as a unit.

These axial features, readily demonstrable in thin sections under polarized light, because of the orientation of the component calcareous fibers, may prove to be of diagnostic value for fossil material. The actual relationships of many gorgonacean fossils reported from the Cretaceous upward may well be clarified by a re-examination of their basic axis structure.

The members of 2 orders stand clearly apart from other octocorals in the structure of their colonies. These are the blue coral (Coenothecalia), which lacks spicules entirely but produces a massive calcareous skeleton similar to those of some hydrozoans; and the sea pens (Pennatulacea), which have evolved a highly specialized type of colony symmetrically developed around a long primary polyp which usually contains a supporting axial rod.

In the Coenothecalia the gastric cavities of the spiculeless polyps occupy the upper ends of tabulate tubes in a massive fibrocrystalline coenosteum. The tubes of the polyps are separated by numerous narrower tubes, also tabulate, occupied by caeca of the solenial system, which extend downward from the surface network.

In the Pennatulacea, the autozooids are symmetrically arranged around the primary oözooid, from whose walls they have budded, with radial orientation, in some, and bilateral in others. In the second instance they may be situated in longitudinal, oblique or transverse rows, and may be partly or wholly united into broad polyp leaves. The axial rod, which lies between the 4 longitudinal canals of the stalk and rhachis, is calcified in a radial pattern similar to that described above for certain holaxonian families.

MORPHOLOGICAL TERMS APPLIED TO OCTOCORALLIA

Many of the terms used in describing the morphological and anatomical features of octocorals are the same as those applied to other Anthozoa and to coelenterates in general. Among terms more closely associated with the Octocorallia some are derived from similarities with everyday objects, being thus readily understandable, but a few border on the ridiculous, especially when translated from one language to another.

GLOSSARY OF MORPHOLOGICAL TERMS

anthocodia. Upper, tentacular part of polyp which in many cases can be retracted within the rind or calycular structure (Fig. 134).

anthostele. Lower, thickened part of polyp's body wall into which the anthocodia may be withdrawn, commonly stiffened by spicules; may be equivalent to calyx.

asulcal. Side of polyp opposite siphonoglyph.

autozooid. Polyp with 8 well-developed tentacles and septa; only kind of polyp in monomorphic species and major type in dimorphic species.

axis. Central supporting structure of Gorgonacea and Pennatulacea; in the former, it may be spicular, consolidated or unconsolidated, or horny, with more or less nonspicular calcareous matter.

axis epithelium. Layer of cells derived from ectoderm that produces the axis of Holaxonia.

axoblast. Individual scleroblast of the axis epithelium.

balloon club. *See* spicule.

bark. Rind of holaxonian Gorgonacea.

bracket. *See* spicule.

buccal scale. *See* spicule.

calyx. Wartlike projecting anthostele (Fig. 134).

capstan. *See* spicule.

cell string. Nonlumenate strand of cells meandering through coenenchyme.

club. *See* spicule.

cnidoblast. Cell that produces nematocyst.

coenenchyme. Colonial spiculiferous mesogloea.

collaret. Transverse, subtentacular ring of spicules (Fig. 134).

cortex. Outer coenenchymal layer of gorgonaceans, especially Scleraxonia; outer, horny layer of holaxonian axis, as opposed to its medulla (central chord or core) (Fig. 134).

crown and points. Transverse collaret with superposed opercular rays.

disc spindle. *See* spicule.

double club. *See* spicule.

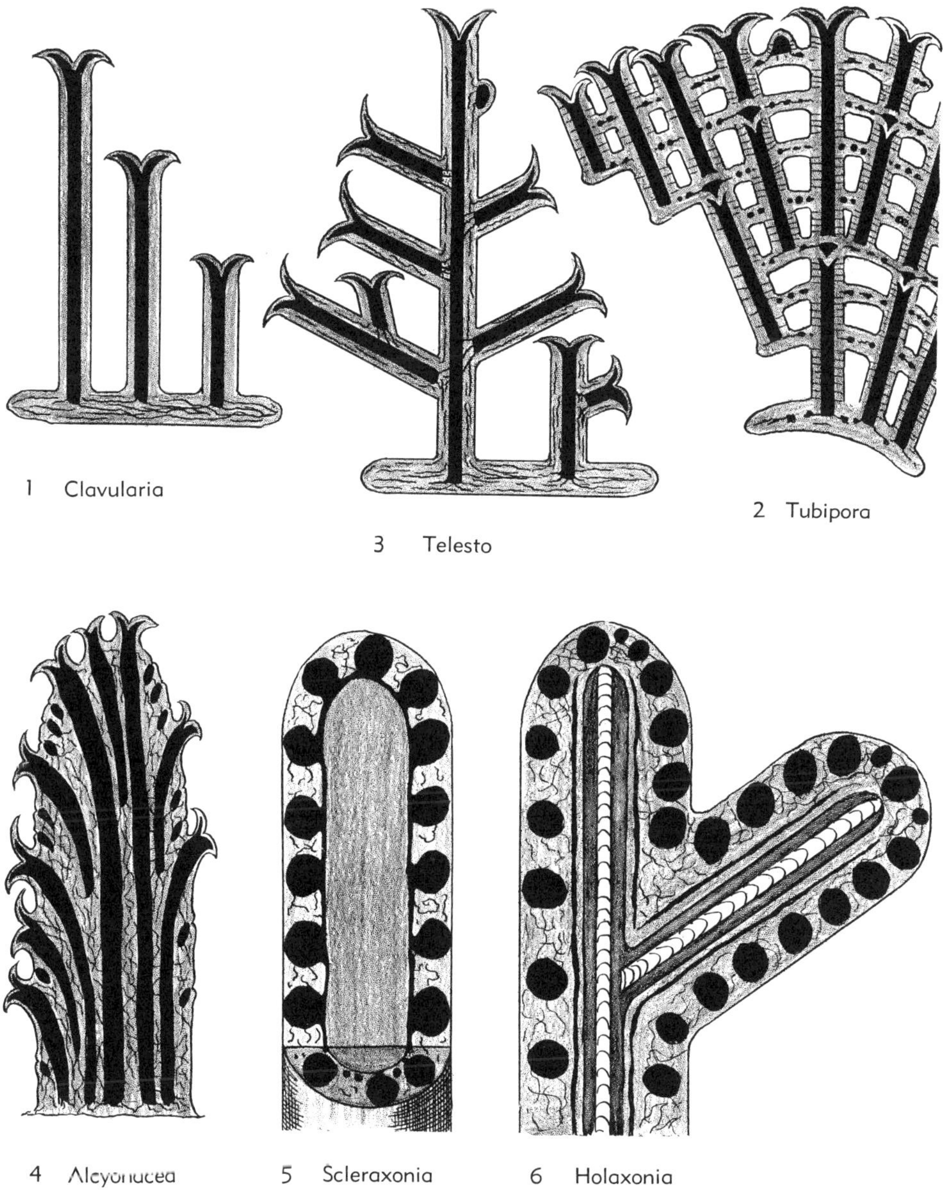

Fig. 135. Colony formation in Octocorallia (26n).

double wheel. *See* spicule.

filament. Thickened, convoluted edge of septum (mesentery); in Octocorallia, filaments of 2 septa opposite siphonoglyph are very long and heavily flagellated, whereas those of remaining 6 are shorter and glandular (Fig. 134).

gastrovascular cavity. Interior space of polyp radially partitioned by septa.

infrabasal scale. *See* spicule.

leaf club. *See* spicule.

loculus. Crescentic or lenticular calcified area or fiber-filled space in holaxonian axis (especially Plexauridae).

medulla. Central zone of scleraxonian stem; rarely, central chord of holaxonian axis.

mesentery. Soft septum.

mesogloea. Jelly-like substance separating the 2 cellular epithelial layers of coelenterates.

nematocyst. "Stinging cell" or "thread cell" formed within cnidoblasts.

operculum. Anthocodial spicular apparatus that more or less closes calyx or protects tentacles in contraction.

oral disc. Distal integument of polyp surrounding mouth, enclosed by ring of tentacles, forms roof over gastrovascular cavity.

pharynx. Tubular passageway between mouth and gastrovascular cavity.

pinnule. Digitate lateral branch of tentacle of octocoral polyps (Fig. 134).

plate. *See* spicule.

polyp. Individual of octocorallian colony (see autozooid).

polyp leaf. United proximal part of adjacent polyps that produces a leaflike expansion, from margin of which anthocodiae project; found in several genera of Pennatulacea and Ainigmaptilidae.

polypary. Colony as a whole.

pseudosepta. Spinelike or toothlike skeletal projections in calicles of Coenothecalia; bear no constant relationship with soft septa of polyps.

radiate. *See* spicule.

rind. Outer, spiculiferous coenenchyme of Holaxonia (Fig. 134).

scale. *See* spicule.

scaphoid. *See* spicule.

sclerite. Calcareous skeletal element of mesogloea, irrespective of form.

scleroblasts. Ectodermal cells of mesogloea that produce calcareous spicules.

septum. Thin radial noncalcareous partition (composed of soft tissue) dividing gastrovascular cavity of polyp (*see* mesentery).

siphonoglyph. Strongly ciliated groove extending down one side of pharynx (Fig. 134).

siphonozooid. Polyp with reduced tentacles or none, and commonly reduced septal filaments; usually much smaller than autozooids.

solenium. Canal lined with gastrodermis penetrating coenenchyme and interconnecting gastric cavities of the polyps (Fig. 134).

spicule. Calcareous skeletal element of mesogloea, irrespective of form; properly, a long, sharp sclerite. Specialized forms include:

- **balloon club.** Spicule with practically smooth, spheroidal head (characteristic of *Eunicella*).
- **bracket.** Same as scaphoid.
- **buccal scale.** Thin flat plate just proximal to operculars (Calyptrophorinae).
- **capstan.** Monaxial rod with 2 whorls of tubercles and terminal tufts.
- **club.** Spicule enlarged at one end, may be strongly sculptured.
- **disc spindle.** Derivative of capstan, with tubercles of 2 whorls fused into discs or wheels.
- **double club.** Bispheroidal, 2-handled sclerite (*Corallium*).
- **double wheel.** Disc spnidle.
- **infrabasal scale.** In primnoids, sclerite situated between basal body scales and scales of rind.
- **leaf club.** Spicule with head ornamented by foliate processes.
- **plate.** Flat sclerite too thick to be called scale.
- **radiate.** Sclerite with processes radiating in one plane (stellate), or in various planes in a more or less symmetrical order.
- **scale.** Thin, flat or nearly flat sclerite.
- **scaphoid.** Spindle more or less distinctly bent in form of a "C" and with warts of convex side reduced.
- **spindle.** Straight or nearly straight monaxial spicule pointed at both ends.
- **thorn club.** Clublike sclerite with processes of head sharp and spiny.
- **thorn scale.** Tabular sclerite with central or marginal projecting process.
- **torch.** Club with foliate processes of head strongly laciniate, resembling burning torch.
- **wart club.** Clublike sclerite with only low, blunt protuberances ornamenting the head.

spindle. *See* spicule.

stolon. Creeping basal expansion, ribbon-like or membranous, from which polyps of Stolonifera and Telestacea arise.

sulcal. Side of polyp nearest siphonoglyph.

sulcus. *See* siphonoglyph.

supporting bundle. Sheaf of supporting spicules below anthocodia of nephthyid polyp.

thorn club. *See* spicule.

thorn scale. *See* spicule.

torch. *See* spicule.

verruca. Calyx; tubercle of sclerite.

wart club. *See* spicule.

zooid. In more recent literature in English, any individual of colony, irrespective of its morphological specializations, thus =polyp; in most German works, =siphonozooid.

BIOLOGY

NUTRITION

Many octocorals are largely holozoic in their nutrition, capturing small, planktonic organisms in large numbers. Small crustaceans and larval mollusks are frequently observed in the gastrovascular cavities of their polyps. The common purple or yellow sea whip, *Leptogorgia viminalis* (PALLAS), of the southeastern coast of the United States, is often found to contain large numbers of lamellibranch larvae; where this gorgonian is common, as from Chesapeake Bay to northwest Florida, it must be able to decimate the populations of larval bivalves that pass among its branches.

On the other hand, a number of other octocorals, notably some Xeniidae and a species of *Clavularia,* have been found to live perfectly well without ever ingesting food; these species contain vast numbers of symbiotic algae, the zooxanthellae, in their gastrodermal layers. They thrive in the presence of adequate sunlight, even in filtered water. In the dark, however, they soon necrose and die, even if amply provided with food, which they seem completely unable to ingest (8). These species with abundant zooxanthellae in their tissues furthermore may have very few nematocysts (in some species I have found none at all) and greatly reduced digestive structures. Conversely, abundant nematocysts and well-developed digestive tissues are usually correlated with a small number or complete absence of zooxanthellae. Evidence as to whether the corals actually digest their zooxanthellae as food is inconclusive, but, in some at least, a sort of reciprocal interchange system has developed and the algae are not eaten. In these, the coelenterates seem able to utilize directly the photosynthetic products that the algae elaborate in excess of their own nutritional requirements, and the algae make use of some of the metabolic byproducts of the animal.

REPRODUCTION

Ordinarily it is the autozooids that are fertile, and the sexes are separate; but in some groups the siphonozooids contain the reproductive tissues, and in some, protogyny is usual.

The reproductive cells, which are derived from the endoderm, develop on the edges of the 6 septa away from the siphonoglyph, where they form conspicuous swellings. Ripe eggs commonly are connected to the septal edge only by a narrow stalk, and are covered by a layer of gastrodermis. The sex products may be discharged into the water, where fertilization and subsequent development take place, or the eggs may be retained after fertilization until late in the planula stage. There is no vestige of a medusoid generation.

ECOLOGY AND DISTRIBUTION

The Octocorallia are widely distributed organisms that occupy a great variety of ecological situations. Nevertheless, there are certain conditions, such as desiccation, heavy sedimentation, and reduced salinity, that they rarely tolerate to any extent. Some species, of course, are able to endure such conditions more or less regularly: the Alcyonacea of tropical Pacific reefs may be regularly exposed by the tides, and the common Atlantic *Leptogorgia viminalis* inhabits inshore waters that may be both muddy and brackish. As a rule, however, most species live below the low-tide mark, where they are safe from desiccation, occurring in clean water with a minimum of constant sedimentation. They extend around the world from Arctic to Antarctic, from the low-tide mark to the abysses. Most species are found on the continental shelf and slope, but a few descend to great depths. The deepest record for the subclass belongs to a pennatulacean, *Umbellula leptocaulis* KÖLLIKER, which has been taken at 2,440 fathoms (4,440 m.). The isidid gorgonian, *Keratoisis profunda* (WRIGHT), claims the deepest record for its order, at 2,300 fathoms (4,200 m.); and species of *Corallium* have been collected from depths down to 2,200 fathoms (4,000 m.) in the Atlantic and 1,314 fathoms (2,400 m.) in the Pacific.

Of the 7 orders, members of only one, the Pennatulacea, are regularly adapted for existence on a soft substrate. They have a stalk, the lower end of which is dilatable, serving to anchor the colony in mud or sand. All other octocorals are typically fixed by a basal attachment to some solid object, although a few species have developed alternative adaptations for holding the colony in place. Notably, some alcyonaceans enclose a lump of mud in a basal expansion, and others produce rootlike, fleshy rhizoids, both effective solutions to the problem of fixation. The Telestidae characteristically form colonies that arise from creeping stolons, but individuals of certain species are able to live unattached on sandy bottoms. Associated with them are found gorgoniids and ellisellids that have made the same adjustment to their habitat. Other holaxonian gorgonaceans, for example certain genera of Isididae, have become adapted to a soft substrate by development of rootlike, calcareous processes in place of the usual incrusting, membranous attachment. Many Chrysogorgiidae and a few Primnoidae likewise can do so, or can attach by the conventional means to solid objects, as circumstances require.

The extent to which octocorals contribute to the formation of reef limestone is not known, but it is certain that they are prolific producers of calcium carbonate. At the Dry Tortugas, the spicule content of living colonies on the reef was estimated to average 5.38 tons per acre; with about 1/5 of the population being destroyed annually (based on a 5-year period), about one ton of calcareous material was made available per acre per year (5). Since this material is in the form of minute spicules, easily transported by currents or dissolved, probably no very high percentage of it actually becomes incorporated into the reef mass. On many tropical Pacific reefs the situation is much different. There, the luxuriant growth of gorgonaceans so characteristic of Caribbean reefs is greatly reduced and in most places completely lacking. The predominant octocorals of the reefs are massive alcyonaceans, which, having an even more densely spicular coenenchyme than the West Indian gorgonaceans, may actually become consolidated with the reef while the upper parts of the colonies are still living (6). No estimates have been made of the calcium carbonate production by the alcyonacean populations of these reefs, but it seems obvious that a higher percentage of it would actually be incorporated into the reef rock even though growth rate and productivity might be substantially lower. Some reefs in the tropical Pacific support a heavy growth of the blue coral, *Heliopora,* which can be a locally significant reef former.

SYMBIONTS AND COMMENSALS

As a result of their sedentary nature, octocorals have fallen victim to a considerable array of epizoic hangers-on, and a number of commensal, symbiotic and parasitic relationships have developed. Some of these associations are distinctly beneficial to the coral host, but most are not obviously so. How many of them are detrimental is difficult to say. The host can be killed, as usually happens when a colony of *Millepora* obtains a foothold on the accidentally exposed axis of a gorgonian. The hydrozoan by its rapid growth speedily pushes the living gorgonian substance farther and farther back until it occupies the entire axis and the gorgonian dies. Such occurrences are entirely fortuitous, however, depending upon the chance contact of a millepore larva in a settling mood with a suitably denuded axis. At any other time, the larva would continue on its way; at any other place on the gorgonian it might be captured and eaten, or at least prevented from settling. The millepore is not dependent upon the octocoral for support; it can form its colony independently elsewhere. In many other cases, however, the dependent partner indeed cannot live elsewhere. A complicated relationship has developed, as in the case of the ovulid gastropods, which seem completely dependent upon their alcyonarian hosts. The exact nature of this dependence is not known; the mollusk may feed directly upon the polyps without ever devouring enough of them to damage the colony, or it may only browse upon foreign materials that collect on the coral colony. In any event, alcyonarians infested with ovulid snails never seem to be unhealthy or damaged in any way. Another interesting mollusk-octocoral relationship is that between *Rapa* and *Sinularia.* The gastropods live imbedded deep in the fleshy tissue of the octocoral, with no direct means of communication with the outside; water currents maintained by the polyps apparently suffice for their respiration, and their nutrition is probably derived in some way from host tissues.

Most, if not all, alcyonarian corals are infested with copepods of the peculiar family Lamippidae. In some cases, the commensals live in the gastrovascular cavities of the individual polyps, but in a great many others they inhabit the colonial gastrodermal canal system, which may be enlarged locally to accommodate them. Most of these copepods have degenerated into small, grublike creatures as a result of their confined habitat. Even more peculiar are the copepods apparently related to *Ophioika*, which have been found in the polyps of *Paramuricea* in the north Atlantic and of *Paracis* in the central Pacific. These are so modified as to bear little resemblance to conventional copepods. Some pycnogonids spend their youth in the polyps of *Chrysogorgia*, but details of their life history are unknown (22). These various arthropods betray their presence in the host only by localized galls or swellings, or by enlargement of individual polyps. Annelid commensals, however, frequently induce more conspicuous external modifications upon their hosts. They cause the polyps of primnoids to enlarge their lateral scales to form archways along the branches, in which the worms live. The polyps of acanthogorgiids, which have only slender fusiform sclerites, respond by sending out fleshy lobes that meet to form similar tunnels along the branches. In octocorals without projecting polyps, polychaetes may inhabit deep grooves in the coenenchyme, and in a few other species they live in the gastrodermal canal system.

Since it is impractical to mention here, even briefly, all of the associations in which alcyonarians are involved, I give below a systematic list of the principal organisms that have been observed to take part in such associations, together with their host corals.

Associations of Octocorallia with Other Organisms

Algae. ZOOXANTHELLAE: widespread in octocorals of warm, shallow waters.

Protozoa. FORAMINIFERA: commonly borne by telestids and other octocorals with an outer cuticle.

Porifera. Sponges may envelop colonies of *Telesto, Anthothela, Parisis* and other octocorals, which seem not to suffer as a result.

Coelenterata. HYDROZOA: *Hydrichthella epigorgia* STECHOW on *Anthoplexaura dimorpha* KÜK., *Millepora* on various Gorgonacea. ANTHOZOA: *Sagartia acanellae* on *Acanella arbuscula, Palythoa* sp. on *Corallium inutile, Parerythropodium coralloides* on *Eunicella verrucosa.*

Ctenophora. *Tjalfiella* and other Platyctenea on various octocorals (11).

Platyhelminthes. A polyclad was found clinging to branches of *Paracis orientalis* dredged in the Marshall Islands.

Brachiopoda. Unidentified species on *Chrysogorgia* (13).

Echinodermata. CRINOIDEA: comatulids often found clinging to gorgonaceans. OPHIUROIDEA: gorgonocephalid *Astrophyton muricatum* living entwined in branches of shallow-water gorgonaceans, especially *Antillogorgia,* in West Indian region; other gorgonocephalids and trichasterids occasionally discovered on branches of deeper-water gorgonaceans as *Schizostella* on *Swiftia* and *Muricea* (W.Indies); *Asteroschema* on *Narella* (Hawaii) and *Nicella* (W.Indies), *Astrogomphus* on *Thesea* (W.Indies), and many others.

Annelida. POLYCHAETA: *Leucodore* (=*Polydora*) in corallum of *Heliopora;* Polynoidae especially frequent upon Primnoidae, but found also on octocorals of other families, as *Harmothoë* on *Arthrogorgia, Narella, Thouarella* and *Corallium; Polyeunoa* on *Acanthogorgia* and *Thouarella; Polynoë* on *Anthomastus, Corallium, Acanthogorgia, Stenella, Acanella, Isidella* and *Pennatula;* Syllidae of the genera *Haplosyllis* and *Typosyllis* recovered from gastrodermal canals of every specimen examined of *Anthogorgia* (Indo-Pac.) and *Acanthacis* (W.Indies) (*fide* M. H. PETTIBONE).

Arthropoda. PYCNOGONIDA: *Ascorhynchus* sp. in polyps of *Chrysogorgia papillosa* (22). COPEPODA: Lamippidae in polyps and in gastrodermal canals of many octocorals; *Ophioika*-like forms in the gastrovascular cavities of *Paramuricea* (Atl.) and *Paracis* (Pac.). CIRRIPEDIA: lepadomorph barnacles sometimes found clinging to gorgonaceans; verrucomorph species observed on Telestidae; balanomorphs commonly forming galls on branches of gorgoniids or imbedding themselves in fleshy forms; Ascothoracica not uncommon on branches of Chrysogorgiidae. ISOPODA: Sphaeromidae found on rind of plexaurids and gorgoniids in W.Indies and Panama; Arcturidae on *Primnoella* in Antarctic. AMPHIPODA: caprellids sometimes found clinging in large numbers to colonies of Gorgoniidae. DECAPODA: anomurans *Orthochela* on *Eugorgia* and *Uroptychus* on *Calyptrophora* and *Chrysogorgia.*

Mollusca. APLACOPHORA: Solenogasters taken from branches of gorgonian corals. GASTROPODA: essentially entire family Ovulidae commensal on Alcyonacea and Gorgonacea; in W.Indies *Cyphoma* common on *Antillogorgia* and *Plexaura*, and *Simnia* upon *Leptogorgia* and *Gorgonia;* coralliophilid snail *Rapa* living imbedded in *Sinularia* (*fide* A. B. BRONSON); small dorid nudibranch common on living *Heliopora*. CEPHALOPODA: eggs of *Sepia* occasionally found attached to *Paramuricea placomus*.

Chordata. PISCES: egg-cases of small sharks attached to branches of gorgonians (13).

METHODS OF STUDY

It is practically impossible to outline methods of study for fossil Octocorallia. Perhaps it may suffice to describe techniques applicable to Recent forms, in order to indicate methods of getting at pertinent characters. From this foundation the paleontologist may then be able to erect a system of procedure appropriate to the material that confronts him.

When one consults a key to the orders of Octocorallia, such as that in DEICHMANN's monograph (7), it immediately becomes obvious that general macroscopical characters are of considerable value: whether or not the colony is solidly attached to a firm surface or only stuck in a muddy sea floor, and the way its component polyps are distributed upon it; whether there is a horny axis or not; whether the colonies are arborescent, lobate, or creeping; whether the gastrovascular cavities of the polyps are long (reaching to base of the colony) or short; whether the polyps are simple or produce branchlike daughters from their side walls; and whether the polyps are of one or more kinds. Many such characters can be observed directly and require no special techniques; some require dissection of fresh material to reveal structures that probably never could be fossilized; and a few require microscopic examination, as, for instance, the distribution of horny material in the coenenchyme.

In modern gorgonaceans the average diameter of the final branches is a useful character, for it seems to remain reasonably constant regardless of the total size of the colony. Also, in the terminal region the canal system and spiculation are not so likely to have suffered modifications attendant upon secondary growth. Therefore it is usually desirable to base conclusions on a study of this area rather than of older parts. As might be expected, in view of this fact, a small specimen is fully as satisfactory for study as a large one, and a fragment may supply all the information necessary for determination. Sad to relate, the same happy state of affairs does not obtain among the sea pens, of which a young specimen can be treacherously misleading, nor among the nephthyid alcyonaceans, in which the variations that occur from base to apex of the colony are important.

Color is usually quite fast in the Octocorallia, and where it is also constant it provides a useful character. In a few species the characteristic life color is due to pigmentation of the soft tissues and fades almost immediately when a specimen is placed in preservative. The greenish, brown, or yellowish coloration of many tropical reef-dwelling species is due to intracellular symbiotic algae, the zooxanthellae and zoochlorellae. The neontologist thus finds color to be of inconsistent value; the paleontologist, however, may find the persistent skeletal coloration of such things as tubipores (red) and heliopores (blue) of some use in the analysis of tropical reef deposits.

Of the 7 orders, the members of 3 (Pennatulacea, Gorgonacea, Trachypsammiacea) produce substantial, fossilizable axial structures. Those of the pennatulids are slender and rodlike, being known to the paleontologist under the generic name *Graphularia*. The axes of sea pens, in general, are quite uniform throughout the group and, to judge by Recent material, it is in most cases completely impossible to distinguish between species or even higher categories on external axial characters alone.

The axial structures of Gorgonacea can often be identified to family or genus with reasonable certainty. The jointed ones, belonging to the Melitodidae, Parisididae and Isididae, are recognizable at once and can be distinguished from one another by means of thin sections which reveal the spicular or nonspicular nature of the calcareous joints. In the Holaxonia, the structure of the central chord or medulla and its relationship to the cortical layer, and the concentric distribution of horny and calcareous substances are significant and must be investigated by sections also. Polarized light may reveal features indistinct or invisible in ordinary illumination, such as the orientation of the calcareous fibers in both axis and spicules.

In general, the calcareous spicules form the most important single character in the determination of octocorals. In certain families (Chrysogorgiidae, Primnoidae), the arrangement of sclerites on the polyps is as important as the shape of the spicules themselves, and in these the zooidal armature must always be studied intact on the polyps, as well as disassociated. In several other families, the organization of the zooidal armature must be considered, but this is somewhat less important than form of the coenenchymal spicules. No specimen of fossil octocoral in which the original arrangement of spicules has been preserved is known to me.

The isolated spicules recovered from marine deposits are probably best handled in a routine fashion as micropaleontological specimens. In order to study the spicules of Recent specimens it is necessary to free them from surrounding tissues. Traditionally, this is done by boiling a small fragment of the colony in a strong solution of caustic soda or potash, which is a cumbersome, messy, and potentially dangerous method. Far more practical is the use of sodium hypochlorite solution, which is readily obtainable as a laundry bleach under several trade names, or as a pharmaceutical. In some places, the commercial solution is known as "antiformin," in others as "Javelle water." It is quick, safe, and simple to use; for routine spicule examinations it is merely necessary to place a fragment of the coral on a slide, drop a small amount of "clorox" on it with a medicine dropper, and wait for a few moments. The bubbles that form can be removed by adding some water and removing excess fluid with bibulous paper. The preparation may then be covered with a glass coverslip and is ready for study. Should it be desirable to make a permanent spicule mount, the sample may be processed in a small vial to facilitate the repeated washings required to remove all traces of reagents. An alcoholic (or distilled water) suspension of spicules is pipetted onto a clean slide and allowed to dry. Any neutral mounting medium may be used, but some of the synthetic resins are better than balsam, which has a refractive index close to that of the spicules themselves.

If one wants to observe the arrangement of opercular elements, it is necessary only to dissect out the anthocodial region and place it in dilute clorox, which will act more slowly. As the tissues are dissolved they swell and become transparent, thus separating and clearly revealing the opercular spicules. Of course, the soft tissue is destroyed completely before long, so the preparation should be watched carefully under a stereoscopic microscope. If it is desirable to keep the spicular arrangement intact for a longer period, the polyp may be placed in a dilute solution of potassium hydroxide for a short time and then removed to water, whereupon the soft tissues swell, separating the spicules, and become transparent, rendering the arrangement of spicules clearly visible. With skill and some practice, such a preparation can be washed, dehydrated, and permanently mounted in balsam.

CLASSIFICATION

HISTORICAL SUMMARY

The classification of the octocorals can be divided historically into three periods: (1) the herbalistic, from ARISTOTLE to LINNÉ, during most of which corals were thought of and classified as plants and referred to in erudite polynomials; (2) the Linnean, from the 10th Edition of the *Systema* to KÖLLIKER's *Icones,* during which it became accepted that corals of all kinds were animals and the Linnean system of nomenclature became established, making possible a crude expression of relationships; and (3) the modern, from KÖLLIKER's *Icones* to the present, during which the relationships of various groups were investigated and systems of classification drawn up by the use of modern methods.

Little need be said about the pre-Linnean, herbalistic period of coral classification, save that it is probably as old as civilization itself and that the octocorals, along with other coelenterates, were universally thought of as plants. Black corals (the axes of Antipatharia and Gorgonacea), precious coral *(Corallium rubrum)*, and the jointed white coral *(Isis hippuris)* were known to the ancients, who valued them for their magical, medicinal, and decorative properties. Hickson (1924, *Introduction to the Study of Recent Corals*, Univ. Manchester) gives an extensive history of the commerce in these corals.

Descriptions and illustrations were, for the most part, poor. The earliest (but not the worst) illustration of a gorgonian that I know of appears in Gesner's *De Rerum Fossilium, Lapidum et Gemmarum*, 1565. The best are probably those in Knorr's *Deliciae Naturae Selectae*. Although the value of the skeletal spicules in the classification of octocorals was not discovered until a hundred years later, these structures were first figured in John Ellis' *Essay on the Natural History of the Corallines*, 1755.

The principal difference between the classification of alcyonarian corals during the years immediately preceding the 10th edition of Linné's *Systema* and that of the hundred years following it is found in the expression of natural relationships made possible by Linnean nomenclature.

In organizing the corals for his *Systema Naturae*, Linné brought together most of the species previously described and arranged them in 6 genera. *Tubipora, Madrepora, Isis, Gorgonia, Alcyonium*, and *Pennatula*. Pallas (1766) and Esper (1788-1830) recognized the same genera, with minor shifting of species and addition of others. New genera were added in profusion by Lamarck (1801, 1816), Lamouroux (1812, 1816), Ehrenberg (1834), Valenciennes (1855), and Milne-Edwards & Haime (1857). Lamarck (1801) began expanding the classification, dividing his order of radiates *(polypes à rayons)* into sections, the 2nd of which included the octocorals, distributed between 2 subdivisions. There was no grouping into families, and it is not likely that Lamarck thought of any category in a familial sense. Lamouroux grouped the genera first into families, later into orders that did not differ, except in name, from his original families. It was Ehrenberg (1828, 1834) who first really elaborated on the classification of zoophytes, creating orders, tribes, and families within a "circle" Anthozoa. Unfortunately, his scheme was marred by its artificial duality, being divided into orders "Phytocorallia" and "Zoocorallia" each with identical tribes (e.g., Octactinia). The families of zoocorallian tribes were distinct from those of the phytocorallian tribes; e.g., the families Isidea and Gorgonina were phytocorallian Octactinia, whereas the Xenina, Tubiporina, Alcyonina, and Pennatulina were zoocorallian Octactinia. Although Ehrenberg was the first to penetrate deeper than superficial appearances and to use the characters of the polyps themselves in devising his classification, his artificial subdivision rendered the resulting system untenable. Therefore, Dana, Milne-Edwards, and other subsequent workers reunited the parts of the Octactinia into a single taxonomic unit variously called Zoophytaria (de Blainville), Alcyonaria (Dana), and Octocorallia (Haeckel).

Valenciennes (1855) was apparently the first to recognize the value of the calcareous spicules in the classification of alcyonarian corals, but he failed to put his observations into actual practice. His *Extrait d'une Monographie de la Famille des Gorgonidées* was only an incomplete outline and served little purpose in advancing knowledge of relationships in the group. Thus it fell to Kölliker to substantiate the suggestions of Valenciennes and put the classification of octocorals on a sounder basis through the publication of his *Icones histiologicae*. Subsequently, Verrill, Saville-Kent, E. P. Wright, and Théophile Studer adopted and expanded the spicular system introduced by Kölliker.

The arrangement of the families and orders of Octocorallia, essentially as we now know it, emerged from the work of A. E. Verrill, with improvements by Hickson, Kükenthal, and others. This system, as emended by Hickson (1930), is basically employed herein. It is admittedly not perfect, and a re-examination of specimens and re-evaluation of characters is certain to bring about many changes. I have already modified the families of Scleraxonia, largely in accord with Verseveldt's researches on the anatomy of several genera, and recognize a 7th order, Trachypsammiacea MONTANARO-GALLITELLI, which is established herein.

PHYLOGENY

In his chapter on geographical distribution of the Gorgonacea, KÜKENTHAL (13) presents a phylogenetic tree of the group based partly upon inferences from his systematic data. Although voluminous and valuable, these data contain erroneous elements, which led KÜKENTHAL to some conclusions that I believe untenable. Moreover, he neglected to reckon fully with paleontological evidence that several groups of octocorals have persisted with very little change through an appreciable part of geological time, and with the consequent implication that many not readily fossilizable octocorals may have existed as long. The distribution of certain modern genera, especially in the tropical American region where we are reasonably sure of Cenozoic paleogeography, furnishes evidence supporting the same point.

KÜKENTHAL's supposition that the 2 species of *Erythropodium* recognized by him, because of their discontinuous distributions in the Atlantic and Indo-Pacific, represent the stocks from which the Atlantic and Indo-Pacific groups of briareids developed independently is quite untenable because its basic premise, that *E. caribbaeorum* and *E. marquesarum* are congeneric, is fallacious. These membranous forms were united in a single genus by KÜKENTHAL in spite of their obvious differences, whereas *Iciligorgia* of the Caribbean is kept separate from *Semperina* and *Solenocaulon* of the Indo-Pacific in spite of their obvious strong similarities. It seems quite clear that the origin of Atlantic and Pacific groups of Scleraxonia antedates the geographical discontinuity of those areas, and that the 2 groups have common elements.

HICKSON (10) has discussed evolution of the Pennatulacea, concluding that the group is derived from an *Anthomastus*-like alcyonacean ancestor. This inference is supported by the peculiar alcyonacean known as *Bathyalcyon,* apparently related to *Anthomastus.* In this odd genus, the colony consists of a monstrous primary autozooid, corresponding to the primary oözooid of the Pennatulacea, with many siphonozooids budding from the solenial system of its thickened outer wall. Although *Bathyalcyon* is an attached form, insofar as known specimens go, its relative, *Anthomastus,* may be either permanently attached to a hard substrate or anchored in a soft bottom by fleshy processes of the lower stalk.

Other workers have derived the sea pens from *Telesto*-like colonies in which secondary autozooids originate from the solenial system of the body wall of the primary polyp. Whatever its origin, the Pennatulacea is certainly a highly specialized group of octocorals.

The interrelationship of the various orders and the origin of the subclass from other Anthozoa may never be fully understood. If the tabulate Coenothecalia have phylogenetic affinity with any of the ancient Tabulata, *Heliopora* is a survivor of a very old lineage. But we will probably never know if any of the older tabulates are octocorals, and the tabulate structure of *Heliopora* may simply represent an independent solution to problems attendant upon upward growth and thus have no phylogenetic significance.

It is pertinent to remark that phylogenetic assumptions must be based on tangible morphological facts, on careful interpretation of geographical distributions, and on paleontological evidence. We must attain a much wider knowledge in these fields before a realistic phylogeny of the Octocorallia can be synthesized. We may never be sure of their origin. A remark by KÜKENTHAL on this subject is apt: "Der Ursprung der Alcyonarien ist in Dunkel gehüllt, und auch die Paläontologie vermag uns nichts darüber auszusagen."

SUPRAGENERIC TAXONOMIC DIVISIONS

The subclass Octocorallia is divided into 7 orders. Of these, one (Pennatulacea) is sharply differentiated from the remaining 5, which are less clearly distinguished one from the other. An examination of extensive material indicates that many genera, as usually regarded, are unrealistically deployed within the existing families, some of which also fail of proper definition. In consequence, I have redefined some families and established or re-established others, and reassigned certain genera, to arrive at a systematic structure useful not only for the collections of the U. S. National Museum but also in further phylogenetic studies on octocorals.

The suprageneric divisions of the Anthozoa Octocorallia, as recognized herein, are listed in the following tabulation, which also records the known geologic range of each. The numbers enclosed in parentheses indicate totals of described genera considered as valid.

Suprageneric Divisions of Octocorallia

Stolonifera *(order)*, polyps arising from a membranous or bandlike creeping stolon, without lateral budding; spicules when present mostly slender, rather spinous spindles (5). *Cret.-Rec.*
- Cornulariidae (1). *Rec.*
- Clavulariidae (3). *Cret.-Rec.*
- Tubiporidae (1). *Rec.*

Telestacea *(order)*, polyps arising from stolons but producing daughter polyps by lateral budding from solenial system; spicules slender and spinous or blunt and tuberculate, inseparably fused together in some (4). *Rec.*
- Telestidae (3). *Rec.*
- Pseudocladochonidae (1). *Rec.*

Alcyonacea *(order)*, polyps proximally imbedded in fleshy, spiculiferous mesogloea, primary individuals extending from base of colony to its top; secondary individuals also extremely elongate; monomorphic or dimorphic (38). *L.Jur.-Rec.*
- Xeniidae (7). *Rec.*
- Alcyoniidae (13). *Rec.*
- Astrospiculariidae (1). *Rec.*
- Nephtheidae (8). *L.Jur.-Rec.*
- Siphonogorgiidae (6). *Rec.*
- Viguieriotidae (3). *Rec.*

Trachypsammiacea *(order)*, polyps surrounded by calcified coenenchymal vesiculae in radiating rays and short, interlocking longitudinal rows, arising from a central longitudinal canal system (1). *Perm.*
- Trachypsammiidae (1). *Perm.*

Coenothecalia *(order)*, massive calcareous skeleton made up of dimorphic, tubular, tabulate corallites comprising large autopores with pseudosepta, and small "siphonopores" (4). *Cret.-Rec.*
- Helioporidae (3). *Cret.-Rec.*
- Ceratoporellidae (1). *Rec.*

Gorgonacea *(order)*, colonies branched treelike, but in some flagelliform, membranous or lobate, the polyps having uniformly short gastrovascular cavities; spicules of diverse form, sculptured with warts or spines (116; 5). *Cret.-Rec.*
- Scleraxonia *(suborder)*, branches consisting of an inner medullar zone composed of horny material and calcareous spicules which may be inseparably fused, and an outer cortical zone with all spicules free, the polyps being imbedded in this zone; monomorphic or dimorphic (26). *Cret.-Rec.*
  - Briareidae (2). *Rec.*
  - Anthothelidae (11). *Rec.*
    - Anthothelinae (1).
    - Semperininae (3).
    - Spongiodermatinae (7).
  - Subergorgiidae (1). *Rec.*
  - Paragorgiidae (2). *Rec.*
  - Coralliidae (2). *Cret.-Rec.*
  - Melithaeidae (6). *Tert.-Rec.*
  - Parisididae (2). *Tert.-Rec.*
- Holaxonia *(suborder)*, branches consisting of an inner axis of horny material which may have in it much calcareous matter but no true spicules (except Keroeididae, which have a chambered medulla unlike Scleraxonia), and a fleshy cortical zone containing spicules and the polyp cavities (90; 5). *Cret.-Rec.*
  - Keroeididae (1). *Rec.*
  - Acanthogorgiidae (3). *Rec.*
  - Paramuriceidae (25). *Rec.*
  - Plexauridae (12). *Rec.*
  - Gorgoniidae (8). *Rec.*
  - Ellisellidae (6). *Cret.-Rec.*
  - Ifalukellidae (2). *Rec.*
  - Chrysogorgiidae (7). *Rec.*
    - Lepidogorgiinae (1).
    - Chrysogorgiinae (5).
    - Chalcogorgiinae (1).
  - Primnoidae (13; 5). *Cret.-Rec.*
    - Primnoeidinae (1). *Rec.*
    - Primnoinae (7;5). *Cret.-Rec.*
    - Calyptrophorinae (3). *Rec.*
    - Callozostrinae (2). *Rec.*
  - Ainigmaptilidae (1). *Rec.*
  - Isididae (12). *Cret.-Rec.*
    - Keratoisidinae (4). *Rec.*
    - Mopseinae (3). *Eoc.-Rec.*
    - Muricellisidinae (1). *Rec.*
    - Isidinae (2). *Cret.-Rec.*
    - Incertae sedis (2). *Cret.*

Pennatulacea *(order)*, free-living colonies consisting of primary oözooid with barren lower part that acts as a stalk and anchor, and upper part which produces secondary individuals singly, in groups, or fused together into bilaterally arranged leaflike processes. Spicules mostly smooth rods, bars or needles, commonly 3-flanged, or small plates or discs. Axial rod usually present (41). *?Sil., Rec.*
- Sessiliflorae *(suborder)* (23). *?Sil., Rec.*
  - Veretillidae (5). *Rec.*
  - Echinoptilidae (2). *Rec.*
  - Renillidae (2). *?Sil., Rec.*
  - Kophobelemnidae (3). *Rec.*

Anthoptilidae (1). *Rec.*
Funiculinidae (1). *Rec.*
Protoptilidae (3). *Rec.*
Stachyptilidae (1). *Rec.*
Scleroptilidae (2). *Rec.*
Chunellidae (2). *Rec.*
Umbellulidae (1). *Rec.*
Subselliflorae *(suborder)* (13). *Cret.-Rec.*
Virgulariidae (7). *Cret.-Rec.*
Virgulariinae (6). *Tert.-Rec.*
Balticininae (1). *?U.Cret., Rec.*
Pennatulidae (3). *Rec.*
Pteroeididae (3). *Tert.-Rec.*
Suborder and family uncertain (5). *Sil.-Cret.*

SYSTEMATIC DESCRIPTIONS

Subclass OCTOCORALLIA Haeckel, 1866

[=Octactinia EHRENBERG, 1828; Zoophytaria DE BLAINVILLE, 1834; Alcyonaria DANA, 1846]. The name Octocorallia is used in preference to prior terms because the 1st, Octactinia, implies relationship with Actiniaria, and was also part of EHRENBERG's unusual dual classification; the 2nd, Zoophytaria, a relict of the botanical associations of these corals, has diverse connotations undesirable in modern usage; and the 3rd, Alcyonaria, although widely used, has been employed for the order Alcyonacea as well as for the subclass as a whole and thus leads to confusion.

Sedentary, colonial Anthozoa with polyps invariably bearing 8 tentacles, almost invariably pinnate; and 8 complete septa (mesenteries); pharynx with one "ventral" siphonoglyph. Skeletal structures usually consisting of calcareous spicules; in many, an axis of more or less calcified, hornlike substance is also present which may be in the form of fused spicules; rarely a massive, fibrocrystalline corallum. *?Sil., Rec.*

Order STOLONIFERA Hickson, 1883

Colonies composed of tall or short, cylindrical or blunt-conical polyps springing from a basal stolon which creeps upon solid objects; stolon ribbon-like, a reticulum of flat bands, or a thin flat sheet. Distal, tentacular part of polyps usually retractile within the stiffened proximal part (anthostele); connecting tubes or platforms may join adjacent polyps. New individuals budded only from stolon or connecting platforms, never from body wall of fully developed polyps. Spicules, when present, are usually slender, spinose rods; stolons and polyps usually invested with a horny external cuticle, which in some forms is the only skeletal structure. Colonies usually not conspicuous, although some may attain considerable size. *Cret.-Rec.*

The value of the horny cuticle in establishing limits of this order needs to be more thoroughly explored. Such a formation occurs in the Telestacea as well as in the Stolonifera, but in neither group are its extent, origin, and formation fully understood. A study of the internal anatomy and histology of both groups may throw light on their proper relationships.

At present, the only features safely utilizable for discrimination are structure of the individual polyps, general character of the stolon, and details of spiculation.

In the Stolonifera the polyps commonly have a well-defined proximal zone stiffened by densely packed spicules, into which the oral part, tentacles and all, may be retracted. The polyps arise from the stolon, which may be a meandering ribbon adhering to the substrate, a complex network of such ribbons, or a membranous sheet, but in none is there much thickening of the mesogloea. The anthostelar tubes of adjacent polyps may be connected by transverse tubes or platforms, which can give rise to new polyps just as the stolon itself is able to do, but daughter polyps are never budded off from primary individuals.

The form of the calcareous spicules is an indispensable character in classification of most stoloniferans. When present, the spicules are almost exclusively long, slender, and sharply thorny; they do not intrude the lower part of the gastric cavities. The arrangement of sclerites in and below the tentacles may be of significance.

TERMIER & TERMIER (23) include the Paleozoic families Syringoporidae MILNE-EDWARDS & HAIME (*Syringopora* GOLDFUSS, 1826), Syringoalcyonidae TERMIER-T. (*Syringoalcyon* TERMIER-T., 1945), and Auloporidae NICHOLSON (*Aulopora* GOLDF., 1826; *Cladochonus* M'COY, 1847; *Ceratopora* GRABAU, 1899; *Vermipora* HALL, 1874) in the Stolonifera. The skeleton in these forms has not been established to be spicular in nature nor to resemble in structure the calcareous skeletons of Recent octocorals. Most paleontologists refer the forms mentioned to the Tabulata, whose relationships are obscure and probably diverse. OKULITCH (20) considers the Syringoporidae, Halysitidae, and Favositidae to be possible Paleozoic alcyonarians.

Family CORNULARIIDAE Dana, 1846

Small stoloniferans with completely separate polyps and no spicules; a delicate, horny cuticle invests stolons and anthosteles. *Rec.*

Cornularia LAMARCK, 1816 [**Tubularia cornucopiae* PALLAS, 1766; SD M.EDW.-H., 1850]. *Rec.*, Medit.-IndoPac.

Family CLAVULARIIDAE Hickson, 1894

Stoloniferans with small to large tubular polyps, armed with spicules which are usually separate but may be inseparably fused, and commonly (?invariably) also a horny cuticle. Spicules generally in the form of spinose spindles arranged longitudinally in body walls and more or less branched sclerites in stolons; tentacles may have spicules like those of polyp body, or small discs, or none at all. *Cret.-Rec.*

Of the 14 genera which HICKSON (10) refers to this family, only 2 seem clearly referable to it; the others fall either as synonyms of these 2 or out of the family altogether.

Clavularia DE BLAINVILLE, 1830 [**C. viridis* QUOY & GAIMARD; SD M.EDW.-H., 1850]. Polyps simple, arising from bandlike or membranous stolons; spicules including simple spinous spindles, rods, or clubs, and a few double forms. *Rec.*, IndoPac.-Atl.——FIG. 136,*1*. **C. viridis*, IndoPac.; *1a*, spindles of body wall (26n); *1b*, discs from tentacles (26n); *1c*, group of polyps (34).

Epiphaxum LONSDALE, 1850 [**E. auloporoides*]. Axis attached throughout, formed of solid fibers; visceral cavities seated in the axis and provided with 8 dentations or blunt lamellae; investing layer unknown. Possibly precursor of *Sarcodictyon*. *Cret.*, Eng.

Sarcodictyon FORBES, 1847 [**S. catenatum*]. Stolons form anastomosing network; anthosteles low and conical; spicules solidly fused together. Anthocodial armature of 8 triangular fields of spindles. *Rec.*, NE.Atl.-C.Pac.

Family TUBIPORIDAE Ehrenberg, 1828

[*nom. correct.* DANA, 1846 (*pro* Tubiporina EHR., 1828)]

Polyps long, forming rigid tubes by solid fusion of sclerites, tubes perforate, partitioned by peculiar double tabulae having funnel-shaped lower and lenticular upper element, nearly parallel, closely placed, joined by transverse platforms containing solenia; new polyps arising from stolons or from platforms, not from walls of old polyps. *Rec.*

For the time being, this family must be taken to include only a single genus with one rather variable species. Its alliance with *Pachyclavularia* and *Hicksonia* (=*Clavularia*) requires verification.

Tubipora LINNÉ, 1758 [**T. musica*]. *Rec.*, IndoPac.——FIG. 136,*2*. **T. musica*, Rec., Carolinel.; *2a*, part of corallum; *2b*, free and partly fused spicules of anthocodiae; *2c*, minute pharyngeal sclerites; *2d*, tentacular sclerites (26n).

Order TELESTACEA Hickson, 1930

Erect, branching colonies arising from bandlike, reticulate or membranous stolons; primary polyp produces daughters from its body wall; sclerites numerous, free, or partially or entirely fused to form rigid tubes; lower part of gastrovascular cavities may be filled in with spiculiferous intrusion tissue. *Rec.*

This order is comprised of 2 families clearly distinguished by their methods of colony formation: (1) Telestidae, which produce colonies by monopodial branching that results from lateral budding of very long axial polyps; and (2) Pseudocladochonidae, with sympodial colonies resulting from lateral budding of successive orders of subordinate polyps all of equal length. The external grooving of the outer wall, anthocodial armature, and structure of the lower part of the zooids are characters of importance in classification. The nature and extent of the external cuticle and cuticular lining of the coelentera are yet to be investigated in detail.

The members of this order are distinguished from stoloniferans by production of daughter polyps from the body wall of the axial or primary polyp, and from alcyonaceans by the absence of thick, spiculiferous mesogloea between the polyps.

Apparently, no Telestacea have been detected in the fossil record, although the species with fused sclerites are surely sturdy enough to allow of preservation. The branching of the Recent *Pseudocladochonus*, which has rigid, though delicate, body tubes, is strikingly like that of the Paleozoic fossil

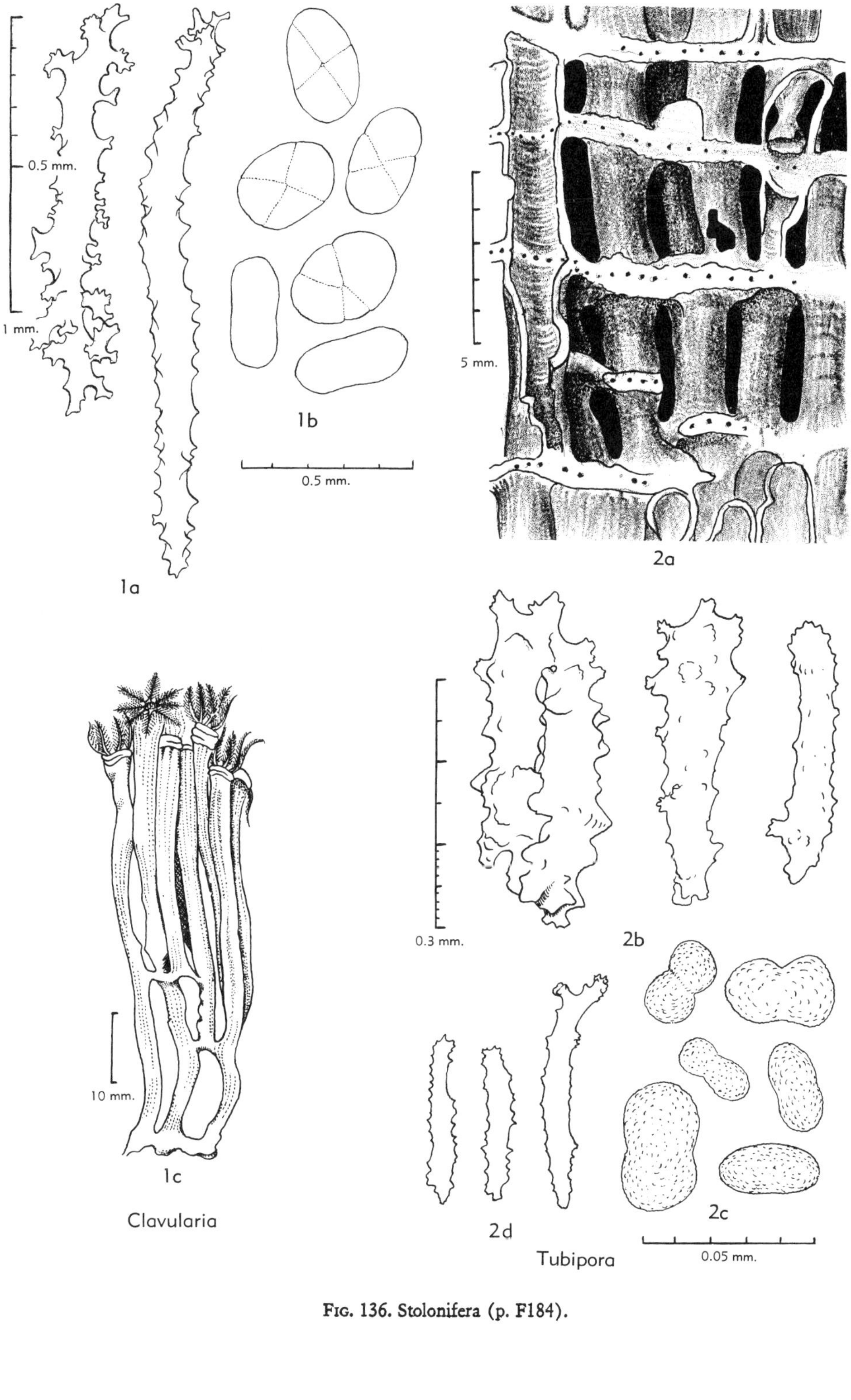

FIG. 136. Stolonifera (p. F184).

Cladochonus, but this feature seems to speak for homeomorphy rather than close phylogenetic relationship.

In modern seas, telestaceans occur widely from the low-tide line to 1,675 fathoms (3,180 m.).

Family TELESTIDAE Milne-Edwards & Haime, 1857

Colonies monopodial, with the primary polyp usually much taller than those of subordinate rank. *Rec.*

Telesto Lamouroux, 1812 [**T. aurantiaca;* SD M. Edw.-H., 1850]. Stolons reticulating or ribbonlike, producing tall axial polyps and shorter lateral polyps; lower parts of gastrovascular cavities not filled in with spiculiferous mesogloeal substance, and body walls not especially thickened. Sclerites long, spinose rods or blunt, coarsely tuberculate ovals, in some species fused solidly together, either in lower parts of polyps or throughout length of anthosteles. *Rec.*, Atl.-IndoPac.——Fig. 137,*2,5*. *T. sanguinea* Deichmann, Rec., Fla.; *2a*, small colony; *2b*, sclerites from body walls; *2c*, sclerites from anthocodiae; *5*, cross section of axial polyp near base, diagrammatic (26n).

Coelogorgia M.Edw.-H., 1857 [**C. palmosa*]. Colonies erect, arborescent, like *Telesto* but stolons not extensively developed, axial polyps very long and laterals very short; body wall of primary polyp greatly thickened by mesogloea perforated with solenia; gastrovascular cavities of axial polyps open to base, not filled in with spiculiferous intrusion tissue. Sclerites are tuberculate spindles and clubs. *Rec.*, Ind.O.——Fig. 137,*3,6*. **C. palmosa*, Rec., E.Afr.; *3a*, part of small colony; *3b*, spicules; *6*, cross section of axial polyp near base, diagrammatic (37).

Telestula Madsen, 1944 [**T. septentrionalis*]. Colonies like *Telesto* but smaller and less profusely branched; stolons bandlike; axial polyps long, sinuous, subordinates shorter; lower part of gastric cavities of both primary and secondary polyps filled in with spiculiferous intrusion tissue, leaving open only 8 canals. Sclerites of body wall usually blunt, coarsely tuberculate oval plates; those of intrusion tissue much-branched irregular forms more or less fused together. *Rec.*, Atl.-IndoPac.——Fig. 137, *1,7*. **T. septentrionalis*, Rec., Iceland; *1a*, small colony lacking secondary polyps (31); *1b*, spicules (31); *7*, cross section of polyp near base, diagrammatic (26n).

Family PSEUDOCLADOCHONIDAE Madsen, 1944

All polyps about equal in length, each giving rise near its distal end to one daughter at an acute angle (where branching occurs, 2 on opposite sides of parent polyp), thus forming zigzag, sympodial colonies; below point of origin of its daughter, each polyp is partially filled in with spiculiferous mesogloeal substance which leaves open 8 canals. *Rec.*

Pseudocladochonus Versluys, 1907 [**P. hicksoni*]. Spicules of body walls much branched, anastomosing, fused to form rigid but delicate tubes; spicules of mesogloeal intrusion tissue larger and less branched. *Rec.*, IndoPac.——Fig. 137,*4,8*. **P. hicksoni*, Rec., Japan; *4a*, part of a colony; *4b*, fused sclerites of body tube; *4c*, fused sclerites of intrusion tissue; *4d*, spicules from pinnules; *4e*, spicules from tentacle backs; *8*, cross section of polyp near base, diagrammatic (*4a-e*, 26n; *8*, 36).

Order ALCYONACEA Lamouroux, 1816

[*nom. transl.* Studer, 1887 (*ex* suborder Alyconacea Verrill, 1865; *nom. transl. et correct. ex* order Alcyoneae Lamx., 1816)] [Although the type genus of this order is *Alcyonium*, present rules do not require incorporation of the stem Alcyoni- in the ordinal name.]

Colonies thick and incrusting, lobate or erect and arborescent, consisting of very long polyps united in more or less extensively developed, usually spiculiferous coenenchyme. Primary polyps extend from base of the colony to the uppermost branches, secondary individuals arising from solenial system of primary polyp walls or of colonial coenenchyme. *L.Jur.-Rec.*

The development of colonial coenenchyme and the extremely long primary polyps distinguish this order from the Telestacea on the one hand and scleraxonian Gorgonacea on the other. The genera and species are recognizable only with difficulty in Recent material, and the indisputable determination of fossil material or isolated spicules seems beyond the realm of possibility.

The Alcyonacea embraces 6 families. In modern seas, representatives occur at all latitudes and in all depths from low tide to several hundred fathoms. Members of the family Alcyoniidae are the predominant reef-dwelling octocorals in the tropical Pacific, in contrast to the situation in the Atlantic, where gorgonaceans dominate the scene and no alcyonaceans at all are to be found in reef habitats. It is thus possible to assign virtually any large (up to 5 mm.), stout, spindle-shaped alcyonarian spicule

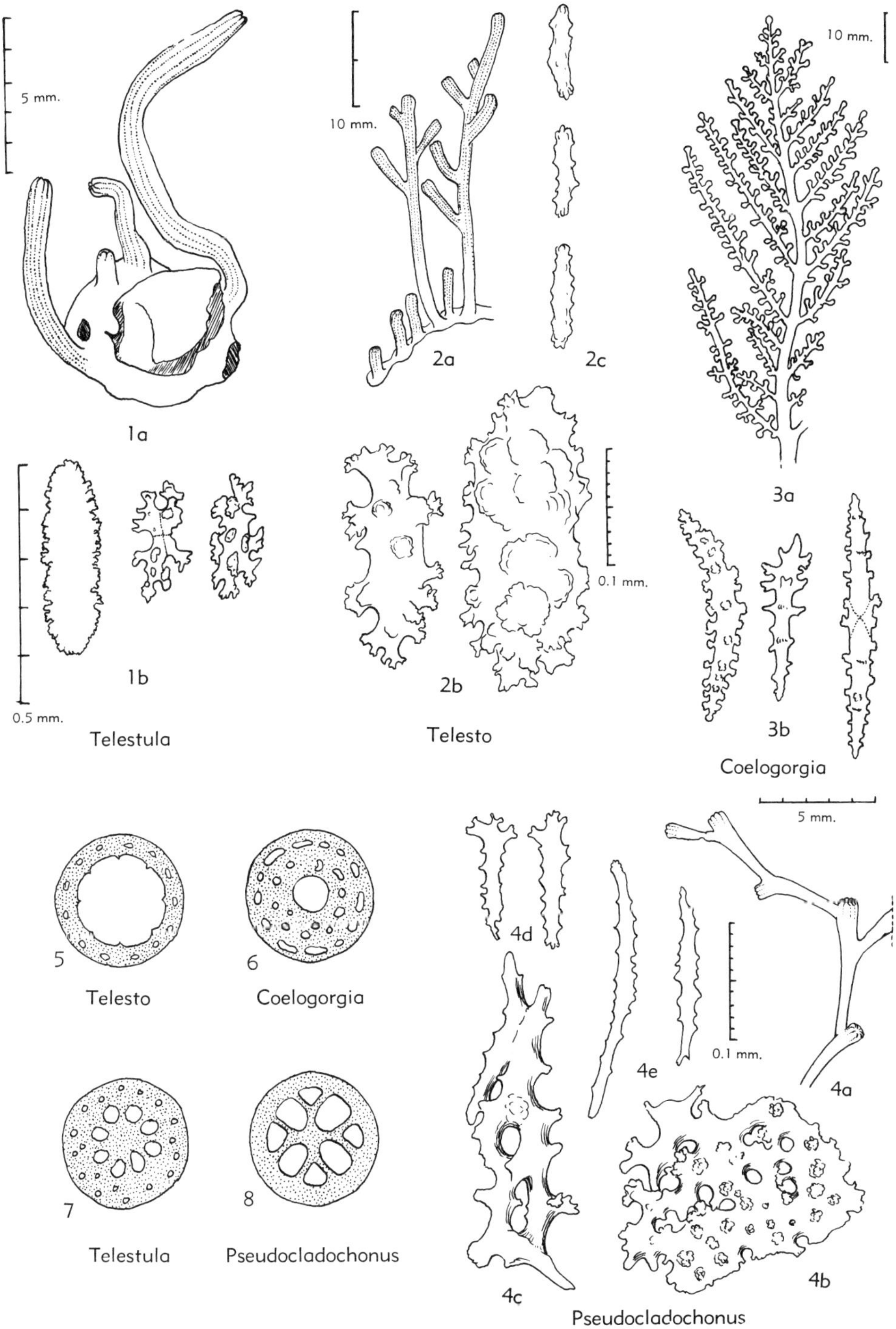

Fig. 137. Telestacea (p. F186).

from tropical Indo-Pacific reef sediments to the family Alcyoniidae.

Family ALCYONIIDAE Lamouroux, 1812

[*nom. correct.* MAY, 1899 (*pro* Alcyonidae DANA, 1846; *nom. correct. pro* family Alcyoneae LAMX., 1812)]

Fleshy or membranous Alcyonacea with mostly spindle-like spicules, some with clubs, crosses, and multiradiate spicules but none with small biscuit-like platelets exclusively. Mostly dimorphic. *Rec.*

A large proportion of the many described genera inevitably fall into synonymy. The following list is neither complete nor critical. The valid type designations have, in the main, never been determined; of those given here, only the monotypes, original designations, and a few subsequent designations should be considered final.

Alcyonium LINNÉ, 1758 [**A. digitatum;* SD M. EDW.-H., 1850] [=*Lobularia* LAMARCK, 1816]. *Rec.,* Atl.——FIG. 138,*4*. **A. digitatum,* Rec., N.Atl.; spicules (26n).

Acrophytum HICKSON, 1900 [**A. claviger*]. *Rec.* S.Afr.

Anthomastus VERRILL, 1878 [**A. grandiflorus*]. *Rec.*——FIG. 138,*1*. **A. grandiflorus; 1a,* spicule of inner stalk (26n); *1b,* surface spicules (26n); *1c,* colony (35).

Bathyalcyon VERSLUYS, 1906 [**B. robustum*]. Single giant autozooid partially retractile within a colonial calyx containing numerous siphonozooids. Spicules similar to those of *Anthomastus. Rec.,* Pac.

Carotalcyon UTINOMI, 1952 [**C. sagamianum*]. *Rec.,* Japan.

Cereopsida STRAND, 1928 [*pro Cereopsis* KENT, 1870 (*non* LATHAM, 1801)] [**Cereopsis bocagii* KENT, 1870]. *Rec.,* E.Atl.

Lobophytum MARENZELLER, 1886 [**L. crassum;* SD MOSER, 1919]. *Rec.,* IndoPac.——FIG. 138,*6*. **L. crassum,* Marshall I.; spicule (26n).

Nidalia GRAY, 1835 [**N. occidentalis*]. *Rec.*——FIG. 138,*5*. **N. occidentalis,* Rec., Fla.; *5a,* spicule in outline; *5b,* tip of same at higher magnification to show tubercles (26n).

Nidaliopsis KÜKENTHAL, 1906 [**N. pygmaea*]. *Rec.,* W.Afr.

Sarcophyton LESSON, 1834 [**S. lobulatum;* SD M. EDW.-H., 1850]. *Rec.,* IndoPac.——FIG. 138,*2*. *S. trocheliophorum* MARENZELLER, Marshall I.; *2a,* clubs of capitulum surface; *2b,* spindle of stalk (26n).

Sinularia MAY, 1898 [**S. brassica*]. *Rec.,* Indo-Pac.

Sphaerella GRAY, 1869 [**Alcyonium tuberculosum* QUOY & GAIMARD] [=*Lobularia* EHRENBERG (*non* LAMARCK); *Cladiella* GRAY; *Microspicularia* MACFAYDEN] (4). *Rec.,* IndoPac.——FIG. 138,*3*. *S. krempfi* HICKSON, Marshall I.; spicules (26n).

Family ASTROSPICULARIIDAE Utinomi, 1951

Fleshy, monomorphic Alcyonacea of bushy habit with sterile stalks and polypiferous terminal lobes, polyps of uniform size, tentacles without pinnules; armature of spicules continued from rind; coenenchyme fleshy, honeycombed with chambers containing several stellate spicules each; solenia penetrating mesogloea between spicule chambers.

Astrospicularia UTINOMI, 1951 [**A. laurae*]. *Rec.,* Formosa.

Family NEPHTHEIDAE Gray, 1862

[*nom. correct.* UTINOMI, 1954 (*pro* Nepthyadae GRAY, 1862)]

Arborescent or lobular colonies with sterile stalks, polyps placed singly or in clusters at tips of ultimate branches. Spiculation usually dense but sparse in some; spicules strongly sculptured spindles, capstans and clubs, in some genera reaching a length of several mm. *L.Jur.-Rec.*

Nephthea AUDOUIN, 1826 [**N. chabrolii;* SD M. EDW.-H., 1850]. *L.Jur.*(Eu.)-*Rec.* (RedSea-Indo-Pac.).

Dendronephthya KÜKENTHAL, 1905 [**Nephthya savignyi* EHRENBERG, 1834; SD UTINOMI, 1954]. *Rec.,* RedSea-IndoPac.

Eunephthya VERRILL, 1869 [**Nephthya thyrsoidea* VERRILL, 1865]. *Rec.,* N.Atl.-S.Afr.-Pac.

Gersemia MARENZELLER, 1877 [=?*Eunephythya* VERRILL, 1869 (*fide* MADSEN, 1944)]

Lemnalia GRAY, 1868 [**L. jukesii*]. *Rec.,* RedSea-IndoPac.

Litophyton FORSKÅL, 1775 [**L. arboreum;* SD KÜKENTHAL, 1903] [=*Ammothea* LAMARCK, 1816]. *Rec.,* RedSea-IndoPac.

Neospongodes KÜKENTHAL, 1903 [**N. atlantica;* SD DEICHMANN, 1936]. *Rec.,* W.Atl.

Stereonephthya KÜKENTHAL, 1905 [**S. papyracea;* SD UTINOMI, 1954]. *Rec.,* W.Pac.

Family SIPHONOGORGIIDAE Kölliker, 1874

Tall slender arborescent colonies, with polyps retractile into low calyces; coenenchyme separating long gastrovascular canals in stem is packed with spicules and forms rigid walls. *Rec.*

Siphonogorgia KÖLLIKER, 1874 [**S. godeffroyi*]. *Rec.,* Indo-Pac.

Agaricoides SIMPSON, 1905 [**A. alcocki*]. *Rec.,* Ceylon.

Cactogorgia SIMPSON, 1907. *Rec.,* Ind.O.

Paranephthya STUDER, 1887 [**P. capitulifera* WRIGHT & STUDER, 1889; SM WRIGHT-S., 1889]. *Rec.,* Philip.

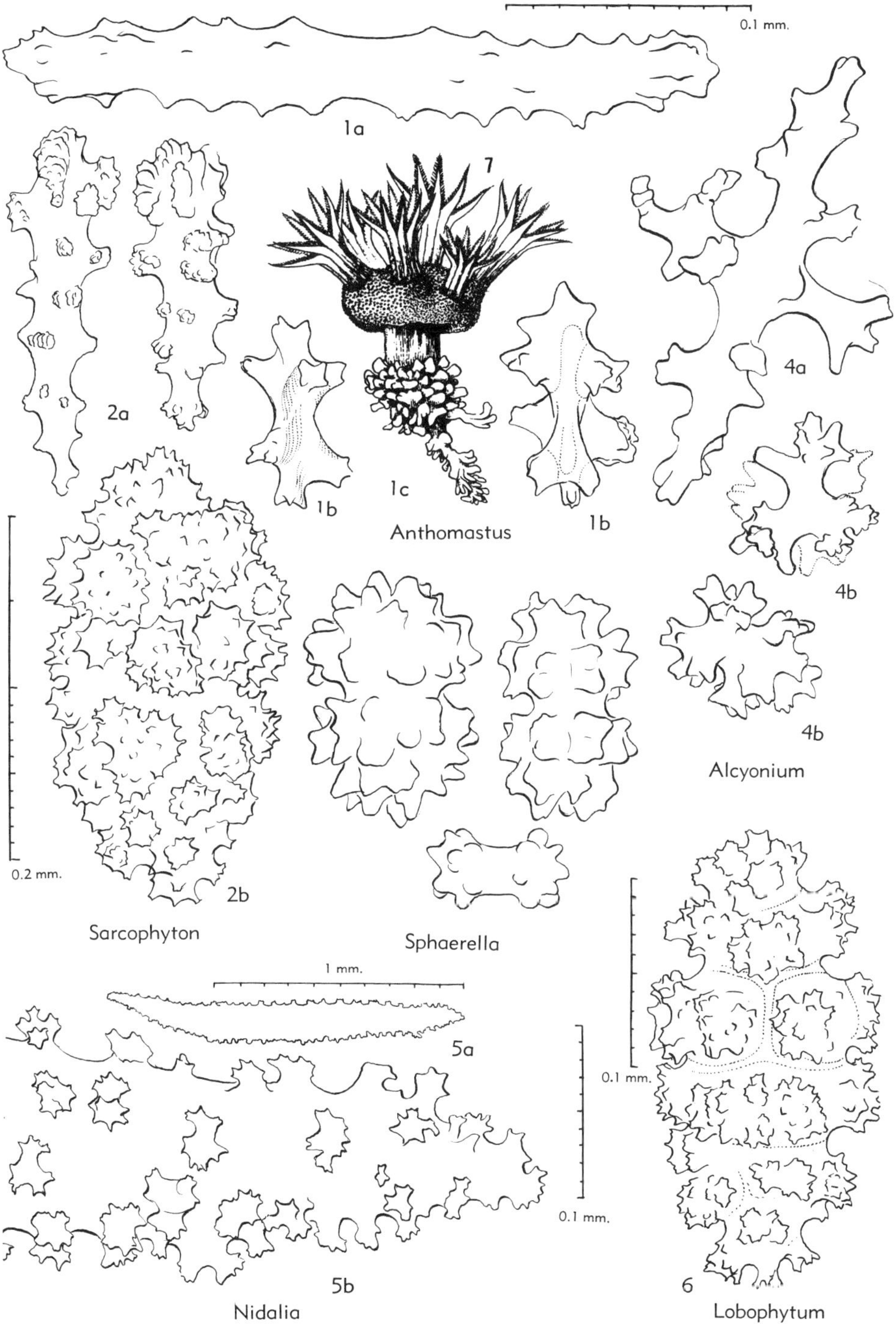

FIG. 138. Alcyonacea: Alcyoniidae (p. F188).

Scleronephthya STUDER, 1887 [**S. pustulosa* WRIGHT & STUDER, 1889; SM WRIGHT-S., 1889]. *Rec.*, Philip.

Stereacanthia THOMSON & HENDERSON, 1906 [**S. indica*]. *Rec.*, AndamanI.

Family VIGUIERIOTIDAE Bayer, 1954

Alcyonacea with polyps united in groups, distally free and bearing lateral polyps, proximally fused into common trunk within which the upper free, finger-like processes may be retracted. Spicules in the form of spindles. *Rec.*

Viguieriotes BAYER, 1954 [*pro Fascicularia* VIGUIER, 1888 (*non* LAMARCK, 1816; *nec* LYELL, 1839; *nec* DYBOWSKI, 1873)] [**Fascicularia radicans* (=*Paralcyonium edwardsii* DE LACAZE-DUTHIERS, 1888)]. *Rec.*, Algeria.

Paralcyonium M.EDW.-H., 1850 [**Alcyonidia elegans* M.EDW., 1835]. *Rec.*, Algeria.

Studeriotes THOMSON & SIMPSON, 1909 [**Studeria mirabilis* THOMSON, 1908]. *Rec.*, AndamanI.

Family XENIIDAE Ehrenberg, 1828

[*nom. correct.* MAY, 1899 (*pro* Xenidae VERRILL, 1866, *pro* Xenina EHR., 1828)]

Polyps soft, fleshy, nonretractile; sclerites minute, biscuit-shaped (8). Some authors have treated the family as a distinct order. *Rec.*

Xenia LAMARCK, 1816 [**X. umbellata;* SD M.EDW.-H., 1850]. *Rec.*, RedSea-IndoPac.

Anthelia LAMARCK, 1816 [**A. glauca*]. *Rec.*, Red Sea-IndoPac.

Ceratocaulon JUNGERSEN, 1892 [**C. wandeli*]. *Rec.*, NorwegianSea-Antarct.

Cespitularia M.EDW.-H., 1850 [**Cornularia multipinnata* QUOY & GAIMARD]. *Rec.*, IndoPac.

Efflatounaria GOHAR, 1939 [**E. tottoni*]. *Rec.*, Austral.

Heteroxenia KÖLLIKER, 1874 [**H. elisabethae*]. *Rec.*, RedSea-IndoPac.

Sympodium EHRENBERG, 1834 [**S. fuliginosum;* SD M.EDW.-H., 1850 (but see GOHAR, 8)]. *Rec.*, RedSea-IndoPac.

TRACHYPSAMMIACEA

By EUGENIA MONTANARO-GALLITELLI[1]

Order TRACHYPSAMMIACEA Montanaro-Gallitelli, nov.

Corallum dendroid, with medullar canal system which gives rise to new corallites; skeletal structures of cortical region oriented radially. *Perm.*

Family TRACHYPSAMMIIDAE Gerth, 1921

[*nom. correct.* LECOMPTE, 1952 (*pro* Trachypsammidae GERTH, 1921)]

Characters of order. *Perm.*

Although the medullar system can in general be compared with that of the Octocorallia, identity of *Trachypsammia*[2] with known forms is not possible because of its combination of a central, medullar canal system, extramedullar, radial coenenchyme, and origin of new corallites from the central medullar canal.

The first description of this unique form was given by GERTH (1921), based upon specimens from the Upper Permian of Timor. He recognized its peculiarity and referred it to the Tabulata, near the Pachyporinae, albeit with some trepidation because of the structure of the extramedullary skeleton and the external sculpture, which he compared with the coenenchyme and sculpture of Tertiary and living Scleractinia of the families Oculinidae, Eupsammiidae, and Dendrophylliidae. GERTH admitted also the possibility that the trachypsammiids could be referred to the Octocorallia, according to some authors, closely related to the Tabulata.

The structure and systematic position of this fossil were discussed subsequently by HEHENWARTER, who excluded any connection with the Pachyporidae, emphasized possible relationship with the Octocorallia as already suggested by GERTH, and related *Trachypsammia* to the Gorgonacea. Recent investigations by MONTANARO-GALLITELLI (16) on topotypes of the original species and on a new species, *T. mediterranea* MONT.-GALL., have permitted a revision of structural interpretations and an effort to verify the relationship with Octocorallia. An affinity with fossil colonial forms like *Thamnopora* is definitely and obviously ruled out.

Although the medullar system can be compared generally with that of the Octocorallia, identity of *Trachpysammia* with known forms is not possible because of its combination of a central medullar canal sys-

[1] Istituto di Geología e Paleontología, Università, Modena, Italia.

[2] Classed in Tabulata by HILL & STUMM (p. F469).

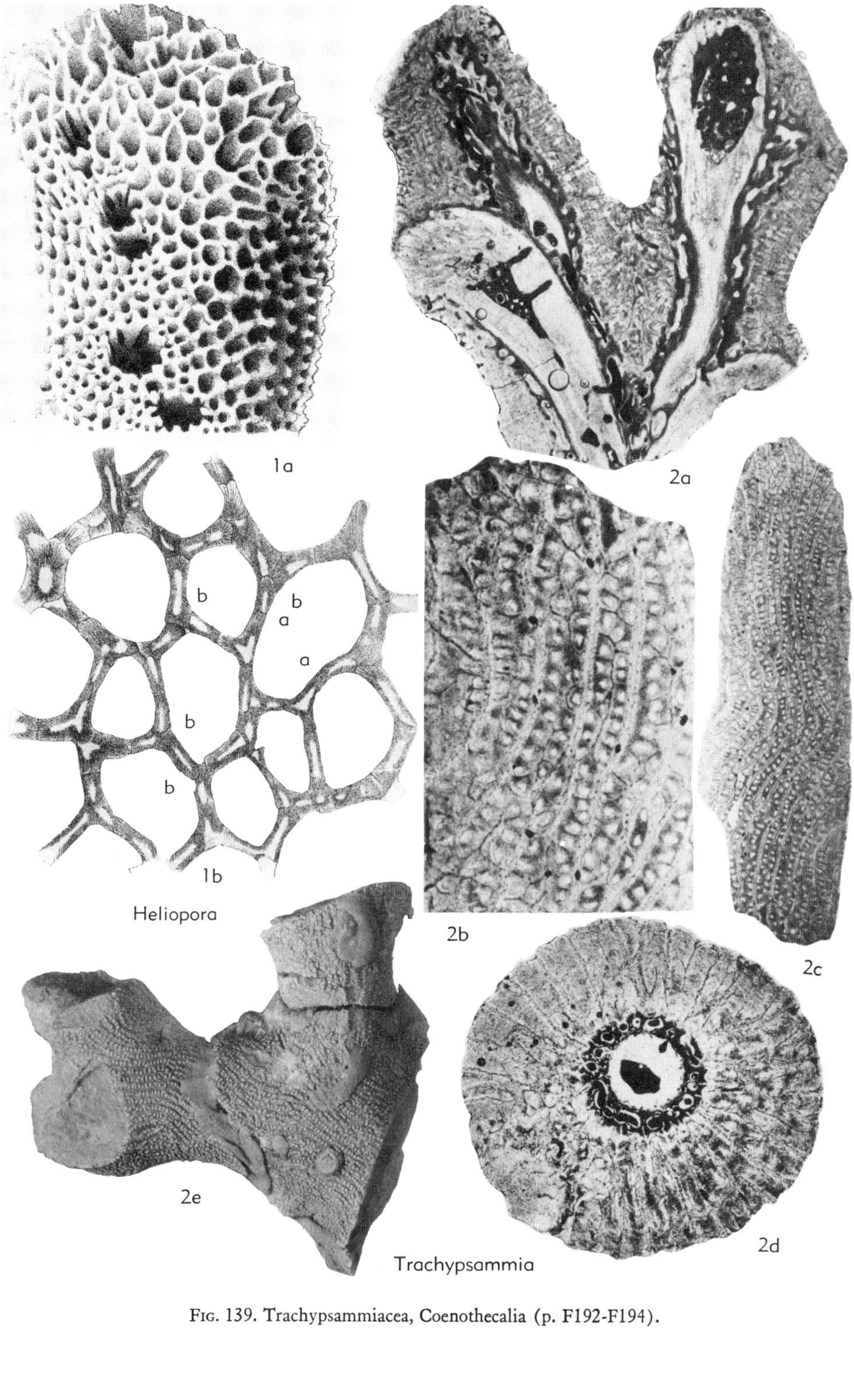

FIG. 139. Trachypsammiacea, Coenothecalia (p. F192-F194).

tem, extramedullar radial coenenchyme, and origin of new corallites from the central medullar canal.

Relationship to the Gorgonacea (e.g., *Gorgonia*), as suggested by HEHENWARTER, is not possible because of the existence of a strongly developed mesoskeleton and the great difference in structure and development of the medullar region, which in *Gorgonia* is an ectodermal product consisting of succeeding cupuliform, overlapping lamellae, concentric layers more or less calcified, or irregular sclerodermites radially oriented in the center and surrounded by secondary calcification arranged either radially or longitudinally.

In comparison with other types of modern Octocorallia, *Trachypsammia* clearly differs from Stolonifera (e.g., *Tubipora*) that have parallel corallites connected only by horizontal, stolonic platforms. However, a similarity can be drawn with certain forms of Telestacea and Alcyonacea, in which young corallites originate at various levels from solenia of the vertical corallite walls of the parent. *Trachypsammia* could be considered, in regard to the medullar system, as an archaic precursor of this structural plan, having a single primary basal polyp, a single central medullar structure with one large central cavity (therefore different from forms like *Briareum*). Some octocorals possess intriguing structural similarities to *Trachypsammia.* It is important to note that Gorgonacea such as *Isis hippuris* LINNE and *Corallium rubrum* (LINNÉ) present a singularly analogous radial arrangement of sclerodermites in the calcareous axis. These differ from *Trachypsammia* by the absence of the central canal system and the exclusion of corallites from the axial skeleton, but these can be found in the gorgonacean genus *Tripalea* (=*Suberia, partim*) (with unconsolidated skeleton) which, moreover, has a very open, vesicular cortex that may show traces of radial arrangement. A comparison of the radial, extramedullar coenenchymatic system with any known coenenchymatic structure of Alcyonacea is extremely problematic and remains an open question. On the other hand, similarity between the radiate structures of *Trachypsammia* with the coenenchymatic structures of the Dendrophylliidae and Eupsammiidae, as suggested by GERTH, fails entirely of homological documentation.

Trachypsammia[1] GERTH, 1921 [**T. dendroides*]. *External characters:* Corallum compound, dendroid, adherent by a broadly cuneiform base. External sculpture of minute tubercles in single or double files, separated by thin ridges axially or sinuously directed, increasing in number and anastomosing in some, chiefly in the vicinity of the corallites. Calyces minute, generally in 2 opposite longitudinal series, less commonly scattered sparsely; slightly projecting, with coronal tubercles and with wall subdivided into 24 to 28 sectors, simulating septal striae. *Structure of coenenchyme:* Medullar region formed of a system of sinuous, labyrinthine canals with frequent anastomoses and ramification (Fig. 139,*2a,d*). Corallites originating in the canalicular region as an axial calcareous process in the fossil skeleton, inserted in the canal system, part of which becomes peripheral to them; one or more canals penetrating also the developing skeleton and forming the axial cavity of a new corallite, which bends centrifugally as it elongates, the central cavity becoming funnel-shaped, and the calcareous column being subdivided into a series of 24 to 28 radial sectors ("septal ridges" of GERTH and HEHENWARTER). In the mature stage the peripheral canals are arranged in a single series around the calicular margin. *Extramedullar tissue:* Transverse, axial and tangential thin sections demonstrate (1) a system of vertical, concentric thin lamellae ("interradial bridges" of HEHENWARTER), which are finely and regularly vesicular (Fig. 139,*2d*) and produce the external sculpture of winding files of tubercles (Fig. 139,*2e*); (2) a system of 28 to 50 longitudinal septum-like lamellae that cross the concentric system, undulating mostly in a vertical direction, and not rarely anastomosing as shown in tangential section (Fig. 139,*2b*); (3) a reticulate system of minute dark points (?pores) on each side of the lamellae, visible at even low power (×5) in both tangential and cross sections; (4) a system of relatively few, large cortical lumina irregularly disposed on the sides of the vertical, radial lamellae. *Dimensions:* fragmentary colonies measure 1.5 to 6 cm. in height, 0.5 to 2.0 cm. in diameter. *U. Perm.*, Timor-Italy.——FIG. 139,2. *T. mediterranea* MONT.-GALL., Sicily; *2a*, axial section of branch tips, ×4; *2b*, tangential section, ×12; *2c*, same, ×4; *2d*, transverse section, ×7; *2e*, fragment of colony, ×3 (*2a-d*, 32; *2e*, 28).

(End of section by MONTANARO-GALLITELLI)

Order COENOTHECALIA Bourne, 1895

Octocorals without spicules but produc-

[1] Classed in Tabulata by HILL & STUMM (p. F469).

ing a massive skeleton of fibrocrystalline aragonite. Corallum penetrated by wide tubes occupied by the polyps and narrow tubes containing blind downgrowths of the solenial system; both kinds of tubes regularly tabulate. Calicles with pseudosepta varying in number, not corresponding with the 8 soft septa. *Cret.-Rec.*

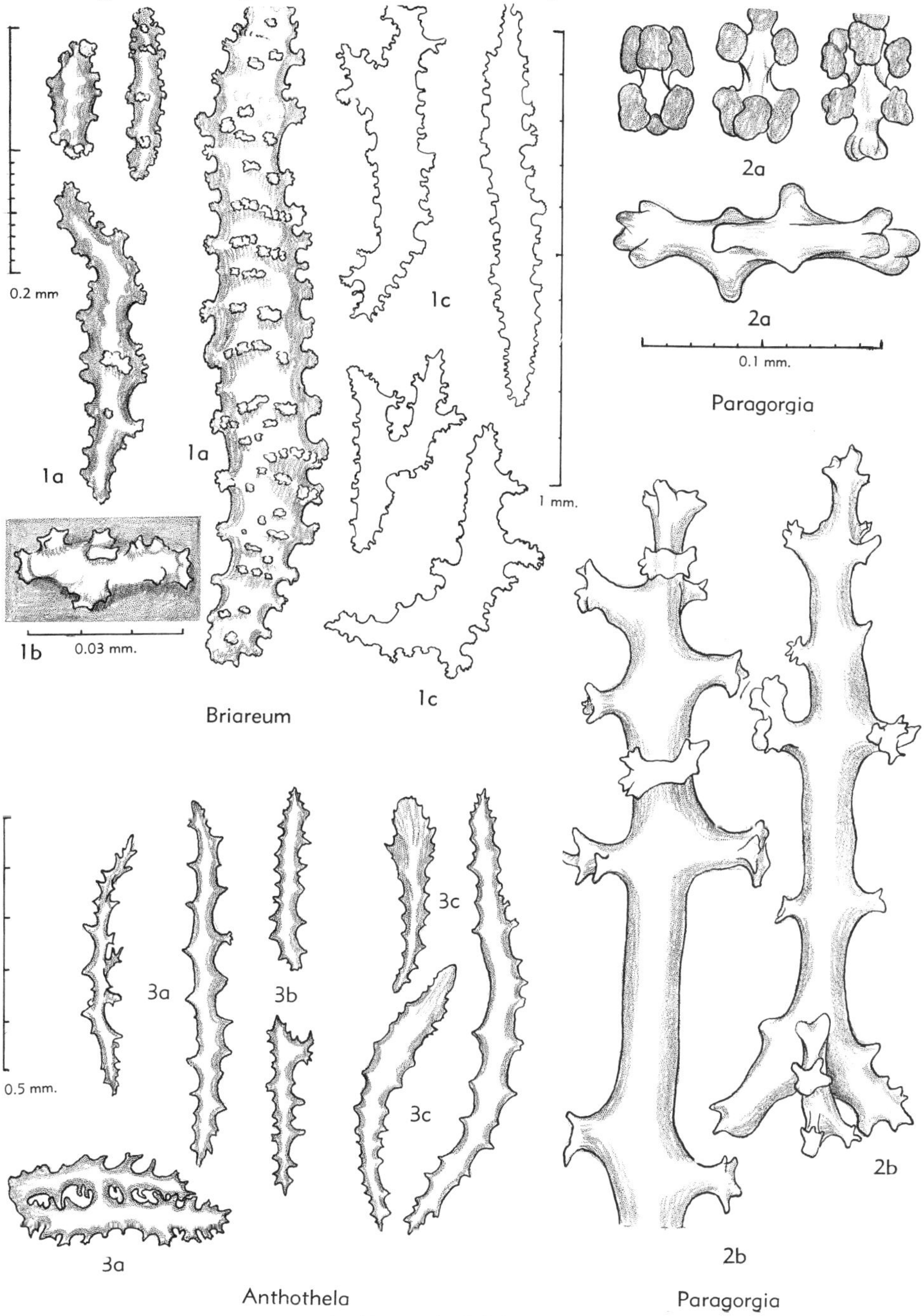

FIG. 140. Gorgonacea: Briareidae, Anthothelidae, Paragorgiidae (p. F194, F197).

A number of fossil genera have at one time or another been placed in this order. Most, however, seem not referable to the Octocorallia and have been removed.

Family HELIOPORIDAE Moseley, 1876

Heliopora DE BLAINVILLE, 1830 [**Millepora coerulea* PALLAS, 1766]. Corallum bright blue in color, massive, vertical, commonly lobate plates with 10 to 16 (usually 15) pseudosepta; an important reef-former in some areas. *Rec.*, IndoPac.——FIG. 139, *1*. **H. coerulea; 1a,* tip of corallum, ×6 (33); *1b,* tangential section of growing edge of corallum; (*a,* junctions of adjacent triradiate figures; *b,* junctions of 3 calcareous lamellae to form triradiate figure) (27).

Octotremacis GREGORY, 1900 [**Polysolenia hochstetteri* REUSS, 1866]. Large, well-developed pseudosepta, 8 in each calicle, in 2 cycles. *Mio.*, Java.

Polytremacis D'ORBIGNY, 1849 [**P. bulbosa*]. Pseudosepta strong, reaching almost to center of calyces, 8 to 20 in number. *U.Cret.-L.Tert.*, Eu.

Family CERATOPORELLIDAE Hickson, 1912

Colonies forming a massive skeleton of crystalline calcium carbonate, in which a few slender spicules are imbedded; no tabulae, tubes closing below by continuous growth of thecal walls; pores monomorphic, small. *Rec.*

Ceratoporella HICKSON, 1912 [**Ceratopora nicholsonii* HICKSON, 1911]. *Rec.*, Cuba.

Order GORGONACEA Lamouroux, 1816

[*nom. transl.* STUDER, 1887 (*ex* suborder Gorgonacea VERRILL, 1866; *nom. transl. et correct. ex* order Gorgonieae LAMX., 1816)] [Although the type genus of this order is *Gorgonia*, present rules do not require incorporation of the stem Gorgoni- in the ordinal name; furthermore, the fact that the originally published ordinal name does incorporate this stem is not a compelling reason for changing Gorgonacea to Gorgoniacea.]

Octocorals with uniformly short gastrovascular cavities; chiefly arborescent, rarely lobate or incrusting colonies producing more or less specialized axial structures, either as a distinct central axis of horny or calcareous material, or both, or as a central medullar zone of calcareous sclerites which are loosely or inseparably bound together by horny or calcareous matter. *Cret.-Rec.*

Suborder SCLERAXONIA Studer, 1887

[=Pseudaxonia (*partim*) VON KOCH, 1878]

Gorgonaceans with central axis, medullar zone or inner layer containing spicules bound together more or less solidly either by horny or calcareous material; outer layer of coenenchyme (cortex) containing proximal portions of gastrovascular chambers of the polyps, endodermal canals and solenia; cortical spicules free and commonly of a different sort from those in medullar region; axial cylinder or medulla may contain canals and solenia but polyp cavities do not penetrate it and it has no cross-chambered central chord. *Cret.-Rec.*

Family BRIAREIDAE J. E. Gray, 1859

Monomorphic Scleraxonia with continuous medulla containing separable spicules, perforated by gastrodermal canals all the way to branch tips, not separated from cortex by a ring of boundary canals. *Rec.*

Briareum DE BLAINVILLE, 1830 [**Briareum gorgonoideum* (=**Alcyonium asbestinum* PALLAS, 1766)]. Sclerites basically spindles with irregular tubercles, very large and sinuous or branched in medulla; white and purple in color. *Rec.*, W.Indies. ——FIG. 140,*1*. **B. asbestinum,* Fla.; *1a,* spicules of cortex; *1b,* sculptural detail of same; *1c,* spicules of medulla in outline (26n).

Solenopodium KÜKENTHAL, 1916 [**Erythropodium stechei;* SD KÜKENTHAL, 1916]. Sclerites like those of *Briareum* but more slender, with tubercles displaying strong tendency toward alignment in regular transverse belts; sclerites of cortex colorless, those of medulla chiefly violet. *Rec.*, RedSea-Malay Arch.-Oceania.

Family ANTHOTHELIDAE Broch, 1916

Monomorphic Scleraxonia with medulla surrounded by longitudinal boundary canals which separate it from cortex; medulla only rarely perforated by gastrodermal solenia in smaller branches and then not extensively as in lower parts of colony. *Rec.*

Subfamily ANTHOTHELINAE Broch, 1916

Cortical sclerites elongate and pointed, anthocodial spindles bent and commonly strongly clubbed; polyps occupying projecting, cylindrical calyces. *Rec.*

Anthothela VERRILL, 1879 [**Briareum grandiflorum* SARS, 1856]. Slender, tortuous, commonly anastomosing branches; no main stems developed. *Rec.*, Atl.-IndoPac.——FIG. 140,*3*. **A. grandiflora,* Grand Banks; *3a,* spicules of medulla; *3b,* spicules of cortex; *3c,* spicules of anthocodia (26n).

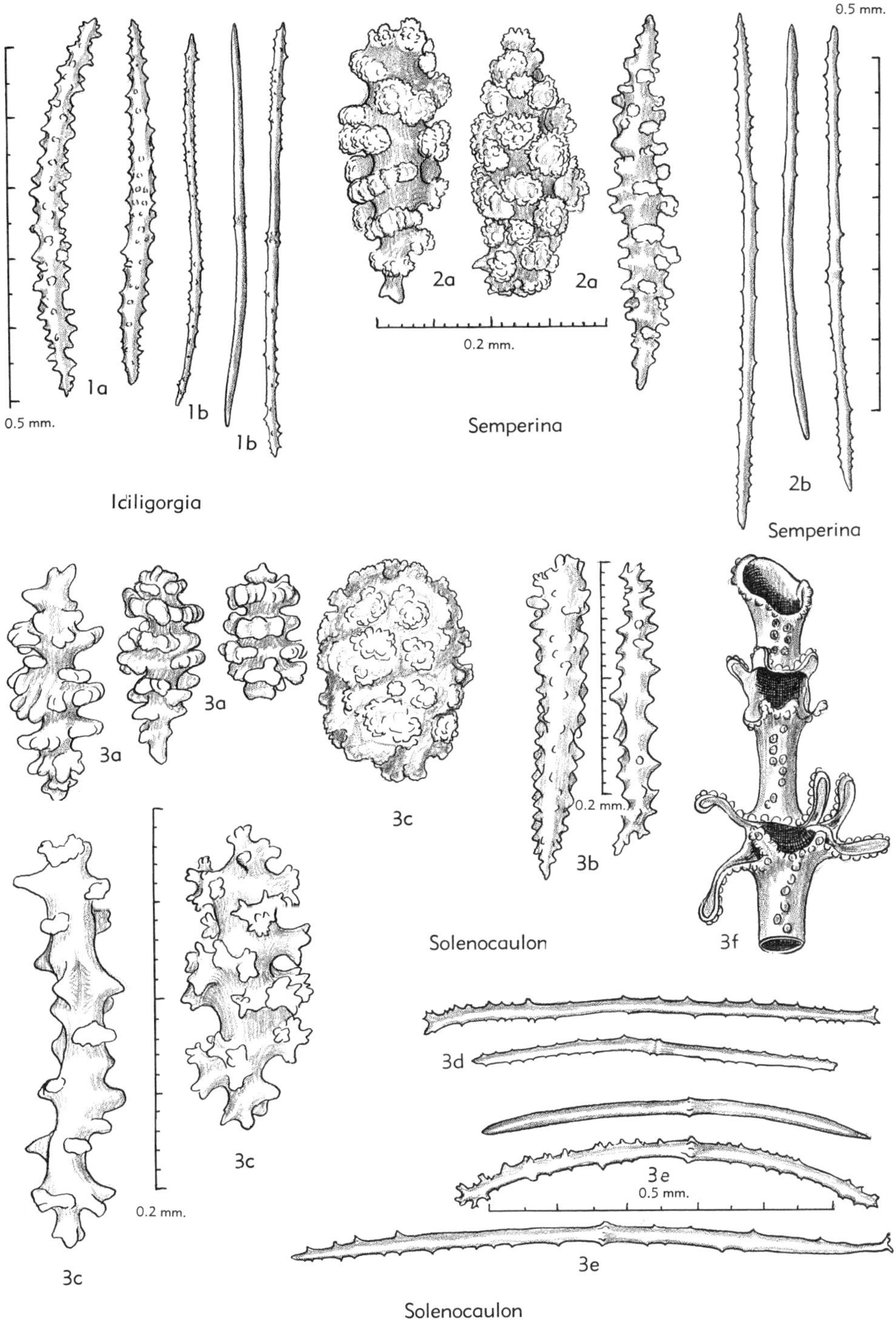

Fig. 141. Gorgonacea: Anthothelidae (p. F196).

Subfamily SEMPERININAE Aurivillius, 1931

[*nom. transl.* BAYER, herein (*ex* Semperinidae AURIVILLIUS, 1931)]

Axial sclerites consisting of long, slender needles; cortical sclerites including thorny or tuberculate spindles but no radiate forms; polyps occupying hemispherical calyces. *Rec.*

Semperina KÖLLIKER, 1870 [**S. rubra*]. Zooids leaving one face of colony free; branches slightly flattened, tips clavate and commonly fistulose; spindles of cortex elongate-fusiform, openly tuberculated; axial needles finely prickly. *Rec.*, Indo-Pac.——FIG. 141,*2*. *S. brunnea* NUTTING, Ind. O.; *2a*, cortical sclerites; *2b*, medullar needles (26n).

Iciligorgia DUCHASSAING, 1870 [**I. schrammi*]. Polyps in 2 longitudinal series along opposite edge of branches; twig tips commonly fistulose, cortical spicules spinous bent spindles; axial spicules slender needles, usually thorny, some smooth. *Rec.*, trop.W.Atl.——FIG. 141,*1*. **I. schrammi*, Fla.; *1a*, cortical spicules; *1b*, medullar spicules (26n).

Solenocaulon GRAY, 1862 [**S. tortuosum*]. Polyps chiefly on small twigs, which may be clavate and terminally fistulose; main stem expanded, gutter-like or closed and tubular; stalk solid, some spatulate below; cortical sclerites consisting of capstans and coarsely tuberculated spindles, with a few spinose rods; medullar sclerites including many very slender needles, all spinose in branchlets, few completely smooth in stalk. *Rec.*, IndoPac.——FIG. 141,*3*. *S. sterroklonium* GERMANOS, Philip.; *3a,b*, spicules of branch cortex; *3c*, spicules of stalk cortex; *3d*, spicules of branch medulla; *3e*, spicules of stalk medulla; *3f*, tip of tubular stem (26n).

Subfamily SPONGIODERMATINAE Aurivillius, 1931

[*nom. transl.* BAYER, herein (*ex* Spongiodermatidae AURIVILLIUS, 1931)]

Medullar sclerites consisting of stout rods, commonly with strong spines or branching processes, cortical sclerites including capstans or radiate forms. *Rec.*

Spongioderma KÖLLIKER, 1870 [**Solanderia verrucosa* MÖBIUS, 1861]. Spicules of cortex mostly 4-, 5- and 6-radiates, coarser than those of *Titanideum;* 2 rings of longitudinal canals. *Rec.*, S.Afr.

Alertigorgia KÜKENTHAL, 1908 [**Iciligorgia orientalis* RIDLEY, 1884] [=*Machaerigorgia* KÜKENTHAL, 1916 (obj.)]. Colonies with long, flat branches, zooids fully retractile and aligned biserially along narrow edges; cortical spicules short oval bodies with closely and irregularly placed tubercles, longer spindles with smaller warts less closely placed, and intermediate forms; spicules of medulla mostly spinous rods, with a few tuberculate forms like those of cortex. *Rec.*, IndoPac.

Callipodium VERRILL, 1869 [**C. pacificum*]. Incrusting species in form of broad bands or sheets; spicules including short belted spindles, 3- and 6-radiates, those of inner zone commonly fused together in a solid layer next to substratum; red or yellow; placed in this family because of spicular characters. *Rec.*, W.Mexico-Panama.——FIG. 142, *2*. **C. pacificum*, Panama; *2a*, cortical sclerites; *2b*, part of the fused basement layer (26n).

Diodogorgia KÜKENTHAL, 1919 [**D. ceratosa*]. Spicules of cortex mostly small 6-radiates in anthosteles, coarsely warted spindles in rind; stout rods with sparse processes predominant in medulla; with double ring of longitudinal canals. *Rec.*, trop. W.Atl.——FIG. 143,*1*. **D. ceratosa*, Carib.; *1a*, rod from medulla; *1b*, cortical spindle; *1c*, longer spindle of inner cortex; *1d*, radiates from polyp walls (26n).

Erythropodium KÖLLIKER, 1865 [**Xaenia caribaeorum* DUCHASSAING & MICHELOTTI, 1860]. Spicules 4- and 6-radiates, those of cortical zone colorless, of medullar region red; discontinuous layer of fused sclerites next to substratum; membranous form placed here because of spicular characters. *Rec.*, trop.W.Atl.——FIG. 142,*3*. **E. caribaeorum*, Fla.; *3a*, colorless cortical sclerites; *3b*, red medullar sclerites (26n).

Titanideum VERRILL, 1863 [**Gorgonia suberosa* ELLIS & SOLANDER, 1786 (*non* PALLAS, 1766) (=*Solanderia frauenfeldtii* KÖLLIKER, 1865)]. Spicules of cortex chiefly pink or red 4- and 5-radiates; of medulla, colorless branched rods; double ring of longitudinal canals, one immediately surrounding medulla, other dividing cortex into thin outer and thick inner layer. *Rec.*, SE. USA.——FIG. 142,*1*. **T. frauenfeldtii*, N.Car., *1a*, radiate spicules or cortex; *1b*, medullar spicules (26n).

Tripalea BAYER, 1955 [**Suberia clavaria* STUDER, 1878]. Spicules of cortex short capstans in thin outer layer, warted and spinous spindles in spongy inner layer; medulla with rods branched at ends; one ring of boundary canals. *Rec.*, E.S.Am.——FIG. 143,*3*. **T. clavaria*, Arg.; *3a*, spicules of inner cortex; *3b*, of outer cortex; *3c*, of medulla (26n).

Family SUBERGORGIIDAE Gray, 1859

[*nom. correct.* BAYER, herein (*pro* Subergorgiadae, GRAY, 1859)]

Arborescent, monomorphic, with medullar region containing spicules partially fused together to form a well-differentiated axis set off from cortex by ring of longitudinal boundary canals of which 2 may be larger than others; no solenia or central chord in medulla; cortical sclerites include short capstans and longer spindles with transverse belts of tubercles, irregularly tuberculated spindles, and disc spindles ("double wheels"); medullar sclerites mostly long, sinuous, commonly branching and anastomosing, without spines or warts; a few shorter rods, smooth except for tiny terminal cluster of tubercles at each end. *Rec.*

Subergorgia GRAY, 1857 [**Gorgonia suberosa* PALLAS, 1766]. *Rec.,* includes some monstrous sea fans of E.Indies-Oceania.——FIG. 143,*2*. *S. mollis* (NUTTING), Rec., MarshallI.; *2a,* disc-spindles, and *2b,* belted spindles of cortex; *2c,* sclerites of medulla (26n).

Family PARAGORGIIDAE Kükenthal, 1916

[*nom. transl.* AURIVILLIUS, 1931 (*ex* Paragorgiinae KÜKENTHAL, 1916)]

Like Briareidae except polyps dimorphic. *Rec.*

Paragorgia M.EDW.-H., 1857 [**Alcyonium arboreum* LINNÉ, 1758]. Massive, treelike colonies with thick branches; sclerites of cortex small 6-, 7- and 8-radiate capstans; in medulla, long rods with branching processes, derived from capstan type. *Rec.,* Arct.-Atl.-Pac.——FIG. 140,*2*. **P. arborea,* Grand Banks; *2a,* cortical capstans and rod; *2b,* long medullar sclerites (26n).

Sibogagorgia STIASNY, 1937 [**S. weberi* (=**Suberia kölli̧keri* NUTTING, 1911, *non* STUDER, 1879)]. Spicules similar to those of *Paragorgia;* boundary canals present. *Rec.,* E.Indies.

Family CORALLIIDAE Lamouroux, 1812

[*nom. correct.* GRAY, 1857 (*pro* Corallieae, LAMX., 1812)]

Dimorphic, with treelike or bushy form and hard unjointed axis consisting of in-

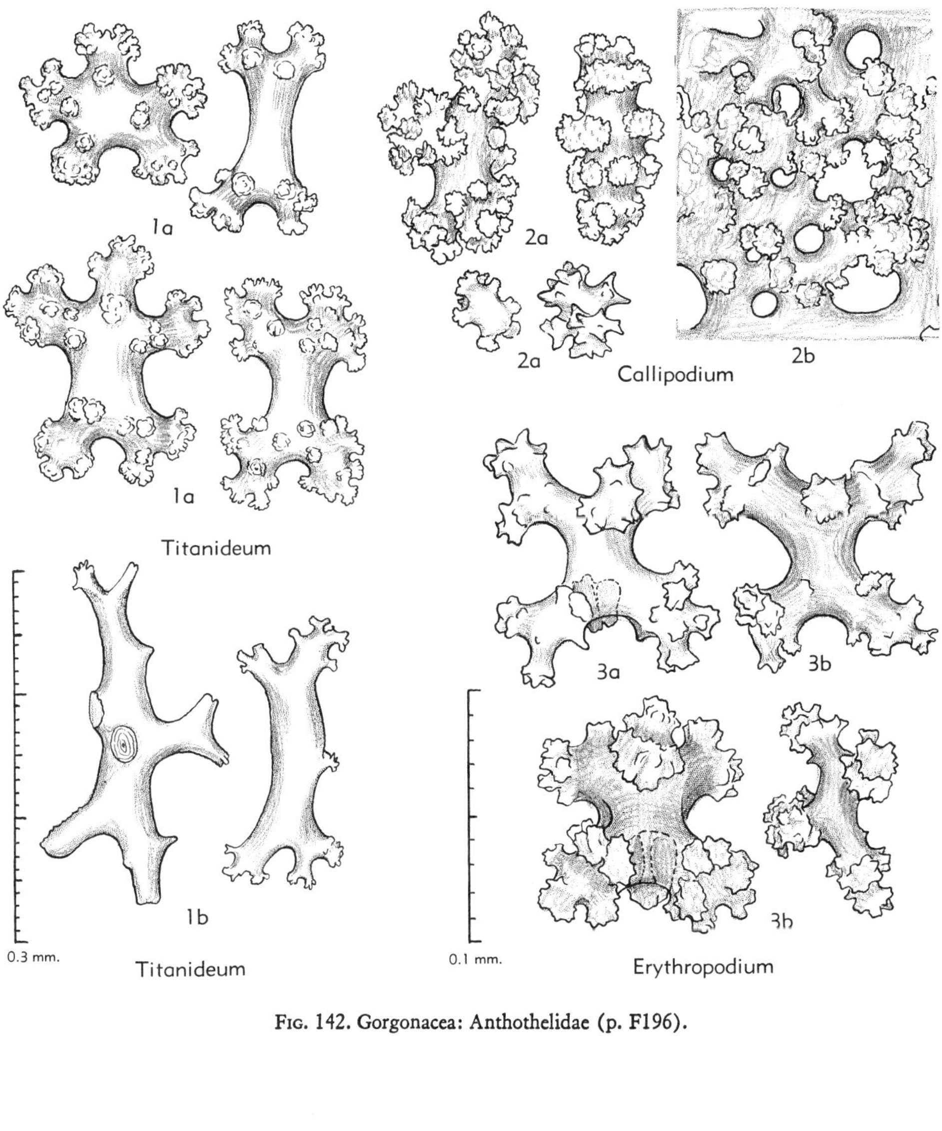

FIG. 142. Gorgonacea: Anthothelidae (p. F196).

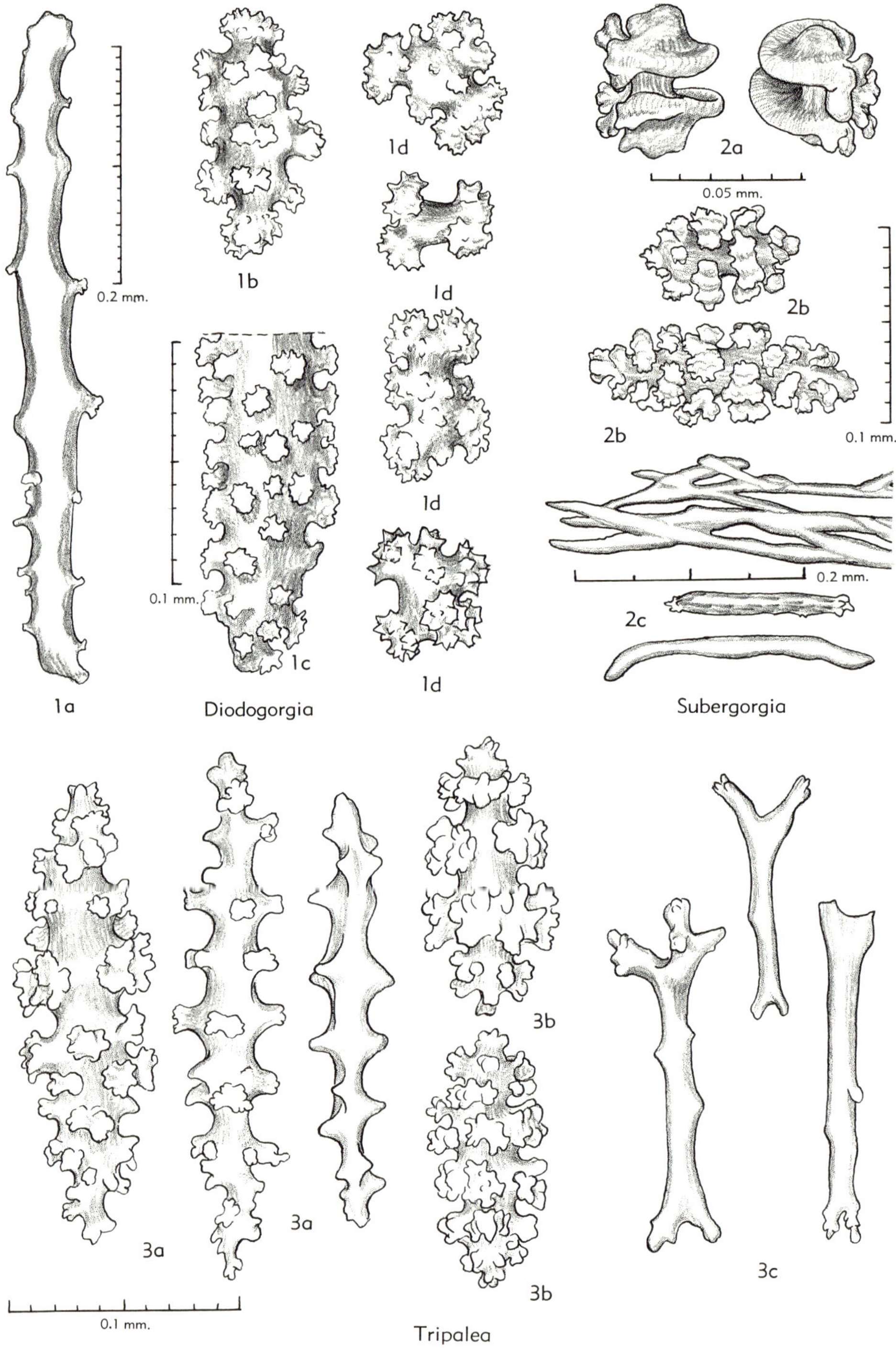

FIG. 143. Gorgonacea: Anthothelidae, Subergorgiidae (p. F196-F197).

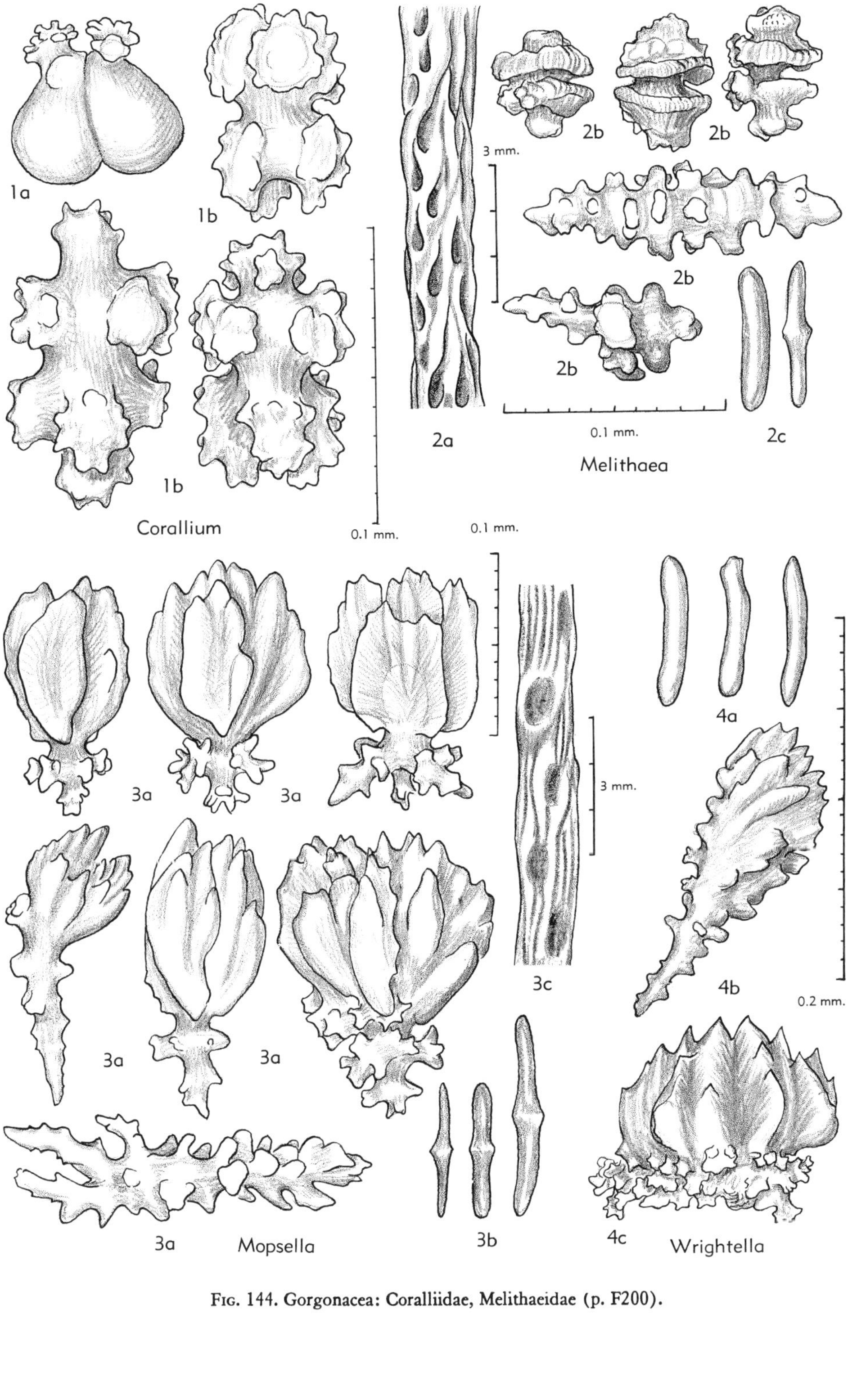

FIG. 144. Gorgonacea: Coralliidae, Melithaeidae (p. F200).

dividual sclerites solidly fused by interstitial calcareous material; polyps occurring on all sides, on one face and laterally, or biserially, and invariably forming distinct calyces; zooids very small, distributed between and upon polyp calyces; rind of varying thickness, containing small 6-, 7- and 8-radiate capstans and commonly abundant double clubs or in some species thorny spindles, or broad plates beset with thorns on outer surface; predominantly pink or red, some yellow or white. *Cret.-Rec.*

Corallium Cuvier, 1798 [**Madrepora rubra* Linné, 1758; SM Lamarck, 1801]. Cortical sclerites including capstans, in some species exclusively, in others with crosses, spindles and highly characteristic double clubs. *Cret.-Rec.*, E.Atl.-Medit.-Ind. O.-E. Indies-Philip.-Japan-Hawaii-L.Calif.——Fig. 144,*1*. *C. borneënse* Bayer, Rec., N.Borneo; *1a*, double club; *1b*, radiate spicules (26n).

Pleurocoralloides Moroff, 1902 [**P. formosum*]. Like *Corallium* but cortical sclerites comprising large spindles and plates instead of capstans and double clubs. *Rec.*, Japan.

Family MELITHAEIDAE Gray, 1870

[=Melitodidae Wright & Studer, 1889 (*partim*)]

Arborescent, monomorphic, with axis composed of alternating hard internodes of inseparably fused calcareous spicules and soft nodes of spongy, spiculiferous horny material; branches arising usually from horny nodes; cortical sclerites strongly sculptured spindles and clubs of various kinds; axial spicules smooth rods, commonly with median transverse raised belt. *Tert.-Rec.*

Melithaea M.Edw.-H., 1857 [*pro Melitaea* Lamouroux, 1812 (*non* Péron & Lesueur, 1810)] [**Isis ocracea* Linné, 1758] [=*Melitodes* Verrill, 1864 (obj.)]. Spicules of cortex very diverse in form, including warty spindles, clubs and disc-spindles, but no predominance of any one kind; axial internodes frequently with sculpture of anastomosing grooves and ridges. *Tert.-Rec.*, RedS.-IndoPac.——Fig. 144,*2*. **M. ocracea*, Rec., FijiI.; *2a*, axial internode; *2b*, spicules of cortex; *2c*, rods from axial node (26n).

Acabaria Gray, 1859 [**A. divaricata*]. Cortical sclerites chiefly long spindles; some species with a few clubs and capstans; axial internodes frequently with sculpture of parallel grooves and ridges. *Rec.*, RedS.-Ind.O.-Japan-Austral.-Oceania.——Fig. 145,*1*. *A. crosslandi* Stiasny; *1a*, spicules of cortex; *1b*, rods of axial nodes; *1c*, surface detail of internode (26n).

Clathraria Gray, 1859 [**C. rubrinodis*]. Cortical spicules including spindles, thorny and leafy clubs; internodes solid, with sculpture of both pits and parallel grooves and ridges. *Rec.*, IndoPac.

Melitella Gray, 1859 [**M. elongata;* SD Hickson, 1937]. Cortical spicules predominantly of peculiar, birotulate type. *Rec.*, IndoPac.

Mopsella Gray, 1857 [**Isis dichotoma* Pallas, 1766]. Colonies erect and flabellate, commonly retiform, some bushy; cortical spicules characteristically including very numerous foliate clubs, some resembling flower buds, others recalling a torch with flame blown aslant; axial internodes with shallow pits and a few grooves. *Rec.*, IndoPac.——Fig. 144,*3*. *M. aurantia* (Esper), Philip.; *3a*, clubs and spindle from cortex; *3b*, rods of axial nodes; *3c*, axial internode (26n).

Wrightella Gray, 1870 [**Isis coccinea* Ellis & Solander, 1786; SD Kükenthal, 1908]. Cortex with surface layer of leafy globes packed closely together. *Rec.*, IndoPac.——Fig. 144,*4*. **W. coccinea*, Mauritius; *4a*, rods of axial nodes; *4b*, club, and *4c*, leaf-globe of cortex (26n).

Family PARISIDIDAE Aurivillius, 1931

[=Trinelladae Gray, 1870]

Arborescent, monomorphic, like Melithaeidae but internodal sclerites tuberculate, nodal sclerites lobate rods, cortical sclerites coarse plates, and ramification proceeds from calcareous internodes. *Tert.-Rec.*

Parisis Verrill, 1864 [**P. fruticosa*]. Large rigid colonies branched in one plane. *Tert.*(Italy-India)-*Rec.* (IndoPac.).——Fig. 145,*2*. **P. fruticosa*, Rec., Sulu Sea; *2a*, cortical sclerites; *2b*, detail of tubercle of large plate from cortex; *2c*, lobate rods of axial nodes; *2d*, part of axis, decorticated; *2e*, end view of internode; *2f*, surface detail of internode (26n).

?**Skleranthelia** Studer, 1878 [**S. musiva*]. Membranous or stoloniferous forms placed in this family (probably incorrectly) because of spicular similarity. *Rec.*, E.Atl. (no other record of Parisididae in Atl.).

Suborder HOLAXONIA Studer, 1887

[=Axifera von Koch, 1878]

Gorgonacea with distinct central axis made up of horny material alone or of horny material more or less heavily per meated with calcareous substance, continuous or with alternating horny and calcareous joints. *Cret.-Rec.*

Calcareous material of the peripheral zone of the axis is in nonspicular form, with single exception of the Keroeididae, a small family in which it takes the form of smooth, glassy, elongated, pointed scales or rods, the axis having a definite chambered central

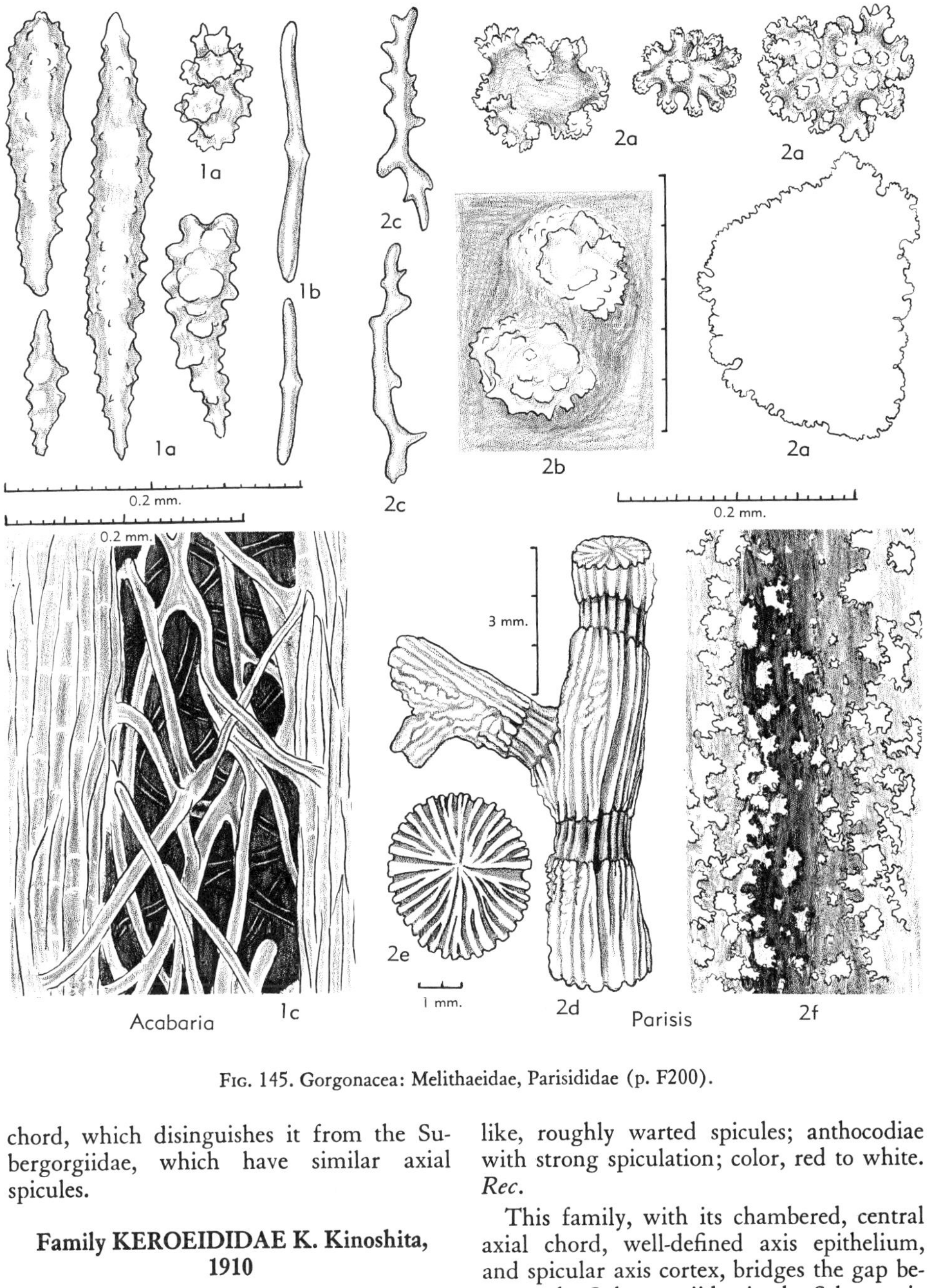

FIG. 145. Gorgonacea: Melithaeidae, Parisididae (p. F200).

chord, which disinguishes it from the Subergorgiidae, which have similar axial spicules.

Family KEROEIDIDAE K. Kinoshita, 1910

[=Suberogorgidae STUDER, 1887 *(partim)*]

Axis with chambered central chord surrounded by peripheral zone of smooth, elongated, pointed scales or rods which are not solidly fused together; coenenchymal cortex (rind) with large fusiform or plate-like, roughly warted spicules; anthocodiae with strong spiculation; color, red to white. *Rec.*

This family, with its chambered, central axial chord, well-defined axis epithelium, and spicular axis cortex, bridges the gap between the Subergorgiidae in the Scleraxonia on one side and the primitive families of the Holaxonia on the other. In spite of the outer, spicular layer of its axis, it must be referred to the Holaxonia.

Keroeides STUDER, 1887 [**K. koreni* WRIGHT &

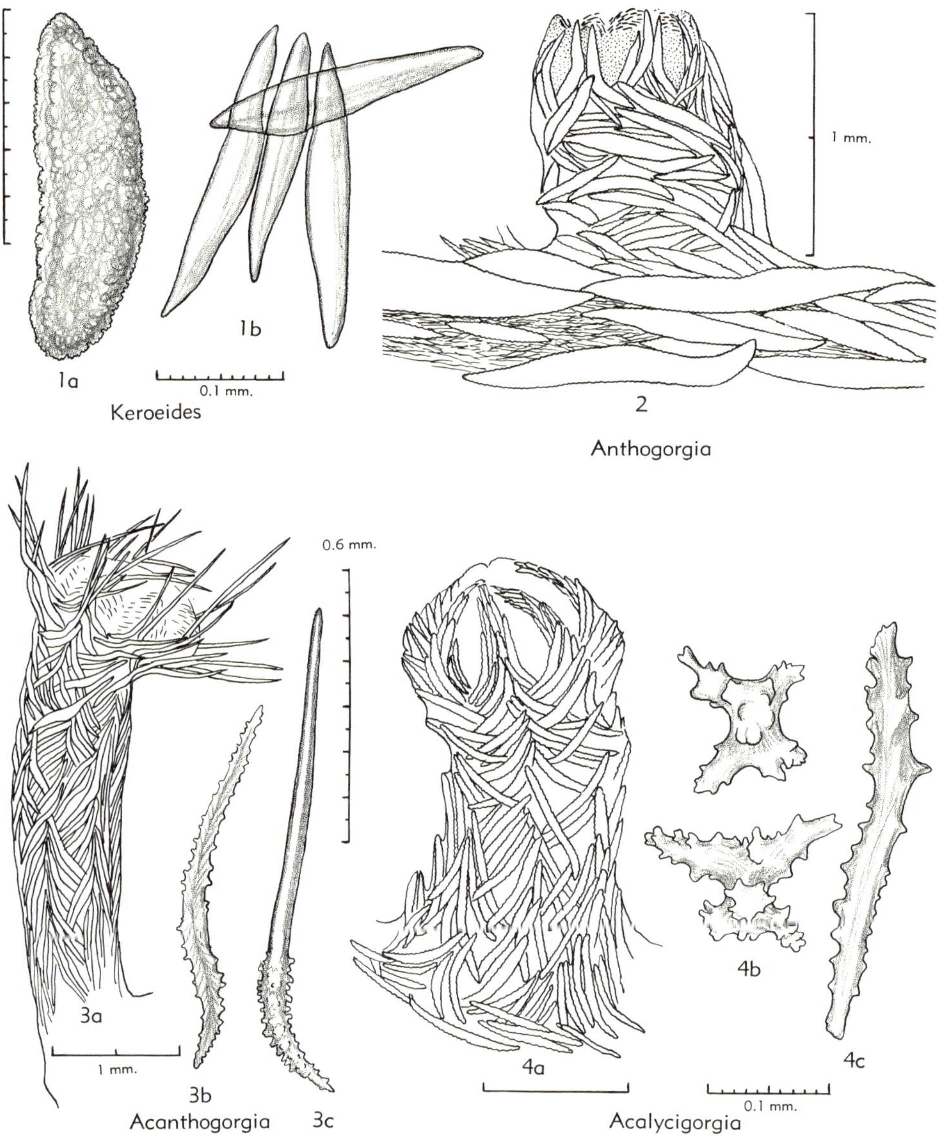

FIG. 146. Gorgonacea: Keroeididae, Acanthogorgiidae (p. F202-F203).

STUDER, 1889; SM WRIGHT-S., 1889]. *Rec.*, Indo-Pac. (Andamans-Hawaii; Japan-New Brit.)-Atl. (Guadalupe).——FIG. 146,*1*. **K. koreni*, Marshall I.; *1a*, coarse spindle of rind; *1b*, spicules of axis (26n).

Family ACANTHOGORGIIDAE Gray, 1859

[*nom. correct.* KÜKENTHAL & GORZAWSKI, 1908 (*pro* Acanthogorgiadae GRAY, 1859)]

Axis purely horny, with wide, chambered central chord; rind thin, polyps not retractile, forming prominent, cylindrical calyces with spicules arranged in 8 *en chevron* fields; calycular spicules continuous with those of tentacular crown, without any intervening spicule-free neck zone or transverse collaret and consequently no clear division between anthocodia and anthostele; spicules fusiform but in stem rind may be radiate. *Rec.*

Species of this family inhabit moderate to considerable depths; various species of *Acanthogorgia* occur in all seas, some thriving in very cold waters. The genera *Anthogorgia* and *Calicorgorgia* are limited to the Indo-Pacific region, so far as now known.

Acanthogorgia GRAY, 1857 [**A. hirsuta* (?=*A. aspera* POURTALÈS, 1867)] [=*Paracanthogorgia* STIASNY, 1943]. Calyces tall and cylindrical, with crown of strongly projecting spinous spindles in tentacle bases; stem rind with spindles or in some with radiates. [Species with rough projecting spines have been separated by STIASNY (1943, 1947) as a distinct genus, *Paracanthogorgia*, which at present cannot be maintained.] *Rec.*, IndoPac.-Atl.——FIG. 146,*3*. *A. aspera* POURTALÈS, W.Atl. (off Ga.); *3a*, calyx; *3b*, bent spindle of calyx wall; *3c*, spinous crown spicule (26n).

Acalycigorgia KÜKENTHAL, 1908 [**A. grandiflora* KÜKENTHAL & GORZAWSKI, 1908; SD KÜKENTHAL & GORZAWSKI, 1908]. Calyces similar to those of *Acanthogorgia* but without the crown of strongly projecting spines. *Rec.*, IndoPac.——FIG. 146,*4*. *A. ceylonensis* (THOMSON & HENDERSON), Timor; *4a*, calyx; *4b*, radiates from stem rind; *4c*, spindles from calyx wall (26n).

Anthogorgia VERRILL, 1868 [**Muricea divaricata* VERRILL, 1865]. Calyces cylindrical, spicules converging *en chevron* as 8 marginal points, beneath which they may be transverse or disposed irregularly. *Rec.*, IndoPac.——FIG. 146,*2*. *A bocki* AURIVILLIUS, 1931, Japan; part of a branch with one calyx (26n).

Family PARAMURICEIDAE Bayer, nov.

Axis horny, with wide, chambered central chord; rind thin, or in some moderately thick, polyps retractile within protruding calyces by virtue of a neck zone with few or no spicules; anthocodial armature of strong points of *en chevron* spicules usually resting on a transverse collaret; coenenchymal sclerites spindles only in some but commonly modified strongly as thorn scales or having other forms, ordinarily best-dedeveloped in or near calyces. *Rec.*

The establishment of this family is necessitated by removal of the genus *Muricea* from this assemblage, formerly called Muriceidae (DUCHASSAING & MICHELOTTI, 1864), which forms a reasonably homogeneous group. It occurs at moderate to considerable depths in all seas.

Paramuricea KÖLLIKER, 1865 [**Gorgonia placomus* LINNÉ, 1758; SD VERRILL, 1883]. Characteristic sclerites are thorn scales with lobed or laciniated edges and a stout outer projection, some long and spinelike, others short and blunt. *Rec.*, Atl.-Medit.-IndoPac.——FIG. 147,*1*. **P. placomus*, Grand Banks; *1a*, contracted polyp in its calyx; *1b*, spicules of inner rind; *1c*, calycular thorn scales (26n).

Acanthacis DEICHMANN, 1936 [**A. scabra*]. Colonies irregularly branched in one plane; rind spicules in one layer, stout spindles or plates; calyces low, armed with stout thorn scales. *Rec.*, W.Indies.——FIG. 148,*1*. *A. austera* DEICHMANN, Porto Rico; *1a*, calyx with surrounding thorn scales; *1b*, thorn scales, isolated (26n).

Anthomuricea STUDER, 1887 [**A. argentea* WRIGHT & STUDER, 1889; SD NUTTING, 1910]. No thorn scales; outer rind with blunt spindles sculptured with rounded, simple warts; inner rind with small, bent rods. Calyces tall, cylindrical, biserial or on all sides, armature in 8 double rows, *en chevron;* opercular spicules in converging rows, commonly including clublike spindles. *Rec.*, Japan-Hawaii-MalayArch.-Patagonia.——FIG. 149,*2*. *A. tenuispina* NUTTING, Hawaii; *2a*, spicules of outer rind; *2b*, of inner rind (26n).

Bebryce PHILIPPI, 1842 [**B. mollis*]. Characteristic sclerites are stellate plates with central boss, which make up inner layer of rind spicules, and smaller, rosette-like bodies, which lie closely packed in surface layer and become asymmetrical toward calycular margins where they project as spinous processes, commonly branched. *Rec.*, Medit.-IndoPac.-Atl.——FIG. 150,*4*. *B. cinerea* DEICHMANN, Fla.; *4a*, sclerites of calycular margin; *4b*, rosette of rind surface; *4c*, stellate plate of inner rind (26n).

Calicogorgia THOMSON & HENDERSON, 1906 [**C. investigatoris;* SD KÜKENTHAL, 1924].

Cyclomuricea NUTTING, 1908 [**C. flabellata*]. Small, fanlike colonies branched in one plane. Calyces shelflike, directed toward one face of colony; anthocodiae not retractile but with narrow neck zone that is inconspicuous in contraction. Rind sclerites clubs and spindles irregularly and asymmetrically sculptured with foliate tubercles; anthocodial spicules resembling flat tentacular rods of Gorgoniidae. *Rec.*——FIG. 149,*3*. **C. flabellata*, Hawaii; *3a*, contracted polyp; *3b*, rind sclerites; *3c*, tentacular rods (26n).

Discogorgia KÜKENTHAL, 1919 [**Placogorgia campanulifera* NUTTING, 1901]. Characteristic sclerites are discoidal or ovoidal plates with more or less lobate edges and sculpture of complicated tubercles; spindles may occur also in some, plates of calycular margin are thorn scales; inner rind contains various radiate forms. *Rec.*, IndoPac.——FIG. 147, *2*. *D. dendritica* (NUTTING), Celebes; *2a*, calycular thorn scales; *2b*, large plate from outer rind; *2c*, radiates from inner rind (26n).

?**Echinogorgia** KÖLLIKER, 1865 [**E. pseudosasappo* (=**Gorgonia sasappo reticulata* ESPER, 1791)]. Characteristic sclerites are leafy clubs, but simple

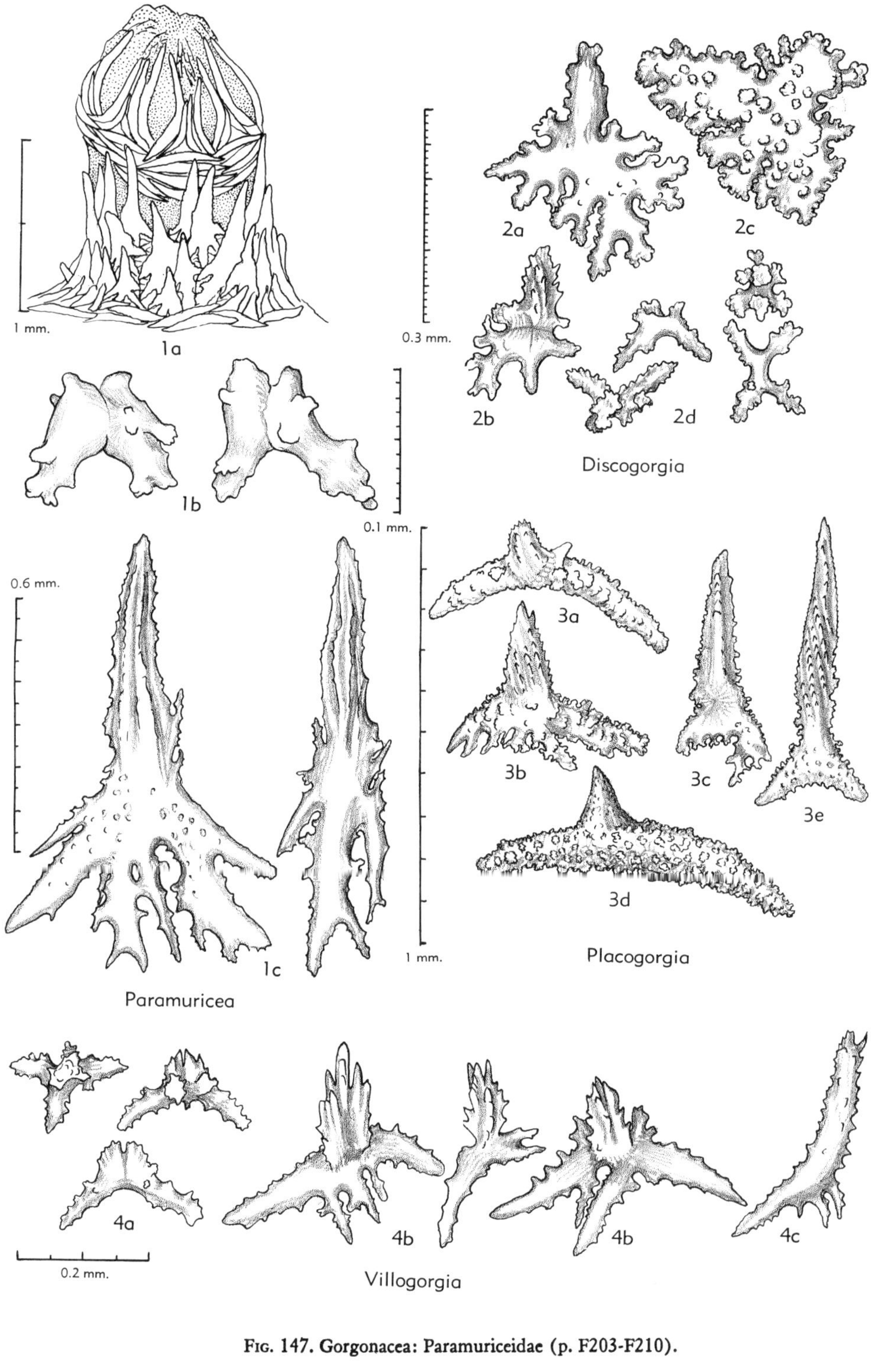

Fig. 147. Gorgonacea: Paramuriceidae (p. F203-F210).

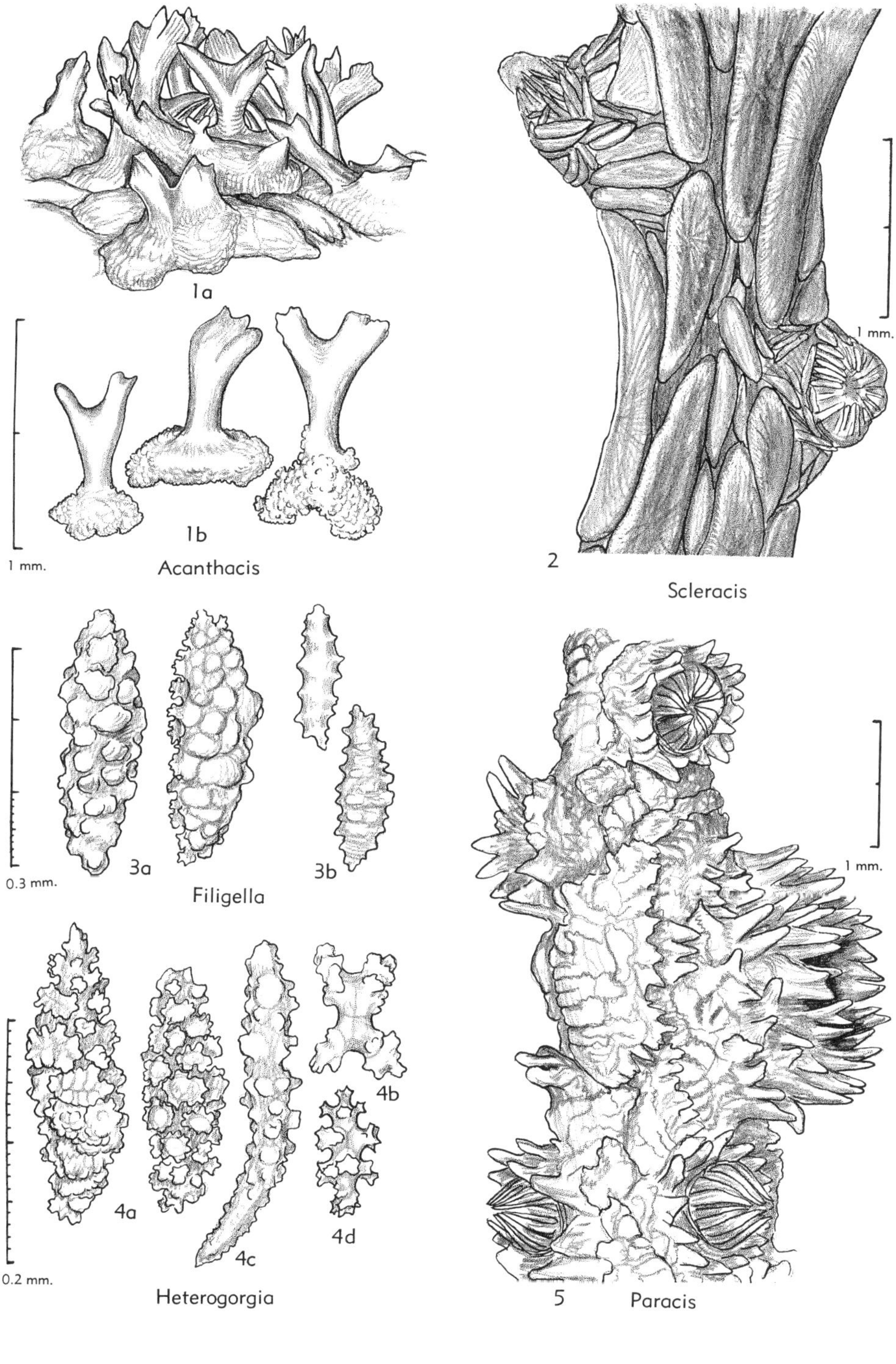

FIG. 148. Gorgonacea: Paramuriceidae (p. F203-F206).

spindles, commonly large and some with high external processes, may also be present at rind surface; rind moderately thick, calyces low and crowded. [This genus may belong in Plexauridae, for it seems close to, if not identical with, *Plexauroides.*] *Rec.,* IndoPac.——Fig. 150,*3. E. flexilis* Thomson & Simpson, 1909; *3a,* leaf clubs of calyces; *3b,* sclerites of inner rind (26n).

Echinomuricea Verrill, 1869 [**Nephthya coccinea* Stimpson, 1855]. Characteristic sclerites are thorn scales with stout, simple spine projecting from branched or lobed basal plate; spindles with or without strong outer processes may be present. Colonies usually branched in one plane and may be reticulate. [?Not separable from *Lissogorgia.*] *Rec.,* IndoPac.-Atl.——Fig. 150,*2. E. coronalis* Germanos, Celebes; *2a,* thorn scales; *2b,* quadriradiate (26n).

Filigella Gray, 1868 [**F. gracilis*]. Slender, unbranched or weakly branched colonies that may be attached to solid objects by basal disc or lie free on soft bottom. Spicules fusiform, some clavate, commonly stout and with coarse tubercles on outer surface. *Rec.,* IndoPac.-Calif.——Fig. 148,*3. F. mitsukurii* K. Kinoshita, Japan; *3a,* spicules from rind surface; *3b,* from deeper rind (26n).

Filigorgia Stiasny, 1937 [**F. riodouroi*]. Similar to *Filigella* but large coenenchymal spindles with tubercles of outer surface tall and spinelike; possibly identical with *Swiftia. Rec.,* W.Afr.

Heterogorgia Verrill, 1868 [**H. verrucosa;* SD Nutting, 1910]. Rind sclerites chiefly irregularly warted spindles, smaller capstans, and a few crosses; anthocodiae with strong armature of stout, bent spindles. *Rec.,* trop.E.Pac.——Fig. 148,*4. H. papillosa* Verrill, L.Calif.; *4a,* spindles of rind; *4b,* quadriradiate; *4c,* bent spindle from opercular radius; *4d,* capstan (26n).

Hypnogorgia Duchassaing & Michelotti, 1864 [**H. pendula*]. Tall, flexible colonies branched in one plane in a loose pinnate manner. Calyces bracket-like, topped by tall opercula composed of long rods, biserial on branches. Spicules long, simple spindles or rods. *Rec.,* W.Indies.

Lepidomuricea Kükenthal, 1919 [**Acanthomuricea ramosa* Thomson & Henderson, 1906] [*non Lepidomuricea* Verrill, 1922 (=*Paramuricea* Kölliker)].

Lissogorgia Verrill, 1864 [**Antipathes flabellum* Verrill (*non* Esper, 1791) =*Lissogorgia eubrandella* Bayer, *nom. nov.*] [=*Eubrandella* Deichmann, 1936 (obj.)]. Colonies flabellate, reticulate. Cortical sclerites including large stellate thorn scales with single, projecting spine, and spindles. *Rec.,* ?Fla.

Menacella Gray, 1870 [**M. reticularis*]. Colonies flabellate and reticulate; rind sclerites including spindles with crenulate or foliate tubercles. *Rec.,* MalayArch.

Muriceides Studer, 1887 [**M. fragilis* Wright & Studer, 1889; SM Wright-S., 1889]. Colonies scantily branched in one plane. Spicules comprising spindles with spinous sculpture, some differentiated as thorn scales. *Rec.,* IndoPac.-E.Atl.

Muricella Verrill, 1869 [**Lissogorgia flexuosa* Verrill, 1866; SD Nutting, 1910]. Colonies branching in one plane, some anastomosing. Calyces prominent, their sclerites usually forming 8 marginal points. Spicules of rind in 2 distinct layers, mostly spindles, some clubs or discoidal forms. Tentacular operculum distinct. *Rec.,* IndoPac.——Fig. 149,*1. M. abnormalis* Nutting, Japan; part of branch with calyces (26n).

Paracis Kükenthal, 1919 [**Acis orientalis* Ridley, 1882]. Colonies usually but not invariably branched in one plane. Calyces distinct, distributed generally or leaving one face of colony free. Rind sclerites in form of large spindles, plates or scales, some with spines. *Rec.,* IndoPac.——Fig. 148,*5.* **P. orientalis,* MarshallI.; part of branch with calyces, large one parasitized by a copepod (26n).

Placogorgia Studer, 1887 [**P. atlantica* Wright & Studer, 1889; SM Wright-S., 1889] [=*Pseudothesea* Kükenthal, 1919]. Colonies branched in one plane. Calyces low, on all sides, armed with thorn scales. Rind with large spindles having strong spines on their outer surfaces. *Rec.,* Atl.-IndoPac.——Fig. 147,*3. P. tenuis* Deichmann, W.Indies; *3a-c,* variously developed thorn scales; *3d,* spinous spindle of rind; *3e,* crutch-shaped central spicule of opercular radius (26n).

Scleracis Kükenthal, 1919 [**S. pumila* (=**Acis guadalupensis* Duchassaing & Michelotti, 1860)]. Colonies branched in one plane, lateral and dichotomous combined. Calyces biserial, alternating. Rind sclerites in 2 layers; in outer, chiefly large warted spindles, somewhat flattened and commonly bent; in inner, mostly straight sparsely warted spindles. *Rec.,* W.Atl.——Fig. 148,*2.* **S. guadalupensis,* Cuba; part of branch with 2 calyces (26n).

Swiftia Duchassaing & Michelotti, 1864 [**Gorgonia exserta* Ellis & Solander, 1786 (*non Thesea exserta* Duch.-M., 1860)] [=*Stenogorgia* Verrill, 1883; *Platycaulos* Wright & Studer, 1889; *Allogorgia* Verrill, 1928]. Colonies with lax branching, chiefly in one plane; polyps scattered or biserial, forming prominent conical calyces, anthocodiae commonly exsert. Rind thin, outer layer filled with spinous rods, spindles, and capstans; inner layer mostly restricted to areas between longitudinal canals, containing only small capstans. Tentacles with stout rods. *Rec.,* IndoPac.-Atl.——Fig. 149,*4.* **S. exserta,* Fla.; *4a,* spindle from tentacles; *4b,* from calyx; *4c,* from outer rind; *4d,* capstans of outer rind; *4e,* small capstans from inner rind between stem canals (26n).

Thesea Duchassaing & Michelotti, 1860 [**T. exserta* Duch-M. (*non Gorgonia exserta* Ellis & Solander, 1786) (=*Thesea guadalupensis* Duch.-M., 1864)]. Colonies usually branched in one plane, branches slender to rather stout; calyces distinct, with 8 marginal teeth formed of simple spindles. Sclerites of rind in 2 layers, outer containing large, spheroidal or platelike bodies with outer faces commonly undulated, inner layer including warted spindles. *Rec.,* W.Atl.——FIG. 149, *5. T. grandiflora* DEICHMANN, Barbados; *5a,* spicules of outer rind; *5b,* of inner (26n).

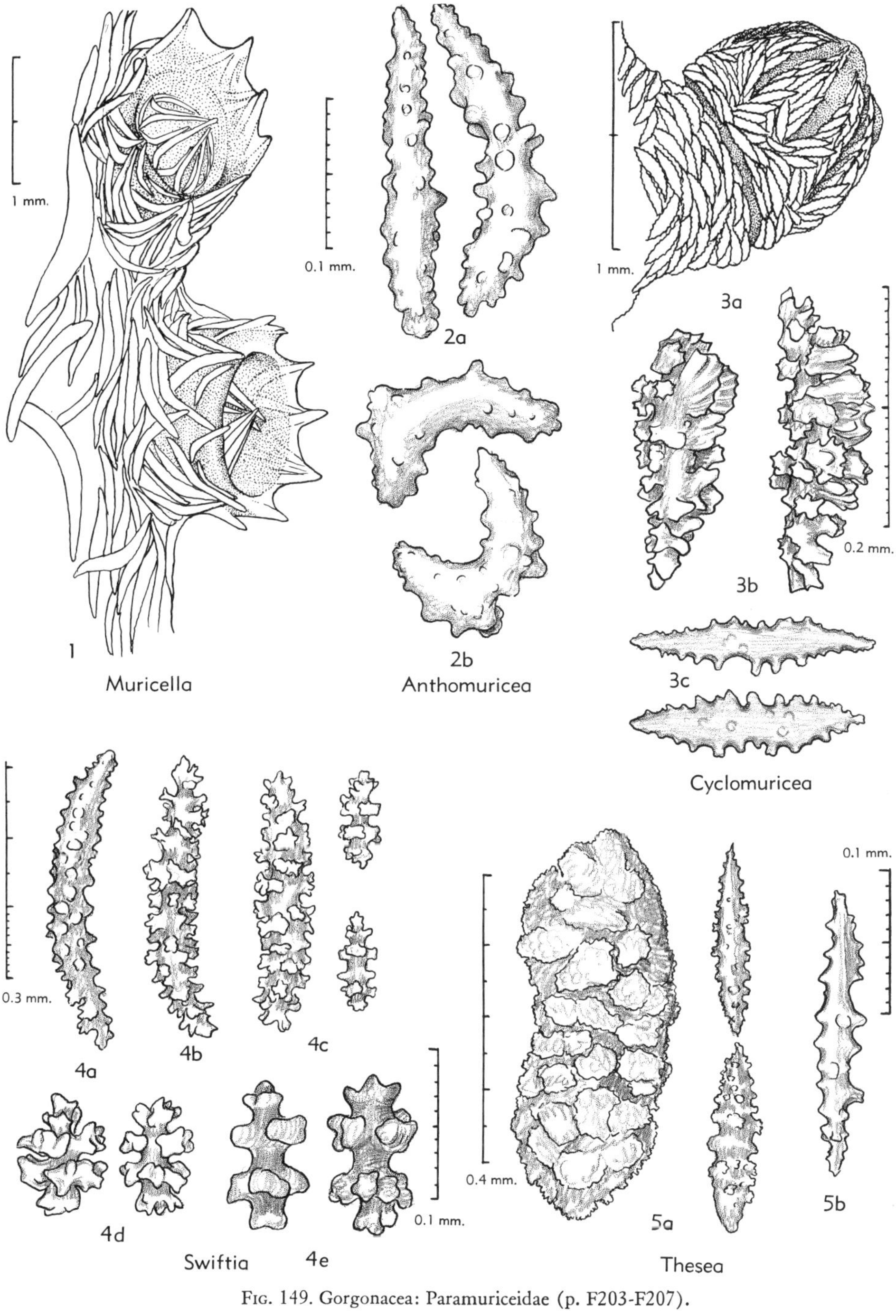

FIG. 149. Gorgonacea: Paramuriceidae (p. F203-F207).

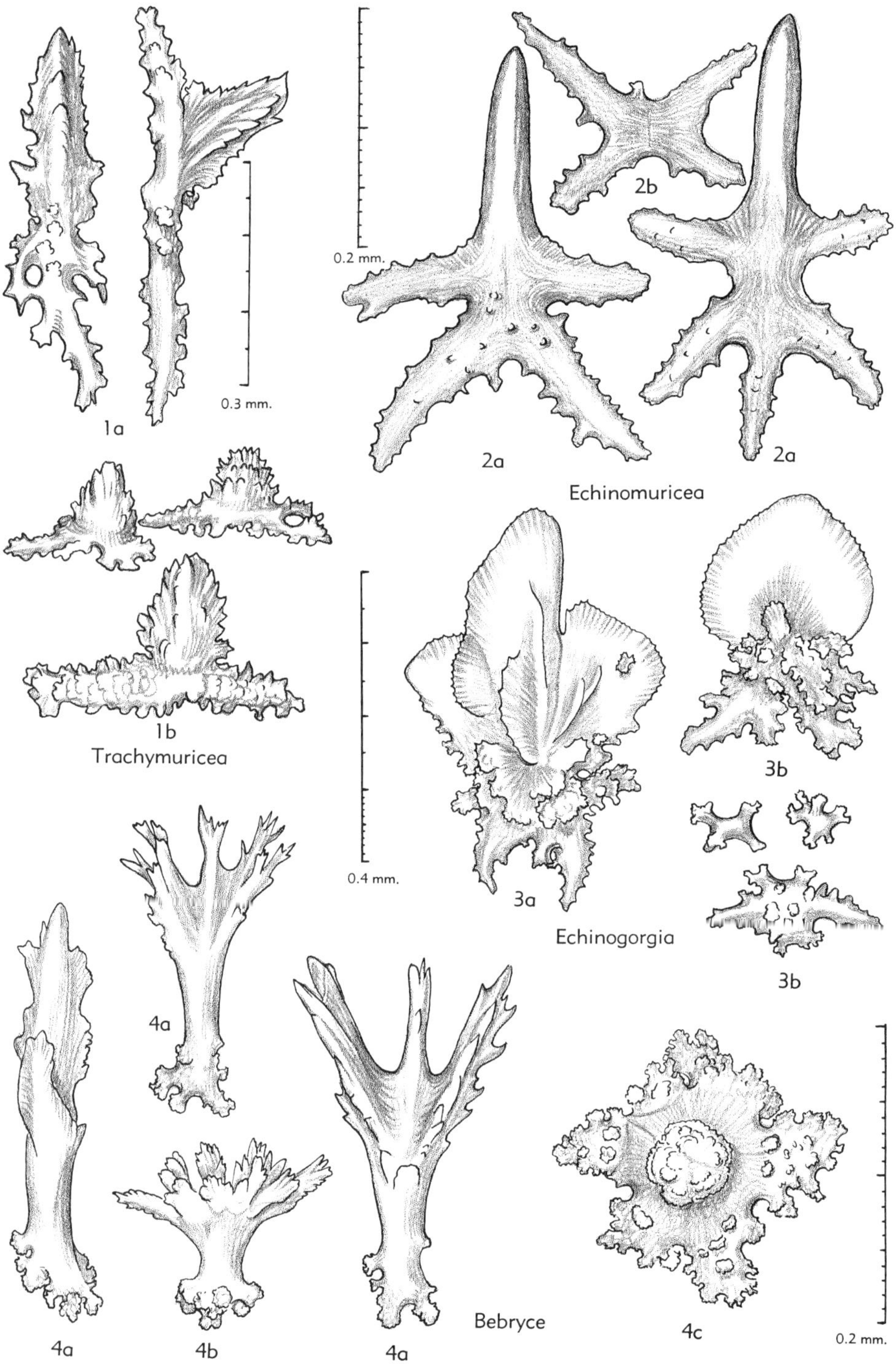

FIG. 150. Gorgonacea: Paramuriceidae (p. F203-F210).

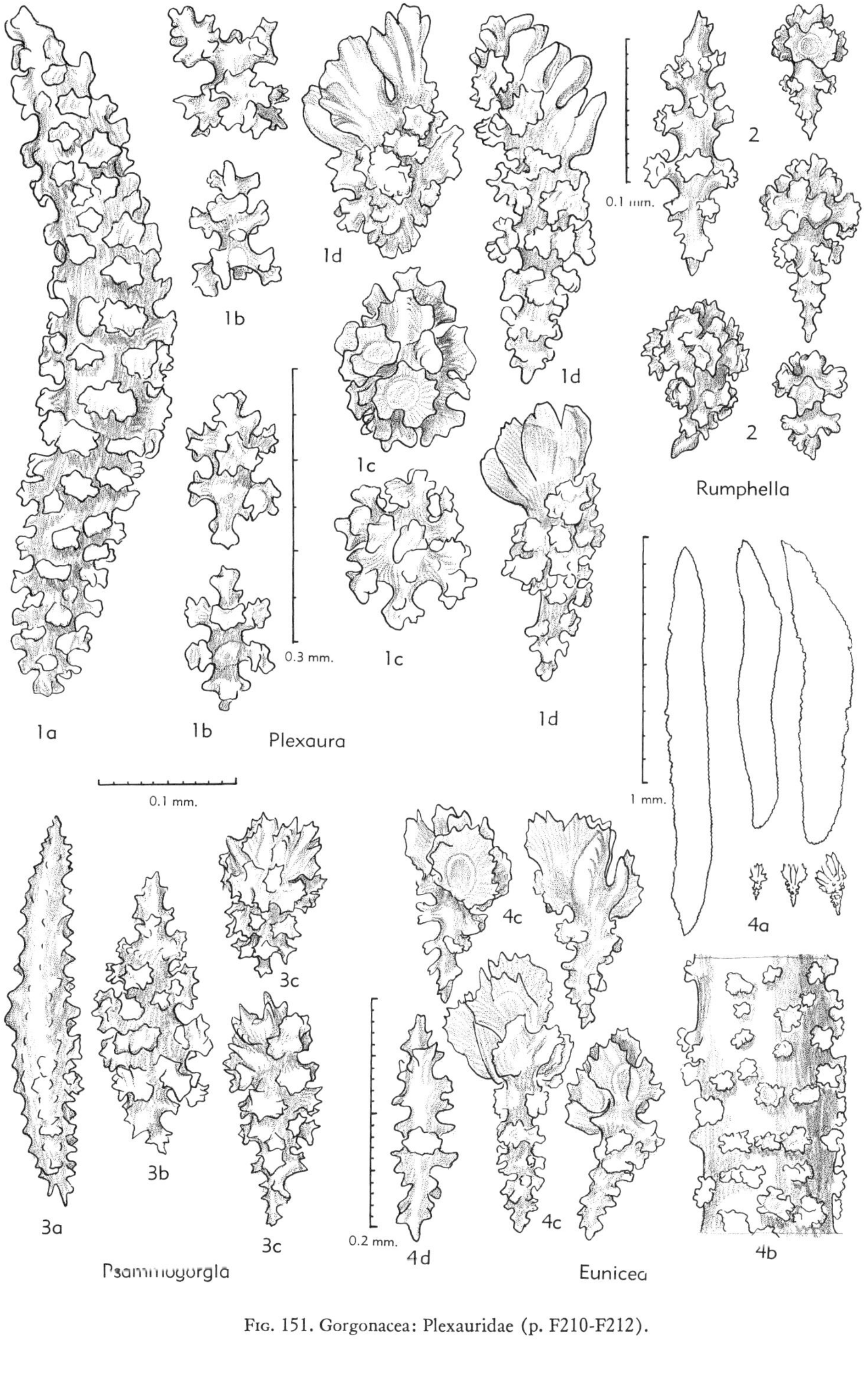

FIG. 151. Gorgonacea: Plexauridae (p. F210-F212).

Trachymuricea DEICHMANN, 1936 [**Acanthogorgia hirta* POURTALÈS, 1867]. Small colonies with large, conical calyces and strong opercula of converging rows of spindles above a high collaret. Rind sclerites are spinous rods with rough projecting process; calycular thorn scales with strong, laciniated projection. *Rec.*, W.Atl.——FIG. 150,*1*. **T. hirta*, Fla.; *1a*, calycular thorn scales; *1b*, spicules of outer rind (26n).

Villogorgia DUCHASSAING & MICHELOTTI, 1860 [**V. nigrescens*] [=*Brandella* GRAY, 1869; *Perisceles* STUDER, 1887; *Acamptogorgia* WRIGHT & STUDER, 1889; *Paracamptogorgia* KÜKENTHAL, 1919]. Colonies branched in one plane; calyces prominent, operculum strong, collaret well-developed. Rind sclerites including specialized thorn scales with spinous, digitate or foliate process. *Rec.*, Atl.-Medit.-IndoPac.——FIG. 147,*4*. **V. nigrescens*, Fla.; *4a*, spicules of stem rind; *4b*, of calyces; *4c*, bent spindle of opercular radius (26n).

Family PLEXAURIDAE Gray, 1859

Colonies of very diverse form but generally with thick branches arising laterally, dichotomously or in some pinnately. Polyps completely retractile or forming distinct calyces into which anthocodiae can be withdrawn. Axis with wide, chambered central chord and peripheral zone of loculated horny material usually containing nonspicular calcareous matter, commonly with tendency toward heavy calcification of base in old colonies. Rind thick, perforated by system of longitudinal canals surrounding axis, which delimits outer and inner (axial sheath) layer of cortex that generally differs in spiculation. Sclerites usually including some form of club, but some with spindles only, oval bodies, or large quadriradiates. *Rec.*

Plexaura LAMOUROUX, 1812 [**Gorgonia homomalla* ESPER, 1792; SD VERRILL, 1912] [=*Plexauropsis* VERRILL, 1907] Stout, treelike colonies with thick branches, bushy but commonly tending to spread in one plane. Rind thick, polyps fully retractile; without prominent calyces. Outer layer of spicules mostly spindles and clubs, colorless or purple; inner layer with spiny spindles, mostly purple, some colorless; in some species also short, multiradiate bodies. *Rec.*, trop.W.Atl.——FIG. 151,*1*. **P. homomalla*, W.Indies; *1a*, superficial spindle; *1b*, 6-, 7-, and 8-radiates of inner rind; *1c*, multiradiate oval bodies of inner rind; *1d*, superficial clubs (26n).

Anthoplexaura KÜKENTHAL, 1908 [**A. dimorpha*]. Bushy colonies with large, completely unarmed polyps retractile within prominent calyces. Rind with oval spindles sculptured with tall irregularly placed warts and small capstans with tubercles in transverse belts; red. *Rec.*, Japan.

Eunicea LAMOUROUX, 1816 [**E. mammosa;* SD BAYER, 1955]. Stout, thick-branched, arborescent colonies resembling *Plexaura* but generally with prominent tubular or shelflike calyces; polyps with anthocodial crown and points. Outer layer of rind containing numerous small clubs, commonly foliate, mostly colorless; deeper parts of outer rind with warty spindles, mostly purple; inner rind with colorless and purple spinous spindles. *Rec.*, trop.W.Atl.——FIG. 151,*4*. **E. mammosa*, Bahamas; *4a*, clubs and spindles of outer rind in outline, showing relative sizes; *4b*, sculptural detail of large spindle; *4c*, surface clubs in detail; *4d*, purple spindle of inner rind (26n).

Eunicella VERRILL, 1869 [**Gorgonia verrucosa* PALLAS, 1766]. Colonies branched scantily or moderately in one plane; calyces prominent or not. Rind with surface layer of characteristic balloon clubs and deeper layer of belted spindles. *Rec.*, Atl.-Medit.-IndoPac.——FIG. 152,*3*. **E. verrucosa*, Spain; *3a*, spindles of inner rind; *3b*, balloon clubs of surface layer (26n).

Euplexaura VERRILL, 1865 [**E. capensis*]. Erect colonies branched in one plane, arising from spreading base that bears polyps and commonly short stems also; polyps fully retractile. Rind with superficial layer of oval, coarsely warted sclerites, and deeper layer of capstans. *Rec.*, IndoPac.——FIG. 153,*2*. *E. erecta* KÜKENTHAL, Japan; *2a*, bent spindle from collaret; *2b*, coarse ovoids from rind (26n).

Hicksonella NUTTING, 1910 [*non* SIMPSON, 1910] [**H. princeps*] [=*Rhabdoplexaura* KÜKENTHAL, 1919 (obj.)]. Thick-branched arborescent colonies; polyps retractile into low calyces. Spiculation of capstans commonly modified as clubs and large rods. *Rec.*, IndoPac.——FIG. 152,*4*. **H. princeps*, Japan; *4a*, large rod; *4b*, capstans; *4c*, spindle; *4d*, clubs (26n).

Muricea LAMOUROUX, 1821 [**M. spicifera*]. Arborescent colonies richly branched, commonly tending to remain in one plane; branches moderately to very thick, with close-set, prickly, tubular or shelf-like calyces on all sides. Sclerites usually fusiform, with strong outer or terminal spines or both; rarely some irregular forms. *Rec.*, trop.W.Atl.-E.Pac.——152,*1* **M. spicifera*, W.Indies; *1a*, outlines of large spindles; *1b-c*, spinous spindles at higher power; *1d*, sculptural detail of *1b* (26n).

Muriceopsis AURIVILLIUS, 1931 [**Gorgonia tuberculata* ESPER, 1791]. Colonies low, shrublike and irregularly branched, or taller and pinnate; calyces low but usually distinct, with small lower lip. Spicules of outer rind mostly foliate clubs and stout spindles with foliate processes on one side. Inner rind contains symmetrically sculptured spindles, usually bright purple. *Rec.*, trop.W.Atl.——

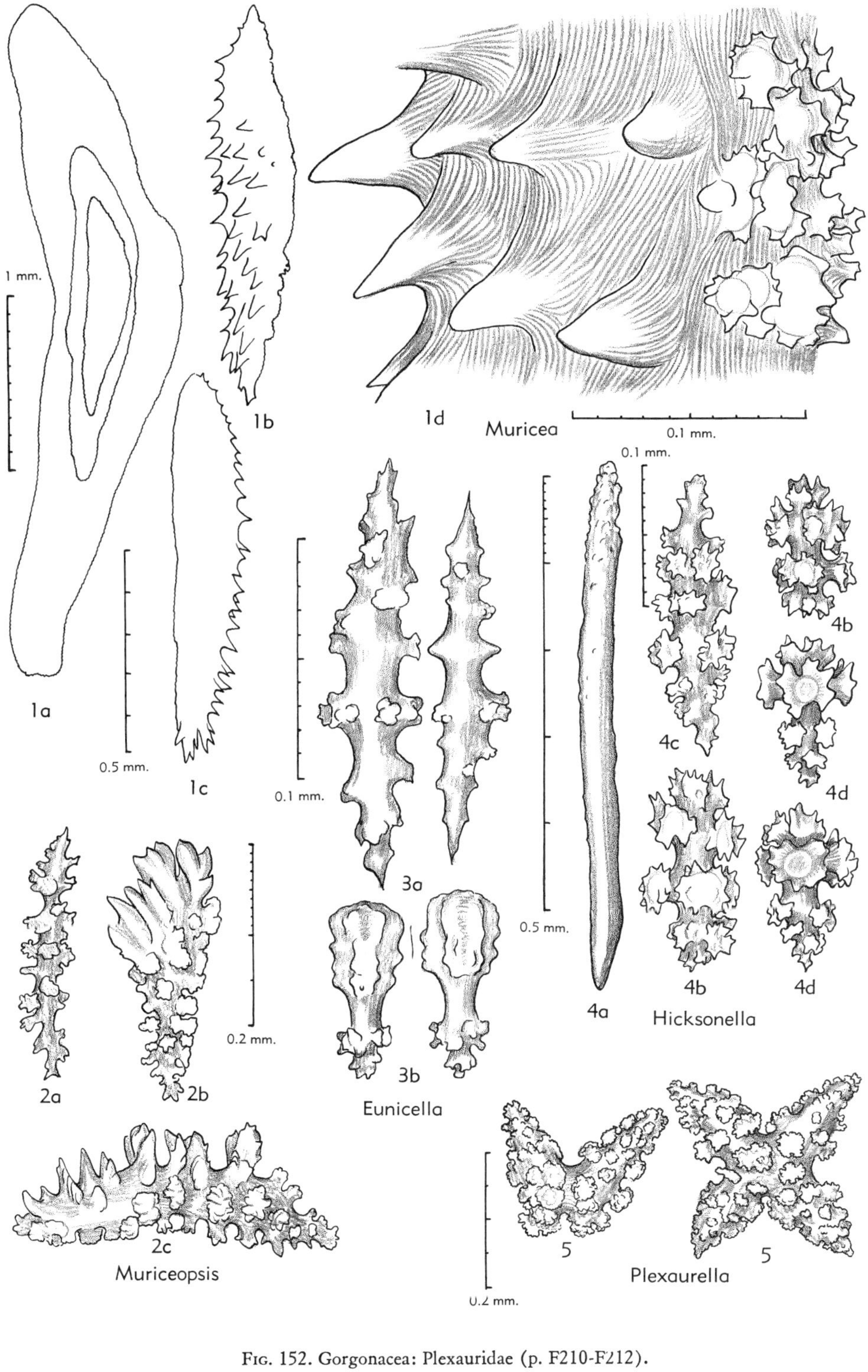

FIG. 152. Gorgonacea: Plexauridae (p. F210-F212).

FIG. 152,*2*. *M. flavida* STIASNY, VirginI.; *2a*, purple spindle of inner rind; *2b*, surface club; *2c*, asymmetrical foliate spindle of outer rind (26n).

Paraplexaura KÜKENTHAL, 1909 [**Plexauroides asper* MOROFF, 1902; SD KÜKENTHAL, 1917]. Branching mostly in one plane; polyps forming distinct calyces on all sides. Rind thick, containing warty and leafy clubs, commonly with very coarse heads, some developed into platelike forms; inner rind with fusiform and radiate sclerites. *Rec.*, IndoPac. ——FIG. 153,*1*. **P. asper*, Japan; leaf clubs and coarse plate (26n).

Plexaurella VALENCIENNES, 1855 [**Gorgonia dichotoma* ESPER, 1791]. Erect, dichotomously branched colonies with thick, stiff branches; polyps on all sides, fully retractile, with only raised margins at most, none with prominent calyces. Characteristic sclerites are quadriradiates; simple spindles may be present. *Rec.*, trop.W.Atl.——FIG. 152,*5*. **P. dichotoma*, W.Indies; spicules (26n).

Psammogorgia VERRILL, 1868 [**Echinogorgia arbuscula* VERRILL, 1866]. Colonies low, bushy, irregularly branched dichotomously; polyps on all sides, calyces low or absent. Rind rather thick, with spindles and aculeate wart clubs; spicules red. *Rec.*, W.Mexico-Panama.——FIG. 151,*3*. **P. arbuscula*, Costa Rica; spicules (26n).

Rumphella BAYER, 1955 [**Plexaura aggregata* NUTTING, 1910]. Erect, bushy colonies with thick, sinuous or nearly straight branches; axis dark, with abundant calcareous deposits, especially near base which is much thickened and stony in old specimens; polyps on all sides, calyces low or absent. Rind spicules as warty spindles, with superficial layer of small wart clubs having whorl of 3 warts as a head. *Rec.*, E.Indies-Oceania.——FIG. 151,*2*. *R. antipathes*, Rec., GilbertI.; spicules (26n).

Family GORGONIIDAE Lamouroux, 1812

[*nom. correct.* JOHNSON, 1862 (*pro* Gorgonidae *auctt.*; *nom. correct. pro* family Gorgonieae LAMX., 1812)]

Axis purely horny with narrow but distinct chambered central chord and cortex little loculated, if at all; polyps fully retractile, some forming low calyces, scattered or biserially disposed. Rind moderately thick, packed with spindles and capstans with regular belts of tubercles, in certain genera modified into disc spindles, scaphoids, or unilaterally spined forms; anthocodial armature weak, in form of crown composed of flat rodlets with scalloped edges, or lacking entirely. Colonies of diverse form, from unbranched to pinnate, closely reticulate or foliate. *Rec.*

A number of species of "*Gorgonia*" are recorded as fossils by early authors but these records are almost certainly erroneous. For a few such references see G. FISCHER DE WALDHEIM, *Bibliographia Palaeonthologica Animalium Systematica* (1834).

The genus of fossils called *Plumalina* HALL, 1858, has been placed in this family by some authors (J. S. WILLIAMS, 24), but it has more resemblance to certain delicate plumose hydroids and probably belongs to that group.

Websteria M.EDW.-H., 1850, also was assigned tentatively by its authors to the Gorgoniidae, but probably it does not belong here.

Gorgonia LINNÉ, 1758 [**G. flabellum*; SD VERRILL, 1868] [=*Rhipidigorgia* VALENCIENNES, 1855 (obj.)]. Spicules in form of strong scaphoids and symmetrical spindles. Branches in one plane, closely anastomosing. *Rec.*, W.Indies.——FIG. 153,*3*. **G. flabellum*, Nassau; *3a*, scaphoid; *3b*, spindle (26n).

Antillogorgia BAYER, 1951 [**Gorgonia acerosa* PALLAS, 1766]. Spicules are slender scaphoids and symmetrical spindles. Branches closely pinnate, not anastomosing. *Rec.*, W.Indies.——FIG. 153,*8*. **A. acerosa*, Key West; *8a*, scaphoids; *8b*, spindle (26n).

Eugorgia VERRILL, 1868 [**Leptogorgia ampla* VERRILL, 1864]. Spicules chiefly disc spindles; ordinary spindles present in small numbers in some species. Branching lateral, in one plane or bushy. *Rec.*, Calif.-Peru.——FIG. 153,*4*. **E. ampla*, Acapulco; *4a*, disc spindles in side, end and oblique views; *4b*, spindle (26n).

Leptogorgia M.EDW.-H., 1857 [**Gorgonia viminalis* PALLAS, 1766; SD VERRILL, 1868]. Spicules are spindles, shorter ones with warts of one side fused like those of disc spindles; long ones symmetrical or with warts on one side simple and conical, elsewhere compound. Colonies little-branched, slender, and whiplike. *Rec.*, W.Atl.——FIG. 153,*7*. **L. viminalis*, W.Fla.; *7a*, short, almost disc spindle, and top view of one disc; *7b*, spindle (26n).

Lophogorgia M.EDW.-H., 1857 [**Gorgonia palma* PALLAS, 1766]. Spicules exclusively symmetrical spindles without unilateral fusion of warts to form discs; anthocodial armature of flattened rods may be present. Branches long and slender, or short and pinnately arranged, in one plane or bushy. *Rec.*, Atl.-Medit.-IndoPac.——FIG. 153,*5*. **L. palma*, S.Afr.; *5a*, spindles of rind; *5b*, rods from crown (26n).

Pacifigorgia BAYER, 1951 [**Gorgonia stenobrochis* VALENCIENNES, 1846]. Spicules as in *Lophogorgia*, commonly long and pointed. Branches in one plane and closely anastomosed to form netlike fans. *Rec.*, GulfCalif.-Panama-Brazil.

Phycogorgia M.EDW.-H., 1850 [**Gorgonia fucata*

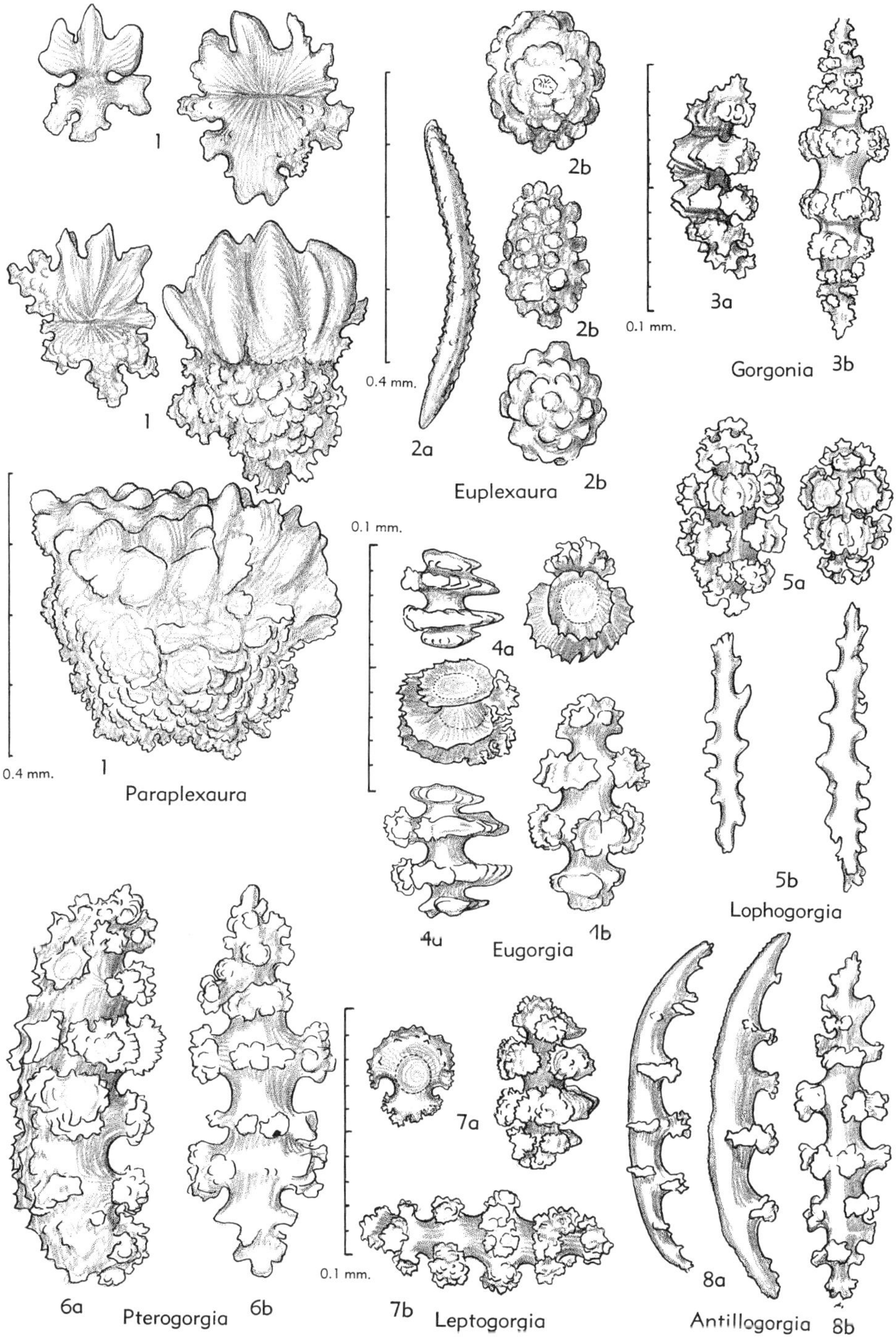

FIG. 153. Gorgonacea: Plexauridae, Gorgoniidae (p. F210-F214).

VALENCIENNES, 1846] Spicules exclusively short, blunt, symmetrical capstans. Branches flattened into broad fronds, with lamellar axis. *Rec.*, Mazatlan-Chile.

Phyllogorgia M.EDW.-H.,, 1850 [**Gorgonia dilatata* ESPER, 1806]. Spicules including scaphoids as well as spindles. Branches flattened into broad fronds, with loosely reticulating axis. *Rec.*, Brazil.

Pterogorgia EHRENBERG, 1834 [**Gorgonia anceps* PALLAS, 1766; SD M.EDW.-H., 1850] [=*Xiphigorgia* M.EDW.-H., 1857 (obj.)]. Spicules are very coarse spindles and scaphoids. Ramification lax, branches rather long; polyps bi- or triserial, retracting into thin, raised coenenchymal flanges. *Rec.*, W.Indies.——FIG. 153,*6*. **P. anceps*, Curaçao; *6a*, scaphoid; *6b*, spindle (26n).

Family ELLISELLIDAE Gray, 1859

[*nom. correct.* BAYER, 1955 (*pro* Elliselladae GRAY, 1859)]

Axis strongly calcified, made up of concentric layers not markedly undulating, central chord not soft and chambered; calcareous material deposited in irregular radial sectors (4). Colonies unbranched, loosely branched, or forming broad, flat fans with or without anastomoses. Spicules small (0.05-0.15 mm.), of characteristic dumbbell shape, some modified into clubs, and longer double spindles, all regularly sculptured with strong, hemispherical tubercles. *Cret.-Rec.*

Ellisella GRAY, 1858 [**Gorgonia elongata* PALLAS, 1766; SD NUTTING, 1910] [=*Scirpearia* KÖLLIKER, 1865, and later authors (*non* CUVIER, 1817, 1830); *Viminella* GRAY, 1870]. Colonies simple or with few long, whiplike branches; calyces biserial or in lateral tracts, usually prominent. Spicules including double spindles approaching 0.2 mm. in length concentrated in calyces, and double heads or dumbbells half as long or less, chiefly in general cortex. *Rec.*, Atl.-Medit.-RedS.-IndoPac.——FIG. 154,*4*. *E. funiculina* (DUCHASSAING & MICHELOTTI), Rec., GulfMexico; dumbbells from calyces and spindles from rind (26n).

Ctenocella VALENCIENNES, 1855 [**Gorgonia pectinata* PALLAS, 1766]. Branching characteristically lyrate. Spicules chiefly stout double heads. *Rec.*, IndoPac.——FIG. 154,*5*. **C. pectinata*, Torres Straits; *5a*, colony (30); *5b*, spicules (26n).

Junceella VALENCIENNES, 1855 [**Gorgonia juncea* PALLAS, 1766; SD NUTTING, 1910]. Colonies simple or dichotomously branched, calyces usually distinct and biserially disposed, some generally scattered. Characteristic sclerites are flattened clubs in outer cortex and dumbbells with strong, conical warts in thin axial sheath layer. *?Eoc.*, *Rec.*, IndoPac.——FIG. 154,*2*; 158,*2*. **J. juncea*, Rec., Philip.; 154,*2a*, dumbbell from axial sheath zone; 154,*2b*, face and edge view of cortical club; 158,*2a*, axis in transverse, 158,*2b* in longitudinal section (left margin of *2b* follows edge of central chord), as seen in polarized light (all 26n).

Nicella GRAY, 1870 [**N. mauritiana* (=**Scirpearia dichotoma* GRAY, 1859)]. Colonies ramified in one plane, branches usually slender and cylindrical; polyps biserial, with hemispherical calyces. Rind with inner layer of double spindles up to twice length of dumbbells of outer layer. *Rec.*, IndoPac.-Atl.——FIG. 154,*3*. *N. guadalupensis* DUCHASSAING & MICHELOTTI, Barbados; spicules (26n).

Riisea DUCHASSAING & MICHELOTTI, 1860 [**R. paniculata*]. Colonies finely divided, with numerous very slender branchlets arranged on main branches in alternating pinnate fashion; core of axis excentric; calyces prominent, clavate, placed at ends of short twigs and along upper side of branchlets. Spiculation much as in *Nicella*, with dumbbells in rind and long, tapered spindles (longer ones flattened) in calyces. *Rec.*, trop.W.Atl.——FIG. 155, *2*; 158,*3*. **R. paniculata*, Porto Rico; 155,*2a*, rods of calyces; 155,*2b*, dumbbells of rind; 155,*2c*, branchlet, 158,*3a,b*, axis in transverse and long. sect. (light zone along left edge of *3b* is axial core), as seen in polarized light (all 26n).

Toeplitzella DEICHMANN, 1936 [**Junceella laevis* VERRILL, 1866]. Like *Ellisella* but no long double spindles. *Rec.*, IndoPac.

Verrucella M.EDW.-H., 1857 [**Gorgonia flexuosa* LAMARCK, 1816 (=**Gorgonia umbraculum* ELLIS & SOLANDER, 1786); SD BAYER, 1955]. Colonies flabellate, some reticulate, branches slender; polyps with distinct verruciform calyces. Rind containing slender double heads and somewhat longer double spindles; twin or triplet forms producing crosses or stars may be more or less numerous. *Cret.(Dan.)-Rec.*, Eu.-RedS.-IndoPac.——FIG. 154,*1*. *V. delicatula* (NUTTING), *Rec.*, Philip.; spicules (26n).

Family IFALUKELLIDAE Bayer, 1955

Arborescent, with strongly calcified axis made up of gently undulating concentric lamellae, calcareous material not oriented in radial sectors, central chord not soft and cross-chambered. Sclerites minute (less than 0.1 mm. in length), flattened oval biscuit-shaped, or double discs, with irregular surfaces, very similar to spicules of *Xenia* and polyps of *Tubipora*. *Rec.*

Ifalukella BAYER, 1955 [**I. yanii*]. Colonies densely bushy, not pinnate; axis spirally ridged, arising from massive calcareous base onto which ridges extend as high, calcareous crests with lobed, in many places strongly laciniated, edges that may give rise to small twigs, which here and there develop into full-sized colonies. Spicules extremely

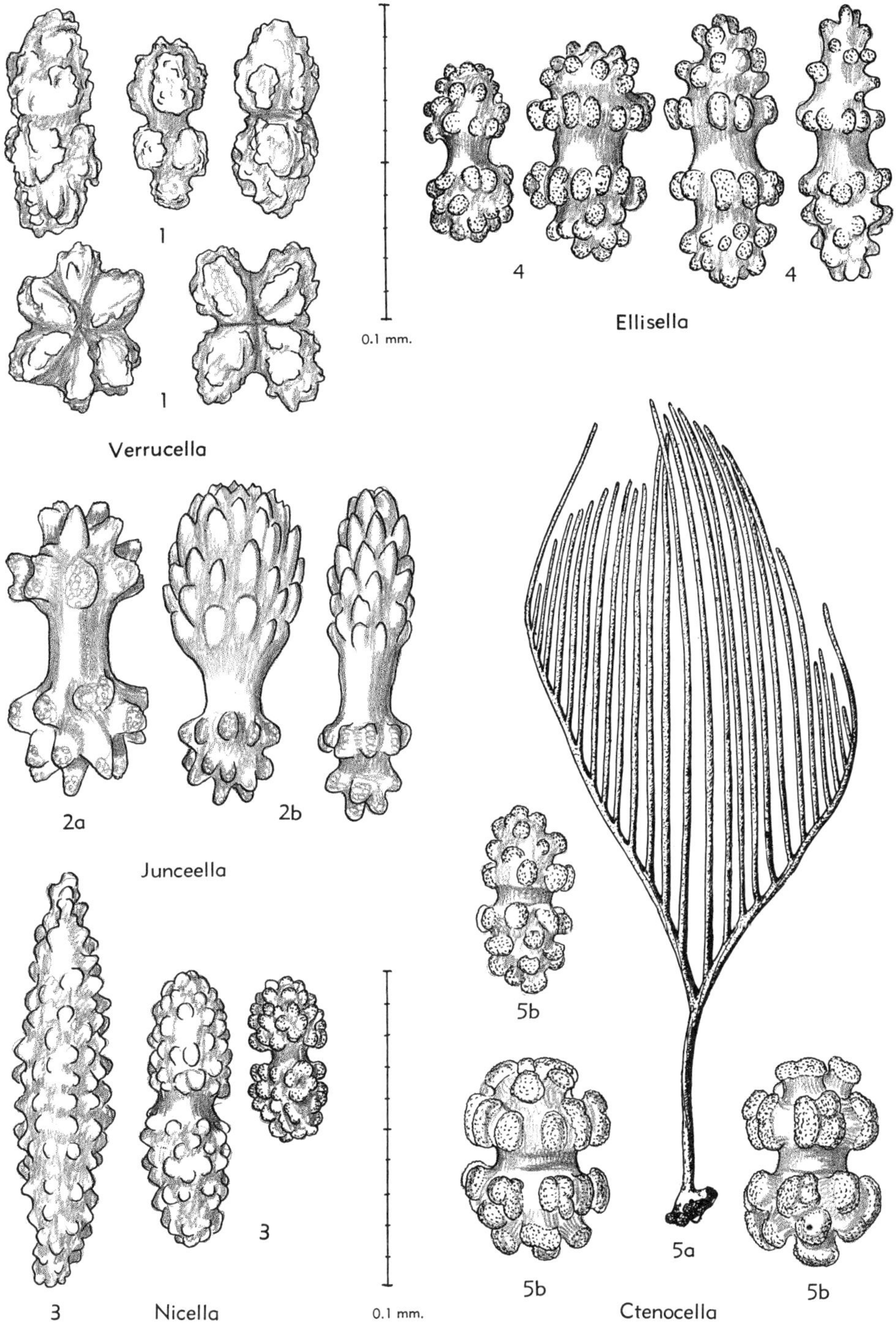

FIG. 154. Gorgonacea: Ellisellidae (p. F214).

scarce, those present being mostly oval rods with some twin forms. Zooxanthellae extraordinarily numerous. Distribution unknown, but type species is frequent at Ifaluk Atoll (Carolines), where it is the one abundant reef-dwelling gorgonian, inhabiting reef fronts beyond surge-channels in 15 or 20 feet of water; apparently quite restricted ecologically. Its peculiar, calcareous base should be easy to recognize if found in reef limestones, and might be a good paleoecological indicator. *Rec.* ——FIG. 156,*2*. **I. yanii; 2a,* spicules (circle represents relative size of zooxanthellae); *2b,* calcareous colonial base (26).

Plumigorgia NUTTING, 1910 [**P. hydroides*]. Colonies delicate, flexible, pinnately branched, arising from calcareous base. Polyps forming calyces or wholly retractile into general coenenchyme. Spicules numerous. *Rec.,* SuluArch.-ChinaS.——FIG. 156,*1*. **P. hydroides;* spicules (26).

Family CHRYSOGORGIIDAE Verrill, 1883

Axis highly calcified, continuous, glossy, iridescent or metallic, made up of smooth concentric lamellae (Fig. 158,*1*), rarely undulating; calcareous matter not oriented in radial sectors; branching may not occur, but more frequently does so in a very regular order, uniserial, spiral or dichotomous. Sclerites are basically rods, commonly sharp and needle-like or flattened and becoming scale- or platelike. The oval or elongate scales of the Chrysogorgiidae show in polarized light brilliant interference colors, concentrically arranged (Fig. 158,*6*), but never a cruciform pattern radiating out from the center. Base may be either a large, branched, calcareous root or a small disc, former being adapted for life on soft bottoms and latter for attachment to rocks or other solid material. *Rec.*

Subfamily LEPIDOGORGIINAE Versluys, 1902

Colonies unbranched; polyps uniserial, without operculum of 8 plates. *Rec.*

Radicipes STEARNS, 1883 [**R. pleurocristatus*] [=*Lepidogorgia* VERRILL, 1884]. Sclerites of polyps mostly fusiform rods; of rind, variable but commonly consisting of small scales. Roots subdivide dichotomously, forming long zigzag processes by dominance of alternate branches of succeeding dichotomies. *Rec.,* Atl.-IndoPac.—FIG. 156,*3*. **R. pleurocristatus,* Japan; *3a,* sclerites; *3b,* sculptural detail of spicule; *3c,* part of branch with calyces; *3d,* the rootlike colonial base (26n).

Subfamily CHRYSOGORGIINAE Verrill, 1883

Colonies with main stem giving rise to branches, which may be either simple or further branched. Polyps with rods or scales but no specialized operculum of 8 plates. *Rec.*

Chrysogorgia DUCHASSAING & MICHELOTTI, 1864 [**C. desbonni*]. Colonies erect, sympodial, branches arranged in a spiral and dichotomously subdivided in one plane or in alternating planes at right angles to each other. *Rec.,* Atl.-IndoPac.——FIG. 155,*3;* 158,*6*. *C. elegans* VERRILL, W.Atl.; 155,*3a,* colonies (25); 155,*3b,* spicules (26n); 158,*6,* scale in polarized light.

Iridogorgia VERRILL, 1883 [**I. pourtalesii*]. Colonies monopodial, main trunk coiled in upright spiral with simple branchlets arising from its outer side in single series. *Rec.,* W.Atl.-Hawaii.——FIG. 157,*4,* **I. pourtalesii,* Dominica; colony (25).

Metallogorgia VERSLUYS, 1902 [**Dasygorgia melanotrichos* WRIGHT & STUDER, 1889]. Colonies monopodial, straight main trunk distally giving off strong lateral branches, which profusely subdivide in a dichotomous pattern. *Rec.,* IndoPac.-Atl.——FIG. 158,*1*. **M. melanotrichos;* axis in thin section, in polarized light; *1a,* cross; *1b,* longitudinal.

Pleurogorgia VERSLUYS, 1902 [**P. plana*]. Main branches with simple branchlets uniserially placed along one side; polyps chiefly on branchlets, uniserial. Sclerites are spindles and very irregular bodies. *Rec.,* IndoPac.——FIG. 157,*5*. **P. plana,* Moluccas; spicules and branching pattern (36).

Trichogorgia HICKSON, 1905 [**T. flexilis*]. Colonies branched mostly in a plane, some with long terminal branches; calyces thickly set on all sides. *Rec.,* S.Afr.-W.Atl.

Subfamily CHALCOGORGIINAE Bayer, 1949

Colonies unbranched, zooids biserial. Sclerites of rind are medially constricted scales, those of zooids irregularly lobed plates, 8 triangular plates forming a distinct operculum. *Rec.*

Chalcogorgia BAYER, 1949 [**C. pellucida*]. *Rec.*——FIG. 155,*1*. **C. pellucida,* Cuba; *1a,* part of stem; *1b,* single calyx; *1c,* colonial base; *1d,* calycular scales; *1e,* opercular scale; *1f,* accessory rods lying between opercular scales (26).

Family PRIMNOIDAE Gray, 1857

[*nom. correct.* VERRILL, 1868 (*pro* Primnoadae GRAY, 1857)]

Axis strongly calcified, unjointed, made up of undulated concentric layers (Fig. 158,*4*), core not a soft, chambered central chord, base a calcareous disc, rarely a branched, rhizoidal structure. Colonies usually profusely branched, rarely flagelliform; polyps

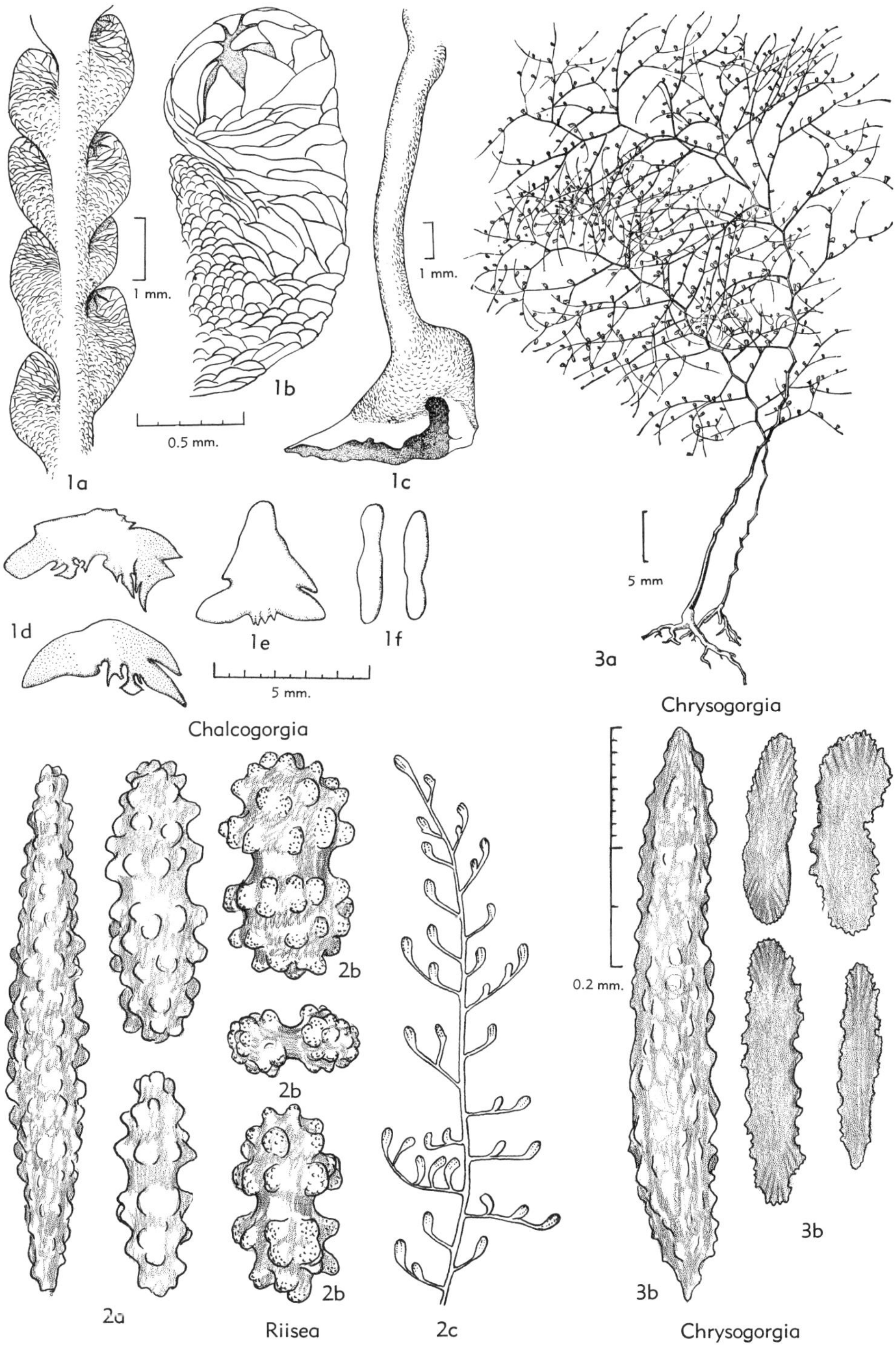

FIG. 155. Gorgonacea: Ellisellidae, Chrysogorgiidae (p. F214-F216).

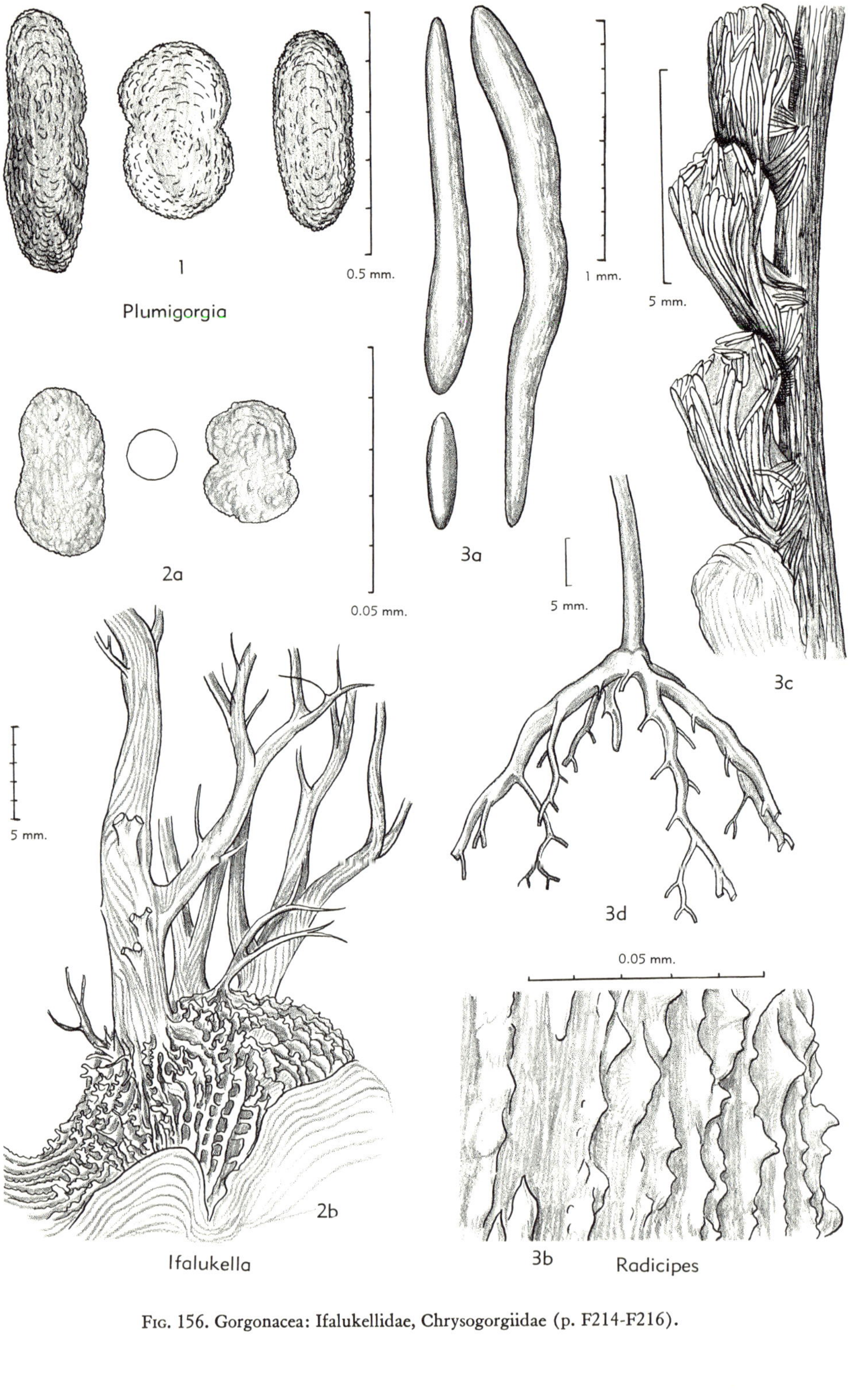

FIG. 156. Gorgonacea: Ifalukellidae, Chrysogorgiidae (p. F214-F216).

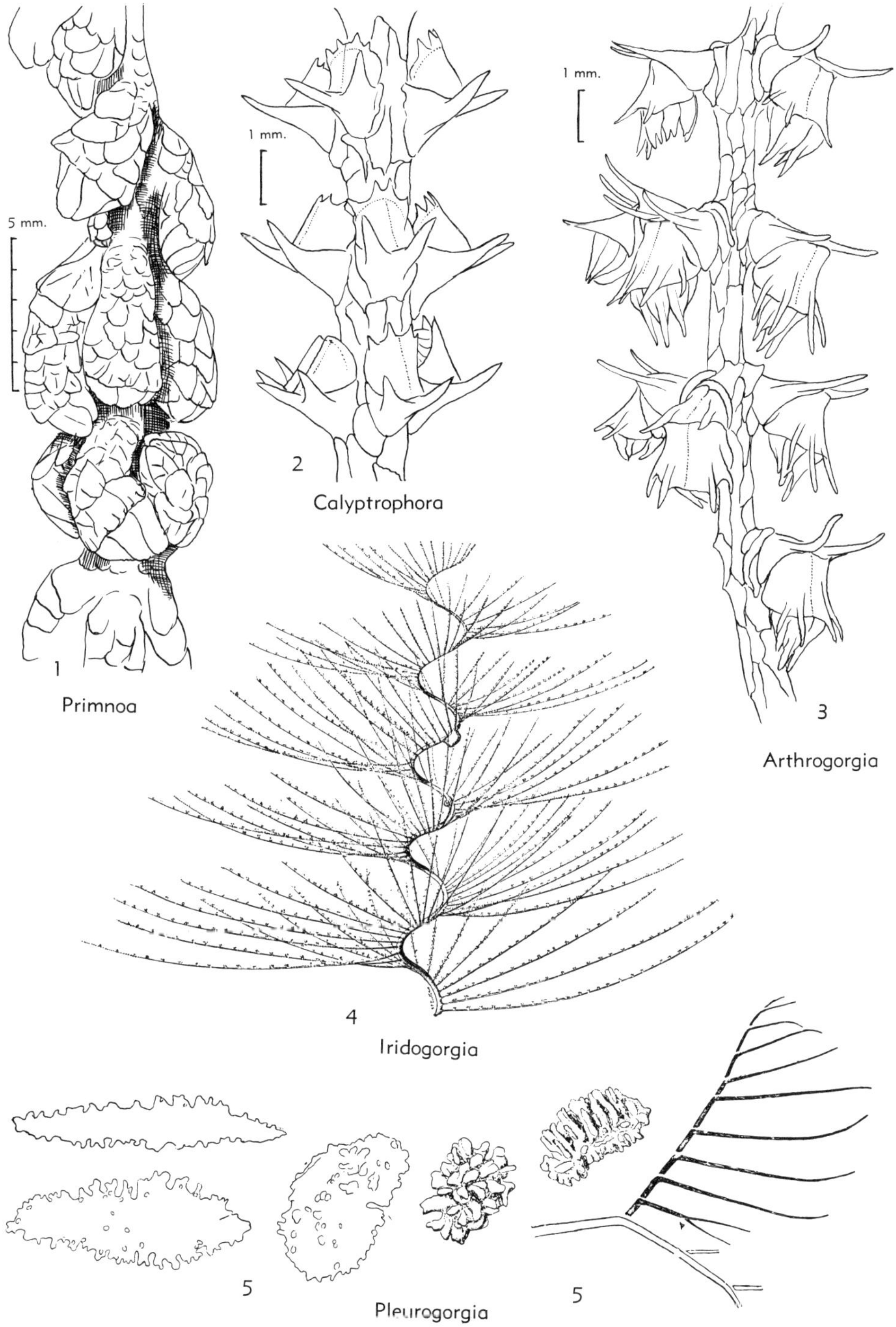

FIG. 157. Gorgonacea: Chrysogorgiidae, Primnoidae (p. F216-F222).

single, in pairs, or regular whorls, heavily armored with calcareous scales (Fig. 158,5) and permanently exsert; in contraction tentacles are infolded and protected by 8 triangular scales making up a distinct operculum, below which scales of polyp body are aligned in 8 rows, some of which may be reduced or missing on adaxial side; rarely (1 sp.) scales not regularly arranged and operculum undifferentiated. In rind, a layer of plates or scales, commonly elongate, some with an inner layer of stellate spicules. *Cret.-Rec.*

The scalelike sclerites of the Primnoidae can always be distinguished by the cruciform pattern of extinction that they show in polarized light (dark field) (Fig. 158,5, *Primnoella*). The scales of chrysogorgiids, isidids and pennatulids never show such a pattern.

Subfamily PRIMNOEIDINAE Studer, 1887

Polyps without differentiated operculum, body scales not in rows. *Rec.*

Primnoeides STUDER, 1887 [**Primnoides sertularoides* WRIGHT & STUDER, 1889; SM WRIGHT-S., 1889]. *Rec.*, Ind.O.

Subfamily PRIMNOINAE Gray, 1857

[*nom. transl.* STUDER, 1887 (*ex* Primnoidae *nom. correct.* VERRILL, 1868, *pro* Primnoadae GRAY, 1857)]

Polyps with well-differentiated operculum, body scales in 8 complete or incomplete rows, some of which may be missing. *Cret.-Rec.*

Primnoa LAMOUROUX, 1812 [**Gorgonia lepadifera* LINNÉ, 1767 (=**Gorgonia reseda* PALLAS, 1766)]. Colonies treelike, dichotomously branched, commonly very large; axis densely calcareous; polyps thickly distributed all around twigs and branches, irregularly oriented but commonly facing downward, especially at twig tips; polyps large (up to 5 mm. long) with 8 opercular and 8 marginal scales, but body scale rows reduced; adaxial surface of polyps without scales. Rind sclerites are scales. *U.Cret.-Rec.*, Arct.——FIG. 157,*1*. **P. reseda* (PALLAS), *Rec.*, Grand Banks; part of branch with polyps (26n).

Callogorgia GRAY, 1858 [**Gorgonia verticillata* PALLAS, 1766]. Colonies usually pinnately branched, some dichotomously, mostly in one plane; axis longitudinally striated, commonly iridescent; polyps in regular whorls, bent inward; adaxial rows of body scales reduced; spicules usually elaborately sculptured externally; cortical sclerites thick and pebble-like or more elongate. *Rec.*, Atl.-Medit.-IndoPac.-Arct.-Antarct.——FIG. 158,*4*; 159,*2*. **C. verticillata*, W.Indies; 158,*4*, cross section of axis in polarized light; 159,*2*, part of branch with 3 whorls of polyps (26n).

?**Eogorgia** HICKSON, 1938 [**E. sullivani*]. Stout, straight primnoid axis assigned by HICKSON to Scleraxonia; not definitely assignable to any subfamily but its size suggests Primnoinae. *Eoc. (Jackson.)*, Miss.

Plumarella GRAY, 1870 [**Gorgonia penna* LAMARCK, 1815]. Branching in one plane, pinnate, closely so in many; polyps biserial, alternate or rarely opposite, never in whorls. All 8 rows of body scales present, adaxials of some with fewer scales. Cortical sclerites of some species scales or warty radiates in lower parts of colony and inner rind. *Rec.*, Atl.-Medit.-IndoPac.——FIG. 159,*7*. **P. pourtalesii* (VERRILL), Fla.; part of branch with polyps (26n).

Primnoella GRAY, 1858 [**Primnoa australasiae* GRAY, 1849]. Colonies branched sparingly or not at all, stems long and whiplike; polyps in whorls, closely appressed to stem. Opercular scales more or less overreached by marginals; abaxial rows of body scales well developed, adaxials reduced or wanting. Superficial sclerites of rind are scales; commonly an inner layer with small, warty sclerites is present. *Rec.*, Atl.-IndoPac.-Arct.-Antarct. ——FIG. 158,*5*; 159,*1*. *P. polita* DEICHMANN, W. Indies; 158,*5*, scale in polarized light; 159,*1*, part of stem with whorls of polyps (26n).

Pseudoplumarella KÜKENTHAL, 1915 [**Plumarella thetis* THOMSON & MACKINNON, 1911]. Similar to *Plumarella*, but polyps may be in whorls. Opercular scales 8, but marginals fewer than 8; fewer than 8 rows of body scales. *Rec.*, Indo-Austral.

Thouarella GRAY, 1870 [**Primnoa antarctica* VALENCIENNES, 1846]. Main stem producing short, slender branches all around, like bottle-brush. Marginal scales overreach operculars; rows of body scales commonly reduced. *Rec.*

T. (Thouarella) [=*Parathouarella* KÜKENTHAL, 1915] Polyps irregularly distributed. Marginal scales with foliate process. *Rec.*

T. (Euthouarella) KÜKENTHAL, 1915 [**Plumarella hilgendorfi* STUDER, 1878; SD BAYER, herein]. Polyps in pairs or whorls of 3 to 4. Marginal scales with median spine. *Rec.*, IndoPac.——FIG. 159,*3*. **T. (E.) hilgendorfi*, Philip.; part of branch with polyps (26n).

T. (Epithouarella) KÜKENTHAL, 1915 [**Thouarella crenelata* KÜKENTHAL, 1907; SD BAYER, herein]. Polyps closely placed all around branches, but not in whorls. Marginal scales without projecting spine; body scales dentate. *Rec.*, Antarct.

Subfamily CALYPTROPHORINAE Gray, 1870

[*nom. transl. et correct.* STUDER, 1887 (*ex* Calyptrophoradae GRAY, 1870)]

Polyps with operculum well differentiated; large body scales reduced to 2 or 3 pairs of plates curved to encircle zooid; adaxial marginal scales retained in some as "adaxial buccal scales." *Rec.*

Calyptrophora GRAY, 1866 [**C. japonica*]. Two pairs of large body scales meeting or nearly meeting adaxially so as to encircle polyp, in some inseparably fused ring-wise; only one pair of infrabasal scales connecting body scales to those of rind; never any vestigial adaxial buccal scales. Branching dichotomous or unilateral pinnate, commonly in one plane, in some flabellate. *Rec.,* Atl.-IndoPac.——FIG. 157,*2*. **C. japonica,* Japan; branch with 3 whorls of polyps (26n).

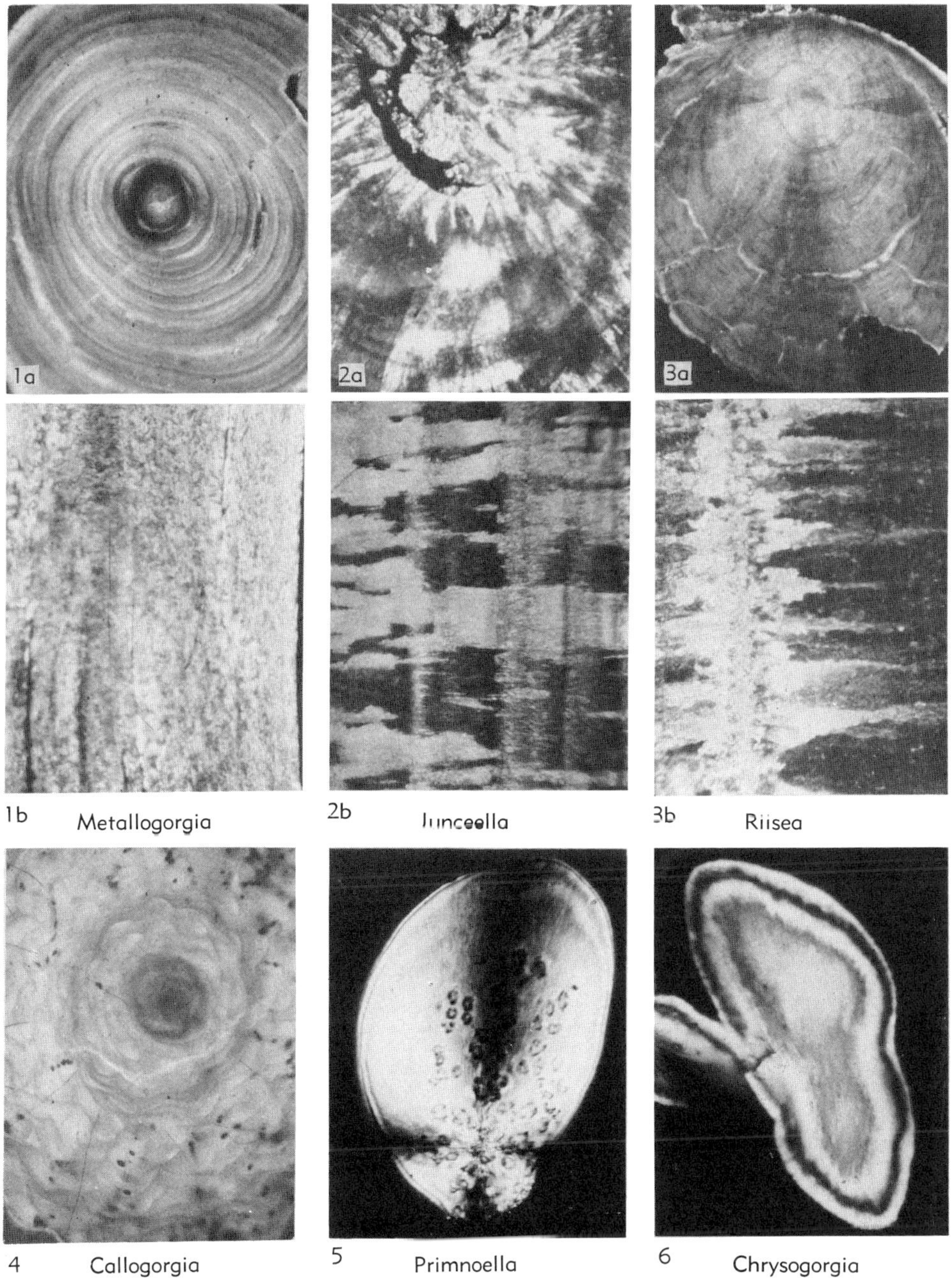

FIG. 158. Gorgonacea: Ellisellidae, Chrysogorgiidae, Primnoidae (p. F214-F220).

Arthrogorgia KÜKENTHAL, 1908 [**A. membranacea* (=**Calyptrophora ijimai* KINOSHITA, 1907)]. Like *Calyptrophora* but more than one pair of infrabasal scales and some vestigial adaxial buccals present. Branching pinnate or dichotomous. *Rec.*, NW. Pac.——FIG. 157,*3*. **A. ijimai,* Japan; part of branch with whorls of polyps (26n).

Narella GRAY, 1870 [**Primnoa regularis* DUCHASSAING & MICHELOTTI, 1860] [=*Stachyodes, Calypterinus* STUDER, 1887]. Three or 4 pairs of large body scales partially encircling polyp but rarely meeting adaxially; adaxial buccal scales commonly present. Branching pinnate, dichotomous, or trichotomous. *Rec.*, Atl.-IndoPac.-Antarct.——FIG. 159, *5*. **N. regularis,* W.Indies; part of branch with 3 whorls of polyps (26n).

Subfamily CALLOZOSTRINAE Studer, 1887

Polyps with operculum well differentiated; polyps in whorls, erect, not bent inward toward stem. *Rec.*

Callozostron WRIGHT, 1885 [**C. mirabile*]. Polyps in close-set whorls of at least 8. Marginal scales with long, projecting spines. *Rec.*, Antarct.

Candidella BAYER, 1954 [*pro Stenella* GRAY, 1870 (*non* GRAY, 1866)] [**Primnoa imbricata* J. Y. JOHNSON, 1862]. Polyps in widely spaced whorls of up to 5.

C. (Candidella). Marginal scales 4. *Rec.*, Atl.-IndoPac.——FIG. 159,*4b*. **C. (C.) imbricata,* W. Indies; part of branch with 3 whorls of polyps (26n).

C. (Parastenella) VERSLUYS, 1906 [**Stenella doederleini* WRIGHT & STUDER, 1889; SD BAYER, herein]. Marginal scales 8. *Rec.*, IndoPac.——FIG. 159,*4a*. **C. (P.) doederleini,* Japan; part of branch with 3 whorls of polyps (26n).

Family AINIGMAPTILIDAE Carlgren, 1943

[*nom. correct.* BAYER, 1950 (*pro* Ainigmaptilonidae CARLGREN, 1943)]

Axis strongly calcified; polyps borne on expanded, unsupported, simple or branched polyp leaves. Spicules are thin scales; operculum of 8 scales. *Rec.*

Ainigmaptilon DEAN, 1926 [**A. haswelli*] [=*Lycurus* MOLANDER, 1929]. Several specimens of various species are provided with expanded funnel-like base, apparently adapted for anchoring colony by enclosing a lump of mud. *Rec.*, Antarct. ——FIG. 159,*6*. *A. edisto* BAYER; *6a*, polyp leaf; *6b*, isolated polyp; *6c*, body scales; *6d*, opercular scale (26).

Family ISIDIDAE Lamouroux, 1812

[*nom. correct.* KÜKENTHAL, 1915 (*pro* Isideae LAMX., 1812)]

Axis composed of alternating purely horny nodes and nonspicular calcareous internodes; base may be either a rootlike calcareous structure for anchoring colony in soft bottom or a basal disc for attachment to hard objects. *Cret.-Rec.*

Subfamily KERATOISIDINAE Gray, 1870

[*nom. transl. et correct.* BAYER, herein (*ex* Keratoisidae GRAY, 1870)] [=Ceratoisidinae STUDER, 1887]

Polyps nonretractile, with spindles, needles or rods placed along septa. *Rec.*

Keratoisis WRIGHT, 1869 [**K. grayi*]. Colonies with branching from calcareous internodes, some unbranched; internodes hollow or solid. Polyps cylindrical, irregularly disposed but with tendency toward biserial arrangement; distal part of polyp body with 8 longitudinal rows of spindles and needles, some projecting beyond tentacles; stellate forms in pharynx. *Rec.*, Atl.-IndoPac.——FIG. 160, *1*. *K. ornata* VERRILL, Grand Banks; section of stem with node, and internode with branch; short, horny node at base of branch is not an invariable feature (26n).

Acanella GRAY, 1870 [**Mopsea arbusculum* JOHNSON, 1862]. Colonies bushy, branched in whorls from horny nodes. Spicules of calyces are mostly spindles and some flattened blunt rods, with fine prickles or low warts. *Rec.*, Atl.-IndoPac.——FIG. 160,*3*. **A. arbuscula,* Grand Banks; *3a*, entire colony (25); *3b*, tip of branch with terminal polyp (26n); *3c*, node with one branch (26n).

Isidella GRAY, 1857 [**Isis elongata* ESPER, 1788]. Colonies branched from nodes dichotomously and in one plane; internodes hollow. Calycular sclerites mostly rodlets which do not project. *Rec.*, E. Atl.-Medit.

Lepidisis VERRILL, 1883 [**L. caryophyllia;* SD KÜKENTHAL, 1915]. Colonies simple or sparsely branched from nodes; internodes hollow. Calycular sclerites projecting needles and elongate scales. *Rec.*, Atl.-?IndoPac.——FIG. 160,*2*. **L. carophyllia,* NW.Atl.; *2a*, polyp; *2b*, colonial base; *2c*, scales, and *2d*, rods from calyces (26n).

Subfamily MOPSEINAE Gray, 1870

[*nom. transl. et correct.* WRIGHT & STUDER, 1889 (*pro* Mopseadae GRAY, 1870)]

Polyps nonretractile, with transversely arranged platelike sclerites; branching from internodes. *Eoc.-Rec.*

Mopsea LAMOUROUX, 1816 [**Isis dichotoma* LINNÉ, 1758; SD M.EDW.-H., 1850]. Colonies branched pinnately or dichotomously in one plane. *Eoc.-Rec.*, MalayArch.-Austral.-Antarct.——FIG. 161,*3*. *M. elongata* ROULE, Rec., Antarct.; 2 internodes (26n).

Peltastisis NUTTING, 1910 [**P. uniserialis*]. Colonies unbranched; polyps with operculum of 8 plates. *Rec.*, MalayArch.

Primnoisis STUDER, 1887 [**Isis antarctica* STUDER, 1878; SD KÜKENTHAL, 1915]. Colonies branched on all sides bottle-brush fashion. *Rec.*, Antarct.-

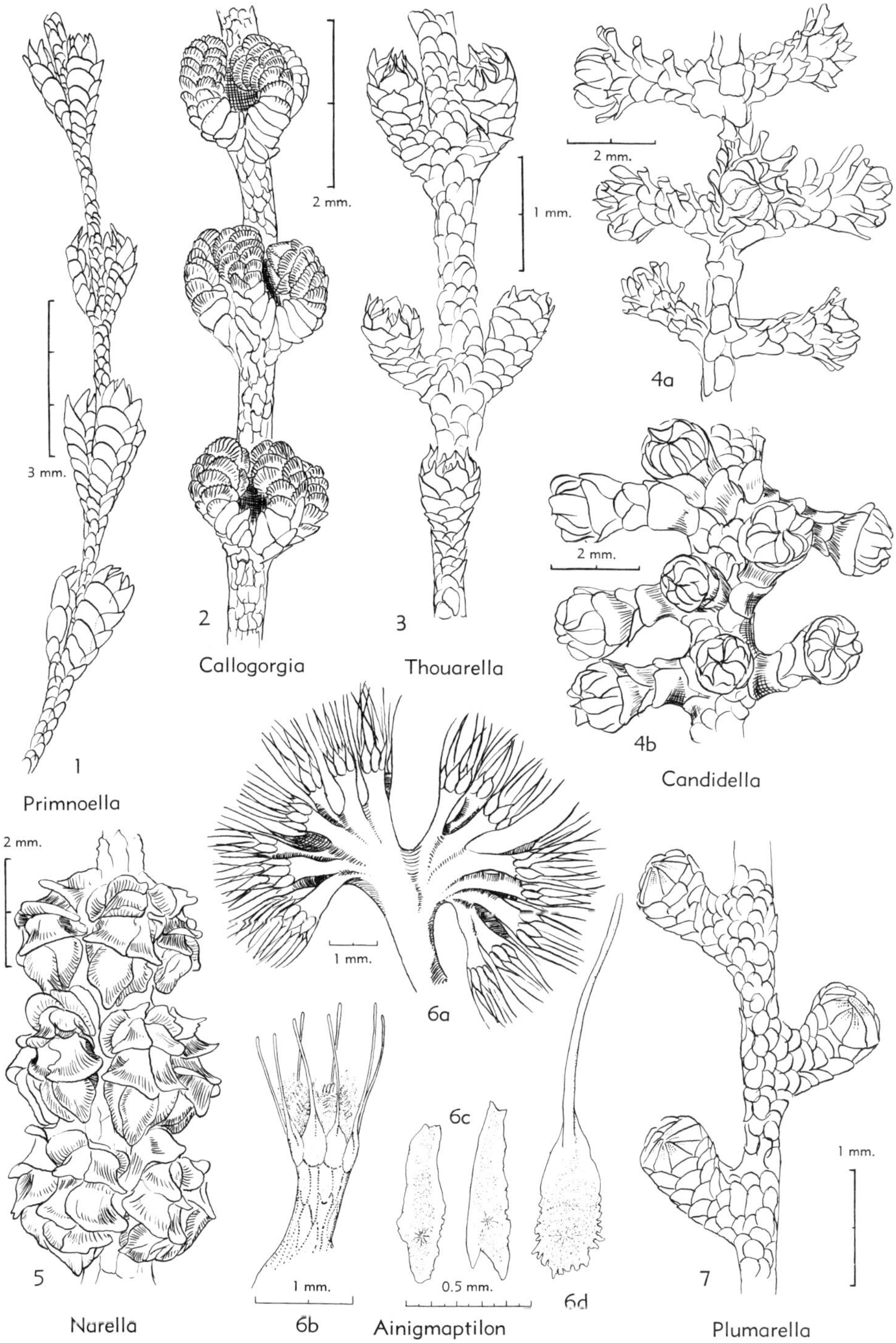

FIG. 159. Gorgonacea: Primnoidae, Ainigmaptilidae (p. F220-F222).

Subantarct.-S.Afr.-Arg.——FIG. 161,2. **P. antarctica*, Marguerite Bay; *2a*, sclerites; *2b*, part of branch partially decorticated to show ramification of axial internode (26n).

Subfamily MURICELLISIDINAE Kükenthal, 1915

Anthocodiae rectractile within distinctly separated calyces, armed with strongly thorned spindles. Branching from internodes, which are hollow. *Rec.*, Japan.

Muricellisis KÜKENTHAL, 1915 [**M. echinata*]. *Rec.*, Japan.

Subfamily ISIDINAE Lamouroux, 1812

[*nom. transl.* STUDER, 1887 (*ex* Isididae *nom. correct.* KÖLLIKER, 1865, *pro* Isideae LAMX., 1812)]

Anthocodiae fully or partially retractile within coenenchyme, armed with small rods having transverse girdles of tubercles, and various radiate forms. *Rec.*

Isis LINNÉ, 1758 [**I. hippuris;* SD M.EDW.-H., 1850]. Branching from internodes; colonies bushy, with thick rind into which polyps are fully retractile. [Various fossil species attributed to this genus almost certainly do not belong here.] *Rec.*, IndoPac.——FIG. 161,*1*. **I. hippuris*, Philip.; *1a*, spicules; *1b*, decorticated axis (26n).

Chelidonisis STUDER, 1890 [**C. aurantiaca*]. Branching from internodes, dichotomous, in one plane. Rind thin, polyps retractile into hemispherical verruciform calyces. *Rec.*, Atl.-S.Afr.

Subfamily UNCERTAIN

Axogaster LONSDALE, 1850 [**A. cretacea*]. Base branched, creeping, rootlike. Calcareous internodes with calycular pits having fewer than 12 pseudosepta; branching from internodes. *Cret.*, Eng.

Moltkia STEENSTRUP, 1846 [**M. isis*]. Colonies spread roughly in a plane, dichotomously branched, anastomosing. Nodes overgrown by the calcareous internodes, completely enclosed. Axis pitted by calyces and sculptured by beaded longitudinal striae which may continue into calycular pits. *U.Cret. (Dan.)*, Denm.

Order PENNATULACEA Verrill, 1865

[*nom. transl.* KÖLLIKER, 1880, as order Pennatulida (*ex* suborder Pennatulacea VERRILL, 1865); *nom. correct.* STUDER, 1887, as order Pennatulacea]

Colonial octocorals unbranched, not firmly attached, consisting of a primary polyp (oözooid) that elongates to produce a barren, proximal stalk, which anchors colony in mud, and a polypiferous distal rhachis from which secondary polyps spring, either directly or from ridgelike or broadly expanded polyp leaves. Gastric cavity of primary polyp divided into 2 primary and 2 secondary longitudinal canals by fleshy partitions at center of which a more or less calcified horny axis usually is produced. Secondary polyps invariably of at least 2 kinds. Spicules in form of smooth or 3-flanged rods or needles, rarely tuberculated; or small scales or plates which, in polarized light, behave optically like plates of Chrysogorgiidae. Axes of pennatulids formed of irregular, prismatic columns of calcareous material radiating outward from axis core, which seems to contain a higher proportion of organic matter, thus resembling closely axes of ellisellid Gorgonacea. *?Sil., Rec.*

Classification of the sea pens depends upon arrangement of the autozooids and siphonozooids, distribution and form of the calcareous spicules, and other features ordinarily not preserved in fossils. External characters of the axis are so variable as to be useless except in a most general way. Consequently, the "species" of *Graphularia*, established on such characters, are of very questionable standing. The differences in patterns of calcification may, however, afford a new means of evaluating these fossils, and need to be more thoroughly investigated.

A number of presumed botanical fossils have recently been assigned to the Renillidae, and it is probable that still more need reinterpretation.

Molds interpreted as fossil pennatulids have lately been reported from both East and West Indian Tertiary deposits (3), and similar remains have been noted in European strata.

Suborder SESSILIFLORAE Kükenthal, 1915

Sea pens with polyps standing separately and arising directly from rhachis without being united near their bases by ridgelike or leaflike structures (12). *?Sil., Rec.*

Family VERETILLIDAE Herklots, 1858

[*nom. correct.* GRAY, 1870 (*pro* Veretilloides HERKLOTS, 1858)]

Stout, commonly clavate colonies without trace of bilaterality; polyps fully retractile, no calyces. Spicules of various types, none 3-flanged. *Rec.*

Veretillum CUVIER, 1798 [**Pennatula cynomorium*

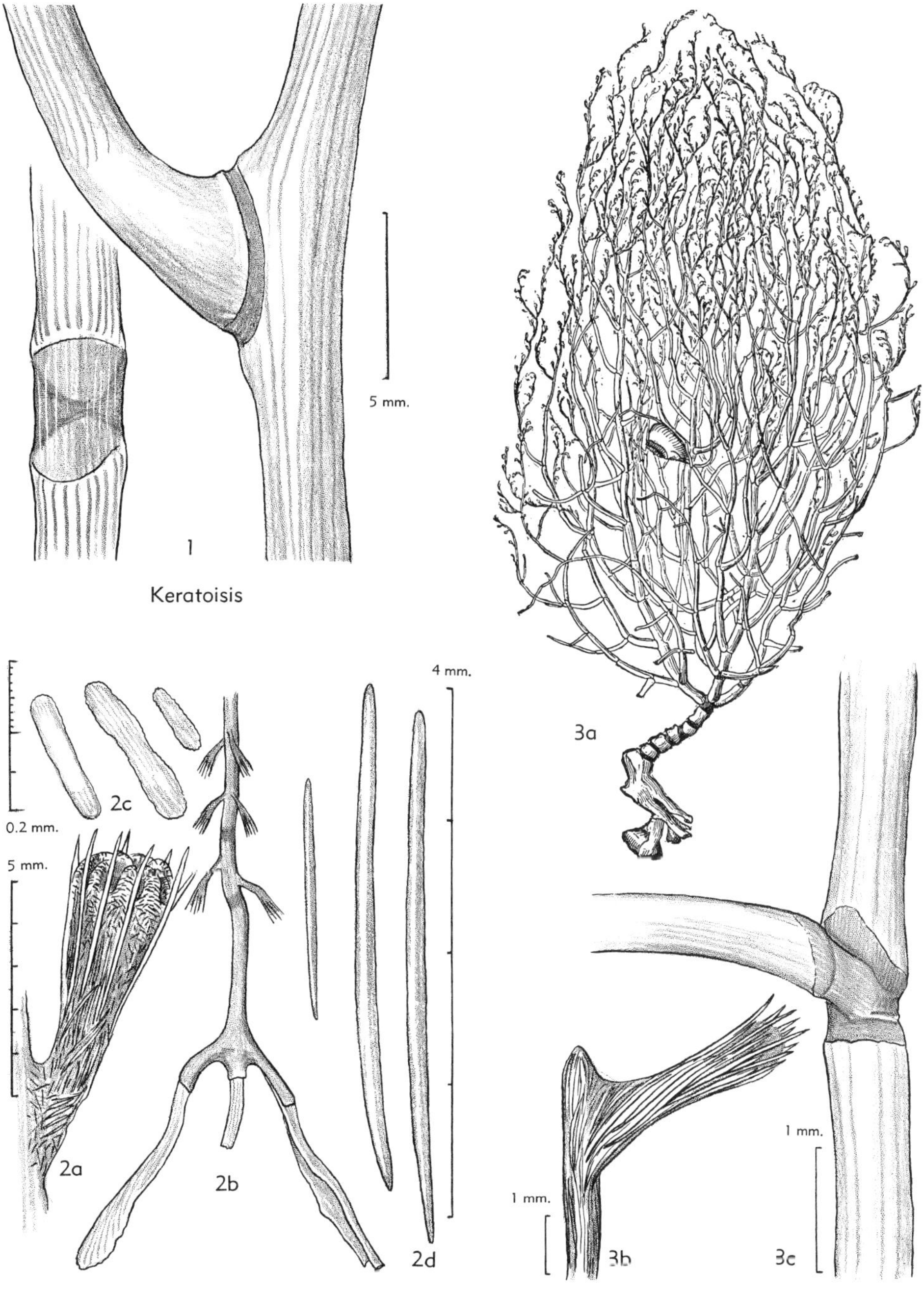

FIG. 160. Gorgonacea, Isididae (p. F222).

PALLAS, 1766; SD M.EDW.-H., 1850]. Spicules are minute oval constricted plates in polyps as well as in rhachis and stalk. *Rec.*, W.Afr.-S.Eu.

Cavernularia M.EDW.-H., 1850 [**C. obesa*]. Spicules consisting of oval or elongated rods. *Rec.*, W.Afr.-Medit.-IndoPac.

Cavernulina KÜKENTHAL & BROCH, 1911 [**C. cylindrica*]. Spicules are rods with bilobed ends. *Rec.*, Ind.O.

Lituaria M.EDW.-H., 1850 [**Pennatula phalloides* PALLAS, 1766]. Spicules are plates, commonly with terminal denticles. *Rec.*, IndoPac.

Policella GRAY, 1870 [**P. australis*]. Spicules are minute, oval bodies in interior of stalk, and small, biscuit-shaped platelets in outer layer; spicules absent in polyps. *Rec.*, IndoPac.

Family ECHINOPTILIDAE Hubrecht, 1885

Colonies cylindrical, bent, lacking axis, with polyps radially disposed but in some showing traces of bilateral orientation; polyps standing in transverse rows, with calyces. Spicules are 3-flanged rods. *Rec.*

Echinoptilum HUBRECHT, 1885 [**E. mackintoshi*]. *Rec.*, Ind.O.

Actinoptilon KÜKENTHAL, 1910 [**A. molle*]. *Rec.*, S.Afr.

Family RENILLIDAE Gray, 1860

[*nom. correct.* GRAY, 1870 (*pro* Renilleae GRAY, 1860)]

Sea pens with slender stalk and oval or reniform foliate rhachis bearing polyps on upper surface only; no axis. Spicules are 3-flanged rods which may be more or less platelike. *?Sil., Rec.*

Renilla LAMARCK, 1816 [**Renilla americana* (=**Pennatula reniformis* PALLAS, 1766)]. *Rec.*, W.Atl.-E.Pac.——FIG. 162,*4*. **R. reniformis*, S. Car.; colony (26n).

?**Alectorurus** SCHIMPER, 1869 [*non* GLOGER, 1842]. *Sil.*

?**Cancellophycus** SAPORTA, 1872 [**Taonurus liasinus* FISCHER-OOSTER, 1858] [=*?Taonurus*]. Colonies flattened, in form of broad, undulating spiral blade, some with accessory blades; spiral may encompass several turns or may be quite lax, blade being formed of successive more or less concentric and arcuate imbricating lamellae joined together at close intervals by oblique cross trabeculae, altogether producing a reticulate structure with meshes consisting of sigmoid lozenges thought to have contained gastric cavities of the individual polyps. In section, the lamellae are seen to contain calcite needles (15 to 40 microns in diameter and about 1 mm. in length), trifoliate in cross section and practically identical with the spicules of various pennatulids (1). *Jur.-Cret.*

?**Spirophyton** HALL, 1863 [**S. typum*] [=*?Taonurus*]. *Sil.*

?**Taonurus** FISCHER-OOSTER, 1858 [**Fucoides brianteus* VILLA]. *Cret.*

Family KOPHOBELEMNIDAE Gray, 1860

[*nom. correct.* KÜKENTHAL, 1915 (*pro* Kophobelemnonieae GRAY, 1860)]

Sea pens with polyps bilaterally oriented on rhachis but with some tendency to radial symmetry; colonies clavate, with axis. *Rec.*

Kophobelemnon ASBJÖRNSEN, 1856 [**K. mülleri* (=**Pennatula stellifera* O. F. MÜLLER, 1776)]. Elongate, clavate colonies with large autozooids, which do not form calyces. Spicules are prismatic rods, shorter ones commonly distinctly tuberculated. Axis slender, rounded quadrangular. *Rec.*, Atl.-Medit.-IndoPac.——FIG. 162,*2*. **K. stelliferum*, Grand Banks; *2a*, spicules (26n); *2b*, entire colony (35).

Mesobelemnon GRAVIER, 1907 [**M. gracile*]. Similar to *Kophobelemnon* but spicules are small bone-shaped rods. Axis rounded in section. *Rec.*, Red Sea.

Sclerobelemnon KÖLLIKER, 1872 [**S. schmeltzei*]. Similar to *Kophobelemnon* but spicules are small plates, smooth or with serrate ends. *Rec.*, IndoPac.

Family ANTHOPTILIDAE Kölliker, 1880

Bilateral sea pens with polyps in transverse or somewhat diagonal rows on 2 sides of rhachis. Sclerites absent except for minute oval bodies in interior of stalk. Axis round or quadrangular with rounded angles. *Rec.*

Anthoptilum KÖLLIKER, 1880 [**Virgularia grandiflora* VERRILL]. *Rec.*, Atl.-IndoPac.

Family FUNICULINIDAE Gray, 1860

[*nom. correct.* GRAY, 1870 (*pro* Funiculineae GRAY, 1860)]

Colonies elongate, slender; autozooids rather small, arranged laterally and ventrally on rhachis, producing distinct calyces with 8 marginal teeth; siphonozooids infrequent. Spicules are prismatic needles. Axis quadrangular. *Rec.*

Funiculina LAMARCK, 1816 [**Pennatula quadrangularis* PALLAS, 1766; SD M.EDW.-H., 1850]. *Rec.*, Atl.-IndoPac.

Family PROTOPTILIDAE Kölliker, 1872

[*nom. correct.* KÖLLIKER, 1880 (*pro* Protoptileae KÖLLIKER, 1872)]

Bilateral sea pens with autozooids longitudinally arranged in one or more lateral rows. Spicules 3-flanged. Axis stout, rounded. *Rec.*

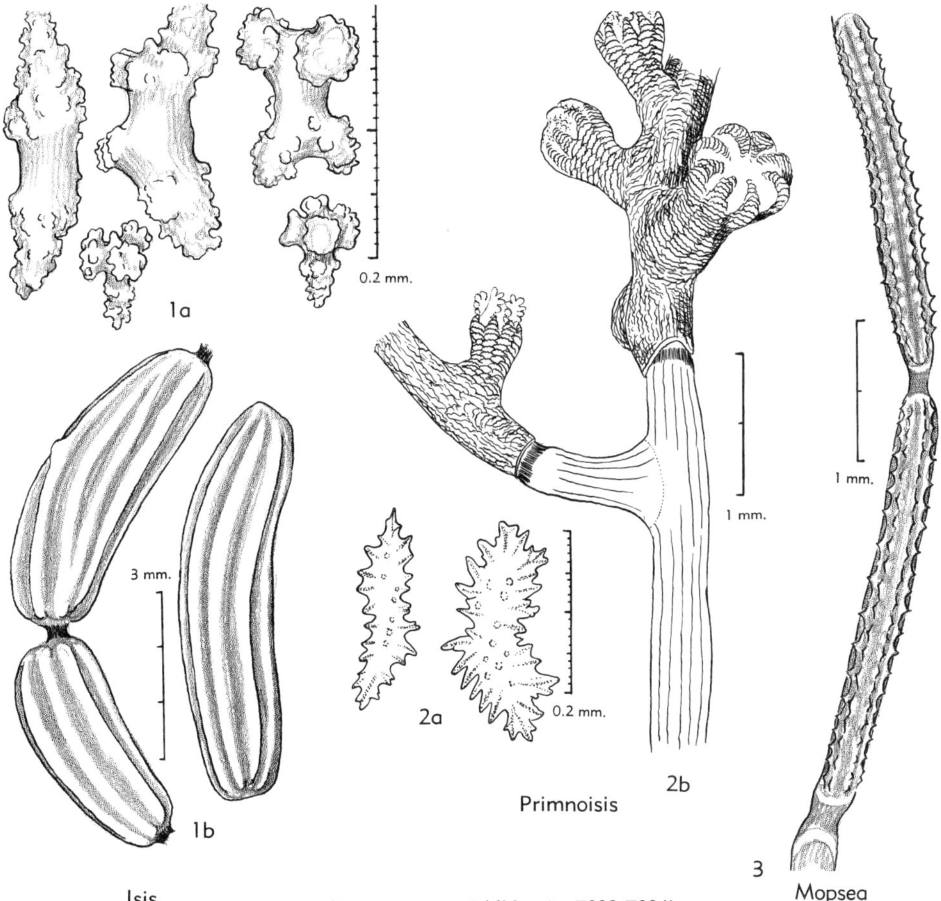

FIG. 161. Gorgonacea: Isididae (p. F222-F224).

Protoptilum KÖLLIKER, 1872 [**P. carpenteri;* SD KÜKENTHAL, 1915]. *Rec.,* N.Atl.-Indo.O.

Distichoptilum VERRILL, 1882 [**D. gracile*]. *Rec.,* Atl.-Pac.

Helicoptilum NUTTING, 1912 [**H. rigidum*]. *Rec.,* Japan.

Family STACHYPTILIDAE Kölliker, 1880

Bilateral colonies with autozooids arranged laterally in transverse rows but not in longitudinal rows also. Autozooids and siphonozooids with well-developed, scale-like calyces. Spicules are 3-flanged needles. *Rec.*

Stachyptilum KÖLLIKER, 1880 [**S. macleari*]. *Rec.,* trop.Pac.

Family SCLEROPTILIDAE Jungersen, 1904

Rhachis elongate, bearing autozooids closely arranged in indistinct whorls; dorsal track free of autozooids; siphonozooids scattered between autozooids. *Rec.*

Scleroptilum KÖLLIKER, 1880 [**S. grandiflorum;* SD BAYER, herein]. With 3-flanged spindles. *Rec.,* Atl.-N.Pac.-Ind.O.

Calibelemnon NUTTING, 1908 [**C. symmetricum*=*Protocaulon indicum* THOMSON & HENDERSON, 1906]. No spicules. Axis quadrangular. *Rec.,* Ind. O.-Hawaii.

Family CHUNELLIDAE Kükenthal, 1902

Rhachis extremely slender; autozooids in widely separated whorls; small siphonozooids on lateral tracts of rhachis, larger ones within autozooid whorls. Axis apparently quadrangular. *Rec.*

Chunella KÜKENTHAL, 1902 [**C. gracillima*]. Terminal autozooid rudimentary or absent. *Rec.*, E. Afr.

Amphiacme KÜKENTHAL, 1903 [**Amphianthus abyssorum* KÜKENTHAL, 1902]. Terminal autozooid well developed, asymmetrical. *Rec.*, E.Afr.

Family UMBELLULIDAE Lindahl, 1874

[*nom. correct.* KÖLLIKER, 1880 (*pro* Umbelluleae LINDAHL, 1874)]

Rhachis slender, elongate, bearing at its apex an umbelliform tuft of large autozooids; siphonozooids situated among autozooids and in groups or rows on barren parts of rhachis. Spicules are 3-flanged needles in polyp walls, rhachis and stalk rind, and small oval bodies in deep layers of stalk. Axis round or quadrangular. *Rec.*

Umbellula CUVIER, 1798 [**Isis encrinus* LINNÉ, 1758]. *Rec.*, cosmop., deep-sea.——FIG. 162,*3*. *U. güntheri* KÖLLIKER, 1880, W.Indies; small colony, about ×1 (25).

Suborder SUBSELLIFLORAE Kükenthal, 1915

Polyps united by their bases, situated in rows on lateral swellings or foliate polyp leaves. *Cret.-Rec.*

Family VIRGULARIIDAE Verrill, 1868

[*nom. correct.* JUNGERSEN, 1904 (*pro* Virgularidae VERRILL, 1868)]

Bilateral, with slender rhachis; autozooids situated in transverse rows and united together by their bases, rhachis beneath them raised into lateral swellings or small leaves. Spicules present as prismatic needles, small biscuit-shaped plates, or entirely absent. Axis stout. *Cret.-Rec.*

Subfamily VIRGULARIINAE Verrill, 1868

[*nom. transl. et correct.* KÜKENTHAL & BROCH, 1911 (*ex* Virgularidae VERRILL, 1868)]

Autozooids form low to prominent polyp leaves. *Cret.-Rec.*

Virgularia LAMARCK, 1816 [**Pennatula mirabilis* MÜLLER, 1776; SD M.EDW.-H., 1850]. Long, slender colonies with lateral rows of low polyp leaves separated by dorsal track and in some also by ventral groove; autozooids retractile; siphonozooids between or on polyp leaves and along dorsal track. Spicules absent except for small, oval platelets in interior of stalk. Axis stout, projecting apically. *Cret.-Rec.*, nearly cosmop.——FIG. 162,*6*. *V. presbytes* BAYER; *6a*, dorsal, lateral, and ventral views of part of rhachis with fully developed polyp leaves (Rec., GulfMexico) (26); *6b*, mold of ventral side of colony (Wilcox.-Claiborn., Trinidad) (29).

Acanthoptilum KÖLLIKER, 1870 [**A. pourtalesii;* SD KÜKENTHAL, 1915]. *Rec.*

Graphularia M.EDW.-H., 1850 [**G. wetherelli*]. Axial rods slightly compressed, narrower sides flattened or slightly concave; central core wide; cortical zone consisting of long, irregular, calcareous radii extending from core to periphery, perforated by radial tubules extending outward from central core. These axes correspond closely in structure with those of modern *Stylatula*, with which they may be congeneric (4). *Eoc.*(London Clay), Eng.

Scytaliopsis GRAVIER, 1906 [**S. djiboutiensis*]. *Rec.*

Scytalium HERKLOTS, 1858 [**S. sarsi*]. *Rec.*

Stylatula VERRILL, 1864 [**S. elongata;* SD KÜKENTHAL, 1915 (?=*Virgularia elongata* GABB, 1863)]. Polyp leaves supported by radiating spicules. *Rec.*, Atl.-E.Pac.

Subfamily BALTICININAE Gray, 1870

[*nom. transl.* BAYER, herein (*ex* Balticinidae GRAY, 1870)] [=Pavonariinae KÜKENTHAL & BROCH, 1911]

Autozooids stand in oblique rows surrounded by ridges or folds of rhachis. *?U. Cret.(Senon.), Rec.*

Balticina GRAY, 1870 [**Virgularia finmarchica* SARS, 1851] [=*Pavonaria* KÖLLIKER, 1869 (*non* SCHWEIGGER, 1820)]. *?U.Cret.(Senon.), Rec.*, N. Atl.-IndoPac.

Family PENNATULIDAE Ehrenberg, 1828

[*nom. correct.* DANA, 1846 (*pro* family Pennatulina EHRENBERG, 1828)]

Bilateral sea pens with well-developed polyp leaves bearing one or more marginal rows of autozooids, which have calyces with marginal teeth formed by spicules; siphonozooids on rhachis, not on leaves. Spicules as minute oval bodies, plates, rods, and prismatic needles. *Rec.*

Pennatula LINNÉ, 1758 [**P. phosphorea;* SD KÜKENTHAL, 1915]. Calyces usually with 8 marginal teeth; spicules as prismatic rods and needles, inconspicuously 3-flanged in stalk, where plates occur also. *Rec.*, cosmop.——FIG. 162,*1*. *P. aculeata* DANIELSSEN, Mass.; *1a*, colony from the side (35); *1b*, spicules (26n).

Leioptilus GRAY, 1860 [**Pennatula fimbriata* HERKLOTS, 1858]. *Rec.*

Ptilosarcus GRAY, 1860 [**Sarcoptilus (Ptilosarcus) gurneyi;* SD BAYER, herein]. Calyces with only 1 or 2 teeth; large plates in interior of stalk. *Rec.*, off W.N.Am.

Family PTEROEIDIDAE Kölliker, 1880

Bilateral sea pens like Pennatulidae but polyp leaves supported by rays of stout

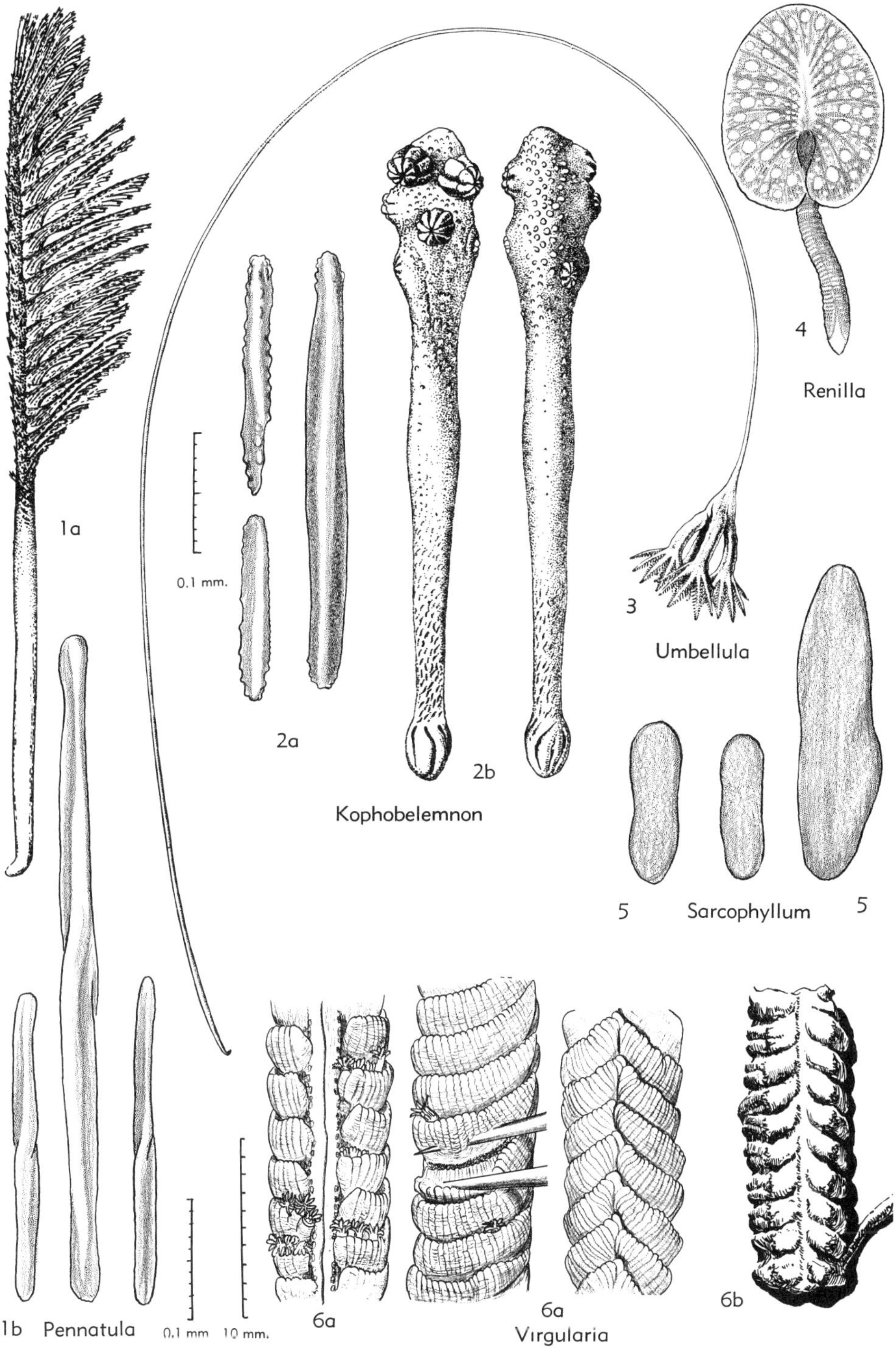

FIG. 162. Pennatulacea (p. F226-F230) (for *Sarcophyllum* read *Sarcoptilus*, p. F230).

spicules; autozooid calyces not formed by spicules; spicules not 3-flanged. *Tert.-Rec.*

Pteroeides HERKLOTS, 1858 [**Penna grisea* BOHADSCH, 1761 (=*Pennatula spinosa* ELLIS, 1764; *Pteroeides spinosum* HERKLOTS, 1858); SD KÜKENTHAL, 1915]. Polyp leaves lack accessory leaves. *Tert.*(Sumatra)-*Rec.*(E.Atl.-Medit.-IndoPac.).

Sarcoptilus GRAY, 1848 [**S. grandis*] [=*Sarcophyllum* KÖLLIKER, 1860]. Supporting rays absent, but spicules as rods and plates, never prismatic. *Rec.*, Austral.-N.Z.——FIG. 162,*5*. **S. grandis;* spicules (26n).

Struthiopteron BROCH, 1910 [**Pteroeides caledonicum* KÖLLIKER, 1869]. Leaves with supporting rays and accessory leaves along dorsal track. *Rec.*, NewCaledonia-Amboina.

Suborder and Family UNCERTAIN

Glyptosceptron BOEHM, 1890. *U.Cret.(Senon.)*, Bavaria.

Palaeosceptron COCCHI, 1870. *U.Cret.*

Pennatulites COCCHI, 1870. *U.Cret.(Senon.)*.

Prographularia FRECH, 1890. Like *Graphularia*, but longitudinally striated and with symmetrical pair of deep grooves. *Trias.*

Protovirgularia M'COY, 1850. *Sil.*

REFERENCES

Publications cited in the following list are judged to be helpful in furnishing additional information concerning the octocorals and in offering a guide to more extensive literature. The index numbers enclosed in parentheses in the column at left are employed in the text for identification of the publications.

Alloiteau, James

(1) 1952, *Sous-classe des Alcyonaria:* in PIVETEAU, JEAN, Traité de Paléontologie, Masson (Paris), tome 1, p. 408-417, 3 figs.

Bayer, F. M.

(2) 1953, *Zoogeography and evolution in the octocorallian family Gorgoniidae:* Bull. Marine Sci. Gulf and Caribbean, v. 3, no. 2, p. 100-119, fig. 1-5.

(3) 1955, *Remarkably preserved fossil sea-pens and their recent counterparts:* Jour. Washington Acad. Sci., v. 45, p. 294-300, fig. 1-2.

(4) 1955, *Contributions to the nomenclature, systematics and morphology of the Octocorallia:* Proc. U.S. Natl. Mus., v. 105, p. 207-220, pl. 1-8.

Cary, L. R.

(5) 1918, *The Gorgonaceae as a factor in the formation of coral reefs:* Carnegie Inst. Washington, Pub. 213, p. 341-362, pl. 100-105.

(6) 1931, *Studies on the coral reef of Tutuila, American Samoa, with especial reference to the Alcyonaria:* Papers from Tortugas Lab., Carnegie Inst. Washington, v. 27, no. 3, p. 53-98, pl. 1-7, fig. 1-14, table 1-14, map 1.

Deichmann, Elisabeth

(7) 1936, *The Alcyonaria of the western part of the Atlantic Ocean:* Mem. Mus. Comp. Zool. Harvard, v. 53, p. 1-317, pl. 1-37.

Gohar, H. A. F.

(8) 1940, *Studies on the Xeniidae of the Red Sea:* Pub. Marine Biol. Sta. Ghardaqa (Red Sea), no. 2, p. 25-118, pl. 1-7.

Hickson, S. J.

(9) 1916, *The Pennatulacea of the Siboga Expedition:* Siboga-Exped. (Leiden) Mon. 14, x+265 p., 10 pl., 45 text-fig., 1 chart.

(10) 1930, *On the classification of the Alcyonaria:* Proc. Zool. Soc. London, 1930, pt. 1, p. 229-252, fig. 1-2.

Hyman, L. H.

(11) 1940, *The Invertebrates: Protozoa through Ctenophora:* McGraw-Hall (New York), xii+726 p., 221 fig.

Kükenthal, Willy

(12) 1915, *Pennatularia:* Das Tierreich, Lief. 43, xv+132 p., 126 fig.

(13) 1919, *Gorgonaria:* Wissenschaft. Ergebn. deutsch. Tiefsee-Expedition auf dem Dampfer "Valdivia" 1898-1899, Band 13, pt. 2, p. i-viii+1-646+i-vi+647-946, pl. 30-89, fig. 1-319.

(14) 1924, *Gorgonaria:* Das Tierreich, Lief. 47, xxviii+478 p., 209 fig.

Milne-Edwards, Henri, & Haime, Jules

(15) 1850, *A monograph of the British fossil corals. Pt. 1, Introduction; corals from the Tertiary and Cretaceous formations:* Palaeontographical Soc., London, lxxxv+71 p., 11 pl.

Montanaro-Gallitelli, Eugenia

(16) 1955, *Il Permiano del Sosio e i suoi coralli:* Palaeontographia Italica, tomo 49, p. 1-98, pl. 1-10, figs. 1-10.

Nielsen, K. B.

(17) 1913, *Moltkia isis, Steenstrup og andre Octocorallia fra Danmarks Kridttidsaflejringer:* Mindeskrift for Japetus Steenstrup, pt. 18, p. 1-20, pl. 1-4, fig. 1-2.

(18) 1917, *Heliopora incrustans nov. sp., with a survey of the Octocorallia in the deposits of the Danian in Denmark:* Meddelelser fra

Dansk geologisk Forening, v. 5, pt. 8, p. 1-13, fig. 1-17.

(19) 1925, *Nogle nye Octocoraller fra Danienet:* Same, v. 6, pt. 28, p. 1-6, fig. 1-3.

Okulitch, V. J.

(20) 1936, *On the genera Heliolites, Tetradium, and Chaetetes:* Am. Jour. Sci., ser. 5, v. 32, p. 361-379, fig. 1.

Počta, Philipp

(21) 1885, *Über fossile Kalkelemente der Alcyoniden und Holothuriden und verwandte recente Formen:* Sitzungsber. Kaiserlichen Akad. Wissenschaft. (math.-naturw. Cl.), Band 92, no. 1, p. 7-12, pl. 1.

Stock, J. H.

(22) 1953, *Contributions to the knowledge of the pycnogonid fauna of the East Indian Archipelago:* Biol. Results of Snellius Exped., XVII, Temminckia, v. 9, p. 276-313, fig. 1-18.

Termier, Geneviève & Termier, Henri

(23) 1950, *Foraminifères, Spongiaires et Coelentérés. Invertébrés de l'Ère Primaire, fasc. 1:* Paléontologie Marocaine, II. Morocco, Service Géol., Notes et Mémoires, no. 73, p. 1-218, pl. 1-51.

Williams, James Steele

(24) 1943, *Stratigraphy and fauna of the Louisiana limestone of Missouri:* U. S. Geol. Survey Prof. Paper 203, iv+133 p., 9 pl.

SOURCES OF ILLUSTRATIONS

Explanation of the use of index numbers for citing sources of illustrations is given in the Editorial Preface.

(25) **Agassiz, Alexander**
(26) **Bayer, F. M.**
(27) **Bourne, G. C.**
(28) **Cooper, G. A.**
(29) **Hogue, R. E.**
(30) **Kükenthal, Willy**
(31) **Madsen, F. J.**
(32) **Montanaro-Gallitelli, Eugenia**
(33) **Moseley, H. N.**
(34) **Thomson, J. A., & Dean, L. M. I.**
(35) **Verrill, A. E.**
(36) **Versluys, Jan**
(37) **Wright, E. P., & Studer, Théophile**

ZOANTHARIA—GENERAL FEATURES

By J. W. WELLS and DOROTHY HILL

Subclass ZOANTHARIA de Blainville, 1830

[=Helianthoïda LATREILLE, 1825; Zoantha BLAINV., 1834; Zoocorallia polyactinia, Phytocorallia polyactinia, Phytocorallia dodecactinia EHRENBERG, 1834; Anthozoa helianthoida JOHNSTON, 1837 (*partim*); Actinaria DANA, 1846; Hexacorallia, Tetracorallia HAECKEL, 1866; Hexactiniae SCHNEIDER & RÖTTEKEN, 1871; Hexacoralla + Rugosa NICHOLSON in LANKESTER, 1877; Zoanthactiniaria + Rugosa VAN BENEDEN, 1898; Actinanthida DELAGE & HÉROUARD, 1901; Dodecacorallia CARLGREN, 1908]

Solitary and colonial anthozoans, with or without calcareous trabecular exoskeleton, characterized by coupled and paired mesenteries, and by insertion of new pairs of mesenteries, generally after the first 6, in 2 (ventrolateral), 4 (lateral and ventrolateral), or all 6 primary exocoeles. *Ord.-Rec.*

Representatives of the Zoantharia are divided into orders according to arrangement of their mesenteries and corresponding septa if a skeleton is present.

Classification of Zoantharia

1. Septa inserted only within "bifurcated" outer ends of 4 original septa (cardinal, counter, and alars) .... HETEROCORALLIA
2. Mesenterial pairs, after first 6, inserted only in ventrolateral exocoeles .... ZOANTHINIARIA
3. Septa inserted in 4 spaces: between cardinal and alar septa, and between alar and counter-lateral septa (?mesenterial pairs inserted in lateral and ventrolateral exocoeles) .... RUGOSA
4. Mesenterial pairs (and septa) inserted in dorsolateral, lateral, and ventrolateral exocoeles .... HEXACTINIARIA
   - A. One tentacle over each exocoele and entocoele; exoskeleton
     - i. Absent .... ACTINIARIA
     - ii. Present .... SCLERACTINIA
   - B. More than one tentacle over each exocoele and entocoele; skeleton absent .... CORALLIMORPHARIA

5. Mesenterial arrangement unknown; septa short, equal, acanthine, generally 12 in number ....TABULATA

Genera and subgenera of the Zoantharia recorded as fossils total 831 and 60, respectively, distributed as follows: Rugosa, 333 genera and 12 subgenera; Heterocorallia, 2 genera and 2 subgenera; Scleractinia, 388 genera and 46 subgenera; and Tabulata, 108 genera. In addition, 3 genera of Actiniaria are doubtfully reported to be represented by fossils.

ZOANTHINIARIA, CORALLIMORPHARIA, AND ACTINIARIA

By J. W. WELLS and DOROTHY HILL

Order ZOANTHINIARIA van Beneden, 1898

[=Actinaria LAMOUROUX, 1824 (*partim*); Zoantharia coriacés DE BLAINVILLE, 1830; Zoanthina coriaces EHRENBERG, 1834; Zoanthidae GRAY, 1840; Zoanthinae MILNE-EDWARDS & HAIME, 1851; Zoanthacea VERRILL, 1865; Zoantharia KLUNZINGER, 1877 (*non* BLAINV., 1830); Zoantheae HERTWIG, 1882; Zoanthida HAECKEL, 1896; Zoanthidea BOURNE, 1900].

Mostly colonial anemones (Figs. 132*f*, 163), with mesenterial pairs beyond the first 6 inserted only in intermesenterial spaces on each side of ventral directive pair. *Rec.*

Includes 5 or 6 genera, most abundant in warm shallow waters. Unknown as fossils.

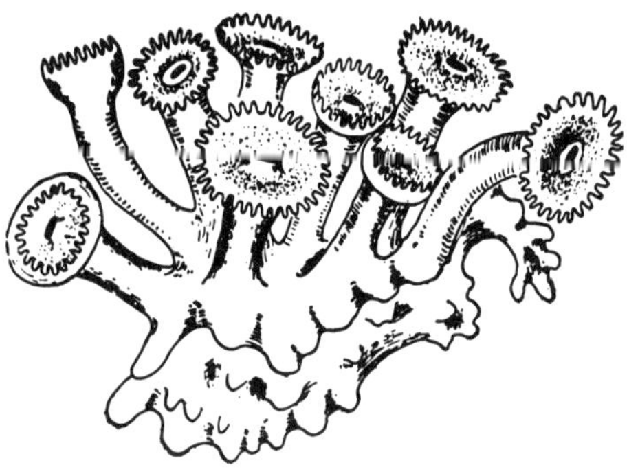

FIG. 163. Zoanthid colony, Rec., Bermuda, ×1 (L. H. HYMAN, *Invertebrates: Protozoa through Ctenophora,* McGraw-Hill, New York).

Order CORALLIMORPHARIA Stephenson, 1937

[=Stichodactylina DELAGE & HÉROUARD, 1901; Asclerocorallia CARLGREN, 1949 (*non* PAX, 1940)]

Solitary or colonial anemones, with principal tentacles cyclically arranged and secondary tentacles developed from the same exo- and entocoeles. No basilar muscles or ciliated mesenterial tracts. Mesenteries inserted cyclically in all 6 primary exocoeles. *Rec.*

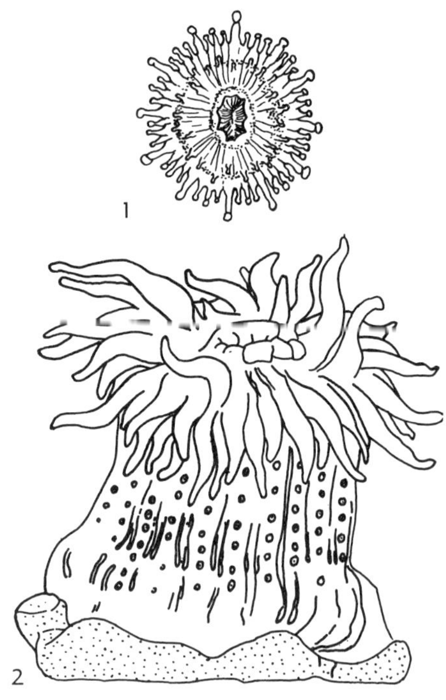

FIG. 164. Corallimorpharia and Actiniaria. *1, Corallimorphus rigidus,* Rec., S.Ind.O., 1,950 fathoms; oral aspect, ×0.5 (R.HERTWIG). *2, Stomphia coccinea,* Rec., Eu.; ×1 (DANIELSSEN).

A small but widely distributed group, including 10 living genera, with no known fossil representatives. It is regarded as coordinate with the Scleractinia, from which it differs by (1) lack of skeleton, (2) tentacular arrangement (Fig. 164,*1*), (3) somewhat more varied nematocysts, and (4) presence of cells in the mesogloea.

Order ACTINIARIA R.Hertwig, 1882

[=Actinina EHRENBERG, 1834; Actiniadae GRAY, 1840; Actinidae DANA, 1846; Zoantharia malacodermata MILNE-EDWARDS & HAIME, 1851 (*partim*); Actinacea VERRILL, 1865; Actinida HAECKEL, 1876; Hexactinia GÖTTE, 1898 (*partim*); Malacactiniae (*partim*), Edwardsidea BOURNE, 1900; Hexactinidae DELAGE & HÉROUARD, 1901 (*partim*)]

Solitary skeletonless anemones (Fig. 164, *2*), with mesenterial pairs beyond the first 6 cyclically inserted in primary exocoeles (or rarely in lateral entocoeles) in dorsoventral succession (Fig. 132*h*). Tentacles cyclically arranged, commonly retractile and only one over each exo- and entocoele. *Rec.*

The small family Ptychodactidae, containing 2 genera, with a number of supposedly primitive features as regards musculature, nematocysts, and mesenterial structures, is considered by CARLGREN to constitute a separate order (Ptychodactiaria).

The suborder Endocoelactaria, comprising a small group in which new mesenterial pairs beyond the first 6 are developed by introduction of pairs oriented like directives in the lateral exocoeles, has been suggested as related to the Rugosa, which differ, if septal and mesenterial equivalence be granted, by having the new mesenteries after the first secondaries apparently introduced only in the ventral sides of the lateral exocoeles (Fig. 132,*h*).

Although about 200 living genera and some 700 species of actiniarians are distinguished, fossil forms are not certainly known; a few genera have been attributed to Actiniaria, however.

?**Mackenzia** WALCOTT, 1911 (**M. costalis*). Originally identified as an holothurian, it has been referred by CLARK to the Edwardsiidae, a primitive actiniarian family. *M.Cam.*(Burgess sh.), B.C.

?**Palaeactis** DOLLFUS, 1875 (**P. vetula*). *M.Ord.(Llandeil.)*, Fr.

?**Palaeactinia** RUEDEMANN, 1934 (**P. halli*). Possibly an actinodendrid. *M.Ord.(Mohawk.)*, N.Y.

RUGOSA

By DOROTHY HILL

CONTENTS

MORPHOLOGY

GENERAL FEATURES OF SKELETON

The Rugosa are an extinct, Paleozoic order of corals forming calcareous skeletons in which the radial plates **(septa)** of each individual skeleton **(corallite)** are generally more noticeable than the transverse plates. The septa are typically of 2 orders, major and minor, alternating in size; in solitary corallites, after the insertion of the first 6 septa, the remaining major septa are inserted at 4 points only. In many Rugosa the only transverse plates are the arched, flat, or sagging **tabulae;** in others, the **tabularium** (space occupied by the tabulae) is surrounded by smaller, commonly arched and inclined plates called **dissepiments,** the marginal zone occupied by them being the **dissepimentarium.** Another type of marginal zone **(marginarium),** common in Rugosa, is a zone of dense calcareous tissue without loculi, a marginal stereozone.

MICROSTRUCTURE

By analogy with Scleractinia, the skeleton is an exoskeleton which must have been laid down by the ectoderm of the base of the polyp. The calcareous tissue is **sclerenchyme,** but use of this term is common only as applied to elements which are notably thickened. The basal ectoderm is considered to have exuded a gel in which $CaCO_3$ crystallized out in needles ("fibers") according to the laws of spherulitic crystallization and at right angles to the secreting ectodermal surface (3). The exudation and crystallization are probably periodic, for growth lamination may be distinguished across the length of the fibers. The fibers of the septa and other radial elements crystallized in the gel secreted in radial upfolds (invaginations) in the base of the polyp, while the transverse elements and those parts of the wall between the peripheral edges of the septa were deposited in the gel

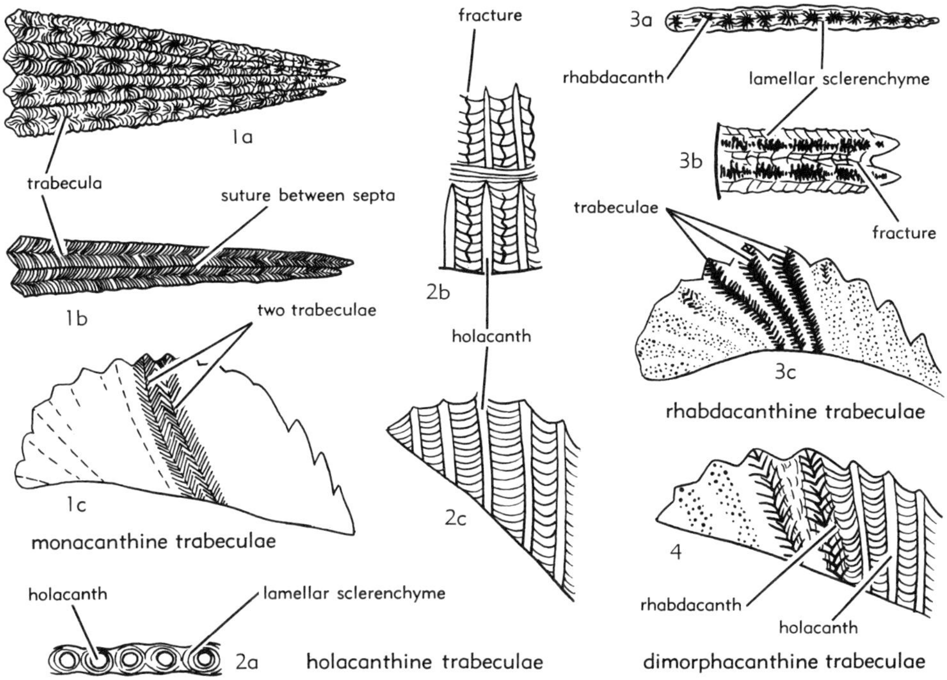

FIG. 165. Rugose coral morphology; types of trabeculae. *1*, Monacanths in *Porpites porpita* (LINNÉ), Sil., Eu.; *1a*, 4 septa in transv. sec. cut at right angles to trabeculae; *1b*, 2 septa in transv. sec. cut obliquely to trabeculae; *1c*, 1 septum in long. sec., structure of only 2 trabeculae being shown; enlarged (85). *2*, Holacanths in *Tryplasma primum* HILL, M.Sil., Eng.; *2a*, transv. sec. cut at right angles to trabeculae; *2b*, transv. sec. cut obliquely to trabeculae; *2c*, long. sec.; enlarged (85). *3*, Rhabdacanths in *Rhabdocyclus porpitoides* LANG & SMITH, Sil., Eng.; *3a*, transv. sec. cut at right angles to trabeculae; *3b*, transv. sec. cut obliquely to trabeculae; *3c*, long. sec. with 3 central trabeculae shown in median long. sec., others in tangent. long. sec.; enlarged (85). *4*, Dimorphacanthine septa in *Tryplasma primum* HILL, M.Sil., Eng.; long. sec., 2 cut medially, 3 trabeculae at right being holacanths; enlarged (85).

exuded by the ectoderm of the unfolded parts.

Skeletal tissue deposited by unfolded ectoderm shows planar control of the spherulitic crystallization, for in tabulae and dissepiments the fibers lie at right angles to the upper and lower surfaces of the plates, the plane of control being the uniform ectodermal surface secreting the gel. When the secretory surface is not uniform, the fibers deposited from a more advanced point may be grouped in a radiating aggregate of the type termed "fascicle" in much scleractinian literature; but such a grouping is by no means a fundamental unit of the coral skeleton; fascicles will not be found where perfect planar control exists.

Tissue formed by the crests of the upfolds shows control by points in, or by the line of, the crest. Puntal control is commoner than linear, for most septa consist of cylindrical or flattened spines called trabeculae (Fig. 165), each apparently formed by continuous deposition from one or more persistent centers of calcification in a conical hollow at the crest of the upfold. The trabeculae of a septum may be separate or contiguous, and their nature and arrangement vary from genus to genus and family to family. A trabecula may be a simple **monacanth** in which the fibers are related to one center only, radiating upward and outward from the axis formed by upward shifting of the center; or its fibers may be related to any number of separate, smaller, transient centers grouped about the main one, thus forming a **rhabdacanth** (17). The fibrous nature of the skeletal elements may be ob-

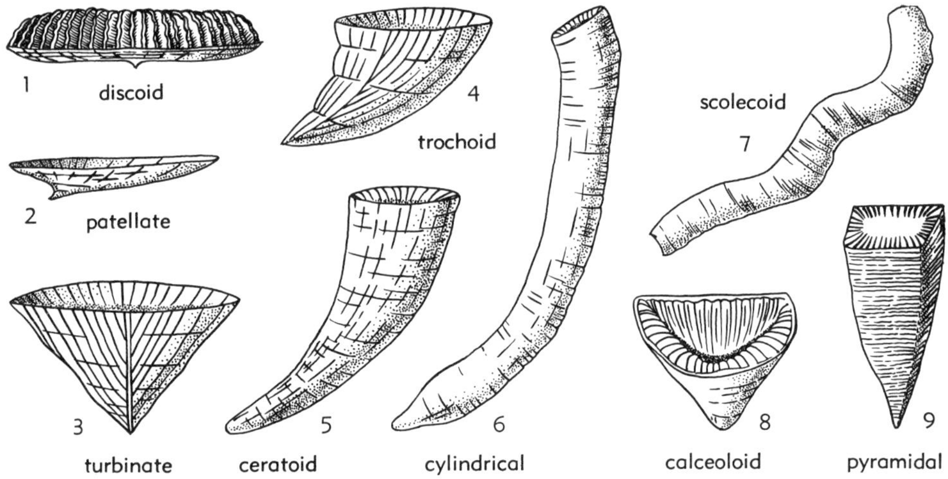

FIG. 166. Rugose coral morphology; types of simple coralla (85). *1, Porpites porpita* (LINNÉ), Sil., Eu.; ×2. *2, Rhabdocyclus fletcheri* (EDW.-H.), Sil., Eng.; ×1.5. *3, Phaulactis* sp., Sil., Eng.; ×1. *4, Zaphrenthis* sp., M.Dev., N.Am.,; ×1. *5, Streptelasma* sp., Sil., Eng.; ×1. *6, Amplexus* sp., L.Carb., Eng.; ×0.3. *7, Helminthidium mirum* LINDSTRÖM, Sil., Eu.; ×1. *8, Calceola sandalina* (LINNÉ), L.Dev.-M.Dev., Eu.-Asia-Afr.-Austral.-N.Am.; ×1. *9, Goniophyllum flletcheri* (EDW.-H.), Sil., Eng.; ×1.

scured either by original failure of the material to crystallize spherulitically or, more usually, by recrystallization before or during fossilization. Some parts of an individual may show the fibrous nature perfectly, while in other similar parts it is obscured. Growth lamellation may be emphasized, either by recrystallization or by original failure of the fibers to come out of the gel spherulitically, and this may be combined with a complete recrystallization to clear calcite of the axial parts of the trabeculae, causing for instance the perfect rhabdacanthine septa of *Tryplasma* to be replaced by a series of seemingly structureless rods of clear calcite, termed **holacanths,** set in **lamellar sclerenchyme.** Such obscurations are as characteristic of genera and families as are the patterns they obscure.

The trabeculae are variously arranged in the septa. In most Rugosa they lie in the median plane of the septum and project upward and inward toward the axis of the corallite; their course is invariably at right angles to the successive platforms of dissepiments. In a few genera, an area of divergence of the septal trabeculae is observed within the dissepimentarium, where the trabeculae of one part of a septum are directed peripherally in the opposite sense to the remainder which are directed axially (e.g., *Acervularia*); an outward direction at right angles to the successive tabulae is taken by the trabeculae in radial plates in any axial structure. In the Phillipsastraeidae with thick septa, the trabeculae diverge from the median plane toward the faces (sides) of the septum. In some genera, trabeculae may develop in other than the median plane of the septum and may then form an open network *(Arachnophyllum).*

In the septa of many genera the fibers all appear to proceed from the mid-plane toward the septal faces at exactly the same angle, so that individual trabeculae cannot be distinguished; such an attitude results from spherulitic crystallization under linear control, the line being the crest of the septal invagination; this is in contrast to trabecular growth resulting from control by a number of separated points along the crest.

The latest fibrous growth lamellae of many thick septa may be seen in parts of some corals to be continuous with those of neighboring tabulae or dissepiments; this indicates a septal invagination so deep that the fibers initiated at its crest are lengthened by its smooth side, which is continuous over change of slope with the unfolded ectoderm between the invaginations.

FORM OF CORALLA

SOLITARY CORALS

Turning from the fine structure of the skeletal elements to megascopic characters of the skeleton, we note that **solitary corallites** may be curved or erect, and the fundamental shape appears to be a reversed cone. Depending on the apical angle of the cone and other characteristics, such as the growth form of the mature region and occurrence of sharp angulations or flattened areas (Fig. 166), the shape may be designated as **ceratoid** (very slenderly conical, horn-shaped), **cylindrical** (nearly straight and of uniform diameter except in the apical region), **scolecoid** (like cylindrical but irregularly bent in the manner of a worm), **trochoid** (regularly expanding from an apical angle of about 40 degrees), **turbinate** (like trochoid but with wider apical angle, about 70 degrees), **patellate** (with still wider apical angle, about 120 degrees), **discoid,** nearly all in a single plane, **calceoloid** (like the tip of a slipper, as in *Calceola*), or **pyramidal** with flattened sides which meet at angles (Fig. 166).

Solitary corallites are laterally and basally sheathed in a thin film of ?granular (?nonfibrous) $CaCO_3$, the **epitheca** (Fig. 167,*3*) lying upon the **corallite wall,** which is not a simple unit. In some corallites, the wall is formed by the peripheral edges of the septa thickening to contiguity; in others, a nontrabeculate fibrous lining bridges the space between each such peripheral edge, the fibers lying radially. This lining may be quite thin, but where the fibers lengthen

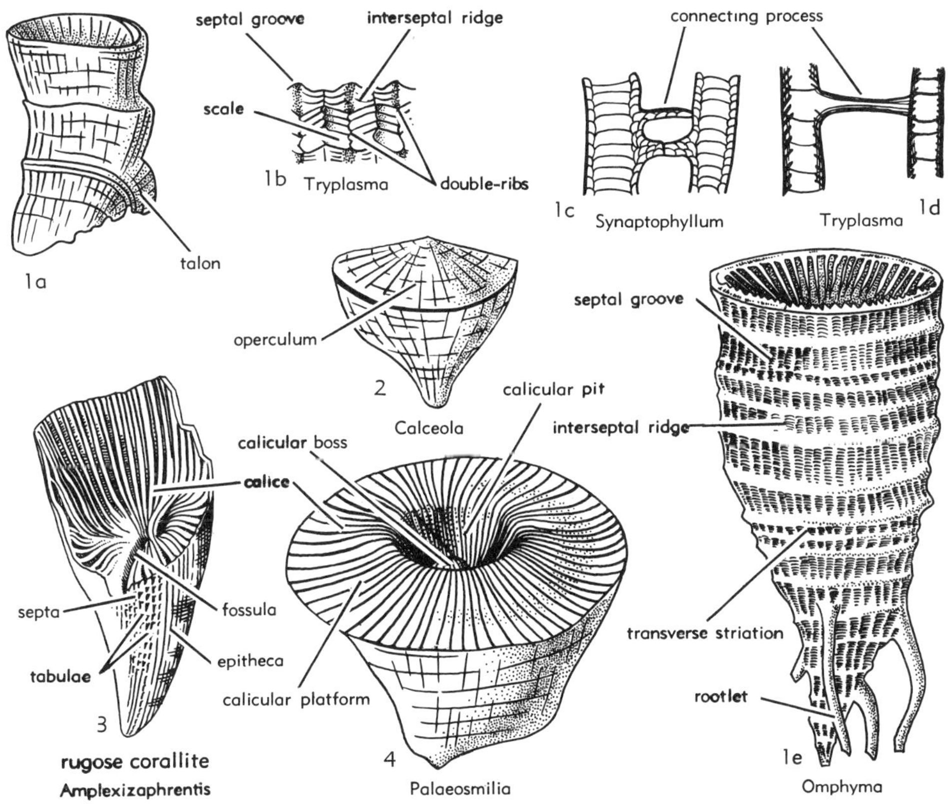

FIG. 167. Rugose coral morphology; epithecal and calical characters (85). *1a, Tryplasma* sp., M.Sil., Eng.; ×1. *1b, Tryplasma lovéni* (EDW.-H.), M.Sil., Eu.; enlarged. *1c, Synaptophyllum simcoense* (BILL.), Dev., N.Am.; ×2. *1d, Tryplasma lonsdalei* ETHERIDGE, M.Sil., N.S.W.; ×2. *1e, Ketophyllum subturbinatum* (ORB.), Sil., Eu.; ×1. *2, Calceola sandalina* (LINNÉ), M.Dev., Eu.-Asia-Afr.-Austral.-N.Am.; ×1. *3, Amplexizaphrentis enniskilleni* (EDW.-H.), L.Carb., Eu.; ×1. *4, Palaeosmilia murchisoni* EDW.-H., L.Carb., Eu.; ×1.

greatly, a marginal stereozone may be formed.

The lateral epithecate surfaces of the corallite are vertically grooved, each such depression corresponding in position to a major or a minor septum and being therefore termed a **septal groove; scales** may be attached to these furrows and the **interseptal ridges** ocurring between them *(Tryplasma)*. Transverse or growth wrinklings of the epitheca may also be noted. Buttressing outgrowths **(talons)** from the cone of attachment and rootlike structures **(radiciform processes** or **rootlets** and **spines)** may assist in anchoring the corallite (Fig. 167, *1a,1e*).

COMPOUND CORALS

The entire skeleton of a compound organism is termed a **corallum.** This term is traditionally used for the skeleton of the whole organism whether compound or solitary. A compound corallum is called **fasciculate** if the corallites are cylindrical and not in contact. Fasciculate coralla may be classified as **dendroid** (irregularly branching), or **phaceloid** (with neighboring corallites more or less parallel, not in contact but joined at some points by connecting processes, Fig. 168,*1*). Compound coralla are described as **massive** if neighboring corallites are in contact and polygonal in section. Massive coralla

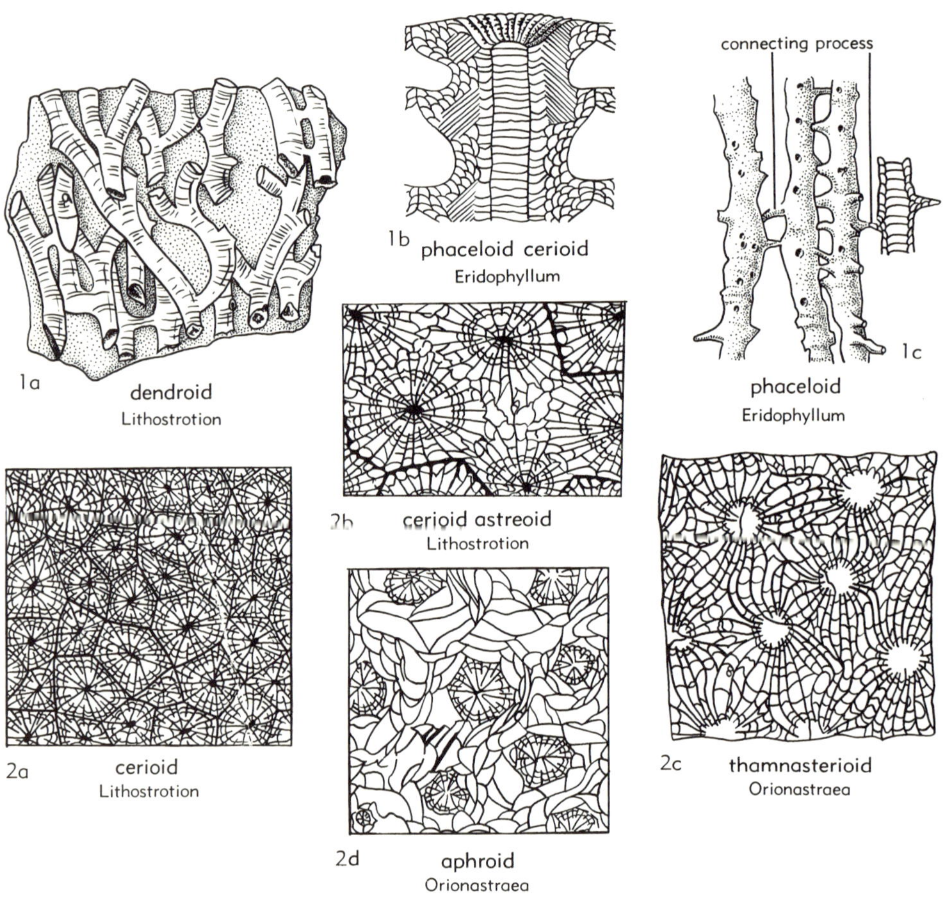

Fig. 168. Rugose coral morphology; types of compound coralla (85). *1a, Lithostrotion irregulare* (Phillips), L.Carb., Eu.; ×1. *1b, Eridophyllum colligatum* (Bill.), Dev., N.Am.; median long. sec., ×1.3. *1c, Eridophyllum simcoense* (Bill.), Dev., N.Am.; corallite in median long. sec., ×1.5. *2a, Lithostrotion arachnoideum* (M'Coy), L.Carb., Eng.; transv. sec., ×1. *2b, Lithostrotion columnare* Etheridge, L.Carb., E.Austral.; transv. sec., ×1.5. *2c, Orionastraea garwoodi* Hudson, L.Carb., Eng.; transv. sec., ×2. *2d, Orionastraea lonsdaleoides* Hill, L.Carb., E.Austral.; transv. sec., ×1.5.

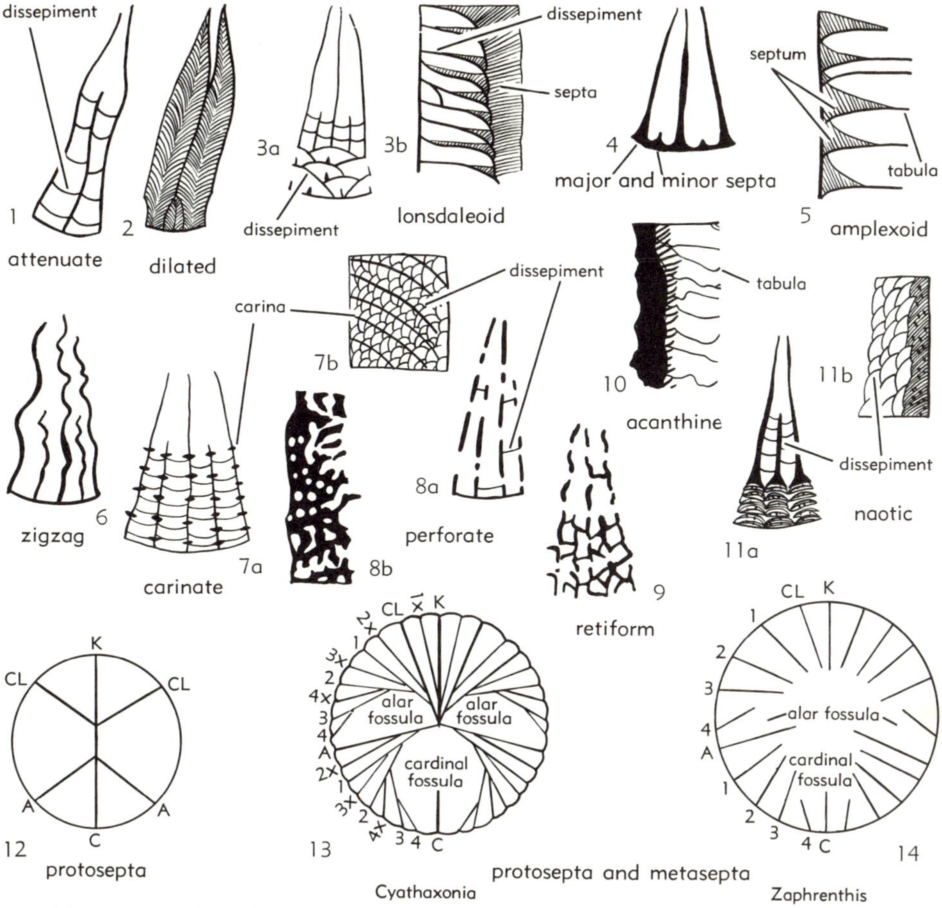

FIG. 169. Rugose coral morphology; types and insertion of septa (85n). *1-11,* Types of septa, diagrammatic, enlarged; *1,2,3a,4,6,7a,8a,9,11a,* transv. secs.; *3b,5,7b,8b,10,11b,* median long. secs.; *12-14* insertion of septa, diagrammatic, enlarged; *12,* showing 6 protosepta; *13, Cyathaxonia,* numerals indicating order of insertion of septa, those marked with "x" staying short (minor septa), others lengthening and becoming major septa; *14, Zaphrenthis,* numerals indicating order of insertion of metasepta, minor septa being inserted later in cycle, after which insertion proceeds as in *Cyathaxonia.* A= alar septum; C= cardinal septum; CL= counter-lateral septum; K= counter septum.

(Fig. 168,*2*) are clothed in a common epitheca. They are distinguished as (1) **cerioid,** if each corallite is defined by a wall; (2) **astreoid,** if the individual corallites lose their walls without reduction of the septa, which usually alternate in neighboring corallites; (3) **thamnasterioid,**[1] if the septa of neighboring corallites are confluent and arranged between the axes of corallites like lines of force between poles in a magnetic field; or (4) **aphroid,** if the septa have withdrawn from the periphery so that those of neighboring corallites are united by dissepiments only. Meandroid coralla are not known in Rugosa.

CALICE

The **calice** is the distal surface of a corallite, and is a mold of the base of the polyp (Fig. 167,*3,4*). Many corallites have an

[1] The structure of corals in which "the septa of neighbouring corallites become confluent" originally was designated by LANG (1923, *Trends in British Carboniferous Corals,* Geol. Assoc., Proc., v. 34, p. 123) as "thamnastraeoid," without reference suggesting that the name was derived from the scleractinian *Thamnastraea* LESAUVAGE, 1832 (=*Thamnasteria* LESAUVAGE, 1823). Accordingly, HILL sees no reason to alter spelling to "thamnasterioid," as is done by WELLS and used in the *Treatise;* also, she prefers "astraeoid" rather than "astreoid" (*Astrea* LAMARCK, 1801), although no genus bears the name *Astraea.*—EDITOR.

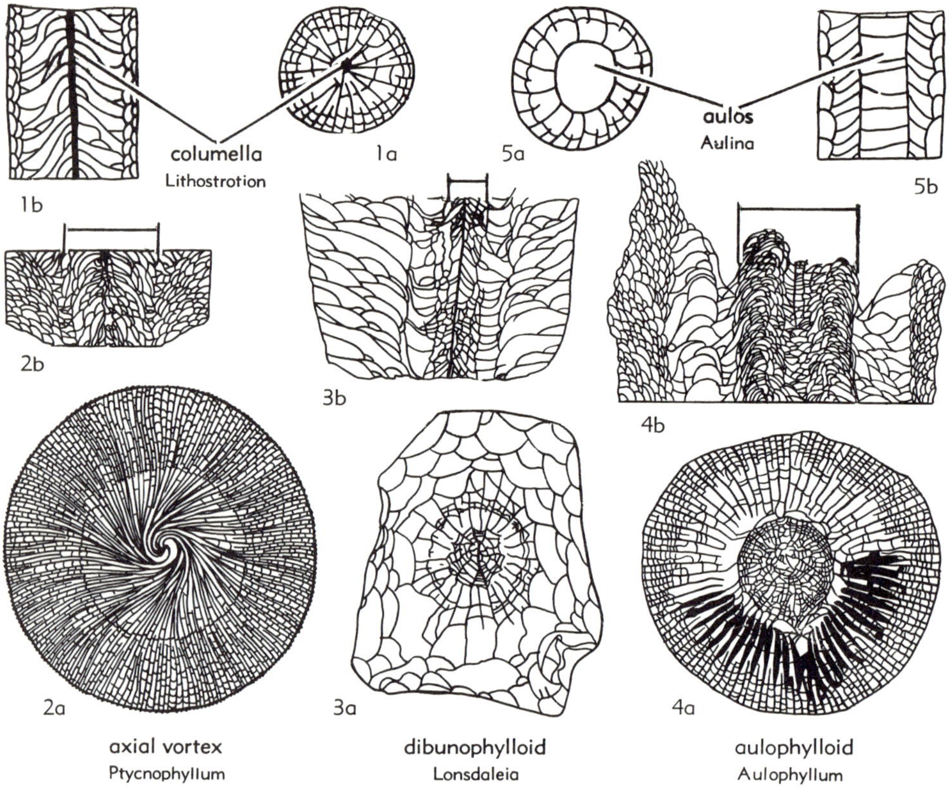

FIG. 170. Rugose coral morphology; axial structures (85). *1, Lithostrotion* sp., L.Carb., Eu.; *1a,b,* transv. and median long. secs., enlarged. *2, Ptychophyllum stokesi* EDW.-H., Sil., Can.; *2a,b,* transv. and median long. secs., ×1. *3, Lonsdaleia laticlavia* SMITH, L.Carb., Eng.; *3a,b,* transv. and median long. secs., ×1.7. *4, Aulophyllum fungites* (FLEMING), L.Carb., Eu., *4a,b,* transv. and median long. secs., ×1.25. *5, Aulina furcata* SMITH, L.Carb., Eng.; *5a,b,* transv. and median long. secs., ×2.

evenly rounded bowl-shaped calice, ranging in depth from a very shallow saucer-like shape to a deep beaker-like form. The calice in some genera, particularly those with marginaria, shows marked differentiation of inner and outer parts, the outer area (which may be nearly plane or everted) constituting a **calicular platform,** while the central part is abruptly depressed **(calicular pit)** or raised in a rounded to sharp-pointed prominence **(calicular boss)** (Fig. 167,*4*). In pyramidal or calceoloid coralla, the calice may be closed by an independent plate or plates, the **operculum** (Fig. 167,*2*), evidently drawn into a position of closure by the retraction of the polyp.

SEPTA

MAJOR SEPTA

The first skeletal elements to be laid down upon the initial deposit from the attached planula, the basal epitheca are 6 of the major septa, termed **protosepta.** These are first the **cardinal septum** and its opposite, the **counter septum,** marking the plane of bilateral symmetry of the corallite, commonly with their axial ends conjoined; then the 2 **alar septa,** one on each side of the cardinal septum, and later the 2 **counter-lateral septa,** between the alar septa and the counter septum. This order of appearance is not, however, invariable. Insertion of later **metasepta** takes place at 4 points only, pinnately, immediately on each side of the cardinal septum, and on the counter side of each alar septum (Fig. 169, *12-14*). The septa inserted at each of these points form a **quadrant;** in some corals more septa are developed in the counter than in the cardinal quadrants, which accordingly are smaller, and septal insertion is then said to be accelerated in the counter quadrants; in other corals, the opposite occurs.

MINOR SEPTA

So-called minor septa are distinguished by their shorter length as compared with major septa and by their placement between the major septa so as to alternate with them. In some genera, the minor septa may be seen to enter almost simultaneously with neighboring major septa, one on the counter side of any major septum before another on its cardinal side. In others, although septal grooves corresponding to the minor septa appear almost simultaneously, as above, with their neighboring major septal grooves, the minor septa themselves are nonexistent, or so small as to escape notice until a later stage when they all appear together right around the corallite, after which new minor septa appear as in the former case. In some Artinskian and later Rugosa, irregularities in the normal septal insertion occur which indicate a transition from the rugose, pinnate type to the hexacoral, cyclic type (42). **Contratingent** minor septa lean on their neighboring major septa on the counter sides; the minor septa on either side of the counter septum may be long. A third **(tertiary)** order of septa is observed in some genera, between major and minor septa (*Iranophyllum,* Fig. 211,*2*).

SPECIAL TYPES

The **septa** are variously developed in the Rugosa, not only in their trabecular constitution as described above but in their shapes (Fig. 169,*1-11*). Generally the septa are laminar, i.e., the trabeculae are closely united without intervening openings to form a continuous sheet; but **acanthine** septa occur, with the trabeculae separate distally or throughout. Septa may be **long** or **short, attenuate, thin,** or commonly **dilated** in whole or in part; in the Zaphrenticae a septum may be thicker within the tabularium than in the dissepimentarium. **Rhopaloid** septa are expanded at their inner edges and are club-shaped in section (*Ufimia,* 177,*12*). Thickening of the septa is effected by elongation of the fibers, accretion lamellae showing that the process is periodic (possibly related to feeding times). Where thickening affects the septa in a particular zone of the coral, a **stereozone** is formed (peripheral stereozone of *Kodonophyllum,* Fig. 172,*2*). A septum may have a number of irregular, upwardly directed, slender, or cylindrical **axial lobes** springing from its axial edge, as in Streptelasmatidae (cf. the "paliform lobes" of Scleractinia). It may be represented in an axial structure by a **septal lamella,** which is a radial axial plate not in continuity with it. Its distal (calicular) edge may show teeth representing the growing tips of trabeculae. The sides of a septum may be smooth, provided with small denticulations, or flanged. The flanges may be parallel to the calicular edge, as in Cyathaxoniicae, or parallel to the trabeculae, of which they are then lateral extensions called **carinae,** or **yard-arm** carinae if those of both sides of the septum are opposite *(Heliophyllum),* or zigzag carinae if they are sub-opposite *(Entelophyllum)* (Fig. 169, *7*).

The septa may not be continuous vertically and radially. Thus **amplexoid** septa are fully developed only on upper surfaces of the tabulae, but above this position they extend progressively a shorter distance from the epitheca, as in *Amplexus* (Fig. 169,*5*). **Lonsdaleoid** septa are discontinuous toward the opposite peripheral edge; they do not extend continuously through the dissepimentarium, but break up the dissepiments for a certain distance only, dying out toward the epitheca in a series of septal crests, each resting on a platform of dissepiments. **Perforate** septa are rare among the Rugosa, but occur, with **retiform** septa (perforate septa connected in an irregular network) in the Calostylidae (Fig. 169,*8,9*). A **naotic** septum is represented in its peripheral parts by a series of closely spaced, dissepiment-like plates, each connected by short rods of $CaCO_3$, each rod representing the axis of an imperfectly developed trabecula (for example, in *Naos,* Fig. 169,*11*).

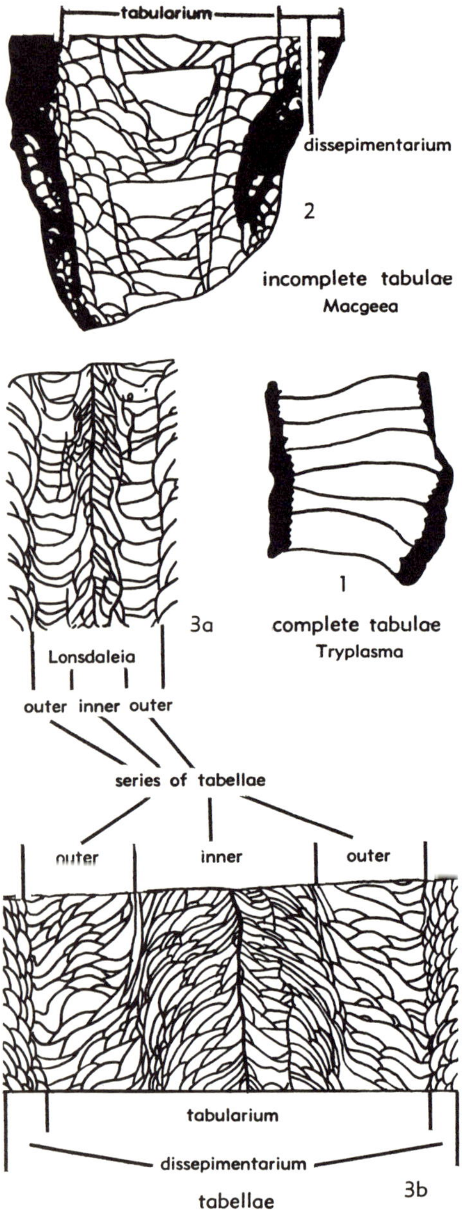

FIG. 171. Rugose coral morphology; types of tabulae (85). *1, Tryplasma* sp., Sil., Eu.; median long. sec., ×1.5. *2, Macgeea* sp., U.Dev., Eu.; median long. sec., ×1.5. *3a, Lonsdaleia laticlavia* SMITH, L.Carb., Eng.; enlarged. *3b, Dibunophyllum bipartitum* (M'COY), L.Carb., Eu.; ×2.

FOSSULAE

Consequent upon the rugose plan of tetrameral, pinnate septal insertion, interseptal loculi of unusual shape and size occur at the points of insertion; these are **fossulae,** the largest being naturally at the cardinal septum, and termed the **cardinal fossula,** or simply the fossula (Fig. 169,*13,14*). The other two are the **alar fossulae.** A **counter fossula** may develop but is not connected with the insertion of new septa. The fossulae may be marked further by quite sharp depression of the tabulae therein. A fossula is **open** if the adjacent septa do not join around its axial end, **closed** if they do; and it may be variously shaped in transverse section, according to the genus or family. In some corals there is a narrowing of the dissepimentarium at the fossula.

AXIAL STRUCTURES

Axial structures commonly are formed from the axial parts of septa reinforced by transverse skeletal elements (Fig. 170); these axial parts tend to be continuous with the peripheral parts only in very young stages. Thus, the axial lobes of many of the Streptelasmatidae form an axial structure. Many genera have a **columella** formed by the vertical prolongation of the axial end of the counter septum (Fig. 170,*1*) which may be swollen by the adherence of vestigial axial ends of the other septa (*Carruthersella,* Fig. 197,*5*). Others have an **axial vortex** formed by the similar twisting of the axial ends of the long major septa (*Ptychophyllum,* Fig. 170,*2*). Still others have an axial structure composed of the axial parts of septa (septal lamellae), usually discontinuous in adult stages from the septa to which they belong and united by transverse skeletal elements; such may be **clisiophylloid,** like a spider web, with a short medial plate derived from the conjoined cardinal and counter septum in the young stages (*Clisiophyllum,* Fig. 195,*1*), **dibunophylloid** (Fig. 170,*3*; 196,*1*), bisected by a medial plate, or **aulophylloid** (Fig. 170,*4*), without a medial plate. An **aulos** (inner tube) may be formed from the similar turning aside of the axial edges of the septa to meet their neighbors, as well as in other ways (Fig. 170,*5*).

TABULAE AND DISSEPIMENTS

Tabulae are the chief horizontal skeletal

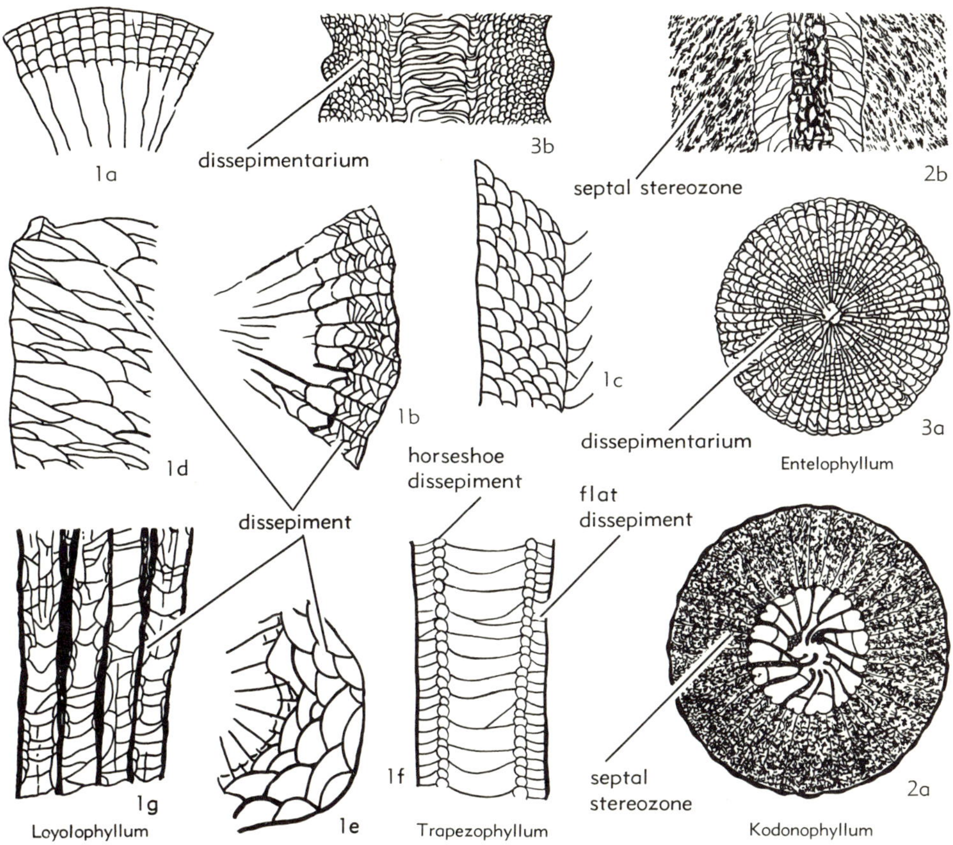

FIG. 172. Rugose coral morphology; dissepiments and marginaria (85). *1a, Caninia subibicina* M'COY, L. Carb., Eu.; part of transv. sec. showing dissepiments arranged in regular concentric pattern; enlarged. *1b, Caninia juddi* (THOMSON), L.Carb., Eu.; part of transv. sec. showing dissepiments arranged in herringbone pattern; enlarged. *1c, Caninia* sp., part of median long. sec. showing many vertical series of dissepiments, corresponding to structure of *1a,b. 1d,e, Lonsdaleia duplicata* (MARTIN), L.Carb., Eu.; parts of long and transv. secs. showing lonsdaleoid dissepiments; enlarged. *1f, Trapezophyllum elegantulum* ETHERIDGE, Dev., Vict.; median long. sec. showing inner series of horseshoe dissepiments and outer series of flat dissepiments, ×3. *1g, Loyolophyllum cresswelli* CHAPMAN, L.Dev., Vict.; median long. sec. showing isolated dissepiments, ×3. *2, Kodonophyllum truncatum* (LINNÉ), Sil., Eu.; *2a,b,* transv. and median long. secs. showing marginarium composed of septal stereozone, ×2. *3, Entelophyllum articulatum* (WAHLENBERG), Sil., Eu.; *3a,b,* transv. and median long. secs. showing marginarium composed of dissepimentarium, ×2.

elements of the Rugosa (Fig. 171). In the Cyathaxoniicae and Streptelasmatidae, they alone are developed; but in many others the space which they occupy **(tabularium)** is surrounded by a **marginarium** (Fig. 172, *2,3*), which is either a peripheral stereozone or a zone of dissepiments comprising a **dissepimentarium.** Each tabula may be a complete floor, or each may be replaced by a number of smaller plates **(tabellae)** which together form a floor and is thus incomplete. They may be domed, horizontal, or inversely conical, or with a median boss or notch.

The development of dissepiments (Fig. 172, *1*) seems not to be possible without minor septa, though many genera may develop minor septa without developing **dissepiments.** They are small, curved or globose plates inclined parallel to the calical slope of the dissepimentarium, each overlapped slightly by its outer neighbor. In a **regular** dissepimentarium each is developed entirely in the interseptal loculus between a major and a minor septum, its side edges being drawn upward at the septal faces; the upward concavity so formed may become an-

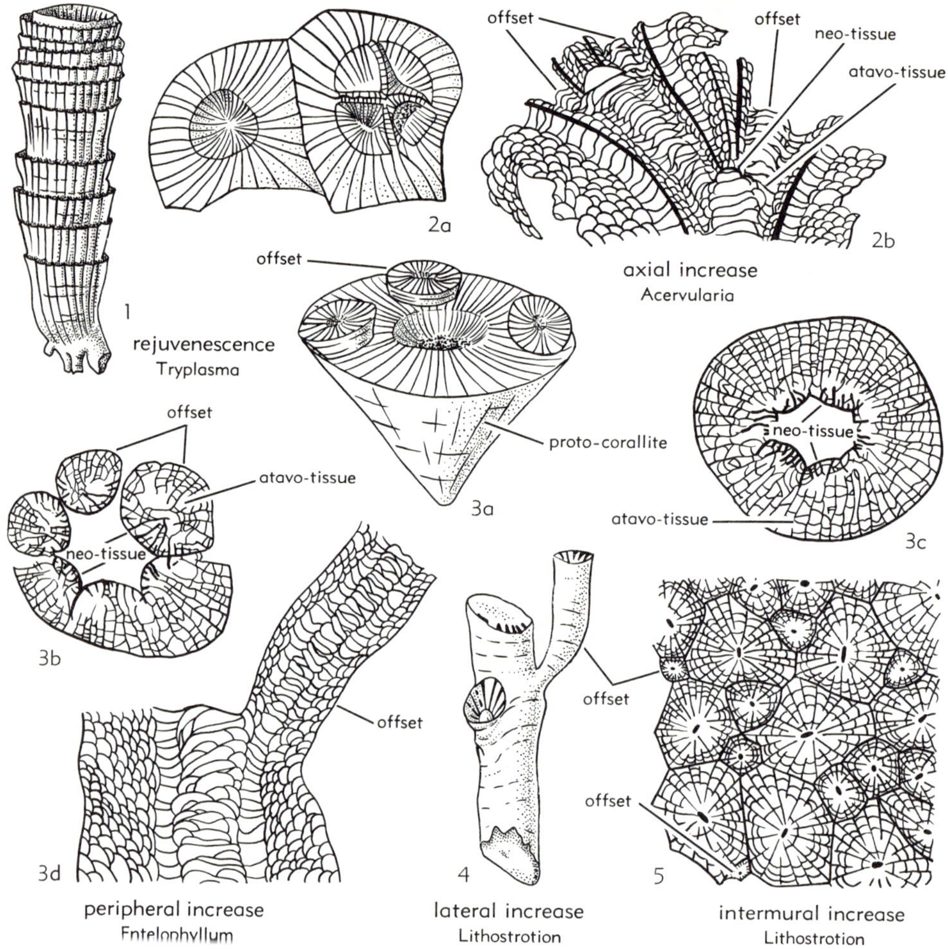

FIG. 173. Rugose coral morphology; rejuvenescence and increase (85). *1, Tryplasma lovéni* (EDW.-H.), Sil., Eu.; ×1. *2, Acervularia ananas* (LINNÉ), M.Sil., Eu.; *2a,* calical view; *2b,* median long. sec.; enlarged. *3, Entelophyllum articulatum* (WAHLENBERG), Sil., Eu.; *3a,* external view; *3b,c,* successive transv. secs.; *3d,* median long. sec.; enlarged. *4, Lithostrotion affine* (FLEMING), L.Carb., Eng.; ×1. *5, Lithostrotion* sp., L.Carb., Scot.; enlarged.

gular. A **herringbone** dissepimentarium forms where the minor septa fail, and the dissepiments between neighboring major septa inosculate. In a **lonsdaleoid** dissepimentarium, large dissepiments with a convex upper surface develop, associated with vertical discontinuities in the septa extending across more than one interseptal loculus. **Horseshoe** dissepiments occur in some phillipsastreids (Fig. 172,*1f*); these are dissepiments in single series placed vertically one above the other, horizontally based and with tall domes. In many Columnariina isolated dissepiments occur, both top and bottom edges being adherent to the wall. **Lateral** (blister-like) dissepiments may be developed similarly along the sides of septa (*Mochlophyllum,* Fig. 218,*1*). In some genera, flat dissepiments occur in association with a stereozone (*Acervularia*).

INCREASE OF CORALLITES

The original corallites of colonies are **protocorallites,** and those produced by **increase** are **offsets.** Three types of increase characterize the Rugosa (Fig. 173). (1) In

axial increase, the old corallite ceases to grow as such; 2 or more offsets are formed, occupying its whole calice, and are built up at the periphery of this old calice by the extension of its septa and epitheca, but at its axis by new septa laid down above its tabularium. (2) In **peripheral increase,** which does not necessarily cause the old corallite to cease growth, small offsets arise from the old calice in its outer septate zone. (3) In **lateral increase** (in branching coralla) or **intermural increase** (in cerioid coralla), the offsets appear to project laterally through or to spring from the wall of the old corallite.

"Rejuvenescence" is often observed (Fig. 173,*1*). A corallite, instead of continuing its growth with increasing or constant diameter, suddenly becomes much constricted, usually leaving a ledge of older calice round the constricted part. It then increases in diameter, and the process may be repeated. During rejuvenescence the corallite recapitulates earlier growth stages in a condensed form.

GLOSSARY OF MORPHOLOGICAL TERMS APPLIED TO CORALS

By Raymond C. Moore, Dorothy Hill, and John W. Wells

For convenience of reference, an alphabetically arranged glossary of morphological terms applied to corals has been prepared, including together terms commonly employed for description of Rugosa, Heterocorallia, Scleractinia, and Tabulata. This is advantageous because many terms are applied to 2 or more of these ordinal divisions of the Zoantharia.

Classification of the terms is indicated typographically, (1) those thought to be most important because most generally used being printed in boldface capital letters (as **SEPTUM**); (2) less universally used terms being printed in uncapitalized letters (as **amplexoid septum**); and (3) least important terms printed in italics (as *cnidocil*).

H=Heterocorallia; R=Rugosa; S=Scleractinia; T=Tabulata

acanthine septum. Type composed of a vertical or steeply inclined series of trabeculae and commonly marked by spinose projections along axially directed margin of septum (*see* other types: amplexoid, dilated, fenestrate, laminar, lonsdaleoid, naotic, retiform, rhopaloid) (R, S, T).

AHERMATYPIC. Not reef-forming.

ALAR FOSSULA. Relatively prominent interseptal space developed in position of an alar septum or adjoining it on side toward counter septum (*see* closed fossula, open fossula, also other types: cardinal, counter) (R).

ALAR SEPTUM (symbol, A). One of 2 protosepta located about midway between cardinal and counter septa, distinguished by insertion of newly formed metasepta on side facing counter septum (*see* other types: cardinal, counter, counter-lateral) (R).

alveolitoid. Type of reclined corallite having a vaulted upper wall and nearly plane lower one parallel to surface of adherence of colony, as in *Alveolites* (T).

ambulacrum. Trough of coenosteum separating collines on surface of some meandroid coralla (S).

amplexoid septum. Type characterized by extreme shortness except where septum is extended axially on distal side of a tabula, as in *Amplexus* (*see* other types: acanthine, dilated, fenestrate, laminar, lonsdaleoid, naotic, retiform, rhopaloid) (R).

anthoblast. *Acrosmilia* stage of ontogeny derived by transverse division from a solitary *Fungia* individual by extratentacular budding (*see* anthocaulus, anthocyathus) (S).

anthocaulus. *Acrosmilia* stage of ontogeny derived by transverse division from a solitary *Fungia* individual of sexual generation (*see* anthoblast, anthocyathus) (S).

anthocyathus. Neanic *Fungia* stage after separation from an anthocaulus or anthoblast (S).

aphroid. Massive corallum like astreoid type but with septa shortened peripherally, adjacent corallites being united by a dissepimental zone (*see* astreoid, cerioid, meandroid, plocoid, thamnasterioid) (R, S).

astreoid. Massive corallum in which septa of each corallite are fully developed but walls between corallites are lacking, septa of adjacent corallites generally in alternating position (*see* aphroid, cerioid, meandroid, plocoid, thamnasterioid) (R).

aulophylloid. Type of axial structure like that termed clisiophylloid but lacking a medial plate, as in *Aulophyllum* (*see* clisiophylloid, dibunophylloid, axial structure, axial vortex) (R).

AULOS. Axial structure consisting of tube commonly formed by abrupt sideward deflection of inner edges of septa and junction of them with neighbors (*see* axial structure) (R).

AXIAL. With reference to corallite oral-aboral axis.

AXIAL INCREASE. Budding of corallites in coralla characterized by appearance of dividing walls between newly formed corallites approximately in position of axis of parent (*see* lateral increase, peripheral increase, intermural increase) (R, T).

axial lobe. Irregular upwardly directed slender extension from axial edge of a septum (R).

AXIAL STRUCTURE. Collective term for various longitudinal structures in axial region of corallite, whether a solid or spongy rodlike columella or an axial vortex (*see* axial vortex, aulos, aulophylloid, clisiophylloid, dibunophylloid) (R, S).

axial vortex. Longitudinal structure in axial region of corallite formed by twisting of inner edges of major septa associated commonly with transverse skeletal elements (*see* axial structure, aulophylloid, clisiophylloid, dibunophylloid) (R).

axis of divergence. Generally vertical or oblique line in septum from which trabeculae incline inward and outward (*see* fan system) (R,S).

basal disc. Aboral fleshy part of coral polyp, typically subcircular in outline (*see* oral disc, basal plate) (S).

basal plate. Thin, initially formed part of corallite from which septa begin to be built upward (S).

calceoloid. Solitary corallite shaped like tip of a pointed slipper, as in *Calceola,* with angulated edges between flattened and rounded sides (*see* cylindrical, discoid, patellate, pyramidal, scolecoid, trochoid, turbinate) (R).

CALICE. Oral surface of corallite, generally bowl-shaped (H, R, S).

calicoblast layer. Part of ectoderm of coral polyp lying against skeleton.

calicular boss. Sharp-pointed protuberance in central part of calice in some corallites (R, S).

calicular pit. Abruptly depressed central part of calice in some corallites, surrounded by a calicular platform (R).

calicular platform. Part of calice floor having a subhorizontal plane or outwardly sloping (everted) form; generally surrounds a calicular pit (R).

canalicula. Minute canal, as in extended mesenterial chambers of some scleractinians.

CARDINAL FOSSULA. Relatively prominent interseptal space developed in position of cardinal septum (*see* closed fossula, open fossula, also other types: alar, counter) (R).

CARDINAL SEPTUM (symbol, C). Protoseptum in plane of bilateral symmetry of a corallite, distinguished from other protosepta by insertion of newly formed metasepta adjacent to it on both sides (*see* other types: alar, counter, counter-lateral) (R).

CARINA. Flangelike elevation on side of septum formed by thickened trabecula (*see* types: yardarm, zigzag) (R).

cateniform. Corallum with corallites united laterally as palisades which appear chainlike in cross section, the palisades commonly forming a network (T).

ceratoid. Very slenderly conical, horn-shaped solitary corallite (R, S).

cerioid. Massive corallum in which walls of adjacent polygonal corallites are closely united (R, S, T).

CIRCUMMURAL BUDDING. Type of polystomodaeal budding with indirectly linked stomodaea arranged around discontinuous collines or monticules of corallum (*see* circumoral, intramural types of polystomodaeal budding) (S).

CIRCUMORAL BUDDING. Type of polystomodaeal budding with directly linked stomodaea arranged concentrically around central parent stomodaeum (*see* circummural, intramural types of polystomodaeal budding) (S).

clisiophylloid. Type of axial structure with short medial plate in cardinal-counter plane, in transverse sections resembling a spider web, as in *Clisiophyllum* (*see* aulophylloid, dibunophylloid, axial structure, axial vortex) (R).

closed fossula. Prominent interseptal space inclosed toward axis of corallite by united edges of septa (*see* fossula, open fossula) (R).

cnidocil. Same as nematocyst.

coenenchymal. Collective term for corallum having common skeletal tissue uniting individual corallites (S, T).

COENENCHYME. Collective term for both coenosteum and coenosarc (S, T).

coenosarc. Common soft tissue connecting coral polyps in a colony.

COENOSTEUM. Skeletal deposits formed between individual corallites of a colony (S).

colline. Protuberant ridge of corallum surface between corallites (*see* monticule).

COLUMELLA. Solid or nonsolid calcareous axial structure formed by various modifications of inner edges of septa; commonly projects into calice in form of a calicular boss (R, S). (*See* axial structure, axial vortex, also various types of columella: trabecular, parietal, spongy, fascicular, styliform, lamellar.)

column. Smooth cylindrical body wall of coral polyp between basal and oral discs (S).

complete mesentery. Defined by attachment of mesentery to stomodaeum (*see* incomplete mesentery).

complete tabula. Type consisting of a single platform, not composed of several small plates joined together (*see* incomplete tabula) (R, S, T).

compound synapticula. Type consisting of broad bars formed by fusion of opposed ridges on adjacent septa (*see* simple synapticula) (S).

compound trabecula. Composed of bundles of sclerodermites (*see* simple trabecula) (S).

connecting tubule. Subhorizontal tubular connection between neighboring corallites in fasciculate coralla (R, T).

contratingent. Minor septum that leans against adjoining major septum on side toward counter septum (R).

CORALLITE. Exoskeleton formed by an individual coral polyp (*see* protocorallite, corallum) (H, R, S, T).

CORALLUM. Exoskeleton of a coral colony or solitary coral (H, R, S, T).

costa. Prolongation of septum on outer side of corallite wall (S).

COUNTER FOSSULA. Relatively prominent interseptal space developed in position of the counter septum (*see* closed fossula, open fossula, also other types: alar, cardinal) (R).

COUNTER-LATERAL SEPTUM (symbol, KL). One of 2 protosepta that adjoin counter septum on either side (*see* other types: alar, cardinal, counter) (R).

COUNTER SEPTUM (symbol, K). Protoseptum opposite cardinal septum in position (see other types: alar, cardinal, counter-lateral) (R).

couple. Corresponding mesenteries on either side of dorsoventral plane (S).

cuneiform. Wedge-shaped corallite (S).

cupolate. Corallite with flat base and highly convex oral surface (S).

cylindrical. Nearly straight solitary corallite with subequal diameter except near base (*see* calceoloid, discoid, patellate, pyramidal, scolecoid, trochoid, turbinate) (H, R, S).

dendroid. Irregularly branching types of fasciculate corallum (*see* foliose, massive, phaceloid, ramose, reptant, umbelliferous) (R, S).

dibunophylloid. Type of axial structure like that termed clisiophylloid but with longer medial plate joining cardinal and counter septa, as in *Dibunophyllum* (*see* aulophylloid, clisiophylloid, axial structure, axial vortex) (R).

dicentric. Type of corallite formed by polyp retaining distomodaeal condition permanently (*see* monocentric, tricentric, polycentric) (S).

dilatated septum. Type partly or wholly thickened (*see* other types: acanthine, amplexoid, fenestrate, laminar, lonsdaleoid, naotic, retiform, rhopaloid) (R).

direct linkage. Type of intratentacular budding characterized by development of 2 or more mouths with stomodaea inside same tentacular ring and with mesenterial strands connecting adjacent stomodaea (*see* indirect linkage, trabecular linkage, lamellar linkage) (S).

DIRECTIVE COUPLE. Pair (couple) of mesenteries in so-called dorsoventral plane of coral polyp characterized by pleats on opposite rather than facing sides of mesenteries.

directive mesentery. One of 2 mesenteries of a directive couple.

discoid. Solitary corallite with button-like form (*see* calceoloid, cylindrical, patellate, pyramidal, scolecoid, trochoid, turbinate) (R, S).

DISSEPIMENT. Small domed plate forming a cyst-like enclosure in peripheral region of a corallite (R, S, T).

DISSEPIMENTARIUM. Peripheral zone of corallite interior occupied by dissepiments (R, S).

distal. Direction away from point of origin of a corallite (*see* proximal) (H, R, T).

DISTOMODAEAL BUDDING. Having 2 stomodaea developed within a common tentacular ring and 2 interstomodaeal couples of mesenteries between original and each new stomodaeum (*see* tristomodaeal, triple stomodaeal, and polystomodaeal types of budding) (S).

dorsal. In oral aspect of polyp, direction away from pleats of first lateral protocnemes (*see* ventral) (S).

ectoderm. Outer layer of oral and basal discs, tentacles, and column wall of coral polyp (*see* endoderm, mesogloea).

EDGE ZONE. Fold of body wall of coral polyp extending over edge of wall.

endoderm. Inner layer of outer body walls of coral polyp and occurring as a double lamina in mesenteries (*see* ectoderm, mesogloea).

endotheca. Collective term for dissepiments inside corallite wall (S).

entocoele. Space within pair of mesenteries (*see* exocoele).

ENTOSEPTUM. Septum developed within a mesenterial entocoele (*see* exoseptum) (S).

entotentacle. Tentacle occupying position over an entocoelous mesenterial chamber (*see* exotentacle).

EPITHECA. Sheath of skeletal tissue laterally surrounding a corallite comprising extension of basal plate (R, S, T).

exocoele. Space between adjacent pairs of mesenteries (*see* entocoele).

EXOSEPTUM. Septum developed within a mesenterial exocoele (*see* entoseptum) (S).

exotentacle. Tentacle occupying position over an exocoelous mesenterial chamber (*see* entotentacle).

EXOTHECA. Collective term for dissepiments outside of corallite wall (S).

extramural. Occurring outside of wall of a corallite.

EXTRATENTACULAR BUDDING. Formation of new coral polyps by invagination of the edge zone or coenosarc outside of ring of tentacles surrounding mouth of parent (*see* intratentacular budding) (S).

FAN SYSTEM. Fan-shaped pattern formed by diverging trabeculae in plane of septum (*see* axis of divergence) (S).

fascicular columella. Axial structure formed by twisted vertical ribbons or rods resembling pali or paliform lobes (*see* other types: lamellar, styliform, trabecular) (S).

fasciculate. Corallum with cylindrical corallites which are somewhat separated from one another but may be joined by connecting tubules (R, S, T).

fenestrate. Regularly perforated septum (S).

flabellate. Fan-shaped corallite or meandroid corallum with single continuous laterally free linear series of corallites (S).

foliaceous. Same as foliose.

foliose. Type of corallum with laminar branches (*see* dendroid, fasciculate, massive, phaceloid, ramose, reptant, umbelliferous) (R, S, T).

FOSSULA. Interseptal space distinguished by its unusual shape and size (*see* types; alar, cardinal, counter, closed, open) (R).

HERMATYPIC. Reef-forming.

herringbone dissepimentarium. Type in which dissepiments between major septa inosculate, minor septa failing (*see* regular dissepimentarium) (R).

holacanth. Trabecula seemingly consisting of a clear rod of calcite, as in septa of *Tryplasma* (R).

horseshoe dissepiment. Type with horizontal base and strongly arched top part, arranged in single vertical series (*see* lateral dissepiment) (R).

hydnophoroid. Type of corallum with corallite centers arranged around protuberant collines or monticules (S).

incomplete mesentery. Defined by lack of attachment to stomodaeum (*see* complete mesentery).

incomplete tabula. Type consisting of several small plates (tabellae) joined together (*see* complete tabula) (R, S, T).

INCREASE. Addition of corallites to colonies by budding (*see* axial increase, lateral increase, intermural increase, peripheral increase) (R, T).

incrusting. Thin coralla adhering to a surface and following its irregularities (S, T).

indirect linkage. Type of intratentacular budding characterized by development of 2 or more mouths with stomodaea inside same tentacular ring and with one or more couples of mesenteries between each pair of neighboring stomodaea (*see* direct linkage, trabecular linkage, lamellar linkage) (S).

INTERMURAL INCREASE. Budding of corallites in cerioid coralla characterized by sideward outgrowth of offsets, initial parts of which become surrounded by growing wall of parent corallite (*see* axial increase, lateral increase, peripheral increase) (R, T).

interseptal ridge. Longitudinal elevation on outer side of corallite wall, corresponding in position to space between a pair of adjacent septa inside of wall (*see* septal groove) (H, R).

intramural. Within column wall of a polyp.

INTRAMURAL BUDDING. Type of polystomodaeal budding with stomodaea directly or indirectly linked in a single linear series (*see* circummural types of polystomodaeal budding) (S).

INTRATENTACULAR BUDDING. Formation of new coral polyps by invagination of oral disc of parent inside ring of tentacles surrounding its mouth (*see* extratentacular budding) (S).

lamellar columella. Platelike axial structure, in rugose corals generally in plane of cardinal and counter septa, in scleractinians oriented parallel with longer axis of calice (*see* other types: fascicular, styliform, trabecular) (R, S).

lamellar linkage. Connection between corallite centers corresponding to direct linkage of stomodaea (*see* direct linkage, indirect linkage, trabecular linkage) (S).

laminar septum. Longitudinal radial partitioning wall in corallite formed of trabeculae that are contiguous throughout their length (H, R, S, T).

lateral dissepiment. Type having blister-like form developed in isolated manner on sides of septa (*see* horseshoe dissepiment) (R).

LATERAL INCREASE. Budding of corallites in fasciculate coralla characterized by sideward outgrowth of offsets (*see* axial increase, peripheral increase, intermural increase) (R, T).

linear series. Corallites arranged in a confluent row enclosed within continuous walls (S).

longitudinal skeletal element. Part of corallite oriented in direction of growth of polyp (H, R, T).

lonsdaleoid septum. Type characterized by discontinuity toward peripheral edge of septum, as in *Lonsdaleia* (*see* other types: acanthine, amplexoid, dilated, fenestrate, laminar, naotic, retiform, rhopaloid) (R).

MAJOR SEPTUM. One of the protosepta or metasepta (*see* minor septum) (H, R, S).

MARGINARIUM. Peripheral part of interior of a corallite distinguishable from tabularium by difference in constituent structures, generally abundant dissepiments or dense deposit of skeletal tissue producing a stereozone (R, T).

MASSIVE. Corallum composed of corallites closely in contact with one another (*see* dendroid, fasciculate, foliose, ramose, reptant, umbelliferous) (R, S, T).

meandroid. Corallum characterized by meandering

rows of confluent corallites with walls only between the rows (*see* aphroid, astreoid, cerioid, plocoid, thamnasteroid) (S, T).

mesenterial filament. Ribbon-like prolongation of inner free margin of a mesentery.

MESENTERY. Fleshy radially disposed lamina attached to inner surface of oral disc and column wall of coral polyp.

mesogloea. Noncellular jelly-like middle layer of outer walls and mesenteries of coral poyps (*see* ectoderm, endoderm).

metacneme. One of the mesenteries developed after appearance of protocnemes (S).

METASEPTUM. One of the main septa of a corallite other than protosepta, generally distinguished by their extension axially much beyond that of minor septa (*see* major septum) (H, R, S).

MINOR SEPTUM. One of the relatively short septa that commonly are inserted between adjacent major septa (R, S).

monacanth. Simple trabecula in which fibers are related to a single center of calcification) (R).

monocentric. Type of corallite formed by a monostomodaeal polyp (*see* dicentric, tricentric, polycentric) (S).

monostomatous. Single-mouthed; refers to solitary coral polyps.

monostomodaeal. Polyp stomodaea each having its own tentacular ring after originating by di- or tristomodaeal budding (S).

monticule. Protuberant portion of corallum surface produced in circummural budding (S).

MURAL PORE. Circular or oval small hole in wall between adjoining corallites, as in some tabulates (T).

naotic septum. Type characterized by development peripherally in a series of closely spaced dissepiment-like plates, as in *Naos* (*see* other types: acanthine, amplexoid, dilated, fenestrate, laminar, lonsdaleoid, retiform, rhopaloid) (R).

NEMATOCYST. Stinging or adhesive body characteristic of cnidarians.

OFFSET. New corallite in corallum formed by "budding" (H, R, S, T).

open fossula. Prominent interseptal space not inclosed toward axis of corallite by united edges of septa (*see* fossula, closed fossula) (R).

OPERCULUM. Lidlike covering of calice in some corallites, formed of one or more independent plates (R).

oral disc. Aboral part of coral polyp above column, its center containing the mouth (*see* basal disc).

orthocneme. Longitudinal muscle sheet produced on surface of a sterigmatocneme, by cleavage and elongation toward axis of coral polyp giving rise to complete mesenteries.

pair. Adjacent mesenteries with pleats facing one another (S).

paliform lobe. Structure closely resembling a palus but formed by detached trabecular offset from inner edge of a septum, appearing in vertical succession and differing from pali in not being formed as a result of substitution (*see* palus) (S).

PALUS (pl., pali). Vertical lamella or pillar developed along inner edge of certain entosepta, comprising remnant part of a pair of exosepta joined at their inner margins (*see* paliform lobe) (S).

paratheca. Corallite wall formed by closely spaced dissepiments (*see* septotheca, synapticulotheca) (S).

parietal columella. Same as trabecular columella (S).

parricidal budding. Production of new polyp from inner surface of fragment of parent consisting of wedge-shaped part split off lengthwise (S).

patellate. Low solitary corallite with sides expanding from apex at angle of about 120 degrees (*see* calceoloid, cylindrical, discoid, pyramidal, scolecoid, trochoid, turbinate) (R, S).

perforate. Type of wall between corallites of some colonies characterized by presence of many irregularly arranged small openings through wall (S, T).

PERIPHERAL INCREASE. Budding of corallites in coralla characterized by offsets that arise in marginarial or coenenchymal tissue (*see* axial increase, lateral increase, intermural increase) (R, T).

peristome. Part of oral disc surrounding mouth of coral polyp.

phaceloid. Fasciculate corallum having subparallel corallites (*see* dendroid, foliose, massive, ramose, umbelliferous) (R, S, T).

PLANULA. Free-swimming larval stage of coral polyp.

pleat. Longitudinal fold of retractile muscle fibers with associated mesogloea on side of a mesentery.

plocoid. Massive corallum in which corallites have separated walls and are united by costae, dissepiments or coenosteum (term originally used with different meaning for Rugosa, but since abandoned) (S).

polycentric. Type of corallite formed by polyp retaining polystomodaeal condition permanently (*see* monocentric, dicentric, tricentric) (S).

POLYSTOMODAEAL BUDDING. Having more than 3 stomodaea developed within a common tentacular ring (*see* distomodaeal, tristomodaeal, triple stomodaeal types of budding and intramural, circumoral, and circummural types of polystomodaeal budding) (S).

Pourtalès plan. Arrangement of septa in some scleractinians (notably dendrophylliids) characterized by much greater development of exosepta than that of entosepta (S).

protocneme. One of mesenteries of first 12 (forming 6 mesenteric couples and pairs) developed in early ontogeny of the coral polyp (*see* metacneme) (S).

protocorallite. First-formed corallite of a colony (R, S, T).

PROTOSEPTUM. One of 6 first-formed septa of a corallite (H, R, S).

proximal. Direction toward point of origin of a corallite (*see* distal) (H, R, T).

pyramidal. Solitary corallite with flattened sides that meet at angles (*see* calceoloid, cylindrical, discoid, patellate, scolecoid, trochoid, turbinate) (R).

QUADRANT. Space in interior of a corallite bounded by cardinal septum and an alar septum or by counter septum and an alar septum (R).

radiciform process. Rootlike epithecal outgrowth of corallite wall, serving for fixation (*see* talon) (R, S, T).

ramose. Branching form of cerioid, plocoid, thamnasterioid, hydnophoroid, or meandroid corallum (*see* dendroid, fasciculate, foliose, massive phaceloid, reptant, umbelliferous (R, S, T).

reclined. Corallite growing and opening obliquely with respect to surface of corallum (*see* alveolitoid) (T).

regular dissepimentarium. Type in which dissepiments are developed only in spaces between major and minor septa (*see* herringbone dissepimentarium (R).

reptant. Corallite with creeping habit, growing attached along one side to some foreign body (*see* dendroid, fasciculate, foliose, massive, phaceloid, ramose, umbelliferous) (R, S, T).

reptoid. Creeping (*see* reptant).

retiform septum. Type of perforate septum composed of an irregular network of skeletal tissue (*see* other types: acanthine, amplexoid, dilated, fenestrate, laminar, lonsdaleoid, naotic, rhopaloid) (R).

rhabdacanth. Trabecula characterized by shifting centers of fibrous growth grouped around a main one (R).

rhopaloid septum. Type characterized by distinctly thickened axial edge, appearing club-shaped in cross section (*see* other types: acanthine, amplexoid, dilated, fenestrate, laminar, lonsdaleoid, naotic, retiform) (R).

rootlet. See radiciform process (R, S).

scale. Small platelike structure attached to septal grooves and interseptal ridges in some corallites, as *Tryplasma* (R).

SCLERENCHYME. Calcareous tissue of corallites; term used especially for notably thickened parts of skeleton (*see* stereome) (R).

sclerodermite. Center of calcification and surrounding cluster of calcareous fibers; apparent primary element in septa (S).

scolecoid. Solitary corallite of cylindrical type but bent irregularly in wormlike manner (*see* calceoloid, cylindrical, discoid, patellate, pyramidal, trochoid, turbinate) (R, S).

SEPTAL CYCLE. All septa belonging to a single stage in ontogeny of corallite as determined by order of appearance of septal groups, 6 protosepta comprising first cycle and later-formed exosepta and entosepta in constantly arranged succession being introduced in sextants (S).

septal dentation. Same as septal tooth (S).

SEPTAL GROOVE. Longitudinal furrow on outer side of corallite wall, corresponding in position to a septum on inner side of wall (*see* interseptal ridge) (H, R).

septal lamella. Radially disposed longitudinal plate in axis of a corallite aligned with a septum but not joined to it (R).

septal tooth. Small projection along upper margin of a septum formed by extension of a trabecula beyond calcareous tissue connecting it with others (S).

septotheca. Corallite wall formed by thickened outer parts of septa along axis of trabecular divergence (*see* paratheca, synapticulotheca) (S).

SEPTUM. Radially disposed longitudinal partition of corallite occurring between or within mesenterial pairs (*see* acanthine, alar, amplexoid, cardinal, counter, counter-lateral, dilated, entoseptum, exoseptum, fenestrate, laminar, lonsdaleoid, major, metaseptum, minor, naotic, retiform, rhopaloid) (H, R, S, T).

simple synapticula. Type formed by union of 2 opposed granulations (*see* compound synapticula) (S).

simple trabecula. Composed of a series of single sclerodermites (*see* compound trabecula) (S).

SOLITARY. Corallite of polyp not forming part of a colony (H, R, S).

solum. Small subhorizontal plate in tubule of coenenchyme, as in *Heliolites* (T).

spongy columella. See trabecular columella (S).

squamula. Small plate projecting subhorizontally in eavelike manner from wall of corallite toward axis, as in *Emmonsia* (T).

STEREOME. More or less dense calcareous skeletal deposit, generally covering and thickening various parts of corallite (S).

STEREOZONE. Area of dense skeletal deposits in a corallite, generally peripheral or subperipheral in position (R, S, T).

sterigmatocneme. Longitudinal muscle sheet appearing as bulge in endoderm of coral polyp's body wall forming base for development of a mesentery.

stomodaeum. Esophagus-like tubular passageway or pharynx leading from mouth of coral polyp to gastrovascular cavity.

styliform columella. Solidly fused and longitudinally projecting axial structure, in scleractinians with entosepta fused to it by secondary stereome (*see* other types: fascicular, lamellar, trabecular) (R, S).

substitution. Replacement of a temporary exoseptum by a permanent entoseptum (S).

synapticula. Small rods or bar connecting opposed faces of adjacent septa and perforating mesenteries between them (*see* types: simple, compound) (S).

synapticulotheca. Corallite wall formed by one or more rings of simple or compound synapticulae along axis of trabecular divergence (*see* paratheca, septotheca) (S).

TABELLA. Small subhorizontally disposed plate in central part of corallite forming part of an incomplete tabula (R, T).

TABULA. Transverse partition of corallite, nearly plane, or upwardly convex or concave, extending to outer walls or occupying only central part of corallite (H, R, S, T).

tabular dissepiment. Somewhat flat plates extending across an entire scleractinian corallite or confined to its axial part (*see* tabula) (S).

TABULARIUM. Axial part of the interior of a corallite in which tabulae are developed (H, R, S, T).

talon. Buttress produced by outgrowth of corallite wall, serving as aid in fixation (*see* radiciform process) (R).

TENTACLE. Movable tubular extension of soft integument rising from oral disc of coral polyp, closed terminally at tip, commonly simple but rarely forked (*see* entotentacle, exotentacle).

thamnasterioid. Massive corallum characterized by absence of corallite walls and by confluent septa that join neighboring corallites together, with pattern of septa resembling lines of force in a magnetic field (*see* aphroid, astreoid, cerioid, meandroid, plocoid; also footnote, p. F239) (R, S).

theca. See wall.

TRABECULA. Pillar of radiating calcareous fibers comprising skeletal element in structure of septum and related components (*see* simple trabecula, compound trabecula, monocanth, rhabdacanth) (H, R, S, T).

trabecular columella. Spongy axial structure formed of trabeculae loosely joined with synapticulae or paliform lobes (*see* other types: fascicular, lamellar, styliform) (S).

trabecular linkage. Connection between corallite centers corresponding to indirect linkage of stomodaea (*see* direct linkage, indirect linkage, lamellar linkage) (S).

trabecular spine. More or less completely isolated trabeculae in plane of a septum; may be vertically discontinuous with interruption by endotheca (S).

transverse division. Formation of new coral polyps by separation of parent by splitting into 2 parts transverse to oral-aboral axis (S).

tricentric. Type of corallite formed by polyp retaining tristomodaeal condition permanently (*see* monocentric, dicentric, polycentric) (S).

triple stomodaeal budding. Same as tristomodaeal budding except that 3 stomodaea invariably form triangle and only one interstomodaeal couple of mesenteries occurs between each pair of neighboring stomodaea (*see* distomodaeal, tristomodaeal, and polystomodaeal budding) (S).

tristomodaeal budding. Having 3 stomodaea developed within a common tentacular ring, occurring in series or forming a triangle, with 2 interstomodaeal couples of mesenteries between original and each new stomodaeum (*see* distomodaeal, triple stomodaeal, and polystomodaeal types of budding) (S).

trochoid. Solitary horn-shaped corallite with sides regularly expanding from apex at angle of about 40 degrees (*see* calceoloid, cylindrical, discoid, patellate, pyramidal, scolecoid, turbinate) (R, S).

turbinate. Solitary horn-shaped corallite with sides expanding from apex at angle of about 70 degrees (*see* calceoloid, cylindrical, discoid, patellate, pyramidal, scolecoid, trochoid) (R, S).

tympanoid. Squat drum-shaped form of corallite (S).

umbelliferous. Corallum having corallites arranged like ribs of an umbrella, growing outward in whorls (T).

ventral. In oral aspect of polyp, direction toward which the pleats of the first couples of lateral protocnemes face (S).

WALL. Skeletal deposit inclosing column of polyp and uniting outer edges of septa; it is variously formed in different corallites (*see* septotheca, paratheca, synapticulotheca) (S).

yard-arm carinae. Oppositely placed carinae that give appearance of yard-arms along a mast to cross sections of septum (*see* zigzag carinae) (R).

zigzag carinae. Not quite oppositely placed carinae on the 2 sides of a septum (*see* yard-arm carinae) (R).

zooxanthella. Symbiotic unicellular yellow-brown protistan in endoderm of hermatypic coral polyps.

CLASSIFICATION

MAIN FEATURES

General agreement has not yet been reached on subdivision of the Rugosa. The rugose coral studies of the last half century, led by STANLEY SMITH, have been concerned very largely with the fixing and precise description of type species and type specimens, and with the description of stratigraphical assemblages, all based on thin sections of coralla as an essential preliminary to systematic classification (33). Much further such work is required before safe conclusions can be drawn, for we are still ignorant of the precise internal structure of the type specimens of many species; and perhaps we would be well advised to wait yet longer before attempting systematic classification into families and superfamilies. Nevertheless, several such attempts, all differing widely, have been made recently, and another is made herein. The classification here proposed has benefited from correspondence with Dr. SMITH, and Dr. M. LECOMPTE, and from study of the published views of WANG (53), STUMM (52) and others, though there are still differences of opinion between us. It does not seem as if any single character, such as septal structure, is the key to subdivision of the Rugosa.

It is considered herein that three suborders of rugose corals may be recognized, each with characteristic marginaria and tabulae; these are termed Streptelasmatina, Columnariina, and Cystiphyllina. The Streptelasmatina may be divided into two superfamilies, the Cyathaxoniicae and Zaphrenticae, and while the former appear to have had a common descent, parts of the latter probably derived from different Cyathaxoniicae, as well as from *Streptelasma* itself. It is not improbable that all Streptelasmatina developed from *Lambeophyllum,* but our knowledge of the morphology of this oldest of Rugosa genera is not precise. The Scleractinia may well have developed from Permian Cyathaxoniicae. In the Streptelasmatina, the tabulae are fundamentally domed; in the Cyathaxoniicae, the marginarium, if developed, is a narrow to moderately wide stereozone, while in the Zaphrenticae it may be a wide septal stereozone or a dissepimentarium in which the dissepiments are small, globose, and interseptal, not disrupting the major septa. The septa of the Zaphrenticae show many different complex arrangements of the trabeculae and they may have lobed axial edges; an axial structure is common, particularly in late forms.

The Columnariina are possibly all descended from *Columnaria alveolata* GOLDF. Their tabulae are horizontal or broad flat-topped domes, or axially sagging; in late forms an axial structure develops. Marginaria, where developed in width, are distinctive; a single series of impersistent dissepiments may form, or more commonly, a wide septal stereozone may be replaced by long lonsdaleoid dissepiments which disrupt the major septa. None have septa with lobed axial edges and typically the septa are thin in the tabularium.

The Cystiphyllina are possibly all derived from Ordovician *Tryplasma;* they are characterized by acanthine septa and are divided into 4 families. The Tryplasmatidae have complete flat tabulae and a marginal stereozone in which the cores of the trabeculae appear set in lamellar sclerenchyme. In the Cystiphyllidae, in which the tabulae are incomplete and inversely conical, the marginarium is a dissepimentarium with discrete trabeculae based on the dissepiments. If WEDEKIND'S (54, 55) biostratigraphical conclusions are correct, in the Devonian these discrete trabeculae united to form the partly laminar septa of the Digonophyllidae. This suborder died out with the Devonian.

A tabular summary of suprageneric divisions recognized in the order Rugosa showing stratigraphic range of each as now known and including record of the number of genera and subgenera recognized in each taxonomic unit follows. The presence of 8 genera in a family, for example, is indicated by "(8)" and occurrence of 4 genera and 8 subgenera in a family or other taxonomic division is shown as "(4; 8)."

Suprageneric Divisions of Rugosa

Rugosa *(order),* septa introduced in 4 quadrants with tendency to produce bilateral symmetry (333; 12). *Ord.-Perm.*

Streptelasmatina *(suborder),* marginarium com-

prising a septal stereozone or dissepimentarium, tabulae mostly domed (218). *Ord.-Perm.*
Cyathaxoniicae *(superfamily)* (77). *Ord.-Perm.*
Metriophyllidae (18). *Ord.-Perm.*
Laccophyllidae (4). *Sil.-Perm.*
Petraiidae (2). *Sil.-Dev.*
Polycoeliidae (21). *?Sil., Dev.-Perm.*
Polycoeliinae (10). *?Sil., Carb.-Perm.*
Plerophyllinae (6). *M.Dev.-Perm.*
Endotheciinae (1). *Perm.*
Tachylasmatinae (4). *Carb.-Perm.*
Hadrophyllidae (8). *L.Dev.-M.Carb.*
Cyathaxoniidae (2). *Carb.-Perm.*
Amplexidae (1). *L.Carb.*
Lophophyllidiidae (5). *Carb.-Perm.*
Timorphyllidae (5). *L.Carb.-L.Perm.*
Hapsiphyllidae (11). *L.Carb.-L.Perm.*
Zaphrenticae *(superfamily)* (141). *Ord.-Perm.*
Streptelasmatidae (24). *M.Ord.-M.Dev.*
Streptelasmatinae (18). *M.Ord.-M.Dev.*
Kodonophyllinae (4). *L.Sil.-L.Dev.*
Acrophyllinae (2). *L.Dev.*
Halliidae (12). *Ord.-Dev.*
Lykophyllinae (8). *Ord.-Sil.*
Halliinae (4). *L.Dev.-M.Dev.*
Arachnophyllidae (8). *Sil.-Dev.*
Arachnophyllinae (6). *Sil.*
Kyphophyllinae (1). *Sil.*
Ptychophyllinae (1). *Sil.*
Acervulariidae (3). *Sil.*
Mycophyllidae (4). *M.Sil.-L.Dev.*
Zaphrentidae (9). *Dev.*
Phillipsastraeidae (21). *M.Sil.-L.Carb.*
Phillipsastraeinae (14). *Sil.-L.Carb.*
Phacellophyllinae (7). *Dev.*
Craspedophyllidae (1). *Dev.*
Lithostrotionidae (13). *L.Carb.-L.Perm.*
Aulophyllidae (24). *Carb.-Perm.*
Aulophyllinae (15). *Carb.-Perm.*
Amygdalophyllinae (7). *L.Carb.*
Yatsengiinae (2). *L.Perm.*
Cyathopsidae (19). *L.Carb.-Perm.*
Calostylidae (3). *Sil.*
Columnariina *(suborder)* (60; 2). *Ord.-Perm.*
Stauriidae (18). *Ord.-Dev., ?L.Carb.*
Spongophyllidae (5). *M.Sil.-U.Dev.*
Chonophyllidae (14). *Sil.-Dev.*
Chonophyllinae (5). *Sil.*
Endophyllinae (8). *?Sil., Dev.*
Blothrophyllinae (1). *M.Dev.*
Ptenophyllidae (10). *Sil.-Dev.*
Stringophyllidae (1). *L.Dev.-M.Dev.*
Lonsdaleiidae (12; 2). *Carb.-Perm.*
Lonsdaleiinae (6). *Carb.*
Waagenophyllinae (6; 2). *L.Perm.*
Cystiphyllina *(suborder)* (30; 10). *Ord.-Dev.*
Tryplasmatidae (9). *Ord.-Dev.*
Cystiphyllidae (7). *Sil.*
Goniophyllidae (6). *Sil.-Dev.*
Digonophyllidae (8; 10). *Dev.*
Zonophyllinae (6). *L.Dev.-M.Dev.*
Digonophyllinae (2; 10). *M.Dev.*
Incertae sedis (25). *Ord.-Perm.*

STRATIGRAPHIC DISTRIBUTION

SUMMARY STATEMENT

The Rugosa are first known in Blackriveran strata of Middle Ordovician age in North America; they immediately spread round the globe, but remained relatively unimportant until the late Early Silurian and Middle Silurian, which were times of active reef-building and rapid evolution of new forms. They are less abundant in Upper Silurian and lower Lower Devonian strata but late Early Devonian and Middle Devonian times once more produced a great wealth of new forms. The Late Devonian was in most parts a lean epoch, but the Early Carboniferous was another time of great activity in coral evolution. Until the Devonian the Rugosa was in general less important in numbers of individuals than the Tabulata, and during the Devonian approximate equality was attained; but the Early Carboniferous accession of coral vitality did not affect the Tabulata, which then became relatively unimportant. Middle and Late Carboniferous times saw corals flourishing in only a few regions; but in the Artinskian there was a fresh access of vigor, this time on a much-lessened scale; by the end of the Artinskian practically all of the Rugosa, save members of the Cyathaxoniicae, had died out; and though these had failed by the end of the Permian, it was apparently from this group that the Mesozoic, Tertiary, and Recent Scleractinia arose (42).

RELATIVE IMPORTANCE OF RUGOSA IN SUCCESSIVE PERIODS

ORDOVICIAN

The Rugosa first appeared in Blackriveran strata in the shallow seas associated with the Appalachian geosyncline and it is quite probable that all three suborders were represented therein, the Streptelasmatina by *Lambeophyllum,* the Columnariina by nonspinose *Columnaria,* and the Cystiphyllina by spinose "*Columnaria.*" They are thus of later appearance than the Tabulata, and the possible relation between Tabulata and Rugosa is to be sought in these early North American strata.

During the Ordovician, the Rugosa are fewer in species and individuals than the Tabulata, and all are without dissepimentaria, the only type of marginarium developed being a rather narrow septal stereozone. Further, the tabulae are complete or represented each by only 2 or 3 tabellae. Streptelasmatina and Stauriidae are common, Tryplasmatidae rare.

SILURIAN

Early in the Silurian the curious Calostylidae appeared, forms with perforate septa, reticulate axial structures, and marginaria unknown in other forms. Perforate septa are common in the Scleractinia; the Calostylidae seem to have tried this mode in the Silurian, but were unsuccessful and left no descendants in the Devonian.

Four bursts of evolutionary activity have occurred in the Rugosa since their appearance in the Middle Ordovician. The first of these was in late Llandovery times, when the three suborders separately evolved dissepimentaria. Thus, in the Streptelasmatina, we find the Lykophyllinae and the Arachnophyllinae with their characteristic small interseptal dissepiments; in the Columnariina, the early endophylloid, *Strombodes contortiseptatum* Dybowski, developed its characteristic lonsdaleoid dissepiments which disrupt the major septa in a septal stereozone; and in the Cystiphyllina, the Cystiphyllidae and Goniophyllidae entered with their cystlike plates and discrete trabeculae. The Middle Silurian saw a continuation of this burst; all these families, together with the Streptelasmatidae, Metriophyllidae, Columnariina, and Tryplasmatidae, which continued on from the Ordovician, developed many new forms; and the 2 characteristically Devonian families, Ptenophyllidae and Phillipsastraeidae, originated, though their representatives remained rare.

Whether this first burst of evolution was promoted by exploitation of the reef habitat is unclear; there may have been some connection. Coral reefs were already known in the Ordovician, though rarely, if we understand by reefs, structures which grew above the level of the surrounding sea floor, maintaining their upward growth against wave-base erosion by the activities of their framework organisms (algae, stromatoporoids, or corals) and by filling in the interstices in the framework by lime flour, shells, echinoderms, and their detritus; but the Middle and Upper Silurian rocks contain many such reefs, in which, however, Rugosa played a part minor to that of Stromatoporoidea, Tabulata, and probably also of algae. In these Silurian reefs, solitary as well as colonial Rugosa occur.

DEVONIAN

The next climax of evolutionary activity is not clearly to be correlated with the assumption of some new morphological feature like the dissepiment in the first climax; most of its vitality of change went into the wealth of families and species produced in the late Early and early Middle Devonian, and cannot as yet be bracketed with climatic or other geographical factors. In the Zaphrenticae, the new subfamily Halliinae and the Zaphrentidae appeared, probably first in America, and the Phillipsastraeidae developed in great fecundity and variability, experimenting with the horseshoe dissepiment and with complex arrangements of trabeculae in the septa. The Streptelasmatidae themselves produced one or 2 genera with the small globose interseptal dissepiments of other families. Cyathaxoniicae diversified a little also. In the Columnariina, the Stauriidae and Endophyllinae were only moderately active, but the Ptenophyllidae reached their acme, experimenting like the new Stringophyllidae in septal arrangement and structure, and dying out in the Late Devonian. In the Cystiphyllina, the Tryplasmatidae waned and were extinct by the late Middle Devonian; but the record shows a great burst of activity in the Digonophyllidae, which, according to Wedekind, experimented in the production of laminar septa from discrete trabeculae. These also became extinct before the end of the Devonian.

CARBONIFEROUS

The transition from Devonian to Carboniferous coincided with the greatest transformation of rugose faunas, the chief feature being the development generally of axial structures in all 3 suborders. The primitive Cyathaxoniicae alone continued the slow evolution of its old families, but one new family, the Timorphyllidae, developed an axial structure. Of the Zaphrenticae, the older families Streptelasmatidae, Halliidae, Zaphrentidae, and Phillipsastraeidae seem to have become extinct, but 2 new families, Lithostrotionidae and the Aulophyllidae, each characterized by axial structure, developed. In the Columnariina, also, the Stauriidae and Endophyllinae became extinct; the new family is the Lonsdaleiidae with axial structures.

PERMIAN

The fourth great change in the Rugosa was the final disappearance, during the Artinskian, of all families except those of the Cyathaxoniicae, which were distinguished by having their protosepta developed more strongly than the metasepta, and by characteristic rhopaloid septa.

The earliest Scleractinia may have developed directly from these Polycoeliidae as surmised by SCHINDEWOLF (42), by the general assumption of dominance by the 6 protosepta, the limiting of the minor septa to 6, and the insertion of tertiary and later septa in the counter quadrants as well as in the cardinal and alar quadrants. The last Rugosa were extinct by the end of the Permian.

ECOLOGY

Like the compound Scleractinia today, compound Rugosa frequently acted as framework organisms in reefs; whether these profuse growths then, as now, occurred only in warm shallow seas is still for consideration; the sedimentary environment of the Paleozoic reefs certainly indicates a depth of formation quite similar to that of today, and, in the Carboniferous, at least, profuse coral growth, and in some places reef growth apparently occurred in warm-water seas (23).

In the Paleozoic too, it seems that small, solitary corals (Cyathaxoniicae) were able to exist in numbers in conditions where the large compound corals could not flourish, the sedimentary environment suggesting deeper seas with less light; while in the Carboniferous at least the large, solitary Rugosa, chiefly Zaphrenticae, appear to have flourished best in intermediate environments (18).

ONTOGENY

Studies on the growth of individual corallites have been practically confined to solitary corals, and have concentrated in the main on the position and order of insertion of the septa. Fundamentally, it seems that 6 protosepta are first inserted, but their order of appearance is not invariable (18); a common mode is cardinal and counter together and conjoined, then the 2 alar septa, and last the 2 counter-lateral septa. Thereafter, major septa of the Rugosa are inserted in only 4 positions, on either side of the cardinal septum and on the counter side of the alar septa; minor septa may appear with these major septa, or may delay and be inserted as a cycle between them.

Another aspect of ontogenetic studies is the indication they may give of phylogeny. CARRUTHERS' (4) work on the Carboniferous cyathaxoniicid *Zaphrentites delanouei* shows that the arrangement of septa characteristic of the adult appears in the pre-adult stage in the ontogeny of its descendent species; but the number of such ancestors reflected in the ontogeny is limited. Individuals of a species share a characteristic ontogeny.

Dissepimentaria do not appear in the earliest stages; nor do complex axial structures.

Rejuvenescence (reversion of the corallites to an earlier ontogenetic stage) is common in Rugosa, and is often associated with a contraction in diameter; sometimes too, the marginarium of the calice is deserted by the polyp at the time of rejuvenescence, and grows again only as the diameter increases.

EVOLUTIONARY TRENDS

(1) The most important of the trends in development of the Rugosa seems to have been the production of a marginarium between the epitheca and tabularium. Many kinds of marginarium were evolved, but they may be grouped into 3 main types:
   i. A septal stereozone, formed by the expansion to contiguity of the peripheral edges of the septa.
   ii. A regular dissepimentarium, in which a series of small highly arched dissepiments with troughlike upper surfaces developed in the loculi between major and minor septa.
   iii. A lonsdaleoid dissepimentarium, in which a series of moderately large blister-like dissepiments disrupt both major and minor septa; in many they seem to have developed within a septal stereozone.

(2) A universal trend is toward the replacement of complete tabulae by tabellae.

(3) Another trend is carried on in a great variety of ways; it is towards complication in arrangement of the trabeculae within the septa, and this affects the Streptelasmatina particularly. In the Cystiphyllina it is a trend by which laminar septa are developed from the discrete trabeculae of the primitive members of the suborder. A manifestation of this trend also may be responsible for the development in many late Paleozoic Rugosa of all the fibers at right angles to the median plane of the septum. A few Cyathaxoniicae are affected by a rhopaloid swelling of the axial edges of the septa.

(4) A trend which is unimportant from the point of view of number of genera affected is seen in the predominance of the protosepta over the remaining septa in some Cyathaxoniicae.

(5) Also unimportant in the number of genera affected, is a 5th trend, toward development of a 3rd order of septa (tertiary septa) between the major and minor septa, at different times, as in the Silurian *Acervularia* and the Artinskian *Iranophyllum*.

(6) Similarly, the development of perforate septa in the Silurian Calostylidae may be recognized as a trend. This and trends 4 and 5 affecting the septa seem to be forerunners of developments characteristic in Scleractinia.

(7) Another trend affecting the septa, which may appear in any lineage, is their withdrawal from the axis, generally combined with masking of their original pinnate arrangement and with their thinning.

(8) A very important development is the axial structure. This forms infrequently before the Early Carboniferous, but it then appears in most families, either as a simple axial rod, continuous or discontinuous with the counter septum, or as a complex of tabellae and septal lamellae. Axial structures may disappear in an apparent reversal of this trend, the septa withdrawing from the axis, and the tabulae flattening.

(9) The formation of compound from solitary coralla may be noted as a trend, culminating in such compaction of the corallites that plocoid coralla are developed. Stratigraphic evidence suggests that solitary corals may evolve from compound, in an apparent reversal of this trend.

SYSTEMATIC DESCRIPTIONS

Order RUGOSA Milne-Edwards & Haime, 1850

[*nom. transl.* NICHOLSON, 1872 (*ex* suborder Rugosa M.EDW.-H., 1850)] [=suborder Stauracea VERRILL, 1865; subclass Tetracorallia HAECKEL, 1866 (*partim*); subclass Tetracoralla HKL., 1870 (*partim*); order Pterocorallia FRECH, 1890; order Tetraseptata GRABAU, 1913 (*non* HAACKE, 1879); Tetracoelia YABE & SUGIYAMA, 1940]

Solitary or compound epithecate corals with septa typically in 2 orders alternating in length but in some with 1 or 3 orders; symmetry bilateral; in solitary corallites, after insertion of the first 6 protosepta with metasepta inserted in 4 positions only, on the counter side of each alar septum and on each side of the cardinal septum. A marginarium, consisting of a dissepimentarium or a peripheral stereozone formed by thickening of the septa, may be developed around the tabularium in the zone of minor septa. Tabulae may be conical, domed, horizontal, sagging (in some with a median trough), or inversely conical; each tabula may be complete, consisting of one plate, or incomplete, consisting of a number of tabellae. An axial structure may develop. *Ord.-Perm.*

Diversity of structure in the Rugosa is great, but the order seems divisible into 3 suborders in which diagnostic characters are shape of the tabulae, type of marginarium, arrangement of the septa, and arrangement of trabeculae within the septa. Nevertheless, a character dominant in one superfamily may appear in another.

Suborder STREPTELASMATINA Wedekind, 1927

[*nom. correct.* HILL, 1954 (*pro* suborder Streptelasmacea WDKD., 1927)] [Includes ?Cyathophylloidea, Zaphrentoidea NICHOLSON, in NICH. & LYDEKKER, 1889]

Solitary or conical Rugosa with marginarium a septal stereozone or dissepimentarium composed typically of small, globose interseptal dissepiments; tabulae typically domed. *Ord.-Perm.*

Superfamily CYATHAXONIICAE Milne-Edwards & Haime, 1850

[*nom. transl.* HILL, herein (*ex* Cyathaxoniidae, *nom. correct.* STUMM, 1949, *pro* Cyathaxonidae M.EDW.-H., 1850)] [Includes Semiplana POČTA, 1902; =Metriophyllacea HUDSON, 1945]

Small solitary corals with marked longitudinal costation; marginarium a very narrow septal stereozone, without dissepiments and with tabulae declined from the axis. Major septa are at first long and grouped in quadrants joining in the axial region to form an aulos (Laccophyllidae) or vertical rodlike columella free from the counter septum throughout growth (Cyathaxoniidae); septa in cardinal quadrants may be thicker and fewer than in counter quadrants; cardinal septum may shorten and counter septum be very long and produced vertically at the axis as a columella (Lophophyllidiidae); septa may withdraw unequally from union at the axis and may have their axial edges swollen (Polycoeliidae); minor septa may be long and contratingent, leaning on or toward adjacent major septa on the counter side; major septa may be flanged parallel to their distal edges but are without axial lobes. *Ord.-Perm.*

Family METRIOPHYLLIDAE Hill, 1939

[=?Lindstroemiidae POČTA, 1902 (*nom. correct.* GRABAU, 1928, *pro* Lindströmiidae POČTA, 1902)]

Small, solitary Rugosa with marginarium a very narrow stereozone. All major septa unite at the axis with axial end of counter septum swollen laterally in some forms; cardinal fossula on the convex side of corallum and false counter fossula opposite; minor septa short; the septa may be flanged parallel to their calical edges. Tabulae distant. *Ord.-Perm.*

Lambeophyllum OKULITCH, 1938 [**Cyathophyllum profundum* CONRAD, 1843]. Corallum with very deep calice; septa in young stages very short, later uniting in quadrants which join the unswollen axial end of counter septum; cardinal septum inconstant; septa not flanged. *M.Ord.*, N.Am.——FIG. 174,*1*. **L. profundum* (CONRAD); *1a,b,* transv. secs., ×2 (62).

Duncanella NICHOLSON, 1874 [**D. borealis*]. Very small, slender, erect, apex without epitheca; major septa straight, unflanged, uniting axially; minor septa absent. *M.Sil.*, N.Am.

Asthenophyllum GRUBBS, 1939 [**A. orthoseptatum*]. Very small; major septa twisting slightly together to form an axial structure; minor septa short, tabulae rare. *M.Sil.*, N.Am.——FIG. 174,*2*. **A. orthoseptatum; 2a,b,* side views, ×2; *2c,d,* diagrams of septa (82).

Amplexiphyllum STUMM, 1949 [**Amplexus hamiltonae* HALL, 1876]. Septa withdrawn from axis except at apex, unflanged; tabulae distant, complete, shallow domes (52). *M.Dev.*, N.Am.——FIG. 174,*3*. **A. hamiltonae* (HALL); *3a,b,* transv. and long. secs., ×1 (127).

Stewartophyllum BUSCH, 1941 [**Amplexus intermittens* HALL, 1876]. Very small, major septa dilated, withdrawing from axis in adult stage; tabulae tall domes; no minor septa (52). *Up.M. Dev.*, N.Am.——FIG. 174,*4*. **S. intermittens* (HALL); *4b,a,c,* side view, transv. and long. secs., ×1 (63).

Metriophyllum M.EDW.-H., 1850 [**M. bouchardi*] [=*Lopholasma* SIMPSON, 1900; *?Paterophyllum* POČTA, 1902; *Lophelasma* LANG-S.-T., 1940 (*nom. van. pro Lopholasma*)]. Straight thick septa bearing horizontal flanges usually with upturned edges; axial edge of counter septum not swollen nor produced into a columella but union of septa at the axis forms a solid axial pillar (46). *M.Dev.-U.Dev.*, N.Am.-Eu.-Austral.——FIG. 174,*5*. **M. bouchardi,* U.Dev., Eu.; *5a-c,* transv. and 2 long. secs., ×1 (117).

Stereolasma SIMPSON, 1900 [**Streptelasma rectum*

HALL, 1876 (*partim*)] [=*Stereoelasma* LANG-S.-T., 1940 (*nom. van. pro Stereolasma*)]. Like *Metriophyllum* but septa not flanged (52). *M.Dev.*, N.Am.——FIG. 174,*7*. **S. rectum* (HALL); *7a,b*, transv. and long. secs., ×2 (115).

Nalivkinella SOSHKINA, 1939 [**N. profunda*]. Major septa flanged, and withdrawn evenly from axis, leaving axial space in which tabulae sag; no aulos; minor septa very short (48). *U.Dev.*, USSR.——FIG. 174,*10*. **N. profunda; 10a-c*, 2 transv. secs. and long. sec., ×2 (122).

Buschophyllum STUMM, 1949 [**Caninia complexa* BUSCH, 1941]. Septa meet at axis in metriophylloid manner in all save adult stage when they withdraw, leaving an axial space; minor septa very short (52). *Up.M.Dev.*, N.Am.

Rotiphyllum HUDSON, 1942 [**Densiphyllum rushianum* VAUGHAN, 1908]. Septa dilated, with extra thickening in inner third of their course, joined at axis forming a dense axial structure; counter septum may be longer than others; minor septa long, contratingent (27). *L.Carb.*, Eu.-N.Am.; *L. Perm.(?Artinsk.)*, N.Z.——FIG. 174,*13*. **R. rushianum* (VAUGHAN) Visé., Eu.; transv. sec., ×2 (132).

Meniscophyllum SIMPSON, 1900 [**M. minutum*] [=?*Heptaphyllum* CLARK, 1924; ?*Caenophyllum* CLARK, 1926]. Like *Rotiphyllum* but with counter and counter-lateral septa thinning and withdrawing a little from axis, and without minor septa (10). *L.Carb.*, Eu.-N.Am.——FIG. 174,*8*. **M. minutum*, Miss., N.Am.; transv. sec., ×2 (115).

Disophyllum TOLMACHEV, 1924 [**D. symmetricum*; SD TOLMACHEV, 1933]. Counter septum and septa of counter quadrants become thin and withdraw slightly from axis in adult stages (18). *L.Carb.*, Sib.——FIG. 174,*9*. **D. symmetricum;* transv. sec., ×2 (131).

Bradyphyllum GRABAU, 1928 [**B. bellicostatum*] [=?*Heterelasma* GRABAU, 1922 (*non* GIRTY, 1908; *nec* LANG-S.-T., 1940, *nom. van. pro Heterolasma* EHLERS, 1919)]. Like *Metriophyllum* but without flanges on septa which withdraw from axis in late stages, the cardinal septum being then very short and the counter septum not longer than others; alar septa on convex and concave sides of corallum (13). *M.Carb.*, China.——FIG. 174,*12*. **B. bellicostatum; 12a,b*, transv. secs., ×3 (81).

Empodesma MOORE & JEFFORDS, 1945 [**E. imulum*]. Calice very oblique, septa dilated except distally, metriophylloid in arrangement only just above tabulae, withdrawing from the axis between tabulae into a calophylloid arrangement with the cardinal, counter, and 2 alar septa longer than others; cardinal and counter septa remain in contact longer (39). *Penn.*, N.Am.——FIG. 174,*11*. **E. imulum*, Tex.; *11a,b*, transv. secs., ×2 (104).

Stereocorypha MOORE & JEFFORDS, 1945 [**S. annectans*]. Like *Metriophyllum* but with fossula on concave side and septa without flanges (39). *Penn.*, N.Am.——FIG. 174,*14*. **S. annectans*, Tex.; *14a,b*, transv. secs., ×2 (104).

Duplophyllum KOKER, 1924 [**D. zaphrentoides* (*non Cyathophyllum? zaphrentoides* ETHERIDGE, 1891)]. Septa in early stages thin toward axis where they join but in adult stages are thickened axially and may withdraw; minor septa very long, contratingent; counter septum not enlarged (29). *Perm.(Artinsk.)*, Timor.——FIG. 174,*15*. **D. zaphrentoides;* transv. sec., ×2 (95).

?**Lindstroemia** NICHOLSON & THOMSON, 1876 [*nom. correct.* GRABAU, 1928 (*pro Lindströmia*)] [**L. columnaris*] [=?*Stereolasma* SIMPSON, 1900]. Small, with axial structure; septal thickening and arrangement as in *Stereolasma* (type specimens missing, 52). *Dev.*, N.Am.——FIG. 174,*6*. **L. columnaris; 6a,b*, transv. and long. secs., ×2 (105).

?**Centrocellulosum** THOMSON, 1883 [**C. densothecum*]. *L.Carb.(Visé)*, Scot.

?**Antiphyllum** SCHINDEWOLF, 1952 [**A. inopinatum*]. Septa not united axially; cardinal septum long and swollen axially; counter septum short. *U.Carb.(Namur.)*, Ger.

Family LACCOPHYLLIDAE Grabau, 1928

[=Syringaxonidae HILL, 1939; Amplexocarininae SOSHKINA in SOSHK.-D.-P., 1941]

Small, solitary Rugosa with axial ends of major septa united at an aulos which divides horizontal inner tabellae from inclined outer tabellae; minor septa contratingent, dissepiments absent. *Sil.-Perm.*

Syringaxon LINDSTRÖM, 1882 [**Cyathaxonia siluriensis* M'COY, 1850] [=*Laccophyllum* SIMPSON, 1900; *Alleynia* POČTA, 1902 (*pro Nicholsonia* POČTA, 1902, *non* KIÄR, 1899; *nec* SCHLÜTER, 1885; *nec* others)]. Septa and aulos considerably thickened (46). *Sil.-Dev.*, Eu.-N.Am.——FIG. 175,*2*. **S. siluriense* (M'COY), Sil., Eu.; *2a,b*, transv. and long. secs., ×1 (117).

Barrandeophyllum POČTA, 1902 [**B. perplexum*] [=*Retiophyllum* POČTA, 1902]. Septa and aulos thin, aulos commonly imperfect; minor septa may not form (40). *M.Dev.-U.Dev.*, Eu.-Austral.——FIG. 175,*1*. **B. perplexum*, M.Dev., Boh.; *1a,b*, transv. and long. secs., ×2 (108).

Trochophyllum M.EDW.-H., 1850 [**T. verneuili*] [=*Permia* STUCKENBERG, 1895; *Crassiphyllum* GROVE, 1935]. Septa and aulos extremely thickened; no minor septa (51). *L.Carb.*, Eu.-Asia-N. Am.——FIG. 175,*4*. *T. cavernulum* (HUDSON), Eng.; *4a,b*, transv. and long. secs., ×2 (89).

Amplexocarinia SOSHKINA, 1928 [**A. muralis*] [=*Amplexicarinia* LANG-S.-T., 1940 (*nom. van.*); ?*Paralleynia* SOSHK., 1936]. Septa and aulos thin (49). *Perm.*, Eu.——FIG. 175,*3*. **A. muralis*, Artinsk., USSR; *3a-c*, transv. secs., ×2 (123).

Family PETRAIIDAE de Koninck, 1872

[*nom. correct.* GRABAU, 1922 (*pro* Petraiadae KON., 1872)] [=Petrainae DYBOWSKI, 1873; Petraiaidae SANFORD, 1939]

Small solitary Rugosa with thin septa typically meeting at the axis, minor septa being inserted alternately with the metasepta; calice very deep. *Sil.-Dev.*

Petraia MÜNSTER, 1839 [**P. decussata* MÜNSTER, 1839 *(partim)*; SD MILLER, 1889 (=**P. radiata* MÜNSTER, 1839, *partim*)]. Septa very thin, minor septa long, contratingent; no septal flanges, tabulae few. *U.Sil.*, Ger.——FIG. 175,6. **P. radiata; 6a,* transv. sec., ×5; *6b,* side view, ×2 (113).

Orthophyllum POČTA, 1902 [**O. bifidum* BARRANDE in POČTA, 1902; SD LANG-S.-T., 1940]. Like *Petraia* but septa somewhat withdrawn from axis, order of septal insertion unknown (40). *L.Dev.-M.Dev.*, Boh.——FIG. 175,5. **O. bifidum,* L.Dev., Boh.; *5a,b,* transv. sec. and side view, ×1 (108).

Family POLYCOELIIDAE Roemer, 1883

[*nom. correct* SCHINDEWOLF, 1952 (*pro* Polycoelidae ROEMER, 1883, *ex Polycoelia* KING, 1849, =*Calophyllum* DANA, 1846,

FIG. 174. Streptelasmatina (Cyathaxoniicae): Metriophyllidae (p. F257-F258).

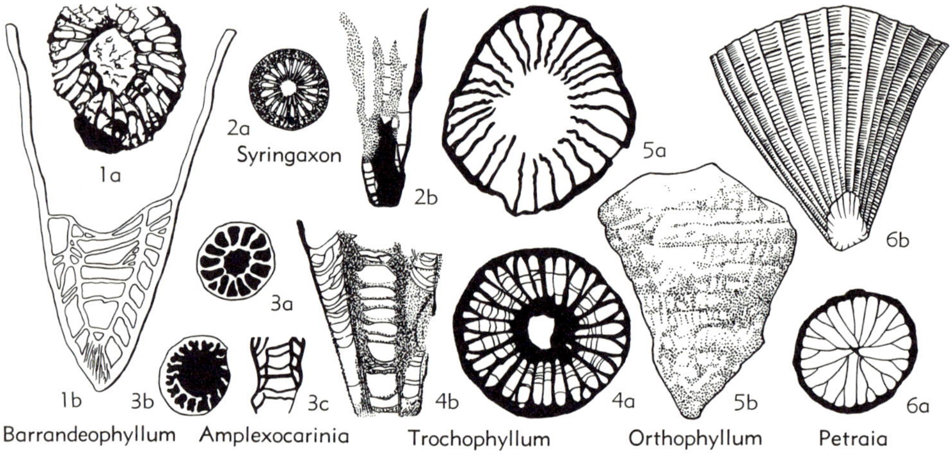

FIG. 175. Streptelasmatina (Cyathaxoniicae): Laccophyllidae, Petraiidae (p. F258-F259).

obj.)] [=Polycaeliens, Polycoeliens DE FROMENTEL, 1861 (invalid vernacular names); Plerophyllidae KOKER, 1924; Polycoelacea LECOMPTE, 1952 (superfam.); Calophyllidae STUMM, 1953]

Small solitary Rugosa in which some or all of the 6 protosepta are longer and thicker than other septa; counter septum not produced vertically into a columella; tabulae are tall domes flattened or slightly concave axially. Dissepiments absent except in highly developed genera. *?Sil., Dev.-Perm.*——FIG. 176, *1-4*. Young stages of subfamilies of Polycoeliidae, ×5 (113).

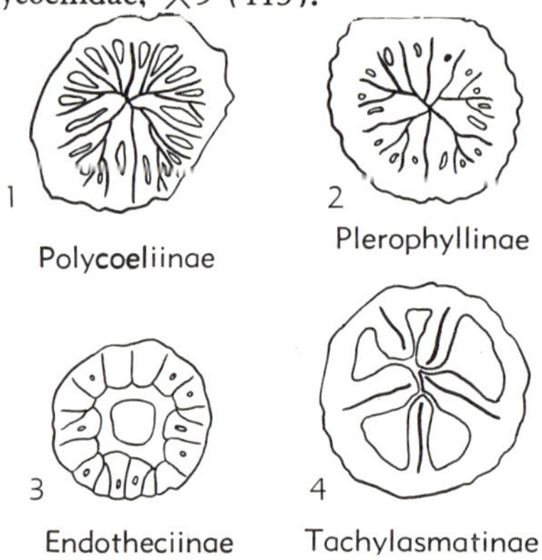

FIG. 176. Streptelasmatina (Cyathaxoniicae): Characters of subfamilies of Polycoeliidae illustrated by transverse sections (85).

Subfamily POLYCOELIINAE Roemer, 1883

[*nom. transl.* SCHINDEWOLF, 1942 (as Polycoelinae) *et correct.*, SCHIND., 1952 (*ex* Polycoelidae ROEMER, 1883)] [=Sochkineophyllinae GRABAU, 1928]

Four protosepta (cardinal, counter and 2 alar) longer and commonly thicker than other septa, especially near axial edges. *?Sil., Carb.-Perm.*

Claviphyllum HUDSON, 1942 [**Cyathopsis? eruca* M'COY, 1851]. Septa meeting axially in young stages; in adult stages all but the rhopaloid counter septum (which may be slightly produced vertically into a low columella) withdrawn from axis, but 2 neighboring lateral septa in each quadrant are outstanding and rhopaloid; cardinal septum short; tabellae globose (27). *L.Carb.*, Eu.——FIG. 177,*5*. **C. eruca* (M'COY); transv. sec., ×2 (89).

Fasciculophyllum THOMSON, 1883 [**F. dybowskii;* SD GREGORY, 1917]. Septa meeting axially in young stages, long, thin, possibly flanged and, except counter septum, only slightly withdrawn in adult stages (27). *L.Carb.*, Eu.——FIG. 177,*2*. **F. dybowskii;* transv. sec., ×2 (130).

Kinkaidia EASTON, 1945 [**K. trigonalis*]. Counter and 2 alar septa dominant, axial edges not markedly rhopaloid; cardinal septum short; metasepta short in young stages (28). *Miss.*, N.Am.——FIG. 177,*1*. **K. trigonalis;* transv. sec., ×3 (72).

Clinophyllum GROVE, 1935 [**Zaphrentis chouteauensis* MILLER, 1891]. Calical floor highly oblique, counter quadrants on longer side of corallum; counter septum dominant, cardinal septum strong, alar septa short; minor septa developed only between counter and counter-lateral septa (10). *Miss.*, N.Am.——FIG. 177,*3*. **C. chouteauense* (MILLER); *3a*, side view, ×1; *3b,c*, transv. and long. secs., ×3 (72).

Calophyllum DANA, 1846 [**C. donatianum* KING, 1850 (=**Turbinolia donatiana* KING, 1848 =**Cyathophyllum profundum* GEINITZ, 1842)] [=*Polycoelia* KING, 1849 (obj.) (*non* DE FROMENTEL, 1860, *nec* FUHRMANN, 1907); *Tetraphyllum* LUDWIG, 1865; *?Phryganophyllum* DE KONINCK, 1872; *Gerthia* GRABAU, 1928; *Pycnocoe-*

lia SCHINDEWOLF, 1952 (*pro Weissermelia* SCHINDEWOLF, 1942, *non* LANG-S.-T., 1940)]. Cardinal, counter and 2 alar septa equally developed, longer and thicker than others; septa united axially in quadrants in young stages (42). *L.Carb.-Perm.*, Eu.-Asia-Austral.——FIG. 177,*4*. **C. profundum* (GEINITZ), Perm., Eu.; *4a,b,* transv. secs., ×3 (113).

Tetralasma SCHINDEWOLF, 1942 [**T. quadriseptata*]. Like *Calophyllum* but with 2 counter-laterals and metasepta extremely short (42). *L.Carb.*, Ger.——FIG. 177,*6*. **T. quadriseptata; 6a,b,* transv. secs., ×3 (113).

Sochkineophyllum GRABAU, 1928 [**Plerophyllum artiense* SOSHKINA, 1925]. In late stages counter septum, 2 alar septa and 1 or 2 metasepta in each quadrant are outstanding, with slightly swollen axial edges; cardinal septum short (42). *L.Carb.-Perm.*, Eu.-Asia.——FIG. 177,*13*. **S. artiense* (SOSHKINA), L.Perm.(Artinsk.), USSR; *13a,b,* transv. and long. secs., ×2 (122).

Prosmilia KOKER, 1924 [**Plerophyllum cyathophylloides* GERTH; SD LANG-S.-T., 1940]. Like *Calophyllum* but with regular, moderately wide dissepimentarium in adult stages (42). *Perm. (Artinsk.)*, Timor.——FIG. 177,*11*. **P. cyathophylloides* (GERTH); *11a,b,* transv. and long. secs., ×2 (95).

?**Anisophyllum** M.EDW.-H., 1850 [**A. agassizi*]. Cardinal? and 2 alar septa taller, thicker and longer than others, meeting at axis; no columella (38). *Sil.*, N.Am.——FIG. 177,*8*. **A. agassizi; 8a,b,* side and calical views, ×1 (73).

?**Lophophrentis** CHI, 1935 [**L. trilobata*]. *Carb.*, China.

Subfamily PLEROPHYLLINAE Koker, 1924

[*nom. transl.* GRABAU, 1928 (*ex* Plerophyllidae KOKER, 1924)]

Five protosepta (cardinal, 2 alar and 2

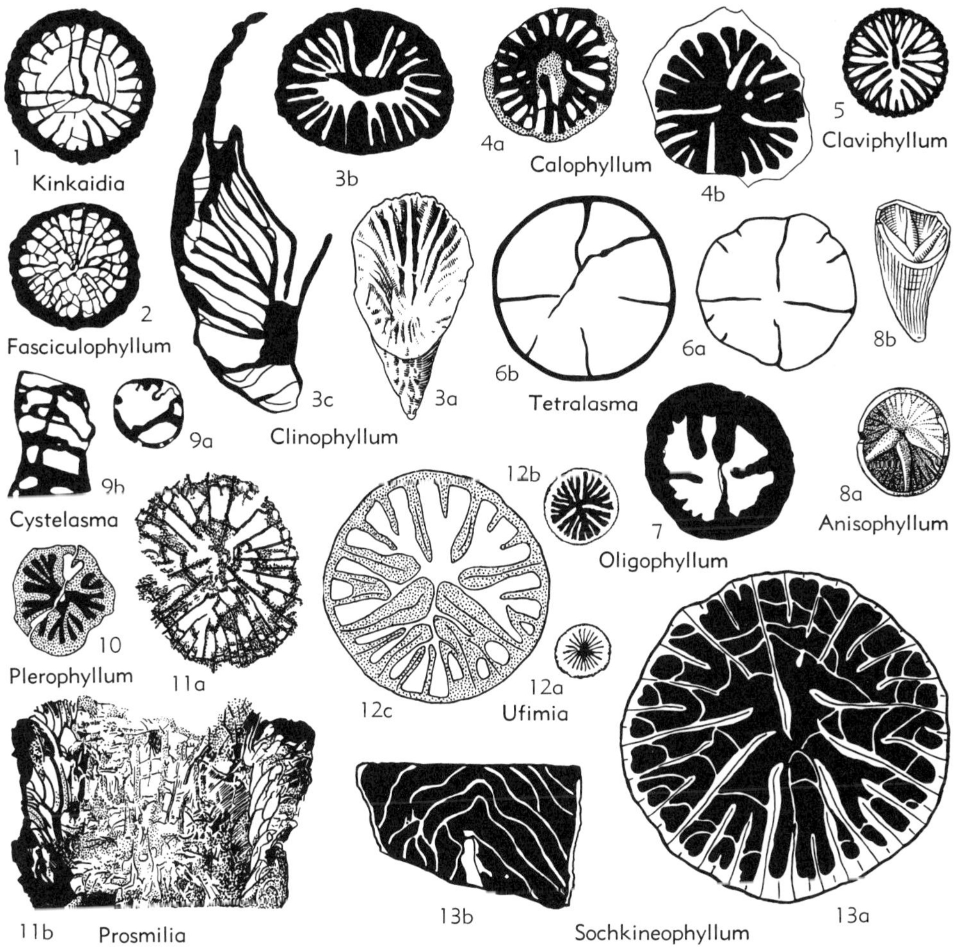

FIG. 177. Streptelasmatina (Cyathaxoniicae): Polycoeliidae (p. F260-F262).

counter-lateral) longer and thicker than other septa; counter septum long in young but shortened in adult stages; septa pinnately arranged, meeting at axis in early stages. No dissepiments or inner wall. *M. Dev.-Perm.*

Oligophyllum POČTA, 1902 [**O. quinqueseptatum*]. Counter-lateral septa club-shaped, dilated more than alar and cardinal septa, which are typically not swollen axially (42). *M.Dev.*, Boh.——FIG. 177,*7*. **O. quinqueseptatum;* transv. sec., ×2 (108).

Plerophyllum HINDE, 1890 [**P. australe;* SD GRABAU, 1928] [=*Timorosmilia* KOKER, 1924]. Of the 5 dominant protosepta the 2 counter-lateral are outstanding (42). *U.Dev.-Perm.*, Eu.-Asia-Austral. ——FIG. 177,*10*. **P. australe,* Perm.(Artinsk.), Austral.; transv. sec., ×2 (113).

Cystelasma MILLER, 1891 [**C. lanesvillense* MILLER, 1892]. With epithecal talons; short cardinal, 2 alar and 2 counter-lateral septa developed, commonly without metasepta; no minor septa; septa may be separated from wall by large globose dissepiments (51). *Miss.*, N.Am.——FIG. 177,*9*. **C. lanesvillense, 9a,b,* transv. and long. secs., ×2 (127).

Ufimia STUCKENBERG, 1895 [**U. carbonaria*] [=*Rhopalolasma* HUDSON, 1936; *Rhopalelasma* LANG-S.-T., 1940]. Like *Plerophyllum* but with cardinal septum short in adult stages, leaving only 4 outstanding protosepta (2 alars, 2 counter-laterals) (42). *U.Dev.-Perm.*, Eu.-Asia.——FIG. 177, *12a,b*. **U. carbonaria,* Carb., USSR; *12a,b,* transv. secs., ×1 (126).——FIG. 177,*12c*. **U. tachyblastum* (HUDSON), type species of *Rhopalolasma,* L.Carb., Eu.; transv. sec., ×3 (89).

Pseudobradyphyllum DOBROLYUBOVA, 1940 [**P. nikitini*]. Septa metriophylloid in arrangement in early stages but withdrawn slightly from axis in adult stages, cardinal and counter being shortest; longest septa are counter-laterals, alars, and metasepta nearest cardinal (7). *U.Carb.(Gshel.)*, USSR. ——FIG. 178,*3*. **P. nikitini; 3a,b,* transv. and long. secs., ×2 (68).

Pleramplexus SCHINDEWOLF, 1940 [**P. similis*]. Like *Plerophyllum* in young stages, but in ephebic stages all septa shorten toward periphery (42). *Perm.(Artinsk.)*, Timor.——FIG. 178,*7*. **P. similis; 7a,b,* transv. and long. secs., ×2 (113).

Subfamily ENDOTHECIINAE, Schindewolf, 1942

[=Endotheciidae (*nom. transl.* LECOMPTE, 1952)]

Like Plerophyllinae but with inner wall formed between septa; no dissepiments. *Perm.*

Endothecium KOKER, 1924 (*non* FRASER, 1935) [**E. apertum;* SD LANG-S.-T., 1940]. Alar septa outstanding, cardinal and counter short; septa withdrawn from axis with inner wall near their axial ends (42). *Perm.(Artinsk.)*, Timor.——FIG. 178,*2*. **E. apertum;* transv. sec., ×1 (95).

Subfamily TACHYLASMATINAE Grabau, 1928

[*nom. correct.* HILL, herein (*pro* Tachylasmainae GRABAU, 1928)] [=Pentaphyllinae SCHINDEWOLF, 1942; Tachyelasmidae HILL, 1948; Tachyelasminae HILL, 1952]

Cardinal, 2 alar and 2 counter-lateral septa outstanding; counter septum may be absent throughout or developed only in late stages; metasepta inserted relatively late and not pinnately; without dissepiments. *Carb.-Perm.*

Cryptophyllum CARRUTHERS, 1919 [**C. hibernicum*] [=*?Pentaphyllum* DE KONINCK, 1872]. Septa not all radial in course (42). *L.Carb.-Perm.*, Eu.-Asia. ——FIG. 178,*5*. **C. hibernicum,* L.Carb., Eu.; transv. sec., ×2 (64).

Pseudocryptophyllum EASTON, 1944 [**P. cavum*] [=*Hexalasma* SOSHKINA, 1928; *Hexelasma* LANG-S.-T., 1940 (*non* HOEK, 1915)]. Six protosepta dominant, may be only septa developed (10). *Miss.*, N.Am.; *Perm.(Artinsk.)*, USSR.——FIG. 178,*6a-c*. **P. cavum* Miss., N.Am.; *6a,b,* transv. secs., ×3; *6c,* long. sec., ×3 (10).——FIG. 178, *6d,e*. *P. primitivum* (SOSHKINA), type species of *Hexalasma,* Artinsk., USSR; *6d,e,* transv. secs., ×4 (123).

Tachylasma GRABAU, 1922 [**T. cha*] [=*Tachyelasma* LANG-S.-T., 1940 *(nom. van.); Prionophyllum* SCHINDEWOLF, 1942]. Cardinal septum short in adult stages (42). *Perm.(Artinsk.)*, Asia-Austral.——FIG. 178,*1*. **T. cha,* China; transv. sec., ×1 (81).

Pentamplexus SCHINDEWOLF, 1940 [**P. simulator*]. Young stages like *Cryptophyllum* with 5 dominant septa; in adult stages all septa shorten toward periphery, leaving wide axial space (42). *Perm. (Artinsk.)*, Timor.——FIG. 178,*4*. **P. simulator;* transv. sec., ×2 (113).

Family HADROPHYLLIDAE Nicholson in Nicholson & Lydekker, 1889

Small, simple, broadly trochoid to button-shaped corals with flattened or recurved calices; septa arranged in quadrants; minor septa short, contratingent. Tabulae and dissepiments absent. *L.Dev.-M.Carb.*

Combophyllum M.EDW.-H., 1850 [**C. osismorum*]. Small, free, discoid; without epitheca; minor septa long, thinner than major septa, with bare edges of septa on lower surface of disc crenulate; fossula well marked, with low cardinal septum (38). *L. Dev.*, Fr.——FIG. 179,*1*. **C. osismorum; 1a-c,* top, base, side, ×2 (73).

Hadrophyllum M.EDW.-H., 1850 [**H. orbignyi*]. Discoid to very widely trochoid; cardinal septum in prominent fossula somewhat contracted near periphery, other septa meeting in wall of fossula, those of counter quadrants radially arranged, others pinnate (52). *L.Dev.-M.Dev.*, N.Am.——FIG. 179, 2. **H. orbignyi*, M.Dev., N.Am.; *2a,b*, top, side, ×1 (73).

Xenocyathellus BASSLER, 1937 [**Homalophyllum thedfordensis* STEWART, 1936]. Very small, calceoloid, flattened, attached on cardinal side; cardinal fossula contracted near periphery, with short cardinal septum; counter quadrants with many fewer septa (52). *L.M.Dev.*, N.Am.——FIG. 179,*5*. **X. thedfordensis* (STEWART); *5a,b*, side, ×1; *5c,d*, transv. secs., ×3 (125).

Microcyclus MEEK & WORTHEN, 1868 (*non* SIMROTH, 1896) [**M. discus*]. Flat, discoid, very small, fossula opening inward into a wide shallow axial depression with slopes formed by contact of axial edges of major septa (52). *M.Dev.*, N.Am.——FIG. 179,*8*. **M. discus; 8a,b*, top, base, ×1 (102).

Dipterophyllum ROEMER, 1883 [**Zaphrentis glans* WHITE, 1862]. Short, turbinate, with counter and cardinal septa shortened so that a deep median trough crosses calice (1). *Miss.*, N.Am.——FIG. 179,*7*. **D. glans* (WHITE); *7a*, top, ×2; *7b*, side, ×1 (60).

Baryphyllum M.EDW.-H., 1850 [**B. verneuilianum*]. Discoid; septa exsert except for small axial epithecate part of base; counter septum short, cardinal septum dominant and very long; septa twist remarkably in adults (1). *Miss.*, N.Am.——FIG. 179, *4*. **B. verneuilianum*, *4a,b*, top, base, ×1 (60).

Gymnophyllum HOWELL, 1945 [**G. wardi*]. Like *Baryphyllum* but with cardinal and counter developed equally with other septa. *Penn.*, N.Am. ——FIG. 179,*3*. **G. wardi; 3a-c*, top, base, side, ×1 (87).

Cumminsia MOORE & JEFFORDS, 1945 [**Hadrophyllum aplatum* CUMMINS, 1891]. Patellate, apex al most central, epitheca without septal grooves; quadrants diverge from inner edge of rather short counter septum; cardinal septum short, fossula long and parallel-sided, expanding outwards near periphery; oldest metasepta are longest in each quadrant (39). *Penn.*, N.Am.——FIG. 179,*6*. **C. aplata* (CUMMINS), Tex.; *6a,b*, top, base, ×1 (104).

Family CYATHAXONIIDAE Milne-Edwards & Haime, 1850

[*nom. correct.* STUMM, 1949 (*pro* Cyathaxonidae M.EDW.-H., 1850)] [=Cyathaxoninae (*nom. transl.* DYBOWSKI, 1873)]

Small, ceratoid rugose corals with tall columella developed independently of the major septa but in contact with them and with long minor septa inserted alternately

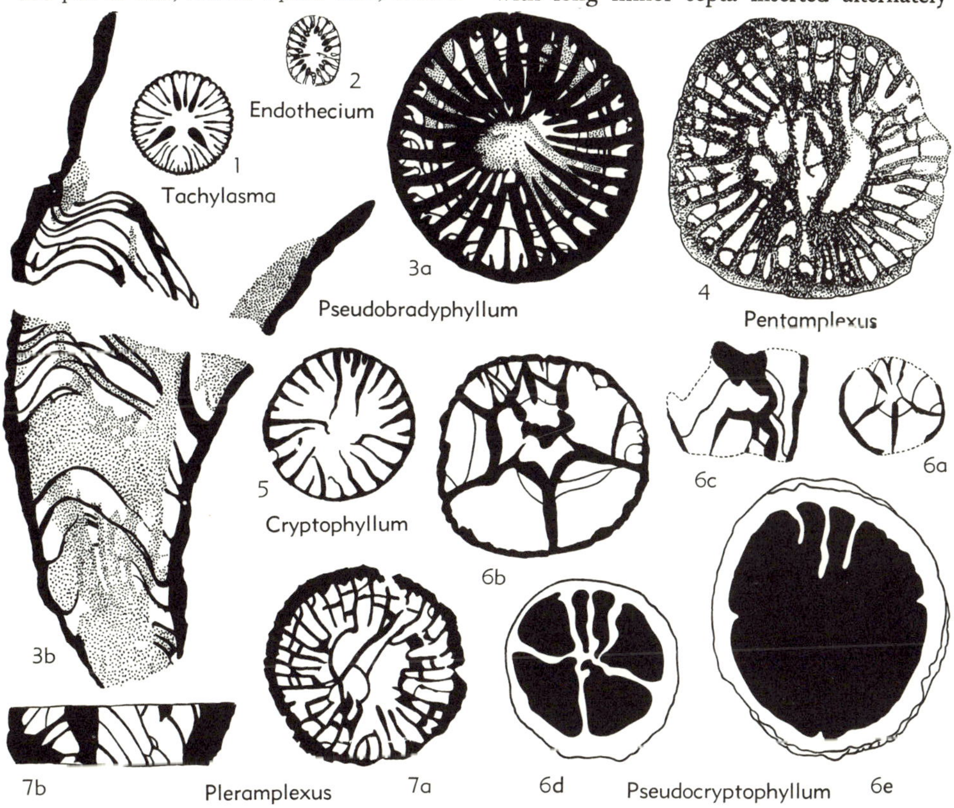

FIG. 178. Streptelasmatina (Cyathaxoniicae): Polycoeliidae (p. F262).

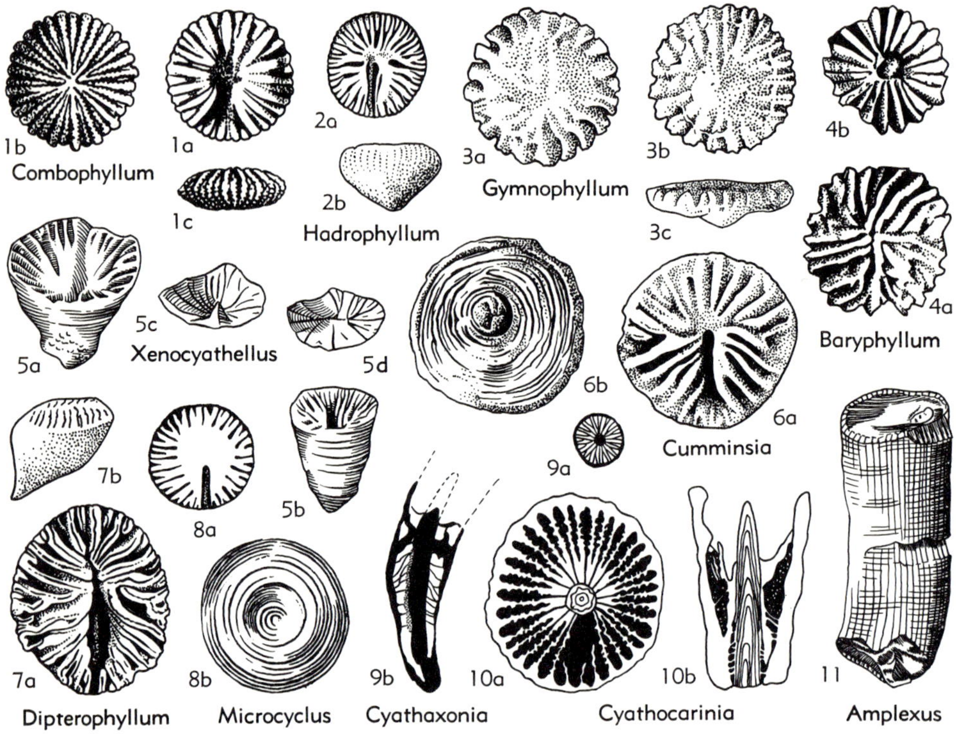

FIG. 179. Streptelasmatina (Cyathaxoniicae): Hadrophyllidae, Cyathaxoniidae, Amplexidae (p. F262-F264).

with the metasepta; complete tabulae inclined outward to the epitheca; dissepiments lacking. *Carb.-Perm.*

Cyathaxonia MICHELIN, 1847 [**C. cornu;* SD M. EDW-H., 1850] [=?*Zaphrentula* BOLKHOVITNOVA, 1915]. Septa without flanges; minor septa long and contratingent (10). *L.Carb.,* N.Am.-Eu.-Asia-Austral.——FIG. 179,*9*. **C. cornu,* L.Carb., Eu.; *9a,b,* transv. and long. secs., ×2 (64).

Cyathocarinia SOSHKINA, 1928 [**C. tuberculata;* SD LANG-S.-T., 1940]. Like *Cyathaxonia* but with sides of septa strongly tuberculated (49). *Perm. (Artinsk.),* USSR.——FIG. 179,*10*. **C. tuberculata; 10a,b,* transv. and long. secs., ×4 (123).

Family AMPLEXIDAE Chapman, 1893

[=Amplexinae (*nom. transl.* WANG, 1947)]

Solitary, without marginarium; tabulae flat, with down-turned edges; major septa thin, continuous vertically only at peripheral edges, but may extend toward axis as low, short ridges developed only on upper surfaces of tabulae; minor septa absent or appearing very late. *L.Carb.*

May possibly be more closely related to Cyathopsidae than to the Cyathaxoniicae.

Amplexus SOWERBY, 1814 [**A. coralloides*]. Long, cylindrical or scolecoid, interseptal ridges absent; tabulae with cardinal and alar fossular depressions. *L.Carb.,* Eu.——FIG. 179,*11*. **A. coralloides;* side, ×0.5 (124).

Family LOPHOPHYLLIDIIDAE Moore & Jeffords, 1945

[=?Lophophyllidae GRABAU, 1928; Lophophyllidiinae, *nom. correct.* WANG, 1950 (*pro* Lophophyllidinae, *nom. transl.* WANG, 1947)]

Small solitary coralla without dissepiments and with conical tabulae; septa long, arranged in quadrants in young stages, meeting enlarged counter septum at the axis; columella formed by swollen, vertically produced axial edge of the counter septum which may be reinforced by vestigial axial ends of other septa; cardinal septum shortened and other septa withdrawn from the axis in adult stages, commonly rhopaloid. *Carb.-Perm.*

Lophocarinophyllum GRABAU, 1922 [**L. acanthiseptum*]. Columella a prolongation of counter septum only, no septal lamellae taking part; septa flanged parallel to their distal edges (28). *Carb.*, China.——FIG. 180,*1*. **L. acanthiseptum; 1a,b*, transv. secs., ×3; *1c*, long. sec., ×3 (91).

Lophophyllidium GRABAU, 1928 [**Cyathaxonia prolifera* MCCHESNEY, 1860 [=*Sinophyllum* GRABAU, 1928; ?*Malonophyllum* OKULITCH & ALBRITTON, 1937]. Columella wide, typically with radial lamellae conjoined to median lamella but not tabellate, may be separated from counter septum in adult stages; axial edges of other septa, except cardinal, may be thickened and fused to one another in a collar around the columella (28). *Penn.-Perm.*, N.Am.-Eu.-Asia.——FIG. 180,*3*. **L. proliferum* (MCCHESNEY), Penn., USA; *3a,b*, transv. secs., ×3; *3c*, long. sec., ×3 (91).

Stereostylus JEFFORDS, 1947 [**S. lenis*]. Columella a simple expansion of axial end of the long counter septum, from which it may separate in adult stages; other septa typically only slightly swollen at axial edges, subequal (28). *Penn.-L.Perm. (Artinsk.)*, N.Am.-USSR-?N.Z.——FIG. 180,*5*. **S. lenis*, Penn., USA; *5a-c*, transv. secs., ×3; *5d*, long. sec., ×3 (91).

Lophamplexus MOORE & JEFFORDS, 1941 [**L. eliasi*]. Like *Stereostylus*, but columella discontinuous or lacking in upper part of corallite. *Penn.-L.Perm.*, N.Am.——FIG. 180,*4*. **L. eliasi*, L.Perm., Kans.; *4a-c*, transv. secs., ×2; *4d*, long. sec., ×2 (91).

Lophotichium MOORE & JEFFORDS, 1945 [**L. vescum*]. Septa long, thin and flanged, meeting long counter septum with its slight rhopaloid swelling axially in all save the late adult stages, when they withdraw, those of the cardinal quadrants re-

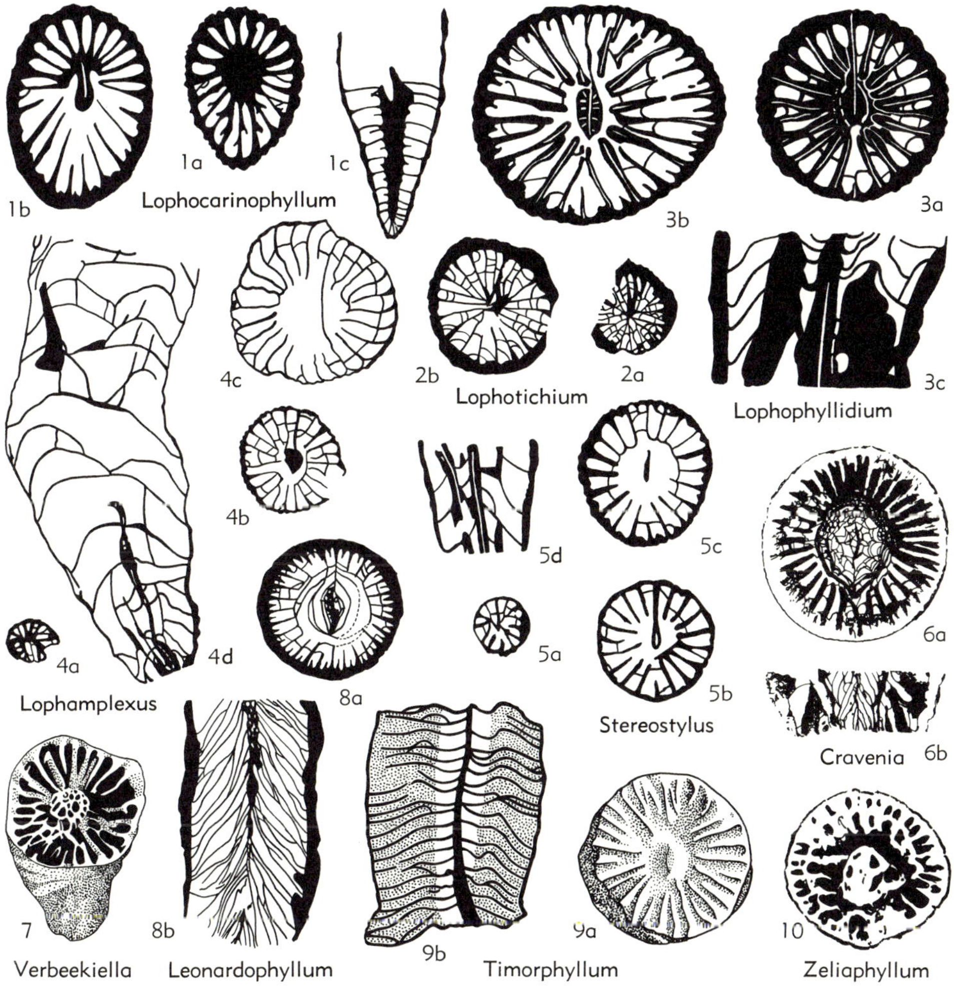

FIG. 180. Streptelasmatina (Cyathaxoniicae): Lophophyllidiidae, Timorphyllidae (p. F265-F267).

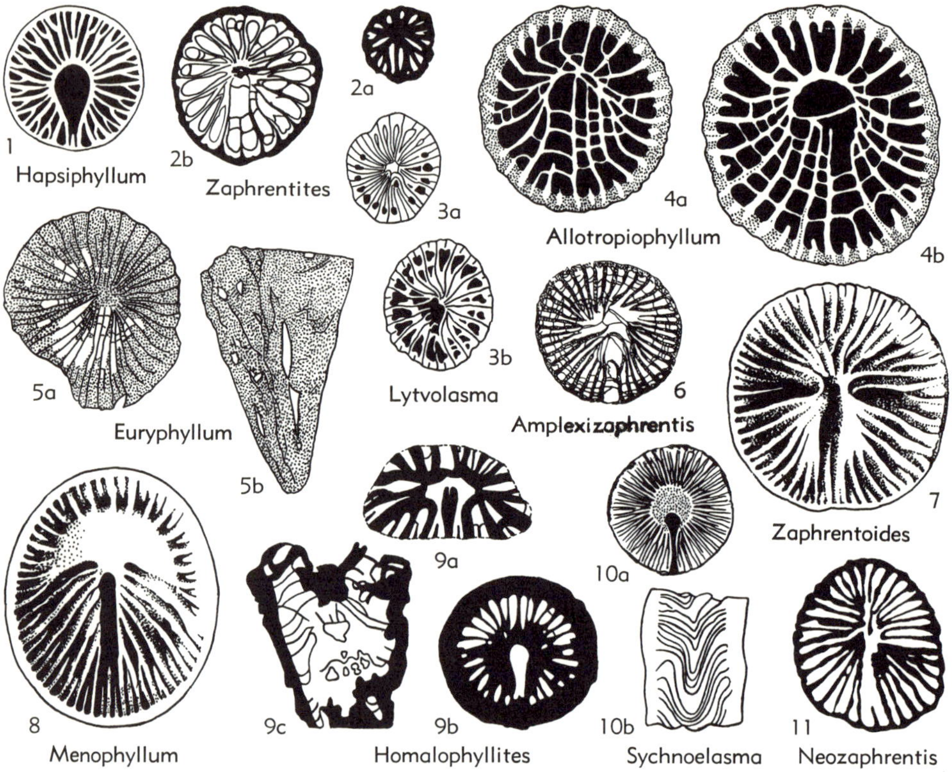

FIG. 181. Streptelasmatina (Cyathaxoniicae): Hapsiphyllidae (p. F267-F268).

treating first, leaving counter septum and its columellar vertical prolongation dominant (39). *Penn.*, N.Am.——FIG. 180,2. **L.* [illegible], Tex.; *2a,b,* transv. secs., ×3 (104).

?**Lophophyllum** M.EDW.-H., 1850 [**L. konincki*]. *L.Carb.(Tournais.),* Belg.

Family TIMORPHYLLIDAE Soshkina in Soshkina, Dobrolyubova & Porfiriev, 1941

[*nom. transl.* HILL, herein (*ex* Timorphyllinae SOSHK. in SOSHK.-D.-P., 1941)]

Solitary coralla, typically without dissepiments; with an axial structure of median lamella, radial lamellae, and tabellae; with narrow peripheral stereozone and conical tabulae. *L.Carb.-L.Perm.(Artinsk.).*

Cravenia HUDSON, 1928 [**C. rhytoides*]. Small; axial structure dibunophylloid; no dissepiments (18). *L.Carb.,* Eng.——FIG. 180,6. **C. rhytoides,* Tournais., Eng.; *6a,b,* transv. and long. secs., ×2 (89).

Zeliaphyllum HERITSCH, 1936 [**Z. suessi*] [=*Zelaeophyllum* LANG-S.-T., 1940]. Axial structure of tabellae and lamellae irregularly interwoven; minor septa short; only one series of dissepiments (14). *L.Perm.(Artinsk.),* Carnic Alps. FIG. 180,*10*. **Z. suessi,* transv. sec., ×3 (84).

Timorphyllum GERTH, 1921 [**T. wanneri*]. Somewhat scolecoid, with smooth epitheca; axial structure simple, comprising a barlike columella typically separate from septa in adult stages, buttressed by sharply upturned narrow axial parts of tabulae, which are otherwise flat but downturned at the edges; no dissepiments (12). *L.Perm.(Artinsk.),* Asia.——FIG. 180,*9*. **T. wanneri,* Timor; *9a,b,* calical view, long. sec., ×1 (79).

Verbeekiella GERTH, 1921 [*pro Verbeekia* PENECKE, 1908 (*non* FRITSCH, 1887; *nec* SILVESTRI, 1908)] [**Verbeekia permica* PENECKE, 1908]. Axial structure large, with well-defined wall separating its tabellae from periaxial tabellae; skeletal dilatation common; radial lamellae numerous, medial plate generally not dominant; no dissepiments (12). *L.Perm.(Artinsk.),* Eu.-Asia-Austral.——FIG. 180,*7*. **V. permica* (PENECKE), Timor; side, ×1 (79).

Leonardophyllum MOORE & JEFFORDS, 1941 [**L. distinctum*]. Axial structure without defined wall; counter septum typically continuous with medial lamella; peripheral stereozone moderately wide; no dissepiments. *L.Perm.(Leonard.)*, N.Am.——FIG. 180,*8*. **L. distinctum*, W.Tex.; *8a,b*, transv. and long. secs., ×2 (104).

Family HAPSIPHYLLIDAE Grabau, 1928

[=Zaphrentoididae SCHINDEWOLF, 1938; Zaphrentoidinae (*nom. transl.* WANG, 1950); Zaphrentoidida (superfam.) (*nom. transl.* SCHIND., 1952)]

Small, solitary, ceratoid or trochoid coralla with fossula bounded laterally by cardinal lateral septa and axially (in younger stages at least) by a wall consisting of fused axial ends of major septa of the counter quadrants. The septa may withdraw from the axis. Tabulae incomplete, conical, with highest point at inner edge of the fossula. No dissepiments. Minor septa present or absent. *L.Carb.-L.Perm.(Artinsk.)*.

Group I

Fossula on the shorter, concave side of the corallum. *L.Carb.-L.Perm.*

Hapsiphyllum SIMPSON, 1900 [**Zaphrentis calcariformis* HALL, 1882 (=*Zaphrentis cassedayi* M. EDW., 1860)] [=*Enallophyllum* GREEN, 1901]. Fossula expanded axially; septa arranged pinnately in cardinal and radially in counter quadrants; minor septa long, contratingent; cardinal septum short in adult stages (10). *Miss.*, N.Am.-Eu.——FIG. 181,*1*. **H. cassedayi* (M.EDW.), N.Am.; transv. sec., ×2 (115).

Zaphrentites HUDSON, 1941 [*non* VON BUBNOFF, 1926, p. 150 (*?lapsus calami*)] [**Zaphrentis parallela* CARRUTHERS, 1910] [*?=Cypellophyllum* TOLMACHEV, 1933 (*pro Craterophyllum* TOLMACHEV, 1931, *non* FOERSTE, 1909, *nec* BARBOUR, 1911)]. Septa may withdraw from fossula, first in cardinal quadrants, then in counter quadrants, so that radial arrangement supersedes earlier pinnate arrangement; minor septa absent or extremely short (26). *L.Carb.*, Eu.-Asia.——FIG. 181,*2*. **Z. parallela* (CARRUTHERS), Eu., *2a,b*, transv. secs., ×3 (64).

Canadiphyllum SUTHERLAND, 1954 [**C. knoxi*]. Septa dilated; fossula parallel-sided, met by septa of cardinal quadrants almost at right angles; alar fossulae almost bisect corallite; septa of counter quadrants almost perpendicular to alar fossula. *Miss.*, B.C.

Allotropiophyllum GRABAU, 1928 [**A. sinense*]. Major septa of counter quadrants and (in some) alar and first metasepta of cardinal quadrants grouped in a narrow crescentic area on the counter (convex) side, other major septa directed to a point or points on its inner side roughly midway between the corallum axis and epitheca; major septa amplexoid in late stages; minor septa absent or very short (13). *L.Carb.-L.Perm.(Artinsk.)*, Eu.-Asia.——FIG. 181,*4*. **A. sinense*, Artinsk., Asia, *4a,b*, transv. secs., ×5 (81).

Amplexizaphrentis VAUGHAN, 1906 [**Zaphrentis bowerbanki* THOMSON, 1883; SD LANG-S.-T., 1940] [=*Triplophyllites* EASTON, 1944; *?Barytichisma* MOORE & JEFFORDS, 1945]. Corallum large, may be subquadrate in section; fossula commonly oblique and septa wavy, arranged inequilaterally; withdrawal from axis may begin first in counter quadrants; minor septa very short; cardinal septum long at first, short later (10). *L.Carb.*, Eu-Asia-N.Am.; *Penn.*, N.Am.——FIG. 181,*6*. *A. curvilinea* THOMSON, L. Carb., Scot.; transv. sec., ×1 (130).

Zaphriphyllum SUTHERLAND, 1954 [**Z. disseptum*]. Like *Amplexizaphrentis* but with normal dissepimentarium. *Miss.*, Can.

Lytvolasma SOSHKINA, 1925 [**L. asymetricum*] [=*Lytvelasma* LANG-S.-T., 1940]. Septa few and dilated, slightly and irregularly withdrawn from axis; fossula shallow, open axially in adult stages when cardinal septum is short (49). *L.Perm.(Artinsk.)*, USSR.——FIG. 181,*3*. **L. asymetricum*; *3a,b*, transv. secs., ×2 (123).

Euryphyllum HILL, 1937 [**E. reidi*]. Calice oblique, septa initially much dilated, dilatation decreasing first midway between periphery and axis; counter septum long and thin. *L.Perm.(Artinsk.)*, Austral.-N.Z.——FIG. 181,*5*. **E. reidi*, Austral.; *5a,b*, transv. and long. secs., ×2 (85).

Group II

Fossula on the longer, convex side of the corallum. *L.Carb.*

Zaphrentoides STUCKENBERG, 1895 [**Zaphrentis griffithi* M.EDW.-H., 1851; SD SCHINDEWOLF, 1938]. Minor septa very short. Type specimen insufficiently known (10). *L.Carb.*, Eu.-N.Am.——FIG. 181,*7*. **Z. griffithi* (M.EDW.-H.), Eu., calical view, ×1 (73).

Menophyllum M.EDW.-H., 1850 [**M. tenuimarginatum*]. Fossula long, narrow, deep and parallel-sided, containing a short cardinal septum in adult stages; major septa of counter quadrants equal, short and withdrawn from axis (38). *L.Carb.*, Fr.——FIG. 181,*8*. **M. tenuimarginatum*, Tournais., Fr.; calical view, ×2 (73).

Homalophyllites EASTON, 1944 [**Lophophyllum calceola* WHITE & WHITFIELD, 1862]. Fossula expanded axially with bounding septa reinforced by thickening; cardinal septum short; convex side of corallum flattened near apex (10). *Miss.*, N.Am.——FIG. 181,*9*. **H. calceola* (WHITE-W.); *9a*, neanic transv. sec., ×6; *9b*, ephebic transv. sec., ×3; *9c*, long. sec., ×3 (72).

Neozaphrentis GROVE, 1935 [**Zaphrentis tenella* MILLER, 1891]. Counter septum longer than others but not swollen axially; cardinal septum short, fossula oblique and variably placed, usually on the convex side; septa not rhopaloid (10). *Miss.*, N.

Am.——FIG. 181,*11*. **N. tenella* (MILLER); transv. sec., X3 (72).

Sychnoelasma LANG-S.-T., 1940 [*pro Verneuilia* STUCKENBERG, 1895 (*non* HALL & CLARKE, 1894)] [**Verneuilia urbanowitschi* STUCK., 1895]. Major septa withdrawn equally from axis, leaving space in which the tabulae are infundibuliform (33). *L.Carb.*, USSR.——FIG. 181,*10*. **S. urbanowitschi* (STUCK.); *10a,b*, transv. and long. secs., X1 (126).

Superfamily ZAPHRENTICAE Milne-Edwards & Haime, 1850

[*nom. transl.* HILL, herein (*ex* Zaphrentidae M.EDW.-H., 1850)] [=Zaphrentoidea, ?Cyathophylloidea NICHOLSON in NICH. & LYDEKKER, 1889; Streptelasmacea WEDEKIND, 1927; Streptelasmaticae HILL, 1954]

Solitary or compound, with marginarium consisting either of a septal stereozone or a regular or herringbone dissepimentarium; among forms with a septal stereozone, the septa bear axial lobes; tabulae conical or domed, rarely sagging. *Ord.-Perm.*

NICHOLSON based his conception of the Zaphrentoidea on the Hapsiphyllidae, rather than on *Zaphrenthis,* although he names *Zaphrenthis ("Zaphrentis")* as the typical genus.

The Cyathopsidae are placed in this superfamily somewhat doubtfully because many of them possess a lonsdaleoid dissepimentarium such as is typical of the Columnariina, but otherwise they are morphologically quite close to the Aulophyllidae. The ontogeny of the early Cyathopsidae suggests that they developed from a member of the Hapsiphyllidae, herein regarded as Cyathaxoniicae.

Family STREPTELASMATIDAE Nicholson in Nicholson & Lydekker, 1889

[*nom. correct.* WEDEKIND, 1927 (*pro* Streptelasmidae NICH. in NICH.-L., 1889)] [=Streptelasmaidae GRABAU, 1922]

Solitary and conical or fasciculate coralla, with short minor septa and marginarium consisting of a septal stereozone; dissepiments typically not developed; major septa lobed axially; early uniform septal dilatation decreasing from the axis outwards; tabulae domed, complete or incomplete. *M.Ord.-M.Dev.*

Subfamily STREPTELASMATINAE Nicholson in Nicholson & Lydekker, 1889

[*nom. transl.* LECOMPTE, 1952 (as Streptelasminae (*ex* Streptelasmatidae, *nom. correct.* WEDEKIND., 1927, *pro* Streptelasmidae NICH. in NICH.-L., 1889)] [Includes Dinophyllidae WANG, 1947; Dalmanophyllinae LECOMPTE, 1952; Dinophyllinae (*nom. transl.* LECOMPTE, 1952)]

Septal stereozone only moderately wide, if present; axial edges of septa and axial lobes typically not forming a conical boss in calice. *M.Ord.-M.Dev.*

Streptelasma HALL, 1847 [**S. corniculum;* SD C. F. ROEMER, 1861] [=*Streptoplasma* HALL, 1847; *Palaeocyathus* FOERSTE, 1888]. Solitary; septa long; axial lobes few, forming a loose, narrow axial structure (20). *M.Ord.-M.Sil.,* N.Am.-Eu.-Asia-Austral.——FIG. 182,*1*. **S. corniculum,* M.Ord., N.Am.; *1a,b,* transv. and long. secs., X1 (66).

Palaeophyllum BILLINGS, 1858 [**P. rugosum*]. Like *Streptelasma* but phaceloid (30). *M.Ord.-U.Ord.,* N.Am.-Eu.——FIG. 182,*2*. **P. rugosum,* ?U.Ord., Can.; *2a,* side of corallum, X1; *2b,c,* transv. and long. secs., X1 (96).

Coelostylis LINDSTRÖM, 1880 [**C. törnquisti* (=*Cyathaxonia? törnquisti* LIND., 1873)]. Solitary; septa with axial lobes; bladelike columella projecting into calice. *M.Ord.,* Swed.——FIG. 182,*8*. **C. toernquisti* (LIND.); long. sec., X1 (100).

Grewingkia DYBOWSKI, 1873 [**Clisiophyllum buceros* EICHW., 1855; SD SHERZER, 1891] [=*Kiaerophyllum* WEDEKIND, 1927]. Solitary; septa short, axial lobes numerous, axial structure wide and present even in young stages (55). *U.Ord.,* Eu.-N.Am.——FIG. 182,*10*. *G. kiaeri* (WDKD.), type species of *Kiaerophyllum,* Norway; *10a,b,* transv. secs., X1.5; *10c,* long. sec., X1.5 (137).

Brachyelasma LANG-S.-T., 1940 [*pro Dybowskia* WEDEKIND, 1927 (*non* DALL, 1876, *nec* others)] [**Dybowskia prima* WDKD., 1927]. Solitary; septa short; axial lobes and axial structure absent in adult stages (55). *U.Ord.,* Eu.——FIG. 182,*5*. **B. primum* (WDKD.), Norway; *5a,b,* transv. and long. secs., X1 (137).

Dalmanophyllum LANG & SMITH, 1939 [**Cyathaxonia dalmani* M.EDW.-H., 1851] [=*Centrotus* LINDSTRÖM in THOMSON & NICHOLSON, 1876 (*non* FABRICIUS, 1803); *Tyria* SCHEFFEN, 1933 (*non* HÜBNER, 1819)]. Solitary or compound; major septa long and thick, with axial lobes joining a bladelike columella produced upward from axial parts of conjoined cardinal and counter septa; minor septa very short, septal stereozone narrow. *L.Sil.-M.Sil.,* Eu.——FIG. 182,*6a,b*. **D. dalmani* (M.EDW.-H.); *6a,b,* transv. and long. secs., X2 (73).——FIG. 182,*6c*. *D. inserta* (SCHEFFEN), type species of *Tyria,* L.Sil., Norway; transv. sec., X5 (112).

Dinophyllum LINDSTRÖM, 1882 [**D. involutum* (=*Clisiophyllum hisingeri* M.EDW.-H., 1851)] [=*Streptophyllum* GRABAU in CHI, 1931]. Solitary; long major septa with few axial lobes but reaching to axis where, with or without convolution of their thin axial ends, they form an axial structure with steeply conical tabulae; minor septa extremely short (37). *L.Sil.-M.Sil.,* Eu.-Asia.——

FIG. 182,*3*. **D. hisingeri* (M.EDW.-H.), M.Sil., Gotl.-Sib.; *3a*, calical view, ×1; *3b*, long. sec., ×1 (100).

Rhegmaphyllum WEDEKIND, 1927 [**Turbinolia turbinata* HISINGER, 1831 (*partim*) (?=*Zaphrentis? conulus* LINDSTRÖM, 1868); SD SOSHKINA, 1937] [=*Regmaphyllum* WDKD., 1927; *Rhegmatophyllum* LANG-S.-T., 1940 (*nom. van.*)]. Solitary; major septa meeting near axis around inner edge of strongly marked fossula; axial lobes few; cardinal septum thin or short (55). *M.Sil.*, Eu.-Asia. ——FIG. 182,*9*. *R. conulus* (LIND.), Gotl.; *9a,b*, secs., ×1; *9c*, calical view of septa, diagrammatic (100).

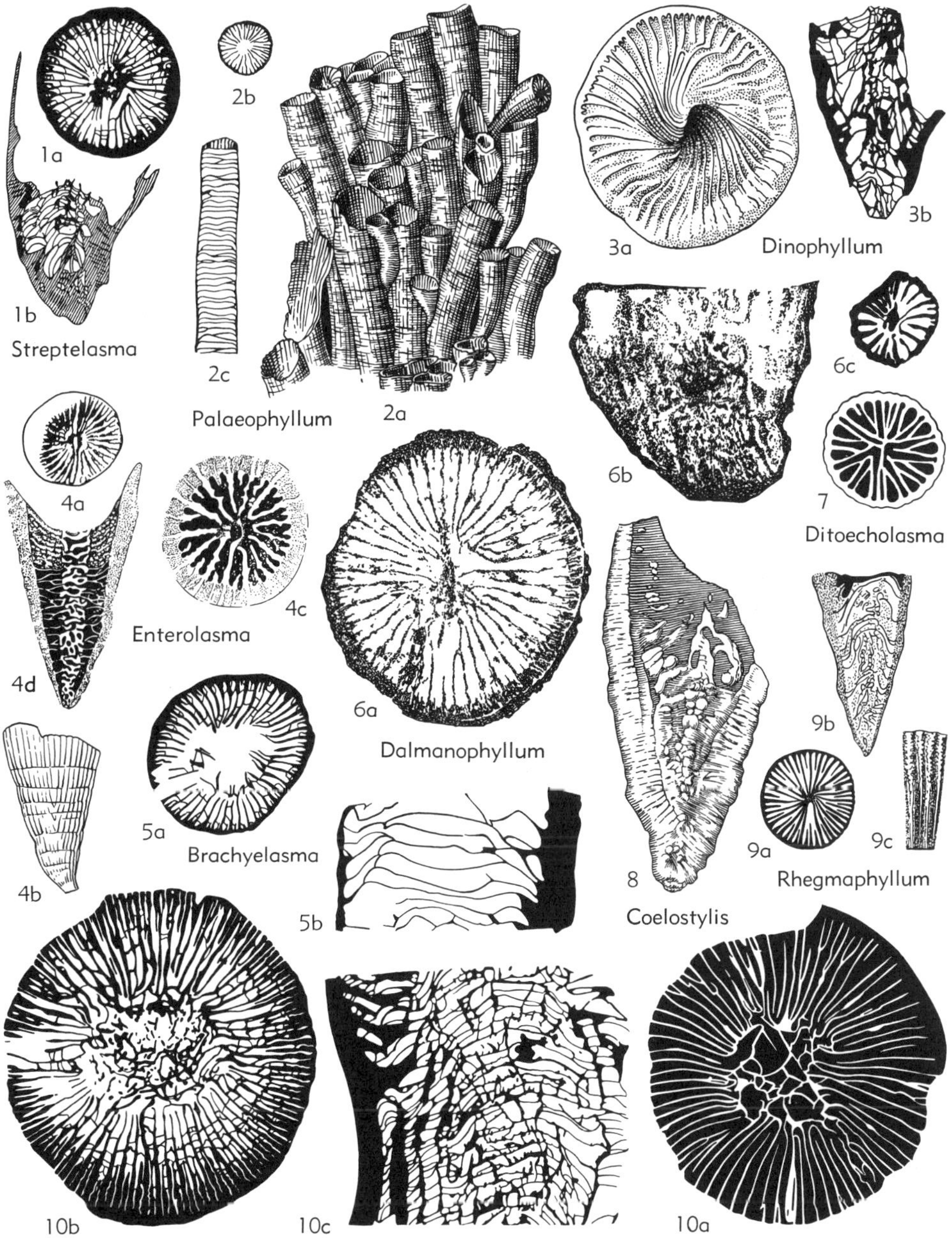

FIG. 182. Streptelasmatina (Zaphrenticae): Streptelasmatidae (p. F268-F270).

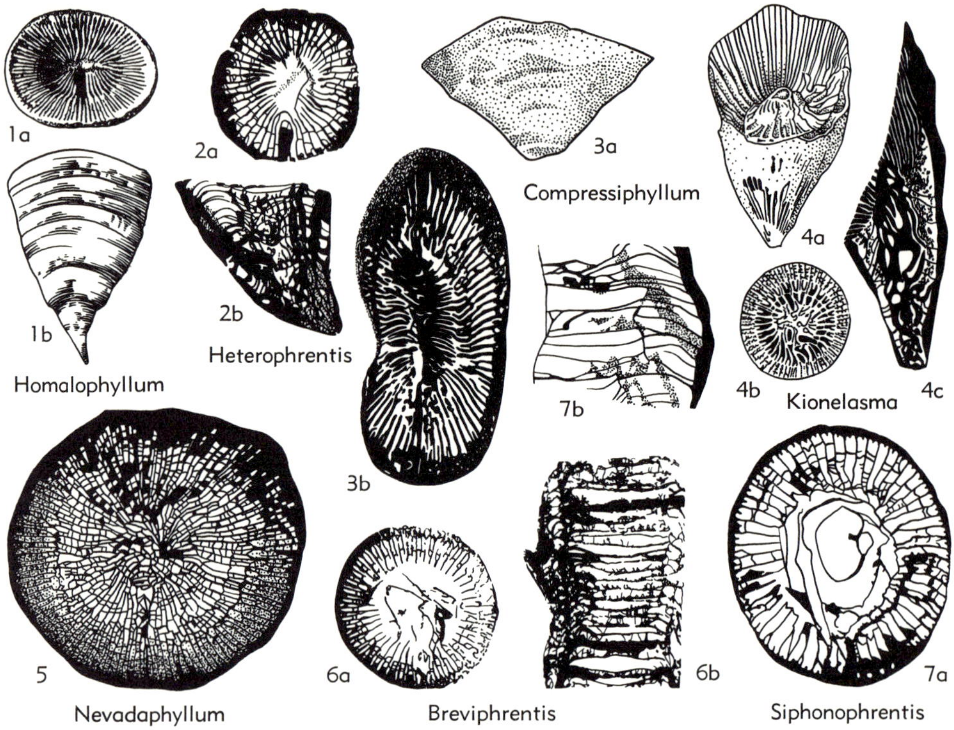

FIG. 183. Streptelasmatina (Zaphrenticae): Streptelasmatidae (p. F270-F271).

Ditoecholasma SIMPSON, 1900 [**Petraia fanningana* SAFFORD, 1869] [=*Ditoechelasma* LANG-S.-T., 1940 *(nom. van.)*]. Solitary, slender; septal stereozone very narrow, septa grouped as in *Petraia,* minor septa long; axial septal lobes forming spongy axial structures; tabulae sagging axially, incomplete. *M.Sil.,* N.Am.——FIG. 182,7. **D. fanninganum* (SAFFORD); transv. sec., ×2 (115).

Enterolasma SIMPSON, 1900 [**Streptelasma (Petraia) stricta* HALL, 1874] [=*Enterelasma* LANG-S.-T., 1940 *(nom. van.)*]. Solitary, small, septal stereozone narrow, septa with sides tuberculate and axial edges lobed; cardinal and counter septa long (44). *L.Dev.,* N.Am.——FIG. 182,4. **E. strictum* (HALL); *4a,b,* calice, side, ×1; *4c,d,* transv. and long. secs., ×2 (127).

Kionelasma SIMPSON, 1900 [**Streptelasma mammiferum* HALL, 1882] [≐*Cionelasma* LANG-S.-T., 1940 *(nom. van.)*]. Solitary, septal stereozone wide; dense axial structure formed from swollen and twisted axial lobes and edges of major septa united by dilating tissue (44). *L.Dev.,* N.Am.——FIG. 183,4. **K. mammiferum* (HALL); *4a,* side, ×0.5; *4b,c,* transv. and long secs., ×1 (127).

Heterophrentis BILLINGS, 1875 [**H. spatiosa* BILL., 1875 (=*Zaphrentis spatiosa* BILL., 1858; *Zaphrentis prolifica* BILL., 1858); SD MILLER, 1889] [=*Triplophyllum* SIMPSON, 1900]. Solitary, large, conical, flattened in lateral parts of counter quadrants, wtih fossula on convex side; septal stereozone very narrow; septa with sparse axial lobes, withdrawn from axis; tabulae shallow domes with axial depression (52). *L.Dev.-M.Dev.,* N.Am.——FIG. 183,2. **H. prolifica* (BILL.); *2a,b,* transv. and long. secs., ×1 (127).

Compressiphyllum STUMM, 1949 [**Zaphrentis compressa* ROMINGER, 1876 (*non* M.EDW.-H., 1860); renamed *Zaphrentis davisana* MILLER, 1889]. Like *Heterophrentis* but compressed laterally with flattened alar regions (52). *L.Dev.,* N.Am.——FIG. 183,3. **C. davisanum* (MILLER); *3a,* side, ×0.5; *3b,* calice, ×1 (127).

Homalophyllum SIMPSON, 1900 [**Zaphrentis ungula* ROMINGER, 1876]. Solitary, small, flattened on the fossular, convex side; with long major and short minor septa (44). *L.Dev.-M.Dev.,* N.Am. ——FIG. 183,1. **H. ungula* (ROMINGER); *1a,b,* calice and side, ×1 (127).

Siphonophrentis O'CONNELL, 1914 [**Caryophyllia gigantea* LESUEUR, 1821]. Solitary, large, subcylindrical with septa withdrawn from axis, leaving wide space with very sparse axial lobes; septal stereozone very narrow; tabulae mesa-shaped, depressed at fossula (52). *L.Dev.-M.Dev.,* N.Am. ——FIG. 183,7. **S. gigantea* (LESUEUR); *7a,b,* transv. and long. secs., ×1 (127).

Breviphrentis STUMM, 1949 [**Amplexus invaginatus* STUMM, 1937]. Solitary, cylindrical, with septa withdrawn from axis and complete only on tops of the tabulae, which become flat-topped domes with down-turned edges; tabulae not depressed at fossula which is on convex side (52). *L.Dev.-M.Dev.*, Nev.——FIG. 183,*6*. **B. invaginata* (STUMM); *6a,b*, transv. and long. secs., ×1 (127).

Nevadaphyllum STUMM, 1937 [**N. masoni*]. Large, solitary; septal stereozone partly replaced by a regular dissepimentarium of small globose interseptal dissepiments; major septa numerous and convolute in the tabularium, where the domed tabulae are incomplete (52). *L.Dev.*, N.Am.——FIG. 183,*5*. **N. masoni;* transv. sec., ×1 (127).

?**Strobilasma** SCHEFFEN, 1933 [**S. dentatum*] [=*Strobilelasma* LANG-S.-T., 1940 *(nom. van.)*]. *L.Sil.*, Norway.

Subfamily KODONOPHYLLINAE Wedekind, 1927

[*nom. transl.* LECOMPTE, 1952 (*ex* Kodonophyllidae WDKD., 1927)]

Marginarium a wide septal stereozone. *L.Sil.-L.Dev.*

Schlotheimophyllum SMITH, 1945 [**Fungites patellatus* SCHLOTHEIM, 1820]. Solitary, large discoid or patellate with reflexed calicular platform; minor septa long, septal stereozone wide; major septa long and twisted at axis into loose axial structure, axial lobes few (44). *L.Sil.-M.Sil.*, Eu. ——FIG. 184,*3*. **S. patellatum* (SCHLOTH.); *3a,b*, transv. and long. secs., ×1 (97).

Kodonophyllum WEDEKIND, 1927 [**Streptelasma milne-edwardsi* DYBOWSKI, 1873 (=*Madrepora truncata* LINNÉ, 1758)] [=*Patrophontes* LANG & SMITH, 1927; *Codonophyllum* LANG-S.-T., 1940 *(nom. van.)*]. Fasciculate or solitary with flat or sloping calicular platform; otherwise like *Schlotheimophyllum* (31). *M.Sil.-U.Sil.*, Eu.——FIG. 184,*1*. **R. truncatum* (LINNÉ); *1a,b*, transv. and long. secs., ×2 (98).

Circophyllum LANG & SMITH, 1939 [*pro Rhysodes* SMITH & TREMBERTH, 1927 (*non* ILLIGER in DALMAN, 1823)] [**Rhysodes samsugnensis* SMITH-T., 1927]. Fasciculate, corallites slender and septal stereozone narrow; major septa long, straight, united axially by secondary thickening. *U.Sil.*, Gotl.

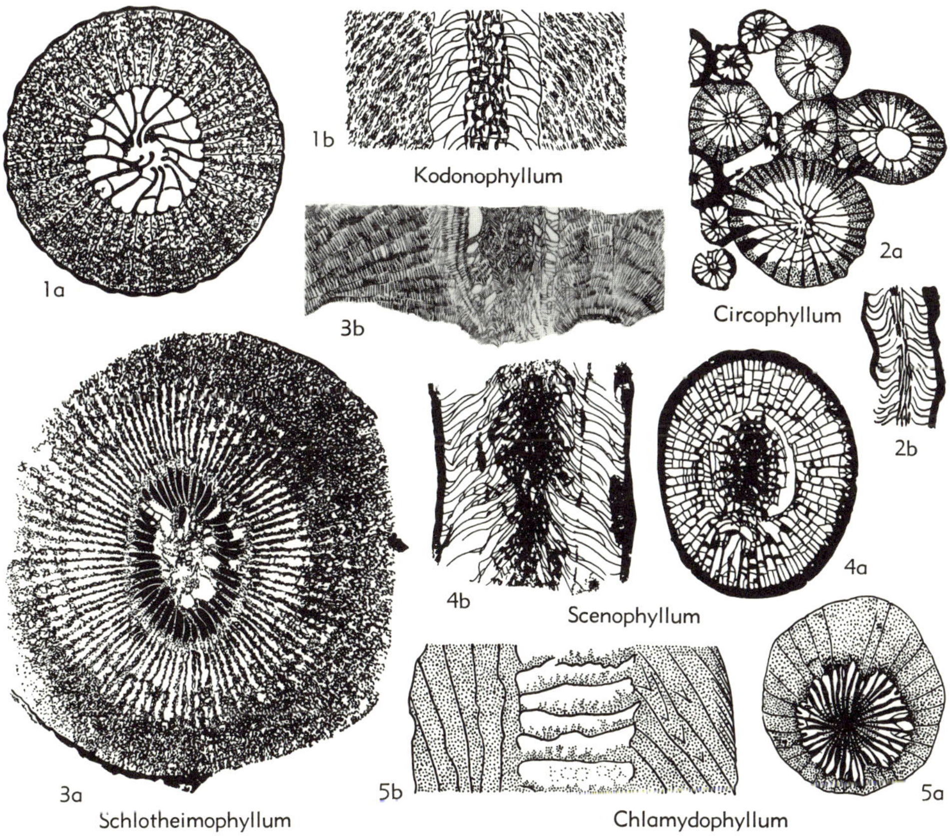

FIG. 184. Streptelasmatina (Zaphrenticae): Streptelasmatidae (p. F271-F272).

——FIG. 184,2. *C. samsugnense (SMITH-T.); 2a, transv. sec., ×2; 2b, long. sec., ×1 (120).

Chlamydophyllum POČTA, 1902 [*C. obscurum]. Solitary, with very wide septal stereozone from which major septa project with moderate thickening into tabularium; major septa may unite at axis in proximal parts of corallum, and on surface of tabulae in distal parts, being then withdrawn between tabulae; axial lobes few and thickened (40). *L.Dev.*, Eu.-Austral.——FIG. 184,5. *C. obscurum, Czech.; 5a,b, transv. and long secs., ×2 (108).

Subfamily ACROPHYLLINAE Stumm, 1949

[*nom. transl.* HILL, herein (*ex* Acrophyllidae STUMM, 1949)]

Marginarium a narrow septal stereozone or dissepimentarium; axial edges of septa and axial lobes forming with tabulae a conical boss in calice (52). *L.Dev.*

Scenophyllum SIMPSON, 1900 [*Zaphrentis conigera ROMINGER, 1876]. Solitary, large, conico-cylindrical, with numerous long major septa with axial lobes based on close, tall, conical tabulae forming an axial structure with a dominant long irregular lamella in its counter-cardinal plane; septal stereozone narrow (52). *L.Dev.*, N.Am.——FIG. 184,4. *S. conigerum (ROMINGER); 4a,b, transv. and long. secs., ×1 (127).

Acrophyllum THOMSON & NICHOLSON, 1876 [*Clisiophyllum oneidaense BILLINGS, 1859]. Like *Scenophyllum* but with marginarium of angulo-concentric dissepiments in which minor septa are suppressed (52). *L.Dev.*, N.Am.

Family HALLIIDAE Chapman, 1893

Solitary coralla typically with elongate cardinal septum; marginarium comprising a narrow septal stereozone or regular dissepimentarium of small, globose, interseptal dissepiments and thin septa; septa dilated and in contact in the tabularium, dilatation decreasing first from the axial edges and in counter quadrants; septa without axial lobes; in curved coralla the deep fossula is on the convex side; tabulae domed or sagging, usually incomplete. *Ord.-Dev.*

Subfamily LYKOPHYLLINAE Wedekind, 1927

[*nom. transl.* HILL, herein (*ex* Lykophyllidae WDKD., 1927)] [=Neocystiphyllidae WDKD., 1927; ?Stratiphyllidae SCHEFFEN, 1933; Pycnactidae HILL, 1940]

Pinnate arrangement of the septa moderate or weak. *Ord.-Sil.*

Holophragma LINDSTRÖM, 1896 [*Hallia calceoloides LIND., 1866 (*partim*)]. Calceoloid, small; long cardinal septum on the flattened side; marginarium a septal stereozone without dissepiments (37). *U.Ord.*, N.Am.; *Sil.*, Gotl.——FIG. 185,1. *H. calceoloides (LIND.), Sil., Gotl.; 1a,b, side views, ×1; 1c, calice, ×2 (100).

Onychophyllum SMITH, 1930 [*O. pringlei]. Small; long cardinal and short counter septa continuous in young stages; septa of counter quadrants short, dilated and in contact; septa of cardinal quadrants longer, thinner and with axial edges bent like a claw toward counter septum; no minor septa or dissepiments (45). *L.Sil.*, Eng.——FIG. 185,3. *O. pringlei; 3a-h, transv. secs. (inverted), ×2 (117).

Pycnactis RYDER, 1926 [*Hyppurites mitratus SCHLOTHEIM (*partim*), 1820]. Solitary, major septa dilated, contiguous and pinnately arranged, cardinal septum long; minor septa extremely short, dissepiments absent except at rim of calice. *Sil.*, Eu.——FIG. 185,5. *P. mitrata (SCHLOTH.); transv. sec., ×3 (137).

Phaulactis RYDER, 1926 [*P. cyathophylloides] [=*Mesactis* RYDER, 1926; *Desmophyllum* (*non* EHRENBERG, 1834), *Lykophyllum, Lykocystiphyllum, Neocystiphyllum* WEDEKIND, 1927; *Semaiophyllum* VOLLBRECHT in WDKD., 1927; *Hercophyllum* JONES, 1936; *Lycophyllum, Lycocystiphyllum, Semaeophyllum* LANG-S.-T. (*nom. van.*), 1940; all synonyms or possibly subgenera of *Phaulactis*]. Solitary, large; dissepimentarium regular and wide, septa unthickened in it but dilated in tabularium, dilatation decreasing in adult; tabulae domed, flat or sagging, complete or incomplete; pinnate arrangement of septa generally not marked, and cardinal septum commonly not elongate (55). *M.Sil.-U.Sil.*, Eu.-N.Am.-Austral.——FIG. 185, 4a-d. *P. tabulata* (WDKD.), type species of *Lykophyllum*, Gotl.; 4a,b, transv. and long. secs., ×2; 4c,d, transv. secs., ×3 (137).——FIG. 185,4e,f. *P. shearsbyi* (SÜSSMILCH), type species of *Hercophyllum*, N.S.W.; 4e,f, transv. and long secs., ×1 (85).

Lamprophyllum WEDEKIND, 1927 [*L. degeeri]. Solitary; marginarium a herringbone dissepimentarium, minor septa suppressed; major septa thin, short; tabulae horizontal, complete and incomplete (55). *M.Sil.-U.Sil.*, Eu.-N.Am.——FIG. 185, 2. *L. degeeri, Eu.; 2a,b, transv. and long. secs., ×1 (137).

?**Pseudocystiphyllum** WANG, 1947 [*P. lini]. Large, subcylindrical; marginarium a lonsdaleoid dissepimentarium; tabulae concave, incomplete, septa thin and in a narrow irregular zone in outer tabularium. *M.Sil.*, China.——FIG. 186,1. *P. lini; 1a,b, transv. and long. secs., ×1 (136).

?**Stratiphyllum** SCHEFFEN, 1933 [*non Stratophyllum* SMYTH, 1933 (=*Ethmoplax* SMYTH, 1933, *nom. van.*; *nec Stratiphyllum* LANG-S.-T., 1940, *nom. van.*)] [*S. cavernosum]. *L.Sil.*, Norway.

?**Phragmophyllum** SCHEFFEN, 1933 [*P. corrivatum]. *L.Sil.*, Norway.

Subfamily HALLIINAE Chapman, 1893

[*nom. transl.* STUMM, 1949 (*ex* Halliidae CHAPMAN, 1893)] [=Papiliophyllinae STUMM, 1949; Aulacophyllidae SOSHKINA, 1951]

Pinnate arrangement of septa well marked. *L.Dev.-M.Dev.*

Hallia M.Edw.-H., 1850 [**H. insignis*]. Cardinal, counter and 2 alar septa differently developed from others, the cardinal very long and in deep fossula on convex side; septa of cardinal quadrants longer than others and directed pinnately (30°) to the cardinal septum; septa of counter quadrants radial, directed at 30° toward alar septa; marginarium a septal stereozone in youth, a regular dissepimentarium later (52). *L.Dev.*, N.Am.——Fig. 186, *5. *H. insignis;* side, ×1 (73).

Aulacophyllum M.Edw.-H., 1850 [**Caninia sulcata* d'Orbigny, 1850] [=*Pinnatophyllum* Grabau, 1922]. Like *Hallia* but cardinal septum extremely short, leaving fossula empty (52). *L.Dev.-M.Dev.*, N.Am.; *M.Dev.*, N.S.W.——Fig. 186,*3a. *A. sulcatum* (Orb.), L.Dev., N.Am.; side, ×1 (106). ——Fig. 186,*3b,c. A. scyphus* (Rominger), type species of *Pinnatophyllum*, M.Dev., N.Am.; *3b,c,* calice, side, ×1 (127).

Odontophyllum Simpson, 1900 [**Aulacophyllum convergens* Hall, 1882]. Like *Aulacophyllum* but with long thin cardinal septum; distal edges of septa denticulate (52). *L.Dev.-M.Dev.*, N.Am.——Fig. 186,*2. *O. convergens* (Hall); *2a,* calice, ×1; *2b,c,* transv. and long. secs., ×1 (127).

Papiliophyllum Stumm, 1937 [**P. elegantulum*] [=*Eurekaphyllum* Stumm, 1937]. Like *Aulacophyllum* but with lonsdaleoid dissepimentarium and septa withdrawing from axis (52). *L.Dev.-*

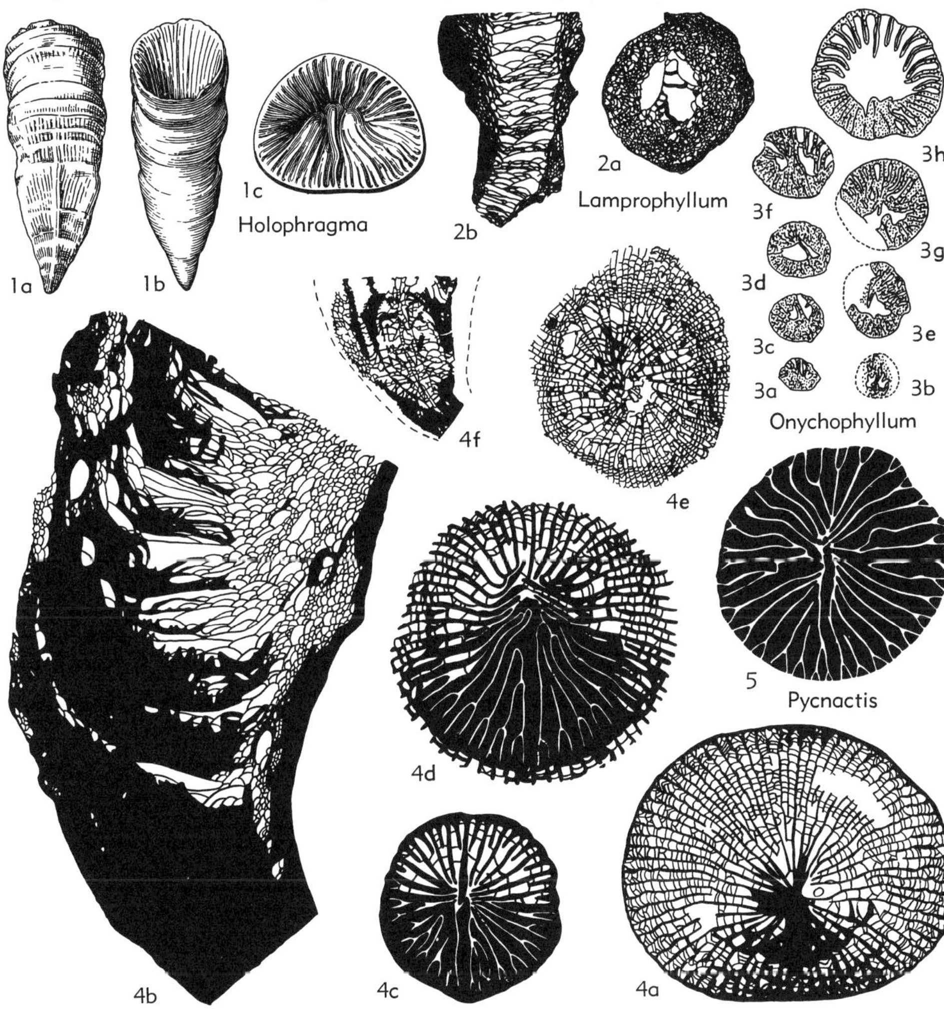

Fig. 185. Streptelasmatina (Zaphrenticae): Halliidae (p. F272).

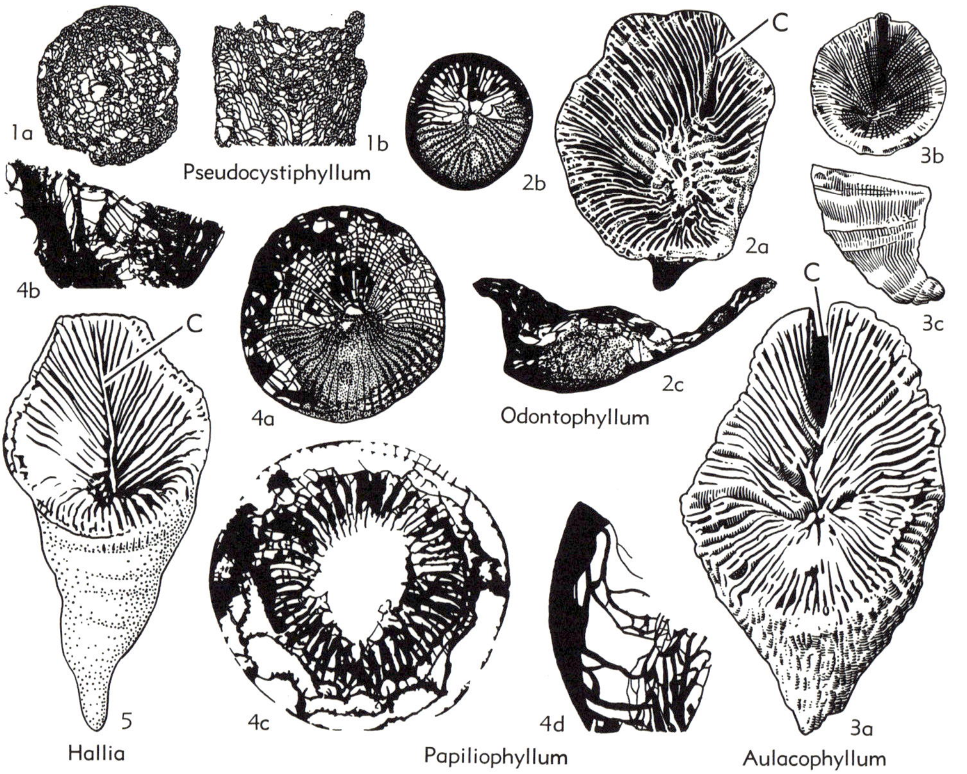

FIG. 186. Streptelasmatina (Zaphrenticae): Halliidae (p. F272-F274).

M.Dev., Nev.——FIG. 186,*4a,b*. **P. elegantulum* L.Dev., Nev.; transv. and long. secs., ×1 (127). ——FIG. 186,*4c,d*. *P. breviseptatum* (STUMM), type species of *Eurekaphyllum*, L.Dev., Nev.; transv. and long. secs., ×1 (127).

Family ARACHNOPHYLLIDAE Dybowski, 1873

Solitary or compound; marginarium a regular dissepimentarium, in some with a few lonsdaleoid dissepiments; major septa either meeting in an axial boss formed by steeply domed tabulae without axial lobes or withdrawn slightly from the axis, where tabular domes flatten or sag axially. *Sil.-Dev.*

Subfamily ARACHNOPHYLLINAE Dybowski, 1873

[*nom. transl.* HILL, herein (*ex* Arachnophyllidae DYB., 1873] [=Entelophyllidae, Entellophyllidae (misprint) HILL, 1940; Entelophyllinae (*nom. transl.* WANG, 1950)]

Septa may be represented by a network of vertical trabeculae or may be carinate; dissepiments small, globose. *Sil.*

Arachnophyllum DANA, 1846 [**Acervularia baltica* SCHWEIGGER *(partim)*, LONSDALE, 1839 (=*Strombodes murchisoni* M.EDW.-H., 1851) SD LANG & SMITH, 1927] [=*Favastraea* ORB., 1850 (*non* LANG-S.-T., 1940, *nom. van pro Favastrea* DE BLAINVILLE, 1834); *Darwinia* DYBOWSKI, 1873 (*non* BATE, 1857; *nec* others); *Arachniophyllum* LANG-S.-T., 1940 (*nom. van. pro Arachnophyllum* DANA, 1846) (*non* SMYTH, 1915)]. Astreoid; tabularium narrow, tabulae steeply domed, incomplete; dissepimentarium very wide, of numerous small dissepiments based horizontally, in which septa develop sporadically; septa thickened and contiguous, or each a network of small trabeculae standing vertically on the dissepiments but not piercing more than 1 or 2 successive dissepimental platforms (31). *L.Sil.-M.Sil.*, Eu.-N.Am.-Austral.——FIG. 187,*3*. **A. murchisoni* (M.EDW.-H.), Eu.; *3a,b*, transv. and long. secs., ×2 (98).

Petrozium SMITH, 1930 [**P. dewari*]. Fasciculate, corallites slender, like *Entelophyllum* but with numerous long septa meeting to form an axial structure; tabulae without axial sag (45). *L.Sil.*, Eng.——FIG. 187,*1*. **P. dewari*; *1a,b*, transv. secs., ×2; *1c*, long. sec., ×2 (117).

Entelophyllum WEDEKIND, 1927 [**Madreporites articulatus* WAHLENBERG, 1821; SD LANG-S.-T., 1940] [=*Xylodes* LANG-S., 1927 (*non* WATERHOUSE, 1876); *Stereoxylodes* WANG, 1947]. Solitary or fasciculate; septa commonly with zigzag carinae, may be thickened, slightly withdrawn from axis; a narrow regular zone of flat tabellae at outer margins of the tabular domes which usually sag axially (55). *M.Sil.-U.Sil.,* Eu.-Asia-N.Am.-E.Austral.——FIG. 187,*2a,b.* **E. articulatum* (WAHLENBERG), Eu.; *2a,b,* transv. and long. secs., ×2 (98).——FIG. 187,*2c. E. pseudodianthus* (WEISSERMEL), type species of *Stereoxylodes*), Eu.-Asia-N.Am.; transv. sec., ×2 (98).

Tenuiphyllum SOSHKINA, 1937 [**T. ornatum*]. Like *Petrozium* but cerioid and with some lonsdaleoid dissepiments (47). *M.Sil.,* Urals.——FIG. 187,*5.* **T. ornatum; 5a,b,* transv. and long. secs., ×2 (122).

Craterophyllum FOERSTE, 1909 [*non* BARBOUR, 1911; *nec* TOLMACHEV, 1931] [**Chonophyllum (Craterophyllum) vulcanius;* SD LANG-S.-T., 1940] [=*Naos* LANG, 1926]. Large, solitary, with inverted calice; septa long, major and minor equal, naotic, with or without narrow interseptal loculi with dissepiments; tabularium narrow, tabulae incomplete. *M.Sil.,* N.Am.——FIG. 188,*3a.* **C. vulcanius;* calice, ×1 (75).——FIG. 188,*3b,c. C. pagoda* (SALTER), type species of *Naos;* transv. and long. secs., ×1 (97).

Weissermelia LANG-S.-T., 1940 [*pro Ptilophyllum* SMITH-T., 1927 (*non* GUÉRIN-MÉNEVILLE, 1845; *non Weissermelia* SCHINDEWOLF, 1942)] [**Ptilophyllum lindströmi* SMITH-T., 1927]. Phaceloid, corallites slender; septa thin, carinate; major reaching axis; tabulae incomplete, axially depressed; dissepimentarium wide, peripheral plates large, interseptal, horizontally based and flattened. *U.Sil.,* Gotl.——FIG. 187,*4.* **W. lindstroemi* (SMITH-T.); *4a,b,* transv. and long. secs., ×3 (120).

Subfamily KYPHOPHYLLINAE Wedekind, 1927

[*nom. transl.* HILL, herein (*ex* Kyphophyllidae WDKD., 1927)]

Solitary; septa not carinate, large lonsdaleoid dissepiments predominant over small globose dissepiments. *Sil.*

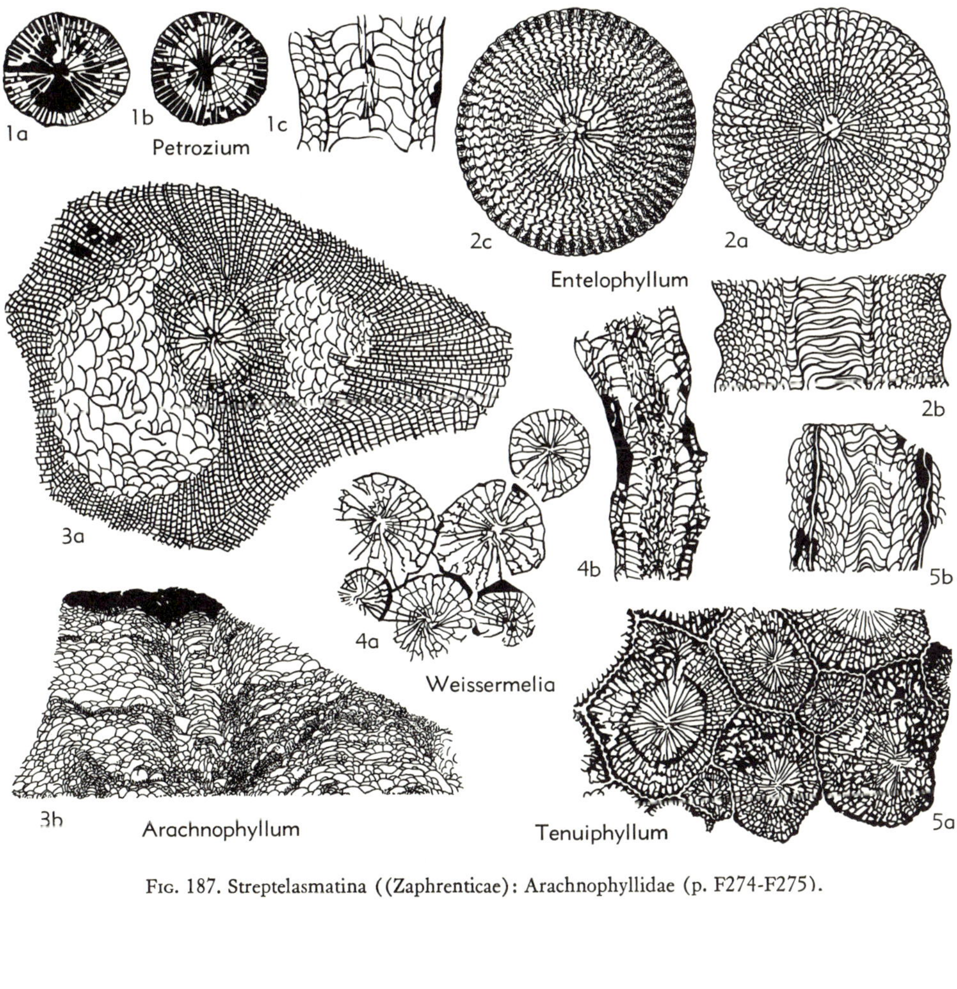

FIG. 187. Streptelasmatina ((Zaphrenticae): Arachnophyllidae (p. F274-F275).

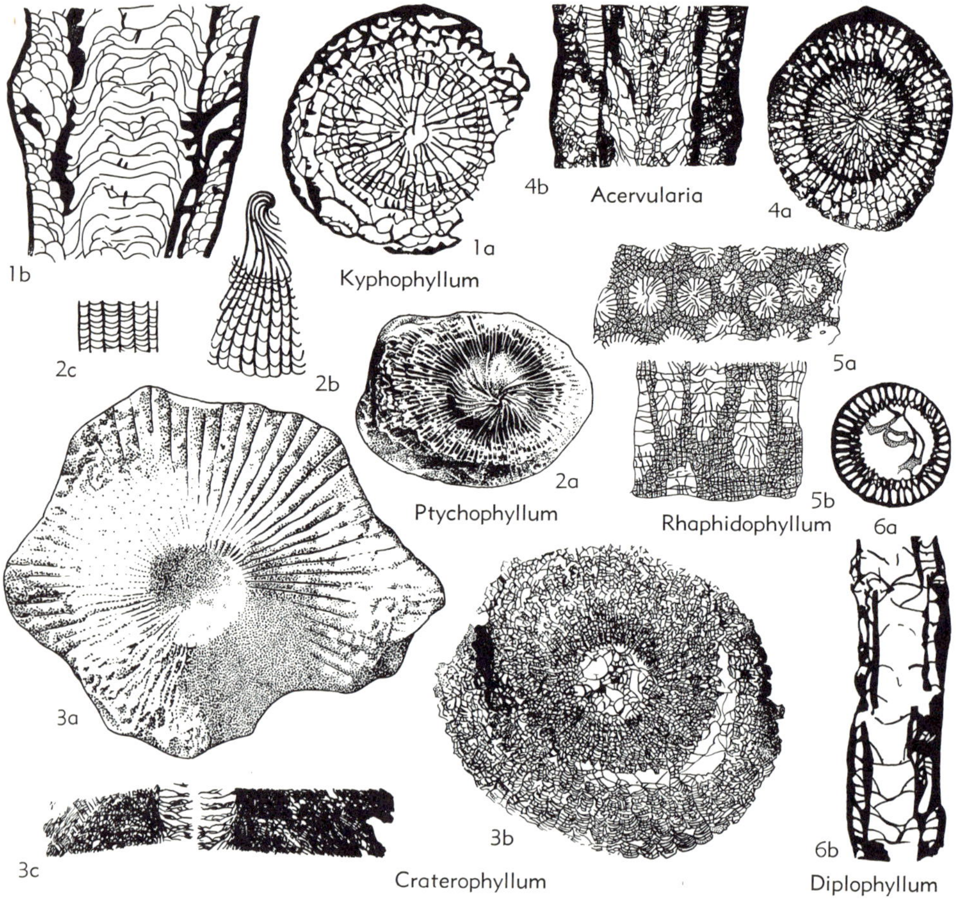

FIG. 188. Streptelasmatina (Zaphrentidae). Arachnophyllidae, Acervulariidae (p. F275-F277).

Kyphophyllum WEDEKIND, 1927 [**K. lindströmi*] [=*Cyphophyllum* LANG-S.-T., 1940 *(nom. van.)*] (55). *M.Sil.*, Gotl.——FIG. 188,*1*. **K. lindstroemi; 1a,b,* transv. and long. secs., ×2 (137).

Subfamily PTYCHOPHYLLINAE Dybowski, 1873

[*nom. transl.* HILL, herein (*ex* Ptychophyllidae DYB., 1873)]

Large, solitary; major septa reaching axis, their axial ends similarly curved; with elongate blister-like dissepiments lining sides of septa. *Sil.*

Ptychophyllum M.EDW.-H., 1850 [**P. stokesi*] [=*Cystiphorolites* MILLER, 1889 (*pro Vesicularia* ROMINGER, 1876, *non* THOMPSON, 1830)]. With wide reflexed dissepimentarium; fossula in tabularium; tabulae domes, incomplete (46). *M.Sil.*, N.Am.——FIG. 188,*2*. **P. stokesi; 2a,* calice, ×0.5; *2b,c,* transv. and long. secs., ×1 (117).

Family ACERVULARIIDAE Lecompte, 1952

[=Acervulariens DE FROMENTEL, 1861 (invalid vernacular name)]

Phaceloid or cerioid with axial, quadripartite increase; marginarium a complex dissepimentarium; septa carinate and dilated near inner ends of long minor septa, forming a wall within the dissepimentarium, at which septal trabeculae diverge; tabulae inversely conical. *Sil.*

Acervularia SCHWEIGGER, 1819 [**A. baltica* (=*Madrepora ananas* LINNÉ, 1758)] [=*Floscularia* EICHWALD, 1829 (*non* CUVIER, 1798, *nec* OKEN, 1815); *Favastrea* DE BLAINVILLE, 1834; *Favastraea* LANG-S.-T., 1940 *(nom. van.)* (*non* D'ORBIGNY, 1850); *Arachnium* KEYSERLING, 1846; *Cyathogonium* CHAPMAN, 1893; *Rhabdophyllum* WEDEKIND, 1927]. Tabulae complete or in part incomplete;

dissepimentarium with 3 zones, an outer comprising 2 or more series of globose plates, a middle consisting of flat plates just outside the inner wall, and an inner composed of globose plates inside this wall and merging into the tabulae; major septa long; third order septa may develop (31). *Sil.*, Eu.—FIG. 188,*4*. **A. ananas* (LINNÉ); *4a,b*, transv. and long. secs., ×2 (117).

Diplophyllum HALL, 1851 (*non* SOSHKINA, 1939) [**D. caespitosum*]. Like *Acervularia* but with only flat dissepiments developed, septa not extending beyond inner wall, and tabulae complete. *M.Sil.*, N.Am.——FIG. 188,*6*. **D. caespitosum*; *6a,b*, transv. and long. secs., ×2 (118).

?**Rhaphidophyllum** LINDSTRÖM, 1882 [**R. constellatum*]. Cerioid; septa each a lamina within the narrow marginarium of small regular dissepiments, but within the wide tabularium each represented by long, separated spines, directed upward and inward; tabulae flat, complete (35). *Sil.*, Sib.——FIG. 188,*5*. **R. constellatum*; *5a,b*, transv. and long. secs., ×2 (100).

Family MYCOPHYLLIDAE Hill, 1940

[*ex Mycophyllum* LANG-S.-T., 1940 (*nom. van. pro Mucophyllum* ETHERIDGE, 1894)] [=Pseudamplexinae STUMM, 1949]

Solitary, with tabularium surrounded by a broad marginarium as wide as the minor septa are long, in which the septa are so dilated as to be in contact; trabeculae rhabdacanthine, but lamellar sclerenchyme not continuous from one septum to another; tabulae horizontal, complete; no dissepiments. *M.Sil.-L.Dev.*

Mucophyllum ETHERIDGE, 1894 [**M. crateroides*] [=*Mycophyllum* LANG-S.-T., 1940 (*nom. van.*)]. Corallites broadly trumpet-shaped, peripheral stereozone expanding at calice into a wide rim that may be everted (20). *M.Sil.-U.Sil.*, Austral.-Gotl.——FIG. 189,*3*. **M. crateroides*, M.Sil., N.S. W.; *3a*, long. sec., ×1; *3c*, transv. sec. septa, ×1; *3b*, tang. long. sec. septa, ×1 (85).

Pseudamplexus WEISSERMEL, 1897 [**Zaphrentis ligeriensis* BARROIS, 1889] [=*Pselophyllum* POČTA,

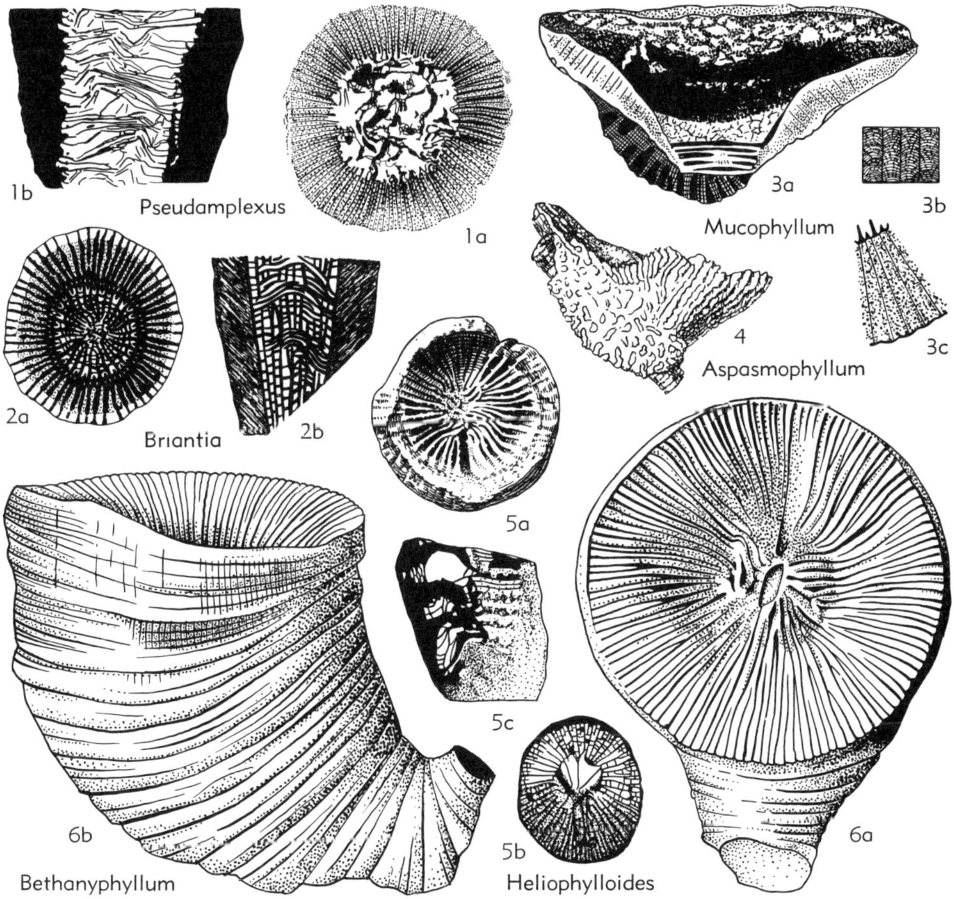

FIG. 189. Streptelasmatina (Zaphrenticae): Mycophyllidae, Zaphrentidae (p. F277-F278).

1902; *Pseliophyllum* LANG-S.-T., 1940 *(nom van.); Pseudomphyma* WEDEKIND, 1927]. Corallites conico-cylindrical, large with wide tabularium (55). *Sil.-L.Dev.*, Eu.-Austral.——FIG. 189,*1*. *P. bohemicus* (POČTA), type species of *Pselophyllum*, L.Dev., Czech.; *1a,b*, transv. and long. secs., ×0.5 (108).

Briantia BARROIS, 1889 [**B. repleta*]. Solitary, marginarium a broad septal stereozone; major septa long and thin in tabularium, almost reaching axis; tabulae broadly domed, complete or incomplete (52). *L.Dev.*, Fr.——FIG. 189,*2*. **B. repleta; 2a,b*, transv. and long. secs., ×1 (59).

Aspasmophyllum C. F. ROEMER, 1880 [**A. crinophilum*]. Corallum very broadly conical adherent by outgrowths from sides and base, usually attached to a crinoid stem (52). *M.Dev.*, Ger.——FIG. 189,*4*. **A. crinophilum;* side, ×1 (110).

Family ZAPHRENTIDAE Milne-Edwards & Haime, 1850

[*nom. correct.* ROEMER, 1883 (*ex* Zaphrentinae M.EDW.-H., 1850, *ex Zaphrentis* HAIME, 1850, *nom. van. pro Zaphrenthis* RAFINESQUE & CLIFFORD, 1820)] [=?Cyathophyllidae DANA, 1846; Zaphrentiniens DE FROMENTEL, 1861 (invalid vernacular name); Heliophyllidae NICHOLSON in NICH. & LYDEKKER, 1889; Zaphrenthidae MOORE & JEFFORDS, 1945; Zaphrentiadae MINATO, 1943; ?Bethanyphyllidae STUMM, 1949]

Solitary or compound, marginarium (when developed) a regular dissepimentarium; major septa either meet in an axial boss formed by steeply domed tabulae, or withdraw from the axis when the tabular domes flatten or sag axially; septa commonly carinate. *Dev.*

This name has been used previously for small solitary corals which herein are placed in the Cyathaxoniicae; the type genus of the family (*Zaphrenthis*) differs from the Cyathaxoniicae and from the conception attached to its name by possessing a dissepimentarium in its calical borders, within which the septa have crossbar carinae; and the type genus leads on, as SMITH (46) has shown, to *Heliophyllum* HALL.

Zaphrenthis RAFINESQUE & CLIFFORD 1820 [**Z. phrygia* RAFINESQUE & CLIFFORD; SD MILLER, 1889 (=*Caryophyllia cornicula* LESUEUR, *partim*, 1821] [=*Zaphrentis* HAIME, 1850 *(nom. van.); Helenterophyllum, Heliophrentis* GRABAU, 1910]. Major septa long, dilated equally in the tabularium where they are in contact axially; a narrow dissepimentarium of single globose dissepiments appears in the calical rim, in which the septa are thin with yardarm carinae (46). *Dev.*, N.Am.——FIG. 190,*3*. **Z. phrygia; 3a,b*, calice, long. sec., ×1 (113).

Heliophylloides STUMM, 1949 [**Cyathophyllum brevicorne* DAVIS, 1887]. Like *Zaphrenthis* but with dissepimentarium better and earlier developed and with major septa commonly withdrawn from the axis, when the tabular domes become flat-topped (52). *L.Dev.*, N.Am.——FIG. 189,*5*. **H. brevicornis* (DAVIS); *5a*, calice, ×1; *5b,c*, transv. and long. secs., ×1 (127).

Heliophyllum HALL in DANA, 1846 [**Strombodes helianthoides? (sic)* GOLDF., HALL, 1843 (=*Heliophyllum halli* M.EDW.-H., 1850) (*non* =*Cyathophyllum helianthoides* GOLDF., 1826)] [=*Heliogonium* CHAPMAN, 1893]. Corallites large, aggregated or weakly compound; septa with yardarm carinae, which may be dilated in the tabularium, when all are thickened equally; fossula marked by narrowing of the wide dissepimentarium and by axial ends of neighboring septa being short and curving around it; tabulae shallowly domed with axial sag or sagging, incomplete (52). *L.Dev.-M.Dev.*, N.Am.; *M.Dev.*, Morocco; *Dev.*, S.Am.——FIG. 190,*4*. **H. halli* M.EDW.-H., N.Am.; *4a,b*, transv. and long. secs., ×1 (127).

Radiophyllum HILL, 1942 [**Entelophyllum arborescens* HILL & JONES, 1940]. Like *Heliophyllum* but septa not carinate, long and thin with lateral blister-like plates; fossula not marked; tabulae shallowly domed, incomplete; subequal, large globose tabellae. *L.Dev.*, E.Austral.——FIG. 190,*1*. **R. arborescens* (HILL & JONES); *1a,b*, transv. and long. secs., ×1 (85).

Keriophyllum WEDEKIND, 1923 [**K. heiligensteini*] [=*Ceriophyllum* LANG-S.-T., 1940 *(nom. van.)*]. Corallites turbinate; septa thin, with zigzag carinae; fossula not marked; tabulae domed, may be complete (52). *M.Dev.*, Eu.——FIG. 190,*2*. **K. heiligensteini*, Eifel; *2a*, transv. sec., ×2; *2b*, long. sec. of dissepimentarium, ×2 (137).

Moravophyllum KETTNEROVA, 1932 [**M. ptenophylloides*]. Like *Heliophyllum* but with very numerous long septa without yardarm carinae, thin in tabularium in adult (52). *Up.M.Dev.*, Eu.——FIG. 190,*5*. **M. ptenophylloides*, Czech.; *5a,b*, transv. and long. secs., ×1 (93).

?Tortophyllum SLOSS, 1939 [**Zaphrentis cystica* WINCHELL, 1866]. Large, solitary, with numerous, long, thin, noncarinate septa, convolute in an axial structure with domed tabulae not sagging axially; regular dissepimentarium wide, narrowing at fossula (52). *Up.M.Dev.*, Mich.——FIG. 190,*6*. **T. cysticum* (WINCHELL); *6a,b*, transv. and long. secs., ×1 (116).

?Cyathophyllum GOLDF., 1826 [**C. dianthus;* SD DANA, 1846]. *M.Dev.*, Ger.(Eifel). GOLDFUSS' 6 figured syntype specimens are probably all different species. M.EDW.-H. (1851) selected the original of pl. 15, fig. 13, as type by excluding the other 5. This has not been sectioned; it may be close to *Heliophyllum*.

?Bethanyphyllum STUMM, 1949 [**Cyathophyllum robustum* HALL, 1876]. Large, curved, fossula on convex side; major and minor septa long, thin. *Up.M.Dev.*, N.Am.——FIG. 189,*6*. **B. robustum* (HALL); *6a,b*, calice, side view, ×1 (83).

Family PHILLIPSASTRAEIDAE C. F. Roemer, 1883

[*ex Phillipsastraea* D'ORBIGNY, 1850 (*nom. van. pro Phillipsastrea* ORB., 1849)] [=Phillipsastraeidea ROEMER, 1883 (misprint); Disphyllidae HILL, 1939; Mictophyllidae HILL, 1940; Phillipsastreidae HILL, 1954]

Compound or solitary; major and minor septa, carinate or dilated, typically reaching epitheca; minor septa long; marginarium a wide dissepimentarium of globose, usually small dissepiments, commonly with one vertical series of plates which are horseshoe-shaped in vertical section, with septal trabeculae diverging from the mid-plane; the trabeculae are slender and when the septum is dilated, they curve out laterally from its median plane in regular zones; tabularium in 2 regions separated at axial edges of major septa which are somewhat withdrawn from the axis; tabellae of axial region horizontal, periaxial ones being inclined either inward or outward. *M.Sil.-L.Carb.* (maximum *Dev.*).

Subfamily PHILLIPSASTRAEINAE C. F. Roemer, 1883

[*nom. transl.* HILL, herein (*ex* Phillipsastraeidae C.F.ROEMER, 1883)] [=Disphyllinae HILL, 1939 (*nom. transl.* WANG, 1948, *ex* Disphyllidae HILL, 1939; includes Breviphyllinae TAYLOR, 1951]

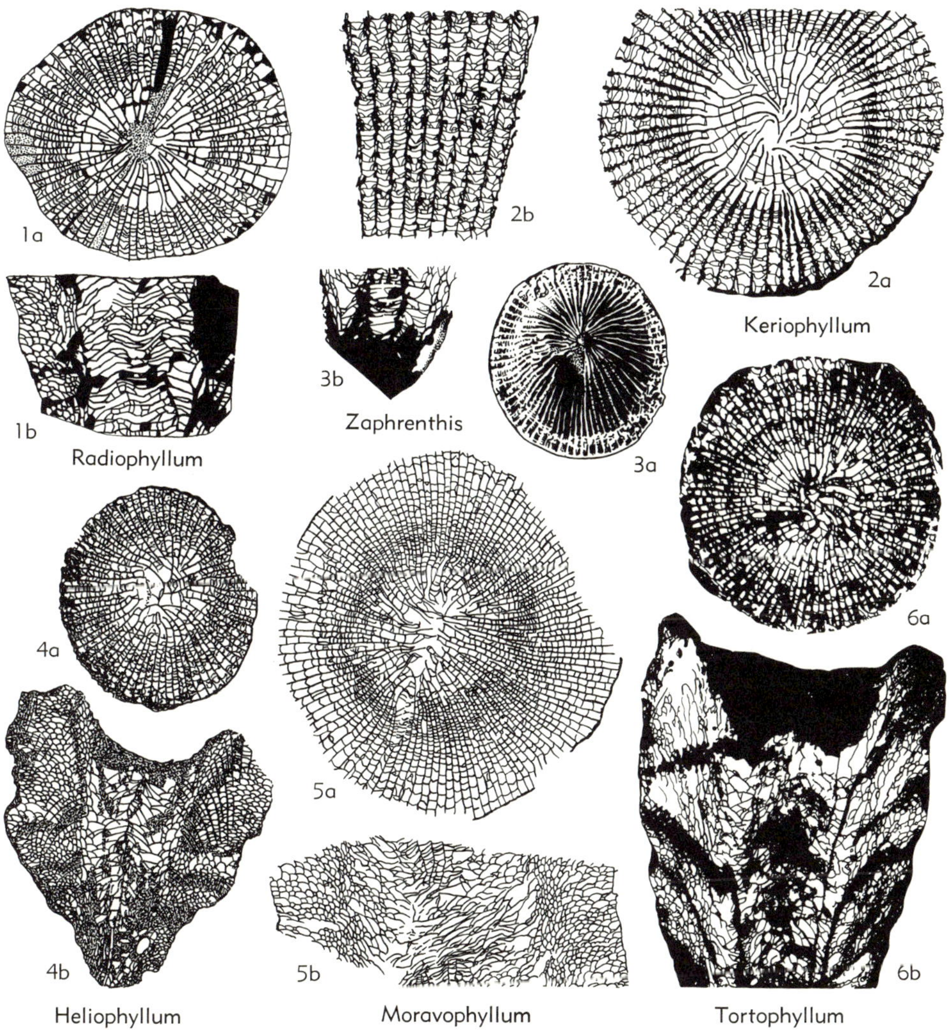

FIG. 190. Streptelasmatina (Zaphrenticae): Zaphrentidae (p. F278).

No aulos or well-defined series of horseshoe dissepiments developed. *Sil.-L.Carb.*

Disphyllum DE FROMENTEL, 1861 [**Cyathophyllum caespitosum* GOLDFUSS, 1826 *(partim)*; SD LANG & SMITH, 1934] [=*Schlüteria* WEDEKIND, 1922 (*non* FRITSCH, 1887; *nec* others); *Pseudostringophyllum, Megaphyllum* SOSHKINA, 1939 (*non* VERHOEFF, 1894); *Ceratinella* SOSHK., 1941 (*non* EMERTON, 1882)]. Phaceloid, with lateral or peripheral increase; septa only slightly dilated, carinate in some; dissepiments in several series of small, equal plates, the innermost being highly inclined; horseshoe dissepiments not developed (32). *M.Sil.*, N.S.W.; *Dev.*, cosmop.——FIG. 191,*10*. **D. caespitosum* (GOLDF.), M.Dev., Eu.; *10a,b*, transv. and long. secs., ×2 (97).

Cylindrophyllum SIMPSON, 1900 (*non* YABE & HAYASAKA, 1915) [**C. elongatum*] [=*Cylindrohelium* GRABAU, 1910; *Spinophyllum* WEDEKIND, 1922]. Like *Disphyllum* but septa with yardarm carinae; tabulae commonly complete (52). *L.Dev.-M.Dev.*, N.Am.; *Up.M.Dev.*, Ger.——FIG. 191,*8*. **C. elongatum, L.Dev.*, N.Am.; long. sec., ×1 (117).

Synaptophyllum SIMPSON, 1900 [**Diphyphyllum arundinaceum* BILLINGS, 1859]. Like *Disphyllum* but corallites commonly connected by lateral prolongations of dissepimentarium; septa thickest near inner margin of dissepimentarium; trabeculate thickening spreads as a coating over the dissepiments (52). *Dev.*, N.Am.——FIG. 191,*6*. **S. arundinaceum* (BILL.); *6a,b*, transv. and long. secs., ×4 (98).

Breviphyllum STUMM, 1949 [**Amplexus lonensis* STUMM, 1937]. Solitary, with short septa; tabulae flat, complete; dissepimentarium regular, narrow (52). *L.Dev.*, N.Am.-Eu.-Austral.——FIG. 191,*5*. **B. lonense* (STUMM); transv. sec., ×1 (127).

Phillipsastrea D'ORBIGNY, 1849 [**Astraea hennahi* LONSDALE, 1840 *(partim)*; SD M.EDW.-H., 1850] [=*Phillipsastraea* ORB., 1850; *Smithia* M.EDW.-H., 1851 (*non* SAUSSURE, 1855; *nec* others); *Streptastrea* SANDBERGER & SANDBERGER, 1856; *Pseudoacervularia* SCHLÜTER, 1881; *Streptastraea* LANG-S.-T., 1940 *(non. van.)*]. Massive, with walls between corallites absent; septa dilated, especially at inner margin of dissepimentarium (32). *Dev.*, Eu.-Asia-Austral.——FIG. 191,*3a*. **P. hennahi* (LONSD.), U.Dev., Eu.; *3a*, transv. sec., ×2 (117).——FIG. 191,*3b*. *P. currani* ETHERIDGE, ?M.Dev., N.S.W.; *3b*, long. sec., ×2 (117).

Billingsastraea GRABAU, 1917 [**Phillipsastrea verneuili* M.EDW-H., 1851] [=?*Asterocycles* VANUXEM, 1842; *Radiastraea* STUMM, 1937]. Like *Phillipsastrea* but septa carinate and attenuate; tabulae flat-topped domes (52). *L.Dev.-M.Dev.*, N.Am.; *M.Dev.*, Eu.——FIG. 191,*7*. **B. verneuili* (M.EDW.-H.); upper surface, ×1 (73).

Hexagonaria GÜRICH, 1896 [**Cyathophyllum hexagonum* GOLDFUSS, 1826; SD LANG-S.-T., 1940] [=*Polyphyllum* DE FROMENTEL, 1861 (*non* BLANCHARD, 1850); *Prismatophyllum* SIMPSON, 1900; *Hexagoniophyllum* GÜRICH, 1909]. Cerioid; septa thin, carinate or dilated, long, in some meeting or intertwining at the axis, when axial tabellae may be arranged in an axial structure, otherwise withdrawn slightly from axis when the axial tabellae are horizontal, periaxial ones being inclined inward or outward; dissepiments small and globose, numerous (52). *Dev.*, cosmop.; *L.Carb.*, China.——FIG. 191,*2*. **H. hexagona* (GOLDF.), M.Dev., Eu.; *2a,b*, transv. and long. secs., ×1 (127).

Haplothecia FRECH, 1885 [**Madrepora filata* SCHLOTHEIM (*partim*, var.α), 1820]. Cerioid, but dividing walls weak; septa with regular close yardarm carinae; trabeculae diverge in mid-region of dissepimentarium (11). *U.Dev.*, Ger.——FIG. 191, *1*. **H. filata* (SCHLOTH.); *1a,b*, transv. and long. secs., ×2 (77).

Ceratophyllum GÜRICH, 1896 [**C. typus*]. Solitary, ceratoid, with complete or incomplete broad flat tabulae and narrow dissepimentarium (52). *M. Dev.*, Ger.

Kunthia SCHLÜTER, 1885 [**K. crateriformis*]. Small, solitary, with calical floor extending almost to apex of cone; septa as in *Temnophyllum* (52). *M.Dev.*, Ger.——FIG. 191,*4*. **K. crateriformis; 4a,b*, calice, side view, ×1 (113).

Mictophyllum LANG & SMITH, 1939 [**M. nobile*]. Large, solitary, with a wide herringbone dissepimentarium in which minor septa are suppressed (46). *M.Dev.*, Austral.; *U.Dev.*, Can.——FIG. 192, *7*. **M. nobile*, U.Dev., Can.; *7a,b*, transv. and long. secs., ×1 (117).

Temnophyllum WALTHER, 1928 [**T. latum;* SD LANG-S.-T., 1940] [=*Diplophyllum* SOSHKINA, 1939 (*non* HALL, 1851); *Temeniophyllum* LANG-S. T., 1940 (*nom. van.*)]. Small, ceratoid, with outer half of septa in the dissepimentarium so dilated as to be in contact; trabeculae slender, curving outward from median plane of septum; dissepiments small and globose, merging with outer tabellae (52). *Up.M.Dev.*, Ger.; *U.Dev.*, USSR.——FIG. 191,*11a*. *T. ornatum* WALTHER, Up.M. Dev., Ger.; *11a*, transv. sec., ×2 (135).——FIG. 191,*11b*. **T. latum*, Up.M.Dev., Ger.; *11b*, long. sec., ×2 (135).

Charactophyllum SIMPSON, 1900 [**Campophyllum nanum* HALL & WHITFIELD, 1872]. Small, solitary, with moderately dilated and carinate septa widely withdrawn from axis; dissepimentarium narrow, formed of steeply inclined plates; trabeculae slender but not divergent (46). *U.Dev.*, N.Am.——FIG. 191,*9*. **C. nanum* (HALL-W.), *9a,b*, transv. and long. secs., ×2 (117).

?**Lyliophyllum** KELUS, 1939 [**L. pulcherrimum*]. Weakly compound, a few conico-cylindrical individuals united by lateral processes; internal structure unknown (52). *M.Dev.*, Pol.

Subfamily PHACELLOPHYLLINAE Wedekind, 1921

[=Pexiphyllidae WALTHER, 1928; Phacellophyllidae (*nom. transl.* KETTNEROVA, 1932); Thamnophyllidae, Peneckiellidae SOSHKINA, 1947; Pachyphyllinae STUMM, 1948]

Without aulos but with a well-defined series of horseshoe dissepiments. *Dev.*

Thamnophyllum PENECKE, 1894 [**T. stachei;* SD LANG & SMITH, 1935]. Dendroid, corallites united

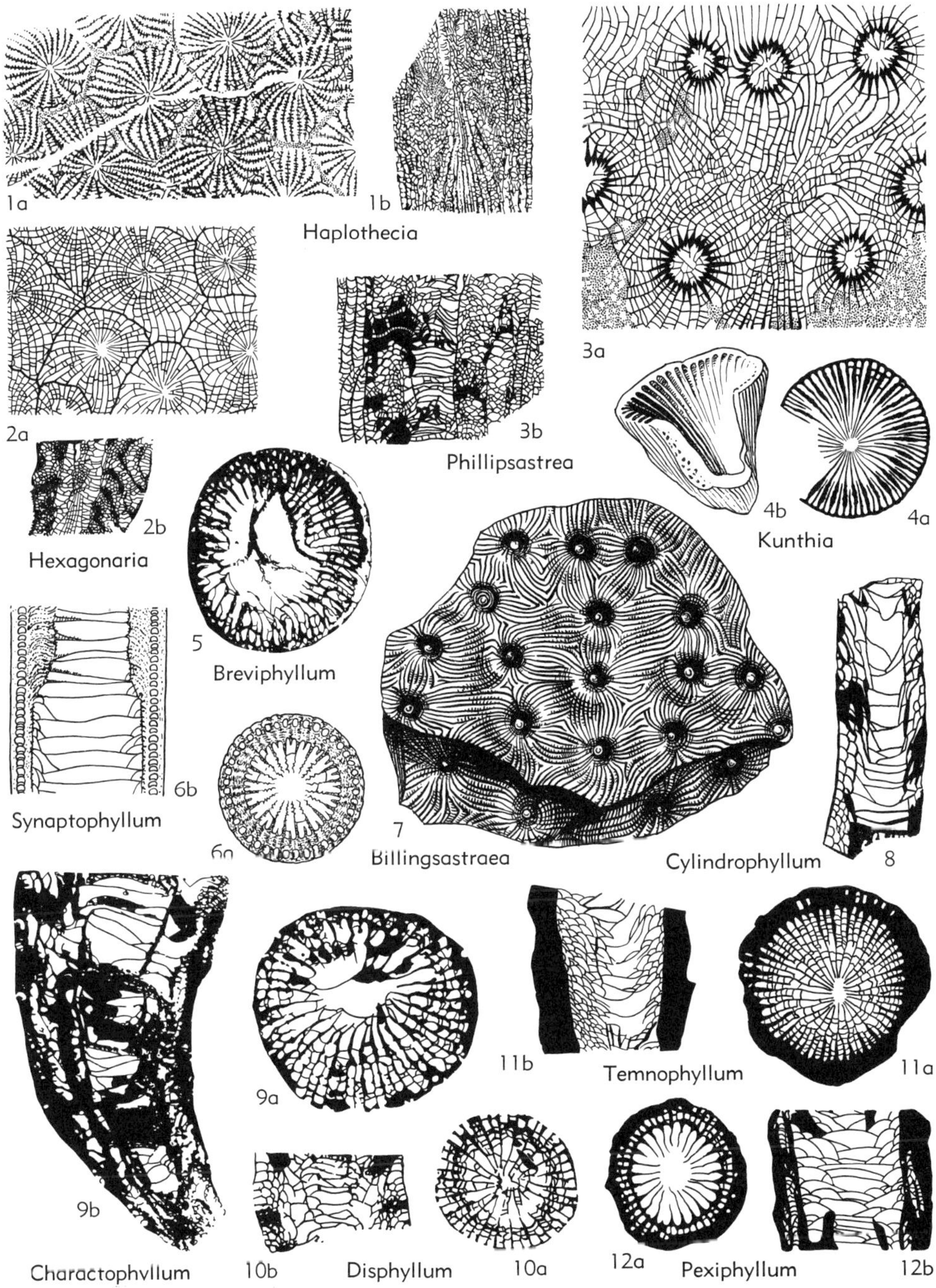

FIG. 191. Streptelasmatina (Zaphrenticae): Phillipsastraeidae (p. F280-F282).

by dissepimental tissue at axils of branches; septa short, dilated, the trabeculate thickening distributed over the very small dissepiments, of which one series comprises horseshoe plates; tabulae commonly complete (32). *L.Dev.-M.Dev.*, Eu.-Austral.——Fig. 192,*4*. **T. stachei*, Aus.; *4a,b*, transv. and long. secs., ×2 (98).

Trapezophyllum Etheridge, 1899 [**Cyathophyllum elegantulum* Dun, 1898]. Cerioid; dissepiments an outer series of flat plates and inner vertical series of horseshoe plates; tabulae complete (19). *L.Dev.-M.Dev.*, E.Austral.——Fig. 192,*8*. *T. coulteri* Hill, L.Dev., N.S.W.; transv. sec., ×2 (85).

Phacellophyllum Gürich, 1909 [**Lithodendron caespitosum* Goldfuss, 1826] [=*Fascicularia* Dybowski, 1873 (*non* Lamarck, 1816, *nec* M. Edw. in Busk, 1859); *Phacelophyllum* Lang-S.-T., 1940 (*nom. van.*)]. Dendroid or phaceloid; branches may be united in their axils by dissepimental tissue; dissepiments in 2 series, an outer of flat plates and an inner vertical series of horseshoe-shaped plates (32). *M.Dev.-U.Dev.*, Eu.——Fig. 192,*6*. **P. caespitosum* (Goldf.), M.Dev., Eu.; *6a,b*, transv. and long. secs., ×2 (98).

Peneckiella Soshkina, 1939 [**Diphyphyllum minus* F. A. Roemer, 1855]. Phaceloid or cerioid; septa short, tabulae commonly in one series and complete; one series of dissepiments, horseshoe-shaped but flattened above (48). *U.Dev.*, Eu.——Fig. 192,*3*. **P. minus* (Roemer); *3a,b*, transv. and long. secs., ×1 (77).

Pachyphyllum M.Edw.-H., 1850 [**P. bouchardi*] [=*Medusaephyllum* F. A. Roemer, 1855]. Massive, may be aphroid, with walls between corallites absent; septa dilated at inner margin of dissepimentarium; innermost dissepiments a vertical series of horseshoe-shaped plates, coated by trabeculate extensions from the dilated septa (52). *U.Dev.*, Eu.-N.Am.——Fig. 192,*5*. *P. ibergense* (F. A. Roemer), type species of *Medusaephyllum*, Ger.; *5a,b*, transv. and long. secs., ×2 (77).

Macgeea Webster, 1889 [**Pachyphyllum solitarium* Hall & Whitfield, 1873; SD Fenton & Fenton, 1924]. Solitary or with a few offsets, epitheca not extending to rim of calice so that peripheral edges of septa are exposed distally; septa somewhat dilated in the moderately wide dissepimentarium, which includes one vertical series of horseshoe-shaped dissepiments; trabeculae slender, laterally curving and divergent from horseshoe dissepiments (46). *U.Dev.*, N.Am.,-Asia-Austral.——Fig. 192,*2*. **M. solitaria* (Hall-W.), N.Am.; *2a*, side, ×1; *2b,c*, transv. and long. secs., ×2 (98).

Pexiphyllum Walther, 1928 [**P. rectum;* SD Lang-S.-T., 1940]. Small, slender, solitary, with septa much dilated in dissepimentarium, which includes one vertical series of horseshoe-shaped plates and other globose plates (52). *U.Dev.*, Eu.——Fig. 191,*12*. **P. rectum*, Ger.; *12a,b*, transv. and long. secs., ×2 (135).

Family CRASPEDOPHYLLIDAE Dybowski, 1873

[=Eridophylliens de Fromentel, 1861 (invalid vernacular name); Crepidophyllidae Grabau in Yü, 1934; Eridophyllinae Stumm, 1949]

Compound, with carinate or dilated septa, minor ones long; marginarium a wide dissepimentarium with small mostly globose plates; carinae curving upward and inward from peripheral bases of septa; an aulos separates horizontal axial tabellae from outwardly declined periaxial tabellae. *Dev.*

Eridophyllum M.Edw.-H., 1850 [**E. seriale*] [=*Craspedophyllum* Dybowski, 1873 (subj.); *Crepidophyllum* Nicholson & Thomson, 1876; *Schistotoecholasma* Stewart, 1938; *Schistotoechelasma* Lang-S.-T., 1940 (*nom. van.*)]. Phaceloid; septa but slightly dilated, carinate in some; dissepiments in several series of small equal plates, the innermost highly inclined; with an aulos (52). *L.Dev.-M.Dev.*, N.Am.; *M.Dev.*, Morocco.——Fig. 192,*1*. **E. seriale; 1a,b,* transv. and long. secs., ×1 (125).

Family LITHOSTROTIONIDAE d'Orbigny, 1851

[*pro* Lithodendroninae M.Edw.-H., 1850, *ex Lithodendron* Phillips, 1836 (*non* Schweigger, 1819) =*Lithostrotion* Fleming, 1828] [=Nematophyllinae M'Coy, 1851; Stylaxiniens de Fromentel, 1861 (invalid vernacular name); Axophylloidae (invalid), Diphyphyllinae Dybowski, 1873; Stylaxinidae Gerth, 1921; Lithostrotiontidae Grabau in Chi, 1931; includes Diphyphyllinae Dybowski, 1873]

Compound Rugosa with an axial structure, generally a columella formed by elongation and vertical prolongation of axial end of counter septum, and conical tabulae; with a regular concentric dissepimentarium in which major and minor septa are continuous; diphymorphs are characteristic, when the axial structure disappears and tabulae flatten (19). *L.Carb.-L.Perm.*

Lithostrotion Fleming, 1828 [**L. striatum* Fleming; SD ICZN Opinion 117 (=*Madrepora vorticalis* Parkinson, 1808)] [=*Lithostrotium* Agassiz, 1846; *Nemaphyllum, Stylaxis, Siphonodendron* M'Coy, 1849; *Lasmocyathus* d'Orbigny, 1849; *Nematophyllum* M.Edw.-H., 1850; *Petalaxis* M. Edw.-H., 1852; *?Fischerina* Stuckenberg, 1904 (*non* Terquem, 1878); *Cystidendron, Cystistrotion* Schindewolf, 1927; *Stylostrotion* Chi, 1935]. Phaceloid or cerioid; typically with columella, long major septa, and large conical tabulae, generally supplemented by outer, smaller and nearly horizontal tabulae; dissepiments absent in very small forms, normal and well developed in large forms; increase nonparricidal; diphymorphs common (18). *L.Carb.*, cosmop.; *M.Carb.*, C.Eu.-Asia.——Fig. 193,*2*. *L. clavaticum* Thomson, L.Carb., Scot.; *2a,b*, long. and transv. secs., ×1 (85).

Orionastraea SMITH, 1916 [**Sarcinula phillipsi* M'COY, 1849; SD SMITH, 1917] [=*?Protolonsdaleiastraea* GORSKY, 1932]. Astreoid, thamnasterioid or aphroid; like *Lithostrotion* but with columella weakly developed or absent; septa withdrawn from axis, and in some from periphery also when dissepiments become lonsdaleoid (18). *L. Carb.*, Eu.-Asia-Austral.——FIG. 193,*3*. *O. phillipsi* (M'COY), Eu.; transv. sec., ×2 (117).

Diphyphyllum LONSDALE, 1845 [**D. concinnum*] [=*Depasophyllum* YÜ, 1934 (*non* GRABAU, 1936); *Donophyllum* FOMICHEV, 1939]. Like fasciculate *Lithostrotion* but with columella failing to develop, and with tabulae domed or flattened axially and downturned peripherally (18). *L.Carb.*, Eu.-Asia-Austral.——FIG. 193,*4*. *D. lateseptatum* (M'COY), Br.I.; *4a,b,* transv. secs., ×2; *4c,d,* long. secs., ×2 (118).

Stylastraea LONSDALE, 1845 [*non* DE FROMENTEL, 1861] [**S. inconferta;* SD MILLER, 1889] [=*Diphystrotion* SMITH & LANG, 1930]. Cerioid, but with columella absent or reduced to spines on successive flat or slightly domed tabulae; minor septa discontinuous through inosculating dissepi-

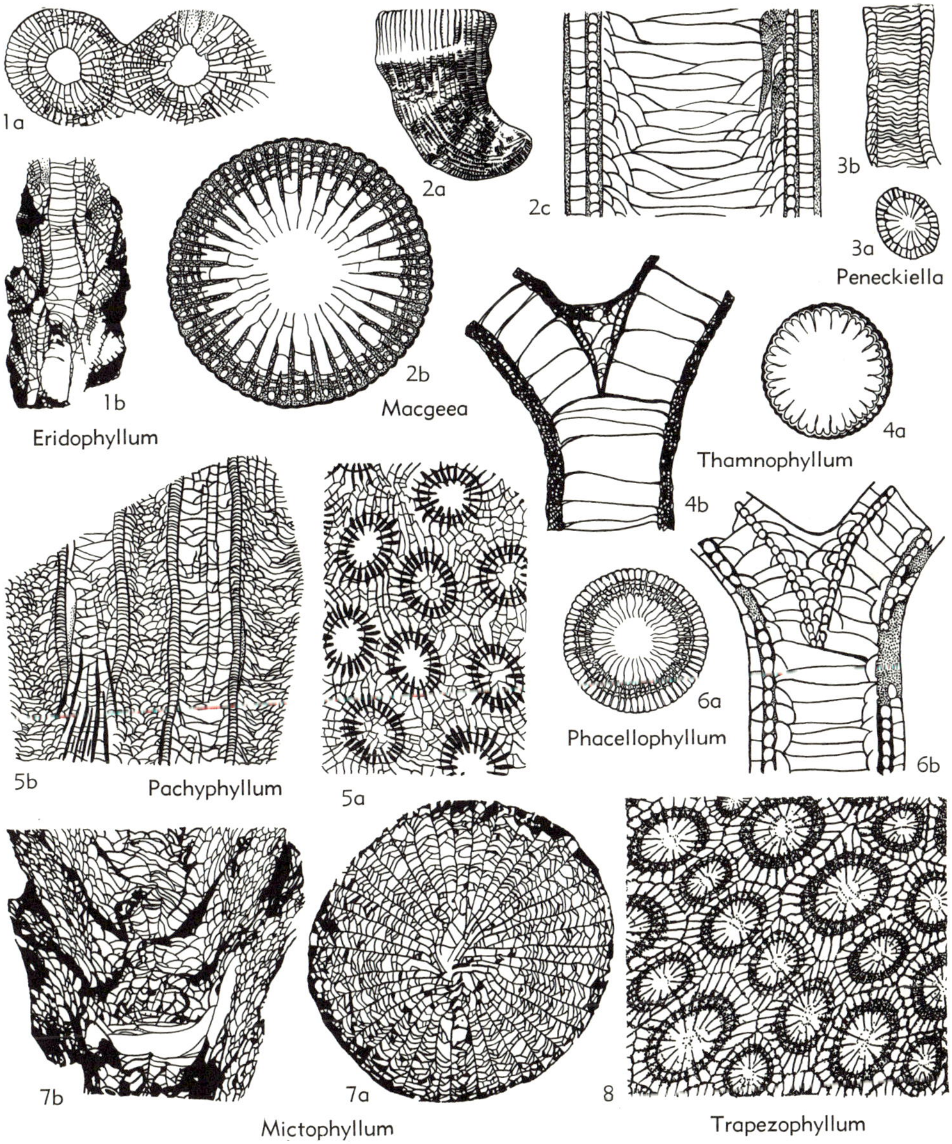

FIG. 192. Streptelasmatina (Zaphrenticae): Phillipsastraeidae, Craspedophyllidae (p. F280-F282).

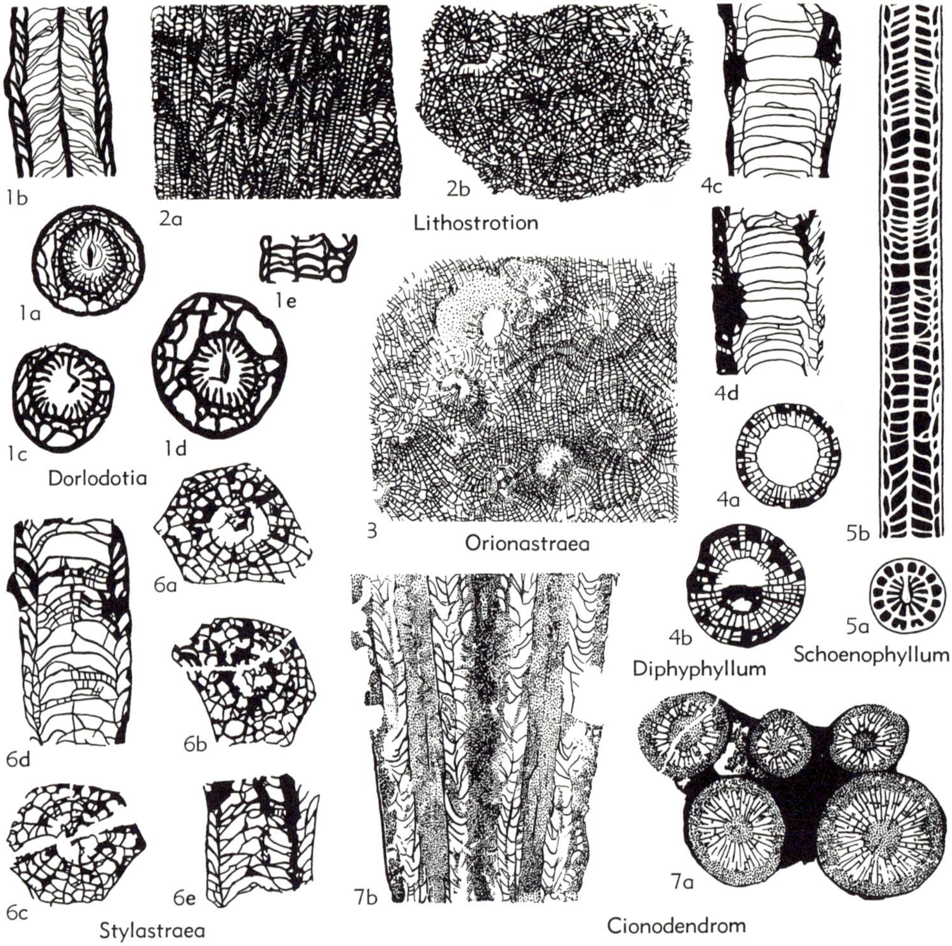

FIG. 193. Streptelasmatina (Zaphrenticae): Lithostrotionidae (p. F282-F284).

ments (18). *L.Carb.*, Eu.——FIG. 193,*6*. **S. inconferta*, USSR; *6a-c*, transv. secs., ×2; *6d,e*, long. secs., ×2 (118).

Schoenophyllum SIMPSON, 1900 [**S. aggregatum*]. Slender, fasciculate, with a single series of dissepiments and columella formed by elongation of counter septum (44). *Miss.*, N.Am.——FIG. 193, *5*. **S. aggregatum*, *5a,b*, transv. and long. secs., ×2 (115).

Cionodendrom BENSON & SMITH, 1923 [**C. columen*]. Fasciculate, like *Lithostrotion* but with a large dense round axial structure formed by axial septal ends fused with columella (15). *L.Carb.*, N.S.W.——FIG. 193,*7*. **C. columen*; *7a,b*, transv. and long. secs., ×3 (61).

Dorlodotia SALÉE, 1920 [**D. briarti*] [=?*Lytvophyllum* DOBROLYUBOVA in SOSHK.-D.-P., 1941]. Fasciculate, like *Lithostrotion* but with lonsdaleoid dissepimentarium (18). *L.Carb.*, Fr.-Ger.-?China; ?*L.Perm.(Artinsk)*, USSR.——FIG. 193,*1a,b*. **D. briarti*, L.Carb., Fr.; *1a,b*, transv. and long. secs., ×1 (111).——FIG. 193,*1c-e*. *L. tschernowi* (SOSHKINA), Artinsk., USSR; *1c,d*, transv. secs., ×2; *1e*, long. sec., ×2 (122).

Nemistium SMITH, 1928 [**N. edmondsi*]. Dendroid, with peripheral, parricidal increase; with narrow, regular dissepimentarium, short major septa and axial structure consisting of an irregular medial plate with few straggling septal lamellae impersistent in axial zone of globose, regularly superposed tabellae, from which the outer tabellae slope downward (18). *L.Carb.(Visé.)*, Eng.-Scot.——FIG. 194,*2*. **N. edmondsi*, Visé., Eng.; *2a*, transv. sec., ×2; *2b,c*, long. secs., ×2 (117).

Aulina SMITH, 1916 [**A. rotiformis*]. Compound, with regular dissepimentarium and aulos but no columella; septa may have zigzag but not yardarm carinae in dissepimentarium (18). *L.Carb.* (D_2-

Namur.), Eu.-Asia.——FIG. 194,*6*. **A. rotiformis,* Namur., Br.I.; *6a,b,* transv. and long. secs., ×2, ×3 (117).

Arachnastraea YABE & HAYASAKA, 1916 [**A. manchurica*] [=*Cystophorastraea* DOBROLYUBOVA, 1935; *Cystiphorastraea* LANG-S.-T., 1940 *(nom. van.)*]. Corallum astreoid, thamnasterioid, or aphroid; septa thin, major septa meeting conjoined cardinal and counter septa which are only faintly thickened to suggest a medial plate; tabulae conical, complete or incomplete; dissepimentarium regular (57). *M.Carb.,* USSR; *?L.Perm.(Artinsk.),* Manch.——FIG. 194,*3b,c.* **A. manchurica* ?Artinsk., Manch.; *3b,c,* long. and transv. secs., ×2 (140).——FIG. 194,*3a,d. A. molli* (STUCKENBERG), type species of *Cystophorastraea,* M.Carb., USSR; *3a,d,* transv. and long. secs., ×2 (68).

Tschussovskenia DOBROLYUBOVA, 1936 [**T. captiosa*]. Like *Nemistium,* but with impersistent dissepimentarium, and axial series of tabellae seldom

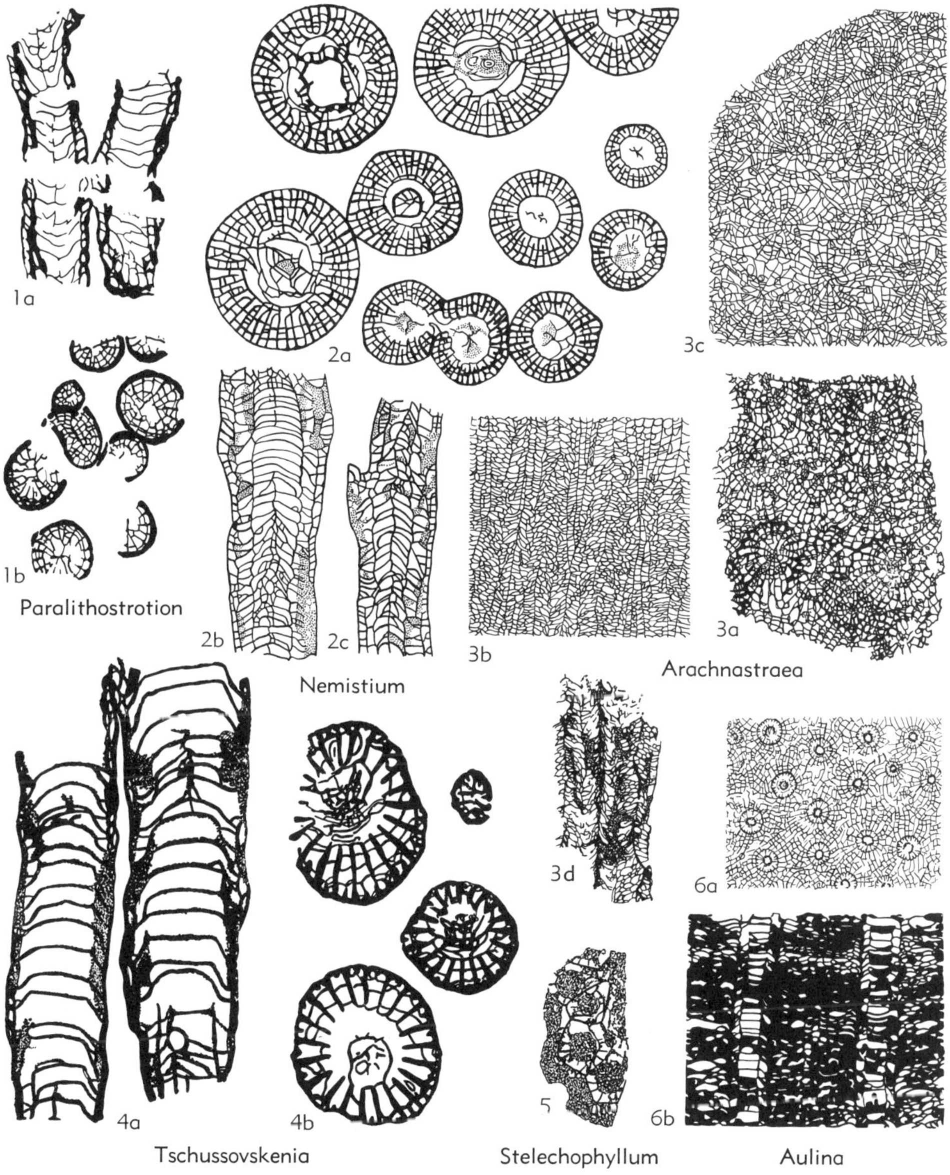

FIG. 194. Streptelasmatina (Zaphrenticae): Lithostrotionidae (p. F284-F286).

differentiated (49). *U.Carb.-L.Perm.(Artinsk.)*, USSR.——FIG. 194,*4*. **T. captiosa; 4a,b,* long. and transv. secs., ×2 (68).

Stelechophyllum TOLMACHEV, 1933 [*pro Stylophyllum* TOLM., 1924 (*non* REUSS, 1854] [**Stylophyllum venukoffi* TOLM., 1924; SD TOLM., 1933]. Cerioid, with long major septa almost meeting at axis, without distinct columella or median plate, and with border of lonsdaleoid dissepiments (18). *L.Carb.(Tournais.)*, W.Sib.——FIG. 194,*5*. **S. venukoffi* (TOLM.); transv. sec., ×1 (131).

?Paralithostrotion GORSKY, 1938 [**P. jermolaevi*]. Fasciculate, with *Lithostrotion*-like columella, single series of dissepiments, rudimentary minor septa, and concave tabulae (53). *Carb.*, N.Zem. ——FIG. 194,*1*. **P. jermolaevi; 1a,b,* long. and transv. secs., ×2 (80).

Family AULOPHYLLIDAE Dybowski, 1873

Simple or less commonly compound Rugosa with numerous septa, a regular dissepimentarium, incomplete conical tabulae, and generally an axial structure. Septa are equally spaced and seldom curved about the small, open cardinal fossula, which is marked by extension of the tabularium into the dissepimentarium; major septa may be dilated in the tabularium, particularly in cardinal quadrants; minor septa may be degenerate. Dissepiments are small and globose, concentric, angulo-concentric, or inosculating, rarely lonsdaleoid. The axial structure normally consists of straight or curved septal lamellae, commonly with a columella or median plate, and an inner series of tabellae. *Carb.-Perm.*

Subfamily AULOPHYLLINAE Dybowski, 1873

[*nom. transl.* HILL, herein (*ex* Aulophyllidae DYBOWSKI, 1873)] [=Rhodophyllesia THOMSON, 1874 (tribe); Aspidophyllacea THOMSON, 1875; Diplocyathophyllidae THOMSON, 1882 (invalid, not based on generic name); Clisiophyllidae NICHOLSON & THOMSON, 1883; Clisaxophyllidae GRABAU in YÜ, 1934, Dibunophyllinae, Koninckophyllinae WANG, 1950]

Tabularial part of fossula short, open and narrowing axially. *Carb.-Perm.*

Clisiophyllum DANA, 1846 [**C. keyserlingi* M'COY, 1849; SD DINGWALL, 1926] [=*Arachniophyllum* SMYTH, 1915 (*non* LANG-S.-T., 1940, *nom. van. pro Arachnophyllum* DANA, 1846); *Clisaxophyllum* GRABAU in CHI, 1931; *Clisiaxophyllum* LANG-S.-T., 1940 (*nom. van*)]. Solitary, with normal minor septa in a regular dissepimentarium; axial structure wide, with thin septal lamellae about half as numerous as major septa, commonly convolute and abutting on a short, thickened median plate (18). *L.Carb.*, Eu.-Asia.——FIG. 195,*1a,b*. **C. keyserlingi* M'COY, Visé.-Namur., Eu.; *1a,b,* transv. and long. secs., ×1 (85).——FIG. 195,*1c,d*. *C. coniseptum* (KEYSERLING) (type species of *Clisaxophyllum*), Carb., USSR; ×1 (94).

Aulophyllum M.EDW.-H., 1850 [**Clisiophyllum prolapsum* M'COY, 1849 (=*Turbinolia fungites* FLEMING, 1828)] [=*Cyclophyllum* DUNCAN & THOMSON, 1867 (*pro Cyclocyathus* DUNCAN & THOMSON, 1867, *non* M.EDW.-H., 1850); *?Setamainella* MINATO, 1943]. Solitary, with dissepimentarium as in *Clisiophyllum;* axial structure a compact, well-defined column, cuspidate in transverse section and built up of closely packed lamellae and tabellae without a median plate (18). *L. Carb.(Visé.)*, Eu.-N.Afr.-?Japan.——FIG. 195,*3a,b*. **A. fungites* (FLEMING), *Visé.*, Eu.-N.Afr.; *3a,b,* transv. and long. secs., ×1 (85).——FIG. 195,*3c*. **S. hayasakai* MINATO, Visé., Japan; *3c,* transv. sec., ×2 (103).

Auloclisia LEWIS, 1927 [**A. mutatum*]. Like *Aulophyllum,* but axial column with dibunophylloid medial plate in young stages (18). *L.Carb.(Visé.)*, W.Eu.——FIG. 195,*2*. **A. mutatum; 2a,b,* transv. secs., ×1; *2c,* long. sec., ×1 (99).

Dibunophyllum THOMSON & NICHOLSON, 1876 [**D. muirheadi* (=*Clisiophyllum bipartitum* M'COY, 1849); SD GREGORY, 1917] [=*Rodophyllum* THOMSON, 1874; *Rhodophyllum, Aspidiophyllum, Kurnatiophyllum* THOMSON, 1875; *Kumatiophyllum* THOMSON, 1876 *(nom. van.), Aspidophyllum* THOMSON-N., 1876; *Cymatiophyllum (nom. van.), Albertia* THOMSON, 1878 (*non* DUJARDIN, 1838, *nec* RONDANI, 1843); *Histiophyllum* THOMSON, 1879; *Centrephyllum* THOMSON, 1880; *Centrophyllum* THOMSON, 1883 *(nom. van.); Cymatophyllum (nom. van.), Centrolamellum* THOMSON, 1901; *?Protodibunophyllum* LISSITZIN, 1925]. Solitary, with inner parts of minor septa degenerate, leaving inner dissepiments inosculating; axial structure variable, typically one-third as wide as corallite, consisting of a median plate and 4 to 8 septal lamellae on either side, less typically, the lamellae may become curved, the median plate disappear, or the bilateral arrangement be lost (18). *L.Carb.*, Eu.-N.Afr.-N.Am.; *M.Carb.*, Carpathians-Japan. ——FIG. 196,*1a*. **D. bipartitum bipartitum* (M'COY) (lectotype of *D. muirheadi*), Eu.; transv. sec., ×1 (85).——FIG. 196,*1b*. *D. bipartitum bipartitum* (M'COY, 1849, lectotype), Eu.; long. sec., ×1 (85).——FIG. 196,*1c,d*. *D. bipartitum konincki* (M.EDW.-H.) (lectotype of *Aspidiophyllum koninckianum* THOMSON, type of this genus), Br.I.; transv. and long. secs., ×1 (85).——FIG. 196,*1e,f*. *D. bipartitum konincki* (M.EDW.-H.) (lectotype of *Kurnatiophyllum concentricum* THOMSON, type of this genus), Br.I.; transv. and long. secs., ×1 (130).——FIG. 196,*1g*. *D. bipartitum craigianum* (THOMSON) (type species of *Rodophyllum*). Visé.-Namur., Br.I.; transv. sec., ×1 (85).

Koninckophyllum THOMSON & NICHOLSON, 1876 [**K. magnificum;* SD THOMSON, 1883] [=*Lophophylloides* STUCKENBERG, 1904; *Eostrotion* VAUGHAN, 1915; *?Neokoninckophyllum* FOMICHEV,

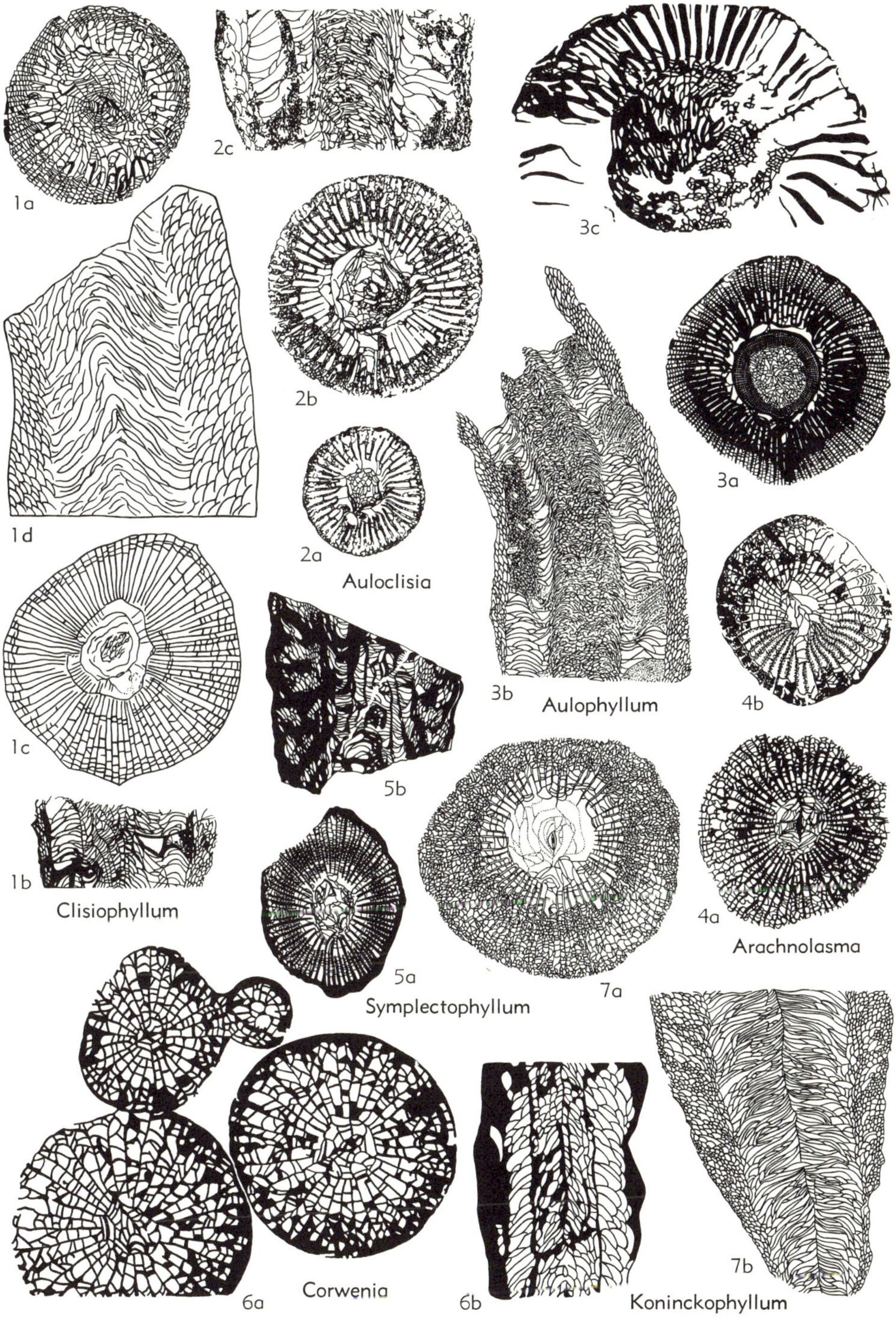

Fig. 195. Streptelasmatina (Zaphrenticae): Aulophyllidae (p. F286-F289).

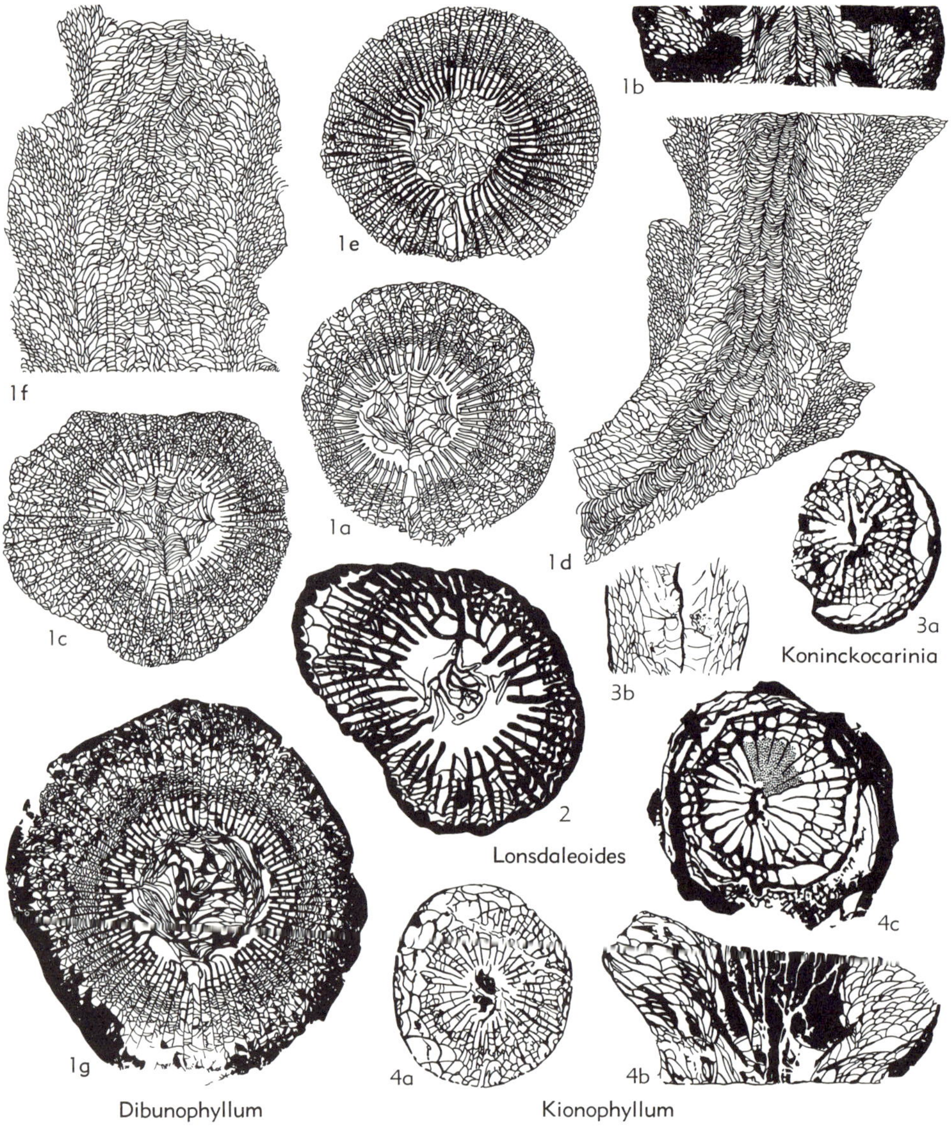

FIG. 196. Streptelasmatina (Zaphrenticae): Aulophyllidae (p. F286-F290).

1939]. Solitary or fasciculate; minor septa may be shortened axially in dissepimentarium; axial structure a columella, which may be supported by a few septal lamellae; tabulae tented and incomplete; if columella is absent, the tabulae flatten and may become complete (18). *Carb.*, Eu.-Asia-N.Am. ——FIG. 195,*7*. **K. magnificum*, Visé.-Namur., Br.I.; transv. and long. secs., ×1 (85).

Arachnolasma GRABAU, 1922 [**Lophophyllum sinense* YABE & HAYASAKA, 1920] [=*Yuanophyllum* YÜ, 1931; *Arachnelasma* LANG-S.-T., 1940 *(nom. van.)*]. Like *Dibunophyllum* but with narrow axial structure, in which the median plate is much thickened and septal lamellae fewer (3 to 5) on each side (58). *L.Carb.*, China; *M.Carb.*, Spitz. ——FIG. 195,*4a*. **A. sinense* (YABE-H.), L.Carb., China; transv. sec., ×1 (140).——FIG. 195,*4b*. *A. kansuense* (YÜ) (type species of *Yuanophyllum*), L.Carb., China; transv. sec., ×1 (143).

Corwenia SMITH & RYDER, 1926 [**Lonsdaleia rugosa* M'COY, 1849]. Fasciculate, with dibunophylloid axial structure and dissepimentarium (18).

L.Carb., Eu.-Asia.——Fig. 195,*6.* **C. rugosa* (M'Coy), Visé., Eu.; *6a,b,* transv. and long. secs., ×2 (119).

Carruthersella Garwood, 1913 [**C. compacta*]. Solitary, small; columella of tightly packed septal lamellae abutting on a medial plate; septa dilated, dissepimentarium narrow, with lonsdaleoid plates peripherally (18). *L.Carb.,* Eng.——Fig. 197,*5.* **C. compacta,* U.Tournais., Eng.; *5a,b,* transv. and long. secs., ×3 (78).

Symplectophyllum Hill, 1934 [**S. mutatum*]. Solitary; axial structure variable, dibunophylloid; skeletal thickening great, not regular; septa naotic peripherally (16). *L.Carb.,* E.Austral.——Fig. 195, *5.* **S. mutatum,* Visé., E.Austral.; *5a,b,* transv. and long. secs., ×1 (85).

Nagatophyllum Ozawa, 1925 [**N. satoi*]. Fasciculate, corallites large; axial structure consisting of median plate surrounded by tightly packed tabellae without septal lamellae; major and minor septa long, naotic peripherally; dissepimentarium wide; tabulae conical. *L.Carb.-M.Carb.,* Japan.——Fig. 197,*3.* **N. satoi; 3a,b,* transv. and long. secs., ×2 (107).

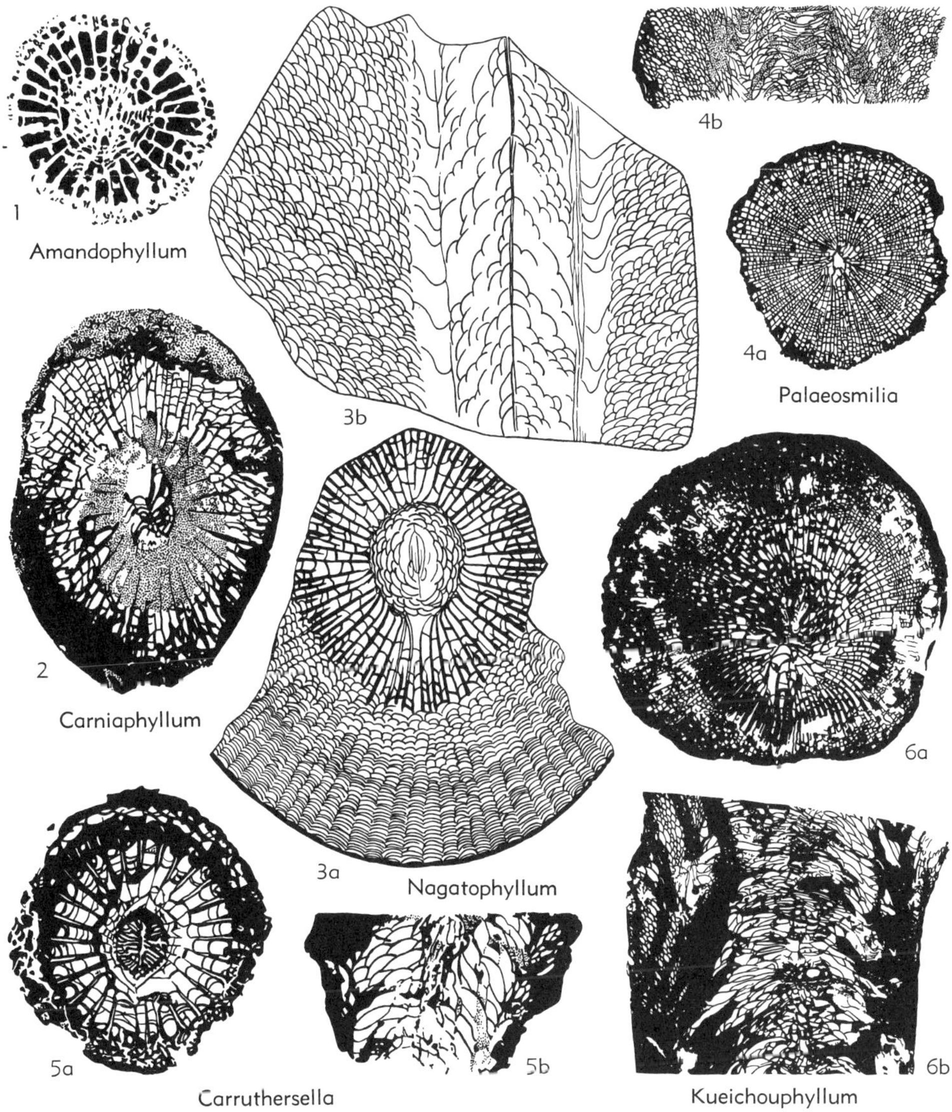

Fig. 197. Streptelasmatina (Zaphrenticae): Aulophyllidae (p. F289-F290).

Lonsdaleoides HERITSCH, 1936 [**L. boswelli*]. Like *Symplectophyllum* but corallum compound and bushy, with large corallites (14). *L.Perm.(Artinsk.)*, Carnic Alps.——FIG. 196,*2*. **L. boswelli;* transv. sec., ×2 (84).

Kionophyllum CHI, 1931 [**K. dibunum*] [=*Geyerophyllum* HERITSCH, 1936; *Cionophyllum* LANG-S.-T., 1940 *(nom. van.)*]. Like *Koninckophyllum* but with lonsdaleoid dissepiments peripherally and septal lamellae irregularly reinforcing columella (5). *M.Carb.*, China; *U.Carb.(Triticites zone)*, Carnic Alps.——FIG. 196,*4a,b*. **K. dibunum* M.Carb., China; *4a,b,* transv. and long. secs., ×1 (65).——FIG. 196,*4c*. **K. carnicum* (HERITSCH) (type species of *Geyerophyllum*), U.Carb. *(Triticites* zone), Carnic Alps; transv. sec., ×2 (84).

Koninckocarinia DOBROLYUBOVA, 1937 [**K. flexuosa*]. Solitary, conico-cylindrical, with one septum produced into an elongated columella not reached by other septa; septa with tuberculate sides; peripheral dissepiments lonsdaleoid, others normal; tabulae flat, with upturned edges (6). *M.Carb.*, USSR.——FIG. 196,*3*. **K. flexuosa; 3a,b,* transv. and long. secs., ×2 (68).

Carniaphyllum HERITSCH, 1936 [**C. gortanii*]. Solitary, with axial structure consisting of long thickened medial plate surrounded by tabellae arranged tent-wise, and rarely crossed by septal lamellae; dissepimentarium regular, wide (14). *M.Carb.*, USSR; *L.U.Carb.(Auernig)*, Carnic Alps.——FIG. 197,*2*. **C. gortanii,* U.Carb.(Auernig), Carnic Alps; transv. sec., ×3 (84).

Amandophyllum HERITSCH, 1941 [**Clisiophyllum carnicum* HERITSCH, 1936]. Solitary, axial structure of 4 or 6 septal lamellae developed on either side of an irregular, unthickened median lamella to which some run parallel; dissepimentarium narrow, dissepiments inosculating; minor septa may be withdrawn to periphery (14). *M.Carb.*, USSR; *L.U.Carb.(?Triticites* zone), Carnic Alps.——FIG. 197,*1*. **A. carnicum* (HERITSCH), L.U.Carb., Carnic Alps; transv. sec., ×3 (84).

Subfamily AMYGDALOPHYLLINAE Grabau in Chi, 1935

[*nom. transl.* HILL, herein (*ex* Amygdalophyllidae GRABAU in CHI, 1935)] [=?Campophyllidae, ?Campophyllinae WEDEKIND, 1921; Palaeosmiliidae HILL, 1940]

Fossula deepened, lengthened or expanded axially in tabularium; septa very numerous. *L.Carb.*

Cyathoclisia DINGWALL, 1926 [**C. tabernaculum*]. Dissepimentarium regular, minor septa normal; axial structure crowded with convolute septal lamellae as numerous as the major septa and commonly continuous with them; median plate small and thickened; tabellae not clearly differentiated by change of slope into inner and outer series; fossula expanding into outer part of tabularium (18). *L.Carb.(Tournais.)*, W.Eu.——FIG. 198,*5*. **C. tabernaculum; 5a,b,* transv. and long. secs., ×1 (67).

Heterocaninia YABE & HAYASAKA, 1920 [**H. tholusitabulata*]. Large, solitary, with loose axial structure of curved septal lamellae about one-sixth as numerous as septa and mostly not continuous with them, usually without distinct medial plate; minor septa absent, but dissepimentarium wide, with inosculating dissepiments; major septa dilated in tabularium in cardinal quadrants; fossula with short cardinal septum (58). *L.Carb.*, China.——FIG. 198,*6*. **H. tholusitabulata; 6a,b,* transv. and long. secs., ×1 (140).

Kueichouphyllum YÜ, 1931 [**K. sinense*] [=*Yabeella* YÜ, 1934; *Kesenella* NAGAO & MINATO, 1941]. Like *Heterocaninia* but with long minor septa developed in the wide dissepimentarium (58). *L.Carb.*, Asia; *M.Carb.*, Spitz.——FIG. 197,*6*. **K. sinense,* L.Carb., Asia; *6a,b,* transv. and long. secs., ×1 (143).

Turbophyllum PARKS, 1951 [**T. multiconum*] [=*Faberophyllum* PARKS, 1951]. Like *Heterocaninia* but with long minor septa and septa not generally dilated in tabularium; axial structure weak, as in *Heterocaninia,* or with sharply tented axial tabellae. *L.Carb.*, N.Am.

Amygdalophyllum DUN & BENSON, 1920 [**A. etheridgei*] [=*Echigophyllum* YABE & HAYASAKA in HAYASAKA, 1924; *?Carinthiaphyllum* HERITSCH, 1936; *Ekvasophyllum* PARKS, 1951]. Solitary, with large columella, almond-shaped in section, made of dilated axial ends of the long major septa welded together with intercalated septal lamellae to form a medial plate; with wide normal dissepimentarium and incomplete domed tabulae having upturned edges (15). *L.Carb.*, E.Austral.-Eu.-Asia; *M.Carb.*, Eu.——FIG. 198,*1*. **A. etheridgei,* Visé., E.Austral.; *1a,b,* transv. and long. secs., ×1 (70).

Palaeosmilia M.EDW.-H., 1848 [**P. murchisoni*] [=*Strephodes* M'COY, 1849; *Palastraea* M'COY, 1851; *Palaeastraea* LANG-S.-T., 1940 *(nom. van.)*; *?Campophyllum* M.EDW.-H., 1850; *Clisiophyllites* LÖWENECK, 1932]. Solitary or compound, with numerous long septa, wide regular dissepimentarium, and fossula narrow at outer part of tabularium, widening slightly in middle of tabularium; major septa reach the axial structure of incomplete, domed tabulae sagging at the axis; no septal lamellae (18). *L.Carb.*, cosmop.——FIG. 197,*4*. **P. murchisoni,* Visé., Eu.; *4a,b,* transv. and long. secs., ×1 (85).

Aphrophyllum SMITH, 1920 [**A. hallense*]. Like *Palaeosmilia,* but with corallites elongated in the counter-cardinal plane, some skeletal dilatation; dissepimentarium may have an irregular lonsdaleoid border (15). *L.Carb.*, E.Austral.——FIG. 198,*2*. **A. hallense,* Visé., E.Austral.; *2a,b,* transv. and long. secs., ×1 (85).

Subfamily YATSENGIINAE Hill, nov.

Corallites slender, septa few, minor septa poorly developed, dissepimentarium very narrow; axial structure dibunophylloid, commonly without regular medial plate. *L. Perm.(Artinsk.)*.

Yatsengia HUANG, 1932 [**Y. asiatica*]. Fasciculate, axial structure surrounded by an outer series of tabellae sloping downward and outward; dissepimentarium commonly not crossed by the poorly developed minor septa, dissepiments inosculating; major septa dilated in tabularium (25). *L.Perm. (Artinsk.)*, Asia.——FIG. 198,3. **Y. asiatica; 3a,b*, transv. and long. secs., ×4 (88).

Protowentzelella PORFIRIEV, 1941 [**P. simplex*]. Like *Yatsengia* but cerioid, axial structure typically weakly developed (49). *L.Perm.(Artinsk.)*, Eu.-Asia.——FIG. 198,4. **P. simplex*, Eu.; *4a,b*, transv. and long. secs., ×2 (123).

Family CYATHOPSIDAE Dybowski, 1873

[*ex Cyathopsis* D'ORBIGNY, 1849 (=*Caninia*, MICHELIN in GERVAIS, 1840)] [=Caniniidae HILL, 1938]

Solitary or fasciculate Rugosa with an open tabular fossula; septa typically dilated

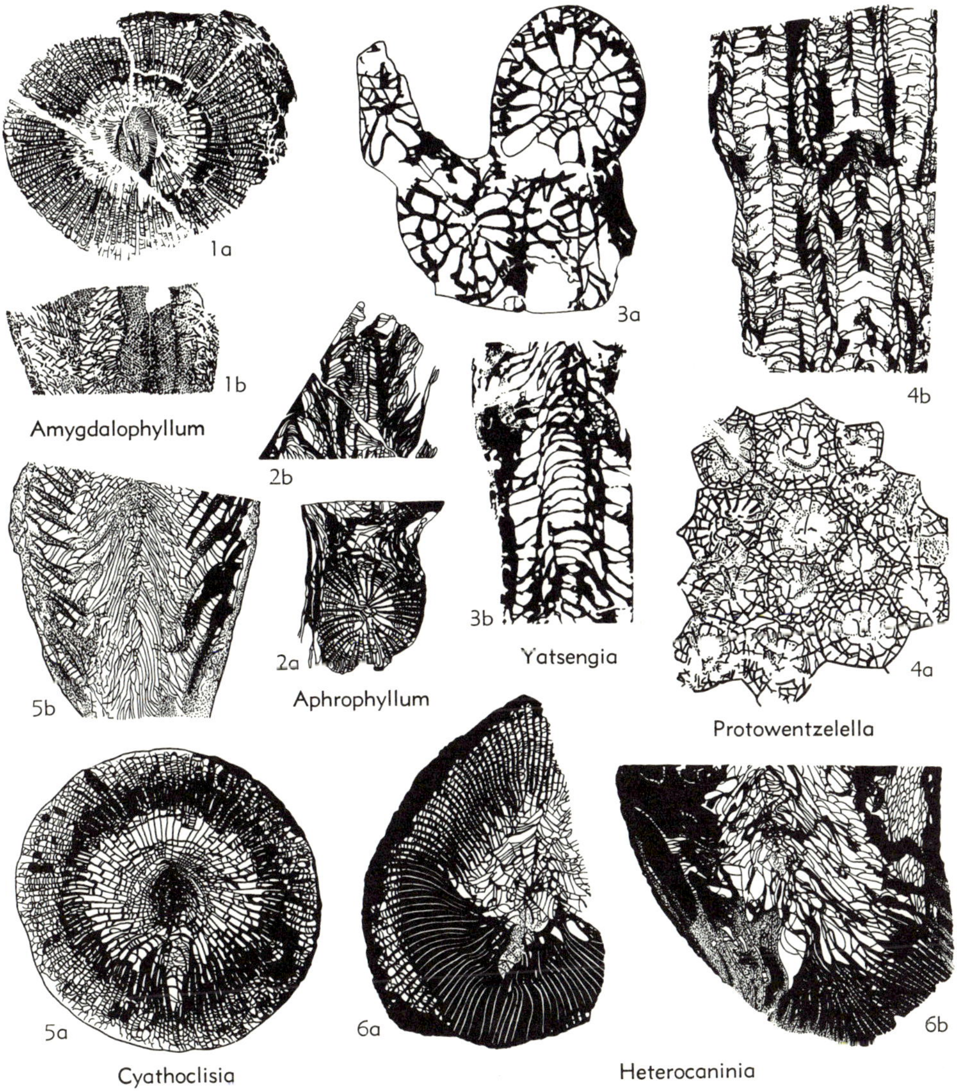

FIG. 198. Streptelasmatina (Zaphrenticae): Aulophyllidae (p. F290-F291).

and amplexoid in the wide tabularium; tabulae complete, domed or flat, with downturned edges; marginarium a regular or (in some) a lonsdaleoid dissepimentarium; cardinal septum short, counter septum commonly elongate. *L.Carb.-Perm.*

The appearance of a lonsdaleoid dissepimentarium in some genera suggests relationship to the Columnariina, but all other characters are indicative of the Zaphrenticae.

Caninia MICHELIN in GERVAIS, 1840 (*non* OWEN, 1846) [**C. cornucopiae*] [=*Cyathopsis* D'ORBIGNY, 1849 (*non* THOMSON, 1883) (obj.); *Pseudozaphrentoides* STUCKENBERG, 1904; *Peetzia* TOLMACHEV, 1924]. Solitary; in youth, the long major septa are slightly sinuous, with lanceolate dilatation in tabularium, particularly in cardinal quadrants; septa become amplexoid and less dilated in adult stages, when a marginarium of concentric, inosculating or lonsdaleoid dissepiments may develop; fossula open, neighboring septa curving about it; tabulae flat with downturned margins (18). *Carb.-Perm.*, N.Am.-Eu.-Asia-Austral.——FIG. 199,*2*. **C. cornucopiae* L.Carb., Eu.; *2a,b*, transv. secs., ×1; *2c*, long. sec., ×1 (64).

Siphonophyllia SCOULER in M'COY, 1844 [**S. cylindrica*] [=*Paleocaninia* LISSITZIN, 1925]. Large, solitary; like *Caninia* but with a wide lonsdaleoid dissepimentarium. *L.Carb.*, Eu.——FIG. 199,*1*. **S. cylindrica*; transv. sec., ×1 (121).

Caninophyllum LEWIS, 1929 [**Cyathophyllum archiaci* M.EDW.-H., 1852] [=?*Neocaninia* LISSITZIN, 1925]. Large, solitary, dissepimentarium wide, with angulate, not lonsdaleoid dissepiments; fossula open; major septa long, not confluent at axis, dilated in cardinal quadrants of tabularium in youth (18). *L.Carb.*, Eu.-Asia; *M.Carb.-U.Carb.*, USSR.——FIG. 199,*7*. **C. archiaci* (M.EDW.-H.), L.Carb., Eu.; *7a,b*, transv. secs., ×1 (99).

Bothrophyllum TRAUTSCHOLD, 1879 [**B. conicum* (FISCHER) TRAUT. (?=*Turbinolia conica* FISCHER, 1830)] [=*Rossophyllum*, *Pseudocaninia* STUCKENBERG, 1888]. Like *Caninophyllum* but with axial ends of longer major setpa commonly joined to an elongate counter septum forming a weak, impersistent axial structure; major septa dilated in tabularium, those of counter quadrants less than those in cardinal quadrants. *L.Carb.*, Eu.-N.Am.-Austral.; *M.Carb.-U.Carb.*, USSR.-China.——FIG. 199,*6a,b*. **B. conicum* (FISCHER), *M.Carb.*, USSR; *6a,b*, transv. secs., ×2 (68).——FIG. 199,*6c,d*. *B. novum* (STUCKENBERG) (type species of *Rossophyllum*), M.Carb., USSR; *6c,d*, transv. and long. secs., ×1 (126).

Lithodrumus GREENE, 1904 [**L. veryi*] [=*Lithodrymus* LANG-S.-T., 1940 *(nom. van.)*]. Solitary, fasciculate, or cerioid; corallites large; dissepimentarium wide, minor septa withdrawn to periphery, leaving inosculating dissepiments; lonsdaleoid dissepiments may occur; major septa withdrawn somewhat from axis; tabulae flat, complete, with downturned edges. *Carb.*, Eu.-Asia-N.Am.

Timania STUCKENBERG, 1895 [**T. schmidti*]. Large, solitary; tabular fossula expanding inward, with ends of major septa in cardinal quadrants twisting sharply across its inner end; counter septum very long, septa in counter quadrants radially arranged; minor septa withdrawn to outer part of dissepimentarium which is not lonsdaleoid (50). *M.Carb.-U.Carb.*, USSR-Spitz.——FIG. 199,*4*. **T. schmidti*, M.Carb., USSR; *4a,b*, transv. and long. secs., ×1 (126).

Gshelia STUCKENBERG, 1888 [**G. rouilleri*]. Like *Timania* but with more evenly developed dissepimentarium in which minor septa extend to the tabularium; counter septum longer than others only in young stages, when it is connected with a platelike columella absent in adult (7). *M.Carb.*, China; *U.Carb.*, USSR.——FIG. 199,*3*. **G. rouilleri*, U.Carb., USSR; *3a,b*, transv. and long. secs., ×1 (68).

?**Bordenia** GREENE, 1901 [**B. zaphrentiformis*]. Solitary, with an epithecal talon; septa amplexoid; no dissepimentarium (51). *L.Carb.*, N.Am.——FIG. 199,*5*. **B. zaphrentiformis*; *5a,b*, calice, nat. long. sec., ×1 (127).

?**Vesiculophyllum** EASTON, 1944 [**Chonophyllum sedaliense* WHITE, 1880]. Peripheral dissepiments irregular, lonsdaleoid, inner dissepiments small, regular; major tabulae sagging deeply and incomplete (10). *Miss.*, N.Am.——FIG. 200,*5*. **V. sedaliense* (WHITE); *5a*, transv. sec., ×3; *5b*, long. sec., ×2 (72).

?**Enygmophyllum** FOMICHEV, 1931 [**E. taidonense*] [=*Aenigmatophyllum* LANG-S.-T., 1940 *(nom. van.)*]. Dissepimentarium narrow, regular; major septa withdrawn from wide axial space in which the tabulae have a wide deep flat-bottomed depression. *L.Carb.*, USSR.——FIG. 200,*6*. **E. taidonense*; *6a,b*, transv. and long. secs., ×1 (76).

?**Uralinia** STUCKENBERG, 1895 [**Heliophyllum multiplex* LUDWIG, 1862; SD LANG-S.-T., 1940] [=?*Pseudouralinia* YÜ, 1931 (?subgenus)]. Dissepimentarium of lonsdaleoid dissepiments only, septa represented only by short ridges at outer edges of tabularium (50). *L.Carb.*, USSR-China.——FIG. 200,*4a*. **U. multiplex* (LUDWIG); transv. sec., ×1 (126).——FIG. 200,*4b,c*. *P. tangpakouensis* (YÜ) (type species of *Pseudouralinia*), China; *4b,c*, transv. and long. secs., ×1 (143).

?**Cystophrentis** YÜ, 1931 [**C. kolaohoensis*] [=*Cystiphrentis* LANG-S.-T., 1940 *(nom. van)*]. Like *Caninia* but with more and longer septa in counter than in cardinal quadrants, and without minor septa (58). *L.Carb.*, China.——FIG. 200,*1*. **C. kolaohoensis*; *1a,b*, transv. secs., ×2 (143).

?**Humboldtia** STUCKENBERG, 1895 [**H. rossica*]. Large, solitary; dissepimentarium lonsdaleoid;

septa of cardinal quadrants few, dilated and grouped around the cardinal and alar fossulae; septa of counter quadrants many, long and grouped about counter fossula; counter septum short; tabulae incomplete (50). *L.Carb.*, USSR-Fr.——FIG. 200,*3*. **H. rossica*, USSR; transv. sec., ×1 (126).

?**Keyserlingophyllum** STUCKENBERG, 1895 [**Cystiphyllum obliquum* KEYSERLING, 1846; SD LANG-S.-T., 1940]. Solitary; dissepimentarium wide in counter, narrow in cardinal quadrants; major septa long and dilated in tabularium but thinning in counter quadrants in adult; septa grouped pinnately about marked cardinal and alar fossulae, radially arranged in counter quadrants; tabulae

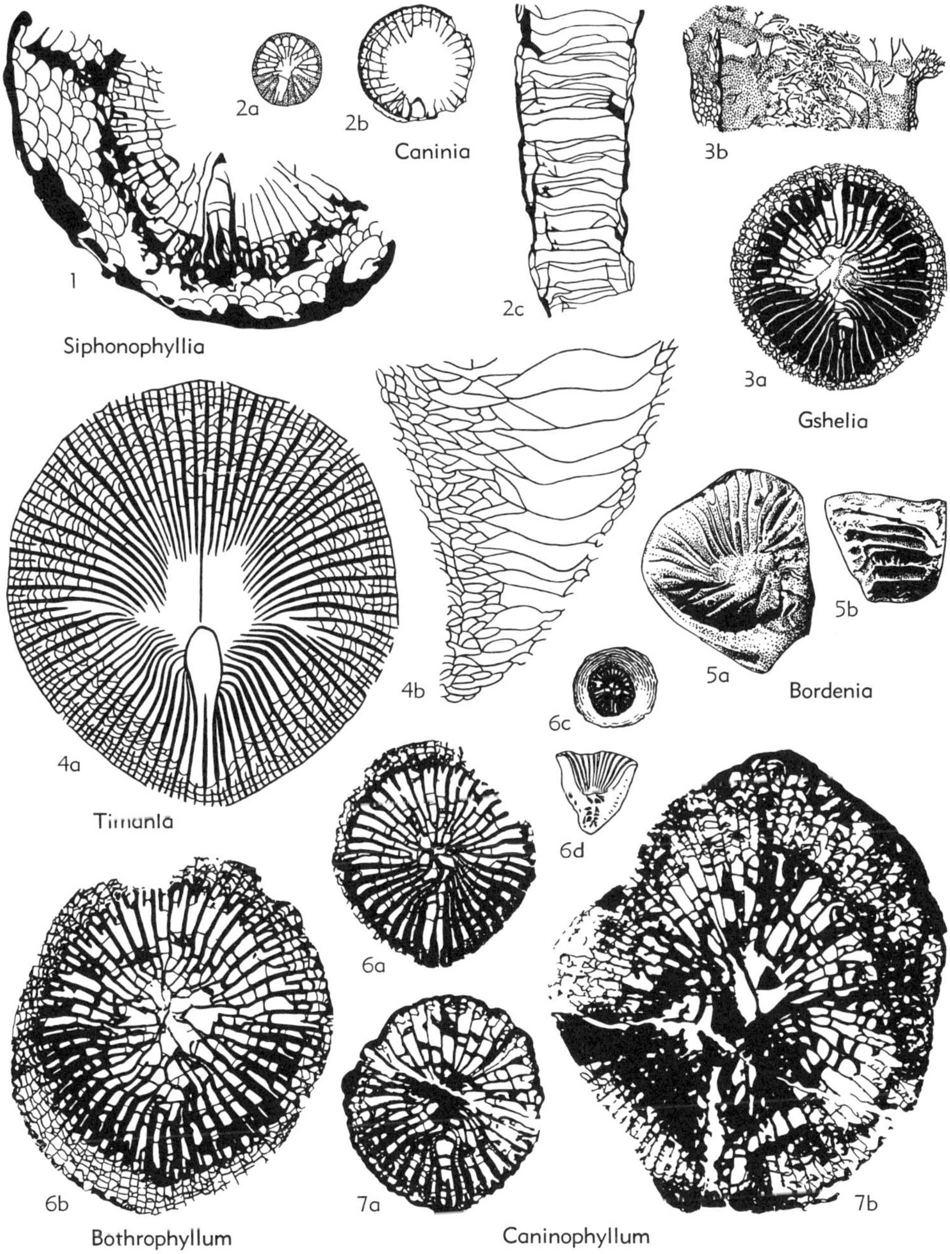

FIG. 199. Streptelasmatina (Zaphrenticae): Cyathopsidae (p. F292).

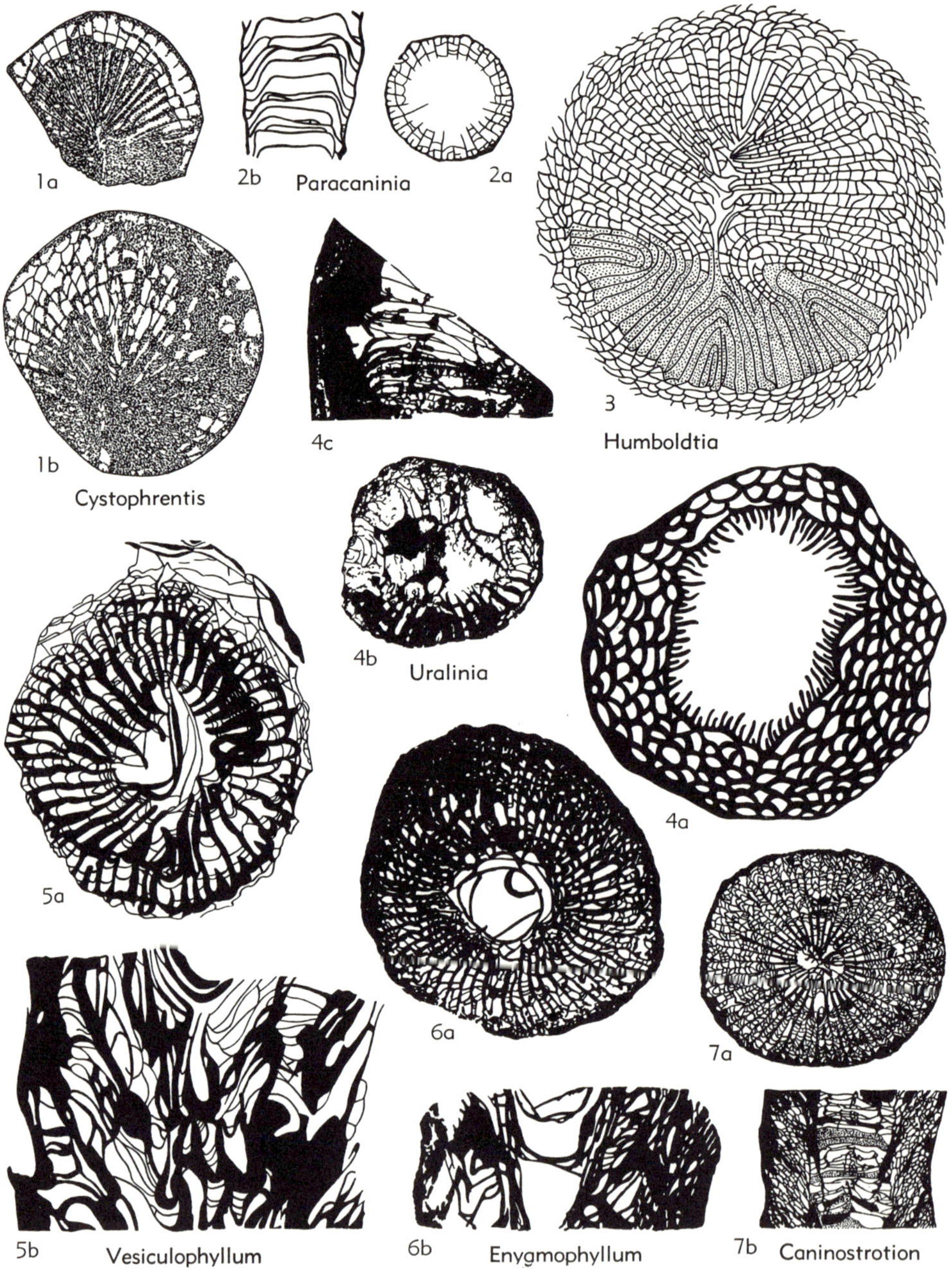

FIG. 200. Streptelasmatina (Zaphrenticae): Cyathopsidae (p. F292-F295).

flat (50). *L.Carb.*, USSR.——FIG. 201,*4*. **K. obliquum* (KEYS.); *4a,b*, transv. and long. secs., ×1 (126).

?**Kakwiphyllum** SUTHERLAND, 1954 [**K. dux*]. Solitary; elongated in counter-cardinal plane; with lonsdaleoid dissepiments peripherally and normal dissepiments within; septa grouped; tabulae incomplete, may rise to axial boss. *Miss.*, Can.

?**Liardiphyllum** SUTHERLAND, 1954 [**L. hagei*]. Solitary; with irregular large, lonsdaleoid dissepiments, a long counter septum and septa of cardinal quadrants dilated. *Miss.*, Can.

?**Paracaninia** Chi, 1937 [**P. sinensis*]. Like *Caninia* but with epithecal spines, and without dissepiments, although short minor septa occur; septa amplexoid in cylindrical parts of corallum. *L.Perm.(Artinsk.)*, China.——Fig. 200,*2*. **P. sinensis; 2a,b,* transv. and long. secs., ×1 (65).

?**Endamplexus** Koker, 1924 [**E. dentatus*] [=*Endoamplexus* Lang-S.-T., 1940 *(nom. van.)*]. Solitary; septa short, thick, irregular, in places separated from epitheca by an irregular lonsdaleoid dissepimentarium (29). *L.Perm.(Artinsk.)*, Timor. ——Fig. 201,*1*. **E. dentatus;* transv. sec., ×1 (95).

?**Caninostrotion** Easton, 1943 [**C. variabilis*]. Large, conical, with peripheral increase; dissepimentarium not completely crossed by minor septa, some dissepiments angulate; major septal ends in tabularium dilated, usually stopping short of axis, but a few may form an impersistent axial structure. May be an aulophyllid. *U.Miss.(Chester.)*, N. Am.; *U.Carb.(Namur.)*, Scot.——Fig. 200,*7*. **C. variabilis,* Chester., Ark.; *7a,b,* transv. and long. secs., ×1 (72).

?**Caninella** Gorsky, 1938 [**C. pulchra*]. Dissepimentarium regular, wide, with lateral dissepiments on septa. *Genus inquirendum;* fig'd. specimens crushed axially. May be an aulophyllid. *L.Carb.*, USSR.——Fig. 201,*3*. **C. pulchra; 3a,b,* transv. and long. secs., ×1 (80).

?**Rylstonia** Hudson & Platt, 1927 [**R. benecom-*

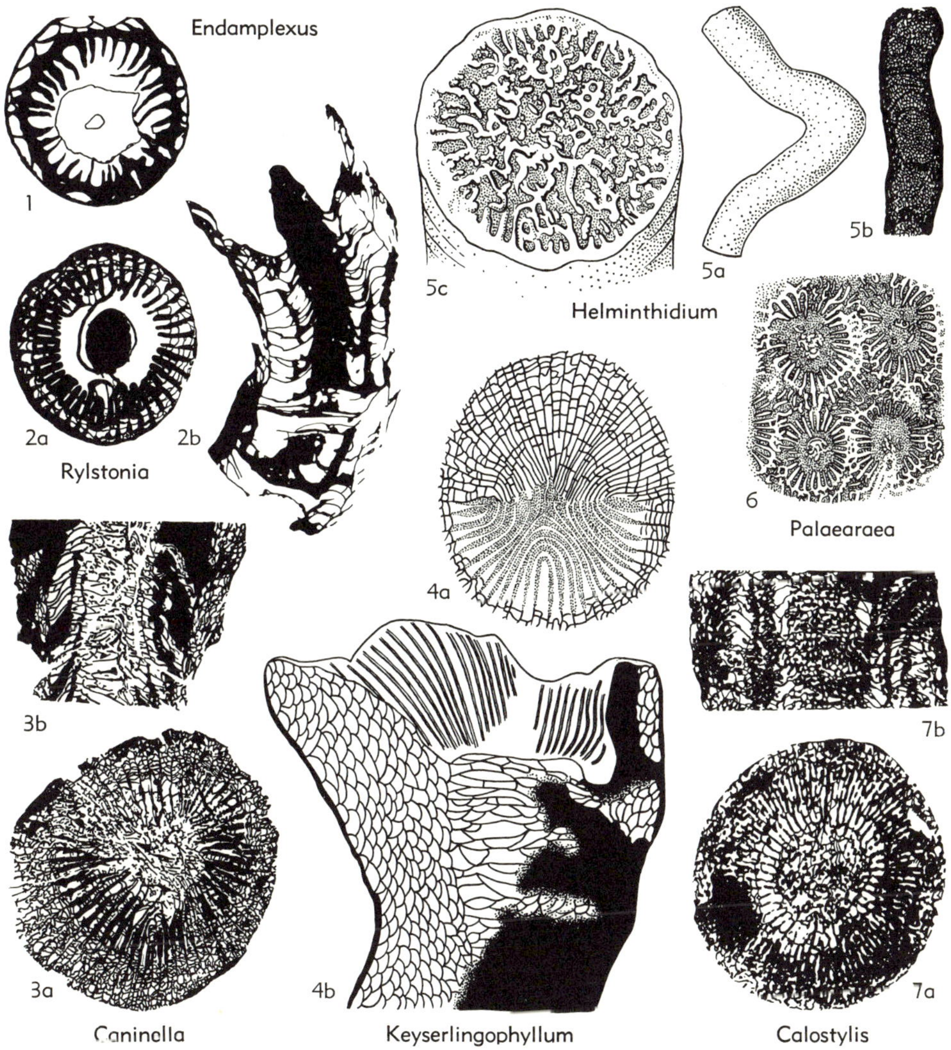

Fig. 201. Streptelasmatina (Zaphrenticae): Cyathopsidae, Calostylidae (p. F293-F296).

pacta] [=*Hettonia* HUDSON & ANDERSON, 1928]. Curved, conico-cylindrical to cylindrical; marginarium narrow, of large concentric dissepiments, minor septa degenerate; major septa dilated and in contact at first, dilatation disappearing first in counter quadrants; an axial structure of dilated median septal lamellae may develop discontinuous from the septa, when the incomplete tabulae are arched axially and incomplete, or the axial structure may disappear when the incomplete tabulae sag. May be an aulophyllid. *L.Carb.(Visé.)*, Eng.——FIG. 201,*2*. *R. fallax* (HUDSON & ANDERSON) (type species of *Hettonia*); *2a,b*, transv. and long. secs., ×1 (90).

Appendix to STREPTELASMATINA

Family CALOSTYLIDAE C. F. Roemer, 1883

Solitary and colonial corals with major and minor septa perforate; axial ends of long major septa lobed and reticulate, forming a spongy axial structure; tabulae domed, slender, complete and distant; dissepiments of unusual type, large, distant, slender and almost horizontally based; epitheca developed in bands which fail to cover the corallum. *Sil.*

The systematic position of this family is doubtful.

Calostylis LINDSTRÖM, 1868 [**C. cribraria* (=*Clisiophyllum denticulatum* KJERULF, 1865)] [=*Hemiphyllum* TOMES, 1887]. Solitary, conical, epithecate in apical region only; with narrow peripheral zone where the highly perforate major and minor septa are connected by irregular septal elements (46). *L.Sil.-M.Sil.*, Eu.-N.Am.——FIG. 201,*7*. **C. denticulata* (KJERULF), M.Sil., Eu.; *7a,b*, transv. and long. secs., ×2 (117).

Helminthidium LINDSTRÖM, 1882 [**H. mirum*]. Solitary, scolecoid; all trace of radial arrangement of the perforate and reticulate septa is lost (46). *M.Sil.-U.Sil.*, Eu.——FIG. 201,*5*. **H. mirum; 5a,b*, side, long. sec., ×1; *5c*, transv. sec., ×4 (100).

Palaearaea LINDSTRÖM, 1882 [**P. lopatini*]. Astreoid; axial structure marked (35). *Sil.*, Sib.——FIG. 201,*6*. **P. lopatini;* transv. sec., ×2 (100).

Suborder COLUMNARIINA Rominger, 1876

[*nom. correct.* HILL, 1954 (*pro* Columnariae ROMINGER, 1876)] [=Columnariacea SOSHKINA, 1947; Stauriacea MOORE, 1952; includes Deuteroseptata GRABAU, 1922 (*partim*)]

Corallum compound or (less commonly) solitary; marginarium absent in oldest forms, but later it develops as a septal stereozone which may be replaced by a lonsdaleoid dissepimentarium, or an incomplete series of elongate dissepiments; septa invariably thin in tabularium, somewhat withdrawn from axis, not lobed axially; tabulae complete and flat or with downturned edges, or sagging medially; late forms develop an axial structure of septal lamellae and conical tabellae. Axial increase may occur. *Ord.-Perm.*

Family STAURIIDAE Milne-Edwards & Haime, 1850

[*nom. correct.* HILL, herein (*pro* Stauridae M.EDW.-H., 1950)] [=Cyathophylloidae, Cyathophylloinae, Densiphyllinae DYBOWSKI, 1873; Columnariae ROMINGER, 1876; ?Fletcherinae ZITTEL, 1876; Columnaridae NICHOLSON, 1879; ?Coelophyllidae ROEMER, 1883; Favistellidae CHAPMAN, 1893; Columnariadae LAMBE, 1901; ?Fletcheriidae WEISSERMEL, 1939; Columnariidae HILL, 1939; Columnariinae STUMM, 1949; ?Pycnostylidae STUMM in SHROCK & TWENHOFEL, 1953]

Fasciculate and cerioid coralla with slender corallites; a marginarium may develop as a narrow peripheral stereozone; a single series of elongate dissepiments, discontinuous vertically may occur between major and minor septa. *Ord.-Dev., ?L.Carb.*

Favistella DANA, 1846 [**Columnaria alveolata* GOLDFUSS, 1826 (VAN CLEVE in WHITE, 1882) (=*F. stellata* HALL, 1847)] [=*Favistella* HALL, 1847]. Cerioid, with major septa reaching almost to axis; minor septa very short; marginarium absent; tabulate with downturned edges (24). *M.Ord.-U.Ord.*, cosmop.——FIG. 202,*5*. **F. alveolata* (GOLDF.), N.Am.; *5a,b*, transv. and long. secs., ×1.5 (98).

Cyathophylloides DYBOWSKI, 1873 [**C. kassariensis;* SD SHERZER, 1891]. Cerioid, with major septa running together in groups axially, and with long minor septa; marginarium absent, tabulae highly domed and incomplete (9). *Ord.*, Est.——FIG. 202, *3*. **C. kassariensis, 3a,b*, transv. and long. secs., ×3 (139).

Densiphyllum DYBOWSKI, 1873 [**D. thomsoni;* SD SHERZER, 1891] [=*Pycnophyllum* LINDSTRÖM, 1873; *Densyphyllum* THOMSON, 1883]. Small, solitary; minor septa buried in wide peripheral septal stereozone; major septa straight and touching at axis; tabulae flat; no dissepiments (9). *Sil.*, Est.——FIG. 202,*2*. **D. thomsoni; 2a,b*, transv. and long. secs., ×2, ×4 (71).

Stauria M.EDW.-H., 1850 [**S. astreiformis* (=*Madrepora favosa* LINNÉ, 1758)] [=*Ceriaster* LINDSTRÖM, 1883]. Like *Favistella* but with axial increase (quadripartite or quinquepartite) and with a few elongate dissepiments at periphery (38). *U.Sil.*, Balt.-China.——FIG. 202,*1*. **S. favosa* (LINNÉ), Gotl.; *1a*, transv. sec., ×2; *1b*, long. sec., ×1 (119).

Loyolophyllum CHAPMAN, 1914 [**Columnaria (Loyolophyllum) cresswelli*]. Like *Favistella* but with shorter septa and a discontinuous vertical series of elongate, in part lonsdaleoid dissepiments and with sagging tabulae; increase intermural (19). *L.Dev.*, Austral.——FIG. 202,*4*. **L. cresswelli*, Vict.; *4a,b*, transv. and long. secs., ×2 (85).

Fasciphyllum SCHLÜTER, 1885 [**F. conglomeratum* (=*Fascicularia? conglomerata* SCHLÜTER, 1880)]. Fasciculate; corallites slender, with peripheral stereozone inside which a single series of elongate dissepiments develops between the thin, long major and short minor septa; tabulae distant and

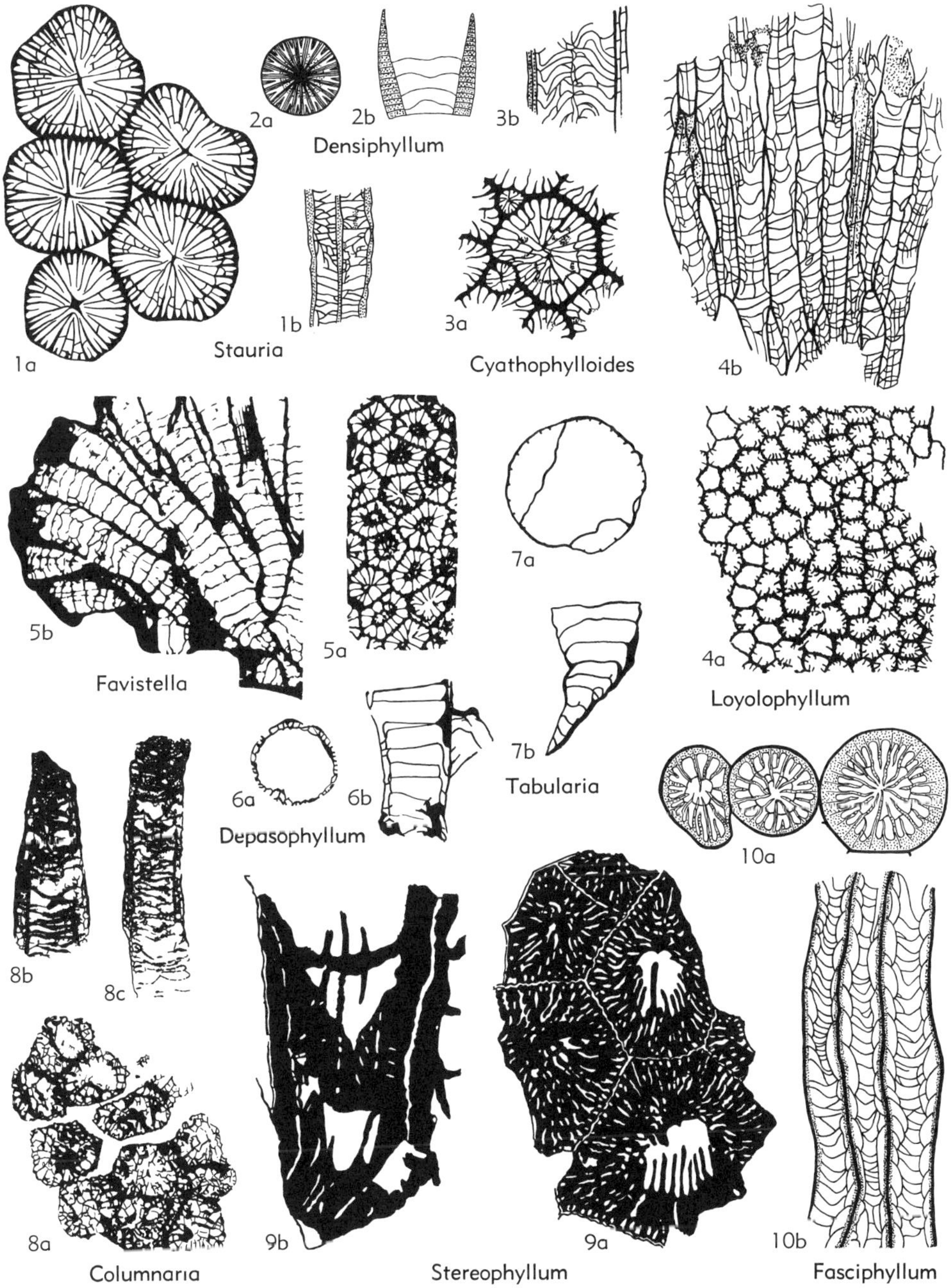

FIG. 202. Columnariina: Stauriidae (p. F296-F298).

sagging. (22). *L.Dev.-M.Dev.*, Eu.-Austral.——FIG. 202,*10*. **F. conglomeratum* (Schlüter), M. Dev., Ger. (Eifel); *10a,* transv. sec., ×4; *10b,* long. sec., ×2 (114).

Vepresiphyllum Etheridge, 1920 [**V. falciforme*]. Cerioid, major septa with lateral and axial upcurved spines, minor septa very short, marginarium absent; tabulae sagging (52). *L.Dev.-M.Dev.*, Austral.

Columnaria Goldfuss, 1826 (*non* Levinson, 1909) [**C. sulcata;* SD M'Coy, 1849] [=*Lithostroma* Rafinesque in Brongniart, 1829]. Cerioid, septa short, dissepiments absent or small, in 1 or 2 series; tabulae complete (19). *M.Dev.*, Eu.——FIG. 202, *8*. **C. sulcata; 8a,* transv. sec., ×1.5; *8b,c,* long. secs., ×1.5 (98).

?**Tabularia** Soshkina, 1937 [**T. turiensis*]. Solitary, cylindroconical, small; septa very short thin lamellae; tabulae complete, flat, with fossular depression and median sag (47). *M.Sil.*, USSR.——FIG. 202,*7*. **T. turiensis; 7a,b,* transv. and long. secs., ×2 (122).

?**Amplexoides** Wang, 1947 [**Amplexus appendiculatus* Lindström, 1883]. Solitary, with major septa as long low ridges on tabulae; minor septa very short; tabulae complete, flat, marginarium a narrow septal stereozone. *M.Sil.*, China.

?**Fletcheria** M.Edw.-H., 1851 [**F. tubifera*] [=*Pycnostylus* Whiteaves, 1884; *?Synamplexus* Grabau, 1922]. Phaceloid, with axial increase typically quadripartite; septa very short, thin; no dissepiments; tabulae flat, complete (38). *M.Sil.*, Eu.-N.Am.-Austral.

?**Stereophyllum** Soshkina, 1937 [*non* Schlüter, 1889; *nec* Grabau, 1917] [**S. massivum*]. Cerioid, with long unequal major septa and long minor septa; marginarium a wide septal stereozone; no dissepiments; tabulae flat, complete (47). *U.Sil.-L.Dev.*, USSR.——FIG. 202,*9*. **S. massivum,* U.Sil., USSR; *9a,b,* transv. and long. secs, ×2 (122).

?**Placophyllum** Simpson, 1900 [**P. tabulatum*]. Phaceloid; corallites without marginarium, tabulae distant, complete, shallowly domed, septal arrangement unknown (52). *L.M.Dev.*, Can.——FIG. 203,*6*. **P. tabulatum;* long. sec., ×1 (115).

?**Depasophyllum** Grabau, 1936 [*non* Yü, 1934] [**D. adnetum*]. Phaceloid; corallites without marginarium, tabulae with very sharply downturned edges, commonly resting on the tabula next below; septa very short, thin; no minor septa (52). *M.Dev.*, USA.——FIG. 202,*6*. **D. adnetum; 6a,b,* transv. and long. secs., ×1 (127).

?**Cyathopaedium** Schlüter, 1889 [*pro Coelophyllum* C. F. Roemer, 1883 (*non* Scudder, 1875; *nec* Schrammen, 1924)] [**Calophyllum paucitabulatum* Schlüter, 1880]. Fasciculate; corallites with distant horizontal tabulae and septa developed only as low nonspinose ridges at periphery and on upper surfaces of tabulae (43). *M.Dev.*, Ger.——FIG. 203,*7*. **C. paucitabulatum* (Schlüter); *7a,* side view of corallum, ×1; *7b,* calice of corallite showing peripheral increase, ×1; *7c,* long. sec., ×1 (114).

?**Fletcherina** Lang-S.-T., 1955 [*pro Yabeia* Lang-S.-T., 1940 (*non Yabeia* Resser & Endo, 1935), *pro Cylindrophyllum* Yabe & Hayasaka, 1915 (*non* Simpson, 1900)] [**Cylindrophyllum simplex* Yabe-H.]. Fasciculate, corallites without septa or marginarium, tabulae complete and horizontal (22). *Dev.*, China; *U.M.Dev.*, Queensl.——FIG. 203,*2*. **Y. simplex* (Yabe-H.), Dev., China; *2a,b,* transv. and long. secs., ×1 (140).

?**Decaphyllum** Frech, 1885 [**D. koeneni*]. Massive; marginarium of each corallite a wide septal stereozone merging with neighbors without bounding wall; counter laterals, alars, and youngest septa neighboring the cardinal septum may be longer and stronger than the rest, meeting at the axis; tabulae flat (11). *U.Dev.*, Ger.——FIG. 203,*1*. **D. koeneni;* surface, ×4 (77).

?**Kwangsiphyllum** Grabau & Yoh in Yoh, 1931 [*pro Syringophyllum* Grabau & Yoh in Yoh, 1929 (*non* M.Edw-H., 1850; *nec* Ulrich in Miller, 1889)] [**Syringophyllum permicum* Grabau & Yoh in Yoh, 1929]. Fasciculate, corallites slender, with connecting tubules; major septa short, minor septa absent or very short, tabulae complete, horizontal; no dissepiments. *L.Carb.*, China.

Family SPONGOPHYLLIDAE Dybowski, 1873

[=Spongophyllinae (*nom. transl.* Wedekind, 1922)]

Septa attenuate, minor septa and commonly major septa discontinuous; major septa carinate in tabularium; dissepimentarium wide, dissepiments lonsdaleoid in part; tabulae close, thin, complete, and sagging (52). *M.Sil.-U.Dev.*

Spongophyllum M.Edw.-H., 1851 [**S. sedgwicki*] [=*?Grabauphyllum* Foerste, 1917]. Cerioid (52). *M.Sil.*, Austral.-?Ohio; *M.Dev.*, cosmop.——FIG. 203,*3*. **S. sedgwicki,* M.Dev., Eu.; *3a,b,* transv. and long. secs., ×2 (127).

Battersbyia M.Edw.-H., 1851 [**B. inaequalis*]. Phaceloid (41). *Dev.*, Eng.

Neomphyma Soshkina, 1937 [**N. originata*]. Solitary, cylindrical (47). *U.Sil.-Dev.*, USSR.——FIG. 203,*4*. **N. originatum,* U.Sil., USSR; *4a,b,* transv. and long. secs., ×4 (122).

Zenophila Hill, 1940 [**Phillipsastraea walli* Etheridge, 1892]. Thamnasterioid or aphroid (20). *M. Sil.*, N.S.W.——FIG. 203,*5*. **Z. walli* (Etheridge); *5a,b,* transv. and long. secs., ×2 (85).

?**Tabellaephyllum** Stumm, 1948 [**T. peculiare*]. Cerioid; septa absent, large, irregular, rather flat-lying dissepiments merging with sagging incomplete tabulae. *U.Dev.*, N.Am. May belong to the Tabulata, subfamily Micheliniinae.——FIG. 203,*8*. **T. peculiare, 8a,b,* transv. and long. secs., ×2 (127).

Family CHONOPHYLLIDAE Holmes, 1887

Solitary or compound Rugosa, commonly with rejuvenescence contractions; corallites large, with marginarium consisting of a septal stereozone broken up by large lonsdaleoid dissepiments; septa long and attenuate in tabularium, tabulae complete, domes or flattened domes. *Sil.-Dev.*

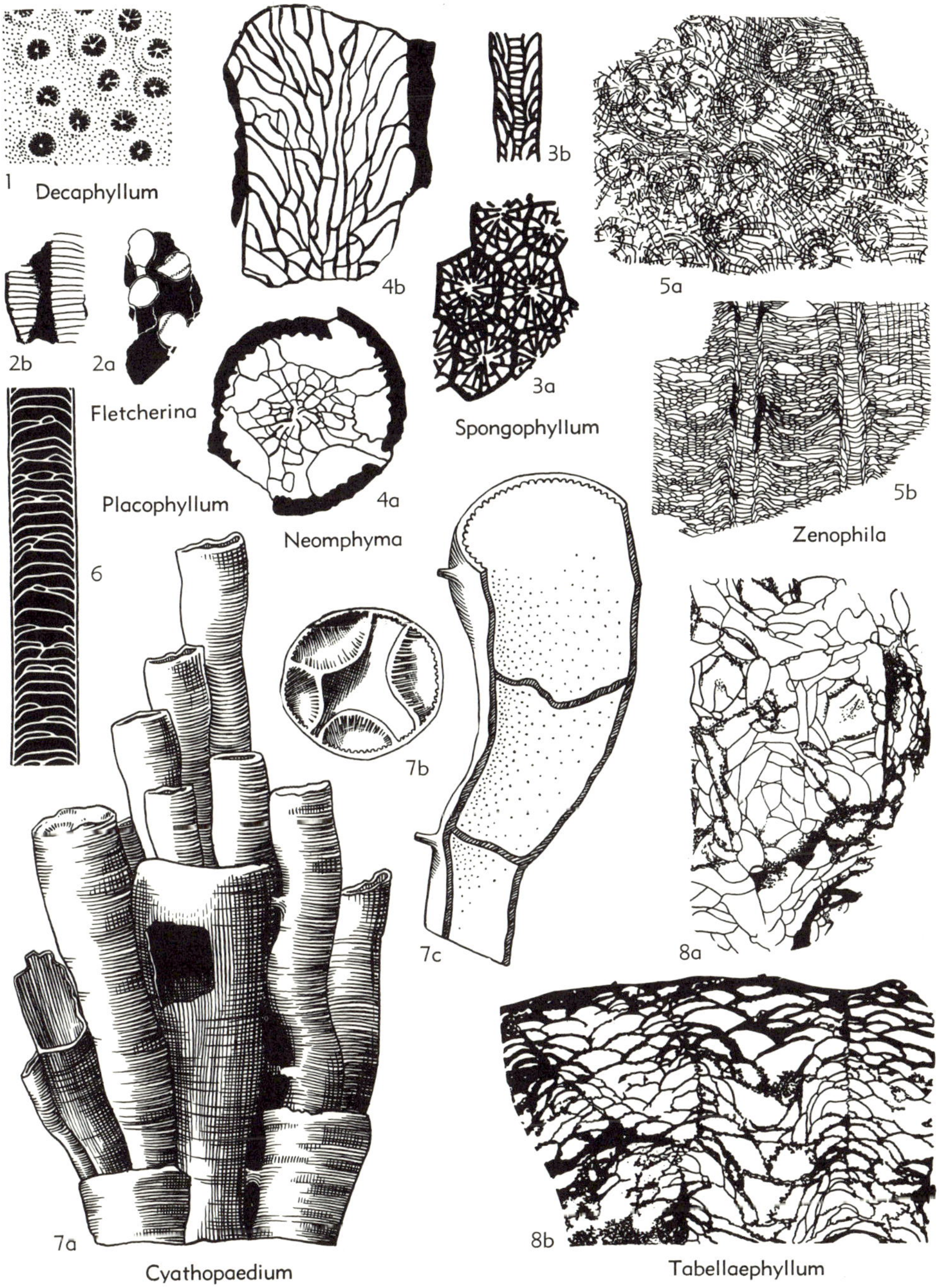

Fig. 203. Columnariina: Stauriidae, Spongophyllidae (p. F298).

Subfamily CHONOPHYLLINAE Holmes, 1887

[*nom. transl.* STUMM, 1949 (*ex* Chonophyllidae HOLMES, 1887)] [=Omphymatidae WEDEKIND, 1927; Ketophyllidae LECOMPTE, 1952)

Solitary or compound; tabulae flat and grouped (55). *Sil.*

Chonophyllum M.EDW.-H., 1850 [**Cyathophyllum perfoliatum* GOLDFUSS in M.EDW.-H., 1850 (*pro Cyathophyllum plicatum* GOLDF., 1826, *partim*)] [=*Omphyma* WEDEKIND, 1927 (*non* RAFINESQUE & CLIFFORD, 1820) The type specimen of *O. verrucosum* RAF. & CLIFFORD, the type species of *Omphyma* RAF. & CLIFFORD cannot be traced; the locality and horizon are uncertain and the description of the species is insufficient for certain identification]. Solitary or compound, septa dilated wedgewise, thickening toward the periphery; tabularium very narrow (46). Many species unrelated to the type have been referred to *Chonophyllum*, giving rise to confused ideas of the family. No vertical section of *C. perfoliatum* has been figured; the transverse section is very like that illustrated by WEDEKIND (55) as *Omphyma flabellatum*, which is known from vertical section and seems closely related to *Ketophyllum*. *M.Sil.-U.Sil.*, Gotl.——FIG. 204,*3a*. **C. perfoliatum* (GOLDF.); transv. secs., ×1 (117).——FIG. 204,*3b*. *C. flabellatum* (WDKD.); long. sec., ×1 (137).

Ketophyllum WEDEKIND, 1927 [**K. elegantulum*] [=*Omphyma auctt.* (*non* RAFINESQUE & CLIFFORD, 1820); ?*Heterolasma* EHLERS, 1919 (=*Heterelasma* LANG-S.-T., 1940 (*nom. van. pro Heterolasma*); *non* GIRTY, 1908; *nec* GRABAU, 1922); *Dokophyllum* WDKD., 1927; *Cetophyllum, Docophyllum* LANG-S.-T., 1940]. Solitary, with flat, grouped tabulae and with long septa, slightly dilated wedgewise in the marginarium, continued over upper surfaces of tabulae as thin, low ridges; fossula marked by depression in tabulae (55). *M. U.Sil.*, Eu.-China.——FIG. 204,*6*. *K. incurvatum* WDKD., M.Sil., Gotl.; *6a,b*, transv. and long. secs., ×1 (137).

?**Lindstroemophyllum** WANG, 1947 [**L. involutum*]. Solitary, with a few very convolute major septa which meet at the axis and are developed axially as low ridges on upper surfaces of tabulae. *M.Sil.*, China.——FIG. 204,*4*. **L. involutum*, Yunnan; *4a,b*, transv. and long. secs., ×1 (136).

?**Mictocystis** ETHERIDGE, 1908 [**M. endophylloides*]. Aphroid very large corallites with wide marginarium of lonsdaleoid dissepiments, and tabularium of flat, close, complete tabulae with downturned edges; major septa developed as long crests on upper surfaces of tabulae only. *Sil.*, Austral.——FIG. 204,*1*. **M. endophylloides*, N.S.W.; surface, ×0.25 (74).

?**Yassia** JONES, 1930 [**Spongophyllum enorme* ETHERIDGE, 1913] [=*Crinophyllum* JONES, 1932]. Very large cerioid corallites with moderately wide marginarium of very large lonsdaleoid dissepiments; tabulae flat, complete; septa very attenuate and discontinuous to absent (20). *M.Sil.*, Austral.——FIG. 205,*2*. **Y. enormis* (ETH.), N.S.W.; *2a,b*, transv. and long. secs., ×1 (92).

Subfamily ENDOPHYLLINAE Torley, 1933

[*nom. transl.* WANG, 1948 (*ex* Endophyllidae TORLEY, 1933)] [=?Pilophyllidae HILL, 1942]

Solitary or compound; tabulae flat-topped domes; major septa convolute in tabularium (22). *?Sil., Dev.*

Sinospongophyllum YOH, 1937 [**S. planotabulatum*]. Solitary, with a few very large lonsdaleoid dissepiments; septa withdrawn from axis; tabulae broad, flat-topped domes (52). *L.Dev.-M.Dev.*, Asia-Austral.——FIG. 204,*5*. **S. planotabulatum*, M.Dev., Kwangsi; *5a,b*, transv. and long. secs., ×1 (142).

Endophyllum M.EDW.-H., 1851 [**E. bowerbanki*; SD SCHLÜTER, 1889] [=*Nicholsonia* SCHLÜTER, 1885 (*non* DAVIS, 1885; *nec* KIÄR, 1899; *nec* POČTA, 1902; *nec* others)]. Subfasciculate, cerioid or aphroid; corallites large, with septal stereozone broken up by large irregular lonsdaleoid dissepiments; major septa long, attenuate, and commonly convolute in tabularium; tabulae complete, domes, mostly flattened (22). *M.Dev.*, Eu.-Asia-E.Austral.——FIG. 205,*3a,b*. *E. abditum* M.EDW-H., Eu.; *3a,b*, transv. and long. secs., ×1 (117).——FIG. 205,*3c*. **E. bowerbanki*, Eu.; long. sec., ×1 (114).

Sanidophyllum ETHERIDGE, 1899 [**S. davidis*]. Alternatingly phaceloid and cerioid, the cerioid parts being periodic thin horizontal expansions of the septal stereozone; no dissepiments (21). *Up. M.Dev.*, Austral.——FIG. 206,*5*. **S. davidis*, N.S.W.; *5a*, side of weathered corallum showing successive levels of tabulae, ×0.25; *5b,c*, secs., ×1 (74).

Tabulophyllum FENTON & FENTON, 1924 [**T. rectum*] [=*Apolythophyllum* WALTHER, 1928] Solitary with talons; septa mostly thin in the wide, irregular lonsdaleoid dissepimentarium, amplexoid; tabularium compressed, tabulae flat-topped domes (52). *U.Dev.*, N.Am.-Eu.-W.Austral.——FIG. 205, *1a,b*. *T. normale* (WALTHER) (type species of *Apolythophyllum*), Ger.; *1a,b*, transv. and long. secs., ×1 (135).——FIG. 205,*1c,d*. **T. rectum* N.Am.; *1c,d*, long. and transv. secs., ×1 (117).

?**Strombodes** SCHWEIGGER, 1819 [*non* GISTL, 1857; *nec* SJÖBRING, 1902] [**Madrepora stellaris* LINNÉ, 1758; SD M'COY, 1849] [=*Strombastraea* EHRENBERG, 1834; *Donacophyllum* DYBOWSKI, 1873]. Alternately phaceloid and cerioid, the cerioid parts being horizontal expansions from the marginarium which is a septal stereozone interrupted by large lonsdaleoid dissepiments; tabulae axially sagging domes, complete or incomplete (46). *L.Sil.-M.Sil.*, Eu.——FIG. 204,*2a-c*. **S. stellaris* (LINNÉ), Gotl.-Oesel; *2a*, side of corallum, ×1 (117); *2b*, transv. sec., ×1 (71); *2c*, long. sec., ×1 (117).——FIG. 204,*2d*. *S. schrenkii* (DYBOWSKI) (lectotype of

Donacophyllum), Balt.; long. sec., ×1 (71).

?Pilophyllum Wedekind, 1927 [**P. keyserlingi*]. Solitary or (less commonly) fasciculate; major septa thin and convolute in tabularium; marginarium a septal stereozone interrupted by lonsdaleoid dissepiments; tabulae incomplete, domed (55). *U.Sil.*, Gotl.——Fig. 206,*3*. **P. keyserlingi; 3a,b*, transv. and long. secs., ×1 (137).

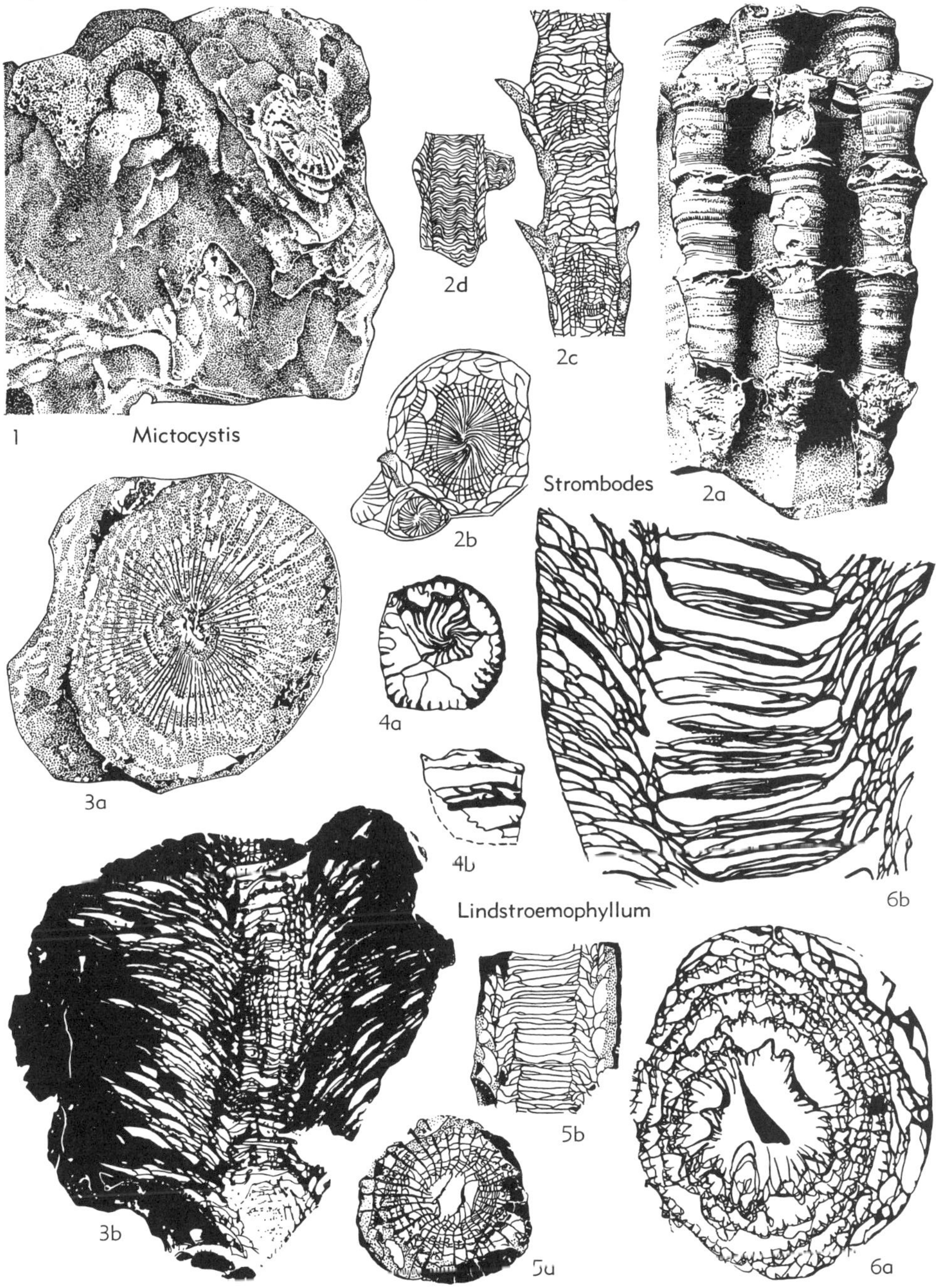

Fig. 204. Columnariina: Chonophyllidae (p. F300).

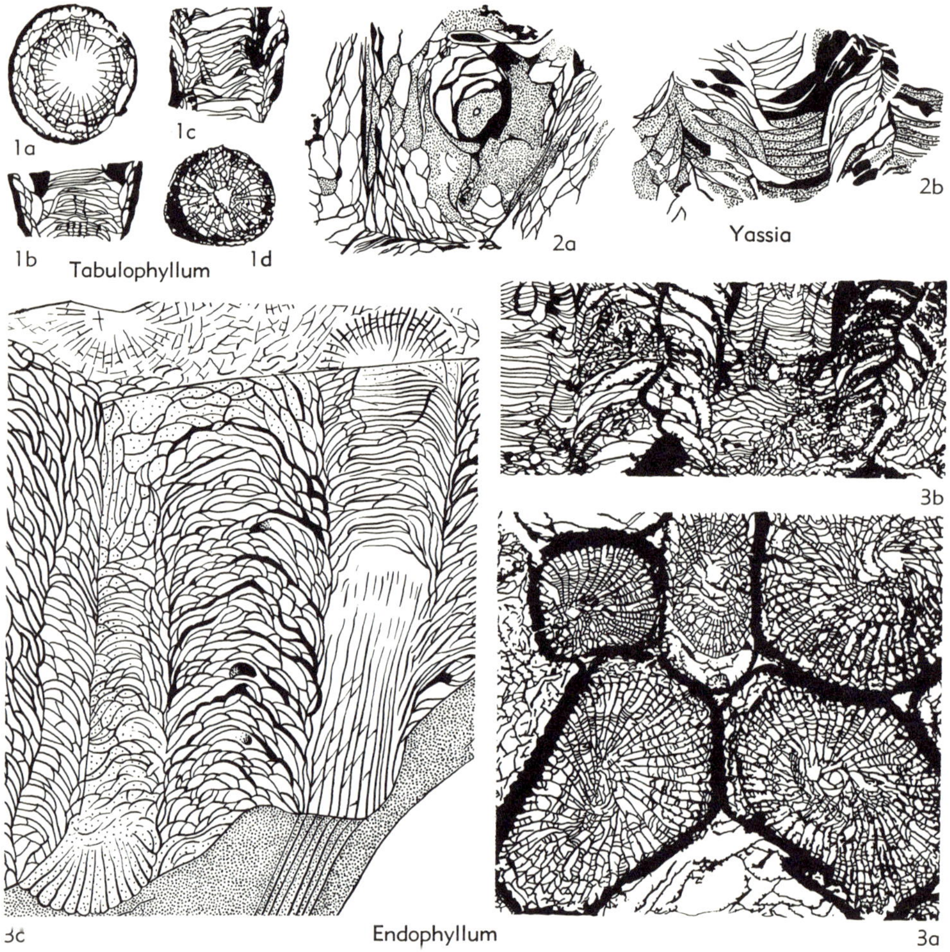

FIG. 205. Columnariina: Chonophyllidae (p. F300).

?**Diversophyllum** SLOSS, 1939 [***Zaphrentis traversensis* WINCHELL, 1866]. Large, solitary; septa long, straight, meeting at axis, attenuate, except in marginarium which is a septal stereozone containing lonsdaleoid dissepiments in adult stages; tabulae flat (52). *M.Dev.-U.Dev.*, N.Am.——FIG. 206,*1*. **D. traversense* (WINCH.), Mich.; *1a,b,* transv. and long. secs., ×1 (116).

?**Iowaphyllum** STUMM, 1949 [***Smithia johanni* HALL & WHITFIELD, 1872]. Astreoid and aphroid; marginarium a septal stereozone much interrupted by large lonsdaleoid dissepiments; septa thin in outer part of narrow tabularium; tabulae steeply domed, forming a false aulos at which axial edges of septa are dilated (52). *M.Dev.* (Czech.)-*U.Dev.* (N.Am.).——FIG. 206,*4*. **I. johanni* (HALL-W.), Iowa; *4a,b,* transv. and long. secs., ×1 (127).

Subfamily BLOTHROPHYLLINAE Stumm, 1949

Solitary; tabulae flat-topped domes with a depression in the well-marked cardinal fossula (52). *L.M.Dev.*

Blothrophyllum BILLINGS, 1859 [**B. decorticatum*]. Large, solitary, with wide lonsdaleoid dissepimentarium; thin straight major septa confined to outer parts of tabularium; tabulae complete flat-topped domes, with down-sinking at the cardinal fossula (52). *L.M.Dev.*, N.Am.——FIG. 206,*2*. **B. decorticatum; 2b,a,* corallite partly long. bisected, transv. sec., ×0.5 (96).

Family PTENOPHYLLIDAE Wedekind, 1923

[*ex Ptenophyllum* WDKD., 1923 (=*Acanthophyllum* DYBOWSKI, 1873, subj.)] [=Stenophyllidae WDKD., 1924; Actinocystidae WDKD., 1927; Acanthophyllidae HILL, 1939; Acanthophyllinae (*nom. transl.* WANG, 1948); Leptoinophyllidae, Leptoinophyllinae, Grypophyllinae STUMM, 1949]

Solitary and compound corals with elongate septa and wide dissepimentarium, lonsdaleoid in early forms but with regular small globose dissepiments in later forms; major septa long, unequal, their axial ends commonly flanged and typically arranged in 4 groups differing in curvature; axial ends of outer septa of each group directed toward longer ends of inner septa; trabeculae so closely spaced as to obscure lines of contact; fossula insignificant; tabulae shallow funnel-shaped, very close and incomplete; tabellae elongate. *Sil.-Dev.*

This family is somewhat doubtfully placed in the Columnariina; its small dissepiments suggest affinities with the Zaphrenticae, but *Spongophylloides,* from which its members seem to stem, has a lonsdaleoid dissepimentarium apparently developed as in the Columnariina from a peripheral septal stereozone. The infundibuliform tabulae also suggest affinities with the Spongophyllidae and the Stringophyllidae. The inconspicuousness of the fossula appears a columnariine rather than a zaphrentid character.

Cymatelasma HILL & BUTLER, 1936 [**C. corniculum*]. Solitary, with marginarium a wide septal stereozone lacking dissepiments; major septa thin and flanged in the tabularium, as in *Spongophylloides. L.Sil.-U.Sil.,* Eng.——FIG. 207,*3*. **C. corniculum,* L.Sil.(U.Llandov.), Eng.; *3a,b,* transv. and long. secs., ×3 (86).

Spongophylloides MEYER, 1881 [**S. schumanni* (=*Cystiphyllum grayi* M.EDW-H., 1851)] [=*Actinocystis* LINDSTRÖM, 1882]. Solitary, with lonsdaleoid dissepimentarium; major septa have lateral flanges parallel to their upper edges (31). *M.Sil.-U.Sil.,* Eu.-N.Am.——FIG. 208,*4*. **S. grayi* (M.EDW.-H.), M.Sil., Eu.; *4a,* transv. sec., ×2 (98); *4b,* long. sec., ×2 (BUTLER).

Acanthophyllum DYBOWSKI, 1873 [**Cyathophyllum heterophyllum* M.EDW.-H., 1851; SD SCHLÜTER, 1889] [=?*Grypophyllum, Mesophylloides* WEDEKIND, 1922; *Ptenophyllum* WDKD., 1923; *Astrophyllum, Rhopalophyllum* WDKD., 1924; *Stenophyllum* AMANSHAUSER in WDKD., 1925 (*non* VERHOEFF, 1897); *Leptoinophyllum* WDKD., 1925;

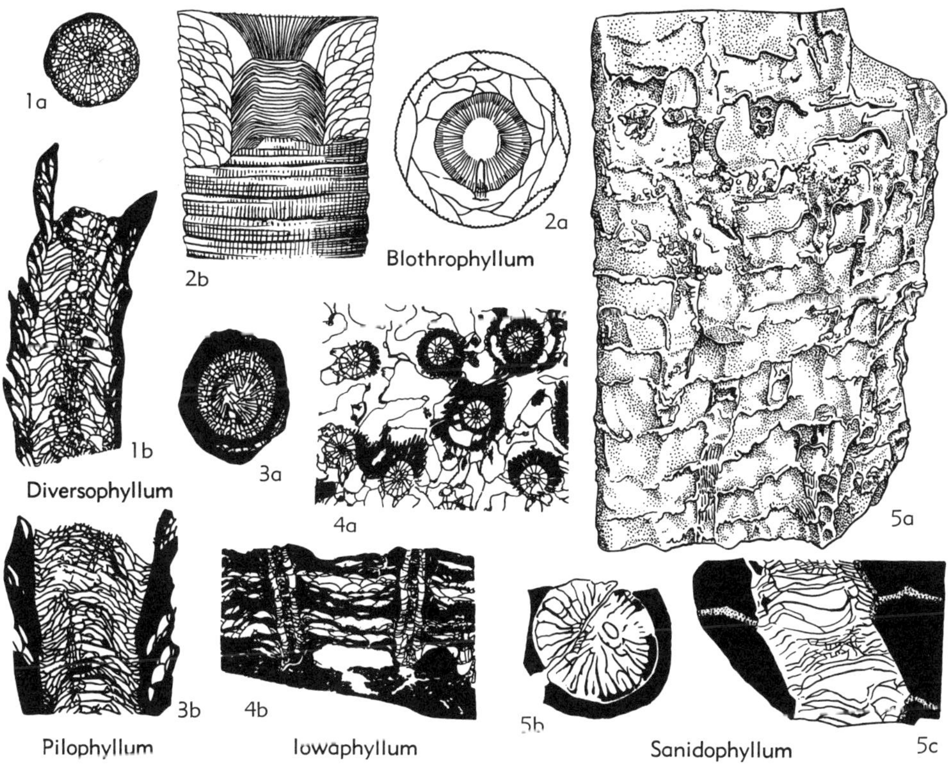

FIG. 206. Columnariina: Chonophyllidae (p. F300-F302).

?*Hooeiphyllum* TAYLOR, 1951]. Solitary; septa commonly with modifications such as lateral cystose dissepiments; major septa dilated more than minor, and parts in the tabularium may be much dilated in young stages (22). *M.Sil.-U.Sil.,* Eu.-N.Am.; *L. Dev.-M.Dev.,* cosmop.——FIG. 207,*1a. Acanthophyllum* sp., Dev., Eu.; long. sec., ×3 (137).——FIG. 207,*1b. A. fibratum* (WDKD.), L.M.Dev., Ger. (Eifel); transv. sec., ×3 (137).——FIG. 207,*1c,d. A.? denckmanni* (WDKD.) (type species of *Grypophyllum*), U.M.Dev., Ger.; *1c,d,* transv. and long. secs., ×2 (137).

Xystriphyllum HILL, 1939 [**Cyathophyllum dunstani* ETHERIDGE, 1911]. Cerioid; corallites as in *Acanthophyllum* but septa thin in tabularium (21). *L.Dev.-M.Dev.,* Eu.-Austral.——FIG. 207,*2.* **X. dunstani* (ETH.), L.M.Dev., Queensl.; *2a,b,* transv. and long. secs., ×2 (85).

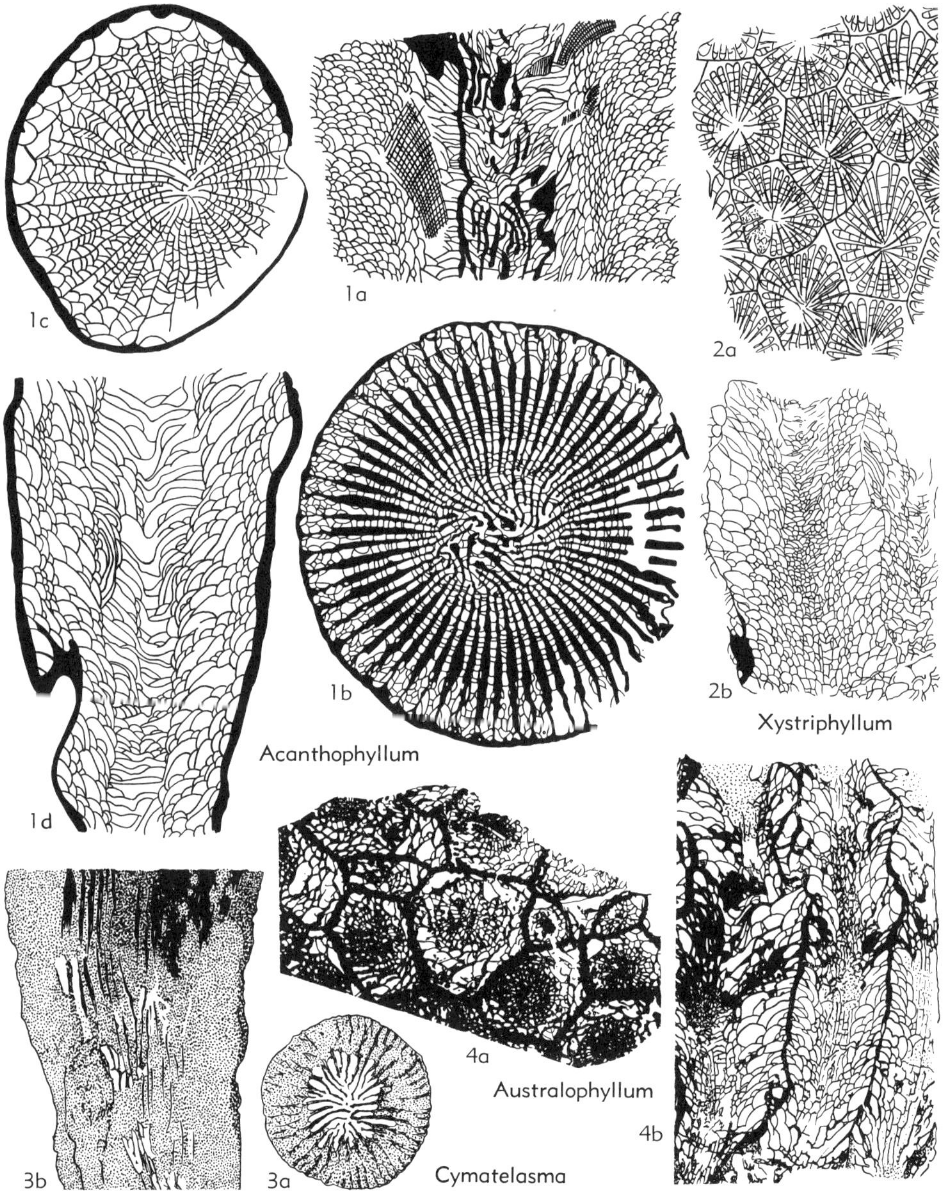

FIG. 207. Columnariina: Ptenophyllidae (p. F303-F306).

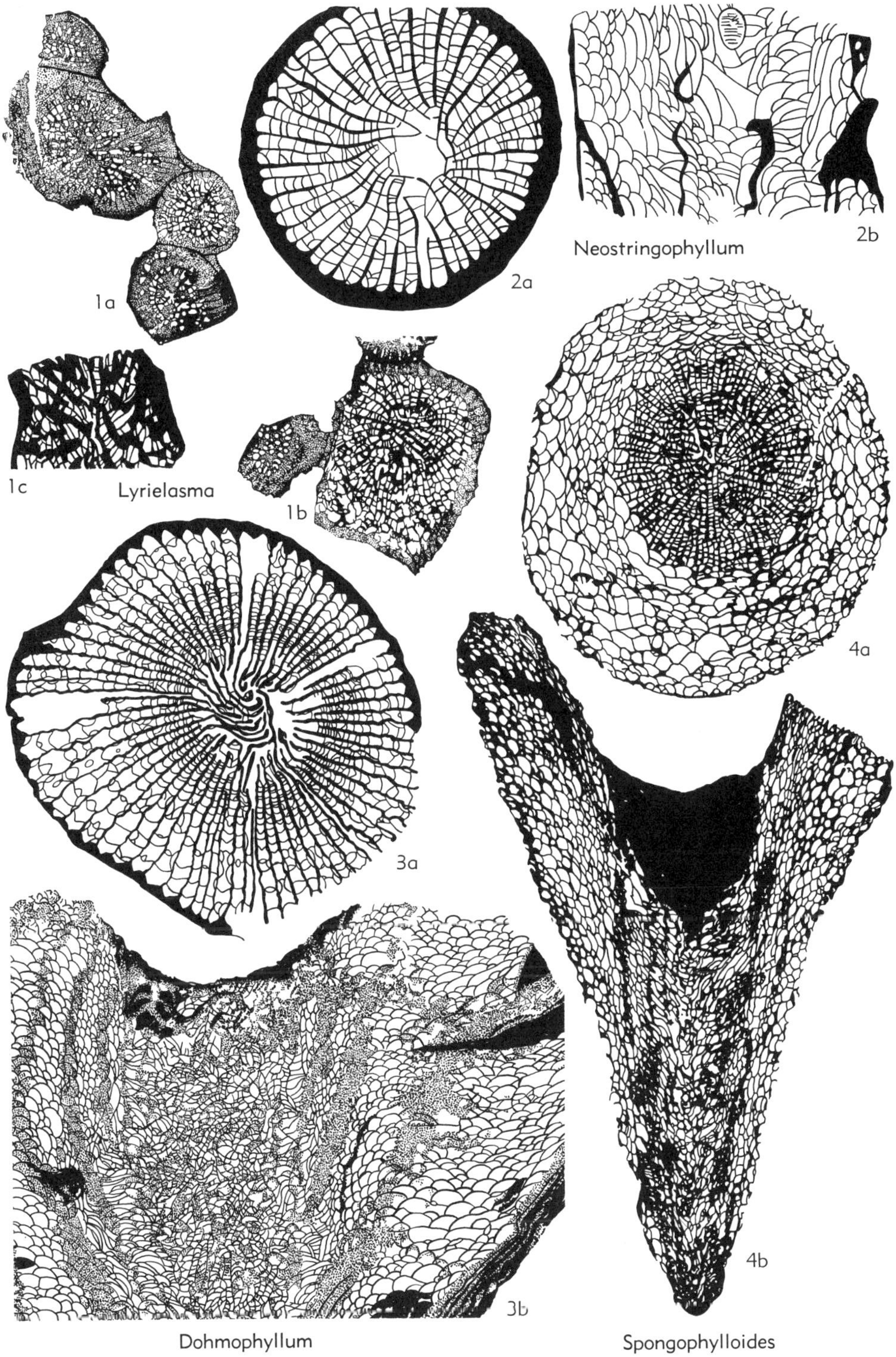

FIG. 208. Columnariina: Ptenophyllidae (p. F303-F306).

Eddastraea HILL, 1942 [*Phillipsastraea grandis* DUN in BENSON, 1918]. Like *Xystriphyllum* but thamnasterioid, with very large corallites (21). *L. Dev.* or *M.Dev.*, E.Austral.-?Ellesm.——FIG. 209,*2*. **E. grandis* (DUN), E.Austral.; *2a,b*, transv. and long. secs., ×2 (85).

Lyrielasma HILL, 1939 [*Cyathophyllum subcaespitosum* CHAPMAN, 1925]. Phaceloid, like *Acanthophyllum* but with septa dilated at their bases, and with their unequal axial ends in a bipartite pattern (19). *L.Dev.-M.Dev.*, E.Austral.-N.Am.——FIG. 208,*1*. **L. subcaespitosum* (CHAPMAN), M.Dev., Vict.; *1a,b*, transv. secs., ×2; *1c*, long. sec., ×2 (85).

Australophyllum STUMM, 1949 [*Spongophyllum cyathophylloides* ETHERIDGE, 1911]. Cerioid, like *Xystriphyllum* but with septa discontinuous peripherally in a partly lonsdaleoid dissepimentarium (52). *L.M.Dev.*, Queensl.——FIG. 207,*4*. **A. cyathophylloides* (ETH.); *4a,b*, transv. and long. secs., ×2 (85).

Neostringophyllum WEDEKIND, 1922 [*N. ultimum*]. Like *Acanthophyllum* but with septa withdrawn from the axis in a wide tabularium (52). *U.M.Dev.*, Ger.——FIG. 208,*2*. **N. ultimum*; *2a,b*, transv. and long. secs., ×1 (137).

Dohmophyllum WEDEKIND, 1923 [*D. involutum*] [=*Trematophyllum* WDKD., 1923; *Sparganophyllum, Pseudoptenophyllum* WDKD., 1925]. Solitary or rarely cerioid, corallites large; septa irregularly carinate or with lateral cystose dissepiments, twisted and united in the tabularium, causing the close tabulae to be irregularly domed, not funnel-shaped 52). *M.Dev.*, Eu.-E.Austral.——FIG. 208,*3a*. **D. involutum*; transv. sec., ×2 (137).——FIG. 208, *3b*. *D. helianthoides* (GOLDFUSS); Ger.(Eifel); long. sec., ×2 (85n).

Pseudochonophyllum SOSHKINA, 1937 [*Chonophyllum pseudohelianthoides* SHERZER, 1892]. Solitary; like *Acanthophyllum* but septa naotic, with very narrow dissepimental alleys (52). *L.Dev.*, Eu.——FIG. 209,*1*. **P. pseudohelianthoides* (SHERZER), Boh.; *1a,b*, transv. and long. secs., ×1 (108).

Family STRINGOPHYLLIDAE Wedekind, 1921

[*nom. transl.* WDKD., 1925 (*ex* Stringophyllinae WDKD., 1921)]

Corallum solitary or phaceloid; septa rather thick, each consisting of a single series of very large monocanths which may be isolated from one another; long major septa bilaterally arranged about the cardinal-counter plane; minor septa tend to be imperfectly developed; dissepimentarium usually lonsdaleoid; tabulae close, complete and sagging, with deepening along median plane (22). *L.Dev.-M.Dev.*

Stringophyllum WEDEKIND, 1922 [*S. normale*; SD WDKD., 1925] [=*Neospongophyllum* WDKD., 1922; *Loipophyllum, Schizophyllum* WDKD., 1925 (*non* VERHOEFF, 1895); *Loepophyllum* LANG-S.-T., 1940 (*nom. van.*); *Vollbrechtophyllum* TAYLOR, 1951 (=*Schizophyllum* WDKD., 1925, obj.); *Sunophyllum* WANG, 1948]. *Up.M.Dev.*, Eu.-E.Austral. ——FIG. 209,*3a,b*. **S. normale*, Eu.; *3a*, transv. sec., ×2; *3b*, long. sec., ×1 (137).——FIG. 209, *3c,d*. *S. variabile* (WDKD.) (type species of *Neospongophyllum*), U.M.Dev., Eu.; *3c,d*, transv. and long. secs., ×2 (137).

Family LONSDALEIIDAE Chapman, 1893

[*nom. correct.* CHI, 1931 (*pro* Lonsdalidae CHAPMAN, 1893)] [=Lonsdaliens DE FROMENTEL, 1861 (invalid vernacular name); Lonsdaleidae GRABAU, 1927]

Compound or rarely solitary Rugosa with an axial structure containing a columella or medial plate and axial tabellae; with lonsdaleoid dissepiments peripherally and with conical tabular floors upturned at margins. *Carb.-Perm.*

Subfamily LONSDALEIINAE Chapman, 1893

[*nom. transl.* HILL, herein (*ex* Lonsdaleiidae, *nom. correct. pro* Lonsdalidae CHAPMAN, 1893)] [=?Axophyllinae M.EDW.-H., 1851; ?Axophylliens DE FROMENTEL, 1861 (invalid vernacular name); ?Axophyllidae (*nom. transl.* DYBOWSKI, 1873); ?Axaphyllidae DYB., 1873 (misprint); Carcinophyllinae HUDSON, 1942; Lithostrotionellidae SHROCK & TWENHOFEL, 1953]

Lonsdaleoid dissepiments predominant; only 2 orders of septa present. *Carb.-L.Perm.* (*Artinsk.*).

Lonsdaleia M'COY, 1849 [*Erismatolithus Madreporites (duplicatus)* MARTIN, 1809, ICZN pend.] [=*Actinocyathus* D'ORBIGNY, 1849 (*non* KENT, 1882); ?*Chonaxis* M.EDW.-H., 1851; *Stylidophyllum* DE FROMENTEL, 1861; *Protolonsdalia* LISSITZIN, 1925; *Protolonsdaleia* LANG-S.-T., 1940 (*nom. van.*)]. Fasciculate or cerioid; corallites with an axial structure of septal lamellae, medial plate, or columella, and conically arranged, shallowly curved tabellae (18). *L.Carb.-M.Carb.*, Eu.-Asia, *L.Carb.*, E.Austral.-N.Am.——FIG. 210,*4*. **L. duplicata* (MARTIN), L.Carb., Eu.; *4a,b*, transv. and long. secs., ×2 (117).

Thysanophyllum NICHOLSON & THOMSON, 1876 [*T. orientale*; SD GREGORY, 1917] [=?*Sublonsdalia* LISSITZIN, 1925; *Sublonsdaleia* LANG-S.-T., 1940 (*nom. van.*)]. Fasciculate or cerioid, with wide lonsdaleoid dissepimentarium; septa withdrawn from axis except for counter septum which may be very long, particularly in young stages; tabulae complete, cones or flat-topped domes (18). *L.Carb.*, Eu.-Asia.——FIG. 210,*3a,b*. **T. orientale*, Scot.; *3a,b*, transv. and long. secs., ×1 (85).——FIG. 210,*3c,d*. *T. ? intermedia* (LISSITZIN) (type species of *Sublonsdalia*), Asia; *3c,d*, transv. and long. secs., ×1 (101).

Lithostrotionella YABE & HAYASAKA, 1915 [**L.*

unica] [=?*Acrocyathus* d'Orbigny, 1849]. Like *Thysanophyllum* but with columella a persistent vertical lath, in some corallites continuous with counter and cardinal septa (57). *Carb.*, N.Am.; *L.Perm.(Artinsk.)*, China.——Fig. 210,*1a*. *L. floriformis* (Orb.), Carb., N.Am.; upper surface, ×0.5 (106).——Fig. 210,*1b,c*. **L. unica*, Artinsk., Yunnan; *1b,c*, transv. and long. secs., ×1 (140).

Sciophyllum Harker & McLaren, 1950 [**S. lambarti*]. Cerioid, without septa and axial structure; dissepimentarium a single series of lonsdaleoid plates; tabulae complete, flat. *Carb.*, Can.

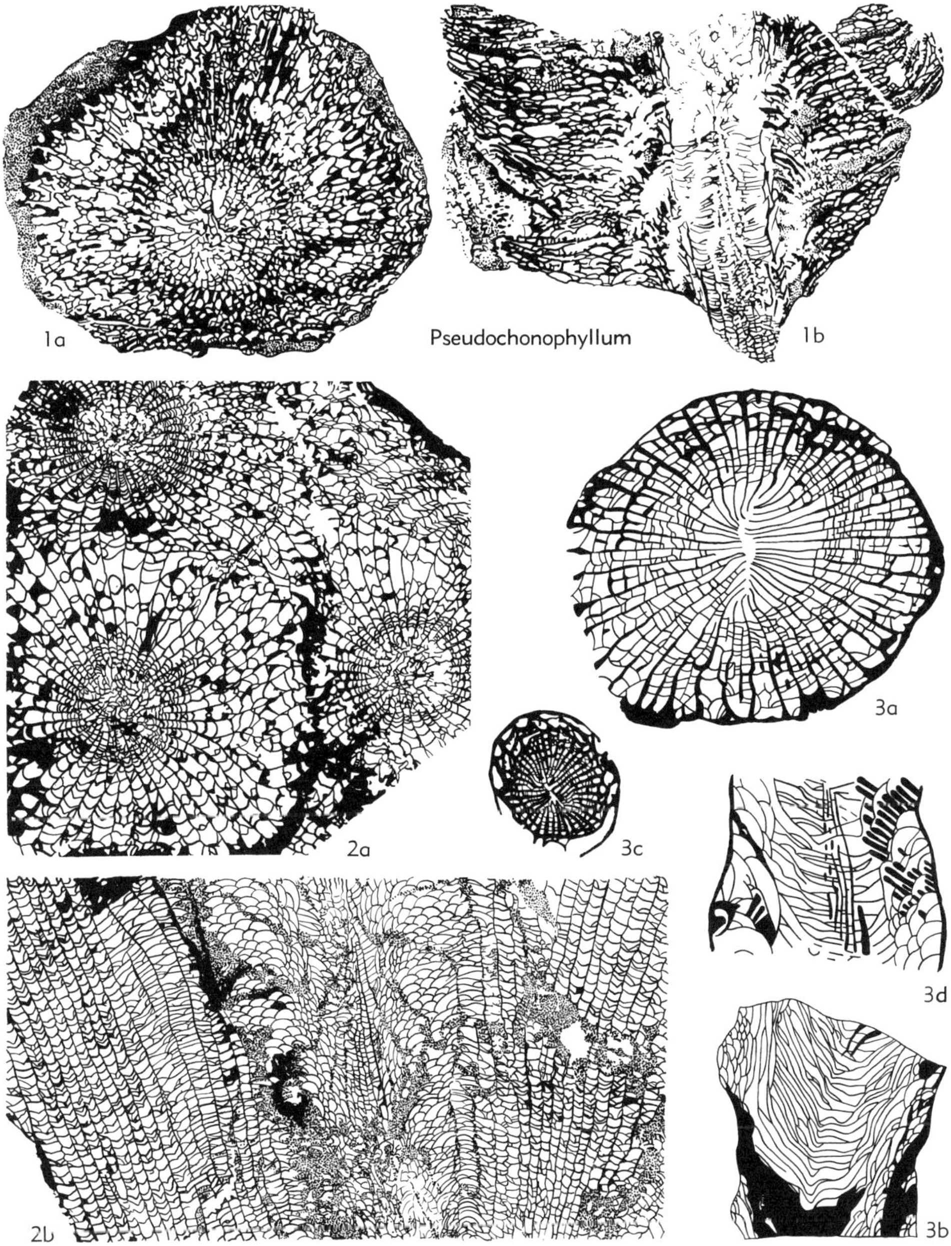

Fig. 209. Columnariina: Ptenophyllidae, Stringophyllidae (p. F306).

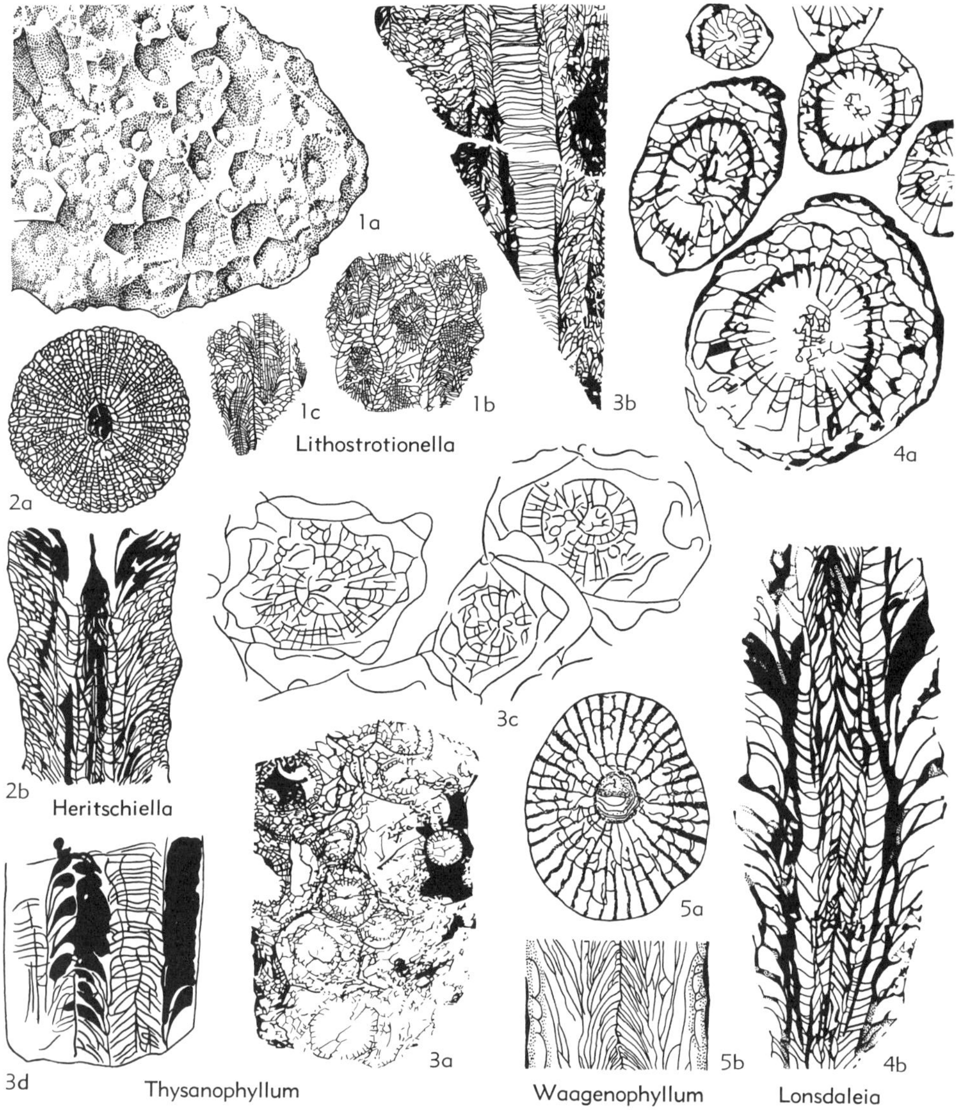

FIG. 210. Columnariina: Lonsdaleiidae (p. F306-F310).

Carcinophyllum THOMSON & NICHOLSON, 1876 [**C. kirsopianum* THOMSON, 1880; SD THOMSON, 1880] [=?*Axophyllum* M.EDW.-H., 1850 (type specimens require restudy); *Agassizia* THOMSON, 1883 (*non* BEHR, 1870; *nec* VETTER, 1881) (*non Agassisia* VALENCIENNES, 1846)]. Solitary; axial structure of irregular anastomosing dilated lamellae, with a median plate; marginarium a peripheral stereozone interrupted by lonsdaleoid dissepiments; tabulae flat or sagging periaxially, conical in the axial structure (18). *L.Carb.*, Eu.——FIG. 211,*4*. **C. kirsopianum* Scot.; *4a,b,* transv. and long. secs., ×1 (85).

Ivanovia DOBROLYUBOVA, 1935 [**I. podolskiensis*] [=*Cystophora* YABE & HAYASAKA, 1916 (*non* NILSSON, 1820); *Cystiphora* LANG-S.-T., 1940 (*nom. van.*) (*non* KIEFFER, 1892)]. Aphroid, with some traces of wall, septa long, dilated in tabularium; axial structure compact, thickened, with a few tabellae arranged in cones; periaxial tabulae sagging (57). *M.Carb.*, USSR; ?*L.Perm.*, Manch.——FIG. 211,*3a,b*. *I. manchurica* DOBR. (type species of *Cystophora*), ?Artinsk., Manch.; *3a,b,* transv. and long. secs., ×2 (85n).——FIG. 211, *3c,d*. **I. podolskiensis*, M.Carb., USSR; *3c,d,* transv. and long. secs., ×1.5 (68).

Subfamily WAAGENOPHYLLINAE Wang, 1950

Dissepimentarium of small interseptal plates; lonsdaleoid dissepiments rare except in *Polythecalis,* third-order septa commonly developed. *L.Perm.*

Iranophyllum Douglas, 1936 [**I. splendens*] Solitary; axial structure of medial plate, septal lamellae, and tented tabellae, surrounded by flat tabulae in the narrow space left free of septa and just outside this by elongate tabellae which slope downward and inward; dissepimentarium not lonsdaleoid, wide, with tertiary as well as minor septa (8). *L.Perm.(?Artinsk.),* Iran-?China.——Fig. 211,2. **I. splendens,* Iran; *2a,b,* transv. and long. secs., ×2 (69).

Waagenophyllum Hayasaka, 1924 [*pro Waagenella* Yabe & Hayasaka, 1915 as *Warganella*) (*non Waagenella* de Koninck, 1883)] [**Lonsdaleia indica* Waagen & Wentzel, 1886; SD Grabau, 1931]. Fasciculate, corallites slender, like *Iranophyllum* internally but without tertiary septa (8). *L.Perm.(Artinsk.),* USSR-Asia-N.Z.——Fig. 210, 5. **W. indica* (Waag.-W.), India; *5a,b,* transv. and long. secs., ×10 (134).

Wentzelella Grabau in Huang, 1932 [**Lonsdaleia salinaria* Waagen & Wentzel, 1886]. Like *Waagenophyllum* but massive (8). *L.Perm. (?Artinsk.),* Asia.

W. (Wentzelella). Cerioid (8). *?Artinsk.,* Asia-N.Z.——Fig. 211,1. **W. (W.) salinaria* (Waagen & Wentzel), India; *1a,* transv. sec., ×4; *1b,* long. sec., ×20 (134).

W. (Wentzelloides) Yabe & Minato, 1944 [**W. (W.) maiyaensis*]. Like *W. (Wentzelella)* but with incomplete fission, so that the corallum is partially meandroid. *Artinsk.,* Japan.——Fig. 212, 3. **W. (W.) maiyaensis; 3a,b,* transv. and long. secs., ×4 (141).

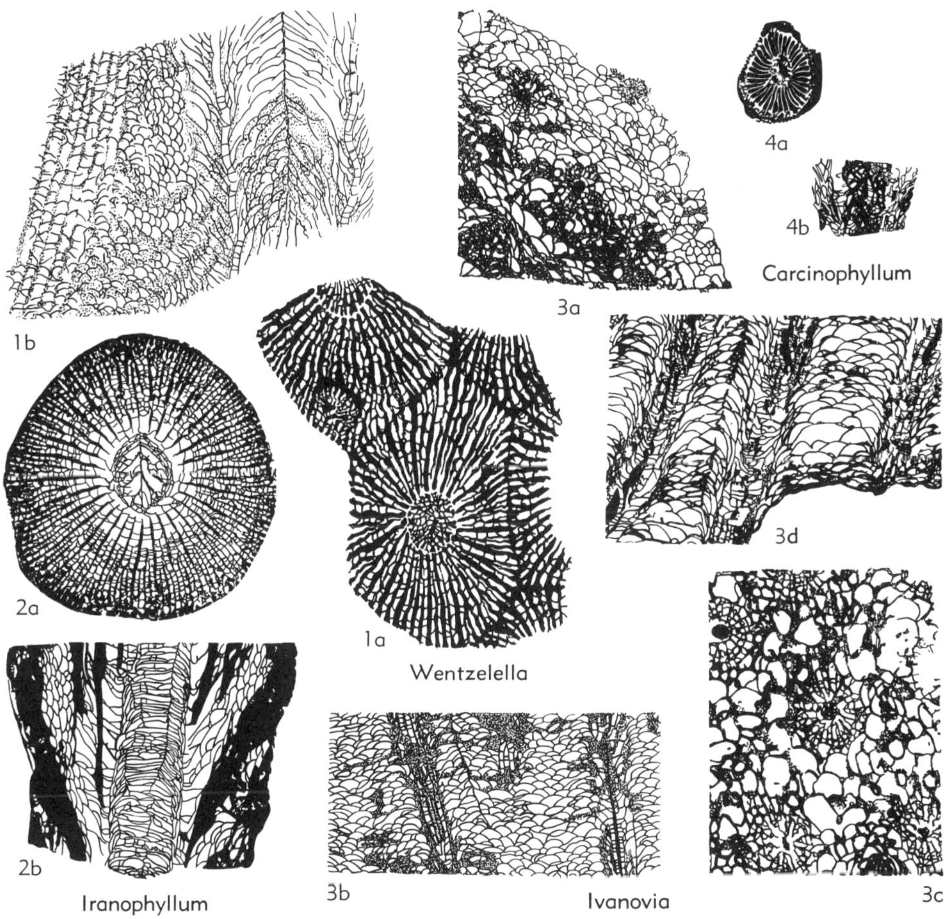

Fig. 211. Columnariina: Lonsdaleiidae (p. F307-F309).

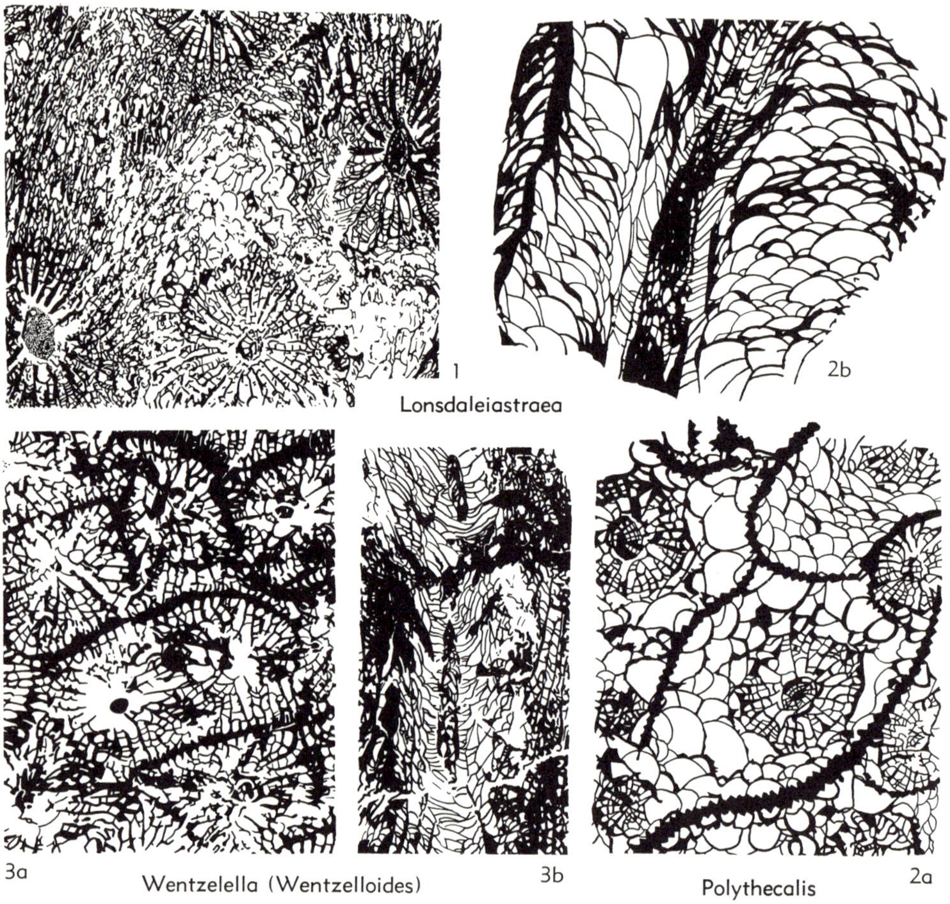

FIG. 212. Columnariina: Lonsdaleiidae (p. F309-F310).

Lonsdaleiastraea GERTH, 1921 [**L. vinassai*]. Thamnasterioid or partly aphroid; like *Iranophyllum* internally, but with some lonsdaleoid dissepiments (12). *L.Perm.(Artinsk.),* Asia.——FIG. 212, *1. *L. vinassai,* Timor; transv. sec., ×2 (79).

Polythecalis YABE & HAYASAKA, 1916 [**P. confluens*]. Cerioid or in part aphroid; like *Iranophyllum* internally but with lonsdaleoid dissepimentarium (25). *L.Perm.(Artinsk.),* Asia.——FIG. 212,2. **P. confluens; 2a,b,* transv. and long. secs., ×4 (88).

Heritschiella MOORE & JEFFORDS, *nom. nov.* [*pro Heritschia* MOORE-J., 1941 (Kans. Geol. Survey Bull. 38, p. 94) (*non* TEPPNER, 1922)] [**Heritschia girtyi* MOORE-J., 1941]. Fasciculate; like *Waagenophyllum* but with thin, numerous septa; tertiary septa present; periaxial zone of tabulae wide, with tabulae flat or outwardly and upwardly inclined. *L.Perm.,* N.Am.-Eu.-Asia.——FIG. 210,2. **H. girtyi* (MOORE-J.), Kans.; *2a,b,* transv. and long. secs., ×2 (104).

Suborder CYSTIPHYLLINA Nicholson in Nicholson & Lydekker, 1889

[*nom. correct.* HILL, herein (*pro* Cystiphylloidea NICH., in NICH.-L., 1889)] [=Lythophyllacea WEDEKIND, 1925; Cystiphyllida, Cystiphyllacea WDKD., 1927; Calceolacea WDKD., 1937; Cystiphylalacea WANG, 1948 (misprint); includes Tryplasmacea LECOMPTE, 1952 (superfam.)]

Solitary or compound Rugosa; septal trabeculae large; marginarium either a stereozone in which cores of trabeculae appear set in a mass of lamellate sclerenchyme, or a dissepimentarium of small globose dissepiments, typically with septa represented by separate trabeculae based on upper surfaces of the dissepiments; tabulae flat and complete, or inversely conical and incomplete. *Ord.-Dev.*

Family TRYPLASMATIDAE Etheridge, 1907

[*nom. correct.* HILL, herein (*pro* Tryplasmidae ETHERIDGE, 1907)] [=?Palaeocyclidae DYBOWSKI, 1873; ?Palaeocyclinae (*nom. transl.* ZITTEL, 1876); ?Cyclinidae CHAPMAN, 1893 (invalid); Pholidophyllidae, Pholidophyllida WEDEKIND, 1927; Acanthocyclidae HILL, 1936; ?Paleocyclidae BASSLER, 1937; Rhabdocyclidae HILL, 1940; ?Porpitidae MOORE & JEFFORDS, 1945)]

Solitary and compound Rugosa with complete, horizontal or inversely conical tabulae, rarely with a narrow median notch; typically without dissepiments, marginarium a stereozone with very numerous septa, each represented by a vertical series of trabeculae bound together by lamellate sclerenchyme but free at their inner ends; major septa may be continued toward axis as small spines on upper surfaces only of tabulae; trabeculae rhabdacanthine or holacanthine. *Ord.-Dev.*

Rhabdocyclus LANG & SMITH, 1939 [*pro Acanthocyclus* DYBOWSKI, 1873 (*non* LUCAS, 1844)] [**Palaeocyclus fletcheri* M.EDW.-H., 1851]. Discoid or patellate, with cone of attachment excentric; without tabulae or dissepiments, with stereozone of rhabdacanthine septa (17). *Sil.,* Eu.-N. Am.——FIG. 213,*1*. **R. fletcheri* M.EDW.-H., M. Sil., Eu.; *1a,b,* calice, long. sec., ×2 (98).

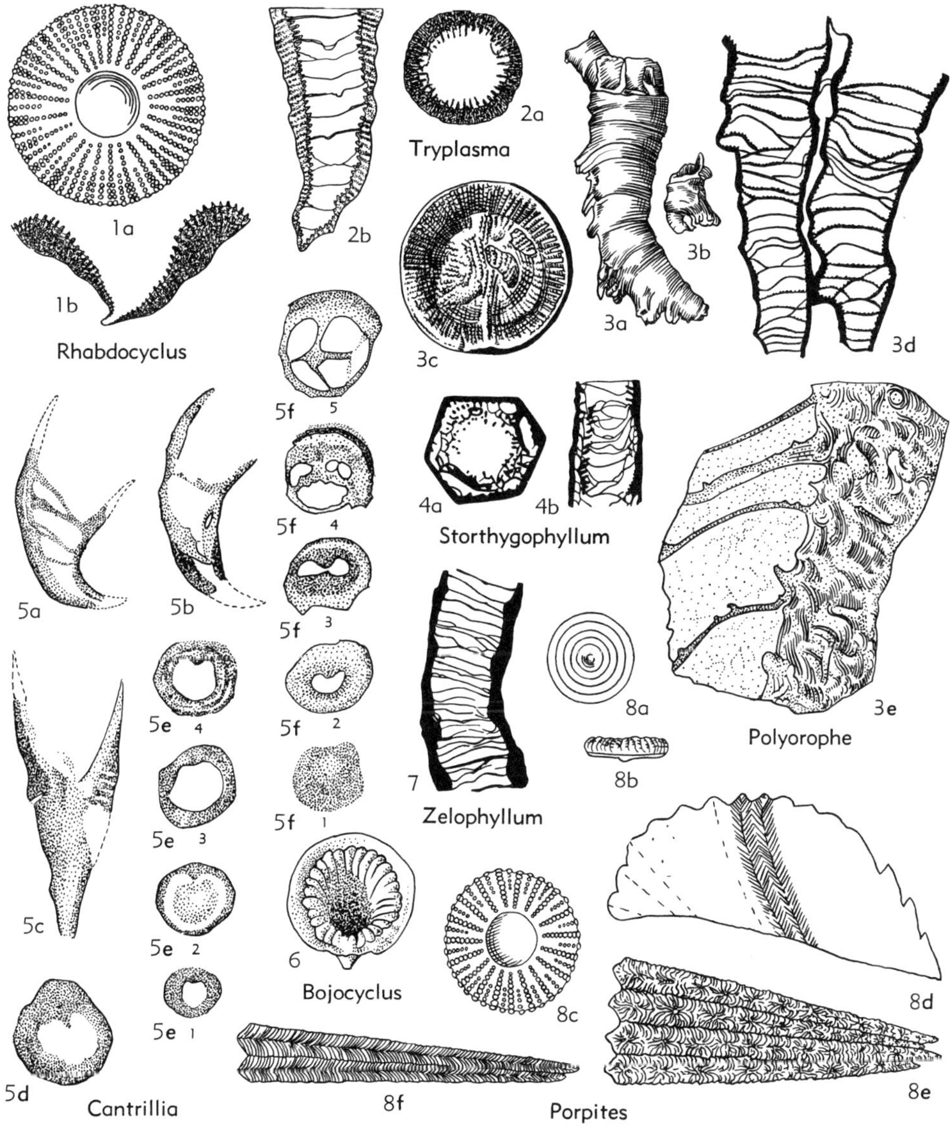

FIG. 213. Cystiphyllina: Tryplasmatidae (p. F311-F312).

Tryplasma LONSDALE, 1845 [**T. aequabile;* SD ETHERIDGE, 1907] [=*Pholidophyllum* LINDSTRÖM, 1871, *Pholadophyllum* LANG-S.-T., 1940 *(nom. van.); Spiniferina* PENECKE, 1894 *(pro Acanthodes* DYBOWSKI, 1873; *non* AGASSIZ, 1833; *nec* DE HAAN, 1833; *nec* BALY, 1864); *Aphyllostylus* WHITEAVES, 1904; *?Tyrrellia* PARKS, 1913 *(non* KOENIKE, 1895); *Stortophyllum* WEDEKIND, 1927; *?Aphyllum* SOSHKINA, 1937 *(non Aphylum* BERGROTH, 1906)]. Solitary, with marked rejuvenescence rims, or fasciculate; minor septal grooves on epitheca weaker than major septal grooves (17). *Sil.-L.Dev.,* Eu.-Asia-N.Am.-Austral.——FIG. 213, *2. T. loveni* (M.EDW.-H.), M.Sil., Eu; *2a,b,* transv. and long. secs., ×1 (85).

Storthygophyllum WEISSERMEL, 1894 [**S. megalocystis (partim)*] [=*Xiphelasma* SMITH & LANG, 1931]. Like *Tryplasma* but cerioid, and with a narrow zone of lonsdaleoid dissepiments. *Sil.,* Gotl. ——FIG. 213,*4. S. tubulatum* (SCHLOTHEIM) (type species of *Xiphelasma*), *4a,b,* transv. and long. secs., ×3 (118).

Cantrillia SMITH, 1930 [**C. prisca*]. Like *Tryplasma* but with infold of peripheral stereozone, and tabulae invested with lamellate sclerenchyme continuous with that of the stereozone (17). *L.Sil.,* Br.I. ——FIG. 213,*5.* **C. prisca,* U.Llandov., Br.I.; *5a-c,* long. secs., ×3; *5d-g,* transv. secs. (incl. 2 ontogenetic series), ×3 (117).

Polyorophe LINDSTRÖM, 1882 [**P. glabra*]. Like solitary *Tryplasma* but with remarkable development of epithecal outgrowths of attachment (36). *L.Sil.-M.Sil.,* Gotl.——FIG. 213,*3.* **P. glabra, 3a,b,* side view of corallites, ×0.5; *3c,* calice, ×1; *3d,e,* long. secs., ×1, ×5 (100).

Zelophyllum WEDEKIND, 1927 [**Z. intermedium*]. Fasciculate, with no free ends of trabeculae projecting beyond narrow peripheral stereozone (55). *M.Sil.,* Gotl.——FIG. 213,*7.* **Z. intermedium;* long. sec., ×1 (137).

?Porpites SCHLOTHEIM, 1820 [*non* GUETTARD, 1770 (*non* Linnean)] [**P. haemisphericus;* (=*Madrepora porpita* LINNÉ, 1767 *partim;* SD LANG-S.-T., 1940] [=*Palaeocyclus* M.EDW.-H., 1849]. Discoid, with central cone of attachment; stereozone of monacanthine septa, no tabulae and no dissepiments (17). *L.Sil.-M.Sil.,* Eu.-N.Am.——FIG. 213,*8.* **P. porpita* (LINNÉ), L.Sil. (U.Llandov.), Eng.; *8a,b,* base, side, ×1; *8c,* top, ×2; *8d-f,* transv. and long. secs. of septa showing trabeculae, enlarged (85).

Kitakamiphyllum HILL, *nom. nov.* [*pro Maia* SUGIYAMA, 1940 (Sci. Rept. Tôhoku Imp. Univ., ser. 2, v. 21, p. 122) *(non* BRISSON, 1760; *nec* LAMARCK, 1801; *nec* REICHENBACH, 1850; *nec* FREDERICKS, 1924)] [**Maia cylindrica* SUGIYAMA, 1940]. Like *Tryplasma* but without trace of septa. *M.Sil.,* Japan.

?Bojocyclus PRANTL, 1939 [**B. bohemicus*]. Discoid with cone of attachment excentric; calice with smooth, aseptate outer border; septa strongly crenulate (52). *M.Dev.,* Czech.——FIG. 213,*6.* **B. bohemicus;* calice, ×2 (109).

Family CYSTIPHYLLIDAE Milne-Edwards & Haime, 1850

[=Cystiphyllinae *(nom. transl.* M'COY, 1851); Araeopomatidae LINDSTRÖM, 1883; Cystiphyllacea WEDEKIND, 1927 (suborder); Holmophyllinae WANG, 1947]

Solitary or fasciculate; septa very numerous, each represented by a vertical series of discrete, typically holacanthine trabeculae commonly set in lamellar sclerenchyme which coats the horizontal skeletal elements; tabulae sagging to inversely conical, incomplete; dissepimentarium wide, dissepiments usually smaller than tabellae. *Sil.*

Cystiphyllum LONSDALE, 1839 [**C. siluriense;* SD M.EDW.-H., 1850] [=*Conophyllum* HALL, 1851]. Solitary; with long septa, each represented only by distant, separate trabeculae developed only on upper surfaces of successive globose dissepiments and tabellae; minor septa very long, dissepimentarium very wide; calicular floor inversely conical, inclination of dissepiments and tabellae being similar (31). *Sil.,* Eu.-Asia-Austral.-N.Y.——FIG. 214,*1. C. cylindricum* LONSD., M.Sil., Eu.; *1a,b,* secs., ×1 (98).

Hedstroemophyllum WEDEKIND, 1927 [*nom. correct.* HILL, herein *(pro Hedströmophyllum* WDKD., 1927)] [**H. articulatum*]. Like *Cystiphyllum* but corallites slender, with trabeculae long enough to pierce several successive dissepimental floors (55). *M.Sil.,* Gotl.——FIG. 214,*3.* **H. articulatum; 3a,* transv. sec., ×3; *3b,* long. sec., ×2 (137).

Holmophyllum WEDEKIND, 1927 [**H. holmi*]. Like *Cystiphyllum* but trabeculae extending from epitheca into tabularium; tabulae complete and flat or sagging (55). *U.Sil.,* Gotl.-Austral.——FIG. 214, *6.* **H. holmi,* Gotl.; *6a,b,* transv. and long. secs., ×3 (137).

Gyalophyllum WEDEKIND, 1927 [**G. angelini*]. Like *Cystiphyllum* but with lamellar sclerenchyme forming a dense zone as wide as the minor septa are long and broken only by dissepiments developed on distant calical floors (55). *U.Sil.,* Gotl. ——FIG. 214,*7.* **G. angelini; 7a,* transv. sec., ×1; *7b,* long. sec., ×2 (137).

Araeopoma LINDSTRÖM, 1883 [**A. prismaticum* (=*Cystiphyllum prismaticum* LIND., 1868)]. Like *Cystiphyllum* but square in section in the erect part of corallum (36). *L.Sil.(U.Llandov.)-M.Sil.,* Gotl. ——FIG. 214,*4.* **A. prismaticum* (LIND.); side view, ×1 (100).

Microplasma DYBOWSKI, 1873 [**M. gotlandicum;* SD WEDEKIND, 1927]. Like *Cystiphyllum* but fasciculate, corallites slender; trabeculae confined to extremely narrow peripheral stereozone; horizontal skeletal elements large, not distinctly divisible into dissepiments and tabellae (9). *M.Sil.-U.Sil.*, Eu.——FIG. 214,2. **M. gotlandicum; 2a,b,* long. secs., ×1; *2c,* transv. sec., ×2 (71).

?**Nipponophyllum** SUGIYAMA, 1940 [**N. giganteum*] [=*Baeophyllum* HILL, 1940]. Fasciculate, corallites slender, with connecting processes formed by outgrowths of dissepimentarium; septa represented by separate trabeculae or by segments in which 2 or 3 trabeculae are in contact; dissepiments large, in 1 or 2 series, inner edges steeply sloping; tabulae concave, complete or supplemented by wide tabellae (20). *M.Sil.*, Japan-E.Austral.——FIG. 214,5. **N. giganteum,* Japan; *5a,* transv. sec., ×1; *5b,c,* long. secs., ×1 (128).

Family GONIOPHYLLIDAE Dybowski, 1873

[=Goniophyllinae (*nom. transl.* ZITTEL, 1876); Calceolidae LINDSTRÖM, 1883 (also Calceolidae ROEMER, 1883); Calceolacea WEDEKIND, 1937]

Corallites semicircular (calceoloid) or square in transverse section; calice with an operculum of 1 or 4 plates of dense sclerenchyme; corallites either with wide dissepimentarium and incomplete, inversely conical tabulae or completely filled with dense sclerenchyme; septa may be confined to a few short plates on flat side of corallum or developed as short plates so thick as almost to preclude dissepiments, and with little difference between major and minor septa. *Sil.-Dev.*

Goniophyllum M.EDW.-H., 1850 [**Turbinolia pyramidalis* HISINGER, 1831]. Solitary, square in section, with operculum of 4 triangular plates; septa

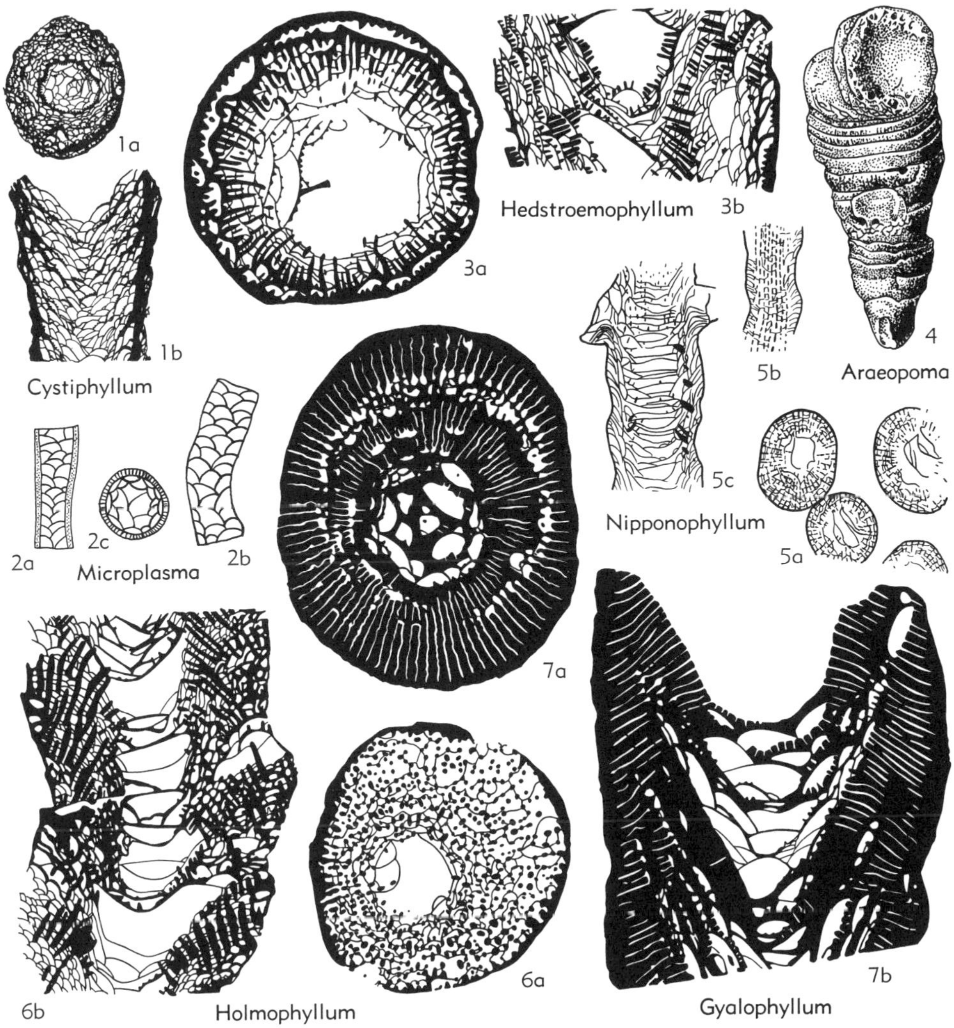

FIG. 214. Cystiphyllina: Cystiphyllidae (p. F312-F313).

thick, mostly contiguous, as long as the dissepimentarium is wide; dissepiments and tabellae numerous, commonly thickened (36). *L.Sil.-M.Sil.*, Eu.——FIG. 215,*2*. **G. pyramidale* (HISINGER); *2a*, corallite, ×1; *2b,c*, secs., ×1, ×2 (100).

Rhizophyllum LINDSTRÖM, 1866 [**Calceola gotlandica* F. A. ROEMER, 1856]. Calceoloid; dissepiments and tabulae little thickened or not at all; septa developed only along flattened side as short plates not distinguishable into major and minor cycles (36). *M.Sil.-L.Dev.*, Eu.-Asia-N.Am.-Austral. ——FIG. 215,*3*. **R. gotlandicum* (ROEMER), U.Sil., Gotl.; *3a,b*, side view, long. sec., ×1 (100).

Teratophyllum LANG-S.-T., 1940 [*pro Platyphyllum* LINDSTRÖM, 1883 (*non* AUDINET-SERVILLE, 1831)] [**Platyphyllum sinense* LIND., 1883]. Calceoloid; septa numerous; dissepiments and tabulae as in *Goniophyllum* (36). *M.Sil.*, China; *L.Dev.*, USSR. ——FIG. 215,*1*. **T. sinense* (LIND.), M.Sil., China; *1a,b*, secs., ×1, enlarged (100).

Rhytidophyllum LINDSTRÖM, 1883 [**R. pusillum*]. Minute, operculate and calceoloid, with faint septa; ?cystiphylloid structure (36). *U.Sil.*, Gotl.——FIG. 215,*5*. **R. pusillum;* side view, ×4 (100).

Calceola LAMARCK, 1799 [*non* SWAINSON, 1840] [**Anomia sandalium* GMELIN, in LINNÉ, 1791 (=*Anomia sandalinum* LINNÉ, 1771)] [=*Calceolina* RAFINESQUE, 1815 (*non* ADAMS, 1863)]. Corallites semicircular in section, with semicircular operculum; completely filled with dense sclerenchyme (36). *L.Dev.(Ems.)-M.Dev.*, Eu.-Asia; *M. Dev.*, Afr.-Austral.-Calif.——FIG. 215,*4*. **C. sandalina* (LINNÉ); *4a*, convex side; *4b,c*, convex side and interior of another specimen; *4d,e,g*, exterior, interior, side of operculum; *4f*, diagram of septal arrangement in calice and interior of operculum (C, cardinal septum; K, counter septum); all ×1 (129).

?**Protaeropoma** TING, 1937 [**P. wedekindi*] [=*Protaraeopoma* LANG-S.-T., 1940 *(nom. van.)*]. *Sil.*, Gotl.

Family DIGONOPHYLLIDAE Wedekind, 1923

[=?Plasmophyllidae, ?Plasmocystidae DYBOWSKI, 1873; ?Plasmophyllinae (*nom. transl.* ZITTEL, 1876); Lythophyllidae, Lythophyllacea WDKD., 1925; Arcophyllidae MARCOV, 1926; Lytophyllidae, Lytophyllinae, Cosmophyllinae WDKD. & VOLLBRECHT, 1931 (invalid); Glossophyllidae KETTNEROVA, 1932; Cystiphylloidae, Cystiphylloinae, Arcophyllinae STUMM, 1949 (=Cystiphylloididae, *nom. correct.* HILL, herein, *ex Cystiphylloides*)]

Solitary coralla with inversely conical calicular floors, wide dissepimentarium merging into tabularium; in some, septal tissue is developed only in successive inverse cones of thickening (septal cones of WEDEKIND), in which individual large trabeculae may be distinguished, or septa may be developed in short, thick, contiguous laminar segments or in part as discrete trabeculae; in others, septa are developed as contiguous laminae, and in these a minor septum near the counter septum is elongated into the axial region. *Dev.*

In this family, if WEDEKIND's (54) analysis is correct, laminar septa are built up from discrete trabeculae and from discontinuous laminar septal segments in which the trabeculae are contiguous and may be indistinguishable. The family thus may have been derived from the earlier family Cystiphyllidae with discrete trabeculae, in which some genera (*Gyalophyllum*) show extreme dilatation affecting tissue in successive inversely conical calicular floors ("septal cones"), just as in the Digonophyllidae. It is possible, however, to regard the "septal cones" as rejuvenescent returns to earlier stages of ontogeny in which the septa were of thick, contiguous laminae, and in this view ancestors of the Digonophyllidae may be outside the Cystiphyllina altogether, though at present no satisfactory ancestor for such an origin can be suggested for them. Therefore, they are herein referred to the Cystiphyllina and placed close to the Cystiphyllidae.

Subfamily ZONOPHYLLINAE Wedekind, 1924

[=?Plasmophyllidae, ?Plasmocystidae DYB., 1873; Lythophyllidae WDKD., 1925; Lytophyllidae, Lytophyllinae WDKD. & VOLLBRECHT, 1931 (invalid); Zonophyllidae (*nom. transl.* WDKD., 1937)]

Solitary or rarely fasciculate coralla with inversely conical calicular floors of globose tabellae and smaller, less globose dissepiments; septa discontinuous, of monacanthine trabeculae (separate in some) developed only in successive zones (calicular floors, "septal cones") of skeletal dilatation; the separate trabeculae do not pierce successive dissepiments; there is no trace of any long counter minor septum, nor of crossbar carinae. *L.Dev.-M.Dev.*

Zonophyllum WEDEKIND, 1924 [**Z. duplicatum;* SD LANG-S.-T., 1940] [=?*Coleophyllum* HALL, 1883; *Legnophyllum, Pseudozonophyllum* WDKD., 1924; *Lythophyllum, Nardophyllum, Paralythophyllum* WDKD., 1925; *Plagiophyllum* WDKD. & VOLLBRECHT, 1931; *Lithophyllum (nom. van. pro Lythophyllum), Paralithophyllum (nom. van. pro Paralythophyllum)* LANG.-S.-T., 1940 (*non Lithophyllum* MÜLLER, 1859); *Wedekindophyllum* STUMM, 1949 *(nom. van. pro Lythophyllum)*]. Solitary, conical or cylindrical; calicular floors erect or oblique inverse cones (54). *L.Dev.-M.Dev.*, Eu.-Asia-N.Am.——FIG. 216-*1a-c*. **Z. duplicatum*,

Ger.; *1a-c,* transv. secs., ×2 (137).——Fig. 216, *1d,e. Z. marginatum* Wdkd. (type species of *Lythophyllum* Wdkd. and *Wedekindophyllum* Stumm), U.M.Dev., Ger.; *1d,e,* transv. and long. secs., ×1 (137).——Fig. 216,*1f,g. Z. pseudoseptatum* (Schulz), U.M.Dev., Ger.; *1f,g,* transv. and long. secs., ×1 (138).——Fig. 270,*1h,i. Z.? romingeri* (Hall) (type species of *Coleophyllum*), L.Dev., N.Am.; *1h,i,* side and calical views of corallite, ×0.5 (83).

Cayugaea Lambe, 1901 [**C. whiteavesiana*]. Subcylindrical without trace of septa; tabulae large, saucer-shaped, complete, distinctly demarcated from dissepimentarium formed of rather steeply sloping plates (52). *L.Dev.,* N.Am.——Fig. 216,*2.* **C. whiteavesiana,* Can.; *2a,b,* transv. and long. secs., ×0.5 (96).

Skoliophyllum Wedekind, 1937 [**Cyathophyllum lamellosum* Goldfuss, 1826; SD Lang-S.-T., 1940] [=*Scoliophyllum* Lang-S.-T., 1940 *(nom. van.)*]. Solitary, with flat calicular floors and repeated rejuvenescence, giving a series of laminae en echelon; practically filled by skeletal dilatation (52). Uppermost *L.M.Dev.,* Ger.——Fig. 216,*3.* **S. lamellosum* (Goldf.); side view, ×1 (137).

?**Bucanophyllum** Ulrich, 1886 [**B. gracile*]. Small, solitary, slender at first, then rapidly expanding. *L.Dev.,* N.Am.

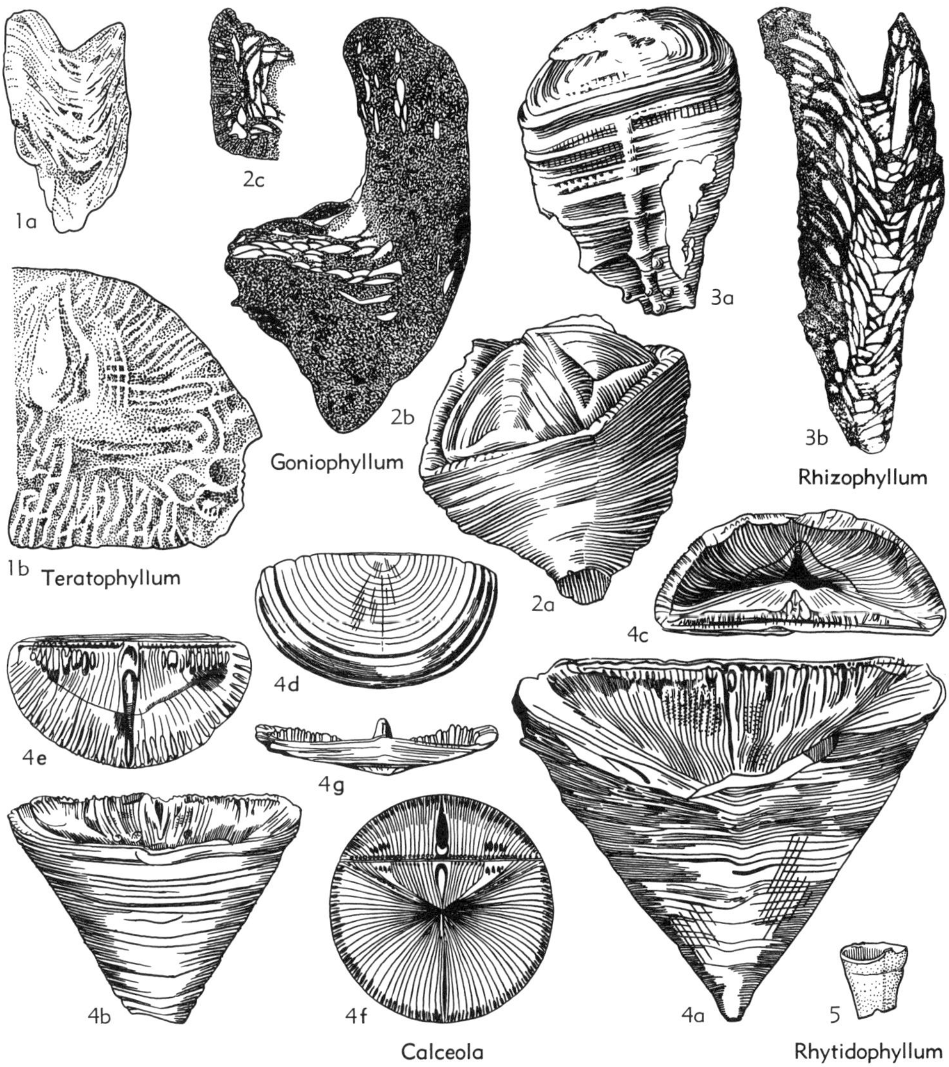

Fig. 215. Cystiphyllina: Goniophyllidae (p. F314).

?**Plasmophyllum** Dybowski, 1873 [**Cyathophyllum goldfussi* M.Edw.-H., 1851] [=*Stereophyllum* Schlüter, 1889 (*non* Grabau, 1917; *nec* Soshkina, 1937)]. Small, curved, conical; typically filled with skeletal dilating tissue except at calical periphery, where lonsdaleoid dissepiments develop (52). *M.Dev.*, Ger.——Fig. 216,6. **P. goldfussi* (M.Edw.-H.); *6a*, calice, ×1; *6b,c*, transv. and long. secs., ×1 (114).

?**Diplochone** Frech, 1886 [**D. striata*]. Solitary;

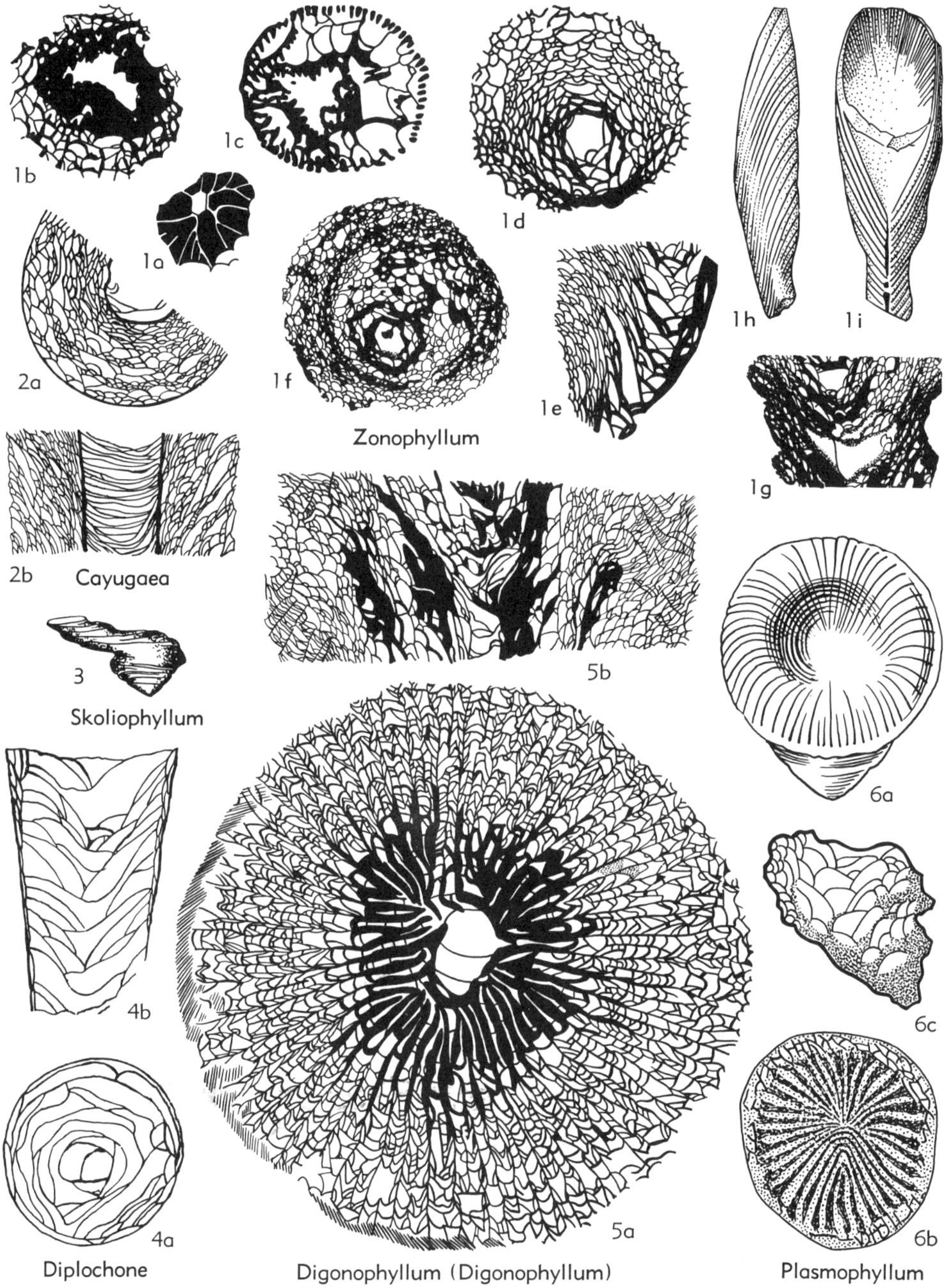

Fig. 216. Cystiphyllina: Digonophyllidae (p. F314-F318).

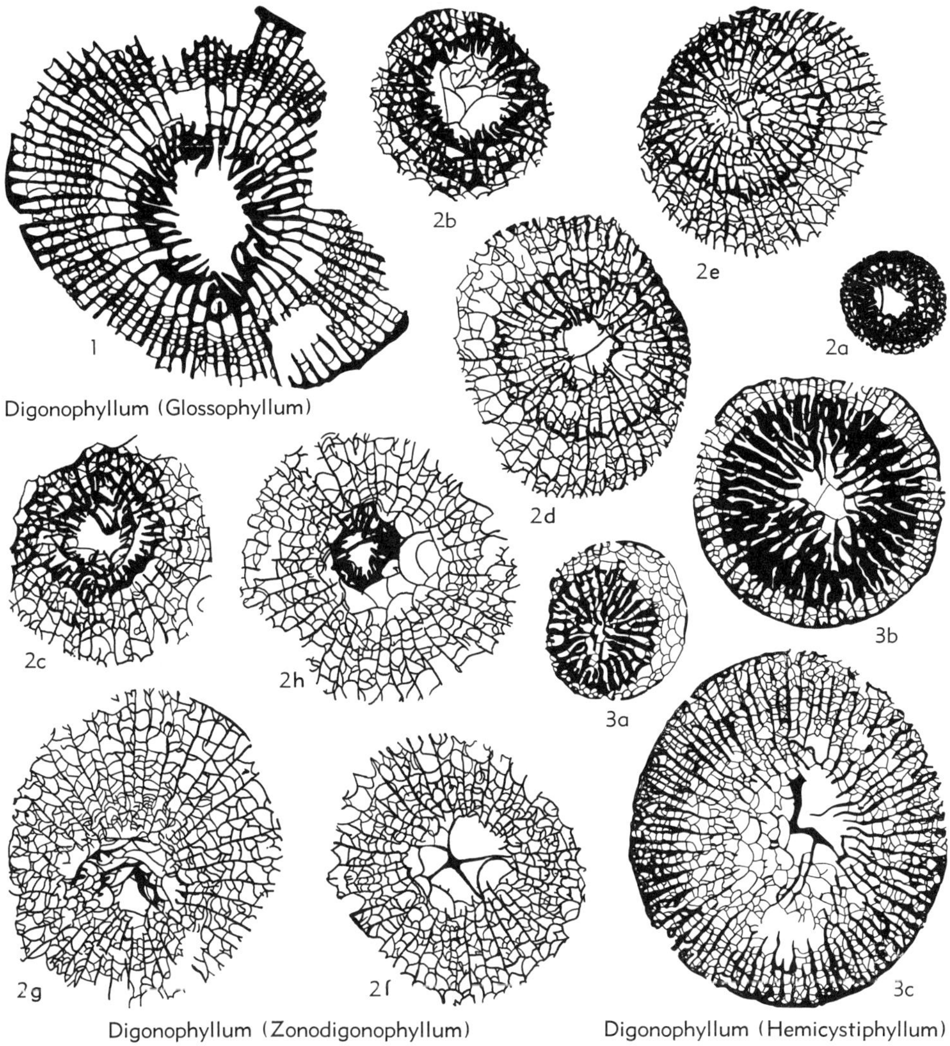

FIG. 217. Cystiphyllina: Digonophyllidae (p. F318).

septa developed only as very low ridges (not spinose) on upper surfaces of tabulae and on inner side of wall, but numerous and rather long; dissepimentarium very narrow, of 1 or 2 series of elongate, highly inclined plates; tabulae complete or incomplete, funnel-shaped (52). *Up.M.Dev.*, Ger.——FIG. 216,*4*. **D. striata; 4a,b,* transv. and long. secs., ×1 (77).

Subfamily DIGONOPHYLLINAE Wedekind, 1923

[=Arcophyllidae MARKOV, 1926; Cosmophyllinae WEDEKIND & VOLLBRECHT, 1931; Glossophyllidae KETTNEROVA, 1932; Cystiphylloidae STUMM, 1949 (*errore pro* Cystiphylloididae); Atelophyllinae, Mochlophyllinae TAYLOR, 1951]

Large solitary Rugosa with inversely conical calicular floors of globose tabellae and small, less globose dissepiments; septa are continuous vertical plates thinnest near periphery, or are represented in dissepimentarium or throughout by septal crests, by separate crossbar carinae, or they are quite absent (especially minor septa); they may be lined by lateral dissepiments; usually one minor septum near counter septum is very long; arrangement of septa in tabularium characteristic; fossula like a keyhole in transverse section, being outlined in cardinal part of tabularium by deepening of the calical floor, by parallelism of the short axial ends of 2 major septa nearest to the short cardinal septum, and by sudden inward expansion of this trough into a circular space left by axial ends of septa in cardinal quadrants. *M.Dev.*

Digonophyllum WEDEKIND, 1923 [**D. schulzi*]. Large; septa typically continuous laminae, but minor septa may be represented by sparse crests only (22). *M.Dev.*, Eu.-Austral.

D. (Digonophyllum). Very large; minor septa continuous vertical plates except near periphery; major septa thick in tabularium, the thickening dying away in inner parts of dissepimentarium; lateral dissepiments line the septa, and crossbar carinae may develop at periphery (22). *L.M.Dev.*, Ger.——FIG. 216,*5*. **D. (D) schulzi; 5a,b,* transv. and long. secs., ×0.25 (133).

D. (Glossophyllum) WEDEKIND, 1924 [**G. dohmi;* SD LANG-S.-T., 1940]. Like *D. (Digonophyllum)* but with well-developed minor septa (54). *L.M. Dev.*, Ger.——FIG. 217,*1*. **D. (G.) dohmi;* transv. sec., ×2 (137).

D. (Zonodigonophyllum) VOLLBRECHT, 1926 [**Z. primum;* SD LANG-S.-T., 1940]. Minor septa represented by rare crests only; major septa typically continuous plates, septa withdrawn from axis; rejuvenescent cones of septal thickening marked (52). *L.M.Dev.*, Ger.——FIG. 217,*2*. **D. (Z.) primum; 2a-h,* series of transv. secs. showing growth changes, ×1 (133).

D. (Hemicystiphyllum) WEDEKIND, 1925 [**H. frechi*]. Minor septa commonly continuous; major septa in adult stage withdrawn from tabularium or represented there by crests (52). *U.M.Dev.*, Ger.——FIG. 217,*3*. **D. (H.) frechi; 3a-c,* transv. secs., ×2 (137).

D. (Mochlophyllum) WEDEKIND, 1923 [**Actinocystis maximus* SCHLÜTER, 1882 (=*Mesophyllum maximum* SCHLÜTER, 1889] [=*Pseudocosmophyllum* WEDEKIND & VOLLBRECHT, 1931]. Exceptionally large; major and minor septa continuous plates except near periphery where they are replaced by separate crossbar carinae; lateral dissepiments line septa; septal ends in tabularium dilated, unequal, not straight, shorter ones impinging on longer; one counter minor septum very long, extending into tabularium; keyhole fossula may be obscured (52). *U.M.Dev.*, Ger.-Austral.——FIG. 218,*1c*. **D. (M.) maximum* (SCHLÜTER), Ger.; transv. sec., ×0.25 (133).——FIG. 218,*1a,b*. *D. (M.) geigeri* WDKD. & VOLLBRECHT (type species of *Pseudocosmophyllum*), Ger.; *1a,b,* transv. and long. secs., ×1 (138).

D. (Enteleiophyllum) WALTHER, 1928 [**E. sundwigense;* SD LANG-S.-T., 1940]. Septa continuous plates except near periphery, with slight dilatation, without crossbar carinae; major septa unequal, not straight in tabularium rarely leaving a rounded axial space at inner end of the fossula (52). *U.M.Dev.*, Ger.——FIG. 218,*2*. **D. (E.) sundwigense;* transv. sec., ×1 (135).

D. (Uralophyllum) SOSHKINA, 1936 [**U. unicum*]. Like *D. (Mochlophyllum)* but with septa dilated only in tabularium, and thread-thin in the inner parts of dissepimentarium (52). *U.M.Dev.*, USSR.——FIG. 218,*3*. **D. (U.) unicum,* Urals; *3a,b,* transv. and long. secs., ×1 (122).

Mesophyllum SCHLÜTER, 1889 [*non* HAHN, 1911] [**M. defectum* SCHLÜTER, 1889 (=*Actinocystis defecta* SCHLÜTER, 1882); SD WDKD., 1925]. Septa represented in dissepimentarium or throughout by septal crests, by separate yardarm carinae, or by rare traces, especially the minor septa; commonly one counter minor septum very long; keyhole fossula characteristic (22). *M.Dev.*, Eu.-Asia-Austral.

M. (Mesophyllum). Septa withdrawn from axis, leaving an axial space, pear-shaped in transverse section, with cardinal fossula at the edge of tabularium; septa unthickened, represented peripherally by discrete yardarm carinae and axially by narrow laminar segments (22). *U.M.Dev.*, Ger.-Austral.-Asia.—FIG. 219,*5*. **M. (M.) defectum* (SCHLÜTER), Ger.; transv. sec., ×2 (137).

M. (Dialytophyllum) AMANSHAUSER in WEDEKIND, 1925 [**D. complicatum*] [=*Bothriophyllum* VOLLBRECHT, 1926]. Like *M. (Mesophyllum)* but with major septa dilated, continuous but partly replaced by lateral dissepiments, and without discrete yardarm carinae peripherally (22). *U.M. Dev.*, Ger.-Austral.——FIG. 219,*4*. **M. (D.) complicatum,* Ger.; transv. sec., ×1 (137).

M. (Hemicosmophyllum) WEDEKIND & VOLLBRECHT, 1931 [**H. limbatum*]. Like *M. (Mesophyllum)* but septa may be dilated in inner parts of dissepimentarium, major septa proceeding as continuous dilated vertical plates to axis, with shorter ones impinging on longer, keyhole fossula obscure (52). Lowermost *U.M.Dev.*, Ger.——FIG. 219,*3*. **M. (H.) limbatum; 3a-c,* transv. secs., ×1 (138).

M. (Lekanophyllum) WEDEKIND, 1923 [**L. punctatum;* SD LANG-S.-T., 1940] [=*Lecanophyllum* LANG-S.-T., 1940]. Like *M. (Mesophyllum)* but with septa represented in young stages by sparse monacanthine trabeculae and in adult by more or less continuous, moderately thick plates (52). Lowermost *U.M.Dev.*, Ger.——FIG. 219,*2a-c*. **M. (L.) punctatum; 2a-c,* ×2 (54).

M. (Arcophyllum) MARKOV, 1926 [**A. typus*] [=*Cosmophyllum* VOLLBRECHT, 1922 (*non* BLANCHARD, 1851)]. Like *M. (Mesophyllum)* but with greater continuity in septa in inner parts of dissepimentarium (52). *U.M.Dev.*, Eu.——FIG. 219,*1*. **M. (A.) typus,* Ger.; enlarged (133).

M. (Atelophyllum) WEDEKIND, 1925 [**A. emsti*]. Like *M. (Arcophyllum)* but with peripheral discrete crossbar carinae rare or absent (52). *U.M. Dev.*, Eu.-Asia.——FIG. 218,*4a,b*. **M. (A.) emsti, 4a,b,* ×1 (137).

M. (Cystiphylloides) CHAPMAN, 1893 [**Cystiphyllum aggregatum* BILLINGS, 1859] [=*Cystiphylloides* YOH, 1937; ?*Cystiplasma* TAYLOR, 1951]. Solitary or weakly aggregate, with septa represented only by radiating striae in calice (52). *U.M.Dev.*, Eu.-N.Am.-Asia.——FIG. 219,6. **M. (C.) aggregatum* (BILL.), Can.; ×1 (BILLINGS).

RUGOSA Incertae Sedis

Akiyosiphyllum YABE & SUGIYAMA, 1942 [**A. stylo-*

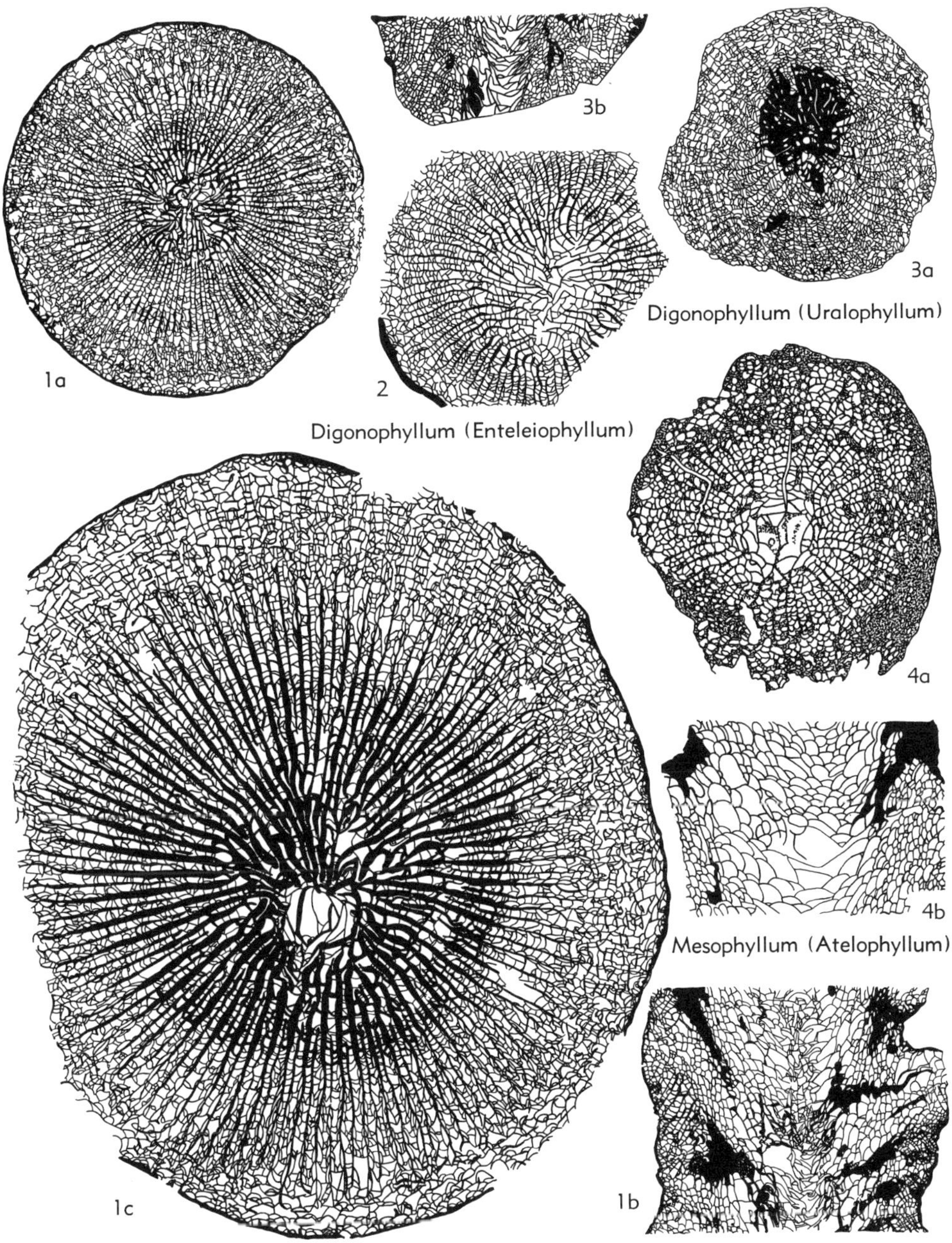

FIG. 218. Cystiphyllina: Digonophyllidae (p. F318).

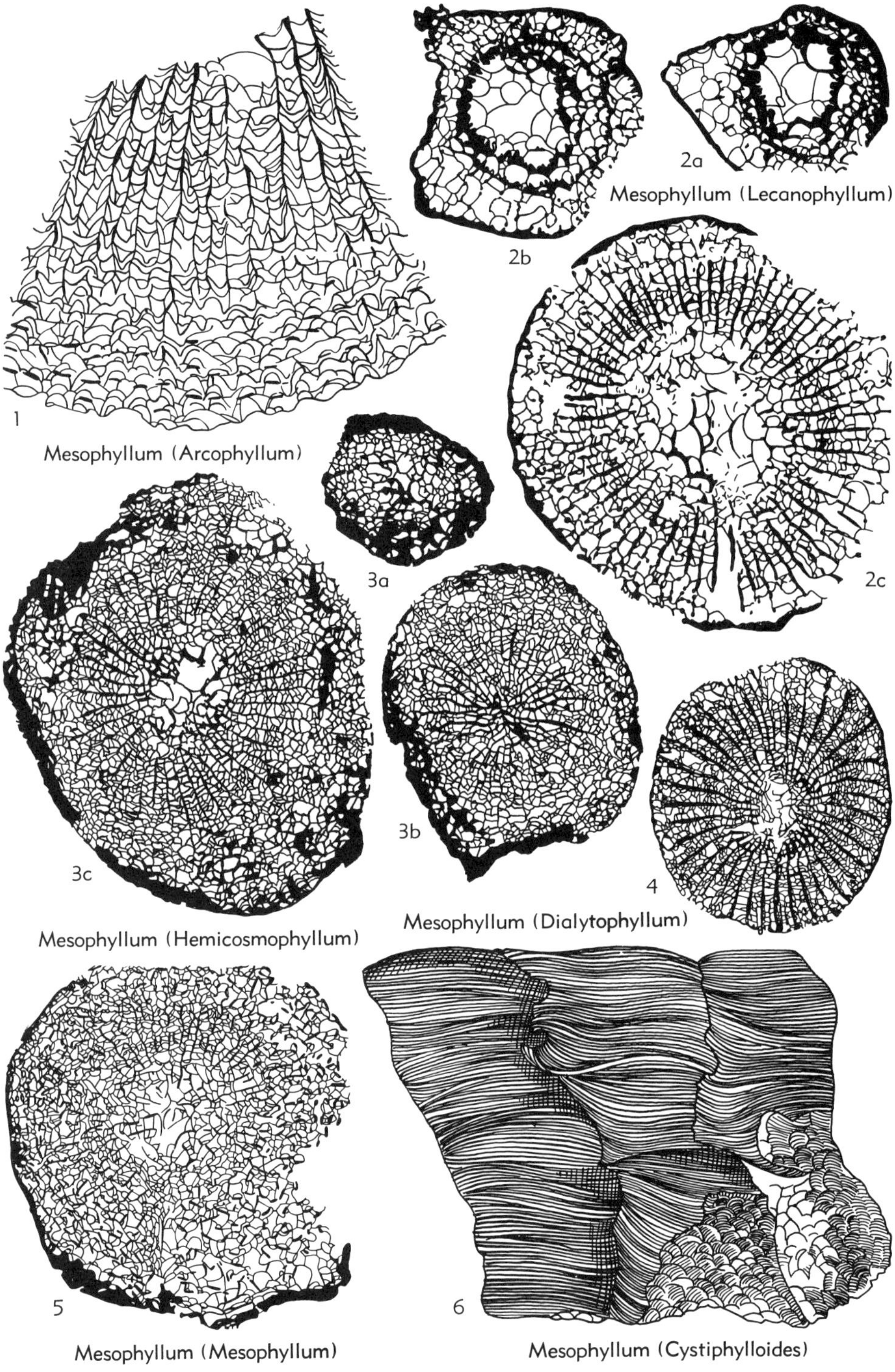

FIG. 219. Cystiphyllina: Digonophyllidae (p. F318-F319) (For *Lecanophyllum* read *Lekanophyllum*).

phorum]. *Perm.*, Japan [not seen].

Astraeophyllum NICHOLSON & HINDE, 1874 [**A. gracile*]. *Sil.*, Can. [types missing].

Axinura CASTELNAU, 1843 [**A. canadensis*]. Drift, *ex Dev.*, N.Am. [holotype lost (=?*Eridophyllum*)].

Barbouria LANG-S.-T., 1940 [*pro Craterophyllum* BARBOUR, 1911; *non* FOERSTE, 1909; *nec* TOLMACHEV, 1931] [**Craterophyllum verticillatum* BARBOUR, 1911]. *Penn.*, Nebr.

Brochiphyllum WEDEKIND, 1923 [*genus caelebs*]. *M.Dev.*, Ger.

Campsactis RAFINESQUE & CLIFFORD, 1820 [**C. canaliculata;* SD LANG-S.-T., 1940]. *Dev.*, Ky. [types lost].

Chusenophyllum TSENG, 1948 [**C. palonoidea*]. *Perm.*, China [not seen].

Cyathaxonella STUCKENBERG, 1895 [**C. gracilis*]. *L.Carb.*, USSR [?Cyathaxoniidae or ?Timorphyllidae].

Cystostylus WHITFIELD, 1880 [**C. typicus*] [=*Cystistylus* LANG-S.-T., 1940 (*nom. van.*)]. *M. Sil.*, USA.

Duncania DE KONINCK, 1872 [**D. simplex*]. *L. Carb.*, Belg.

Edaphophyllum SIMPSON, 1900 [**Cystiphyllum bipartitum* HALL, 1882]. *L.Dev.(Onond.)*, N.Am. [cf. *Coleophyllum;* types lost].

Elasmophyllum HALL, 1882 [**E. attenuatum*]. *L. Dev.(Onond.)*, N.Am. [types lost].

Exostega RAFINESQUE & CLIFFORD, 1820 [**E. tecta;* SD LANG-S.-T., 1940]. *Dev.*, Ky. [types lost].

Gangamophyllum GORSKY, 1938 [**G. boreale*]. *Carb.*, N.Zem.

Huangophyllum TSENG, 1948 [**H. symmetricum*]. *Perm.*, China [not seen].

Kenophyllum DYBOWSKI, 1873 [**K. subcylindricum*] [=*Cenophyllum* RYE, 1875]. *Ord.* (Z_2), Est.

Omphyma RAFINESQUE & CLIFFORD, 1820 [**O. verrucosum*]. *?Miss.*, Ky. [types lost].

Peripaedium EHRENBERG, 1834 [**Cyathophyllum turbinatum* GOLDF., 1826]. M.Dev., Ger.

Polydilasma HALL, 1851 [**P. turbinatum*] [=*Polydiselasma* LANG-S.-T., 1940 (*nom. van.*)]. *M.Sil.*, N.Y. [types missing].

?Prisciturben KUNTH, 1870 [**P. densitextum*]. *Sil.*, Swed. (*See* p. F133.)

Pterorrhiza EHRENBERG, 1834 [**Cyathophyllum marginatum* GOLDF., 1826; SD LANG-S.-T., 1940]. *M.Dev.*, Ger. [type missing].

Rhipidophyllum SANDBERGER, 1889 [**R. vulgare*]. *L.Dev.*, Ger. [not seen].

Sphaerophyllum WEDEKIND, 1923 [*genus caelebs*]. *L.M.Dev.*, Ger.

Stegophyllum SCHEFFEN, 1933 [**S densum*]. *U. Ord.*, Norway.

Sugiyamaella YABE & MINATO, 1944 [**S. carbonarium*]. *Perm.*, Japan [not seen].

REFERENCES

Publications cited in the following list may be found most helpful in furnishing additional information concerning the Rugosa and Heterocorallia and in offering a guide to more extensive literature. The index numbers enclosed in parentheses in the column at left are employed in the text for identification of the publications.

Bassler, R. S.

(1) 1937, *The Paleozoic rugose coral family Paleocyclidae:* Jour. Paleont., v. 11, p. 189-201, pl. 30-32.

(2) 1950, *Faunal lists and descriptions of Paleozoic corals:* Geol. Soc. America, Mem. 44, 315 p., 20 pl.

Bryan, W. H., & Hill, Dorothy

(3) 1941, *Spherulitic crystallisation as a mechanism of skeletal growth in the hexacorals:* Proc. Roy. Soc. Queensland, v. 52, p. 78-91.

Carruthers, R. G.

(4) 1910, *On the evolution of Zaphrentis delanouei in Lower Carboniferous times:* Quart. Jour. Geol. Soc., v. 66, p. 523-538, pl. 36-37.

Chi, Y. S.

(5) 1931, *Weiningian (Middle Carboniferous) corals of China:* Palaeont. Sinica, ser. B, v. 12, fasc. 5, p. 1-70, pl. 1-5.

Dobrolyubova, T. A.

(6) 1937, *Simple corals of the Myatshkovo and Podolsk horizons of the Middle Carboniferous of the Moscow Basin:* Trav. Inst. Paleozool. Acad Sci. U.R.S.S., v. 6, livr. 3, p. 1-92, pl. 1-23.

(7) 1940, *The Rugosa corals of the Upper Carboniferous of the Moscow Basin:* Same, v. 9, livr. 3, p. 1-88, pl. 1-25.

Douglas, J. A.

(8) 1936, *A Permo-Carboniferous fauna from southwest Persia (Iran):* Palaeont. Indica, n.ser., v. 22, Mem. 6, p. 1-59, pl. 1-5.

Dybowski, W. N.

(9) 1873-74, *Monographie der Zoantharia sclerodermata rugosa aus der Silur-formation Estlands, Nord-Livlands und der Insel Gotland:* (Dorpat), p. 257-532, pl. 1-5.

Easton, W. H.

(10) 1944, *Corals from the Chouteau and related formations of the Mississippi Valley region:* Ill. State Geol. Survey, Rept. Inv. 97, p. 1-91, pl. 1-17.

Frech, Fritz

(11) 1885, *Die Korallenfauna des Oberdevons in*

Deutschland: Zeitschr. deutsch. geol. Gesell., 1885, p. 21-130, pl. 1-11.

Gerth, H.

(12) 1921, *Die Anthozoën der Dyas von Timor:* Paläont. von Timor, Lief 9, Abt. 16, p. 67-147, pl. 145-150.

Grabau, A. W.

(13) 1928, *Palaeozoic corals of China, Pt. 1, Tetraseptata, II, Second contribution to our knowledge of the streptelasmoid corals of China and adjacent territories:* Palaeont. Sinica, ser. B, v. 2, pt. 2, p. 1-175, pl. 1-6.

Heritsch, F.

(14) 1936, *Korallen der Moskauer-, Gshel- und Schwagerinen-stufe der Karnischen Alpen:* Palaeontograph., Band 83, Abt. A, p. 99-162, pl. 14-18, fig. 1-6 (Stuttgart).

Hill, Dorothy

(15) 1934, *The Lower Carboniferous corals of Australia:* Proc. Roy. Soc. Queensl., v. 45, p. 63-115, pl. 7-11.

(16) 1935, *British terminology for rugose corals:* Geol. Mag., v. 72, p. 481-519, fig. 1-21.

(17) 1936, *The British Silurian rugose corals with acanthine septa:* Philos. Trans. Roy. Soc. London, ser. B, v. 226, p. 189-217, pl. 29-30.

(18) 1938-41, *The Carboniferous rugose corals of Scotland:* Palaeontograph. Soc. Mon. (London), 213 p., 11 pl.

(19) 1939, *The Devonian rugose corals of Lilydale and Loyola, Victoria:* Proc. Roy. Soc. Vict., n. ser., v. 51, p. 219-256, pl. 13-16.

(20) 1940, *The Silurian Rugosa of the Yass-Bowning district, N.S.W.:* Proc. Linn. Soc. N.S.W., v. 65, p. 388-420, pl. 11-13.

(21) 1942, *The Devonian rugose corals of the Tamworth district, N.S.W.:* Jour. Proc. Roy. Soc. N.S.W., v. 76, p. 142-164, pl. 2-4.

(22) 1942a, *The Middle Devonian rugose corals of Queensland, III, Burdekin Downs, Fanning River and Reid Gap, North Queensland:* Proc. Roy. Soc. Queensl., v. 53, p. 229-268, pl. 5-11.

(23) 1948, *The distribution and sequence of Carboniferous coral faunas:* Geol. Mag., v. 85, p. 121-148.

(24) 1951, *The Ordovician corals:* Proc. Roy. Soc. Queensl., v. 62, p. 1-27.

Huang, T. K.

(25) 1932, *Permian corals of southern China:* Palaeont. Sinica, ser. B, v. 8, fasc. 2, p. 1-163, pl. 1-16.

Hudson, R. G. S.

(26) 1941, *On the Carboniferous corals, Zaphrentis carruthersi, sp. nov., from the Mirk Fell beds and its relation to the Z. delanouei species group:* Proc. Yorks. Geol. Soc., v. 24, p. 290-311, pl. 21-22.

(27) 1942, *Fasciculophyllum Thomson and other genera of the "Zaphrentis" omaliusi group of Carboniferous corals:* Geol. Mag., v. 79, p. 257-263.

Jeffords, R. M.

(28) 1947, *Pennsylvanian lophophyllidid corals:* Univ. Kansas Paleont. Contr., Coelenterata, art. 1, 84 p., 28 pl.

Koker, E. M. J.

(29) 1924, *Anthozoa uit het Perm van het eiland Timor, I, Zaphrentidae, Plerophyllidae, Cystiphyllidae, Amphiastraeidae:* Jaarb. Mijnw. Nederl. Oost-Indië, v. 51 (1922), p. 1-50, pl. 1-11.

Lambe, L. M.

(30) 1899-1901, *A revision of the genera and species of Canadian Palaeozoic corals:* Can. Geol. Survey, Contr. Can. Paleont., v. 4, p. 1-197, pl. 1-18.

Lang, W. D. & Smith, Stanley

(31) 1927, *A critical revision of the rugose corals described by W. Lonsdale in Murchison's "Silurian System":* Quart. Jour. Geol. Soc. (London), v. 83, p. 448-491, pl. 34-37.

(32) 1935, *Cyathophyllum caespitosum Goldfuss and other Devonian corals considered in a revision of that species:* Same, v. 91, p. 538-590, pl. 35-37.

———, ———, & Thomas, H. D.

(33) 1940, *Index of Palaeozoic coral genera:* Brit. Mus. (Nat. Hist.) (London), 231 p.

Le Maitre, Dorothée

(34) 1947, *Le récif coralligene de Ouihalane:* Service géol. Maroc, Div. Géol. et Mines, v. 67, 112 p., 24 pl.

Lindström, G.

(35) 1882, *Silurische Korallen aus Nord-Russland und Sibirien:* Bih. k. Svenska Vetenskapsakad., Handl. 6 (18), p. 1-23, pl. 1.

(36) 1883, *Om de Palaeozoiska formationernas operkelbarande koraller:* Same, v. 7, no. 4, p. 1-112, pl. 1-9.

(37) 1896, *Beschreibung einiger Obersilurischer Korallen aus der Insel Gotland:* Same, v. 21, no. 7, p. 1-50, pl. 1-8.

Milne-Edwards, H., & Haime, Jules

(38) 1851, *Monographie des polypiers fossiles des terrains palaeozoiques:* Arch. Mus. Hist. Nat. (Paris), tome 5, p. 1-502, pl. 1-20.

Moore, R. C., & Jeffords, R. M.

(39) 1945, *Description of Lower Pennsylvanian corals from Texas and adjacent regions:* Univ. Tex. Pub. 4401, p. 77-208, pl. 14, fig. 1-214.

Počta, P.

(40) 1902, *Anthozoaires et alcyonaires:* in BARRANDE, J., Systême silurien du centre de la Bohême (Prag), tome 8, pt. 2, p. 1-347, pl. 20-118.

Schindewolf, O. H.

(41) 1941, *Zur Kenntnis der Heterophylliden,*

einer eigentümlichen palaeozoischen Korallengruppe: Paläont. Zeitschr., Band 22, p. 213-306, pl. 9-16.

(42) 1942, *Zur Kenntnis der Polycoelien und Plerophyllen:* Abh. Reichsamt. Bodenf., Neue Folge, Heft 204, 324 p., 36 pl.

Schlüter, C.

(43) 1889, *Anthozoen des rheinischen Mittel-Devon:* Abh. geol. Spezialkarte Preuss. Thuring. Staat: Band 8, pt. 4, p. 259-465, pl. 1-16.

Simpson, G. B.

(44) 1900, *Preliminary descriptions of new genera of Palaeozoic rugose corals:* Bull. N.Y. State Mus., v. 8 (39), p. 199-222.

Smith, Stanley

(45) 1930, *Some Valentian corals from Shropshire and Montgomeryshire:* Quart. Jour. Geol. Soc. (London), v. 86, p. 291-330, pl. 26-29.

(46) 1945, *Upper Devonian corals of the Mackenzie River region:* Geol. Soc. America, Spec. Paper 59, 126 p., 35 pl.

Soshkina, E. D.

(47) 1937, *Corals of the Upper Silurian and Lower Devonian of the eastern and western slopes of the Urals:* Trav. Inst. Paléozool. Acad. Sci. U.R.S.S., v. 6, livr. 4, 112 p., 21 pl.

(48) 1939, *Upper Devonian corals Rugosa of the Ural:* Same, v. 9, livr. 2, 88 p., 14 pl.

———, Dobrolyubova, E., & Porfiriev, G.

(49) 1941, *The Permian rugose corals of the European part of the U.S.S.R.:* Paleont. of U.S.S.R. (Moscow), v. 5, pt. 3, fasc. 1, 304 p., 63 pl.

Stuckenberg, A.

(50) 1895, *Korallen und Bryozoen der Steinkohlen Ablagerungen des Ural und des Timan:* Mém. Com. géol. St. Petersbourg, n. ser., v. 10, pt. 3, 244 p., 24 pl.

Stumm, E. C.

(51) 1948, *A revision of some Mississippian tetracoral genera:* Jour. Paleont., v. 22, p. 68-74, pl. 17.

(52) 1949, *Revision of the families and genera of the Devonian tetracorals:* Geol. Soc. America, Mem. 40, 92 p. 25 pl.

Wang, H. C.

(53) 1950, *A revision of the Zoantharia Rugosa in the light of their minute skeletal structures:* Philos. Trans. Roy. Soc. London, ser. B, v. 234, p. 175-246, pl. 4-9.

Wedekind, R.

(54) 1924-25, *Das Mitteldevon der Eifel, eine biostratigraphische Studie; Teil 1, Die Tetrakorallen des unteren Mitteldevon:* Schr. Gesell. Beford. gesamten Naturw. Marburg, Band 14, pt. 3, p. 1-91; *Teil 2,* Band 14, pt. 4, p. 1-85.

(55) 1927, *Die Zoantharia Rugosa von Gotland:* Sver. Geol. Undersök., v. 19, 94 p., 30 pl.

Weissermel, W.

(56) 1937, *Coelenterata: Anthozoa, Hydrozoa, Scyphozoa* (Review): Fortschritte Paläont., Band 1, p. 84-96.

Yabe, H., & Hayasaka, I.

(57) 1915-16, *Palaeozoic corals from Japan, Korea and China:* Jour. Geol. Soc. Tokyo, v. 22, p. 55-70, 79-109, 127-142; v. 23, p. 57-75.

Yü, C. C.

(58) 1934, *Lower Carboniferous corals of China:* Palaeont. Sinica, ser. B, v. 12, fasc. 3, 211 p., 24 pl.

SOURCES OF ILLUSTRATIONS

Explanation of the use of index numbers for citing sources of illustrations is given in the Editorial Preface.

(59) **Barrois, C.**
(60) **Bassler, R. S.**
(61) **Benson, W. N., & Smith, S.**
(62) **Brown, T. C.**
(63) **Busch, D. A.**
(64) **Carruthers, R. G.**
(65) **Chi, Y. S.**
(66) **Cox, I.**
(67) **Dingwall, J. M. M.**
(68) **Dobrolyubova, T. A.**
(69) **Douglas, J. A.**
(70) **Dun, W. S., & Benson, W. N.**
(71) **Dybowski, W. N.**
(72) **Easton, W. H.**
(73) **Edwards, H. Milne, & Haime, Jules**
(74) **Etheridge, R.**
(75) **Foerste, A. F.**
(76) **Fomichev, V. D.**
(77) **Frech, F.**
(78) **Garwood, E. J.**
(79) **Gerth, H.**
(80) **Gorsky, J.**
(81) **Grabau, A. W.**
(82) **Grubbs, D. M.**
(83) **Hall, James**
(84) **Heritsch, F.**
(85) **Hill, Dorothy**
(86) **——— & Butler, A. J.**
(87) **Howell, B. F.**
(88) **Huang, T. K.**
(89) **Hudson, R. G. S.**
(90) **——— & Anderson, F. W.**

(91) Jeffords, R. M.
(92) Jones, O. A.
(93) Kettnerova, M.
(94) Keyserling, A.
(95) Koker, E. M. J.
(96) Lambe, L. M.
(97) Lang, W. D.
(98) ——— & Smith, Stanley
(99) Lewis, H. P.
(100) Lindström, G.
(101) Lissitzin, K. I.
(102) Meek, F. B., & Worthen, A. H.
(103) Minato, M.
(104) Moore, R. C., & Jeffords, R. M.
(105) Nicholson, H. A., & Etheridge, R.
(106) Orbigny, A. D. d'
(107) Ozawa, Y.
(108) Počta, P.
(109) Prantl, F.
(110) Roemer, C. F.
(111) Salée, A.
(112) Scheffen, W.
(113) Schindewolf, O. H.
(114) Schlüter, C.
(115) Simpson, G. B.
(116) Sloss, L. L.
(117) Smith, Stanley
(118) ——— & Lang, W. D.
(119) ——— & Ryder, T. A.
(120) ——— & Tremberth, R.
(121) Smyth, L. B.
(122) Soshkina, E. D.
(123) ———, Dobrolyubova, E., & Porfiriev, G.
(124) Sowerby, J.
(125) Stewart, G. A.
(126) Stuckenberg, A.
(127) Stumm, E. C.
(128) Sugiyama, T.
(129) Termier, Henri, & Termier, Geneviève
(130) Thomson, J.
(131) Tolmachev, I. P.
(132) Vaughan, A.
(133) Vollbrecht, E.
(134) Waagen, W., & Wentzel, J.
(135) Walther, C.
(136) Wang, H. C.
(137) Wedekind, R.
(138) ——— & Vollbrecht, E.
(139) Weissermel, W.
(140) Yabe, H., & Hayasaka, I.
(141) ——— & Minato, M.
(142) Yoh, S. S.
(143) Yü, C. C.

HETEROCORALLIA

By Dorothy Hill

MORPHOLOGY

The heterocorals are a small, short-lived (Carboniferous) order of very elongate solitary corals. They possess **septa** (longitudinal radial plates) and **tabulae** (transverse plates) but may lack an external sheath (**epitheca**) and their narrow wall is formed by thickening and apposition of the steeply sloping edges of their tabulae. The method of insertion of their septa is characteristic. Material so far collected consists of fragments only, which show little change in diameter from one end to another. There are but rare indications of branching and the corallites may be slightly twisted, or may show sudden changes in direction of growth.

The skeleton is a framework of radial longitudinal and transverse skeletal elements; each consisting of fibers of $CaCO_3$ at right angles to the growth lamellation, which is weakly apparent. An external sheath or epitheca is absent, according to Schindewolf (41), but Hill (18) thought it to be present in one species (*Heterophyllia phillipsi*).

The only radial skeletal elements are the septa, plates in which the fibers are arranged approximately at right angles to the median plane. The outer edges of septa typically project, so that the corallum is fluted rather than cylindrical, and they may have a linear series of denticulations which may be hooked. The inner edges of septa are scarcely anywhere free, but each typically abuts on to another septum, the abutted septum commonly diverging outwardly from its radial course at the point of contratingency, so as to give an appearance of forking. This attachment of inner edges of the septa is one of the distinctive features of the order. The septa are flexuous, and serial sections show that the roles of apparent abutter and abutted may be interchangeable.

Owing to the fragmentary nature of the material, and the very slow change in diameter with consequently slow rate of appear-

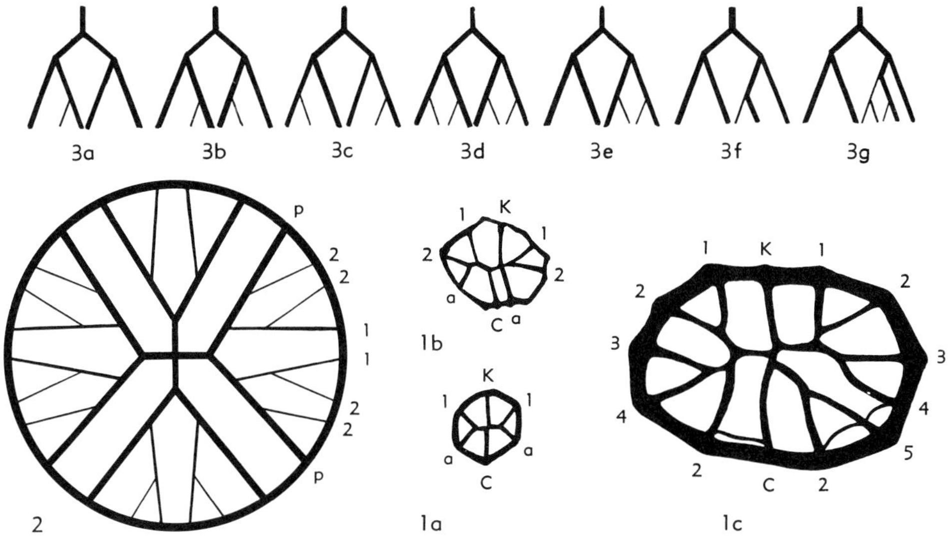

FIG. 220. Arrangement of septa in Heterocorallia. *1,* Development of septa in *Heterophyllia kitakamiensis* YABE & SUGIYAMA, L.Carb., Japan, shown by transverse sections of juvenile part of corallite *(1a,b)* and ephebic part *(1c)*, ×15; C, cardinal septum; K, counter septum; a, accessory (alar) septa; p, protoseptum; 1-5, metasepta in order of appearance (140). *2,* Diagram of septal arrangement in *Heterophyllia;* p, protoseptum; 1, 2, metasepta in order of appearance (113). *3,* Diagrams showing insertion of 2nd-order metasepta (light-weight lines) in a quadrant as illustrated by various heterocorals, heavy lines indicating protosepta and 1st-order metasepta; the ideal regular plan *(3d)* is commonly not realized (113).

ance of new septa, the **insertion of septa** has been the subject of deductive rather than inductive studies. Thus, YABE & SUGIYAMA (1940) consider that 6 protosepta were present, as in Rugosa and Scleractinia, and that new septa (metasepta) were thereafter pinnately inserted, as in Rugosa, not cyclically as in Scleractinia, but that this pinnate insertion occurred only in 2 positions, as against 4 positions in Rugosa (Fig. 220,*1a-c*).

SCHINDEWOLF (1941), however, included in his studies serial sections from fragmentary coralla, and his strikingly different conclusions may be outlined as follows. There are 4 original septa **(protosepta).** Each of these may split into 2 near the periphery; new septa are inserted only in the 4 spaces formed between the 2 split portions of each such septum, thus in "quadrants," and do not arise in the original 4 spaces ("fossulae") between the protosepta (Fig. 220,*2*). Of these 4 insertion spaces, 2 opposing ones, regarded as lateral, show far more newly formed septa than the remaining 2, of which one, the "upper" (oriented in SCHINDEWOLF's drawings in the position he usually assigns to the cardinal septum) is more backward than the other. The new septa develop in groups of 8 irregularly, appearing neither pinnately nor cyclically, the various possible arrangements being as indicated in Fig. 220,*3*. There may be 16 in the group derived from "splitting" of each lateral septum, as against 2 from the cardinal, and 4 from the counter septum. In SCHINDEWOLF's view, the new septa are inserted in the original endocoeles, leaving the original 4 exocoeles undivided.

The only transverse skeletal elements are **tabulae.** These are domed (or in SCHINDEWOLF's opinion saucered) floors with steeply sloping edges. Each is formed of fibrous $CaCO_3$, the fibers growing on one surface only, which in HILL's view is the upper surface. Each floor consists of a series of segments developed in the loculi between septa, the segments being discontinuous through the septa, but on the same level in neighboring loculi. Each segment is curved both from axis of corallite to periphery, and from one septum to its neighbor. In HILL's view, the latter curve is concave on the growing side, as in similar tabulae or dissepimental segments in Rugosa; but according to inter-

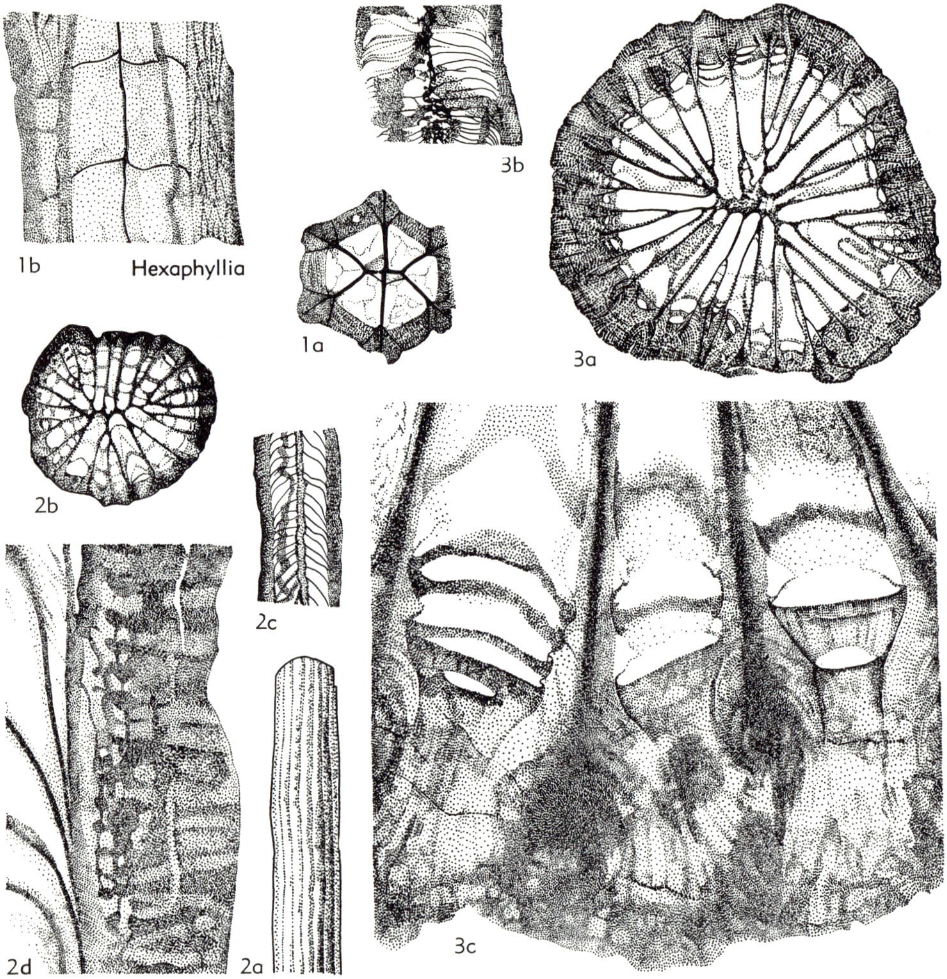

FIG. 221. Heterophylliidae (p. F327).

pretation of the tabulae as saucered, not domed, the curve from septum to septum is domed, not saucered, and is thus the opposite of the condition in all other corals. Scattered incomplete tabulae developed in one or only a few interseptal loculi indicate by their manner of adhesion that the tabulae are domes, not saucers.

In their steeply sloping edges, the tabulae, thickened by outward growth of their fibers, are in contact in a zone of dense tissue; the septa also become thickened at their peripheral edges in this zone, and the combined result is a **stereozone,** called "heterotheca" by SCHINDEWOLF. In this stereozone, trabeculae may develop from the tendency for fibers to grow grouped about straight or curved axes perpendicular to the outer surface of the corallum. This is the only type of marginal zone developed in the Heterocorallia; dissepiments are observed nowhere (Fig. 221,*3*).

CLASSIFICATION

The classification of the Heterocorallia as Zoantharia seems justified by their possession of septa and tabulae of fibrous $CaCO_3$; but the grouping, contratingency, and flexuousness of the septa, and their manner of septal insertion as known at present, clearly distinguishes them from Tabulata, Rugosa, and Scleractinia. They consist of one small family only, divisible into 2 genera.

DISTRIBUTION

Their stratigraphic distribution is limited, both in time and space, for they are known only in the Visean and Namurian divisions of the Carboniferous system in Europe and Asia. At most localities, both genera occur together, and in view of the fragmentary nature of the material, KUNTH's suggestion that the more primitive genus, *Hexaphyllia,* may be merely the young stage of *Heterophyllia,* is not unreasonable. Because of their fragmentary nature, also, we have direct knowledge only of later parts of their ontogeny. It is not possible at present to indicate ancestry for the order, which seems to have left no descendants in post-Namurian time.

They are found in calcareous shales and in pure limestones, and SCHINDEWOLF (41) suggests that their elongate nature, and the hooks on outer septal edges of some, indicate a pseudo-planktonic existence attached to seaweeds.

SYSTEMATIC DESCRIPTIONS

Order HETEROCORALLIA Schindewolf, 1941

[as suborder; =Dicoelia YABE & SUGIYAMA, 1940 (subdivision of Tetracoralla)]

Elongate coralla originally with 4 septa conjoined axially and with new septa formed in attachment to these so that the 4 original interseptal loculi remain undivided; marginarium a narrow trabeculate stereozone; tabulae complete, domes with steeply sloping to vertical edges. *Carb.*

Family HETEROPHYLLIIDAE Dybowski, 1873

[*nom. transl.* HILL, 1940 (*ex* Heterophyllinae DYBOWSKI, 1873); Heterophyllidae YABE & SUGIYAMA, 1940]

Hexaphyllia STUCKENBERG, 1904 [**H. prismatica*]. Slender, with only 6 septa (41). *Carb.(Visé.-Namur.)* Eu.-Asia.——FIG. 221,*1*. *H. mirabilis* (DUNCAN), Eu.; *1a,b,* transv. and long. secs., ×10 (113).

Heterophyllia M'COY, 1849 [*non* D'ORBIGNY, 1849] [**H. grandis;* SD M.EDW.-H., 1850]. With numerous septa. *Carb.(Visé.-Namur.),* Eu.-Asia.

H. (Heterophyllia). With all 4 original interseptal loculi separated by new septa (41). *Carb.(Visé.-Namur.),* Eu.-Asia.——FIG. 221,*3*. **H. (H.) grandis* (M'COY), Eu.; *3a,c,* transv. secs., ×4, ×20; *3b,* long. sec., ×2 (113).

H. (Heterophylloides) SCHINDEWOLF, 1941 [**H. (H.) reducta*]. With 2 of the original interseptal loculi not separated by new septa or separated by but one new septum (41). *Carb.(Visé.-Namur.),* Eu.——FIG. 221,*2*. **H. (H.) reducta; 2a,* exterior of corallite, ×1.5; *2b,* transv. sec., ×4; *2c,d,* long. secs., ×2, ×20 (113).

REFERENCES AND SOURCES OF ILLUSTRATIONS

See preceding section on Rugosa (p. F321, F323)

SCLERACTINIA

By John W. Wells

CONTENTS

GENERAL CHARACTERS

The order Scleractinia is the major division of the Zoantharia which includes nearly all post-Paleozoic fossil and Recent corals. It includes solitary and colonial forms with polyps less specialized than those of the Actiniaria and resembling more closely those of the Corallimorpharia, a highly specialized and probably closely parallel group. The Scleractinia are distinguished by a calcareous external skeleton consisting essentially of radial partitions or septa situated between the mesenteries and secreted by the ectodermal body layer within upward infoldings of the basal part of the polyp wall, together with a more or less developed external sheathing and variously developed attendant supporting structures. In ontogeny the septa are developed following the pattern of the mesenteries: after the establishment of the first 6 septa, others normally are inserted between those previously formed in successive groups of 6, 12, 24, and so on.

As in all other anthozoans, scleractinian corals are exclusively marine in habit. They are commonest in warm, clear, shallow waters of the tropical zone, but some are adjusted to shallow or deep, cold water and may be found in all latitudes. They have little or no tolerance for other than normal salinity.

The oldest known Scleractinia are a few shallow-water forms of primitive aspect from the Middle Triassic, and like the majority of other invertebrates, were essentially modernized by the beginning of Cenozoic time. Scleractinians obviously are allied so closely to the soft-bodied actiniarian and corallimorph anemones that either the 3 groups had a common origin, or as is more likely, the scleractinians were a late Paleozoic or early Mesozoic development from the same stock as the corallimorphs.

Morphological terminology applied to the scleractinian corals is explained in following sections on anatomy of the polyps and morphology of the skeleton, most special terms being called to notice by printing the name in boldface type where it first appears. In addition, an alphabetically arranged glossary of morphological terms used in description of Scleractinia, Rugosa, Heterocorallia, and Tabulata is given on pages F245 to F251.

For valuable criticism and suggestions relating especially to the organization of general discussion of the Scleractinia, it is appropriate to express acknowledgment and appreciation to Dr. H. Dighton Thomas, of the British Museum (Natural History) and the *Treatise* Editor, Dr. Raymond C. Moore.

ANATOMY OF POLYPS

Scleractinian polyps resemble those of the simpler types of actiniarians (Fig. 222). Alive, they are variously colored and assume different aspects as they are fully expanded or contracted. The expanded polyp shows 2 distinct regions: the smooth, cylindrical **column,** terminated above by the horizontal **oral disc,** below by the **basal disc.** In the

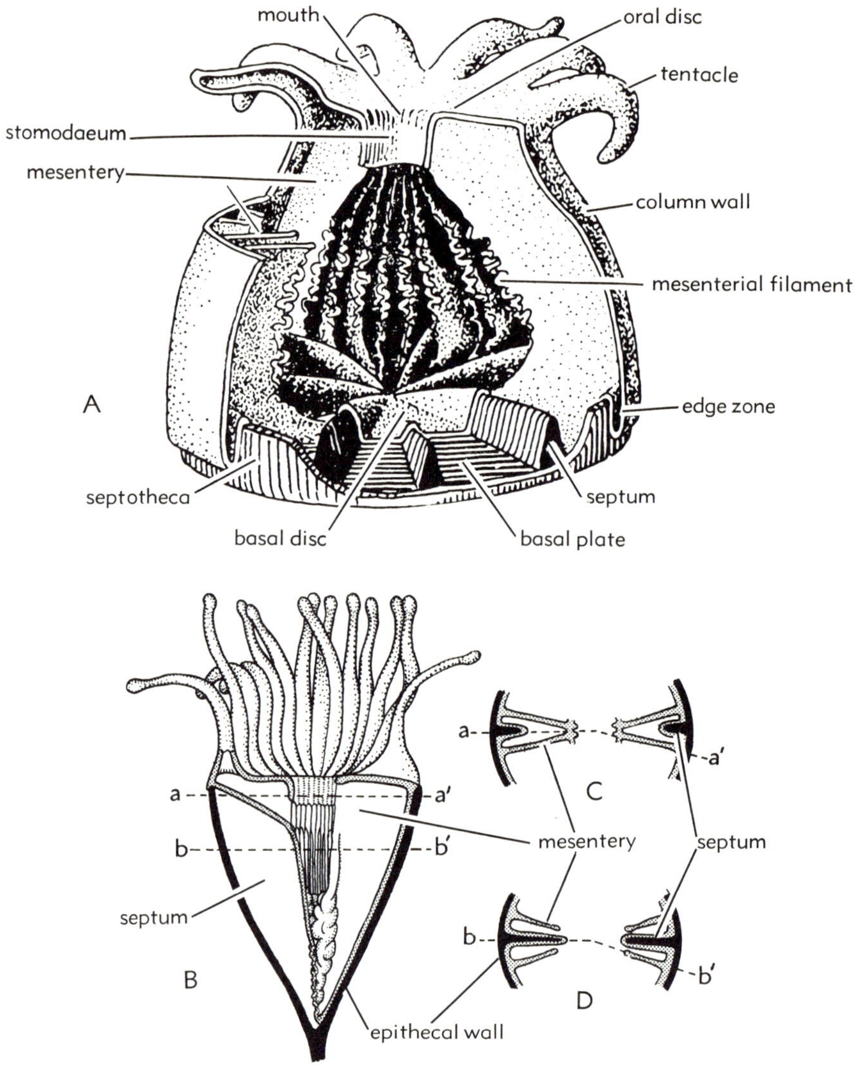

FIG. 222. Relations of polyp and skeleton in scleractinian corals. *A*, Young caryophylliid, with edge zone (mesenteries not shown between septa in foreground) (185). *B*. *Flabellum*, lacking edge zone. *C,D*, Transverse sections of *Flabellum* along line *a-a′* and *b-b′* (151).

center of the oral disc is the **mouth,** a hole leading to the polyp interior. The oral-disc region around the mouth is the **peristome,** peripherally bounded by one or more rings of **tentacles.** The **stomodaeum** is an esophagus-like connection between the mouth and the interior gastrovascular cavity. The latter is hollow and subdivided by 2 series of vertical radial partitions. One series is formed by the **mesenteries,** fleshy plates attached to the inner surface of the oral disc and column wall. Their upper parts may or may not be connected to the stomodaeum, and their lower edges commonly are free. The other series of partitions comprises the septal invaginations, which are inturnings of the basal disc occupied by the calcareous **septa,** and which alternate in position with the mesenteries or pairs of mesenteries.

BODY LAYERS

Microscopically the column wall, oral

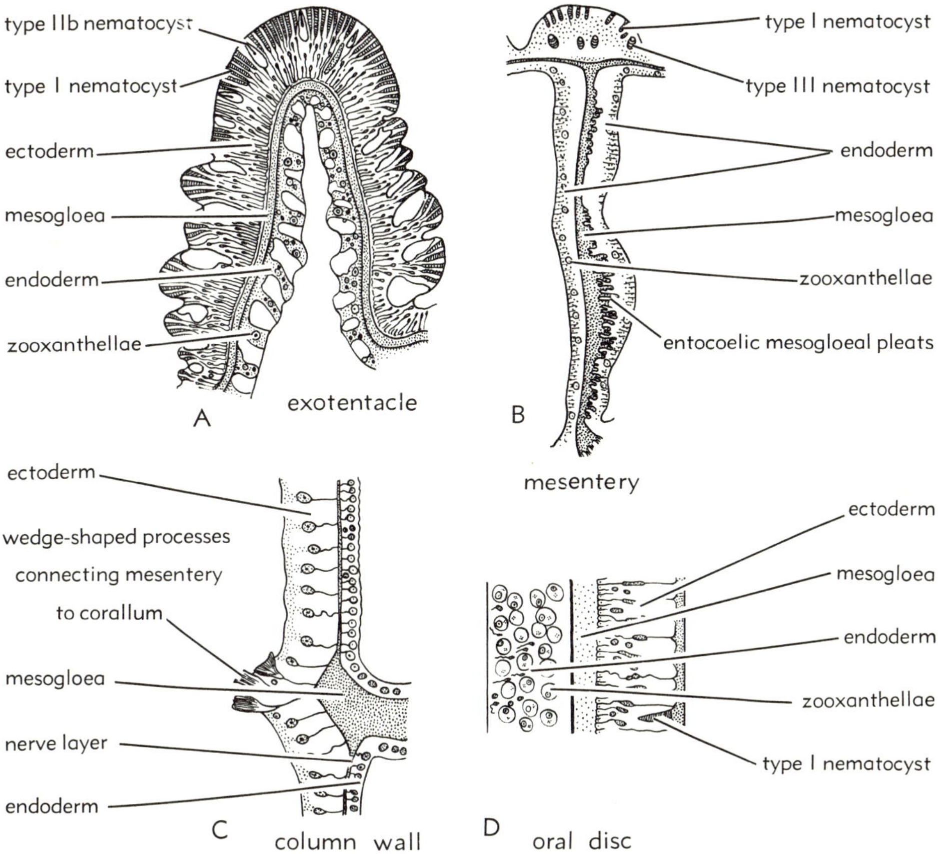

FIG. 223. Body layers of scleractinian polyps (83). *A,* Longitudinal section of exotentacle of *Favia doreyensis* M.EDW.-H., Maldives, ×90. *B,* Transverse section of mesentery of *Favia bertholleti* M.EDW.-H., Ceylon, ×42. *C,* Transverse section near upper part of column wall of *Favites pentagonus* (ESPER), Maldives, ×315. *D,* Transverse section of part of oral disc of *Favia hululensis* GARDINER, Maldives, ×37.

disc, tentacles, and basal disc (but not the mesenteries), are composed of 3 distinct layers of tissue: **ectoderm, mesogloea,** and **endoderm** (Fig. 223). The outer layer of ectoderm is externally ciliated and contains nuclei, granular and mucous vacuoles, and nematocyst capsules. The middle, supporting lamina of mesogloea is nearly homogeneous, noncellular, nucleated, and jelly-like. The inner layer of endoderm consists of glandular and muscular elements, and granular vacuoles. The endoderm of most reef-dwelling corals contains large numbers of symbiotic, unicellular, yellow-brown, dinoflagellate algae, termed **zooxanthellae.**

The ectoderm consists of 2 histologically distinct regions: (1) the exposed surface of the polyp: oral disc, tentacles, and outer areas known as **edge zone,** if they merely extend over the corallite wall, and **coenosarc,** if they are continuous between polyps of colonial coralla; and (2) the **calicoblast layer,** comprising whatever parts of the ectoderm that lie directly against the skeleton. The calicoblast layer, nonexistent in the larval stage, is differentiated after fixation of the larva from the ectoderm of the basal disc antecedent to development of the skeleton which it secretes.

MESENTERIES

NATURE AND KINDS

The radial partitions of the gastrovascular cavity, the mesenteries, are the most vital part of anthozoan polyps. With their filaments, they are the organs of digestion, absorption, and excretion, and are the site of development of the gonads; their arrangement and order of development are the distinguishing feature of the Anthozoa. Mesenteries are laminar in appearance and consist of 2 layers of endoderm between which is a sheet of mesogloea (Fig. 223). They are attached above to the under surface of the oral disc and along their outer margins to the inner surface of the hard parts by wedge-shaped mesogloeal processes. Those that extend to and join the inner surface of the stomodaeum are said to be **complete mesenteries;** those that do not reach the stomodaeum are **incomplete mesenteries.** A short distance below the stomodaeum the inner free margin of the mesentery may be drawn out into long ribbon-like prolongations **(mesenterial filaments)** which lie coiled in the gastrovascular cavity.

On each side of the mesogloea, as seen in the cross section of a mesentery, is a longitudinal layer of retractor muscle fiber. Invariably one of these muscle layers is better developed than the other and the mesogloea beneath it is extended as a series of longitudinal folds or **pleats** (Fig. 223). The position of these pleats is constant in all mesenteries of the same relative position and is useful in orientation of the polyp. Most of the mesenteries in a polyp are disposed radially with respect to an axis passing through the stomodaeum; they are arranged in **pairs** in which the mesogloeal pleats of each member face those of the other (Fig. 224). The space between the mesenteries of such a pair is an **entocoele,** and the space between neighboring pairs is an **exocoele.**

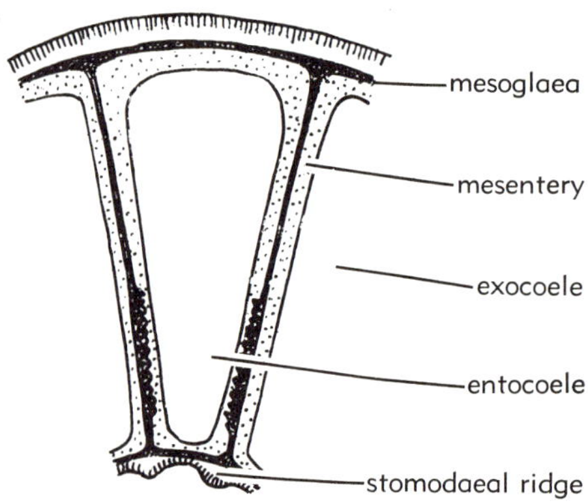

Fig. 224. Generalized transverse section of part of polyp, showing relations of normal mesenterial couple (151).

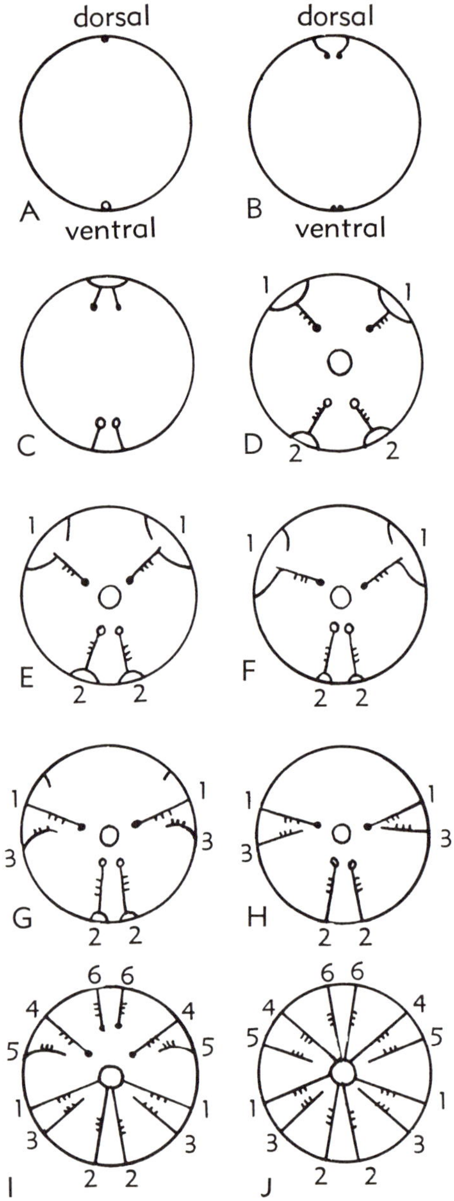

Fig. 225. Development of first cycle of mesenteries (protocnemes) in *Pocillopora* and *Seriatopora, 1,2, 4,6* (=orthocnemes); *3,5* (=sterigmatocnemes). Position of muscular pleats indicated by short lines perpendicular to mesenteries (76).

In all solitary **(monostomatous)** polyps and in each of the new polyps of colonial corals developed by budding outside the tentacular ring, the mesenteries of 2 oppos-

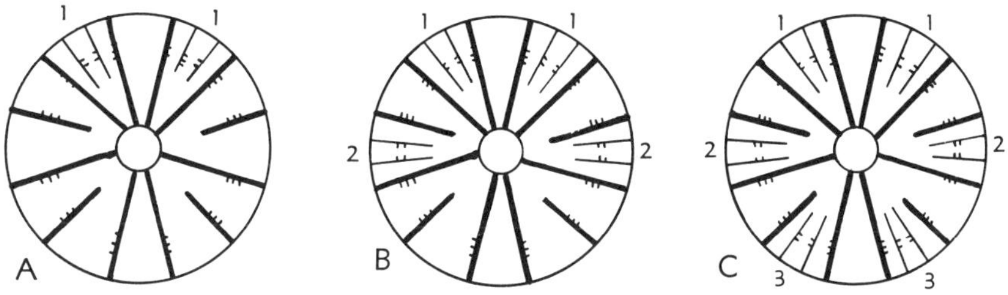

FIG. 226. Development of second cycle of mesenteries (metacnemes) in *Manicina; A,B,C,* successive stages. Second cycle of mesenteries (metacnemes) indicated in order of appearance by *1,2,3* (26).

ing pairs in a so-called dorsoventral plane have their pleating reversed; these are **directive pairs** or **directive mesenteries.** In some forms, the tentacles lying over these directive pairs and skeletal septa within are enlarged. Where colony-formation is by budding within the tentacular ring, the founder polyp has 2 directive pairs, but upon "division" of the polyp, the pairs are separated and the vast majority of the polyps in such a colony have no directive pairs, although theoretically each original pair is in some polyp somewhere in the colony. In such cases it is convenient to regard the polyps as lacking directive mesenteries.

DEVELOPMENT

Before the appearance of mouth opening or tentacles in the larva **(planula),** 2 opposing bulges consisting of longitudinal muscular sheets appear in the endoderm, and define a polarity designated as dorsoventral (Fig. 225*a*). These sheets, called **sterigmatocnemes,** are substrates on which the mesenteries develop. The dorsal bulge precedes the ventral and undergoes cleavage in which the 2 halves spread apart from each other (Fig. 225*b-d*). The ventral bulge does the same, but later on. Another sheet, termed **orthocneme,** develops on the surface of each sterigmatocneme, undergoes cleavage, and elongates toward the axis. These new sheets are predecessors of complete mesenteries.

After the mouth and stomodaeum have appeared (Fig. 225*d*), musculature and pleats appear on the orthocnemes, the ventral couple of which becomes the ventral pair of directive mesenteries, and the dorsal pair becomes the first pair of complete mesenteries. Thus far, development of the first 2 pairs of orthocnemes has been similar, the ventral pair being slightly retarded, but subsequently the sterigmatocnemes of the ventral pair retrogress and disappear (Fig. 225*e-h*). The laterodorsal part of the 2 dorsal sterigmatocnemes also disappears (Fig. 225 *f-h*), but thickening of the ventral part produces the third couple of mesenteries which form pairs with the dorsal orthocnemes (Fig. 225*g,h*). Repetition of this sequence in the dorsal sector (Fig. 225*i,j*) gives rise to another couple of mesenteries before the dorsal directive pair appears. When the first 6 mesenteric pairs (comprising 12 **protocnemes**) have developed (Fig. 225*j*) the embryonic period is terminated. Mesenteries developed subsequently are called **metacnemes.**

The sequence just described seems to represent the primitive order of appearance. In most groups of scleractinians development of orthocnemes in the dorsal sector is greatly accelerated, and they may be well formed before the appearance of the ventral directives.

The developmental order of the metacnemes (Fig. 226) follows the dorsoventral sequence established by the protocnemes. The second cycle of 6 metacnemic pairs appears after fixation of the planula, one pair at a time within the exocoeles of the protocnemic pairs. The first are introduced in the spaces between a dorsal directive and the first lateral protocnemes (Fig. 226*A*), the next between the first and second lateral pairs of protocnemes (Fig. 226*B*), and so on (Fig. 226*C*). The 12 pairs of third-cycle metacnemes arise in 2 series of 6 couples each (Fig. 227), the first in the exocoeles on the dorsal side of the second-cycle couples (Fig. 227*A*), and in the same order. After the appearance of the first series, the second series arises in the ventral exocoeles (Fig. 227*B,C*).

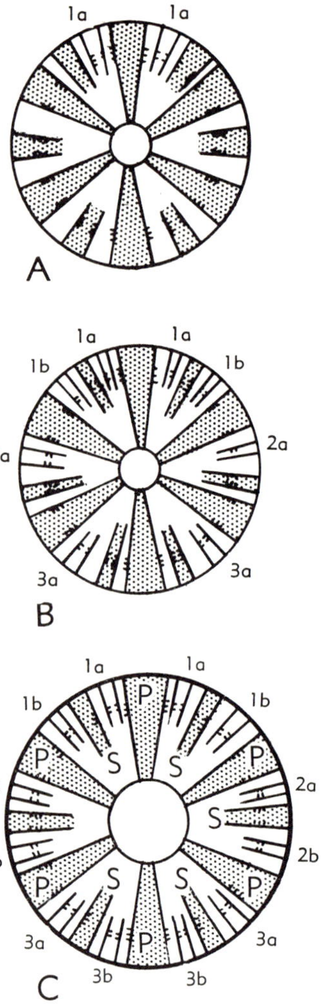

FIG. 227. Development of 3rd cycle of mesenteries (metacnemes) in *Manicina.* First cycle (protocnemes) *(P)* and 2nd cycle (metacnemes) *(S)* indicated by shading between mesentery pairs; order of appearance of 3rd-cycle metacnemes indicated by sequences *1a,2a,3a* (1st series) and *1b,2b,3b* (2nd series) (26).

The dorsoventral succession in each of 2 moieties of mesenterial insertion, which is characteristic for the Scleractinia, Corallimorpharia, and most of the Actiniaria, is basically bilateral, as in the Rugosa, not radial. But the sites of insertion and order contrast with the Rugosa, as inferred from the septal succession in the latter, where a dorsoventral (or ?ventrodorsal) succession occurred in each of 4 quadrants defined by the cardinal-counter and alar pairs of septa, after the establishment of the protocnemes and the 6 primary entocoelic septa; or, to put it differently, in the Scleractinia additions of metacnemes (and septa) occur in all 6 of the primary exocoeles, and in only 4 of the 6 primary exocoeles in the Rugosa.

STOMODAEUM

The stomodaeum connects the mouth with the gastrovascular cavity (Fig. 222). It is generally rather short and compressed laterally with respect to the directive mesenteries. The complete mesenteries are attached to its inner surface and over the points of attachment the ectoderm may be raised into vertical ridges (Fig. 224). In some groups, these stomodaeal ridges are lacking and the stomodaeum is smooth (Astrocoeniina, Agariciicae, Poriticae). Since free-swimming larvae lack stomodaeal ridges, their absence in adult forms may be a primitive character, or it may simply be correlated with size of the polyps. In polyps with directive couples of mesenteries, 2 vertical grooves in the stomodaeum face these couples.

TENTACLES

The tentacles arise from the oral disc as out-pushings of the mesenterial chambers below—**entotentacles** from entocoeles, and **exotentacles** from exocoeles. They are arranged in one or more rings around the mouth, from the edge of which they are separated by the peristome, or they may be spread over much of the entire oral disc. They are mostly simple, formed by a single tubular extension, but in *Siderastrea* the entotentacles are bifurcated near their tips. The tips of the tentacles are swollen or knobbed with concentrated nematocyst batteries, or obtuse with scattered batteries. In most corals, a longitudinal ectodermal musculature is capable of retracting the tentacle, but in some forms *(Stephanophyllia, Fungiacyathus, Euphyllia)*, the tentacles are nonretractile. In a few genera of the Agariciidae, the tentacles are very small or even rudimentary. The function of the tentacles is tactile and prehensile, to detect and seize food particles and move them toward the mouth.

NEMATOCYSTS

Stinging or adhesive structures, termed

nematocysts or cnidocils, are found in all scleractinian polyps. They are ectodermal, differing in shape and structure in different species or in different parts of the same polyp. Four main types are recognized, of which some have a stinging function, whereas others may be merely adhesive. The combination of nematocyst types found in scleractinian polyps is similar to that in the corallimorph anemones, and much less varied than in the Actiniaria.

EDGE ZONE AND COENOSARC

The **edge zone** is the part of a polyp that lies outside the wall of a solitary coral (Figs. 222, 228*A*). It is simply a horizontal outfold of the column wall overflowing the corallite wall, and contains a continuation of the gastrovascular cavity. In solitary corals, only edge zone is developed; in colonies the edge zone becomes coextensive between the polyps, constituting the **coenosarc** (Fig. 228*B*).

The calicoblast layer is present on the under surface of edge zone or coenosarc and lays down calcareous deposits called **costae,** which are essentially prolongations of the septa outside the column wall, and other extramural skeletal deposits such as dissepiments and stereome, which are collectively termed **coenosteum** when deposited by coenosarc. **Coenenchyme** is a term often applied to either coenosteum or coenosarc, or to both together; and properly speaking, it refers to the latter combination, a connotation unnecessary in scleractinian terminology.

Edge zone and coenosarc typically contain extensions of the mesenteries, hence the strict homology between intra- and extramural skeletal structures, but in some groups (acroporids, dendrophylliids), the extended mesenterial chambers are modified into a network of canaliculae. In corals with perforate skeletal wall *(Acropora, Porites)*, the interior of the coenosarc and the intramural gastrovascular cavity of the polyps are directly connected through the wall as well as over it. In corals with epitheca, the edge zone is narrow and extends down to the edge of the upgrowing epitheca *(Caryophyllia)*, or is absent *(Flabellum)*.

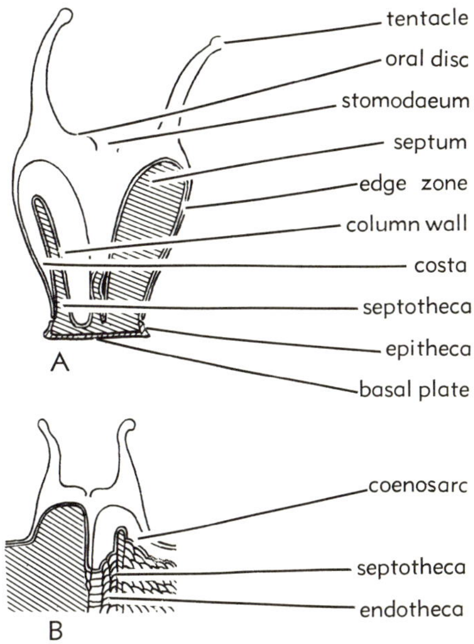

FIG. 228. Edge zone and coenosarc: diagrammatic vertical sections. *A*, Solitary caryophylliid (151). *B*, Colonial plocoid faviid (151).

MORPHOLOGY OF THE SKELETON

The scleractinian skeleton (Fig. 229) consists of a thin **basal plate** from which rise radiating vertical partitions called **septa,** and attendant structures. An upward continuation of the basal plate, marking the limit of the oral disc or edge zone, is termed **epitheca.** The column wall of the polyp is enclosed by a limiting calcareous deposit (**wall** or **theca**) consisting of different structures in different corals. **Costae** are extensions of the septa beyond the wall into the edge zone of coenosarc.

The entire skeleton deposited by a single polyp or by a colony of polyps is a **corallum.** Individual polyp skeletons in a colony are **corallites.** The upper or open end of a corallum or corallite is the **calice** (pl., **calices**), usually concave and deepest in the center, and occupied in life by the polyp. Extrathecal structures supporting the edge zone and coenosarc are collectively termed **coenosteum. Dissepiments** are thin, curved or tabular sheets in the spaces between septa or costae that support the lower surface of the soft parts and cut off the void left by upward growth (Fig. 230). **Endotheca** com-

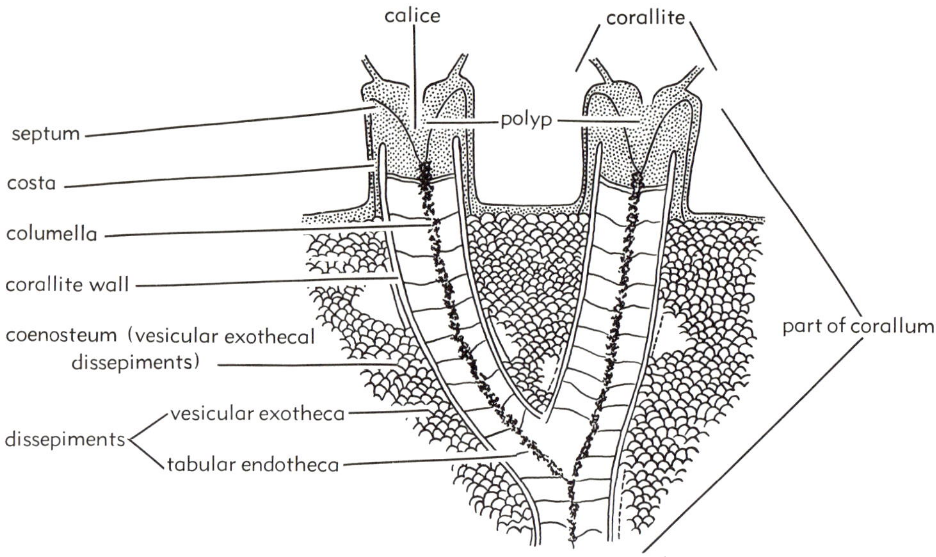

FIG. 229. General relations of polyp and corallum in *Galaxea* (98).

prises dissepiments inside the wall of a corallum or corallite, **exotheca** those outside the wall. A central axial structure, originating from various modifications of the inner ends of the septa, is the **columella.** Septa and thecal structures are commonly thickened secondarily by a deposit of **stereome,** similar in structure to epitheca.

COMPOSITION AND ORIGIN

The exact way in which the skeleton is deposited by the polyp is not thoroughly understood, but it has been supposed that it is carried out within certain specialized cells of the calicoblast layer of the body wall ectoderm (intraprotoplasmic), and that it constitutes an exoskeleton. The calcareous tissue of the exoskeleton consists of minute crystalline fibers or needles of aragonite ($CaCO_3$), about 2 microns in diameter and

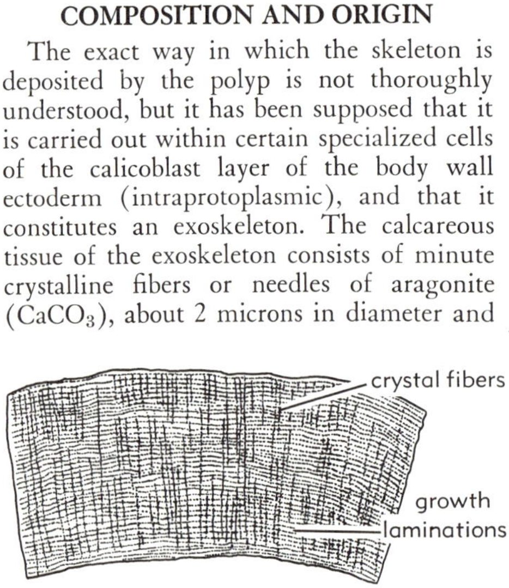

FIG. 230. Structure of scleractinian dissepiment. Vertical section of dissepiment of *Galaxea, ca.* ×50 (98).

each an orthorhombic crystal, arranged normal to the secreting surface of the ectoderm. In the basal plate, dissepiments, and stereome, all of which are deposited from unfolded, even-surfaced ectoderm, the skeletal tissue appears as dense, fibrous sheets marked only by growth laminations (Fig. 230).

When septa and structures derived from them are examined in thin section under considerable magnification (Fig. 231), they are seen to contain numerous dark spots or **"centers of calcification,"** from each of which fibrous crystals radiate toward those of neighboring centers. The centers of calcification and their clusters of fibers are termed **sclerodermites,** and are considered for practical purposes as the primary units of the septa. However, sclerodermites do not differ fundamentally from the fibrous sheets of the basal plate and its related structures. They are secreted in the upfoldings of the basal disc of the polyp by persistent areas of calcification in hollows, often very small and close together along the upper ectodermal surface of the upfold. The dark "centers of calcification" of the sclerodermites may be an optical effect or possibly patches of residual organic matter in the center toward which the fibrous crystals have grown. The remaining ectodermal surface of the upfold does not lack the ability to secrete crystalline

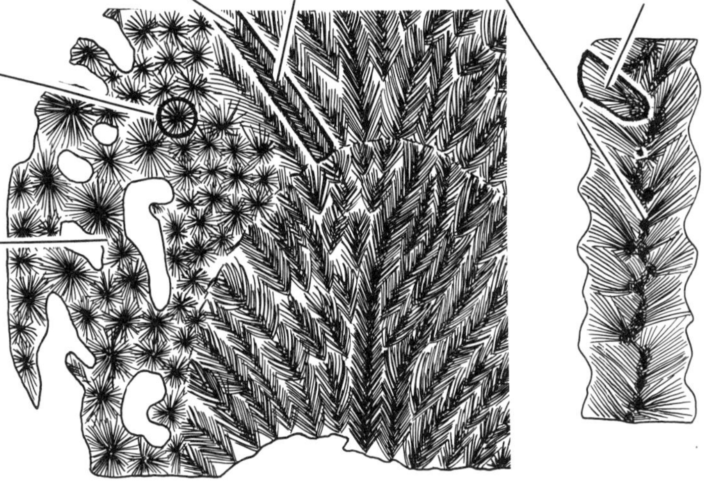

FIG. 231. Scleractinian septal structure. *A*, Radial section of septum and columella of *Galaxea, ca.* ×50 (98). *B*, Tangential section of septum of *Galaxea, ca.* ×50 (98).

fibers, sheets of which (stereome) may afterwards be laid over the sclerodermites, thickening and strengthening the septa.

SEPTA

GENERAL NATURE

Relative to the polyp, the septa lie in the ento- and exocoeles, their function being that of support and separation of the mesenteries. In development, they are the first skeletal structures to appear after the deposition of the basal plate. All other skeletal parts are subsequent to the septa and of secondary importance.

The sclerodermites are more or less persistent vertically in the plane of the upfold from the basal disc of the polyp, and develop spines or rods called **trabeculae** (Fig. 231). A septum is a palisade of trabeculae. Depending on the development of the sclerodermites along the crest of the upfold, the trabeculae vary widely in different scleractinian groups. They are **simple trabeculae,** if composed of series of single sclerodermites (Fig. 232), or **compound trabeculae,** if composed of bundles of sclerodermites (Fig. 233). Sclerodermites are rarely arranged in a uniformly continuous vertical series; commonly they incline outward from the axis of the trabecula and emerge from the septal plane as striae or granulations. They may emerge at the same level in adjacent trabeculae and fuse laterally in form ridges parallel to the upper or growing margin of the septum. Trabeculae are rarely parallel to one another, but are so arranged that they are increasingly inclined outward from an **axis of divergence** which may be vertical, inclined, or even horizontal (Fig. 234). The pattern thus produced constitutes a **fan system.** Septa may consist of more than one fan system lying in the septal plane (Fig. 232*B*). If the trabeculae are closely united without intervening spaces to form a continuous sheet (Fig. 232), the septum is **laminar** and in cross section the centers of calcification may be discrete or so close together as to form an apparently continuous dark line. If trabeculae are more or less loosely connected, they form a **fenestrate septum** (Fig. 223), in which **pores** or **perforations** appear more or less regularly between adjacent trabeculae. Such perforations may become filled secondarily. In a few primitive genera of stylophyllids and astrocoeniids, the trabeculae may be more or less completely isolated from each other in the septal plane as **trabecular spines** (Fig. 235). These spines may also be discontinuous vertically and interrupted by endotheca, as if septal secretion began anew after each period of growth. This structure is homologous with the acanthine septal type of some Rugosa.

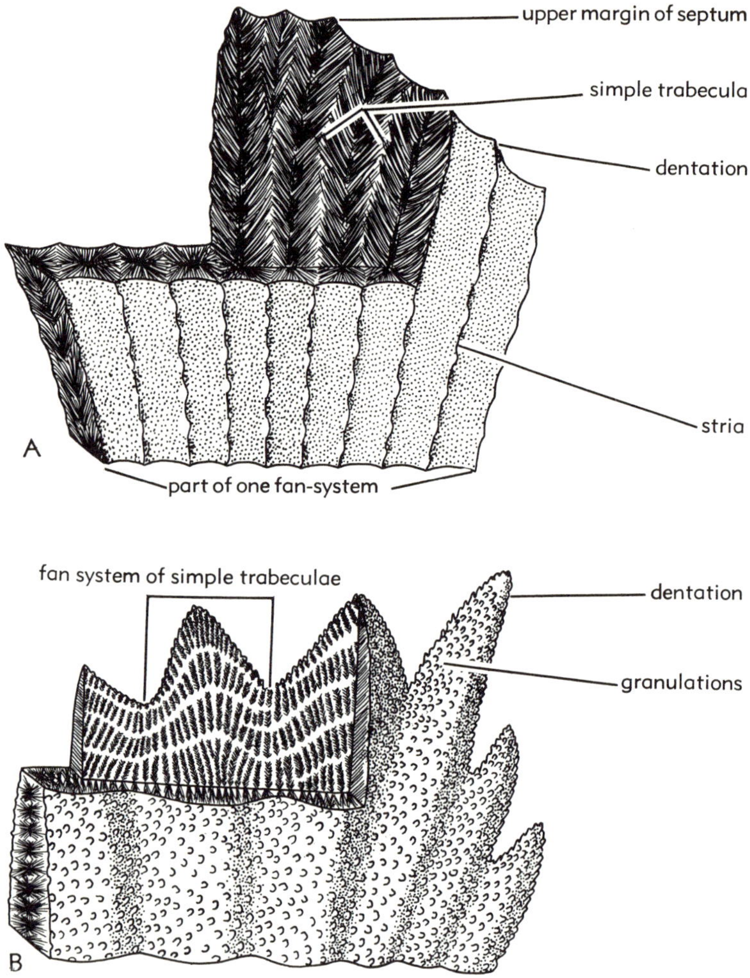

FIG. 232. Trabecular structure of scleractinian septa. *A*, Single fan system of simple trabeculae in laminar septum of *Galaxea* (98). *B*, Multiple fan systems of simple trabeculae in laminar septum of *Mussa* (98).

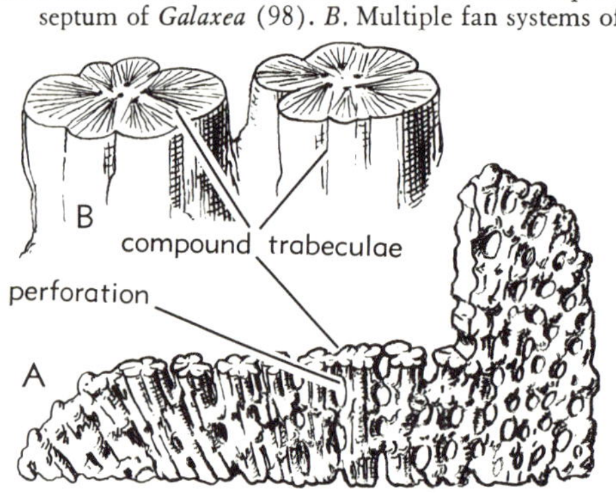

FIG. 233. Trabecular structure of scleractinian septum. *A*, Single fan system of compound trabeculae in fenestrate septum of *Cyclolites* (105). *B*, Compound trabeculae enlarged (105).

Septal teeth or **dentations** are formed along the upper margin of the septum if the trabecular axes project beyond the connective tissue or if the trabeculae are too far apart for intervening spaces to be filled (Fig. 236). They reflect the underlying structure of the septum. Granular or bead-like teeth are found in scleractinians having the septa composed mostly of simple trabeculae arranged in one fenestrate fan system (Astrocoeniidae, Calamophylliidae, Agariciidae, Thamnasteriidae, Microsolenidae). Nearly smooth, acute teeth occur (Montlivaltiidae, Faviidae) where the septa are composed of simple trabeculae in 1 or 2

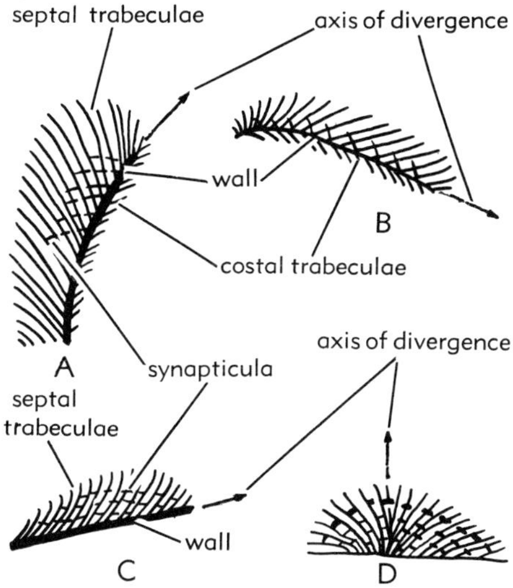

FIG. 234. Axis of trabecular divergence in fungiid genera (*A, Acrosmilia; B, Fungia; C, Cyclolitopsis; D, Cyclolites*) (151).

laminar fan systems. Dentations like the last but larger and tending to be irregularly lacerate or hirsute, are common in groups (Pectiniidae, Merulinidae) in which the septa are laminar but with compound trabeculae. In the stylophyllid corals of large size (*Oppelismilia*), the compound trabeculae emerge as blunt, lobulate teeth. In the Fungiidae, trabeculae are compound, forming small beadlike dentations in the fenestrate primitive forms and early stages of advanced types, and large, hirsute, acute to lobulate teeth in later stages. Where the septa are formed by a number of small fan systems, as in the Mussidae, each system may form a single large, lobulate, conical, or triangular tooth. Even septa which apparently have quite smooth upper margins are very minutely dentate (Caryophylliidae).

The septal structure, as briefly outlined above, varies regularly from group to group within the Scleractinia. This variation of a fundamental structure forms the principal basis for the classification of the order into suborders, superfamilies, and families (Fig. 237).

ARRANGEMENT AND DEVELOPMENT

In general, the septal order of appearance and arrangement corresponds to that of the mesenteries, as is to be expected. Septa developed within mesenterial entocoeles are **entosepta,** those in the exocoeles are **exosepta.** The first-formed septa, called **protosepta,** are the 6 located in the entocoeles of the first cycle of mesenterial pairs. Unlike the protocnemes, however, the protosepta appear simultaneously rather than in succession in basal invaginations of the prosepta, the latter being high vertical endodermal ridges which appear on fixation of the polyp. Other cycles of septa, collectively termed **metasepta,** are subsequently introduced one after the other. If only one cycle of mesenteries (protocnemes) has developed when the 6 protosepta appear, the second cycle of septa appears in the exocoeles as exosepta. If 2 cycles of mesenteries (protocnemes and first-cycle metacnemes) have developed, both the protosepta and second-cycle metasepta appear nearly simultaneously and both are entoseptal. The first-cycle protosepta develop simultaneously, and the second-cycle ento- or exosepta apparently also originate simultaneously, but in some forms the order of appearance of the second cycle is arranged in a dorsoventral order following that of the metacnemes, emphasizing the basic bilaterality of the scleractinian polyp and corallum. This order of succession in sextants constitutes the most obvious difference between the Scleractinia and the Rugosa. In the latter, the metasepta develop in a dorsoventral (?ventrodorsal) direction in each of 4 quadrants of insertion.

The succession of septal cycles after the first 2 may be complicated by the first appearance of the second- or third-cycle exosepta in the positions ultimately occupied by the entosepta (Fig. 238). This **substitution**

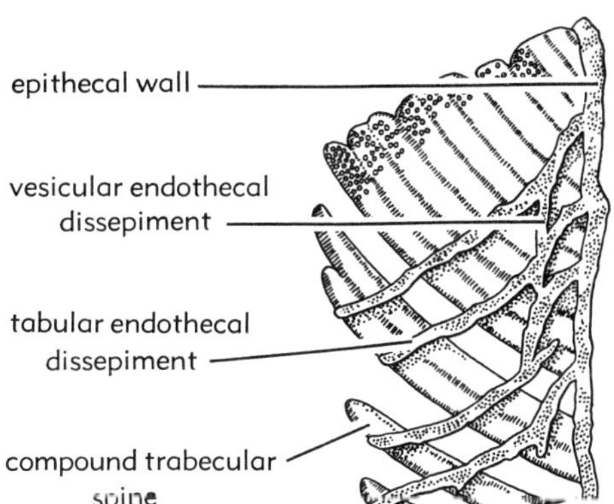

FIG. 235. Trabecular structure of scleractinian septum. Trabecular spines in *Oppelismilia*, ×2 (172).

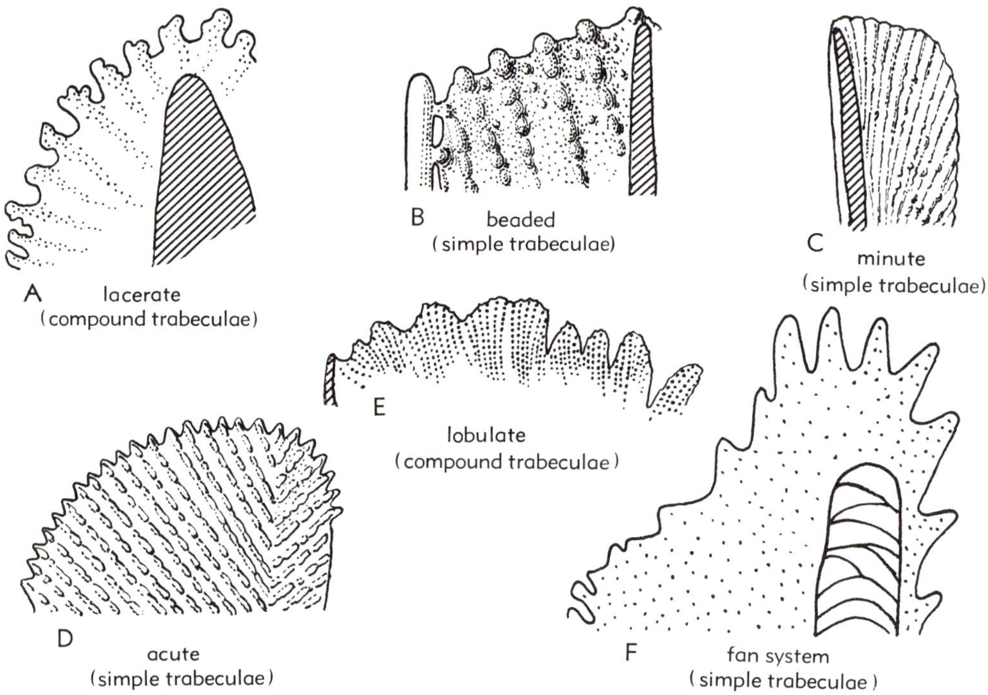

FIG. 236. Septal margins. *A*, Lacerate teeth (compound trabeculae), *Merulina*, ×8 (172). *B*, Beaded teeth (simple trabeculae), *Actinastrea*, ×12 (172). *C*, Minute teeth (simple trabeculae), *Caryophyllia*, ×8 (151). *D*, Acute teeth (simple trabeculae), *Montlivaltia*, ×4 (151). *E*. Lobulate teeth (compound trabeculae), *Oppelismilia*, ×8 (151). *F*, Fan system teeth (simple trabeculae), *Acanthophyllia*, ×1.5 (172).

of a permanent entoseptum for a temporary exoseptum is brought about by bifurcation of the outer (with respect to the axis of the corallum and polyp) end of the exoseptum as it grows upward, by coincident development of a mesenterial pair between the exoseptal branches, and by the growth of a new entoseptum within the entocoele formed by the new mesenteries. The 2 exosepta formed by branching of the original exoseptum may remain united proximally, fuse to the new entoseptum, or become free. This process

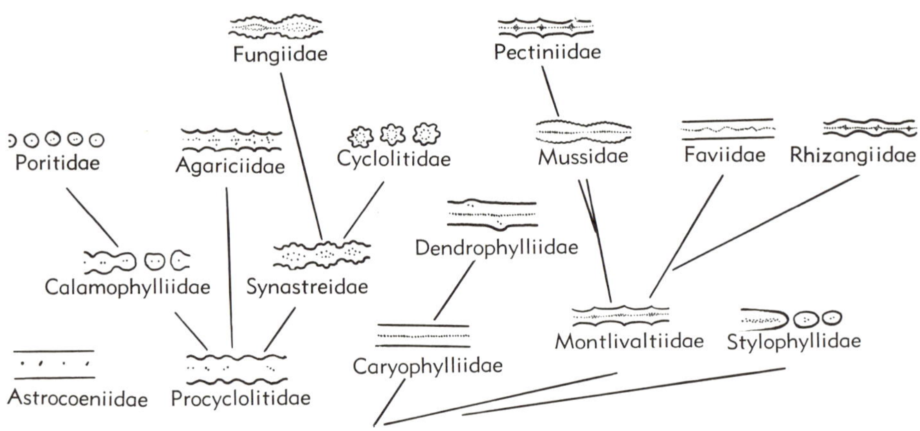

FIG. 237. Principal types of septal structure and probable relationships of important scleractinian groups. The figures, which are highly schematic and not to scale, represent horizontal sections of septa, and the fine dots indicate grouping of trabecular centers (151).

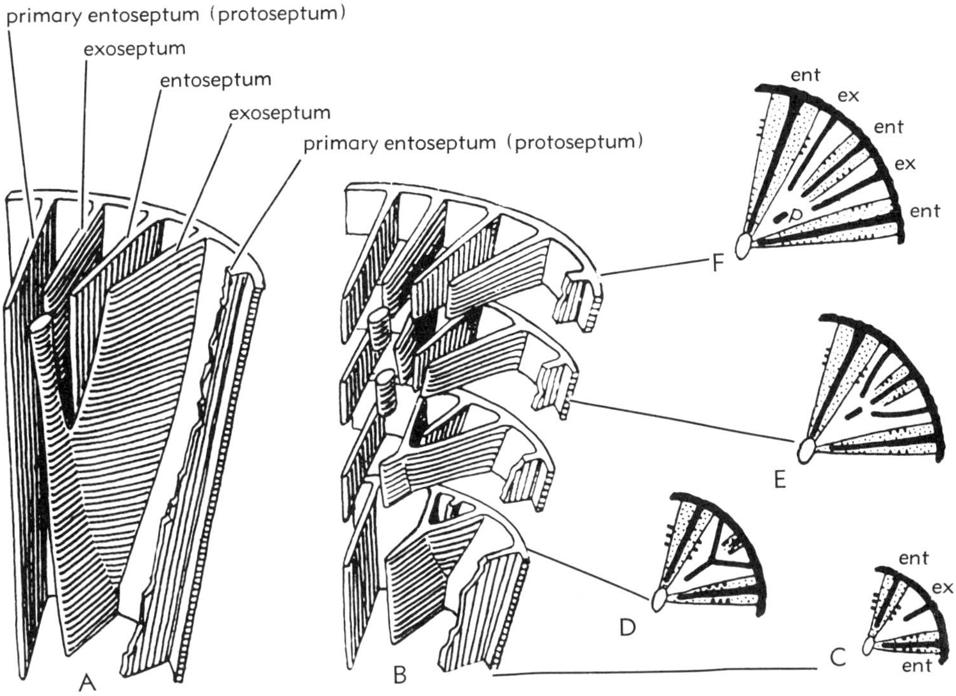

FIG. 238. Septal substitution and origin of pali. *A*, Sextant of solitary scleractinian (primary entoseptum to right is cut away). *B*, Same sextant with horizontal layers cut away. *C,D,E,F*, Sections of corallum and polyp at various levels (entocoeles indicated by stippling) (185).

may be repeated with the formation of new entosepta and doubling of the exosepta. In some groups (notably the dendrophylliids), the development of the exosepta is commonly much greater than that of the entosepta, so that, except for the first 1 or 2 cycles, they appear subordinate to the exosepta: the **Pourtalès plan** (Fig. 239). If septa of a higher cycle unite with those of a lower, it may be assumed that substitution has occurred, but if their inner ends are free, substitution may or may not have taken place, generally not. But in some genera no exosepta are ever formed, and in a few others even the first cycle is exocoelic.

The primary plan of the septa, like that of the mesenteries, is hexameral and seemingly radial (Fig. 240*A*). The 6 protosepta, appearing simultaneously, form the **first cycle.** The **second cycle** also consists of 6 septa inserted in spaces between the first. The **third cycle** includes 12 septa, the **fourth** 24, the **fifth** 48, and so on. Only very exceptionally are found other than traces of more than 8 cycles (768 septa, if complete). Septa of the second and third cycles commonly do not appear until all septa of the previous cycle have been formed. This general rule holds true for many scleractinians, but exceptions are common. Thus, in some forms, septa of equal size and apparent rank may be octameral or decameral in arrangement: 8/8/16/32, or 10/10/20. In such corals, certain mesenterial couples of the normal second cycle have attained a level of de-

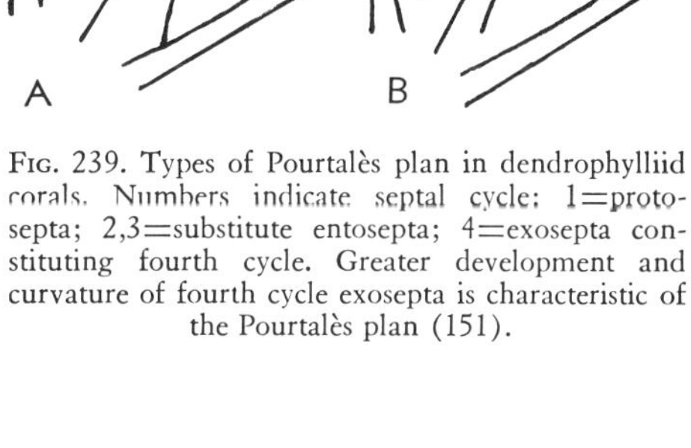

FIG. 239. Types of Pourtalès plan in dendrophylliid corals. Numbers indicate septal cycle: 1=protosepta; 2,3=substitute entosepta; 4=exosepta constituting fourth cycle. Greater development and curvature of fourth cycle exosepta is characteristic of the Pourtalès plan (151).

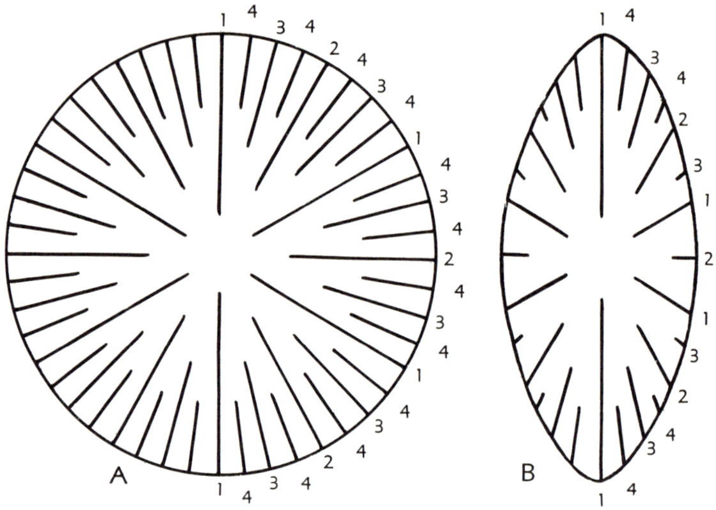

FIG. 240. Septal cycles in scleractinian corals. *A*, Normal cyclical order (numbers indicate cycles) (175). *B*, Cyclical order in calice elongated in plane of directive septa and mesenteries (higher cycles accelerated in end sextants; lower cycles retarded in lateral sextants) (175).

velopment equal to those of the first cycle, the complement of the second-cycle mesenteries, being filled by couples belonging to the third normal cycle.

The rule that septa of one cycle must be complete before those of the next cycle appear does not hold for septa of cycles higher than the third, especially in forms having an elongate calice (Fig. 240*B*). Higher-cycle septa tend to be accelerated and to appear in order from positions nearest the 2 opposing protosepta in the entocoeles of the directive mesenterial couples *(Flabellum, Notophyllia, Fungia)* and to be retarded in the transverse sextants.

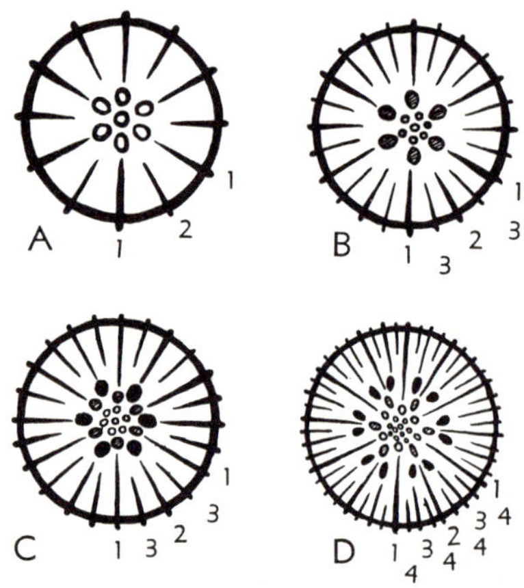

FIG. 241. Development of pali and paliform lobes in *Trochocyathus rawsoni* POURTALÈS, off South Africa (55). *A*, Early stage with central columellar tubercle (unshaded) resulting from fusion of ends of primary entosepta *(1)*, surrounded by 6 paliform lobes (shaded) offset from entosepta. *B*, Later stage with a 2nd ring of 6 paliform lobes (shaded) before primary entosepta *(1)* [1st ring of paliform lobes (unshaded) now forming columellar tubercles]. *C*, Still later stage with 1st crown of 6 pali (black) resulting from substitution of 2nd cycle entosepta *(2)* for former 2nd cycle exosepta which have doubled and now form 3rd cycle *(3)*. *D*, Adult stage, with 2nd crown of 12 pali, in pairs of chevrons, resulting from substitution of 3rd cycle entosepta *(3)* for former 3rd cycle exosepta now doubled and forming the 4th cycle *(4)*.

SPECIAL STRUCTURES OF SEPTAL ORIGIN

PALI AND PALIFORM LOBES

Vertical lamellae or pillars along the inner edges of certain entosepta, especially in the Caryophylliidae (Figs. 238, 241), are called **pali.** They are structurally identical with the septa and secondary to them, and are the internal portions of exosepta which remain after substitution, as shown by their attachment basally to the united inner ends of exosepta. In forms of *Caryophyllia* with 48 septa (in 4 cycles), for example, the first 2 cycles (6/6) are entoseptal, the third (12) is originally exoseptal but by substitution becomes entoseptal, the doubled exosepta forming the fourth and last cycle (24), and the inner or proximal ends of the third exoseptal cycle are left in the form of a circle or **crown** of pali adjacent and opposite to the inner edges of the permanent third-cycle entosepta. Thus, when pali are present, the number of crowns equals the number of cycles of exosepta less one.

Paliform lobes resemble pali, and are often difficult to differentiate in some genera. They are more or less detached trabecular offsets from the inner edges of some septa, appear in vertical succession, and apparently are not the result of septal substitution. Paliform lobes from the first 1 or 2 cycles of septa may mingle axially to form a columella (Fig. 241).

COLUMELLA

An axial structure found in many corals is termed **columella.** A **trabecular (parietal** or **spongy) columella** is formed by the intermingling of loose trabeculae, synapticulae, or paliform lobes from the inner septal margins of lower-cycle septa. It may be a weakly developed and loose, or a compact, spongy mass. A **fascicular columella** appears as a group of twisted vertical ribbons or rods similar to pali and paliform lobes in origin, that is, composed of remnant inner portions of exosepta, or offsets from the first 1 or 2 cycles. In some corals these elements may be solidly fused together as a single columnar rod, a **styliform columella,** free at its upper end with the entosepta secondarily fused to it by stereome. A **lamellar columella** is a plate, commonly free in its upper part, lying lengthwise in the longer axis of elongate calices. In calicular series resulting from intratentacular budding, a lamellar columella may be continuous or discontinuous between calicular centers. It originates by elongation of the dorsal directive protoseptum into the axis of the corallite. Later it may become separated and appear free. In no case does the columella originate independently of the septa, but in some forms its development may be accelerated and it may appear nearly at the same time as the protosepta.

SYNAPTICULAE

Rods or bars connecting opposed faces of adjacent septa and perforating mesenteries between them are termed **synapticulae** (Fig. 242). They first develop as large granulations formed by horizontal growth of strongly diverging trabecular elements, and may include 1 or 2 extra sclerodermites. **Simple synapticulae** are rods formed by 2 opposed granulations, as in *Siderastrea;* **compound synapticulae** are broad bars formed by the fusion of opposing ridges on 2 adjacent septa, parallel to the upper septal margin, as in *Fungia.* Synapticulae are found in nearly every scleractinian, especially in inner and outer parts of septa, and are particularly characteristic of forms with fenestrate septa (Fungiina).

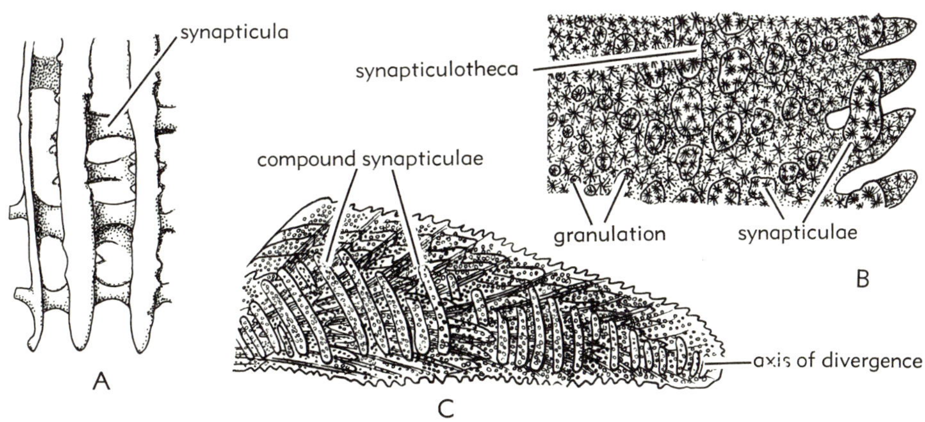

FIG. 242. Synapticulae. *A,* Simple synapticulae between septa of *Micrabacia* (30). *B,* Simple synapticulae in *Siderastrea:* longitudinal section parallel with septal plane (98). *C,* Compound synapticulae in *Fungia:* longitudinal section parallel with septal plane (98).

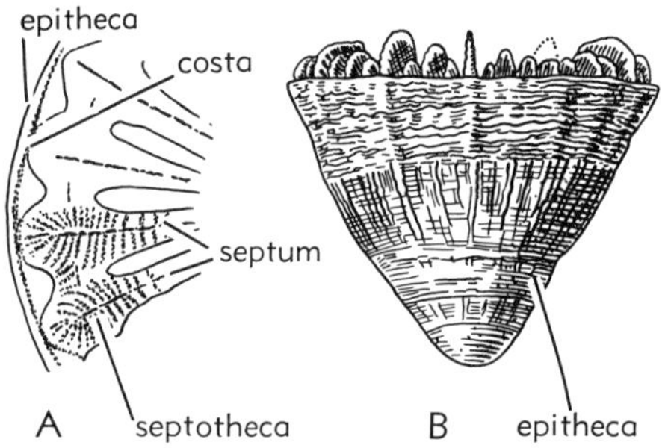

FIG. 243. Epitheca of *Trochocyathus hyatti* VAUGHAN, Paleoc.(Midway.), Ala. *A*, Schematic transverse section of mural region, ×8 (144). *B*, Lateral aspect of corallite, showing growth-wrinkled epitheca overlying costae, ×3 (144).

OTHER STRUCTURES

BASAL PLATE, EPITHECA, AND ROOTLETS

The first-formed part of the skeleton is the **basal plate,** which is deposited by the basal disc shortly after fixation of the larva, before the appearance of any septa. It is very thin, nearly transparent, firmly adherent to the substrate, and is commonly thickened by a later deposit of stereome.

As the polyp grows upward from the basal plate, the margin of the latter turns upward, forming a cup. This extension of the basal plate is the **epitheca,** secreted by the lower or outer edge of the calicoblast layer of the column wall where it is flexed upward, or at the lower margin of the edge zone where this is present. It is thin and externally marked only by growth lines. In corals lacking an edge zone, the epitheca unites the outer ends of the septa and keeps pace with their upward growth, forming the outer part of the epithecal wall. In forms with an edge zone, if this extends to the base, the epitheca may never be developed or it may lag considerably behind the septa and wall. If the edge zone extends only part way to the base, the epitheca extends from it to the base, closely covering the costae (Fig. 243).

Rootlets occur in a few genera, such as *Monomyces,* without an edge zone. Temporary extrusions of a tongue of edge zone over the margin of the calice toward the substrate may develop, depositing a layer of epitheca around themselves, then degenerating and leaving a hollow tube as an added element of stability. Such hollow rootlets are structurally distinct from the solid costal spines serving the same function, which are found in some genera provided with an edge zone.

DISSEPIMENTS

These are the partitions by which the polyp is cut off from the lower part of the corallum which it no longer occupies by reason of upward growth. Just how they are formed is not known exactly, but they seem to be responses to strains and shifts of the basal disc and column wall of the polyp resulting from secretion of vertical skeletal elements such as septa and walls. As skeletal deposition takes place along the tops of septa and wall, the soft parts are drawn upward, developing tensions sufficient to pull away the basal disc of the polyp from contact with the basal plate, either as a whole (whence tabular dissepiments) or as blister-like upward bulges (whence vesicular dissepiments). After the development of such bulges new skeletal tissue is deposited on the ectodermal surface. Dissepiments thus are successive, locally constructed basal discs.

Two types of dissepimental structures are **tabular** and **vesicular** (Fig. 235). The former are more or less flat plates extending across the whole corallite or confined to the axial part. Vesicular dissepiments appear as small, curved, overlapping plates inclined downward and inward from the corallite boundary or wall, and appear vesicular in section. Dissepiments developed between septa inside the corallite are **endothecal;** those outside the corallite, between costae or costal elements, are **exothecal** (Fig. 229).

STEREOME

The surfaces of the septa and other structures may be thickened secondarily by a closely adherent layer of variable thickness with smooth or lightly granular surface, the **stereome.** Structurally this is identical with the basal plate and dissepiments, and is especially common in corals having laminar septa, as the Faviina and Caryophylliina.

COENOSTEUM

In colonial corals, the corallites are united to each other by structures deposited by the coenosarc and referred to generally as **coenosteum.** Where the coenosarc contains extrathecal extensions of the mesenteries, the coenosteum is made up of costae, which

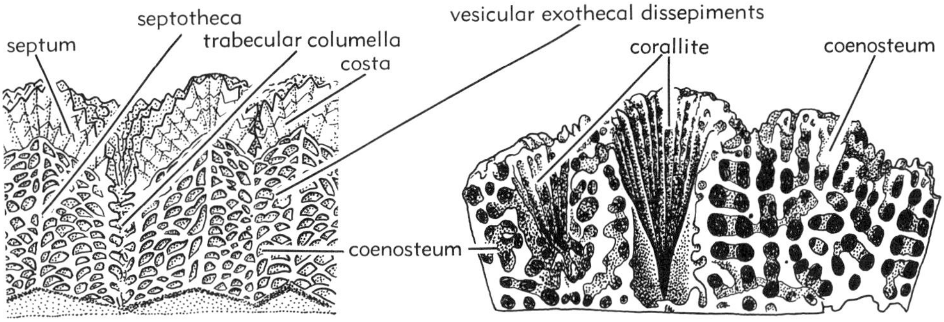

FIG. 244. Coenosteum. *A, Montastrea:* coenosteum composed of costae and vesicular exothecal dissepiments, in vertical section (98). *B, Astreopora:* porous coenosteum, in vertical section (57).

may be confluent between corallites, and tabular or vesicular exothecal dissepiments (Fig. 244). Where the extrathecal mesenteries are not developed, with consequent reduction or absence of costae, the coenosteum consists wholly of dissepimental tissue (Fig. 229). In such groups as the Poritidae, Acroporidae, and Dendrophylliidae, the lower part of the coenosarc is modified into a tangle of canaliculae with ramifying vertical and lateral connections. The coenosteum laid down around this soft tissue consists of sclerodermites united horizontally and vertically by rods in more or less distinct laminae (Fig. 244); or, if the canaliculae do not extend downward, as in the Pocilloporidae, the coenosteum is either a solid or vesicular granulated or spinulose deposit lacking sclerodermites.

Modifications of coenosteal structure parallel those of the septa, for increasing porosity of the septa is accompanied by porosity of

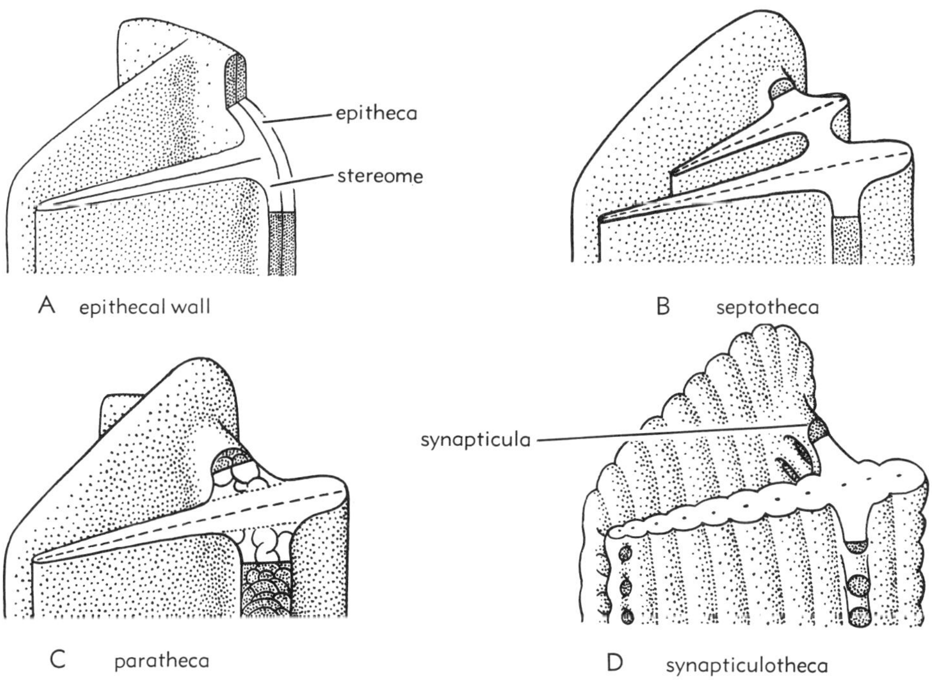

FIG. 245. Wall structures (198n). *A,* Epithecal wall. *B,* Septotheca. *C,* Paratheca. *D,* Synapticulotheca.

the coenosteum, and reduction of the extrathecal mesenterial continuations to canaliculae results in porous coenosteum.

WALL

The skeletal **wall** or **theca** encloses the column wall of the polyp, and unites and secures the outer ends of the septa. It is entirely secondary in origin and its function is assumed by different structures in different groups. In many primitive (*Pinacophyllum,* procyclolitids, stylophyllids), and a few neotenic (flabellids, guyniids) scleractinians, the epitheca encircles the outer ends of the septa, with an internal thickening of stereome (Fig. 245*A*). Such a wall is found in the earliest stages of nearly all corals, but generally is superseded during ontogeny by other types of mural structure. In forms with an edge zone, where the development of epitheca is limited or even prevented, another type of wall, the **septotheca,** is developed by thickening of the outer part of the septa along the axis of trabecular divergence, with or without additional sclerodermites (Fig. 245*B*). A septotheca is solid, forming a partition between the septa and costae. In some corals having well-developed dissepiments, these may push upwards so as to reinforce the septotheca, or even replace it as the wall (Fig. 245*C*). In the latter instance the wall is termed **paratheca.** Similar to septotheca but not solid, is a **synapticulotheca,** formed by one or more rings of simple or compound synapticulae along the axis of trabecular divergence (Fig. 245*D*). Such a wall enables direct communication through the wall as well as over it between the gastrovascular cavity and its extension into the edge zone or coenosarc.

REPRODUCTION

SEXUAL REPRODUCTION

Scleractinian polyps are hermaphroditic, unisexual, or sterile. In hermaphroditic forms, ovaria and spermaria are closely associated in the same polyp, commonly on the same mesentery. Dioecious forms are common, the ovaria usually bearing but one large egg at a time.

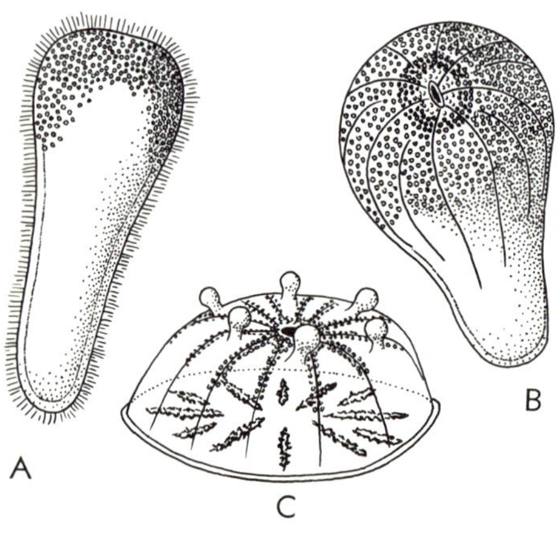

Fig. 246. Planulae and early stages of *Siderastrea radians, ca.* ×25 (28). *A,* Larva immediately on extrusion. *B,* Second-day larva just before settling, with six pairs of mesenteries (protocnemes). *C,* Polyp a few days after fixation. First two cycles of ento- and exosepta beginning to develop.

Viviparity is the rule. Before extrusion, the larvae are free in the gastrovascular cavity of the parent, and are ejected through the mouth singly or in batches up to a dozen or more at a time. The breeding season for many reef corals seems to have a roughly lunar periodicity, but some species have an almost continuous season, and still others have a prolonged annual season.

The young larva or **planula** (Fig. 246*A*) is spheroid, pyriform, or elongate-cylindrical, ranging in length from 1 to 3 mm., externally ciliated, and able to swim very shortly after extrusion. At first, the oral aperture is not visible, but soon appears as a minute opening at or near one end. Initially almost opaque, the planula later becomes more transparent, and the mesenterial attachments to the wall can be seen, generally with 6 orthocnemes developed (Fig. 246*B*).

Planulae may swim about sometimes for weeks, but settling on a substrate usually occurs within a few days. Fixation is by the aboral end, which flattens and broadens as the basal disc and column wall become differentiated. The first complete cycle of mesenteries appears within a few days after fixation when the entocoeles of the protocnemic mesenteries have been formed, and at about the same time the first cycle of septa is developed (Fig. 246*C*).

ASEXUAL REPRODUCTION AND COLONY FORMATION

To be discussed here are the various modes by which new polyps or stomodaea[1] may be produced from other polyps, either as separate, **solitary individuals,** or as members of **colonies** with or without organic continuity between parent and daughter polyps. The many modes by which this is accomplished and the different conditions to which they lead are one of the most remarkable aspects of the Scleractinia; they have great systematic importance because the form of the colonies is largely determined by them.

New stomodaea or polyps arise (1) by invagination, either of the oral disc inside the tentacular ring **(intratentacular budding),** or of the edge zone or coenosarc outside the tentacular ring **(extratentacular budding),** into the coelenteric cavity; (2) by union of the marginal filaments of the broader subsidiary mesenteries, with openings in the oral disc **(intratentacular),** or in the edge zone or coenosarc **(extratentacular),** and (3) by **transverse division** of the polyp. For practical purposes the first 2 generally have essentially the same result, and it is more convenient to recognize 3 main modes: (1) intratentacular budding ("fission"), (2) extratentacular budding ("budding"), and (3) transverse division.

INTRATENTACULAR BUDDING

The development of 2 or more mouths with stomodaea within the same tentacular ring, (a) with one or more couples of mesenteries between each pair of neighboring stomodaea **(indirect linkage),** or (b) with neighboring stomodaea connected by mesenterial strands **(direct linkage),** characterizes intratentacular budding. Indirect linkage of stomodaea, which concerns soft parts, is reflected in the hard parts by **trabecular linkage** between corallite centers (Fig. 247*A*); in direct linkage of stomodaea, corallite centers are joined by lamellar septal plates **(lamellar linkage)** (Fig. 247*B*).

[1] A "polyp" includes a mouth and stomodaeum enclosed by a tentacular ring or rings, but in many scleractinian polyps intratentacular budding ends with 2 or more mouths and stomodaea enclosed within the same tentacular ring. It is not yet settled whether this latter condition (di- to polystomodaeal) constitutes one "polyp" or as many polyps as there are mouths and stomodaea within a common tentacular ring.

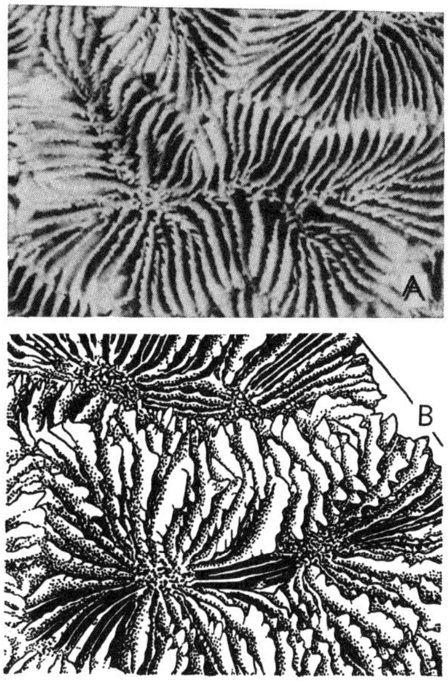

FIG. 247. Linkages after intratentacular budding. *A,* Trabecular linkage of centers in *Isophyllia* (88). *B,* Lamellar linkage in *Symphyllia* (88).

Modes of Intratentacular Budding

1. *Distomodaeal budding,* with 2 stomodaea developed within a common tentacular ring and 2 interstomodaeal couples of mesenteries between original and each new stomodaeum (Fig. 248*A*; 249,*2*).
2. *Tristomodaeal budding,* like distomodaeal budding except that 3 stomodaea occur in series or forming a triangle (Fig. 249,*1,3,6*).
3. *Triple stomodaeal budding,* like tristomodaeal budding except that the 3 stomodaea invariably form a triangle and only one interstomodaeal couple of mesenteries occurs between every 2 neighboring stomodaea.
4. *Polystomodaeal budding,* with more than 3 stomodaea developed within a common tentacular ring; includes 3 types, as follows.
   a. *Intramural budding,* with stomodaea directly or indirectly linked in a single linear series; the polystomatous polyps may be continuous or discontinuous throughout a colony and exhibit lateral branching or terminal forking (Fig. 248*B;* 249,*4*).
   b. *Circumoral budding,* with directly linked stomodaea arranged concentrically around central parent stomodaeum (Fig. 248*C*).
   c. *Circummural budding,* with indirectly linked stomodaea arranged around discontinuous collines or monticules of corallum (Fig. 248*D*).

The several basic modes of budding indicated in the tabular outline just given may

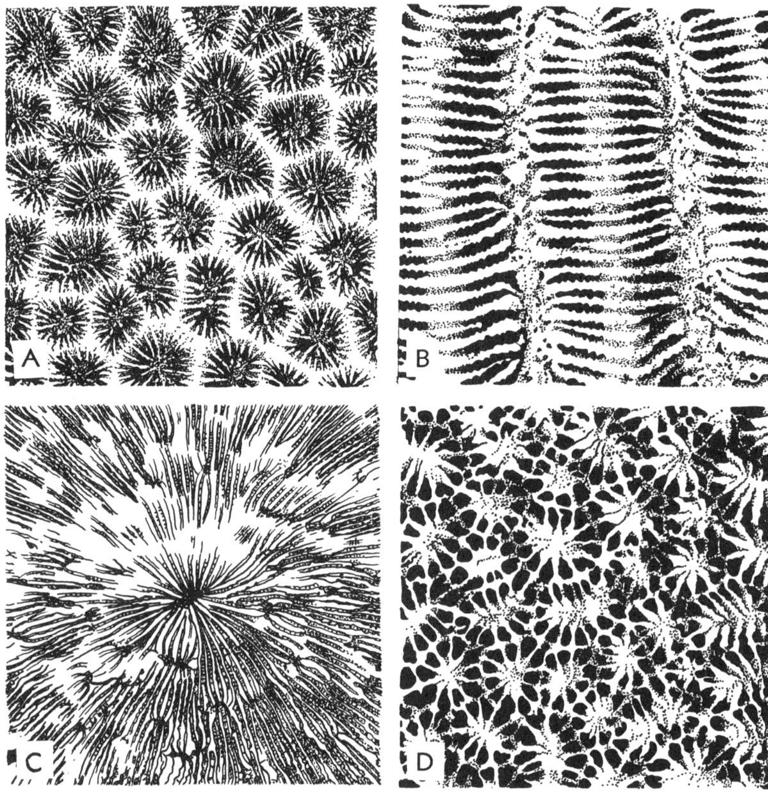

FIG. 248. Intratentacular budding. *A,* Distomodaeal budding in *Goniastrea retiformis* (LAMARCK), Rec., Rotuma, ×3 (83). *B,* Intramural budding with indirect linkage in *Diploria strigosa* (DANA), Rec., W. Indies, ×3 (88). *C,* Circumoral budding in *Halomitra philippinensis* STUDER, Rec., Jaluit, ×1 (165). *D,* Circummural budding in *Hydnophora microconos* (LAMARCK), Rec., Maldives, ×3 (88).

be permanent or only temporary. After di- or tristomodaeal budding, the new stomodaea may remain within the same tentacular ring or they may become individualized as separate polyps, each with its own tentacular ring. In the latter case, the **monostomodaeal** condition is said to be permanent, and the corallites are **monocentric.** If the permanent condition is di- or tristomodaeal, the corallites are **di-** or **tricentric,** and when the polystomodaeal condition is permanent the corallum is **polycentric.** These conditions, and likewise conditions of linkage, are generally the same within the limits of a genus.

In many genera, the mode of colony formation changes as the colony grows: at first it may be circumoral, followed by intramural or di- or tristomodaeal budding. When polystomodaeal budding occurs, individualization of polyps or corallites rarely takes place.

EXTRATENTACULAR BUDDING

The presence of only one stomodaeum within a tentacular ring, combined with development of new stomodaea from edge zone or coenosarc outside the tentacular ring of the parent, characterizes extratentacular budding (Fig. 250). Significant also is the absence of interstomodaeal mesenterial couples or mesenterial strands between stomodaea of neighboring polyps, as well as the absence of polystomodaeal polyps or polycentric corallites.

Solitary corals normally lacking an edge zone, as *Culicia,* or with a very narrow edge zone, as *Rhizangia,* may develop reptoid colonies by the outpushing of a temporary tongue of edge zone on which a new extratentacular bud is formed. Such stoloniferous expansions may lay down a calcareous base, or degenerate and sever organic continuity.

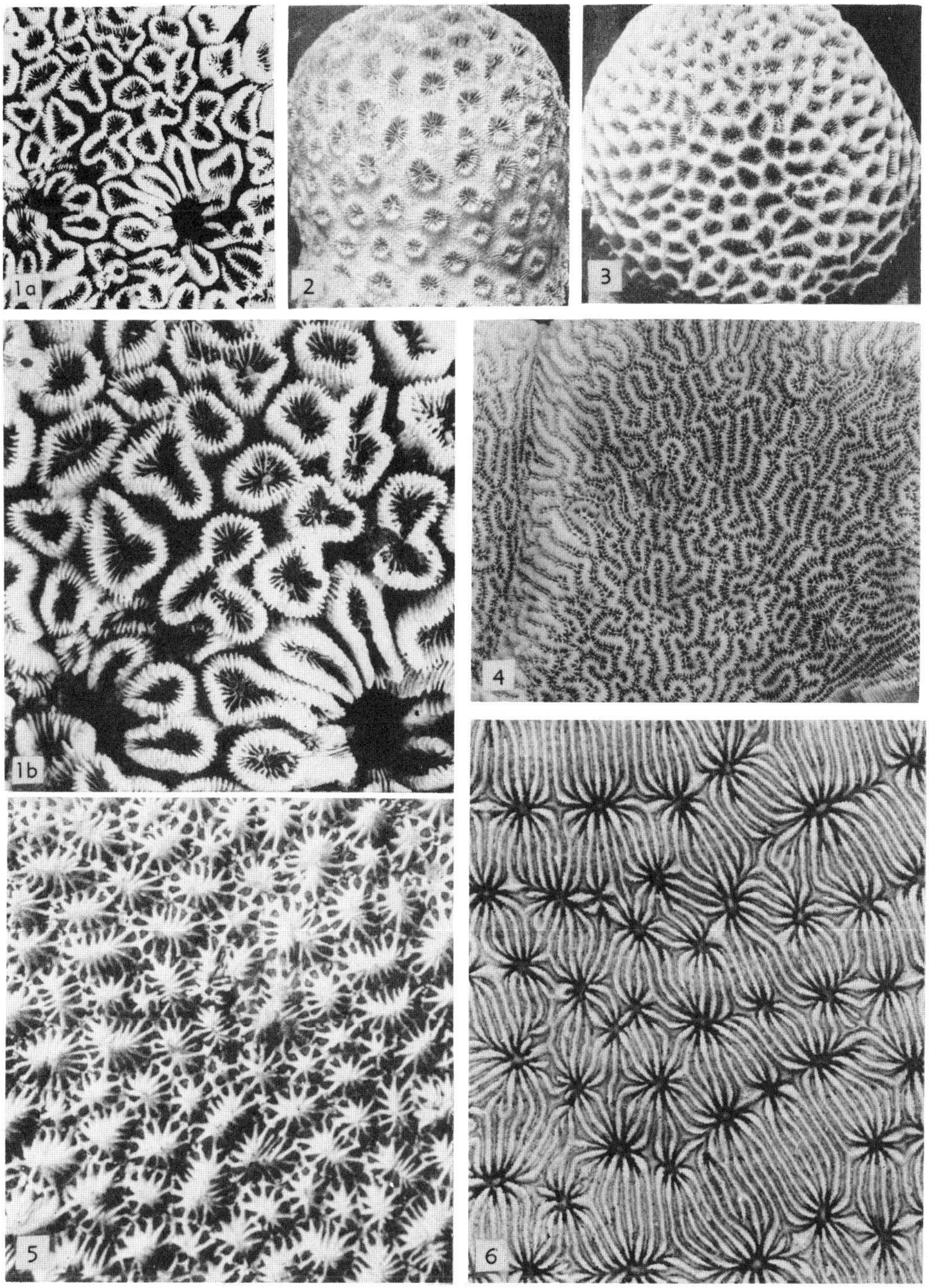

FIG. 249. Recent hermatypic scleractinian corals. *1, Caulastrea tumida,* Japan, showing phaceloid mono- to tristomodaeal intratentacular budding; *1a,b,* ×0.5, ×1 (YABE & SUGIYAMA). *2, Favia speciosa,* Fiji, showing plocoid mono- to distomodaeal intratentacular budding, ×0.5 (VAUGHAN). *3, Goniastrea pectinata,* Fiji, showing cerioid mono- to tristomodaeal intratentacular budding, ×0.5 (VAUGHAN). *4, Leptoria phrygia,* Bonin I., showing meandroid polystomodaeal intramural intratentacular budding, ×0.8 (WELLS). *5, Hydnophora microconos,* Murray I., Great Barrier Reef, showing hydnophoroid circummural intratentacular budding, ×2 (VAUGHAN). *6, Pavona crassa,* Ryukyu I., showing thamnasterioid mono- to tristomodaeal intratentacular budding, ×4 (WELLS).

FIG. 250. Extratentacular budding in *Diploastrea heliopora* (LAMARCK), Rec., Minikoi, ×2 (83).

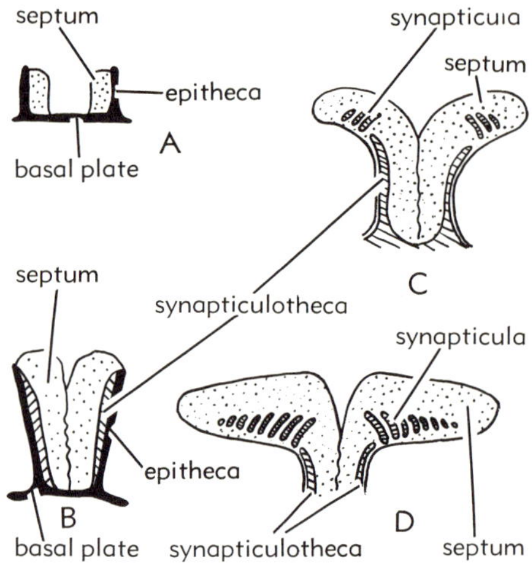

FIG. 251. Transverse division in *Fungia* (schematic vertical sections) (151). *A*, Early post-embryonic stage (anthocaulus) with epithecal wall. *B*, Later anthocaulus stage, with synapticulotheca developing inside epithecal wall. *C*, *Acrosmilia* stage. *D*, Detached anthocyathus.

TRANSVERSE DIVISION

Transverse fission with complete severance of connection has been observed in only a few scleractinians.

Fungia may reproduce by transverse division of a fixed conical solitary polyp and corallum, the **anthocaulus** or **anthoblast** (*Acrosmilia* stage of ontogeny) (Fig. 251). An anthocaulus is the result of sexual generation, an anthoblast, of extratentacular budding. **Anthocyathi** (neanic *Fungia* stage) are the buds separated from anthocauli or anthoblasts. This is a special form of transverse division, differing from the usual concept of fission in that the anthocaulus or anthoblast apparently never attains the adult *Fungia* condition and either dies or remains in the *Acrosmilia* stage and produces more anthocyathi (successive transverse division). Other genera related to *Fungia* show the same mode of division, but in some the anthocyathi do not break away from the anthocaulus and a fixed adult corallum results. Genera of extinct groups, such as the Cyclolitidae, may have had this mode of increase.

REGENERATION

Genera of the Fungiidae and a few others have remarkable ability to regenerate more or less complete adult coralla from fragments. The minimum amount of polyp and corallum necessary for successful regeneration is a segment between and including 2 adjacent entosepta of the first cycle, about one-sixth of most solitary scleractinians.

In several genera, such as *Dasmosmilia*, the conical corallite shows a strong tendency to split lengthwise into wedge-shaped fragments, either as a result of injury or an inherent structural weakness. Each fragment is able to produce one or more buds on its inner surface **(parricidal budding).**

MORPHOGENESIS OF CORALLUM

SOLITARY CORALLA

The form of the corallum depends on the character of the polyp, the relative growth rates of its parts, and in colonial forms, on the mode of asexual increase.

The basic corallum, as suggested by the earliest stages of living forms, is a disc formed by the horizontal basal plate fixed to the substratum (Fig. 252). In development, this is followed immediately by a narrow band of epitheca forming a shallow cup, after which the first septa appear. The adult corallum of some groups (Flabellidae, Guyniidae) is of this simple type, but in most it is followed by development of more advanced types of wall and the epitheca is to be regarded as an early post-embryonic structure of diminishing importance. Ability to form epitheca, however, is always retained by the edge zone and it frequently develops when the edge zone is retracted as upward growth proceeds. Some genera in which the corallum is wholly covered by the polyp, invariably lack all traces of epitheca (turbinolians).

The commonest form of the solitary corallum is conical or **turbinate,** with a basal angle of 60° to 80° (Fig. 252). If the angle is about 40°, the corallum is termed **trochoid,** and if the angle is only about 20° the corallum is said to be **ceratoid.** These essentially conical coralla result from differing rates of vertical and peripheral growth, vertical growth generally being faster than peripheral. If peripheral enlargement ceases in the adult stage, and vertical growth continues, the resulting corallum is elongate and **cylindrical** if regular, or **scolecoid** if irregular. In a few forms peripheral growth ceases at a very early stage and the corallum resembles a short, squat plug, or drum, the **tympanoid** form. **Cupolate** coralla, with flat base and highly convex oral surface, are the result of nearly equal rates of peripheral and vertical growth. **Discoid** coralla, with horizontal wall and flat or slightly concave or convex oral surface, develop if peripheral growth is much faster than vertical. **Patellate** coralla, which are broadly flattened conical in form, are the result of slight acceleration of peripheral over vertical growth. **Cuneiform** (wedge-shaped) and **flabellate** (fan-shaped) coralla develop if the calice is elongated in the direction of the directive entosepta. If the polyp continues to grow in the gerontic stage, the skeletal structures generally show a reduction in size from their maxima and the calices become smaller.

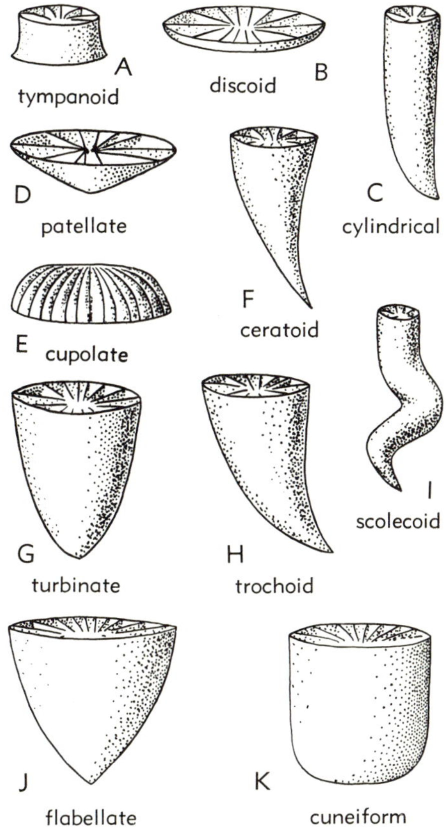

FIG. 252. Forms of solitary coralla.

COLONIAL CORALLA

The form of a colonial corallum depends to a great extent on the mode of colony formation. It may be **reptoid** (creeping), with stolon-like expansions of edge zone from which new corallites are budded; **dendroid,** with spreading branches; **phaceloid** (bushy), with parallel or nearly parallel, laterally free corallites forming tufts; **plocoid** (Fig. 250), with more or less cylindrical corallites having distinct walls and united to each other by confluent or nonconfluent costae and dissepiments; **cerioid** (Fig. 248*A*), with closely appressed prismatic corallites united directly to each other by fused walls; **aphroid,** found only in a few amphiastreids, with septa withdrawn or separated from the wall by a zone of dissepiments; **thamnasterioid** (Fig. 249,*6*), with corallites lacking definite walls and with directly confluent septa or septocostae; **hydnophoroid** (Fig. 248*D*), with corallite centers arranged around protuberant **collines** or **monticules,** the result of circummural budding; or **meandroid** (Fig. 248*B*; 249,*4*), with corallites forming **linear series** or groups of series within the same walls. In meandroid coralla the exsert edges of the septa and walls may form a ridge or colline on each side of the series. Collines of adjacent series may be fused to form a single colline (Fig. 248*B*), or separated by a trough of coenosteum, an **ambulacrum.** Meandroid coralla with a single continuous slightly contorted or straight laterally free linear series are termed **flabellate.**

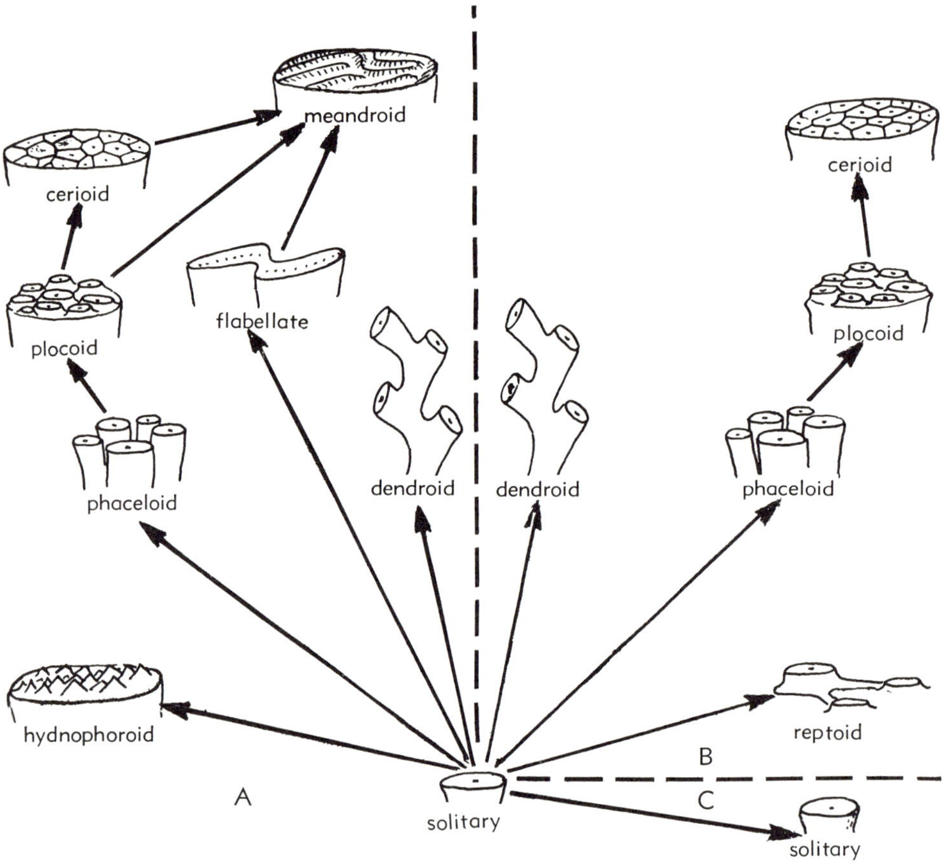

FIG. 253. Morphogenetic trends in colony formation (198mod.). *A,* Intratentacular budding; *B,* Extratentacular budding; *C,* Transverse division.

Cerioid, plocoid, thamnasterioid, hydnophoroid, and meandroid coralla are **massive** in habit if they form more or less thick masses or heads; **ramose** if branching; **foliaceous** if they form thin, expanding, overlapping sheets with a small basal attachment, like Savoy cabbages; and **incrusting** if they form thin sheets adherent to and following the substrate. Reptoid coralla result only from extratentacular budding; thamnasterioid, hydnophoroid, meandroid, and flabellate, only from intratentacular budding; and dendroid, phaceloid, cerioid, and plocoid from either mode.

Colonial coralla may also be produced by the fixation of a number of planulae of the same species in such close proximity that they subsequently fuse together.

MORPHOGENETIC TRENDS

Most scleractinian families show trends from solitary to various types of complex colonial coralla, especially if the principal mode of colony formation is by intratentacular budding, as partly summarized in Fig. 253. In addition to the trends there shown, it should be noted also that trends from solitary to thamnasterioid occur, and from plocoid and cerioid to thamnasterioid likewise. Thus, from a solitary progenitor some 13 colonial forms can arise by various modes of colony formation, 6 of these forms having 4 distinct habits apiece, a total of at least 31 principal growth forms of specific to generic value.

ECOLOGY

Scleractinian corals may be divided into two ecological groups: (1) **hermatypic** (reef) corals (Figs. 254-256), characterized by the presence of vast numbers of symbiotic, unicellular, dinoflagellates or zooxanthellae in their endodermal tissues, and (2) **ahermatypic** ("deep-sea") corals, which lack zooxanthellae and are less restricted environmentally than reef corals (Fig. 257). Seemingly, the hermatypic corals are closely dependent on their contained zooxanthellae for general well-being and vigorous growth, and their restricted habitat is largely controlled by the ecologic requirements of the symbionts.

CONTROLLING FACTORS

DEPTH

Hermatypic corals live between the surface and a maximum depth of about 90 m., but most of them occur in depths less than 50 m., and the most luxuriant growth is attained where the water is less than 20 m. deep. Ahermatypic corals live at all depths down to a maximum of nearly 6,000 m., but mostly between the surface and 500 m.

TEMPERATURE

A few hermatypic corals can live in temperatures as low as 15°C. but most are found in waters above 18°C. and they flourish best between 25° and 29°C.; the maximum endurable temperature is about 36°C. Ahermatypic corals live within a range from average reef temperatures to a minimum of about −1.1°C., but maximum development is found between 4.5° and 10°C.

SALINITY

The salinity tolerated by scleractinians lies between 27 and 40 parts per thousand, but they live best in salinities at or near the ocean normal of 36 ppm.

LIGHT

Strong sunlight is essential for vigorous growth of hermatypic corals, whereas ahermatypic corals live in partial to total darkness.

WATER MOVEMENT

Circulation of water is necessary, both to insure adequate supplies of nutrients (largely zooplankton) and oxygen, and to remove sediment. Corals rarely survive or live in areas where sedimentation is rapid.

The duration of the free-swimming planula stage of sexual increase is frequently a matter of days or weeks, and the transportation of planulae by currents is responsible for the wide distribution of many species, whereas shortness of the planula stage in other species accounts for their absence in some regions, even where the environment is suitable, as adjacent to Hawaii and Bermuda.

SUBSTRATUM

Coral planulae can settle only on a firm substratum such as bed rock, other corals, shells and skeletal parts of other sedentary organisms, loose blocks and smaller stable particles down to a few millimeters in size. In general, fine sand, silt, or mud bottoms are inimical to coral development unless there are scattered larger, clean particles and sedimentation is slow.

GROWTH RATE

The growth rate of corals differs in different areas and is greater in places where the annual average water temperatures are moderately higher. It also varies according to the structure of the skeleton, being slower in types with dense coralla, and faster in types with light, porous skeletons. In the tropical Indo-Pacific region, where ecologic conditions are more or less optimum, the annual growth increment of hermatypic corals, expressed in percentage gain in weight of coralla weighing 100 to 200 grams, ranges from as little as 20 per cent in solid coralla such as *Leptastrea,* to as much as 1,200 per cent in coralla with highly porous light skeletons, such as *Porites.* Annual increments in overall height range from 5 mm. (with 20 per cent weight increment) to 82 mm. (with 78 per cent weight increment).

FORM AND SIZE OF COLONIES

To some extent hermatypic corals, whether massive, foliaceous, ramose, or incrusting, owe their growth form to their situation on the reefs with respect to water movements.

The development of enormous colonies

consisting of millions of polyps and corallites, surpassing anything found amongst Paleozoic corals, is attributable to the development of (1) the porous, fast-growing skeletons of many forms and (2) the symbiosis of polyps and zooxanthellae.

PALEOECOLOGY

Certain ecologic factors controlling the distribution and development of corals, such as salinity, substratal requirements and sedimentation, certainly have been significant since the first appearance of the Scleractinia. Probably they were similarly operative on earlier occupants (rugose and tabulate corals) of the same ecologic niche. How long the single most important ecologic factor among the living hermatypic corals, the symbiosis with zooxanthellae, has existed, is uncertain, since it has no direct reflection in the skeletal structure and only an indirect effect on growth form. It is a characteristic and invariable condition only in a few families, and most families include both hermatypic and ahermatypic genera. The condition certainly was a gradual development, at first ecologically insignificant, but of increasing importance as zooxanthellae-coral symbiotic lineages became more and more mutually interdependent. Coral groups with established dinoflagellate symbiosis became further limited not only by factors controlling growth of the zooxanthellae, such as temperature and light intensity, which change with depth, but also of effects of resultant vastly increased colonial habit. In spite of these rather severe restrictions, which effectively limit hermatypic corals to the tropical surface temperature zone of the seas, this ecologic group has developed to an extent unknown in the Paleozoic corals and thus construction of true coral reefs became possible.

The earliest scleractinians, of Triassic age, were, by analogy with living forms, reef types. Symbiosis wtih zooxanthellae may have existed among Triassic corals, for they thrived best under conditions similar to those governing modern hermatypic forms, but presumably the relationship was not yet very significant and therefore development of vast reef structures was not possible. In the Jurassic, ahermatypic caryophylliids first appear in nonreef environments and a slow spread of corals into cooler, darker water began. In later Jurassic and Cretaceous time, ahermatypic corals representing other families are found less frequently in reef environments and more and more commonly in areas interpreted as marginal to the then tropical zone and beyond, and in areas of deeper and deeper water. Whereas modern ahermatypic corals at present equal or even surpass reef-builders in number of genera, less than a third of these are colonial in habit. The small colonies, many of which are delicately dendroid (an effect of sediment control in regions below wave base), can develop extensive banks and thickets only in a few places where food supplies are high, such as along the edge of the continental shelf off western Europe. Ahermatypic corals represent amongst the scleractinians the deployment away from the ancient, ancestral, and most fitted hermatypic environment, into all other possible environments so characteristic of the later history of successful organic groups.

STRATIGRAPHIC AND GEOGRAPHIC DISTRIBUTION

TRIASSIC

Scleractinian corals have not yet been found in rocks older than Middle Triassic. The oldest are found in lower Middle Triassic (Anisian-Muschelkalk) deposits of Germany, the southern Alps, Corsica, and Sicily; they include representatives of the Thamnasteriidae, Astrocoeniidae, Stylinidae, Procyclolitidae, and Montlivaltiidae, all of which seem to be hermatypic forms that lived on banks and in patches rather than forming true reefs. The late Middle Triassic (Ladinian) faunas of the same region are very similar.

The Late Triassic (Carnian, Norian, Rhaetian) witnessed world-wide extension of the earlier scleractinian faunas and considerable increase in their variety, including the first appearance of the Stylophyllidae in the Carnian. Species of the Montlivaltiidae, Procyclolitidae, Stylophyllidae, and Thamnasteriidae formed scattered reef patches during the Norian in southern and southeastern Europe, southeastern Asia, and

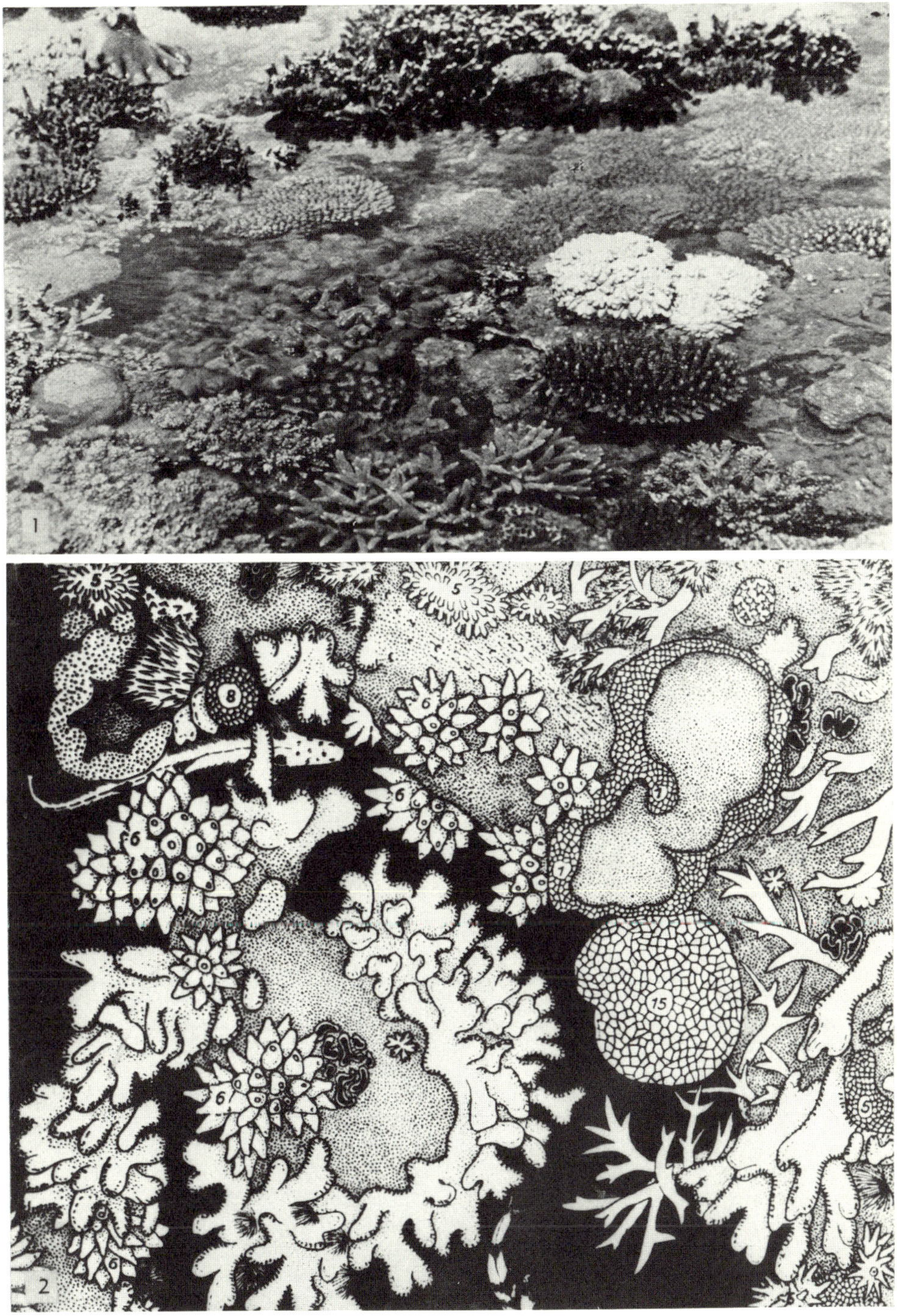

FIG. 254. Recent hermatypic scleractinian and alcyonarian corals. *1,* Calm-water reef tract at low tide, Low Isles, Great Barrier Reef, Australia (STEPHENSON). *2,* Map of coral colonies in area about 6 by 7 feet on part of Yonge Reef, Great Barrier Reef, Australia (MANTON).

Malaysia. These, with the exception of the Stylophyllidae, had already reached western North America in the Carnian.

The maximum range of the Late Triassic patch reefs and banks was from 60°N. to 10°S., greater than that of succeeding periods, but it is possible that hermatypic corals had not so close an adjustment to the tropical zone as they had later.

JURASSIC

The Early Jurassic (Liassic) corals were not greatly different from those of the Late Triassic. The families were practically the same; a few genera died out and a few new ones appeared. They were still of hermatypic type, but no important reefs are known. The Pinacophyllinae disappeared, and the Procyclolitidae lost most of their more primitive types as genera with more advanced septal structure developed. The Stylophyllidae culminated and became extinct by the Middle Jurassic. The most important addition was the appearance of the first caryophylliids in the Toarcian of northern Europe, marking the beginning of the distinction between reef and nonreef corals.

The Middle Jurassic (Aalenian-Bajocian-Bathonian) was a turning point in scleractinian evolution. From the Bajocian, diversification and multiplication of families and genera increased, with the center of this acceleration seemingly located in the western part of the Tethys. New groups were the Microsolenidae, Amphiastreidae, Calamophylliidae, and Synastreidae. In Bathonian time, conditions suitable for reef types extended eastward in the Tethys and southward along the eastern border of Africa. New Bathonian groups were the Stylininae, Parasmiliinae, and Rhipidogyridae. Few reefs were built, but patches and banks were common.

The Upper Jurassic was an epoch of extensive reef development, especially in the Tethys and its extensions into northern Europe, and as far east as Japan. Important new groups were the Actinacididae, Haplaraeidae, and Faviidae.

The early Late Jurassic (Callovian-Oxfordian) was much like the late Liassic in its lack of large faunas, but beginning in Lusitanian time corals are a most important element of the Tethyan faunas, even though they are almost absent in the Americas. Especially typical are the Rhipidogyridae, Amphiastreidae, *Enallhelia,* and *Tiaradendron.* The northernmost reef patches of this time are in England at 54°30′N., in Germany at 54°N., and in Japan at 38°N. Scattered occurrences of reef types are known as far north as 58° in Scotland. The most southerly are about 5°S. in eastern Africa. The reef-coral belt had about the same extent as today, but the zone as a whole was nearly 20° farther north.

CRETACEOUS

Little or no reef development is observed in earliest Cretaceous rocks and the corals of the Valanginian and Hauterivian are Jurassic in aspect; they are found mainly in Tunisia, Portugal, France, Italy, Carpathians, eastern Balkans, and the Crimea, where they lived on banks under conditions somewhat less than optimum.

In the Barremian and Aptian, another world-wide epoch of reef building began (Urgonian phase). Sizable reefs are known in France, Switzerland, Capri, Tunisia, Algeria, Catalonia, western Balkans, Bulgaria, Rumania, the Crimea, Kenya-Tanganyika, Japan, Venezuela, Mexico, and Texas. Coral banks existed as far south as Neuquen, Argentina (37°S.), and in Europe north to the Isle of Wight (50°45′N.). New families and subfamilies include the Cyclolitidae, Placosmiliinae, Montastreinae, Siderastreidae, Rhizangiidae, Micrabaciidae, and Desmophyllinae. The Rhipidogyridae died out, and the Thecocyathinae and Amphiastreidae declined.

Little reef development occurred during the Middle Cretaceous (Albian-Cenomanian), while ahermatypic corals living in the varied environments of neritic and bathyal zones became increasingly abundant. Albian deep- or cold-water coral faunas are known from France, Switzerland, England, Texas, and Antarctica (Graham Land), reef-coral faunas from Texas, northern Mexico, England, France, and central India. Cenomanian hermatypic faunas occur in Devon, France, northern Italy, Westphalia, Czechoslovakia, Bavaria, Albania, Greece, western Balkans, Belgium, Algeria, Egypt, Asia Minor, and southern India. They are similar to those of the Urgonian phase but few reefs are known. Cenomanian ahermatypic corals are widespread from Antarctica to England,

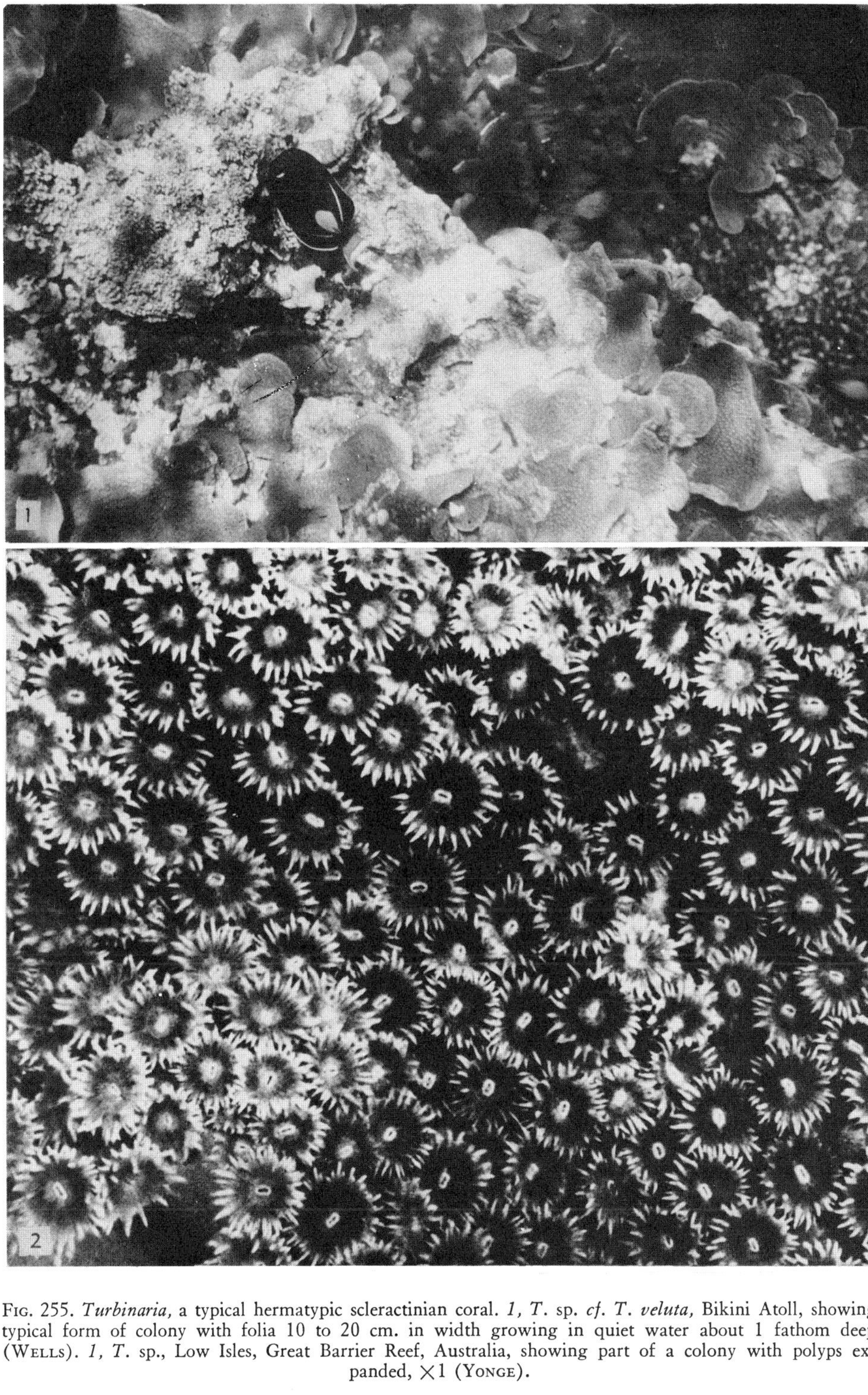

FIG. 255. *Turbinaria*, a typical hermatypic scleractinian coral. *1, T.* sp. *cf. T. veluta*, Bikini Atoll, showing typical form of colony with folia 10 to 20 cm. in width growing in quiet water about 1 fathom deep (WELLS). *1, T.* sp., Low Isles, Great Barrier Reef, Australia, showing part of a colony with polyps expanded, ×1 (YONGE).

Poland to Texas, and in Mexico, Madagascar, and India.

The Late Cretaceous was yet another time of reef building, but less so than the Late Jurassic or Early Cretaceous. The most important reefs are in Austria, Carinthia, southern France, Tunisia, Serbia, Switzerland, Czechoslovakia, Catalonia, and India. By the latest Cretaceous (Campanian-Maastrichtian) the Tethyan reef-coral fauna extended westward as far as the West Indies and Mexico and eastward to Malaysia, and the extent of the reef zone at this time was from about the equator to 50°N., although most of the reefs lie between 30° and 47°N.

By the Late Cretaceous, ahermatypic corals had secured a firm foothold in nearly every part of the littoral, neritic, and bathyal environments, and occur in Late Cretaceous beds of England, northern France, Catalonia, northern Germany, Scandinavia, Bavaria, Volga Basin, Poland, northern Africa, southeastern Africa, Arabia, India, United States, Chile, and Antarctica.

By the end of the Cretaceous, the modern aspect of both hermatypic and ahermatypic coral faunas became evident; this is marked by the appearance of early representatives of the Pocilloporidae, Poritidae, Turbinoliinae, and Dendrophylliidae, and disappearance of the Microsolenidae and Amphiastreidae.

TERTIARY

During the Paleocene and Eocene epochs, conditions favorable for extensive formation of reefs seemingly were absent everywhere, both hermatypic and ahermatypic types living sporadically in the Tethys, the coastal plains of the United States, the West Indies, and a few other spots. In the eastern Tethys, the Indian and Malaysian faunas began to become distinct from those of the central Tethys, as did those of the West Indies.

At the close of Eocene times most of the older elements of coral faunas had become extinct or were rapidly dying out; these include the Placosmiliinae, Montlivaltiinae, Haplaraeidae, Cyclolitidae, Actinacididae, Calamophylliidae, and Cyathophorinae. Dominant groups at this time were nearly the same as those of the present day: Eusmiliinae, Dichocoeniinae, Mussidae, Pocilloporidae, Poritidae, Acroporidae, Agariciidae, Siderastreidae, Fungiidae, and Faviidae. Ahermatypic groups, especially the Caryophylliidae, Dendrophylliidae, Flabellidae, and Rhizangiidae, were developing rapidly.

In the Oligocene, fringing reefs were built in the southeastern United States, West Indies, Mexico, northern South America, northern Italy, and India, and in the latest Oligocene in the East Indies and Japan. In Europe, reef-coral banks occurred in Bavaria, southeastern France, and the Balkans. During the late Oligocene, the European elements of the West Indian fauna began to die out, leaving Miocene faunas composed of essentially the same genera as those of today.

In Miocene time the final severance of connection between the Mediterranean and Indian Ocean areas eliminated the eastward route by which reef corals had migrated from the central Tethys to the East Indies, and from this epoch originate the 2 great modern reef-coral provinces, the Caribbean and Indo-Pacific.

During the later Tertiary, the Indo-Pacific coral faunas, already richer than those of the Caribbean, expanded greatly. Especially significant is the enormous development of the Pocilloporidae, Poritidae, and Acroporidae, which had been relatively unimportant in the early Tertiary, but now accounting for more than two-thirds of the present reef fauna.

The relic hermatypic fauna of the European Miocene consisted of a few genera related to both Indo-Pacific and Caribbean faunas, but few reefs were built and adverse conditions soon wiped out reef corals, leaving only a European Pliocene fauna of ahermatypic genera which is similar to that of the Mediterranean of the present day.

The contraction of the tropical zone in Pliocene time shifted the northern limit of reef coral growth southward to a maximum of 35°N., the southern boundary remaining nearly the same, and with little change of faunal composition.

Development of ahermatypic corals (Fig. 258) in the later Tertiary shows no sharp differentiation into faunal provinces. Because of their adaptability to extremes of temperature and depth, in strong contrast to reef corals, many genera and even species are cosmopolitan in distribution. Of the 11 families and subfamilies which include ahermatypic forms, all had appeared by the

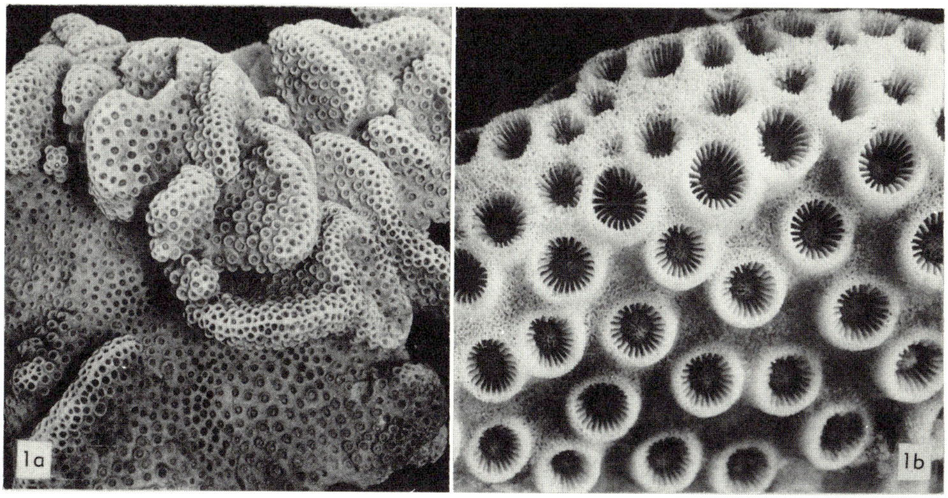

FIG. 256. *Turbinaria,* a typical hermatypic scleractinian coral. *1, T. peltata,* Formosa; *1a,* part of corallum, ×0.3; *1b,* calices, ×3 (YABE & SUGIYAMA).

Eocene. None is peculiar to any single environment, although the Rhizangiidae as a whole prefer rocky bottoms in the littoral and shallow neritic zones, whereas the Turbinoliinae, Anthemiphylliinae, and Micrabaciidae, which include only free-living solitary types, occur on sandy or silty bottoms at depths below wave base. Several families, such as the Micrabaciidae and Flabellidae, which were cosmopolitan in the early Tertiary, are now mostly confined to the Indo-Pacific-Antarctic region, whereas the Parasmiliinae are more abundant in the North Atlantic and West Indies. Several genera, such as *Conocyathus* and *Platytrochus,* which originated in the Tethyan or West Indian areas, now survive in the Australian area.

PLEISTOCENE AND RECENT

Pleistocene conditions had little noticeable effect on hermatypic corals, except possibly for a slight reduction in their numbers, and the faunal effect was an accentuation of porous types, such as the Poritidae and Acroporidae. Known Pleistocene coral faunas show almost no difference between those of the Pliocene and the present day.

In general, Recent hermatypic corals are found in the shallow waters of oceans and seas in the subtropical and tropical zones within an earth-encircling band from about 35°N. to 32°S. These are the extreme limits of reef-coral growth but reefs are only well developed within somewhat narrower limits, around oceanic islands away from influxes of quantities of terrigenous sediments, and in places along the eastern sides of continental masses.

Two large provincial areas of hermatypic corals are recognizable within the tropical zone: (1) *Caribbean* and (2) *Indo-Pacific.* The Caribbean province lies within the area bounded by the southeastern coast of the United States, eastern coast of Mexico and Central America, and northern coast of South America eastward to the Windward Islands, between latitudes 27°30′N. and 11°30′N. In this area, however, corals do not occur everywhere, even where the water is sufficiently shallow. They are not found along the margin of the Gulf of Mexico from north of Cape Roxo around to Tampa, Fla., and are very sparse from the latter place southward to Cape Sable. Along the South American coast they are found only around a few islands off Colombia and Venezuela. They are most abundant in the West Indies away from large land masses, and in the Bahamas and Florida Keys, where fringing reefs are fairly well developed.

The Caribbean reef fauna consists of about 20 genera and 36 species of the Acroporidae, Faviidae, Mussidae, Poritidae, Eusmiliidae, Meandrininae, Dichocoeniinae,

FIG. 258. *1, Sphenotrochus auritus* POURTALÈS, Plio., Venez.; *1a,b,* ×13 (WELLS).

Siderastreidae, Agariciidae, and Astrocoeniidae. Distinctive is the absence of any Pocilloporidae (except *Madracis*), and Fungiidae and the genera *Montipora, Astreopora, Goniopora, Alveopora, Hydnophora,* and *Pavona,* although all are known in the Caribbean middle Tertiary.

Outposts of the Caribbean fauna occur on the Bermuda reefs (32°N.), off the Brazilian coast between Alcantara (2°S.) and Rio de Janeiro (23°S.), and in the Gulf of Guinea, West Africa.

The Indo-Pacific province includes the Red Sea, Gulf of Aden, Persian Gulf, Gulf of Oman, Indian Ocean south to 26°S., the Australasian region, China Sea, Tropical Pacific, Bay of Panama, and the lower part of the Gulf of California. Nowhere in this vast area are any strongly marked faunal differences found, with exception of the Panamanian region which is depauperate rather than different. Coral growth in general is far more luxuriant and vigorous than that of the Caribbean, and in favorable places the annual growth rate is several times that of Caribbean species. The total number of species within the entire area is not less than 500, representing about 80 genera and subgenera, and the number of species living on a single reef tract ranges from 50 to as many as 200. As a whole, contrasting with the Caribbean, the fauna is notable for the protean development of the Acroporidae, of which the genera *Acropora, Astreopora,* and *Montipora* have at least 250 species, compared with 3 only of *Acropora* in the Caribbean; the Poritidae, with about 50 species of *Porites, Goniopora,* and *Alveopora,* compared with 3 only of Caribbean *Porites;* and the Seriatoporidae, with some 25 species of *Pocillopora, Stylophora,* and *Seriatopora,* all unrepresented in the Caribbean. A few genera have only local distribution and subprovincial areas can be broadly outlined.

The Indo-Pacific hermatypic coral fauna shows more differentiation bathymetrically than geographically. This variation as yet is only partially studied in a few localities, but 2 broad ecologic associations are evident: (1) surface reefs to a depth of 15 to 20 m., limited downward by normal wave base, and (2) quiet waters of lagoons and seaward slopes of reefs below wave base to the extreme lower limit of hermatypic coral growth, about 90 m. A number of forms are confined to the second association, but they are very widely distributed; examples are *Anacropora, Leptoseris, Cycloseris, Madracis,* and delicately branching and thinly foliaceous species of common surface-reef genera.

The ahermatypic corals are not readily separable into distinct faunal provinces, partly because of their commonly wide, seemingly erratic distribution, and partly because of lack of adequate information concerning certain areas. In general, their distribution varies with depth: the deeper the normal habitat of a species the greater is its geographic range, and forms occurring below about 500 m. are mostly cosmopolitan. Most species occur within definite bathymetric zones; also, more species occur in tropical than in temperate or cold seas. Corals inhabiting the littoral and shallower parts of the neritic zones along the coasts of great land masses generally are sparse both in variety and individuals, and in the tropics

FIG. 257. Recent ahermatypic scleractinian corals. *1*, *Cyathelia axillaris*, Amboina; *1a*, corallum, ×0.5; *1b*, calice, ×3.5 (BEDOT). *2*, *Paracyathus stearnsi*, off Calif., depth 110 m.; *2a,b*, ×2 (DURHAM & BARNARD). *3*, *Cycloseris vaughani*, Hawaii; *3a,b*, ×1, ×1.2 (BOSCHMA). *4*, *Caryophyllia clavus*, off Azores, depth 1,900 m.; *4a,b*, ×1 (GRAVIER). *5*, *Stephanophyllia formosissima*, Kii I., E.Indies; *5a,b*, ×1 (WELLS).

away from large land masses these zones are even more thinly populated by ahermatypic corals.

The greatest development of ahermatypic corals occurs near and down the edges of the continental shelves and the equivalent bathymetric zone around oceanic islands in depths from 175 to about 800 m., below the light floor of the ocean, in temperatures of 4° to 21°C. The family and subfamily groups consisting wholly of or including these corals are shown in the following tabulation.

Bathymetric Distribution of Ahermatypic Corals

Littoral-neritic
- Rhizangiidae

Littoral-bathyal
- Oculinidae
- Dendrophylliidae

Littoral-abyssal
- Caryophylliinae
- Desmophyllinae
- Flabellidae

Neritic-bathyal
- Anthemiphylliidae
- Parasmiliinae
- Guyniidae

Neritic-abyssal
- Fungiidae
- Micrabaciidae
- Turbinoliinae

EVOLUTION

GENERAL DISCUSSION

The major source of information on the evolution of scleractinian corals, in the absence of information on the polyps of extinct forms, is the structure of the septa (Fig. 237). The latter are the most important skeletal structures and differ from suborder to suborder, family to family, in progressive and orderly fashion. Wall structures are of some value, but since the same type of wall may appear in analogous genera of widely separated families, or 2 different types may appear in the same corallum or even in the same corallite, they are of secondary significance (Fig. 259). Structures independent of the septa, such as dissepiments, epitheca, and basal plate, are of little value. Evolution within small, compact groups, such as the Rhipidogyridae, is best shown by morphogenetic trends (Fig. 260).

A large question in the evolution of anthozoan corals is the relation of scleractinians to rugosans: whether the scleractinians were filiated from the rugosans or whether they were independently derived from skeletonless anemone stocks—the scleractinians from the same stock that led to the living corallimorphs and actiniarians and the rugosans from earlier stocks leading to the zoanthids. Chronologically and to some extent ecologically, the scleractinians succeeded the rugose corals and many have wished to view them as redivived rugosans (Fig. 261). If this were so, it would seem that the ancestral stock should be found among the Late Paleozoic rugose corals, perhaps in the Permian calophyllids (plerophyllids), the specialized end-forms of a long-lived line of conservative solitary forms. To derive the Scleractinia from these it would have to be assumed that there was (1) a change in septal structure from laminar to subfenestrate, (2) a change in the pattern of mesenterial and septal insertion, and (3) a possible change in skeletal substance from calcite to aragonite. Further, this involves acceptance of a monophyletic derivation by sudden mutation on the ordinal level (typostrophism) affecting some early ontogenetic stage (proterogenesis) of the Late Permian ancestor. The Scleractinia, however, like the Rugosa, are seemingly not a monophyletic group. At the earliest recorded appearance of both orders, each already included widely divergent groups at the subordinal level, such as the columnariids in the Rugosa and the astrocoeniids in the Scleractinia. Supposed survival of aspects of the rugosan mode of septal insertion in scleractinians, indicative of transitional characters, seems more apparent than real. In the absence of any known corals in latest Permian and Early Triassic or other evidence clearly linking the known early scleractinians with Paleozoic forms, no definite conclusion as to origin is now warranted. At present it seems most probable that the Rugosa and Scleractinia were developed at different times from different lines or stocks of the ancient anemones.

ASTROCOENIINA

The isolated suborder Astrocoeniina, of exclusively colonial habit, is first known from the Middle Triassic, where it is represented by the Pinacophyllinae. This little group, with spinelike septal trabeculae, was either an early offshoot from the astrocoeniid line, or represented the primitive type of the suborder, but a considerable previous history

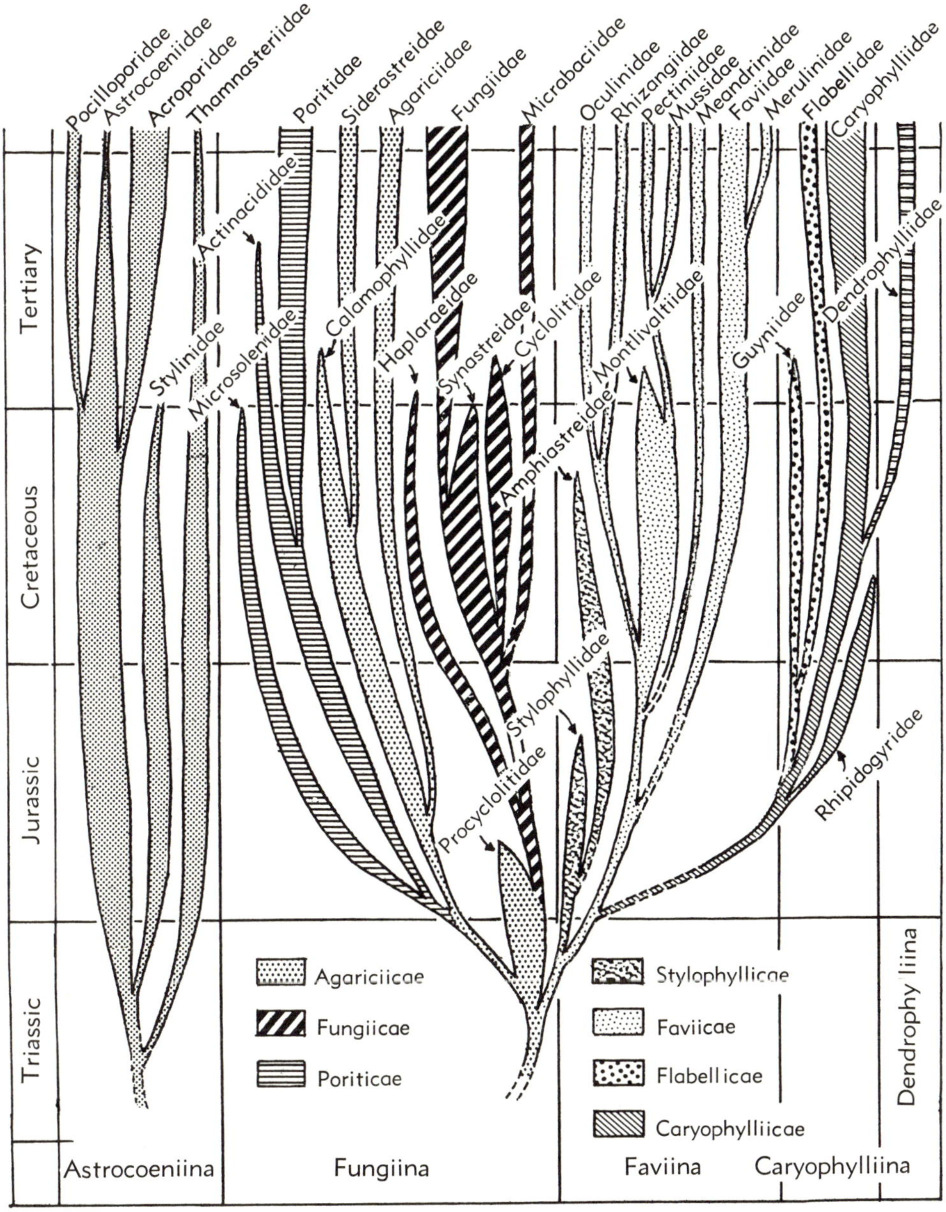

FIG. 259. Evolutionary pattern of the scleractinian suborders and families (198 mod.). [The author chooses to recognize superfamily *Faviicae* in preference to *Oculinicae,* which has priority, despite Copenhagen Decisions (1953).—ED.]

must be inferred, for no solitary astrocoeniids are yet known. The Astrocoeniinae, with more compact septa and closely united corallites, appeared in the Late Triassic, and within a relatively short time were worldwide in distribution. No marked change is apparent until the late Middle Cretaceous when the development of a more or less solid coenosteum between the corallites, the result of expansion of coenosarc, led to the Pocilloporidae. The Acroporidae came in the Late Cretaceous and were the result of

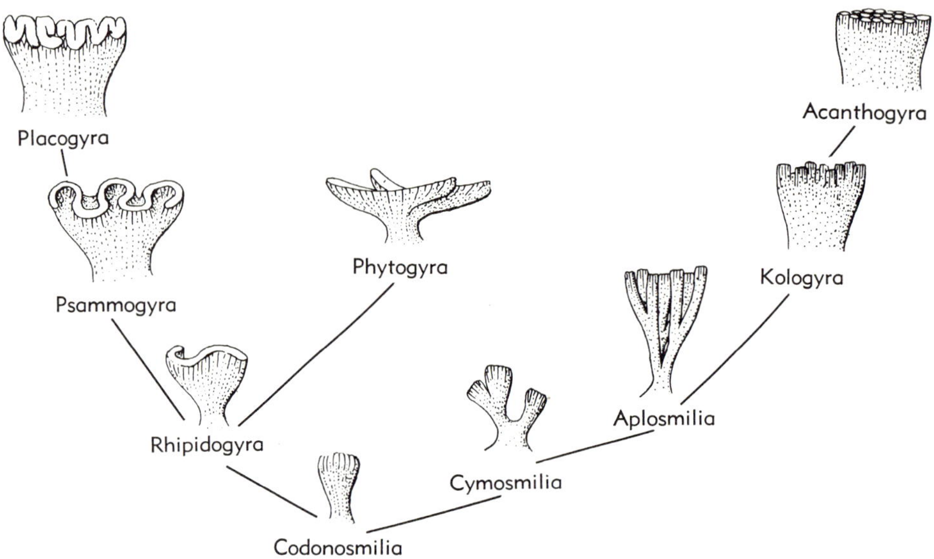

FIG. 260. Evolution in the Rhipidogyridae, a middle Mesozoic family (198mod.).

great compaction of the first-cycle septal trabeculae to form thin, laminar, smooth-margined but commonly reduced septa, accompanied by development of porous, canaliculate coenosteum. This group, on the basis of their light, very rapidly growing skeleton, and enormous numbers of small polyps forming huge colonies, is perhaps the most successful of all scleractinians.

The main astrocoeniid line weakened in the Tertiary, and there are now only 3 or 4 living species of *Astrocoenia,* its recent derivative *Stylocoeniella,* and *Stephanocoenia.*

The Thamnasteriidae, which very early were derived from astrocoeniids by development of the intratentacular mode of colony formation and septa supported by synapticulae, appeared in the Middle Triassic and persisted with little modification or diversification until the close of the Cretaceous. Late in the Cretaceous, however, canalicular modification of the extratentacular extensions of the mesenteries, similar to that which took place in the poritids, gave rise to *Ahrdorffia* and a few Tertiary and living forms such as *Psammocora,* without achieving any great success.

The Stylinidae, a distinct family of uncertain relations, were colonial corals with small corallites and relatively few, smooth-margined septa, suggesting an astrocoeniid connection. The structure of the earliest and most primitive of the Cyathophorinae, *Cassianastraea* of the Middle Triassic, further strengthens such a supposition. The Stylininae appeared in the Late Triassic, the Euheliinae in the Middle Jurassic, and the whole group was practically extinct and without issue by the end of the Mesozoic.

FUNGIINA

The suborder Fungiina consists of 3 superfamily groupings, Agariciicae, Fungiicae, and Poriticae, derived from the Procyclolitidae, an early Mesozoic central agariciid family. One of the earliest members of this family, *Triadophyllum* of the Middle Triassic, is a prototypic form with rather simply constructed septa in which each trabecula consists of a single vertical series of sclerodermites scarcely diverging from the vertical plane and closely placed side by side to form a continuous laminar sheet. Progressively, however, in later members of the family, the trabeculae were less closely united laterally in the septal plane and the sclerodermites were inclined more and more away from the plane, appearing on the face as granulations or ridges, so as to make fenestrate septa. The earlier stages of Late Triassic genera, such as *Margarophyl-*

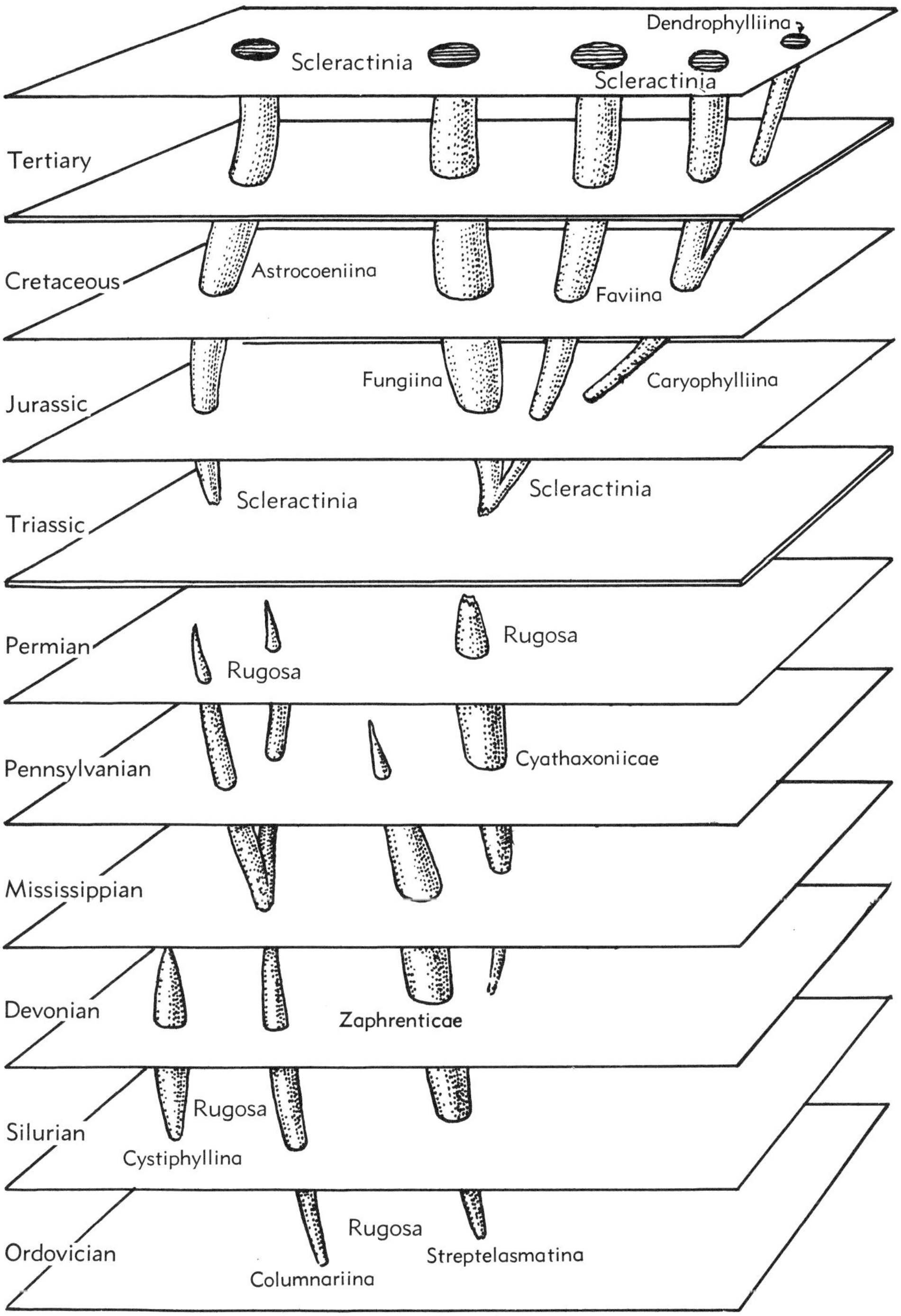

FIG. 261. Chronologic distribution of major groups of scleractinian and rugose corals (198mod.).

lia and Conophyllia, have the laminar *Triadophyllum*-type of septum and more fenestrate septa in later stages.

Early in the Jurassic, the procyclolitids diverged along 2 lines. In one, the septa were composed of mostly simple trabeculae, tending to become increasingly fenestrate and perforate. In the earliest family (Calamophylliidae) of this line, the septa have only scattered perforations. The Agariciidae, first known in the Middle Cretaceous *(Trochoseris)*, originated from calamophylliids with very compact septa, resulting from such close growth of the trabeculae that they were staggered in the septal plane. From the Early Jurassic calamophylliids developed the Microsolenidae with highly perforate, confluent septa. In the Middle Jurassic, mesenterial continuations beyond the tentacular ring began to break down *(Actinaraea)* into canaliculate gastro-vascular prolongations, resulting in the disappearance of definite septocostae outside the calice (Actinacididae). From the actinacidids in Late Cretaceous time came the modern, highly successful Poritidae, in which the extracalicular coenosteum diminished as the corallites became smaller and more numerous, and the number of septal trabeculae and septa diminished. Late Jurassic offshoots from the calamophylliids were the Siderastreidae, in which extratentacular mesenteries nearly disappeared as common synapticular walls developed between corallites, and compound trabeculae developed, coarsening the septa.

From the other procyclolitid line arose the Fungiicae of probable *Thecoseris* ancestry, in which the number of vertical series of sclerodermites in each trabecula increased from 1 or 2 to 4, 8, or even more, resulting in compound trabeculae whose emergent edges appeared as spinose or granulose teeth instead of the small, nearly smooth, beadlike teeth of the Agariciidae and Poriticae. The earliest in this line of development were the Synastreidae of the Middle Jurassic-Cretaceous, with nearly imperforate septa. Early offshoots in the Middle Jurassic were the Haplaraeidae with large, irregularly spaced septal perforations. A later one comprised the Cyclolitidae, Cretaceous-Early Tertiary ecologic predecessors of the Fungiidae, with regular septal perforations and broad flat-based coralla adapted for lying loose on soft or shifting substrates.

In the Middle Cretaceous, the axis of trabecular divergence, which had remained essentially vertical in the cyclolitids, became inclined towards the horizontal in some synastreids, resulting in coralla shaped like those of the cyclolitids but lacking epitheca. The septal structure in earlier forms was more or less perforate (as it still is in early developmental stages of modern fungiids) but later it became more compact. These were the earliest Fungiidae *(Cycloseris)*. Reduction of the solid septotheca to a more or less perforate synapticulotheca gave direct connection between oral and aboral surfaces, allowing development early in the Tertiary of large coralla characteristic of *Fungia* and its derivatives. Living fungiids pass through a very brief procyclolitid stage with epithecal wall, followed by a trochoid acrosmiliid stage and a short cycloserid stage during the anthocaulus and anthocyathus periods of ontogeny.

FAVIINA

The suborder Faviina had its origin in Triassic procyclolitids from which there are 2 possible lines of development: (1) through the Stylophyllidae to the Montlivaltiidae, or (2) directly through the Montlivaltiidae from which the stylophyllids may have been early derivatives. The Stylophyllidae first appeared in the Middle Triassic and were marked by septa having compound trabeculae different from those of the fungiids, by retention of the simple epithecal type of wall, and by more or less nonexsert septa with trabeculae either completely united vertically or discontinuous and appearing as more or less free spines, but never uniting as a fenestrate sheet. Later forms showed a trend towards the laminar septum. Abundant and widely distributed in the Late Triassic and Liassic, they declined and disappeared by Middle Jurassic time. The family Amphiastreidae, with septa similar to those of advanced stylophyllids and a tendency towards the aphroid condition in the corallites, are provisionally associated here. Many genera showed a greater development of one or more adjoining (dorsal?) septa, leading to the development of a lamellar columella in some later forms. The group as a whole was not a success—edge zone was never developed—and after attaining wide distribution in Late Jurassic and Early Cretaceous time, disappeared before the end of the Cretaceous.

The earliest montlivaltiids, of Late Triassic age, are not easily distinguished from procyclolitids with compact septa, but they soon developed an edge zone leading to septothecal or parathecal walls. The characteristic laminar septa of the faviids, with sharp, conical dentations, resulted from closely spaced trabeculae with sclerodermites laterally diverging only slightly from the septal plane. By Mid-Jurassic, the Montlivaltiidae were highly developed.

Some of the earliest derivatives of the montlivaltiids, the Faviidae, appearing in the Late Jurassic and increasing steadily to the present day, owe much of their success to broadening of the edge zone with consequent reduction of epitheca, and thorough exploitation of various modes of colony formation. The earlier forms were intratentacular in mode of colony formation (Faviinae), but early in the Cretaceous the extratentacular mode became dominant in the Montastreinae, which are artificially separated from the Faviinae solely on this distinction. The Faviidae will eventually be more properly reorganized on the basis of subfamilial lineages. Late in the Tertiary, the Merulinidae appeared, characterized by small compound trabeculae produced by very irregular divergence of the sclerodermites. The Trachyphylliinae and Agathiphylliinae are minor offshoots from the Faviidae.

The Mussidae, first definitely known early in Tertiary time, show the culmination of septal development in the main montlivaltiid line, with large polyps, and septa formed by a number of fan systems of trabeculae, each system forming a large dentation. An early Tertiary group was the Pectiniidae, with irregular, ragged, reduced dentations.

The Rhizangiidae are first known in the Early Cretaceous and are now widely distributed in the littoral and neritic zones. The earliest form, *Arctangia,* like the progenitors of other faviid groups, was small and solitary, with irregular septal dentations resulting from irregular divergence of sclerodermites, probably developed from some early montlivaltiid. In many later rhizangiid genera the entosepta are nearly smooth-margined, but the exosepta retain the more primitive, coarse teeth.

The Anthemiphylliidae as yet have no paleontological history, but the septal structure suggests a rhizangiid ancestry.

The Meandrinidae, which are distinguished by the minutely dentate septal margins, were seemingly derived from montlivaltiids late in the Jurassic. The group includes 2 short, but well-defined lineages of hermatypic corals with lamellar columella dating from the Cretaceous. The *Pachygyra* lineage lasted only into the Eocene; the *Meandrina* lineage still survives. The Dichocoeniinae are a very small group dating from the Late Cretaceous, distinguished by development of a heavy coenosteum.

CARYOPHYLLIINA

The earliest Caryophylliina, the most successful of all scleractinian groups in adaptation to extremes of environment and prolific in generic differentiation, are 2 genera which constitute the Thecocyathinae. *Thecocyathus,* of the Early Jurassic, had an epithecal wall. By the Middle Jurassic, *Discocyathus* had developed an edge zone and the beginnings of a septotheca. The first Caryophylliinae came in the Middle Jurassic, living in a neritic, nonreef environment. By the commencement of the Tertiary, some genera had become adjusted to the littoral zone, and on the other hand, penetration of the deep waters of the bathyal zone had begun.

The Turbinoliinae, first known in the Late Cretaceous, are a curious group of handsome little corals especially adapted to substrates affording few or very small sites for larval fixation. Except in the very earliest ontogenetic stages, these exclusively solitary corals are unattached, the edge zone completely covering the exterior of the corallum which lies loose on the bottom or partly imbedded in it.

Earlier in the Cretaceous appeared the small group of the Desmophyllinae, caryophyllians in which septal substitution was scarcely developed, with little or no development of columella and pali. The Parasmiliinae came later in the Jurassic from caryophyllians in which the relatively great vertical extent of the corallum provoked the development of dissepiments. The hermatypic Eusmiliinae are parasmilian derivatives.

Septal substitution had diminished in an even earlier offset from the caryophyllians during the Middle Jurassic, the Rhipidogyridae, a compact lineage extinct by the Early Cretaceous.

The ahermatypic Flabellidae, dating from the Cretaceous, contain a few neotenic forms which seem to have been derived from caryophyllians by arrested development of the edge zone, resulting in a wholly epithecal wall. The somewhat similar Guyniidae, first appearing in the Middle Jurassic, may represent a still earlier neotenic caryophyllian offshoot.

DENDROPHYLLIINA

The dendrophylliids were the last of the scleractinian suborders to evolve. About the middle of the Cretaceous, the sclerodermites of the septa of some caryophyllians became very irregular in development, diverging strongly sideward from the plane of the normally compact, laminar septum, especially in the vicinity of the wall with locally opening perforations. This brought about a thickening of the outer parts of the septa, and the septotheca gave way to an irregularly porous synapticulotheca. In this new group, the Dendrophylliidae, the substitution of entosepta for exosepta, a common occurrence in scleractinians, became very marked, and the Pourtalès plan of septal arrangement is characteristic of the suborder.

SUMMARY OF MAIN EVOLUTIONARY TRENDS

(1) *Increase or decrease in compactness of septa.* Increase in compactness of some procyclolitids led to the laminar septa of the Faviina and Caryophylliina. Irregular development of sclerodermites in the latter led in turn to the porous septa of the Dendrophylliina. Decrease in compactness in other procyclolitids led to the fenestrate septa of the Fungiina. A trend from simple to compound trabeculae can be observed in some family groups, and from many to few septa in others.

(2) *Suppression of primitive epithecal wall.* The development of an edge zone with septotheca, paratheca, or synapticulotheca, and coenosteal structures, accompanied by suppression of the primitive epithecal wall, is regularly observed in the Fungiina, Faviina, and Caryophylliina. Reduction of extramesenterial extensions to canalicular mazes, with modification of coenosteal structure, is found especially in some lines of the Astrocoeniina and Fungiina.

(3) *Development of colonial habit.* This is particularly notable in hermatypic corals, which display many modes of colony formation, many of which are significant by making possible large, efficient colonies.

Broadly speaking, the classification used here is based on the following features, in order of decreasing taxonomic importance:

(1) Structure of septa and ancillary structures such as wall, synapticulae, and costae; and nature of polyps: *suborders, superfamilies,* and some *families.*

(2) Primary mode of colony formation, and relative development and structure of extrathecal structures: *families.*

(3) Secondary modes of colony formation, involving shape of colonial coralla and lineages of trends, habit of colonial coralla, shape of solitary coralla, secondary septal structures such as pali and columella, relative development and form of dissepiments: some *families, subfamilies, genera,* and *subgenera.*

(4) Dimensions of corallites, number, arrangement and structural details of septa and costae, habit of colonial coralla and shape of solitary coralla, and features of the coenosteum: *species* and *subspecies, varieties, formae,* etc.

CLASSIFICATION

A tabular summary of suprageneric divisions recognized in the order Scleractinia, showing stratigraphic range of each as now known and including record of the number of genera and subgenera recognized in each taxonomic unit, follows. The presence of 10 genera in a family, for example, is indicated by "(10)" and occurrence of 6 genera and 2 subgenera in a family or other taxonomic division is shown as "(6; 2)."

Suprageneric Divisions of Scleractinia

Scleractinia *(order)* (450; 57). *M.Trias.-Rec.*

Astrocoeniina *(suborder)* (46; 14). *M.Trias.-Rec.*

Astrocoeniidae (13). *M.Trias.-Rec.*
Pinacophyllinae (2). *M.Trias.-U.Trias.*
Astrocoeniinae (11). *U.Trias.-Rec.*
Thamnasteriidae (4; 5). *M.Trias.-Rec.*
Pocilloporidae (4). *U.Cret.-Rec.*
Acroporidae (6; 2) *U.Cret.-Rec.*
Stylinidae (19; 7). *M.Trias.-Eoc.*
Cyathophorinae (6; 3). *M.Trias.-Eoc.*
Stylininae (8; 4). *U.Trias.-Cret.*
Euheliinae (5). *M.Jur.-L.Cret.*
Fungiina *(suborder)* (108; 12). *M.Trias.-Rec.*
Agariciicae *(superfamily)* (51; 10). *M.Trias.-Rec.*
Procyclolitidae (10). *M.Trias.-M.Jur.*
Calamophylliidae (23). *M.Jur.-Cret.*
Agariciidae (12; 5). *M.Cret.-Rec.*
Siderastreidae (6; 5). *Cret.-Rec.*
Fungiicae *(superfamily)* (35). *M.Jur.-Rec.*
Synastreidae (7). *M.Jur.-Rec.*
Cyclolitidae (5). *Cret.-Eoc.*
Haplaraeidae (9). *U.Jur.-Eoc., ?Oligo.-?Mio.*
Fungiidae (11). *M.Cret.-Rec.*
Micrabaciidae (3). *Cret.-Rec.*
Poriticae *(superfamily)* (22; 2). *Jur.-Rec.*
Microsolenidae (14). *Jur.-Cret.*
Actinacididae (4). *U.Jur.-Oligo.*
Poritidae (4; 2). *M.Cret.-Rec.*
Faviina *(suborder)* (178; 13). *M.Trias.-Rec.*
Stylophyllicae *(superfamily)* (28). *M.Trias.-Cret.*
Stylophyllidae (7). *M.Trias.-L.Jur.*
Amphiastreidae (21). *M.Jur.-Cret.*
Faviicae *(superfamily)* (150; 11). *M.Trias.-Rec.*
Montlivaltiidae (20). *M.Trias.-Eoc.*
Montlivaltiinae (15). *M.Trias.-Cret.*
Placosmiliinae (5). *U.Jur.-Eoc.*
Faviidae (58, 4). *M.Jur.-Rec.*
Faviinae (30; 2). *U.Jur.-Rec.*
Montastreinae (19, 2). *M.Jur.-Rec.*
Agathiphylliinae (4). *Eoc.-Mio.*
Trachyphylliinae (5). *Oligo.-Rec.*
Rhizangiidae (14; 3). *L.Cret.-Rec.*
Oculinidae (15; 2). *Cret.-Rec.*
Oculininae (11; 2). *Cret.-Rec.*
Galaxeinae (4). *Mio.-Rec.*
Meandrinidae (14; 2). *Cret.-Rec.*
Meandrininae (9). *Cret.-Rec.*
Dichocoeniinae (5; 2). *U.Cret.-Rec.*
Merulinidae (4). *Rec.*
Mussidae (17; 2). *?U.Jur., Eoc.-Rec.*
Pectiniidae (6). *Oligo.-Rec.*
Anthemiphylliidae (2). *Rec.*
Caryophylliina *(suborder)* (93; 18). *Jur.-Rec.*
Caryophylliicae *(superfamily)* (79; 19). *Jur.-Rec.*
Caryophylliidae (73; 19). *Jur.-Rec.*
Thecocyathinae (2). *Jur.-M.Cret.*
Caryophylliinae (27; 12). *U.Jur.-Rec.*
Turbinoliinae (16; 6). *U.Cret.-Rec.*
Desmophyllinae (8). *Cret.-Rec.*
Parasmiliinae (14). *U.Jur.-Rec.*
Eusmiliinae (6). *Eoc.-Rec.*
Rhipidogyridae (9). *M.Jur.-L.Cret.*
Flabellicae *(superfamily)* (12). *M.Jur.-Rec.*
Flabellidae (6). *Cret.-Rec.*
Guyniidae (6). *M.Jur.-Eoc.*
Dendrophylliina *(suborder)* (24; 2). *U.Cret.-Rec.*
Dendrophylliidae (24; 2). *U.Cret.-Rec.*

SYSTEMATIC DESCRIPTIONS

Order SCLERACTINIA Bourne, 1900

[Zoantharia aporosa et perforata M.EDW.-H., 1851; Madreporaria M.EDW.-H., 1857 *(partim)*; Hexacorallia HAECKEL, 1866 *(partim)*; Hexacorallidae DELAGE & HÉROUARD, 1901; Cyclocorallia SCHINDEWOLF, 1942]

Solitary or colonial Zoantharia with calcareous external skeleton secreted by the ectodermal body layer, consisting essentially of radial partitions or septa, which are intermesenterial in position and formed within upward infoldings of basal part of polyp column wall, and attendant supporting structures: basal plate, epitheca, dissepiments, synapticulae, and mural structures. Septa developed in ontogeny following pattern of mesenteries: additional septa after first 6 being inserted in all 6 primary mesenterial exocoeles in successive cycles of 6, 12, 24, 48, and so on, in dorsoventral order. *M. Trias.-Rec.*

The classification followed here is a modification of that proposed by VAUGHAN & WELLS (1943), and the 5 suborders are separable by characters stated in the following key.

Key for Recognition of Scleractinian Suborders

1. Septa composed of relatively few (up to 8) simple trabeculae; mostly colonial, with small corallites (diam. 1 to 3 mm.); polyps rarely with more than 2 cycles of tentacles, and lacking stomodaeal ridges .... ASTROCOENIINA
2. Septa composed of numerous trabeculae; solitary or colonial, with corallites generally larger in diameter than 2 mm.; polyps commonly with more than 2 cycles of tentacles, and stomodaea generally ridged.
   A. Synapticulae present
      i. Septa basically fenestrate, more or less perforate .... FUNGIINA

ii. Septa basically laminar, irregularly perforate .... DENDROPHYLLIINA

B. Synapticulae rarely present

i. Septal margins dentate .... FAVIINA

ii. Septal margins smooth .... CARYOPHYLLIINA

Suborder ASTROCOENIINA Vaughan & Wells, 1943

[*nom. transl.* WELLS, herein (*ex* Astrocoeniida VAUGHAN & WELLS, 1943)]

Colonial, rarely solitary: corallites small. Septa formed by relatively few (up to 6 or 8) simple or compound trabeculae, appearing as simple spines to solid laminae, inclined in series from wall rather than in divergent fan systems, dentate to practically smooth marginally; sclerodermites regularly continuous or irregularly divergent along trabecular axes. Polyps small, with smooth stomodaea, rarely with more than 12 tentacles arranged in a single ring. *M.Trias.-Rec.*

Family ASTROCOENIIDAE Koby, 1890

Colonial; hermatypic; phaceloid to cerioid by extratentacular budding. Corallite walls septothecal (except *Pinacophyllum*). Septa exsert (except *Pinacophyllum*), composed of one series of simple trabecular spines projecting inward and upward from the wall, those of lower cycles fusing in the septal plane to form nearly smooth- or beaded-margined laminae. Columella absent or styliform and continuous. Endothecal dissepiments tabular; exothecal dissepiments tabular when developed. *M.Trias.-Rec.*

Subfamily PINACOPHYLLINAE Vaughan & Wells, 1943

Primitive astrocoeniids with septa formed by distinct trabecular spines; columella absent. *M.Trias.-U.Trias.*

Pinacophyllum FRECH, 1890 [**Amplexus? parallelus* FRECH, 1889; SD DIENER, 1921]. Small phaceloid colonies. Corallites with wall of epitheca internally thickened by stereome. Septa not exsert. *U.Trias.*, Eu.——FIG. 262,*2*. **P. parallelum* FRECH, U.Trias. (Nor.), Aus.; *2a,b*, ×1, ×2 (50);*2c*, ×10, long. sec. (151).

Koilocoenia DUNCAN, 1884 [**Phyllocoenia decipiens* LAUBE, 1865] [=*Coelocoenia (nom. van.)*, *?Cyathocoenia* VOLZ, 1896 (*non* DUNCAN, 1867)]. Small, plocoid to cerioid colonies. Corallite walls septothecal, costae nonconfluent. Septa exsert. *M. Trias.-U.Trias.*, Eu.——FIG. 262,*4*. **K. decipiens* (LAUBE), U.Trias., Aus.; *4a*, ×2 (155); *4b*, transv. sec., ×5 (155); *4c*, long. sec., ×7 (151).

Subfamily ASTROCOENIINAE Koby, 1890

[*nom. transl.* FELIX, 1898 (*ex* Astrocoeniidae KOBY, 1890)]

Astrocoeniids with septa of lower cycles laminar. Columella generally present, continuous. *U.Trias.-Rec.*

Actinastrea ORB., 1849 [**Actinastrea goldfussi* (=*Astrea geminata* GOLDFUSS, 1826 *(partim)*] [=*Aplosastrea (partim)*, *Enallocoenia* ORB., 1849; *Cyathocoenia* DUNCAN, 1867 (*non* VOLZ, 1896); *Stelidioseris* TOMES, 1893; *?Toechastraea* VOLZ, 1896; *Praestephanocoenia* OPPENHEIM, 1930; *Platastrocoenia* GREGORY, 1930; *Araiocoenia* ALLOITEAU, 1948; *?Septastraeopsis* ALLOITEAU, 1954; *Astrocoenia auctt.* (*non.* M.EDW.-H., 1848). Cerioid; ramose, massive, or incrusting; by extratentacular budding. Columella generally well developed, styliform. *U.Trias.-Rec.*, cosmop.——FIG. 262,*6a,b*. *A. ramosa* (M.EDW.-H.), U.Cret.(Turon.), Gosau, Aus.; long. transv. secs., ×25 (42).——FIG. 262, *6c*. *A. matheyi* (KOBY), U.Jur.(Sequan.), Switz.; lateral aspect of septum, ×2 (71).——FIG. 262,*6d*. *A. guadalupae* (ROEMER), L.Cret.(Alb.), Tex.; ×5 (157).

Stylocoeniella YABE & SUGIYAMA, 1935 [**Stylocoenia hanzawai* YABE-S., 1933 (=*Stylophora armata* EHR., 1834)]. Like *Actinastrea* but with striate, columniform projections arising at junctions of adjacent calices. *Eoc.-Rec.*, USA-W.Indies-IndoPac. ——FIG. 262,*7a,b*. **S. armata* (EHR.), Rec., Bikini, Marshall I.; *7a*, ×2; *7b*, ×5 (163).

Platycoenia VAUGHAN, 1900 [**P. jacksonensis*]. Like *Actinastrea* but plocoid with echinulate coenosteum. *Eoc.-Mio.*, USA-Eu.

Isastrocoenia GREGORY, 1900 [**I. kachensis*]. Like *Actinastrea* but subplocoid with some intratentacular budding. Columella absent or very weak. *M. Jur.*, India-E.Afr.——FIG. 262,*3*. **I. kachensis*, M.Jur.(Bath.), Cutch; transv. sec., ×6 (65).

Columactinastraea ALLOITEAU, 1952 [**C. rennensis*] [=*Stephanastraea* FROMENTEL, 1886 (*non* ÉTALLON, 1862)]. Like *Actinastrea* but with pali before first 2 septal cycles in one crown. *Cret.*, Eu.

?Stephanastrea ÉTALLON, 1862 [**S. ramulifera*]. Like *Columactinastraea* but with pali in 2 crowns. *U.Jur.-Cret.*, Eu.

Stylocoenia M.EDW.-H., 1849 [**Astrea emarciata* LAMARCK, 1816] [=*Aplosastrea (partim)*, *Triphyllocoenia* ORB., 1849]. Like *Stylocoeniella* but much lighter corallum and thin laminar septa with acute dentations. *Eoc.-Mio.*, Eu.-W.Indies. ——FIG. 262,*8*. **S. emarciata* (LAM.), M.Eoc. (Lut.), Paris Basin, Fr.; *8a*, ×1; *8b*, ×5 (95).

Astrocoenia M.EDW.-H., 1848 [**Astrea numisma* DEFRANCE, 1826] [=*Goniocoenia* ORB., 1849]. Like *Stylocoenia* but lacking intercorallite pillars or styliform columella. *Eoc.-Mio.*, Eu.

Lithostrotionoides ALLOITEAU, 1952 [**L. tessieri*] Like *Astrocoenia* but walls very thin and zigzag in cross section; columella sublamellar. *U.Cret. (Maastr.)*, Senegal.

Stephanocoenia M.Edw.-H., 1848 [**Astrea intersepta* Lamarck, 1816 (*non Madrepora intersepta* Esper, 1719) (=**S. michelini* M.Edw.-H., 1848] [*Antillastraea* Duncan, 1884]. Plocoid to subcerioid, massive. Costae short. Septa with minutely dentate margins, one crown of 12 pali before first 2 septal cycles. Columella styliform. *Cret.-Rec.*, Eu.-USA-W.Indies.——Fig. 262,*1*. **S. michelini* M.Edw.-H., Rec., W.Indies, ×5 (95).

Columastrea Orb., 1849 [**Astrea striata* Goldf., 1827] [=*Columellastrea* Orb., 1850; *Columnastrea* M.Edw.-H., 1851, ?*Psammocoenia* M.Edw.-H., 1857; *Haldonia* Duncan, 1879; ?*Stylinopsis* Chevalier, 1954]. Like *Stephanocoenia*, but with one crown of 6 pali before second cycle. *U.Jur.-U.Cret.*, Eu.-USA.——Fig. 262,*5*. *C. striata* (M.Edw.-H.), U.Cret.(Turon.), Gosau, Aus.; transv. sec., ×12 (42).

Family THAMNASTERIIDAE Vaughan & Wells, 1943

Colonial, hermatypic; massive, ramose, or incrusting, growth by intratentacular mono- to tristomodaeal budding. Corallite walls absent or weakly defined by synapticular rings.

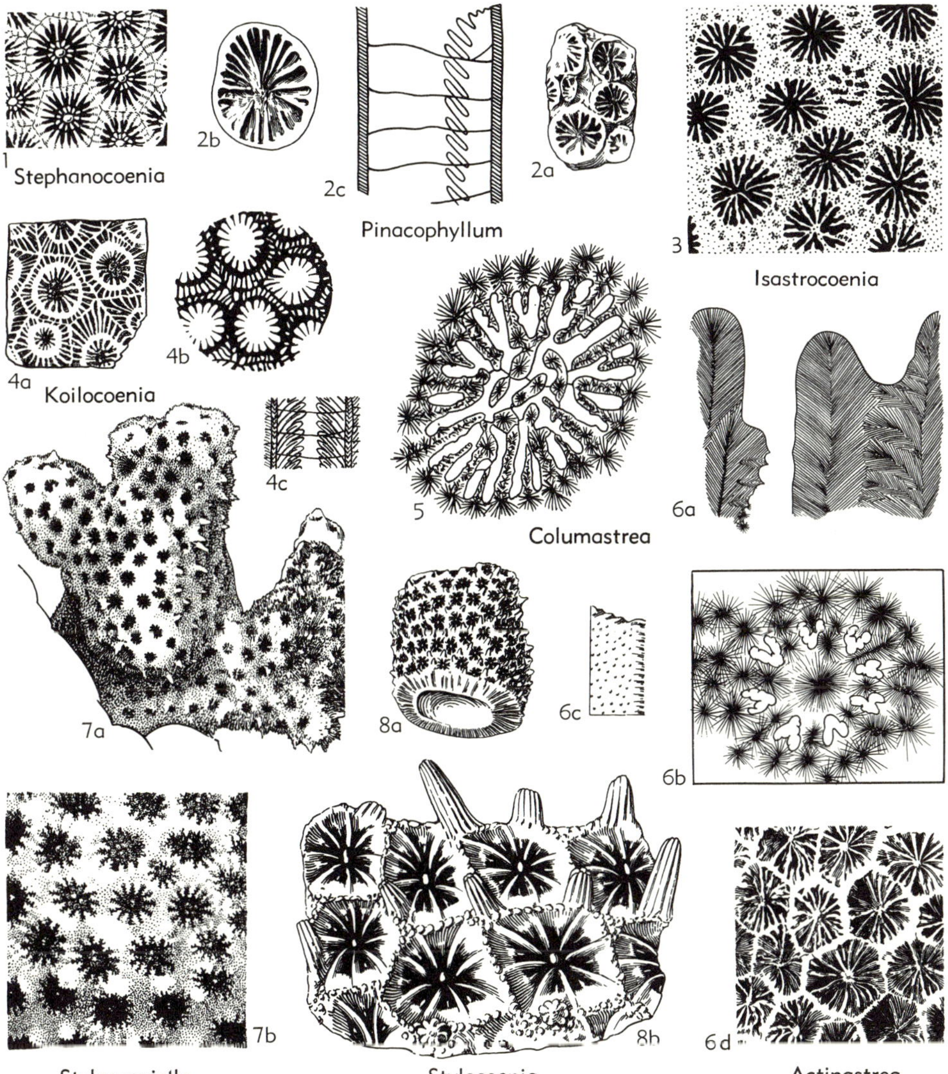

Fig. 262. Astrocoeniina: Astrocoeniidae (p. F370-F371).

Septa confluent between centers, similar in structure to those of the astrocoeniids, mostly of simple trabeculae with sclerodermites diverging laterally and producing granulations or continuous ridges, more or less porous, with beaded margins. Columella styliform or absent. Dissepiments tabular. *M.Trias.-Rec.*

Thamnasteria Lesauvage, 1823 [**T. lamourouxi* (=*Astrea dendroidea* Lamarck, 1821)] [=*Thamnastraea* Lesauvage, 1832; *Centrastrea, Dactylastrea, Dactylocoenia, Dendrastrea* Orb., 1849; *Morphastrea* Orb., 1850; *Holocoenia* M.Edw.-H., 1861; *?Thamnoseris* Fromentel, 1861; *Stylomaeandra* From., 1873; *Mesomorpha* Pratz, 1883; *Astraeofungia* Alloiteau, 1952]. Massive, ramose, or incrusting colonies. Corallite centers closely united by confluent, short, septocostae. Columella styliform. Septa with rows of granulations or discontinuous ridges parallel to margins. *M.Trias.-M. Cret.*, Eu.-Asia-Afr.-N.Am.-S.Am.

T. (Thamnasteria). *M.Trias.-M.Cret.*, cosmop.——Fig. 263,*1*. *T. (T.) rectilamellosa* (Winkler), U.Trias.(Nor.), Gosau, Aus.; *1a*, ×1; *1b*, transv. sec., ×5 (50).

T. (Astraeomorpha) Reuss, 1854 [**A. crassisepta*]. Septal ridges strong and continuous. *U.Trias.*, Eu.

Stereocaenia Alloiteau, 1952 [**Holocoenia collinaria* Fromentel, 1867]. Like *Thamnasteria* but septa thickened by development of irregular multiple series of large trabeculae. *L.Cret.*, Eu.——Fig. 263,*2*. **S. collinaria* (From.), L.Cret. (Neocom.), Yonne, Fr.; transv. sec., ×35 (172).

Ahrdorffia Trauth, 1911 [**Porites stellulata* Reuss, 1854]. Nodular or ramose. Corallites small, separated by ramifying, uniting septocostae. Columella a single trabecular pillar. *U.Cret.*, Eu.——Fig. 263,*5*. **A. stellulata* (Reuss), U.Cret.(Turon.), Gosau, Aus.; ×10 (112).

Psammocora Dana, 1846 [**Pavona obtusangula* Lamarck, 1816; SD M.Edw.-H., 1850]. Massive, foliaceous, ramose, columnar, or incrusting. Corallites small, structures as in *Ahrdorffia*. Collines low and rounded, enclosing several centers or series of centers. *Mio.-Rec.*, W.Indies-IndoPac.

P. (Psammocora), *Mio.-Rec.*, W.Indies-IndoPac.——Fig. 263,*3a*. *P. (P.) folium* Umbgrove, Rec., Indon.; ×5 (138).——Fig. 263,*3b*. *P. (P.) nierstraszi* Horst, Rec., Indon.; ×5 (141).

P. (Stephanaria) Verrill, 1867 [*pro Stephanocora* Verrill, 1866 (*non* Ehr., 1834)] [**Stephanocora stellata* Verrill, 1866]. Intercorallite areas very narrow and low, the collines enclosing one center or series, or absent. *Rec.*, IndoPac.

P. (Plesioseris) Duncan, 1884 [**Maeandroseris australiae* Rousseau, 1854]. Centers or series separated by high, acute, collines. *Rec.*, IndoPac.

Family POCILLOPORIDAE Gray, 1842

Plocoid, generally ramose, mostly hermatypic, colony formation by extratentacular budding. Septa rarely more than 2 cycles, reduced to narrow laminae or striae, even to spines. Columella styliform, vertically discontinuous. Coenosteum solid or vesicular. *U.Cret.-Rec.*

Stylophora Schweigger, 1819 [**Madrepora pistillata* Esper, 1797; SD M.Edw.-H., 1850] [=*Sideropora* Blainv., 1830; *Anthopora* Gray, 1835; *Enallastrea* Orb., 1849; *Phyllopora* T.-Woods, 1879; *Stylomadrepora* Oppenheim, 1923; *Palauastrea* Yabe & Sugiyama, 1941; *Stylacropora* Kühn, 1948]. Ramose to submassive. Corallites tending to spiral irregularly around branches, set in solid, spinose coenosteum. Septa of first cycle uniting with styliform columella. *Eoc.-Rec.*, Eu.-N.Am.-S.Am.-IndoPac.——Fig. 263,*8*. *S. mordax* Dana, Rec., Fanning I., Pac.; *8a*, ×0.5; *8b*, ×10 (148).

Seriatopora Lamarck, 1816 [**S. subulata;* SD M. Edw.-H., 1850]. Like *Stylophora* but corallites arranged in longitudinal series along branches. *Mio.-Rec.*, IndoPac.——Fig. 263,*7*. *S. hystrix* Dana, Rec., Fiji; ×0.5 (148).

Pocillopora Lamarck, 1818 [**P. acuta;* SD M.Edw.-H., 1850]. Ramose, rarely massive or incrusting. Calices commonly borne on short protuberances (verrucae) from the branches. Septa rudimentary, represented by striae or spines. Endothecal dissepiments solid or tabular. Columella a low boss. *Eoc.-Rec.*, W.Indies-IndoPac.——Fig. 263,*6a*. *P. elegans* Dana, Rec., Banda, E.Indies; ×0.5 (129).——Fig. 263,*6b*. *P. eydouxi* M.Edw.-H., Rec., Torres Strait, Austral.; ×10 (148).

Madracis M.Edw-H., 1849 [**M. asperula*] [=*Axhelia* M.Edw.-H., 1849; *Axohelia* M.Edw.-H., 1857; *Reussia* Duchaissing & Michelotti, 1860; *Pentalophora* Saville-Kent, 1871; *?Stylopsammia* Oppenheim, 1930]. Ramose to submassive. Septa well developed with smooth margins, higher cycles reduced to spines, arranged in groups of 8 or 10. Columella styliform, prominent. *U.Cret.-Rec.*, ?Eu.-IndoPac.-N.Am.-W.Indies.——Fig. 263,*4a*. **M. asperula*, Rec., 15 m. depth, Porto Rico; ×10 (195).——Fig. 263,*4b*. *M. mirabilis* Duch. & Mich., Rec., Porto Rico; ×1 (195).

Family ACROPORIDAE Verrill, 1902

Massive or ramose colonies by extratentacular budding; hermatypic. Corallites small, synapticulothecate, pseudocostate, slightly differentiated from coenosteum. Septa nonexsert, in 2 cycles, formed by simple spiniform trabeculae projecting inward and upward from vertical mural trabeculae, commonly fusing to form laminae. Columella absent or trabecular and weak. Dissepiments thin and tabular when developed. Coenosteum extensive, light reticulate, flaky, generally spinose or striate on surface. *U.Cret.-Rec.*

Acropora Oken, 1815 [**Millepora muricata* Linné, 1758; SD Verrill, 1902] [=*Madrepora auctt.* (*non* Linné, 1758), *Heteropora* Ehr., 1834; *Isopora* Studer, 1878; *Conocyathus, Distichocyathus, Eumadrepora, Lepidocyathus, Odontocyathus, Polystachys, Rhabdocyathus, Trachylopora, Tylopora*

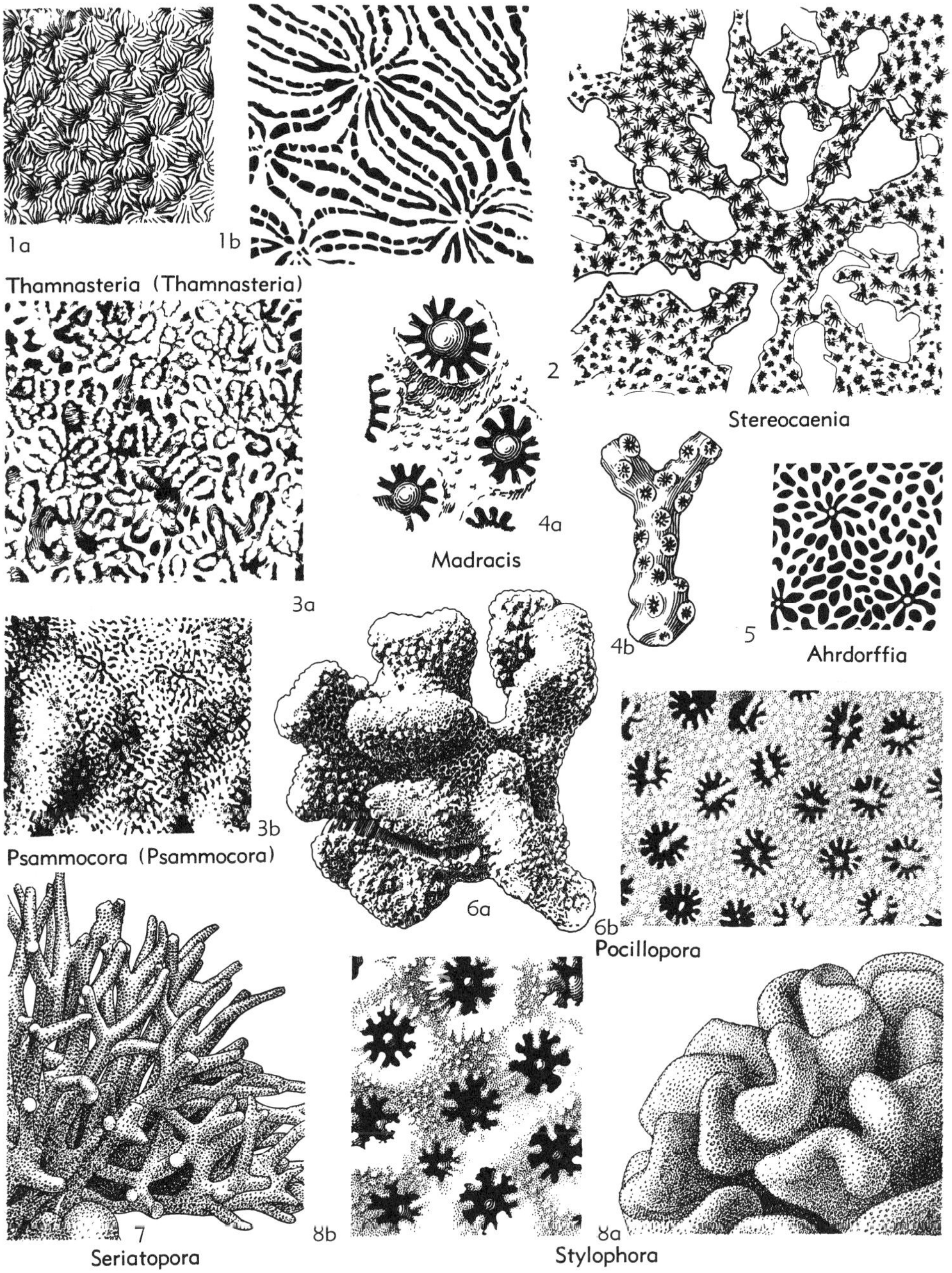

Fig. 263. Astrocoeniina: Thamnasteriidae, Pocilloporidae (p. F372).

Brook, 1893]. Ramose, rarely massive or incrusting; branches with an axial or leading corallite larger than the more numerous radial corallites budded from it, united by light, reticulate, spinose or pseudocostate coenosteum. Columella and dissepiments absent. *Eoc.-Rec.,* Eu.-W.Indies-N.Am.-IndoPac. The most important and protean genus of hermatypic corals, with more than 200 living species representing about 40 per cent of living scleractinians; not yet satisfactorily divided subgenerically.——Fig. 264,*2a. A. formosa* (Dana), Rec., Formosa; ×0.2.——**Fig. 264,*2b. A. tubicinaria*** (Dana), Bikini, Marshall I.; ×4.

Dendracis M.Edw.-H., 1849 [**Madrepora gervillei* Defrance, 1823]. Like *Acropora* but without persistent axial corallite, with coenosteum solid below. *Eoc.-Mio.,* Eu.-Afr.-E.Indies-W.Indies.——Fig. 264,*3. D. mammillosa* Reuss, M.Olig.(Rupel.), Mte.Grumi, Italy; ×2 (116).

Astreopora Blainv., 1830 [**Astrea myriophthalma* Lamarck; SD M.Edw.-H., 1850] [=*Phyllopora* Ehr., 1834; *Araeacis* M.Edw.-H., 1849; *Polysolenia* Reuss, 1867; *Astraeacis* Quenst., 1880; *Octotremacis* Gregory, 1899; *Annulopora, Palaestraeopora* Kühn & Andrusov, 1937]. Massive or subramose; no axial corallites. Coenosteum reticular, formed by outwardly inclined trabeculae, with spinose surface. Dissepiments tabular. Corallite walls solid. *U.Cret.-Rec.,* Eu.-W.Indies-IndoPac.

A. (Astreopora). *U.Cret.-Rec.*——Fig. 264,*1. A. listeri* Bernard, Rec., Bikini, Marshall I.; *1a,* ×0.4; *1b,* ×2.

A. (Cebuphyllia) Yabe & Sugiyama, 1941 [**C. chitanii*]. Like *A. (Astreopora),* but with well-developed trabecular columella and thick septa. *?Plio.,* Philip.——Fig. 264,*4. *A. (C.) chitanii,* transv. sec., ×3 (200).

?Gravieropsammia Filliozat, 1910 [**G. cornucopiae*]. Solitary, free, ceratoid. Corallite wall imperforate, with interseptal costae; a solitary *Astreopora* (?). *M.Eoc.,* Fr.

Montipora Quoy & Gaimard in Blainv., 1830 [**M. verrucosa* (*non Madrepora verrucosa* Lamarck, 1816); SD M.Edw.-H., 1850 (=*Manopora foveolata* Dana, 1846)] [=*Manopora* Dana, 1846]. Submassive, foliaceous, ramose or incrusting; no axial corallite. Corallite walls porous. Columella feeble or absent. Coenosteum reticular with strong vertical trabeculae, thin horizontal connections, surface spinulose or hirsute. No dissepiments. *Eoc.-Rec.,* IndoPac.——Fig. 265,*1a. M. foliosa* (Pallas), Rec., Cocos Keeling; ×0.1 (148).——Fig. 265,*1b. M. turgescens* Bernard, Rec., Bikini, Marshall I.; ×12.

Anacropora Ridley, 1884 [**A. forbesi*]. Like *Monti-*

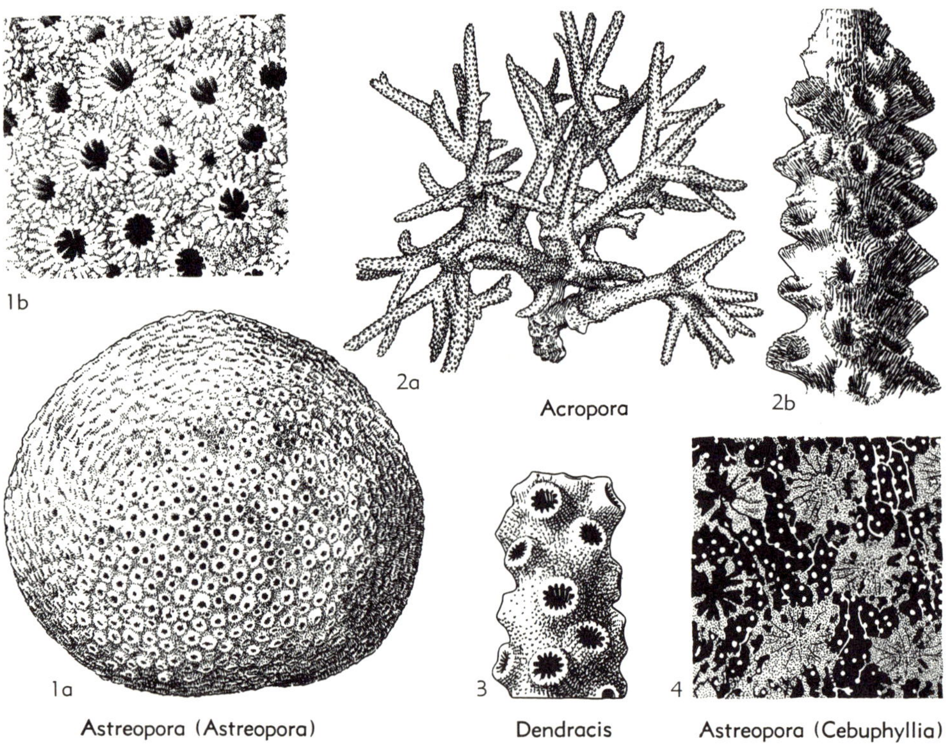

Fig. 264. Astrocoeniina: Acroporidae (p. F373-F374).

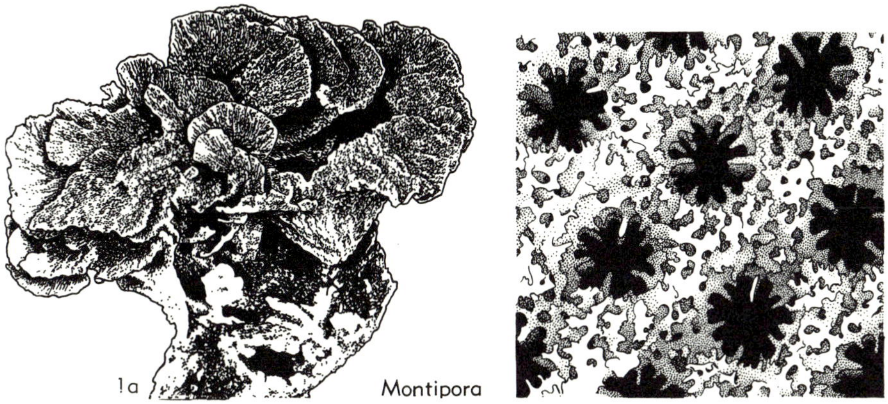

FIG. 265. Astrocoeniina: Acroporidae (p. F374).

pora but invariably ramose, with dense coenosteum. *Rec.,* IndoPac.

Family STYLINIDAE d'Orbigny, 1851

Colonial, hermatypic; colony formation by intra- and extratentacular budding. Septotheca present, generally thickened by stereome. Septa composed of a single fan system of simple trabeculae, upper margins smooth to beaded, laterally smooth or finely granulated. Endothecal dissepiments tabular when developed, rarely vesicular. *M.Trias.-Eoc.*

Subfamily CYATHOPHORINAE Vaughan & Wells, 1943

Ramose or plocoid, rarely cerioid; colony formation by extratentacular budding. Septa rarely more than 2 cycles, slightly exsert. No columella. Endothecal dissepiments thin and tabular, or absent. Coenosteum vesicular or tabular, surface spinose. *M.Trias.-Eoc.*

Cassianastraea VOLZ, 1897 [**Stylina reussi* LAUBE, 1865]. Small clumps consisting of ramose branches with crowded corallites lacking costae, united by some vesicular coenosteum. No endothecal dissepiments or columella. *U.Trias.,* Aus.——FIG. 266,*1.* **C. reussi* (LAUBE), Carn., Aus.; *1a,* ×2; *1b,* transv. sec., ×4 (155).

Procyathophora WEISSERMEL, 1928 [**Cyathophora fürstenbergensis* ECK, 1880]. Massive, plocoid; corallites noncostate, united by spinose tabular coenosteum. Septa rudimentary. Endothecal dissepiments tabular. *M. Trias.,* Ger.——FIG. 266,*11.* **P. fürstenbergensis,* Baden; *11a-c,* surf., transv. and long. secs., ×2 (156).

Cyathophora MICHELIN, 1843 [**C. richardi* (=*Astrea bourgeti* DEFRANCE, 1826)] [=*Pentacoenia* ORB., 1850; *Depaphyllum, Bathycoenia, Scyphocoenia* TOMES, 1883; *Orbignycaenia* ALLOITEAU, 1948]. Massive, plocoid. Costate corallites united by more or less costate tabular coenosteum. Septa well developed but rarely extending to corallite axis. No columella. Endothecal dissepiments tabular. *Jur.-Cret.,* Eu.-N.Am.-S.Am.

C. (Cyathophora). *Jur.-Cret.*——FIG. 266,*2. C. haysensis* WELLS, L.Cret.(U.Apt.), Tex.; *2a,b,* long. and transv. secs., ×2 (157).

C. (Holocystis) LONSDALE, 1849 [**H. elegans*] [=*Tetracoenia* ORB., 1850; *Cyathocoenia* FROMENTEL, 1884 (*non* DUNCAN, 1867)]. Like *C. (Cyathophora),* but with vesicular coenosteum and tetramerally arranged septa. *L.Cret.,* Eu.

C. (Cyathophoropsis) ALLOITEAU, 1947 [**C. hupei*]. Reputed to differ from *C. (Cyathophora)* by having intercalicular pillars as in *Stylocoenia* and *Stylocoeniella. L.Cret.(Gargas.),* Sp.

Heterocoenia M.EDW-H., 1848]**Lithodendron exiguum* MICHELIN, 1847] [=*Crinopora, Cyclocoenia* ORB., 1849; *Baryhelia* M.EDW.-H., 1857; *Hexasmilia* FROMENTEL, 1873; *Trinacis, Bacillastraea* QUENST., 1881; *?Pachycaenia, Heterocaeniopsis* ALLOITEAU, 1952]. Plocoid; massive, foliaceous, or ramose; corallites noncostate, united by extensive vesicular or nearly dense coenosteum. Septa rarely more than 2 cycles, trimerally arranged; some may be reduced to rudimentary spines. Columella absent or weakly developed *(Heterocaeniopsis). Cret.,* Eu.-N.Am.-Japan.——FIG. 266,*6. H. provincialis* (MICH.), U.Cret.(Turon.), Fr.; *6a,* ×1; *6b,* nat. long. sec., ×1.5 (181).

Miyakosmilia EGUCHI, 1936 [**M. densa*]. Like *Heterocoenia,* but corallites with low costae. *L. Cret.,* Japan.

Ewaldocoenia OPPENHEIM, 1921 [**E. hawelkai*]. Like *Heterocoenia,* but cerioid. *U.Eoc.,* Eu.

Subfamily STYLININAE d'Orbigny, 1851

[*nom. transl.* VERRILL, 1864 (*ex* Stylinidae ORB., 1851)]

Phaceloid, plocoid, meandroid, or rarely hydnophoroid; colony formation by extra-

or intratentacular budding. Septa exsert. Columella styliform, lamellar, or absent. Endothecal dissepiments subtabular or vesicular. Coenosteum, where developed, subtabular or vesicular. *U.Trias.-Cret.*

Aplophyllia Orb., 1849 [**Lithodendron dichotomum* Michelin, 1843 (*non* Goldf., 1827) (=**A. orbignyi* M.Edw.-H., 1851]. Phaceloid clumps by extratentacular budding. Costae distinct only near calicular margins, covered below by dense granular stereome. Columella feeble, trabecular. No coenosteum. *M.Jur.-U.Jur.*, Eu.

Stylosmilia M.Edw.H., 1848 [**S. michelini*]. Like *Aplophyllia* but with strong styliform columella. *M.Jur.-L.Cret.*, Eu.-S.Am.——Fig. 266,*7b-d*. **S. michelini*, U.Jur.(Raur.), Switz.; *7b-d*, ×2 (71).——Fig. 266,*7a*. *S. suevica* Becker, U.Jur. (Kimm.), Ger.; ×1 (7).

Stylina Lamarck, 1816 [**S. echinulata*] [=*Fascicularia* Lam., 1816; *Branchastrea* Blainv., 1830; *Gemmastrea* Blainv., 1834; *Stylopora* M'Coy, 1848; *?Adelocoenia, Conocoenia, ?Convexastrea, Cryptocoenia, ?Decacoenia, ?Dendrocoenia, Lobocoenia, ?Tremacoenia* Orb., 1849; *Acanthocoenia, ?Octocoenia, Pseudocoenia* Orb., 1850; *Platysmilia* Toula, 1889; *Dendroseris* Gregory, 1929]. Plocoid; massive or ramose; corallites protuberant, united by costate, subtabular coenosteum. Columella styliform or absent. Septa variously arranged (hexamerally, octamerally, etc.). *U.Trias.-L.Cret.*, Eu.-AsiaM.-Afr.-N.Am.-S.Am.

S. (Stylina). *U.Trias.-L.Cret.*——Fig. 266,*8*. *S. girodi* Étallon, U.Jur.(Raur.), Switz.; calices, ×1 (71).

S. (Heliocoenia) Étallon, 1859 [**H. variabilis*] [=*Dendrohelia* Étallon, 1860; *Pleurostylina, Psammohelia, ?Stylohelia* Fromentel, 1861]. Like *S. (Stylina)* but coenosteum noncostate, surface granulate. *U.Jur.-L.Cret.*, Eu.——Fig. 266,*9*. *S. (Heliocoenia) corallina* Koby, U.Jur. (Raur.), Switz.; long. sec., ×3 (71).

Elasmocoenia M.Edw.-H., 1851 [**Oculina explanata* Michelin, 1845; SD Wells, 1936]. Thick laminae produced by marginal extratentacular and some intratentacular budding. Corallites strongly costate, united by extensive vesicular exothecal dissepiments. No columella. *M.Cret.-U.Cret.*, Eu.

Myriophyllia Orb., 1849 [**Meandrina rastellina* Michelin, 1843] [=*Eugyra* Fromentel, 1857; *Ktenodema* Schafhäutl, 1877; *??Diplothecophyllia* Alloiteau, 1952]. Meandroid colonies produced by intratentacular polystomodaeal budding; series united directly by walls. Septa short. Columella lamellar, continuous, discontinuous, or absent. *U.Jur.-L.Cret.*, Eu.-Afr.-N.Am.-S.Am.——Fig. 266,*5*. *M. angustata* (Orb.), U.Jur.(Lusit.), Port.; ×1 (73).

Diplogyra Eguchi, 1936 [**D. lamellosa*]. Like *Myriophyllia* but with series separated by ambulacra. *L.Cret.*, Japan.

Felixigyra Prever, 1909 [**F. deangelisi*] [=*Eohydnophora* Yabe & Eguchi, 1936]. Hydnophoroid; series continuous; collines short, discontinuous. Columella weak, usually not developed. *L.Cret.-M.Cret.*, Eurasia-Tex.

?Haimesastraea Vaughan, 1900 [**H. conferta*]. Plocoid; massive or ramose colonies produced by extratentacular budding; corallites small, united by dense, costate coenosteum. Septa marked by strong lateral divergence of trabeculae from septal plane, alternately or oppositely, producing thin ridges or carinae parallel to margins. Columella trabecular. *L.Eoc.-M.Eoc.*, N.Am.-S.Am.

H. (Haimesastraea). *L.Eoc.-M.Eoc.*——Fig. 266, *10*. **H. conferta*, L.Eoc. (Wilcox.), Ala.; *10a*, ×0.5; *10b*, ×5 (144).

H. (Peruviastrea) Vaughan, 1922 [**P. peruviana*]. Like *H. (Haimesastraea)* but columella compressed, substyliform. *M.Eoc.*, Peru.

Subfamily EUHELIINAE de Fromentel, 1861

Dendroid colonies formed by extratentacular budding, generally in one plane. Coenosteum dense, surface nearly smooth or finely granulated. Hermatypic. *M.Jur.-L. Cret.*

Euhelia M.Edw.-H., 1850 [**Oculina gemmata* Michelin, 1843]. Colonies small; buds produced simultaneously on opposite sides of branches in pairs. Costae concealed below by stereome. Columella absent or weak. *M.Jur.-U.Jur.* Eu.

Tiaradendron Quenst., 1857 [**Lobophyllia germinans* Quenst., 1852]. Larger than *Euhelia*, with calices tending to alternate, and with 2 opposing septa (?directives) larger and lying in the plane of the branch. *U.Jur.*, Eu.-Afr.——Fig. 266,*4*. **T. germinans*, U.Jur.(Kimm.), Württ., Ger.; ×0.6 (110).

?Pseudogatheria Euchi, 1951 [**P. hiraigaensis*]. Fasciculate tufts produced by irregular budding at right angles. Columella formed by 2 to 3 trabecular papillae. *L.Cret.*, Japan.

Enallhelia M.Edw.-H., 1849 [**Lithodendron compressum* Goldf., 1829; SD M.Edw.-H., 1850]. Branches generally anastamosing with tendency to lie in one plane; calices alternately produced. Costae short, becoming covered by granular stereome. Columella styliform, well developed. *M.Jur.-L.Cret.*, Eurasia-S.Am.——Fig. 266,*3*. *E. tubulosa* Becker, U.Jur.(Kimm.), Württ., Ger.; *3a*, ×0.6; *3b*, ×3.

Stylangia Fromentel, 1857 [**S. neocomiensis*] [=*Prohelia* From., 1857]. Like *Enallhelia* but calices produced from only one side of branches. *L.Cret.*, Eu.-N.Am.-S.Am.

Suborder FUNGIINA Verrill, 1865

[*nom. correct.* Wells, herein (*ex* Fungiida Duncan, 1884; *ex* Fungacea Verrill, 1865)]

Solitary and colonial. Septa fenestrate, formed by simple or compound trabeculae united by simple or compound synapticulae, margins beaded or dentate. *M.Trias.-Rec.*

Superfamily AGARICIICAE Gray, 1847

[*nom. correct.* WELLS, herein (*ex* Agaricioidae VAUGHAN & WELLS, 1943, *nom. transl. ex* Agariciidae GRAY, 1847)]

Solitary and colonial. Septa fundamentally fenestrate, irregularly porous, formed by one (rarely more) fan system of mostly simple trabeculae, united by simple synapticulae, margins beaded, commonly laminar in later stages. *M.Trias.-Rec.*

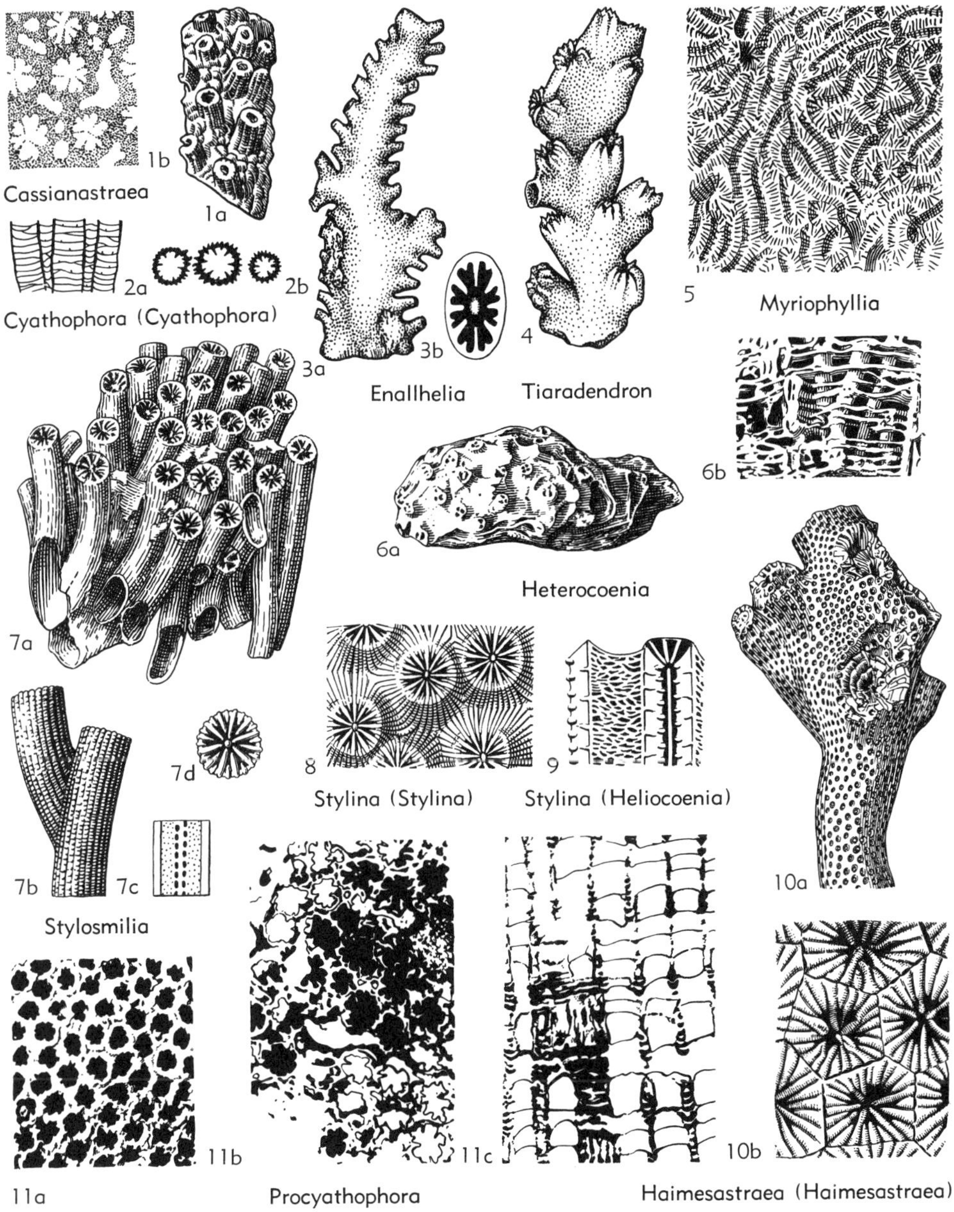

FIG. 266. Astrocoeniina: Stylinidae (p. F375-F376).

Family PROCYCLOLITIDAE Vaughan & Wells, 1943

Solitary and colonial; hermatypic. Corallite wall epithecal. Septa primarily formed by one fan system of simple trabeculae; compound trabeculae in advanced forms; fenestrate structure not strongly developed. Synapticulae simple, uncommon. Endothecal dissepiments well developed. Columella trabecular or absent, rarely otherwise. *M.Trias.-M.Jur.*

Triadophyllum WEISSERMEL, 1925 [**T. posthumum*]. Solitary, trochoid, free in ephebic stage. Septa few, slightly exsert, margins beaded, laterally striate or spinose. No columella. *M.Trias.*, Eu.——FIG. 267,*1*. **T. posthumum*, Anis., Ger.; *1a,b*, ×1 (156).

Gigantostylis FRECH, 1890 [**G. epigonus*]. Like *Triadophyllum* but one septum forms an elongate styliform columella. *U.Trias.*, Aus.——FIG. 267,*6*. **G. epigonus*, Nor., Aus.; *6a,b*, ×1 (50).

Margarophyllia VOLZ, 1896 [**Montlivaltia capitata* MÜNSTER, 1841]. Solitary, subcylindrical, fixed. Septa faintly fenestrate in later stages. *U.Trias.*, Eu.——FIG. 267,*4*. *M. crenata* (MÜNSTER), Carn., Aus.; *4a*, ×1; *4b*, transv. sec., ×10 (155).

Margarosmilia VOLZ, 1896 [**Cyathophyllum confluens* MÜNSTER, 1841; SD DIENER, 1921]. Like *Margarophyllia* but small phaceloid colonies formed by intratentacular budding. *M.Trias.-U.Trias.*, Eu.——FIG. 267,*3a*. *M. septanectens* (LORETZ), M.Trias.(Ladin.), Aus.; transv. sec., ×15 (204).——FIG. 267,*3b*. *M. zieteni* (KLIPSTEIN), M.Trias. (Ladin.), Aus.; ×1 (155).

Craspedophyllia VOLZ, 1896 [**C. cristata* (=*Omphalophyllia boletiformis* LAUBE, 1865, *non Montlivaltia boletiformis* MÜNSTER, 1841); SD VAUGHAN, 1905]. Like *Margarophyllia* but with septal structure as in *Triadophyllum*. Columella trabecular. *U.Trias.*, Eu.

Conophyllia ORB., 1849 [**Montlivaltia granulosa* MÜNSTER, 1841] [=*Omphalophyllia* LAUBE, 1865]. Like *Craspedophyllia* in early stages but septa developing more distinct fenestrate structure. *M.Trias.-U.Trias.*, Eurasia.——FIG. 267,*5*. *C. boletiformis* (MÜNSTER), M.Trias.(Ladin.), Aus.; *5a,b*, transv. and long. secs., ×2 (155).

Procyclolites FRECH, 1890 [**P. triadicus*] [=*Myriophyllia* VOLZ, 1896 (*non* ORB., 1849)]. Solitary or with some polystomodaeal intratentacular budding, fixed, subcylindrical to subturbinate. Septa laminar and imperforate with tendency toward porous fenestrate condition with simple (Trias. species) and compound (Trias.-Jur. species) trabeculae, commonly with strong lateral ridges. *U.Trias.-L.Jur.*, Eurasia.——FIG. 267,*2*. **P. triadicus*, U.Trias. (Nor.), Aus.; *2a*, corallum, ×1; *2b*, long. sec., ×2 (50).

Thecoseris FROMENTEL & FERRY, 1869 [**T. patellata;* SD FROM., 1870]. Like *Procyclolites* but compound trabeculae well developed; septa porous, united by simple trabeculae. *L.Jur.-M.Jur.*, Eu.

Tricycloseris TOMES, 1878 [**T. anningi*]. Like *Thecoseris* but small thamnasterioid colonies formed by incomplete mono- to tristomodaeal intratentacular budding. *L.Jur.*, Eng.

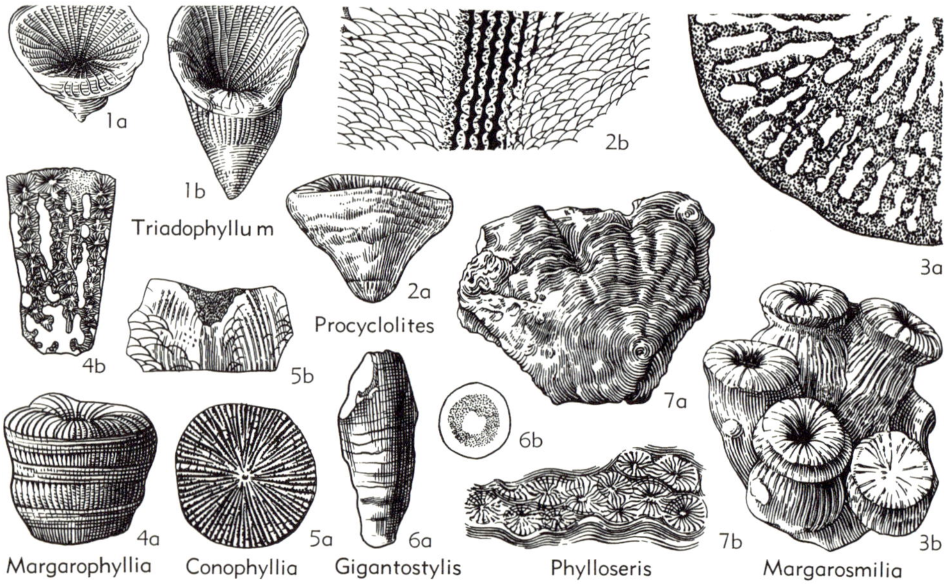

FIG. 267. Fungiina (Agariciicae): Procyclolitidae (p. F378-F379).

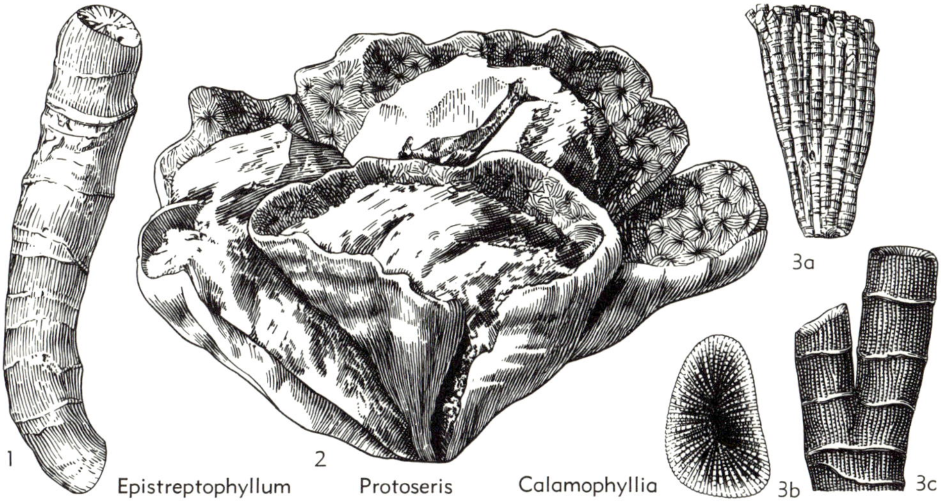

FIG. 268. Fungiina (Agariciicae): Calamophylliidae (p. F379-380).

Phylloseris TOMES, 1882 [**P. rugosa*]. Structures as in *Thecoseris* and *Procyclolites* but corallites smaller, forming foliaceous colonies by ?extratentacular budding. *M.Jur.*, Eng.——FIG 267,7. **P. rugosa*, Baj., Eng.; *7a*, ×0.6; *7b*, ×2 (131).

Family CALAMOPHYLLIIDAE Vaughan & Wells, 1943

Solitary and colonial; hermatypic. Colony formation mostly by intratentacular budding, linkages generally lamellar. Wall synapticulothecal, tending to be thickened secondarily and solid, costate. Septa fenestrate, formed by one fan system of simple trabeculae, axis of divergence vertical or slightly inclined and coinciding with the wall, more or less perforate internally, upper margins beaded, laterally granulated or ridged. Synapticulae sparse except near walls. Endothecal dissepiments thin or absent. Exothecal dissepiments and coenosteum present in some forms. Columella trabecular, feeble or well developed. *M.Jur.-Cret.*

Cyclastraea ALLOITEAU, 1952 [**Cyclolites spinosa* FROMENTEL, 1864]. Solitary, cupolate, with basal epitheca. Septa imperforate with strong dentations and vertical carinae. *M.Cret.(Cenom.)*, Fr.

Epistreptophyllum MILASCHEWITSCH, 1875 [**E. commune*; SD WELLS, 1936] [=*Phegmatoseris* MILASCHEWITSCH, 1876; *Phragmatoseris* DUNCAN, 1884; *Lithoseris* KOBY, 1886; *Frechia, Metethmos, Protethmos* GREGORY, 1900; *Calamosmilia* KOBY, 1904]. Solitary, turbinate to cylindrical. Epitheca more or less developed. Dissepiments present. *M. Jur.-M.Cret.*, Eurasia-N.Am.——FIG. 268,*1*. *E. typum* (KOBY), U.Jur.(Sequan.), Port.; ×1 (73).

Gonioseris DUNCAN, 1872 [**G. angulata*]. Solitary, free, hexagonal-pyramido-cupolate with flat base. Basal costae prominent, parallel, extending inward at right angles from sides of each basal segment. *M.Jur.*, Eng.

?**Antilloseris** VAUGHAN, 1905 [**Turbinoseris eocaenica* DUNCAN, 1873]. Solitary, turbinate, subceratoid or subpatellate, generally compressed. No columella or dissepiments. *Eoc.*, W.Indies-N.Am.

Dermosmilia KOBY, 1884 [**D. divergens;* SD WELLS, 1936[[=*Calamophylliopsis* ALLOITEAU, 1952]. Dendroid colonies produced by complete monostomodaeal intratentacular budding, corallites united basally. *U.Jur.-L.Cret.*, Eu.——FIG. 269,*2*. *D. crassa* (ÉTALLON), U.Jur.(Raur.), Switz.; ×0.3 (71).

Sematethmos GREGORY, 1900 [**S. sinuosa*]. Low colonies formed by circumoral budding with 1 or 2 concentric series of centers. *M.Jur.-U.Jur.*, Eurasia.

Protoseris M.EDW.-H., 1851 [**P. waltoni*] [=*Illinodendron* QUENST., 1880]. Colony formation at first circumoral followed by polystomodaeal budding, several series of centers formed between walls with common forking and formation of collines, resulting in foliaceous or contorted fronds. *U.Jur.*, Eu.——FIG. 268,*2*. **P. waltoni*, Coral., Eng.; ×0.6 (94).

?**Synhelia** M.EDW-H., 1849 [**Lithodendron gibbosum* GOLDF., 1829]. Slender ramose, like *Oculina* in form, corallites superficial, with confluent septocostae. Structures apparently calamophylliid but secondarily thickened by heavy stereome. *Cret.*, Eu.

Viminohelia GEYER, 1954 [**V. seminuda*]. Like *Synhelia* but espalier-like in form, with calices on one side of corallum. *U.Jur.*, Eu.

Calamophyllia BLAINV., 1830 [**C. striata;* SD M. EDW.-H., 1850] [=*Rhabdophyllia* M.EDW.-H., 1851; *Calamoseris* ALLOITEAU, 1952]. Phaceloid clumps produced by complete polystomodaeal budding, centers permanently monocentric. *M.Jur.-Cret.,* Eurasia-Afr.-N.Am.-W.Indies.——FIG. 268,*3. C. stokesii* M.EDW.-H., U.Jur.(Raur.), Eng.; *3a,* ×0.3; *3b,* ×2; *3c,* ×1 (94).

Latomeandra M.EDW.-H., 1849 [**Lithodendron plicatum* GOLDF., 1827] [=*Latimaeandra* M.EDW.-H., 1857; *Chorisastrea* FROMENTEL, 1861; *Thorisastrea* ÉTALLON, 1864; *Jugodendron, Plicodendron,* QUENST., 1880]. Like *Calamophyllia* but corallites mostly di- or tricentric. *M.Jur.-L.Cret.,* Eu.——FIG. 269,*6. *L. plicatum,* U.Jur.(Kimm.), Ger.; ×1 (7).

Gyrodendron QUENST., 1880 [**G. lobatum;* SD WELLS, 1936]. Like *Latomeandra* but polycentric condition mostly permanent, corallum fasciculate, formed by clusters of polycentric corallites. *U.Jur.,* Eu.——FIG. 269,*4. *G. lobatum,* Kimm., Ger.; ×1 (7).

Comophyllia ORB., 1849 [**C. elegans*] [=*Dimorphophyllia* BECKER, 1875; *Archaeoseris* GREGORY, 1900]. Circumoral followed by intramural budding, with irregularly radiating collines between groups of serial centers. *U.Jur.,* Eu.——FIG. 269,*3. C. polymorpha* (KOBY), U.Jur.(Raur.), Port.; ×1 (73)

Rabdastrea ÉTALLON, 1859 [**R. jurensis;* SD WELLS, 1936]. Submassive subplocoid colonies, with mono- to tricentric corallites closely united nearly to the calices by costae and exothecal dissepiments. *U.Jur.,* Eu.

Baryphyllia FROMENTEL, 1857 [**Barysmilia gregaria* ORB., 1850; SD WELLS, 1936]. Like *Rabdastrea* but corallites monocentric. *Cret.,* Eu.

Ovalastrea ORB., 1849 [**Astrea caryophylloides* GOLDF., 1827] [=*Ebrayia* FERRY, 1870; *Favoidioseris* WELLS, 1933; *Favioseris* WELLS, 1934; *Ambiguastraea* ALLOITEAU, 1952]. Plocoid colonies by mono- to tristomodaeal budding, corallites mono- to tricentric, united by costae and exothecal dissepiments. *M.Jur.-Cret.,* Eurasia-Afr.-N.Am.-W.Indies.——FIG. 270,*2a. O. bihinensis* (GREGORY), U.Jur. (Kimm.), Br.Somali.; transv. sec., ×2 (130).——FIG. 270,*2b. *O. caryophylloides,* U.Jur.(Kimm.), Ger.; ×1 (130).

Isastrea M.EDW-H., 1851 [**Astrea helianthoides* GOLDF., 1827; SD QUENST., 1857] [=*Desmastraea* FROMENTEL, 1865; *Placastrea* STOLICZKA, 1873; *Platastrea* TOMES, 1886]. Laminar to massive cerioid colonies; corallite walls septothecal to parathecal, but commonly partly absent with septa confluent; corallites mono- or dicentric. *M.Jur.-Cret.,* Eu.-Afr.-N.Am.——FIG. 269,*1a. I. limitata* (LAMX.), M.Jur.(Bath.), Eng.; ×2 (94).——FIG. 269,*1b. I. bonanomii* (KOBY), U.Jur.(Raur.), Switz.; lat. aspect of septum, ×4 (71).

Periseris FERRY, 1870 [**Agaricia elegantula* ORB., 1850] [=?*Fungiastraea* ALLOITEAU, 1952]. Like *Isastrea* but thamnasterioid, corallite walls absent at surface, parathecal below; calices superficial. Septa laterally carinate. *M.Jur.-Cret.,* Eu.-N.Am.-S.Am.——FIG. 269,*5. P. irregularis* (FELIX), L. Cret.(Apt.), Tex.; transv. sec., ×2 (157).

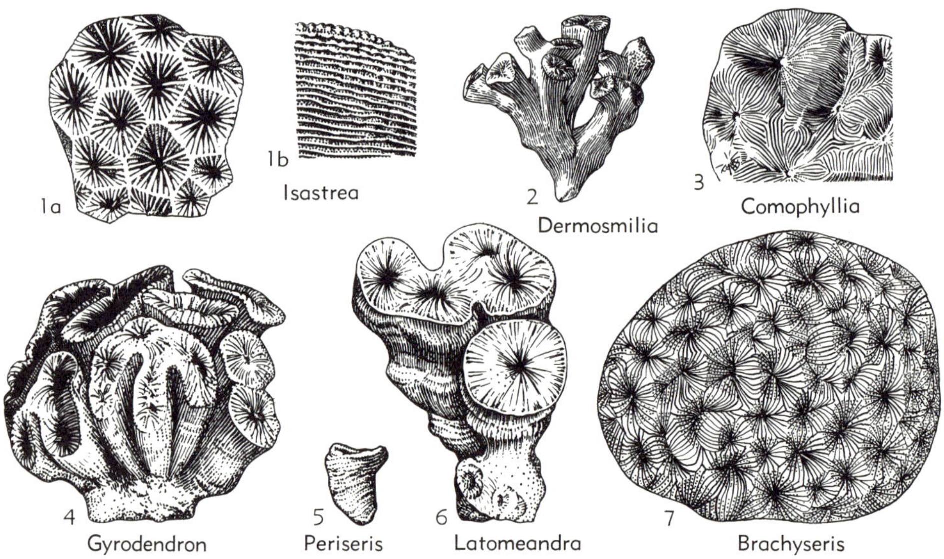

FIG. 269. Fungiina (Agariciicae): Calamophylliidae (p. F379-F381).

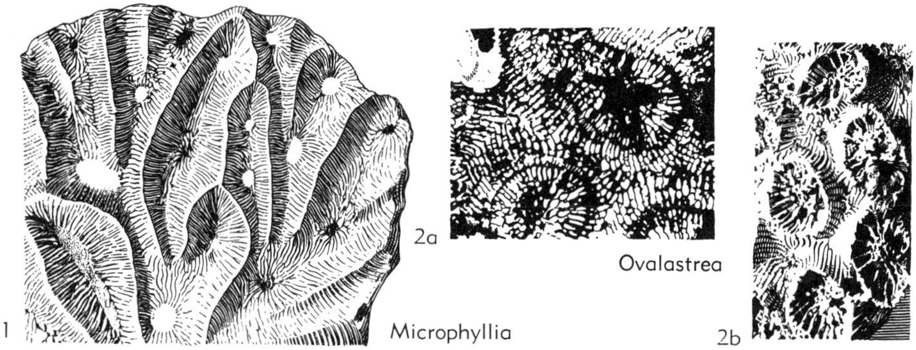

FIG. 270. Fungiina (Agariciicae): Calamophyllidae (p. F380-F381).

Ferrya FROMENTEL in FERRY, 1870 [*F. burgundiae]. Like *Ovalastrea* but meandroid, the calicular series separated by broad ambulacra. *M.Jur.*, Eu.

Stibastrea ÉTALLON, 1859 [*S. edwardsi]. Meandroid, derived from *Latomeandra* by incomplete separation of corallites with intramural budding. Ambulacra narrow, costate. *U.Jur.*, Eu.

Stiboria ÉTALLON, 1862 [*S. suprajurensis]. Like *Stibastrea* but ambulacra granulated, noncostate. *U.Jur.*, Eu.

Microphyllia ORB., 1849 [*Meandrina soemmeringii MÜNSTER in GOLDF., 1829] [=*Axophyllia* ORB., 1849; *Comastrea* FROMENTEL, 1862; ?*Koilomorpha* ALLOITEAU, 1952]. Like *Stibastrea* but lacking ambulacra; collines subacute; series in single rows of centers in the valleys, or less commonly several rows or groups. *U.Jur.*, Eurasia-Afr.-S.Am.——FIG. 270,*1*. **M. soemmeringii*, U.Jur.(Kimm.), Ger.; ×1 (7).

Brachyseris ALLOITEAU, 1947 [*Latimeandra morchella REUSS, 1854]. Like *Microphyllia*, but series short (1 to 3 centers), surrounded by acute collines; centers with trabecular linkage. *Cret.*, Eu. ——FIG. 269,*7*. **B. morchella*, U.Cret.(Turon.), Gosau, Aus.; ×1 (112).

Family AGARICIIDAE Gray, 1847

Solitary and colonial; hermatypic. Colony formation mainly by intratentacular budding. Wall synapticulothecal, usually becoming solid, or absent. Septa formed by one fan system of simple trabeculae, rarely porous, margins beaded, directly confluent between centers, united by some compound synapticulae. Endothecal dissepiments mostly absent. Columella trabecular or absent. *M.Cret.-Rec.*

Trochoseris M.EDW.-H., 1849 [*Anthophyllum distortum MICHELIN, 1844] [=*Gyroseris* REUSS, 1854; *Palaeoseris* DUNCAN, 1870; *Trochoseropsis* SÖHLE, 1897]. Solitary, turbinate or trochoid, fixed. Columella papillose or weak. *M.Cret.-Oligo.*, Eurasia-W.Indies.

Vaughanoseris WELLS, 1934 [*V. catadupensis]. Solitary, discoid-patellate, free. Well-developed trabecular columella. *U.Cret.*, W.Indies.——FIG. 272,*1*. **V. catadupensis*, Camp., Jamaica; *1a-c*, ×0.5 (198).

Elliptoseris DUNCAN, 1880. [*E aperta]. Like *Trochoseris* but lacking columella. *Eoc.*, India.

Brachyphyllia REUSS, 1854 [*B. dormitzeri; SD REUSS, 1864]. Small subplocoid colonies formed by extratentacular budding. *U.Cret.*, Eu.——FIG. 271,*2*. *B. depressa* REUSS, Turon., Gosau, Aus.; *2a,b*, ×1 (112).

Cyathoseris M.EDW.-H., 1849 [*Pavonia infundibuliformis BLAINV., 1830] [=*Dimorphophyllia* REUSS, 1864; ?*Spinellia* D'ACHIARDI, 1867; *Pterastrea* DUNCAN, 1880; *Pratzia* DUNCAN, 1884; *Mycetoseris* REIS, 1889; *Axoseris* OPPENHEIM, 1901; *Kühnophyllia* WELLS, 1937]. Like *Trochoseris* but expanding colonies formed by circumoral, followed by marginal or intramural budding with terminal forking. Collines more or less radiating. *U.Cret.-Mio.*, Eurasia.——FIG. 271,*1a,b*. *C. irradians* (REUSS), M.Oligo.(Rupel.), Italy; *1a*, ×0.6; *1b*, ×3 (116).——FIG. 271,*1c*. *C. centrifuga* (REUSS), M.Oligo.(Rupel.), Italy; ×0.6 (116).——FIG. 271, *1d,e*. *C. applanata* REUSS, M.Oligo.(Rupel.), Italy; ×0.6 (116).——FIG. 271,*1f*. *C. oxylopha* (REUSS), M.Oligo.(Rupel.), Italy; ×0.6 (116).

Hydnophorabacia D'ACHIARDI, 1875 [*H. variabilis]. Hydnophoroid; colony formation by circummural budding. *Eoc.*, Eu.

Agaricia LAMARCK, 1801 [*Madrepora undata ELLIS & SOLANDER, 1786; SD M.EDW.-H., 1850] [=*Undaria* OKEN, 1815; *Mycedia* DANA, 1846]. Colony formation by circumoral budding followed by marginal budding, forming foliaceous unifacial or bifacial fronds or submassive coralla. Collines discontinuous, enclosing several centers. Corallite walls well developed, calices inclined slightly toward margin. *Mio.-Rec.*, W.Indies.——FIG. 273,*2*. *A. agaricites* (PALLAS), Rec., Fla.; ×1 (151).

Heterogyra REUSS, 1868 [*H. lobata]. Colony for-

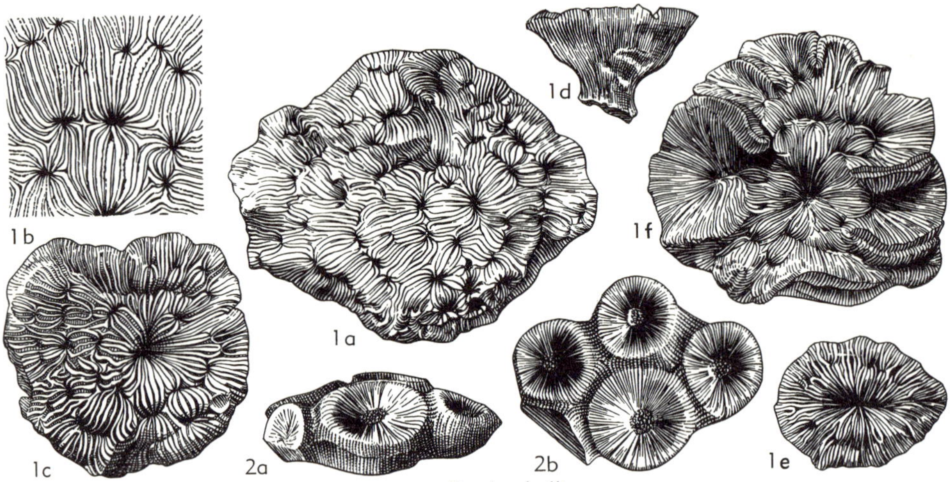

FIG. 271. Fungiina (Agariciicae): Agariciidae (p. F381).

mation by intramural budding with lateral branching and terminal forking; series generally discontinuous, laterally free or closely united with acute collines. *Oligo.*, Eu.——FIG. 273,*1*. **H. lobata*, Rupel., Italy; *1a,b*, ×0.6 (116).

Pavona LAMARCK, 1801 [**Madrepora cristata* ELLIS & SOLANDER, 1786 (=*M. cactus* FORSKÅL, 1775); SD VAUGHAN, 1918] [=*Pavonia* LAM., 1816; *Lophoseris* M.EDW.-H., 1849; *Pseudastraea* REUSS, 1864; *Reussastraea* D'ACHIARDI, 1875; *Heteroseris* FERRY, 1870]. Massive to foliaceous with bifacial fronds. Corallite walls absent. Collines radiating, discontinuous. *Oligo.-Rec.*, Eu.-W.Indies-IndoPac.

P. (Pavona). *Oligo.-Rec.*——FIG. 273,*3*. *P. frondifera* LAM., Rec., Jaluit, Marshall I.; *3a*, ×0.6; *3b*, ×4.

P. (Polyastra) EHR., 1834 [**Polyastra venosa*] [=*Tichoseris* QUELCH, 1884; *Asteroseris* VERRILL, 1902; (*non* FROMENTEL, 1867)]. Like *Pavona* but centers in short or submeandrine series, bounded by well-developed, acute walls. Corallum massive or incrusting. *Rec.*, IndoPac.

P. (Pseudocolumnastraea) YABE & SUGIYAMA, 1933 [**Pseudocolumnastraea yamanarii*]. Like *Pavona* but corallites bounded by distinct walls, appearing plocoid; no collines. *Rec.*, IndoPac.

Leptoseris M.EDW.-H., 1849 [**L. fragilis*] [=*Haloseris, Helioseris* M.EDW.-H., 1849; *Domoseris, Cylloseris* QUELCH, 1886; *Folioseris* REHBERG, 1892; *Agariciella* MA, 1937]. Circumoral followed by marginal budding, forming thin, unifacial folia, crateriform to digitate. Centers superficial, protuberant in some species; collines poorly developed. *Oligo.-Rec.*, W.Indies-IndoPac.——FIG. 274,*1*. *L. hawaiiensis* VAUGHAN, Rec., depth 45 m., Hawaii; *1a,b*, ×0.6 (146).

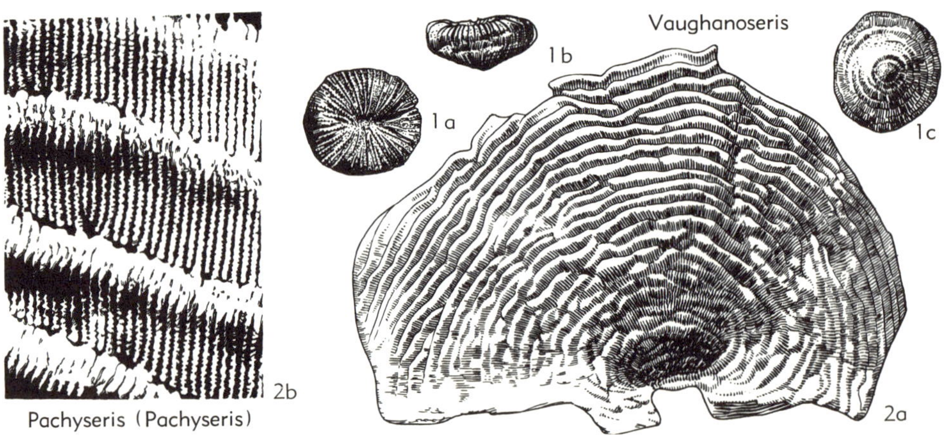

FIG. 272. Fungiina (Agariciicae): Agariciidae (p. F381-F383).

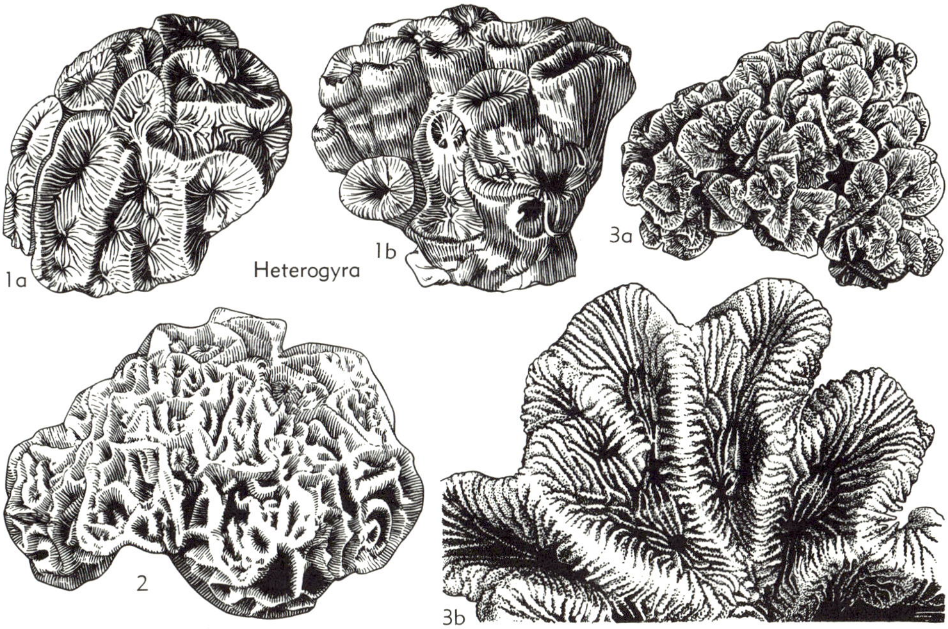

FIG. 273. Fungiina (Agariciicae): Agariciidae (p. F381-F382).

Coeloseris VAUGHAN, 1918 [*C. mayeri*]. Corallum cerioid; colony formation by monostomodaeal budding. Columella absent, axial space open. *Rec.*, IndoPac.——FIG. 275,*1*. **C. mayeri*, Rec., Torres Strait, Austral.; *1a*, ×2; *1b*, transv. sec., ×10 (148).

Pachyseris M.EDW.-H., 1849 [*Agaricia rugosa* LAMARCK, 1801]. Foliaceous, unifacial or bifacial. Centers in concentric rows between long collines parallel to margin, indistinct, linked by columellar laminae or trabeculae. Walls well developed in collines. *Mio.-Rec.*, IndoPac.

P. (Pachyseris), *Mio.-Rec.*——FIG. 272,*2*. *P. (P.) speciosa* DANA, Rec., E.Indies; *2a*, ×0.6; *2b*, ×5 (148).

P. (Pavonaraea) UMBGROVE, 1946 [**Pavonaraea irregularis*]. Like *Pachyseris* but centers distinct. *Plio.*, Java.——FIG. 275,*2*. **P. (P.) irregularis*; ×0.5 (140).

Family SIDERASTREIDAE Vaughan & Wells, 1943

Colonial, rarely solitary; hermatypic. Colony formation by intra- or extratentacular budding. Synapticulothecate. Septa composed of one fan system of small, simple or compound trabeculae, laterally strongly granulated, more or less porous, margins beaded or dentate, laterally united by simple synapticulae. Columella composed of one or more papillary trabeculae. Endothecal dissepiments present. *Cret.-Rec.*

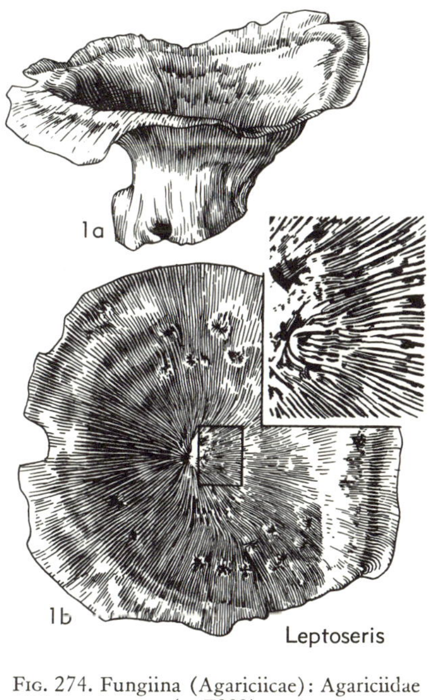

FIG. 274. Fungiina (Agariciicae): Agariciidae (p. F382).

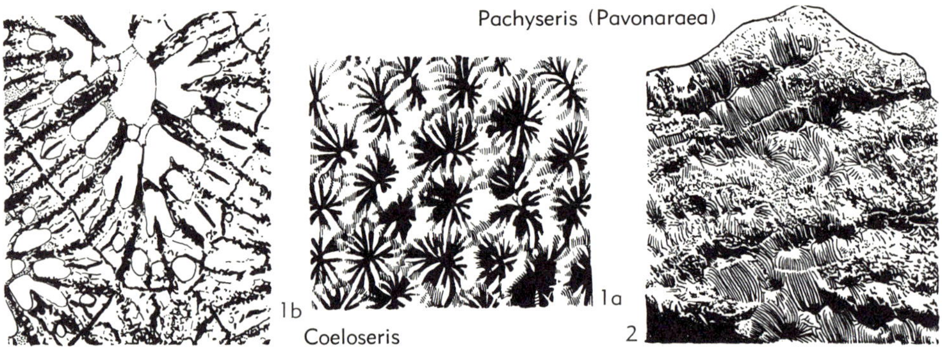

FIG. 275. Fungiina (Agariciicae): Agariciidae (p. F383).

Siderastrea BLAINV., 1830 [**Madrepora radians* PALLAS, 1766] [=*Siderina* DANA, 1846]. Massive, ramose or incrusting, cerioid colonies formed by extratentacular budding. Corallite walls well defined, formed by several synapticular rings. *Cret.-Rec.,* Eu.-Afr.-N.Am.-S.Am.-W.Indies-Red Sea-W. Ind.O.

S. (Siderastrea), *Cret.-Rec.*——FIG. 276,*1a. S. (S.) radians* (PALLAS), Rec., Fla.; ×1 (151).——FIG. 276,*1b. S. (S.) italica* (DEFRANCE), Mio.,

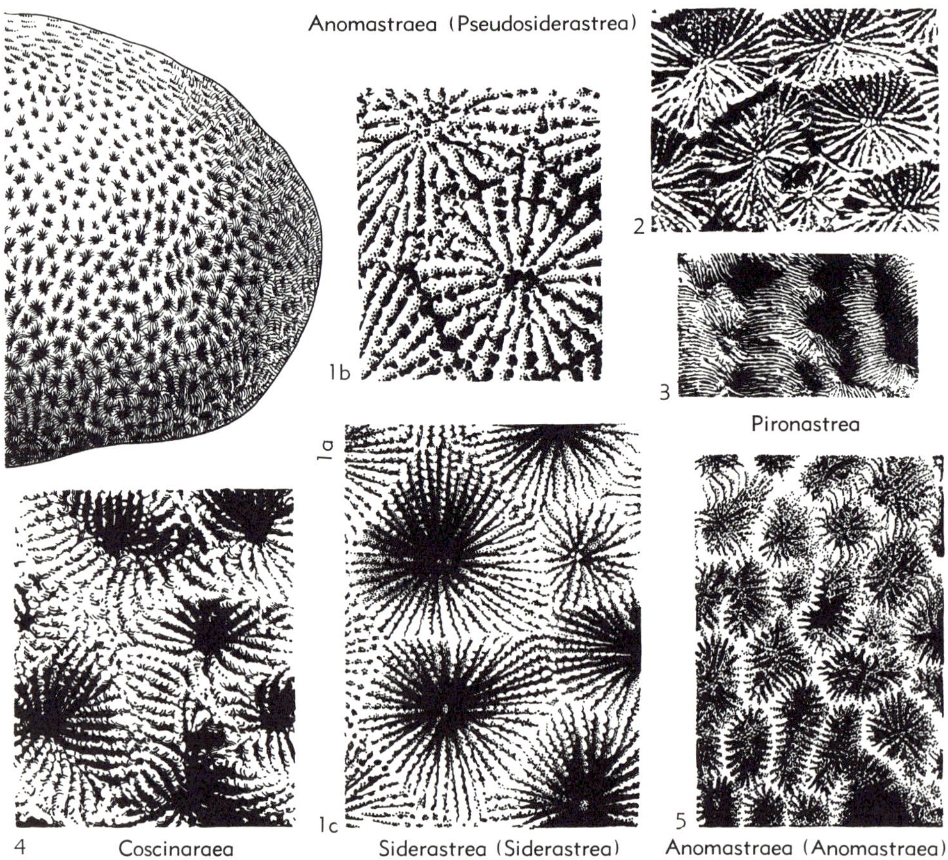

FIG. 276. Fungiina (Agariciidae): Siderastreidae (p. F384-F385).

Poland; transv. sec., ×8 (190).——FIG. 276,*1c*. *S. (S.) siderea* (ELLIS & SOLANDER), Rec., Bahamas; ×6 (149).

S. (Siderofungia) REIS, 1889 [**Columnastraea bella* REUSS, 1869]. Like *S. (Siderastrea)* but corallite walls indistinct; septa confluent, corallum thamnasterioid. *Cret.-Oligo.*, Eurasia-N.Am.

S. (Stephanomorpha) VAUGHAN, 1900 [**Stephanomorpha monticuliformis*]. Like *E. (Siderastrea)* but inner ring of mural synapticulae strongly developed, giving corallum a plocoid aspect. *Eoc.*, N.Am.

Sideroseris WELLS, 1945 [**S. durhami*]. Like *S. (Siderastrea)* but solitary, *Eoc.*, Barbados.

Pironastrea D'ACHIARDI, 1875 [**P. discoides*]. Like *S. (Siderastrea)* but intratentacular circumoral budding forms explanate corallum with centers in concentric rings separated by low concentric collines. *Eoc.-Oligo.*, Eu.-W.Indies.——FIG. 276,*3*. *P. anguillensis* VAUGHAN, Oligo., W.Indies; ×2 (149).

Maeandroseris ROUSSEAU, 1854 [**M. bottae;* SD WELLS, 1936]. Like *Pironastrea* but circumoral followed by intramural budding forms irregularly radiating series separated by collines. *Pleisto.*, Red Sea.

Anomastraea VON MARENZELLER, 1901 [**A. irregularis*] [=?*Thalamocaeniopsis* ALLOITEAU, 1954]. Like *S. (Siderastrea)* but colony formation by intratentacular budding. Corallite wall nearly solid, formed by a single synapticular ring. Septa partly perforate, nonexsert. *?L.Cret.*, N.Afr. (*Thalamocaeniopsis*). *Rec.*, W.Ind.O.-W.Pac.

A. (Anomastraea).——FIG. 276,*5*. **A. (A.) irregularis*, Pt.St.Johns, Pondoland; ×4 (176).

A. (Pseudosiderastrea) YABE & SUGIYAMA, 1935 [**Pseudosiderastrea tayamai*]. Like *A. (Anomastraea)* but colony formation by extratentacular budding; septa nearly imperforate. W.Pac.——FIG. 276,*2*. **A. (P.) tayamai*, Aru I., ×4 (200).

Coscinaraea M.EDW.-H., 1848 [**C. bottae* (=*Astrea monile* FORSKÅL, 1775)] [=*Coscinastrea* M.EDW.-H., 1848]. Massive, subspherical, columniform, or incrusting colonies by intratentacular mono- to tristomodaeal budding. Synapticulotheca formed by several rings of synapticulae forming low collines. Septa perforate, formed mostly by compound trabeculae. *Pleisto.-Rec.*, IndoPac.——FIG. 276,*4*. **C. monile* (FORSKÅL), Rec., E.Afr.; ×4 (195).

Superfamily FUNGIICAE Dana, 1846

[*nom. correct.* WELLS, herein (*ex* Fungioidae VAUGHAN & WELLS, 1943, *nom. transl. ex* Fungiidae DANA, 1846)]

Solitary and colonial; colony formation commonly by intratentacular budding. Corallite wall synapticulothecal, secondarily septothecal. Septa formed by compound trabeculae, united by mostly compound synapticulae, irregularly porous in early genera, more regularly porous in later ones, strongly dentate on margins. *M.Jur.-Rec.*

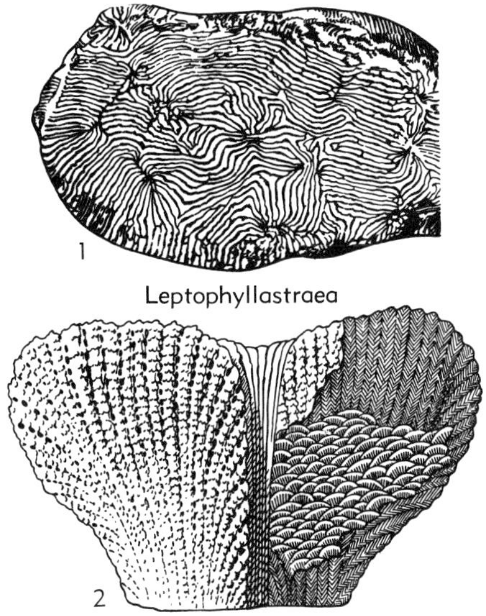

FIG. 277. Fungiina (Fungiicae): Synastreidae (p. F385).

Family SYNASTREIDAE Alloiteau, 1952

[=Leptophylliidae VAUGHAN, 1905]

Solitary and colonial; hermatypic. Synapticulothecate, wall more or less perforate near calice, solid basally, costate. Septa exsert, axis of divergence coinciding with wall, vertical or inclined outward; perforate inwardly and near upper margins, strongly beaded marginally. Endothecal dissepiments well developed. Columella trabecular. *M.Jur.-Rec.*

Acrosmilia ORB., 1849 [**Turbinolia cernua* MICHELIN, 1846 (*non* GOLDF., 1826) (=*Trochosmilia crassa* M.EDW.-H., 1850] [=*Polyphyllia* ORB., 1849 (*non* QUOY & GAIMARD, 1830); *Leptophyllia* REUSS, 1854; *Placoseris* FROMENTEL, 1863; *Turbinoseris* DUNCAN, 1870; *Septophyllia* QUENST., 1881; *?Carantoseris, ?Proleptophyllia* ALLOITEAU, 1952]. Solitary, turbinate, with weak columella. *M.Jur.-Cret.*, Eu.-N.Am.——FIG. 277,*2*. *A. clavata* (REUSS), U.Cret.(Turon.), Aus.(Gosau); vert. sec., ×2 (105).

Leptophyllastraea OPPENHEIM, 1930 [**L. irregularis*]. Like *Acrosmilia* but small colonies produced by irregular mono- to tristomodaeal budding, corallites remaining closely united. *U.Cret.*, Eu.——FIG. 277,*1*. **L. irregularis*, U.Cret.(Turon.), Aus. (Gosau); calicular surface, ×1 (100).

Felixastraea OPPENHEIM, 1930 [**Cyathoseris zitteli* FELIX, 1903]. Like *Acrosmilia* but thamnasterioid, colony formation by polystomodaeal intramural budding with terminal forking; centers with lamellar linkages. Series free laterally near summits,

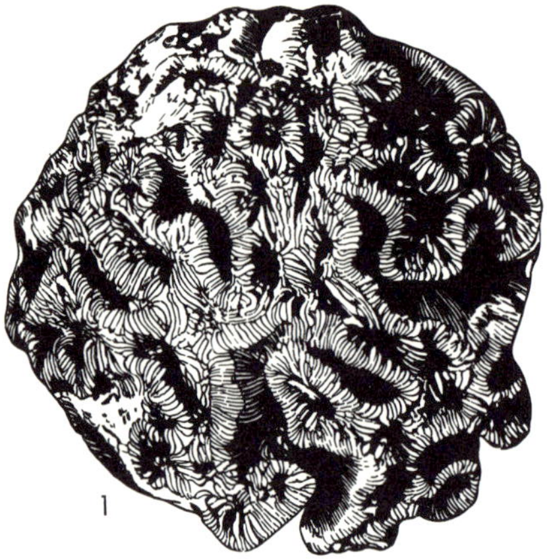

Fig. 278. Fungiina (Fungiicae): Synastreidae (p. F385).

with narrow ambulacra. *U.Cret.*, Eu.——Fig. 278,*1*. **F. zitteli*, U.Cret.(Turon.), Aus.(Gosau); ×1 (100).

Edwardsoseris Alloiteau, 1946 [**Latomeandra corrugata* M.Edw.-H., 1849] [=*Kobymeandra* Alloiteau, 1954]. Like *Felixastraea* but lacking ambulacra. *M.Jur.-Cret.*, Eu.——Fig. 279,*1*. **E. corrugata*, M.Jur., Switz.; *1a,b*, ×0.7; *1c*, lat. aspect of septum, ×4 (71).

Synastrea M.Edw.-H., 1848 [**Astrea agaricites* Goldf., 1826] [=*Thamnastrea auctt.* (*non Thamnastraea* Lesauvage, 1832, =*Thamnasteria* Lesauvage, 1823); *Adelastrea* Reuss, 1854; *Mesoseris* Ferry, 1870; *Aphragmastraea* Solomko, 1888]. Thamnasterioid, colony formation by incomplete polystomodaeal budding, corallites united to summits; calices flush. *M.Jur.-Cret.*, Eu.-N.Am.-W. Indies.——Fig. 279,*4a*. *S. raristella* (Reuss), U. Cret.(Turon.), Aus.(Gosau); ×0.5 (112).——Fig. 279,*4b*. *S. arachnoides* (Parkinson), U.Jur. (Raur.), Switz.; lat. aspect of septum, ×4 (71).

Dimorphoseris Duncan, 1872 [**D. oolitica*] [=*Kobya* Gregory, 1900]. Thamnasterioid, colony formation by circumoral budding; homeomorphic with *Dimorphastrea*, etc. *M.Jur.-Cret.*, Eurasia.——Fig. 279,*3*. *D. waehneri* (Felix), U. Cret.(Turon.), Aus.(Gosau); ×1 (100).

Crateroseris Tomes, 1883 [**C. fungiformis*]. Like *Dimorphoseris* but calices protuberant as bosses (bourrelets); homeomorphic with *Polyphylloseris*. *U.Jur.*, Eu.——Fig. 279,*2*. **C. fungiformis*, Sequan., Eng.; ×0.5 (194).

Family CYCLOLITIDAE d'Orbigny, 1851

Solitary and colonial, free in ephebic stage, subdiscoid, patellate, or cupolate; hermatypic. Colony formation by circumoral or circummural intratentacular budding. Corallite wall synapticulothecate, epithecate. Septa as in Synastreidae, perforations generally filled secondarily, axis of trabecular divergence inclined outwards. Endothecal dissepiments and columella weak or absent. Homeomorphic with Fungiidae. *Cret.-Eoc.*

Cyclolites Lamarck, 1801 [**C. ellipticus*] [=?*Porpites* Schloth., 1820; *Cyclolithas* Ehr., 1834; *Funginella* Orb., 1849; *Orbitolina* Orb., 1850; *Episeris* Fromentel, 1861; *Zittelofungia* Duncan, 1884; ?*Cunnolites* Gagnebin, 1772, *Plesiocunnolites* Alloiteau, 1952]. Solitary, cupolate; septa mostly perforate, axis of divergence nearly vertical. *Cret.-Eoc.*, Eurasia-N.Afr.-W.Indies.——Fig. 280, *4a*. *C.* sp. restored long. sec., ×2 (105).——Fig. 280,*4b,c*. **C. ellipticus*, U.Cret.(Turon.), Aus. (Gosau); ×0.5 (112).

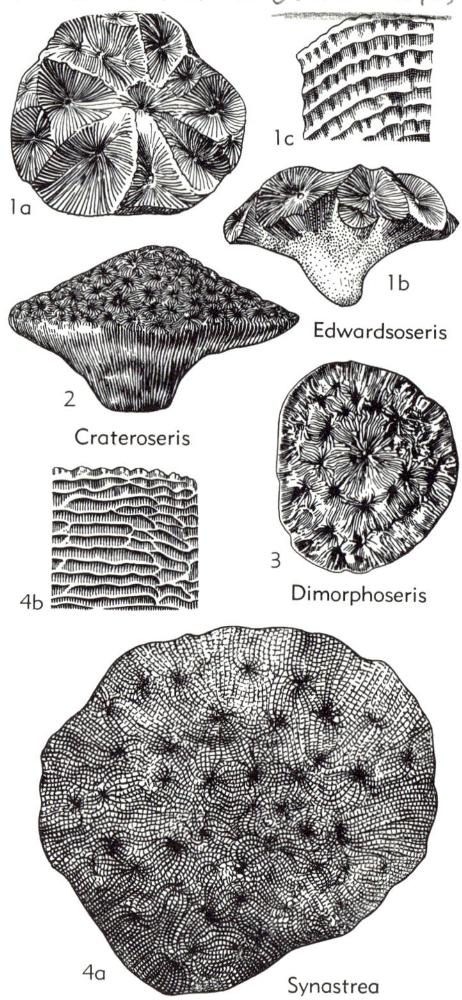

Fig. 279. Fungiina (Fungiicae): Synastreidae (p. F386).

Cyclolitopsis Reuss, 1873 [**Cyclolites patera* d'Achiardi, 1867]. Solitary, patellate, fixed basally in early stages. Septa nearly imperforate, axis of divergence outwardly inclined. *Eoc.*, Eu.

Paracycloseris Wells, 1934 [**P. elizabethae*]. Like *Cyclolites* but discoid or patellate, with well-developed columella. *U.Cret.*, W.Indies-Mex.——Fig. 280,*2*. **P. elizabethae* Campan., Jamaica; *2a,b*, ×1 (198).

Aspidastraea Kühn, 1933 [**A. orientalis*]. Like *Cyclolites* but small colonies formed by circumoral budding. Homeomorphic with *Halomitra* and *Genabacia*. *U.Cret.*, SW.Asia.——Fig. 280,*3*. **A. orientalis*, Senon., Iran; ×1 (78).

Aspidiscus König, 1825 [**A. shawi* (=*Cyclolites cristata* Lamarck, 1801)] [=*Cyclophyllia* M.Edw.-H., 1848; *Helladastraea* Avnimelech, 1947]. Small, cupolate, hydnophoroid colonies produced by circummural budding. *M.Cret.*, Eu.-AsiaM.-N. Afr.——Fig. 280,*7*. *A. felixi* Renz, Cenom., Greece; *7a*, lat. aspect, ×1; *7b*, transv. sec., ×2 (188).

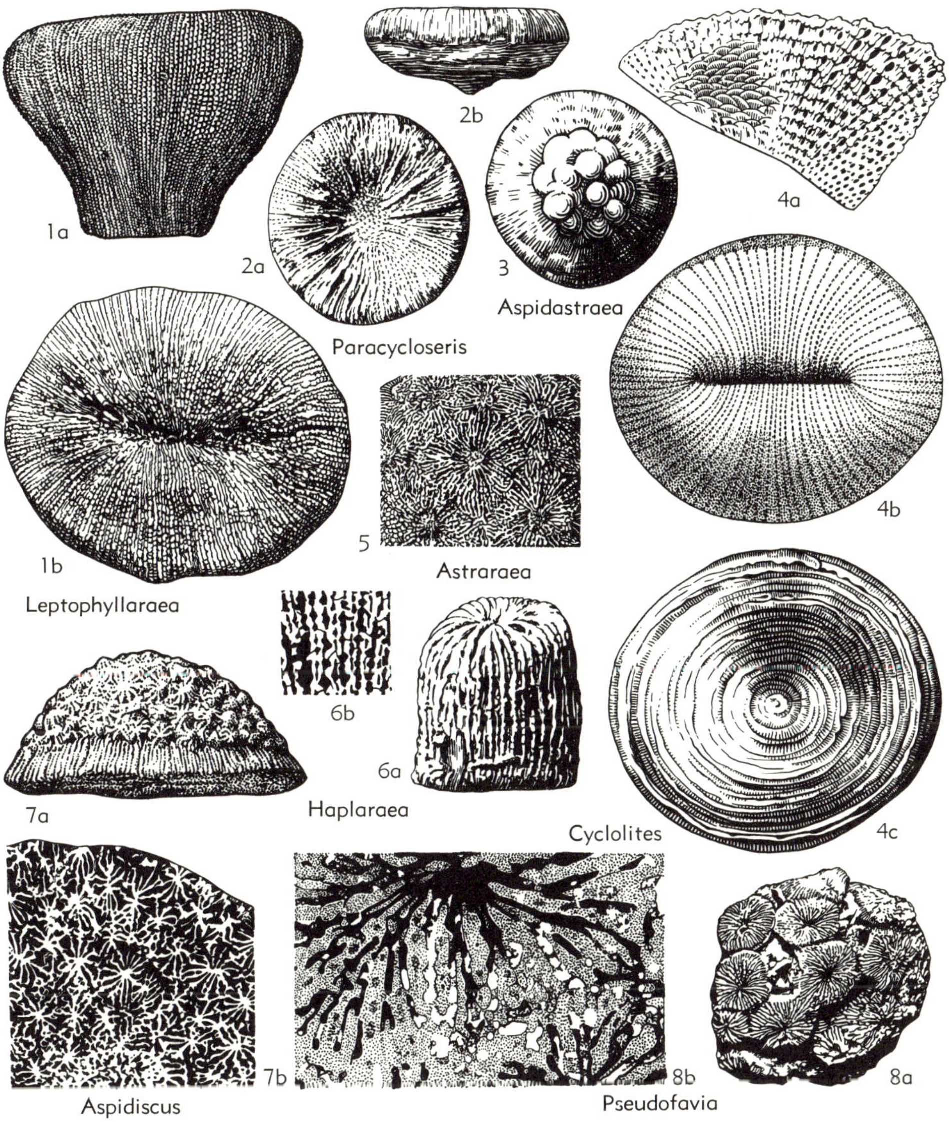

Fig. 280. Fungiina (Fungiicae): Cyclolitidae, Haplaraeidae (p. F386-F388).

Family HAPLARAEIDAE Vaughan & Wells, 1943

Solitary and colonial; hermatypic. Colony formation by intra- and extratentacular budding. Wall synapticulothecate, porous, poorly defined, costate. Septa exsert, formed by one fan system of compound trabeculae, vertical axis of divergence, irregularly porous and thick. Dissepiments thin. Columella trabecular and weak, or absent. *U.Jur.-Eoc., ?Oligo.-?Mio.*

Haplaraea MILASCHEWITSCH, 1876 [**H. elegans*]. Solitary, cylindrical, fixed, epitheca thin. *U.Jur.-Cret.,* Eu.——FIG. 280,*6. H. pratzi* FELIX, U.Cret. (Turon.), Aus.(Gosau); *6a,* ×1; *6b,* transv. septal sec., ×2 (42).

Leptophyllaraea ALLOITEAU, 1952 [**Leptophyllia granulata* FROMENTEL, 1863]. Like *Haplaraea* but turbinate and lacking epitheca. *M.Cret.-U.Cret.,* Eu.——FIG. 280,*1. L. conclavina* (OPPENHEIM), U.Cret.(Turon.), Aus.(Gosau); *1a,b,* ×1 (100).

Physoseris VAUGHAN, 1905 [**Trochoseris insignis* DUNCAN, 1873]. Like *Haplaraea* but lacking epitheca and with lower part of wall parathecal. *Eoc.,* W.Indies.

Diplaraea MILASCHEWITSCH, 1876 [**D. arbuscula*]. Like *Haplaraea* but phaceloid colonies formed by intratentacular budding, corallites monocentric. *U.Jur.-L.Cret,* Eu.-S.Am.

Pseudofavia OPPENHEIM, 1930 [*pro Parastraea* REUSS, 1854 (*non* M.EDW.-H., 1848)] [**Parastraea grandiflora* REUSS, 1854] [=*Cretastraea* KÜHN, 1930]. Like *Haplaraea* but massive plocoid heads formed by extratentacular budding. Corallites closely united by short costae. *U.Cret.,* Eu.——FIG. 280, *8. *P. grandiflora* (Turon.), Aus.(Gosau); *8a,* ×0.5; *8b,* transv. sec., ×2.5 (100).

Trechmannaria WELLS, 1935 [**T. montanaroae*] [=*?Stiboriopsis* VAUGHAN, 1899]. Like *Pseudofavia* but colony formation by intratentacular polystomodaeal budding, with short meandrine series; centers linked by porous lamellae. *U.Cret.,* W. Indies.

Summiktaraea ALLOITEAU, 1952 [**Meandrastrea reticulata* ORB., 1850]. Like *Trechmannaria* but series long and contorted. *U.Cret.,* Eu.

Astraraea FELIX, 1900 [**Thamnastrea media* SOWERBY, 1832; SD WELLS, 1936]. Like *Pseudofavia* but corallites much smaller, colony formation by intratentacular mono- to tristomodaeal budding; corallites permanently monocentric. *U.Cret.,* Eu.——FIG. 280,*5. A. columellata* OPPENHEIM, Turon., Aus.(Gosau); ×1 (100).

?**Confusastraraea** GERTH, 1933 [**Confusastraea obsoleta* GERTH, 1921] [=*?Polyaraea* FRITSCH, 1878]. Small plocoid colonies formed by intratentacular budding, with large corallites; septa very porous. *?Eoc., Mio.,* E.Indies.

Family FUNGIIDAE Dana, 1846

Solitary and colonial, fixed or free, mostly hermatypic. Solitary corallites discoidal or elongate-oval, aboral surface flat or concave. Colonies discoidal or elongate-oval, free or foliaceous and fixed; colony formation by incomplete intratentacular polystomodaeal budding. Wall synapticulothecate, commonly secondarily septothecal or thickened. Septa numerous, fenestrate in early stages, perforate or solid, composed of a single fan system of compound trabeculae producing simple or compound marginal dentations, laterally united by stout compound synapticulae (Fig. 242*C*). Axis of trabecular divergence horizontal in ephebic stage. Costae continuous or broken up into spinose projections. Columella trabecular, feeble. Dissepiments absent; epitheca only in early anthoblast stage. Homeomorphic with Cyclolitidae. *M.Cret.-Rec.,* Eurasia-IndoPac.

Cycloseris M.EDW.-H., 1849 [**Fungia cyclolites* LAMARCK, 1801] [=*Diaseris* M.EDW.-H., 1849; *Actinoseris* ORB., 1849; *Asteroseris* FROMENTEL, 1867; *Microseris* FROM., 1870; *Javanoseris* GERTH, 1922; *?Funginellastraea* ALLOITEAU, 1952]. Solitary, small discoid to cupolate. Wall imperforate, with well-defined costae. Lower-cycle septa imperforate with acute dentations; higher cycles more or less perforate; cyclolitid in structure. *M.Cret.-Rec.,* Eurasia-IndoPac., depth range, 0-411 m.——FIG. 281,*1. C. fragilis* ALCOCK, Rec,. Sulu I., 37 m.; ×0.5 (175).

?**Discotrochus** M.EDW.-H., 1848 [**D. orbignyanus*]. Like *Cycloseris* but smaller, with relatively few septa (4 cycles), beaded on margins; ahermatypic. *Eoc.,* USA; *Oligo.-Mio.,* ?Russia, ?Italy.——FIG. 281,*2. *D. orbignyanus,* M.Eoc.(Claib.), USA; *2a-c,* ×2 (151); *2d,* nat. long. sec., ×3 (144).

Lithophyllon REHBERG, 1892 [**L. undulatum* (=*Leptoseris edwardsi* ROUSSEAU, 1850)]. Like *Cycloseris* in structure but forming fixed, unifacial, foliaceous colonies by marginal budding. *Oligo.-Rec.,* Eu.-IndoPac.

Fungia LAMARCK, 1801 [*pro Fungites* CUVIER, 1798 (*non* GLEDITSCH, 1765)] [**Madrepora fungites* LINNÉ, 1758; SD LEUCKART, 1841] [=*Haliglossa* EHR., 1834 *(partim); Ctenactis, Lobactis, Pleuractis* VERRILL, 1864]. Solitary, discoid, or elongate-oval, flat or convex, free. Wall perforate in ephebic stage. Costae mostly reduced to rows of spines. Septa imperforate except highest cycles, strongly dentate. Contains 6 species groups: (1) *F. echinata* group, with triangular septal dentations, costal spines tall and spinose; (2) *F. scutaria* group, with rounded septal dentations and horizontal lateral ridges, costal spines small, beaded; (3) *F. repanda* group, with subtriangular septal dentations and irregular lateral ridges, costal spines beaded, commonly branching; (4) *F. danai* group, with triangular spinulose septal dentations, costal spines club-shaped, finely spinulose; (5) *F. actiniformis* group, with lobulate, finely spinulose septal teeth, costal spines short, minutely spinulose; (6) *F. fungites* group with triangular, acute, nearly smooth septal teeth, costal spines tall, nearly smooth. *Mio.-Rec.,* IndoPac.——FIG. 281,*4*. *F. scutaria* LAM., Rec., Hawaii; *4a,b,* ×0.5 (146).

Herpolitha ESCHSCHOLTZ, 1826 [**Madrepora limax* ESPER, 1797; SD M.EDW.-H., 1850] [=*Herpetolithus* LEUCKART, 1841; *Herpetolitha* M.EDW.-H., 1860]. Colonial, elongate-oval, convex, free. Colony formation by linear polystomodaeal budding forming a deep axial furrow with a few lateral circumoral centers. Septal and costal structure as in *Fungites scutaria* group. *Rec.,* IndoPac.——FIG. 281,*6*. **H. limax* (ESPER), Jaluit, Marshall I.; ×0.4 (198).

Polyphyllia QUOY & GAIMARD, 1830 [**P. pelvis* (=*Fungia talpina* LAMARCK, 1816)] [=*Lithactina* LESSON, 1832; *Cryptabacia* M.EDW.-H., 1849]. Similar to *Herpolitha* in early stages but secondary centers numerous and nearly equal in size to those of axial furrow. Structures as in *Fungites scutaria* group. *Rec.,* Pac.——FIG. 281,*5*. **P. talpina* (LAM.), E.Indies; ×0.2 (128).

Halomitra DANA, 1846 [**Madrepora pileus* LINNÉ, 1758]. Subcircular, convex, free, lacking axial furrow. Colony formation by circumoral followed by marginal budding. Structures as in *Fungia fungites* group. *Rec.,* IndoPac.——FIG. 282,*1*. **H. pileus,* E.Indies; ×0.2 (193). (Also Fig. 248*C*.)

Parahalomitra WELLS, 1937 [*pro Doederleinia* GARDINIER, 1909 (*non* STEINDACHNER, 1883)] [**Halomitra irregularis* GARDINER, 1898] [=*?Sandalolitha* QUELCH, 1884]. Like *Halomitra* but structures as in *Fungia repanda* group. *Rec.,* IndoPac.

Podabacia M.EDW.-H., 1849 [**Agaricia cyathoides* VALENCIENNES in M.EDW.-H., 1849 (=*Madrepora crustacea* PALLAS, 1766)]. Like *Parahalomitra* and

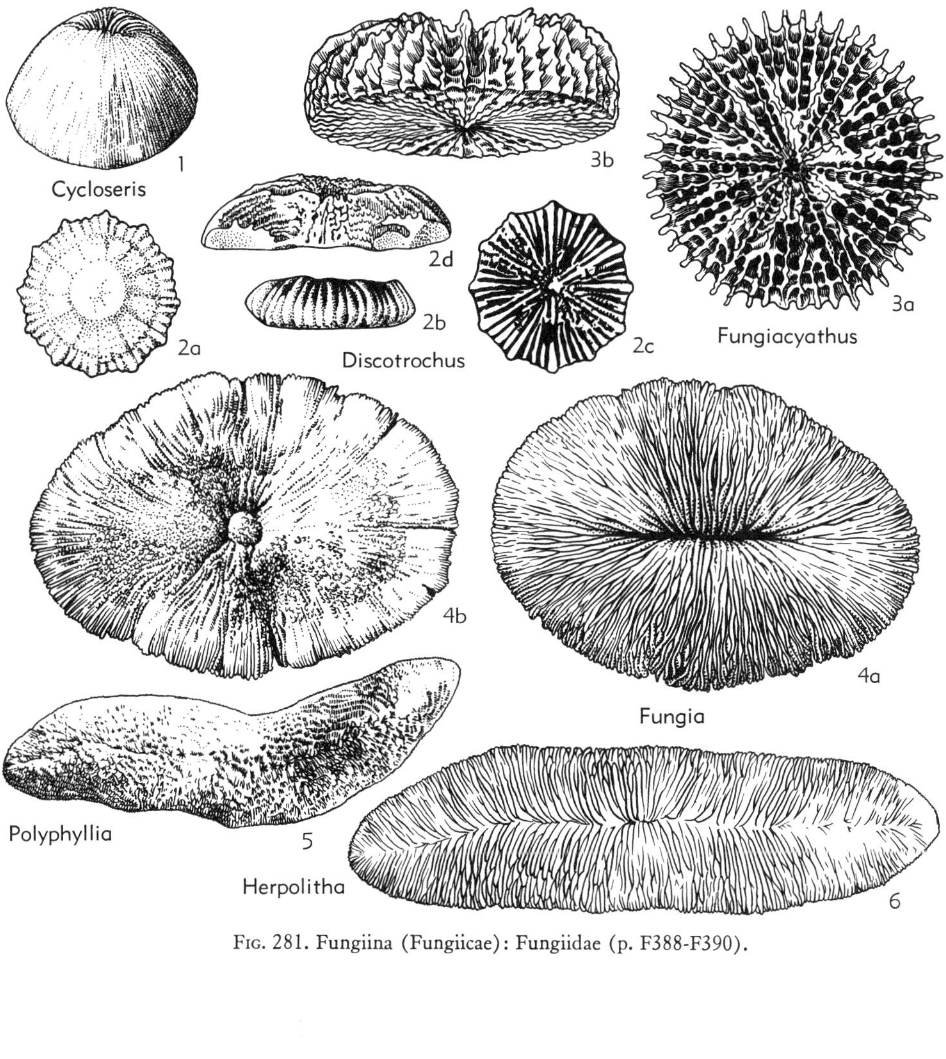

FIG. 281. Fungiina (Fungiicae): Fungiidae (p. F388-F390).

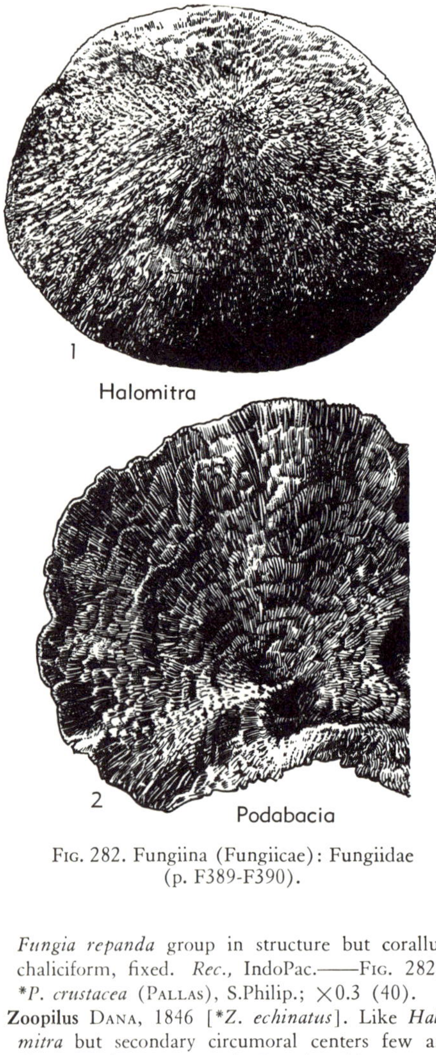

FIG. 282. Fungiina (Fungiicae): Fungiidae (p. F389-F390).

Fungia repanda group in structure but corallum chaliciform, fixed. *Rec.*, IndoPac.——FIG. 282,*2*. **P. crustacea* (PALLAS), S.Philip.; ×0.3 (40).

Zoopilus DANA, 1846 [**Z. echinatus*]. Like *Halomitra* but secondary circumoral centers few and distant, and structures as in *Fungia echinata* group. *Rec.*, Pac.

?**Fungiacyathus** SARS, 1872 [**F. fragilis* (=*Fungia symmetrica* POURTALÈS, 1871)] [=*Bathyactis* MOSELEY, 1881]. Solitary, discoidal or cupolate, free; ahermatypic. Costae thin, acute, spinose. Septa thin, dentations irregular and acute, laterally braced by thin, perforate plates (synapticulae?) extending upward and outward from the horizontal septotheca and by thin septal striae. *Tert.-Rec.*, cosmop., depth range 58-5,868 m.——FIG. 281,*3*. **F. symmetrica*, Rec., off Tristan da Cunha, 3,700 m.; *3a,b*, ×2 (97).

Family MICRABACIIDAE Vaughan, 1905

Solitary, cupolate, free; ahermatypic. No corallite wall, the costae alternating in position with the septa, beaded or smooth, fixed to septa by regularly spaced simple synapticulae. Septa composed of one fan system of simple trabeculae, perforate, margins dentate. Dissepiments absent. Columella trabecular. *Cret.-Rec.*

Micrabacia M.EDW.-H., 1849 [**Fungia coronula* GOLDF., 1827] [=*Discopsammia* ORB., 1849; *Stephanoseris* FROMENTEL, 1863 (*non* M.EDW.-H., 1851); *Cyclabacia* BÖLSCHE, 1866; *Diafungia* DUNCAN, 1884]. Small, up to 15 mm. in diameter. Septal margins nearly vertical peripherally, with small, transverse dentations. *Cret.-Rec.*, cosmop., depth range 91-328 m.——FIG. 283,*1a-c*. *M. mineolensis* STEPHENSON, U.Cret., Tex.; *1a-c*, ×4 (191).——FIG. 283,*1d*. *M. rotatilis* STEPHENSON, U.Cret., Md.; nat. long. sec., ×4 (191).

Stephanophyllia MICHELIN, 1841 [**Fungia elegans* BRONN, 1837; SD M.EDW.-H., 1850] [=*Letepsammia, Stephanopsammia* YABE & EGUCHI, 1932]. Larger than *Micrabacia*, with septal margins forming peripheral shelf. *Eoc.-Rec.*, Eurasia-W.Pac., depth range 73-814 m.——FIG. 283,*3*. *S. imperialis* MICH., Mio.(Vindob.), Baden; *3a,b*, ×1 (113). (Also Fig. 257,*5*.)

Leptopenus MOSELEY, 1881 [**L. discus*; SD WELLS, 1936]. Like *Stephanophyllia* but with very porous, delicate structures. Costae smooth, acute; septal dentations reduced to a few long spines. *Rec.*, S. Atl.-S.IndoPac., depth range 682-4,115 m.——FIG. 283,*2*. *L. hypocoelus* MOSELEY, 4,000 m., off Chile; *2a,b*, ×2 (97).

Superfamily PORITICAE Gray, 1842

[*nom. correct.* WELLS, herein (*ex* Poritoidae VAUGHAN & WELLS, 1943, *nom. transl. ex* Poritidae GRAY, 1842)]

Colonial, rarely solitary. Septa formed by simple trabeculae; sclerodermites diverging at more or less regular intervals and fusing in plane of septum, forming a fenestrate, porous meshwork united laterally by simple synapticulae. *Jur.-Rec.*

Family MICROSOLENIDAE Koby, 1890

Solitary and colonial; hermatypic. Colony formation by intra- and extratentacular budding. Corallite walls absent or marked by synapticular rings. Septa thin, formed by one fan system of simple trabeculae, fenestrate, with abundant, regularly arranged perforations. Synapticulae simple, abundant between septa. Dissepiments feebly developed. Columella trabecular, feeble. *Jur.-Cret.*

Chomatoseris THOMAS, 1936 [*pro* **Orbulites* SCHWEIGGER, 1819 (*non* LAMARCK, 1801)] [**Madrepora porpites* SMITH, 1816] [=*Anabacia* auctt. (*non* M.EDW.-H., 1849)]. Solitary, cupolate,

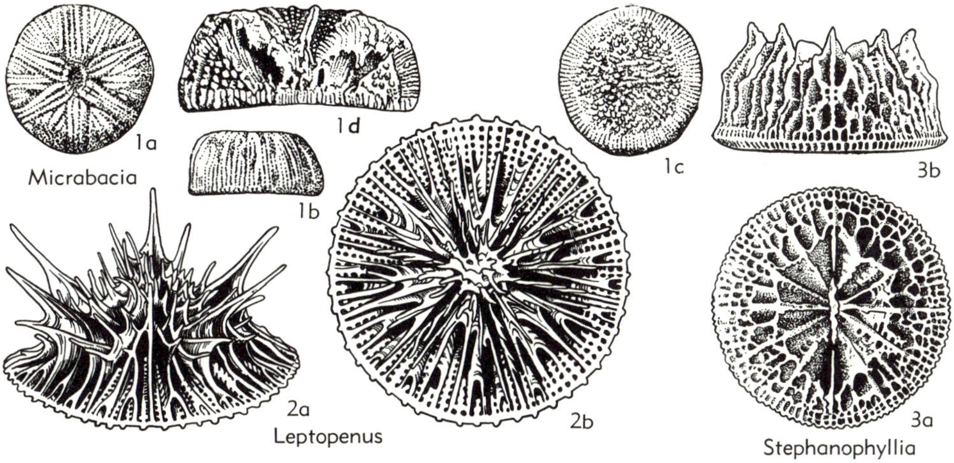

FIG. 283. Fungiina (Fungiicae): Micrabaciidae (p. F390).

free. *L.Jur.-M.Jur.*, Eurasia.——FIG. 285,*3*. *C. complanata* (DEFRANCE), M.Jur.(Bath.), Fr.; transv. sec., ×5 (199).

Semeloseris THOMAS, 1935 [**Tricycloseris limax* TOMES, 1883]. Like *Chomatoseris* but small subthamnasterioid colonies with several centers produced by polystomodaeal budding. *M.Jur.*, Eng.

Genabacia M.EDW.-H., 1849 [**Fungia stellatifera* D'ARCHIAC, 1843]. Like *Chomatoseris* but small colonies produced by circumoral budding, with 1 or 2 rings of secondary centers. Homeomorphic with *Aspidastraea*. *M.Jur.*, Eu.——FIG. 284,*3*. **G. stellatifera*, Bath., Fr.; *3a,b*, ×1 (186).

Trocharea ÉTALLON, 1864 [**T. actiniformis*] [=*Trochoplegma* GREGORY, 1900; *Trochophlegma* VAUGHAN, 1905; *Paracyclolites* WELLS, 1933]. Solitary, patellate to trochoid, or subcylindrical, fixed. *M.Jur.-Cret.*, Eurasia-N.Am.——FIG. 284,*1*. *T. bakerae* WELLS, M.Cret(Cenom.), Tex.; *1a,b*, ×1; *1c*, ×2 (158).

Mycetaraea PRATZ, 1882 [**Thamnastraea? dimorpha* BÖLSCHE, 1866]. Small, thamnasterioid colonies formed by intratentacular mono- to tristomodaeal budding, corallites permanently di- or tricentric. *U.Jur.*, Eu.

Disaraea FROMENTEL, 1861 [**D. cotteaui;* SD WELLS, 1936]. Phaceloid colonies formed by extratentacular budding; corallites cylindrical with stout epitheca. *U.Jur.*, Eu.

Dermoseris KOBY, 1887 [**D. schardti;* SD WELLS,

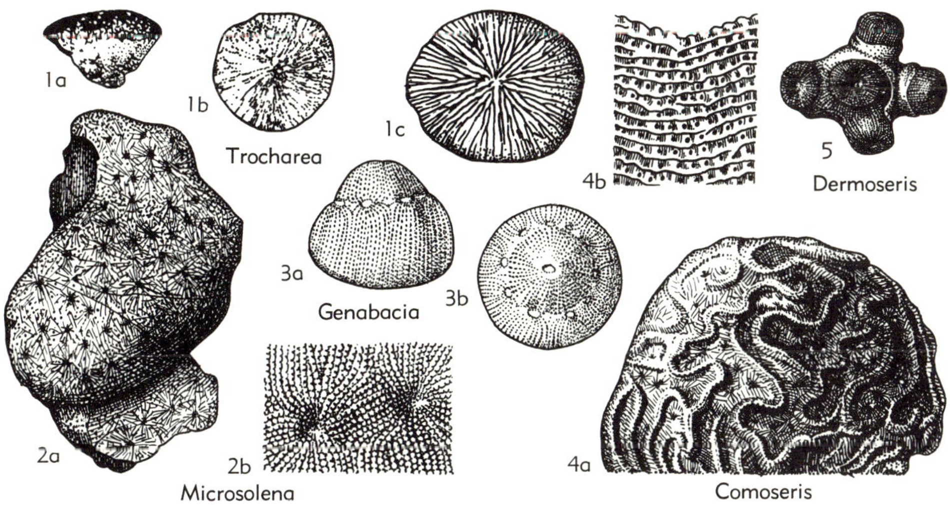

FIG. 284. Fungiina (Poriticae): Microsolenidae (p. F391-F392).

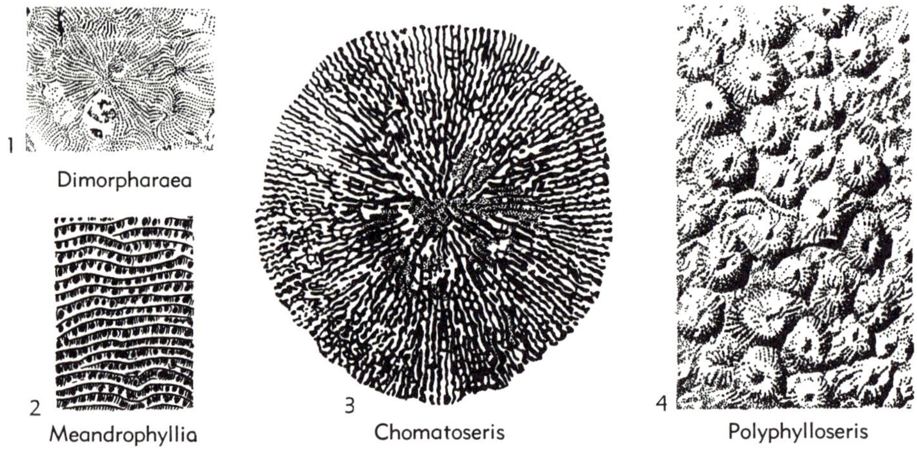

FIG. 285. Fungiina (Poriticae): Microsolenidae (p. F390-F392).

1936]. Like *Disaraea* but colony formation by intratentacular budding, corallites permanently monocentric. *U.Jur.*, Eu.——FIG. 284,*5*. *D. corymbosa* KOBY, Sequan., Port.; ×1 (73).

Dactylaraea ORB., 1849 [**D. truncata*]. Phaceloid tufts of small epithecate branches each with 1 to 3 centers formed by intratentacular polystomodaeal budding. *U.Jur.*, Eu.

Microsolena LAMOUROUX, 1821 [**M. porosa*] [=?*Turbinolopsis* LAMX., 1821; *Turbinastrea* BLAINV., 1830; *Trochopsis* EHR., 1834; *Polyphyllastrea* ORB., 1849; *Microsaraea* KOBY, 1890; *Camptodocis* DIETRICH, 1926]. Thamnasterioid, massive or explanate colonies formed by intratentacular marginal budding. Calices superficial. *M.Jur.-Cret.*, Eurasia-Afr.-N.Am.-S.Am.——FIG. 284,*2*. *M. fromenteli* KOBY, U.Jur.(Raur.), Switz.; *2a*, ×0.5; *2b*, ×2 (71).

Dimorpharaea FROMENTEL, 1861 [**Microsolena koechlini* M.EDW.-H., 1860; SD GREGORY, 1900] [=?*Anomophyllum* ROEMER, 1836]. Like *Microsolena* but colony formation by circumoral with some marginal budding. Homeomorphic with *Dimorphoseris* and *Dimorphastrea*. *M.Jur.-Cret.*, Eurasia-Afr.-N.Am.——FIG. 285,*1*. *D. manchacaensis* WELLS, M.Cret.(Cenom.), Tex.; transv. sec., ×2 (158).

Polyphylloseris FROMENTEL, 1857 [**Polyphyllastrea convexa* ORB., 1849; SD WELLS, 1936] [=*Mastophyllia* FELIX, 1891]. Like *Microsolena* but calices elevated on bosses or bourrelets. *U.Jur.-Cret.*, Eurasia-N.Am.——FIG. 285,*4*. *P. simondsi* WELLS, L.Cret.(Apt.), Tex.; ×1 (157).

Meandrophyllia ORB., 1849 [**Meandrina lotharinga* MICHELIN, 1843] [=*Meandraraea* ÉTALLON, 1859; *Latimaeandraraea* FROMENTEL, 1861; *Michelinaraea* ALLOITEAU, 1952]. Massive or columniform colonies formed by intratentacular intramural budding; collines continuous; valleys discontinuous. *M.Jur.-Cret.*, Eurasia-Afr.-N.Am.-S.Am.——FIG. 285,*2*. *M. gressleyi* (ÉTAL.), U. Jur.(Raur.), Switz.; lat. aspect of septa, ×4 (71).

Comoseris ORB., 1849 [**Pavonia maeandrinoides* MICHELIN, 1843] [=*Oroseris* M.EDW.-H., 1851; *Commoseris* FROMENTEL, 1865]. Like *Meandrophyllia* but commonly foliaceous, with several series of centers between collines. *M.Jur.-U.Jur.*, Eu.——FIG. 284,*4*. **C. maeandrinoides*, U.Jur.(Sequan.), Port.; *4a*, ×1 (73); Switz.; lat. aspect of septum, *4b*, ×5 (71).

Gosaviaraea OPPENHEIM, 1930 [**G. camerina*]. Like *Microsolena* but septa irregularly porous with some distal bifurcation. *U.Cret.*, Eu.

Family ACTINACIDIDAE Vaughan & Wells, 1943

Colonial; hermatypic. Colony formation by intra- and extratentacular budding. Synapticulothecate. Septa relatively few, formed by one fan system of 3 to 5 nearly vertical simple trabeculae, pores common, laterally united by simple synapticulae, innermost trabeculae commonly differentiated as paliform lobes. Distinct septocostae scarcely distinguishable, or absent where distal ends of septa bifurcate and become lost in coenosteum formed by discontinuous vertical trabeculae united horizontally by simple trabeculae. Columella absent or formed by one or more trabecular pillars. Endothecal dissepiments thin and subtabular. *U.Jur.-Oligo.*

Actinaraea ORB., 1849 [**Agaricia granulata* MÜNSTER in GOLDF., 1833]. Colony formation by intratentacular budding. Calicular margins definite. Septocostae present but somewhat vague. No palar trabeculae. *U.Jur.*, Eu.——FIG. 286,*3*. **A. granulata*, U.Jur.(Kimm.), Ger.; ×3 (7).

Dendraraea ORB., 1849 [**Alveopora racemosa* MICHELIN, 1843] [=*Thamnaraea* ÉTALLON, 1864; *Spongitamnia* GREGORIO, 1884]. Colony formation by intratentacular budding, producing ramose or submassive colonies. Structures similar to *Actinaraea* but septocostae much reduced and coenosteum distinctly layered. *M.Jur.-L.Cret.*, Eurasia-S.Am.

Elephantaria OPPENHEIM, 1930 [**E. lindstroemi*]. Submassive colonies produced by extratentacular budding. Septocostae very reduced. Columellar trabeculae developed. *U.Cret.*, Eu.-W.Indies.——FIG. 286,*2*. **E. lindstroemi*, U.Cret.(Turon.), Aus. (Gosau); transv. sec., ×5 (100).

Actinacis ORB., 1849 [**A. martiniana*] [=*Neostroma* TORNQUIST, 1901]. Submassive to ramose; colony formation by extratentacular budding. Septa commonly in 3 cycles with one crown of 6 to 8 palar trabeculae. One columellar trabecula. *M. Cret.-Oligo.*, Eurasia-N.Am.-S.Am.-W.Indies-Afr. ——FIG. 286,*1a*. *A.* sp., U.Cret.(Coni.), Corbières, Fr.; long sec., ×30 (5).——FIG. 286,*1b*. **A. martiniana*, U.Cret.(Turon.), Aus.(Gosau); transv. sec., ×10 (151).——FIG. 286,*1c*. *A. elegans* REUSS, U.Cret.(Turon.), Aus.(Gosau); ×1 (100).

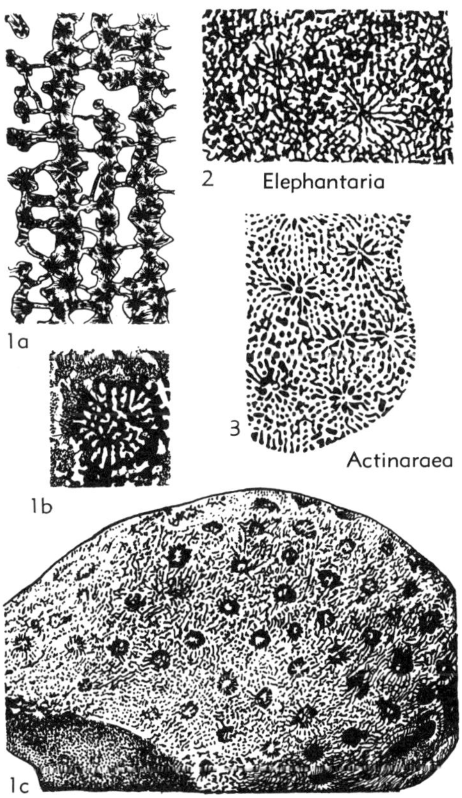

FIG. 286. Fungiina (Poriticae): Actinacididae (p. F392-F393).

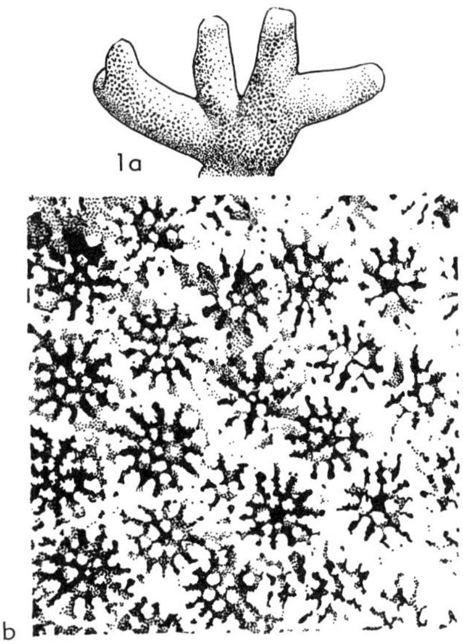

FIG. 287. Fungiina (Poriticae): Poritidae (p. F393).

Family PORITIDAE Gray, 1842

Colonial, hermatypic. Colony formation by extratentacular budding. Corallites mostly united closely without coenosteum, limited by one or more synapticular rings. Septa (except *Alveopora*) formed by 3 to 8 nearly vertical trabeculae, loosely united, with more or less regular perforations. Innermost trabeculae of certain septa differentiated as pali. A single columellar trabecula. *M.Cret.-Rec.*

Goniopora BLAINV., 1830 [**G. pedunculata*] [=*Goniaraea* ORB., 1849; *Litharaea, Rhodaraea* M.EDW.-H., 1849; *Tichopora* QUELCH, 1886]. Massive, columniform, or ramose, rarely incrusting. Septa generally in 3 cycles, formed by 4 to 8 trabeculae. *M.Cret.-Rec.*, Eu.-Pac.——FIG. 289,*3* (131). *G. duofaciata* Thiel, Rec., Philip.; ×5 (128).

Porites LINK, 1807 [**P. polymorphus* (=*Madrepora porites* PALLAS, 1766, *partim*)] [=?*Stylaraea* M. EDW.-H., 1851; *Cosmoporites, Neoporites* DUCHAISSING & MICHELOTTI, 1860]. Massive, ramose, or incrusting. Corallites smaller (to 2 mm.) than *Goniopora*, with only 2 septal cycles. Septa formed by 3 to 4 trabeculae. One of the most important hermatypic coral genera, second only to *Acropora*. *Eoc.-Rec.*, cosmop.

P. (Porites), *Eoc.-Rec.*——FIG. 287,*1*. **P. (P.) porites*, Rec., W.Indies; *1a*, ×0.3; *1b*, ×10 (151).

P. (Synaraea) VERRILL, 1864 [**Porites erosa* DANA, 1846; SD VAUGHAN, 1919] [=*Napopora* QUELCH, 1884]. Like *P. (Porites)* but calices superficial,

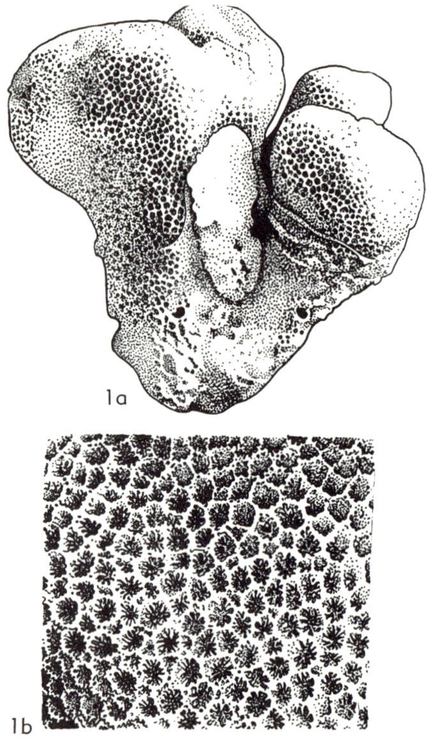

FIG. 288. Fungiina (Poriticae): Poritidae (p. F393-F394).

separated by finely porous coenosteum, commonly raised in ridges. *Oligo.-Rec.*, W.Indies-IndoPac.——FIG. 289,*2*. *P. (S.) horizontalata* HOFFMEISTER, Rec., Bikini, Marshall I.; ×5 (198).

Dictyaraea REUSS, 1867 [**D. micrantha;* SD GREGORY, 1925]. Ramose with slender branches, cerioid. Septa on plan of *Porites* but generally very irregular and much thickened secondarily with obliteration of basic plan. *Mio.-Plio.*, W.Pac.——FIG. 289, *1*. **D. micrantha*, Mio., Java; *1a*, ×1; *1b*, ×5 (115).

Alveopora BLAINV., 1830 [**Madrepora daedalea* FORSKÅL, 1775; SD WELLS, 1936] [=*Porastrea* M.EDW.-H., 1848; *Favositipora* SAVILLE-KENT, 1870; ?*Diechoraea* T. WOODS, 1879]. Massive or ramose. Septa in 1 to 3 cycles, represented by nearly horizontal spines projecting inward from mural trabeculae. *Eoc.-Rec.*, Eu.-IndoPac.——FIG. 288,*1*. *A. verrilliana* DANA, Rec., Japan; *1a*, ×0.4; *1b*, ×2 (198).

Suborder FAVIINA Vaughan & Wells, 1943

[*nom. correct.* WELLS, herein (*ex* Faviida VAUGHAN & WELLS, 1943)]

Solitary and colonial. Corallite wall epithecal, septothecal, or parathecal. Septa formed by one or more fan systems of simple or compound trabeculae, ranging from isolated spines to imperforate laminae, margins more or less regularly dentate. Dissepiments well developed. Synapticulae very rare. *M. Trias.-Rec.*

Superfamily STYLOPHYLLICAE Volz, 1896

[*nom. transl.* WELLS, herein (*ex* Stylophyllidae, VOLZ, 1896)]

Solitary and colonial; colony formation by extratentacular (parricidal or intracalicular) or intratentacular budding. Wall epithecal, secondarily thickened internally. Septa essentially laminar, ranging from isolated trabecular spines to solid laminae, commonly discontinuous vertically. Columella absent or represented by elongated septum. Endothecal dissepiments generally bizonal: peripherally vesicular, and centrally tabular. *M. Trias.-Cret.*

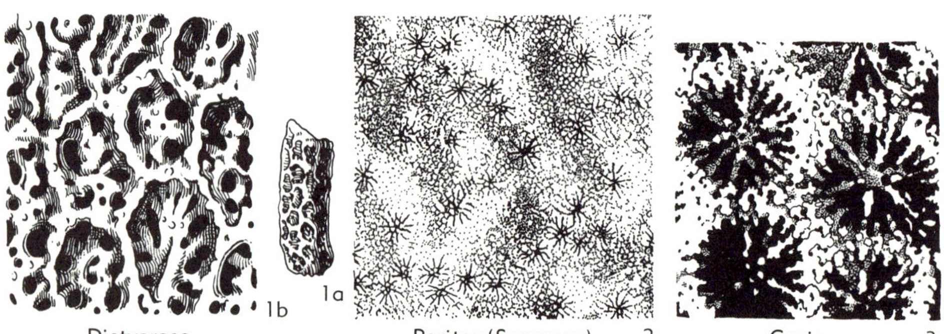

FIG. 289. Fungiina (Poriticae): Poritidae (p. F393-F394).

Family STYLOPHYLLIDAE Volz, 1896

Solitary and colonial, hermatypic. Septa scarcely exsert, formed by simple or compound trabecular spines inclined inward from periphery, isolated or laterally united as vertically discontinuous laminae, interrupted by dissepiments. No columella. *M. Trias.-L.Jur.*

Stylophyllum Reuss, 1854 [**S. polyacanthum*]. Solitary, or small phaceloid or cerioid colonies formed by extratentacular intracalicular budding. Corallites free laterally. Septal spines rarely fused into laminae. Endothecal dissepiments tabular. *U.Trias.,* Eu.——Fig. 290,*2a,b*. **S. polyacanthum,* U.Trias.(Nor.), Aus.; transv. and long. secs., ×1 (112).——Fig. 290,*2c,d*. *S. paradoxum* Frech, U.Trias.(Nor.), Aus.; ×1 (50).

Oppelismilia Duncan, 1867 [**O. gemmans*] [=?*Epismilia* Fromentel, 1861; *Stylophyllopsis* Frech, 1890; *Mollukkia* Jaworski, 1915; ?*Epismiliopsis, Plesiophyllum* Alloiteau, 1952]. Like *Stylophyllum* but the trabecular septal spines tend to fuse and form subcompact septal laminae. *U. Trias.-Jur.,* Eurasia-E.Indies-N.Am.-S.Am.——Fig. 290,*1a,b*. *O. rudis* (Emmrich), U.Trias. (Rhaet.), Aus.; ×1 (50).——Fig. 290,*1c*. *O. rugosa* (Duncan & Wright), L.Jur., Eng.; transv. sec., ×5 (192).——Fig. 290,*1d*. *O. mucronata* (Duncan), U.Jur., Eng.; septal margin, ×3 (192).

Coccophyllum Reuss, 1865 [**C. sturi*]. Like *Stylophyllum* but cerioid. Endothecal dissepiments mostly tabular. *U.Trias.,* Eu.——Fig. 290,*3*. **C. sturi,* U.Trias.(Nor.), Aus.; *3a,b,* surface and long secs., ×1 (50).

Maeandrostylis Frech, 1890 [**Stylophyllum (M.) irregulare*]. Like *Oppelismilia* but small corallites forming small, submassive colonies by intratentacular intramural budding, in short series. *U. Trias.,* Eu.

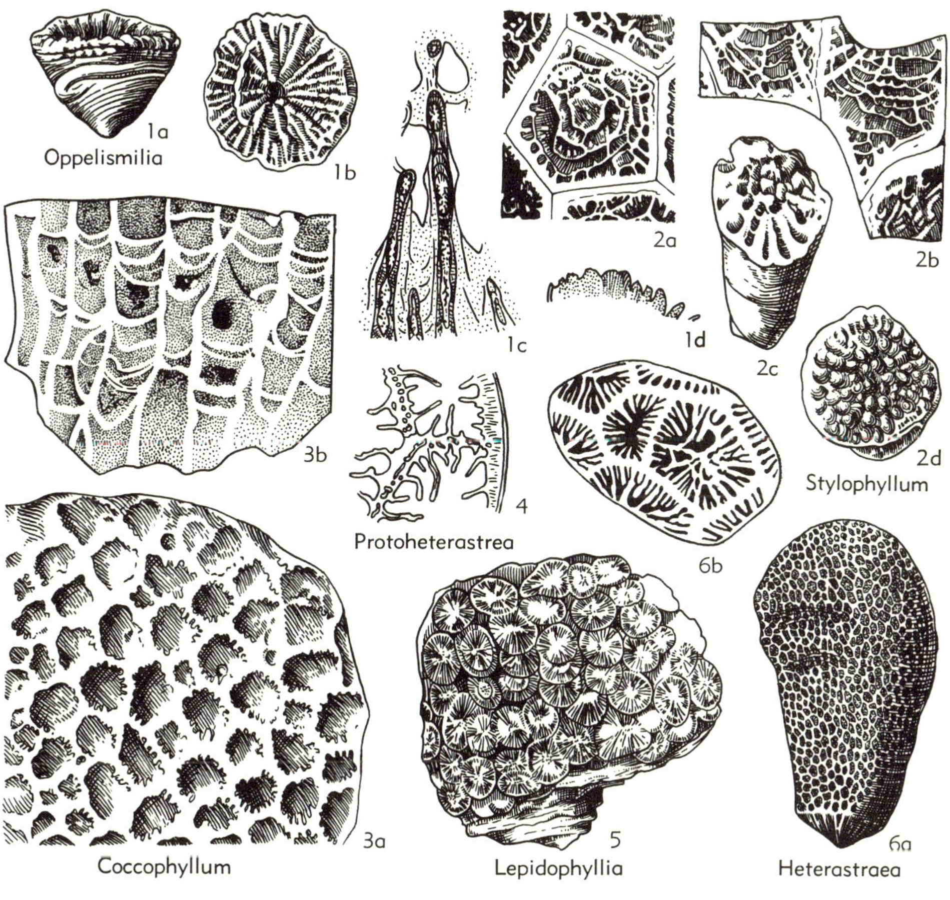

Fig. 290. Faviina (Stylophyllicae): Stylophyllidae (p. F395-F396).

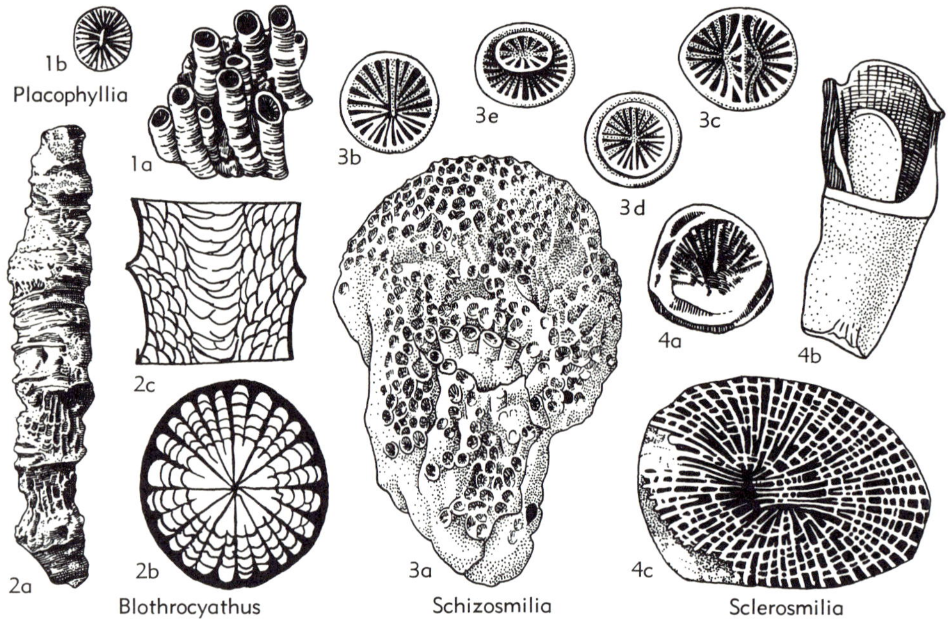

FIG. 291. Faviina (Stylophyllicae): Amphiastreidae (p. F396-F398).

Lepidophyllia DUNCAN, 1868 [**L. hebridensis;* SD WELLS, 1936]. Like *Oppelismilia* but phaceloid; massive, subglobose colonies with small corallites, formed by extratentacular budding. *L.Jur.*, Br.I.-S.Am.——FIG. 290,*5*. **L. hebridensis,* L.Jur. (Sinem.), Hebrides; ×0.5 (30).

Protoheterastrea WELLS, 1937 [*pro Hexastraea* VOLZ, 1896 (*non* SISMONDA, 1871)] [**Hexastraea leonhardi* VOLZ, 1896]. Small, dendroid or phaceloid colonies by intratentacular budding; corallites monocentric. Septa composed of inwardly inclined spines commonly fusing in subcompact laminae. Endothecal dissepiments mostly tabular, but may be vesicular peripherally. *M.Trias.*, Eu.——FIG. 290,*4*. *P. fritschi* (VOLZ), Aus.; transv. sec., ×10 (155).

Heterastraea TOMES, 1888 [**Isastrea tomesi* DUNCAN, 1868; SD VAUGHAN & WELLS, 1943]. Like *Protoheterastrea* but cerioid, with some di- or tristomodaeal budding. *U.Trias.-L.Jur.*, Eu.——FIG. 290,*6*. *H. haimei* (DUNCAN), L.Jur.(Hett.), Eng.; *6a*, ×0.5; *6b*, ×3 (30).

Family AMPHIASTREIDAE Ogilvie, 1896

Solitary and colonial, hermatypic. Septa nonexsert, formed by one scarcely diverging system of relatively few trabeculae, laminar, with nearly smooth or beaded margins, commonly interrupted vertically. Columella absent, but one larger septum generally projects into the axial space. In early forms, an outer endothecal zone of dissepiments is absent or weakly developed. In central members of the family, outer ends of septa may be withdrawn from the corallite wall (aphroid condition). In later types, bilaterality produced by columellar septum or septal group is lost. *M.Jur.-Cret.*

Ceratocoenia TOMES, 1884 [**C. elongata*]. Solitary, elongate, ceratoid or cylindrical. Septa few, radial, laminar. Endothecal dissepiments tabular. *M.Jur.*, Eu.——FIG. 292,*5*. **C. elongata,* M.Jur.(Bath.), Fr., ×0.5 (194).

Discocoenia TOMES, 1884 [**D. bononiensis*]. Like *Ceratocoenia* but patellate. *M.Jur.*, Eu.

Budaia WELLS, 1933 [**B. travisensis*]. Like *Ceratocoenia* but with 2 endothecal zones, an outer vesicular and an inner tabular. *L.Cret.-M.Cret.*, N.Am.-S.Am.——FIG. 292,*3*. **B. travisensis,* M.Cret. (Cenom.), Tex.; *3a,b,* ×1 (158).

Aulastraeopora PREVER, 1909 [**A. deangelisi;* SD WELLS, 1936]. Like *Budaia* but phaceloid or cerioid. *M.Cret.*, Eu.

Cheilosmilia KOBY, 1888 [**C. microstoma*]. Solitary, elongate, ceratoid. Septa rudimentary. *U.Jur.*, Eu.——FIG. 292,*4*. **C. microstoma,* U.Jur.(Raur.), Switz., *4a,b,* ×0.5 (71).

Sclerosmilia KOBY, 1888 [**S. rugosa;* SD WELLS, 1936] [=*Lingulosmilia* KOBY, 1888; *Opisthophyllum* OGILVIE, 1897]. Solitary, elongate, cylindrical. Major septum well developed. *U.Jur.*, Eu.——FIG.

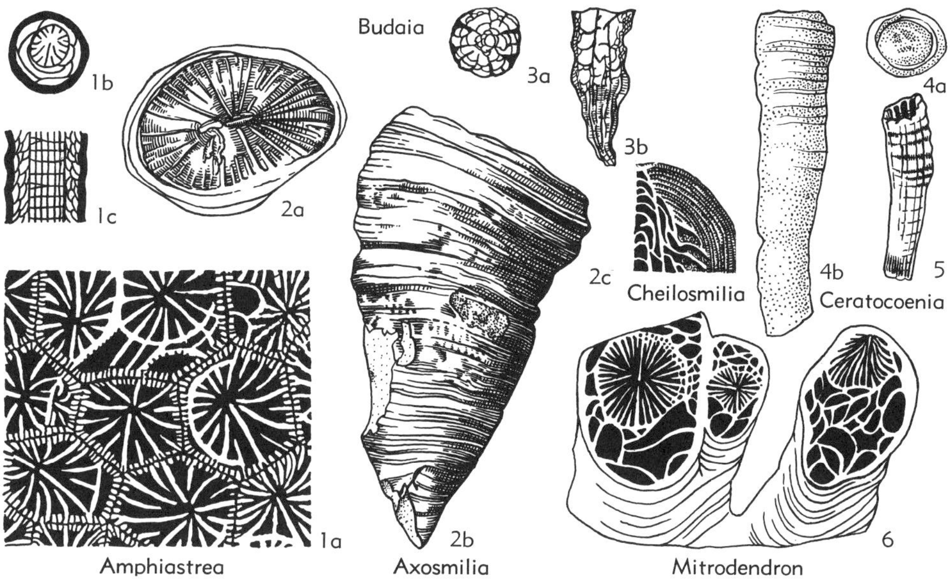

FIG. 292. Faviina (Stylophyllicae): Amphiastreidae (p. F396-F398).

291,*4a,b*. *S. excavata* (KOBY), U.Jur.(Sequan.), Switz.; ×1 (71).——FIG. 291,*4c*. *S. vesiculare* (OGILVIE), U.Jur.(Portl.), Czech.; ×1 (99).

Donacosmilia FROMENTEL, 1861 [**D. corallina;* SD WELLS, 1936] [=?*Pleurophyllia* FROM., 1861; *Pseudothecosmilia* KOBY, 1888; ?*Pseudostylina* ALLOITEAU, 1939; ?*Pseudopisthophyllum* GEYER, 1955]. Like *Sclerosmilia* but forming phaceloid colonies *U.Jur.*, Eu.-AsiaM.

Schizosmilia KOBY, 1888 [**S. excelsa;* SD WELLS, 1936]. Like *Donacosmilia* but budding apparently intratentacular. *U.Jur.*, Eu.——FIG. 291,*3*. **S. excelsa,* U.Jur.(Sequan.), Switz.; *3a,* ×0.5; *3b-e,* ×3 (71).

Polymorphastraea KOBY, 1907 [**P. variabilis*]. Solitary, ceratoid, but producing 4 smaller corallites parricidally in calice; structures as in *Sclerosmilia. M.Jur.*, Eu.

Mitrodendron QUENST., 1880 [**Lithodendron mitratum* QUENST., 1858] [=*Aulastraea* OGILVIE, 1897; *Amphiaulastrea* GEYER, 1955]. Columniform, composed of corallite groups enclosed by common epitheca as result of incomplete separation after intensive budding. Broad outer zone of vesicular endothecal dissepiments. *U.Jur.*, Eu.——FIG. 292, *6*. *M. schaferi* (OGILVIE), U.Jur.(Portl.), Czech.; ×1 (99).

Amphiastrea ÉTALLON, 1859 [**A. basaltiformis*] [=*Diplocoenia* DUNCAN, 1867; *Diplothecastraea* DUNCAN, 1884; *Connectastrea* KOBY, 1904; *Metaulastraea* DIETRICH, 1926]. Like *Donacosmilia* and *Sclerosmilia* but massive, cerioid. Outer vesicular endothecal zone narrow. *M.Jur.-M.Cret.*, Eurasia-Afr.-N.Am.——FIG. 292,*1a*. *A. gracilis* KOBY, U.Jur.(Portl.), Czech., transv. sec., ×4 (99).——FIG. 292,*1b,c*. *A. waltheri* DE ANGELIS, L.Cret.(Apt.), Italy; transv. and long. secs., ×2 (173).

Oyonnaxastraea ALLOITEAU, 1952 [**O. schlumbergeri*]. Like *Amphiastrea* but wall thickened and major septum axially swollen. *U.Jur.*, Eu.

Latusastrea ORB., 1849 [**Explanaria alveolaris* GOLDF., 1831] [=*Pleurocoenia* ORB., 1849; *Thecidiosmilia* KOBY, 1888]. Colonial, foliaceous or incrusting; plocoid. Calices opening towards growing margin like pockets. *U.Jur.-L.Cret.*, Eu. ——FIG. 293,*1*. **L. alveolaris,* U.Jur.(Kimm.), Ger.; ×1 (7).

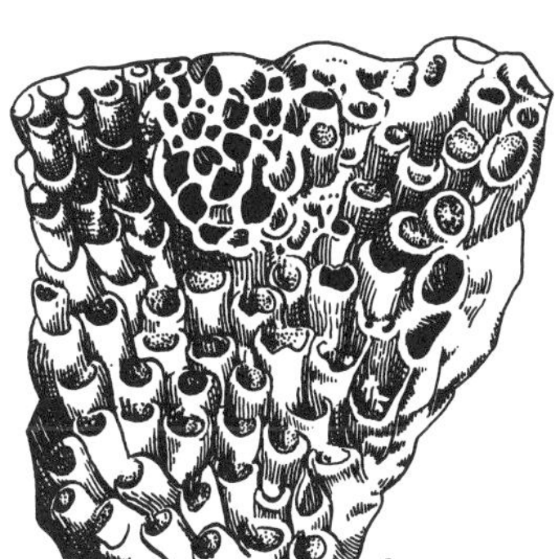

FIG. 293. Faviina (Stylophyllicae): Amphiastreidae (p. F397).

?**Thamnosmilia** Tomes, 1886 [**T. annulata*]. Dendroid, formed by lateral budding as in *Donacosmilia*. Columella trabecular. *M.Jur.*, Eng.

?**Lochmaeosmilia** Wells, 1943 [*pro Eunomia* Lamx., 1821 (*non* Rafinesque, 1815; *nec* Hübner, 1818)] [**Stylosmilia trapeziformis* Gregory, 1900]. Phaceloid, corallites very small, elongate, cylindrical or subprismatic. Endothecal dissepiments tabular (possibly a stylinid). *M.Jur.-U.Cret.*, Eurasia-Afr.

Selenegyra Ogilvie, 1897 [**S. geikiei*]. Phaceloid, colony formation as in *Schizosmilia;* corallites lunate or crescentiform in cross section. *U.Jur.*, Eu.

Axosmilia M.Edw.-H., 1848 [**Caryophyllia extinctorum* Michelin, 1840] [=*Pleurosmilia* Fromentel, 1856; *Blastosmilia* Étallon, 1859; *Trismilia* From., 1861; *Axiphyllum* Quenst., 1880]. Solitary, turbinate, or subcylindrical. Lamellar columella formed by enlarged major septum. Endothecal dissepiments vesicular. *M.Jur.-M.Cret.*, Eu.-Afr.-N.Am.——Fig. 292,*2a,b*. *A. cuneata* (Koby), U.Jur.(Sequan.), Port.; ×1 (71).——Fig. 292,*2c*. *A. pumila* (Koby), U.Jur. (Sequan.), Switz.; septum and wall, ×1 (71).

Columnaphyllia Geyer, 1955 [**C. tithonica*]. Like *Axosmilia* but forming phaceloid colonies. *U.Jur.*, Eu.

Blothrocyathus Wells, 1932 [**B. harrisi*]. Solitary, cylindrical, with relatively few septa, no columella; 2 zones of endothecal dissepiments. *L.Cret.*, N.Am. ——Fig. 291,*2*. **B. harrisi*, L.Cret.(Apt.), Tex.; *2a*, ×0.25, *2b,c*, transv. and long. secs., ×0.5 (157).

?**Plesiosmilia** Milaschewitsch, 1876 [**P. turbinata;* SD Wells, 1936]. Like *Axosmilia* but septa slightly exsert; wall parathecal. *U.Jur.*, Eu.

Placophyllia Orb., 1849 [**Lithodendron dianthus* Goldf., 1827]. Like *Plesiosmilia* but forming small phaceloid colonies. *M.Jur.-U.Jur.*, Eu.——Fig. 291, *1*. *P. rugosa* Becker, U.Jur.(Kimm.), Ger.; *1a*, ×0.5; *1b*, ×2 (7).

Superfamily FAVIICAE Gregory, 1900[1]

[*nom. transl.* Wells, herein (*ex* Faviidae Gregory, 1900)]

Solitary and colonial. Wall septothecal or parathecal. Septa laminar with very rare perforations, margins dentate. *M.Trias.-Rec.*

Family MONTLIVALTIIDAE Dietrich, 1926

Solitary and colonial, hermatypic. Colony formation by various plans of complete and incomplete intratentacular budding, lamellar linkages between corallite centers. Epitheca well developed. Septa exsert, formed by one fan system of large, mostly simple trabeculae, with regular conical dentations and lateral striae or granulations. Endothecal dissepiments abundant. *M.Trias.-Eoc.*

Subfamily MONTLIVALTIINAE Dietrich, 1926

[*nom. transl.* Vaughan & Wells, 1943 (*ex.* Montlivaltiidae Dietrich, 1926)]

Montlivaltiids with columella absent or trabecular. *M.Trias.-Cret.*

Montlivaltia Lamouroux, 1821 [**M. caryophyllata*] [=*Thecophyllia* M.Edw.-H., 1848; *Amblophyllia, Lasmophyllia, Perismilia* Orb., 1849; *?Cyathophyllia* Fromentel, 1865; *Coenotheca* Quenst., 1881; *Montlivaltiopsis, ?Paramontlivaltia, Stereophyllia, Trochophyllia* Alloiteau, 1952]. Solitary, cupolate, trochoid to subcylindrical, usually free in ephebic stage. Columella generally weak. *M.Trias.-Cret.*, cosmop.——Fig. 294,*2a,b*. *M. nattheimensis* Milaschewitz, U.Jur.(Kimm.), Ger.; ×0.5 (7). ——Fig. 294,*2c*. *M. matheyi* Koby, U.Jur.(Raur.), Fr.; lat. aspect of septum and wall, ×0.5 (71).

?**Ellipsosmilia** Orb., 1849 [**Montlivaltia? cornucopia* M.Edw.-H., 1849] [=*Veliphyllum* Quenst., 1880]. Like *Montlivaltia* but trabeculae inclined only inwards with no line of divergence and with irregular septal dentation (?may be a stylophyllid). *Cret.*, Eu.

Dimorphosmilia Tomes, 1902 [**D. eboracensis*]. Thamnasterioid; formed by circumoral followed by intramural budding. Radiating collines. *M.Jur.*, Eu.

Dimorphastrea Orb., 1850 [**D. grandiflora;* SD Gregory, 1900] [=*Dimorphocoenia* Fromentel, 1857]. Like *Dimorphosmilia* but colony formation by circumoral budding only; no collines. *U.Jur.-M.Cret.*, Eurasia-Afr.——Fig. 294,*5*. *D. crassisepta* Orb., L.Cret.(Urgon.), Fr.; ×2 (72).

Thecosmilia M.Edw.-H., 1848 [**Lithodendron trichotomum* Goldf., 1826] [=*Hymenophyllia* M.Edw.-H., 1851; *Bavarosmilia* Kühn, 1942]. Phaceloid colonies formed by polystomodaeal budding, mono- to tristomodaeal conditions permanent. Polyphyletic. *M.Trias.-Cret.*, cosmop.——Fig. 294,*7a*. *T. annularis* (Fleming), U.Jur., Eng.; ×0.5 (94).——Fig. 294,*7b*. *T. costata* Fromentel, U.Jur.(Raur.), Switz.; lat. aspect of septum and wall, ×2 (71).

Palaeastraea Kühn, 1936 [**Phyllocoenia grandissima* Frech, 1890] [=*Thigmastrea* Wells, 1937]. Massive, plocoid, corallites mono- to tricentric, derived from Triassic Thecosmilias by incomplete separation of corallites. *M.Trias.-U.Trias.*, Eu.-N.Am.——Fig. 294,*1*. **P. grandissima*, U.Trias. (Nor.), Aus., transv. sec., ×1 (50).

Elysastraea Laube, 1864 [*pro Clausastrea* Orb., 1949 (*non* M.Edw.-H. 1849)] [**E. fischeri*]. Like *Palaeastraea* but cerioid. *M.Trias.-L.Jur.*, Eurasia-N.Am.

Margarastraea Frech, 1896 [**Isastrea (Margarastraea) klipsteini*]. Like *Elysastraea* but meandroid.

[1] The name Faviicae, which is accepted by the author for this superfamily, is not authorized by the International Rules, since it is based on a family group other than the first-published such taxon included in the assemblage (Oculinidae Gray, 1847, or Meandrinidae Gray, 1847). Accordingly, the author has been urged to submit an application to ICZN for validation of Faviicae by action under plenary powers.—Editor.

Series discontinuous, united directly by fused walls; collines acute. *U.Trias.*, Eu.

Phyllogyra Tomes, 1882 [**Symphyllia etheridgei* Duncan, 1874; SD Wells, 1936]. Meandroid, derived from Middle Jurassic Thecosmilias by polystomodaeal budding, commonly linear. Series discontinuous, closely joined by costae and exothecal dissepiments; collines rounded. *M.Jur.*, Eng.

Complexastrea Orb., 1849 [**C. subburgundiae* (=*Astrea rustica* Defrance, 1826] [=*Confusastrea* Orb., 1849]. Massive, plocoid, derived from Middle Jurassic Thecosmilias. Corallites monocentric, with parathecal wall. *M.Jur.-M.Cret.*, Eu.——Fig. 294,*4a*. **C. rustica*, U.Jur.(Raur.), Switz.; lat. aspect of septum and wall, ×2 (71).——Fig. 294,*4b*. *C. magnifica* (Tomes), M.Jur.(Bath.), Eng.; ×0.5 (131).

Meandrastrea Orb., 1849 [**Astrea pseudomeandrina* Michelin, 1841]. Massive, submeandroid colonies formed by polystomodaeal budding; series short, some corallites monocentric. *U.Cret.*, Eu.

Latiphyllia Fromentel, 1861 [**Lobophyllia requienii* Michelin, 1841; SD Wells, 1936] [=*Glyphephyllia* From., 1873]. Flabello-meandroid, colony formation by polystomodaeal linear budding; series free laterally. *U.Jur.-Cret.*, Eurasia-Madag.——Fig. 294,*6*. *L. suevica* (Quenst.), U. Jur.(Kimm.), Ger.; ×0.5 (7).

Fromentellia Ferry, 1863 [**F. fabryana;* SD Wells, 1936]. Like *Latiphyllia* but series closely united to summits by costae and exothecal dissepiments, separated by ambulacra. *M.Jur.-U.Jur.*, Eu.

Lasmosmilia Orb., 1849 [**Lobophyllia lobata* Michelin, 1846 (=*L. lobata* Blainv., 1830)]. Massive, plocoid, corallites united nearly to summits by costae and exothecal dissepiments. *M.Cret.-U.Cret.*, Eu.

Mycetophyllopsis Oppenheim, 1930 [**Mycetophyllia antiqua* Reuss, 1854]. Meandroid, series directly united over low common paratheca. *U.Cret.*, Eu. ——Fig. 294,*3*. **M. antiqua*, U.Cret.(Turon.), Aus.(Gosau); ×0.5 (112).

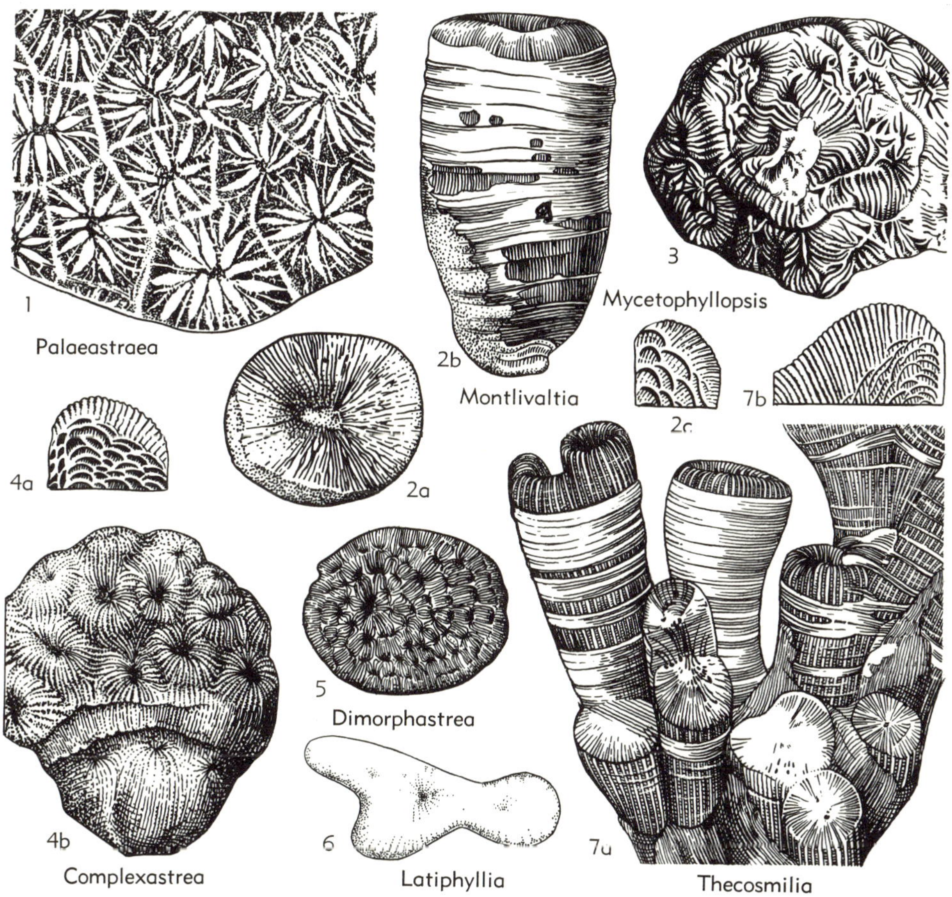

Fig. 294. Faviina (Faviicae): Montlivaltiidae (p. F398-F399).

Subfamily PLACOSMILIINAE Alloiteau, 1952

Montlivaltiids with well-developed lamellar columella. *U.Jur.-Eoc.*

Peplosmilia M.Edw.-H., 1850 [**P. austeni*] [=*Plesiosmilia* Koby, 1884 (*non* Sismonda, 1871)]. Solitary, subcylindrical, fixed. *U.Jur.-M. Cret.,* Eu.——Fig. 296,*3*. **P. austeni,* M.Cret. (Cenom.), Eng.; *3a,b,* ×0.5 (94).

Elasmophyllia d'Achiardi, 1875 [**E. gigantea*]. Phaceloid; colony formation by polystomodaeal budding, corallites mono- to tricentric. *Cret.-Eoc.,* Eu.-N.Am.-S.Am.——Fig. 295,*2*. *E. deformis* (Reuss), U.Cret.(Turon.), Aus.(Gosau); *2a,b,* ×1 (100).

Placosmilia M.Edw.-H., 1848 [**Turbinolia cymbula* Michelin, 1846; SD M.Edw.-H., 1850] [=*Lasmogyra* Orb., 1849; *Epigyra* Ferry, 1870; *Firia* Gregorio, 1894]. Flabellate, with a linear contorted series. *U.Cret.-Eoc.,* Eu.——Fig. 298,*2*. *P. tortuosa* (Felix), U.Cret(Turon.), Aus.(Gosau); transv. sec., ×1 (42).

Astrogyra Felix, 1900 [**Gyrosmilia edwardsi* Reuss, 1854]. Like *Lasmogyra* but with forked series united nearly to summits by costae and exothecal dissepiments; narrow ambulacra. *U.Cret.,* Eu.——Fig. 295,*4*. **A. edwardsi,* U.Cret.(Turon.), Aus.(Gosau); ×05 (100).

Taxogyra Wells, 1937 [*pro Heterophyllia* Orb., 1849 (*non* M'Coy, 1849)] [**Meandrina macroreina* Michelin, 1847]. Like *Astrogyra* but lacking ambulacra; series closely united to summits. *U. Cret.,* Eu.

Family FAVIIDAE Gregory, 1900

Solitary and colonial, mostly hermatypic. Colony formation by extratentacular or various plans of intratentacular budding. Septothecate or parathecate, rarely partially synapticulothecate. Septa exsert, laminar, formed by 1 or 2 fan systems of simple (compound in some later forms) trabeculae, more or less regularly dentate marginally. Paliform lobes formed by inner fan system commonly developed. Columella trabecular or laminar, rarely styliform or absent. *M. Jur.-Rec.*

Subfamily FAVIINAE Gregory, 1900

[*nom. transl.* Vaughan & Wells, 1943 (*ex* Faviidae Gregory, 1900)]

Solitary and colonial faviids, hermatypic. Colony formation by intratentacular budding (rare extratentacular budding). Septa formed by one fan system of simple trabeculae, commonly with small inner fan system. *U.Jur.-Rec.*

?Cladophyllia M.Edw.-H., 1851 [**Lithodendron*

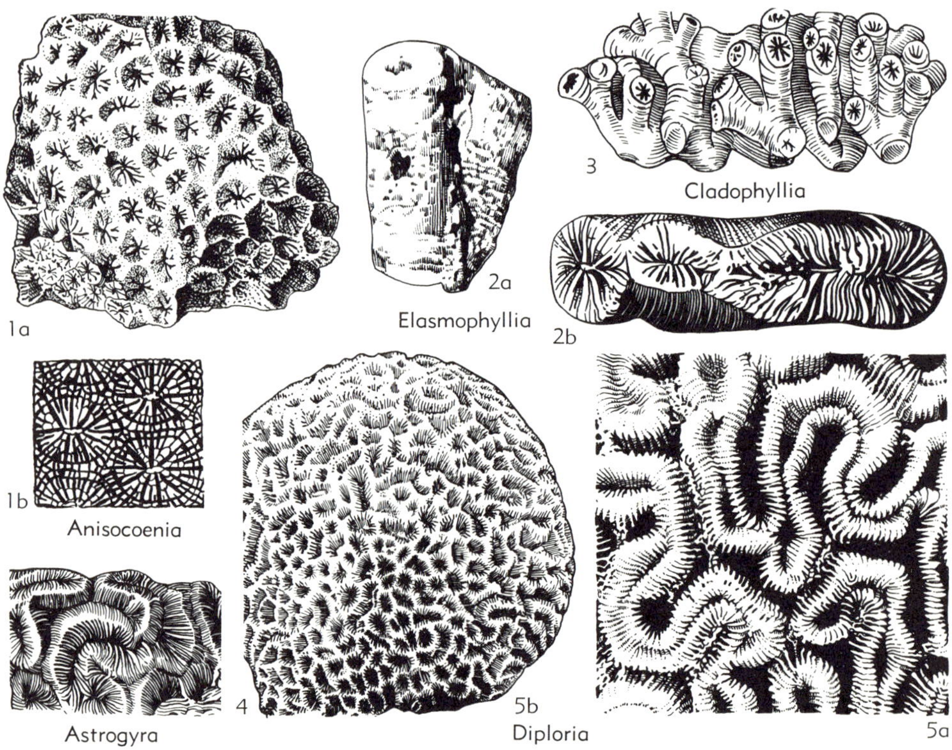

Fig. 295. Faviina (Faviicae): Montlivaltiidae, Faviidae (p. F400-F402).

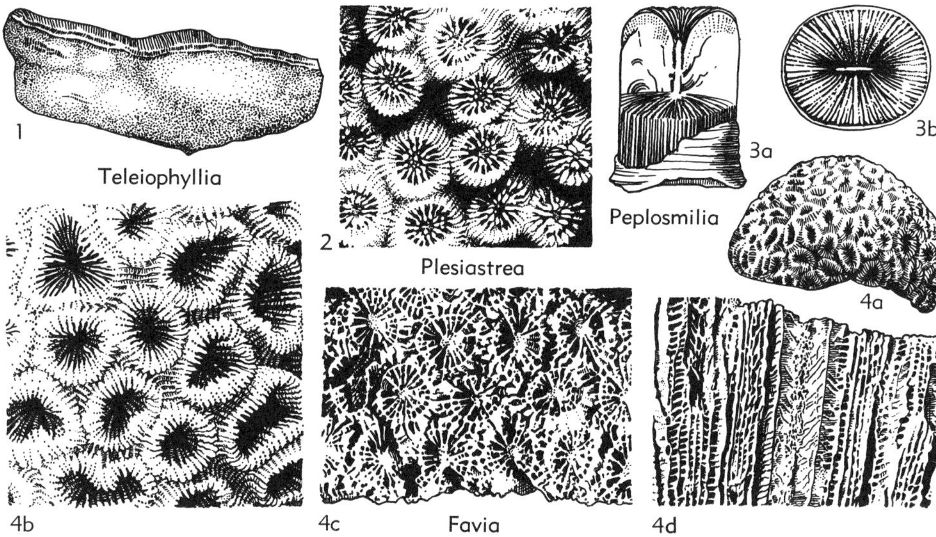

FIG. 296. Faviina (Faviicae): Montlivaltiidae, Faviidae (p. F402).

dichotomum GOLDF., 1829; SD WELLS, 1933]. Phaceloid tufts. Corallite walls epithecal; columella feeble, trabecular. *U.Jur.-M.Cret.,* Eu.-N.Am.-S. Am.——FIG. 295,*3*. *C. babeana* (ORB.), M.Jur. (Bath.), Eng.; ×1 (94).

Indosmilia GERTH, 1933 [**I. rembangensis*]. Solitary, trochoid, lacking columella. *Oligo.-Mio.,* Eu.-E.Indies.

Petrophylliella FELIX, 1925 [*pro Petrophyllia* FELIX, 1885 (*non* CONRAD, 1855)] [**Caryophyllia grumi* CATULLO, 1847]. Solitary, ceratoid, with well-developed trabecular columella. *Oligo.,* Eu.——FIG. 300,*5*. **P. grumi,* M.Oligo., Italy; *5a,* ×0.5; *5b,* ×1 (116).

Grumiphyllia WELLS, 1937 [*pro Grumia* OPPENHEIM, 1899 (*non* ALPHÉRAKY, 1892)] [**Grumia diploctenium* OPPENHEIM, 1899]. Solitary (or ?colonial), flabellate, lacking columella. Homeomorphic with *Diploctenium. Oligo.,* Eu.

Laterophyllia KÜHN, 1933 [=*L. turriformis*]. Forming tall colonies by budding from edge zone, buds so produced in turn budding intratentacularly to form short straight series. Columella lamellar. *Mio.,* Asia.

Caulastrea DANA, 1846 [**C. furcata*] [=*Dasyphyllia* M.EDW.-H., 1848; *Cricotheca* QUENST., 1881]. Phaceloid colonies produced by mono- to tristomodaeal budding, permanent condition mono- to tricentric. Columella feeble. *Eoc.-Rec.,* Eu.-Asia-IndoPac.——FIG. 297,*1*. **C. furcata,* Rec., Fiji; ×0.5 (88).

Desmocladia REUSS, 1874 [**D. septifera*]. Like *Caulastrea* but with short series partially united by some exothecal dissepiments. *Oligo.,* Eurasia.

Astraeosmilia ORTMANN, 1892 [**A. connata*]. Like *Caulastrea* but subplocoid. *Rec.,* E.Afr.

Bikiniastrea WELLS, 1954 [**B. laddi*]. Like *Caulastrea* but distomodaeal budding; corallum subdendroid, the corallites anastomosing, with some coenosteum. Corallite wall parathecal. *Rec.,* Pac.

Barabattoia YABE & SUGIYAMA, 1941 [**B. mirabilis*]. Subplocoid, with corallites free at summits; colony formation by intra- and extratentacular budding. Intermediate between *Bikiniastrea* and *Plesiastrea. Rec.,* Pac.

Plesiastrea M.EDW.-H., 1848 [*pro Astrea* LAMARCK, 1801 (*non* BROWN, 1789; *nec* BOLTEN, 1798)] [**Astrea versipora* LAM., 1816] [=*Heliastrea* M. EDW.-H., 1857; *?Plesiastreopsis* CHEVALIER, 1954]. Plocoid, corallites closely united nearly to summits; colony formation mostly by extratentacular budding. *Mio.-Rec.,* Eu.-IndoPac.——FIG. 296,*2*. **P. versipora,* Rec., Austral.(Torres Strait); ×2 (148).

FIG. 297. Faviina (Faviicae): Faviidae (p. F401).

Favia Oken, 1815 [**Madrepora fragum* Esper, 1795; SD M.Edw.-H., 1848] [=*Dipsastraea* Blainv., 1830; *Fissicella* Dana, 1846; *Parastrea, Phymastrea* M.Edw.-H., 1848; *Ellipsocoenia, Thalamocoenia* Orb., 1850; *Stegiastrea* Meneghini, 1857; *Clypeofavia* Angelis, 1895; *?Pseudaulina* Alloiteau, 1954]. Plocoid, massive, foliaceous, or incrusting colonies formed by mono- to tristomodaeal budding, corallites permanently monocentric. Vesicular endo- and exothecal dissepiments. Columella trabecular, spongy. *Cret.-Rec.*, cosmop. ——Fig. 296,*4a*. **F. fragum*, Rec., Fla.; ×0.5 (199).——Fig. 296,*4b-d*. *F. speciosa* (Dana), Rec., E.Indies; Cocos-Keeling; ×1 (148). (Also Fig. 249,*2*.)

Barbadiastrea Wells, 1945 [**B. favioides*]. Like *Favia* but with lamellar columella. *Eoc.*, W.Indies.

Nefocoenia Oppenheim, 1930 [*pro Phyllastraea* Fromentel, 1877 (*non* Dana, 1846)] [**Areacis lobata* Reuss, 1854] [=*Proplesiastraea* Oppenheim, 1930]. Like *Favia* but with spinose coenosteum and short, thin, lamellar columella. *U.Cret.*, Eu.——Fig. 298,*3*. **N. lobata*, U.Cret.(Turon.), Aus.(Gosau); transv. sec., ×8 (42).

Diploria M.Edw.-H., 1848 [**Meandrina cerebriformis* Lamarck, 1816 (=*Madrepora labyrinthiformis* Linné, 1758)] [=*Maeandrella* Oppenheim, 1930 (*non* Perna, 1903)]. Like *Favia* but meandroid; colony formation by polystomodaeal intramural budding with lateral branching and terminal forking; series long with thick collines, separated by ambulacra in some species. Columella continuous, trabecular. *U.Cret.-Rec.*, Eu.-N.Am.-W.Indies.——Fig. 295,*5a*. **D. labyrinthiformis*, Rec., Fla.; ×1 (151).——Fig. 295,*5b*. *D. strigosa* (Dana), Rec., Fla.; ×0.5 (151). (Also Fig. 248*B*.)

Favites Link, 1807 [**F. astrinus* (=*Madrepora abdita* Ellis & Solander, 1786; SD Vaughan, 1901)] [=*Astrophyllia* Ehr., 1834; *Cellastraea* Blainv., 1834; *Aphrastrea, Prionastrea* M.Edw.-H., 1848; *Clausastrea* M.Edw.-H., 1849; *Plerastrea* M. Edw.-H., 1851; *Metastraea* M.Edw.-H., 1857; *Dyctioastraea, Halysiastraea* Sismonda, 1871; *Ellasmoastraea, Plesiophyllia* Michelotti in Sismonda, 1871; *?Narcissastraea* Pratz, 1883; *Allastraea* Gregory, 1900]. Like *Favia* but cerioid. *Eoc.-Rec.*, Eu.-W.Indies-IndoPac.——Fig. 299,*2a,b*. **F. abdita*, Rec., Fiji; *2a*, ×1; *2b*, ×2 (148)——Fig. 299,*2c*. *F. affinis* Klunzinger, Rec., Red Sea; lat. aspect of septum, ×3 (151).

Oulophyllia M.Edw.-H., 1848 [**Meandrina crispa* Lamarck, 1816] [=*Ulophyllia* M.Edw.-H., 1857]. Like *Favites* but meandroid; colony formation by intramural polystomodaeal budding. Series short, discontinuous, separated by simple, acute, parathecal collines. Centers linked by trabeculae. *M. Oligo.-Rec.*, Eu.-IndoPac.——Fig. 300,*6*. *O. aspera* Quelch, Rec., Pac.; ×1 (88).

Goniastrea M.Edw.-H., 1848 [**Astrea retiformis* Lamarck, 1816] [=*?Coelastrea* Verrill, 1866]. Cerioid or submeandroid; colony formation by mono- to tristomodaeal budding, corallites permanently monocentric. Septa with scarcely dentate paliform lobes. Columella feeble. *Eoc.-Rec.*, N.Am.-W.Indies-IndoPac.——Fig. 298,*6*. *G. pectinata* (Ehr.), Rec., Austral.(Torres Strait); ×2 (148). (Also Figs. 248*A*, 249,*3*.)

Anisocoenia Reuss, 1867 [**A. crassisepta*] [=*Favoidea* Reuss, 1867]. Like *Goniastrea* but lacking columella and with reduced paliform lobes. *Mio.*, E.Indies.——Fig. 295,*1*. **A. crassisepta*, Mio., Java; *1a*, ×1; *1b*, ×2 (115).

Lamellastraea Duncan, 1867 [**L. smythi*]. Like *Goniastrea* but with short, solid, lamellar columella. *Oligo.*, W.Indies.——Fig. 298,*1*. **L. smythi*, M.Oligo., Antigua; ×2.5 (198).

Ogilviastraea Oppenheim, 1930 [*pro Placohelia* Felix, 1903 (*non* Počta, 1887)] [**Placohelia bigemmis* Felix, 1903]. Small subplocoid colonies formed by mono- or distomodaeal budding, permanent condition monocentric. Coenosteum extensive, solid. Columella elongate, commonly appearing lamellar. *U.Cret.*, Eu.——Fig. 300,*3*. **O. bigemmis*, U.Cret.(Turon.), Aus.(Gosau); *3a*, ×1; *3b*, ×3 (42).

?Diplocoenia Fromentel, 1857 [**D. mirabilis*]. Cerioid or subplocoid; colony formation by mono- or distomodaeal budding, permanent condition monocentric. Endothecal dissepiments vesicular, forming false "inner wall." Columella substyliform, fixed to several larger septa. *U.Jur.-L.Cret.*, Eu.-Afr.

Platygyra Ehr., 1834 [**P. lamellina;* SD Brueggemann, 1879] [=*Coeloria, Astroria* M.Edw.-H., 1848; *Brachymeandrina* Duncan, 1884; *Miria* Gregorio, 1894]. Meandroid; colony formation by linear intramural polystomodaeal budding with lateral branching and terminal forking. Collines narrow, septothecal. Larger septa commonly with small internal paliform lobes. Columella continuous, trabecular. *Eoc.-Rec.*, Eu.-IndoPac.——Fig. 298,*4*. **P. lamellina*, Rec., Austral.(Torres Strait); *4a*, ×1; *4b*, ×4, lat. aspect of septum; *4c*, ×4 (148).

Leptoria M.Edw.-H., 1848 [**Meandrina phrygia* Lamarck, 1816 (=*Madrepora phrygia* Ellis & Solander, 1786)] [=*Cycloria* Reuss, 1854]. Meandroid; colony formation by intramural polystomodaeal budding. Collines simple, thin, septothecal. Columella thin, lamellar, continuous or not. *U.Cret.-Rec.*, Eu.-IndoPac.

L. (Leptoria). *U.Cret.-Rec.*——Fig. 299,*3*. *L. (L.) gracilis* (Dana), Rec., Fiji; *3a*, ×1; *3b*, ×3 (148). (Also Fig. 249,*4*.)

L. (Dictuophyllia) Blainv., 1830 [**Meandrina reticulata* Goldf., 1827] [=*?Anisoria* Vidal, 1917]. Like *L. (Leptoria)* but collines separated by costate ambulacra. *U.Cret.-Oligo.*, Eu.-W. Indies-N.Am.——Fig. 298,*5*. *L. (D.) crassolamellosa* (M.Edw.-H.), U.Cret.(Turon.), Aus. (Gosau); ×0.5 (112).

Hydnophora Fischer, 1807 [**H. demidovii*

(=*Madrepora exesa* PALLAS, 1766)] [=*Monticularia* LAMARCK, 1816; *Actinhelia, Monticulastraea* DUNCAN, 1880; *Monticulina* SAVILLE-KENT, 1893; *Hydnophorella* DELAGE & HÉROUARD, 1901; *Staminocoenia* GREGORY, 1930; *Hydnophoraraea* OPPENHEIM, 1930]. Hydnophoroid; colony formation by circummural polystomodaeal budding. Collines discontinuous, short, conical. Columella trabecular to lamellar, discontinuous. *Cret.-Rec.,* Eurasia-W. Indies-S.Am.-IndoPac.——FIG. 300,*1*. *H. microconos* (LAM.), Rec., Austral.(Torres Strait); *1a,* ×1; *1b,* ×3 (148). (Also Figs. 248*D*, 249,*5*.)

Thysanus DUNCAN, 1863 [**T. excentricus*]. Flabellate; colony formation by intramural budding proceeding away from parent center in one direction. Larger septa with small internal lobes. Columella trabecular below, sublamellar and interrupted above. *Mio.-Plio.,* W.Indies-N.Am.——FIG. 300,*2*. **T. excentricus,* Mio., Jamaica; *2a,b,* ×1 (195).

Teleiophyllia DUNCAN, 1864 [**T. grandis*]. Like *Thysanus* but budding proceeds in both directions from parent center. *Mio.,* W.Indies.——FIG. 296,*1*. **T. grandis,* Dominican Republic; ×0.5 (29).

Manicina EHR., 1834 [**Madrepora areolata* LINNÉ, 1758; SD M.EDW.-H., 1850] [=*Podasteria* EHR., 1834]. Like *Teleiophyllia* but with forking and branching of series and development of ambulacra. *Mio.-Rec.,* W.Indies-Fla.——FIG. 299,*1*. **M. areolata,* Rec., Fla.; ×0.75 (201).

Colpophyllia M.EDW.-H., 1848 [**Meandrina gyrosa* LAMARCK, 1816 (=*Madrepora gyrosa* ELLIS & SOLANDER, 1786, =*Madrepora natans* MÜLLER, 1775)] [=*Hydnophyllia* REIS, 1889]. Meandroid; colony formation by intramural budding with terminal forking. Collines discontinuous, rarely with ambulacra. Series commonly discontinuous, linkages lamellar. Septa with very small or no internal lobes. *Eoc.-Rec.,* Eu.-W.Indies.——FIG. 300,*4*. *C. amaranthus* (MÜLLER), Rec., Barbados; ×1 (88).

Montigyra MATTHAI, 1928 [**M. kenti*]. Subhydnophoroid; colony formation by intramural and circummural budding. Collines discontinuous; series continuous with lamellar linkages. *Rec.,* Ind.O.

Subfamily MONTASTREINAE Vaughan & Wells, 1943

Colonial, hermatypic faviids. Colony for-

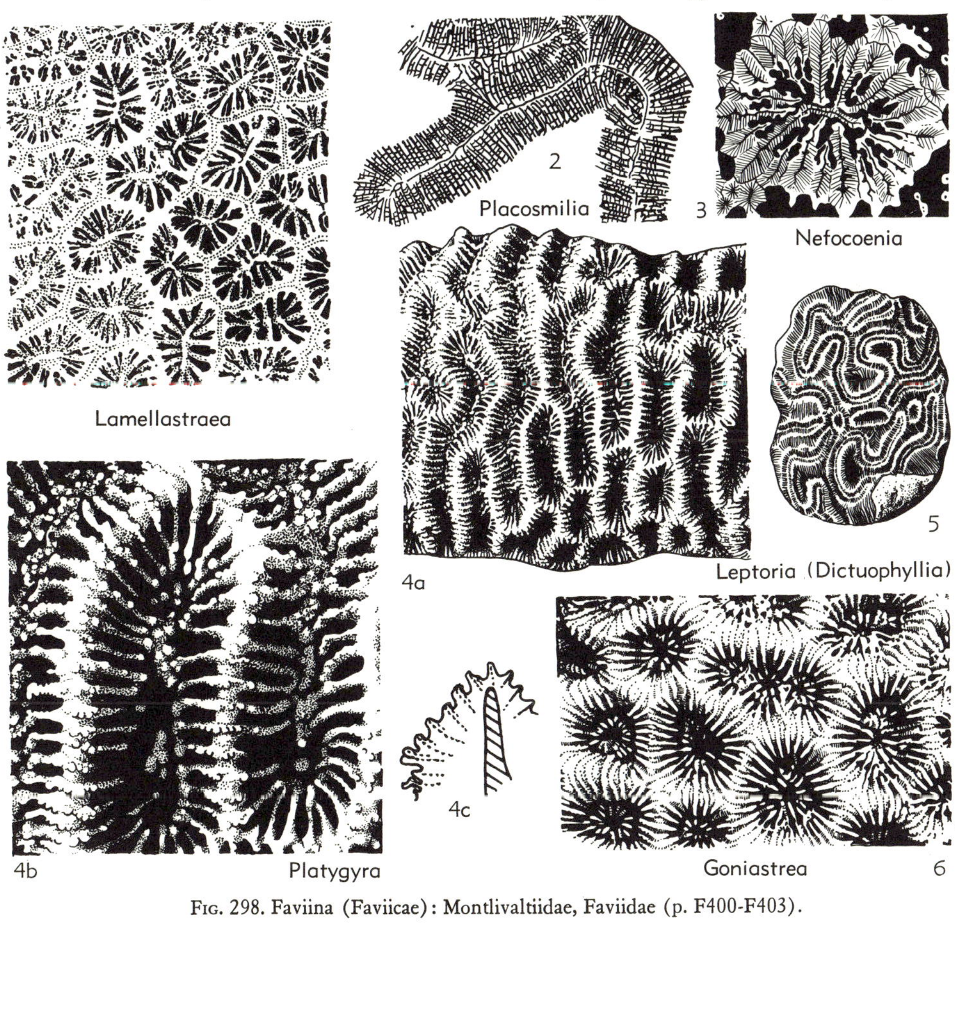

FIG. 298. Faviina (Faviicae): Montlivaltiidae, Faviidae (p. F400-F403).

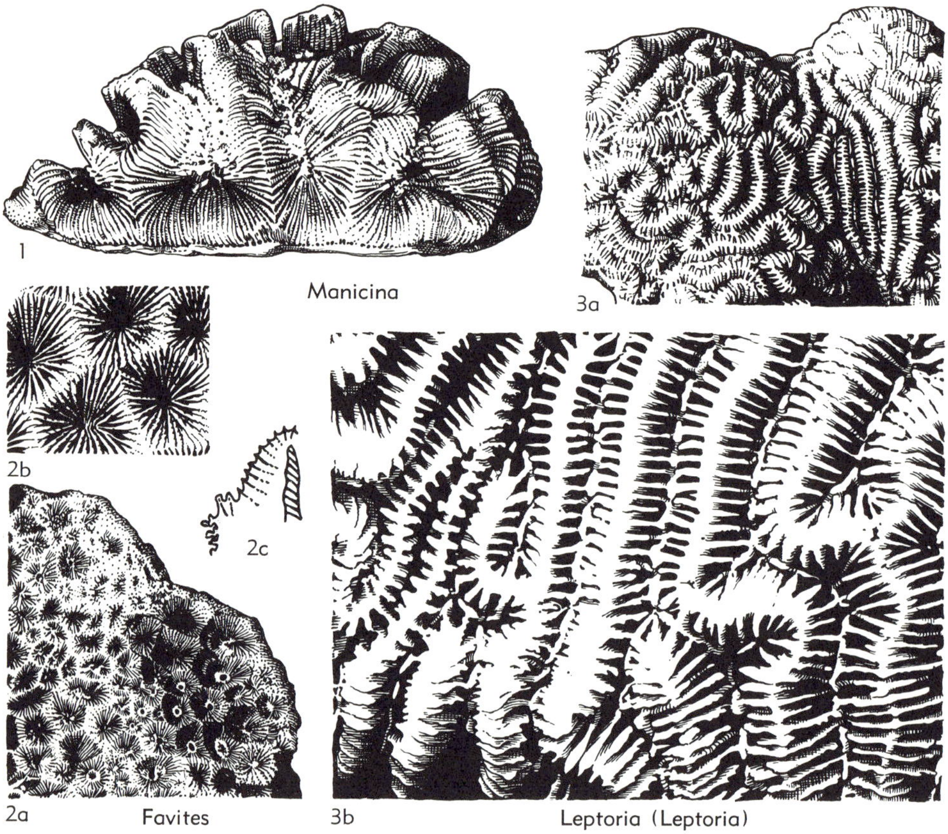

FIG. 299. Faviina (Faviicae): Faviidae (p. F402).

mation by extratentacular budding (with a few exceptions). Septa formed by one fan system of mostly simple trabeculae. *M.Jur.-Rec.*

Goniocora M.EDW.-H., 1851 [**Lithodendron sociale* ROEMER, 1836] [=*?Thamnocoenia* TOMES, 1885]. Phaceloid. Columella styliform. No paliform lobes. *M.Jur.-U.Jur.*, Eu.——FIG. 301,*4*. **G. socialis*, U.Jur.(Raur.), Eng.; *4a*, ×1.5; *4b*, ×3 (94).

Rhabdocora FROMENTEL, 1873 [**R. cretacea*]. Like *Goniocora* but with strong lamellar columella. *M.Cret.-U.Cret.*, Eu.

Cladocora EHR., 1834 [**Madrepora caespitosa* LINNÉ, 1767; SD M.EDW.-H., 1850] [=*Procladocora* ALLOITEAU, 1952]. Phaceloid. Paliform lobes opposite all but last cycle of septa, merging with papillose columella. *U.Cret.-Rec.*, Eu.-Medit.-Atl.-W.Indies-N.Am.-E.Pac.

C. (Cladocora). *U.Cret.-Rec.*——FIG 302,*1*. *C. arbuscula* (LESUEUR), Rec., Fla.; *1a*, ×0.75; *1b*, ×5 (198).

C. (Dendrocora) DUNCAN, 1876 [**D. fissipara*]. Like *Cladocora* but colony formation mostly by intratentacular budding. *Rec.*, W.Afr.

Stylocora REUSS, 1871 [**S. exilis*]. Like *Cladocora* but columella styliform; paliform lobes reduced. *Mio.*, Eu.

Pleurocora M.EDW.-H., 1848 [**Lithodendron gemmans* MICHELIN, 1847] [=*Stylocora* FROMENTEL, 1873 (*non* REUSS, 1871); *?Phyllohelia* ALLOITEAU, 1952]. Like *Cladocora* but subplocoid, subdendroid, or explanate, with coenosteum between corallites formed by costae and exothecal dissepiments. *M.Cret.-U.Cret.*, Eu.-N.Am.——FIG. 301,*1*. *P. angelisi* FELIX, U.Cret.(Turon.), Carp., ×2 (177).

Psammiophora FROMENTEL, 1870 [**P. cenomana*]. Corallum compressed, thin lamina with calices on both sides. Corallites protuberant, united by smooth coenosteum. Paliform lobes and papillose columella. *M.Cret.-U.Cret.*, Eu.

Montastrea BLAINV., 1830 [**Astrea guettardi* DEFRANCE, 1826; SD SMITH & LANG, 1935] [=*?Astroites* WALCH, 1775; *Orbicella* DANA, 1846; *Phyllocoenia* M.EDW.-H., 1848; *Actinocoenia* ORB.,

1849; *Cellulastraea* BLANCKENHORN, 1890; *Hydnophoropsis* SÖHLE, 1899; *Phyllocoeniella* FELIX, 1926; *Prodiploastrea* WELLS, 1934; *Neocoenia* HACKEMESSER, 1936; *Aquitanastrea, Heliastreopsis, Phyllocaeniopsis, ?Provinciastrea* CHEVALIER, 1954]. Massive, incrusting, or subfoliaceous, plocoid colonies. Septothecate. Septal margins regularly dentate. Columella trabecular. *U.Jur.-Rec.,* Eu.-W. Indies-W.Afr.-Brazil-N.Am.——FIG. 301,*5a.* **M. annularis* (ELLIS & SOLANDER), Rec., P.R., ×3 (195).——FIG. 301,*5b. M. corollaris* (REUSS), U. Cret.(Turon.), Aus.(Gosau); transv. sec., ×5 (42).

Diploastrea MATTHAI, 1914 [**Astrea heliopora* LAMARCK, 1816] [=?*Thegioastraea* SISMONDA, 1871]. Plocoid colonies. Walls mostly septothecate, but partially synapticulothecate and porous at level of calices. Septa formed by compound trabeculae (especially in later forms); dentations relatively large; columella well developed. *Cret.-Rec.,* Eu.-W.Indies-N.Am.-S.Am.-IndoPac.——FIG. 302,*4a. D. harrisi* WELLS, L.Cret.(Apt.), Tex.; transv. sec., ×3 (157).——FIG. 302,*4b,c.* **D. heliopora,* Rec., Fr.Somali.; *4b,* ×1; *4c,* ×3 (195).

?**Oulastrea** M.EDW.-H., 1848 [**Astrea crispata* LAMARCK, 1816] [=*Ulastrea* M.EDW.-H., 1857]. Like *Montastrea* and *Diploastrea* but wall more porous and septal structures smaller. Paliform lobes opposite larger septa, merging with papillose columella (may be an agariciid). *Rec.,* IndoPac.——FIG. 301,*2.* **O. crispata,* Rec., Japan; *2a,* ×1; *2b,* ×3 (164).

Antiguastrea VAUGHAN, 1919 [*pro Heterastraea* REIS, 1889 (*non* TOMES, 1888)] [**Astraea cellulosa* DUNCAN, 1863]. Like *Montastrea* but subcerioid, and with thin lamellar columella. *U.Cret.-Oligo.,* Eu.-N.Am.-W.Indies-S.Am.——FIG. 303,*3.* **A. cellulosa,* U.Oligo., USA; ×2 (149).

Placocoenia ORB., 1849 [**Astrea macrophthalma* GOLDF., 1829]. Like *Antiguastrea* but plocoid, with short, stout, lamellar columella. *Cret.,* Eu.-N.Am.——FIG. 302,*2. P. orbignyana* REUSS, U. Cret.(Turon.), Aus.(Gosau); transv. sec., ×4 (42).

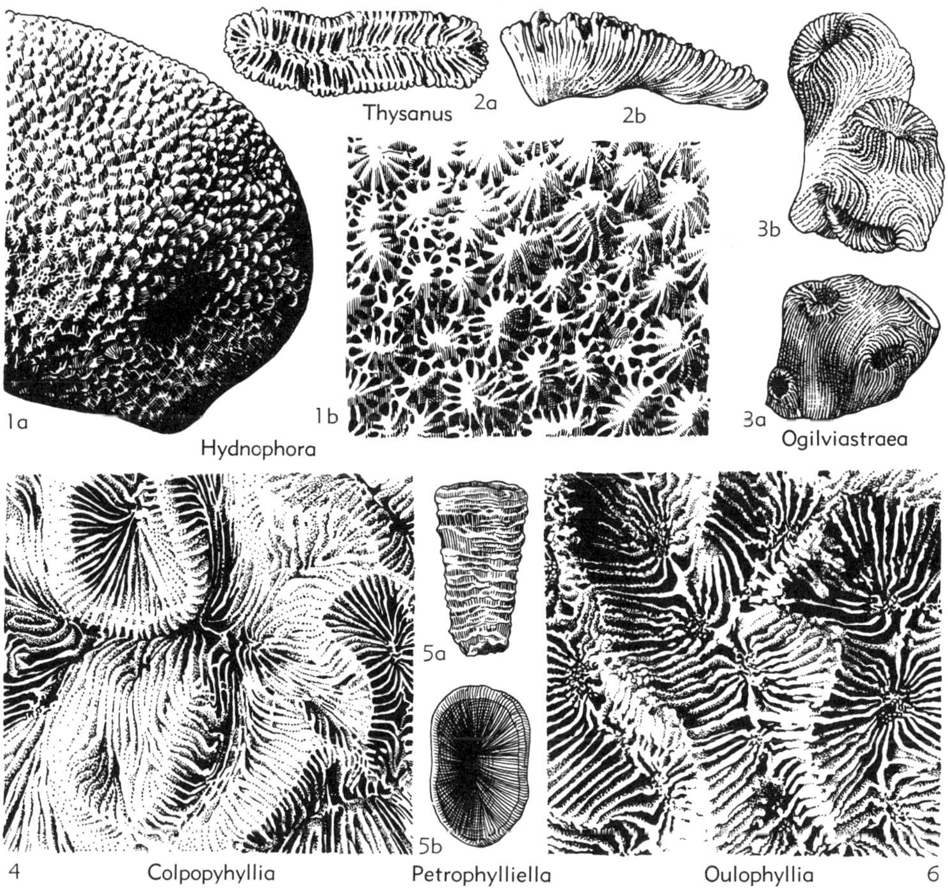

FIG. 300. Faviina (Faviicae): Faviidae (p. F401-F404).

Leptastrea M.Edw.-H., 1848 [**L. roissyana;* SD M.Edw.-H., 1850] [=*Baryastrea* M.Edw.-H., 1848]. Like *Montastrea* but subcerioid, with dense coenosteum. Columella generally papillose. Septa minutely dentate. *M.Oligo.-Rec.,* IndoPac.——Fig. 303,*4a. L. bottae* M.Edw.-H., Austral.(Torres Strait); ×4 (194).

Solenastrea M.Edw.-H., 1848 [**Astrea turonensis* Michelin, 1847; SD M.Edw.-H., 1850] [=*D'Achiardia* Duncan, 1884]. Like *Montastrea* but coenosteum vesicular, nearly devoid of costae, with blistered surface. Paliform lobes opposite first 2 septal cycles. *Oligo.-Rec.,* Eurasia-W.Indies-N. Am.——Fig. 303,*4b. S. bournoni* M.Edw.-H., Rec., W.Indies; ×4 (195).

Cyphastrea M.Edw.-H., 1848 [**Astrea microphthalma* Lamarck, 1816] [=*Chypastraea* Sismonda, 1871; *Quelchia* Duncan, 1886]. Like *Montastrea* but costae rarely extending over coenosteum, the surface of which is spinose. Columella trabecular, spongy. *Oligo.-Rec.,* W.Indies-Indo-Pac.——Fig. 301,*3. *C. microphthalma,* Rec., Cocos-Keeling; ×4 (148).

Multicolumnastraea Vaughan, 1899 [**Heliastraea cyathiformis* Duncan, 1865]. Like *Cyphastrea* but columella formed by 3 or 4 trabecular pillars. *U. Cret.,* W.Indies-Mex.——Fig. 302,*3. *M. cyathiformis,* U.Cret.(Camp.), Jamaica; transv. sec., ×5 (151).

Agathelia Reuss, 1854 [**A. asperella*]. Like *Cyphastrea* but much intratentacular budding; coenosteum dense with minutely granulated surface. *M.Cret.-U.Cret.,* Eu.-Tex.——Fig. 303,*2. *A. asperella,* U. Cret.(Turon.), Aus.(Gosau); *2a,* ×1 (112); *2b,* trans. sec., ×5; *2c,* long. sec., ×5 (42).

Placophora Fromentel, 1879 [**P. neocomiensis*]. Like *Cyphastrea* but with short, lamellar columella. *L.Cret.,* Eu.

Tarbellastraea Alloiteau, 1950 [**Astrea ellisiana* Defrance, 1826]. Like *Cyphastrea* but with parathecal wall and lamellar or sublamellar columella. *Mio.,* Eu.

Echinopora Lamarck, 1816 [**Madrepora lamellosa* Esper, 1797] [=*Echinastrea* Blainv., 1830; *Stephanocora* Ehr., 1834; *Acanthopora* Verrill, 1864 (*non* Orb., 1849); *Acanthelia* Wells, 1937]. Submassive to foliaceous or ramose. Coenosteum dense or vesicular; costae reduced to spines. Corallite walls defined but may be perforate near level of calices. Columella spongy. *Mio.-Rec.,* IndoPac. ——Fig. 303,*1. *E. lamellosa,* Rec., Banda; ×2 (128).

Subfamily AGATHIPHYLLIINAE Vaughan & Wells, 1943

Solitary and colonial, hermatypic. Colony formation by extratentacular budding. Corallite wall synapticulothecate near calices, septothecate or parathecate below. Septa of lower cycles composed of 2 fan systems of

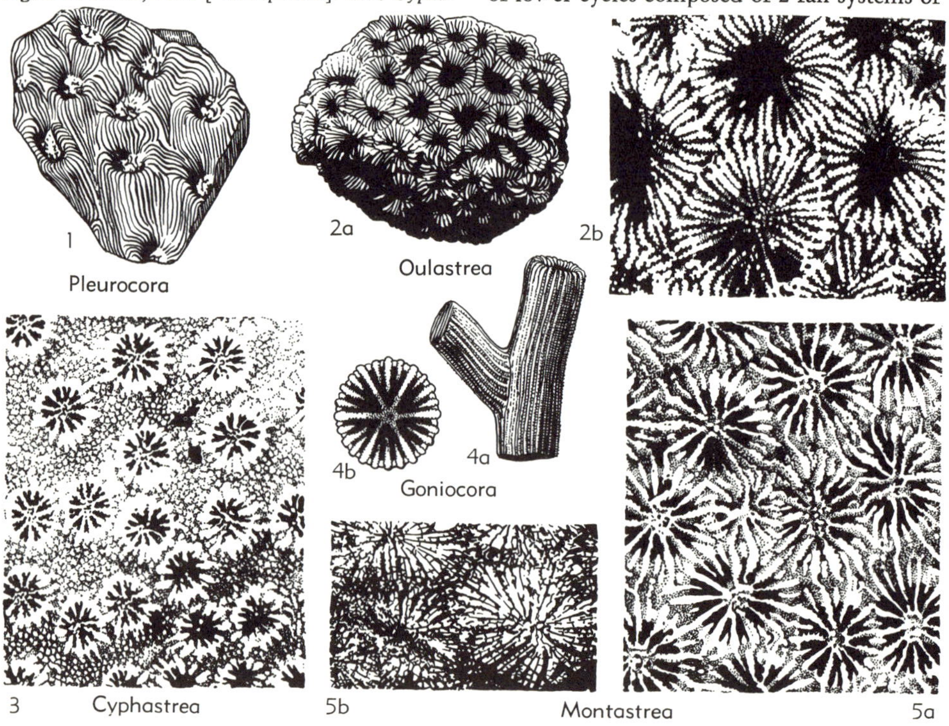

Fig. 301. Faviina (Faviicae): Faviidae (p. F404-F406).

mostly compound trabeculae, the inner fan system forming prominent internal septal lobes; higher cycles of one fan system and more or less perforate internally. *Eoc.-Mio.*

Pattalophyllia d'Achiardi, 1867 [**Turbinolia subinflata* Catullo, 1856 (=*Caryophyllia pseudocalvimonti* Catullo, 1856)]. Solitary, turbinate, fixed by small base. Wall septothecal and synapticulothecal. Columella trabecular, well developed. *Eoc.-Oligo.*, Eu.

?Ceratophyllia Fritsch, 1878 [**C. flabelloides;* SD Felix, 1925]. Like *Pattalophyllia* but ceratoid, and lacking columella. *Eoc.*, E.Indies.

Cricocyathus Quenst., 1881 [*pro Leptaxis* Reuss, 1868 (*non* Lowe, 1852)] [**Cyathophyllia annulata* Reuss, 1868] [=*Stephanosmilia* Reuss, 1874 (*non* Fromentel, 1862)]. Like *Pattalophyllia* but with broad base and synapticuloparathecal wall. *Eoc.-Mio.*, Eu.——Fig. 304,*1*. **C. annulata*, M.Oligo. (Rupel.), Italy; *1a*, ×0.6; *1b*, ×1 (116).

Agathiphyllia Reuss, 1864 [**A. explanata;* SD Vaughan, 1919] [=*Cyathomorpha* Reuss, 1868]. Like *Cricocyathus* but forming submassive plocoid colonies by budding from edge zone. *Oligo.*, Eu.-W.Indies-S.Am.-N.Am.——Fig. 304,*2c*. **A. explanata* Reuss, M.Oligo.(Rupel.), Italy; lat. aspect of minor septum, ×3 (116).——Fig. 304, *2a,b*. *A. umbellata* Reuss, M.Oligo.(Rupel.), Italy; *2a*, ×0.6; *2b*, ×1 (116).

Subfamily TRACHYPHYLLIINAE Wells, nov.

Solitary and colonial, hermatypic. Colony formation by intratentacular polystomodaeal intramural budding. Wall septothecal to parathecal. Septa formed by 2 fan systems of simple trabeculae; inner fan system forming prominent lobes. *Oligo.-Rec.*

Antillophyllia Vaughan, 1932 [**Antillia lonsdaleii* Duncan, 1864]. Solitary, free in ephebic stage, trochoid, compressed or bilobate in cross section. Columella trabecular, spongy. *Oligo.-Mio.*, W. Indies.——Fig. 305,*3*. *A. sawkinsi* (Vaughan), Mio., Trinidad; *3a,b*, ×1 (150).

Indophyllia Gerth, 1921 [**I. cylindrica*]. Like *Antillophyllia* but discoid to subcylindrical. *Oligo.-Mio.*, E.Indies.——Fig. 305,*1*. **I. cylindrica*, Mio., Borneo; *1a*, basal aspect ×1; *1b*, nat. long. sec., ×1.33 (57).

Trachyphyllia M.Edw.-H., 1848 [**Turbinolia geoffroyi* Audouin, 1826; SD M.Edw.-H., 1850]. Like *Antillophyllia* but colonial, flabello-meandroid. Series short, free laterally. Trabecular linkages between centers. *Mio.-Rec.*, IndoPac.——Fig. 305,*2*. **T. geoffroyi*, Rec., Austral.; *2a*, ×0.5; *2b*, lat. aspect of septum, ×2 (88).

Callogyra Verrill, 1902 [**C. formosa*]. Like *Trachyphyllia* but series united laterally to summits. *Rec.*, Pac.

?Moseleya Quelch, 1884 [**M. latistellata*]. Cerioid,

Fig. 302. Faviina (Faviicae): Faviidae (p. F404-F406).

colony formation by extra- and intratentacular budding (stomodaeal diversion and oral disc invagination). Parathecate. Columella trabecular, small and spongy. *Rec.*, N.Austral., Great Barrier Reef.——Fig. 305,*4*. **M. latistellata;* ×0.5 (109).

Family RHIZANGIIDAE d'Orbigny, 1851

[=Astrangiidae Verrill, 1869]

Colonial, ahermatypic. Colony formation by extratentacular budding from edge zone or stolon-like expansions of edge zone, polyps remaining organically connected or not, colonies commonly consisting of scattered corallites with no apparent connection, or united basally by coenosteum, or they form compact masses. Corallites small and low. Septa composed of one fan system of simple or compound trabeculae; irregular divergence of sclerodermites producing scattered lateral granulations and more or less irregular marginal dentations. Columella trabecular, rarely solid or absent. Endothecal dissepiments thin. *L.Cret.-Rec.*

Arctangia Wells, 1937 [**Thecocyathus nathorsti* Lindström, 1900]. Solitary, subturbinate, fixed by small base. Septa dentate; columella papillary. *L. Cret.*, Arct.——Fig. 306,*3*. **A. nathorsti*, L.Cret. (Neocom.), King Charles I., Arct.; *3a*, ×1; *3b,c*, ×4 (183).

Rhizangia M.Edw.-H., 1848 [**Astrea brevissima* Deshayes, 1834]. Tympanoid, reptoid; all septa dentate; columella a single tubercle. *U.Cret.-Mio.*, Eu.——Fig. 307,*1*. *R. michelini* Reuss, U.Cret. (Turon.), Aus.(Gosau); *1a*, ×1; *1b*, ×2 (42).

Culicia Dana, 1846 [**C. stellata;* SD Wells, 1936] [=*Angia* M.Edw.-H., 1848; *Cylicia* M.Edw.-H., 1857]. Tympanoid, reptoid, basal connections temporary. First-cycle septa generally weakly dentate or lobate. Columella weak. *Pleisto.-Rec.*, IndoPac., depth range 0-366 m.——Fig. 307,*3*. *C.* sp., Rec., Austral., 55 m.; ×2 (180).

Cryptangia M.Edw.-H., 1848 [**C. woodii*]. Corallites subtrochoid, immersed in mass of bryozoan. Septa dentate; columella papillary. *Mio.-Plio.*, Eu.——Fig. 306,*4*. **C. woodii*, Plio., Eng.(Norf.); ×3 (120).

Astrangia M.Edw.-H., 1848 [**A. michelini;* SD M.

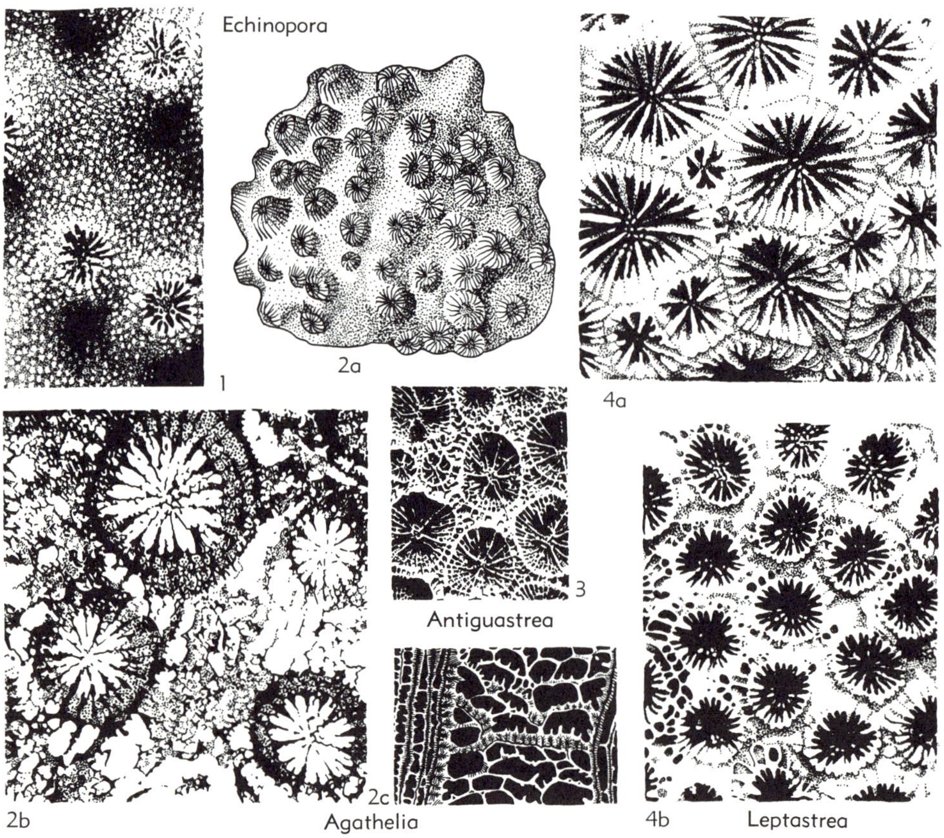

Fig. 303. Faviina (Faviicae): Faviidae (p. F406-F407).

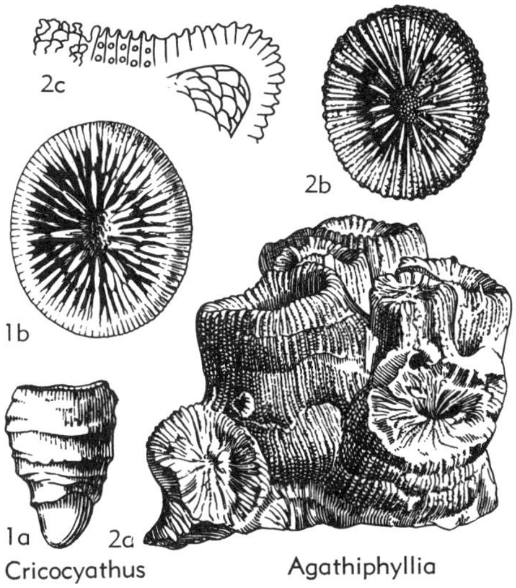

FIG. 304. Faviina (Faviicae): Faviidae (p. F407).

EDW.-H., 1850] [=*Stellangia* DUCHAISSING & MICHELOTTI, 1860; *Gombertangia* OPPENHEIM, 1899]. Incrusting, subplocoid, corallites united basally by thin coenosteum. Septa dentate; columella papillary. *M.Cret.-Rec.*, Eu.-N.Am.-S.Am.-W. Indies-Austral., shallow water.

A. (Astrangia). *Eoc.-Rec.*——FIG. 306,5. *A. rathbuni* VAUGHAN, Rec., Braz.; ×2 (195).

A. (Coenangia) VERRILL, 1869 [**Coenangia conferta;* SD VAUGHAN & WELLS, 1943]. Like *A. (Astrangia)* but cerioid. *M.Cret.-Rec.*, N.Am.-C. Am., shallow water.——FIG. 306,2. **A. (C.) conferta*, Rec., GulfCalif.; ×1 (36).

A. (Hoplangia) GOSSE, 1860 [**H. durotrix*]. Like *A. (Astrangia)* but septal margins nearly smooth. *Rec.*, Eu.

Oulangia M.EDW.-H., 1848 [**O. stokesiana;* SD M. EDW.-H., 1850] [=*Ulangia* M.EDW.-H., 1857]. Small reptoid colonies, the corallites united by thin coenosteum. Septa of first 2 or 3 cycles obscurely dentate and exsert. Columellar papillae merging with inner septal dentations. *Pleisto.-Rec.*, IndoPac., shallow water.——FIG. 307,2. **O. stokesiana*, Rec., Philip.; *2a*, ×1; *2b*, ×2.5 (93).

Phyllangia M.EDW.-H., 1848 [**P. americana;* SD M.EDW.-H., 1850] [=*Syndepas* LYMAN, 1857]. Incrusting, habit like *Oulangia* but with more extensive coenosteum. Lower-cycle septa obscurely dentate; columella weak. *Mio.-Rec.*, Eu.-W.Afr.-N.Am.-S.Am., depth range 0-100 m.——FIG. 307, *5*. **P. americana*, Rec., W.Indies; ×1 (198).

Colangia POURTALÈS, 1871 [**C. immersa*]. Rounded, incrusting, plocoid; corallites united basally by heavy coenosteum. Lower-cycle septa obscurely dentate, highly exsert; paliform lobes before third cycle. Columella a thin, interrupted lamella, trabecular below. *Rec.*, W.Indies.

Cladangia M.EDW.-H., 1851 [**Astrea semispherica* DEFRANCE, 1826]. Like *Astrangia* but massive or subramose, plocoid; corallites united by striate or

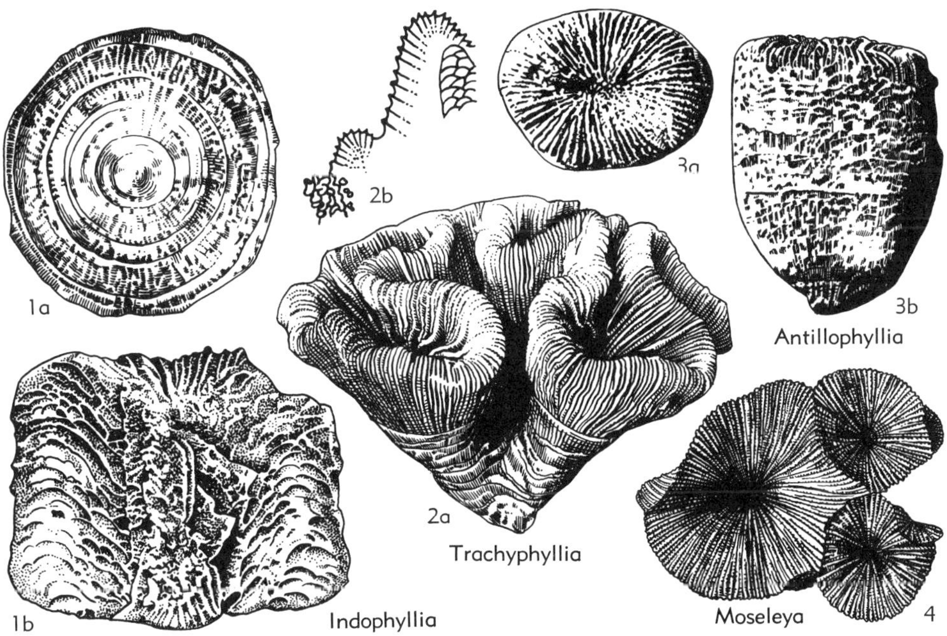

FIG. 305. Faviina (Faviicae): Faviidae (p. F408).

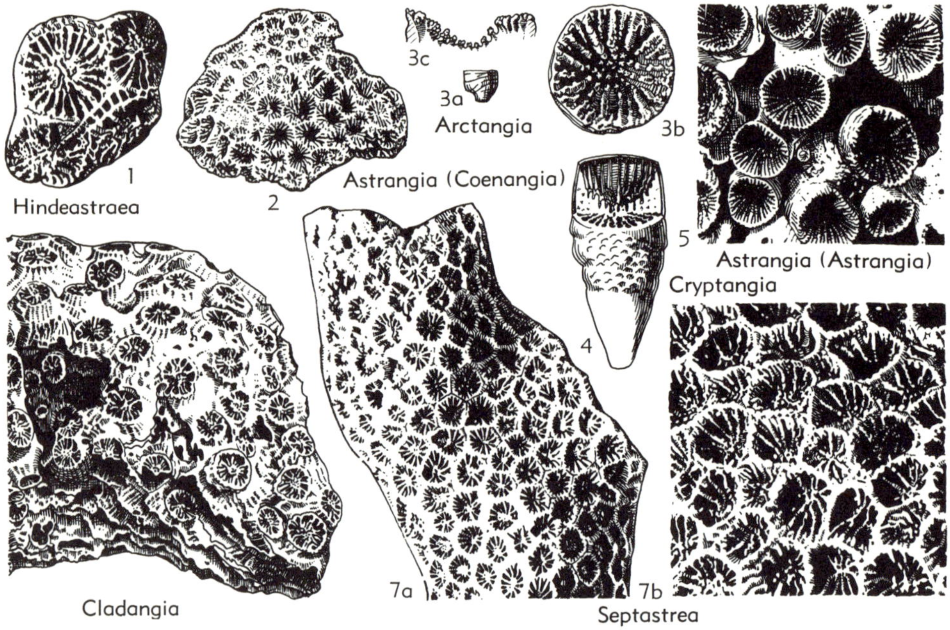

FIG. 306. Faviina (Faviicae): Faviidae (p. F409-F410).

spinose coenosteum. *Eoc.-Rec.*, Eurasia.——FIG. 306,*6*. *C. exusta* LÜTKEN, Rec., India; ×1 (184).

Reussangia WELLS, 1937 [*pro Stylangia* REUSS, 1874 (*non* FROMENTEL, 1857)] [=**Stylangia elegans* REUSS, 1874]. Like *Cladangia* but columella a single tubercle as in *Rhizangia*. Coenosteum costate. *Eoc.-Oligo.*, Eu.-C.Am.——FIG. 307,*4*. **R. elegans*, M.Eoc.(Lut.), Italy; *4a*, ×1; *4b*, ×2 (116).

Hindeastraea WHITE, 1888 [**H. discoidea*]. Ramose colonies formed by extratentacular budding; cerioid or subplocoid. Septa dentate; columella feeble, trabecular. *U.Cret.*, Tex.——FIG. 306,*1*. **H. discoidea*, Navarro, Tex.; ×2 (180).

Septastrea ORB., 1849 [**S. subramosa (nom. nud.)* (=*S. forbesi* M.EDW.-H., 1849; =*Astrea marylandica* CONRAD, 1841)] [=?*Hexastraea* SISMONDA, 1871; *Glyphastraea* DUNCAN, 1887]. Like *A. (Coenangia)* but forming large, ramose colonies. Corallites rarely with more than 2 cycles of septa. Columella generally appearing solid. *Mio.-Plio.*, Eu.-N.Am.-C.Am.——FIG. 306,*7*. *S. crassa* (TUOMEY & HOLMES), Piio., N.Car.; *7a*, ×1; *7b*, ×2 (151).

Platyhelia TENISON-WOODS, 1880 [**P. distans*]. Like *Septastrea* but with more than 2 cycles of septa and paliform lobes or lobate dentations before 1st 2 or 3 cycles; columella papillary. *U.Oligo.-Mio.*, N.Z.

?**Podoseris** DUNCAN, 1869 [**P. mammiliformis;* SD DUNCAN, 1889]. Solitary, tympanoid. Septa dentate, exsert; columella papillose. *M.Cret.*, Eng.

Family OCULINIDAE Gray, 1847

Colonial; colony formation by extratentacular (or rarely intratentacular) budding. Corallites externally thickened by extensive, noncostate, granulated or smooth, dense (rarely vesicular) coenosteum. Septa exsert, formed by one fan system of simple trabeculae, margins minutely dentate, laterally granulose or spinose. Pali generally developed. Columella papillose, trabecular, or absent. Endothecal dissepiments, when developed, subtabular, thin, or replaced by stereome. *Cret.-Rec.*

Subfamily OCULININAE Gray, 1847

[*nom. transl.* VAUGHAN & WELLS, 1943 (*ex.* Oculinidae GRAY, 1847)]

Mostly ahermatypic oculinids forming dendroid colonies; corallites united basally by dense coenosteum. *Cret.-Rec.*

Oculina LAMARCK, 1816 [**O. virginea* (*non Madrepora virginea* LINNÉ, 1758; *nec* ELLIS & SOLANDER, 1786) (=*O. diffusa* LAM., 1816; SD M.EDW.-H., 1850)] [=*Dentipora* BLAINV., 1830; *Trymhelia* M.EDW.-H., 1849; *Trymohelia* M.EDW.-H., 1857; *Coelohelia* VAUGHAN, 1900]. Ramose colonies formed by alternate budding, corallites tending to spiral around branches; ahermatypic and hermatypic. Coenosteum dense, striated. Pali before first

2 cycles in an irregular crown. Columella papillose. *Cret.-Rec.*, Eu.-W.Indies-N.Am.-Austral., depth range 0-91 m.

O. (Oculina). *Cret.-Rec.*——Fig. 308,*4*. **O. (O.) diffusa*, Rec., Fla.; ×3 (151).

O. (Schizoculina) Wells, 1937 [**O. fissipara* M. Edw.-H., 1850]. Like *O. (Oculina)* but colony formation mostly by intratentacular monostomodaeal budding. *Rec.*, W.Afr.

Archohelia Vaughan, 1919 [**A. limonensis*]. Like *Oculina* but with persistent axial corallite as in *Acropora*. *M.Cret.-Plio.*, N.Am.-C.Am.-W.Indies.——Fig. 309,*2*. *A. vicksburgensis* (Conrad), Oligo., Miss.; *2a*, ×0.5; *2b*, ×1.5 (144).

Sclerhelia M.Edw.-H., 1850 [**Madrepora hirtella* Pallas, 1766] [=*Sclerohelia* M.Edw.-H., 1857]. Like *Oculina* but pali in one crown opposite 2nd cycle; columella papillose. *Rec.*, S.Atl.-IndoPac., depth range 110-400 m.——Fig. 309,*1*. *S. formosa* (Alcock), Maldive I., 400 m.; *1a*, ×0.5; *1b*, ×4 (171).

Bathelia Moseley, 1881 [**B. candida*]. Regular and alternate budding from edge zone near calices. Septa apparently entire marginally. Pali in one crown opposite 3rd cycle. Columella of curled laths. *Rec.*, S.Atl., depth range 1,097 m.——Fig. 308,*1*. **B. candida*, 1,097 m.; *1a*, ×0.35; *1b*, ×1.5 (97).

Haplohelia Reuss, 1864 [**H. gracilis*]. Budding from calicular margin, singly and usually from same side. Pali in 2 crowns opposite 1st 2 cycles. Columella papillose. *Oligo.*, Ger.——Fig 310,*2*. **H. gracilis; 2a*, ×1; *2b*, ×3 (114).

Cyathelia M.Edw.-H., 1849 [**Madrepora axillaris* Ellis & Solander, 1786] [=*Cyathohelia* M.Edw.-H., 1857]. Regular alternate budding. Pali opposite 1st 3 cycles in 2 crowns. Columella of twisted trabecular rods. *Rec.*, IndoPac., depth range 119-1,509 m.——Fig. 308,*2*. **C. axillaris* (Ellis-S.), Amboina; *2a*, ×0.5; *2b*, ×6 (174).

Astrhelia M.Edw.-H., 1849 [**Madrepora palmata* Goldf., 1829] [=*Astrelia* Orb., 1849; *Astrohelia* M.Edw.-H., 1857]. Subramose with stout anastomosing branches. Corallites united by dense, smooth coenosteum, slightly protuberant. Septal dentations lacerate; pali absent; columella trabecular. *Mio.*, N.Am.——Fig. 310,*1*. **A. palmata* (Goldf.), Md.; *1a*, ×0.5; *1b*, ×1.25 (195).

Diplhelia M.Edw.-H., 1850 [**Oculina raristella* Defrance, 1825] [=*Diplohelia* M.Edw.-H., 1857]. Dendroid with coalescing branches. Paliform lobes opposite 1st 2 cycles of dentate septa.

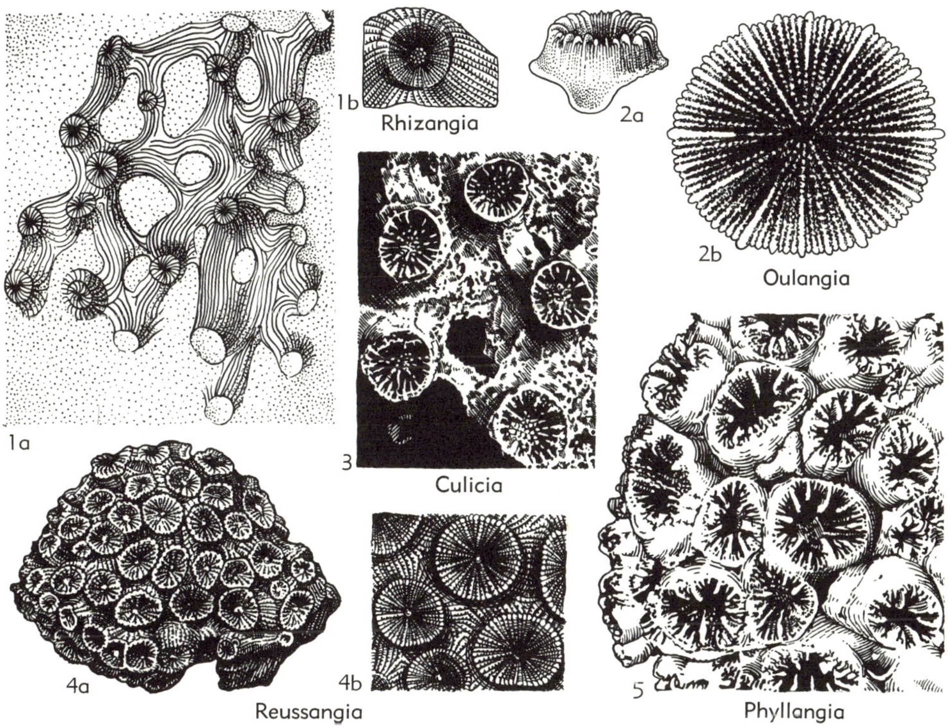

Fig. 307. Faviina (Faviicae): Rhizangiidae (p. F409-F410).

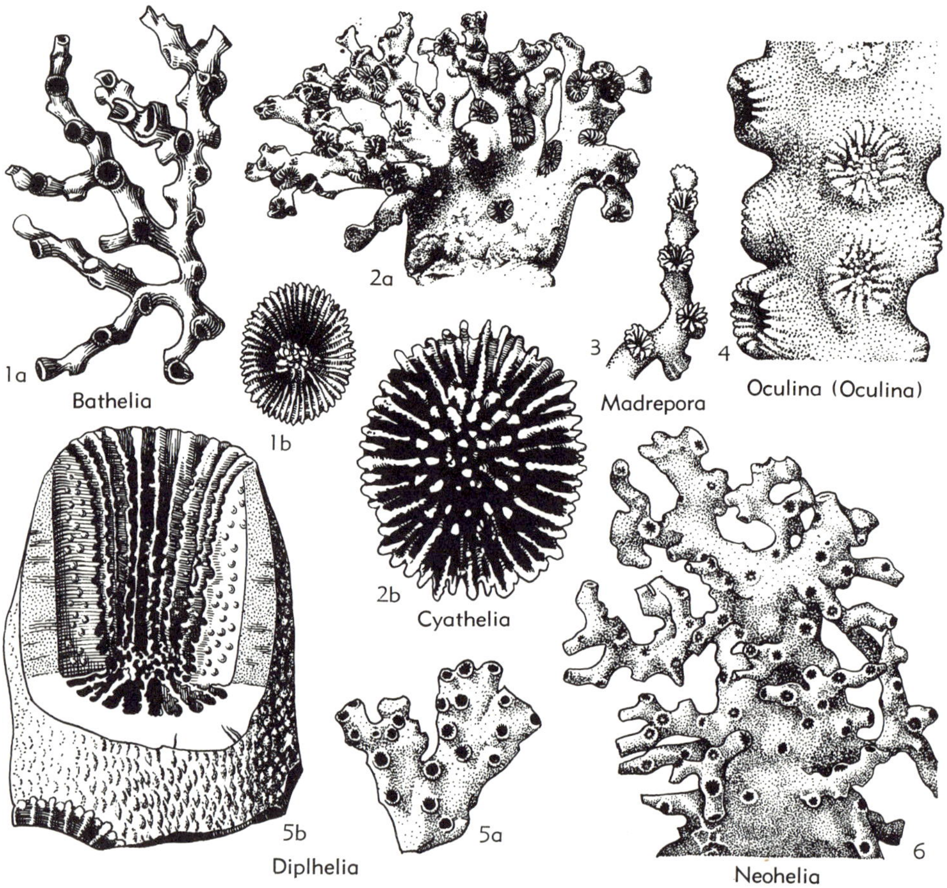

FIG. 308. Faviina (Faviicae): Oculinidae (p. F411-F413).

Eoc., Eu.——FIG. 308,*5*. *D. papillosa* M.EDW.-H., L.Eoc.(Lond.), Isle of Wight; *5a*, ×0.5; *5b*, ×6 (94).

Madrepora LINNÉ, 1758 [**M. oculata;* SD VERRILL, 1902] [=*Amphelia* M.EDW.-H., 1849; *Amphihelia* M.EDW.-H., 1857]. Dendroid colonies with more or less regular, alternate budding. Coenosteum dense, corallites filled internally by stereome. Pali absent. Columella spongy or absent. *Eoc.-Rec.*, cosmop., depth range 183-1,554 m.——FIG. 308,*3*. *M. galapagensis* VAUGHAN, Rec., Galapagos I., 575 m.; ×1 (195).

Diblasus LONSDALE, 1850 [**D. grevensis*]. Like *Madrepora* but colony formation irregular, mostly by basal budding, branches irregularly dividing. Corallites with little or no internal stereome. *U. Cret.*, Eu.

Neohelia MOSELEY, 1881 [**N. porcellana*]. Subdendroid, with dense coenosteum. Septa nonexsert, apparently nondentate. Columella absent. *Rec.*, Pac., depth range 91-115 m.——FIG. 308,*6*. **N. porcellana*, New Hebrides; ×1 (97).

Subfamily GALAXEINAE Vaughan & Wells, 1943

Hermatypic oculinids, ramose or plocoid. Corallites united by vesicular coenosteum. Septa highly exsert. *Mio.-Rec.*

Galaxea OKEN, 1815 [**Madrepora fascicularis* LINNÉ, 1767; SD VAUGHAN, 1918]. Massive, plocoid colonies formed by budding from broad edge zone of coenosarc. Corallites costate, cornute to cylindrical, united basally by extensive, vesicular, spinose, noncostate coenosteum. Columella absent or weak. Endothecal dissepiments weak and thin. *Mio.-Rec.*, N.Am.-IndoPac.——FIG. 311,*2*. **G. fascicularis* (LINNÉ), Rec., Austral.(Torres Strait); *2a*, ×2 (194); *2b*, Amboina, long. sec., ×2 (174).

Simplastrea UMBGROVE, 1939 [**S. vesicularis*]. Like *Galaxea* but corallites united to calicular margins by coenosteum. Columella trabecular. *Rec.*, E.

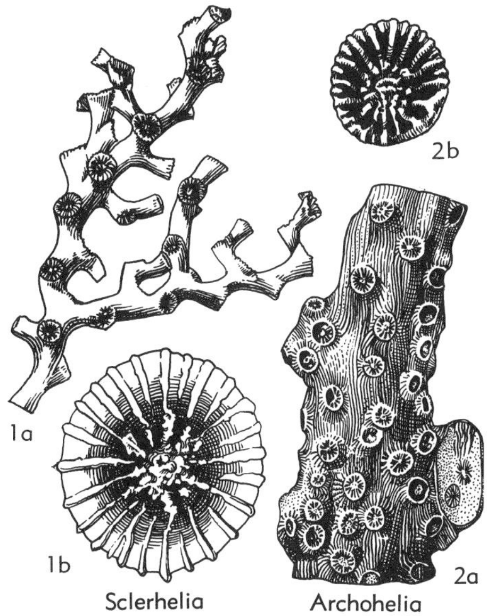

FIG. 309. Faviina (Faviicae): Oculinidae (p. F411).

Indies.——FIG. 311,*1*. **S. vesicularis,* Batavia Bay; ×2 (138).

Acrhelia M.EDW.-H., 1849 [**Oculina horrescens* DANA, 1846] [=*Acrohelia* M.EDW.-H., 1857]. Ramose, with corallites tending to spiral around branches. Calices protuberant with very exsert septa. Costae rudimentary. Coenosteum mostly solid, vesicular in lower parts of corallum. *Rec.,* Pac.——FIG. 312,*1*. **A. horrescens* (DANA), Austral.(Torres Strait); ×1 (148).

Bantamia YABE & EGUCHI, 1943 [**B. gerthii*]. Like *Acrhelia* but columella developed. *Mio.,* Java.

Family MEANDRINIDAE Gray, 1847

Solitary and colonial; colony formation by intratentacular budding. Wall septothecal or rarely parathecal, costate. Septa formed by one fan system of simple trabeculae, exsert, margins minutely dentate. Columella lamellar or trabecular. Endothecal dissepiments

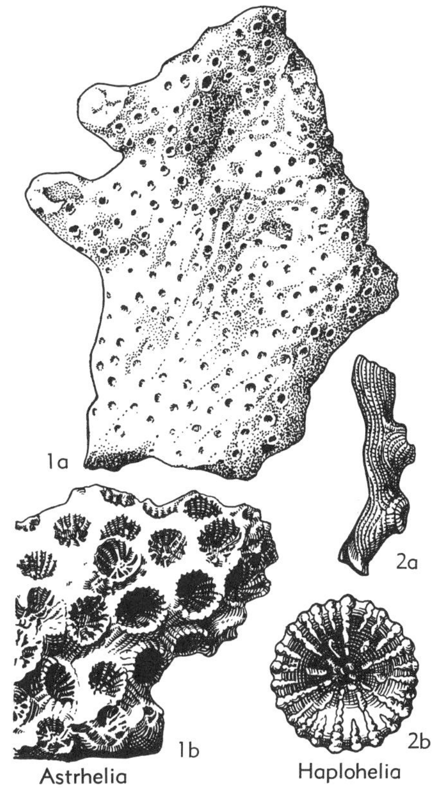

FIG. 310. Faviina (Faviicae): Oculinidae (p. F412).

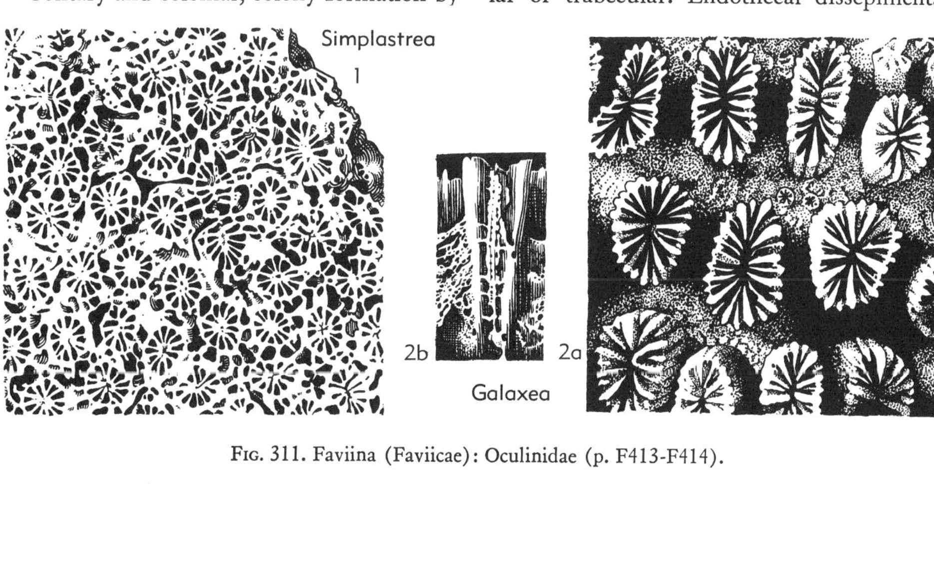

FIG. 311. Faviina (Faviicae): Oculinidae (p. F413-F414).

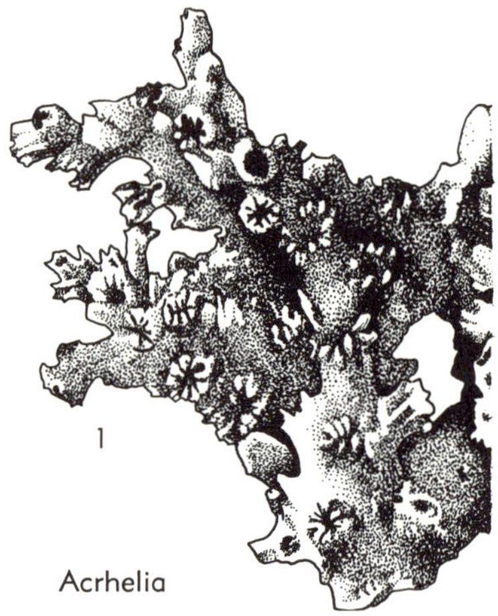

FIG. 312. Faviina (Faviicae): Oculinidae (p. F414)

well developed; exothecal dissepiments in some forms. *Cret.-Rec.*

Subfamily MEANDRININAE Gray, 1847

[*nom. transl.* VAUGHAN & WELLS, 1943 (*ex* Meandrinidae GRAY, 1847)]

Solitary and colonial, apparently most hermatypic. Colony formation by intratentacular polystomodaeal intramural budding. Septothecate or parathecate, costate. Septa laminar, exsert, with very small simple trabeculae. Columella a thin lamella, generally continuous, deep in calice or series. *Cret.-Rec.*

Aulosmilia ALLOITEAU, 1952 [**Trochosmilia archiaci* FROMENTEL, 1862] [=*Placosmilia auctt.* (*non* M. EDW.-H., 1848); *Phragmosmilia* ALLOITEAU, 1952]. Solitary, compressed or flabellate, with simple costae. *M.Cret.-Eoc.*, Eu.-N.Am.-W.Indies.——FIG. 313,*4*. *A. cuneiformis* (M.EDW.-H.), U.Cret. (Turon.), Aus.(Gosau); *4a*, ×0.5 (112); *4b*, transv. sec., ×0.5; *4c,d*, long. secs., ×1, ×6 (42).

Flabellosmilia OPPENHEIM, 1930 [**Flabellum bisinuatum* REUSS, 1854; SD VAUGHAN & WELLS, 1943]. Solitary, flabellate, with alate opposing end costae. *U.Cret.*, Eu.——FIG. 314,*1*. *F. subcarinata* (REUSS), Turon., Aus.(Gosau); *1a*, ×1; *1b*, transv. sec., ×2 (100).

Nefophyllia WELLS, 1937 [*pro Platysmilia* FELIX, 1899 (*non* TOULA, 1889)] [**Placosmilia angusta* REUSS, 1854]. Phaceloid colonies formed by extratentacular budding. *U.Cret.*, Eu.——FIG. 313,*5*. *N. multicincta* (REUSS), Turon, Aus.(Gosau); ×0.5 (100).

Phyllosmilia FROMENTEL, 1862 [**Turbinolia basochesi* DEFRANCE, 1828]. Colonial, flabellate, with continuous end costae, others dividing. *U.Cret.-Eoc.*, Eu.——FIG. 313,*2*. *P. transiens* FELIX, U. Cret.(Turon.), Aus.(Gosau); *2a,b*, ×0.5 (42).

Diploctenium GOLDF., 1827 [**D. cordatum;* SD WELLS, 1936]. Like *Phyllosmilia* but series curved

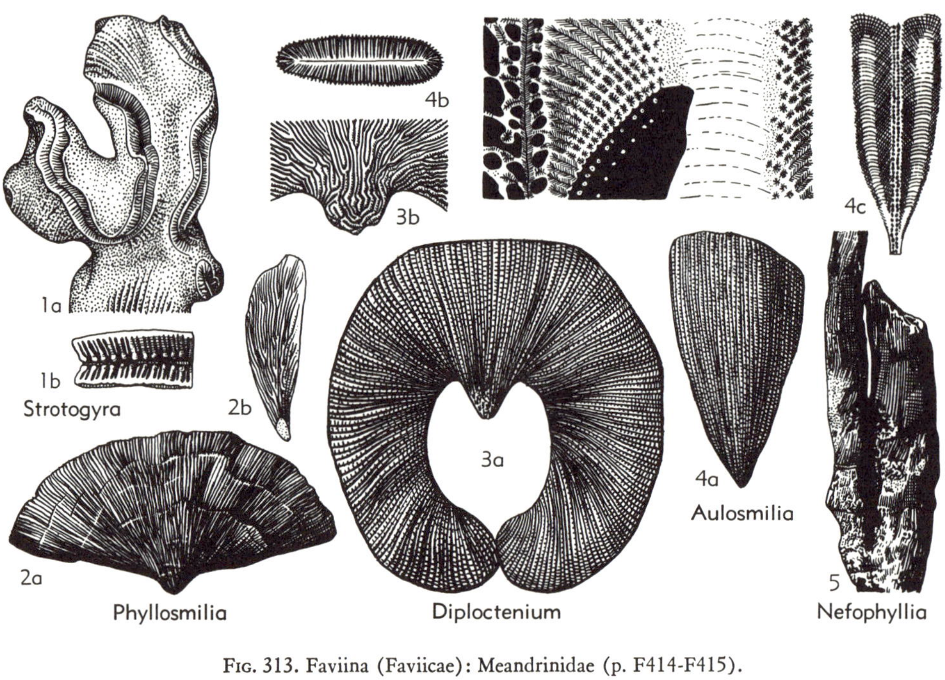

FIG. 313. Faviina (Faviicae): Meandrinidae (p. F414-F415).

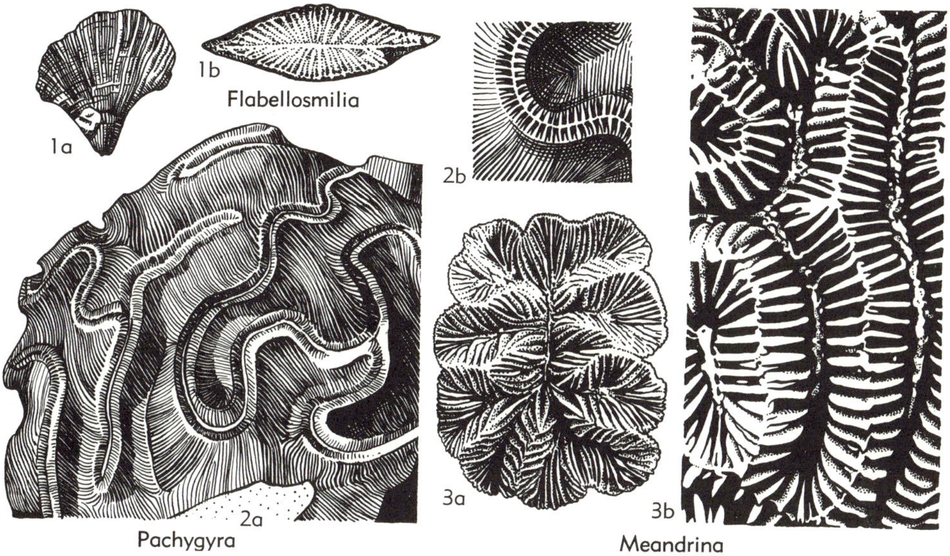

FIG. 314. Faviina (Faviicae): Meandrinidae (p. F415).

toward base, not horizontal. *M.Cret.-U.Cret.*, Eu.-N.Afr.-W.Indies.——FIG. 313,*3*. *D. haidingeri* REUSS, U.Cret.(Turon.), Aus.(Gosau); *3a*, ×0.5 (112); *3b*, detail of base, ×1 (100).

Strotogyra WELLS, 1937 [**Rhipidogyra undulata* REUSS, 1854]. Like *Phyllosmilia* and *Diploctenium* but corallum contorted, formed by one continuous, laterally free, series. *U.Cret.-Eoc.*, Eu.——FIG. 313, *1*. *S. savii* (D'ACHIARDI), M.Eoc.(Lut.), Italy; *1a*, ×1; *1b*, ×2 (116).

Pachygyra M.EDW.-H., 1848 [**Lobophyllia labyrinthica* MICHELIN, 1846]. Meandroid, massive, formed by one or a series united to summits by costae and exothecal dissepiments which form ambulacra. Columella discontinuous. *U.Cret.-Eoc.*, Eu.——FIG. 314,*2*. *P. princeps* REUSS, U.Cret. (Turon.), Aus.(Gosau); *2a*, ×0.5; *2b*, ×1 (112).

Meandrina LAMARCK, 1801 [**M. pectinata* (=*Madrepora maeandrites* LINNÉ, 1758)] [=*Maeandra* OKEN, 1815; *Placocyathus* M.EDW.-H., 1848; *Ctenophyllia* DANA, 1846; *Sphenophyllia* MOSELEY, 1881; *?Thecophyllia* DAINELLI, 1915 (*non* M.EDW.-H., 1848)]. Meandroid; valleys continuous or discontinuous, the series united directly by septothecae without ambulacra. Columella generally discontinuous. *?Eoc., Rec.*, Eu.-W.Indies-Braz.——FIG. 314,*3a*. *M. braziliensis* (M.EDW.-H.), Rec., Braz.; ×0.5 (88).——FIG 314,*3b*. **M. maeandrites* (LINNÉ), Rec., Bahamas; ×1 (88).

Ctenella MATTHAI, 1928 [**C. chagius*]. Like *Meandrina* but columella continuous. *?Rec.*, Ind.O.

Subfamily DICHOCOENIINAE Vaughan & Wells, 1943

Meandroid, colonial, hermatypic meandrinids with trabecular columella and well-developed, heavy coenosteum between corallites. *U.Cret.-Rec.*

Barysmilia M.EDW.-H., 1848 [**Dendrophyllia brevicaulis* MICHELIN, 1841] [=*Stenosmilia* FROMENTEL, 1870; *?Axophyllia* DAINELLI, 1915 (*non* ORB., 1849)]. Massive, plocoid heads formed by intratentacular polystomodaeal budding, permanent conditions mono- to tricentric. Corallites protuberant, united by costate and vesicular to solid coenosteum. Columella trabecular, commonly appearing lamellar at surface. *U.Cret.-Eoc.*, Eu.

Dichocoenia M.EDW.-H., 1848 [**D. stokesi*]. Like *Barysmilia* but with some meandroid series, noncostate ambulacra, and granulose coenosteum. *U. Cret.-Rec.*, W.Indies-N.Am.

D. (Dichocoenia). *U.Cret.-Rec.*——FIG. 315,*1a*. *D. trechmanni* WELLS, U.Cret.(Camp.), Jamaica; long. sec., ×1 (151).——FIG. 315,*1b*. **D. stokesi*, Rec., Bahamas; ×1 (195).

D. (Psilogyra) FELIX, 1903 [**Psilogyra telleri*]. Like *D. (Dichocoenia)* but without columella. *U.Cret.*, Eu.——FIG. 316,*1*. **D. (P.) telleri*, Turon., Aus.(Gosau); ×1 (100).

Dendrogyra EHR., 1834 [**D. cylindrus*; SD M.EDW.-H., 1850] [=*Gyrophyllia* ORB., 1849; *Meandroria* ALLOITEAU, 1952]. Corallite series long, united directly by thickened septothecae. Columella trabecular and solid, a discontinuous lamella, or substyliform. *Eoc.-Rec.*, Eu.-W.Indies-N.Am.——FIG. 315,*2*. **D. cylindrus*, Rec., Bahamas; *2a*, ×0.08; *2b*, ×3 (88).

Orbignygyra ALLOITEAU, 1952 [**Diploria neptuni*

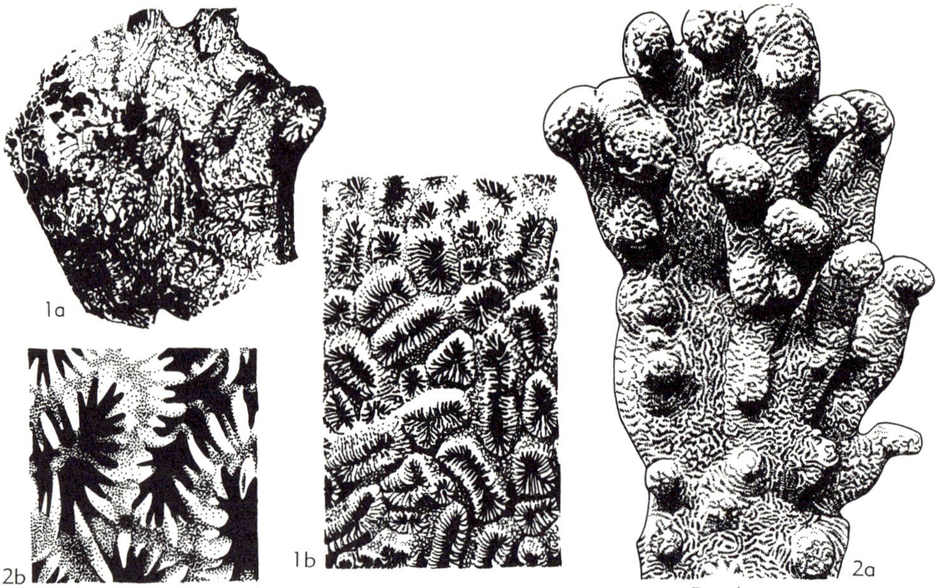

FIG. 315. Faviina (Faviicae): Meandrinidae (p. F416).

ORB., 1850]. Like *Dendrogyra* but series separated by ambulacra. *U.Cret.*, Eu.

Scalariogyra GERTH, 1923 [**S. escharoides*]. Like *Dendrogyra* but inner ends of septa with strong internal swellings or paliform lobes. *Oligo.*, Borneo.

Family MERULINIDAE Verrill, 1866

Colonial, hermatypic. Colony formation by intratentacular polystomodaeal budding; centers linked by trabeculae. Septa of one fan system of compound trabeculae, regularly dentate at first, later with scattered, spinose, ragged, very irregular dentations.

FIG. 316. Faviina (Faviicae): Meandrinidae (p. F416).

Columella feeble or absent. Dissepiments sparse. *Rec.*

Boninastraea YABE & SUGIYAMA, 1935 [**B. boninensis*]. Cerioid; colony formation by polystomodaeal budding, permanent condition of corallites di- to tricentric. No columella. *Rec.*, Pac.——FIG. 317,*3*. **B. boninensis*, Ogasawara I.; *3a*, ×0.5 (164); *3b*, septum, ×2 (151).

Merulina EHR., 1834 [**Madrepora ampliata* ELLIS & SOLANDER, 1786]. Foliaceous, thin, meandroid; colony formation by intramural budding, series relatively straight, spreading fanwise by lateral branching and terminal forking. Columella trabecular, generally fused into a continuous mass. *Rec.*, IndoPac.——FIG. 317,*2a,b*. *M. studeri* BEDOT, Amboina; ×0.25, ×2 (174).——FIG. 317,*2c*. **M. ampliata*, Pac.; septum, ×2 (151).

Clavarina VERRILL, 1864 [**Merulina scrabicula* DANA, 1846] Like *Merulina* but ramose, the series discontinuous with circumscribed corallites. Columella rudimentary. *Rec.*, Pac.

Scapophyllia M.EDW.-H., 1848 [**S. cylindrica*]. Like *Merulina* but massive-columniform, corallite series sinuous, continuous. *Rec.*, W.Pac.——FIG. 317,*1*. **S. cylindrica*, Philip.; septum, ×2 (151).

Family MUSSIDAE Ortmann, 1890

Solitary and colonial, hermatypic. Colony formation by intratentacular budding; centers linked by lamellae or trabeculae. Septothecate or parathecate. Septa entocoelic, formed by several fan systems of large, simple trabeculae, each fan system producing a lobulate dentation. Endothecal dissepiments well developed. Columella trabecular. *?U. Jur., Eoc.-Rec.*

?**Palaeomussa** ALLOITEAU, 1952 [**Montlivaltia minor* FROMENTEL, 1864]. Small, solitary, turbinate, with trabecular columella. Septa composed of a few (4 or 5) irregular fan systems. *U.Jur.*, Eu.

Circophyllia M.EDW.-H., 1848 [**Anthophyllum truncatum* GOLDF., 1826] [=*Felixopsammia* FILLIOZAT, 1910]. Solitary, turbinate or trochoid, fixed or free. Septothecate, costate. Septa with one broad peripheral fan system and several smaller inner ones. *Eoc.*, Eu.——FIG. 318,*6*. **C. truncatum* (GOLDF.), M.Eoc.(Lut.), Fr.; lat. aspects of septum ×1 (151).

?**Trochosmilia** M.EDW.-H., 1848 [**Turbinolia cornicula* MICHELIN, 1846; SD GREGORY, 1900] [=*?Feddenia* DUNCAN, 1880]. Solitary, turbinate or trochoid, nonepithecate, septothecate. Septa vertically discontinuous, formed by isolated trabeculae and trabecular bundles, thickened by stereome, with large dentations. Exo- and endothecal dissepiments developed. *Eoc.*, Eurasia.

Mussismilia ORTMANN, 1890 [**Mussa harttii* VERRILL, 1867] [=*Protomussa* MATTHAI, 1928]. Phaceloid; colony formation by di- or tristomodaeal budding. Parathecate. Septal dentations small, more or less uniform but ragged. *Rec.*, Braz.——FIG. 318,*2a*. **M. harttii;* long. sec., ×1 (151).——FIG. 318,*2b*. *M. braziliensis* (VERRILL); long. sec., ×1 (151).

Syzygophyllia REUSS, 1860 [**S. brevis*]. Solitary or in small phaceloid clumps produced by distomodaeal budding. Centers with lamellar linkage. *Mio*, Eu.——FIG. 319,*3*. **S. brevis,* Vindob., Czech.; *3a,* ×1; *3b,* long. sec., ×1 (189).

Homophyllia BRUEGGEMANN, 1877 [**Caryophyllia australis* M.EDW.-H., 1849]. Solitary, broadly turbinate; or with 1 or 2 secondary centers with lamellar linkage. Septal dentations numerous, generally triangular. *U.Oligo.-Rec.*, Austral.

Acanthastrea M.EDW.-H., 1848 [**A. spinosa* (=*A. hirsuta* M.EDW.-H., 1848; =*Astraea echinata* DANA, 1846)]. Low, cerioid (rarely plocoid) colonies with monocentric corallites formed mostly by marginal budding. *Mio.-Rec.*, IndoPac.——FIG. 318,*8*. **A. echinata* (DANA), Rec., Fiji; *8a,* ×2; *8b,* long. sec., ×2; *8c,* septum, ×1 (148).

Acanthophyllia WELLS, 1937 [**Caryophyllia deshayesiana* MICHELIN, 1850] [=*?Apostrophyllum* THOMAS, 1935]. Solitary, turbinate, free or fixed by small base. Parathecate with spinose costae. Septal dentations large and lobulate. Columella spongy. *?Oligo., Mio.-Rec.*, Eu.-IndoPac.——FIG. 319,*1a,b*. *A. ampla* (REUSS), Mio.(Vindob.), Rumania; ×0.5 (117).——FIG. 319,*1c*. **A. deshayesiana,* Rec., Philip.; long. sec., ×1 (151).

Antillia DUNCAN, 1863 [**A. dentata;* SD FROMENTEL, 1867] [=*Smilophyllia* FROM. & FERRY, 1865]. Like *Acanthophyllia* but septothecate and septa with larger peripheral fan system. *Eoc.-Mio.*, W.Indies.——FIG. 318,*4a,b*. *A. dentata,* Mio., Dominican Rep.; ×0.7 (29).——FIG. 318,*4c*. *A. gregorii* VAUGHAN, Mio., Dominican Rep.; long. sec., ×1 (151).

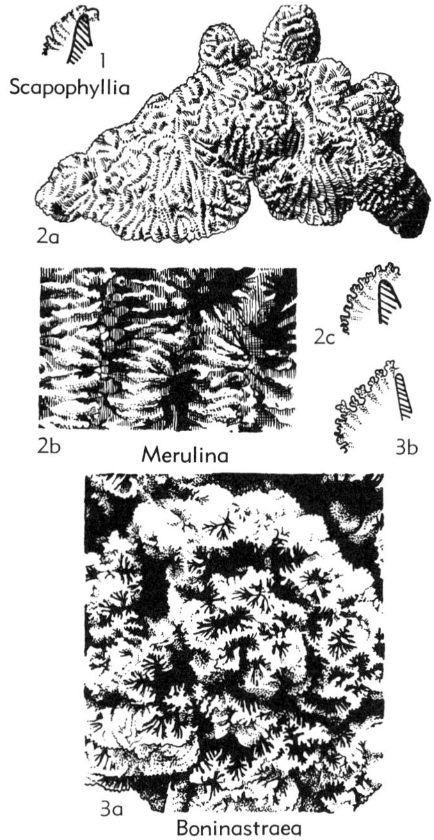

FIG. 317. Faviina (Faviicae): Merulinidae (p. F416-F417).

Cyclomussa WELLS, 1941 [**C. concinna*]. Solitary, cupolate to low cylindrical, free, with nearly flat base and highly exsert septa. Septo- and parathecate. *Oligo.*, Peru.——FIG. 319,*2*. **C. concinna; 2a-c,* basal, lat., oral aspects, ×1; *2d,* long. sec., ×1 (198).

Leptomussa D'ACHIARDI, 1867 [**L. variabilis*]. Solitary, fixed, ceratoid, septothecate. Septa with large fan systems, each bearing dentations. Columella absent. *Oligo.*, Italy.——FIG. 318,*7*. **L. variabilis,* Oligo., Italy, long. sec. of septum, ×1 (116).

Lobophyllia BLAINV., 1830 [**Madrepora corymbosa* FORSKÅL, 1775; SD MATTHAI, 1928] [=*Cynarina* BRUEGGEMANN, 1877; *Rhodocyathus* BOURNE, 1905; *Protolobophyllia* YABE & SUGIYAMA, 1935]. Meandroid-phaceloid clumps formed by intramural polystomodaeal budding, series laterally free. Centers with lamellar linkage. *Rec.*, IndoPac.

L. (Lobophyllia). *Rec.*——FIG. 318,*3*. *L. echinata* (DANA), Amboina; ×0.4 (174).

L. (Palauphyllia) YABE, SUGIYAMA & EGUCHI, 1936

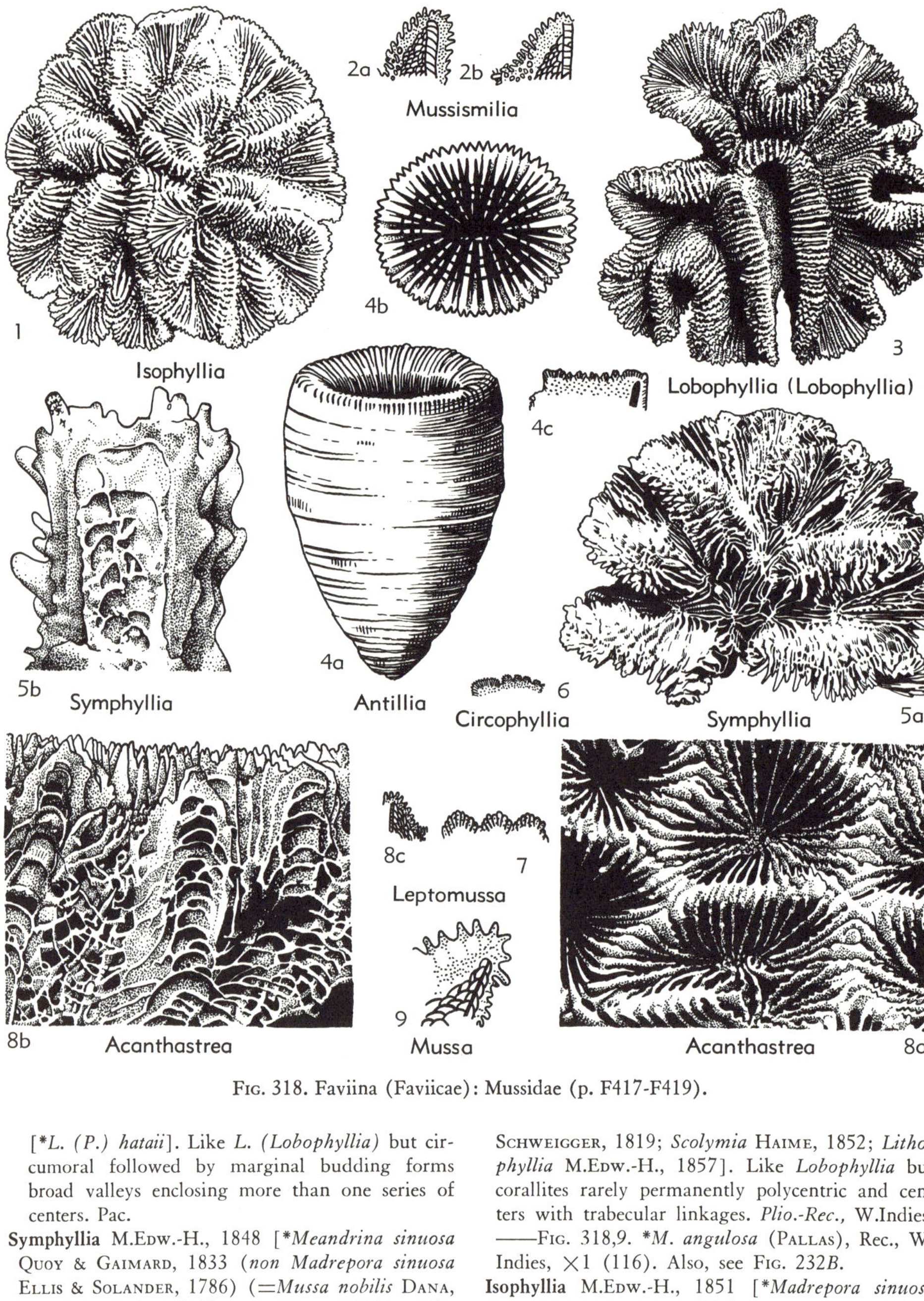

FIG. 318. Faviina (Faviicae): Mussidae (p. F417-F419).

[**L. (P.) hataii*]. Like *L. (Lobophyllia)* but circumoral followed by marginal budding forms broad valleys enclosing more than one series of centers. Pac.

Symphyllia M.EDW.-H., 1848 [**Meandrina sinuosa* QUOY & GAIMARD, 1833 (*non Madrepora sinuosa* ELLIS & SOLANDER, 1786) (=*Mussa nobilis* DANA, 1846)] [=*Sclerophyllia* KLUNZINGER, 1879]. Like *Lobophyllia* but series directly united by walls to summits; no ambulacra. *Rec.*, IndoPac.——FIG. 318,*5*. *S. indica* M.EDW.-H., Amboina; *5a*, ×0.5; *5b*, ×2 (174).

Mussa OKEN, 1815 [**Madrepora angulosa* PALLAS, 1766; SD VAUGHAN, 1918] [=*Lithodendron* SCHWEIGGER, 1819; *Scolymia* HAIME, 1852; *Lithophyllia* M.EDW.-H., 1857]. Like *Lobophyllia* but corallites rarely permanently polycentric and centers with trabecular linkages. *Plio.-Rec.*, W.Indies. ——FIG. 318,*9*. **M. angulosa* (PALLAS), Rec., W. Indies, ×1 (116). Also, see FIG. 232*B*.

Isophyllia M.EDW.-H., 1851 [**Madrepora sinuosa* ELLIS & SOLANDER, 1786; SD MATTHAI, 1928 (=*Oulophyllia? spinosa* M.EDW.-H., 1849)]. Colony formation by circumoral followed by intramural budding, the series later discontinuous, closely united laterally. Trabecular linkages. *Rec.*, W.Indies.——FIG. 318,*1*. **I. sinuosaa* ELLIS-S., Bermuda; ×0.5 (88).

Isophyllastraea MATTHAI, 1928 [**Astraea rigida* DANA, 1846]. Like *Isophyllia* but circumoral budding is followed by separation of buds to form mono- to tricentric corallites. *Rec.*, W.Indies.

Mycetophyllia M.EDW.-H., 1848 [**M. lamarckiana*]. Like *Isophyllia* but several series of centers enclosed between collines. Lamellar linkages. *Mio.-Rec.*, W.Indies.——FIG. 319,*4*. **M. lamarckiana*, Rec., W.Indies; ×1 (88).

Family PECTINIIDAE Vaughan & Wells, 1943

Solitary and colonial, hermatypic. Colony formation by intratentacular polystomodaeal budding; polyps and corallites organically united throughout; linkages by interstomodaeal mesenteries and thin lamellae. Corallum fixed, explanate to foliaceous, rarely submassive, with light structures. Corallites lacking definite walls; coenosteum abundant to practically absent. Septa irregularly dentate, formed by one fan system of compound trabeculae which may produce spinose dentations of mussoid or merulinid type, or dentations may be almost completely reduced. Columella trabecular, feeble or absent. *Oligo.-Rec.*

Fungophyllia GERTH, 1923 [**F. monstrosa;* SD VAUGHAN & WELLS, 1943]. Solitary, with expanded calice whose oral surface is nearly flat. Dentations mussoid. *Oligo.-Mio.*, E.Indies-W.Pac.-W.Indies.——FIG. 321,*1*. **F. monstrosa*, Mio., Borneo; *1a,b*, ×0.5 (58).

Echinophyllia KLUNZINGER, 1879 [**Madrepora aspera* ELLIS & SOLANDER, 1786; SD WELLS, 1936] [=*Oxyphyllia* YABE & EGUCHI, 1935]. Colonial; colony formation in earlier stages by circumoral budding, followed by irregular polystomodaeal budding, forming spreading explanate coralla. Calices parallel to plane of frond, united by vesicular coenosteum and confluent septocostae. *Mio.-Rec.*, Eu.-IndoPac.——FIG. 320,*2*. **E. aspera*, Rec., Japan; ×1 (164).

Oxypora SAVILLE-KENT, 1871 [*pro Trachypora* VERRILL, 1864 (*non* M.EDW.-H., 1851)] [*Trachypora lacera* VERRILL, 1864; SD WELLS, 1936]. Like *Echinophyllia* but folia very thin, nonvesicular, perforated at points of insertion of new septa. Calices superficial. Septa with simple to trifurcate dentations. *Rec.*, Pac.——FIG. 320,*1*. **O. lacera*, Palau I.; *1a,b*, ×1 (164).

Mycedium OKEN, 1815 [**Madrepora elephantotus* PALLAS, 1766; SD VERRILL, 1902] [=*Phyllastraea* DANA, 1846]. Like *Echinophyllia* but forming contorted folia with protuberant, nariform, outwardly inclined calices. *Rec.*, Pac.——FIG. 320,*3a*. *M. tenuicostatum* VERRILL, Singapore; ×1 (152).——FIG. 320,*3b*. *M. tubifex* (DANA), Mergui I.; ×1 (85).

Physophyllia DUNCAN, 1884 [**P. ayleni* WELLS,

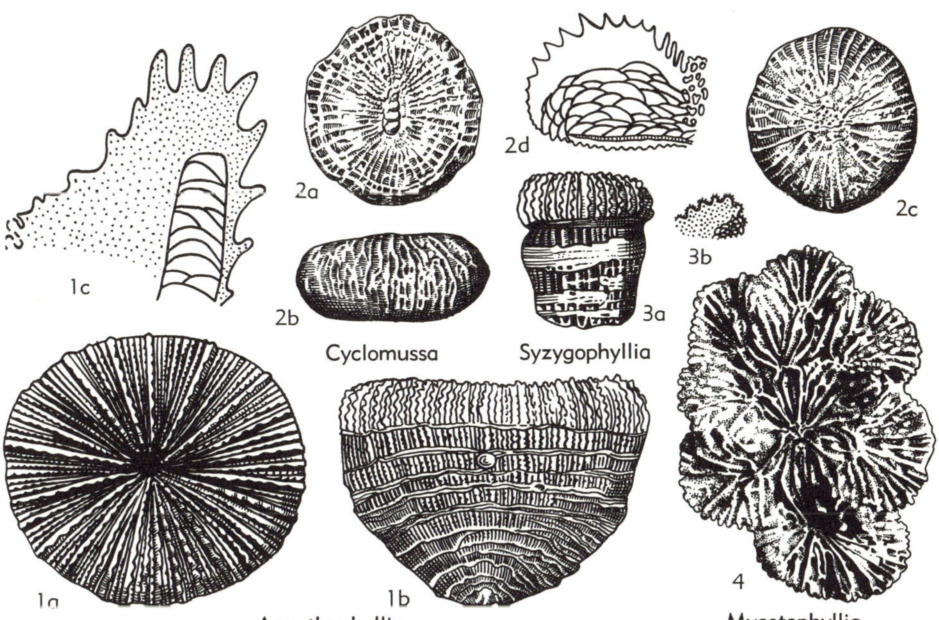

FIG. 319. Faviina (Faviicae): Mussidae (p. F417-F419).

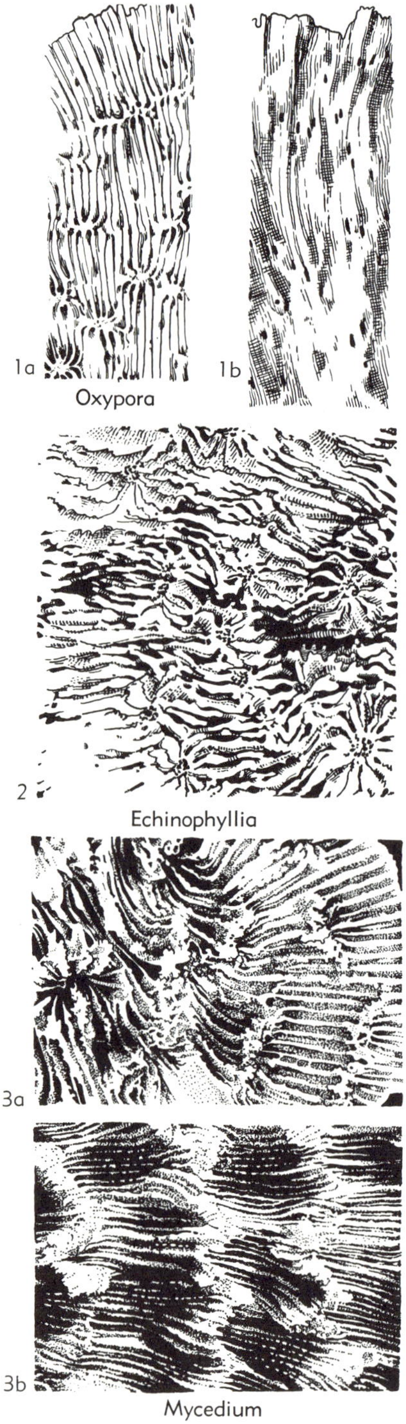

FIG 320. Faviina (Faviicae): Pectiniidae (p. F419-F421).

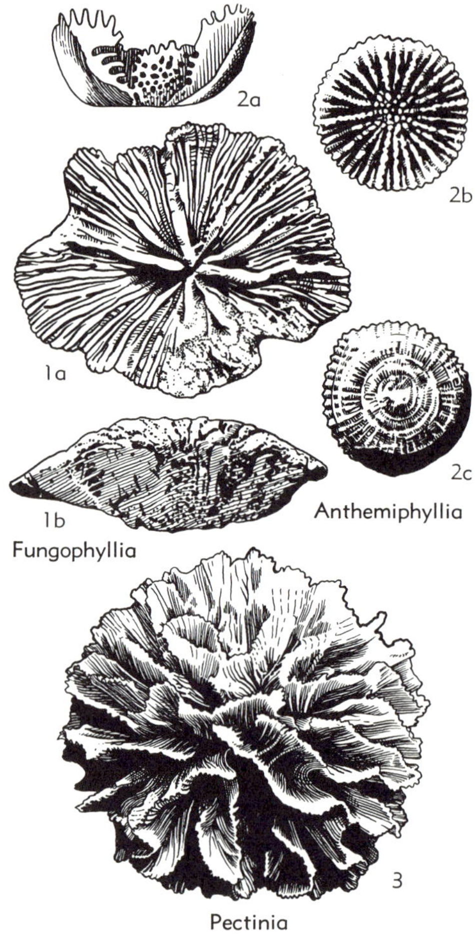

FIG. 321. Faviina (Faviicae): Pectiniidae, Anthemiphylliidae (p. F419-F421).

1934; SD WELLS, 1934]. Like *Echinophyllia* but abundant vesicular coenosteum rising between centers forming low collines. Septal dentations small, spinose, distant. *Rec.,* Pac.

Pectinia OKEN, 1815 [**Madrepora lactuca* PALLAS, 1766; SD VAUGHAN, 1901] [=**Tridacophyllia* BLAINV., 1830; *?Trydacnophyllia* PREVER, 1921]. Colony formation by circumoral followed by intramural budding, with submeandrine series separated by high, thin, acute collines bearing secondary centers. *Rec.,* IndoPac.——FIG. 321,*3*. **P. lactuca,* Amboina; ×0.3 (174).

Family ANTHEMIPHYLLIIDAE Vaughan, 1907

Solitary, free, patellate; ahermatypic. Septothecate, costate. Septa exsert, formed by one fan system of simple trabeculae produc-

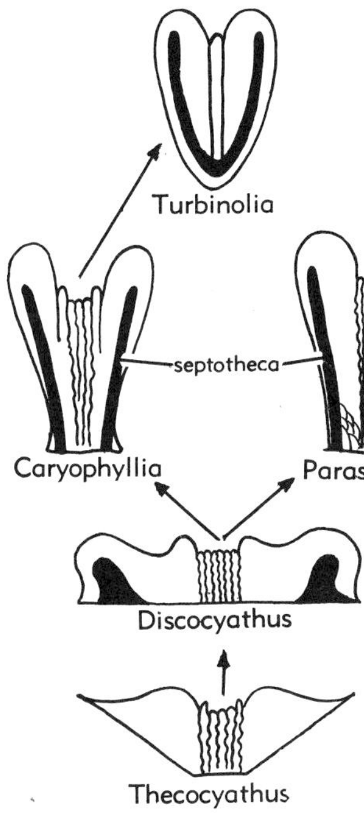

FIG. 321A. Structure of Carophylliicae shown by diagrammatic longitudinal sections (198n).

ing strong dentations. Columella trabecular or absent. Dissepiments absent. *Rec.*

Anthemiphyllia POURTALÈS, 1878 [**A. patera*]. Patellate, with smooth or granulated costae distinct to center of base. Columella trabecular, papillose on surface. *Rec.*, W.Indies-Pac., depth range 183-732 m.——FIG. 321,*2a*. **A. patera,* off Havana, 450-732 m.; long. sec., ×3 (187).——FIG. 321,*2b,c*. *A. hawaiiensis* VAUGHAN, Hawaii; ×3 (146).

Bathytrochus GRAVIER, 1915 [**B. hexagonus*]. Like *Anthemiphyllia* but lacking columella. *Rec.*, Atl., depth range 4,023 m.

Suborder CARYOPHYLLIINA Vaughan & Wells, 1943

[*nom. correct.* WELLS, herein (*pro* Caryophylliida VAUGHAN & WELLS, 1943)]

Solitary and colonial. Septa laminar, with smooth or nearly smooth margins, composed of one fan system of small, simple trabeculae. *Jur.-Rec.*

Superfamily CARYOPHYLLIICAE Gray, 1847

[*nom. correct.* WELLS, herein (*ex* Caryophyllioidae VAUGHAN & WELLS, 1943, *nom. transl. ex* Caryophyllidae GRAY, 1847)]

Caryophylliina with septothecal wall; rarely parathecal (epithecal in *Thecocyathus*). *Jur.-Rec.*

Family CARYOPHYLLIIDAE Gray, 1847

Solitary and colonial. Colony formation by extratentacular (rarely intratentacular) budding, forming phaceloid or dendroid colonies. Costae commonly covered by stereome or epitheca. Septa exsert. Columella formed by curled trabecular laths, solid, spongy or absent. Pali or paliform lobes common. Endothecal dissepiments developed in some groups (Fig. 321*A*) (199). *Jur.-Rec.*

Subfamily THECOCYATHINAE Vaughan & Wells, 1943

Solitary, discoidal or turbinate, free; ahermatypic. Wall epithecal or partly septothecal. *Jur.-M.Cret.*

Thecocyathus M.EDW.-H., 1848 [**Cyathophyllum tintinnabulum* GOLDF., 1826; SD M.EDW.-H., 1850]. Turbinate, wall epithecal, some internal stereome. Septa slightly exsert; pali opposite all but last cycle. Columella papillose. *Jur.*, Eu.——FIG. 322,*2*. *T. mactrus* (GOLDF.), L.Jur.(Toarc.), Fr.; *2a-c*, ×1.3 (53).

Discocyathus M.EDW.-H., 1848 [**Cyclolites eudesii* MICHELIN, 1840] [=*Brachycyathus* M.EDW.-H., 1848; *Cyclocyathus* M.EDW.-H., 1850]. Discoid; septotheca developed vertically near periphery of epithecal base, with granulated costae. Pali opposite 3rd septal cycle. Columella of several pillars, commonly fused to form lamella. *M.Jur.-M.Cret.*, Eurasia.——FIG. 322,*1*. **D. eudesii* (MICH.), M. Jur.(Baj.), Fr.; *1a-c*, ×1.3 (53).

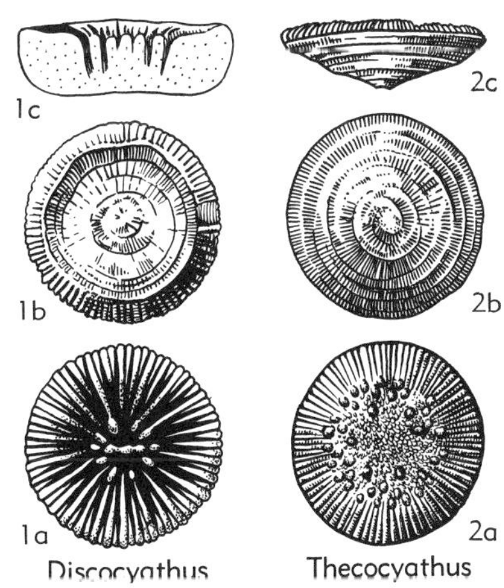

FIG. 322. Caryophylliina (Caryophylliicae): Caryophylliidae (p. F421).

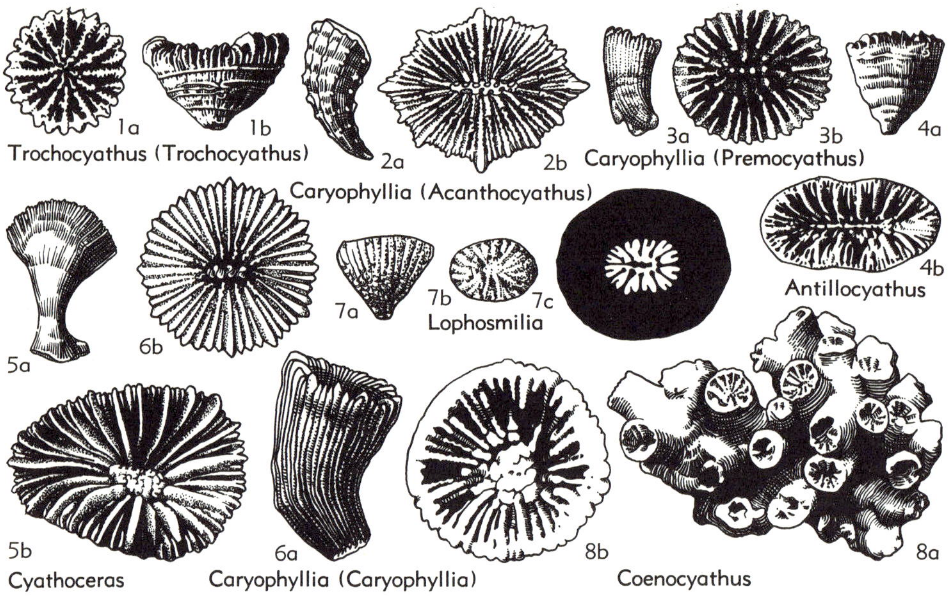

FIG. 323. Caryophylliina (Caryophylliicae): Caryophylliidae (p. F422-F423).

Subfamily CARYOPHYLLIINAE Gray, 1847

[*nom. transl.* M.EDW.-H., 1857 (*ex* Caryophyllidae GRAY, 1847)]

Mostly solitary, fixed or free; ahermatypic. Phaceloid or dendroid colonies formed by extratentacular budding from edge zone. Septothecate, with strong costae. Endothecal dissepiments rare. *U.Jur.-Rec.*

Caryophyllia LAMARCK, 1801 [**Madrepora cyathus* ELLIS & SOLANDER, 1786; SD BRODERIP, 1828] [=*Anthophyllea* LINK, 1807; *Anthophyllum* SCHWEIGGER, 1819; *Cyathina* EHR., 1834; *Amblocyathus* ORB., 1849; *Paterocyathus* DUCHASSAING & MICHELOTTI, 1860; *Ceratocyathus* SEGUENZA, 1863; *Blastocyathus* REUSS, 1864; *Goniocyathus* YABE & EGUCHI, 1932]. Solitary, turbinate to subcylindrical, fixed or free. Pali opposite third cycle in one crown (or before 2nd group of septa where hexameral symmetry is lost). Columella fascicular, formed by twisted trabecular laths. *U.Jur.-Rec.*, cosmop., depth range 0-2,743 m.

C. (Caryophyllia). *U.Jur.-Rec.*——FIG. 323,*6*. *C. (C.) bukowskii* JÜSSEN, Plio., Rhodes; *6a*, ×1; *6b*, ×1.5 (182).

C. (Acanthocyathus) M.EDW.-H., 1848 [**A. grayi;* SD M.EDW.-H., 1850]. Like *C. (Caryophyllia)* but with large costal spines, best developed on costae in plane of longer calicular axis. *Mio.-Rec.*, Eu.-IndoPac., depth range 68-732 m.——FIG. 323,*2*. *C. (A.) transylvanica* REUSS, Mio., Rumania; *2a*, ×1; *2b*, ×2 (117).

C. (Premocyathus) YABE & EGUCHI, 1942 [**Premocyathus compressus*]. Like *C. (Caryophyllia)* but compressed, with end costa of convex side expanded. *Plio.-Rec.*, Pac., depth range 34-658 m.——FIG. 323,*3*. **C. (P.) compressa*, Rec., Philip.; *3a*, ×1; *3b*, ×3 (165).

Antillocyathus WELLS, 1937 [**Placocyathus maoensis* VAUGHAN, 1925]. Like *C. (Caryophyllia)* but with more compressed calice and sublamellar columella. *Mio.*, W.Indies.——FIG. 323,*4*. **A. maoensis* (VAUGHAN), Mio., Dominican Rep.; *4a*, ×1; *4b*, ×2 (150).

Coenocyathus M.EDW.-H., 1848 [**C. cylindricus;* SD M.EDW.-H., 1850] [=*Bathangia* KEFERSTEIN, 1859]. Like *C. (Caryophyllia)* but small subphaceloid colonies formed by extratentacular budding. *Oligo.-Rec.*, Eu.-W.Indies-Calif., depth range 0-732 m.——FIG. 323,*8*. *C. bowersi* VAUGHAN, Rec., Calif.; *8a*, ×1; *8b*, ×4 (195).

Dendrocyathus ALLOITEAU, 1952 [**Dactylosmilia cenomana* ORB., 1850]. Like *Coenocyathus* but with small lamellar columella. *M.Cret.*, Eu.

Cyathoceras MOSELEY, 1881 [**C. cornu;* SD FAUSTINO, 1927]. Like *C. (Caryophyllia)* but lacking pali or paliform lobes. *Mio.-Rec.*, Eu.-W.Indies-S. Atl.-IndoPac., depth range 300-1,372 m.——FIG. 323,*5*. *C. diomedeae* VAUGHAN, Rec., Hawaii, 300-330 m.; *5a*, ×0.5; *5b*, ×1 (146).

Oxysmilia DUCHASSAING, 1870 [**Lophosmilia rotundifolia* M.EDW.-H., 1849]. Like *Cyathoceras* but columella fused in solid elongate mass. *Rec.*, W. Indies, depth range 73-293 m.

Lophosmilia M.EDW.-H., 1848 [**Caryophyllia cenomana* MICHELIN, 1845] [=*Actinosmilia* ORB., 1849; *Thalassiotrochus* M.EDW.-H., 1861; *Cenomanosmilia* ALLOITEAU, 1949]. Like *Oxysmilia* but columella a thin, lamellar sheet deep in calice.

M.Cret.-Rec., Eu.-N.Am.-S.Am.-Pac., depth range 183-366 m.——FIG. 323,*7*. **L. texana* (VAUGHAN), U.Cret.(Cenom.), Tex.; *7a,b*, ×1; *7c*, transv. sec., septotheca thickened by stereome, ×4 (158).

Bathycyathus M.EDW.-H., 1848 [**B. chilensis;* SD WELLS, 1936] [=*Steriphonotrochus* VAUGHAN, 1900]. Like *C. (Caryophyllia)* but pali not distinct from columellar laths. *Cret.-Rec.*, Eu.-N.Am.-W. Indies-N.Atl.-SE.Pac., depth range 55-165 m.

?**Nomlandia** DURHAM & BARNARD, 1952 [**N. californica*]. Said to be like *Bathycyathus* but discoidal and lacking wall. *Rec.*, 90 m., Calif.

Trochocyathus M.EDW.-H., 1848 [**Turbinolia mitrata* GOLDF., 1827; SD M.EDW.-H., 1850]. Solitary, turbinate to ceratoid, fixed or free. Pali opposite all but last cycle in 2 crowns. Columella fascicular, spongy, or crispate *M.Jur.-Rec.*, cosmop., depth range 32-1,573 m.

T. (Trochocyathus). *M.Jur.-Rec.*——FIG. 323,*1*. *T. (T.) uber* VAUGHAN & POPENOE, L.Eoc.(Midway.), Tex.; *1a,b*, ×4 (196).

T. (Aplocyathus) ORB., 1849 [**Turbinolia armata* MICHELOTTI, 1838]. Like *T. (Trochocyathus)* but bowl-shaped, with strong basal costal spines. *Eoc.-Rec.*, Eu.-N.Am.-S.Am.-E.Pac., depth range 27-366 m.——FIG. 325,*1*. **T. (A.) armata*, Mio., Italy; *1a*, basal aspect, ×1; *1b*, lat., ×1 (92).

T. (Platycyathus) FROMENTEL, 1863 [**Trochocyathus terquemi* M.EDW.-H., 1857; SD WELLS, 1933] [=?*Ecmesus* PHILIPPI, 1841; *Hemicyathus* SEGUENZA, 1864; ?*Blanfordia* DUNCAN, 1880; *Blanfordiola* STRAND, 1928]. Like *T. (Trochocyathus)* but thin, discoidal, free. *Cret.-Rec.*, Eu.-N. Am.-Pac., depth range 366-549 m.——FIG. 324, *2*. *T. (P.) lunulitiformis* (CONRAD), U.Eoc.(Jackson.), Miss.; *2a,b*, ×2 (144).

Tethocyathus KÜHN, 1933 [**Thecocyathus microphyllus* REUSS, 1871] [=*Thecocyathus auctt.* (*non* M.EDW.-H., 1848)]. Like *Trochocyathus* but edge zone narrow; epitheca extensive. *Eoc.-Rec.*, Eu.-Iran-N.Am.-W.Indies-Ind.O.-E.Indies, depth range 13-732 m.——FIG. 324,*1*. *T. cylindraceus* (POURTALÈS), Rec., W.Indies; *1a,b*, ×1.5 (104).

Stylocyathus ORB., 1849 [**S. dentalinus*]. Like *Trochocyathus* but pali in one irregular crown opposite first 2 cycles. Columella trabecular, appearing substyliform. *M.Cret.*, Eu.

Axocyathus ALLOITEAU, 1952 [**Ceratotrochus exiguis* FROMENTEL, 1861]. Like *Stylocyathus* but with paratheca. Columella fascicular, appearing papillose on surface. *U.Cret.*, Eu.

Deltocyathus M.EDW.-H., 1848 [**Turbinolia italicus* MICHELOTTI, 1838]. Solitary, discoid to patellate, free. Palpi opposite but all last cycle, forming deltas. Columella papillose on surface. *Eoc.-Rec.*, cosmop., depth range 13-4,480 m.

D. (Deltocyathus). *Eoc.-Rec.*——FIG. 324,*6*. **D. (D.) italicus* (MICHELOTTI), Mio.(Vindob.), Aus.; *6a,b*, ×1 (117).

D. (Levipalifer) VAUGHAN, 1900 [**Levipalifer orientalis* (=*D. (L.) vaughani* YABE & EGUCHI, 1937 (*non D. orientalis* DUNCAN, 1876)]. Like *D. (Deltocyathus)* but with pali before all cycles. *Rec.*, IndoPac., depth range 80-549 m.——FIG.

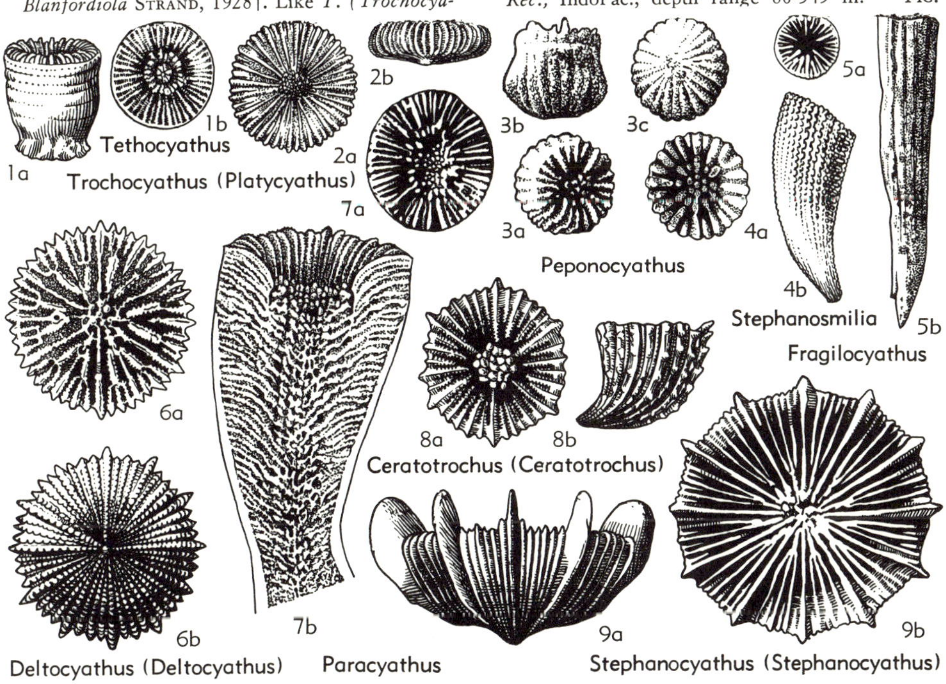

FIG. 324. Caryophylliina (Caryophylliicae): Caryophylliidae (p. F423-F426).

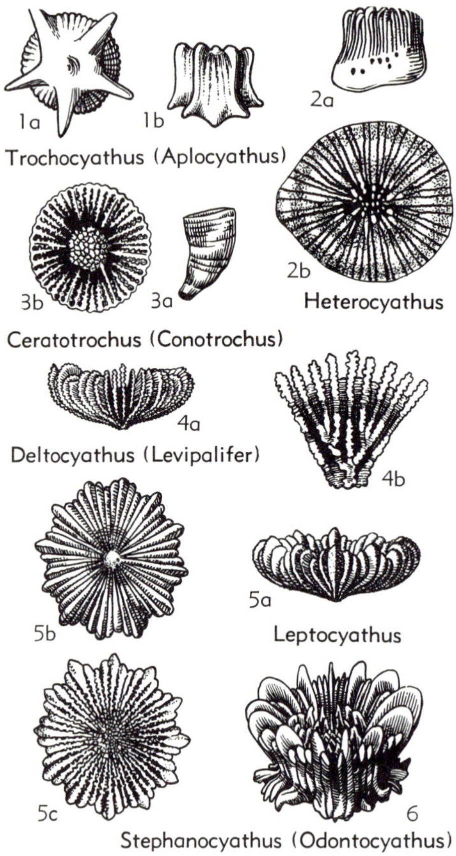

FIG. 325. Caryophylliina (Caryophylliicae): Caryophylliidae (p. F423-F424).

325,*4*. **D.* (*L.*) *vaughani*, Japan; *4a*, ×1; *4b*, ×1.5 (195).

Paracyathus M.EDW.-H., 1848 [**P. procumbens;* SD M.EDW.-H., 1850]. Solitary, turbinate, fixed. Pali opposite all but last cycle, merging with columellar papillae. *Eoc.-Rec.*, cosmop., depth range 13-1,472 m.——FIG. 324,*7b*. **P. procumbens*, M.Eoc.(Lut.), Fr.; long. sec., ×2 (93).——FIG. 324,*7a*. *Paracyathus* sp., Fla.; ×2 (151).

Polycyathus DUNCAN, 1876 [**P. atlanticus*] [=*Agelecyathus* DUNCAN, 1876]. Like *Paracyathus* but low colonies formed by extratentacular budding from edge zone. *Rec.*, S.Atl.-RedSea-Persian Gulf, depth range 20-73 m.

Deltocyathoides YABE & EGUCHI, 1932 [**D. japonicum*]. Like *Paracyathus* but patellate or subturbinate; free. *Rec.*, Japan, depth range 91-229 m.

Leptocyathus M.EDW.-H., 1850 [**L. elegans*]. Solitary, discoidal, free. Pali opposite all cycles, merging with columellar papillae. *M.Cret.-Eoc.*, Eng. ——FIG. 325,*5*. **L. elegans*, L.Eoc.(Lond.), Eng.; *5a-c*, ×1.5 (94).

Heterocyathus M.EDW.-H., 1848 [**H. aequicostatus;* SD M.EDW.-H., 1850] [=*Stephanoseris* M.EDW.-H., 1851; *Brachytrochus* DUNCAN, 1876 (*non* REUSS, 1864); *Brachytrossatus* STRAND, 1928]. Solitary, fixed in neanic stage to small univalve later completely covered by corallite base except for openings communicating with habitation of commensal sipunculid. Costae extending over base as granulations. Pali opposite all cycles; papillose columella. *Plio.-Rec.*, IndoPac., depth range 11-549 m. ——FIG. 325,*2*. *H. rousseanus* M.EDW.-H., Rec., Zanzibar; *2a*, ×1; *2b*, ×2 (93).

Ceratotrochus M.EDW.-H., 1848 [**Turbinolia multiserialis* MICHELOTTI, 1838; SD M.EDW.-H., 1850] [=*Crispatotrochus* T.-WOODS, 1878]. Solitary, trochoid, fixed or free. Pali absent. Columella large and papillose. *M.Cret.-Rec.*, cosmop., depth range 27-732 m.

C. (Ceratotrochus). *M.Cret.-Rec.*——FIG. 324,*8*. *C.* (*C.*) *multispinosus* M.EDW.-H., Mio.(Vindob.), Aus.; *8a,b*, ×1 (117).

C. (Conotrochus) SEGUENZA, 1864 [**C. typus*] [=*Pleurocyathus* MOSELEY, 1881 (*non* KEFERSTEIN, 1859); *Phloeocyathus* ALCOCK, 1902]. Like *C.* (*Ceratotrochus*) but epithecate, with septa almost nonexsert. *Eoc.-Rec.*, Eu.-N.Atl.-E.Indies, depth range 137-1,005 m.——FIG. 325,*3*. **C.* (*C.*) *typus*. Mio.(Vindob.), Aus.; *3a*, ×1; *3b*, ×2 (117).

Lochmaeotrochus ALCOCK, 1902 [**L. oculeus*]. Like *C.* (*Conotrochus*) but forming small phaceloid or subdendroid colonies. *Rec.*, E.Indies, depth range 366-530 m.

Stephanosmilia FROMENTEL, 1862 [**S. perlata*]. Like *Conotrochus* but small columella and with pali in 2 crowns opposite first 2 cycles. *L.Cret.*, Eu.——FIG. 324,*4*. **S. perlata*, L.Cret.(Apt.), Fr.; *4a,b*, ×3 (52).

Stephanocyathus SEGUENZA, 1864 [**S. elegans;* SD WELLS, 1936] [=*Sabinotrochus* DUNCAN, 1873; *Stephanotrochus* MOSELEY, 1881]. Solitary, patellate, free, with strong costae. Pali opposite 1 or 2 cycles or absent. Columella trabecular, papillose on surface. *Eoc.-Rec.*, cosmop., depth range 141-2,195 m.

S. (Stephanocyathus). *Eoc.-Rec.*——FIG. 324,*9*. *S.* (*S.*) *diadema nobilis* (MOSELEY), Rec., off Azores; *9a,b*, ×1 (64).

S. (Odontocyathus) MOSELEY, 1881 [**Platytrochus coronatus* POURTALÈS, 1868]. Like *S.* (*Stephanocyathus*) but with basal part of lower cycles of costae produced into stout spines. *Mio.-Rec.*, W. Indies-Atl.-Pac., depth range 366-1,575 m.—— FIG. 325,*6*. **S.* (*O.*) *coronatus* (POURTALÈS), Rec., VirginI., 750 m.; ×1 (97).

Vaughanella GRAVIER, 1915 [**V. concinna*]. Like *Stephanocyathus* but corallum turbinate or subtrochoid. *Rec.*, N.Atl., depth range 1,097-1,829 m.

Fragilocyathus YABE & EGUCHI, 1932 [**F. conotrochoides* (=*Flabellum matricidum* SAVILLE-KENT, 1871)]. Ceratoid, free. Wall finely costate. Longitudinal parricidal budding. No columella or pali. *Rec.*, Japan, depth range 84-289 m.——FIG. 324,5. **F. matricidum* (SAVILLE-KENT); *5a,b*, ×0.67 (199).

Aulocyathus MARENZELLER, 1904 [**A. juvenescens*]. Like *Fragilocyathus* but with trabecular columella. *Rec.*, E.Afr., depth range 366-430 m.

Stylotrochus FROMENTEL, 1861 [**S. arcuatus*]. Ceratoid, free. Pali irregular; columella sublamellar. *U.Cret.*, Eu.

Subfamily TURBINOLIINAE Milne-Edwards & Haime, 1848

Solitary, free, trochoid, cuneiform, or conical, completely invested by polyp; intercostal grooves deeply incised from calicular margin to points of origin of costae; ahermatypic. *U.Cret.-Rec.*

Bothrophoria FELIX, 1909 [**B. ornata*]. Trochoid, slightly compressed. Wall regularly perforated between costae. No pali; columella spongy. *U.Cret.*, Antarct.——FIG. 329,1. *B. ornata*, Senon., Antarct.; *1a,b*, ×2 (177).

Conocyathus ORB., [**C. sulcata*] [=*Stylocyathus* REUSS, 1855 (*non* ORB., 1849); *Pleurocyathus* KEFERSTEIN, 1859]. Like *Bothrophoria* but not compressed, and with paliform lobes forming one crown opposite second cycle. *Eoc.-Rec.*, Eu.-Afr.-IndoPac., depth range 73 m.——FIG. 326,1. *C. zelandiae* DUNCAN, Rec., Persian Gulf; *1a,b*, ×5 (178).

Trematotrochus T.-WOODS, 1879 [**Conocyathus fenestratus* T.-WOODS, 1878]. Like *Conocyathus* but septa and costae with hispid granulations; paliform lobes reduced or even absent. *U.Oligo.-Rec.*, Austral.-N.Z., depth range 27-457 m.——FIG. 327,2. *T. hedleyi* Dennant, Rec., off Pt. Jackson, 457 m.; *2a,b*, ×5 (21).

Turbinolia LAMARCK, 1816 [**T. sulcata*; SD M.EDW.-H., 1850] [=*Diplactis, Heteractis* CHALMAS, 1895]. Trochoid-conical; wall perforated or externally pitted between costae. Interseptal costae in some species. Styliform or slightly compressed columella formed by fusion of primary septa. *Eoc.-Oligo.*, Eu.-W.Afr.-N.Am.-S.Am.

T (Turbinolia). *Eoc.-Oligo.*——FIG. 326,2a. **T. (T.) sulcata* LAM., M.Eoc.(Auvers.), Eng.; ×5 (94).——FIG. 326,2b. *T. (T.) dixoni* M.EDW.-H., M.Eoc.(Auvers.), Eng.; ×5 (94).

T. (Batotrochus) WELLS, 1937 [**Turbinolia corbicula* POURTALÈS, 1879]. Like *T. (Turbinolia)* but septa and costae with thick-set spinose granulations. *Rec.*, Fla., depth range 183-567 m.

Notocyathus T.-WOODS, 1880 [**Caryophyllia viola* DUNCAN, 1864; SD FELIX, 1927] [=*Nototrochus* DUNCAN, 1884; *Citharocyathus* ALCOCK, 1902]. Trochoid-conical to cuneiform; wall imperforate. Paliform lobes opposite all but last cycle, commonly forming deltas opposite second and third. Columella papillose to pseudolamellar. *U.Eoc-Rec.*, W.Pac.-Austral.

N. (Notocyathus). *U.Oligo.-Rec.*, depth range 68-583 m.——FIG. 329,6. *N. (N.) conicus* (ALCOCK), Rec., Sulu Sea, 550 m.; *6a,b*, ×5 (4).

N. (Paradeltocyathus) YABE & EGUCHI, 1937 [**Deltocyathus orientalis* DUNCAN, 1876]. Like *N. (Notocyathus)* but with bowl-shaped corallum. *U.Eoc.-Rec.*, IndoPac.-N.Atl., depth range 55-835 m.

Sphenotrochus M.EDW.-H., 1848 [**Turbinolia crispa* LAMARCK, 1816; SD M.EDW.-H., 1850]. Cuneiform, with strong crispate costae or costae more or less reduced to granulations. Columella pseudolamellar. Pali absent. *Eoc.-Rec.*, cosmop.

S. (Sphenotrochus). *Eoc.-Rec.*, depth range 21-274 m.——FIG. 327,4. *S. (S.) intermedius* (MÜNSTER), Plio., Eng.; *4a,b*, ×5 (94). (Also Fig. 248.)

S. (Eusthenotrochus) WELLS, 1935 [**S. (E.) moseri* (=*S. gilchristi* GARDINER, 1904)]. Like *S. (Sphenotrochus)* but costae wholly reduced to crispate granulations. *Eoc.-Rec.*, Eu.-W.Indies-Ind. O., depth range 78-142 m.——FIG. 326,3. **S. (E.) gilchristi*, Rec., southeast Africa, 125 m.; *3a,b*, ×1 (198).

Idiotrochus WELLS, 1936 [**Sphenotrochus emarciatus* DUNCAN, 1865]. Like *S. (Sphenotrochus)* but the costae alternate in position with septa, and with pali in one crown opposite first 2 cycles. *Oligo.-Rec.*, W.Pac.-Austral.——FIG. 329,4. **I. emarciatus*, L.Mio.(Balcomb.), SAustral.; *4a,b*, ×5 (198).

Dunocyathus T.-WOODS, 1878 [**D. parasiticus* T.-WOODS, 1878 (=*Deltocyathus rotaeformis* T.-WOODS, 1878)]. Like *Idiotrochus* but patellate to discoidal or turbinate with base overgrown by

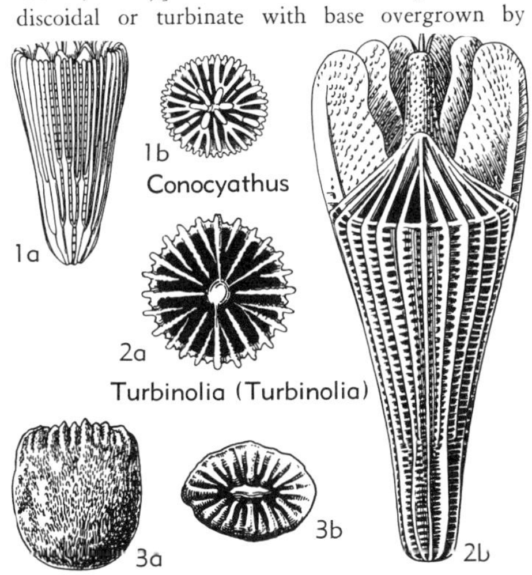

FIG. 326. Caryophylliina (Caryophylliicae): Caryophylliidae (p. F425).

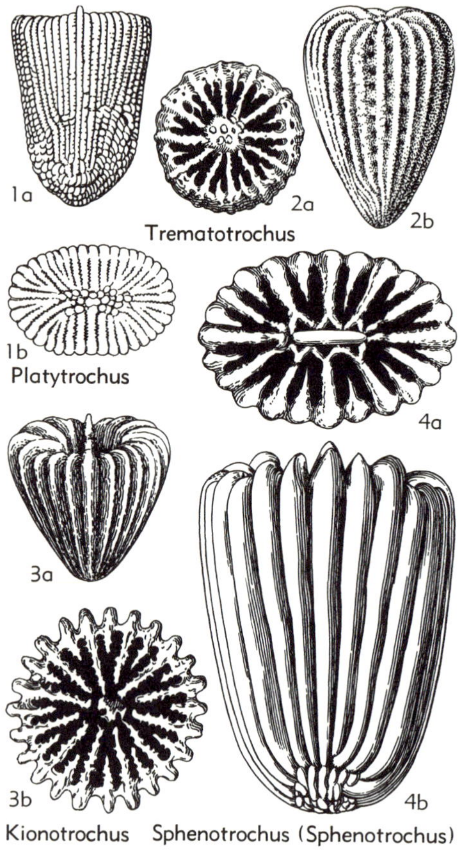

FIG. 327. Caryophylliina (Caryophylliicae): Caryophylliidae (p. F425-F426).

bryozoan *(Bipora)*. *Rec.*, Austral., depth range, 82-457 m.

Platytrochus M.EDW.-H., 1848 [**Turbinolia stokesi* LEA, 1833; SD M.EDW.-H., 1850]. Like *Sphenotrochus* but with spongy columella, and generally with alate, thickened end costae. *U.Cret.-Rec.*, N. Am.-Austral., depth range 27-130 m.——FIG. 327, *1*. *P. claibornensis* GREGORIO, M.Eoc.(Claib.), Ala.; *1a,b*, ×5 (144).

Koilotrochus T.-WOODS, 1878 [**Smilotrochus? vacuus*] [=*Aldrichia* VAUGHAN, 1900 (*non* COQUILLET, 1894); *Aldrichiella* VAUGHAN, 1903]. Like *Platytrochus* but elongate-cornutiform, with normal end costae. *Eoc.-Mio.*, N.Am.-Austral.——FIG. 329,*5*. *K. elegans* (VAUGHAN), U.Eoc.(Jackson.), Miss.; *5a*, ×5; *5b*, ×10 (144).

Dominicotrochus WELLS, 1937 [**Smilotrochus? dominicensis* VAUGHAN, 1925]. Like *Platytrochus* but lacking columella. *Mio.*, W.Indies.

Kionotrochus DENNANT, 1906 [**K. suteri*]. Subtrochoid, bowl-shaped, or patellate. Costae broad and spinose. Columella commonly solid. Pali absent. *Rec.*, Eu.-Pac., depth range 68-470 m.——FIG. 327,*3*. **K. suteri*, Rec., N.Z.; *3a,b*, ×5 (21).

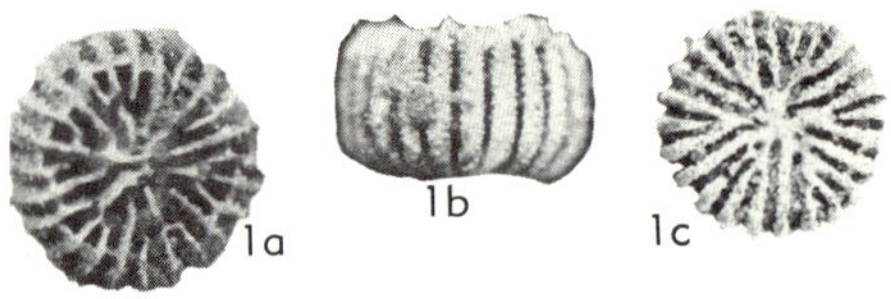

FIG. 328. Caryophylliina (Caryophylliicae): *Cylindrophyllia*.

Cylindrophyllia YABE & EGUCHI, 1937 [**C. minima*]. Tympanoid, generally with 3 cycles of septa and single crown of 6 thin pali before 2nd cycle; columella papillary. *Oligo.-Rec.*, Eu.-Pac., depth range 260 m.——FIG. 328,*1*. **C. minima*, M.Oligo. (Waitakian), N.Z.; *1a-c*, calicular, lateral, and basal views, ×6.7 (198).

Peponocyathus GRAVIER, 1915 [**P. variabilis*]. Globular or bottle-shaped; pali opposite all but last cycle of septa in 2 crowns; columella papillary. *Oligo.-Rec.*, Atl.-W.Indies-E.Pac.-Austral., depth range 110-1,097 m.——FIG. 324,*3*. **P. variabilis*, Rec., Azores, 350-850 m.; *3a-c*, ×5 (64).

Tropidocyathus M.EDW.-H., 1848 [**Flabellum lessoni* MICHELOTTI, 1842] [=*Cyathotrochus* BOURNE, 1905]. Like *Platytrochus* but larger, with strong alate costal extensions, and with pali opposite all but last cycle. *Mio.-Rec.*, IndoPac., depth range 11-480 m.——FIG. 329,*2*. **T. lessoni*, Rec., E. Indies; *2a,b*, ×2 (171).

Holcotrochus DENNANT, 1902 [**H. scriptus*]. Cuneiform, with very broad, spinose or hispid costae. Septa thick, few, equal. Columella trabecular. *Mio.-Rec.*, Austral., depth range 10-41 m.——FIG. 329,*3*. **H. scriptus*, L.Mio.(Balcomb.), S.Austral.; *3a,b*, ×5 (21).

Subfamily DESMOPHYLLINAE Vaughan & Wells, 1943

Solitary and colonial; ahermatypic. Colony formation by intra- and extratentacular budding. Endothecal dissepiments deep and sparse. Coenosteum dense where developed. Columella, pali, and epitheca absent. *Cret.-Rec.*

Desmophyllum EHR., 1834 [**D. dianthus* (*non Madrepora dianthus* ESPER, 1797) (=*D. cristagalli* M.EDW.-H., 1848; SD M.EDW.-H., 1850)] [=*Coelosmilia* M.EDW.-H., 1850; *Psammosmilia* FROMENTEL, 1867; *Thalamophyllia* DUCHASSAING, 1870; *Javania* DUNCAN, 1876]. Solitary, trochoid, fixed. Costae well developed near calice, obsolete or covered by stereome basally. Some sparse endothecal dissepiments. *M.Cret.-Rec.*, cosmop., depth range 0-2,286 m.——FIG. 330,*4*. **D. cristagalli*, Rec., off Cape Breton; ×0.5 (93).

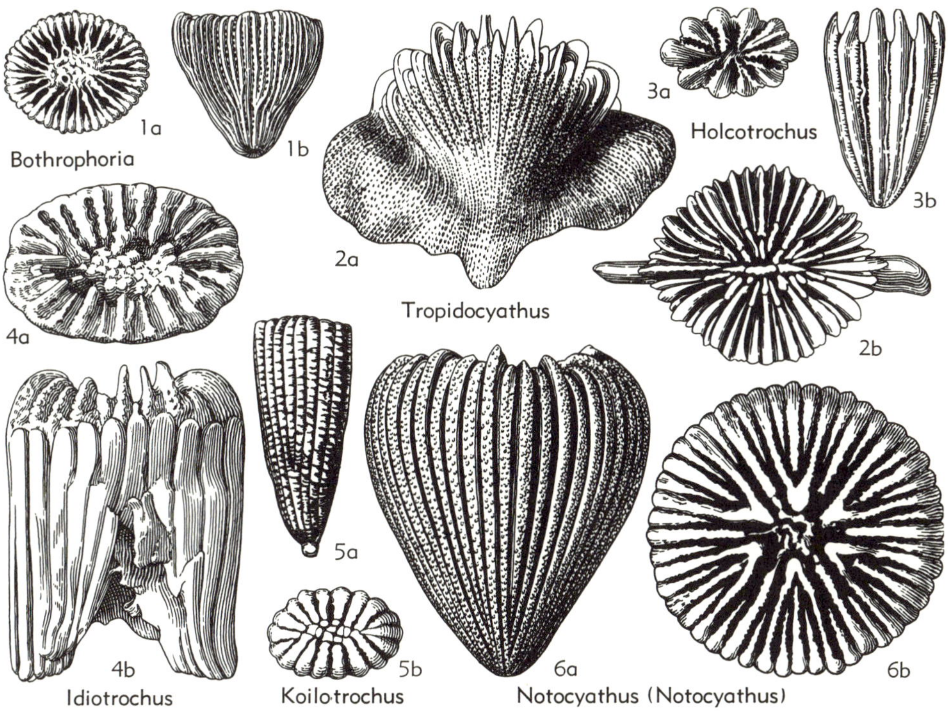

FIG. 329. Caryophylliina (Caryophylliicae): Caryophylliidae (p. F425-F426).

Gemmulatrotrochus DUNCAN, 1878 [**G. simplex*] [=*Microcyathus* DÖDERLEIN, 1913 (*non* HINDE, 1896)]. Like *Desmophyllum* but forming small clumps mostly by extratentacular budding. *Rec.*, Medit.-Eng.Channel, shallow water.

Dactylotrochus WELLS, 1954 [**Tridacophyllia cervicornis* MOSELEY, 1881]. Like *Desmophyllum* but with slender, finger-like extensions of wall and septa flaring outwards from calicular margin. *Rec.*, Pac., depth range 75-135 m.

Lophelia M.EDW.-H., 1849 [**Madrepora prolifera* PALLAS, 1766; SD M.EDW.-H., 1850] [=*Lophohelia* M.EDW.-H., 1857; *Scolangia* T.-WOODS, 1880]. Like *Desmophyllum* but forming large dendroid

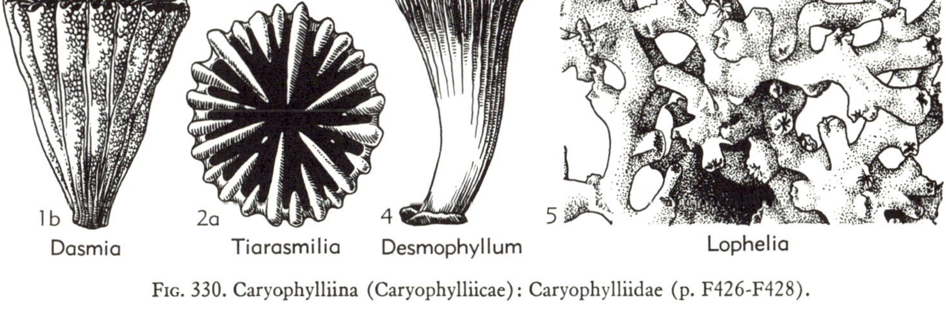

FIG. 330. Caryophylliina (Caryophylliicae): Caryophylliidae (p. F426-F428).

colonies by unequal intratentacular monostomatous budding. Coenosteum dense. Costae feebly developed. Sparse tabular endothecal dissepiments. *Oligo.-Rec.*, N.Atl.-Medit.-N.Z., depth range 64-1,975 m.——FIG. 330,*5*. **L. prolifera*, Rec., off S. Car., 500-800 m.; ×0.5 (151).

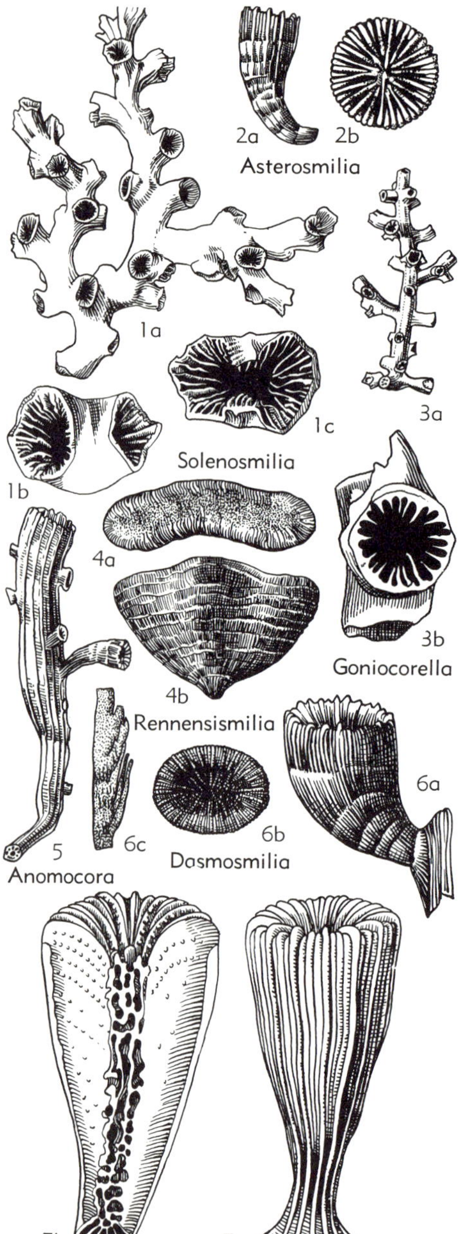

FIG. 331. Caryophylliina (Caryophylliicae): Caryophylliidae (p. F428-F429).

Tiarasmilia WELLS, 1932 [**T. casteri*]. Like *Desmophyllum* but septa with strong vertical striations or lateral ridges. *L.Cret.*, Tex.——FIG. 330,*2*. **T. casteri*, L.Cret.(U.Apt.), Tex.; *2a*, calice, *2b*, cutaway diagram of 3 septa, ×2 (157).

Stelloria ORB., 1849 [**S. elegans* (=*Anthophyllum sulcatum* MICHELOTTI, 1845; SD WELLS, 1936)]. Low and turbinate, fixed by broad base. Costae apparently very low or obsolete. Septa with strong vertical striations or carinae projecting nearly halfway to adjacent septa. *M.Cret.*, Eu.

Dasmia M.EDW.-H., 1848 [**D. sowerbyi*]. Solitary, free in ephebic stage. Septa of first cycle normal; those of higher cycles divided vertically without corresponding costae. *L.Eoc.*, Eng.——FIG. 330,*1*. **D. sowerbyi*, Lond., Eng.; *1a,b*, ×3 (94).

Dasmiopsis OPPENHEIM, 1930 [**Trochocyathus lamellicostatus* REUSS, 1854]. Solitary, cornute, free. Costae of first 2 cycles normal; those of higher cycles bent toward those of lower cycles and covered by stereome, giving appearance of only 12 costae. *U.Cret.*, Aus.——FIG. 330,*3*. **D. lamellicostatus* (REUSS), Turon., Aus.(Gosau); *3a,b*, ×1 (100).

Subfamily PARASMILIINAE Vaughan & Wells, 1943

Solitary and colonial, ahermatypic. Colony formation by intra- and extratentacular budding. Coenosteum, epitheca, and stereome rarely developed. Endothecal dissepiments sparse. Columella trabecular or absent. *U. Jur.-Rec.*

Parasmilia M.EDW.-H., 1848 [**Madrepora centralis* MANTELL, 1822] [=*Cylicosmilia* M.EDW.-H., 1848; *Cyclosmilia* ORB., 1849; *Monocarya* LONSDALE, 1850; *?Plesioparasmilia, Strobilosmilia* ALLOITEAU, 1952]. Solitary, trochoid, fixed. Columella spongy. Endothecal dissepiments developed but deep in corallum. *L.Cret.-Rec.*, cosmop., depth range 313-366 m.——FIG. 331,*7*. **P. centralis*, U. Cret.(Senon.), Eng.; *7a,b*, corallum and long. sec. (endotheca not shown), ×2 (94).

Edwardsosmilia ALLOITEAU, 1949 [**Trochosmilia faujasi* M.EDW.-H., 1849]. Like *Parasmilia* but columella compressed; costae obsolete basally. *U. Cret.*, Eu.

Smilotrochus M.EDW.-H., 1851 [**Trochosmilia tuberosa* M.EDW.-H., 1850] [=*Blagrovia* DUNCAN, 1880]. Like *Parasmilia* but columella absent. *M. Cret.-Eoc.*, Eurasia.

Rennensismilia ALLOITEAU, 1952 [**Turbinolia didyma* GOLDF., 1826]. Like *Smilotrochus* but compressed, with endothecal dissepiments well developed peripherally. Includes many species attributed to *Trochosmilia*. *U.Cret.*, Eu.——FIG. 331,*4*. *R. didymophila* (FELIX), U.Cret.(Turon.), Aus. (Gosau); *4a,b*, ×0.5 (42).

Meandrosmilia ALLOITEAU, 1952 [**Trochosmilia flabellum* FROMENTEL, 1863]. Like *Rennensismilia* but larger, flabellate in shape. *U.Cret.*, Eu.

Coenosmilia POURTALÈS, 1874. [**C. arbuscula*]. Like *Parasmilia* but producing small phaceloid colonies by extratentacular budding from edge zone below calice. *Rec.*, W.Indies, depth range 183 m.

Dungulia OPPENHEIM, 1930 [**Coelosmilia milneri* GREGORY, 1898] [=*Coelosmilia auctt.* (*non* M. EDW.-H., 1850)]. Like *Parasmilia* but columella and endothecal dissepiments feebly developed, with relatively few septa. *U.Jur.*, *?Rec.*, Eurasia-Afr.-N.Am.

Caryosmilia WANNER, 1902 [**C. granosa*]. Like *Dungulia* but with paliform lobes opposite third cycle. *U.Cret.*, N.Afr.

Anomocora STUDER, 1878 [**Coelosmilia fecunda* POURTALÈS, 1871] [=*Blastosmilia* DUNCAN, 1878 (*non* ÉTALLON, 1859); *Pourtalosmilia* DUNCAN, 1884]. Like *Dungulia* but with tendency toward budding from edge zone with loss of organic connection. *Rec.*, W.Indies-Atl.-Ind.O., depth range 91-576 m.——FIG. 331,*5*. **A. fecunda*, Fla., 125-275 m.; ×1 (104).

Solenosmilia DUNCAN, 1873 [**S. variabilis*]. Dendroid or subphaceloid colonies produced by intratentacular budding. Corallites cylindrical with some granular stereome. Septa relatively few; columella very weak; endothecal dissepiments tabular. *Rec.*, Atl.-IndoPac., depth range 457-1,280 m.——FIG. 331,*1*. *S. jeffreyi* ALCOCK, off Travancore, 840 m.; *1a*, ×0.5; *1b,c*, ×2 (171).

Dendrosmilia M.EDW.-H., 1848 [**D. duvaliana*]. Like *Solenosmilia* but colony formation by extratentacular budding, regular and alternate. *Eoc.-Oligo.*, Eu.

Asterosmilia DUNCAN, 1867 [**Trochocyathus abnormalis* DUNCAN, 1865; SD VAUGHAN, 1919]. Solitary, trochoid-ceratoid, generally free. Paliform lobes opposite next to last 1 or 2 cycles. Columella lamellar at surface, trabecular below. *Oligo.-Rec.*, W.Indies, depth range 137-183 m.——FIG. 331,*2*. *A. prolifera* POURTALÈS, Rec., W.Indies; *2a*, ×1; *2b*, ×1.5 (104).

Dasmosmilia POURTALÈS, 1880 [**Trochosmilia lymani* POURTALÈS, 1871; SD WELLS, 1933]. Turbinate or trochoid, commonly increasing by parricidal budding. Paliform lobes before all but last cycle. Columella formed by mingling of inner lobes. *U.Cret.-Rec.*, N.Am.-W.Indies-RedSea, depth range 110-448 m.——FIG. 331,*6*. **D. lymani*, Rec., Fla., 125-275 m.; *6a,b*, lat. and calicular aspects, ×1; *6c*, septum with paliform lobes, ×1 (104).

Goniocorella YABE & EGUCHI, 1932 [**Pourtalosmilia dumosa* ALCOCK, 1902]. Dendroid; colony formation by extratentacular budding at right angles. Corallites cylindrical, faintly costate. Columella absent; endothecal dissepiments tabular and sparse.

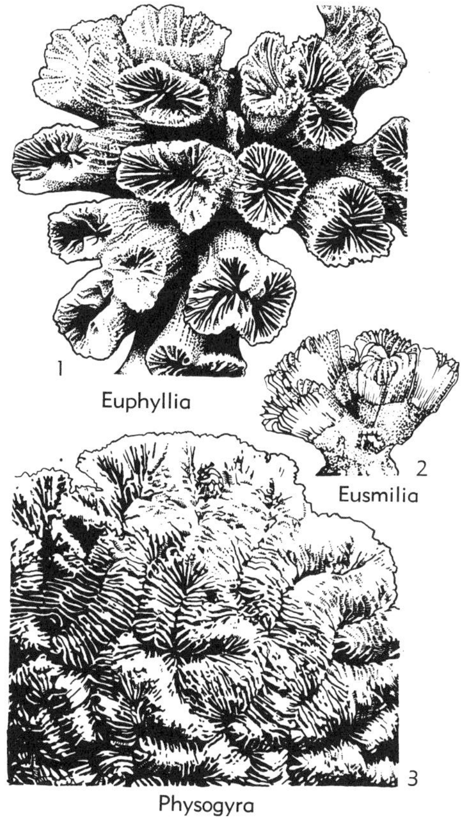

FIG. 332. Caryophylliina (Caryophylliicae): Caryophylliidae (p. F429-F430).

Rec., W.Pac., depth range 100-457 m.——FIG. 331,*3*. **G. dumosa*, E.Indies, 450-457 m.; *3a*, ×0.5; *3b*, ×4 (4).

Subfamily EUSMILIINAE Milne-Edwards & Haime, 1857

Colonial caryophylliids, hermatypic. Colony formation by intratentacular intramural di- to polystomodaeal budding. Corallite walls septothecal or parathecal. Costae low, distant. Septa exsert, mostly entocoelic. Columella trabecular, lamellar, or absent. Endothecal dissepiments vesicular. *Eoc.-Rec.*

Euphyllia DANA, 1846 [**Caryophyllia glabrescens* CHAMISSO & EYSENHARDT, 1821; SD VAUGHAN, 1918] [=*Leptosmilia* M.EDW.-H., 1848; *Plocophyllia* REUSS, 1868]. Phaceloflabellate, with laterally free series. Walls septothecal. Columella absent. *Eoc.-Rec.*, Eurasia-IndoPac.——FIG. 332,*1*. **E. glabrescens*, Rec., Fiji; ×0.25 (148).

Plerogyra M.EDW.-H., 1848 [**P. laxa* M.EDW.-H., 1849; SD M.EDW.-H., 1850 (=*Euphyllia sinuosa* DANA, 1846)]. Like *Euphyllia* but series united basally and more or less laterally by cellular coenosteum; walls parathecal. *Rec.*, IndoPac.——FIG. 333, *1. P. sinuosa*, E.Indies; ×0.25 (88).

Physogyra QUELCH, 1884 [**P. aperta* (=*Plerogyra lichtensteini* M.EDW.-H., 1851)]. Like *Plerogyra* but series closely united by walls; endothecal dissepiments highly vesicular. *Rec.*, IndoPac.——FIG. 332,*3*. *P. somaliensis* VAUGHAN, Fr.Somali.; ×0.5 (195).

Gyrosmilia M.EDW.-H., 1851 [**Manicina interrupta* EHR., 1834]. Like *Physogyra* but with proportionally more septa and mostly septothecal walls. *Rec.*, RedSea.

Progyrosmilia WELLS, 1937 [*pro Coelocoenia* GERTH, 1923 (*non* VOLZ, 1896)] [**Coelocoenia torulosa* GERTH, 1923]. Like *Gyrosmilia* but with shorter series separated by costate ambulacra. *Mio.*, E.Indies.——FIG. 333,*2*. **P. torulosa*, Borneo; *2a*, ×1; *2b*, long. sec., ×1.75 (58).

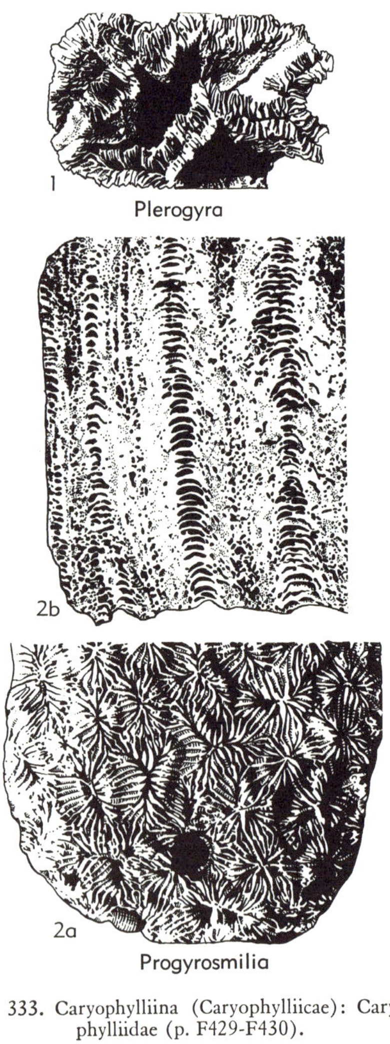

FIG. 333. Caryophylliina (Caryophylliicae): Caryophylliidae (p. F429-F430).

Eusmilia M.EDW.-H., 1848 [**Madrepora fastigiata* PALLAS, 1766]. Like *Euphyllia* but corallites smaller, mono- to tristomatous, with septa more exsert and spinose costae. *Oligo.-Rec.*, W.Indies-N. Am.——FIG. 332,*2*. **E. fastigiata*, Rec., Fla.; ×0.5 (151).

Family RHIPIDOGYRIDAE Koby, 1904

Solitary and colonial, fixed; hermatypic. Colony formation by intratentacular budding (except *Cymosmilia*). Corallites united or externally thickened by solid coenosteum with granulated surface, concealing costae except near calices. Septa relatively few, exsert, thick, nondentate. Columella lamellar, thin, continuous, deep. Endothecal dissepiments thin and sparse. Epitheca absent. *M. Jur.-L.Cret.* (Fig. 260).

APLOSMILIA GROUP

Colony formation (except *Cymosmilia*) by complete mono- or distomodaeal budding.

?**Codonosmilia** KOBY, 1888 [**C. elegans*]. Solitary or forming low colonies by monostomodaeal budding. Presence of columella in this genus not yet established. *M.Jur.-U.Jur.*, Eurasia.——FIG. 334,*4*. **C. elegans*, U.Jur.(Bath.), Fr.; *4a,b*, ×1 (71).

Cymosmilia KOBY, 1894 [**C. conferta*]. Low colonies produced by extratentacular budding from edge zone. *U.Jur.*, Eu.——FIG. 334,*3*. *C. conferta*, U.Jur.(Raur.), Fr.; *3a*, ×0.5; *3b*, ×1 (71).

Aplosmilia ORB., 1849 [**Lobophyllia semisulcata* MICHELOTTI, 1843]. Phaceloid colonies formed by complete mono- or distomodaeal budding. *U.Jur.-L.Cret.*, Eu.-AsiaM.——FIG. 334,*6a*. *A. rugosa* KOBY, U.Jur.(Raur.), Switz.; ×0.25 (71).——FIG. 334,*6b,c*. **A. semisulcata*, U.Jur.(Sequan.), Fr.; long. secs. parallel to and across columella, ×1.5 (71).

Kologyra WELLS, 1937 [**Pachygyra tuberosa* ORB., 1850]. Like *Aplosmilia* but subplocoid with incomplete separation of corallites which may form short series. *U.Jur.*, Eu.

Acanthogyra OGILVIE, 1897 [**A. columnaris*]. Like *Aplosmilia* but cerioid. *U.Jur.*, Eurasia.——FIG. 335,*1*. *A. multiformis* OGILVIE, Portl., Czech., transv. sec., ×1 (99).

RHIPIDOGYRA GROUP

Colony formation by incomplete, linear, polystomodaeal budding.

Rhipidogyra M.EDW.-H., 1848 [**Lobophyllia flabellum* MICHELIN, 1843; SD M.EDW.-H., 1850] [=*Stylogyra* ORB., 1849; *Stenogyra* FROMENTEL,

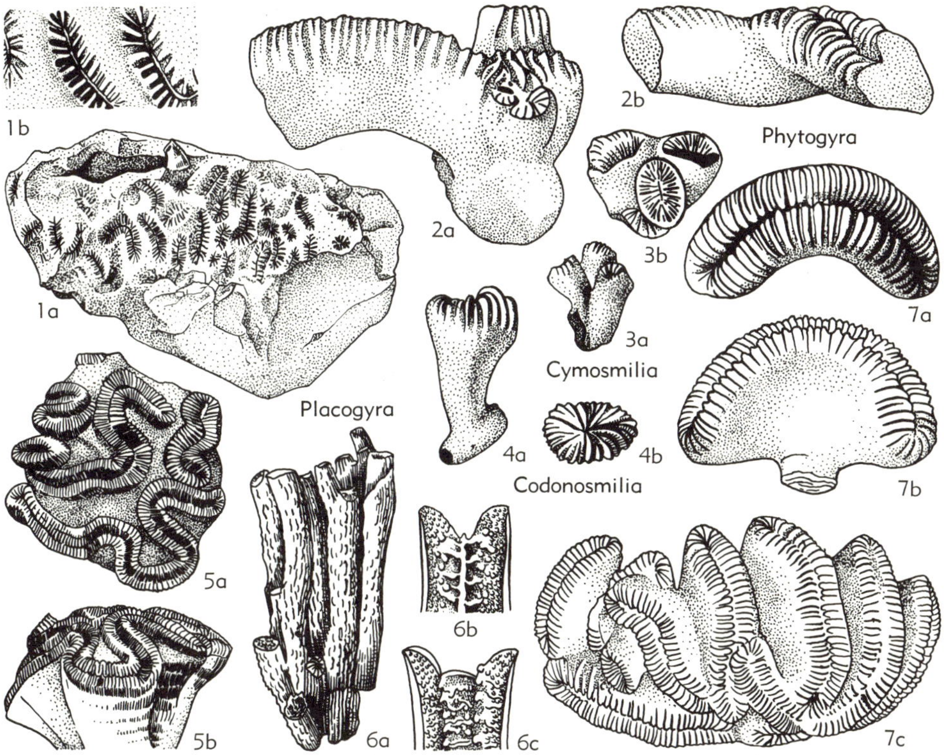

FIG. 334. Caryophylliina (Caryophylliicae): Rhipidogyridae (p. F430-F431).

1861; *Sclerogyra* FERRY, 1870; *Rhypidomontivoltia* GREGORIO, 1884; *Fromentelligyra* ALLOITEAU, 1952; *?Rhipidosmilia* GEYER, 1955]. Flabelliform, corallites in laterally free linear series, commonly contorted. *U.Jur.*, Eu.——FIG. 334,*7a,b*. *R. jaccardi* KOBY, Raur., Fr.; ×0.5 (71).——FIG. 334, *7c*. *R. percrassa* ÉTALLON, Raur., Switz.; ×0.4 (71).

FIG. 335. Caryophylliina (Caryophylliicae): Rhipidogyridae (p. F430).

Phytogyra ORB., 1849 [**P. magnifica*]. Like *Rhipidogyra* but with low horizontal branches with terminal forking. *U.Jur.*, Eu.——FIG. 334,*2*. **P. magnifica,* Raur., Switz.; *2a,b,* ×0.5 (71).

Psammogyra FROMENTEL, 1862 [**Pachygyra cottaldina* ORB., 1850]. Massive, consisting of a single much-contorted meandrine series united laterally by coenosteum. *U.Jur.*, Eu.——FIG. 334,*5*. *P. caudata* ÉTALLON, Sequan., Switz.; *5a,b,* ×0.5 (71).

Placogyra KOBY, 1904 [**P. felixi*]. Several contorted series joined directly by their walls. *U.Jur.*, Eu.——FIG. 334,*1*. **P. felixi,* Lusit., Port.; *1a,* ×0.5; *1b,* ×1 (73).

Superfamily FLABELLICAE Bourne, 1905

[*nom. correct.* WELLS, herein (*ex* Flabelloidae VAUGHAN & WELLS, 1943, *nom. transl.* Flabellidae BOURNE, 1905)]

Caryophylliina with epithecal wall. *M. Jur.-Rec.*

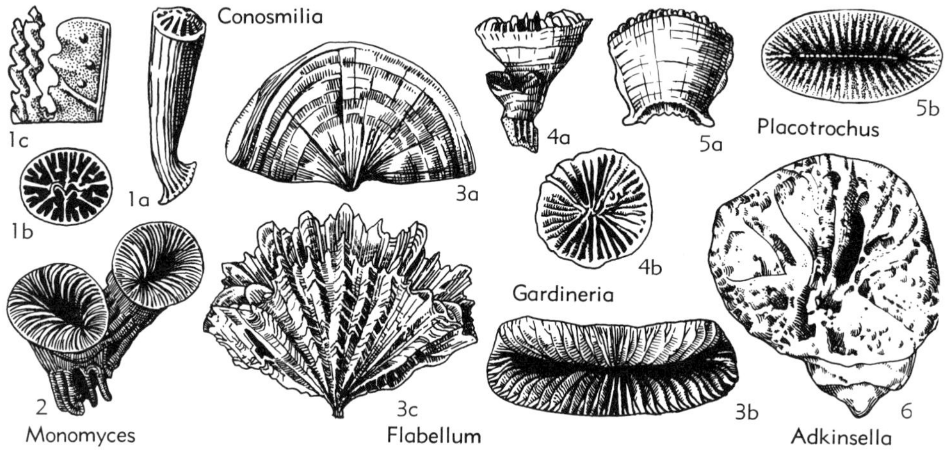

FIG. 336. Caryophylliina (Flabellicae): Flabellidae (p. F432).

Family FLABELLIDAE Bourne, 1905

Solitary, fixed or free; ahermatypic; neotenic. Wall epithecal, thickened internally by stereome. Septa nonexsert, formed by one fan system of simple trabeculae, with smooth margins, and laterally smooth or granulated. Pali and dissepiments absent. Columella present or absent. *Cret.-Rec.*

Flabellum LESSON, 1831 [**F. pavoninum*] [=*Phyllodes* PHILIPPI, 1841; *Blastotrochus* M.EDW.-H., 1848; ?*Ulocyathus* SARS, 1856; *Vasillum* T.-WOODS, 1879; *Lithomyces* PHILLIPI, 1887; *Placotrochides* ALCOCK, 1902]. Cuneiform to compressed turbinate, free. Septa numerous. Columella absent or very feeble. *Eoc.-Rec.*, cosmop., depth range 3-3,183 m.——FIG. 336,*3a,b*. **F. pavoninum*, Rec., Hawaii, 375 m.; ×0.5 (146).——FIG. 336,*3c*. *F. alabastrum* MOSELEY, Rec., off Azores, 1,750 m.; ×0.5 (97).

Placotrochus M.EDW.-H., 1848 [**P. laevis*] [=*Microtrochus* TENISON-WOODS, 1880]. Like *Flabellum* but with thin, irregular, lamellar columella. *Eoc.-Rec.*, IndoPac., depth range 265 m. ——FIG. 336,*5*. **P. laevis*, Rec., Philip.; *5a*, ×1; *5b*, ×1.4 (93).

Monomyces EHR., 1834 [**M. anthophyllum;* SD M. EDW.-H., 1850] [=*Rhizotrochus* M.EDW.-H., 1848; *Coelocyathus* SARS, 1856; *Heterotrochus* M. EDW.-H., 1857; *Biflabellum* DÖDERLEIN, 1913]. Turbinate, cornute or compressed, fixed, with basal or lateral rootlets. Columella absent. *U.Cret.-Rec.*, Eu.-Medit.-IndoPac., depth range 27-1,047 m.——FIG. 336,*2*. *M. typus* (M.EDW.-H.), Rec. Singapore; ×0.5 (93).

Gardineria VAUGHAN, 1907 [*pro Haplophyllia* POURTALÈS, 1868 (*non Aplophyllia* ORB., 1849; *nec Haplophyllia* FROMENTEL, 1865)] [**G. hawaiiensis*] [=*Duncania* POURTALÈS, 1874 (*non* KON., 1872)]. Turbinate, fixed, commonly with basal rootlets. Wall internally thickened by stereome. Relatively few septa. Columella trabecular, well developed. *Rec.*, W.Indies-S.Atl.-Pac., depth range 91-592 m.——FIG. 336,*4*. **G. hawaiiensis*, Hawaii, 525-575 m.; *4a,b*, ×0.5 (146).

Conosmilia DUNCAN, 1865 [**C. elegans*]. Trochoid-ceratoid, fixed. Columella formed by 1 or 2 crispate laths. *U.Eoc.-Mio.*, Austral.——FIG. 336,*1*. *C. anomala* DUNCAN, L.Mio.(Balcomb.); *1a*, ×1; *1b*, ×2; *1c*, septum and columella, ×6 (22).

Adkinsella WELLS, 1933 [**A. edwardsensis*]. Like *Gardineria* but lacking columella. *L.Cret.*, Tex.——FIG. 110,*6*. **A. edwardsensis*, Alb., Tex.; ×1 (158).

Family GUYNIIDAE Hickson, 1910

Solitary, fixed or free, ceratoid to scolecoid; ahermatypic. Wall epithecal, with 1 or 2 series of pores between each septal pair, later filled by thin deposit. Septa nonexsert, relatively few, margins smooth, laterally smooth or finely granulated. Pali present or absent. Columella of trabecular laths, spongy, or absent. Endothecal dissepiments sparse or absent. *M.Jur.-Rec.*

Onchotrochus DUNCAN, 1870 [**O. serpentinus*]. Scolecoid, lacking columella and pali. *M.Cret.-U. Cret.*, Eu.——FIG. 337,*4*. **O. serpentinus*, U.Cret. (Senon.), Eng.; *4a*, ×1; *4b*, ×5 (30).

Guynia DUNCAN, 1873 [**G. annulata*] [=*Gwynia* POURTALÈS, 1874; *Bistylia* T.-WOODS, 1878; *Pyrophyllia* HICKSON, 1910]. Ceratoid, free or laterally attached. Wall imbricated. Columella a single crispate trabecular lath. Pali absent. *U.Eoc.-Rec.*, W.Indies-Medit.-PersianGulf-Austral., depth range 170-658 m.——FIG. 337,*3*. *G. inflata* (HICKSON), Gulf of Oman, 300 m.; *3a*, ×5; *3b*, ×6 (179).

Microsmilia KOBY, 1888 [**Anthophyllum erguelense* THURMANN, 1851; SD VAUGHAN, 1905]. Ceratoid, fixed. Longitudinal parricidal budding. Columella formed by several twisted trabecular laths. *U.Jur.*, Switz.——FIG. 337,*5*. **M. erguelensis*, Oxf.; *5a*, ×1; *5b-d*, 3 regenerated corallites, ×1; *5e*, long. sec., ×3; *5f*, wall, ×5 (71).

Stenocyathus POURTALÈS, 1871 [**Coenocyathus vermiformis* POURTALÈS, 1868]. Ceratoid, with imbricated wall. Columella formed by 1 or 2 crispate trabecular laths. Pali crispate, in one crown of 6. *U.Cret.-Rec.*, W.Indies-Medit.-Atl.-N.Am.-Austral., depth range 155-1,097 m.——FIG. 337,*2*. **S. vermiformis*, Rec., off Azores, 1,150 m.; *2a,b*, ×3 (64).

Cyathosmilia T.-WOODS, 1878 [**C. laticostata*]. Ceratoid, fixed. Pali in 2 crowns opposite first 2 cycles. Columella weak. *U.Eoc.-L.Oligo.*, Austral.

Schizocyathus POURTALÈS, 1874 [**S. fissilis*]. Ceratoid, fixed. Longitudinal parricidal budding. One crown of pali opposite second cycle. Columella absent. *Eoc.-Rec.*, Atl.-W.Indies-Austral., depth range 102-1,445 m.——FIG. 337,*1*. **S. fissilis*, Rec., off Canary I., 625-900 m.; corallum split lengthwise, regenerated, ×3 (64).

Suborder DENDROPHYLLIINA Vaughan & Wells, 1943

[*nom. correct.* WELLS, herein (*ex* Dendrophylliida VAUGHAN & WELLS, 1943)]

Solitary and colonial. Wall synapticulothecal, irregularly porous. Septa primarily laminar as in Caryophylliina but generally secondarily thickened, more or less porous, with smooth or slightly dentate margins. *U. Cret.-Rec.*

Family DENDROPHYLLIIDAE Gray, 1847

Solitary and colonial, mostly ahermatypic. Colony formation by intra- and extratentacular budding. Wall formed by trabecular outer ends of septa and simple but very irregular synapticulae, irregularly porous, usually thick, irregularly costate or covered by reduced costal granulations. Porous, layered coenosteum in some colonial forms. Septa composed of one fan system of simple trabeculae, but trabeculae tend to be very irregular, commonly not united closely in plane of septum and vertically discontinuous with sclerodermites bending outward from septal plane, especially at periphery

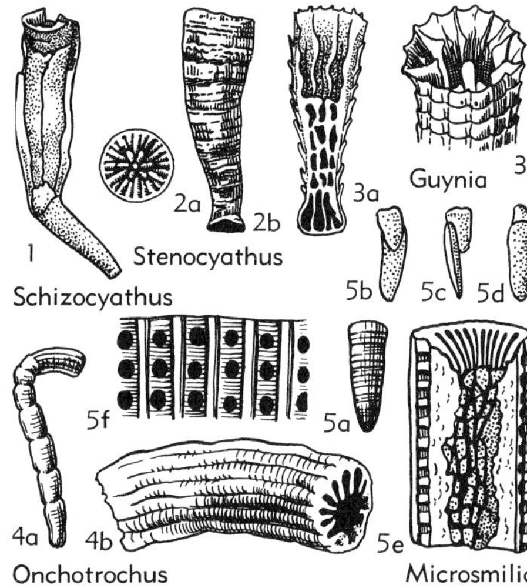

FIG. 337. Caryophylliina (Flabellicae): Guyniidae (p. F432-F433).

and near columella. Septa strongly granulated laterally, mostly smooth marginally except peripherally and centrally where irregular dentations occur, or wholly weakly dentate. Septa inserted following POURTALÈS plan (Fig. 239), at least in early stages. Columella trabecular and spongy, or absent. Endothecal dissepiments thin and poorly developed. *U.Cret.-Rec.*

Palaeopsammia WANNER, 1902 [**P. multiformis*]. Solitary, ceratoid to patellate, free in ephebic stage, externally epithecate. Costae corresponding to septa. POURTALÈS plan not apparent. Columella spongy. *U.Cret.*, N.Afr.——FIG. 338,*3*. **P. multiformis*, Dan., Libyan Desert; *3a*, ×1; *3b*, ×2 (197).

Areopsammia DIETRICH, 1917 [**A. maestrichtensis* (=*Cyclolites alacca* MORREN, 1828)] [=*Porosmilia* UMBGROVE, 1925 (*non* FROMENTEL, 1860)]. Solitary, turbinate, nonepithecate. Costae distinct only near calice, reduced basally to granulations. POURTALÈS plan not apparent. Columella weak. *U.Cret.*, Eu.

Wadeopsammia WELLS, 1933 [**Trochosmilia nodosa* WADE, 1926]. Solitary, small, turbinate, free, base with tube of commensal sipunculid. Wall thin; costae reduced to granulations. Septa arranged following POURTALÈS plan. Columella large and solid. *U.Cret.*, N.Am.——FIG. 338,*6*. **W. nodosa*, Navarro, Tex.; ×2 (158).

Balanophyllia WOOD., 1884 [**B. calyculus*] [=*Osteodes* CONRAD, 1855; *Leptopsammia* M.EDW.-H., 1848; *?Clonotrochus* SCHAFHÄUTL, 1863; *Rhodopsammia* SEMPER, 1872]. Solitary, trochoid, fixed by broad or pedunculate base. Costae well developed,

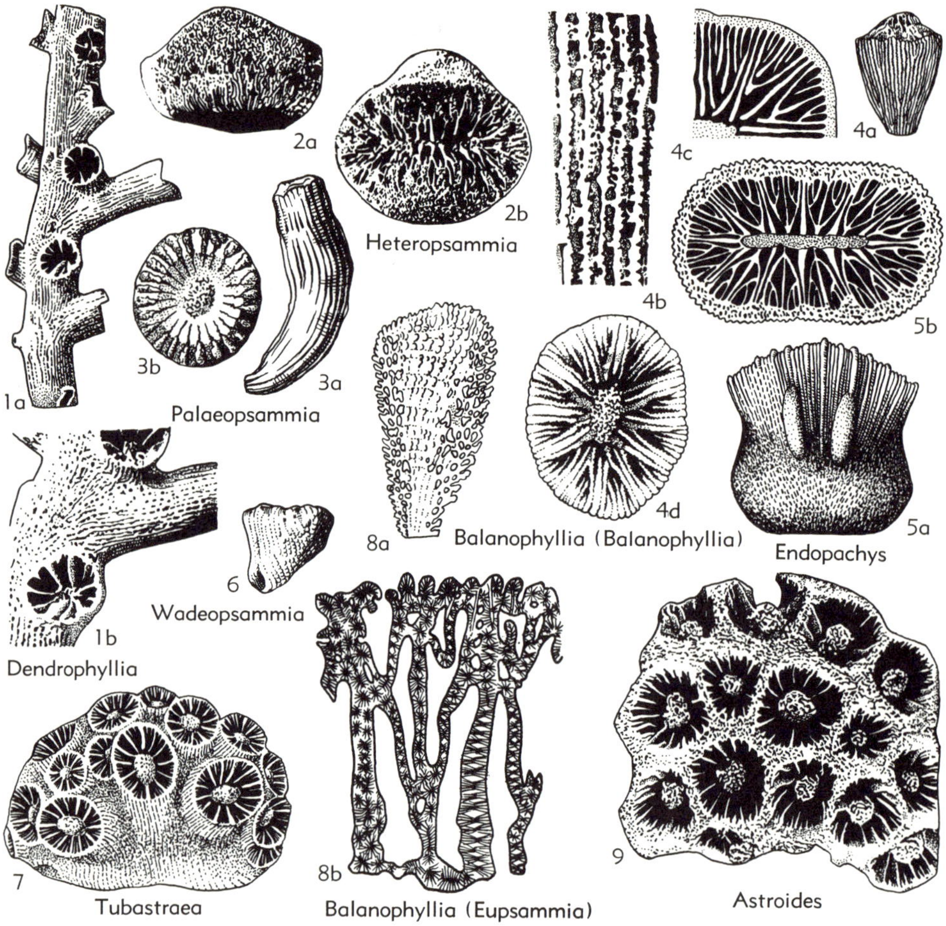

FIG. 338. Dendrophylliina: Dendrophylliidae (p. F433-F436).

corresponding to septa. Septa following POURTALÈS plan. Columella well developed, spongy. *Eoc.-Rec.*, cosmop.

B. (Balanophyllia). *Eoc.-Rec.*, depth range 0-1,100 m.——FIG. 338,*4a,b*. *B. (B.) irrorata* (CONRAD), M.Eoc.(Claib.), Tex.; *4a*, ×1; *4b*, costae, ×5 (144).——FIG. 338,*4c*. *B. (B.) inauris* VAUGHAN, L.Eoc., N.J., transv. sec., ×3 (144).——FIG. 338, *4d*. *B. (B.) italica* MICHELOTTI, Rec., Italy; calice, ×2 (25).

B. (Eupsammia) M.EDW.-H., 1848 [**Madrepora trochiformis* PALLAS, 1766; SD M.EDW.-H., 1850]. Like *B. (Balanophyllia)*, but free in ephebic stage. *U.Cret.-Rec.*, Eu.-IndoPac.-N.Am., depth range *ca.* 55 m.——FIG. 338,*8*. **B. (E.) trochiformis*, M.Eoc.(Lut.), Paris Basin; *8a*, lat. aspect of septum, ×0.7; *8b*, transv. sec., ×5 (98).

Endopachys LONSDALE, 1845 [**Turbinolia maclurii* LEA, 1833; SD M.EDW.-H., 1850] [=*Rhectopsammia* VAUGHAN, 1900]. Like *B. (Balanophyllia)* but cuneiform, free, with alate costal extensions extending outward along longer axis near base. Costae generally distinct around calice. *Eoc.-Rec.*, N.Am.-IndoPac., depth range 37-604 m.——FIG. 338,*5*. **E. maclurii*, M.Eoc.(Claib.), USA; *5a*, ×1.5; *5b*, ×2 (93).

Notophyllia DENNANT, 1899 [**N. semivestita*]. Like *Endopachys* but more elongate, costal extensions reduced, septa appearing normal. Columella elongate, semi-solid, even laminar. Costae represented by granulations. *Mio.-Rec.*, Austral., depth range 37-457 m.——FIG. 339,*8*. *N. etheridgei* HOFFMEISTER, Rec., S.Austral., 37-40 m.; *8a,b*, ×2 (180).

Rhizopsammia VERRILL, 1869 [**R. pulchra*]. Like *B. (Balanophyllia)* but forming small colonies by extratentacular stoloniferous budding as in *Culicia*, the corallites generally losing organic connection. *Rec.*, IndoPac., depth range 0-113 m.

Cladopsammia Lacaze-Duthiers, 1897 [**C. rolandi*]. Like *B. (Balanophyllia)* but in small phaceloid colonies formed by extratentacular budding. *Rec.*, Medit., depth range 18 m.

Dendrophyllia Blainv., 1830 [**Madrepora ramea* Linné, 1758; SD M.Edw.-H., 1850] [=*Brasseyia* Wright, 1882]. Like *B. (Balanophyllia)* but forming dendroid colonies by extratentacular budding from edge zone. *Eoc.-Rec.*, cosmop., depth range 0-1,372 m.——Fig. 338,*1*. *D. micrantha* (Ehr.), Rec., Philip.; *1a*, ×1; *1b*, ×2 (40).

Lobopsammia M.Edw.-H., 1848 [**Lithodendron cariosum* Goldf., 1827; SD M.Edw.-H., 1850] [=*Placopsammia* Reuss, 1859]. Like *Dendrophyllia* but colonies small, formed by mono- or distomodaeal intratentacular budding, corallites permanently monocentric. *Eoc.-Oligo.*, Eu.——Fig. 339,*7*. **L. cariosa*, M.Eoc.(Lut.), Fr.; ×1 (92).

Reussopsammia Wells, 1937 [**Stereopsammia granulosa* Reuss, 1864]. Like *Lobopsammia* but columella very feeble or absent; Pourtalès plan not apparent. *Oligo.*, Ger.——Fig. 339,*3*. **R. granulosa*; *3a*, ×1; *3b*, ×2; *3c*, ×4 (114).

Stichopsammia Felix, 1885 [**S. gyrosa*; SD Felix, 1925]. Like *Lobopsammia* but polycentric condition permanent, forming flabellate series. *Oligo.*, Eu.

Endopsammia M.Edw.-H., 1848 [**E. philippinensis*]. Solitary, subcylindrical, fixed by broad base. Wall thin, costate. Septa thin, following Pourtalès plan only in early stages. Columella weak. *Rec.*, Medit.-S.Atl.-IndoPac., depth range 5-91 m.——Fig. 339,*2*. *E. microcardia* (Döderlein), Naples; ×2 (25).

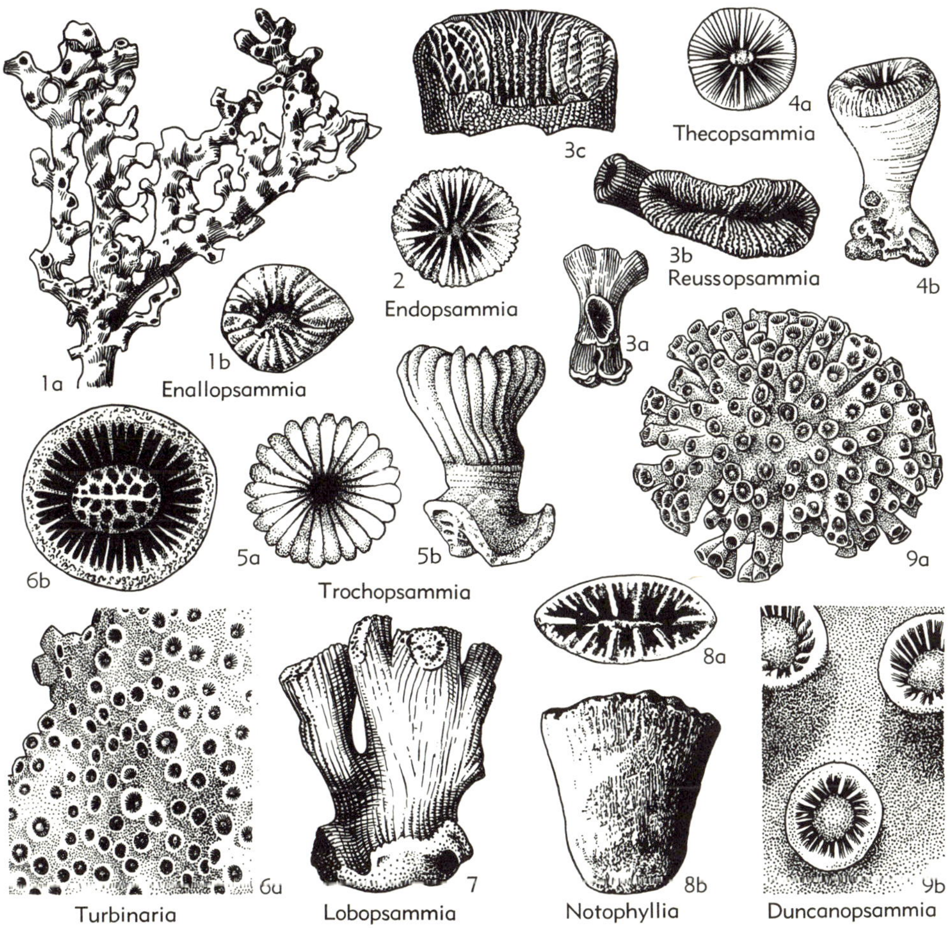

Fig. 339. Dendrophylliina: Dendrophylliidae (p. F434-F436).

Tubastraea LESSON, 1834 [**T. coccinea* (=*Lobophyllia aurea* QUOY & GAIMARD, 1833)] [=*Coenopsammia* M.EDW.-H., 1848; *Astropsammia, Pachypsammia* VERRILL, 1869]. Like *Endopsammia* but in small plocoid clumps formed by extratentacular budding from edge zone; corallites generally united by feeble costae and some coenosteum. Columella mostly feeble. *Rec.*, cosmop., depth range 0-1,463 m.——FIG. 338,*7*. *T. tenuilamellosa* (M.EDW.-H.), Panama; ×1 (93).

Astroides QUOY & GAIMARD, 1827 [**A. luteus* (=*Madrepora calycularis* PALLAS, 1766)] [=*Astreoides* BLAINV., 1830; *Astroitis* DANA, 1846]. Like *Endopsammia* but extratentacular budding from proliferations of edge zone produces small, submassive colonies. Corallites united by noncostate coenosteum; costae reduced to granulations. Columella large, spongy. *Rec.*, Medit., depth range 0-73 m.——FIG. 338,*9*. *A. calycularis*, Naples; ×2 (25).

Stereopsammia M.EDW.-H., 1850 [**S. humilis*] [=*?Bosnopsammia* OPPENHEIM, 1908]. Like *Tubastraea* but much smaller, and columella very weak or absent. *Eoc.*, Eu.

Trochopsammia POURTALÈS, 1878 [**T. infundibulum*]. Solitary, turbinate, fixed, with some epitheca. Costae thick, spongy, equal over thin wall. Septa showing no trace of POURTALÈS plan. Columella absent. *Rec.*, W.Indies, depth range 403-1,490 m.——FIG. 339,*5*. **T. infundibulum*, Fla., 1,100 m.; *5a,b*, ×2 (151).

Enallopsammia MICHELOTTI, 1871 [**Coenopsammia scillae* SEGUENZA, 1864] [=*Anisopsammia* VON MARENZELLER, 1904]. Dendroid colonies formed by extratentacular budding, the calices generally on one side of branches which tend to anastomose in one plane. Septa following POURTALÈS plan only in early stages. Columella weak. *Mio.-Rec.*, Eu.-W. Indies-IndoPac., depth range 293-1,966 m.——FIG. 339,*1*. *E. rostrata* (POURTALÈS), Rec., Morocco, 1,966 m.; *1a*, ×0.4; *1b*, ×2 (64).

Thecopsammia POURTALÈS, 1868 [**T. socialis*]. Solitary, turbinate, fixed, costae reduced to spinose granulations. Epitheca stout, extending halfway to calice. Septa following POURTALÈS plan only in early stages. Columella weak. *Rec.*, W.Indies-N. Atl.-Pac., depth range 0-1,097 m.——FIG. 339,*4*. **T. socialis*, Fla., 375-500 m.; *4a,b*, ×1 (187).

Bathypsammia VON MARENZELLER, 1906 [**Thecopsammia tintinnabulum* POURTALÈS, 1868]. Like *Thecopsammia* but POURTALÈS plan apparent; columella well developed. *Rec.*, W.Atl., depth range 220-329 m.

Psammoseris M.EDW.-H., 1851 [**Heterocyathus hemisphericus* GRAY, 1850] [=*Spongiocyathus* FOLKESON, 1919]. Solitary, subdiscoidal, free, base with tube of commensal sipunculid. Costae thick and spongy, distinct to center of base. Septa thick and spongy. Columella well developed. *Rec.*, E. Indies, depth range 11-44 m.

Heteropsammia M.EDW.-H., 1848 [**H. michelini*]. Solitary, or colonial with 2 or 3 calices resulting from mono- or distomodaeal intratentacular budding, free, base enclosing tube of commensal sipunculid with several lateral opening through corallum. Wall thick and spongy; costae reduced to multiple rows of crispate granulations, becoming labyrinthine over base. Septa following POURTALÈS plan, lower cycles thick and spongy. Columella well developed. *Pleisto.-Rec.*, IndoPac., depth range 11-192 m.——FIG. 338,*2*. **H. michelini*, Pleisto., Formosa; *2a,b*, ×2 (199).

Duncanopsammia WELLS, 1936 [**Dendrophyllia axifuga* M.EDW.-H., 1848]. Broad, pedunculate, subdendroid, hermatypic colonies formed by extratentacular budding from lower margins of corallites. Corallites basally united by coenosteum; costae reduced to rows of crispate granulations extending over coenosteum. Septa following POURTALÈS plan. Columella spongy, deep in calices. *Mio.-Rec.*, Austral.——FIG. 339,*9*. **D. axifuga*, Rec., 5.5 m.; *9a*, ×0.25; *9b*, ×2 (198).

Turbinaria OKEN, 1815 [**Madrepora crater* PALLAS, 1766] [=*Explanaria* LAMARCK, 1816; *Gemmipora* BLAINV., 1830; *Turbinacis* QUENST., 1880]. Hermatypic; large explanate or crateriform, contorted foliaceous colonies; corallites united nearly to summits by extensive coenosteum. POURTALÈS plan apparent only in early stages. Columella well developed. *Oligo.-Rec.*, Eu.-IndoPac.——FIG. 339,*6*. *T. peltata* (ESPER), Rec., E.Indies; *6a*, ×0.5; *6b*, ×6 (40). (Also Figs 255, 256.)

GENERA OF UNCERTAIN SYSTEMATIC POSITION

Andemantastraea ALLOITEAU, 1951 [**Synastraea consobrina* ORB., 1850]. *M.Jur.*, Fr.

Carcicocaenia ALLOITEAU, 1954 [**C. pfenderae*]. *M.Cret.*, Fr.

Cenomanina ALLOITEAU, 1952 [**Rhabdocora exiguis* FROMENTEL, 1873]. *M.Cret.*, Fr.

Ceratosmilia ALLOITEAU, 1952 [**C. arnaudi*]. Horizon and locality not indicated.

Corbariastraea ALLOITEAU, 1952 [**C. rennensis*]. *U.Cret.*, Fr.

Cyclophyllopsis ALLOITEAU, 1952 [**Cyclolites aptiensis* FROMENTEL, 1870]. *L.Cret.*, Fr.

Dactylosmilia ORB., 1849 [**D. carantonensis*]. *M. Cret.*, Fr.

Epitrochus FROMENTEL, 1861 [**E. primus*]. *M.Cret.*, Fr.

Eusmiliopsis ALLOITEAU, 1952 [**Amblophyllia cretacea* ORB., 1849]. *M.Cret.*, Fr.

Ficariastraea ALLOITEAU, 1952 [**Phyllastraea hippuritiformis* FROMENTEL, 1857]. *U.Cret.*, Fr.

Glenaraea POČTA, 1887 [**G. cretacea*]. *M.Cret.*, Czech.

Lamellofungia ALLOITEAU, 1952 [**L. rennensis*]. *U.Cret.*, Fr.

Morchellastraea ALLOITEAU, 1941 [**Astrea coniformis* MICHELIN, 1843]. *U.Cret.*, Fr.

Pleuropodia DENNANT, 1903 [**P. otwayensis*]. *Mio.*, S.Austral.

Pseudoseris ALLOITEAU, 1952 [**P. senessi*]. *U.Cret.*, Fr.

Siniastraea AVNIMELECH, 1947 [**Aspidiscus labyrinthica* HOPPE, 1922]. *U.Jur.*, Sinai.

Subtilicyathus ZUFFARDI-COMERCI, 1924 [**S. zanoni*]. *Mio.*, Libya.

Trigerastraea ALLOITEAU, 1952 [**Isastraea trigeri* FROMENTEL, 1857]. *M.Cret.*, Fr.

Tubicora LATHAM, 1929 [**T. somaliensis* (=*Calamophyllia aylmeri* GREGORY, 1900)]. *Eoc.*, Somali.

Turbinacis GREGORY, 1900 [**T. erythraensis*] [*non Turbinacis* QUENST., 1880]. Oligo., Somali.

Valloria VIDAL, 1874 [**V. egozcuei*]. *U.Cret.*, Sp.

UNRECOGNIZABLE GENERA

Allocoenia ÉTALLON, 1859 [**A. furcata*]. *M.Jur.*, Fr.

Anabacia M.EDW.-H., 1849 [**A. parkinsoni* (=*Madrepora porpita* PARKINSON, 1808)]. ?Age, ?loc.

Archicoeniopsammia KOLOSVARY, 1949 [**A. hungarica*]. *Eoc.*, Hung.

Brevismilia BÖLSCHE, 1866 [**Anthophyllum conicum* F. A. ROEMER, 1836]. *L.Cret.*, Ger. (Brunswick).

Coenastraea ÉTALLON, 1859 [**C. martis*]. *M.Jur.*, Fr.

Corofolius GREGORIO, 1930 [**C. mirus*]. *Trias.*, Sicily.

Digitium GREGORIO, 1930 [**D. taxum*]. *Trias.*, Sicily.

Discoseris GÜMBEL, 1861 [**D. rhaetica*]. *U.Trias.*, Bavaria.

Isocora ÉTALLON, 1864 [**I. thurmanni*]. *M.Jur.*, Switz.

Kraterostrobilos CRICKMAY, 1930 [**K. bathys*]. *L.Jur.*, B.C.

Lepiconus STOPPANI, 1857 [**L. bassii*]. *U.Trias.*, Italy.

Mirmidia GREGORIO, 1884 [**Meandrina? (Mirmidia) serafina*]. *U.Jur.*, Sicily.

Stylastraea FROMENTEL, 1860 [**S. sinemuriensis*]. [*non Stylastraea* LONSDALE, 1845]. *L.Jur.*, Fr.

Pyxidophyllum STOPPANI, 1861 [**P. edwardsii*]. *U.Trias.*, Italy.

Petrophyllia CONRAD, 1855 [**P. arkansasensis*]. *Tert.*, Ark.

Trochosmilietta KOLOSVARY, 1949 [**Trochosmilia comonensis* D'ACHIARDI, 1875]. *M.Eoc.*, Italy.

INVALID GENERIC NAMES

The following names were proposed by ALLOITEAU (1952), without description, illustration, or indication of valid type species: *Clavismilia, Columnocoenia, Cylindrocyathus, Delphinastraea, Dermosmiliopsis, Diplothecophyllia, Ellipsoidastraea, Hemiporites, Metasmilia, Palaeosmilia* (*non* M.EDW.-H., 1848), *Parasmiliopsis, Phyllotrochus, Placocaeniopsis, Polyastropsis, Polystephanastraea, Saltastraea, Thecoseriopsis,* and *Uxacalcaraea.*

REFERENCES

Publications cited in the following list are considered to be most helpful in furnishing additional information concerning the Scleractinia and in offering a guide to more extensive literature. The index numbers enclosed in parentheses in the column at the left are employed in the text for identification of the publications.

Abe, N.

(1) 1937, *Ecological survey of Iwayama Bay, Palao:* Palao Trop. Biol. Sta. Studies, no. 2, p. 219-324, fig. 1-42, tab. 1-25.

Achiardi, A. d'

(2) 1866-68, *Coralli fossili de terreno nummulitico dell 'Alpi Veneti:* Soc. Ital. Sci. Nat., Mem., v. 2 (1866), 53 p., 5 pl.; v. 4 (1868), 31 p., 8 pl.

(3) 1875-76, *Coralli eocenici dei Friuli:* Soc. Toscana Sci. Nat. (Pisa), Atti, v. 1, p. 70-86, 115-124, pl. 1-2, 6-7 (1875); p. 147-221, pl. 8-19 (1876).

Alcock, A. W.

(4) 1902, *Report on the deep-sea Madreporaria of the Siboga Expedition:* Siboga-Expeditie, Mon. 16a, 55 p., 5 pl.

Alloiteau, James

(5) 1952, *Madréporaires post-Paléozoïques:* in Piveteau, J., Traité de Paléontologie, Masson (Paris), tome 1, p. 539-684, pl. 1-10, fig. 1-130.

Atoda, K.

(6) 1951, *The larval and postlarval development of the reef-building corals, III, Acropora brüggemanni (Brook):* Jour. Morphology (Philadelphia), v. 89, p. 1-16, pl. 1, fig. 1-9.

Becker, E., & Milaschewitsch, C.

(7) 1875-76, *Die Korallen der Nattheimer Schichten:* Palaeontographica (Stuttgart), Band 21, p. 121-204, pl. 36-45 (1875); p. 205-244, pl. 45-51 (1876) (by Milaschewitsch).

Bernard, H. M.
(8) 1896, *The genus Turbinaria; the genus Astraeopora:* Brit. Mus. (Nat. Hist.) Cat. Madreporarian Corals (London), v. 2, 106 p., 33 pl.
(9) 1897, *The genus Montipora; the genus Anacropora:* Same, v. 3, 192 p., 34 pl.
(10) 1903, *The genus Goniopora:* Same, v. 4, 206 p., 14 pl.
(11) 1905, *Porites of the Indo-Pacific region:* Same, v. 5, 303 p., 35 pl.
(12) 1906, *Porites of the Atlantic and West Indies, with the European fossil forms:* Same, v. 6, 173 p., 17 pl.

Boschma, Hilbrand
(13) 1925, *Madreporaria I, Fungiidae:* Dansk Naturh. Foren. Meddel. (København), v. 79, p. 185-259, pl. 5-11.

Bourne, G. C.
(14) 1887, *On the anatomy of Mussa and Euphyllia and the morphology of the madreporarian skeleton:* Quart. Jour. Micros. Sci. (London), new ser., v. 28, p. 21-52, pl. 3-4.
(15) 1899, *Studies on the structure and formation of the calcareous skeleton of the Anthozoa:* Same, v. 41, p. 499-547, pl. 40-43.

Brook, G.
(16) 1893, *The genus Madrepora:* Brit. Mus. (Nat. Hist.) Cat. Madreporarian Corals (London), v. 1, 212 p., 35 pl.

Catullo, T. A.
(17) 1856, *Dei terreno di sedimento superiore delle Venezie e dei fossili bryozoari, antozoari, e spongiari:* (Padua), 88 p., 19 pl.

Crossland, C.
(18) 1931, *The reduced building power and other variations in the astraean corals of Tahiti:* Zool. Soc. London, Proc., p. 351-392, pl. 1-22.
(19) 1952, *Madreporaria, Hydrocorallineae, Heliopora:* Brit. Mus. (Nat. Hist.) Great Barrier Reef Expedition, 1928-29, Sci. Repts. (London), v. 6, no. 3, 257 p., 56 pl.

Dana, J. D.
(20) 1846, *Zoophytes:* U. S. Explor. Exped. (Washington), v. 7, 740 p., 61 pl. (1849).

Dennant, J.
(21) 1899-1904, *Descriptions of new species of corals from the Australian Tertiaries:* Roy. Soc. South Australia (Adelaide), Trans., v. 23 (1899), p. 112-122, 281-287, pl. 2-3, 9-10; v. 25 (1901), p. 48-53, pl. 2; v. 26 (1902), p. 1-6, 255-264, pl. 1, 5-6; v. 27 (1903), p. 208-215, pl. 1-2; v. 28 (1904), p. 52-76, pl. 22-25.

Diener, Carl
(22) 1921, *Cnidaria triadica:* Fossilium Catalogus ('s Gravenhage), I, pars 13, 46 p.

Dietrich, W. O.
(23) 1926, *Steinkorallen des Malms und Unterkreide im südlichen deutsch-Ostafrika:* Palaeontographica (Stuttgart), Supp., Band 7, p. 40-102, pl. 5-14.

Döderlein, L.
(24) 1902, *Die Korallen-Gattung Fungia:* Senckenberg. naturf. Gesell. (Frankfurt-a-M.), Abh., Band 27, 162 p., 25 pl.
(25) 1913, *Die Steinkorallen am die Golf von Neapel:* Zool. Sta. Neapel Mitt., Band 21, no. 5, p. 105-152, pl. 7-9.

Duerden, J. E.
(26) 1900, *The order of appearance of the mesenteries and septa in the Madreporaria:* Johns Hopkins Univ. Circ., v. 19, p. 47-53, fig. 1-12.
(27) 1902, *West Indian madreporarian polyps:* Natl. Acad. Sci. (Washington), Mem., v. 8, no. 7, p. 403-648, pl. 1-25, fig. 1-18.
(28) 1904, *The coral Siderastraea radians and its postlarval development:* Carnegie Inst. Washington Pub. no. 20, 130 p., 11 pl., 13 fig.

Duncan, P. M.
(29) 1863-68, *On the fossil corals of the West Indian islands:* Geol. Soc. London Quart. Jour., v. 19 (1863), p. 406-458, pl. 13-16; v. 20 (1864), p. 20-44, 358-374; pl. 2-5, v. 24 (1868), p. 9-33, pl. 1, 2.
(30) 1866-72, *A monograph of the British fossil corals:* Palaeontograph. Soc. (London), pt. 1, 66 p., 10 pl. (1866); pt. 2, p. 1-26, pl. 1-9 (1869), p. 27-46, pl. 10-15 (1870); pt. 3, 24 p., 7 pl. (1873); pt. 4, 43 p., 11 pl. (1867).
(31) 1873, *A description of the Madreporaria dredged up during the expedition of H.M.S. Porcupine in 1869 and 1870:* Zool. Soc. London Trans., v. 8 (1873), p. 303-344, pl. 39-49; v. 10 (1878), p. 235-250, pl. 43-45.
(32) 1870, *On the fossil corals of the Australian Tertiary deposits:* Geol. Soc. London Quart. Jour., v. 26, p. 284-318, pl. 19-21, fig. 1-2.
(33) 1880, *Sind fossil corals and Alcyonaria:* Palaeontologia Indica, ser. 14, v. 1, 110 p., 28 pl.
(34) 1884, *A revision of the families and genera of the sclerodermic Zoantharia, Ed. & H., or Madreporaria (M. rugosa excepted):* Linnean Soc. London Jour., Zool., v. 18, p. 1-204.

Durham, J. W.
(35) 1942, *Eocene and Oligocene coral faunas of Washington:* Jour. Paleont., v. 16, p. 84-104, pl. 15-17, fig. 1.
(36) 1947, *Corals from the Gulf of California and the North Pacific coast of America:* Geol. Soc. America, Mem., v. 20, 68 p., 14 pl., 2 fig.

——— & Barnard, J. L.

(37) 1952, *Stony corals of the Eastern Pacific collected by the Valero III and Valero IV:* Allan Hancock Pacific Expeditions, v. 16, 110 p., 16 pl.

Edmondson, C. H.

(38) 1928, *The ecology of an Hawaiian coral reef:* B. P. Bishop Mus. (Honolulu), Bull. 45, 64 p., 6 fig.

Eguchi, M.

(39) 1951, *Mesozoic hexacorals from Japan:* Tohoku Univ. Sci. Repts. (Sendai), ser. 2, v. 24, 95 p., 28 pl.

Faustino, L. A.

(40) 1927, *Recent Madreporaria of the Philippine Islands:* Philippine Dept. Agric. and Nat. Resources, Bur. Sci., Mon. 22, 310 p., 100 pl.

Felix, J.

(41) 1891, *Versteinerungen aus der mexicanischen Jura- und Kreideformation:* Palaeontographica (Stuttgart), Band 37, p. 140-194, pl. 22-30.

(42) 1903, *Die Anthozoen der Gosauschichten in den Ostalpen:* Palaeontographica (Stuttgart), Band 49, p. 163-359, pl. 17-25, fig. 1-67.

(43) 1914, *Anthozoa palaeocretacea:* Fossilium Catalogus, I, pars 5, p. 1-84.

(44) 1914a, *Anthozoa cenomanica:* Same, pars 6, p. 85-143.

(45) 1914b, *Anthozoa neocretacea:* Same, pars 7, p. 144-273.

(46) 1925, *Anthozoa eocaenica et oligocaenica:* Same, pars 28, p. 1-296.

(47) 1927, *Anthozoa miocaenica:* Same, pars 35, p. 297-488.

(48) 1929, *Anthozoa pliocaenica et pleistocaenica:* Same, pars 44, p. 489-668.

Fowler, G. H.

(49) 1885-90, *The anatomy of the Madreporaria:* Quart. Jour. Micros. Sci. (London), v. 25 (1885), p. 577-597, pl. 40-42; v. 27 (1886), p. 1-16, pl. 1; v. 28 (1887), p. 1-19, pl. 1-2; v. 29 (1888), p. 415-430, pl. 32-33; v. 30 (1890), p. 405-422, pl. 28.

Frech, F.

(50) 1890, *Die Korallenfauna der Trias. I. Die Korallen der juvavischen Triasprovinz:* Palaeontographica (Stuttgart), Band 37, p. 1-116, pl. 1-21.

Fromentel, L.-E. G. de

(51) 1861, *Introduction à l'étude des polypiers fossiles:* (Paris), 357 p.

(52) 1862-87, *Zoophytes:* Paléontologie française, Terrain crétacé (Paris), tome 8, 624 p., 192 pl.

———, & Ferry, H. B. A. T. de

(53) 1865-69, *Zoophytes:* Paléontologie française, Terrain jurassique (Paris), 240 p., 60 pl.

Gardiner, J. S.

(54) 1903-06, *The fauna and geography of the Maldive and Laccadive Archipelagoes:* (Cambridge) v. 2, p. 755-790, pl. 59-64 (1903); p. 933-957, pl. 89-93 (1906).

(55) 1904, *The turbinolid corals of South Africa, with notes on their anatomy and variation:* Cape of Good Hope Dept. Agric. Marine Inv. South Africa (Capetown), v. 3, p. 95-129, pl. 1-3.

(56) 1909, *Family Fungidae, with a revision of its genera and species and an account of their geographical distribution:* Linnean Soc. London Trans., ser. 2, Zool., v. 12, p. 257-290, pl. 33-39.

Gerth, H.

(57) 1921, *Anthozoa:* in MARTIN, K., Die Fossilen von Java, Geol. Reichsmus. Leiden Sammlung, new ser., v. 1, p. 387-445, pl. 55-57.

(58) 1923, *Die Anthozoenfauna des Jungtertiärs von Borneo:* Geol. Reichsmus. Leiden Sammlung, ser. 1, v. 10, p. 37-136, pl. 1-9.

(59) 1928, *Beiträge zur Kenntnis der mesozoischen Korallenfaunen von Südamerika:* Leidsche Geol. Meded. v. 3, p. 1-15, pl. 1-2.

(60) 1931, *Palaeontologie von niederländisch Ost-Indien. Coelenterata:* Same, v. 5, p. 120-151.

(61) 1933, *Neue Beiträge zur Kenntnis der Korallenfauna des Tertiär von Java. I. Die Korallen des Eocän und des älteren Neogen:* Dienst Mijnbouw Nederlandsch-Indië Wetensch. Meded. (Batavia), no. 25, 45 p., 5 pl.

Geyer, O. F.

(62) 1954, *Die oberjurassische Korallenfauna von Württemberg:* Palaeontographica (Stuttgart), Band 104A, p. 121-220, pl. 9-16.

(62a) 1955, *Beiträge zur Korallenfauna des Stramberger Tithon:* Palaeont. Zeitschr., Band 29, p. 177-216, pl. 22-26, 2 fig.

Goldfuss, A.

(63) 1826-29, *Petrefacta Germaniae:* (Düsseldorf) v. 1, Divisio Prima, p. 1-114, pl. 1-38.

Gravier, C.

(64) 1920; *Madréporaires provenant des campagnes des Yachts Princesse-Alice et Hirondelle:* Campagnes Scientifiques Albert 1er Monaco Résultats (Monaco), fasc. 55, 119 p., 16 pl.

Gregory, J. W.

(65) 1900, *The corals. Jurassic fauna of Cutch:* Palaeontologia Indica (Calcutta), ser. 9, v. 2, pt. 2, p. 1-195 pl. 2a-27.

Hackemesser, M.

(66) 1936, *Eine kretazische Korallenfauna aus Mittelgriechenland:* Palaeontographica (Stuttgart), Band 84, Abt. A, p. 1-97, pl. 1-8.

Hickson, S. J.

(67) 1924, *An introduction to the study of recent*

corals: Manchester Univ. Pub., Biol. Ser. no. 4, 257 p., 110 fig.

Hoffmeister, J. E.

(68) 1925, *Some corals from American Samoa and the Fiji Islands:* Carnegie Inst. Washington Pub. 343, 89 p., 23 pl.

Karakasch, N. I.

(69) 1907, *Le Crétacé inférieur de la Crimée et sa faune:* Soc. Imp. Naturalistes St.-Petersbourg Travaux, tome 32, 484 p., 28 pl., 10 fig.

Klunzinger, C. B.

(70) 1877-79, *Die Korallthiere des rothen Meeres:* (Berlin), 286 p., 28 pl.

Koby, F.

(71) 1881-95, *Monographie des polypiers jurassiques de la Suisse:* Soc. Paléont. Suisse, Mém. (Genève): tome 7 (1881), p. 1-60, pl. 1-12; tome 8 (1882), p. 61-108, pl. 13-30; tome 10 (1884), p. 109-148, pl. 31-42; tome 11 (1885), p. 149-212, pl. 43-63; tome 12 (1886), p. 213-304, pl. 64-89; tome 13 (1887), p. 305-352, pl. 90-99; tome 14 (1888), p. 353-400, pl. 100-109; tome 15 (1889), p. 401-456, pl. 110-121; tome 16 (1890), p. 457-582, pl. 122-130; suppl., tome 21 (1895), p. 1-20, pl. 1-4.

(72) 1896-98, *Monographie des polypiers crétacés de la Suisse:* Soc. Paléont. Suisse, Mém. (Genève), tome 22 (1896), p. 1-128, pl. 1-8; tome 23 (1897), p. 29-62, pl. 9-16; tome 24 (1898), p. 63-100, pl. 17-22.

(73) 1905, *Polypiers du jurassique supérieur (Description de la faune jurassique du Portugal):* Comité Service Géol. Portugal (Lisbõa), 167 p., 30 pl.

Koch, G. von

(74) 1896, *Das Skelett der Steinkorallen, eine morphologische Studie:* Festschrift C. Gegenbaur, Band 2, p. 249-276, pl. 1.

(75) 1897, *Entwicklung von Caryophyllia cyathus:* Zool. Sta. Neapel Mitt., Band 12, p. 755-772, pl. 34, fig. 1-21.

Krempf, A.

(76) 1919, *Développement des ébauches musculaires chez la larve d'un Anthozoaire (Pocillopora caespitosa Dana) au stade primitif de la symétrie tetraradiaire:* Acad. Sci. Paris Comptes-Rendus, tome 169, p. 929-932.

(77) 1920, *Sur les principles modalités du développement et des relations de l'orthosepta et du sterigmatosepta dans l'ensemble du groupe des Anthozoaires:* Same, tome 170, p. 136-139.

Kühn, O.

(78) 1933, *Das Becken von Isfahan-Saidabad und seine altmiocäne Korallenfauna:* Palaeontographica (Stuttgart), Band 79, Abt. A, p. 143-218, pl. 17-19, fig. 1-8.

Lacaze-Duthiers, H. de

(79) 1873, *Développement des coralliaires. Deuxième mémoir: actinaires des polypiers:* Archives Zoologie Expérimentale Générale (Paris), tome 2, p. 269-348, pl. 12-15.

(80) 1897, *Faune de Golfe du Lion. Coralliaires. Zoanthaires sclérodermés (Deuxième mémoir):* Same, tome 5, p. 1-249, pl. 1, 2.

Manton, S. M., & Stephenson, T. A.

(81) 1935, *Ecological surveys of coral reefs:* Brit. Mus. (Nat. Hist.) Great Barrier Reef Expedition 1928-29 Sci. Repts. (London), v. 3, no. 10, p. 273-312, pl. 1-16, fig. 1-2.

Marenzeller, E. von

(82) 1904, *Riffkorallen:* Zool. Ergebnisse Exped. S.M.S. *Pola,* Nr. 26, K. Akad. Wiss. Wien Math.-Naturwiss. Kl. Denkschr., Band 80, p. 27-97, pl. 1-29.

Matthai, G.

(83) 1914, *A revision of the Recent colonial Astraeidae possessing distinct corallites:* Linnean Soc. London, Trans., ser. 2 (Zool.), v. 17, p. 1-140, pl. 1-38.

(84) 1923, *Histology of the soft parts of astraeid corals:* Quart. Jour. Micros. Sci. (London), new ser., v. 67, p. 101-122, pl. 7-8.

(85) 1924, *Report on the madreporarian corals in the collection of the Indian Museum, Calcutta. Pt. 1:* Indian Museum Mem., v. 8, p. 1-59, pl. 1-11.

(86) 1926, *Colony formation in astraeid corals:* Roy. Soc. London Philos. Trans., v. 214B, p. 313-356, pl. 24-26.

(87) 1926, *Colony formation in Isophyllia, Symphyllia, Manicina, and Merulina:* Same, p. 356-367, pl. 27-28.

(88) 1928, *A monograph of the recent meandroid Astraeidae:* Brit. Mus. (Nat. Hist.) Cat. Madreporarian Corals (London), v. 7, p. 1-288, pl. 1-72.

(89) 1948, *On the mode of growth of the skeleton in fungid corals:* Roy. Soc. London Philos. Trans., v. 233B, p. 177-195, pl. 3-14, figs.

(90) 1948, *Skeletal variation in two large coralla from Tahiti, one of Pavona varians (Verrill) and another of Psammocora haimiana M. E. & H.:* Same, p. 197-199, pl. 15-16.

(91) 1948, *Colony formation in fungid corals. I. Pavona, Echinophyllia, Leptoseris, and Psammocora:* Same, p. 201-231, pl. 17, fig. 1.

Michelin, H.

(92) 1840-47, *Iconographie zoophytologique: descriptions par localités et terrains des polypiers fossiles de France et pays environnants:* (Paris), 348 p., 79 pl.

Milne-Edwards, Henri, & Haime, Jules

(93) 1848-50, *Recherches sur les polypes:* Annales Sci. Nat. (Paris), ser. 3, tome 9 (1848), p. 37-89, 211-344, pl. 4-6, 7-10; tome 10

(1848), p. 65-114, 209-320, pl. 1, 5-9; tome 11 (1849), p. 233-312; tome 12 (1849), p. 95-197; tome 13 (1850), p. 63-110, pl. 3-4; tome 15 (1850), p. 73-144; tome 16 (1850), p. 21-70.

(94) 1850-54, *Monograph of the British fossil corals:* Palaeontograph. Soc. (London), lxxxv +290 p., 72 pl.

(95) 1857-60, *Histoire naturelle des coralliaires:* (Paris), tome 1 (1857), viii+326 p.; tome 2 (1857), 633 p.; tome 3 (1860), 560 p.; atlas (1857), 31 pls.

Montanaro, Eugenia

(96) 1933, *Faune attinologiche plio-pleistoceniche dell'Africa orientale:* Palaeontographica Italica (Siena), new ser. tomo 3, p. 141-209, pl. 18-22.

Moseley, H. N.

(97) 1881, *On the deep-sea Madreporaria:* Voyage H.M.S. *Challenger* Repts. Sci. Results, Zool. (London), v. 2, pt. 7, p. 127-208, pl. 1-16.

Ogilvie, M. M.

(98) 1896, *Microscopic and systematic study of madreporarian types of corals:* Roy. Soc. London Philos. Trans., v. 187B, p. 83-345, fig. 1-75.

(99) 1897, *Die Korallen der Stramberger Schichten:* Palaeontographica (Stuttgart), Suppl., Bd. 2, pt. 7, p. 73-282, pl. 7-18.

Oppenheim, P.

(100) 1930, *Die Anthozoen der Gosauschichten in der Ostalpen:* (Berlin), 604 p., 48 pl.

Ortmann, A.

(101) 1888, *Studien über Systematik und geographische Verbreitung der Steinkorallen:* Zool. Jahrbücher (Jena), Abt. Systematik, Band 3, p. 143-188, pl. 6.

(102) 1889, *Beobachtungen an Steinkorallen von der Südküste Ceylons:* Same, Band 4, p. 493-590, pl. 11-18.

Počta, P.

(103) 1887, *Die Anthozoen der böhmischen Kreideformation:* K. Böhm. Gesell. Wiss. (Praha) Abh., Band 7, p. 1-60, pl. 1-2.

Pourtalès, L. F. de

(104) 1871, *Deep-sea corals:* Harvard Coll. Mus. Comp. Zool., Illus. Cat., no. 4, p. 1-93, pl. 1-8.

Pratz, E.

(105) 1882, *Über die verwandtschaftlichen Beziehungen einiger Korallengattungen mit hauptsachlicher berücksichtigung ihrer Septal-Struktur:* Palaeontographica (Stuttgart), Band 29, p. 81-122, pl. 1.

Prever, P. L.

(106) 1909, *Anthozoa della fauna cenomaniana dei calcari i scogliera:* in Parona, C. F., La fauna coralligena del Cretaceo dei Monte d'Ocre nell'Abruzzo Aquilano, Cart. Ital. Com. Geol. Mem., tomo 5, p. 51-147, pl. 1-15.

(107) 1921-22, *I coralli oligocenici di Sassello nell'Appennino Ligure. Pte. I. Corallari a calci confluenti:* Palaeontographica Italica (Siena), tomo 27 (1921), p. 53-100, pl. 7-15; tomo 28 (1922), p. 1-40, pl. 1-7.

Quayle, E. H.

(108) 1932, *Fossil corals of the genus Turbinolia from the Eocene of California:* San Diego Soc. Nat. Hist., Trans., v. 7, p. 91-110, pl. 6, figs. 1-12.

Quelch, J. J.

(109) 1886, *Report on the reef corals:* Voyage H.M.S. *Challenger* Repts. Sci. Results, Zool. (London), v. 16, no. 3, 203 p., 12 pl.

Quenstedt, F. A.

(110) 1878-81, *Petrefactenkunde Deutschlands:* (Tübingen), Band 6, Abt. 1, (Korallen), 1093 p., 42 pl.

Reis, O. M.

(111) 1889, *Korallen der Reiter Schichten:* Geogn. Jahresh. (München), Band 11, p. 91-162, pl. 1-4.

Reuss, A. E.

(112) 1854, *Beiträge zur Characteristik der Kreideschichten in den Ostalpen besonders im Gosauthale und am Wolfgangsee:* Akad. Wiss. Wien, Math.-naturwiss. Kl., Denkschr., Band 7, p. 1-157, pl. 1-31.

(113) 1864, *Über Anthozoen und Bryozoen des Mainzer Tertiärbeckens:* Same, Sitzungsber., Band 50, p. 197-210, pl. 1-3.

(114) 1864, *Zur Fauna der deutschen Oberoligocäns, Anthozoen:* Same, Sitzungsber., Band 50, p. 614-690, pl. 6-15.

(115) 1867, *Über fossile Korallen von der Insel Java: Novara*-Expeditie (Wien), Geol. Teil, Band 2, p. 165-185, pl. 1-3.

(116) 1868-72, *Paläontologische Studien über die älter Tertiärschichten der Alpen:* Akad. Wiss. Wien Math.-naturwiss. Kl., Denkschr., Band 28 (1868), p. 129-184, pl. 1-16; Band 29 (1869), p. 215-298, pl. 17-36; Band 33 (1872), p. 1-60, pl. 37-56.

(117) 1871, *Die fossilen Korallen des Österreichisch-ungarnischen Miocäns:* Same, Denkschr., Band 31, p. 197-270, pl. 1-21.

Roemer, F. A.

(118) 1863, *Die Polyparien des norddeutschen Tertiär-Gebirges:* Palaeontographica (Stuttgart), Band 9, p. 199-245, pl. 35-39.

Różkowska, M. D.

(119) 1932, *Korale mioceńskie Polski:* Polnischen Geol. Gesell. (Cracow), Jahrg. 8, p. 1-75, pl. 2-7.

Saville-Kent, W.

(120) 1893, *The Great Barrier Reef of Australia:* (London), xii+387 p., 48 pl., 16 color pl.

Seguenza, G.

(121) 1863-64, *Disquisizioni paleontologiche intorno ai corallarii fossili delle rocce terziarie del distreto di Messina:* Accad. Sci. Torino, Mem, ser. 2, tomo 21, p. 399-458, pl. 1-5 (1863); p. 459-560, pl. 6-15 (1864).

Sismonda, E., & Michelotti, G.

(122) 1871, *Materiaux pour servir à la paléontologie de terrain tertiaire du Piémont. II. Animaux. Types protozoaires et coélentérés:* Same, tomo 25, p. 257-362, pl. 1-10.

Solomko, E.

(123) 1888, *Die Jura- und Kreidekorallen der Krim:* Zap. Imp. Peterb. Min. Obshch., ser. 2, v. 24, p. 67-231, pl. 1-8.

Stephenson, T. A., & Stephenson, A.

(124) 1933, *Growth and asexual reproduction in corals:* Brit. Mus. (Nat. Hist.) Great Barrier Reef Expedition 1928-29 Sci. Repts. (London), v. 3, no. 7, p. 167-217, pl. 1-10, figs. 1-14.

Stoliczka, F.

(125) 1873, *Cretaceous fauna of Southern India. The corals or Anthozoa:* Palaeontologia Indica (Calcutta), v. 4, p. 133-190, pl. 18-28.

Tenison-Woods, J. E.

(126) 1878, *On the extra-tropical corals of Australia:* Linnaean Soc. New South Wales Proc. (Sydney), v. 2, p. 292-341, pl. 4-6.

(127) 1880, *Corals and Bryozoa of the Neozoic Period in New Zealand:* Paleontology of New Zealand (Wellington), pt. 4, p. 1-34, pls. 1-4.

Thiel, M. E.

(128) 1928, *Madreporaria:* in Beiträge zur Kenntnis der Meeresfauna Westafrikas (Hamburg), Band 3, Lief. 6, p. 253-350, pl. 1-5.

(129) 1932, *Madreporaria; zugleich ein Versuch einer vergleichenden Oekologie der gefunden Formen [Résultats scientifiques du voyage aux Indes Orientales Nederlandaises . . .]:* Mus. Roy. Hist. Nat. Belg. (Bruxelles), Mém., hors sér., v. 2, fasc. 12, p. 1-177, pl. 1-21.

Thomas, H. D.

(130) 1935, *Jurassic corals and Hydrozoa, together with a redescription of Astraea caryophylloides Goldfuss:* Geology and palaeontology of British Somaliland (London), v. 3, p. 23-39, pl. 2-5.

Tomes, R. F.

(131) 1882, *On the Madreporaria of the Inferior Oolite of the neighbourhood of Cheltenham and Gloucester:* Geol. Soc. London Quart. Jour., v. 38, p. 409-450, pl. 18.

(132) 1884, *A comparative and critical revision of the Madreporaria of the White Lias of the middle and western counties of England, and of those of the conglomerate at the base of the South-Wales Lias:* Same, v. 40, p. 353-375, pl. 19.

(133) 1885, *On some new or imperfectly known Madreporaria from the Great Oolite of the counties of Oxford, Gloucester, and Somerset:* Same, v. 41, p. 170-190, pl. 5.

(134) 1886, *On some new or imperfectly known Madreporaria from the Inferior Oolite of Oxfordshire, Gloucestershire, and Dorsetshire:* Geol. Mag. (London), ser. 3, v. 3, p. 1-23, pl. 10; p. 385-398, 443-452.

Trauth, F.

(135) 1911, *Die oberkretazische Korallenfauna von Klogsdorf in Mähren:* Mähr. Landesmus. Zeitschr. (Brünn), Band 11, p. 1-104, pl. 1-4.

Umbgrove, J. H. F.

(136) 1925, *De Anthozoa uit het Maastrichtsche Tufkrift:* Leidsche Geol. Meded., deel 1, p. 83-126, pl. 8-11.

(137) 1929, *Anthozoen van N. O. Borneo:* Dienst Mijnbouw Nederlandsche-Indië (Batavia), no. 9, p. 47-78, pl. 1-5, figs. 1-2.

(138) 1939, *Madreporaria from the Bay of Batavia:* Rijksmus. Natuurlijke Historie Leiden Zool. Meded., v. 22, p. 1-64, pl. 1-18, 1 map.

(139) 1940, *Madreporaria from the Togian Reefs (Gulf of Tomini, North Celebes):* Same, v. 22, p. 265-310, pl. 21-35, figs. 1-3.

(140) 1946, *Corals from a lower Pliocene patch reef in central Java:* Jour. Paleont. (Tulsa), v. 20, p. 521-542, pl. 77-82.

Van der Horst, C. J.

(141) 1921, *Madreporaria of the Siboga Expedition. Pt. 2. Madreporaria Fungida: Siboga-*Expeditie (Leiden), Mon. 16b, p. 53-98, pl. 1-6.

(142) 1922, *Madreporaria: Agariciidae [No. 9, Percy Sladen Trust Exped.]:* Linnean Soc. London Trans., Zool., ser. 2, v. 18, p. 417-429, pl. 31-32.

(143) 1922a, *Madreporaria of the Siboga Expedition. Pt. 3. Eupsammidae: Siboga-*Expeditie (Leiden), Mon. 16c, p. 99-127, pl. 7-8.

Vaughan, T. W.

(144) 1900, *Eocene and lower Oligocene coral faunas of the United States with a few doubtfully Cretaceous species:* U. S. Geol. Survey, Mon. 39, p. 1-263, pls. 1-24.

(145) 1905, *A critical review of the literature of the simple genera of the Madreporaria Fungida, with a tentative classification:* U. S. Natl. Mus., Proc., v. 28, p. 371-424.

(146) 1907, *Recent Madreporaria of the Hawaiian Islands and Laysan:* Same, Bull., v. 59, no. 9, p. 1-427, pl. 1-96, fig. 1.

(147) 1913, *Hexacoralla and Rugosa:* in ZITTEL-EASTMAN, Textbook of palaeontology, (London), p. 74-109, figs. 94-172.

(148) 1918, *Some shoal-water corals from Murray Island, Cocos-Keeling Islands, and Fanning Island:* Carnegie Inst. Washington Pub. 213, p. 51-234, pl. 20-93.

(149) 1919, *Fossil corals from Central America, Cuba, Porto Rico, with an account of the American Tertiary, Pleistocene, and Recent coral reefs:* U. S. Natl. Mus., Bull., v. 103, p. 189-524, pl. 68-152.

———, & Hoffmeister, J. E.

(150) 1926, *Miocene corals from Trinidad:* Carnegie Inst. Washington Pub. 344, p. 105-134, pl. 1-7.

———, & Wells, J. W.

(151) 1943, *Revision of the suborders, families, and genera of the Scleractinia:* Geol. Soc. America, Spec. Paper 44, xv+363 p., 51 pl., 39 fig., 3 tab.

Verrill, A. E.

(152) 1902, *Variations and nomenclature of Bermudian, West Indian, and Brazilian reef corals, with notes on various Indo-Pacific corals:* Conn. Acad., Trans. (New Haven), v. 11, p. 63-168, pl. 1-26.

(153) 1902, *Notes on corals of the genus Acropora (Madrepora Lam.), with new descriptions and figures of types, and of several new species:* Same, p. 207-266, pl. 36-36F.

Verwey, J.

(154) 1931, *The depth of coral reefs in relation to their oxygen consumption and the penetration of light in the water:* Treubia (Batavia), v. 13, p. 169-198, fig. 1-2, tab. 1-10.

Volz, W.

(155) 1896, *Die Korallenfauna der Trias monographisch bearbeitet. II. Die Korallen der Schichten von St.-Cassian in Süd-Tirol:* Palaeontographica (Stuttgart), Band 43, p. 1-124, pl. 1-12, fig. 1-49.

Weissermel, W.

(156) 1925-28, *Die Korallen des deutschen Muschelkalks. I. Unterer Muschelkalk:* Preuss. geol. Landesanst. Jahrb. (Berlin), Bd. 46 (1925), p. 1-33, pl. 1-2; Band 49 (1928), p. 224-238, pl. 20-21.

Wells, J. W.

(157) 1932, *Corals of the Trinity group of the Comanchean of central Texas:* Jour. Paleont., v. 6, p. 225-256, pl. 30-39.

(158) 1933, *Corals of the Cretaceous of the Atlantic and Gulf coastal plains and western interior of the United States:* Bull. Am. Paleont. (Ithaca), v. 18, p. 85-288, pl. 14-29.

(159) 1934, *Some fossil corals from the West Indies:* U. S. Natl. Mus., Proc., v. 83, p. 71-110, pl. 1-5.

(160) 1941, *Cretaceous and Eogene corals from Peru:* Bull. Am. Paleont. (Ithaca), v. 26, p. 304-326, pl. 44-46.

(161) 1945, *West Indian Eocene and Miocene corals:* Geol. Soc. America, Mem., v. 9, 25 p., 3 pl.

(162) 1954, *Recent corals of the Marshall Islands:* U. S. Geol. Survey, Prof. Paper 260-I, p. 385-486, pl. 94-187, fig. 119-122, 4 tab.

Yabe, H., & Sugiyama, T.

(163) 1935, *Geological and geographical distribution of reef corals in Japan:* Jour. Paleont., v. 9, p. 183-217, pl. 21.

———, Sugiyama, T., & Eguchi, M.

(164) 1936, *Recent reef-building corals from Japan and the South Sea islands under the Japanese mandate:* Tohoku Imp. Univ. Sci. Repts. (Sendai), ser. 2, spec. v. 1, p. 1-66, pl. 1-59.

(165) 1941, *Recent reef-building corals from Japan and the South Sea islands under the Japanese mandate. II:* Same, spec. v. 2, p. 67-91, pl. 59-104.

———, & Eguchi, M.

(166) 1942, *Fossil and Recent simple corals from Japan:* Same, v. 22, p. 105-178, pl. 9-12.

Yonge, C. M.

(167) 1930, *Physiology of reef corals:* Brit. Mus. (Nat. Hist.) Great Barrier Reef Expedition 1928-29 Sci. Repts. (London), v. 1, p. 13-57, 2 pl.

(168) 1940, *The biology of reef-building corals:* Same, v. 1, no. 13, p. 353-391, pl. 1-6.

———, & Nicholls, A. G.

(169) 1930-31, *Physiology of reef corals:* Same, v. 1, p. 59-91, 1 pl., 6 fig. (1930); p. 135-211, 5 pl. (1931).

———, Yonge, M.J., & Nicholls, A.G.

(170) 1932, *Physiology of reef corals:* Same, v. 1, p. 213-251, 4 fig.

SOURCES OF ILLUSTRATIONS

Explanation of the use of index numbers for citing sources of illustrations is given in the Editorial Preface.

(171) **Alcock, A. W.**
(172) **Alloiteau, James**
(173) **Angelis d'Ossat, G. de**
(174) **Bedot, M.**
(175) **Boschma, Hilbrand**
(176) **Crossland, C.**
(177) **Felix, J.**
(178) **Harrison, R. M.**
(179) **Hickson, S. J.**
(180) **Hoffmeister, J. E.**

(181) Janensch, W.
(182) Jüssen, E.
(183) Lindström, G.
(184) Lütken, C.
(185) Moore, R. C., Lalicker, C. G., & Fischer, A. G.
(186) Pictet, F.-J.
(187) Pourtalès, L. F. de
(188) Renz, C.
(189) Reuss, A. E.
(190) Rózkowska, M. D.
(191) Stephenson, L. W.
(192) Straw, S. H.
(193) Studer, T.
(194) Tomes, R. F.
(195) Vaughan, T. W.
(196) ———, & Popenoe, W. P.
(197) Wanner, Johannes
(198) Wells, J. W.
(199) Yabe, H., & Eguchi, M.
(200) ———, & Sugiyama, T.
(201) Yonge, C. M.

TABULATA

By Dorothy Hill and Erwin C. Stumm

CONTENTS

MORPHOLOGY

The Tabulata are an extinct, almost invariably Paleozoic order of corals characterized by their exclusively colonial mode of growth and secretion of a calcareous exoskeleton of slender tubes crossed by many transverse partitions called tabulae. Relative prominence of these tabulae and inconspicuousness or even absence of radially disposed longitudinal partitions (septa) are features which suggested the name of the group.

Morphological terms applied to tabulate corals are distinguished in introductory parts of the text devoted to this group by printing in boldface type. In addition, they are included in an alphabetically arranged glossary of morphological terms used in describing Rugosa, Heterocorallia, Scleractinia, and Tabulata (p. F245-F251).

FORM OF CORALLA

The form of the complete colonial exoskeleton **(corallum)** varies widely, depending mainly on the shape and arrangement of the exoskeletons **(corallites)** built by the constituent individual polyps. Thus, the corallum may be a laminar expansion, a domed, hemispherical (Fig. 340,*6*), spherical, nodular, or irregular body, or it may be slenderly or coarsely branching, the branches being cylindrical (ramose, Fig. 340,*3*) or flattened (foliose, Fig. 352,*4a*) and either free or joining to form a network; some colonies develop only as thin incrustations attached to foreign objects. Maximum dimensions of coralla range from a few millimeters to 2 m. or more; the diameter of corallites ranges from approximately 0.2 to 20 mm.

In the families Chaetetidae, Syringophyllidae, Heliolitidae, and Favositidae, all the corallites of the corallum are commonly in contact throughout their length; if the corallite walls are prismatic by reason of mutual pressure, the corallum is said to be cerioid (Fig. 340,*7*); if opposite walls fail to develop in rows of corallites it is meandroid (Fig. 340,*5*); if dividing walls are absent and a common tissue (coenenchyme) unites the corallites, the corallum is coenenchymal (Fig. 340,*6*). Corallites in these coralla may be erect or curved but opening at right angles to their axes; or reclined, open ing obliquely. Reclined corallites are generally alveolitoid (Fig. 340,*1*) with a vaulted upper wall and a lower wall parallel to the surface of adherence.

Cateniform coralla have their corallites united laterally in palisades generally one corallite thick, the palisades forming a network (Fig. 341,*9*). These characterize the Halysitidae.

The Auloporidae have a number of types of corallum, of which the fasciculate is commonest, the corallites being cylindrical, parallel, and distant, but united by connecting tubules (Fig. 341,*7*); in a second type, coenenchymal expansions alternate with fasciculate zones (Fig. 340,*4*). In this family, too, the proximal part of a corallum may be an attached (reptant) ring or reticulum (Fig. 340,*2*) and distal parts may consist of cylindrical or trumpet-shaped corallites which are either fasciculate or with erect or reptant branches arranged like umbrella ribs in whorls (umbelliferous) (Fig. 341,*8*) or zigzag (Fig. 341,*6*).

STRUCTURAL ELEMENTS

CALICE, EPITHECA, AND WALL

The distal surface of each corallite is the **calice** (Fig. 341,*2*). In all fasciculate and in some cerioid coralla, each corallite is enclosed laterally in a sheath (**epitheca,** Fig. 341,*2*) of ?granular $CaCO_3$; in many cerioid, meandroid, and coenenchymal coralla, a common epitheca covers all but the calical surface of the corallum. The epitheca typically is without the longitudinal ridges and furrows of the Rugosa, but growth rings are common; some Palaeacinae have superficial epithecal scales, as in some Rugosa; *Michelinia* has radiciform processes consisting of rootlike epithecal outgrowths.

The **corallite wall** (on which lies the epitheca in corallites of fasciculate coralla) is not a simple structure; in some corals it is formed by the peripheral edges of the septa expanding to contiguity; in others, the space between the peripheral edges of the septa is bridged by calcareous tissue consisting of "fibers" (needle-like crystals) directed inwards towards the axes of the corallite; this tissue may show growth lamellation across the fibers.

TRANSVERSE STRUCTURES

Corallites in coralla without coenenchyme have only one type of skeletal element transverse to their length, the **tabulae** —floors which generally are subhorizontal (Fig. 342,*1b;* 345,*4*) but in some are shallow domes or saucers, or in others, funnel-shaped (infundibuliform, Fig. 341,*1*). Each tabula is typically complete, i.e., made of one plate only (Fig. 345,*4*), but in some the tabulae may be incomplete, composed of several small plates (**tabellae,** as in *Michelinia,* Fig. 341,*3*). The **tabularium** is the area in which tabulae are developed and it may occupy the whole corallite.

In some cerioid forms, a **marginarium** may develop between the tabularium and the periphery of each corallite, generally as a dense zone of skeletal tissue without loculi forming a stereozone (Fig. 342,*2*); this is generally a septal stereozone formed by thickening of the peripheral parts of the septa to such extent that they become contiguous, but in some it may be formed in part by the thickening of wall segments between peripheral ends of the septa.

The coenenchyme of coenenchymal coralla is a common marginarium. In coenenchyme, transverse skeletal elements of various types may be found: **dissepiments,** which are small domed plates usually not inclined (as in *Propora,* Fig. 342,*1*); **sola,** which are small horizontal or shallowly saucered plates crossing coenenchymal tubules, as in *Heliolites* (Fig. 342,*3*); or extensions of the tabulae, as in *Calapoecia* (Fig. 346,*5*) and *Sarcinula* (Fig. 346,*6b*).

LONGITUDINAL STRUCTURES

Among skeletal elements oriented in the direction of growth of the corallite (longitudinal skeletal elements), the **septa** are most important and are radially arranged in the outer parts of the tabularium; they are short (extending but a little toward the axis from the periphery) and being equal are presumably of one order; in many they are 12 in number. Studies are required on their order of insertion, though one may not find orderly insertion in offsets owing to the manner of increase. Each septum is typically acanthine (spinose), i.e., each consists of a vertical series of thorns (trabeculae, Fig. 342,*3*) with acutely or obtusely pointed apices, in some separate throughout their length, in others contiguous at their bases at the periphery; in a few genera the septal trabeculae are contiguous throughout their lengths, forming a laminar septum. The trabeculae are generally in single series, the axis of each in the mid-plane of the septum, but in some (Fig. 343,*1*) the axes may diverge from the mid-plane; they are directed toward the axis of the corallite, typically upward but in some horizontally and rarely (*Proheliolites,* Fig. 348,6) downward. In some Favositinae (*Emmonsia*), certain of the trabeculae may expand laterally to form eavelike plates (**squamulae**, Fig. 343,*3*) projecting horizontally toward the axis of the corallite.

Axial structures are exceptional in Tabulata, and where formed, consist merely of vertical, separate trabeculae based on the tabulae. The longitudinal skeletal elements of the coenenchyme are trabeculae, generally separate (Fig. 342,*1*), though in some they are contiguous in linear series to form laminae, of which some may be continuations of the septa, while others may form walls enclosing prismatic or cylindrical spaces or tubules (Fig. 342,*3*). The trabeculae are typically vertical (parallel to the direction of growth of the corallum) and may be thin or stout. In the Palaeacinae, the coenenchyme consists of trabeculae contiguous throughout, except for irregular canal-like spaces which may connect neighboring tabularia or open at the surface.

The coenenchyme indicates the presence of a coenosarc or common soft tissue; but direct communication between the polyps of neighboring corallites seems to have been possible also by means other than the coenosarc. Thus, in the Syringoporinae, connecting tubules (Fig. 341,*7*) join neighboring corallites and may pierce both walls. In the Favositidae and in cerioid Syringoporinae (*Roemeria*), regular or irregular systems of circular or oval **mural pores** (holes through the common walls) are found (Fig. 343,*2*). In Micheliniinae and Palaeacinae, the walls may be perforate (Fig. 343,*4*) in the same irregular manner as in the perforate Scleractinia and the calostylid Rugosa.

MICROSTRUCTURE OF SKELETON

Each of the skeletal elements (septum, tabula, dissepiment, solum or wall segment) consists of needle-like crystals (fibers) of $CaCO_3$, though whether originally of aragonite or calcite is not known. In the transverse skeletal elements these are arranged at right angles to the upper and lower surfaces of the plate, but in the septa they are directed upward and outward from the axes of the constituent trabeculae. In all elements, skeletal thickening is effected by a lengthening of the fibers from their upper ends, and growth lamellation is at right angles to the fibers; in some types of recrystallization during fossilization, the fibrosity disappears and lamellation is emphasized.

The relation of the entire exoskeleton to the soft parts must have been similar to that in Scleractinia. No fossil impressions of the soft parts of Tabulata are known. As in Rugosa, the fibers of Tabulata were formed by spherulitic crystallization, presumably in a gel exuded from the basal ectoderm. Crystallization in the transverse skeletal elements is by planar control, and in the

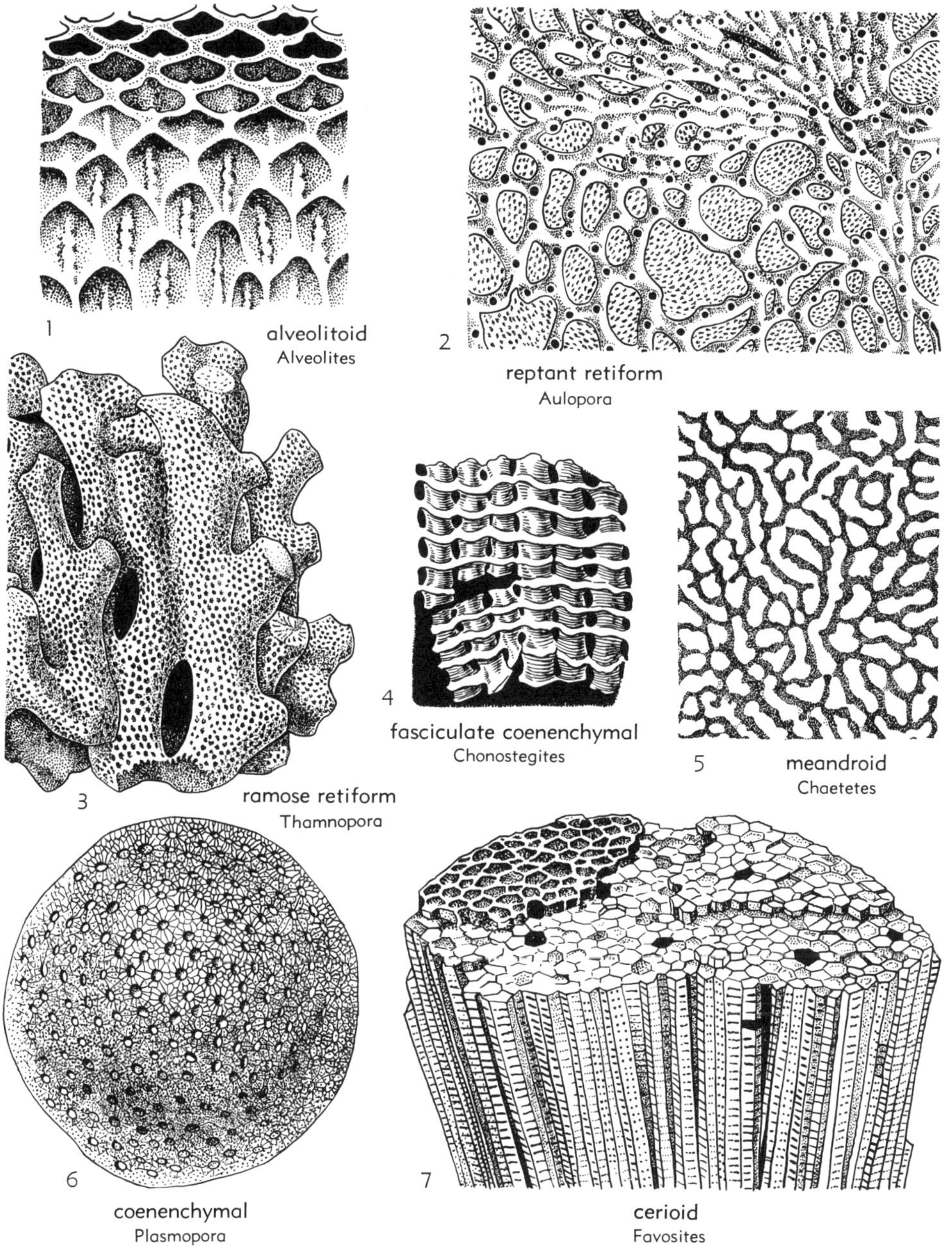

FIG. 340. Types of tabulate coralla. *1, Alveolites suborbicularis* LAMARCK, M.Dev., Ger.; part of surface showing form of nearly recumbent (alveolitoid) corallites with large septal spine in middle of lower wall, enlarged (55). *2, Aulopora serpens* GOLDF., M.Dev., Ger.; upper surface of corallum showing circular calices of branching tubes (reptant on another coral), ×1 (55). *3, Thamnopora reticulata* BLAINV., M.Dev., Ger.; surface of corallum, ×1 (55). *4, Chonostegites clappi* M.EDW.-H., M.Dev., Ohio; side of weathered corallum showing regularly spaced flat coenenchymal extensions connecting the cylindrical corallites, ×1 (70). *5, Chaetetes septosus* FLEMING, L.Carb., Belg.; transv. sec. ×8 (76). *6, Plasmopora petalliformis* (LONSDALE), M.Sil., Eu.; surface of corallum, ×1 (68). *7, Favosites gothlandicus* LAMARCK, Sil. Gotl.; corallum broken along walls of prismatic tubes, ×1 (55).

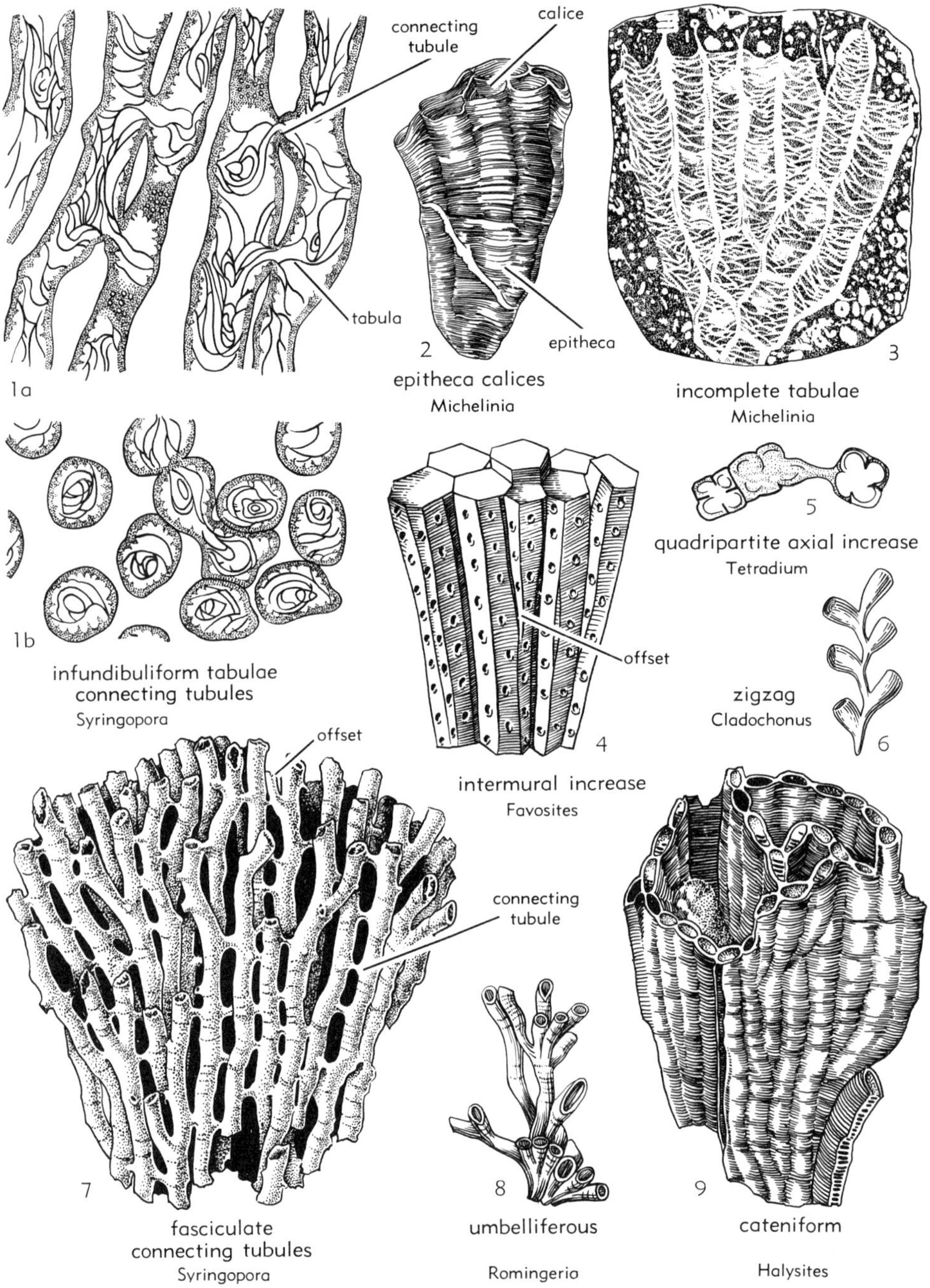

FIG. 341. Types of coralla, increase and tabulae in tabulate corals. *1, Syringopora reticulata* GOLDF., L.Carb., Eng.; long. and transv. secs., showing infundibuliform tabulae and connecting tubules between corallites, ×5 (70). *2,3, Michelinia tenuisepta* (PHILLIPS), L.Carb., Belg.; *2,* side of corallum, ×1 (68); *3,* long. sec., ×1 (68). *4, Thamnopora polymorpha* (GOLDF.), M.Dev., Ger.; part of corallum broken along walls of prismatic tubes, showing prominent mural pores, enlarged (55). *5, Tetradium oklahomense* BASSLER, M. Ord., Okla.; transv. sec. of 3 corallites, ×12 (45). *6, Cladochonus brevicollis* (M'COY), L.Carb., Eng.; side of corallum, ×1 (66). *7, Syringopora ramulosa* GOLDF., L.Carb., Belg.; side of corallum, ×1 (55). *8, Romingeria umbellifera* (BILL.), M.Dev., Ont.; side of corallum, ×1 (70). *9, Halysites labyrinthicus* (GOLDF.), M.Sil., Can.; corallum, ×1 (55).

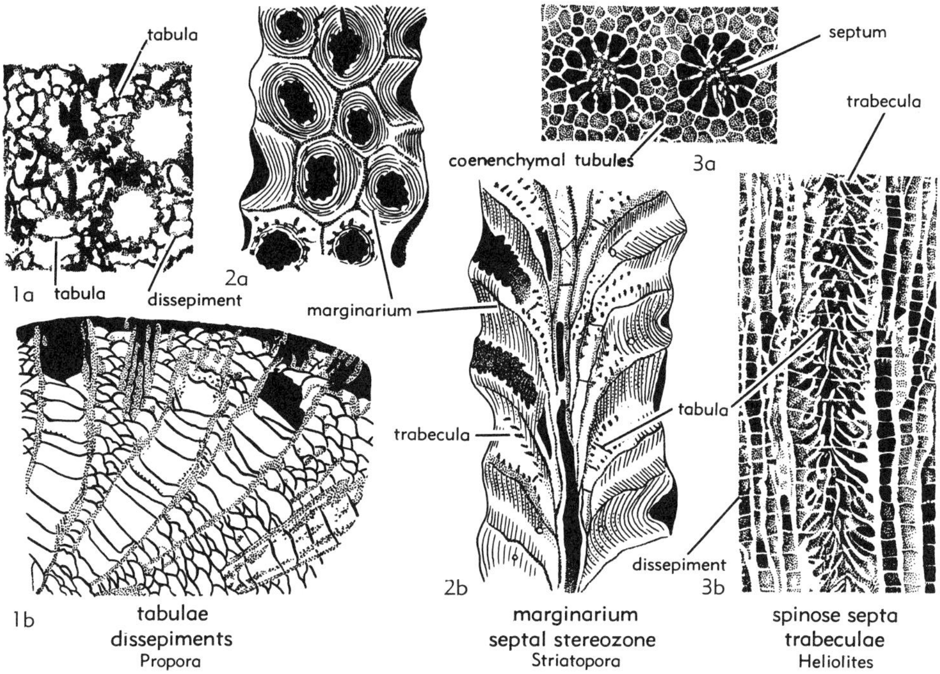

FIG. 342. Morphological features of tabulate corals. *1, Propora speciosa* (BILLINGS), Sil., Eu.; *1a,b,* transv. and long. secs., ×4 (65). *2, Striatopora ornata* (ROMINGER), M.Dev., N.Y.; *2a,b,* surface and long. sec., ×7 (70). *3, Heliolites barrandei* PENECKE, U.Sil.-L.Dev., Eu.; *3a,b,* transv. and long. secs., ×12 (65).

septa by puntal control from a series of points along the crests of invaginations in the base of the polyp, a trabecula being formed from each point.

INCREASE OF CORALLA

The growth or increase of coralla may have started from one corallite only, or possibly from a group of corallites secreted by simultaneously settling planulae (the free-swimming larval stages). New corallites formed during increase are **offsets.**

Three chief types of increase are distinguishable:

(1) **Axial,** in which one or more new dividing walls grow across the axis of a corallite so that small offsets form in the place of the one corallite, whose peripheral tissues continue to grow as part of the offsets; where 2 offsets are so formed the increase is bipartite, and this may be equal or unequal; where 4 offsets are formed, the increase is quadripartite (Fig. 341,*5*).

(2) **Peripheral** (or coenenchymal), where offsets arise in the coenenchyme.

(3) **Lateral** (in fasciculate coralla, Fig. 341, *7*) or **intermural** (in cerioid coralla, Fig. 341,*4*), where the offsets arise laterally, the initial portion becoming surrounded by the growing wall of the older corallite.

CLASSIFICATION

The order has had many changes in its systematics. MILNE-EDWARDS & HAIME included among Tabulata several genera which subsequently have been assigned to other orders, subclasses, or phyla, as for example, *Pocillopora* and *Seriatopora* (scleractinians); *Millepora* (hydrocoralline); *Labechia* (stromatoporoid); *Heliopora* (octocorallian) and *Fistulipora, Stenopora* (polyzoans), etc. As the real affinities of these misfits were discovered, it came to be thought that there was no such zoological entity as the Tabulata, and the remaining genera were distributed among the other Anthozoa, chiefly in the Scleractinia and Octocorallia, but sometimes in specially named orders; the earlier moves were reviewed by NICHOLSON (29); later essays are those of SARDESON (32), GERTH (12), OKULITCH (30), and TERMIER & TERMIER (36). WEISSERMEL (37), in reviewing OKULITCH's conclusions, gave sound reasons for concluding that the Tabulata as now understood and herein defined are a natural unit.

The 6 families forming the order are clearly distinguishable from the Rugosa, Scleractinia, and Octocorallia, and should retain MILNE-EDWARDS & HAIME's name Tabulata. They differ from the Octocorallia (except *Heliopora*) and resemble the Rugosa and Scleractinia in having skeletons of fibrous $CaCO_3$ (not spicular, as in Octocorallia), arranged in trabeculae in the septa and in sheets in the transverse plates. They differ from both Rugosa and Scleractinia in having septa of one series only, while in many the number of septa is 12. They are invariably compound, the corallites are almost universally slender, and they tend to communicate with one another by mural pores, connecting tubules, irregularly perforate wall tissue, or coenenchyme.

Subdivision of the Tabulata is relatively simple, the 6 families recognized having distinctive features. Thus, the most primitive (Chaetetidae) are aseptate and without mural pores; the Favositidae are septate, with mural pores, a few of its members having 12 septa; the Heliolitidae all have 12 septa and are coenenchymal; the Halysitidae have a characteristic cateniform growth and 12 septa; the Auloporidae commonly have funnel-shaped tabulae and connecting tubes which become mural pores in cerioid forms; the Syringophyllidae have mural pores but develop a characteristic coenenchyme.

The aseptate aporose condition of the Chaetetidae might perhaps be thought discrepant from the other Tabulata, but in Devonian forms in which the wall is thickened, the tissue is found to be fibrous as in the other corals.

A tabular summary of suprageneric divisions recognized in the Tabulata, showing stratigraphic range of each as now known and including record of the number of genera (figures enclosed by parentheses) in each taxonomic unit, follows.

Suprageneric Divisions of Tabulata

Tabulata *(order)*, exclusively colonial, corallites mostly small, characterized by presence of tabulae but septa weak or absent (108). *Ord.-Perm., ?Trias.-?Eoc.*
- Chaetetidae *(family)* (8). *M.Ord.-Perm., ?Trias.-?Eoc.*
  - Lichenariinae *(subfamily)* (1). *M.Ord.*
  - Tetradiinae (2). *M.Ord.-U.Ord.*
  - Chaetetinae (5). *Ord.-Perm., ?Trias.-?Eoc.*
- Syringophyllidae (7). *M.Ord.-L.Sil.*
  - Billingsariinae (4). *M.Ord.-U.Ord.*
  - Syringophyllinae (3). *M.Ord.-L.Sil.*
- Heliolitidae (18). *M.Ord.-U.Dev.*
  - Coccoseridinae (5). *M.Ord.-L.Sil.*
  - Palaeoporitinae (3). *M.Ord.-U.Ord.*
  - Plasmoporinae (5). *M.Ord.-U.Sil.*
  - Proheliolitinae (1). *U.Ord.-L.Sil.*
  - Heliolitinae (4). *L.Sil.-U.Dev.*
- Favositidae (40). *U.Ord.-Perm., ?Trias.*
  - Favositinae (16). *U.Ord.-U.Dev.*
  - Theciinae (2). *Sil.*
  - Pachyporinae (7). *Sil.-Perm.*
  - Alveolitinae (3). *Sil.-Dev.*
  - Micheliniinae (4). *Dev.-Perm.*
  - Palaeacinae (8). *L.Dev.-Perm.*
- Halysitidae (3). *Ord.-Sil.*
- Auloporidae (24). *Sil.-Perm.*
  - Auloporinae (15). *Sil.-Perm.*
  - Syringoporinae (9). *Sil.-Perm.*
- Incertae sedis (8). *Sil.-Carb.*

STRATIGRAPHIC DISTRIBUTION

GENERAL SUMMARY

The Tabulata are first known in the Chazyan, early Middle Ordovician strata of the shallow seas associated with the Appalachian geosyncline of eastern North America, but by Trentonian time, they were cosmopolitan. They apparently preceded the Rugosa, first known in Blackriveran strata. In numbers of individuals they exceeded the Rugosa throughout the Ordovician and Silurian, but were about equal in the Devonian until late in Givetian time, when their numbers markedly decreased; by early Carboniferous times they were subordinate to the Rugosa, and retained this relative unimportance until the extinction of the Rugosa at the end of the Paleozoic Era. In families, genera, and species, they outnumbered the Rugosa only in the Ordovician. Probably only one family of tabulate corals survived the Paleozoic, being represented by uncommon examples until its extinction in the

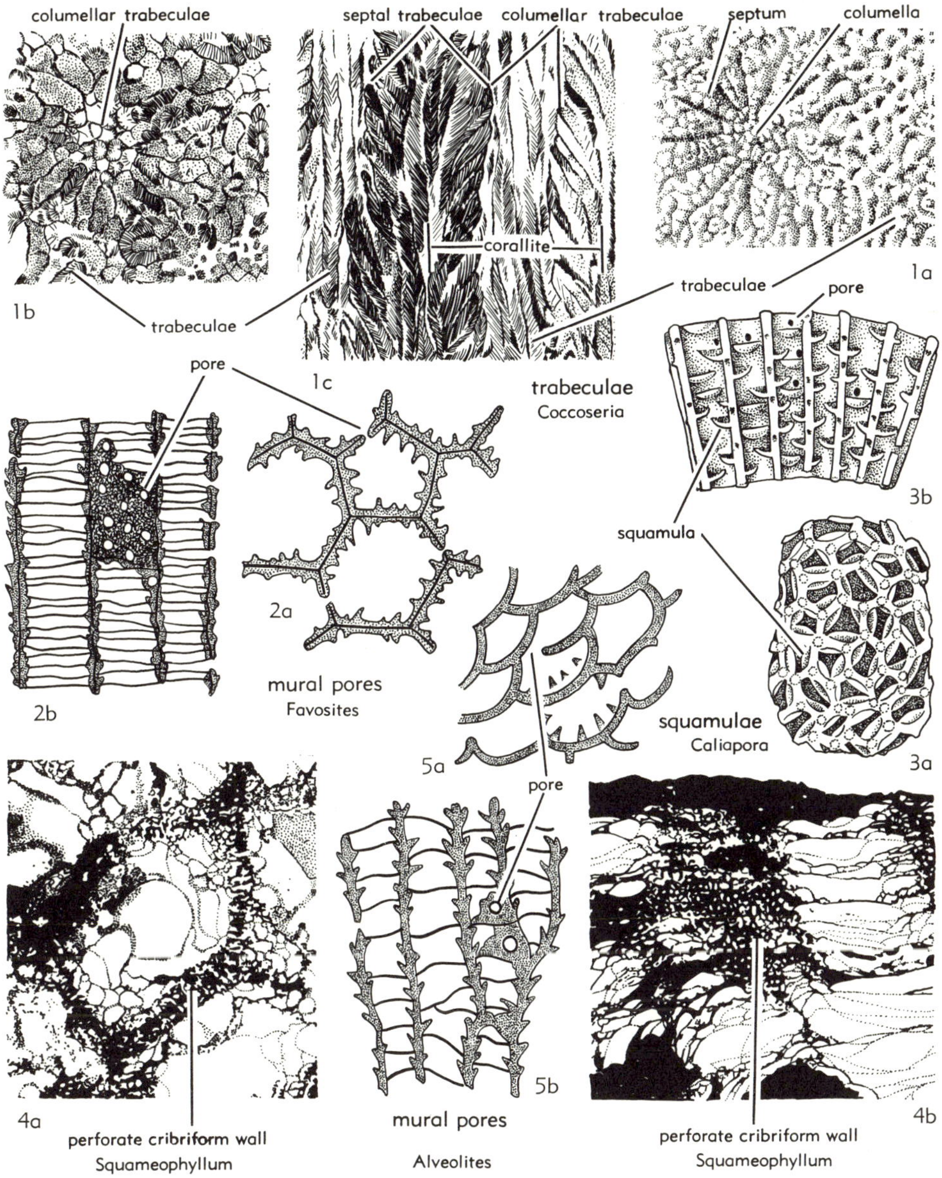

FIG. 343. Morphological features of tabulate corals. *1, Coccoseris ungerni* EICHW., U.Ord., Est.; *1a,* part of distal surface; *1b,c,* transv. and long secs.; all ×10 (65). *2, Favosites* sp., Dev., Queensl.; *2a,* transv. sec. in plane which happens to coincide with position of some mural pores, also showing short spinelike septa; *2b,* long. sec. showing closely spaced tabulae and mural pores in walls both parallel and normal to plane of the section, ×6 (70). *3, Caliapora battersbyi* (M.EDW.-H.), M.Dev., Eu.; *3a,b,* transv. and long. secs. showing shelflike projections from the wall which are termed squamulae, ×6 (74). *4, Squameophyllum antiquum* (M'COY), L.Carb., Ire.; *4a,b,* transv. and long. secs. showing perforate walls and irregular tabulae, enlarged (76). *5, Alveolites labechei* M.EDW.-H., Sil., Eu.; *5a,b,* transv. and long. secs. showing spinose septa and mural pores, ×10 (70).

Eocene. Bursts of fecundity and evolutional activity are less notable in the Tabulata than in the Rugosa.

RELATIVE IMPORTANCE OF FAMILIES IN SUCCESSIVE EPOCHS

ORDOVICIAN

The families appearing in the Chazyan of North America were the Chaetetidae (*Tetradium, Lichenaria*), Syringophyllidae (*Billingsaria*), and Auloporidae (*Eofletcheria*), the chaetetids being perhaps commonest. During Blackriveran (zone of *Nemagraptus gracilis*) time, these Tabulata seem still confined to the strata of the Appalachian seas of North America, with the exception of *Tetradium* in possible Blackriveran equivalents on Bear Island south of Spitzbergen. But by Trentonian time (zones of *Climacograptus peltifer* to *C. linearis,* inclusive) they occurred also in the Caledonian seas of northern Europe and the Himalayan seas of central Asia, and the first Heliolitidae (*Protaraea, Propora*) and Halysitidae (*Halysites*) had entered. The Favositidae are not certainly known until *Palaeofavosites* appeared in the Upper Ordovician, in equivalents of the zone of *Dicellograptus anceps.* In the Upper Ordovician too, the Heliolitidae replaced the Chaetetidae as the dominant family of Tabulata, many genera, particularly of Coccoseridinae, Palaeoporitinae, and Plasmoporinae appearing. The Syringophyllidae were practically extinct by the end of the Ordovician.

SILURIAN

During the Silurian the Chaetetidae were quite unimportant, being represented by a few Chaetetinae only, and the dominant family was perhaps the Favositidae. The ramose Pachyporinae and reclined Alveolitinae were less numerous than the hemispherical or spherical Favositinae, of which *Favosites* itself was the most important, with its pores in the faces rather than the angles of the walls, as in the mainly Ordovician *Palaeofavosites.* In the still important Heliolitidae, the Plasmoporinae and Heliolitinae took the place of the Coccoseridinae, Palaeoporitinae, and Proheliolitinae. The Halysitidae were only slightly more important than the Auloporidae, but whereas the latter continued throughout the Paleozoic, the former became extinct at the end of the Silurian or within the Gedinnian.

DEVONIAN

Devonian faunas are characterizd by only 4 of the families of Tabulata. The Favositidae remained dominant, though they waned in the Upper Devonian. The Heliolitidae were represented by only 2 subfamilies, the Heliolitinae and Plasmoporinae, which continued from the Silurian; the last heliolitid, *Heliolites* itself, became extinct in the early Late Devonian. The Auloporidae and Chaetetidae retained minor importance. Of the Favositidae, the Favositinae and Pachyporinae were very important until late in the Devonian, ramose *Favosites* and the Pachyporinae outnumbering the hemispherical coralla. Many of the Favositinae developed squamulae in the late Early and Middle Devonian. By Late Devonian time, the hemispherical Favositinae were practically absent, and the Alveolitinae and Pachyporinae were the dominant types, the former becoming extinct before the end of the period. The Micheliniinae and Palaeacinae entered in the Early Devonian with *Pleurodictyum* and *Cleistopora,* respectively, but except in the Early Devonian, when *Pleurodictyum* was common, these 2 subfamilies were unimportant.

CARBONIFEROUS AND PERMIAN

In Early Carboniferous time, only the Auloporidae, Chaetetidae, and Favositidae remained, in that order of importance. Few genera and species were involved, but the number of individuals of *Syringopora* and *Chaetetes* was sometimes great. The Favositinae were poorly represented by small squamulate forms and the Micheliniinae were relatively important in numbers of individuals; the Palaeacinae showed considerable variability and a number of these perforate genera have been distinguished; the Pachyporinae were very rare.

The Moscovian, Artinskian, and later Paleozoic faunas are similar to those of the Lower Carboniferous, though there was a slight increase in the evolutional activity of the Auloporidae and Pachyporinae and a decrease in that of the Chaetetidae, the Micheliniinae, and the Palaeacinae.

POST-PALEOZOIC

Most post-Paleozoic species referred to the Tabulata have been removed subsequently; of a number of Mesozoic and Eocene species now recorded under the Chaetetidae, some may be coralline algae, while Triassic species referred to the Favositidae deserve further research.

ECOLOGY

The Tabulata frequently acted as framework organisms in the reefs of the Paleozoic, and in this role were sometimes more important than either the Rugosa or the Stromatoporoidea. But they also occurred in other than reef habitats; thus *Pleurodictyum* is particularly characteristic of the Siegen facies of the Lower Devonian. In general, it may be said that the largest coralla characterize the reef facies and that in deeper- and colder-water facies one finds small or slenderly branched coralla. Like the Rugosa, they are usually found in beds with a considerable lime content and are rare or absent in noncalcareous sandstones and mudstones and in greywackes.

ONTOGENY

Since members of the order are invariably compound, ontogenetic studies have been practically confined to elucidating the position of development of new corallites, and because *Pleurodictyum* has perhaps the largest corallites known in the Tabulata, it has been most studied. The order of insertion of septa in Tabulata is a neglected field of research, just as neglected as in the young corallites of compound Rugosa; one would like to know whether the apparently single order of septa was inserted pinnately in 4 positions, as in the Rugosa, or not, with or without masking by skipping of stages in offsets.

EVOLUTIONARY TRENDS

(1) The production of a marginarium was perhaps the most important of the evolutionary trends in Tabulata, as in Rugosa. There are 2 classes:

(A) In cerioid coralla, where the septa of neighboring corallites are divided by a plane of divergence of their trabeculae, the only marginarium found is the septal stereozone (e.g., *Striatopora*); a dissepimentarium is unknown. Since minor septa are not developed, the width of the septal stereozone does not depend on the length of the minor septa, as in Rugosa.

(B) In coralla where the dividing walls between corallites disappear, the marginarium is a common one, the coenenchyme; different types of coenenchyme develop:

(i) A compact coenenchyme where the common tissue consists (a) of vertical, very short, tightly packed trabeculae only (Coccoseridinae), or (b) of lamellar sclerenchyme in which the original fibrosity is lost in emphasis of the growth lamellation (*Thamnopora lamellicornis,* Fig. 351,*3c-e*).

(ii) A proporoid coenenchyme, where isolated trabeculae, stout or thin, are based on slightly domed dissepiments.

(iii) A heliolitoid coenenchyme in which the trabeculae are organized into vertical walls forming tubules which are crossed by small transverse plates or sola.

(iv) A halysitoid coenenchyme; very narrow vertical tubular spaces crossed by small transverse plates and without vertical skeletal elements may develop between neighboring corallites of cateniform coralla.

(v) A calapoecioid coenenchyme formed by extensions from the tabulae and septa which intersect to enclose horizontal tubular spaces.

(vi) A palaeacoid coenenchyme, i.e., a trabeculate but irregularly tunneled and perforate coenenchyme, as in *Palaeacis*.

(2) The development of intercorallite communication, apart from the coenenchyme, is a trend distinctive of the Tabulata; one finds:

(A) Mural pores (Favositidae, Billingsariinae).

(B) Connecting tubes between the cylindrical corallites of fasciculate Syringoporinae, which in the cerioid *Roemeria* become mural pores.

(C) Irregular perforation of the walls and septa, common in Scleractinia and calostylid Rugosa, and present in the Palaeacinae and Palaeoporitinae.

(3) The development of corallites of reclined, not vertical habit is seen in the Alveolitinae; the upper wall of the reclined corallites becomes vaulted.

Several other trends occur, but less commonly, and some of these which are minor in the Tabulata are common in the Rugosa. Thus, the common Rugosa trend to replacement of tabulae by tabellae is seen in Tabulata only in the Micheliniinae, one of the later subfamilies of Favositidae. The trend to complication of the arrangement of trabeculae within the septa may occur exceptionally in the Tabulata, as in some early Coccoseridinae where the individual trabeculae of each septum diverge from the medial plane; the development of squamulae in Favositinae in the Devonian (*Emmonsia*) and Lower Carboniferous may be due to a specialization of this trend. Laminar septa, where the axes of the trabeculae are so closely spaced in the septum that one cannot distinguish the fibers of the different trabeculae, are rare in Tabulata, but are quite common in Rugosa. The production of an axial structure is also very rare in Tabulata, though common in late Rugosa and in Scleractinia; its early appearance with vertical axial trabeculae in *Billingsaria* and a few Heliolitidae was abortive. The appearance of alternation in size of septa in the early syringophyllid *Nyctopora* seems not to have been due to the trend to develop more than one order of septa which distinguishes both Rugosa and Scleractinia; nor is there any clear trend to the predominance of 6 protosepta in Tabulata, characteristic in Scleractinia and appearing in late Rugosa, though a 12-equiseptate condition is found in many.

SYSTEMATIC DESCRIPTIONS

Order TABULATA Milne-Edwards & Haime, 1850

[as Zoantharia Tabulata (suborder) M.Edw.-H., 1850] [=Aseptata Grabau, 1913; Trichokorallen, Chätokorallen Weissermel, 1927; Chaetokorallen Weissermel, 1937; Trichocorallia Weissermel, 1939] [Includes Tubulosa (section) M. Edw.-H., 1851; Schizocoralla (subclass), Heliolitina (order) Okulitch, 1936; Heliolitida (section) Jones & Hill, 1940; Multisolenida (order) Fritz, 1950]

Corallum compound, with very slender corallites; septa short, equal, in many genera 12 in number, each typically a vertical series of spines; walls with pores in many; tabulae complete or funnel-shaped in some; extratabularial tissue a coenenchyme in many. *Ord.-Perm., ?Trias.-?Eoc.*

Family CHAETETIDAE Milne-Edwards & Haime, 1850

[*nom. transl.* Nicholson, 1877 (*ex* Chaetetinae M.Edw.-H., 1850)] [=Chaetetiniens de Fromentel, 1861 (invalid vernacular name); Chaetetiden Roemer, 1883; Chaetidae Počta, 1902; Tetradina, Chaetetina (orders) Okulitch, 1936]

Massive coralla composed of extremely slender aseptate corallites with imperforate walls and complete tabulae. *M.Ord.-Perm., ?Trias.-?Eoc.*

Subfamily LICHENARIINAE Okulitch, 1936

[*nom. transl. et correct.* Hill, 1951 (*ex* Lichenaridae Okulitch, 1936)]

Increase unequal, offsets arising near the margins of the corallites. *M.Ord.*

Lichenaria Winchell & Schuchert, 1895 [**L. typa*] [=*Lamottia* Raymond, 1924]. Corallites erect, prismatic (18a). *M.Ord.,* N.Am.-Scot.——Fig. 344,*1*. **L. typa*, N.Am.; *1a*, surface, ×1; *1b*, calices on weathered surface, ×6; *1c,d*, transv. and long. secs., ×4 (45).

Subfamily TETRADIINAE Nicholson, 1879

[*nom. transl.* Hill, 1951 (*ex* Tetradiidae Nich., 1879)] [=Paleoalveolitidae, Tetradidae Okulitch, 1935; Tetradina (order) Ok., 1936]

Increase quadripartite, incomplete in many. *M.Ord.-U.Ord.*

Tetradium Dana, 1848 [*non* Schmidt, 1874] [**T. fibratum* Safford, 1856; SD Safford, 1856] [=*Prismostylus* Okulitch, 1935; *Prismatostylus* Lang-S.-T., 1940 *(nom. van.)*]. Corallites erect, cylindrical or prismatic (18a). *M.Ord.-U.Ord.,* N. Am.-Eu.-N.S.W.-Tasm.——Fig. 344,*2*. **T. fibratum, M.Ord.,* N.Am.; *2a,b,d*, transv. secs., ×2, ×4, ×6; *2c*, long. sec., ×4 (45). Other species, Fig. 341,*5*.

Paleoalveolites Okulitch, 1935 [**Tetradium carterense* Bassler, 1932] [=*Kentlandia* Shrock, 1937; *Palaeoalveolites* Lang-S.-T., 1940 *(nom. van.)*]. Ramose; axial corallites prismatic and erect, peripheral corallites inclined, with upper walls vaulted (18a). *M.Ord.,* N.Am.——Fig. 344,*3*. **P. carterensis* (Bassler); *3a,b*, transv. secs., ×4, ×8; *3c*, long. sec., ×4 (45).

Subfamily CHAETETINAE Milne-Edwards & Haime, 1850

Increase bipartite, may be incomplete giving meandroid coralla. *Ord.-Perm., ?Trias.-?Eoc.*

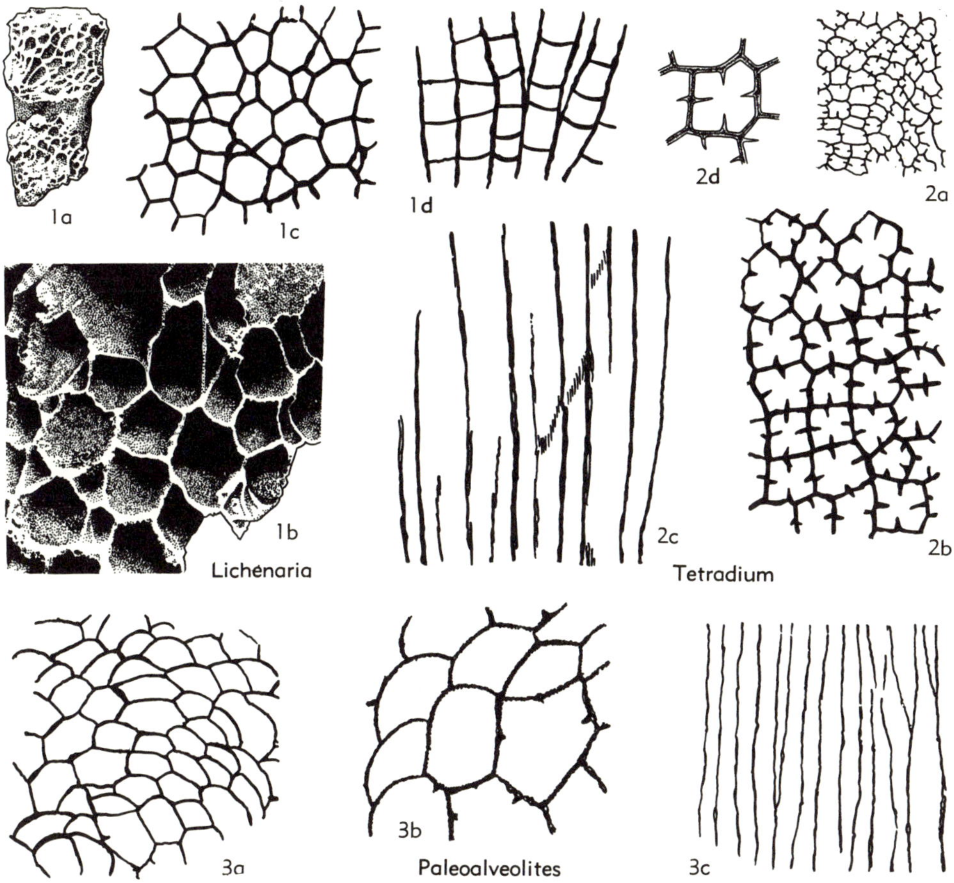

FIG. 344. Chaetetidae (p. F454).

Chaetetes FISCHER in EICHWALD, 1829 [**C. cylindraceus;* SD OAKLEY, 1936] [=*?Chaetetopsis* NEUMAYR, 1890; *Chaetetipora* STRUVE, 1898; *?Carnegiea* GIRTY, 1913 (*pro Carnegia* GIRTY, 1907, *non* HOLLAND, 1896); *Chaetetides* STRAND, 1928; *Boswellia, ?Chaetetella, ?Chaetetiporella, ?Moscovia, ?Spongiothecopora* SOKOLOV, 1939; *?Septachaetetes* RIOS & ALMELA, 1944; *?Fistulimurina* SOKOLOV, 1947]. Massive or incrusting (18a). *Ord.-Perm.,* cosmop.; *?Trias.-?Eoc.,* Eu.——FIG. 340,*5;* 345,*1. C. ?septosus* FLEMING, L.Carb.(Visé.), Eu.; 340,*5,* transv. sec., ×6 (76); 345,*1a,b,* transv. and long. secs., ×4 (58n).

Dania M.EDW.-H., 1849 [**D. huronica*]. Like *Chaetetes* but with tabulae on same level in contiguous corallites and unequally spaced (28). *Sil.,* N.Am.——FIG. 345,*2. *D. huronica, 2a,* weathered top surface, showing form of corallites in transv. sec., ×4; *2b,* weathered side of corallum, showing accordant tabulae, ×1 (68).

Hattonia JONES, 1927 [**H. etheridgei*]. Like *Chaetetes* but with 1 to 3 tabulae occurring in groups which are regularly spaced and on the same level in contiguous corallites. *M.Sil.,* N.S.W.-USSR.——FIG. 345,*3. *H. etheridgei,* N.S.W.; *3a,b,* transv. and long. secs., ×4 (60).

Litophyllum ETHERIDGE, 1899 [**Amplexopora konincki* ETHERIDGE & FOORD, 1884]. Like *Chaetetes* but with walls a little thickened so that tabularium is rounded rather than polygonal. *Up.M.Dev.,* E.Austral.——FIG. 345,*4. *L. konincki* (ETH.-F.), Queensl.; sec., ×4 (58n).

Pachytheca SCHLÜTER, 1885 [*non* HOOKER, 1861; *nec* CANU, 1913] [**P. stellimicans*]. Like *Litophyllum* but with walls so thickened as almost to fill the corallite (33). *M.Dev.,* Ger.——FIG 345,*5. *P. stellimicans,* Eifel; *5a,b,* transv. and long. secs., ×9 (74).

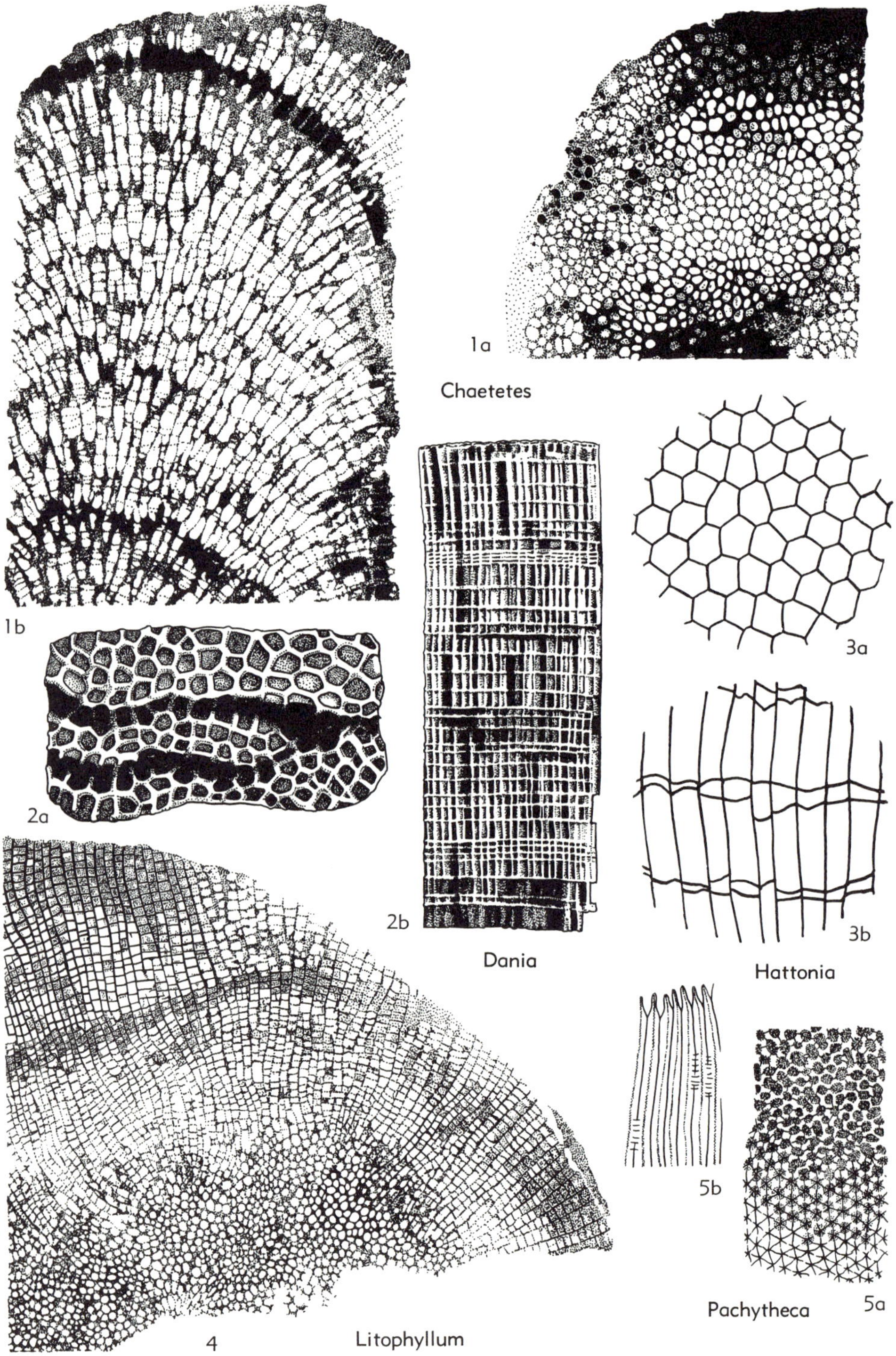

Fig. 345. Chaetetidae (p. F455).

Family SYRINGOPHYLLIDAE Počta, 1902

[*ex Syringophyllum* MILNE-EDWARDS & HAIME, 1850 (=*Sarcinula* LAMARCK, 1816, subj.)] [=Syringophylliden ROEMER, 1883 (invalid vernacular name); Lyoporidae KIÄR, 1930; Billingsaridae OKULITCH, 1936; Calapoeciidae HILL, 1951; Columnoporidae LECOMPTE, 1952]

Massive; septa short, thick, spinose, and typically equal in number in any species; mural pores interseptal and arranged in horizontal rows, each just above a tabula; coenenchyme formed by extensions from the tabulae and septa which intersect to enclose horizontal tabular spaces (18a). *M.Ord.-L.Sil.*

Subfamily BILLINGSARIINAE Okulitch, 1936

[*nom. transl. et correct.* HILL, 1955 (*ex* Billingsaridae OKULITCH, 1936)] [=Nyctoporinae HILL, 1951]

Coenenchyme absent, septa 16, may alternate in size (18a). *M.Ord.-U.Ord.*

Billingsaria OKULITCH, 1936 [**Columnaria parva* BILLINGS, 1859]. Septal trabeculae dilated wedgewise to form a thick wall; vertical trabeculae may

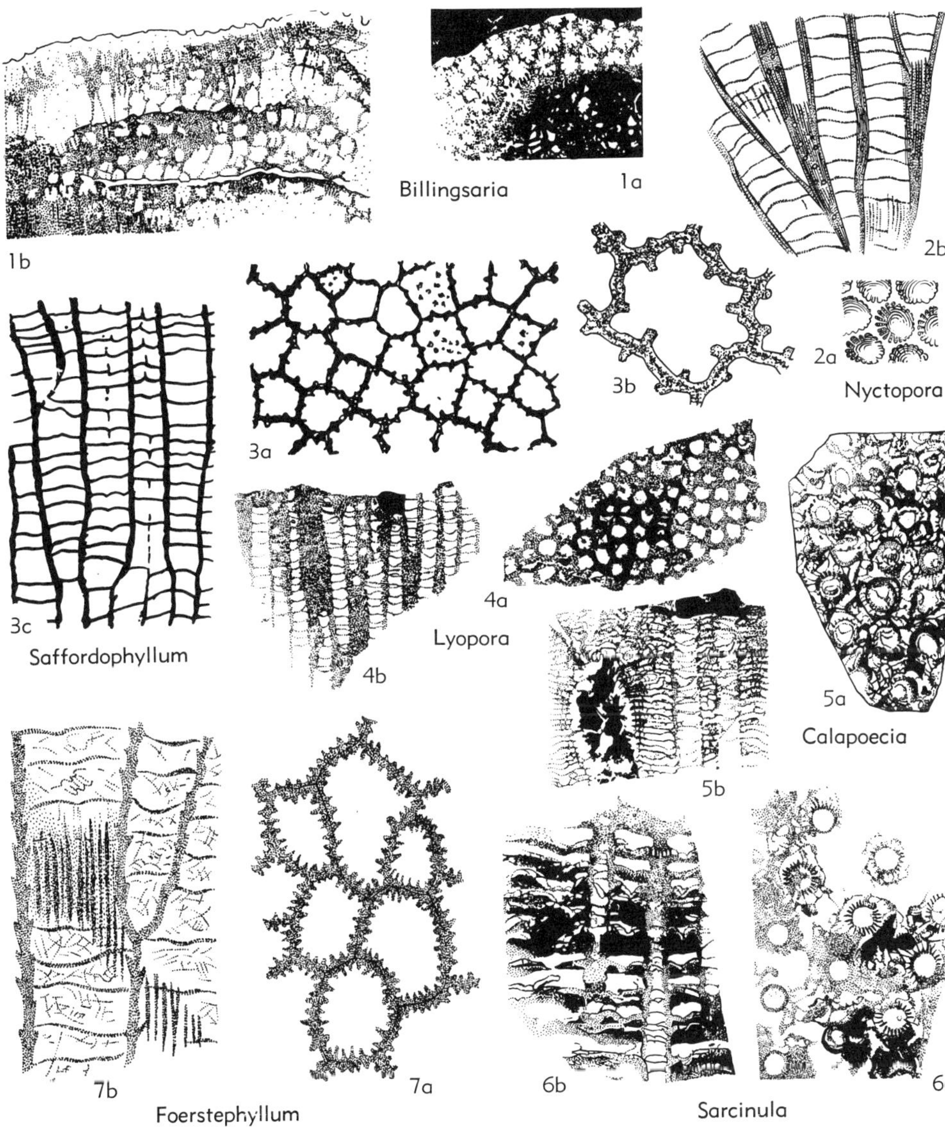

FIG. 346. Syringophyllidae (p. F457-F458).

develop at axis (18a). *M.Ord.*, N.Am.-Austral.——FIG. 346,*1*. **B. parva* (BILL.), N.Am.; *1a,b*, transv. and long. secs., X4 (72).

Nyctopora NICHOLSON, 1879 [**N. billingsi* (*pro Columnaria goldfussi* NICH., 1875, *non* BILL., 1858)]. Walls and septa thin (18a). *M.Ord.-U. Ord.*, N.Am.-Eu.-Austral.; *M.Ord.*, Spiti.——FIG. 346,*2*. **N. billingsi*, M.Ord., N.Am.; *2a*, top surface showing calices, X5; *2b*, long. sec., X5 (70).

Saffordophyllum BASSLER, 1950 [**S. deckeri* (*pro Lichenaria carterensis* DECKER, 1931, *non Columnaria carterensis* SAFFORD, 1869)]. Septa very short, equal, 8, each proceeding from an inward undulation of the wall; tabulae close (4). *M.Ord.*, N.Am.——FIG. 346,*3*. **S. deckeri*; *3a,b*, transv. secs., X4, X10; *3c*, long. sec., X4 (45).

?**Foerstephyllum** BASSLER, 1950 [**Columnaria? halli* NICHOLSON, 1879]. Septa numerous, short, thick, of one order only (4). *M.Ord.-U.Ord.*, N.Am.——FIG. 346,*7*. *F. halli* (NICH.), M.Ord., N.Am.; *7a,b*, transv. and long. secs., X4 (70).

Subfamily SYRINGOPHYLLINAE Počta, 1902

[*nom. transl.* HILL, 1955 (*ex* Syringophyllidae POČTA, 1902)] [=Lyoporidae KIÄR, 1930; Calapoeciinae HILL, 1951; Columnoporidae LECOMPTE, 1952]

Cerioid or with coenenchyme; septa 20 to 24, equal (18a). *M.Ord.-L.Sil.*

Lyopora NICHOLSON & ETHERIDGE, 1878 [**Palaeopora? favosa* M'COY, 1850] [=*Liopora* LANG-S.-T., 1940 (*nom. van.*) (*non* GIRTY, 1915)]. Cerioid, without coenenchyme (18a). *M.Ord.*, Eu.-N.Am. FIG. 346,*4*. **L. favosa* (M'COY), Eu.; *4a,b*, transv. and long. secs., X1 (47).

Calapoecia BILLINGS, 1865 [**C. anticostiensis*; SD LINDSTRÖM, 1883] [=*Columnopora* NICHOLSON, 1874; *Houghtonia* ROMINGER, 1876]. Coenenchyme more or less well developed (18a). *M.Ord.-L.Sil.*, N.Am.-Scand.——FIG. 346,*5*. **C. canadensis* BILL. *anticostiensis*, U.Ord., N.Am.; *5a,b*, transv. and long. secs., X1 (47).

Sarcinula LAMARCK, 1816 [**Madrepora organum* LINNÉ, 1758 (*partim*); SD DANA, 1846] [=*Syringophyllum* M.EDW.-H., 1850 (*non* ULR. in MILLER, 1889, *nec* GRABAU & YOH, 1929)]. Corallum alternatingly fasciculate and with wide coenenchymal platforms (18a). *M.Ord.-L.Sil.*, N. Eu.——FIG. 346,*6*. **S. organum* (LINNÉ), M.Ord.-L.Sil., N.Eu.; *6a,b*, transv. and long. secs., X1 (47).

Family HELIOLITIDAE Lindström, 1876

[*pro* Palaeoporidae M'COY, 1851, *ex Palaeopora* M'COY, 1849 (=*Heliolites* DANA, 1846, obj.); *pro* Heliolithidae LIND., 1873, *ex Heliolithes* LIND., 1873 (=*Heliolites* DANA, 1846, obj.)] [=Palaeoporidae M'COY, 1851; Heliolithidae LIND., 1873; Heliolitiens DOLLFUSS, 1875 (invalid vernacular name); Heliolithinae WENTZEL, 1895; Heliolithoidea (order) FRECH, 1897; Heliolitina (order) OKULITCH, 1936; Heliolitacea (suborder) WEDEKIND, 1937; Heliolitida (section) JONES & HILL, 1940)]

Massive coralla with slender tabularia separated by coenenchyme; each tabularium with 12 equal spinose septa and with complete tabulae (18a). *M.Ord.-U.Dev.*

Subfamily COCCOSERIDINAE Kiär, 1899

[*nom. correct.* HILL & STUMM, herein (*pro* Coccoserinae KIÄR, 1899)] [=Pycnolithinae, Coccoseridae LINDSTRÖM, 1899; Protaraeinae KIÄR, 1904; Protaraeidae LECOMPTE, 1952]

Incrusting, discoid, nodular or branching; coenenchyme tubular; trabeculae greatly thickened; septal trabeculae curving upwards and inwards to become vertical in axial parts of tabularia; coenenchymal trabeculae vertical, spaced so as to form vertical tubules (18a). *M.Ord.-L.Sil.*

Protaraea M.EDW.-H., 1851 [**Porites? vetustus* HALL, 1847; SD MILLER, 1889] [=*Diplastraea* EICHW., 1854; *Tumularia* ROBINSON, 1916 (*pro Stylaraea* SEEBACH, 1866, *non* M.EDW.-H., 1851)]. Trabeculae very stout, coenenchyme narrow or absent (18a). *M.Ord.-L.Sil.*, N.Am.-Eu.-Tasm.——FIG. 347,*1*. **P. vetusta* (HALL) M.Ord., N.Am.; *1a*, side view of corallum, X1; *1b*, calices, X10 (68).

Coccoseris EICHWALD, 1855 [**C. ungerni* (=*Lophoseris ungerni* EICHW., 1855); SD LANG-S.-T., 1940). Coenenchyme wide, trabeculae so stout that all spaces are filled (18a). *U.Ord.-L.Sil.*, Eu.——FIG. 347,*2*. **C. ungerni*, U.Ord., Est.; *2a*, part of surface, X1; *2b,c*, transv. and long. secs., X10 (65).

Acidolites LANG-S.-T., 1940 [*pro Acantholithus* LINDSTRÖM, 1899 (*non* STIMPSON, 1858)] [**Acantholithus lateseptatus* LIND., 1899]. Coenenchyme wide, trabeculae relatively slender, leaving spaces crossed by dissepiments (18a). *U.Ord.-L.Sil.*, Eu.-Tasm.——FIG. 347,*3*. **A. lateseptatus* (LIND.), ?L.Sil., Gotl.; *3a*, part of surface, X1; *3b-d*, transv. and long. secs., X8 (65).

Pycnolithus LINDSTRÖM, 1899 [**P. bifidus*]. Discoid; coenenchymal trabeculae so thick as to fill all spaces; septa very short, leaving open tabularia crossed by tabulae (26). *L.Sil.*, Gotl.——FIG. 347,*4*. **P. bifidus*, L.Sil., Gotl.; *4a,c*, transv. and long. secs., X4; *4b*, part of surface showing calices, X4 (65).

Urceopora EICHWALD, 1855, **Stylidium** EICH., 1855 (*non* DALL in BARTSCH, 1907) are possibly coccoseridines but thin sections are required.

Subfamily PALAEOPORITINAE Kiär, 1899

Like Coccoseridinae, but septa and walls of tabularia and coenenchymal tubules perforate (22). *M.Ord.-U.Ord.*

Trochiscolithus KIÄR, 1904 [**T. micraster*]. Branching, axis of branch with slender trabeculae forming partially perforate septa and walls; outer parts of branch with all spaces filled by dilatation of the trabeculae (18a). *M.Ord.-U.Ord.*, Eu.——FIG. 347, *5*. **T. micraster*, Ord., Norway; *5a*, transv. sec., X4; *5b*, part of surface, X10 (62).

Protrochiscolithus Troedsson, 1928 [**P. kiaeri*]. Coenenchyme narrow; trabeculae slender, forming partially perforate septa and walls, and leaving tabularial and coenenchymal spaces (18a). *U.Ord.*, Arct.N.Am.——Fig. 347,7. **P. kiaeri; 7a,* part of surface, ×1; *7b,c,* transv. and long secs., ×10 (79).

Palaeoporites Kiär, 1899 [**P. estonicus*]. Nodular;

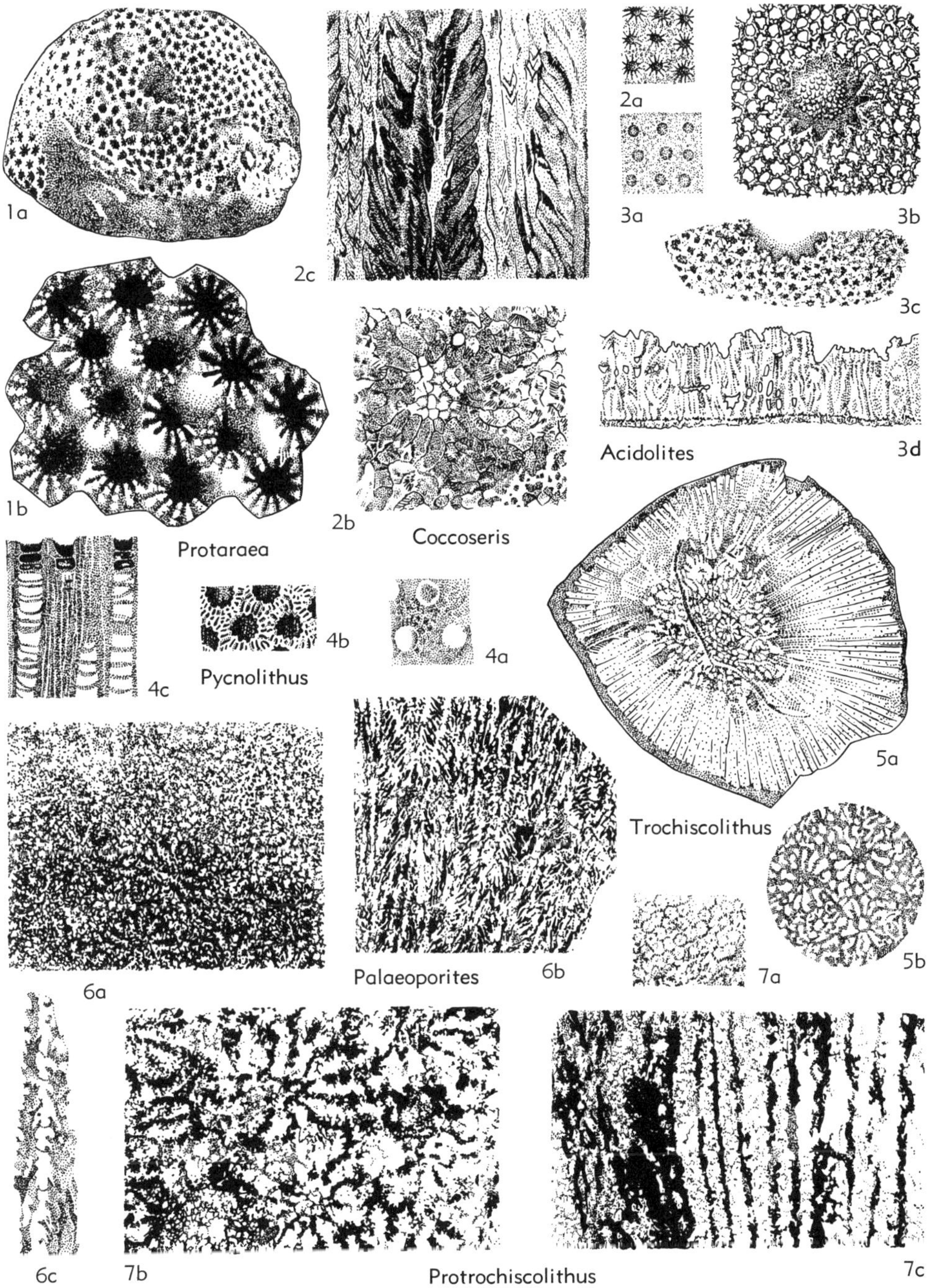

Fig. 347. Heliolitidae (p. F458-F460).

coenenchyme wide; trabeculae moderately thick, leaving spaces in tabularia and coenenchyme, and forming perforate walls and septa; septal trabeculae curving outwards from median plane of septum (22). *L.Sil.* (F_2), Balt.——FIG. 347,*6*. **P. estonicus; 6a,b,* transv. and long. secs., ×4; *6c,* detail of oblique sec. showing septal trabeculae, ×10 (62).

Subfamily PLASMOPORINAE Wentzel, 1895

[*nom. transl.* KIÄR, 1899 (*ex* Plasmoporidae WENTZEL, 1895)] [=Plasmoporinae LINDSTRÖM, 1899 (tribe); Proplasmoporinae KIÄR, 1904 (invalid); Proporinae HILL, 1951]

Coenenchyme of domed plates and isolated trabeculae or with trabeculae uniting to form discontinuously walled tubuli. *M.Ord.-U.Sil.*

Propora M.EDW.-H., 1849 [**Porites tubulatus* LONSDALE, 1839 *(partim)*] [=*Lyellia* M.EDW.-H., 1851; *Pinacopora* NICHOLSON & ETHERIDGE, 1878; ?*Nicholsonia* KIÄR, 1899 (*non* DAVIS, 1885; *nec* SCHLÜTER, 1885; *nec* POČTA, 1902; *nec* others); *Cavella* STECHOW, 1922 (*pro Calvinia* SAVAGE, 1913, *non* NUTTING, 1900); *Koreanopora* OZAKI, 1934]. Coenenchyme of domed plates and isolated trabeculae not more thickened in surface zones than in inner or axial zones of the corallum (18a). *M.Ord.-U.Sil.,* cosmop.——FIG. 348,*1*. **P. tubulata* (LONSDALE), Sil., Eu.; *1a,* part of surface, ×4; *1b,c,* transv. and long. secs., ×4 (65).

Plasmoporella KIÄR, 1897 [**P. convexotabulata* forma *typica* KIÄR, 1899] [=*Camptolithus* LINDSTRÖM, 1899]. Like *Propora* but with tabulae domed and complete or incomplete; trabeculae slender (18a). *U.Ord.-M.Sil.,* Eu.-N.Am.-Tasm.-Asia.——FIG. 348,*2*. **P. convexotabulata,* U.Ord., Norway; *2a,b,* transv. and long. secs., ×4 (62).

Diploëpora QUENST., 1879 [**Heliolites grayi* M. EDW.-H., 1851]. Branching; axially like *Propora* but with surface zones of corallum with great trabecular thickening (22). *M.Sil.,* Eu.-E.Austral. ——FIG. 348,*3*. **D. grayi* (M.EDW.-H.), Eu.; long. sec., ×4 (62).

Plasmopora M.EDW.-H., 1849 [**Porites petalliformis* LONSDALE, 1839]. Coenenchyme tubular, with an aureole around each tabularium of 12 tubules whose radial walls are continuous with the septa (22). *L.Sil.-M.Dev.,* Eu.-N.Am.-E.Austral.-Asia. ——FIG. 340,*6*; 348,*5*. **P. petalliformis* (LONSD.), M.Sil., Eu.; 340,*6,* surface of corallum, ×1 (68); 348,*5a,* part of surface, ×1; 348,*5b,c,* transv. and long. secs., ×4 (65).

?**Cyrtophyllum** LINDSTRÖM, 1882 [**C. densum*]. Coenenchyme tubular, all tubules with discontinuous walls, more than 12 in aureole. *L.Sil.,* Sib. ——FIG. 348,*4*. **C. densum; 4a,b,* transv. and long. secs., ×1.5 (65).

Subfamily PROHELIOLITINAE Kiär, 1899

Tabularia in contact and polygonal or separated in part of their circumference by 1, 2, or 3 small coenenchymal tubuli, each of which may widen to form a normal tabularium; with 12 septa, each composed of a single series of downwardly directed separate spines (18a). *U.Ord.-L.Sil.*

Proheliolites KIÄR, 1897 [*Heliolites dubius* SCHMIDT, 1858].——FIG. 348,*6*. **P. dubius* (SCHMIDT), Eu., *6a,b,* transv. and long. secs., ×4 (62).

Subfamily HELIOLITINAE Lindström, 1876

[*nom. transl.* KIÄR, 1899 (*ex* Heliolitidae LIND., 1876)] [=Palaeoporidae M'COY, 1851; Heliolithinae WENTZEL, 1895]

Coenenchyme of tubuli with complete thin walls, with more than 12 tubuli bounding each tabularium. *L.Sil.-U.Dev.*

Heliolites DANA, 1846 [**Astraea porosa* GOLDF., 1826] [=*Palaeopora* M'COY, 1849 (obj.); ?*Lonsdalia* ORB., 1849 (*non Lonsdaleia* M'COY, 1849); *Geoporites* ORB., 1850 (obj.); *Heliolithes* LINDSTRÖM, 1873 (obj.); *Pachycanalicula* WENTZEL, 1895; *Helioplasma* KETTNEROVA, 1933]. Tabularia polygonal or circular in transverse section, septa spinose (26). *L.Sil.-Up.M.Dev.,* cosmop.——FIG. 348,*7*. **H. porosa* (GOLDF.), M.Dev., cosmop.; *7a,b,* transv. and long. secs., ×5 (65). Other species, FIG. 342,*3*.

Stelliporella WENTZEL, 1895 [**S. lamellata*]. Tabularia stellate in transverse section, septa laminar, interfingering at the axis; walls thin (26). *Sil.,* Eu.——FIG. 348,*8*. **S. lamellata,* Czech.; *8a,b,* transv. and long. secs., ×4 (81).

Cosmiolithus LINDSTRÖM, 1899 [**C. ornatus;* SD LANG.-S.-T., 1940]. Like *Stelliporella* but with thickened walls and septa, and with coenenchymal tubuli of 2 sizes, the smaller around the larger (26). *Sil.,* Gotl.——FIG. 348,*9*. **C. ornatus; 9a,* transv. sec., ×4; *9b,* long. sec., ×8 (65).

Paeckelmannopora WEISSERMEL, 1939 [**P. macrophthalma*]. Tabularial walls thin, in 12 longitudinal waves, without septal spines; coenenchymal tubules small, thin-walled, irregular and unequal. *L.Dev.,* AsiaM.——FIG. 348,*10*. **P. macrophthalma; 10a,b,* transv. and long. secs., ×4 (80).

Family FAVOSITIDAE Dana, 1846

[=Favositacea WEDEKIND, 1937] [Includes Multisolenida FRITZ, 1950]

Massive, typically without coenenchyme; corallites slender, with mural pores; septa short, equal, spinose, variable in number; tabulae complete (18a). *U.Ord.-Perm.,* ?*Trias.*

Subfamily FAVOSITINAE Dana, 1846

[=Multisolenida (order), Multisolenidae (=Multisoleniidae, *nom. correct.,* herein) FRITZ, 1950; Emmonsiinae, Scolioporinae LECOMPTE, 1952]

Corallites prismatic, thin-walled, perforated by mural pores only (18a). *U.Ord.-U. Dev.*

Palaeofavosites TWENHOFEL, 1914 [**Favosites asper* ORB., 1850]. Mural pores predominantly at angles of walls (18a). *U.Ord.-U.Sil.*, Eu.-N.Am.——FIG. 349,*1*. **P. asper* (ORB.), Sil., Eu.; *1a,b*, transv. and long. sec., ×4 (60).

Favosites LAMARCK, 1816 [**F. gothlandicus;* SD M.EDW-H., 1850] [=*Calamopora* GOLDF., 1829; *Astrocerium* HALL, 1851; *?Brignus* GREGORIO, 1930; *?Sapporipora* OZAKI, 1934; *Eufavosites* RUSHKIN, 1936]. Mural pores predominantly near middle of walls (18a). *U.Ord.-M.Dev.*, cosmop.; *?Trias.*, Sicily.——FIG. 340,*7;* 349,*3*. **F. gothlandicus*, Sil.-L.Dev., cosmop.; 340,*7*, corallum,

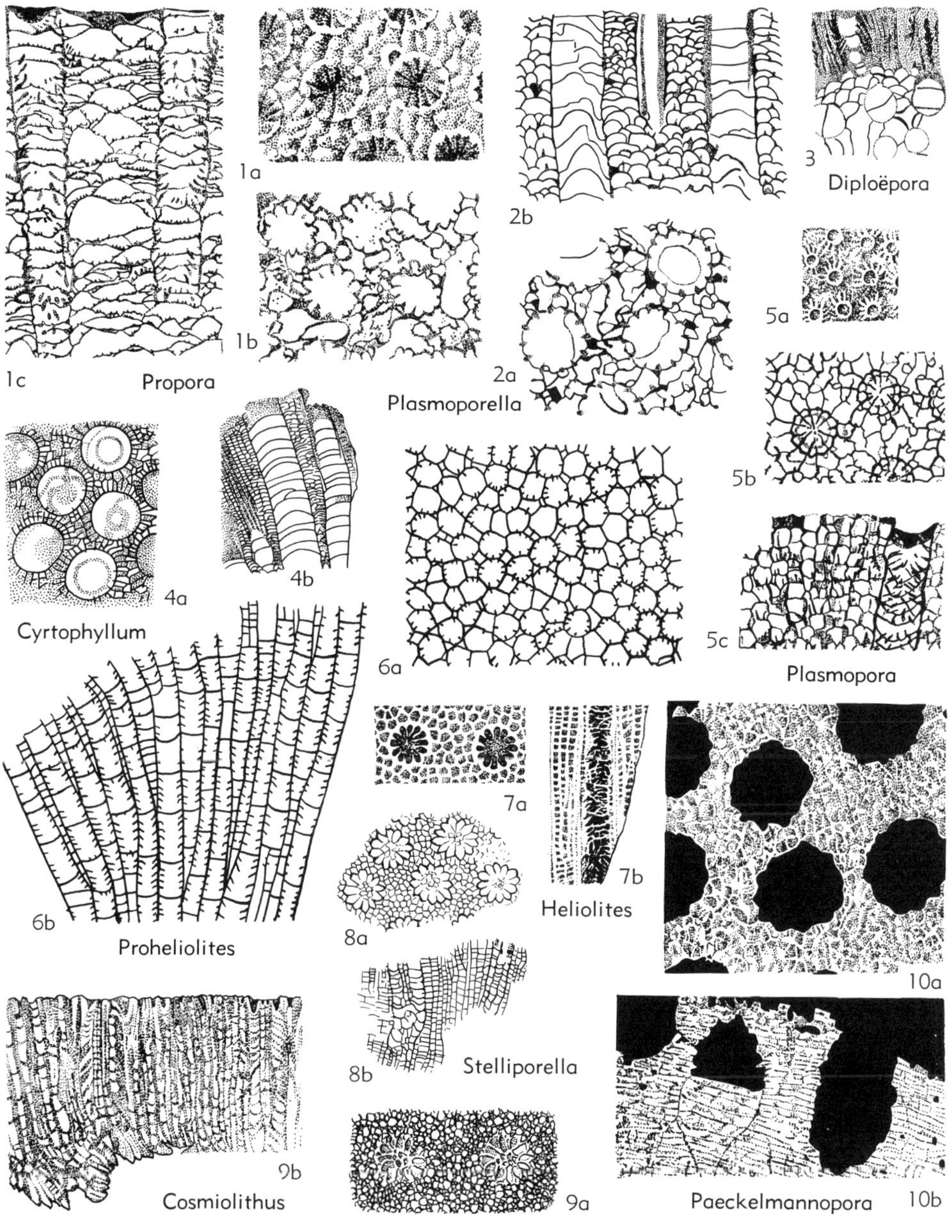

FIG. 348. Heliolitidae (p. F460).

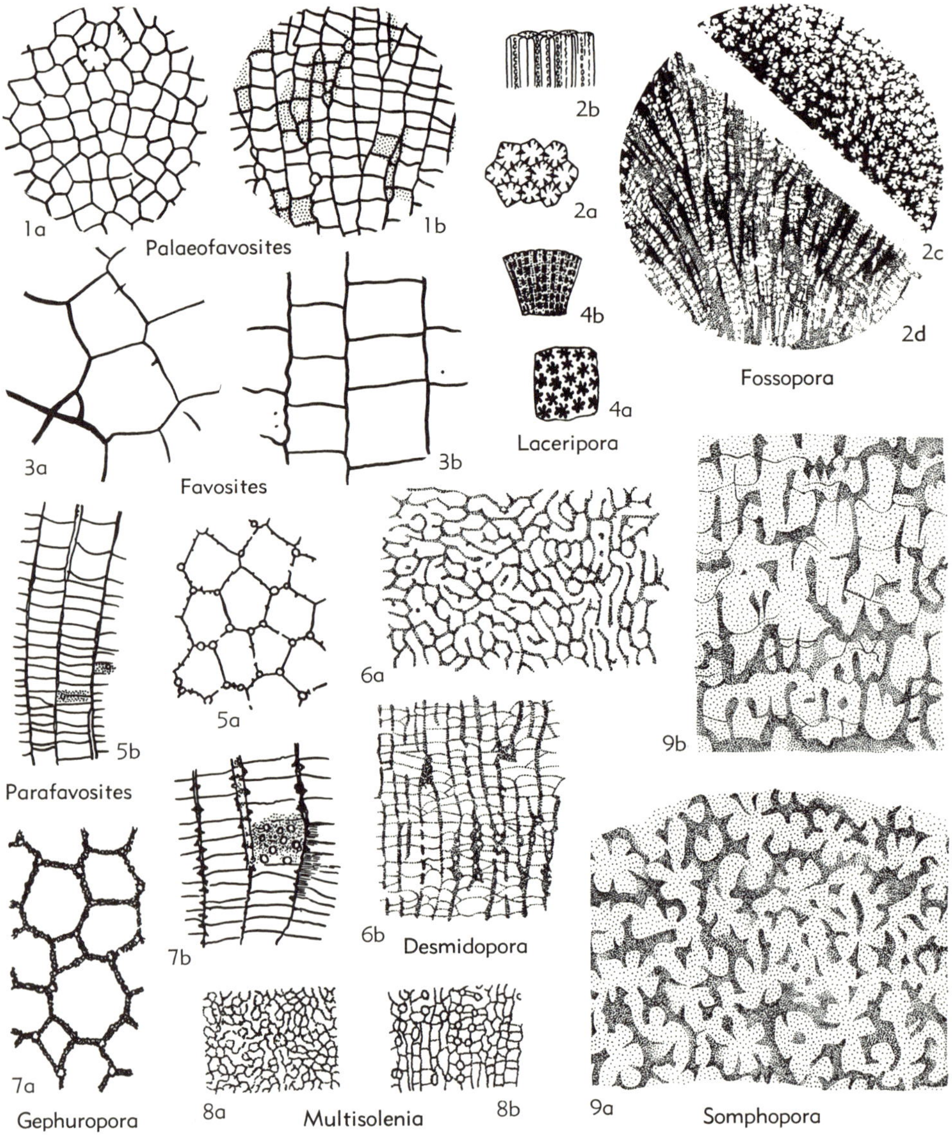

FIG. 349. Favositidae (p. F461-F464).

×1; 349,*3a,b,* transv. and long. secs., ×4 (60). Other species, FIG. 340,*7;* 341,*4.*

Parafavosites ORLOV, 1930 [**P. ferganensis;* SD LANG-S.-T., 1940] [=*Paralleloporella* STRAND, 1934 (*pro Parallelopora* HOLTEDAHL, 1914, *non* BARGATZKY, 1881)]. Like *Favosites* but with vertical aseptate and atabulate tubules expanding the walls either in the angles or faces (3). *Sil.,* Arct. Can.-Asia.——FIG. 349,*5.* **P. ferganensis,* U.Sil., Ferghana; *5a,b,* transv. and long. secs., ×4 (45).

Gephuropora ETHERIDGE, 1920 [**Favosites (?Columnopora) duni* ETH., 1920]. Like *Parafavosites* but tubules may be tabulate (3). *L.M.Dev.,* E. Austral.-Belg.——FIG. 349,*7.* **G. duni* (ETH.); *7a,b,* transv. and long. secs., ×4 (45).

Fossopora ETHERIDGE, 1903 [**F. wellingtonensis*] [=*?Boreaster* LAMBE, 1906; *Fossipora* LANG-S.-T., 1940]. Mural pores oval, one vertical series on each face; septa 6, long, laminar peripherally, spinose axially (8). *Sil.* or *Dev.,* N.S.W.; *Sil.,* Arct.

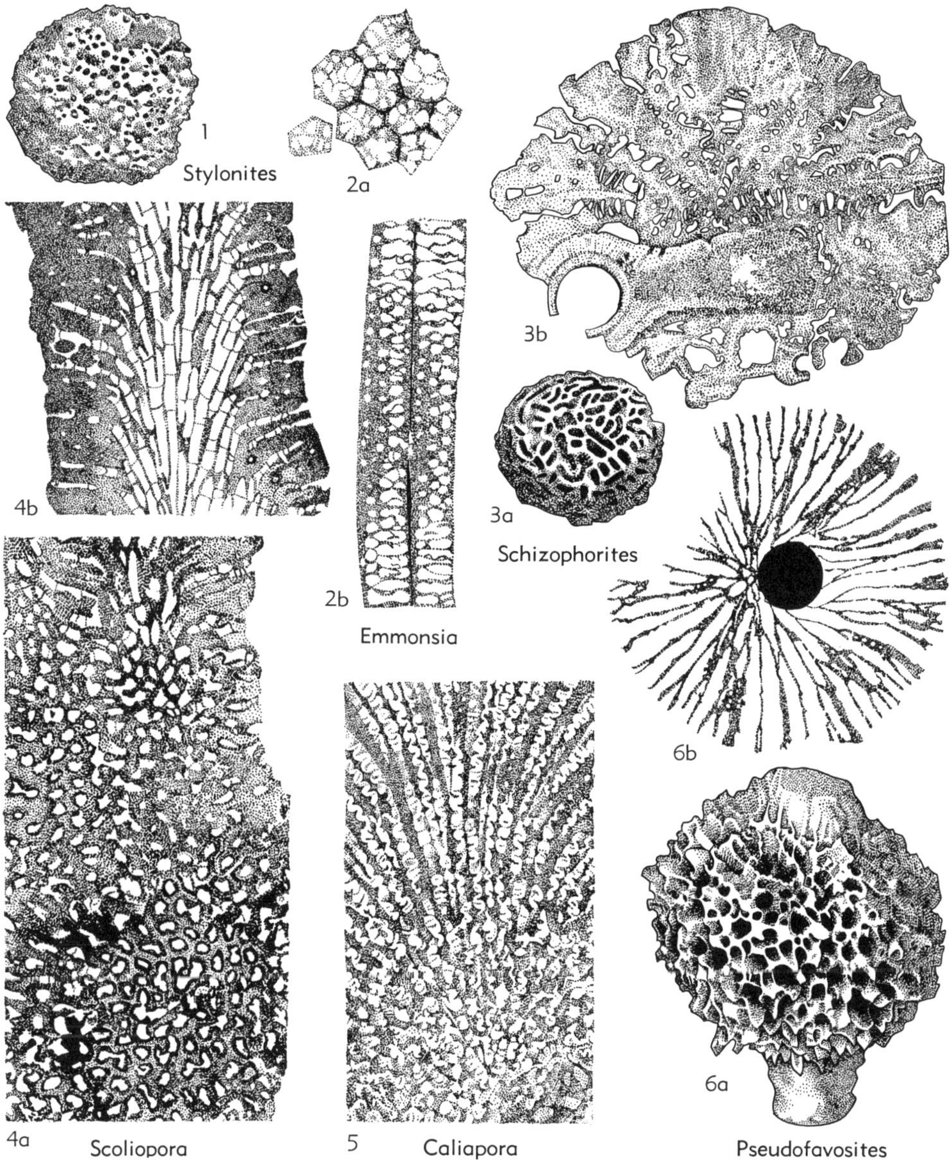

Fig. 350. Favositidae (p. F464).

Can.——Fig. 349,*2c,d*. **F. wellingtonensis,* Sil. or Dev., N.S.W.; *2c,d,* transv. and long. secs., ×4 (49).——Fig. 349,*2a,b*. *F. lowi* (Lambe) (type species of *Boreaster*), Sil., Arct.Can.; *2a,b,* transv. and long. secs., ×4 (45).

Somphopora Lindström, 1883 [**S. daedalea*]. Mural pores numerous, septa 6, spinose (8). *Sil.*, China.——Fig. 349,*9*. **S. daedalea,* Szechuan; *9a,b,* transv. and long. secs., ×4 (65).

Laceripora Eichwald, 1854 [**L. cribrosa*] [=*Laceropora* Lang-S.-T., 1940 *(nom. van.)*]. Mural pores very numerous, septa laminar, not more than 6; tabulae on same level in contiguous corallites throughout corallum (8). *?Sil.,* Est.——Fig. 349,*4*. **L. cribrosa; 4a,b,* transv. and long. secs., ×4 (48).

Desmidopora Nicholson, 1886 [**D. alveolaris*] [=*?Nodulipora* Lindström, 1873]. Meandroid; increase bipartite; mural pores at angles of wall as well as mid-wall (11). *Sil.,* Eng.——Fig. 349,*6*.

D. alveolaris, Wenlock., Eng.; *6a,b,* transv. and long. secs., ×4 (70).

Multisolenia Fritz, 1937 [*M. tortuosa*] [=*Polysolenia* Weissermel, 1939 (*non* Ehr., 1860; *nec* Reuss, 1866)]. Like *Desmidopora,* but with walls waved near the angles, so that rounded projections, circular in vertical section, alternate in neighboring corallites (11). *Sil.,* Can.-USSR.——Fig. 349, *8.* **M. tortuosa,* Lockport., Can.; *8a,b,* transv. and long. secs., ×4 (52).

Scoliopora Lang-S.-T., 1940 [*pro Plagiopora* Gürich, 1896 (*non* MacGill., 1895)]. [**Alveolites denticulatus* M.Edw.-H., 1851]. Branching or lamellar, increase bipartite; calices elongated transversely, opening normal to surface; walls thickened distally, mural pores numerous, septal spines few, tabulae close in peripheral regions (25). *M.Dev.,* Eu.-E.Austral.; *U.Dev.,* Eu.——Fig. 350,*4.* **S. denticulata* (M.Edw.-H.), M.Dev.-U.Dev., Eu.; *4a,b,* ×4 (64).

Emmonsia M.Edw.-H., 1851 [**Favosites emmonsii* Rominger, 1876 (=*F. hemispherica* Yandell & Shumard, M.Edw.-H. *partim*); SD Roemer, 1883]. Massive or tuberose; corallites prismatic, squamulae project toward axis over the mural pores; tabulae thin, dependent from inner ends of squamulae. *L.Dev.-M.Dev.,* N.Am.——Fig. 350,2. **E. emmonsii* (Rom.), Onond., N.Y.; *2a,b,* transv. and long. secs., ×4 (58n).

Caliapora Schlüter, 1889 [**Alveolites battersbyi* M.Edw.-H., 1851]. Like *Emmonsia* but squamulae with concave upper surfaces (25). *M.Dev.,* Eu., Austral.(Vict.).——Fig. 343,*3;* 350,*5.* **C. battersbyi,* Givet., NW.Eu.; 343,*3a,b,* transv. and long. secs., ×6; 350,*5,* long. sec., ×4 (64).

Schizophorites Gerth, 1921 [**S. dubiosus*]. Small, adherent, spherical; corallites with rectangular calices, thick walls, large ragged septal spines; sparse mural pores and numerous thin tabulae (13). *Perm.,* Timor.——Fig. 350,*3.* **S. dubiosus; 3a,* surface, ×1; *3b,* long. sec., ×4 (54).

Pseudofavosites Gerth, 1921 [**P. stylifer*]. Massive or incrusting; corallites as in *Favosites* but with angles of walls produced distally into processes, and apparently without tabulae (13). *Perm.,* Timor.——Fig. 350,*6.* **P. stylifer; 6a,* surface, ×1; *6b,* long. sec., ×1 (54).

Stylonites Gerth, 1921 [*non* Fries, 1848] [**S. porosus*]. Spherical, with distal processes from angles of walls, corallites with a columella and 8 spinose septa (13). *Perm.,* Timor.——Fig. 350,*1.* **S. porosus;* surface, ×1 (54).

?**Plasmadictyon** Wilson, 1926 [**P. irregulare*]. *U. Ord.,* W.Can.

Subfamily THECIINAE Milne-Edwards & Haime, 1850

[*nom. transl. et correct.* Hill & Stumm, herein (*ex* Thecidae M.Edw.-H., 1850)]

Corallites very slender, prismatic; septa typically 12, equal; with or without trabeculate coenenchyme; mural pores may be long tunnels. *Sil.*

Thecia M.Edw.-H., 1849 [**Porites expatiatus* Lonsdale, 1839 (=**Agaricia swinderniana* Goldf., 1829] [=*Angopora* Jones, 1936 (*pro Laminopora* Jones, 1930, *non* Michelin, 1842)]. Septa laminar, thick and in contact peripherally but free and spinose axially; peripheral trabeculae vertical; mural pores few (1). *Sil.,* Eu.-N.Am.——Fig. 351, *1a,b.* **T. swinderniana* (Goldf.), Wenlock., Eng.; *1a,b,* transv. and long. secs., ×4 (58n).——Fig. 351,*1c,d. T. hisingeri* (Jones) (type species of *Angopora*), Sil., Gotl.; *1c,d,* transv. and long. secs., ×4 (60).

Romingerella Amsden, 1949 [**Thecia major* Rominger, 1876]. Like *Thecia* but trabeculae diverging from planes of contact of corallites (1). *Sil.,* N.Am.——Fig. 351,*2.* **R. major* (Rom.), Brownsport, Tenn.; *2a,* surface, ×5; *2b,* transv. sec., ×4 (43).

Subfamily PACHYPORINAE Gerth, 1921

[*nom transl.* Hill & Stumm, herein (*ex* Pachyporidae Gerth, 1921, *ex Pachypora* Lindström, 1873, =*Thamnopora* Steininger, 1831 subj.)] [=?Trachyporinae Waagen & Wentzel, 1886; ?Trachyporidae Sardeson, 1896; Thamnoporinae Hill, 1954]

Tuberose or branching; walls of corallites thickened, with tunnel-like mural pores; calices opening normal to surface; septa spinose, tabulae thin, complete. *Sil.-Perm.*

Thamnopora Steininger, 1831 [*non* Hall, 1883] [**T. madreporacea* (=**Alveolites cervicornis* Blainv., 1830); SD Lang-S.-T., 1940] [=*Pachypora* Lindström, 1873]. Massive, tuberose or ramose; walls thickened, increasing distally; septal spines poorly developed, mural pores numerous, tabulae thin (25). *Sil.-Perm.,* cosmop.——Fig. 351, *3a,b.* **T. cervicornis* (Blainv.), M.Dev. (Givet.), Eu.; *3a,b,* transv. and long. secs., ×4 (64).——Fig. 351,*3c-e. T. lamellicornis* (Lind.) (type species of *Pachypora*), Sil., Gotl.; *3c,* surface, ×1; *3d,e,* transv. and long. secs., ×4 (65).

Striatopora Hall, 1851 [**S. flexuosa*] [=*Cyathopora* Owen, 1844 (*nom. null. pro Cyathophora* Michelin, 1843); *?Thamnoptychia* Hall, 1876]. Ramose; walls thickened distally only; septa 12, with numerous trabeculae. *Sil.,* cosmop.; *Dev.-Perm.,* N.Am.——Fig. 352,*5.* **S. flexuosa*, M.Sil., N.Y.; *5a,b,* transv. and long. secs., ×4 (58n). Other species, Fig. 342,2.

Acaciapora Moore & Jeffords, 1945 [**Michelinia subcylindrica* Mather, 1915]. Like *Thamnopora* but with squamulae. *Penn.,* Tex.-Okla.——Fig. 352,*2.* **A. subcylindrica* (Mather), Tex.; *2a,* surface, ×1; *2b,c,* transv. and long. secs., ×4 (69).

Heterocoenites Gerth, 1921 [**H. variabilis;* SD Lang.-S.-T., 1940]. Like ramose *Thamnopora* but with greatest thickening at axes of the branches (13). *Perm.,* Timor.——Fig. 352,*3.* **H. variabilis; 3a,* surface, ×1; *3b,* long. sec., ×2 (54).

?**Trachypora** M.Edw-H., 1851 [*non* Verrill, 1864] [**T. davidsoni*]. *U.Dev.*, Eu.

?**Taouzia** Termier, 1948 [**T. chouberti*] (36). *M. Dev.*, Morocco.

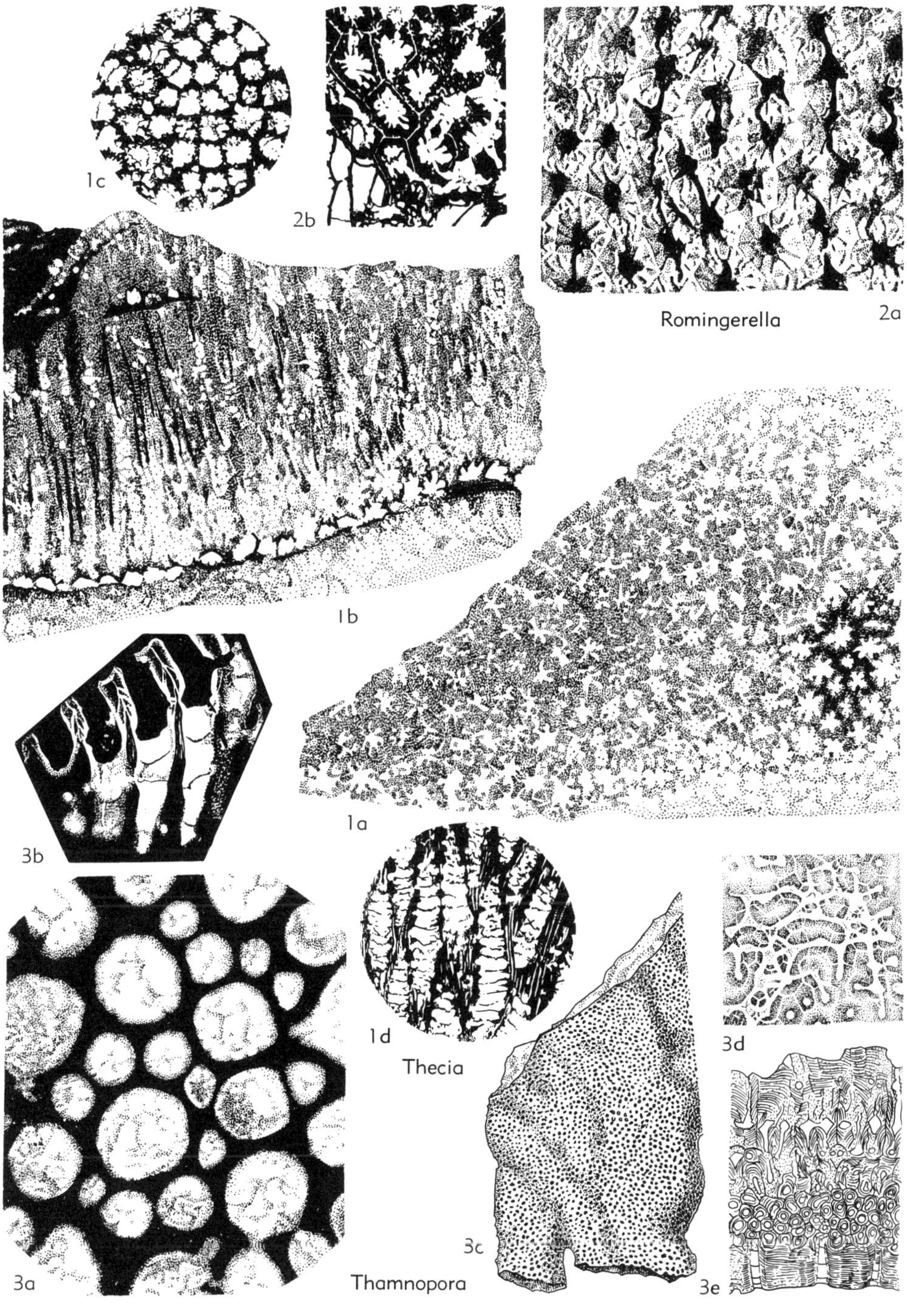

Fig. 351. Favositidae (p. F464).

Subfamily ALVEOLITINAE Duncan, 1872

[*nom. transl.* WAAGEN & WENTZEL, 1886 (*ex* Alveolitidae DUNCAN, 1872)] [=Coenitidae SARDESON, 1896]

Massive, lamellar or branching; corallites inclined, calices opening obliquely, corallites not prismatic but with upper wall vaulted. *Sil.-Dev.*

Alveolites LAMARCK, 1801 [*non* DEFRANCE, 1816] [**A. suborbicularis;* SD NICHOLSON & ETHERIDGE, 1877] [=*Billingsia* DE KONINCK, 1876 (*non* WALCOTT, 1886; *nec* FORD, 1886); ?*Platyaxum* DAVIS, 1887]. Massive, incrusting or ramose; lower or adherent wall parallel to outer (upper) surfaces of earlier corallites; commonly one larger septal spine in the middle of the lower wall; mural pores large, at bases of upper walls (25). *Sil.-Dev.*, cosmop.——FIG. 352,*7*. **A. suborbicularis* M.Dev.-U.Dev., Eu.-Asia-Austral.; *7a,b,* transv. and long. secs., ×4 (64). Other species, FIG. 340,*1*.

Planalveolites LANG & SMITH, 1939 [**Alveolites fougti* M.EDW.-H., 1851]. Like *Alveolites* but thin and flat, the large almost horizontal corallites with very oblique calices, with lower wall produced beyond the upper (1). *Sil.*, Eu.-N.Am.——FIG. 352,*1*. **P. fougti* (M.EDW.-H.), U.Sil., Gotl.; weathered surface, ×1 (68).

Coenites EICHWALD, 1829 [**C. juniperinus;* SD MILLER, 1897 (*species inquirenda*)] [=*Limaria* STEININGER, 1831 (*non* LINK, 1807, *nec* RAFINESQUE, 1815); *Cladopora* HALL, 1851; ?*Dictyostroma* NICHOLSON, 1875 (=*Dictyostoma* SPENCER, 1883); ?*Milleria* DAVIS, 1887 (*non* HARTMAN, 1830, *nec* HERRICH-SCHAEFFER, 1859, *nec* HERING, 1922); ?*Vetofistula* ETHERIDGE, 1917]. Foliar, ramose or massive; corallites short, inclined with crescentic calices; walls thickened except proximally; tabulae few, mural pores rare, septa may be represented by 3 processes in the calice (25). *Sil.-Dev.*, Eu.-E.Austral.-N.Am.——FIG. 352, *4*. *C. escharoides* (STEININGER), M.Dev. (Couvin.), NW.Eu.; *4a,* surface, ×1; *4b,* long. sec., ×4 (64).

Subfamily MICHELINIINAE Waagen & Wentzel, 1886

[=Beaumontidae CHAPMAN, 1893; Michelinidae GERTH, 1921; Pleurodictyidae SARDESON, 1896; Michelininae POČTA, 1902]

Discoid or hemispherical, with large corallites, large mural pores, septal spines or ridges, and with or without tabulae. *Dev.-Perm.*

Pleurodictyum GOLDFUSS, 1829 [**P. problematicum*] [=?*Procteria* DAVIS, 1887 (*non* WERNER, 1924); ?*Dendrozoum* FUCHS, 1915]. Discoid or hemispherical; corallites large, thick-walled, with irregularly distributed tunnel-like mural pores, septal spines, and few or very thin complete tabulae; may grow around a worm tube. *L.Dev.*, cosmop.——FIG. 352,*6a,b*. **P. problematicum,* Ger.; *6a,* mold, ×1 (55); *6b,* calice enlarged (67).——FIG. 352,*6c,d*. *P. michelinoidea* (DAVIS) (type species of *Procteria*); *6c,* surface, ×1; *6d,* long. sec., ×2 (78).

Antholites DAVIS, 1887 [**A. speciosus*]. Like *Pleurodictyum* but incrusting, with pores arranged in regular vertical series, with septal ridges rather than spines, and without tabulae (7). *M.Dev.*, N.Am.——FIG. 353,*1*. **A. speciosus,* Hamilton, Ohio; *1a,* top surface, ×1; *1b,* side, ×2 (78).

Michelinia DE KONINCK, 1841 [*non* DUJARDIN & HUPÉ, 1862] [**Calamopora tenuisepta* PHILLIPS, 1836 (=*Michelinia tenuisepta* KON., 1841); SD M.EDW.-H., 1850] [=?*Conopoterium* WINCHELL, 1865; *Eumichelinia, Protomichelinia, Michelinopora* YABE & HAYASAKA, 1915]. Like *Pleurodictyum* but with numerous, incomplete, convex tabulae; may be thin-walled. *U.Dev.-Perm.*, cosmop.——FIG. 341,*2,3;* 353,*2a,b*. **M. tenuisepta* (PHILLIPS), L.Carb., Eu.; 341,*2,* side of corallum, ×1 (63); 341,*3,* long. sec., ×1 (63); 353,*2a,b,* top and side views, ×0.7 (63).——FIG. 353,*2c,d*. *M. effusa* (WINCHELL) (type species of *Conopoterium*), L.Carb., Mich.; *2c,* surface, ×1; *2d,* long. sec., ×2 (82).

Beaumontia M.EDW.-H., 1851 [*non* EUDES-DESLONGCHAMPS, 1856, *nec* DAVID, 1928] [**Columnaria laxa* M'COY, 1849; SD LANG-S.-T., 1940] [=?*Rhizopora* KON., 1871]. Like *Michelinia,* but corallum phaceloid in part. *L.Carb.*, Eng.——FIG. 353,*5*. **B. laxa* (M'COY); *5,* side of weathered corallum, ×1 (66).

Subfamily PALAEACINAE Počta, 1902

[*nom. transl.* HILL & STUMM, herein (*ex* Palaeacidae POČTA, 1902)] [=Palaeaciden C.F.ROEMER, 1883 (invalid vernacular name); =Leptoporidae MILLER, 1892 (invalid); Trachypsammidae GERTH, 1921;[1] Cleistoporidae EASTON, 1944; Trachypsammiidae,[1] Vaughaniidae LECOMPTE, 1952]

Discoid or cuneiform; walls and septa perforate and retiform; tabulae thin, few, distant. *L.Dev.-Perm.*

Cleistopora NICHOLSON, 1888 [**Michelinia geometrica* M.EDW.-H., 1851]. Small, discoid, adherent; corallites with dense walls pierced by mural pores, short, lined by a reticulate development from the perforate septa, and crossed by thin, distant tabulae (76). *L.Dev.*, Fr.——FIG. 353,*3*. **C. geometrica* (M.EDW.-H.), Fr.; *3a,* surface, ×1; *3b,* long. sec., ×4 (76).

Araeopora NICHOLSON & ETHERIDGE, 1879 [**A. australis*]. Massive, with very slender tall, prismatic corallites, septa and walls cribriform, tabulae few, thin. ?*Dev.*, N.Queensl.——FIG. 353,*4*. **A. australis, 4a,b,* transv. and long. secs., ×4 (71).

Squameophyllum SMYTH, 1933 [**S. spumans*] [=*Leptopora* WINCHELL, 1863 (*non* ORB., 1849); *Ethmoplax* SMYTH, 1939 (*pro Stratophyllum* SMYTH, 1933, *non Stratiphyllum* SCHEFFEN, 1933; *nec* LANG-S.-T., 1940); ?*Helioalcyon* TERMIER,

[1] Although included by HILL & STUMM in the Tabulata, the Trachypsammiidae are classed as Octocorallia by BAYER and MONTANARO-GALLITELLI (see p. F182, F190).

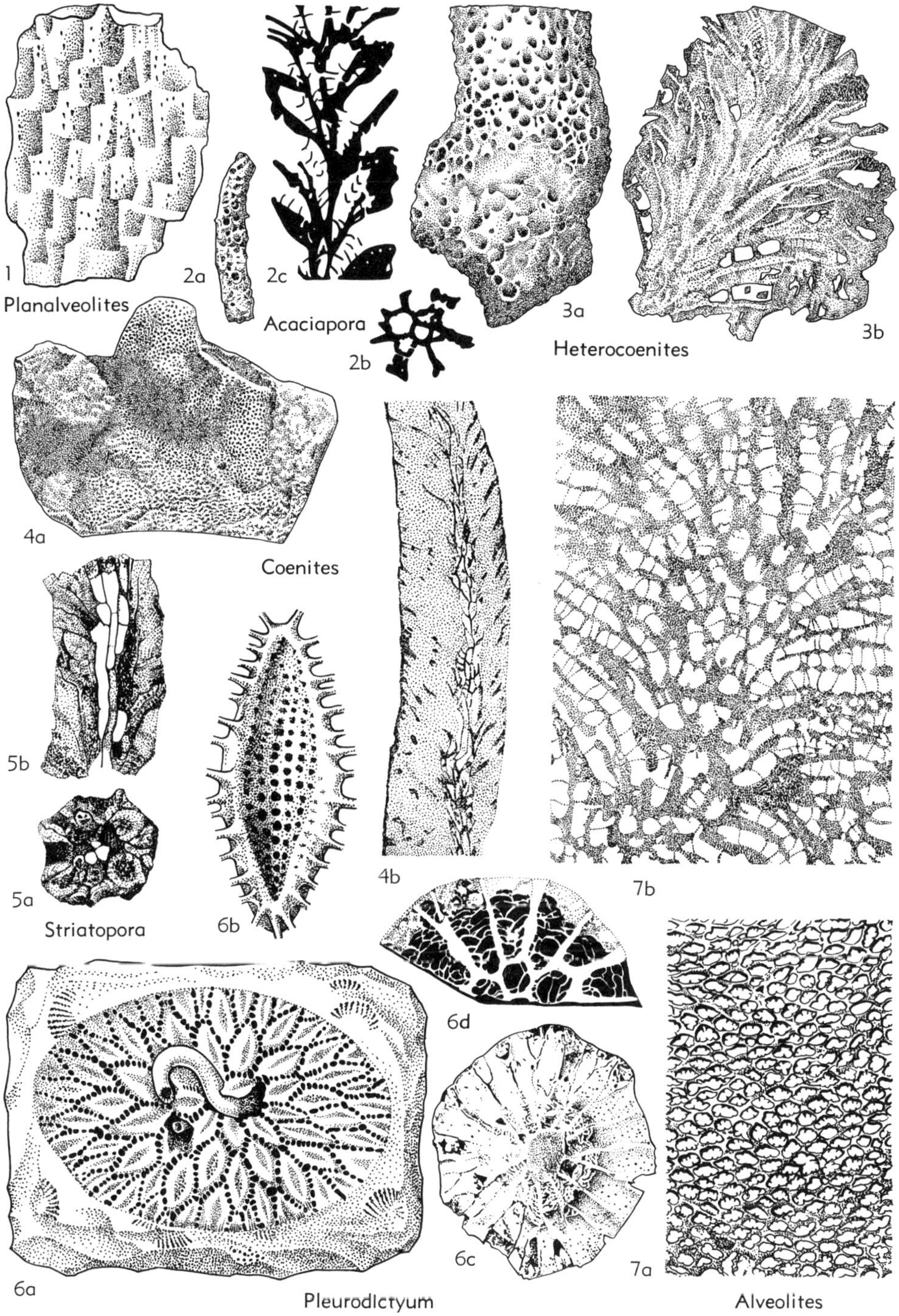

Fig. 352. Favositidae (p. F464-F466).

1945]. Like *Cleistopora* but with epithecal scales and with both walls and septa cribriform; where tabulae develop, the septa are discontinuous from one to the next, being developed only on their upper surfaces. *L.Carb.*, Eu.-N.Am.——FIG. 353, *6a,b*. **S. spumans*, Tournais., Belg.; *6a*, surface, ×1; *6b*, sec., ×4 (76).——FIG. 353,*6c-e*. *S. tenue* (SMYTH) (type species *Stratophyllum*), L.Carb., Belg.; *6c,d*, upper and lower surfaces, ×1; *6e*, sec., ×4 (76). Other species, FIG. 343,*4*.

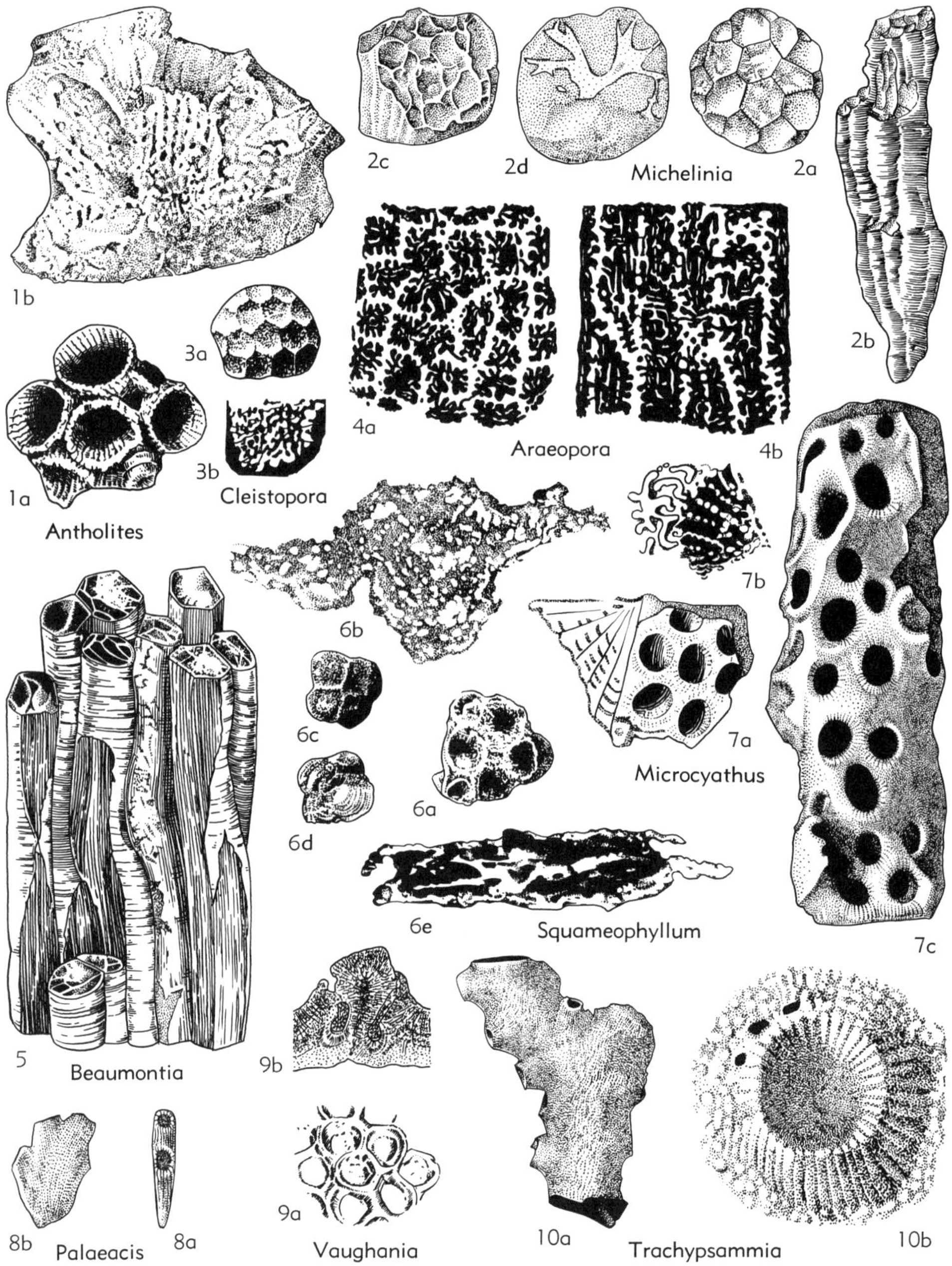

FIG. 353. Favositidae (p. F466-F469).

Vaughania Garwood, 1913 [**V. cleistoporoides*]. Discoid, free; atabulate; walls and floor of dense fibrous tissue in which neither septa, septal loculi, nor tabulae can be distinguished; each corallite with a ring canal around the base where the wall joins the floor; ring canals of neighboring corallites connected by mural pores. *L.Carb.*, Eng.——Fig. 353,*9*. **V. cleistoporoides*, NW.Eng.; *9a*, surface, ×1; *9b*, long. sec., ×10 (53).

Microcyathus Hinde, 1896 [*non* Döderlein, 1913] [**Hydnopora? cyclostoma* Phillips, 1836] [=*?Dictyopora* Gerth, 1921 (*non* Steininger, 1849, *nec* MacGill., 1869)]. Adherent, discoid with calices on upper surface; coenenchyme trabeculate, ridged, and irregularly canaliculate; calices lined with nontrabeculate aperforate fibrous tissue (73). *L.Carb.*, Eu.-N.Am.; *?Perm.*, Timor.——Fig. 353,*7a,b*. **M. cyclostoma* (Phillips), L.Carb., Eng.(Yorks.); *7a*, surface, ×1; *7b*, sec., enlarged (73).——Fig. 353,*7c*. *M. incrustans* (Gerth) (type species of *Dictyopora*), Perm., Timor; surface, ×1.5 (54).

Palaeacis Haime in M.Edw., 1857 [**P. cuneiformis*] [=*Sphenopoterium* Meek & Worthen, 1860]. Wedge-shaped, adherent in young stages; with round calices along narrow sides; flat sides of trabeculate, ridged, and irregularly canaliculate coenenchyme; calices lined with atrabeculate fibrous tissue pierced by perpendicular pores; septa cribriform; atabulate (59). *L.Carb.*, Eu.-N.Am.-Austral.; *Penn.*, N.Am.——Fig. 353,*8*. **P. cuneiformis*, L.Carb., N.Am.; *8a,b*, top and side views, ×1 (59).

Trachypsammia[1] Gerth, 1921 [**T. dendroides*]. Like *Palaeacis* but with tall, somewhat flattened branches, calices on the outer edges (13). *Perm.*, Timor.——Fig. 353,*10*. **T. dendroides*, *10a*, surface, ×1; *10b*, calice, ×4 (54).

?**Yavorskia** Fomichev, 1931 [**Y. barsasensis*]. *L. Carb.*, Sib.(Kouznetzk).

?**Oculinella** Yakovlev, 1939 [**O. gerthi*]. *L.Perm.* (*Artinsk.*), USSR.

Family HALYSITIDAE Milne-Edwards & Haime, 1850

[*nom. transl.* Duncan, 1872 (*ex* Halysitinae M.Edw.-H., 1850)] [=Halysitiniens de Fromentel, 1861 (invalid vernacular name); Halisitinae de Koninck, 1872]

Phaceloid coralla, composed of cylindrical, oval, or subpolygonal corallites united with one another along 2 or 3 sides producing anastomosing chainlike networks. Walls of corallites imperforate. Microcorallites present between corallites in some forms. Tabulae horizontal or arched, more closely set in microcorallites. Some forms have 12 vertical rows of septal spines in corallites. *Ord.-Sil.*

Halysites Fischer, 1828 [*nom. conserv.* (ICZN pend.) *pro Alyssites* Fischer, 1813] [**Tubipora catenularia* Linné, 1767]. Corallites long, cylindrical or oval, united along entire length in manner forming uniserial (rarely biserial or triserial) palisades which interconnect so that transverse sections appear as an anastomosing network of chains. Adjacent normal corallites in a series are separated by a rounded or subpolygonal microcorallite, both types of corallites having imperforate walls. Septa absent or represented by 12 vertical rows of spines; tabulae horizontal or gently arched, more closely set in microcorallites (23, 29). *Ord.-Sil.*, N.Am.-Eu.-Asia-Afr.-Austral.——Fig. 354,*1*. **H. catenularia* (Linné), M.Sil., Gotl., Eng.; *1a*, exterior, ×1; *1b,c*, transv. and long. secs., ×5 (*1a*, 75; *1b,c*, 70). Other species, Fig. 341,*9*.

Catenipora Lamarck, 1816 [**C. escharoides*; SD Thomas & Smith, 1954]. Differs from *Halysites* in lacking microcorallites. *Ord.-Sil.*, cosmop.

Labyrinthites Lambe, 1906 [**L. chidlensis*]. Massive; corallites slender, subpolygonal, connected along entire length with one or more neighbors, giving rise to meandering series of tubes. *Differs* from *Halysites* in form of normal corallites and absence of microcorallites. Septa apparently absent; tabulae numerous, horizontal (24). *?Ord., Sil.*, N.Am.——Fig. 354,*2*. **L. chidlensis*, ?U.Ord., Arct.N.Am.; transv. sec., ×30 (24).

Arcturia Wilson, 1931 [**A. complexa*]. Like *Labyrinthites* but corallites connect by vertical series of horizontal tubes instead of along entire side (38). *?Ord., Sil.*, N.Am.——Fig. 354,*3*. **A. complexa*, drift, *ex* Ord. or Sil., Can.; *3a*, distal surface, ×4; *3b*, side, ×4 (83).

Family AULOPORIDAE Milne-Edwards & Haime, 1851

[=Pyrgiens de Fromentel, 1961 (invalid vernacular name); Moniloporidae Grabau, 1899; Cladochonidae Hill, 1942; Chonostegitidae Lecompte, 1952]

Coralla compound, erect or repent and incrusting; corallites tubular, cylindrical or trumpet-shaped, increasing by lateral gemmation, in some forms connected by transverse stolons; walls solid, covered by transversely wrinkled epitheca. Septa represented by peripheral ridges or vertical rows of spines, lacking in some forms. Tabulae horizontal or distally concave, closely or widely spaced, broken up into tabellae in some forms, rare or absent in others. *?Sil., Dev.-Perm.*

Subfamily AULOPORINAE Milne-Edwards & Haime, 1851

[*nom. transl.* Hill & Stumm, herein (*ex* Auloporidae M.Edw.-H., 1851)]

Small repent or erect coralla, typically with trumpet-shaped corallites that increase by lateral gemmation producing uniserial, biserial, multiserial, ramose, or web-shaped growth patterns; transverse stolons absent.

[1] Trachypsammia is classed as an octocoral by Montanaro-Gallitelli (see p. F192).

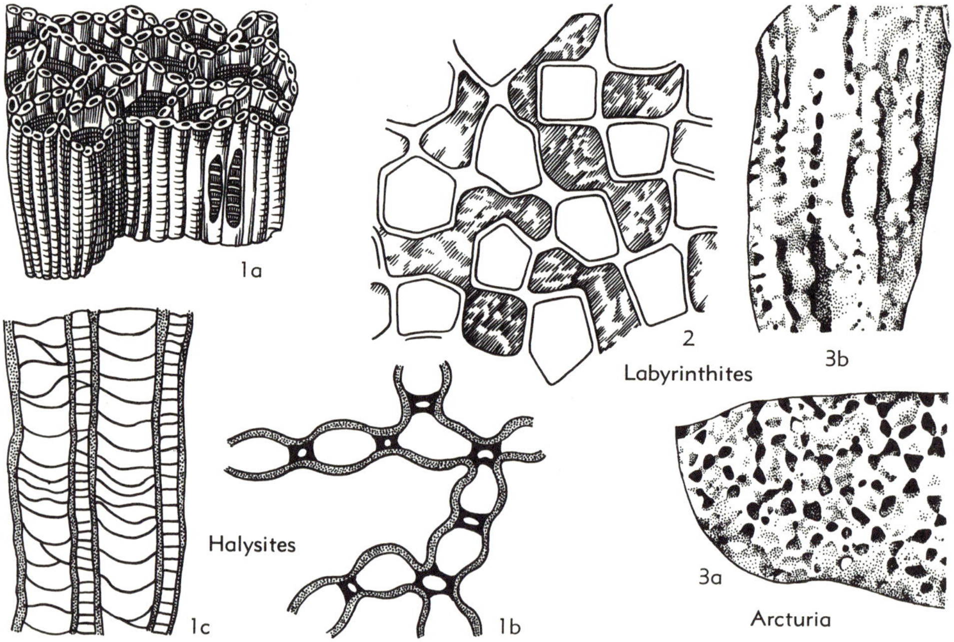

FIG. 354. Halysitidae (p. F469).

Septa represented by peripheral ridges or rows of spinules, absent in some forms. Tabulae widely spaced or lacking. *?Sil.-Dev.-Perm.*

Auloporella GRUBBS, 1939 [**A. typa*]. Like *Aulopora* but has secondary inner chambers, producing a double wall; young corallites retain communication with parent chambers. Tabulae and septa apparently lacking (16). *Sil.,* N.Am.——FIG. 355, *6. *A. typa,* M.Sil., Ill.; *6a,* long. sec., ×2.5; *6b,* side, ×2.5 (56).

Aulozoa GRUBBS, 1939 [**A. constricta*]. Corallum attached, composed of tortuous constricted tubes of nearly uniform diameter; 1 to 6 new tubes may be added within a short space of gemmation at any position about parent, all tubes having the same general direction of growth, tending to bend slightly upward near the apertural end; young tubes communicating with parent through a small subcircular opening at their points of origin. Apertures circular; tabulae apparently lacking; septa rudimentary (16). *Sil.,* N.Am.——FIG. 355,*3. *A. constricta,* M.Sil., Ill.; side, ×3 (56).

Romingeria NICHOLSON, 1879 [*pro Quenstedtia* ROMINGER, 1876 (*non* MORRIS & LYCETT, 1854)] [**Aulopora umbellifera* BILLINGS, 1859]. Resembling *Aulopora* but only attached basally; corallites cylindrical, annulated, multiplying by lateral gemmation, and typically producing new tubes in umbellate whorls or verticels at short intervals. Tabulae complete, remote, apparently not distinctly infundibuliform; septa represented by vertical rows of spinules (23, 29, 31). *Sil.-Dev.,* N. Am.——FIG. 341,*8;* 355,*12. *A. umbellifera* (BILL.), M.Dev., Ont.; 341,*8,* side, ×1 (46); 355, *12a,b,* proximal and side views, ×1 (46).

Bainbridgia BALL & GROVE, 1940 [**B. typicalis*]. Corallum flattened cylindrical; corallites short, uniformly spaced, rising from opposite sides in regular alternating succession. Faint septal ridges in some corallites; tabulae absent (2). *Sil.,* N.Am. ——FIG. 355,*11. *B. typicalis,* M.Sil., Mo.; *11a,* side, ×6; *11b,* long. sec., ×6 (44).

Diorychopora DAVIS, 1887 [**D. tenuis*]. Similar to *Aulopora* except with much smaller corallites arranged in biserial pattern. *Sil.,* N.Am.——FIG. 356, *5. *D. tenuis; 5a,* side, ×1; *5b,* long. sec., ×1 (78n).

Aulocaulis FENTON & FENTON, 1937 [**Aulopora expansa* FENTON-F., 1924]. Growth prostrate as in *Aulopora;* corallites thin-walled, smooth or indistinctly wrinkled, narrowly tubular from base to calice, then expanding abruptly to form circular, vertically directed calices with diameters 1.5 to 4 times that of tubular portions. Tabulae few or absent; septal ridges in calices. An indistinct dorsal ridge on most or all corallites of some species (10). *Dev.,* N.Am.——FIG. 355,*2. *A. expansa,* U.Dev., Iowa; ×1 (50).

Aulocystis SCHLÜTER, 1885 [**A. cornigera*] [=*Drymopora* DAVIS, 1887; *Ceratopora* GRABAU, 1899]. Like *Aulopora* in external form, increasing by lateral gemmation, erect or prostrate, rarely attached above the base; corallites subcylindrical to trumpet-shaped. Septa represented by ridges or rows of spinous trabeculae projecting from inner walls. Calices deep, funnel-shaped, thin-walled, continued downwards in a narrowing tube, formed by thickening of walls through addition internally of coarse, steeply inclined cysts (7, 14, 33). *Dev.*, N.Am.-Eu.-Asia-Austral.——FIG. 355,*5*. **A. cornigera,* M.Dev., Ger.; *5a,* typical corallum, ×1; *5b,* distal view of weathered corallite, ×3; *5c,* long. sec. showing cysts, ×3 (74).

Chonostegites M.EDW.-H., 1851 [**C. clappi*]

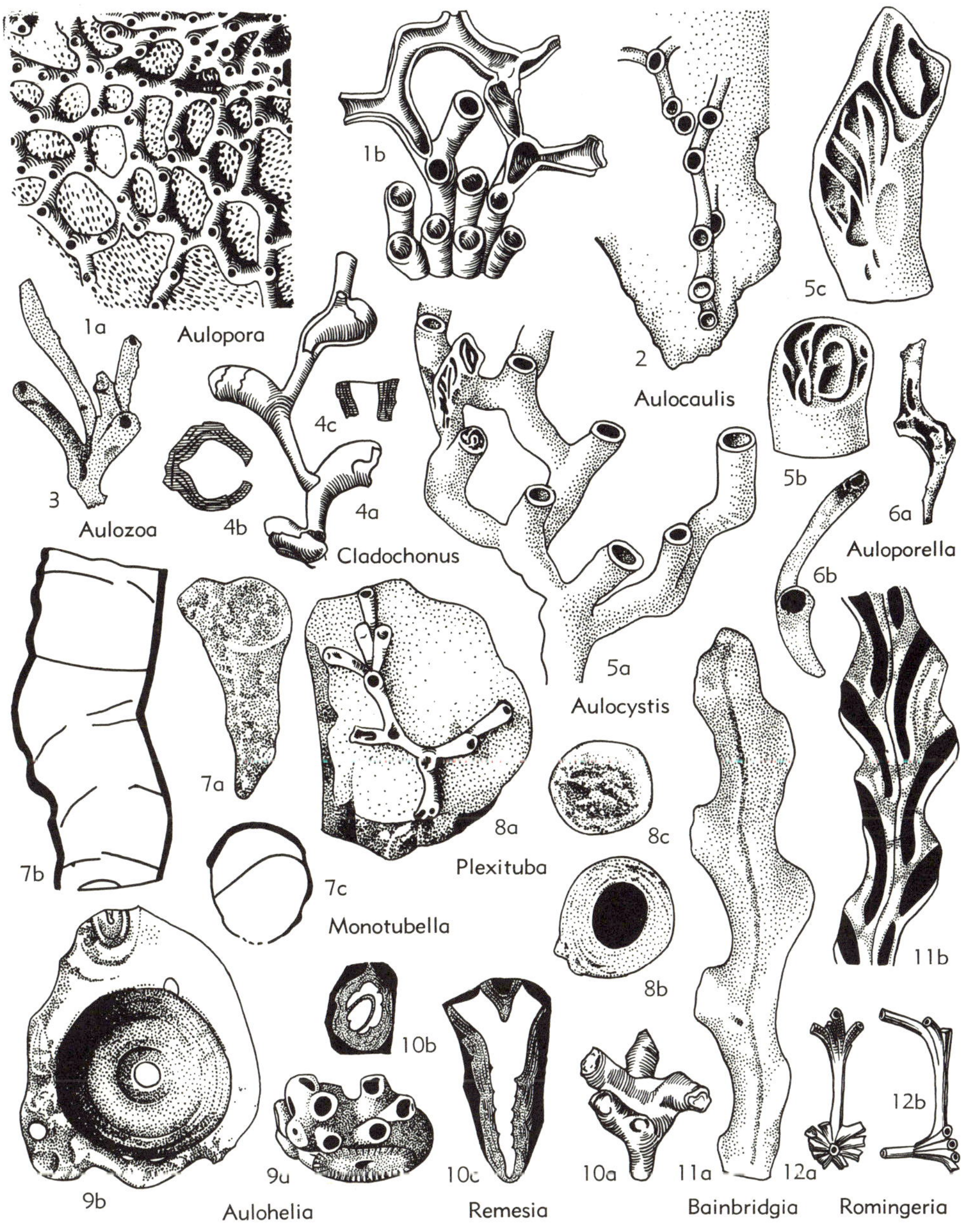

FIG. 355. Auloporidae (p. F470-F472).

[=*Haimophyllum* BILLINGS, 1859]. Coralla fasciculate, with cylindrical corallites connected at regular intervals by disc-shaped expansions. Septa represented by peripheral ridges or spines; tabulae numerous, closely set, broken up into tabellae in some (5, 9). *Dev.*, N.Am.——FIG. 340,*4*; 356,*3*. **C. clappi* M.Dev. (drift), Ohio; 340,*4*, side, ×1; 356,*3a*, side, ×1; 356,*3b*, distal surf., ×1; 356,*3c*, long. sec., ×2 (50).

Remesia KETTNER, 1937 [**R. tubulosa*]. Tubular subcylindrical corallites budding irregularly to form a dendritic colony; walls thick, composed of concentric layers of sclerenchyme, with rudimentary septal spines on inner surface. Calices circular, not enlarged. Tabulae very rare, or absent (21). *Dev.*, Eu.——FIG. 355,*10*. **R. tubulosa*, M.Dev., Czech.; *10a*, side, ×1; *10b,c*, transv. and long. secs., ×3 (61).

Aulopora GOLDFUSS, 1829 [**A. serpens* (*non Tubipora serpens* LINNÉ, 1758, =*Aulopora repens* M.EDW.-H., 1851); SD M.EDW.-H., 1850]. Repent network of trumpet-shaped corallites with circular, obliquely set calices; shape of coralla ranging from a loose, subhexagonal network to biserial or multiserial rows of laterally contiguous corallites. Small septal spines or faint septal ridges in some species; tabulae widely spaced, horizontal, arched or inclined, rudimentary or absent (10, 29). *Dev.*, N.Am.-Eu.-Asia-Austral.-Afr.——FIG. 340,*2*; 355,*1*. **A. serpens*, M.Dev.; 340,*2*, corallum, ×1 (55); 355,*1a*, Ger.; corallum incrusting on *Alveolites*, ×1; *1b*, part enlarged (55).

Plexituba STAINBROOK, 1946 [**P. contexta*]. Like *Aulopora* but with greatly thickened walls composed of fine, concentric lamellae and having internal cysts similar to those of *Aulocystis*. Septal ridges or spines and tabulae absent (34). *Dev.*, N.Am.——FIG. 355,*8*. **P. contexta*, U.Dev., Iowa; *8a*, side, ×1; *8b,c*, transv. secs. showing wall lamellae and edges of cysts, ×5 (77).

Cladochonus M'COY, 1847 [**C. tenuicollis*; SD M.EDW.-H., 1850] [=*Pyrgia* M.EDW.-H., 1851; *Monilopora* NICHOLSON & ETHERIDGE, 1879]. Proximal corallites in reptant ring from which free branches arise; individual corallites trumpet- or pipe-shaped, in contact only at points of origin, each giving rise to another by lateral increase through wall of the expanded calice; each with a thick peripheral stereozone of laminar or reticulate sclerenchyme. Septal spines and tabulae lacking in the narrow lumen, but septal ridges may appear in the calices (19). *Dev.-Perm.*, N.Am.-Eu.-Asia-E.Indies-Austral.——FIG. 355,*4*. **C. tenuicollis*, L.Carb., N.S.W.; *4a*, side, ×1; *4b,c*, transv. and long. secs., ×2 (58). Other species, FIG. 341,*6*.

Monotubella YAKOVLEV, 1939 [**M. permiensis*; SD herein]. Small simple coralla or rarely with one lateral bud; cylindrical or trumpet-shaped, erect, attached by basal extremity. Calices bell-shaped. Interiors with widely spaced, horizontal tabulae; traces of rudimentary septa in some (40). *Perm.*, Asia.——FIG. 355,*7*. **M. permiensis*, L.Perm., USSR; *7a*, side, ×2.5; *7b,c*, transv. and long. secs., much enlarged (85).

Aulohelia GERTH, 1921 [**A. irregularis*; SD LANG-S.-T., 1940]. Typically incrusting crinoid stems. Calices tubular, with thin septal ridges. Interiors almost completely filled with sclerenchyme, leaving narrow axial canals connecting adjacent corallites. Tabulae apparently absent (13). *Perm.*, Timor.——FIG. 355,*9*. **A. irregularis*; *9a*, corallum attached to crinoid columnal, ×1; *9b*, sec. showing parts of axial canals, ×2 (54).

Pseudoromingeria YABE & SUGIYAMA, 1941 [**P. kotoi*]. Coralla fasciculate, composed of many subcylindrical, irregularly twisted corallites multiplying by lateral gemmation. Septa represented by vertical rows of short, blunt spines; tabulae distally concave, crowded in some corallites, widely spaced in others (39). *Perm.*, China.——FIG. 356,*1*. **P. kotoi*; *1a*, sec. showing fasciculate growth, ×1; *1b,c*, transv. and long. secs., ×5 (84).

Subfamily SYRINGOPORINAE Nicholson, 1879

[=Syringoporiens, Thecostegitiniens DE FROMENTEL, 1861 (invalid vernacular names); Syringolitinae WAAGEN & WENTZEL, 1886 (=Syringolitidae, *nom. transl.* LECOMPTE, 1952)]

Large, erect dendroid or fasciculate coralla composed of cylindrical corallites connected by transverse stolons. Septa represented by peripheral ridges or vertical rows of spines, lacking in some forms. Tabulae typically closely set, numerous, horizontal, infundibuliform, or broken up into tabellae. *Sil.-Perm.*

Syringopora GOLDFUSS, 1826 [**S. ramulosa*; SD M.EDW.-H., 1850] [=*Harmodites* FISCHER, 1828; ?*Kazania* STUCKENBERG, 1895; *Kueichowpora* CHI, 1933]. Closely or loosely set cylindrical corallites connected in most species by hollow, transverse stolons. Septa lacking or represented by 12 vertical rows of small spinules; tabulae closely set, deeply depressed axially, coalesced in some species to form a continuous axial tube (6, 29). *Sil.-Penn.*, N.Am.-Eu.-Asia-Austral.-Afr.——FIG. 341,*7*; 356,*2*. **S. ramulosa*, L.Carb., Eng.; 341,*7*, side of corallum, ×1 (68); 356,*2a*, side, ×1; 356,*2b*, distal surf., ×4; 356,*2c*, long. sec., ×4 (68). Other species, FIG. 341,*1*.

Reuschia KIÄR, 1930 [**R. aperta*]. Phaceloid colonies composed of small, thick-walled cylindrical corallites. Walls indented on inner side to form 10 to 12 short, thick septal ridges; tabulae apparently absent (22). *U.Ord.*, Eu.——FIG. 357,*5*. **R. aperta* Norway; *5a,b*, transv. and long. secs. showing thick walls with septal ridges and tubular axial area (lectotype), ×2 (63).

Eofletcheria BASSLER, 1950 [**Columnaria incerta*

Billings, 1859]. Small, phaceloid, thick-walled colonies with poorly developed, intermittently spaced, complete or incomplete tabulae and no septa. Corallites connected by short, horizontal, syringoporoid tubules (4). *M.Ord.*, N.Am.——Fig. 357,*2*. **E. incerta* (Bill.), Chazy., Que.; *2a,b,* long. and transv. secs. showing tubula, tabulae and thick walls, ×10 (72).

Syringolites Hinde, 1879 [**S. huronensis*]. *Sil.*, Mich.

Cannapora Hall, 1852 [**C. junciformis*]. Like *Syringopora* but with closely set corallites connected by stolons at distinct intervals. Interiors of corallites with 12 vertical rows of long septal spines; tabulae horizontal, closely or widely spaced (23, 31). *Sil.*, N.Am.——Fig. 356,*6*. **C. junciformis*, M.Sil., N.Y.; *6a*, side, ×1; *6b*, distal surf., ×2 (56).

Thecostegites M.Edw.-H., 1849 [**Harmodites bouchardi* Michelin, 1846]. Differs from *Syringo-*

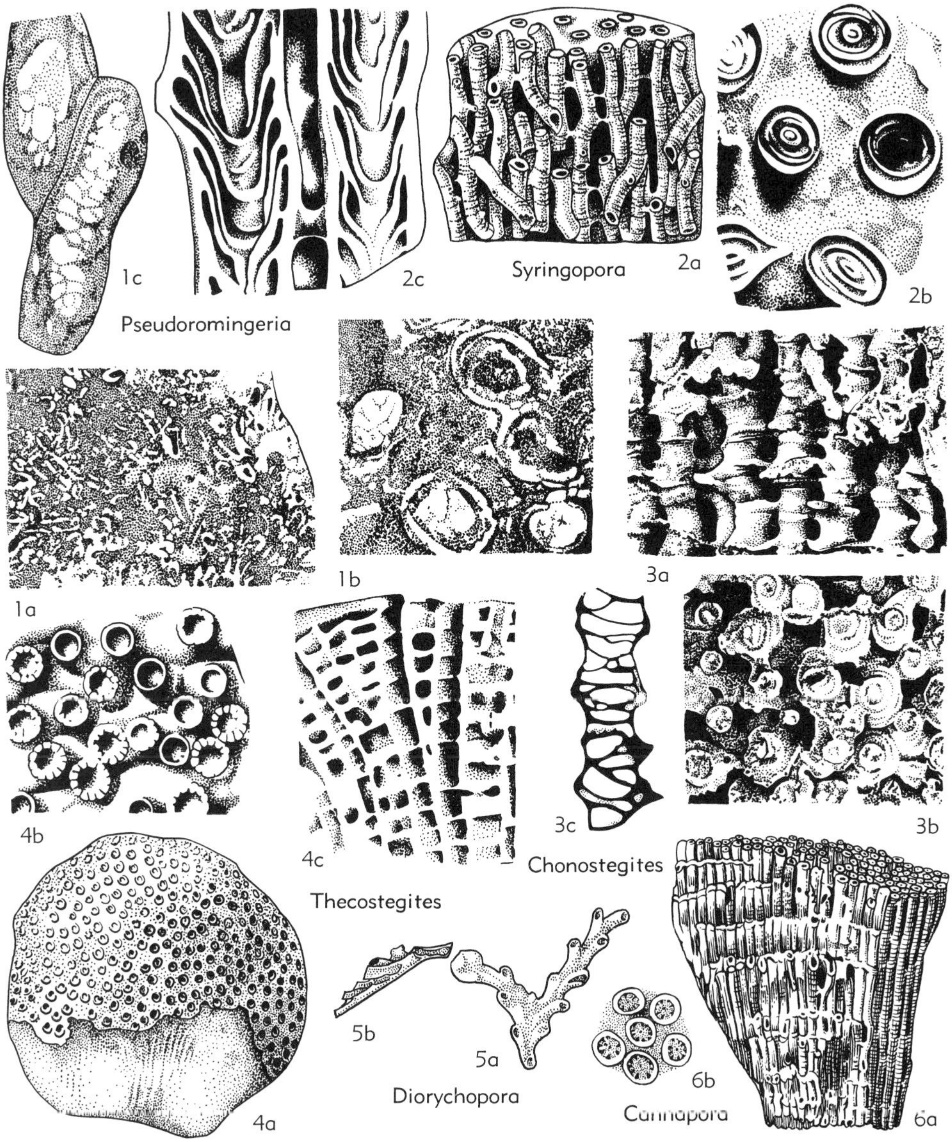

Fig. 356. Auloporidae (p. F470-F474).

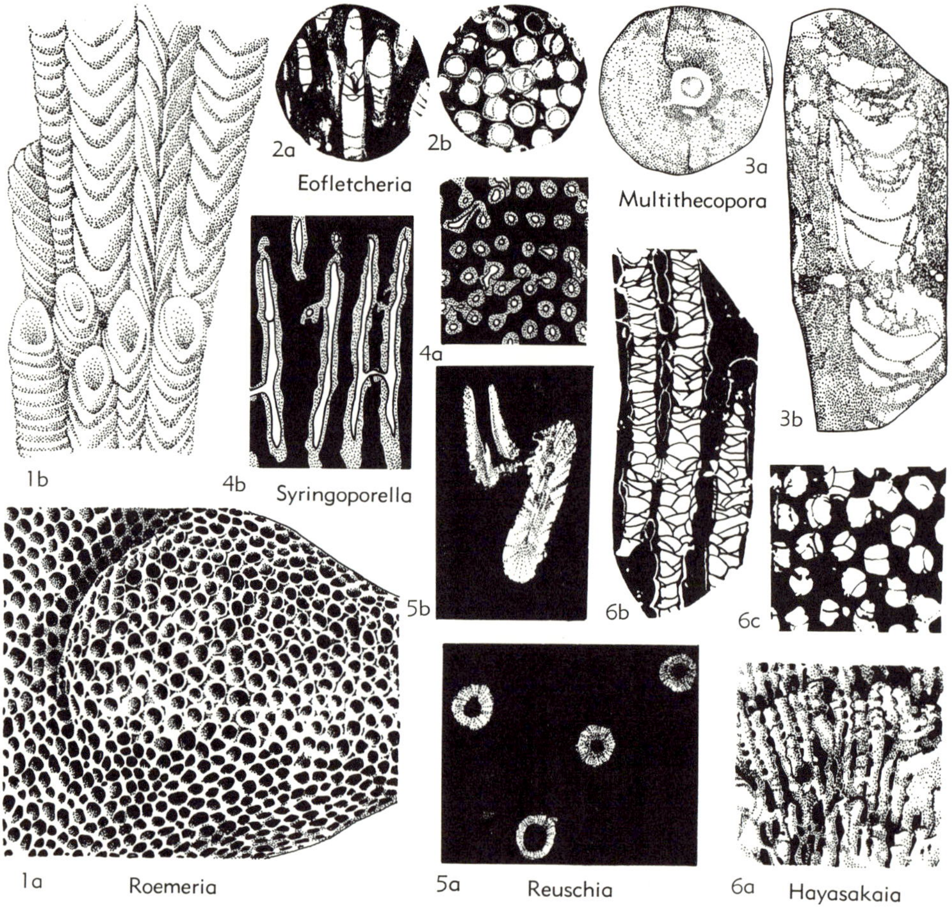

Fig. 357. Auloporidae (p. F472-F474).

pora in having vertical septal ridges and closely set horizontal tabulae (28). *Dev.*, N.Am.-Eu.——Fig. 356,*4*. **T. bouchardi* (Mich.), M.Dev., Fr.; *4a*, corallum, ×1; *4b*, distal surf., ×3; *4c*, long. sec., ×3 (68).

Roemeria M.Edw.-H., 1851 [**Calamopora infundibulifera* Goldfuss, 1829]. Subcerioid to subphaceloid, with peripheral septal spines and complete, funnel-shaped tabulae. Neighboring corallites connected by short syringoporoid tubules, or by pores where corallites adjoin (1). *Dev.*, Eu.-Austral.——Fig. 357,*1*. *R. infundibulifera* (Goldf.), M.Dev., Ger. (Eifel); *1a*, distal view (holotype), ×1; *1b*, long. sec. showing tabulae, ×3 (55).

Syringoporella Kettner, 1937 [**Syringopora moravica* Roemer, 1883]. Externally like *Syringopora* but with much smaller, very thick-walled corallites; connecting stolons short, commonly barrel-shaped, constricted at junctions with corallites. No septal spines or ridges; tabulae horizontal, very remote, lacking in some corallites (21). *Dev.*, Eu. ——Fig. 357,*4*. **S. moravica* (Roemer), M.Dev., Czech.; *4a,b*, transv. and long. secs., ×3 (61).

Multithecopora Yoh, 1927 [**M. penchiensis*]. Like *Syringopora* but corallites with greatly thickened walls, composed of concentric layers of sclerenchyme, occupying four-fifths of the interior. Septa lacking; tabulae thin, closely set, distally concave (41). *U.Carb.*, China.——Fig. 357,*3*. **M. penchiensis*, *3a,b*, transv. and long. secs., ×15 (86).

Hayasakaia Lang-S.-T., 1940 [*pro Tetrapora* Yabe & Hayasaka, 1915 (*non* Quenst., 1857)] [**Tetrapora elegantula* Yabe-H., 1915]. Differs from *Syringopora* in having quadrate corallites connected by closely set short stolons arranged in vertical rows on flattened sides. Tabulae close-spaced, horizontal or convex; no septal spines (20, 41). *Perm.*, China.——Fig. 357,*6*. **H. elegantula* (Yabe-H.); *6a*, side, ×1; *6b,c*, long. and transv. secs., ×2 (87).

Order TABULATA Incertae Sedis

Asteriophyllum PORFIRIEV, 1937 [**A. aenigmaticum*]. *Dev.*, E.Urals (not seen).

Čelechopora PRADÁČOVÁ, 1938 [**C. robustus*]. *Dev.* Czech.

Cylindripora EICHWALD, 1829 [*non* YABE & HAYASAKA, 1915] [**C. serpuloides*] [=*Cylindropora* LANG-S.-T., 1940]. Drift, *ex Sil.*, Est. (possibly bryozoan).

Dendropora MICHELIN, 1846 [**D. explicita*]. *?Dev.*, Fr. (possibly bryozoan).

Dictyopora STEININGER, 1849 [*non* MACGILLIVRAY, 1869; *nec* GERTH, 1921] [**D. reticulata*]. *Dev.*, Ger. (possibly bryozoan).

Donetzites DAMPEL, 1940 [**D. milliporides*]. *Carb.*, Donetz (not seen).

Latepora RAFINESQUE, 1819 [**L. alba*]. *Dev.*, N. Am. May be conspecific with *Michelinia convexa* ORB., 1850 (types missing).

Linipora TROOST, 1840 [**L. rotunda*] [=*Linopora* LANG-S.-T., 1940 *(nom. van)*]. *Sil.*(Brownsport), Tenn. (not recognizable).

Protopora GREENE, 1904 [**(Romingeria cystoides)* GRABAU in GREENE, 1901]. *L.Carb.*, N.Am.

Salpingium SMYTH, 1928 [**S. palinorsum*]. *L.Carb.*, Ire.

Tetraporella SOKOLOV, 1947 [**Labyrinthites? monticuliporoides* TROEDSSON, 1928]. *U.Ord.*, Greenl.

Troedssonites SOKOLOV, 1947 [**Syringopora conspirata* TROEDSSON, 1928]. *U.Ord.*, Greenl.

Vaughanites PAUL, 1937 [*non* WOODRING, 1928] [**Syringopora favositoides* VAUGHAN, 1915]. *L. Carb.*, Eu.

REFERENCES

Publications cited in the following list will be found helpful in furnishing additional information concerning the Tabulata and in offering a guide to more extensive literature. The index numbers enclosed in parentheses in the column at left are employed in the text for identification of the publications.

Amsden, T. W.

(1) 1949, *Stratigraphy and paleontology of the Brownsport formation (Silurian) of western Tennessee:* Peabody Mus. Nat. Hist. (New Haven), Bull. 5, 138 p., 34 pl.

Ball, J. R., & Grove, B. H.

(2) 1940, *New species of corals from the Bainbridge limestone of southeastern Missouri:* Am. Midland Naturalist, v. 24, p. 382-404, 4 pl.

Bassler, R. S.

(3) 1944, *Parafavosites and similar tabulate corals:* Jour. Paleont., v. 18, p. 42-49, fig. 1-29.

(4) 1950, *Faunal lists and descriptions of Paleozoic corals:* Geol. Soc. America, Mem. 44, 315 p., 20 pl.

Billings, Elkanah

(5) 1859, *On the fossil corals of the Devonian rocks of Canada West:* Canad. Jour. Industry Sci. Arts. n.ser., v. 4, p. 97-140, text-figs.

Buehler, E. J.

(5a) 1955, *The morphology and taxonomy of the Halysitidae:* Peabody Mus. Nat. Hist., Yale Univ., Bull. 8, 79 p., 12 pl.

Chi, Y. S.

(6) 1933, *Lower Carboniferous Syringoporas of China:* Palaeont. Sinica, ser. B, v. 12, fasc. 4, v+48 p., 7 pl.

Davis, W. J.

(7) 1887, *Kentucky fossil corals—a monograph of the fossil corals of the Silurian and Devonian rocks of Kentucky, Part 2:* Ky. Geol. Survey, xiii+4 p., 139 pl.

Etheridge, R.

(8) 1903, *Fossopora, a new genus of Palaeozoic perforate corals:* Records Austral. Mus., v. 5, p. 16-19, pl. 1-2.

Fenton, C. L., & Fenton, M. A.

(9) 1936, *The "tabulate" corals of Hall's "Illustrations of Devonian fossils":* Annals Carnegie Mus., v. 25, p. 17-38, pl. 1-8.

(10) 1937, *Aulocaulis, a new genus of auloporoid corals:* Am. Midland Naturalist, v. 18, p. 119-120, pl. 6.

Fritz, M. A.

(11) 1939, *Two unique Silurian corals:* Jour. Paleont., v. 13, p. 512-513, pl. 59.

Gerth, H.

(12) 1908, *Beiträge zur Phylogenie der Tubocorallier:* Zeitschr. indukt. Abstamm.- u. Vererb., Band 1, pt. 1, p. 1-62.

(13) 1921, *Die Anthozoën der Dyas von Timor:* Paläont. von Timor, Abh. 11, p. 67-147, pl. 145-150.

Grabau, A. W.

(14) 1899, *Moniloporidae, a new family of Palaeozoic corals:* Proc. Boston Soc. Nat. Hist., v. 28, p. 409-424, pl. 1-4.

(15) 1936, *Early Permian fossils of China, Part 2, Fauna of the Maping limestone of Kwangsi and Kweichow:* Palaeont. Sinica, ser. B, v. 8, fasc. 4, 441 p., 31 pl.

Grubbs, D. M.

(16) 1939, *Fauna of the Niagaran nodules of the Chicago area:* Jour. Paleont., v. 13, p. 543-560, pl. 61-62.

Hall, James

(17) 1852, *Natural history of New York, Part 6, Palaeontology of New York:* (Albany), v. 2, viii+363 p., 85 pl.

Hill, Dorothy

(18) 1950, *Middle Devonian corals from the Buchan district, Victoria:* Proc. Roy. Soc. Vict., v. 62, pt. 2, p. 158-164, pl. 5-9.

(18a) 1951, *The Ordovician corals:* Proc. Roy. Soc. Queensl., v. 62, p. 1-27.

———, & Smyth, L. B.

(19) 1938, *On the identity of Monilopora Nicholson and Etheridge, 1879, with Cladochonus McCoy, 1847:* Proc. Roy. Irish Acad., v. 45 (B), pt. 6, p. 125-138, pl. 22-23.

Huang, T. K.

(20) 1932, *Permian corals of southern China:* Palaeont. Sinica, ser. B, v. 8, fasc. 2, v+163 p., 16 pl.

Kettner, Radim

(21) 1934, *Paleontologické studie z čelechovického devonu, část 5, O některých Alcyonariích [Paleontological studies of the Devonian of Čelechovice (Moravia); pt. 5, On some alcyonairians]:* Praha Univ. Karlova, ústavu geol. paleont., no. 18, 15 p., 13 fig.

Kiär, J.

(22) 1904, *Revision der mittelsilurischen Heliolitiden und neue Beiträge zur Stammesgeschichte derselben:* Skr. Vidensk.-Selsk. Christiania, Math.-naturv. Kl., no. 10, p. 1-58.

Lambe, L. M.

(23) 1901, *A revision of the genera and species of Canadian Palaeozoic corals, The Madreporaria Aporosa and the Madreporaria Rugosa:* Geol. Survey Can., Contr. Canad. Palaeont., v. 4, pt. 2, p. 97-197, pl. 1-18.

(24) 1906, *Notes on the fossil corals collected by Mr. A. P. Low at Beechey Island, Southampton Island, and Cape Chidley in 1904:* Rept. (Can.) Dom. Govt. Exped. to Hudson Bay and Arctic Islands on board D. G. S. Neptune, 1903-04, App. 4, p. 322-328.

Lecompte, Marius

(25) 1939, *Les tabulés du Devonien moyen et supérieur du bord sud du bassin de Dinant:* Mus. Roy. Hist. Nat. Belg., Mém. 90, 229 p., 23 pl.

Lindström, G.

(26) 1899, *Remarks on the Heliolitidae:* K. Svenska Vetenskapsakad. Handl., v. 32, pt. 1, 140 p., 12 pl.

Milne-Edwards, H., & Haime, Jules

(27) 1850, *A monograph of the British fossil corals, Part 1, Introduction:* Palaeontograph. Soc. (London), Mon., lxxxv+71 p., 11 pl.

(28) 1851, *Monographie des polypiers fossiles des terrains palaeozoiques:* Arch. Mus. Hist. Nat. (Paris), v. 5, 502 p., 20 pl.

Nicholson, H. A.

(29) 1879, *On the structure and affinities of the tabulate corals of the Palaeozoic period:* (Edinburgh, London), xii+342 p., 15 pl.

Okulitch, V. J.

(30) 1936, *On the genera Heliolites, Tetradium, and Chaetetes:* Am. Jour. Sci., 5 ser., v. 32, p. 361-379, fig. 1.

Rominger, Carl

(31) 1876, *Palaeontology, Fossil corals:* Mich. Geol. Survey, v. 3, pt. 2, 161 p., 55 pl.

Ross, M. H.

(31a) 1953, *The Favositidae of the Hamilton group (Middle Devonian of New York):* Buffalo Soc. Nat. Sci., Bull., v. 21, no. 2, p. 37-89, pl. 12-27.

Sardeson, F. W.

(32) 1896, *Beziehungen der fossilen Tabulaten zu den Alcyonarien:* Neues Jahrb. Mineralogie, Geologie u. Paläontologie, Abt. B, Band 10, p. 249-362.

Schlüter, C.

(33) 1889, *Anthozoen des rheinischen Mittel-Devon:* Abh. geol. Spezialkarte Preuss. Thüring. Staat., Band 8, Heft 4, p. 259-465, pl. 1-16.

Stainbrook, M. A.

(34) 1946, *Corals of the Independence shale of Iowa:* Jour. Paleont., v. 20, p. 401-427, pl. 57-61.

Stumm, E. C.

(35) 1949-50, *Type invertebrate fossils of North America (Devonian): Tabulata:* Wagner Free Inst. (Philadelphia), cards 1-114 (Auloporidae) (1949); 115-260 (Favositidae) (1949); 261-405 (Favositidae) (1950).

Termier, Henri, & Termier, Geneviève

(36) 1950, *Invertébrés de l'ère primaire, Fasc. 1, Foraminifères, spongiaires et coelentérés:* Palaeont. Marocaine, Hermann (Paris), 218 p., 51 pl.

Thomas, H. D., & Smith, Stanley

(36a) 1954, *The Genus Halysites Fischer von Waldheim:* Ann. Mag. Nat. Hist., ser. 7, v. 12, p. 765-774, pl. 20-22.

Weissermel, W.

(37) 1937, *Coelenterata: Anthozoa, Hydrozoa, Scyphozoa [Review]:* Fortschritte Paläont., Band 1, p. 84-96.

Wilson, A. E.

(38) 1931, *Notes on the Baffinland fossils:* Trans. Roy. Soc. Canada, ser. 3, v. 25, sec. 4, p. 285-308, 5 pl.

Yabe, Hisakatsu, & Sugiyama, Toshio

(39) 1941, *Pseudoromingeria, a new genus of auloporoids from Japan:* Proc. Imp. Acad. Tokyo, v. 17, p. 379-382, fig. 1-4.

Yakovlev, N. N.

(40) 1939, *Nouveaux genres de coraux Tabulata*

du Permien inférieur de l'Oural et du bassin du Donetz: Comptes Rendus (Doklady) Acad. Sci. U.S.S.R. (Moscow), v. 24, no. 6, p. 629-632, fig. 1-2.

Yoh, S. S.

(41) 1927, *On a new genus of syringoporoid coral from the Carboniferous of Chihli and Fengtien provinces:* Bull. Geol. Soc. China, v. 5, p. 291-293, pl. 1.

———, & Huang, T. K.

(42) 1932, *The coral fauna of the Chihsia limestone of the lower Yangtse Valley:* Palaeont. Sinica, ser. B, v. 8, fasc. 1, 72 p., 10 pl.

SOURCES OF ILLUSTRATIONS

Explanation of the use of index numbers for citing sources of illustrations is given in the Editorial Preface.

(43) **Amsden, T. W.**
(44) **Ball, J. R., & Grove, B. H.**
(45) **Bassler, R. S.**
(46) **Billings, Elkanah**
(47) **Cox, I.**
(48) **Eichwald, C. E. von**
(49) **Etheridge, R.**
(50) **Fenton, C. L., & Fenton, M. A.**
(51) **Fleming, J.**
(52) **Fritz, M. A.**
(53) **Garwood, E. J.**
(54) **Gerth, H.**
(55) **Goldfuss, G. A.**
(56) **Grubbs, D. M.**
(57) **Hall, James**
(58) **Hill, Dorothy**
(58a) **———, & Smyth, L. B.**
(59) **Hinde, G. J.**
(60) **Jones, O. A.**
(61) **Kettner, Radim**
(62) **Kiär, J.**
(63) **Koninck, L. G. de**
(64) **Lecompte, Marius**
(65) **Lindström, G.**
(66) **M'Coy F.**
(67) **Michelin, J. L. H.**
(68) **Milne-Edwards, Henri, & Haime, Jules**
(69) **Moore, R. C., & Jeffords, R. M.**
(70) **Nicholson, H. A.**
(71) **———, & Etheridge, R.**
(72) **Okulitch, V. J.**
(73) **Phillips, John**
(74) **Schlüter, C.**
(75) **Schmidt, F.**
(76) **Smyth, L. B.**
(77) **Stainbrook, M. A.**
(78) **Stumm, E. C.**
(79) **Troedsson, G.**
(80) **Weissermel, W.**
(81) **Wentzel, Joseph**
(82) **Williams, H. S.**
(83) **Wilson, A. E.**
(84) **Yabe, H., & Sugiyama, T.**
(85) **Yakovlev, N. N.**
(86) **Yoh, S. S.**
(87) **———, & Huang, T. K.**

ZOANTHARIA INCERTAE SEDIS

By Dorothy Hill

ZOANTHARIA INCERTAE SEDIS

The following generic names which have been applied to Zoantharia cannot be assigned satisfactorily to any division of the subclass, and some of them may not be zoantharians or Anthozoa.

Cylicopora Steininger, 1849 [**C. fasciculata*]. *M. Dev.*, Ger. (not seen).

Discophyllum Hall, 1847 [**D. peltatum*]. *Ord.*, N.Y. (probably plant).

Favositella Mansuy, 1912 [*non* Eth.-F., 1884] [**F. columnaris*]. *U.Dev.*, Yunnan (may not be coral).

Kazania Stuckenberg, 1895 [**K. elegantissima;* SD Lang-S.-T., 1940]. *U.Carb.* Urals (may not be coral).

Khmeria Mansuy, 1914 [**K. problematica*]. *Perm.*, Indo-China (an operculate coelenterate, probably not a rugose coral).

Lamellopora Owen, 1844 [**L. infundibularia*] [=*Lamellipora* Lang-S.-T., 1940]. *M.Sil.*, N.Am. (may be stromatoporoid).

Mezenia Stuckenberg, 1895 [*non* Simon, 1897] [**M. rozeni*]. *Carb.*, Timan (probably not coral).

Mortieria de Koninck, 1841 [**M. vertebralis*]. *L. Carb.*, Eu. (possibly sponge).

Patinula Eichwald, 1829 [**P. lithuana*] (may not be coral).

Siphonaxis Dybowski, 1873 [**S. tubiferus*]. Drift (undeterminable).

Spongarium Lonsdale, 1839 [*non* M'Coy, 1851] [**S. edwardsii*]. *Sil.*, Eng. (may not be coral).

Sycidium Sandberger, 1849 [*non* Haeckel, 1870] [**S. reticulatum*]. *M.Dev.*, Ger. (probably plant).

CTENOPHORA

By JOHN W. WELLS and DOROTHY HILL

The Ctenophora are an assemblage of marine coelenterates which lack nematocysts and in various ways differ sufficiently from other assemblages to warrant classification as an independent subphylum. Fossil ctenophores are very doubtful or lacking.

Subphylum CTENOPHORA Eschscholtz, 1829

[=Ciliobranches, Ciliogrades DE BLAINVILLE, 1830; Acalephes cténophores MILNE-EDWARDS & HAIME, 1850; Ctenarea DELAGE & HÉROUARD, 1901; Ctenophoraria POCHE, 1911; Acnidaria PAX, 1924]

Ctenophores ("comb-jellies") are transparent, pelagic, mobile, marine, coelenterated animals, mostly globular in shape (Fig. 358). They are characterized by body layers of endoderm, mesogloea, and ectoderm, possession of biradial symmetry, a gastrovascular cavity and stomodaeum, and position of the gonads. From the Cnidaria they are differentiated by the presence of colloblasts (lasso cells) instead of nematocysts, by locomotion almost wholly by means of ciliary movement cencentrated in 8 longitudinal bands, enormous thickening of the mesogloea, presence of a statocyst or sensitive organ at the aboral end, and solid tentacles with axial mesogloea. *Rec.*

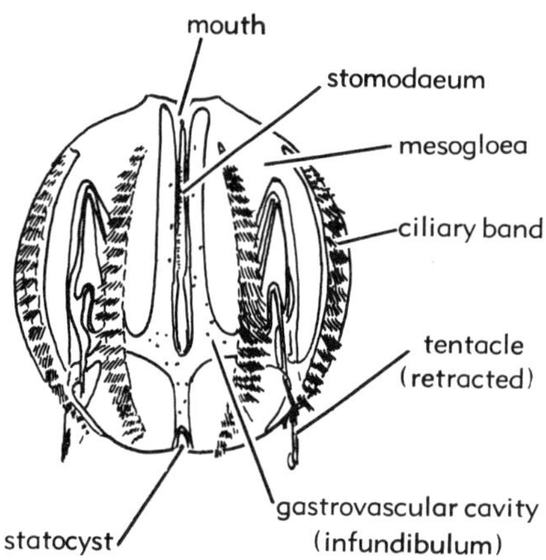

FIG. 358. Morphology of Ctenophora illustrated by *Pleurobrachia rhododactyla,* Rec., Atl., ×1 (AGASSIZ).

No fossil ctenophores are known certainly, their preservation being most improbable, although 2 genera have been suggested as belonging here:

Rangea GÜRICH, 1930 [**R. schneiderhohni*]. Said to resemble *Beroë. L.Paleoz. (Otavi fm., Kuibis quartzite),* S.Afr.

Xenusion POMPECKJ, 1927 [**X. auerswaldae*]. *Cam.* (drift), Ger.

REFERENCES

Gurich, G.

(1) 1930, *Die bishang altesten Spuren von Organismen in Südafrika:* XV Int. Geol. Congr., Comptes Rendus, v. 2, p. 670-680, 5 figs.

Hyman, L. H.

(2) 1940, *The invertebrates: Protozoa through Ctenophora:* McGraw-Hill (New York), xii +726 p., 221 fig. (Ctenophora, p. 662-696, fig. 209-221).

INDEX

Names included in the following index are classified typographically as follows: (1) Roman capital letters are used for suprafamilial taxonomic units which are recognized as valid in classification; (2) italic capital letters are employed for suprafamilial categories which are considered to be junior synonyms of valid names; (3) morphological terms and generic and family names accepted as valid are printed in roman type; and (4) generic and family names classed as invalid, including junior homonyms and synonyms, are printed in italics. Page numbers printed in boldface type (as **F257**) indicate the location of systematic descriptions.